Andrew D. Miall

The Geology of Fluvial Deposits

Sedimentary Facies, Basin Analysis, and Petroleum Geology

4th corrected printing

Springer
Berlin
Heidelberg
New York
Barcelona
Budapest
Hong Kong
London
Milan
Paris
Santa Clara
Singapore
Tokyo

Andrew D. Miall

The Geology of Fluvial Deposits

Sedimentary Facies, Basin Analysis, and Petroleum Geology

4th corrected printing
With 504 Figures and 30 Tables

Springer

Professor Dr. Andrew D. Miall
University of Toronto
Department of Geology
22 Russell Street
5S 3B1 Toronto
Canada

ISBN-10 3-540-59186-9 Springer Berlin Heidelberg New York
ISBN-13 978-3-540-59186-3 Springer Berlin Heidelberg New York

4th printing 2006

Springer is a part of Springer Science+Business Media
springer.com

Printed in Germany

Cover design: Künkel+Lopka, Ilvesheim
Typesetting: Thomson Press (India), Ltd., New Delhi
Production: Almas Schimmel
Printing and binding: Strauss Offsetdruck, Mörlenbach

SPIN: 11751502 Printed on acid-free paper 30/3141/as 5 4 3

For Charlene,
Christopher,
and Sarah

Preface

Fluvial deposits represent the preserved record of one of the major nonmarine environments. They accumulate in large and small intermontane valleys, in the broad valleys of trunk rivers, in the wedges of alluvial fans flanking areas of uplift, in the outwash plains fronting melting glaciers, and in coastal plains. The nature of alluvial assemblages – their lithofacies composition, vertical stratigraphic record, and architecture – reflect an interplay of many processes, from the wandering of individual channels across a floodplain, to the long-term effects of uplift and subsidence. Fluvial deposits are a sensitive indicator of tectonic processes, and also carry subtle signatures of the climate at the time of deposition. They are the hosts for many petroleum and mineral deposits. This book is about all these subjects.

The first part of the book, following a historical introduction, constructs the stratigraphic framework of fluvial deposits, step by step, starting with lithofacies, combining these into architectural elements and other facies associations, and then showing how these, in turn, combine to represent distinctive fluvial styles. Next, the discussion turns to problems of correlation and the building of large-scale stratigraphic frameworks. These basin-scale constructions form the basis for a discussion of causes and processes, including autogenic processes of channel shifting and cyclicity, and the larger questions of allogenic (tectonic, eustatic, and climatic) sedimentary controls and the development of our ideas about nonmarine sequence stratigraphy.

The final chapters address issues of concern to petroleum geologists. The geometry of reservoirs is discussed, and primary reservoir heterogeneities are evaluated from the point of view of facies architecture and porosity-permeability characteristics. Lastly, the stratigraphic and tectonic distribution of petroleum reservoirs is analyzed, leading to a stratigraphic-tectonic classification of oil and gas fields in fluvial deposits, and a description of selected case examples.

The book is intended for advanced students, researchers, and professionals. Extensive references have been made to the published literature, and hundreds of examples and diagrams have been provided. It is to be hoped that the methods and classifications proposed here will assist in future research work and resource production.

Acknowledgments

My earliest work on fluvial sedimentology was carried out as part of my doctoral thesis research in the Canadian Arctic Islands, under the stimulating leadership of the late Brian Rust. I was able to continue research in this areas during the 1970s, while undertaking regional basin studies for the Geological Survey of Canada. I am grateful to the enlightened management of the Survey at the time, for encouraging this "theoretical" interest, particularly Don Stott, the Director of the Institute of Sedimentary and Petroleum Geology at Calgary for most of my time there.

Since my move to the University of Toronto in 1979, my research has been supported by the Natural Sciences and Engineering Research Council of Canada. I have also, at times, received support from the Petroleum Research Fund of the American Chemical Society, and from various petroleum companies, including British Petroleum (USA), Imperial Oil, Union Pacific, Petro-Canada, Petrel-Robertson and Wascana. Some of the compilation of case studies and mapping methods and, in particular, the sections dealing with petroleum geology, formed part of a report prepared for the Japan National Oil Corporation in 1991, and I am very grateful to the Corporation for permission to include this material in the book.

Most of the chapters in the first half of the book were drafted at University of Oxford while I was on sabatical leave there in 1991–1992. I am grateful to John Dewey, Chair of the Department, and to Phil Allen, my host, for making my stay there a memorable one. Later work on the book was completed at my home base, and I appreciate the flexibility I was afforded by my Chair, John Westgate, in rearranging my teaching timetable to provide blocks of time for work on the final chapters and for the production stage of the book.

Many of my students at the University of Toronto have become stimulating and challenging colleagues, and I am grateful to them for the mutual learning experience. In particular, Phil Fralick, David Eberth, Greg Nadon, Jun Cowan, Mike Bromley, Paul Godin, Mark Stephens, Shuji Yoshida, Andrew Willis, and Octavian Catuneanu, have given me much to think about. Most of them are quoted in this book.

I am grateful to various colleagues for reading parts of the book (including some material that was original prepared for review articles and subsequently incorporated into the book instead). Their input has been most valuable. They include Kevin Biddle, Lars Clemmensen, Bob Dalrymple, Bill Galloway, Martin Gibling, John Horne, John Hubert, Gary Kocurek, Mike Leeder, Jim Lowell, Peter McCabe, Judith Totman Parrish, Norm Smith, Finn Surlyk, Jim Tucker, and Paul Wright. Mike Leeder drew my attention to the early understanding of the power of rivers contained in works by Plato and Homer.

Photographs have been contributed by Kevin Crowley, Miles Hayes, Lloyd Homer (and the Institute of Geological and Nuclear Sciences Ltd., Lower Hutt, New Zealand. My thanks to Dale Leckie for the introduction), Tomasz Jerzykiewicz, Ray Levey, Grant Mossop, Sarah Prosser, Derald Smith, and Norm Smith. The field photographs from the Canadian Arctic were taken by myself, nearly all of them while I was employed by the Geological Survey of Canada, Calgary, and the Survey is hereby gratefully acknowledged as the source of these illustrations. Jim Dixon assisted with the preparation of some of these. All other photographs are from my own research collection. Chris Fielding provided good copies of some of his diagrams. Some original or redrawn line drawings were done by

Subash Shanbhag. Finally, I must again thank my wife Charlene, and my children Christopher and Sarah, for all their support, and for putting up with yet another book. This fluvial enterprise owes much to Charlene's field assistance, companionship, professional advice, support, and love over the years.

Contents

Chapter 1

Introduction

1.1 Scope and Purpose of Book

Fluvial deposits represent the preserved record of one of the major nonmarine environments. They accumulate in large and small mountain-girt valleys, in the broad valleys of trunk rivers, in the wedges of alluvial fans flanking areas of uplift, in the outwash plains fronting melting glaciers, and in coastal plains. The composition of fluvial detritus depends on the geology of the source areas from which the sediment was derived, and the climatic influences to which the detritus was subjected before final burial. The nature of the fluvial assemblage – its lithofacies composition, vertical stratigraphic record and architecture – reflects an interplay of many processes, from the wandering of individual channels across a floodplain, to the long-term effects of uplift and subsidence. Fluvial deposits are a sensitive indicator of tectonic processes, and also carry subtle signatures of the climate at the time of deposition. They are the hosts for many petroleum and mineral deposits. This book is about all these subjects.

Why such a book now? The study of fluvial deposits has a lengthy history, as documented in some detail in Chap. 2. Up until the early 1960s such study constituted part of the broader study of "stratigraphy". During the 1960s and 1970s, the new subdiscipline called "sedimentology" came of age, one of the major focuses of which was the development of our ideas about facies and intrabasinal, autogenic depositional processes, leading to the concept of the process-response model and to generation of suites of facies models. Since the early 1980s, sedimentology itself has undergone something of a revolution, and the term "basin analysis" is a more accurate description of what many sedimentary geologists now do. During this most recent phase, a renewed interest in extrabasinal (allogenic) processes has led to the realization that fluvial basin architecture is much more complicated than just a record of the meandering and switching of river channels. Stimulated primarily by developments in the interpretation of regional seismic-reflection data, geologists are learning to appreciate the importance of regional and global controls on basin architecture. The sedimentological effects of orogenic uplift, of basin subsidence, and of base-level change brought about by these processes, and by eustasy, are profound and pervasive. They influence the entire architecture of a fluvial succession, ranging from such small-scale characteristics as the vertical facies sequence in a drill core, to the large-scale stratigraphy of a basin-fill complex. In this sense, fluvial sedimentology has now been incorporated into the larger field of sequence stratigraphy, which has revolutionized sedimentology and basin analysis since the 1980s.

At the same time, two other areas of research have contributed significantly to the development of the science. Firstly, observational and experimental geomorphology have added much to our understanding of sedimentological processes. In recent years, many geomorphological studies have been stimulated by the needs of geologists to better understand the rock record. In particular, much geomorphic work has been carried out in order to explore the effects of geological processes, such as tectonism, acting over geologically significant lengths of time. Experimental geomorphology, simulating natural processes in laboratory flumes, has provided many valuable insights, for example, into the effects of changing the sediment load, or the regional slope, or the base level of a fluvial system. At the time of writing this book, interaction between sequence stratigraphers and geomorphologists was beginning to recognize the considerable complexity of the relationship between climate change, base-level change, and the fluvial response, and a renewed interest was developing in the incised valleys and terrace deposits of rivers formed during the Late Cenozoic cycle of glacial and interglacial phases.

Secondly, numerical models of geological processes are providing a powerful means of quantitatively testing complex, multivariate processes. The effects of base-level change and tectonism on basin stratigraphy have now been modeled by many workers, and investigators are now exploring such additional details as erosion rates, sediment transfer and sedimentary facies.

This complex interplay of processes and products, and the modern concepts that are evolving from our improved methods of observation and analysis, require description, and it is the perceived need for a synthesis of the welter of recent literature on the subject that has stimulated the writing of this book.

In carrying out the literature search for the sections in this book on petroleum geology, it became clear that there is still a vast amount of work to do to find and exploit oil and gas fields in the frontier areas of the world. Despite a growing public belief that most of the world's petroleum has been found, thousands of geologists and geophysicists employed by scores of companies and government organizations are still devoting their careers to carrying out facies studies, stratigraphic correlation, seismic facies analysis, sequence analysis, regional tectonic studies, and production modeling in obscure corners of the earth, in order to track down oil and gas pools and develop them for production (e.g., Ashton 1992; Flint and Bryant 1993). This is as true for those pools that are found to be trapped in fluvial reservoirs as for any other. This book is addressed, in part, to these individuals.

Most of the writer's experience has been in the area of facies analysis, regional basin studies, and petroleum exploration and development, thus these topics form the main thrust of the book. The reader will find very little here about diagenesis or geochemistry. It is to be hoped that the book will provde a useful source of data, analogs, references, and ideas for those carrying out regional basin studies, as well as for those engaged in petroleum exploration and development, and research into groundwater and effluent disposal.

1.2 Data Sources

Research in the area of fluvial sedimentology and basin analysis has appeared primarily in a half dozen geological journals, notably *Sedimentology, Sedimentary Geology, Journal of Sedimentary Petrology* (now the *Journal of Sedimentary Research*), *Journal of the Geological Society* (*London*), *Geological Society of America Bulletin*, and *American Association of Petroleum Geologists Bulletin*. Much important data have, of course, appeared elsewhere, such as in regional journals and the publications of national geological survey organizations. Reference is made in this book to many of these papers.

One of the problems with the study of rivers and their deposits is that they have been examined from different standpoints by geologists, geographers, and engineers, with consequent difficulties of communication of results and ideas. One of the best ways to address this communication problem is to bring the specialties together, and this was a principal reason for convening the first international symposium on fluvial sedimentology, which was held in Calgary in 1977. Since then, there have been four other symposia, and a sixth is in the planning stage. There have also been several other symposia at international meetings that have been devoted in part or entirely to fluvial deposits. The most important of these events are listed below, with references to the sponsors and the resulting publications.

1. First international symposium on fluvial sedimentology. Canadian Society of Petroleum Geologists, Calgary, October 1977 (Miall 1978a).
2. Recent and ancient nonmarine depositional environments. Rocky Mountain Section, Society of Economic Paleontologists and Mineralogists, Casper, Wyoming, June 1979 (Ethridge and Flores 1981).
3. Modern and ancient fluvial systems: sedimentology and processes (Second International Symposium). Sponsored by the International Association of Sedimentologists, Keele, UK, September 1981 (Collinson and Lewin 1983).
4. Fluvial sedimentation and related tectonic framework in western North America. Cordilleran Section, Geological Society of America, Anaheim, California, April 1982 (Nilsen 1984).
5. Rudites formed by unidirectional flow. Symposium, Eleventh Sedimentology Congress, Hamilton, Canada, August 1982 (Koster and Steel 1984).
6. Modern and ancient fluvial sedimentation. General Session, Eleventh International Congress on Sedimentology, Hamilton, Canada, August 1982 (no proceedings publication).
7. Third international fluvial sedimentology conference, Society of Economic Paleontologists and Mineralogists, Fort Collins, Colorado, August, 1985 (Ethridge et al. 1987).
8. The three-dimensional facies architecture of terrigenous clastic sediments, and its implications for hydrocarbon discovery and recovery. Society of Economic Paleontologists and Mineralogists,

Research Symposium at Annual Meeting, San Antonio, April 1989 (Miall and Tyler 1991).

9. Fourth international conference on fluvial sedimentology, Barcelona, International Association of Sedimentologists, October 1989 (Marzo and Puigdefábregas 1993).
10. Braided rivers: form, process and economic applications; special two-day meeting, Geological Society of London, May 1992 (Best and Bristow 1993).
11. Fifth International Conference on Fluvial Sedimentology, Brisbane, July 1993 (Fielding 1993a).
12. Alluvial Basin. European Research Conference, Lunteren, The Netherlands, September 1994 (no proceedings).

These proceedings volumes represent a substantial body of literature and an essential basis for a fluvial library, to which must be added various books, course notes, and review articles. Summaries of progress in research were provided in several of the edited volumes (Miall 1983a, 1987; Fielding 1993b; Bristow and Best 1993). Important review articles include those by Cant (1982), Rust and Koster (1984), Walker and Cant (1984), Bridge (1985, 1993b), Collinson (1986), Miall (1992a), Zaitlin et al. (1994), and Shanley and McCabe (1994). Books that contain valuable syntheses of fluvial sedimentology include those by Turner (1980), Galloway and Hobday (1983), Nilsen (1985), and Rachocki and Church (1990), although none of these use the architectural approach that forms the basis of this book. Short courses by Miall (1981a), Flores (1983a), and Flores et al. (1985) provide useful summaries of fluvial sedimentary processes, but are now largely out of date.

A noticeable characteristic of the First International Symposium on Fluvial Sedimentology (Miall 1978a) was the presentation of several detailed attempts to build suites of facies models (Friend 1978; Miall 1978c; Rust 1978b; Cant 1978; Boothroyd and Nummedal 1978; Heward 1978a), and several other papers providing critiques of the facies model concept (Jackson 1978; Collinson 1978). None of the subsequent symposia have yielded the same focused examination of the facies model idea. The paper by Friend (1983) at the Keele meeting, providing a classification of channel architecture, is the only post-1977 symposium paper to attempt a generalized, theoretical examination of fluvial facies models. Several nonsymposium papers providing critiques of the ideas in this area include Miall (1980, 1985) and Bridge (1985). Many symposium papers and individual journal articles have contained proposals for new facies models or variants on existing ones. Bridge (1985, 1993b) developed a different approach, building theoretical models from flume experiments and dynamic studies of modern rivers.

All the symposia have included large numbers of case studies. Many of these, especially those presented at the second international meeting in Keele (Collinson and Lewin 1983), focused on facies analysis. Contrasting with this body of work is the symposium organized by Nilsen (1984), which set out to examine the regional effects of contemporary tectonism on fluvial systems and the sedimentary record. The 1993 Brisbane symposium saw the presentation of much work in the field of environmental sedimentology.

The most recent developments in fluvial basin analysis are as follows:

1. Firstly, there is a focus on the detailed, formal documentation of three-dimensional facies architecture, stimulated by the pioneering work of Allen (1983a), Ramos and Sopeña (1983), and Ramos et al. (1986). A symposium focusing on this type of work in fluvial and other clastic environments was held at San Antonio in 1989, and has resulted in the production of a large-format atlas, in which several outcrop and subsurface case studies are documened by large-scale cross sections (Miall and Tyler 1991). The two most recent fluvial symposia also devoted considerable attention to alluvial architecture (Marzo and Puigdefábregas 1993; Fielding 1993a), and publications dealing with reservoir heterogeneity are also giving much attention to this subject (Ashton 1992; Flint and Bryant 1993).
2. Secondly, there is increasing interest in bringing together our knowledge of the physics of sedimentation and the physics of basin subsidence, using the methods of quantitative modeling and computer simulation. The book by Cross (1990) is a landmark synthesis in this area, and contains much of importance to the fluvial sedimentologist. Other work of this type is referred to elsewhere in this book, particularly papers by J. Bridge, C. Paola, and P. Heller.
3. Observational and experimental geomorphology has resulted in several important books and many individual papers, particularly by S.A. Schumm (Schumm 1977; Schumm et al. 1987), and his students and coworkers, and these are also cited in this book.
4. Lastly, there are developments involving the larger sphere of sequence stratigraphy, requiring

the researcher to examine the fluvial environment within the context of the nonmarine-marine transition, and the effects of extrabasinal controls. Such books as those by Nummedal et al. (1987), Wilgus et al. (1988), and Van Wagoner et al. (1990, 1991) provide a vital source of data and ideas, and are discussed at length in this book. The geomorphological work referred to above has been particularly useful in this regard (e.g., Schumm 1993; Wescott 1993). A valuable review of the subject was provided by Shanley and McCabe (1994).

Chapter 2

Historical Background

2.1 Introduction

It is difficult (and probably meaningless) to choose the moment at which modern studies of fluvial sedimentology could be said to have begun. Probably the monumental work of Fisk (1944, 1947) on the Mississippi river and delta was the most important single advance. Several developments of fundamental importance took place in the late 1950s and early 1960s. Flume and fieldwork on the origin of bedforms and sedimentary structures culminated in a major structure classification by Allen (1963a) and statements on the flow-regime concepts of bedform origin (Simons and Richardson 1961; Simons et al. 1965; Harms and Fahnestock 1965). Facies models concepts began to appear in the late 1950s, and had a major impact on geologists following the development of the fining-upward cyclothem model by Allen (1963b, 1964) and Bernard et al. (1962). Modern research into stratigraphic traps for petroleum essentially began at about the same time. The Shell Development and Esso groups at Houston, USA, in particular, devoted considerable energies to researching the sedimentary products of modern clastic and carbonate environments. They were interested in improving our knowledge of the geometry of porous bodies in the subsurface, and developing techniques for diagnosing sedimentary environments from such data as wireline logs. Nanz (1954), a member of the Shell group, was probably the first (at least, in the published record) to identify fining-upward fluvial deposits in the subsurface, but they were not clearly understood as the products of fluvial channel-fill and point-bar migration processes for several years, until the work of Bernard et al. (1962), who were also members of the Shell group.

It is this history of the development of ideas that is the subject of this chapter. Some of the areas of interest to fluvial sedimentologists have been the subject of other historical reviews. Some aspects of fluvial geomorphology which bear on fluvial sedimentology were summarized by Schumm (1972a). Jopling (1975) discussed early research on fluvioglacial deposits. Middleton (1965, 1977) provided brief histories of the work carried out on bedforms and sedimentary structures. Pettijohn (1962) and Potter and Pettijohn (1963, 1977) reviewed the subject of paleocurrent analysis. Allen (1965a) described the state of knowledge of alluvial sedimentology as it was in the mid-1960s, but this was not an historically oriented paper. The main historical developments up to the time of the first fluvial conference were described by Miall (1978b), and later developments were detailed by Miall (1987), and it is from these two articles that this chapter is adapted. The intent of this chapter is to outline research progress up to the 1980s. Modern developments are dealt with in the detailed descriptive chapters later in this book.

2.2 Early Developments in the Study of Fluvial Sediments

2.2.1 From the Ancient Greeks to Playfair

Rivers and their deposits did not become subjects of specialized study until the nineteenth century. However, many of the early naturalists and philosophers made notes on this topic in the context of wide-ranging discussions on the origins of the earth and its physical features. The principles of uniformitarianism were not widely accepted until the time of Lyell, but long before this it was realized that rivers could transport and deposit large quantities of sediment, so that correct geological intepretations of Recent alluvium were quickly arrived at. Application of the same ideas to consolidated and structurally deformed deposits took a little longer. The following is a brief review of some of the early writings on the physical geology of rivers.

Plato, in a passage from Critias described the destruction of fertile lands by fluvial erosion and transportation: "... the soil which has kept breaking away from the high lands during these ages and these disasters, forms no pile of sediment worth mentioning, as in other regions, but keeps sliding away ceaselessly and disappearing in the deep." Homer, in *The Iliad*, described the battle of Achilles with the river god Scamander. The river is described as in full flood, and the power of erosion and transportation is graphically conveyed:

"He filled all his channels with foaming cataracts, and roaring like a bull he flung up on dry land the innumerable bodies of Achilles' victims that had choked him, protecting the survivors by hiding them in the deep and ample pools that beautified his course.

The river god sweeps Achilles off his feet, an action likened to a gardener digging an irrigation channel.

Mattock in hand, he clears obstructions from the trench; the water starts flowing; it sweeps the pebbles out of the way; and in a moment it runs singing down the slope and has outstripped its guide."

Later, Scamander calls to Simoïs:

"Dear brother, let us unite to overpower this man ... fill your channels with water from the springs, replenish all your mountain streams, raise a great surge and send it down, seething with logs and boulders. ... [the armour of Achilles] shall lie deep in the slime beneath my flood; and as for him, I'll roll him in the sand and pile up shingle high above him." (Rieu 1953)

The tendency for rivers to overflow and form thick blankets of floodplain deposits has been viewed with mixed feelings by those living in the valley of the Nile since prehistoric times. One of the first to reach correct conclusions about this process was the Greek philosopher Herodotus (born 484 B.C.) who concluded that "Egypt is the gift of the river". Aristotle (384–322 B.C.) was also aware of the depositional activity of the Nile and, in addition, he noted the rapid silting up of navigable river channels around the shores of the Black Sea and on the Bosphorous. These observations were an improvement on that of Thales of Miletus (636–546 B.C.) who thought the presence of new deltaic deposits projecting into the sea showed that water could change into earth. The Meander River in Turkey was mentioned by both Herodotus and Strabo, and its name became used as a general term for highly sinuous rivers (Russell 1954).

Leonardo da Vinci (1452–1519) observed the gravel deposits of the River Arno and wrote in his notebooks "which deposits are still to be seen welded together and forming one concrete mass of various kinds of stones from different localities." Elsewhere the river deposited mud containing shells, which da Vinci used as evidence to dispute the origin of fossils as products of the Flood.

Agricola, in his great treatise *De re metallica*, published in 1556, made some observations on placer gold deposits. He recognized that "gold is not generated in the rivers and streams ... but is torn away from the veins and stringers and settled in the sands of torrents and water courses ..." He commented on a prevailing theory of his time concerning the action of the sun in drawing out the metallic material, which led to the idea that the orientation of veins or gold-bearing streams was critical in their economic exploitation:

"... a river, they say, or a stream, is most productive of fine and coarse grains of gold when it comes from the east and flows to the west, and when it washes against the foot of mountains which are situated in the north, and when it has a level plain toward the south or west." (Hoover and Hoover 1950, p. 75).

Agricola devoted many pages to the extraction of gold and tin from placers, and provided us with a picture of an early fluvial sedimentologist at work with his gravels (Fig. 2.1).

The German physician and geographer Bernhard Varenius (1622–1750) published the first text on physical geography in 1734. He wrote "Of the Changes on the Terraqueous Globe, viz. of Water into Land or Land into Water", and listed a series of "propositions" to explain such changes. Part of Proposition V states, regarding rivers:

"If their water bring down a great deal of Earth, Sand and Gravel out of the high Places, and leave it upon the low, in process of Time these will become as high as the other, from whence the Water flows: or when they leave this Filth in a certain Place on one side of the Channel, it hems in and raises part of the Channel which becomes dry Land."

Robert Hooke (1635–1703), the English physicist and mathematician, wrote about earthquakes changing the level of strata, and "that water counteracts these effects" by processes of erosion and sedimentation:

"We have multitudes of instances of the wasting of the tops of Hills, and of the filling or increasing the Plains or lower Grounds, of Rivers continually carrying along with them great quantities of Sand, Mud, or other Substances from higher to lower places ... Egypt as lying very low and yearly overflow'd, is inlarg'd by the sediments of the Nile, especially towards that part where the Nile falls into the Mediterranean. The Gulph of Venice is almost choked with the sand of the Po. The Mouth of the Thames is grown very shallow by the continual supply of Sand brought down with the Stream."

Nikolaus Steno (1638–1687) recognized the use of certain fossils as environmental indicators, proposing, for example, that the presence of a terrestrial

A—River. B—Weir. C—Gate. D—Area. E—Meadow. F—Fence. G—Ditch.

Fig. 2.1. Early interest in fluvial sedimentation: extraction of tin from placer gravels. From "De Re Metallica" by Georgius Agricola, 1556. (Hoover and Hoover 1950)

fauna, rushes, grasses and tree stems indicated deposition in fresh water.

In 1770, the French naturalist Jeanne Etienne Guettard (1715–1786) published a memoir "On the degradation of mountains effected in our time by heavy rains, rivers and the sea" in which he noted the petrographic variations in the detritus of different river basins. He was aware of the way in which a river basin acts as a long-term, but nevertheless temporary, site of sediment storage in its journey to the sea, and observed how the thickness of alluvium increases toward the river mouth. Guettard interpreted the "*poudingues*" (English: puddingstone, an early term for conglomerate) of the basin as fluviatile in origin, comparing them to the modern gravels carried by the River Seine.

The origin of valleys by the erosive action of streams was first clearly taught from actual examples by Desmarest in a memoir published in 1774, even though others before him, including Targioni-Tozzetti, De Saussure and Guettard, had put forward similar conclusions in earlier publications.

The theory of uniformitarianism is attributed to James Hutton (1726–1797), although many of the ideas that contributed to his theory had appeared before his time. Hutton's first paper on this subject appeared as an extended abstract in 1785 (Bailey 1967), but his principal work was published 10 years later as a four-part book (Hutton 1795). His main contribution to sedimentary geology was the principle that most of the features we see in ancient rocks can be explained by processes taking place at the present day, and that apparently imperceptible processes can have profound effects if they continue long enough. Hutton was also, of course, the chief protagonist of the Plutonist theories of earth history, in opposition to Abraham Gottlab Werner (1749–1817), the principal Neptunist. According to Werner, alluvial deposits were part of the youngest of his four layers of water-laid rocks, immediately below the volcanics, and were always to be found in the same stratigraphic position.

Hutton's language is complex, and his uniformitarianist theories were better explained by his friend and colleague Playfair (1802), from whom the following observations on rivers and their deposits are culled. He suggested that the thickness of alluvial gravels may be used to estimate the depth of erosion in the source area, but concluded: "whether data precise enough could be found to give weight to such a computation must be left for future enquiry to determine." Modern basin modeling techniques are only now coming to grips with this problem, nearly 200 years later.

Fluvial facies and textural studies could be said to have begun with Playfair's following statement:

"It is a fact very generally observed, that where the valleys among primitive mountains open into large plains, the gravel of those plains consists of stones, evidently derived from the mountains. The nearer that any spot is to the mountains, the larger are the gravel stones, and the less rounded is their figure; and, as the distance increases, this gravel, which often forms a stratum nearly level, is covered with a thicker bed of earth or vegetable soil ... The reason of this gradation is evident; the farther the stones have travelled, and the more rubbing they have endured, the smaller they grow, the more regular is the figure they assume, and the greater quantity of that finer detritus which constitutes the soil. The washing of the rains and rivers is here obvious."

These observations state in a more succinct form ideas expressed at length by Hutton (1795, Part II, Chap. IV), although Hutton doubted that a high degree of clast roundness could be attained except in the sea. All these conclusions contrast strongly with those of Finch (1823), Conybeare and Philips (1824), Hayden (1821), and others, all of whom attributed "diluvial gravels" to the Deluge (see Schumm 1972, p. 16), and ideas about the fluvial origin of valleys and alluvial deposits continued to be resisted for many years, because they conflicted with these biblical teachings.

2.2.2 From Lyell to Davis

Charles Lyell's *Principles of Geology*, the first volume of which was published in 1830, is undoubtedly amongst the two or three most important books on geology ever published. The intent of his book was to establish uniformitarianism as the guiding principle for geological interpretation, following Hutton's teachings, and to discredit catastrophism and diluvialism. The immediate success and wide-ranging influence of the book were due in large part to the wealth of field observations that Lyell was able to bring to bear on his thesis, based on his travels throughout Britain, France, and Italy. Von Hoff (1822–1824) had published a treatise earlier on the theme "investigation of the changes that have taken place in the earth's surface conformation since historic times, and the application which can be made of such knowledge in investigating revolutions beyond the domain of history". This work provided copious documentation of recent geological activity around the shores of the Mediterranean Sea and the Black Sea, such as alluvial deposition and changes in sea level, but the work was based mainly on literature research, Von Hoff not having the means for travel available to Lyell, and his treatise did not receive the recognition it deserved (Zittel 1901, p. 188). Lyell acknowledged that much of the information that he used in his *Principles* came from Von Hoff, "but he helped me not to my scientific view of causes" (letter quoted in Bailey 1962, p. 78; Wilson 1972, p. 276).

Among his more important pieces of fieldwork, Lyell observed Tertiary fluvial deposits interbedded with lacustrine sediments and lava flows in the Auvergne, France, and Tertiary fluvial gravels containing beds of clay with marine fossils near Nice (Wilson 1972, Chap. 7). These clearly demonstrated to Lyell the long and complex geological histories of these regions, explainable in terms of processes acting at the present day over long periods of time, but quite incompatible with catastrophist hypotheses.

Five chapters in the first volume of the *Principles* (1st edition) are devoted to rivers and deltas. Lyell described the way in which river meanders gradually enlarge themselves and develop neck cutoffs by wearing away at their banks (Fig. 2.2):

"When the tortuous flexures of a river are extremely great, the aberration from the direct line of descent is often restored by the river cutting through the isthmus which separates two neighboring curves."

Lyell thought that meanders were initiated as deflections caused by the walls of the valley or by variations in the thickness or hardness of the alluvial valley deposits. He quoted earlier work of Flint on the Mississippi River to the effect that opposite each

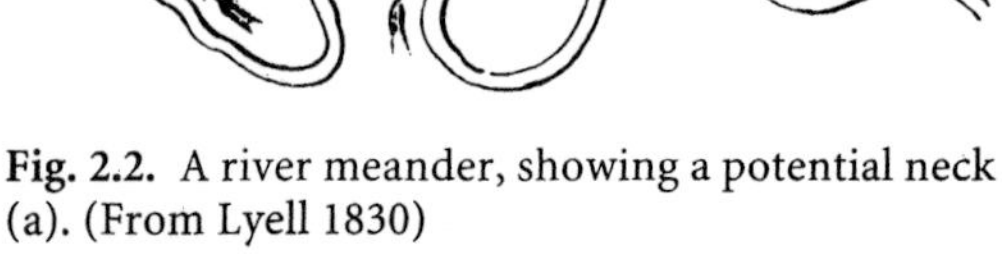

Fig. 2.2. A river meander, showing a potential neck cutoff (a). (From Lyell 1830)

curve "there is always a sand-bar, answering, in the convexity of its form, the concavity of the bend", and described the development of levees, which slope back and become finer grained away from the channel.

Observations of many floods are quoted as testament to the power of rivers to erode and transport large quantities of coarse debris (Playfair was aware of this 30 years earlier), and historical evidence pertaining to the burial of Roman buildings and the isolation of ancient ports far inland is used to demonstrate the fact that rivers can build up thick and extensive alluvial tracts in a few hundred years. The velocity gradient of river channels (slowest near the bed) was already known, and some basic data on sediment entrainment had been obtained:

A velocity of 3 in/s is ascertained to be sufficient to tear up fine clay; 6 in/s, fine sand; 12 in/s, fine gravel; and 3 ft/s, stones of the size of an egg.

Lyell provided what is probably the earliest illustration of fluvial cross-bedding, or "false" bedding (Fig. 2.3), observed in a cutbank in the River Rhône at the confluence with the Arve. He explained the origin of the varying cross-bedding angle as follows:

"Those layers must have accumulated one on the other by lateral apposition, probably when one of the rivers was very gradually increasing or diminishing in velocity, so that the point of greatest retardation caused by their conflicting currents shifted slowly, allowing the sediment to be thrown down in successive layers on a sloping bank. The same phenomenon is exhibited in older strata of all ages."

Nowadays we can recognize a reactivation surface in Lyell's illustration.

Volume 2 contains little of relevance to this book. However, Volume 3 of the Principles, published in 1833 after Lyell's further travels to Spain, Germany, and Switzerland, deals primarily with stratigraphic geology, in particular that of the Tertiary, for which Lyell proposed a detailed subdivision, including the terms Eocene, Miocene, and Pliocene. Amongst the nonmarine deposits described are the alluvium, or "loess" of the Rhine, and the Swiss "molasse". He observed eolian sand dunes evolving near Calais and used his observations in interpreting ripple marks and other cross-bedding structures (Fig. 2.4):

"If a bank [ripple] has a steep side, it may grow by the successive apposition of thin strata thrown down upon its slanting side, and the removal of matter from the top may proceed simultaneously with its lateral extension. The same curent may borrow from the top what it gives to the sides ... Each ridge [ripple] had one side slightly inclined, and the other steep, the lee side being always steep, as bc, de [Fig. 2.4], the windward side a gentle slope, as ab, cd ... We think we shall not strain analogy too far if we suppose the same laws to govern the subaqueous and subaerial phenomena ... We may refer to a drawing given in the first volume [Fig. 2.3], to show the analogy of the arrangement of the ... strata just considered, to that exhibited by deposits formed in the channel of rivers."

Lyell was thus aware of the hydraulic significance of "false" stratification. For example, he illustrated what we would call herringbone cross-bedding, and attributed the opposite dips to "the set of tides and currents in opposite directions". However, the first clear enunciation of the principles of paleocurrent analysis was left to Sorby.

In 1841 and again in 1845, Lyell visited North America; the second visit included some time spent examining the Mississippi River and delta. Both these trips provided him with much new information for revisions of his *Principles*, which subsequently ran to 11 editions (the last in 1872).

Lyell's contributions to geology have been assessed at length elsewhere (Bailey 1962; Wilson 1972), but it is useful here to repeat Middleton's (1978) comment that, although Lyell made many

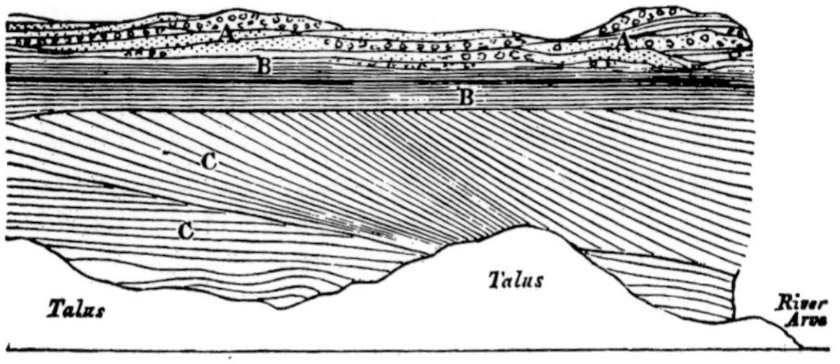

Fig. 2.3. Cross-stratified sand (*C*) overlain by laminated sand (*B*) and interbedded sand and gravel (*A*). (From Lyell 1830)

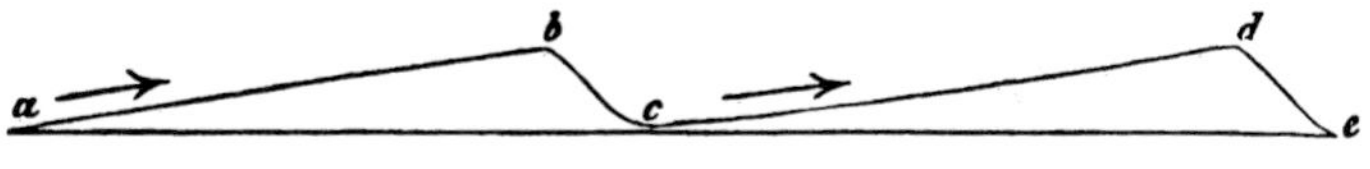

Fig. 2.4. Development of an asymmetric ripple. (Lyell 1830)

observations that we would regard as being sedimentological in nature, his primary concern was with establishing the working methods of historical geology, and it would not be correct to regard him as a founder of sedimentology. It is now suggested by historians of geology that Lyell overemphasized the importance of slow, steady processes in his attempts to discredit biblical ideas. Catastrophism, that is to say, the rare occurrence of catastrophic events, including violent floods, may in fact be very important processes in building the stratigraphic record (Ager 1993).

The next important figure in the study of fluvial sediments is Sorby, whose earliest contribution on this topic constitutes the first paper on paleocurrent analysis (Sorby 1852; available, along with other papers by this great innovator, in a collection edited by Summerson 1976). The principles were stated as follows:

"... By observing the direction of these ripple marks the line in which the current moved may be known ... [when ripple drift or false bedding] is observed in progress in modern sand drifts ... it will be perceived that the line of the dip of the talus [foreset] is not constantly in the true direction of the current on account of the formation of complicated deltoid deposits, in which the line of the dip of their sloping termination varies very considerably on each side of that of the current. Its mean direction, however, coincides with it; and hence, if a number of properly placed observations be made, their mean gives a result very closely agreeing with the true line of the current."

Sorby first studied paleocurrent patterns in the Carboniferous coal measures near Edinburgh. He classified the sandstones into four types: level bedded, ripple laminated (single ripple train), ripple drifted, and drift bedded (large-scale cross-bedding) and referred to them using the first known set of facies symbols. He recognized that structure size is partly dependent on flow velocity.

In a paper published only 7 years later, Sorby (1859) was able to boast "I must now have, in my notebooks, not less than twenty thousand recorded observations" of paleocurrent directions, a record that few have equaled. Most were never published. He went on to say "In various papers ... I have explained many of my deductions, and I have shown that many peculiarities of physical geography at former epochs may be learned from a knowledge of the directions of the currents in various localities". He lamented that others had not taken up his methods of facies and paleocurrent analysis, and could not know that such methods would not be widely used again until the 1950s, over 90 years later.

Sorby, in his 1859 paper, also made the first tentative attempts at understanding bedform hydraulics. He stated " ... when strata are deposited under the influence of a current, the character of the resulting structure must depend on the depth of the water, the velocity of the current, the nature of the deposits and the rate of deposition". He thought that it should be possible to deduce the rate of deposition of ripple-drift cross-bedding from lamina thickness which "indicates the excess of material deposited on the sheltered side of ripples over that washed up again from the exposed side, during the time required for each ripple to advance a distance equal to its own length", and he carried out hydraulic experiments in a stream passing through the grounds of his house (see Sect. 2.3.5).

Sorby's contributions to the study of sedimentary structures appear to have remained unappreciated and largely unknown until exhumed by Pettijohn (1962) and reinterpreted by Allen (1963b). They are barely mentioned in Higham's (1963) biography, and Sorby has generally been much better known for his work in establishing the use of the microscope in petrographic studies and in metallurgy. However, Sorby's accumulation of much data that never became published is, regrettably, a familiar tendency amongst present-day geologists.

Several advances in the study of sedimentary structures were made by other naturalists in this period. For example, Jamieson (1860) first recognized clast imbrication, while studying fluvioglacial gravels in Scotland, and Reade (1884) described examples of ripple lamination in British fluvioglacial deposits. Gilbert (1884, 1899) described large-scale symmetrical ripples from some of the Paleozoic marine sandstones of New York State and stated (1899) "previous to 1882, ideas as to the origin of water-made sand-ripples were crude and unsatisfactory", but quoted several papers that had appeared since that date that had improved theoretical knowledge. He proceeded to supply an explanation of oscillation ripples and current ripples which, for the latter, included Darwin's (1883) discovery of the separation eddy which forms over ripple crests. "An eddy or vortex is created in the lee of the prominence, and the return current of this vortex checks travelling particles, causing a growth of the prominence on its downstream side" (Fig. 2.5). Gilbert (1899) provided a classification of ripple-drift cross-bedding (Fig. 2.6) which was, in effect, only replaced in the early 1960s.

The first investigations into the nature and quantity of sediment carried by rivers were those of Everest (1832), who measured the suspended load of the River Ganges under a variety of flow conditions. Humphreys and Abbot (1861), working for the US

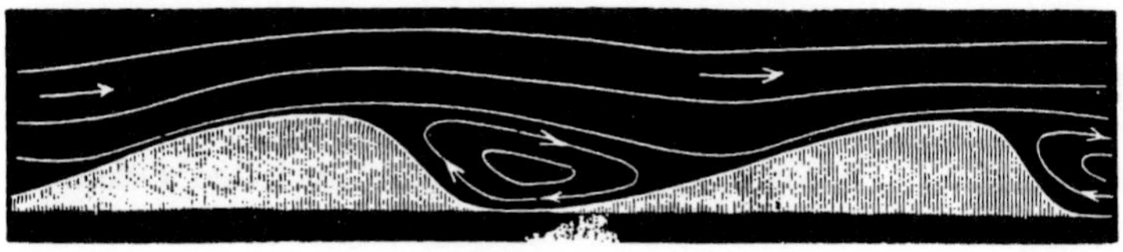

Fig. 2.5. Flow lines over an asymmetric ripple, showing separation eddies. (Darwin 1883)

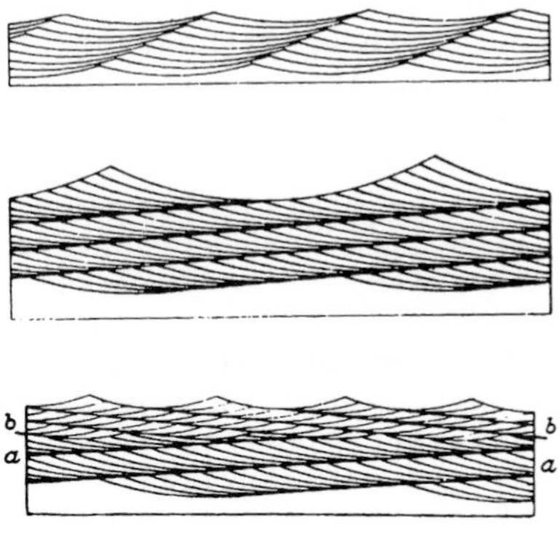

Fig. 2.6. Three types of ripple marks. (Gilbert 1899)

Army Corps of Topographic Engineers, carried out detailed studies of the sediment load in the Mississippi River, and their work was widely quoted by later writers (e.g., Dana, Geikie) as testament to the enormous quantities of detritus that can be moved by fluvial transport. Here is a brief extract from their report:

"... the sediment of the Mississippi is to the water, by weight, nearly as 1 to 1500, and, by bulk, nearly as to 1 to 2900; provided long periods of time be considered ... 812,500,000,000 pounds of sedimentary matter, constituting one square mile of deposit 241 feet in depth [0.19 km^3], are yearly transported in a state of suspension into the gulf ... No exact measurement of the amount of the annual contributions to the gulf from [bedload transport] can be made, but from the yearly rate of progress of the bars into the gulf, it appears to be about 750,000,000 cubic feet, which would cover a square mile about 27 feet deep [0.021 km^3]."

Drew (1873) provided the first complete description of alluvial fans and their deposits, based on extensive field observations in the upper Indus Basin. A wealth of topographic details is given, and the process of fan sedimentation by distributaries wandering laterally over their own deposits is well described: "the lateral changes of position of the depositing stream, and the partial growth of each layer, are denoted by false bedding". Drew described fan-head trenching and discussed the concept of the graded stream. From the point of view of this historical review, his most interesting contribution is as follows:

"We should now consider the relationship of the fans to the alluvium proper of the main valleys.

If the level of this latter alluvium is remaining, on the whole, stationary (the main river neither deepening its bed nor raising it), and if the fan is undergoing increase, then the fan-stuff will just extend over the alluvium, gradually encroach on it in area; and, stratigraphically, rest upon it. But I think it is a more usual case that the river alluvium has been increasing contemporaneously with the increase of fan-stuff, that the river and the ravine-stream both were raising their beds; then there will have been an interstratification of the two deposits at the fan edges as they were at successive epochs, a lapping for a short distance of one set of alluvial beds over the other.

I have sometimes observed in section an interstratification such as to suggest the above origin - beds of well-rounded materials and of sand among the less-worn fan-stuff; for indeed the latter, being nearer its source, is seldom thoroughly rounded."

Gilbert (1880) published what rapidly became a classic of geomorphological literature in his description of the Henry Mountains, Utah. He dealt with fluvial processes extensively, including the growth of alluvial fans and the development of meanders and cutoffs, and provided a detailed discussion of the balance between discharge, slope, and load in a river, showing how they tended towards an equilibrium, or graded, condition. However, from the point of view of fluvial sedimentology, as distinct from geomorphology, little of what he said was an advance on Drew's (1873) work.

The importance of flash flood sedimentation was recognized following work in the arid American west. McGee (1897) described sheet floods 4 km wide and recorded flood deposits that extended for 16 km across strike, and 30 km down depositional dip.

Summaries of the state of the art at this time are provided in two classic textbooks by Dana (1862) and Geikie (1882). Much is written in each about the erosive power of rivers, and details are provided regarding the slope, discharge, and sediment load of selected rivers around the world; the work of Humphreys and Abbot on the Mississippi is quoted extensively. More precise information than before regarding sediment entrainment is given, and the meandering and avulsive habits of river channels are described. However, even as late as the third editions of these books, published in 1880 and 1893, respectively, little attention is paid to fluvial deposits themselves, and no mention at all is made of Sorby's paleocurrent and paleogeographic analyses. In dealing with ancient deposits, the authors' preoccupations are mainly with stratigraphic documentation. However, some glimmerings of fluvial sedimentology do creep in. Dana wrote "in the alluvial deposits from a flood the little layer is begun with a relatively rapid rate of deposition and finished with a slower, as the flood declines, and hence its upper and lower portions will differ as to coarseness and density". Here, we have the first recognition of the cyclic nature of flood deposits. Dana also provided a good description of the shape and development of transverse sand bars, based on the work of General G.K. Warren, another of the US Army topographers, on the Mississippi River. Geikie wrote:

"The deposit of alluvium on river beds is characteristically shown by the accumulation of sand or shingle at the concave side of each sharp bend of a river course. While the main upper current is making a more rapid sweep around the opposite bank, undercurrents pass across to the inner side of the curve and drop their freight of loose detritus which, when laid bare in dry weather, forms the familiar sand-bank or shingle beach."

This is an eminently clear description of the helical flow patterns that occur on river meanders, and of the point-bar deposits that result. It remains only to provide the idea of decreasing competency and sediment sorting up the point-bar slope to arrive at the fining-upward model. However, this was not to come for another 80 years.

In spite of the general acceptance of the principle of uniformitarianism, these excellent observations on modern fluvial environments were not, at this time, being applied to interpretations of ancient rock successions. Copious descriptions of the lithology, fauna, and flora of such nonmarine units as the Old Red Sandstone were given by Dana and Geikie, but environmental and paleogeographic interpretations were limited to general statements concerning the deposition of such rocks in lakes or "inland seas". Their general nonmarine nature was recognized, but further interpretations did not proceed beyond such statements as this, from Dana, concerning the Triassic of the Appalachians:

"The mud-cracks, raindrop-impressions and footprints – these show, wherever they occur, that the layer was for the time a half-emerged mud-flat or sand-flat; and, as they extend through much of the rock, there is evidence that the layers in general were not formed in deep water ... The occurrence ... of coarse conglomerate, some of the stones of which are very large and of a coarse kind of oblique lamination [cross-bedding?] in much of the rock, is evidence that some of the beds were deposited by a flood of waters pouring violently down this valley."

Dana also thought that floating ice was "concerned in part of the deposition". A detailed report on these same rocks by the geographer and geomorphologist W.M. Davis (1898a) included a detailed discussion of the various possible environments in which they could have formed, including tidal, lacustrine, fluvial, glacial, or eolian setings. It is a fine example of sedimentological reasoning, in which modern analogs are clearly appealed to. For example:

"The Pre-Triassic peneplain might have been warped so as to alter the action of the quiescent old rivers that had before flowed across it ... Such a change would set the streams to eroding in their steepened courses, and to depositing where their load increased above their ability of transportation ... The heavy accumulation of river-borne waste on the broad plains of California, or of the Indo-Gangetic depression all agree in testifying that rivers may form extensive stratified deposits, and that the deposits may be fine as well as coarse. They are characteristically crossbedded and variable, and they may frequently contain rain-pitted or sun-cracked layers."

However, Davis was a cautious individual. He concluded:

"Penck has suggested the name 'continental' for deposits formed on land areas, whether in lakes, by rivers, by winds, under the creeping action of waste slopes, or under all these conditions combined. This term seems more applicable than any other to the Triassic deposits of Connecticut. It withdraws them from necessary association with a marine origin, for which there is no sufficient evidence, and at the same time it avoids what is today an impossible task – that of assigning a particular origin to one or another member of the formation."

The last sentence is revealing, for it shows that application of the growing body of knowledge about modern environments to sedimentological problems was still very much in its infancy. Davis was somewhat more self-confident in his fluvial interpretations of the Tertiary strata of the Rocky Mountains, pubished only 2 years later (Davis 1900).

At about this time, Davis (1898b, 1899) also made a major contribution to fluvial geomorphology with his classification of rivers according to their stage of development in the erosion cycle as youthful, mature, and old, and he recognized and described rivers of braided and meandering character.

A rather more advanced level of interpretation of fluvial sediments was given earlier by Medlicott and Blanford (1879), but their observations were not widely applied for some years. They described the Siwalik Formation of the Himalayas, compared it with the "Molasse" of Switzerland and with modern alluvium, as follows:

"In the upper Siwaliks conglomerates prevail largely; they are often made up of coarsest shingle, precisely like that in the beds of the great Himalayan torrents. Brown clays occur often with the conglomerate, and sometimes almost entirely replace it. This clay, even when tilted to the vertical is indistinguishable in hand specimens from that of the recent plains deposit; and no doubt it was formed in a similar manner, as alluvium. The sandstone, too, of this zone is exactly like the sand forming the banks of the great rivers, but in a more or less consolidated condition."

This appeal to a specific modern analog is a good example of an early application of actualistic principles to a sedimentological problem.

2.3 Growth of Present-Day Concepts, up to 1977

The First International Symposium on Fluvial Sedimentology was held in Calgary, in October 1977. This event marks a watershed in the development of ideas about fluvial sedimentary structures and facies, because it was the first time that specialists in a variety of related areas were brought together to deal with this focused theme. The publication that resulted (Miall 1978a) is the only memoir produced by the Canadian Society of Petroleum Geologists that has ever been reprinted as a result of demand. This section deals with developments up to that time, and is adapted from a review prepared for publication in the proceedings volume (Miall 1978b).

2.3.1 Increasing Specialization of the Twentieth Century

Until the end of the nineteenth century, it was possible to be both a general geologist and a geomorphologist. Middleton (1978) regarded Walther (1893–1894) as the first "to draw together the many scattered observations on modern sediments and to document the actualistic method as the basis for a specific science of sedimentary rocks". At about this time, several separate themes gradually began to appear in the study of rivers, fluvial processes, and their deposits. Each theme became a specialized subject, with its own body of workers, its own purposes and applications and, eventually, its own literature. To bridge these divergences has become an increasingly difficult objective for modern research workers. Likewise it is a complex task to trace the various threads that lead to our present understanding of fluvial deposits. The following are the most important areas of specialized study that were actively evolving into the late 1970s.

1. Descriptive fluvial geomorphology: the physical characteristics of rivers, alluvial fans, floodplains, their evolution and their variations from source to mouth.
2. Quantitative fluvial geomorphology: the interrelationships between width, depth, slope, discharge, sinuosity, and sediment load (hydraulic geometry).
3. Sediment transport studies: investigations of sediment entrainment, transport rates, and downstream textural changes.
4. Bedforms: their external form and internal structure, and their relationship to depth, velocity, sediment grain size, and other parameters. Orientation studies, in the form of paleocurrent analysis.
5. Facies studies: recognition of the various discrete lithologic components of a fluvial deposit, with their associated sedimentary structures, and investigation of the vertical (cyclic) and lateral relationships between them. Generalizations in the form of facies models.
6. Paleohydraulics: attempts to construct quantitative descriptions of vanished rivers from evidence contained in their deposits. This area of study is the most recent of the six, and draws on all the other facets of the science of fluviology listed above.

In succeeding sections, an attempt will be made to trace the most significant developments in each of these branches of study down to the present day.

2.3.2 Descriptive Fluvial Geomorphology

Rivers and drainage networks have been classified in four principal ways by geologists and geomorphologists:

1. The cycle-of-erosion classification (youth, maturity, old age) of Davis (1898b, 1899).
2. The structural-control classification (consequent, subsequent, obsequent, antecedent, superimposed, etc.) developed by Powell (1875) and Davis (1898b, 1899; see also Johnson 1932).
3. The morphological classification (braided, meandering, anastomosed, straight) which emerged slowly, mainly during this century.

4. A process-response classification that relates channel morphology to discharge and sediment type.

The first two classifications were the earliest, reflecting the preoccupation of nineteenth-century geologists with the cycle of erosion, landscape evolution, and the power of rivers to remove and transport large quantitites of debris (Dana 1850a,b; Davis 1899).

The morphological study and classification of rivers do not appear to have started with the work of any single individual. The terms "braided", "meandering", and "anastomosing", though poorly defined, were all in use in the nineteenth century (e.g., see Davis 1898b, Chap. IX). Knowledge of the geomorphology of alluvial fans and meandering rivers was well advanced by the turn of the century, as quotes from Drew (1873), Geikie (1882), and others, earlier in this review, have indicated. The work of Gilbert (1880) on this subject has also been much quoted. The concept of the graded stream was beginning to be understood, and Powell (1875) introduced the concept of "base level" in downward erosion.

The Mississippi River has always loomed large in the minds of alluvial sedimentologists and geomorphologists. Nowhere is this more obvious, in contemporary literature, than in the book by Chamberlin and Salisbury (1909), in which detailed maps of the meanders, cutoffs, and bayous copied from charts of the US Geological Survey and the Mississippi River Commission are provided. Figures 2.7 and 2.8, taken from this book, show that the surface processes of meander migration in a large, highly sinuous stream, were well understood by this time.

Davis (1898b) was probably the first to recognize clearly that there is another distinctive type of channel pattern, which he called braided. He illustrated this type using the Platte River. Chamberlin and Salisbury (1909) used the same river as an example of what they termed an anastomosing stream. The term had earlier been used by Jackson (1834) and Peale (1879). The early recognition of the distinctiveness of the Platte, and its more recent prominence in sedimentological studies as a result of N.D. Smith's (1970, 1971) work, undoubtedly qualifies the Platte as the type example of a braided river. Davis (1898b, p. 243) stated:

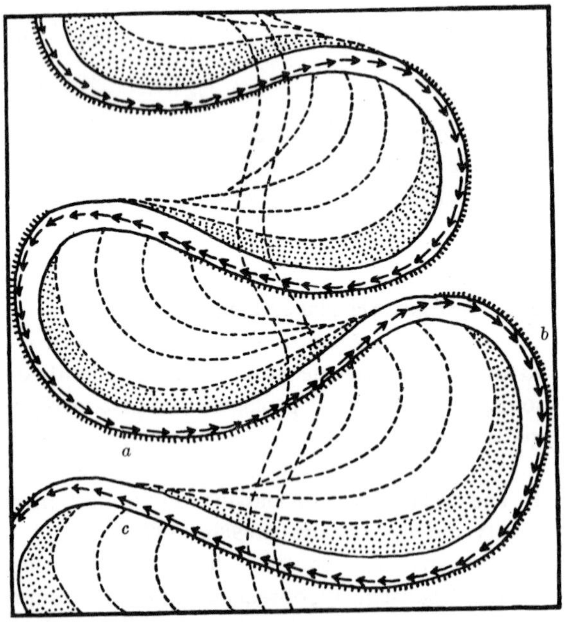

Fig. 2.7. Development of meanders in a high-sinuosity river, showing location of cut-bank erosion (*ticks on outside of curves*) and point bars (*stipple*); based on knowledge of the Mississippi River. (Chamberlin and Salisbury 1909)

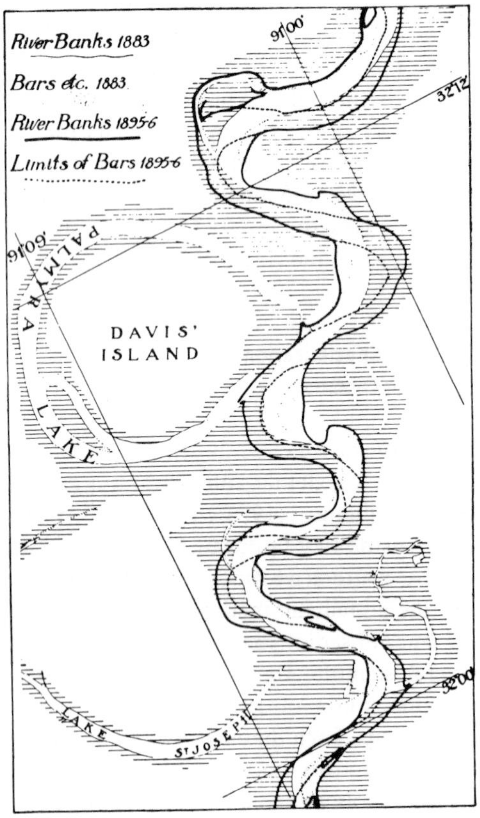

Fig. 2.8. "Meanders and cut-offs (ox-bow lakes) in the Mississippi Valley a little below Vicksburg. The figure also shows the migration of meanders downstream and their tendency to increase themselves". (Chamberlin and Salisbury 1909)

"If a stream has a large load of coarse rock waste, its graded flood plain must be relatively steep (a descent of from 5 to 20 ft or more a mile). In this case the stream does not turn far aside from a direct course along the flood plain; but is constantly embarassed by the formation of bars and islands of gravel and sand, splitting its current into many shifting channels."

Chamberlin and Salisbury (1909) stated:

"In general, anything which greatly increases the load of a stream near its head is likely to cause deposition, and so the development of a flood plain, at some point farther down the valley. Streams which are actively aggrading their valleys are likely to anastomose. This results from the filling of the channels until they are too small to accomodate all the water. The latter then break out of the channel at few or many points. The new channels thus established suffer the same fate."

Nowadays it is recommended that the term anastomosing be restricted to rivers characterized by a network of stable channels of low to high sinuosity, and that the term braided should be used for the type of river described by Davis (1898b) and Chamberlin and Salisbury (1909), which show unstable, low-sinuosity channel patterns, a wide, shallow cross section, and a large bedload (Miall 1977).

The fourth approach to river classification is the one of most interest to fluvial sedimentologists. Its development overlaps with that of the morphological classification. The link between channel cross section and sediment type was first suggested by Griffith (1927), who observed that natural streams with heavy bedload tend to flow in broad, shallow channels, and concluded that a river must tend to adopt the cross-section shape which gives it the maximum sediment-carrying capacity. Actual classification began with the work of Melton (1936). This was an important geomorphological contribution, in which aerial photographs were first used extensively to illustrate fluvial landforms, but it was largely ignored by geologists (with the exception of Fisk). The reason may be that, although the paper contains numerous useful observations regarding channels, bars, and floodplains, the classification proposed by Melton is complex and uses terminology that is difficult to interpret. The word "plain" is used to refer to the bar-covered channel (as in braidplain, meander-plain) and, in a different context, its use implies the entire width of the alluvial valley. Melton's primary subdivision is into "single-crest" and "double-crest" streams. Double-crest streams are those which "possess two phases of geological activity. For example, many streams build meanders during high water, and also deposit an alluvial cover on the flood-plain surface during extreme stages". Single-crest streams are those which lack this second depositional stage. Melton distinguished between vertical-accretion and lateral-accretion deposits; he discussed the difference between braided and meandering rivers and described some of the characteristics of intermediate morphological types; but all this information was assigned a secondary importance in his classification scheme.

An important advance came with the extensive series of flume experiments carried out by Friedkin (1945), who investigated the effects of altering discharge, bed materials, and slope on meander patterns and the stability of bars. Friedkin also produced braided channel patterns by increasing the load and slope, and concluded that an important contributary factor in the development of braided patterns was easily erodible banks. This work was much quoted by Fisk (1947) in his description of the Mississippi alluvial deposits.

Leopold and Wolman (1957, 1960) used both field and flume experiments to study bar and channel behavior. They demonstrated that channel patterns form a continuum resulting from variations in discharge, load, and slope, and showed how channel patterns change downstream in response to variations in these controls.

Schumm (1963) divided rivers into three types: bedload-, suspended-load-, and mixed-load-streams, and showed how load characteristics are related to sinuosity and channel cross-section shape, and to the tendencies toward vertical or lateral erosion and sedimentation within the channel. Several workers, notably Galloway (1981), later used this classification as a basis for a basic subdivision of fluvial styles (Sect. 2.4.4).

Galay et al. (1973) and Mollard (1973) carried these studies further, showing, in particular, the wide range of variability that can exist in channel plan and cross-section shapes and the sediment characteristics associated with them. Research is needed to relate these features in more detail to the sedimentary record, there being a particular need for quantitative studies. Rust (1978a) developed a new braiding index and proposed coupling this with a measure of sinuosity in an inclusive channel classification.

The implications of much of this recent work for studies of ancient fluvial sediments have yet to be explored in detail. (See Sect. 2.3.6 for additional discussion of this area, where its relevance for facies analysis is explored.)

2.3.3 Quantitative Fluvial Geomorphology

An understanding of the quantitative aspects of fluvial geomorphology developed simultaneously with the work carried out on descriptive fluvial geomorphology that was discussed in the previous section. The reason for separating these two aspects of the work is that they have received very different degrees of attention by clastic sedimentologists. Those studying ancient fluvial sediments have generally made some attempts, with varying degrees of sophistication, to relate the features of their deposits to the morphology and behavior of modern rivers, but quantification of these relationships has been attempted only recently.

In this section, the development of some of the principal ideas concerning quantitative aspects of river behavior will be described briefly, emphasizing those aspects most used by geologists.

Early attempts to relate velocity and discharge to channel morphology were carried out by engineers for the purpose of designing irrigation canals in India, Egypt, and arid parts of the United States. Kennedy (1895) published the first important work in this field, relating velocity to flow depth and width in a series of empirical graphs derived from work in India. Contributions by many later workers were summarized by Lane (1935), who showed that empirical equations were not necessarily of universal applicability, depending on a variety of factors that earlier workers, working within the confines of a limited geographic area, had not previously recognized. Lane listed the following factors that may enter into a determination of channel shapes: (a) hydraulic factors (slope, roughness, hydraulic radius or depth, mean velocity, velocity distribution, and temperature); (b) channel shape (width, depth, and side slopes); (c) nature of material transported (size, shape, specific gravity, dispersion, quantity, and bank and subgrade materials); and (d) miscellaneous (alignment, uniformity of flow, and aging).

"A graded stream responds to a change in conditions in accordance with Le Chatelier's general law: – 'if a stress is brought to bear on a system in equilibrium, a reaction occurs, displacing the equilibrium in a direction that tends to absorb the effect of the stress'".

This is the primary conclusion of Mackin (1948) in an important geomorphological paper on the concept of the graded stream – the first to deal with this subject in depth since Gilbert's (1880) pioneer work. Mackin discussed the effects on a river of changes in discharge, load, slope, base level, and other parameters. His discussion is qualitative, but in setting out the logical cause-and-effect nature of river behavior he provided an essential rationale for the detailed empirical work which was in progress at that time on rivers in the United States.

Mackin provided a warning to those engaged in the study of fluvial sediments to the effect:

"(1) that deposits formed by or associated with aggrading streams differ markedly from the loads carried by them; (2) that distinguishing between channel and overbank deposits is the first essential step in interpreting modern valley fills or ancient fluviatile sediments; and (3) that even after this distinction is made it is virtually impossible to work directly from the grade sizes represented in the channel deposits to the characteristics of the depositing streams because there is no simple relationship between the deposits of an aggrading stream and such partly independent factors as slope, discharge, channel characteristics, velocity, and load. We cannot proceed directly from laboratory-determined laws relating to stream transportation processes to interpretations of ancient stream deposits. An alternative and promising route of attack on the problem is via study of deposits now being formed by natural streams of many types to determine whether the sum total of all characteristics of given deposits is uniquely related to the particular modern streams by which they are being formed, and to proceed thence to an understanding of ancient streams by comparison of their deposits with deposits of modern streams of known character."

In this statement, Mackin succinctly outlined the problems we are still tackling in the area of paleohydraulics, and foreshadowed what was to become the most powerful tool in the hands of sedimentologists: facies analysis of modern and ancient sediments.

Lane's work was continued by Leopold and Maddock (1953), who drew on "the mass of data on streamflow collected over a period of seventy years representing rivers all over the United States". This included concurrent measurements of mean velocity, width, shape and area of cross section, discharge, and suspended-sediment concentrations. They provided a series of graphs showing the relationships between these parameters for specific rivers, and arrived at some general relationships that theoretically could be applied to a variety of settings by adjusting the values of various exponents and coefficients. This work was continued by the same group of workers for some years (Leopold and Wolman 1957; Leopold et al. 1964), and additional data on the suspended-sediment load, the relationship of meander wavelength to channel width and radius of curvature, flood magnitude and recurrence interval, and the relationship between discharge and drainage area were published. A later paper (Langbein and Leopold 1966) showed that:

"... meanders are the result of erosion-deposition processes tending toward the most stable form in which the variability of certain essential properties is minimized. This minimization involves the adjustment of the planimetric geometry and the hydraulic factors of depth, velocity, and local slope. The planimetric geometry of a meander is that of a random walk whose most frequent form minimizes the sum of the squares of the changes in direction in each successive unit length. The direction angles are then sine functions of channel distance."

This theoretical conclusion has been used in models of river meanders and in attempts to reconstruct the sinuosity of paleostreams from the record of preserved bedforms.

From the geological viewpoint, quantitative geomorphology has been brought to a relatively advanced state by the work of Schumm, in a variety of papers culminating in a review article (Schumm 1972b). Schumm provided a family of empirical equations relating various hydraulic parameters to features that could be studied and measured in ancient fluvial deposits. Leeder (1973) provided a useful supplement to this, dealing with channel dimensions, and Ethridge and Schumm (1978) provided a review of the methodology. There remains the problem, mentioned earlier, regarding the applicability of these relationships to rivers flowing under different climatic regimes. Even in that classic river, the Mississippi, the concept of the graded river and all its quantitative ramifications were not fully appreciated for many years, as shown by the history of the many man-made neck cutoffs made in the river since 1931, the success of which appears to have ranged from the dubious to the disastrous (Winkley 1977).

2.3.4 Sediment Transport and Textural Studies

The physics of particle movement is of importance to fluvial sedimentologists for a variety of reasons: (1) for an understanding of the gross variation in grain-size and other textural parameters in fluvial deposits; (2) as a basis to the understanding of bedform generation and flow regime concepts; and (3) in providing an explanation for density sorting of clastic grains, of importance in petrological studies, particularly those dealing with economic placer deposits.

Early studies were concerned with empirical measures of the current velocities required to move material of different sizes and with the total volume of sediment transported under various flow conditions. Grabau (1913a) summarized earlier work on this topic, which included the use of hazelnuts, walnuts, peas, pigeons' eggs, and hens' eggs as comparative grain-size standards. However, this picturesque approach to quantification was quickly lost by later workers. Grabau (1913a) also summarized early work dealing with particle sorting and roundness, it being understood that the latter depended on the size, specific gravity, and hardness of the grains, their distance of transport, and the agents by which they had been transported. Sorby (1908), in a classic paper on quantification in geological studies, reported on his investigations into particle settling velocities, and angle of repose. He was the first to attempt to use experimental laboratory studies in investigations of geological problems (see also Sect. 2.3.5), but he appears to have been been ignorant of the many advances made by this time in fluid hydraulics.

An important, widely quoted paper by Gilbert (1914) reported an extensive series of flume investigations into the transport of sediment by streams, the stimulus for which arose from "an investigation of problems occasioned by the overloading of certain California rivers with waste from hydraulic mines". Gilbert was concerned mainly with investigating stream capacity. He examined modes of transport and concluded:

"Some particles of the bed load slide; many roll; the multitide make short skips or leaps, the process being called saltation. Saltation grades into suspension."

Gilbert varied the grade and quantity of debris, width, and discharge, and measured the "slope to which the stream automatically adjusted its bed so as to enable the current to transport the load". He also investigated the velocity required to move particles of different sizes and concluded that competence varied between the 3.2 and 4.0 power of mean velocity, depending on which experimental conditions, discharge, slope, and depth, were varied. Gilbert's (1914) paper was an important bridge between the disciplines in that he brought many of the techniques of fluid hydraulics to bear on a subject of considerable geological significance.

Several theoretical developments followed during the 1920s, including the boundary layer theory, developed mainly by aeronautical engineers, recognition of the importance of bed roughness, and the suggestion by Jeffreys (1929) that particle movement depends not on velocity or drag force, but upon lift induced by the velocity gradient or the rate of shear between adjacent fluid laminae. Turbulent flow in open channels, pipes, and in air was investigated, and the change from laminar to turbulent flow with increased velocity was documented with the use of the Reynolds number (Reynold's classic work in this field dates back to 1883). In 1936, Shields provided the first satisfactory theory regarding the initiation of bedload movement on a flat bed, in which the bottom shear stress or shear velocity was shown to be the critical parameter. Hjulström (1935) used average velocity to predict the critical conditions for the beginning of sediment movement. This approach has been used recently in paleohydraulic reconstructions, but is less accurate. Rubey (1938) reexamined Gilbert's (1914) flume data in the light of these theoretical developments, although he does not seem to have been aware of Shield's (1936) contributions. Rubey was able to explain reasonably well the movement of material of medium sand to pebble grade, but stated that "a satisfactory theory of the force required to start movement of fine sand and silt must await additional observations".

Theoretical developments on the calculation of total load are incomplete. Rouse (1939), and later Vanoni (1946), used the concept of momentum diffusion to predict concentrations of suspended sediment. The earliest successful work on bedload tranportation was that of Du Boys (1879), who deduced that the rate of movement of bed load is proportional to the product of the shear stress and the difference between shear stress and the critical shear stress for the initiation of movement. This physical model has been widely used. However, it fails to take into account bed roughness (grain roughness and form roughness – that induced by the presence of bedforms), a problem first tackled by Einstein (1950).

Developments in fluvial hydraulics up to the mid-1950s were summarized by Leliavsky (1955). Geologists made little use of this field until Sundborg's (1956) major paper on the River Klarälven was published. In this publication, Sundborg applied available theory regarding flow dynamics, erosion, and entrainment of sediment to a study of the bedload and suspended load of a specific river. The geomorphology of the river, and the texture and structure of its deposits were also described in great detail. The paper thus represented a major bridge between the disciplines of hydraulics, geomorphology, and geology, and it was widely quoted during the 1960s and 1970s.

This brief summary has attempted to cover most of the more fundamental developments in the theory of sediment transport. A more complete discussion is beyond the scope of this chapter, and the reader is referred to Bush and to Briggs and Middleton (in Middleton 1965) for further historical details. Recent reviews of the topic have been given by Church and Gilbert (1975), Middleton and Southard (1977), Shen (1978), and Fielding (1993b).

Subaerial debris flows contribute much coarse, poorly sorted sediment to alluvial fan surfaces, as first documented in detail by Blackwelder (1928). He referred to these catastrophic events as mudflows, but data compiled by Bull (1964) and Lustig (1965) showed that their actual mud content may be less than 10%. Observations on modern floods by Chawner (1935) and Sharp and Nobles (1953) demonstrated the transporting power of debris flows and the conditions under which such events occur, notably infrequent but torrential rainfall on unvegetated upland regions, where sufficient time has been allowed for the accumulation of a mass of loose debris. More recent studies of the mechanics of debris flows have been published by Middleton and Hampton (1969), Statham (1976), and Rodine and Johnson (1976). Other special processes operating on alluvial fans, including the tendency of flows to lose competence by infiltration, leaving "sieve deposits", were described by Hooke (1967).

Textural studies, that is, investigations of grain-size parameters (such as mean, sorting, skewness, and kurtosis) and other grain descriptions, including roundness, shape, and sphericity, have long fascinated clastic sedimentologists. Since the work of Udden (1914), the hope has been that simple laboratory measurements on a suite of samples (better still, a single sample) would reveal their depositional environment, and the literature on grain-size analysis is now vast. Modern work on the subject began with the grain-size scale for pebble- to clay-sized material erected by Wentworth (1922; based on Udden 1914) and the investigation of the logarithmic phi scale for grain-size description by Krumbein (1934). Attempts to use sorting parameters as indicators of mode of deposition began with the work of Inman (1949, 1952), and Folk and Ward (1957) proposed methods for describing the sorting of samples of clastic rocks using percentile values of various grain-size intervals derived from the graphs of the cumulative grain-size distributions. These methods were reviewed by Folk (1966). Many attempts have been made to define the sorting characteristics of sand deposits formed in various environments using graphic (Passega 1957; Friedman 1961, 1967; Spencer 1963; Visher 1965a, 1969) and statistical (Sahu 1964;

Klovan 1966) methods. None of these methods appears to be completely satisfactory for discriminating the various depositional environments, although the method of factor analysis (Klovan 1966) and the system of graphic analysis proposed by Visher (1965, 1969a) appear to be the most consistent. Most of this work has been strictly empirical in nature.

Attempts to relate grain-size distribution to sediment tranport mechanics were made by Middleton (1976) and Sagoe and Visher (1977). Middleton (1976) showed that the size break between the traction and intermittent suspension subpopulations in a river bed depends on the dominant shear velocity of the flow, a fact of some potential paleohydraulic significance.

Methods for the study of particle roundness, sphericity, and shape (a more comprehensive description than sphericity) were provided by Krumbein (1941) and Sneed and Folk (1958). More recent developments were summarized by Pettijohn et al. (1972). Roundness increases downstream due to abrasion; mean and maximum grain size decrease in the same direction. These parameters are therefore useful indicators of paleoslope and have been used in many studies of fluvial sediments too numerous to mention (see Pettijohn 1962, for an early review). However, their use demands much field measurement and data processing, and, as paleoslope indicators, they are less flexible tools than paleocurrent analysis. An additional problem is that of inherited textural characteristics – sand which has been recycled from earlier deposits may retain size and shape characteristics which obscure those imprinted by the later environment. For these reasons, grain size and shape studies largely went out of favor during the 1980s, and further developments are not dealt with in detail in this book.

2.3.5 Bedforms and Paleocurrents

It was realized early in sedimentological studies that the morphology of individual hydrodynamic sedimentary structures, and their assemblage and relative arrangement, might yield clues as to the depositional environment under which they were formed, and that their scale was somehow related to the energy level during deposition. Some of the conclusions of Lyell, Sorby, and others along these lines have been quoted above. Other early work included that of Spurr (1894a,b) who discussed the strength and constancy (directional consistency) of currents required to produce ripple marks, and Darwin (1883) who demonstrated the existence of separation eddies in the lee of ripple crests (Fig. 2.5) using a rotating flume. Cornish (1899) observed the formation of antidunes in natural systems and Hunt (1882) studied the growth of ripples on tidal flats. Hobbs (1906) observed cross-bedding with a consistent dip orientation and used this as part of his evidence for interpreting the enclosing strata as "torrential" in origin (see Sect. 2.3.6).

Sorby carried out experimental work on the "effects of current on sand". He explained (in his major 1908 paper, published shortly after his death):

"Fifty-nine years ago, when I was living at Woodbourne, a country-house on the east side of Sheffield, there was at the bottom of the small park a brook entirely under my control. In order to investigate a number of questions, I constructed a place for experiment with some self-registering appliances. I could easily regulate and measure the depth and velocity of the current within certain limits."

The work carried out here and elsewhere, plus careful observations of sedimentary structures in ancient rocks, led Sorby to divide sandstone into four types:

"1. Thinly or thickly bedded rock, without ripples or drift bedding, and showing little or no graining of the surface in the line of the current ... This could be explained by supposing that the water was at considerable depth, and the material mainly deposited from above, not drifted along the bottom where the velocity of the current was much less than 6 inches per second.
2. Thinly bedded rock, with well marked graining in the surface in the line of the current, indicating a mean velocity up to about 6 inches per second, but showing few or no ripple marks.
3. More or less thick masses of rock almost entirely made up of ripple drift. This must have been when the velocity of the current was something like a foot per second ...
4. What I have called drift bedding in numerous published papers ... The velocity of the current is indicated by the nature of the sand; and probably further experiments would enable us to learn the approximate depth, which was probably small, since an increase of a very few feet made so great a difference in the strength of the current."

As pointed out by Allen (1963b), sedimentary structures in types 2 to 4 would now be termed primary current lineation, small-scale cross-lamination, and large-scale cross-stratification, respectively. The order in which Sorby described the sandstone types was one of increasing flow velocity – which was the beginning of the flow regime idea, although, as Allen (1963b) stated, Sorby was mistaken in that type 2 is now known to form at higher flow velocities (or shallower depths) than any of the other three.

The term dune was in use by Swiss workers before the turn of the century, and had been proposed for water-formed bedforms on the basis of their similarity to eolian dunes. The term was introduced to American workers by Gilbert (1914) in preference to the term sand wave, also in current use (e.g., Hider 1882), and this started a nomenclature confusion that is only now being resolved (Ashley 1990). The term antidune was first coined by Gilbert (1914), and the sequence dune, smooth bed, antidune, with increased flow velocity, was documented by Owens (1908).

Gilbert's detailed flume experiments confirmed the results of Sorby's pioneer work, and also reproduced the bedform sequence described by Owens (1908). The mechanism of grain and water movement was observed in detail. It was recognized that dune spacing depends on depth and velocity (Gilbert quoted earlier workers who had made similar observations), and the significance of grain size was realized (although not explored in detail). Similar experimental work was carried out by Hahmann (1912) but has not been widely quoted by English-speaking workers.

Several detailed descriptive papers on cross-bedding structures appeared at about this time, most of which attempted to develop criteria by which such structures could be used as environmental indicators (Grabau 1907; King 1916; Kindle 1917; Bucher 1919). Epry (1913) stated of ripple marks that "no one seems until now, to have definitely ascertained their cause", and "ripple marks are the work of the tide and of that alone", but he obviously had not been keeping up with the literature. Bucher provided data concerning wavelength, amplitude, "horizontal form index" ("the degree of asymmetry may be expressed by the ratio of the horizontal length of the lee-side to that of the stoss-side") and "vertical form index" (wavelength/amplitude). He proposed the terms rhomboid and linguoid for the principal three-dimensional ripple forms, and illustrated their morphology and the flow lines passing over them (Fig. 2.9). Oscillation and current ripples were clearly distinguished on the basis of morphology, origin, and environmental significance.

Bucher emphsized the usefulness of ripples as current-direction indicators:

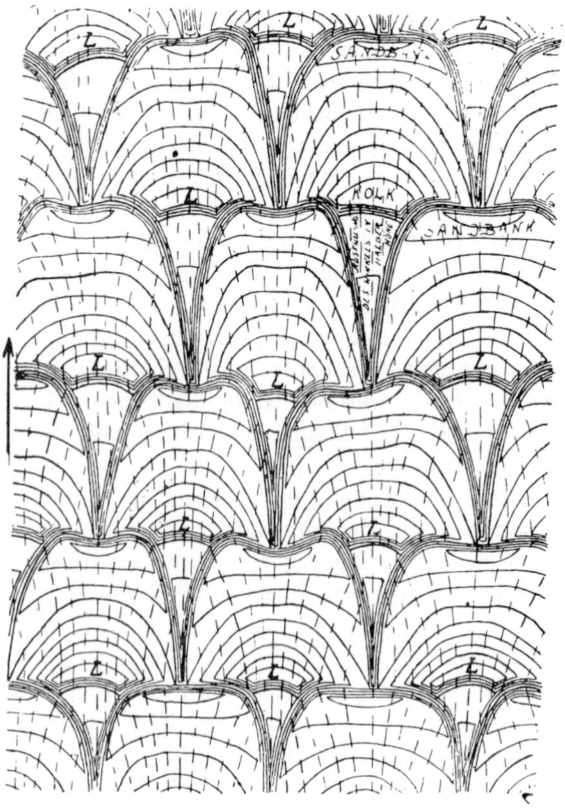

Fig. 2.9. Contour map representing the ideal form of linguoid current-ripples. L Lowest point; *dotted lines* lines of flow of water. (From Bucher 1919; copied from earlier work by Blasius)

"While in marine sediments the trend of ripples seems of little value to the paleogeographer, its study may yield important results in fluviatile deposits, the true nature and wide distribution of which among the sediments of the past we are just beginning to realize ... In the deposits of a river obviously a great majority of all ripples should be found facing approximately in the direction of flow of the river. Locally, of course, they can be found facing even an opposite direction, as along parts of meanders or under the influence of local eddies. Such cases can not, however, seriously affect the average direction obtained from numerous determinations."

This was contrasted with the more varied ripple patterns to be expected in marine environments, for which Bucher provided what are probaby the first published current rose diagrams. Bucher discussed the initiation of ripples on a flat bed, a process that puzzled experimenters for much of this century. Darwin (1883) thought that discontinuous water motion was necessary, whereas Gilbert (1914) thought that a preexistent water rhythm was the cause, and Bucher assumed the need for a small obstacle to flow. The process is now understood to be the result of bed defects produced by the turbulent burst-sweep cycle (see review by Leeder 1983).

A useful review of cross-bedding structures was provided by Twenhofel (1932; parts of this section of Twenhofel's book were contributed by Kindle and Bucher). A description of the development of sandbars (probably what are now termed 2-D and 3-D dunes) was provided:

"Sandbars move over bottoms of water as plateau-like areas with steep slopes on the advancing sides. They may range in height from less than an inch to 10 or more feet, most of them being a foot or less. The sand is rolled over the top, which usually undergoes some erosion. Reaching the edge of the plateau, the sands roll down the slope at the front. The direction of the front slope is usually not constant and the inclinations have a similar range in direction ... Each bed thus formed has its cross-lamination in one direction, which may or may not be the same as that of the beds below and above."

Apart from Sorby's work, the first systematic mapping of cross-bedding orientation in fluvial deposits appears to have been that of Rubey and Bass (1925). According to Pettijohn (1962), "Brinkman (1933) was, perhaps, the first of the 'modern' workers to have a clear concept of the objectives of paleocurrent research and to develop methods to fulfill these aims."

"... through measured observations and statistical analysis a simple facies characteristic – cross-bedding – might allow a number of basic, reliable paleogeographic conclusions to be drawn."

Important advances in the statistical treatment of paleocurrent measurements were made by Reiche (1938), including the use of stereonets for plotting poles to cross-bed foresets, and vector methods for data reduction. McKee (1938, 1939) described the sedimentary structures of the Colorado River in the Grand Canyon, and proposed a classification of ripple cross-lamination; Knight (1929) described festoon cross-bedding.

Except for the German work, led by Brinkman (1933), the pace of research in this area was slow, especially compared to the explosion of paleocurrent studies which took place in the 1950s. Before this time, the attention of most sedimentologists was held by petrographic studies, as revealed by the focus of the first edition of Pettijohn's (1949) textbook, and by the title of the first professional sedimentological journal, the *Journal of Sedimentary Petrology*, founded in 1931. As Pettijohn (1962) has pointed out, there was much misunderstanding as to the origins of many sedimentary structures which, in the case of cross-bedding, was not fully resolved until the various structures were reproduced in exhaustive flume experiments, beginning in the 1950s. The work of Fisk (1944, 1947) stimulated a renewed interest in alluvial sedimentation after World War II, particularly amongst oil companies, who increasingly realized the value of studies of modern environments, and the importance of statistics in geological applications also gained general recognition. These developments all came together in the early 1950s to start the surge in knowledge of fluvial processes that has occurred since the first fluvial facies model was published by Bernard et al. in 1962 (see Sect. 2.3.6).

A contribution by McKee and Weir (1953) focused attention on the internal arrangement and contact relationships of hydrodynamic sedimentary structures, and provided a structure classification that became widely used. Their recognition of the importance of bounding surfaces between cross-bed sets was an important step forward that has been built upon in modern architectural classifications of fluvial deposits (see Sect. 2.4.3.1, Chap. 4). Statistical techniques began to be applied to paleocurrent measurements on a systematic basis, including the use of perfected vector methods of calculating mean and dispersion (Curray 1956), moving average methods for clarifying regional trends (Potter 1955; Pelletier 1958), and the use of analysis-of-variance methods to unravel the sources of dispersion in samples of different outcrop scale (Olson and Potter 1954). Tanner (1955) discussed, with a variety of examples based on his own fieldwork, the use of paleocurrents in basin analysis. A much expanded classification of hydrodynamic sedimentary structures was proposed by Allen (1963a), and this remains the most comprehensive treatment of the subject, although other classifications have been proposed since, especially by workers focusing on specific depositional environments. The use of paleocurrent measurements in basin analysis and regional paleogeographic reconstructions was demonstrated in an innovative case study by Pryor (1960) and reviewed in two important state-of-the-art publications by Pettijohn (1962) and Potter and Pettijohn (1963; revised edition, 1977), and the latter authors produced an illustrated manual of sedimentary structures for use in fieldwork, a year later (Pettijohn and Potter 1964). Another useful book on this topic was published by Conybeare and Crook (1968).

A renewed interest in flume work produced some results of fundamental importance in the early

1960s. Various workers experimented with the production of cross-bedding (McKee 1957; Brush 1958; Jopling 1963, 1965), but the most important developments were those of Simons and Richardson (1961), in which the flow-regime concept was established. This work later appeared (with C.F. Nordin as an additional co-author) in a collection of papers on the hydrodynamics of primary sedimentary strucures edited by Middleton (1965). Much of the work of earlier years on fluid mechanics and sediment transport was brought to bear on sedimentological problems in this book. The statement of the flow-regime concept and demonstrations in the field and the flume of the relationship between bedforms and sedimentary structures were the two principal features of the book, and it had a major influence on sedimentologists over the next decade. The development of ideas on the hydraulic interpretation of sedimentary structures up to the mid-1970s was described in detail by Middleton (1977).

Parallel to the North American work, mentioned above, was that of Allen, who reviewed much engineering work on bedforms and carried out his own flume experiments on the flow patterns associated with the various ripple types. Most of this work is contained in his books (Allen 1968, 1970a, 1984). One of the most widely used parts of this work was the diagram showing the control of flow velocity and sediment grain size on the stability of the principal bedforms.

Descriptive papers on bedforms and bars in modern sandy rivers published prior to the First International Symposium on Fluvial Sedimentology (held in Calgary in 1977) include those of Carey and Keller (1957), Harms et al. (1963), Harms and Fahnestock (1965), Coleman (1969), Collinson (1970), N.D. Smith (1970, 1971), Williams (1971), and Jackson (1976a). All provided much useful data regarding morphology and internal structures; orientation measurements were provided by some, and most of these papers have been widely used in interpretations of the ancient record.

Paleocurrent variance in alluvial systems has been discussed by a few workers. Potter and Pettijohn (1977, p. 108–111) summarized the early research in this area. An important concept was introduced by Allen (1966, 1967), who classified sedimentary structures into a hierarchy of sizes and attempted to relate this to the variability shown by paleocurrent measurements. Miall (1974, 1976) developed these ideas further (Fig. 2.10). Collinson (1971a) and Schwartz (1978) showed how paleocurrent dispersion varies with discharge. Jackson (1975) showed the time-dependent nature of the elements in the bedform hierarchy, and Allen (1973, 1974a) discussed the problems associated with bedform lag – their slowness to respond to changes in flow conditions and the implications for hydraulic interpretation (the concept of lag had been introduced by Kennedy 1963). Southard (1971) and Harms et al. (1975, 1982) demonstrated the importance of a third parameter, depth, in controlling bedform morphology, and the recognition of this important point has allowed a much greater degree of precision in the prediction of bedform morphology from flow conditions. Harms et al. (1975) also recorded the recognition of a new "bar" or "sand wave" bed phase, generally developed between the ripple and dune phase. Such features had been first recognized in experiments by French workers in 1963, and were subsequently recorded in the natural state in intertidal environments (see Middleton 1977).

Pebbly rivers were studied by Johansson (1963), Williams and Rust (1969), Rust (1975), N.D. Smith (1974), and Gustavson (1974, 1978). Measurement and interpretation of clast imbrication have been an important part of this work. Hein and Walker (1977) proposed a model for the evolution of bars in gravel rivers, and Shaw and Kellerhals (1977) and Koster (1978) discussed the interpretation of antidunes in fluvial gravels.

2.3.6 Fluvial Facies Models

2.3.6.1 From Hobbs to Fisk

In 1917, Grabau wrote:

"From the very first, students of sediments have been in the main, marinists, though when the organic evidence pointed unmistakably to the presence of fresh waters they resorted to a lacustrine variant. Even today the interpretation of sediments as of other than marine or lacustrine origin meets with a good deal of skepticism in some circles, and the fluviatilists are not much in favour, especially when they intrude upon territory hitherto preemptied by the marinists ... Fortunately however, American geologists are the most tolerant of men, and even the most startling idea gets a hearing if they are supported by fact and are the result of logical reasoning."

Some of the early attempts at fluvial interpretation have been quoted earlier in this section. That by Medlicott and Blanford (1879) was well ahead of its time, probably because exposures of the Siwalik sediments are located close to the great rivers of the Indo-Gangetic plain – an inescapable modern ana-

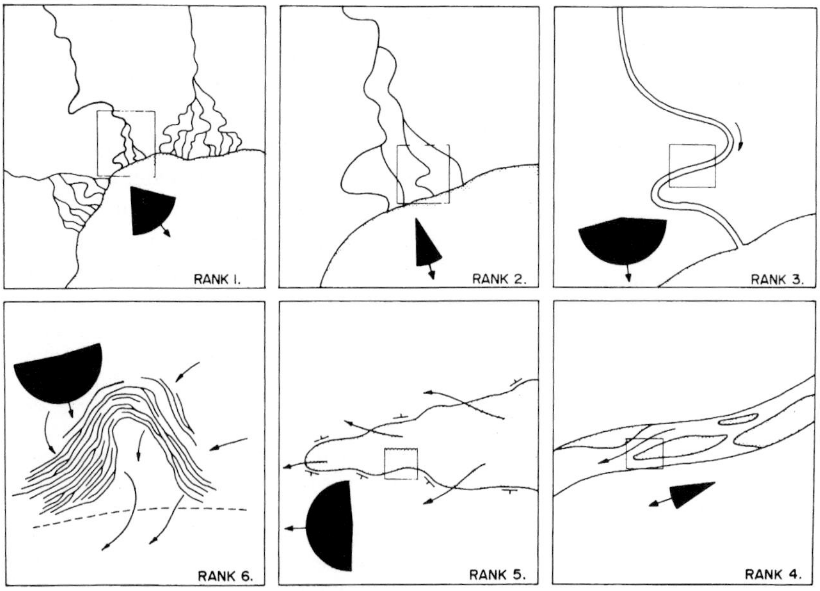

Fig. 2.10. The bedform hierarchy, illustrated for a fluvial-deltaic system. Current rose diagrams illustrate directional variability for the structure of each rank, which must be added to that of the next larger structures to determine the total variability of the structures of each rank. *Small squares* show area enlarged for each lower rank. The structures are: *rank 1* fluvial-deltaic system; *2* channels within a delta; *3* channel reaches; *4* lateral and midchannel bars; *5* cross-bedding flanking a bar; *6* ripple crests on a bar surface, oriented around a sheet run-off gulley. (Miall 1974)

log. A similar spatial proximity led to an interpretation by Hobbs (1906) of various Tertiary sandstone and conglomerate deposits of Spain and Italy as "torrential" in origin, by comparison with modern flood deposits he observed in the process of formation in the same area. Hobbs quoted Italian workers who had reached the same conclusion in 1895, and concluded that such deposits probably are very common in the ancient record. Here are some of Hobb's sedimentological criteria:

"It should be characterized by included lenticular areas of coarser or finer material and by layers of fine material in sharply defined films and plates which reveal its bedding ... Locally, at least, it may show a type of crossbedding ... which, though often observed in ancient sandstone is not adequately explained by the changing currents along a marine shore ... The dominance of ripple marks and the paucity of marine fossils just where marine life should have been most abundant are facts difficult to explain on the theory of marine origin for the great sandstone formations."

Kindle (1911) discussed cross-bedding and fossils as evidence of continental origin, and Grabau (1906) used gross stratigraphic relationships as indicators of environment, suggesting that a "non-marine progressive overlap" (what we would term progradation, with interfingering of marine and nonmarine facies) or an unconformable overlap "away from the sources of supply of the material" (as in an alluvial fan) were good evidence of continental origin.

Virtually nobody was carrying out detailed work in alluvial environments in the early part of the century. A major exception was Trowbridge (1911) who made a careful study of the geomorphology and deposits of some modern alluvial fans in California, "to discuss the causes and processes involved in their deposition" with the objective of deducing "certain criteria whereby materials so deposited may be distinguished from other deposits such as those of lakes and seas, even after cementation has taken place". The criteria so deduced are similar to those of Hobbs (1906), to whom Trowbridge does not refer. Fenneman (1906) used the terms lateral accretion and vertical accretion, in reference to what we now term point-bar and overbank deposits, respectively, in a discussion of the origin of floodplains in modern streams. Melton (1936) made use of this sedimentological concept, but the first geologists to fully exploit it were Happ et al. (1940) and Fisk (1944, 1947), to whom reference is made later.

Barrell's (1912) classic paper on the "criteria for the recognition of ancient delta deposits" provided a detailed discussion of the differences between marine, fluvial and eolian deposits and is, in fact, a sophisticated approach to integrated facies studies following Walther's (1893–1894) teachings, which belies Grabau's pessimistic statement published 5 years later (quoted above). The paper had a profound impact on the development of sedimentology. As regards fluvial sedimentation Barrell stated (in part):

"In rivers where sands are being deposited the channel is subject to meandering. It cuts laterally into the banks and scours down into older floodplain deposits. The sands of the abandoned channels cut across the bedding of the floodplain deposits on the convex sides of the meanders and on the concave side are interlaminated with them."

This last statement is, of course, incorrect. Barrell continued:

"The river works across the floodplain and buries channel structures widely in the fluviatile deposits. The river bars work regularly downstream, being continually cut out above and deposited below ... Gravel tends to be concentrated along such channel bottoms. Lateral discontinuity of the sandstone lenses is also a feature, the ancient channel deposits forming a meshwork."

Barrell was also aware of another style of fluvial sedimentation:

"Effects of sheet-flood deposition: Many aggrading streams overloaded with sand exhibit at low stage shallow braided channels within the main channel. At higher water the main channel may likewise form a braided system, and at highest water the whole floodplain may be covered by a shallow moving water body ... It is such conditions which seem to be required to produce the great depths of regularly bedded and widely extended sandstones which mark certain continental deposits ... They succeed each other without inter-lamination of clays and commonly show neither structure nor fossils. False bedding oblique to the even regular bedding is occasionally observed, but the homogeneous material conceals its frequent presence. Ripple and current marks, however, are rare or absent."

A more recent description of this type of sheet-flood deposition in Bijou Creek, Colorado, by McKee et al. (1967), formed the basis for one of the facies models of (Miall 1977).

Detailed criteria such as those described by Barrell (1912) were used by Grabau (1913b) and Barrell (1913, 1914) in their lengthy descriptions of the Paleozoic deltas of the Appalachians. The alternation of sandstone and shale units in these deposits was noted, and Barrell (1913, p. 458) stated of the alternations in the Catskill Formation:

"A gray or olive sandstone member is commonly sharply delimited at bottom, but at the top grades first into maroon argillaceous sandstone, and this in turn into red sandy shale."

Elsewhere Barrell (1913, p. 466) attributed the alternation of fine and coarse units to "the lateral shifting of distributaries, the red shales representing flood plain areas temporarily removed from the presence of currents". Here, then, is the first inkling of the fining-upward cycle and its origin, although Barrell was not aware of the general significance of this observed sequence. This seems to have first occurred to Dixon (1921, p. 32) in his studies of the South Wales coalfield:

"The relations of each sandstone-band (b) to the marls (c and a) above and below are, typically, as follows:

(c) Red marls; passing down into
(b) Flaggy sandstones, the upper red, the lower green; at the base of the whole band one or more conglomeratic cornstone; eroding
(a) Marl, green immediately below (b) to a depth which varies between a mere skin and several feet, but, in each case, is fairly uniform; often a purple band below; lower still the marl is red and passes down into another sandstone band, which repeats the features of (b).

This sequence is repeated interminably and in all parts of the series."

Other contributions made at this time include a study of the fine banding in Tertiary fluvial sediments in Burma by Stamp (1925), who attributed it to the effects of seasonal floods, and a study of the origins and transportation processes of gravel, including fluvial gravel, by Barrell (1925).

A good summary of the state of knowledge of fluvial sedimentology is included in Twenhofel's (1932) great treatise. Alluvial-fan, valley-flat, and fluviodeltaic environments are distinguished, and the general facies descriptions are much as we would use today, except Twenhofel thought that "channel deposits are entirely ephemeral" and that the distribution of the various lithologies is "extremely erratic".

The next important contributions were those of Happ et al. (1940) and Fisk (1944, 1947). Based on their work in modern rivers, Happ et al. divided fluvial sediments into six types: channel fills, vertical-accretion deposits, floodplain splays, colluvial deposits (hill wash), lateral-accretion deposits, and channel lag deposits. They thought that "the last two types are of little importance under pre-modern conditions". Good descriptions of the facies were given, and four facies associations were proposed: normal floodplain or valley-flat, alluvial-fan, valley-plug, and delta. Their description of the valley-flat association includes the first block diagram of a fluvial facies model (Fig. 2.11) which clearly shows

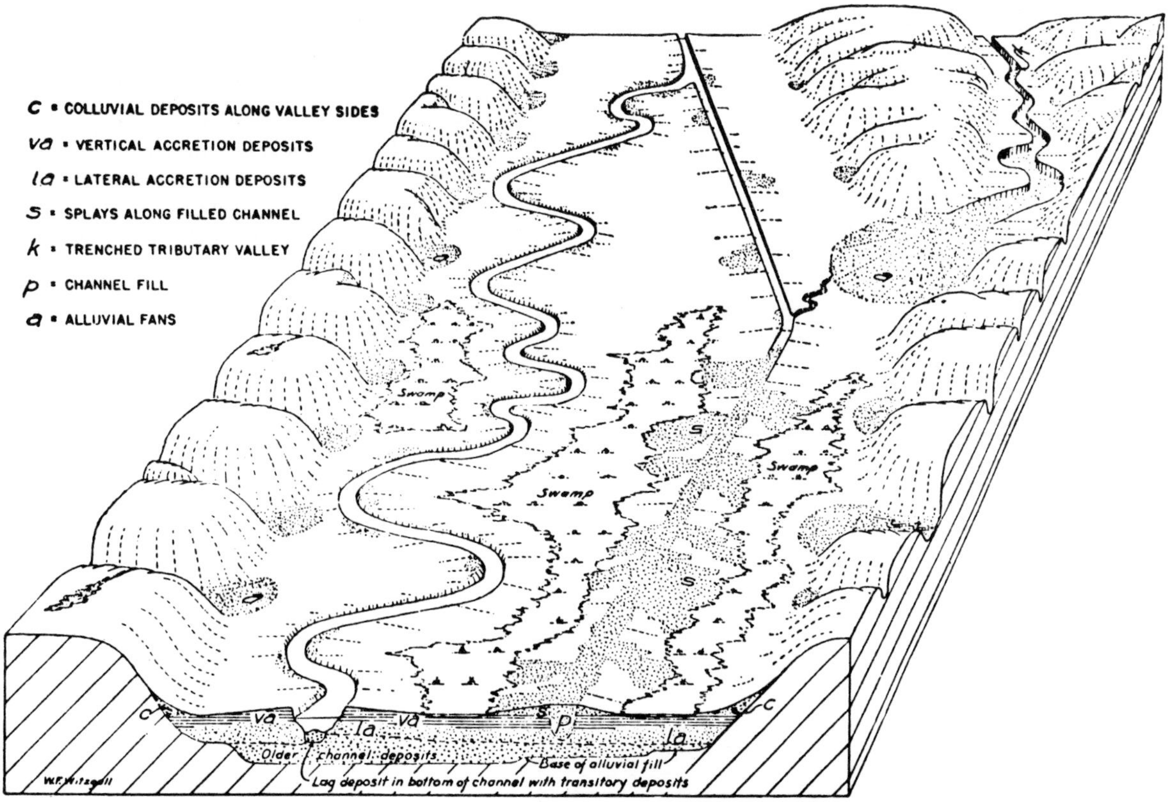

Fig. 2.11. A fluvial block-diagram model: "diagram illustrating typical relations of the various types of deposits in the valley accumulation". (Happ et al. 1940)

the fining-upward nature of meandering stream deposits, but this is nowhere explicitly stated. For example:

"The modern deposits of vertical accretion usually lie upon older deposits of similar origin, which in turn are typically underlain by sandy channel deposits that were accumulated when the channel occupied other positions than at present. These older channel sands may represent lateral accretion on the inside of gradually shifting channel bends, or they may represent filling of former channels abandoned by avulsions. More recent deposits of lateral accretion, in generally crescentic shapes along the inside of stream bends, may still be exposed at the surface but are typically lower than the surface of the adjacent older floodplain which has been covered by deposits of vertical accretion."

Happ et al.'s (1940) work was descended from that of Melton (1936; quoted earlier) and Mackin (1937), who discussed lateral accretion in pebbly rivers and illustrated diagrammatically the development of the now-familiar lateral-accretion surfaces (Fig. 2.12). This paper was incorporated into the geological literature on facies models by Allen (1964) but, apart from this, it has rarely been referred to in the sedimentological literature. This is in marked contrast to Fisk's (1944, 1947) work, which undoubtedly constitutes the most widely quoted source in alluvial sedimentology. Some Russian work (e.g., Shantzer 1951; Botvinkina et al. 1954) rivals that of Fisk in its scope but, being in Russian, it had little influence on the Western "school" of sedimentology (again, except for Allen's work).

Fisk's two classic reports (1944, 1947) contributed a wealth of detail on the Mississippi fluvial and deltaic plain. The postglacial history of the Mississippi was documented in detail, paleochannels were mapped, the sediments were described, and the processes of fluvial sedimentation, including meander migration and cutoff, and the formation of clay plugs, levees, and splay deposits were all discussed in detail. The reports were based on detailed fieldwork, 16000 boreholes, and examination of aerial photos. They remain as one of the most detailed studies of an alluvial system ever undertaken, and this is the most valuable aspect of Fisk's contribution. Most of the main fluvial sedimentary processes and products were fairly well understood in a qualitative way by the 1940s, and Fisk filled out this knowledge with the necessary detailed documentation.

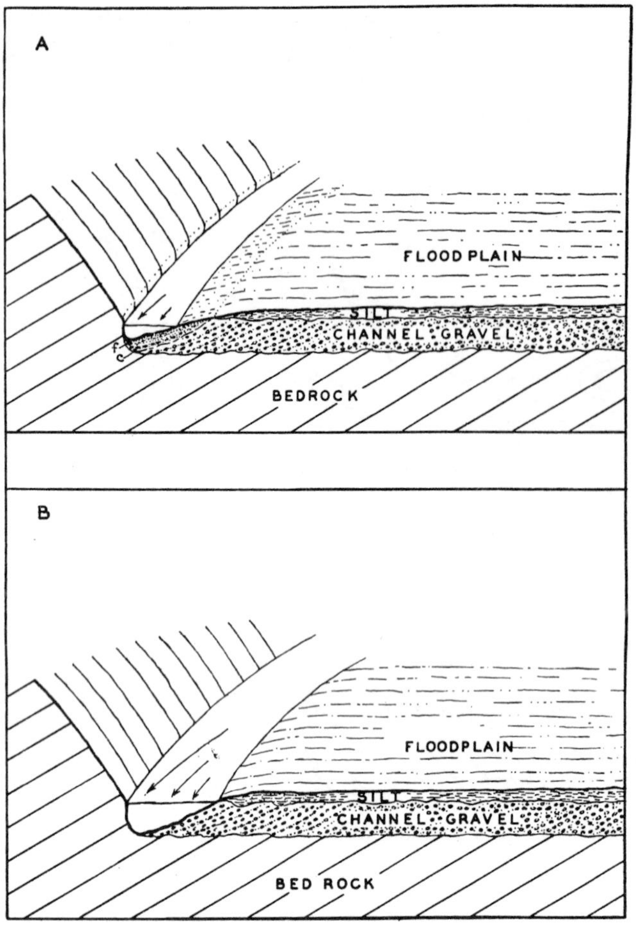

Fig. 2.12. Formation of the Shoshone valley floor. **A** Low water stage *f*, fine detritus, laid down in the stream channel during periods of normal flow; *c* coarse detritus, laid down in the channel at the end of a period of high water. **B** High water stage. (From Mackin 1937). This figure illustrates the principle of lateral accretion, and shows the accumulation of heterogeneous lithologic units in the point bar, preserving large-scale, low-angle cross-bedding surfaces – the structure later termed epsilon cross-bedding by Allen (1963a)

Fisk's main sedimentological contribution was to demonstrate the magnitude and complexity of the deposits of a major river, and to demonstrate the response of that river to fluctuations in base level, in the form of variations in channel morphology and sedimentary facies. This work is still relevant to studies of channel avulsion (Chap. 10) and sequence stratigraphy (Chap. 13). Fisk's contributions to deltaic sedimentology were also of fundamental importance, in that they represented the first major advance in the study of modern deltas since Gilbert's work on lacustrine deltas in the later part of the nineteenth century. In terms of the development of scientific thought, Fisk provided an authoritative demonstration of the value of combining detailed sedimentological analysis with a close examination of sedimentary processes in a modern environment. This was essential groundwork for the development of actualistic facies models, which became a major trend of sedimentological progress in the 1960s, following an innovative conference on the facies models concept in 1958 (Potter 1959).

2.3.6.2 Meandering River Deposits: Development of Modern Facies Model Concepts

A complete description of the characteristics of fining-upward cycles was provided by Bersier (1948) based on his work in the Alpine molasse (Fig. 2.13). The following is a free translation of his description of the typical cycle:

"The most marked characteristic is the base of the sandstone beds, which overlie the shale with a sharp contact, without transition. This is where the grain size is coarsest, with minor gravel and reworked shale pebbles. The hard, sandy bed sits on the surface of the shale with its impressions (sole structures?), borings and tracks. The basal surface is not planar, but undulatory and irregular. The deposit is preceded by an erosion surface in the subjacent shale by which the beds are cut out.

Higher in the sandstone bed the grain size decreases. At the same time fine argillaecous beds appear mixed with the sand. One passes progressively from the sandstone-molasse to argillaceous, immature sandstone, then to fine-sandy shale, with intermediate gradations. The cycle is always terminated at the top by fine-grained units, colored shale and calcareous mud or lithified fresh-water lime-

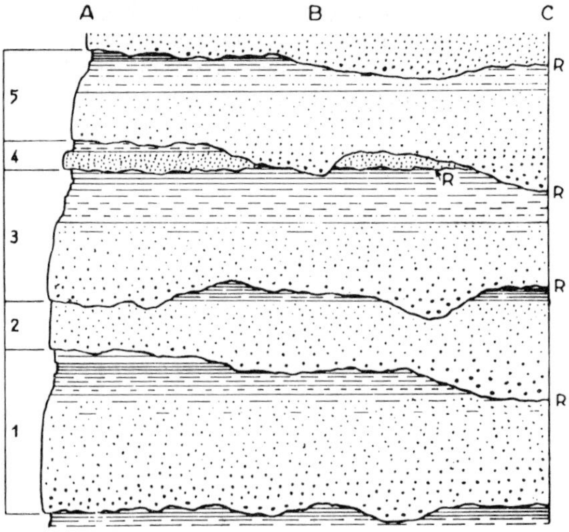

Fig. 2.13. A suite of typical fining-upward fluvial cyclothems in the Alpine molasse. *R* Erosion surface. (Bersier 1948)

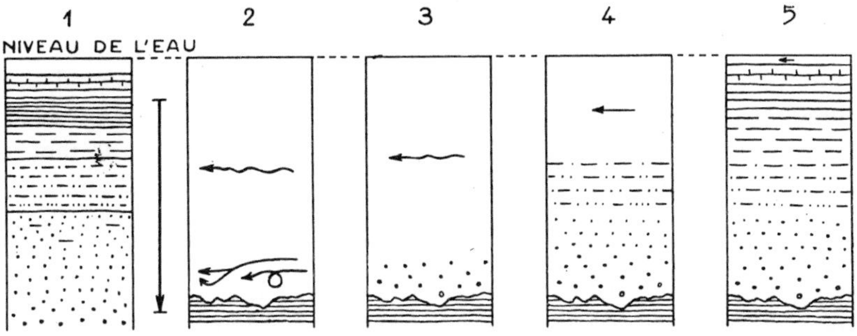

Fig. 2.14. Development of a molasse cyclothem. (Bersier 1948). See text for explanation

stone. A new erosion surface, with a recurrence of coarse material marks the end of one cycle and the beginning of the next."

Bersier thought that these cycles were tectonic in origin, in keeping with prevailing theories of the time that had been developed to explain, among other things, the Carboniferous coal-bearing cycles of the United States. Figure 2.14 is his diagram showing the development of the cycles, and the following is a translation of his figure caption:

"Development of a molasse cycle by subsidence. The length of the horizontal arrows is proportional to the strength of the currents. 1. Completion of underlying cycle; 2. subsidence (vertical arrow) with simultaneous or subsequent erosion by turbulent currents developed in the deeper channel; 3. sedimentation of coarse detritus commences; 4. deposition of fine sand, filling the channel, and slowing of currents by diminution of water depth; 5. deposition of fine mud at top. In beds of Chattian age fresh water limestone and coal."

Later Bersier (1958) provided further details on the molasse cyclothems, including their variability and thickness, and proposed a different explanation of their origin. He ascribed them to vertical aggradation in channels which shift in position on a delta plain, using the modern Mississippi delta as his analog. Allen (1962) proposed a similar mechanism for some Old Red Sandstone (Devonian) cycles in England, suggesting that the upward fining was due to a decrease in slope and current strength as the channel became clogged with sediment. These proposals dispensed with the need for rhythmic subsidence as a cause, but there was still no general comprehension by geologists of the sedimentary mechanism of lateral accretion. This idea first appeared in the geological literature as a result of Wright's (1959) attempt to explain laterally-persistent cross-bedding in fluvial deposits. Wright quoted Van Straaten (1954), who had "described deposition of material on the inside curves of beach gullies", and Dunbar and Rodgers (1957) who "suggested that some wide lenses of cross-stratified sediments may have been deposited as point bars on the inside curves of a meandering stream". Wright does not seem to have been aware of the work on alluvial sediments by Fenneman (1906), Melton (1936), Mackin (1937), or

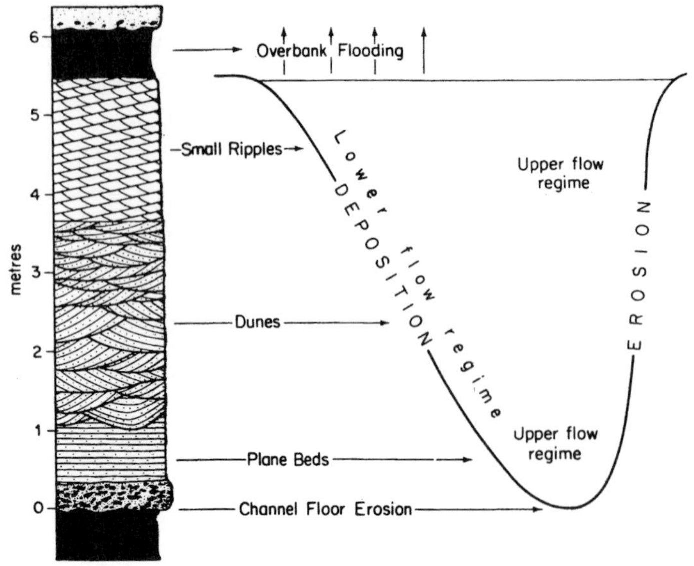

Fig. 2.15. Point-bar model, based on observations in tidal creeks, and applied to the Devonian Old Red Sandstone of Britain (Allen 1963b). Reprinted by permission of Kluwer Academic Publishers

Happ et al. (1940), and he confused point-bar accretion surfaces with cross-bedding. Sundborg (1956) described growth of point bars in the River Klarälven and illustrated what we now recognize to be lateral-accretion surfaces, but his description focuses on texture, and on smaller-scale bedforms.

The lateral-accretion concept, as currently conceived, was first proposed for modern estuarine channel-fill sediments by Oomkens and Terwindt (1960), and later was combined with the fluvial fining-upward cyclothem into a complete process-response facies model by Allen (1963b, 1964), who was working on the Old Red Sandstone of southern Britain (Fig. 2.15). More or less simultaneously, Bernard et al. (1962) and Bernard and Major (1963) were studying Recent and Pleistocene fluvial deposits (Fig. 2.16) as part of a major sedimentological research effort by petroleum geologists in the United States, particularly those of the Shell Development Group, based at Houston. They had identified upward-fining cyclic fluvial deposits in the subsurface some years previously (Nanz 1954) but the origin of such sequences was not discussed in detail at that time (at least, not in print in the public domain).

Allen's (1963b) depositional model is quite different from his earlier (1962) interpretation. He discussed the origins of the various facies present in a fluvial cyclothem and showed that their vertical arrangement is

"... consistent with offlap deposition on the channel bar (?point bar) of a laterally migrating stream. As can be observed on stream channel and point bars today ... both the intensity and speed of flow tend to decline as the bar is ascended from the deepest part of the river bed to the top of the channel."

In a later paper, Allen (1964) documented the lateral-accretion model in detail. This paper was an important bridge between the disciplines in that it drew on an extensive review of geological and geomorphological literature, including European, Russian, and North American work. Elsewhere Allen (1963a) described a type of large-scale, low-angle cross-bedding, termed "epsilon cross-stratification", and stated that:

"Wright's (1959) interesting suggestion that point bar construction generally may explain cross-stratified units where the cross-strata are laid down parallel to the surface of the bar, has not been borne out by a study of the literature on point-bar deposits."

Mackin (1937, Fig. 2) had, in fact, illustrated what Allen termed epsilon cross-bedding based on his work in the Big Horn Basin (Fig. 2.12), as had Sundborg (1956, Fig. 47), but its significance for interpreting the ancient record was missed. But with the new fluvial facies model available, students of ancient rocks rapidly began to find examples of point-bar deposits containing large-scale, low-angle epsilon cross-bedding, thereby providing abundant evidence of lateral accretion as the primary depositional mechanism (Allen 1965b; Moody-Stuart 1966).

Fluvial and deltaic sedimentary processes and their deposits were described in a detailed review paper by Allen (1965a), which summed up the state of knowledge as it existed in the mid-1960s. The influence of this paper has been far-reaching, and next to Fisk's work it became one of the most widely quoted work yet published on fluvial sedimentology. In it, Allen provided, among other things, a series of block diagrams showing various theoretical archi-

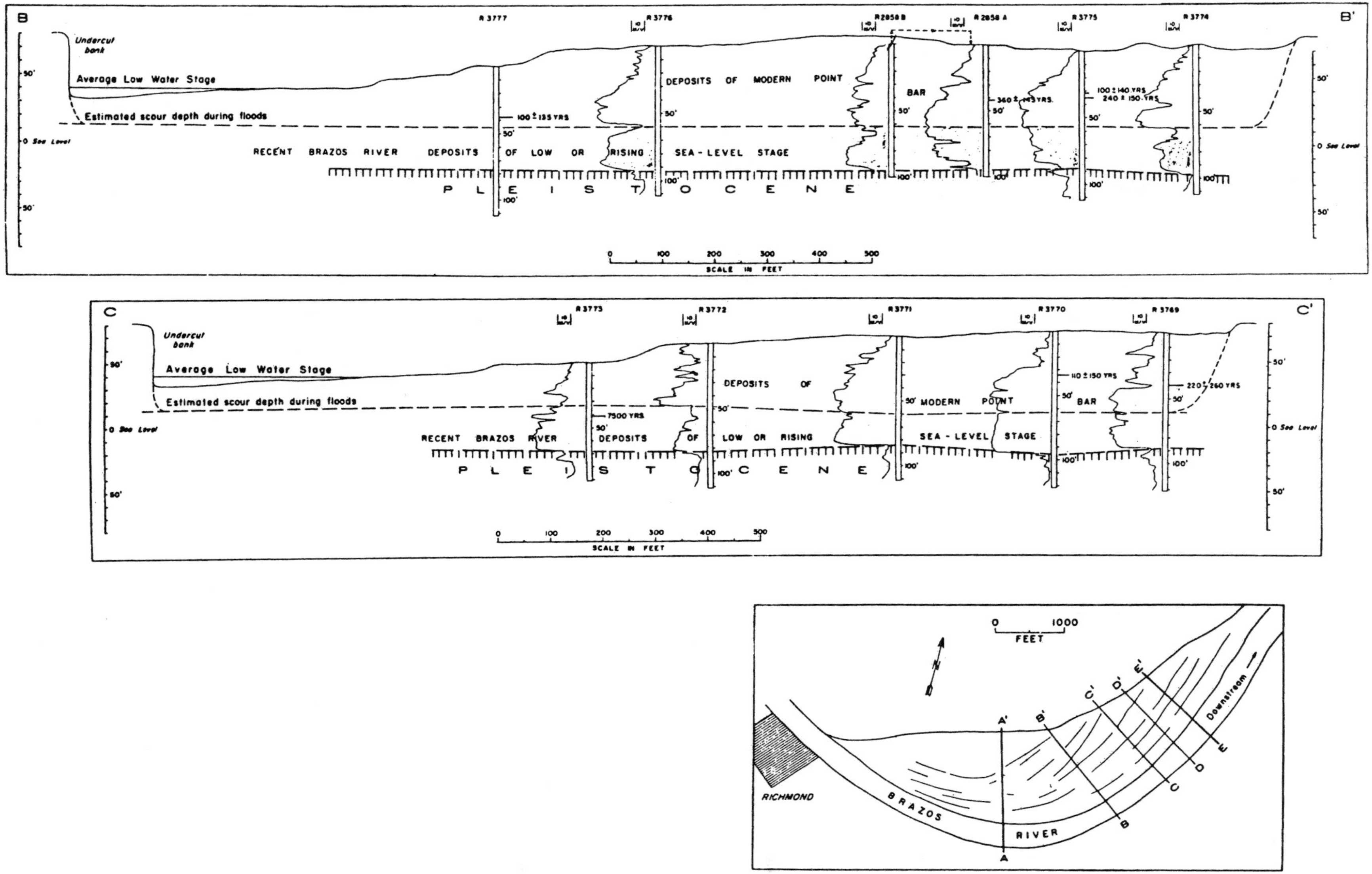

Fig. 2.16. Log responses from borings in Recent point-bar deposits, Brazos River, Texas. (Bernard et al. 1962)

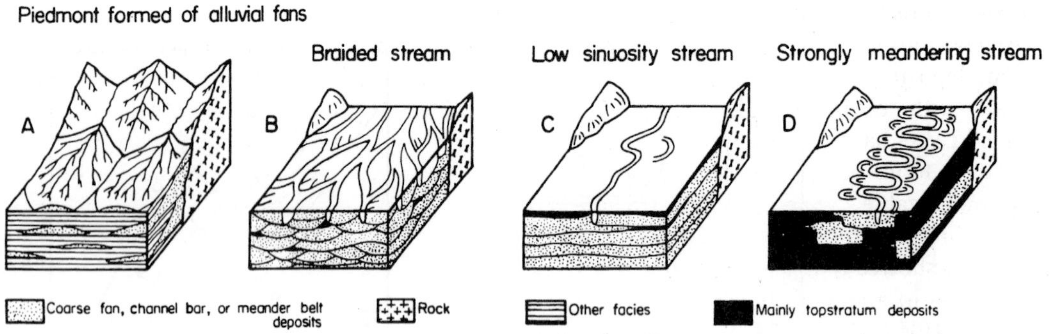

Fig. 2.17A–D. The hypothetical architectural models of Allen (1965a), as redrawn by Friend (1983)

tectural models (Fig. 2.17) which have done much to stimulate the generation of detailed regional paleogeographic reconstructions by later workers.

A useful paper by Beerbower (1964) divided cycles into those of "autocyclic" origin, generated by intrabasinal sedimentary processes, such as channel migration, and those of "allocyclic" or extrabasinal origin, including such processes as tectonism, eustasy, and climate change. Nowadays the terms autogenic and allogenic are preferred, because the processes are not always strictly cyclic in nature.

The importance of Allen's work on the fining-upward cycle was emphasized by the work of Visher (1965a), and it was also seen that this cycle was one of a range of possible stratigraphic variations that could be developed in marine and continental sedimentary environments, so that Visher's (1965b) other, more generalized paper (which actually appeared first), firmly established vertical-profile models as one of the most useful tools available for studying ancient clastic deposits. In subsurface studies it remains a widely used technique, aided by the fact that cyclic character can almost invariably be guessed at from geophysical logs alone, without detailed core or sample studies (Pirson 1970). Problems with the vertical-profile technique, and newer methods of analysis, are discussed in Chap. 9.

The most definitive paper on the fining-upward cycle was that by Allen (1970b), in which statistical studies of a suite of Devonian cycles in Britain and North America were described, and hydraulic interpretations were provided. Little work on the study of modern meandering river analogs was carried out until McGowen and Garner (1970) described some point bars in the southern United States, Bluck (1971) studied the River Endrick in Scotland, and Jackson (1976b) analyzed a reach of the Wabash River (Indiana-Illinois). All of these workers documented the variability that can occur in point-bar deposits and showed that the point-bar cycle may be much more complex than commonly envisaged. Excellent examples of ancient point-bar deposits were described by Puigdefábregas (1973), Puigdefábregas and Van Vliet (1978), Nijman and Puigdefábregas (1978), and Nami and Leeder (1978). Jackson (1978) related depositional characteristics to load and discharge variations, and proposed several new facies associations. These form the basis for the facies-model classification of high-sinuosity rivers adopted by Miall (1985) and elaborated in Chap. 8.

2.3.6.3 Braided Rivers

The distinctiveness of the braided channel pattern was recognized in the nineteenth century, but not until 1962 was an attempt made to relate this pattern to depositional behavior and to recognize a distinctive sedimentary style, different from that of the meandering river. In that year Doeglas (1962) published his analysis of some braided rivers in southern France. Ore (1964) discussed criteria for distinguishing the deposits of meandering and braided streams, and Moody-Stuart (1966) suggested differences in sedimentary facies that might be generated by low-, as distinct from high-, sinuosity rivers (not necessarily the same distinction as that of Ore, as some of Moody-Stuart's criteria may apply to single-channel straight streams).

By the early 1970s the widespread distribution of the braided environment was made apparent by the work of McKee et al. (1967), Coleman (1969), Williams and Rust (1969), and N.D. Smith (1970, 1971), who described, respectively, flash-flood deposits of a small ephemeral stream; the sedimentology of the Brahmaputra – the first work of its kind on one of the world's giant rivers; a gravel-dominated mountain river; and a sandy river of the

American Great Plains. All these papers have now been extensively quoted in studies of ancient braided stream deposits.

By contrast with meandering-river deposits, those of braided rivers have generally been thought of as somewhat random in depositional character, and lacking recognizable cyclicity. Application of the statistical technique of Markov chain analysis to bedding sequences showed that this is not always the case (Miall 1973; Cant and Walker 1976) and, in a review of the braided river environment, Miall (1977) developed four facies models for braided rivers, one of which was strongly cyclic. Generation of these models was aided by the development of a simple lithofacies coding system, which permitted unified description of braided river deposits, including those already described in the literature. This method has been widely adopted, and forms the basis for lithofacies descriptions in this book (Chap. 5).

Cyclic braided deposits were assigned by Miall (1977) to his Donjek model, but subsequent work showed that this name would be best confined to gravel-dominated cyclic deposits, as in the original description of the Donjek sediments (Williams and Rust 1969; Rust 1972). Cant and Walker (1976) and Cant (1978) described a sand-dominated cycle, which has been erected as a fifth model, and termed the South Saskatchewan type (Miall 1978c). Rust (1978b) discussed a slightly different grouping of braided facies associations using his own field examples. Modern work on braided rivers and their deposits is focusing on the details of the internal architecture of their deposits (Sects. 2.4.3, 2.4.4.2, Chap. 8).

2.3.6.4 Alluvial Fans

Because of their distinctive morphology, alluvial fans have always been recognizable as discrete components of any alluvial system, and sedimentologists have attempted to carry this distinction into their interpretations of the ancient record. Several authors have distinguished an alluvial-fan facies from a braided-river facies, which is misleading, because much of the sedimentation that takes place on alluvial fans is brought about by the activity of braided streams. The most distinctive components of alluvial fan environments are debris-flow deposits, although not all the depositional systems termed fans necessarily contain this facies (Miall 1978c; Rust 1978b).

Modern alluvial fans were studied by Tolman (1909), who introduced the term "bajada" for the laterally amalgamated fans that occur at the foot of mountain slopes in many basins. The work of Trowbridge (1911) has been referred to earlier. Lawson (1913) coined the word "fanglomerate", and Blackwelder (1928) first appreciated the importance of mudflows in alluvial-fan deposits. More recent, detailed work on the geology of alluvial-fan deposits by Blissenbach (1954) has been much quoted. Other descriptions were given by Bull (1972) and Boothroyd and Ashley (1975). Ancient fan deposits were described by, amongst others, Nilsen (1968) and Miall (1970a), both of whom used textural parameters and clast composition to map fan outlines and transport directions. Heward (1978a) provided a review of depositional processes on alluvial fans and described some Carboniferous examples in Spain (Heward 1978b). The first of these papers contains an authoritative review of the autogenic and allogenic mechanisms of cycle development in alluvial fans, and has become the standard work on the subject. Boothroyd and Nummedal (1978) documented the depositional facies occurring on humid fans.

Difficulties with the generalized concept of an alluvial-fan depositional model were already apparent by the late 1970s. The work quoted in the preceding paragraphs refers exclusively to gravel-dominated fans, but Friend (1978) described some ancient cone-shaped sandstone units that he compared to the deposits of modern "terminal" fans in arid regions of India (Mukerji 1976). The giant fan of the Kosi River was described by Gole and Chitale (1966), and its deposits have been discussed by Singh et al. (1993). This depositional system, so much greater than the small fans normally conjured up by the term alluvial fan, grades from boulder-dominated in proximal regions to a silt and fine sand system at the toe, 60 km downslope. Later workers became uneasy with the inclusion of fluvioglacial outwash plains in the alluvial-fan category, even though many of them are distributary systems. Important "fan" papers, such as those by Boothroyd and Ashley (1975) and Boothroyd and Nummedal (1978) would be excluded, according to this approach. The current controversy in this area is discussed in Sect. 8.3.

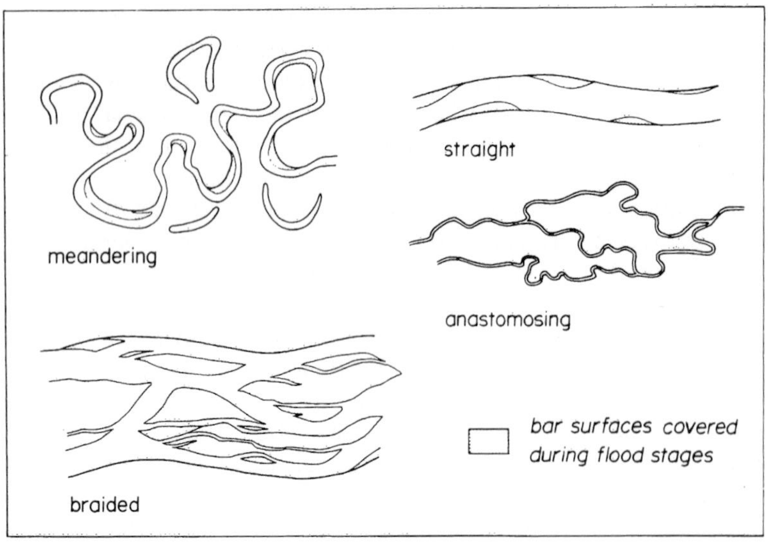

Fig. 2.18. The four basic fluvial channel styles. (Miall 1977)

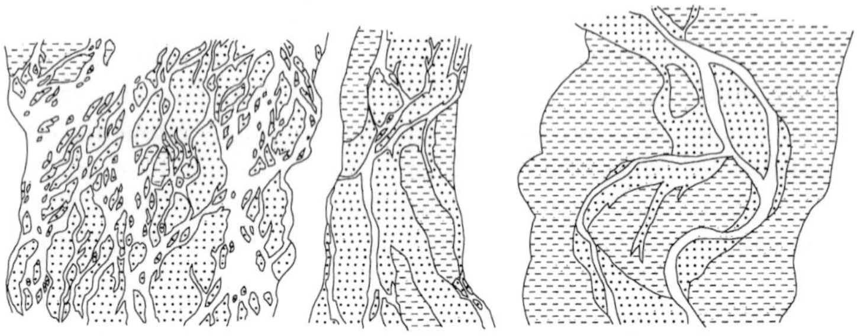

Fig. 2.19. Three typical gravel-bed braided rivers in Yukon Territory, Canada, drawn at the same scale. Note the major differences in fluvial style

2.3.6.5 Other Facies Models

Geomorphological work in the 1970s (e.g., Schumm 1972a,b; Galay et al. 1973; Mollard 1973) emphasized the great variability in fluvial morphology, and this revealed the inadequacy of the simplistic twofold subdivision of fluvial patterns into braided and meandering styles that had become commonplace by that time. Miall (1977) used a fourfold channel classification (braided, meandering, anastomosed, straight; Fig. 2.18), and Rust (1978a) showed how this classification could be based on parameters that are quantifiable, at least in modern rivers, namely sinuosity and a braiding index. However, this classification has never been entirely satisfactory as a basis for facies modeling. The terms braided and meandering refer to different parameters, and it is possible for a river to be defined by both terms. For example, note that the three rivers in Fig. 2.19 are all braided, but differ considerably in their sinuosity. The Slims River is a typical braided river, multichannel, and of low sinuosity. The Donjek, by contrast, has only one or two active channels, and is of moderate to high sinuosity. Point bars are present on some meander bends. In terms of lithofacies characteristics (lithofacies types and vertical profiles), the three rivers would be indistinguishable, yet their fluvial architecture would be clearly quite different.

The sedimentological ramifications of this geomorphological variability are still far from being fully understood. Moody-Stuart (1966) discussed possible depositional products of straight channels, and D.G. Smith (1973) recognized the distinctiveness of what has now become, almost entirely as a result of his research, the standard "anastomosed" fluvial model, although his work did not become widely available until the end of the decade (D.G. Smith and N.D. Smith 1980). D.G. Smith (1973) proposed a quite different definition of "anastomosing" from that used by Miall (1977) and Rust (1978b), whose ideas were based on those of Schumm (1963). Schumm proposed the use of the term for a type of arid fluvial system in Australia, whereas those anastomosed rivers described by D.G. Smith are perennial streams in a humid-temperate climate. The differences in sedimentological style between these two types of rivers have yet to be clearly and definitively evaluated, although some common factors relating to low slope and cohesive banks appear to be emerging (see Chap. 8). The Australian anastomosed rivers are now regarded as something of a special case (Sect. 8.2.10). Baker (1978a) and Taylor and Woodyer (1978) described modern fluvial systems in, respectively, the Amazon basin and New South Wales (Australia) that obey very few of the the "norms" of depositional behavior, as they were understood in the late 1970s. McKee et al. (1967) described the deposits of a flash flood in a small ephemeral stream, Bijou Creek in Colorado, and this work was incorporated into the braided fluvial models of Miall (1977), establishing a category of fluvial depositional styles that is now recognized to be of considerable importance in the ancient record (Sect. 8.2.16). The study of ephemeral streams in Australia by Williams (1971) also played an important part in the development of the ephemeral-stream model (Sect. 8.2.15).

In the 1970s, difficulties were also appearing in the application of geomorphological models to the ancient record. For example, it was shown that under certain circumstances sequences of apparently braided origin could be formed in a single-channel, high-sinuosity fluvial environment (Jackson 1978; Nijman and Puigdefábregas 1978). Collinson (1978) pointed out that not all sandstone units in meandering-stream successions are point-bar deposits, but some are the product of overbank sheet floods.

Jackson (1978) provided a devastating critique of the conventions being used at that time to identify fluvial style. Successions of "braided" and "meandering" origin may look very similar (Fig. 2.20). Indeed, for some years it was thought that the two main fluvial styles were those of the braided and the meandering river, and that these styles could readily be distinguished in vertical profiles. For example, in the first two editions of the well-known *Facies Models* book (Walker and Cant 1979, 1984) these two styles were contrasted, with suggestions given as to how they might be distinguished (Fig. 2.21). Walker and Cant (1979) suggested that the proportions of "vertical accretion" deposits (bar-top and floodplain sediments) would be greater in the meandering style, and the assemblage of sedimentary structures different. If these criteria were correct, they would, indeed, be invaluable for subsurface petroleum geology work, and there would have been little need for further research. However, already by the late 1970s, at the research level, it was realized

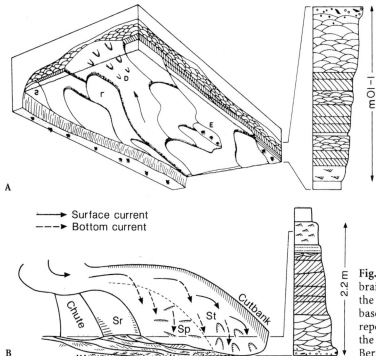

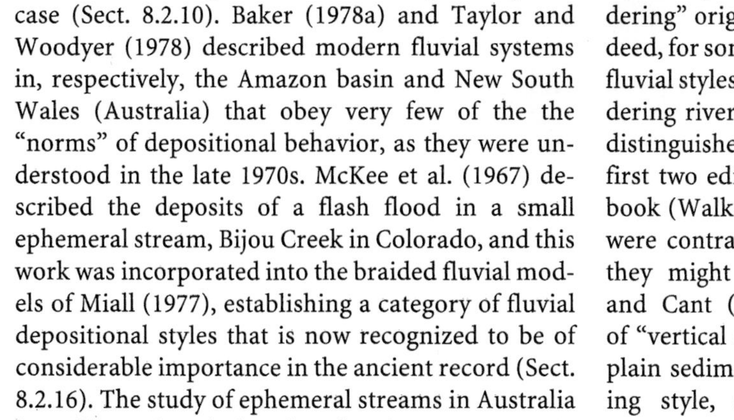

A

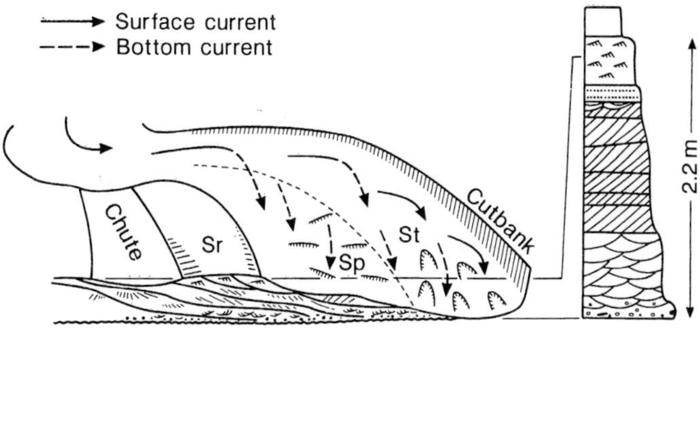

B

Fig. 2.20. Basic facies models for sand-bed braided and meandering rivers, as developed by the late 1970s. That for the braided river (**A**) is based on work in the South Saskatchewan River, reported by Cant and Walker (1978), and that for the meandering river (**B**) is essentially that of Bernard et al. (1962) and Allen (1963a)

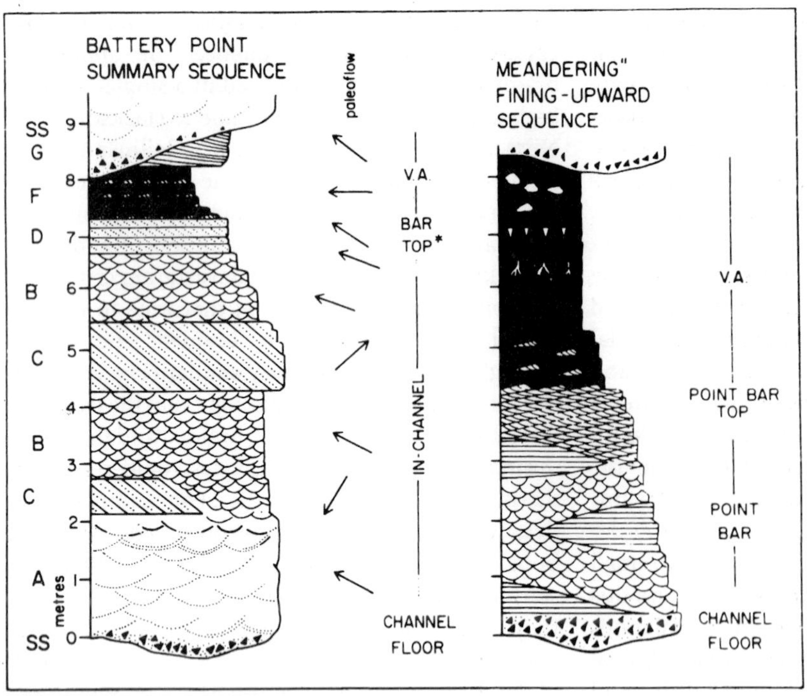

Fig. 2.21. Comparison of model vertical profiles for braided and meandering river deposits. (Walker 1979)

that the braided versus meandering separation was overly simplistic.

2.3.7 Fluvial Architecture

The term "fluvial architecture" was used by J.R.L. Allen (at the First International Symposium on Fluvial Sedimentology in his keynote address) to encompass the geometry and internal arrangement of channel and overbank deposits in a fluvial sequence. Most of the work up to the time of the Calgary symposium had concentrated on vertical profiles, and less attention had been paid to the way in which the various elements in a fluvial sequence stack on one another, although this is obviously of critical importance with regard to reservoir and orebody continuity, and the analysis of present-day and ancient groundwater systems. Allen's (1965a) architectural block diagrams, referred to earlier, were a theoretical approach to this problem, following the preliminary work of Happ et al. (1940). In a later paper, Allen (1974b) emphasized the importance of pedogenic carbonate units as indicators of stability and nonsedimentation in an alluvial plain, and used them in building a variety of hypothetical models showing the variations to be expected in the deposits of a meandering stream system as a result of the interplay of various autogenic and allogenic mechanisms. Again, as with several of Allen's earlier papers, this work contained the seeds of what has evolved into a significant research topic in recent years, that of sequence stratigraphy. As discussed in Chap. 13, one of Allen's models contains many of the elements of a modern sequence model for nonmarine deposits.

In another early and tentative experiment to explore controls on sand-body architecture and the likelihood of exploration holes encountering sand bodies in the subsurface, Leeder (1978) provided a simulation model of subsurface channel distribution. This work was stimulated by the discovery of petroleum in fluvial reservoirs in the North Sea (M.R. Leeder, pers. comm., 1995), and formed the beginning of attempts to develop numerical models of channel architecture, as described in Sect. 10.4.

Knowledge of the three-dimensional (or even two-dimensional) architecture of ancient fluvial deposits was limited at this time. Campbell (1976) provided a detailed cross section through a Jurassic fluvial deposit of possible braided origin, and this became much quoted in later studies because it appeared to demonstrate large-scale lateral shifting of the channels in a fluvial distributary system, as in giant modern fans. Analogies with the Kosi fan have commonly been drawn (e.g., Collinson 1986). Unfortunately, Campbell's reconstruction has now been shown to be seriously flawed (Sect. 10.3.2).

Other work included a study of the geometry of sand units on a delta plain by Ferm and Cavaroc (1968). Horne and Ferm (1976) described the architecture of some Carboniferous coal-bearing rocks in the central Appalachians, based on large highway and railroad exposures. Cant (1978) erected a hypothetical model for the two-dimensional geometry of a braided system of South Saskatchewan type. Nami and Leeder (1978) described the geometry of a well-exposed meandering river deposit. Friend (1978) commented on gross architectural features of several ancient fluvial systems.

Vegetation is an architectural control of considerable importance, affecting channel morphology and shifting behavior through increasing bank stability (D.G. Smith 1976) and, by its presence or absence in headwater catchment areas, playing a large part in determining the nature of the river hydrograph and the sediment supply. The geomorphic effects of vegetation were recognized in the nineteenth century (e.g., Surell 1841, 1870), but Schumm (1968a) was the first to clearly point out its paleoclimatic implications. He suggested that rivers of pre- and post-Devonian age were probably fundamentally different because of the first appearance in the Devonian of an extensive cover of land vegetation. Long (1978) discussed the implications of this difference for studies of the Proterozoic fluvial record, and Cotter (1978) examined the Paleozoic record of the Appalachians.

On all scales ranging from the local to the continental, a knowledge of the geometry and dispersal patterns of alluvial deposits can lead to a better understanding of contemporary tectonics, as several papers published in the 1970s were beginning to demonstrate (e.g., Steel and Aasheim 1978; McLean and Jerzykiewicz 1978; Friend 1978; Potter 1978; Miall 1978d).

2.3.8 Paleohydraulics

The study of fluvial facies in vertical profiles or in two or three dimensions is an essentially qualitative approach to the reconstruction of past river systems, even where statistical techniques are used to provide quantitative information regarding facies assemblages (Sect. 10.4). Paleohydraulics is the study of the quantitative relationships between the hydraulic parameters of a river (depth, width, slope, discharge, sediment type, etc.) and its preserved deposits.

The study of paleohydraulics is fraught with difficulties, and some fluviologists feel that our knowledge of the depositional and preservational processes of rivers is still so inadequate that numerical estimates of past hydraulic parameters verge on the fictional. However, some progress has undoubtedly been made, and continued attempts at paleohydraulic reconstruction can only serve to encourage fluvial sedimentologists to structure their future research into modern rivers so as to ask the right questions, in the hope that some of the difficulties can gradually be overcome.

There are essentially two approaches to paleohydraulics, the engineering approach, which uses theoretical relationships and empirical data regarding sediment transport mechanics and bedform generation to reconstruct the depositional conditions for individual beds or sedimentary structures, and the geomorphological approach, which relates empirical data about the morphology of modern rivers to some of the gross features of their deposits. Each method has its limitations – the engineering approach concerns itself primarily with instantaneous depositional conditions of specific bedding units, the overall, long-term significance of which may be hard to judge because of our lack of knowledge of facies preservability in fluvial environments. And, as Allen (1973, 1974a) has shown, bedforms are slow to react to changes in flow conditions and are therefore inaccurate hydraulic indicators. The geomorphological approach is potentially more useful, in that it deals with long-term, statistical averages, but suffers, at present, from a gross lack of data from all but a limited range of modern climatic, discharge, and sediment-load conditions. Even at the time of writing this book sedimentological concepts regarding rivers are derived from very few, possibly atypical, modern examples, most of which are located in populated or otherwise accessible, relatively temperate climates. The range of different conditions represented by ancient fluvial deposits of all ages has barely been touched, and research which integrates sedimentology with hydraulics is sparse. We are hampered, also, by the influence on modern rivers of the rapid climatic and sea-level changes during and following the Quaternary glaciation, and an uncertainty as to how this distorts our perceptions of rivers at times of greater geological stability. Recent work demonstrating the complexity of the effects of climate and sea-level change is discussed in Chaps. 12 and 13.

Developments in sediment transport mechanics, bedform classification, flow regime concepts, and quantitative geomorphology, on which the study of paleohydraulics rests, have been summarized earlier in this review. The remainder of this section contains a brief discussion of some of the published applications of such knowledge to geological problems.

The fining-upward point-bar cycle has been the most widely used paleohydraulic indicator, almost since the genesis of the cycle as the product of lateral accretion was first deduced in the early 1960s. The discovery of "epsilon" cross-bedding and its interpretation as superimposed point-bar accretion slopes helped to confirm this interpretation. Allen (1965b) was the first to realize the paleohydraulic significance of these cycles, in his study of the Old Red Sandstone of Anglesey, North Wales:

"It appears that the streams which deposited the alluvial facies were already substantial rivers on reaching Anglesey. Each epsilon cross-stratified unit has been interpreted as a point-bar deposit. The thickness of such units therefore corresponds to the stream channel depth at the bankfull state (Wolman and Leopold 1957, pp. 92, 95), and the mean unit thickness is 6 ft 3 in. Leopold and Maddock (1953, Appendix A) gave data leading to relationships between stream width, mean depth and drainage basin area, and Hack (1957, p. 63) later presented an equation connecting stream drainage area and stream length. When the mean unit thickness is substituted for the mean channel depth in these relationships, it becomes statistically very unlikely that the streams were less than 40 miles long, including meanders, and 70 ft wide. Statistically the streams were most likely to have been 400 miles long and 300 ft wide. Needless to say, these figures do not represent predictions, for very many factors combine to determine stream geometry, but are intended merely to convey the probable order of importance of the streams which deposited the Old Red Sandstone of Anglesey."

Schumm (1968b) studied the present and prior courses of the Murrumbidgee River, Australia, and used the empirical geomorphological relationships he had been developing for American and Australian rivers to determine their various paleohydraulic parameters. Sedimentological information, such as sedimentary structures and facies relationships, were not used in this work.

Eicher (1969) attempted to estimate river size, discharge, and slope for a Cretaceous fluvial system in Colorado. His only input datum consisted of stream length, derived from regional paleogeographic reconstruction, whence the other hydraulic parameters were estimated using empirical geomorphological relationships. The building of an edifice of interpretation from one or two items of data by using one estimated parameter as input for the next estimate is one of the principal weaknesses of the paleohydraulic method.

Allen (1970b) used knowledge of bedform hydraulics and flow patterns in meander bends to deduce a generalized paleohydraulic model for a suite of Devonian fining-upward cycles. No attempt was made to proceed from this model to generalizations regarding discharge or drainage area.

Cotter (1971) combined Allen's (1965b, 1970b) observations (in part quoted above) with Schumm's family of equations to produce a paleohydraulic interpretation of a Cretaceous fluvial unit in Utah, including estimates of river length, drainage area, discharge, sinuosity, and slope. Cotter was aware of the possible sources of error in his estimates deriving from differences in climate and vegetation in the past, and emphasized that "the results obtained are only reasonable estimates that must fit the nature of the environment interpreted in other ways."

Similar exercises to those quoted above were used by Friend and Moody-Stuart (1972), Padgett and Ehrlich (1976), Cant and Walker (1976), and Miall (1976) in paleohydraulic reconstructions performed on fluvial units in Spitzbergen, Morocco, Atlantic and Arctic Canada. The routes through the various equations differ in each case; for example, Friend and Moody-Stuart (1972) and Cant and Walker (1976) based much of their work on flow velocity and power deductions derived from sedimentary structures and grain size. Miall (1976) introduced a sinuosity estimation based on paleocurrent measurements.

Baker (1973, 1978b) studied the catastrophic Lake Missoula flood and provided a paleohydraulic reconstruction of some giant bedforms. Leeder (1973) compiled data concerning epsilon cross-bedding, and suggested some refinements of the methods for using this structure and cycle thickness as paleohydraulic indicators. Bridge (1975) used much of the earlier work on bedform hydraulics and flow patterns in a computer simulation of sedimentation in meandering streams. Allen (1977) and Shaw and Kellerhals (1977) examined ripples and antidunes, respectively, as paleohydraulic indicators.

Several papers in Miall (1978a) reported field studies of paleohydraulics, and this section of the book also includes a major review of paleohydraulic methodologies by Ethridge and Schumm (1978). This collection of papers appears to have brought the subject of paleohydraulics to some kind of plateau. Very little has been published in this area since the late 1970s, perhaps because the review by Ethridge and Schumm (1978) revealed the weaknesses and inadequacies of existing paleohydraulic methods. Recent research has taken a quite different form, that of computer modeling of fluvial systems, as discussed in Sect. 10.4.

2.4 Growth of Present-Day Concepts, 1978–1988

The decade 1978–1988 included two fluvial symposia, that at Keele in 1981 (Collinson and Lewin 1983), and the third symposium, at Fort Collins, in 1985 (Ethridge et al. 1987). This period marked the culmination of efforts to define fluvial styles from vertical profiles alone, and the growth of formal architectural methods for lithosome description. Section 2.4 is adapted and expanded from Miall (1980) and the review prepared by Miall (1987) for the Fort Collins meeting. Some post-1988 developments are referred to for the sake of completeness, but the details of most modern studies are given in the appropriate descriptive chapters later in this book.

2.4.1 Bedforms and Sedimentary Structures

During this period, the definitive compilation of bedforms and sedimentary structures by J.R.L. Allen was published. This first appeared as a two-volume work in 1982, and was republished as a single-volume edition 2 years later (Allen 1984). The book contains an exhaustive literature review, and a wealth of data dealing with the experimental production of bedforms in flumes.

Advances in the understanding of the fluid dynamics of bedform generation were discussed by Jackson (1976c), Allen (1983b, 1984), and Leeder (1983). Flow visualization studies revealed important details about the turbulence of traction currents. It was demonstrated that a form of turbulence termed the "burst-sweep cycle" is of critical importance in generating ripples and dunes (Fig. 2.22). At the base of a flow there is an inner layer consisting of streaks of low- and high-speed fluid organized into

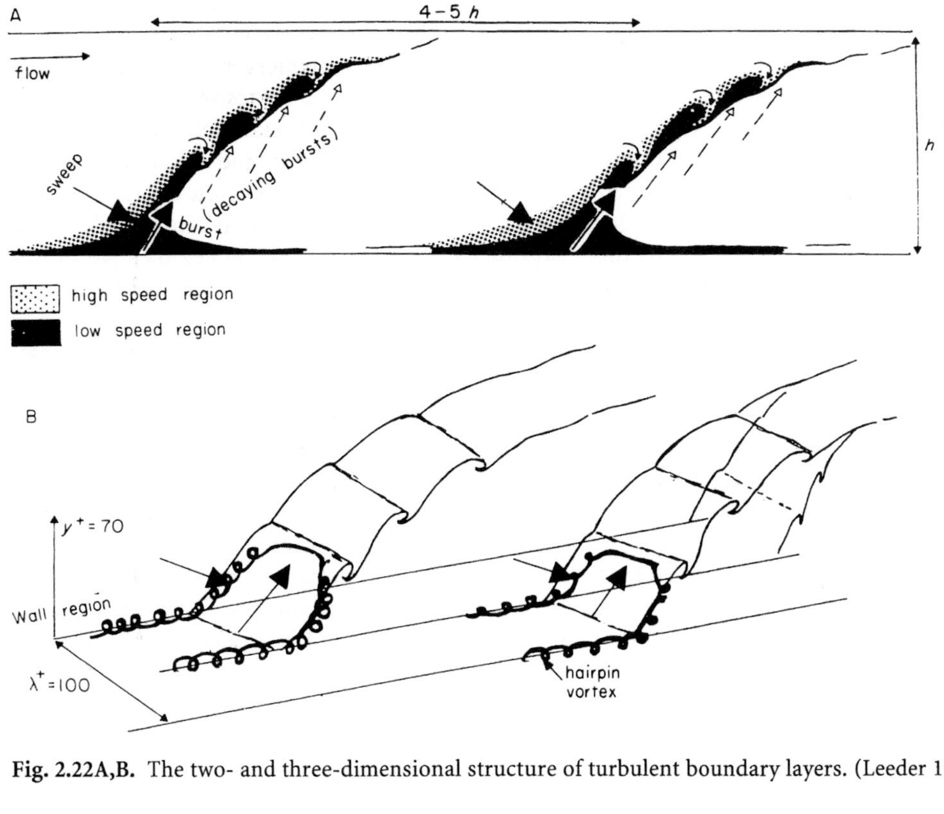

Fig. 2.22A,B. The two- and three-dimensional structure of turbulent boundary layers. (Leeder 1983)

small vortices and subject to periodic disruption and uplift. This passes up into an outer layer containing large breakup vortices that rise steeply to the surface, where they produce boils. These large vortices develop mainly over the lee of dunes, where the pressure gradient encourages bursting. It has been found that the bursts occur, on average, every five water depths in the flow direction. As Leeder (1983, p. 7) stated, "the feedback effect between flow and bedform is clearly of great importance."

Leeder (1983) summarized progress in the application of these data on turbulence to the theory of bedform generation. Ripples are controlled by flow-separation dynamics, and are not influenced by the outer layer. They develop on hydraulically smooth boundaries by the action of flow separation and reattachment. Their form is insensitive to water depth, and only loosely related to bed shear stress. With increased shear stress turbulent vortices become larger but more varied in scale and speed, leading to the generation of a wider range of heights, wavelengths, and forward ripple speeds. Leeder (1983) suggested that the larger ripples cannibalize the smaller ones, leading to enhanced scour in the ripple lee, and still further increases in ripple height.

Dunes, the characteristic bedform in the higher-velocity ranges of subcritical flow, show strong correlation of height and wavelength with water depth (Allen 1968; Jackson 1976a), suggesting that their form is controlled by the outer turbulent layer. It seems likely that bursts develop from large ripples as the ripples grow in size, and that these then dominate the structure of the flow and become the principal mechanism in bedform production (Leeder 1983). Sediment that is lifted into suspension by the turbulent bursts travel downstream a maximum of about five flow depths before deposition takes place.

Continued bursting, erosion and deposition will eventually cause the bedforms to adjust to a relationship between bedform wavelength, height, and flow depth. A stable dune bed will gradually develop in which macroseparation as well as burst macroturbulence play an important role in determining dune morphology and magnitude (Leeder 1983).

The transition to upper-stage plane beds may take place because the high grain concentration that develops at high flow speeds blankets and inhibits turbulence development. Also, the frequency of turbulent sweeps may overcome the scouring effect of leeside separation eddies. Saunderson and Lockett (1983) reported a series of flume experiments carried out to examine this transition. They found that a mixture of flat-bed and dune-like conditions occurred. The decreasing scale and erosive power of separation eddies at higher flow speeds leads to smaller scour pockets and the draping of sand over rounded dune crests, rather than the production of angle-of-repose foreset lamination. The result is the generation of "humpback" dunes with low-angle, sigmoidal cross-bedding.

Considerable problems of classification and nomenclature were apparent by the late 1970s, as discussed by Miall (1977) and N.D. Smith (1978). Leeder (1983) offered a tentative genetic classification of bedforms according to the dominant physical process controlling their size limits and stability (Fig. 2.23). However, the problem of nomenclature was not satisfactorily resolved until an SEPM Research Symposium was held in 1987, the results of which were collated by Ashley (1990). This paper is discussed in a later section. Some other recent research results have been reviewed by Fielding (1993b).

2.4.2 The Decline and Fall of the Vertical Profile

The first fluvial symposium in 1977 was held at a time when a great deal of new information on fluvial facies was being published. At this time the concept of the simplified facies model was much in vogue (Walker 1979). Such models were commonly presented in simplified, abbreviated form, in terms of a vertical stratigraphic profile, which sets out a typical sequence of lithologies with their accompanying biofacies and sedimentary structures (e.g., Visher 1965a,b). This practice developed from the application of Walther's law, which states that only those facies that can be found forming side by side in nature can occur in contact with one another in vertical succession, unless the succession contains internal erosion surfaces (Middleton 1973). The vertical profiles are cyclic, in the sense that they are repeated several or many times in succession, generally with some internal variation. The model may also contain information about lateral variability, particularly with reference to the position of a shoreline or sediment source area and the direction of prevailing air or water currents. Most models were based on a combination of observations from modern environments and the ancient rock record.

In subsurface analysis, application of Walther's law enables us to interpret lateral facies relationships from their vertical sequence as expressed in a facies model. Several textbooks appeared during this period which provided catalogues of vertical profiles as

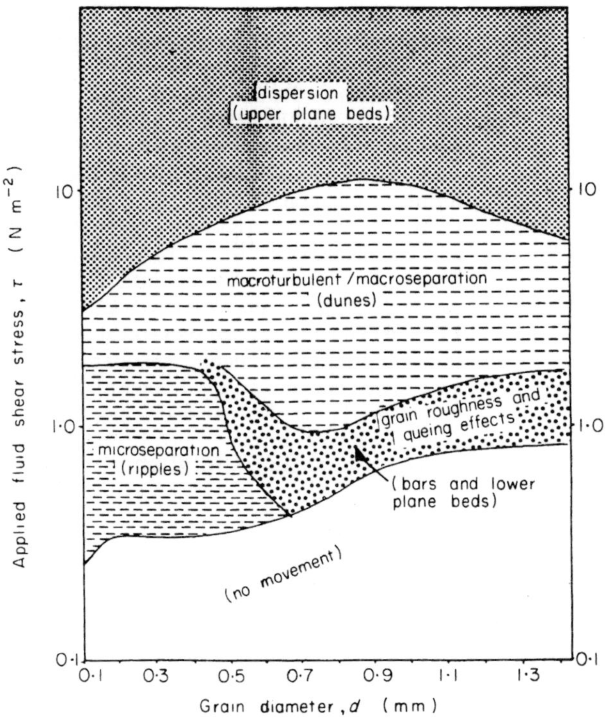

Fig. 2.23. Bedform phase diagram showing how bedforms may be classified according to the physical processes which control their size limits and stability. (Leeder 1983)

a basis for subsurface environmental interpretations (e.g., Pirson 1977; Berg 1986). Commonly, these were identified not on the basis of lithologic logs in cores, but from the profiles displayed by electric logs – the classic "funnel-shaped" and "bell-shaped" gamma-ray, spontaneous potential, and resistivity logs that had first been identified during the 1950s (e.g., Nanz 1954).

The value of a facies model in providing a synthesis of a diverse array of facts is offset by the danger that uncritical use of the model may lead to a loss of information or to misinterpretation, because it is tempting to observe strata in terms of a preconceived model. Objective field measurements commonly are difficult to make, particularly in areas of poor exposure. One of the trends in sedimentological research in the 1970s was to quantify observations of lithological successions in order to enable model sequences or cycles to be defined statistically. Markov chain analysis was commonly used for this purpose (e.g., Allen 1970b; Miall 1973, 1977; Cant and Walker 1976). Although this approach allows more precision in defining the model, it can cause valuable observational detail to be ignored. Collinson (1978, p. 579) made this point about the Battery Point Formation, a highly varied, low-sinuosity-fluvial deposit from which Cant and Walker (1976) extracted a single "summary" cyclic sequence (Fig. 2.24a).

Amongst the problems that had been recognized with the Markov chain method was the fact that it contained no information on the nature of facies contacts. This is, of course, of critical importance, following Walther's law. Cant and Walker (1976) attempted to address the problem by erecting a separate facies state for scour surfaces that, in their field case, were typically accompanied by a distinctive poorly-sorted sandstone. Miall and Gibling (1978) took a different approach, documenting the number of erosional and gradational contacts between each facies state and developing a "contact matrix" that showed the preponderance of each type of contact in tabular form. Although this was a systematic approach to the problem, the method has not been used by other workers because attention at this time began to turn away from vertical profiles. As shown by Godin (1991), cyclic relationships can be developed at several nested levels within fluvial systems, and simple one-step Markov methods are not suitable for analyzing these complexities (Chaps. 8, 10).

A somewhat different approach was taken by Friend et al. (1976), who measured many lengthy sections through the Devonian nonmarine sediments of East Greenland. The sections were divided into 10-m segments, and the abundance of a range of facies attributes (e.g., grain size, sedimentary structures) was determined for each segment. The data were then subjected to factor analysis to determine

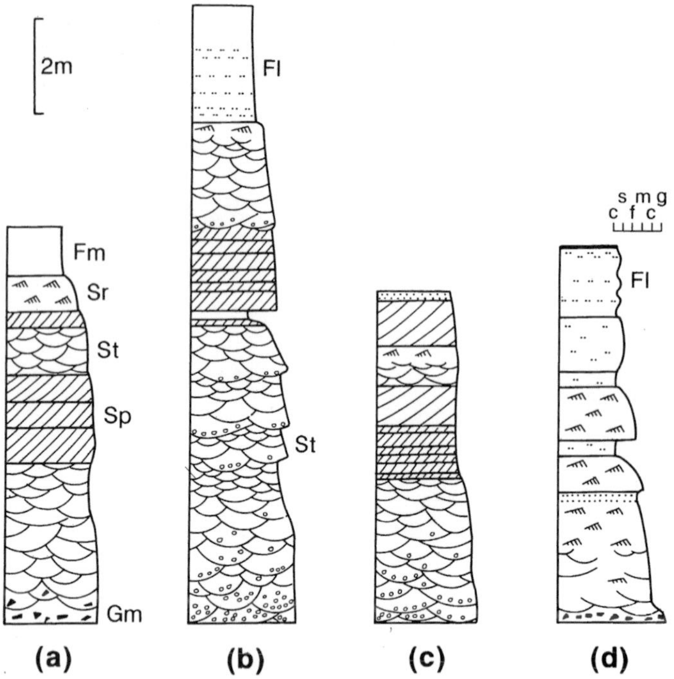

Fig. 2.24. Comparison of four vertical profiles, showing how similar fining-upward cycles may be produced in several different ways. **a** Model cyclic sequence; Battery Point Formation (Devonian), Quebec, probably formed by vertical bar aggradation in a low-sinuosity (braided) river (Cant and Walker 1976). **b** Sequence formed by lateral accretion on a point bar; Castisent Sandstone (Eocene), Spain (Nijman and Puigdefábregas 1978). **c** Sequence formed by lateral accretion within a point bar; modern Amite River, Louisiana (McGowen and Garner 1970). **d** Sequence formed by vertical aggradation and progressive channel abandonment on an alluvial fan, under conditions of tectonic quiescence (allogenic control); Upper Carboniferous coal measures, northern Spain (Heward 1978a). Diagram from Miall (1980)

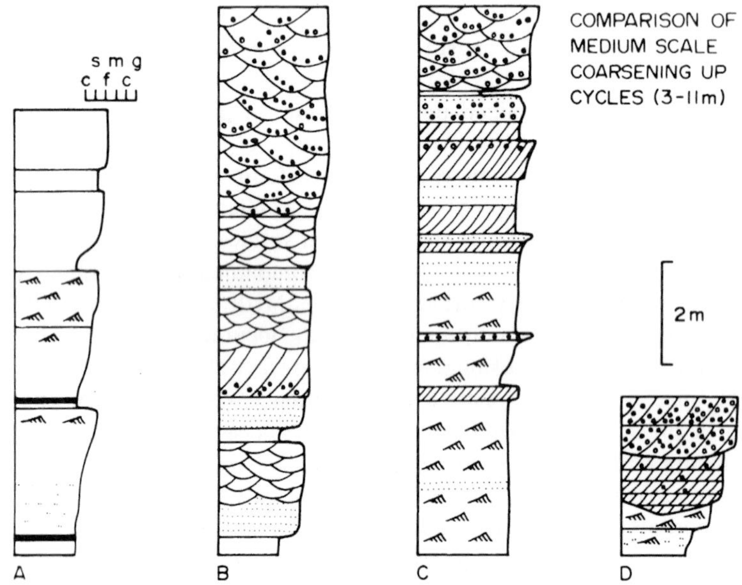

Fig. 2.25. A Progradation and vertical aggradation in distal part of an alluvial fan, Upper Carboniferous coal measures, northern Spain (Heward 1978). **B** Progradation and vertical aggradation in proximal part of an alluvial fan, Devonian of Hornelen Basin, Norway (Steel and Aasheim 1978). **C** Vertical aggradation on a sandur plain in front of an advancing glacier. Sequence is capped by till (not shown); Pleistocene of Denmark (C. Heinberg, pers. comm., 1979). **D** Cycle formed by bar progradation or by fill of a channel deepened by progressively melting ice in a glacial retreat phase; Pleistocene of Ontario (Costello and Walker 1972). Diagram from Miall (1980)

associations ("reaction groups") between the various facies attributes, and the segments were compared using cluster analysis of the reaction groups, in order to define lithofacies associations. These could then be expressed graphically as lithologic logs of typical segments from each association. The results revealed a wide range of sedimentary styles, not necessarily cyclic in nature. Although quantitatively rigorous, this method suffers from the same limitations as all other techniques that rely on the vertical profile – it is not concerned with three-, or even two-dimensional facies architecture.

The proliferation of facies studies led Dott and Bourgeois (1983) to remark that by the early 1980s fluvial facies models had "multipled like rabbits so that every real-world example now seems to require a new model. Such proliferation defeats the whole purpose of the conceptual model by encouraging

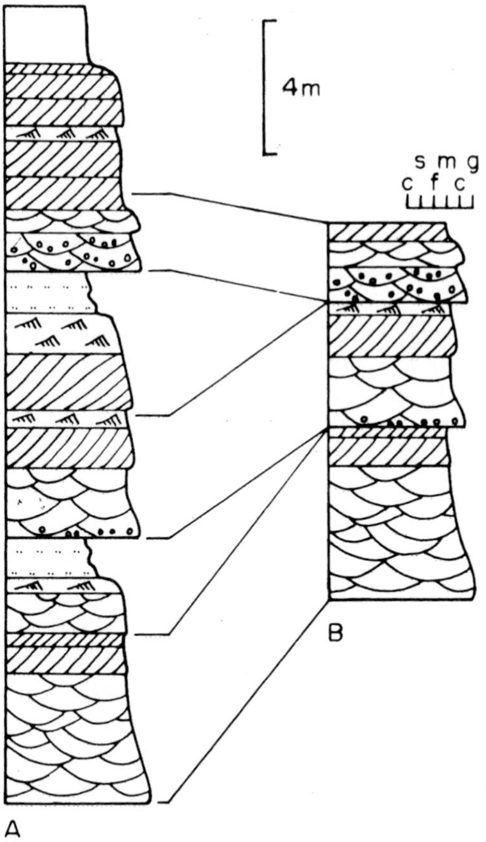

Fig. 2.26. Stacking of three hypothetical point-bar successions under conditions of **A** high and **B** low subsidence rate relative to aggradation rate. In **A**, complete successions are preserved intact, together with overlying floodplain units. In **B**, upper point-bar and floodplain units have been removed by deep scour, and the resulting succession is very similar to those produced in some low-sinuosity multiple-channel rivers, particularly those of South Saskatchewan type. (Miall 1980)

excessive pigeon-holing, which obscures rather than reveals whatever unity may exist among the variants".

There are two main kinds of problem with the vertical profile as a precise analytic tool: (1) in some cases, allogenic and autogenic processes can lead to the same vertical succession; (2) similar vertical profiles can be generated in rivers of different channel pattern by different autogenic processes. These points were explored at length by Miall (1980) from whom Figs. 2.24, 2.25, and 2.26 were taken.

The cycle shown in Fig. 2.24a may be of autogenic or allogenic origin, or a combination of both. It probably represents the filling and progressive abandonment of a relatively deep, braided channel system, but it is unclear whether cyclic repetition of this type of sequence requires tectonic triggering of channel-avulsion events (an allogenic process) or whether it can be accomplished by autogenic lateral channel migration. Information on lateral variability of the section (not available in this case) may supply some clues.

The cycles in Fig. 2.24b,c are unquestionably autogenic in origin. Figure 2.24b shows autogenicity on two nested levels. The smaller cycles may be seasonal in origin or represent infrequent floods, whereas the overall cycle represents a migrating point bar, as does Fig. 2.24c. Figure 2.24d, which is similar to the others in thickness and fining-upward character, represents the deposits of shallow alluvial-fan distributaries in an environment dominated by tectonic controls. The basin in which it formed was a tectonically active one, with differential movement between basin and source area adequate to create and maintain a set of alluvial fans at the basin margin.

The question of scale must also be considered when comparing the columns in Fig. 2.24. Diagrams A, B, and C represent relatively large rivers, with channel depths of several meters. Diagram D represents the deposits of fan distributaries, which probably were much shallower. Autogenic cycles, related to such causes as channel migration and bankfull floods, would be on a correspondingly smaller scale. Several thin fining-upward cycles, of probable autogenic origin, can be discerned in diagram D, nested within the overall fining-upward succession.

The coarsening-upward cycles of Fig. 2.25A,B resulted from rapid progadation following pulses of relative uplift on the source areas of alluvial fans in two different basins; they are therefore of tectonic origin. They are similar to the succession of Fig. 2.25C, in terms of an upward increase in grain size and scale of sedimentary structures. Cycle C, however, represents a climatic effect. During periods of glacial advance, environments in fluvioglacial rivers are shifted progressively downstream, so that at a given point the fluvial setting changes from distal to proximal. The result is a coarsening-upward succession. The example illustrated is from a Pleistocene succession. Other examples were quoted by Miall (1980).

The question of preservation is also of importance in interpreting vertical profiles. Consider Fig. 2.26, which shows the effects of variation in subsidence rates in a basin undergoing deposition by high-sinuosity streams. The main sedimentary process is the formation of coarse channel-belt deposits by lateral accretion. During periods of rapid subsidence, these will be rapidly buried and preserved, largely intact (Fig. 2.26A), but during intervals of slow subsidence most of the floodplain deposits and the upper, finer part of the point-bar succession, may be planed off, and the result would be multistory coarse units that preserve only the evidence of high-energy channel-floor sedimentation (Fig. 2.26B). This part of the succession may well be interpreted as the product of a braided stream system, because of the relatively high proportion of sandy channel deposits, and the paucity of floodplain fines. The fluctuation in fluvial style might be attributed to climatic or tectonic causes. Allen (1978, p. 145) argued that in a basin undergoing a slowly decreasing subsidence rate this effect could be the cause of large-scale coarsening-upward (or at least sandier-upward) cycles (cycles similar to those in Fig. 2.26B overlying those more like 2.24a), whereas such cycles are commonly attributed to progradation caused by increased differential movement, such as the coarsening-upward successions of Steel and Aasheim (1978), Steel et al. (1977), and Heward (1978a). The stacking and separation of channel bodies and preservation of intervening fines are discussed in Chap. 9, and the interpretation of allogenic cyclicity is discussed further in Sect. 11.3.6.

2.4.3 Fluvial Architecture

2.4.3.1 Architectural Scale and the Bounding-Surface Concept

As noted by Allen (1983a, p. 249):

> "The idea that sandstone bodies are divisible internally into 'packets' of genetically related strata by an hierarchically ordered set of bedding contacts has been exploited sedimentologically for many years, although not always in an explicit manner. For example, McKee and Weir (1953) distinguished the hierarchy of the stratum, the set of strata, and the coset of sets of strata, bedding contacts being used implicitly to separate these entities."

In order to trace the development of these ideas, it is necessary to reach back to the 1960s. Allen (1966) showed that flow fields in such environments as rivers and deltas could be classified into a hierarchical order. His hierarchy was designed as an aid to the interpretation of variance in paleocurrent data collected over various areal scales, from the individual bed to large outcrops or outcrop groups. The hierarchy consists of five categories, small-scale ripples, large-scale ripples, dunes, channels, and the integrated system, meaning the sum of the variances over the four scales. Miall (1974) added the scale of the entire river system to this idea, and compiled some data illustrating the validity of the concept (Fig. 2.10).

Brookfield (1977) discussed the concept of an eolian bedform hierarchy and tabulated the characteristics of four orders of eolian bedform elements: draas, dunes, aerodynamic ripples, and impact ripples. These four orders occur simultaneously, superimposed on each other. Brookfield showed that this superimposition resulted in the formation of three types of internal bounding surfaces. His first-order surfaces are major, laterally extensive, flat-lying, or convex-up bedding planes between draas. Second-order surfaces are low to moderately dipping surfaces bounding sets of cross-strata formed by the passage of dunes across draas. Third-order surfaces are reactivation surfaces bounding bundles of laminae within cross-bed sets and are caused by localized changes in wind direction or velocity. (mesoforms to microforms).

Similar hierarchies of internal bounding surfaces have been recognized in some subaqueous bedforms. A particularly useful approach was that of Jackson (1975). Building on Russian work, he defined three classes of bedforms: microforms (e.g., ripples), mesoforms (e.g., dunes, sand waves, produced by "dynamic events", such as floods), and macroforms (longer-term geomorphic products, such as point bars, sand flats, eolian draas). Reactivation surfaces (Collinson 1970) may develop in bedforms as a result of changes in stage or flow direction or under conditions of constant stage, where bedforms randomly interact (McCabe and Jones 1977). They may also separate individual smaller-scale elements within a larger unit, such as the component packages of mesoforms within a macroform. They have been documented in small-scale (Collinson 1970) and large-scale fluvial cross-bed sets (Jones and McCabe 1980; Fig. 2.27) and in tidal sand waves (Allen 1980). In the latter case, the sand waves rest on horizontal surfaces analogous to Brookfield's first-order surfaces and contain second-order reactivation surfaces that relate to individual tidal cycles. Third-order surfaces bound individual cross-bed sets.

Brookfield's (1977) development of the relationship between the time duration of a depositional event, the physical scale of the depositional product, and the geometry of the resulting lithosome was a major step forward that has been of considerable use in the analysis of eolian deposits. Brookfield (1977), Gradzinski et al. (1979), and Kocurek (1981) showed how these ideas could be applied to the interpretation of ancient eolian deposits. Kocurek (1988) has

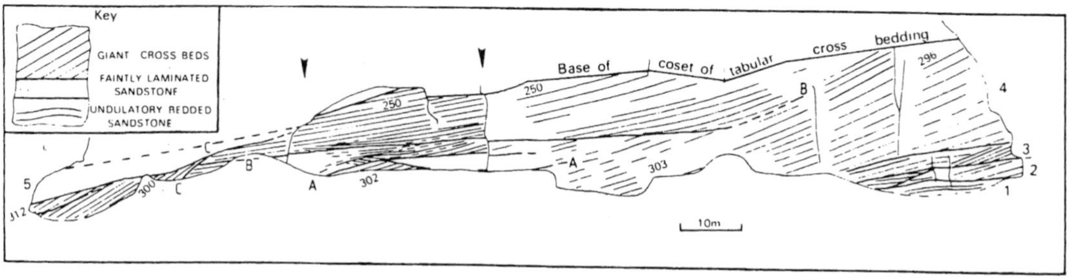

Fig. 2.27. Giant cross-bedding in a Carboniferous fluvial-deltaic sandstone. Surfaces A, B, and C are reactivation surfaces within cross-bed set 4. (Jones and McCabe 1980)

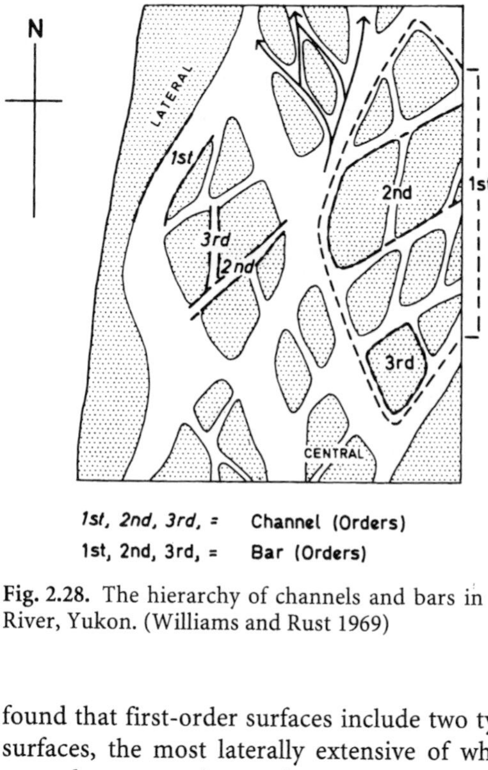

Fig. 2.28. The hierarchy of channels and bars in Donjek River, Yukon. (Williams and Rust 1969)

found that first-order surfaces include two types of surfaces, the most laterally extensive of which he termed super-surfaces. Characterization of eolian deposits therefore now requires a fourfold hierarchy of bounding surfaces.

Several workers have attempted to develop a breakdown of the range of physical scales present in fluvial deposits. Williams and Rust (1969) proposed an ordering of the scales of channels and bars in the modern Donjek River, Yukon (Fig. 2.28). Campbell (1976), in an analysis of the Westwater Canyon Member of the Morrison Formation in New Mexico, recognized several scales of fluvial sequence that occurred in tabular channel-fill sandstone bodies of a range of dimensions (although his work has now been shown to be largely incorrect: Sect. 10.3.2). Jones and McCabe (1980) described three types of reactivation surfaces occurring within sets of giant cross-bedding, and related them to changes in bedform orientation and to stage changes in the river. A similar type of analysis was performed by Haszeldine (1983a,b) on the bounding surfaces within a Carboniferous sand flat deposit. Bridge and Diemer (1983) and Bridge and Gordon (1985) referred to major and minor bounding surfaces within Paleozoic fluvial sequences (Fig. 2.29). The major surfaces are typically horizontal and planar or slightly concave up and enclose tabular sheets representing channel-fill successions.

Allen's (1983a) study of the Devonian Brownstones of the Welsh Borders represents the first explicit attempt to formalize the concept of a hierarchy of bounding surfaces in fluvial deposits (Fig. 2.30) and makes reference to Brookfield's work in eolian strata as a point of comparison. Allen described three types of bounding surfaces. He reversed the order of numbering from that used by Brookfield (1977), such that the surfaces with the highest number are the most laterally extensive. No reason was offered for this reversal, but the result is an open-ended numbering scheme that can readily accomodate developments in our understanding of larger-scale depositional units, as discussed below. First-order contacts, in Allen's scheme, are set boundaries, in the sense of McKee and Weir (1953). Second-order contacts "bound clusters of sedimentation units of the kinds delineated by first-order contacts." They are comparable to the coset boundaries of McKee and Weir (1953), except that more than one type of lithofacies may comprise a cluster. Allen (1983a) stated that "these groupings, here termed complexes, comprise sedimentation units that are genetically related by facies and/or paleocurrent direction." Many of the complexes in the Brownstones are macroforms, in the sense defined by Jackson (1975). Third-order surfaces are comparable to the major surfaces of Bridge and Diemer (1983). No direct relationship is implied between

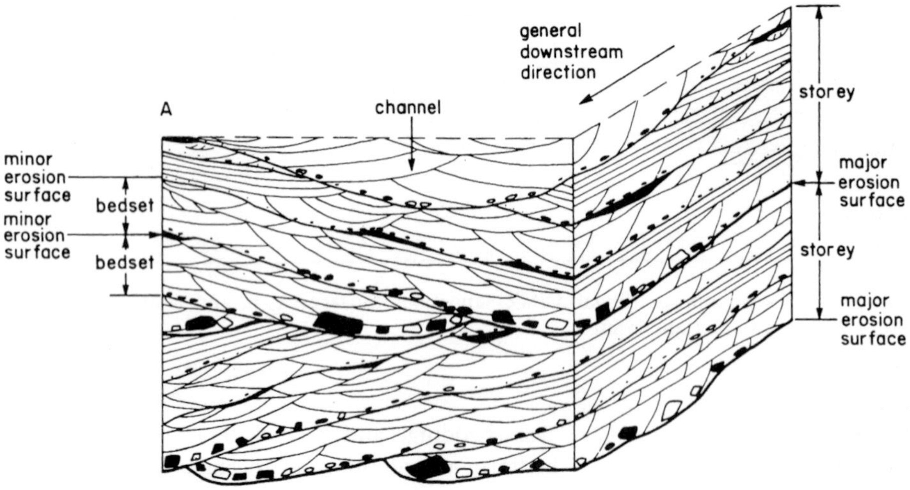

Fig. 2.29. Schematic geometry of a Devonian-Carboniferous channel sand body, Ireland, showing architectural subdivisions proposed by Bridge and Diemer (1983)

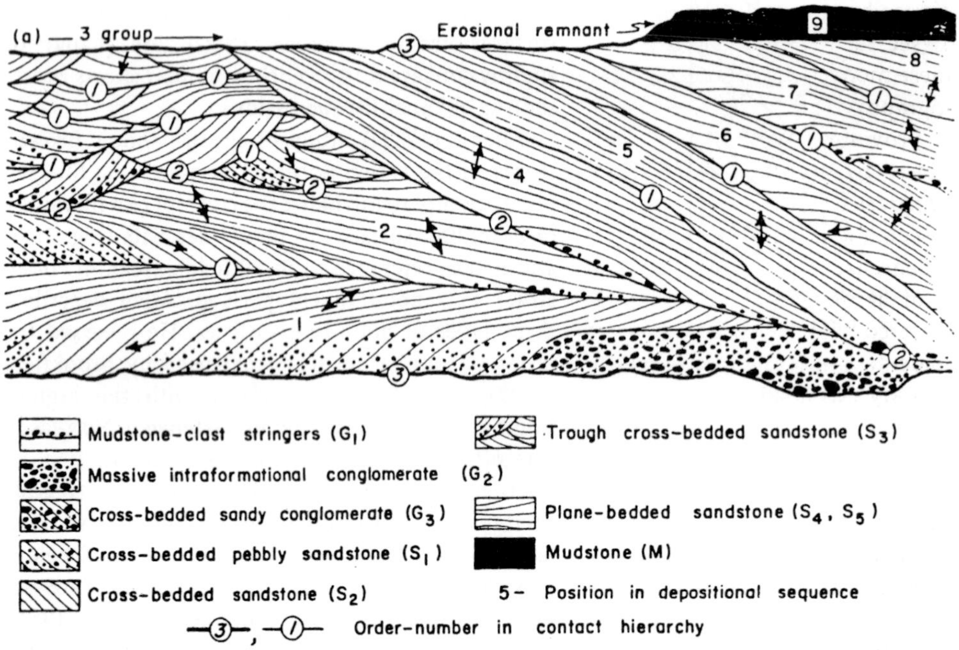

Fig. 2.30. Schematic summary of the lithofacies and bounding surfaces in the Devonian Brownstones of the Welsh Borderlands. (Allen 1983a)

Allen's three orders of surfaces and those of Brookfield because of the different hydraulic behavior and depositional patterns of eolian and aqueous currents.

Miall used Allen's (1983a) classification of bounding surfaces in his first synthesis of architectural-element analysis (Miall 1985), but soon found, in actual field studies that it resulted in the grouping together of some dissimilar types of surfaces, resulting in a loss of significant information on macroform geometry. A more elaborate and comprehensive classification was developed as a result (Miall 1988a,b), consisting of a sixfold hierarchy of lithosomes and bounding surfaces. This is the system used in this book (with additions) and is described in Chap. 4.

2.4.3.2 Alluvial Basin Architecture

The next step beyond a reconstruction of the channel style of a river system is an attempt to reconstruct the basin architecture. As noted by Friend (1983), an influential set of diagrams was that published by Allen (1965a) and reproduced here as Fig. 2.17. These diagrams show a variety of styles of alluvial stratigraphy that reflect the patterns of channel aggradation and migration of various river types. A great deal has been learned about fluvial styles since these diagrams were published.

In order to reconstruct alluvial stratigraphy it is necessary to carry out regional surface or subsurface correlation of outcrops or well sections. Correlation is commonly a problem in fluvial deposits because of rapid lateral facies changes, a lack of regional marker beds, and numerous internal erosion surfaces (channels and interfluve surfaces). Four types of medium- to long-distance correlations can be carried out under certain conditions, however, by focusing on floodplain deposits (see Sect. 9.5 for a more extensive discussion of these modern mapping methods). First, magnetostratigraphy is proving to be extremely valuable for providing local chronostratigraphic datum planes, particularly in the more laterally extensive and possibly less episodically preserved floodplain sequences. Behrensmeyer and Tauxe (1982) and Behrensmeyer (1987) provided excellent examples of architectural interpretation based on this type of control. They were able to distinguish and relate to each other interbedded trunk-river and tributary deposits and to calculate local sedimentation rates.

Second, the same kind of refined lateral control may be possible using tuff beds. These are rarely preserved in channel deposits because of the high transport energy, but may form widespread marker horizons on the floodplain. Allen and Williams (1982) provided an example of a unit less than 30 m thick that can be subdivided into seven allostratigraphic units (these are "chronosomes", to use the term suggested by Schultz 1982), permitting a very detailed reconstruction of architectural evolution.

Paleosols are the third type of floodplain marker horizon. These can be distinguished from each other on the basis of petrological and geochemical signatures, and have the added advantage of yielding a considerable amount of evidence for local environmental conditions, sedimentation rates, and local flooding and avulsion frequencies (Bown and Kraus 1981, 1987; Kraus 1987). They may even be detectable in seismic-reflection data. Kraus (1987) suggested that channel migration patterns could be analyzed by studying paleosol maturity. The "pedofacies" of the paleosols matures with time, indicating the extent of exposure and distance from channel sediment source (Fig. 2.31). These concepts are discussed further in Chaps. 7 and 10. Fourthly, coal seams have a similar local utility for local lithostratigraphic correlation.

Very few studies of carefully documented regional alluvial architecture have been published. Those by Behrensmeyer and Tauxe (1982) and Behrensmeyer (1987) have been referred to above, and one set of reconstructions is illustrated in Fig. 2.32. Campbell (1976) and Blakey and Gubitosa (1984) attempted detailed, large-scale architectural reconstructions. That by Blakey and Gubitosa (1984) described paleovalley-fill styles, and examined the causes of architectural variation, such as regional changes in subsidence patterns. Campbell (1976) provided a reconstruction of what has been interpreted as a large-scale, braided, alluvial distributary system, in which he defined large-scale channel belts and several scales of channel filling. However, Cowan (1991) has shown that most of Campbell's (1976) outcrop units were based on weathering characteristics that reflect diagenetic oxidation and groundwater leaching, not primary depositional architecture (Sect. 10.3.2).

A useful approach to channel classification was developed by Friend et al. (1979) and Friend (1983). Friend (1983) defined channels as "elongate depressions in the alluvial surface, with more or less clearly defined margins or banks between which the river flow is restricted for most of the year". In many rivers, there are channels and other types of "hollow" of more than one size (tributaries, crevasse channels, chute channels, scour hollows), all of which will show variations in size and morphology. A careful analysis of these channels is an essential component of a facies analysis, and Friend (1983) provided three diagrams to assist in this work (Figs. 2.33, 2.34, 2.35).

Note that this classification is strictly descriptive. No attempt need be made, until the final interpretation, to assign terms such as "braided" or "anastomosed" to the deposits. Sheet flood deposits are, strictly speaking, unchannelized by definition. They are characteristic of many distal braid-plain deposits, as discussed in Chap. 8. Fixed channels are typical of anastomosed rivers. They typically are described as "ribbon" sandstones, and have width/depth ratios of less than 15. Some ribbons have "wings" of sandstone extending laterally from the top of the channel margin. These are probably levee

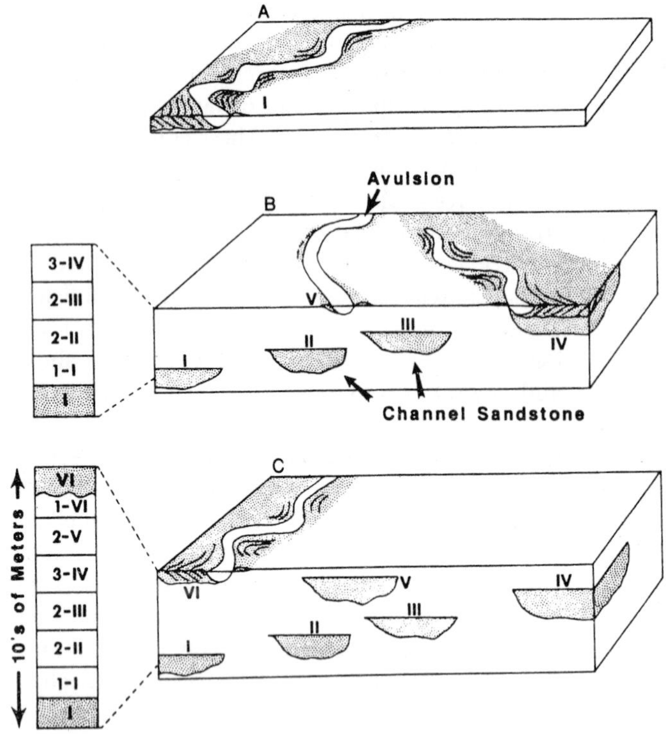

Fig. 2.31. Development of compound pedofacies sequence. *Arabic numerals* indicate developmental stage (maturity) of paleosol. *Roman numerals* indicate depositional sequence of channels. Note initial development of increasingly mature pedofacies at *left side* of area as channels migrate laterally away to *right* (**A**, **B**). Less mature paleosols cap this succession as the channel migrates back to the *left* (**C**). (Kraus 1987)

deposits. Mobile-channel belt deposits include braided and meandering channels, and the various types of channel deposits intermediate between these well-known end members. The deposits are sheet-like in geometry, with width/depth ratios exceeding 15, and commonly greater than 100.

In the 1980s, several important studies documented in considerable detail the internal architecture of ancient complex channel-fill and bar deposits. Okolo (1983) and Hopkins (1985) described some well-exposed channels filled by vertical aggradation. Allen and Matter (1982) and Okolo (1983) described other, more complex, channel fills. Other writers, including Allen (1983a), Haszeldine (1983a,b), Kirk (1983), and Ramos and Sopeña (1983) provided detailed two- and three-dimensional cross sections of large compound bars, revealing a range of complex structures enclosed by and containing major bounding surfaces. These studies provided the first data on the internal geometry and facies architecture of compound bars comparable to the modern side bars of Collinson (1970) and the sand flats of Cant and Walker (1978). These are macroforms (Jackson 1975), comparable to, and gradational with, point bars, in terms of their scale, complexity, and depositional longevity. The discovery of the truly three-dimensional nature of these deposits was one of the most significant contributions to fluvial studies of this period.

Miall (1985) attempted to synthesize these new data into an improved method of field facies analysis and classification, termed "architectural-element analysis". He showed how eight basic elements could be combined to form a variety of fluvial styles, but emphasized the limitation of the fixed-model approach. To some extent, every river and every ancient fluvial succession is unique. Miall (1985) suggested, therefore, that modeling might be simpler at the level of the component elements, which did seem to show some constant characteristics through rivers of all kinds.

Channel morphology changes downstream in response to changes in valley slope, sediment load, bank materials, climate, or tectonic regime, and the same controls may cause changes through time in the morphology of a particular river reach (Schumm 1969; Burnett and Schumm 1983; Peterson 1984; Carson 1984a,b,c). It is therefore unwise to assume that fluvial style will remain constant throughout a given stratigraphic unit.

Modern techniques for basin correlation in nonmarine strata are discussed in Chap. 9. Current studies of alluvial architecture are at the center of a debate in sequence stratigraphy regarding the interpretations of base-level change and tectonism that can be made from alluvial architecture. This is discussed at length in Chaps. 11 and 13.

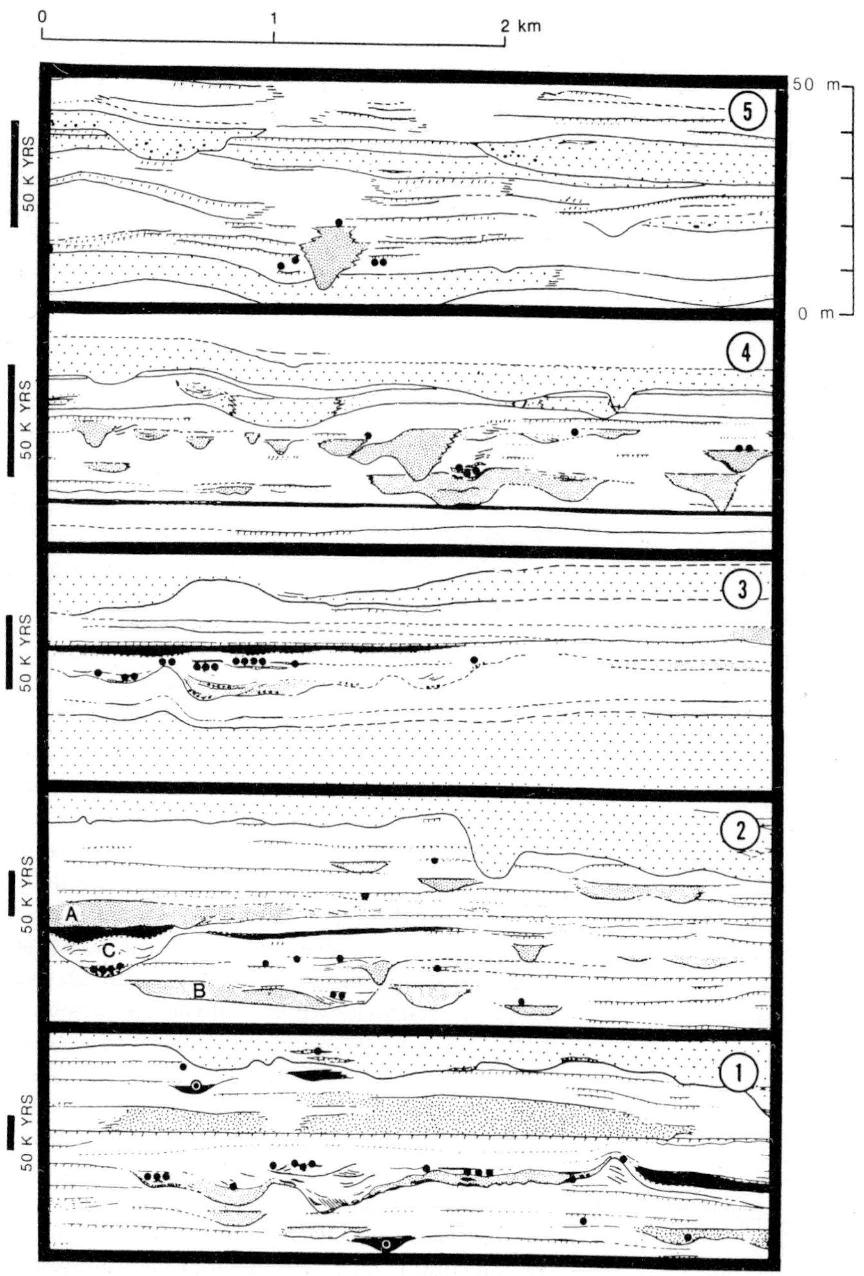

Fig. 2.32. Five lateral architectural sections through the upper Miocene Siwalik Group of northern Pakistan. Laterally continuous pedogenic horizons and magnetic reversal stratigraphy were used for correlation purposes, and permit estimates of sedimentation rate (shown at *left*). *Coarse stippling* Major channels; *fine stippling* minor channels; *black dots* vertebrate localities. (Behrensmeyer 1987)

2.4.4 Fluvial Styles

Some of the developments in facies studies that occurred during the 1970s and 1980s (and more recently) are summarized in this section. Many useful illustrations were published by the authors to whom reference is made below. Some of these are reproduced in Chaps. 5–8, where the modern framework of fluvial facies analysis is formally documented.

The gradual extension of fluvial research into less well-studied regions of the earth, including the Amazon basin, southern Africa, and Australia, has led to a considerable broadening of our ideas regarding fluvial style. Baker (1978a) showed that many of our

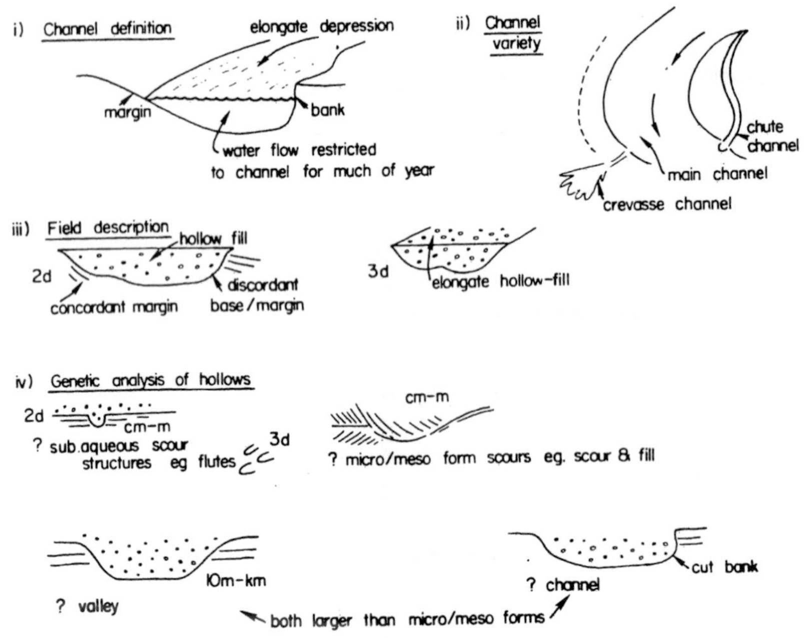

Fig. 2.33. Channel definition and recognition. (Friend 1983)

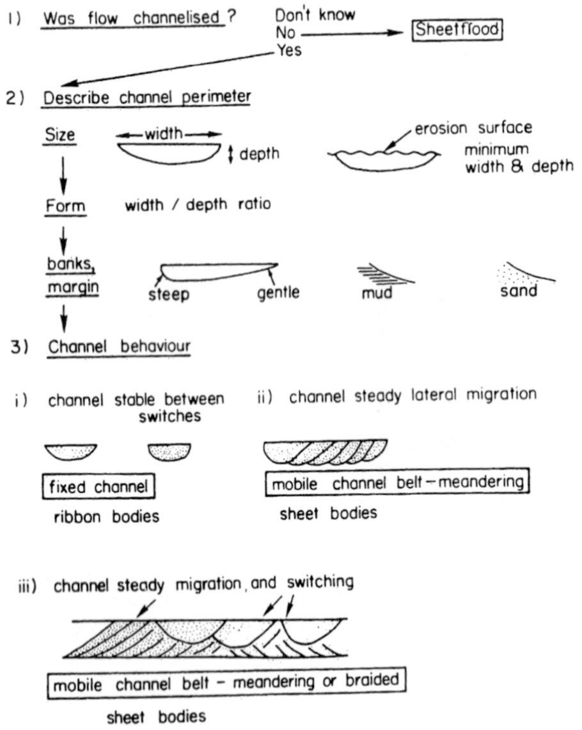

Fig. 2.34. Key for classification of channel behavior. (Friend 1983)

conceptions concerning fluvial geomorphology are appropriate only under certain climatic conditions, particularly temperate climates, and that rivers can show quite different patterns of sinuosity, channel shape, and sediment type under tropical conditions, and where rivers flow from one climatic zone to another. The distinctive climate of Australia has led to the development of several unusual, or even unique, fluvial styles within that continent. Some of these were reviewed by Fielding (1993b), in a volume where much new work on these rivers was presented (Fielding 1993a). A useful classification of fluvial floodplains, reflecting a geomorphological perspective, was provided by Nanson and Croke (1992), and is discussed in Chaps. 7 and 8.

There has been considerable debate regarding the classification of alluvial fans, especially those that prograde directly into standing bodies of water (fan deltas). It is now generally accepted that alluvial fans vary widely in size and may include a wide variety of facies assemblages. Recent work on this subject is discussed in Sect. 8.3.

Schumm's (1963) classification of rivers into bed-load, mixed-load, and suspended-load types has been very influential. A recent manifestation of this classification is given as Fig. 8.6. It was used as a basis for a classification of fluvial styles by Galloway (1981) in his analysis of Cenozoic fluvial systems of the Gulf Coast of the United States (Fig. 2.36).

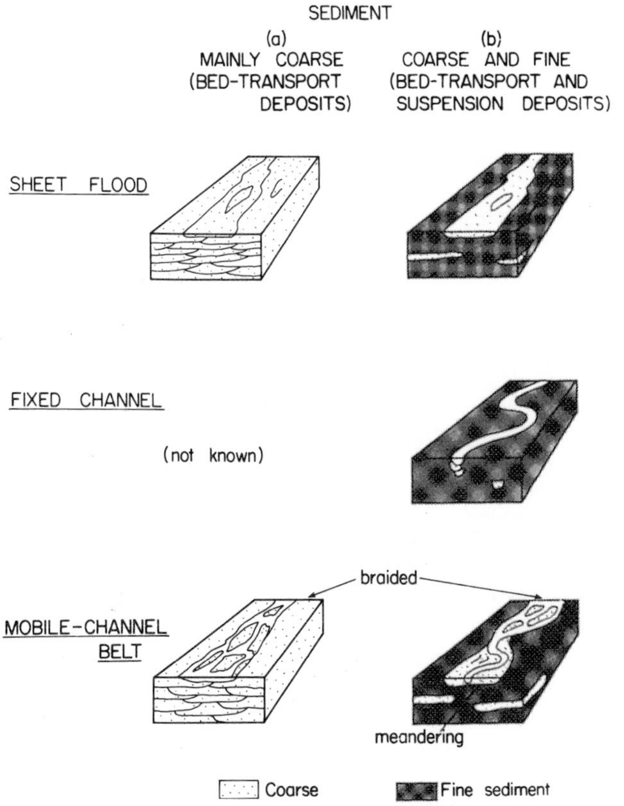

Fig. 2.35. Model of distinctive types of alluvial architecture. (Friend 1983)

2.4.4.1 *High-Sinuosity Rivers*

Jackson's (1978) critique of facies models was a major contribution to fluvial studies. Preliminary work by him led to a tentative classification of high-sinuosity rivers (those with a sinuosity >1.5) into five classes, based mainly on the grain size of the sediment load. Several of these categories were poorly known at the time of writing, but various studies of modern rivers and ancient deposits have confirmed, with some modifications, Jackson's original work. These developments have been integrated into the architectural studies of Miall (1985, 1988a,b) and are discussed in Chap. 8. Only the main points are noted below.

It must be emphasized that high-sinuosity reaches can occur in almost any fluvial setting. Such reaches tend to migrate laterally, developing point bars with the well-known lateral-accretion internal structure (epsilon cross-bedding of Allen 1963a). The occurrence of this structure is therefore not the simple diagnostic criterion of fluvial style that it is often assumed to be, but must be evaluated in context with the other associated facies.

Gravel-bed, high-sinuosity rivers are not common, but our knowledge of them improved considerably during the decade up to 1988. Modern examples have been studied by Gustavson (1978) and Forbes (1983). Point-bar structures are difficult to define in the deposits because of the complexities of small- to medium-scale bedforms and minor channels. Well-exposed ancient examples of gravel-bed rivers were described by Ori (1979) and Ramos and Sopeña (1983). These examples contain gravel lateral-accretion deposits, indicating that the rivers contained sinuous channel reaches.

Rivers with medium to coarse sand or pebbly sand beds have been studied by several workers. Such rivers are similar to the classic examples described by Bernard and Major (1963), McGowen and Garner (1970), Puigdefábregas (1973), and Jackson (1976a,b), and more recent work has not introduced any major modifications to these earlier models. Jackson (1976a) described the development of the helical flow pattern in meander bends of the Wabash River, Illinois. He showed that there is typically a transition zone in the upstream part of most meander bends, in which the helical flow is reversing itself from the previous bend. In this reach, the typical point-bar sorting process does not occur, and fining-upward successions are not developed.

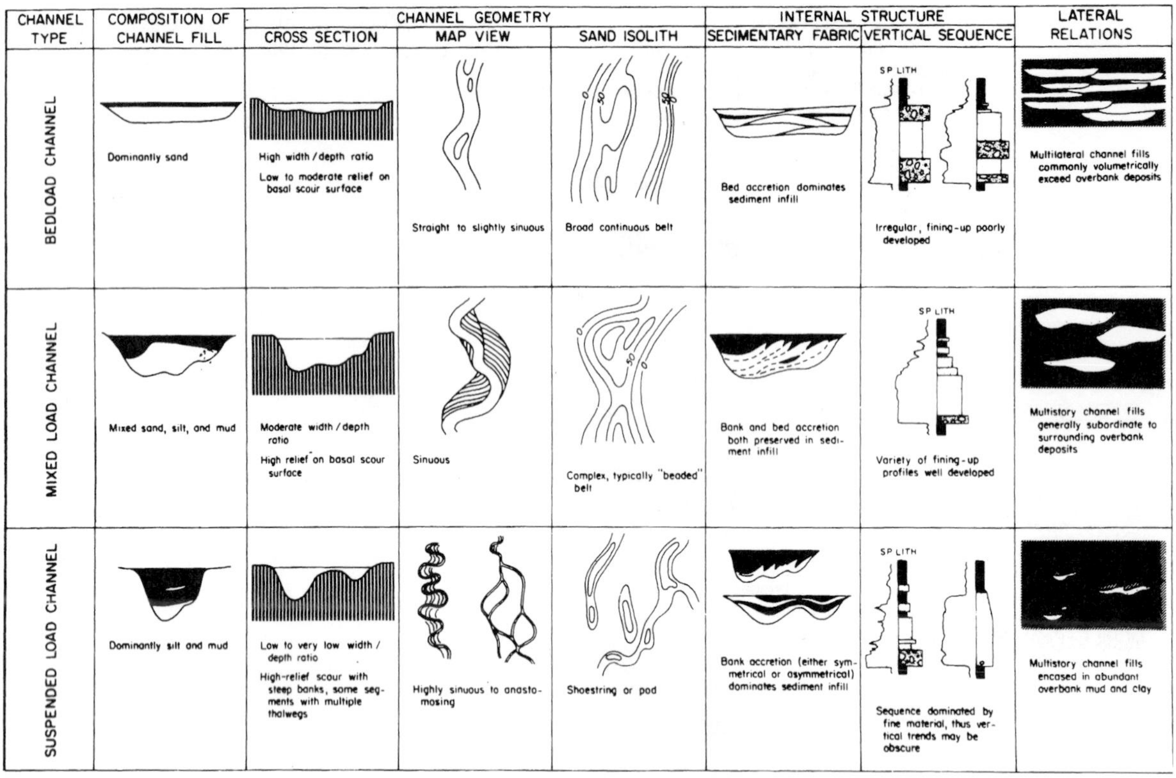

Fig. 2.36. Classification of fluvial styles, following the fluvial-geomorphological classification of Schumm (1963), as given in the left-hand column. (Galloway 1981)

Modern, fine-grained rivers with a fine-grained sediment load have been described by Nanson (1980) and Jackson (1981), and ancient examples by Miall (1979a) and Stewart (1983). Point bars consist of simple epsilon sets with banks or benches formed as a result of flow separation eddies on tight bends (Nanson 1980; Nanson and Page 1983; Fig. 2.37). Subordinate bedforms on the epsilon foreset are small or absent. Point-bar dip is steep (as much as 25°). D.G. Smith (1987) reviewed fine-grained point bars, including those deposited in tidally influenced, estuarine environments, and developed three facies models.

Thomas et al. (1987) added a useful new acronym, IHS, which stand for inclined heterolithic stratification (Fig. 2.38). Observations on a variety of modern meandering rivers in western Canada, and several ancient (Cretaceous) deposits in the Alberta Basin, led to the recognition of a distinctive style of lateral accretion, characterized by alternating fine- and coarse-grained beds. Tidal influence is suspected in many cases, as in estuarine distributary channels.

2.4.4.2 Low-Sinuosity Rivers

Few studies have been carried out on modern, multiple-channel, low-sinuosity (braided) rivers during the decade 1978–1988. Blodgett and Stanley (1980) and Crowley (1983) reexamined the Platte River, long regarded as the classic sandy-braided river, following the work of N.D. Smith (1970, 1971, 1972). Both of the papers focus on the linguoid "bars" which characterize much of the lower Platte River. Crowley (1983) demonstrated that in at least some cases these bedforms form part of much larger structures 200–400 m in length and 0.7–1.5 m high, which he interpreted as macroforms. These ideas are reexamined in Chap. 6. Other depositional processes were described by Bluck (1979, 1980). He demonstrated that coarse bar deposits commonly develop preferentially in the upstream part of bends and may migrate downstream. This process produces coarsening-upward sequences, the reverse of that usually expected in fluvial deposits.

The other major sandy-braided fluvial model is that developed for the South Saskatchewan River by Cant and Walker (1978), and applied by them to an earlier study of the Battery Point Sandstone, Quebec,

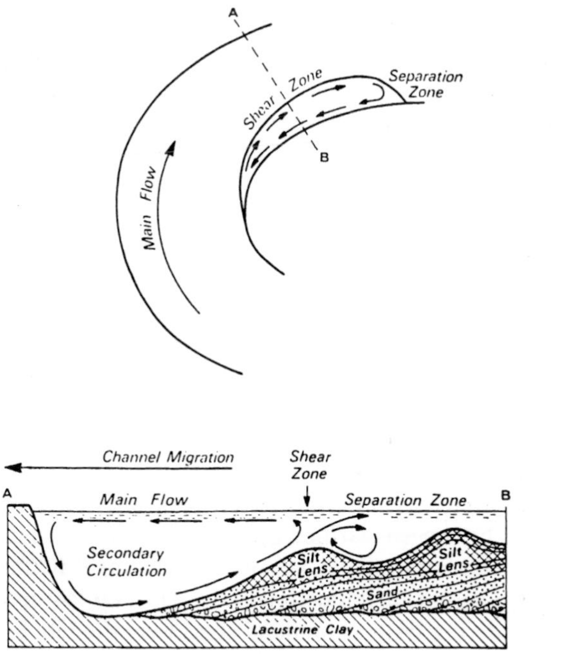

Fig. 2.37. The formation of a flow-separation zone and silt banks on a point bar. (Nanson 1980)

by Cant and Walker (1976). This model is illustrated in Fig. 2.20, and was used by Miall (1978c), together with that of the Platte River, as the two major classes of sandy-braided river for the purposes of defining vertical-profile facies models. Cant and Walker (1978) described the development of "sand flats" by the process of nucleation of small sand bars, which then grow by lateral and downstream capture and turning of migrating transverse and linguoid bedforms. However, in their work, they did not consider the internal architecture of the sand flats, and their diagrams imply that the structure is tabular. Subsequent architectural work by Miall (1988a,b,c), discussed in detail in Chap. 6, shows that the sand flats contain significant internal three-dimensional structure.

N.D. Smith and D.G. Smith (1984) described the William River, a suspended-load stream showing steady discharge that becomes characteristically braided, with linguoid bedforms and sand flats similar to those of the Platte River, where abundant sand bedload is added by eolian processes.

Bristow (1987) carried out a reconnaissance of the Brahmaptura River, one of the world's giant braided rivers. Earlier work on this river by Coleman (1969) was of considerable significance in establishing the style of bedform generation in deep, vigorous

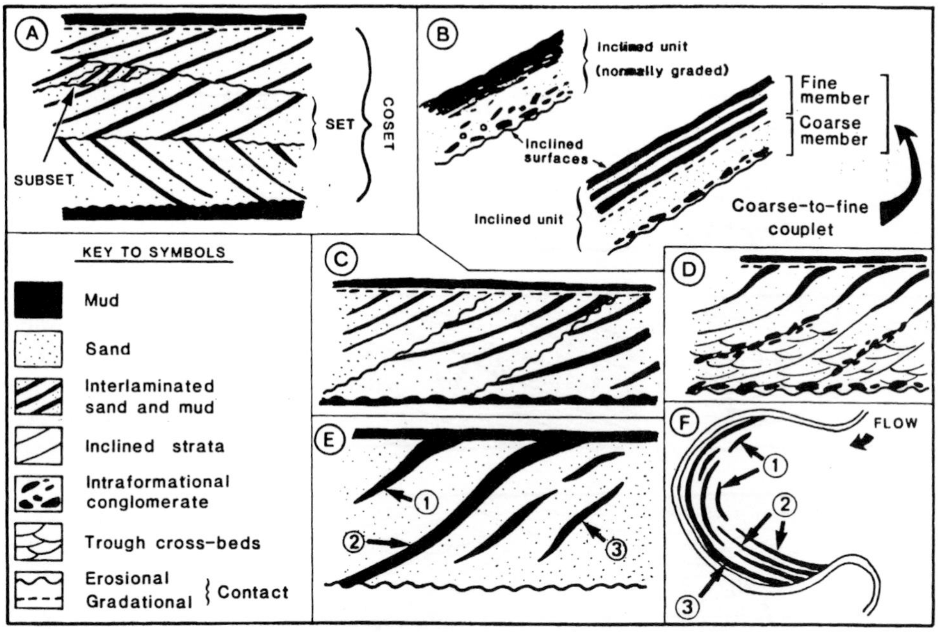

Fig. 2.38. Nomenclature for the description of inclined heterolithic stratification (IHS). **A** Vertically stacked sets, each individually may be 1–30 m thick; **B** individual IHS units, comprising graded couplets 1 cm to several meters in thickness; **C** imbricate sets; **D** composite sets; IHS set showing the three types of IHS geometry; E dip section, showing three types of geometry of discontinuous lamination (*1–3*); **F** plan view of bar showing three possible types of along-strike continuity (*1–3*). (Thomas et al. 1987)

channels. Bristow's (1987) main contribution was to document the growth of the large bar forms (macroforms), both bank-attached, lateral and point bars, and midchannel bars. Lateral and downstream accretion was demonstrated to be of considerable importance. He also demonstrated that the giant "sand waves" described by Coleman (1969) have downstream accretion faces that dip at only 4°. They are, therefore, not the simple avalanche-face bedforms that many authors have assumed them to be, but are probably complex macroforms, comparable to the downstream-accretion deposits described in Chap. 6.

The first attempts at a systematic documentation of the architectural details of a sandstone succession were those of Allen (1983a) and Ramos et al. (1986). Their classifications of architectural elements are shown in Figs. 2.39 and 2.40.

2.4.4.3 Anastomosed Rivers

A facies model for anastomosed rivers was first brought to the general attention of sedimentologists by D.G. Smith and N.D. Smith (1980). The term "anastomosed" is used "for an interconnected network of low-gradient, relatively deep and narrow, straight to sinous channels with stable banks composed of fine-grained sediment (silt/clay) and vegetation ... Separating the channels are floodplains consisting of vegetated islands, natural levees, and wetlands. Occasionally, crevasse channels and splay deposits occur in wetland areas" (D.G. Smith and N.D. Smith 1980). Channels have low width/depth ratios (< 15) and, for the most part, aggrade vertically. They may correspond to the ribbon channel bodies of Friend et al. (1979) and Friend (1983).

D.G. Smith (1983) focused on anastomosed rivers in humid climates. Rust (1981), Rust and Legun (1983), and Nanson and Rust (1985) showed that a similar fluvial style could occur in arid regions, but it has now been shown that these rivers are probably anomalous (Sect. 8.2.10).

Several misconceptions regarding anastomosed rivers have been clarified by later work in this environment. For example, because the channels do not sweep across their floodplains, as do braided and meandering systems, wetland areas were thought by D.G. Smith and N.D. Smith (1980) to be prime sites for long-term peat accumulation, and several coal deposits were initially interpreted as the product of anastomosed fluvial systems (D.G. Smith and Putnam 1980; Leblanc Smith and Eriksson 1979; Cairncross 1980; Flores and Hanley 1984). However, it is now realized that the process of channel crevassing and avulsion, which is the normal style of

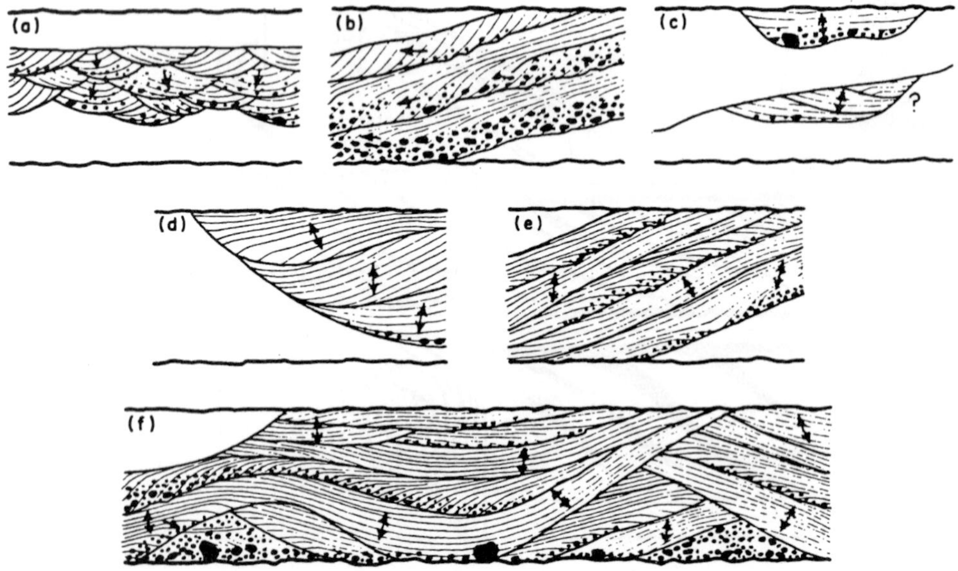

Fig. 2.39. The main types of depositional features in the Brownstones of the Welsh Borderlands. **a** Tabular layers of trough cross-bedding; **b** assemblages of down-climbing (forward accreting) bar units; **c** minor channels; **d** major channels; **e** lateral-accretion units; **f** symmetrical complexes (sand shoals) of lateral-accretion units with gravel cores. (Allen 1983a)

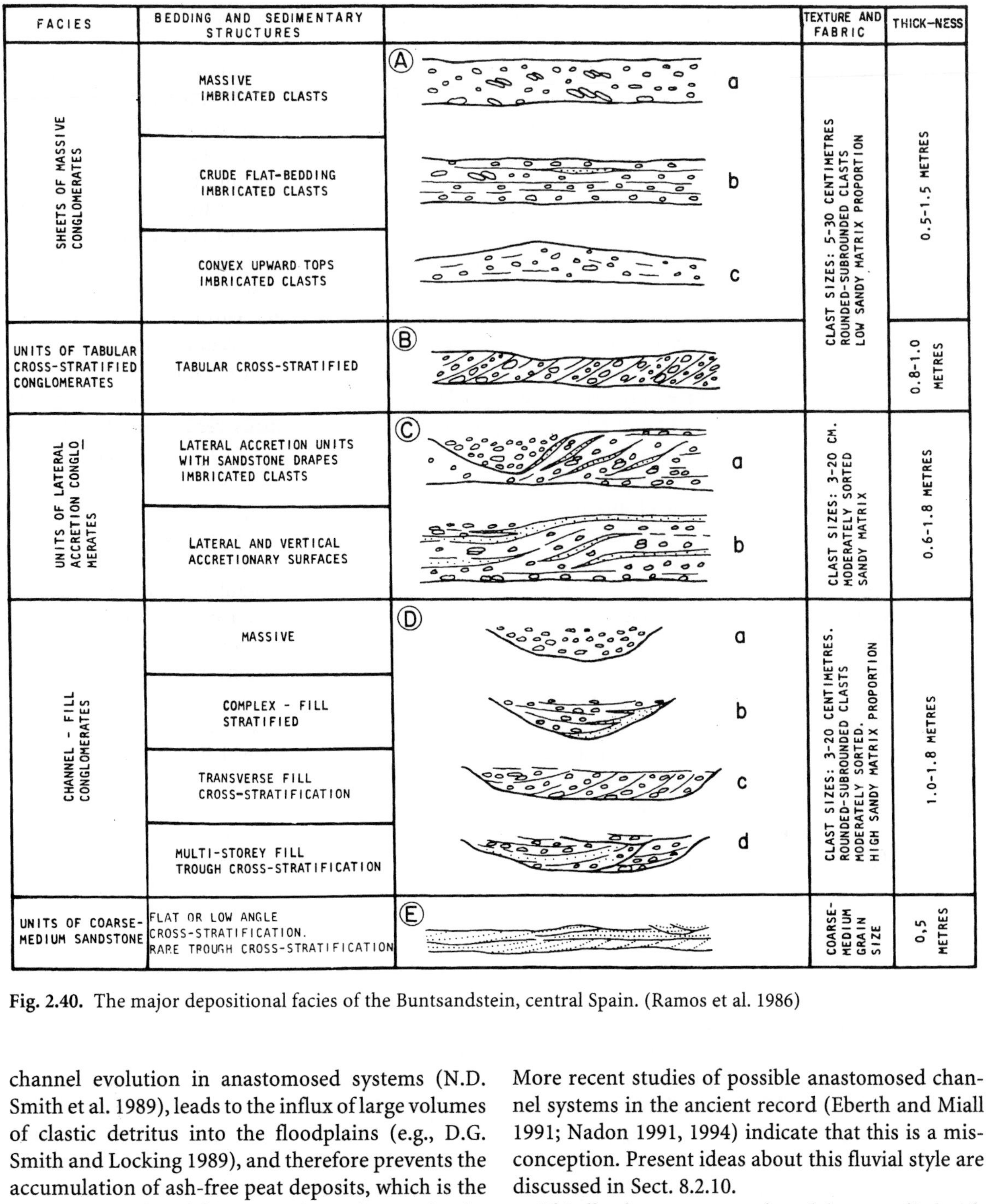
FACIES
BEDDING AND SEDIMENTARY STRUCTURES
TEXTURE AND FABRIC
THICK-NESS
SHEETS OF MASSIVE CONGLOMERATES
MASSIVE IMBRICATED CLASTS
CRUDE FLAT-BEDDING IMBRICATED CLASTS
CONVEX UPWARD TOPS IMBRICATED CLASTS
A
a
b
c
CLAST SIZES: 5-30 CENTIMETRES ROUNDED-SUBROUNDED CLASTS LOW SANDY MATRIX PROPORTION
0.5-1.5 METRES
UNITS OF TABULAR CROSS-STRATIFIED CONGLOMERATES
TABULAR CROSS-STRATIFIED
B
0.8-1.0 METRES
UNITS OF LATERAL ACCRETION CONGLOMERATES
LATERAL ACCRETION UNITS WITH SANDSTONE DRAPES IMBRICATED CLASTS
LATERAL AND VERTICAL ACCRETIONARY SURFACES
C
a
b
CLAST SIZES: 3-20 CM. MODERATELY SORTED SANDY MATRIX
0.6-1.8 METRES
CHANNEL - FILL CONGLOMERATES
MASSIVE
COMPLEX - FILL STRATIFIED
TRANSVERSE FILL CROSS-STRATIFICATION
MULTI-STOREY FILL TROUGH CROSS-STRATIFICATION
D
a
b
c
d
CLAST SIZES: 3-20 CENTIMETRES. ROUNDED-SUBROUNDED CLASTS MODERATELY SORTED. HIGH SANDY MATRIX PROPORTION
1.0-1.8 METRES
UNITS OF COARSE-MEDIUM SANDSTONE
FLAT OR LOW ANGLE CROSS-STRATIFICATION. RARE TROUGH CROSS-STRATIFICATION
E
COARSE-MEDIUM GRAIN SIZE
0.5 METRES

Fig. 2.40. The major depositional facies of the Buntsandstein, central Spain. (Ramos et al. 1986)

channel evolution in anastomosed systems (N.D. Smith et al. 1989), leads to the influx of large volumes of clastic detritus into the floodplains (e.g., D.G. Smith and Locking 1989), and therefore prevents the accumulation of ash-free peat deposits, which is the prerequisite of coal formation (McCabe 1984). The coal deposits referenced above will therefore need reevaluation in the light of this improved understanding.

Secondly, the original facies model of D.G. Smith and N.D. Smith (1980) overemphasized the idea of vertical-channel aggradation, to the extent of proposing that "walls" of channel-fill sand or gravel could accumulate beneath an aggrading channel. More recent studies of possible anastomosed channel systems in the ancient record (Eberth and Miall 1991; Nadon 1991, 1994) indicate that this is a misconception. Present ideas about this fluvial style are discussed in Sect. 8.2.10.

Thirdly, the anastomosed model was applied with great enthusiasm to the interpretation of some channelized petroleum-bearing sandstone bodies in the subsurface, notably the Mannville Sandstones (Cretaceous) of the Lloydminster area, Alberta (Putnam and Oliver 1980; D.G. Smith and Putnam 1980). This interpretation was controversial, in part because the modern analogs of the channel sandstones undergoing interpretation seemed quite inappropriate, being small rivers in Banff National Park, in narrow mountain valleys undergoing degradation, quite different from the large channels within the broad, aggrading alluvial plain of the Alberta foreland basin (as revealed by sandstone body geometries). Putnam (1993) has noted one of the major difficulties in relating surface facies studies to subsurface interpretations, and that is that "sand" bodies interpreted from wireline logs may include crevasse and levee deposits, which, combined with the width of the actual channel deposit, would indicate much wider sand-body dimensions than that for the channel alone, as observed in studies of modern rivers. This controversy, and an alternative interpretation, are discussed further in Sect. 9.5.3.

Nanson and Croke (1992) indicated that cohesive banks and low stream power are the key characteristics that lead to the development of anastomosed channels. N.D. Smith et al. (1989) suggested that the anastomosed fluvial pattern may be but a transitional style that develops during a long-term process of channel avulsion within a meandering system. However, this has yet to be confirmed by later work. The reversion to a single, stable channel observed by these authors may be a reflection of the stable, cratonic location of the river they studied. Work by other researchers in steadily-subsiding basins (e.g., D.G. Smith 1986; Shuster and Steidtmann 1987; Kirschbaum and McCabe 1992; Nadon 1994) suggests that the anastomosed pattern may be dynamically stable for significant periods of geological time.

2.4.4.4 *Ephemeral Rivers*

Miall (1977) proposed using the flood deposits of Bijou Creek, Colorado, as the type example of an ephemeral stream, based on the work of McKee et al. (1967). Subsequently, it has been shown that some of the details of this model need modification, notably the presumed predominance of the upper flow-regime plane-bed lithofacies (flat-bedded sandstone) has been shown to be incorrect (F.G. Ethridge, pers. comm., 1985).

Important new details of facies associations and bed geometries in ephemeral streams were presented by Tunbridge (1981, 1984), Stear (1983), and Wells (1983), based on well-exposed ancient examples. Tunbridge (1984) and Stear (1983) showed that point-bar deposits, with lateral-accretion cross-bedding, could develop into ephemeral streams. Stear (1983) illustrated channel sandstones showing both the ribbon and sheet geometries of Friend (1983).

Deposits of some recent floods have been described by Lucchitta and Suneson (1981), Sneh (1983), and Stear (1985), and include useful data on sediment characteristics. Parkash et al. (1983) described the sedimentology of a modern low-gradient "terminal alluvial fan," on which flow decreases downstream to zero at the distal edges, due to evaporation and infiltration. Some ancient deposits that were possibly formed in this type of environment were described by Friend (1978).

Detailed facies and architectural work has now led to the recognition of several subtly different fluvial styles that develop under arid, ephemeral conditions. These are described at length in Chap. 8.

2.4.4.5 *Large Rivers*

Few thorough sedimentological studies of any modern giant river have been undertaken since the seminal contribution by Coleman (1969) on the Brahmaputra. Bristow (1987, 1993a) reexamined this river, as discussed in Sect. 2.4.4.2 (and in more detail in Sect. 6.6).

Recognition of the deposits of giant rivers (including delta distributaries) requires identification of bar complexes or channel scour surfaces with appropriately large relief. A few recent studies have achieved this. Jones and McCabe (1980) described complex cross-bed sets 40 m thick (Fig. 2.27) which they attributed to deposition on alternate bars in a deep deltaic distributary channel. Several studies of the Athabasca Oil Sands, Alberta (Mossop and Flach 1983; Flach and Mossop, 1985) documented point bars 25 m thick, also interpreted as the deposits of distributary channels, probably in a tidally influenced environment. These deposits are discussed further in Sect. 8.2.9. The deep fluvial channels in the Morrison Formation, which were reconstructed by Campbell (1976), have now been shown to be a result of misinterpretation of the outcrop record (Cowan 1991; see Sects. 2.4.3.2, 10.3.2).

2.4.4.6 *Floodplain Environments*

The predominantly fine-grained deposits of the floodplain have generally received much less detailed study than the coarse "framework" gravel and sand units of the channels. Several commmentators noted this at the first fluvial symposium. The major exception was comprised of those geologists concerned with the stratigraphy and sedimentology of coal (e.g., Horne et al. 1978; Flores 1981, 1984; Fielding 1984; Rahmani and Flores 1984; McCabe 1984).

One of the most distinctive characteristics of the anastomosed fluvial environment is the presence of large, stable floodplain areas, typically wetlands with large crevasse splays. Some of the early references to this model are noted in Sect. 2.4.4.3, although most of these studies are concerned with the composition and geometry of the channel and crevasse-splay units rather than the fine-grained deposits of the floodplain. Similarly, Bridge (1984), in a paper on facies sequences in overbank environments of high-sinuosity rivers, dealt exclusively with the types of fining-upward and coarsening-upward sandstone-dominated successions found in levee and crevasse-splay deposits.

Several studies of coal-bearing successions subdivided floodplain deposits into a series of distinct facies assemblages. For example, Ethridge et al. (1981) recognized crevasse-splay, levee, abandoned-channel, well-drained swamp, poorly drained swamp, and lacustrine assemblages. A similar range of facies was described by Gersib and McCabe (1984). Both of these studies focused on vertical profiles, and on the building of generalized facies models, but did not deal in detail with floodplain architecture. Wing (1984) subdivided plant-bearing floodplain fines into two main facies assemblages, and was able to construct a separate depositonal model for each, based on lithofacies, plant assemblages, and unit geometry. Tabular units were interpreted as the deposits of gradually aggrading swamps, whereas lenticular bodies represent the fill of abandoned channels. This is one of the few studies to have exploited the floodplain deposits themselves for their potential to yield architectural information. As discussed in Chap. 9, floodplain sequences may hold the key for unlocking a vast amount of architectural detail, but few workers have made use of this potential.

Coal is commonly interpreted as an integral part of the floodplain succession in fluvial and delta-plain environments. McCabe (1984) pointed out, however, that in an active clastic environment characterized by crevassing and avulsion, floodplains receive so much detrital input that high-quality, low-ash coals are unlikely to form. Raised peat swamps may develop where rainfall permits an elevated water table to be maintained as rapid peat accumulation outpaces channelized clastic aggradation. The relief of these swamps prevents detrital influx, and constrains channel migration and avulsion (Flores 1984; McCabe 1984).

The use of floodplain deposits in reconstructing fluvial architecture is discussed in Sect. 2.4.3.2. A modern classification of floodplains by Nanson and Croke (1993) is presented in Chap. 8 (Table 8.4) and is discussed in Chaps. 7 and 8.

2.5 Conclusions

Although sedimentology and basin analysis have always been concerned with the geology of rocks in three dimensions, during the 1970s the main focus was on the elaboration of facies models, relying heavily on the use of the vertical profile. The weaknesses in this method were apparent by the end of that decade, and several unrelated attempts to systematize architectural methods were made during the 1980s. It is only since the late 1980s that we have had the tools to carry out systematic three-dimensional investigations of fluvial strata, using the architectural techniques described in later chapters of this book.

Bersier, in 1948, attributed fluvial fining-upward cycles to a tectonic cause. However, the 1960s and 1970s were a time when sedimentological studies were dominated by a fascination with autogenic processes. Much of the fascination of the facies-model revolution was the discovery that appeals to such awkward mechanisms as pulsating tectonism were not necessary. In the case of fluvial sedimentology, this led to the development of the point-bar model and, later, an unraveling of the complexities of braided and anastomosed rivers. To some exent, however, we are coming full circle. The latest stratigraphic revolution – that of sequence stratigraphy – is returning us to some old ideas about allogenic causes. The field of sequence stratigraphy, currently one of the most exciting areas of research in stratigraphy, evolved from new ideas about basin development that derived from developments in the analysis of seismic-reflection data during the 1970s. These burst onto the geological scene in 1977 with the publication of AAPG Memoir 26 (Payton 1977), but the implications of this research for fluvial sedimentology only became apparent with the work of Posamentier and Vail (1988) and Posamentier et al. (1988), at the end of the period dealt with in this chapter. The ideas in these latter papers are still in the category of current research, and are discussed in Chaps. 11 and 13. It is being shown that many clastic successions consist of sequences that are basin-wide or regional in extent, and are developed by allogenic mechanisms such as source-area tectonism, changes in rates of basin subsidence, basin tilting, or eustasy. As discussed later in this book, many fluvial successions contain fining- and coarsening-upward sequences of regional extent. Correlation and mapping of these sequences have led to a renewed interest in vertical profiles, but have emphasized the need for architectural methods to discriminate the various cycles and sequences that have similar vertical profiles, but are of several different origins.

Miall (1978b) noted the paucity of studies of modern rivers, and the consequently limited scope of the data base on which fluvial sedimentology rested. Little has changed in the intervening years. There is still considerable room for detailed sedimentological studies of the facies composition and architecture of modern rivers, and their relationship to flow patterns and discharge characteristics. As noted by N.D. Smith (pers. comm., 1995) many of the fluvial styles and facies models now used by sedimentologists (Chap. 8) are based largely on studies of the ancient record.

Several areas of research that were active up to the end of the 1980s, and are discussed at length in this chapter, are not dealt with in detail elsewhere in this book. Grain-size analysis, once thought to hold great promise as a basin-analysis tool (Sect. 2.3.4), is a far more cumbersome technique for field research than facies analysis, requiring considerable laboratory followup work, and the results have been shown to be ambiguous or plain wrong. Paleohydraulic analysis reached a kind of plateau of development with the work of Ethridge and Schumm (1978), and has been pursued by very few workers since that time, presumably because of the pitfalls and imprecisions in the method that were revealed by that review. However, some of the methods of paleohydraulic analysis have been adapted for the investigation of sandbody geometry in the subsurface, and are described in Chap. 10.

Chapter 3

Concepts of Scale

3.1 Time Scales and Physical Scales in Sedimentation

It has become a geological truism that many sedimentary units accumulate as a result of short intervals of rapid sedimentation separated by long intervals of time when little or no sediment is deposited (Ager 1981, 1993). It is also now widely realized that rates of sedimentation measured in modern depositional environments or the ancient record vary in proportion to the time scale over which they are measured. Sadler (1981) documented this in detail, and showed that measured sedimentation rates vary by 11 orders of magnitude, from 10^{-4} to 10^{7} m/ka. This wide variation reflects the increasing number and length of intervals of nondeposition or erosion factored into the measurements as the length of the measured stratigraphic record increases. Breaks in the record include such events as the nondeposition or erosion that takes place in front of an advancing bedform (a few seconds to minutes), the nondeposition due to drying out at ebb tide (a few hours), up to the major regional unconformity generated by orogeny (millions of years). The variation in sedimentation rate also reflects the variation in actual rates of continuous accumulation, from the rapid sand flow or grain-fall accumulation of a cross-bed foreset lamina (time measured in seconds), and the dumping of graded beds from a turbidity current (time measured in hours to days), to the slow pelagic fill of an oceanic abyssal plain (undisturbed in places for hundreds or thousands of years, or more). There clearly exists a wide variety of time scales of sedimentary processes (Fig. 3.1).

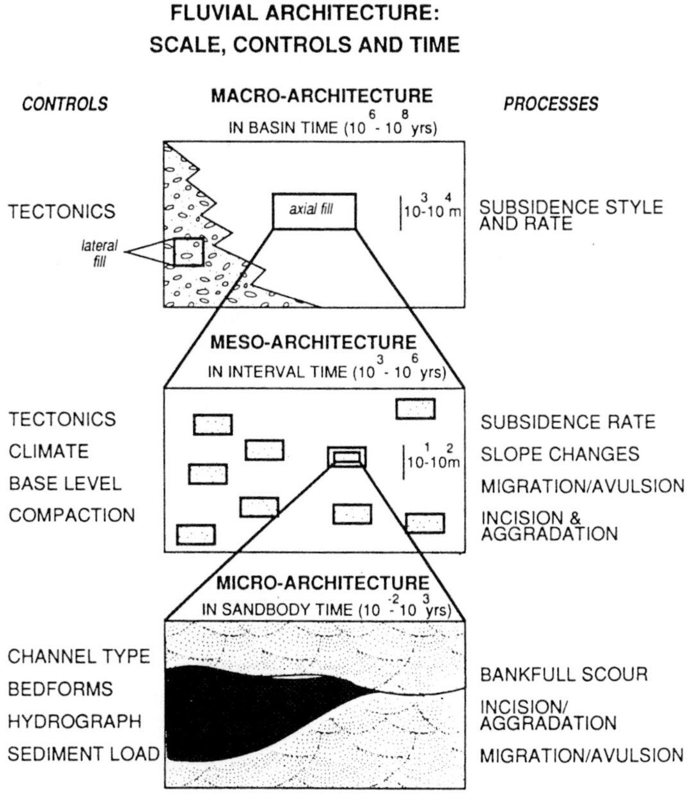

Fig. 3.1. Hierarchies of scale and time in fluvial deposits. (Leeder 1993)

There also exists a hierarchy of physical scales, which the same two examples illustrate – the cross-bed foreset at one extreme to the basin-fill at the other extreme (Fig. 3.1). At least 14 orders of magnitude are represented, from the few square centimeters in area of the smallest scale of ripple foreset, to the tens of thousands of square kilometers of a major sedimentary basin. This chapter, which is based on an earlier review by Miall (1991a), is a systematic exploration of these wide ranges in scales.

The ways by which earth scientists study sedimentary processes and the resultant depositional products vary according to the scale of interest. Bedforms in flumes are studied during experimental runs of, at most, few days duration. Nonmarine and marginal-marine sediments and processes have been much analyzed in modern environments, using studies of surface processes, and by sampling the sediments themselves in trenches and shallow cores. The use of old maps and aerial photographs extends the record as far back as about 100 years, and ^{14}C dates may enable stratigraphic records of the last few tens of thousands of years to be calibrated. Many sedimentological studies draw on geomorphological work on landforms and Recent sediments. However, such work is hampered by the specific, and possibly nongeneralizable nature of the Recent record, such as the Holocene deglaciation, climatic change, and rapid rise of global sea levels. Stratigraphic studies typically deal with much longer time periods, as represented by the deposits of basin fills, which may have taken hundreds of thousands to millions of years to accumulate. Intermediate scales, represented by such major depositional elements as large channels and bars, delta lobes, draas, coastal barriers and shelf sand ridges, which may represent thousands to tens of thousands of years of accumulation, are particularly difficult to document in the ancient record and to analyze in modern environments. The time scales of the relevant sedimentary processes are difficult to resolve, and the physical scale of the deposits falls between the normal size of large outcrops and the well spacing or the scale of geophysical resolution in the subsurface. Yet, it is this scale of deposit that is of particular interest to economic geologists, representing as it does the scale of many stratigraphic petroleum reservoirs and their internal heterogeneities.

Geomorphologists have devoted considerable attention to the problem of time scales and their effects on analysis and prediction (Cullingford et al. 1980; Hickin 1983; Schumm 1985a). As Hickin (1983, p. 61) has stated, "time-scale selection largely determines the questions that we can ask." Schumm (1985a) showed that the significance of an event diminishes as the time-scale increases. Thus, an individual volcanic eruption, a spectacular geological event at the time of its occurrence (a "megaevent", to use Schumm's term), diminishes in geological importance as the millenia go by and other eruptions take place, until eventually, after perhaps millions of years, all evidence of the eruption is lost (it becomes a "nonevent") as a result of erosion or burial of the rocks and landforms formed by the eruption. Events that seem random in the short term (such as turbidity-current events) may assume a regular episodicity, or even cylicity, with definable recurrence intervals, if studied over a long enough time scale. Many events occur only when some critical threshold has been passed, such as the buildup of deposists on a depositional slope leading to gravitational instability and failure. In several essays, Schumm (1977, 1979, 1985a, 1988; Schumm and Brakenridge 1987) has discussed the concept of "geomorphic thresholds" and their impact on sedimentary processes. Such thresholds reflect both autogenic and allogenic processes, and are characterized by a wide range of time scales (Fig. 3.2) and scales of cyclicity (Fig. 3.3). Schumm (1985a) also provided a table which illustrates the magnitude of various geological events as related to the time scale over which they are considered (Table 3.1).

The incorporation of hierarchical scale concepts into fluvial studies requires an architectural approach. Earlier approaches to the architectural study of fluvial deposits are described in Chap. 2, notably the work of J.R.L. Allen and A. Ramos and his colleagues. The main classification used in this book is described in Chap. 4. The current explosion of interest in sequence stratigraphy represents an increasing interest in large-scale stratigraphic architecture, and its dependence on such allogenic controls as tectonics and sea-level change. This work is referred to briefly here, but the main discussion is presented in Chap. 13. The applications of the concepts to the study of reservoir heterogeneities are discussed in Chap. 14.

A review of the depositional processes in clastic environments, and the ways by which various workers have systematized them in depositional hierarchies, has revealed some common themes, although it is premature to propose a unified hierarchy of depositional elements for use in the description and classification of all clastic rocks. Most within-basin (autogenic) or external (allogenic) sedimentary controls have durations of a rather constant and predict-

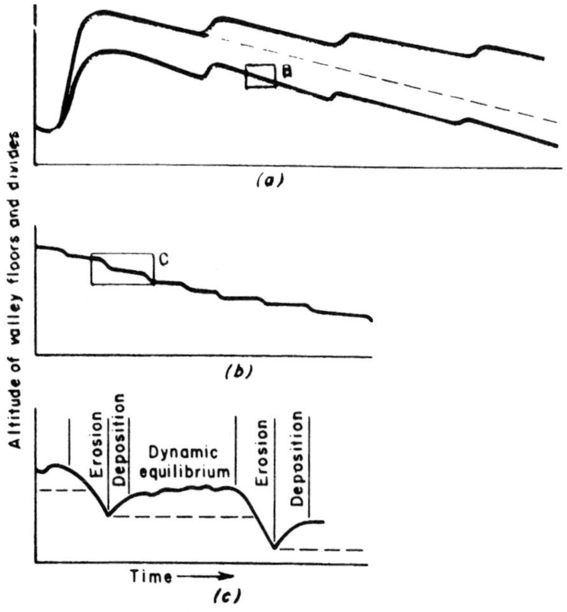

Fig. 3.2a–c. The various time scales of geomorphic processes. **a** The erosion cycle, as envisioned by W.M. Davis in the nineteenth century. The *lower line* indicates the elevation of the valley floor, the *upper line* that of drainage divides. Initial uplift is followed by degradational lowering and episodic pulses of isostatic uplift in response to erosional unroofing. Total elapsed time is in the order of 10^{7-8} years for a major drainage basin, with minor uplift events occurring on the scale of 10^{6-7} years (corresponding to the tectonic cyclothems of Blair and Bilodeau 1988). *Box labeled B* is enlarged in **b** In detail, the valley floor shows an episodicity on a smaller time scale (in the range of 10^{2-3} years) as a result of the periodic storage and flushing of sediment from bars and floodplain deposits, for example by avulsion events. *Box labeled C* is enlarged in **c** The episodicity of **b** is shown here in greater detail. (Diagram from Schumm 1977)

able order of magnitude, permitting a grouping of sedimentary units according to the length of depositional time that they represent. The sedimentary results vary from environment to environment because of the various combinations of processes that occur. Thus, a comparison between subaqueous unimodal flow (e.g., fluvial systems), subaqueous oscillatory flow (tidal environments), and subaerial flow (eolian environments) is not straightforward, however, the assignment of common groupings facilitates clearer thinking about depositional controls. As a simple example, the recognition of "super surfaces" in eolian systems enabled Kocurek (1988) to speculate about the potential mapping of migrating ergs, the detailed correlation of ergs with marine units, the application of sequence-stratigraphy concepts to nonmarine systems, and the documentation

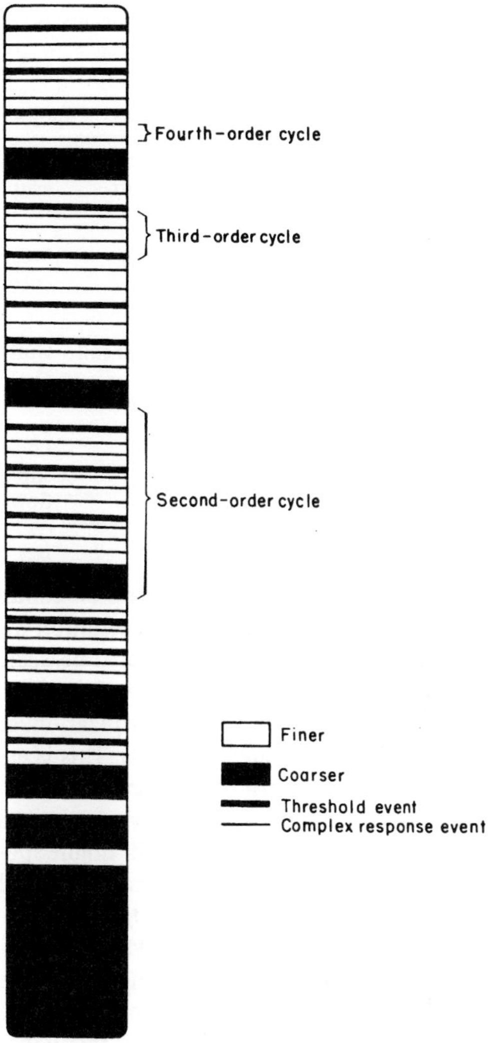

Fig. 3.3. The hierarchy of cycles of sedimentation, based on geomorphic concepts of the complex and episodic response of fluvial systems to autogenic and allogenic forcing. The primary cycle is the entire succession, reflecting the gradual diminution of sediment grade following initial uplift (corresponding to the "erosion cycle" curve of Fig. 3.2a; group 11 of Table 9.1). Second-order geomorphic cycles reflect isostatic adjustments (tectonic cyclothems) or major climate change (the kinks in the curves of Fig. 3.2a; corresponding to group 10 of Tables 3.1, 4.2, 9.1). Third-order geomorphic cycles are those exceeding geomorphic thresholds, leading to periods of "metastable equilibrium" and periods of rapid change and adjustment (the events shown in Fig. 3.2b). These processes occur over various time scales (groups 8 and 9). Fourth-order cycles are related to episodic erosion, and to the complex response of the fluvial system to any of the above changes (group 7?). Fifth-order cycles are related to seasonal and other major hydrological events, such as the "100-year flood" (groups 5, 6). (Schumm 1977)

Table 3.1. Perceived importance of geologic events as related to size and age. (After Schumm 1985a)

Relative magnitude of event	1 day	1 year	10 yr	102 yr	103 yr	105 yr	106 yr	108 yr
Megaevent	Local soil slip or flow	Gully	Meander cutoff	Volcanic eruption	Terrace formation	Continental glaciation	Major folding, faulting	Mountain building
Mesoevent	Rill	Local soil slip or flow	Gully	Meander cutoff	Volcanic eruption	Terrace formation	Continental glaciation	Major folding, faulting
Microevent	Sand grain movement	Rill	Local soil slip or flow	Gully	Meander cutoff	Volcanic eruption	Terrace formation	Continental glaciation
Nonevent	–	Sand grain movement	Rill	Local soil slip or flow	Gully	Meander cutoff	Volcanic eruption	Terrace formation

of regional climatic and tectonic controls on eolian deposits. The eventual availability of a common hierarchy of architectural units based on time scale and physical scale would facilitate more of this type of correlation of events and cross-fertilization of ideas between contrasting environments. This is especially important in an era of compartmentalization of knowledge, where sedimentary geologists studying different aspects of a basin fill might specialize in such diverse areas as (for example) fluid hydraulics, facies analysis, geochemistry, paleoclimatology, regional tectonics, or geophysical subsidence models, yet have very little common language to facilitate exchange of ideas and concepts. The classification and grouping discussed in this paper are offered as a first step toward a "grand unified theory" of architectural classification.

Architectural scale concepts are also important in the petroleum industry. Reservoir heterogeneities are present at at least four scales (Fig. 3.4) that can be related to the scale hierarchy described here. The analysis of each scale of heterogeneity requires different techniques, as described throughout the rest of this book.

3.2 The Grouping of Architectural Units in Clastic Rocks According to Depositional Time Scale

A tentative classification and grouping of architectural units into ten classes spanning at least 12 orders of magnitude of time scale is shown in Table 3.2. Deposits in most clastic environments can be assigned a grouping based primarily on the time scale represented by the deposit, or by the total time elapsed during the formation of the deposit and its bounding surfaces (which may be considerably greater). The grouping here is based primarily on the latter criterion – the recurrence interval of the depositional process. In some cases, such as sediment-gravity flows, there is a marked contrast between the duration of a depositional event and its recurrence interval. The latter may vary by more than an order of magnitude within a basin or between basins, but quantifying such variations may be difficult or impossible using present techniques. Because of such variation, the grouping of some types of deposit in the discussion that follows is somewhat arbitrary.

In a recent attempt to systematize our knowledge of large-scale subaqueous bedforms, Ashley (1990) suggested that the scale of a bedform may simply be a reflection of the physical space, volume of sediment, and elapsed time available for its formation within a flowing system. Dunes, sand waves, and other two- and three-dimensional bedforms, which have gone by various names in the past, may all be part of a genetically related continuum. Whereas this argues against any attempt to impose an arbitrary hierarchical size classification on hydrodynamic sedimentary structures, the fact remains that the timing of depositional and erosional events within trains of bedforms is strongly controlled by rhythmic or episodic events on various time scales (diurnal, seasonal, etc.). Packages of cross-bedded strata and their enclosing bounding surfaces are, therefore, amenable to a hierarchical classification based on depositional recurrence interval, as discussed in this book.

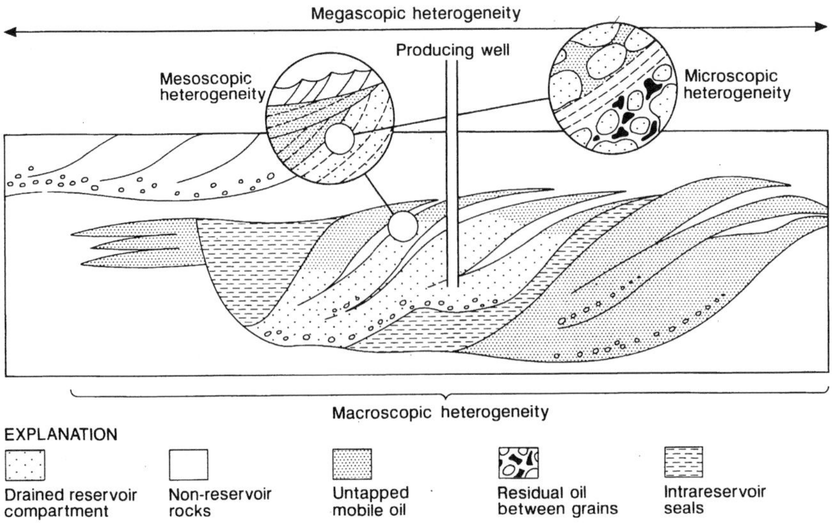

Fig. 3.4. Scales of heterogeneity in a meander-belt reservoir sandstone. Untapped mobile oil is oil trapped by heterogeneities that require special mapping techniques to document (Sects. 9.5, 10.4). In terms of the sediment "groups" described in this book (Table 3.2), *megascopic heterogeneity* corresponds to groups 9–11, *macroscopic heterogeneity* is equivalent to groups 7 and 8, *mesoscopic heterogeneity* refers to groups 4–6 and *microscopic heterogeneity* corresponds to groups 1–3. (Ambrose et al. 1991)

The rankings that have been assigned to the types of bounding surface that enclose the deposits of a given group are shown in square brackets for the fluvial/deltaic, eolian and coastal/estuarine columns in Table 3.2. The bounding surfaces have typical geometric shapes, areal extents and lithofacies associations, and can be used to define hierarchies of depositional units or architectural elements. The rankings are tentative, as discussed in the next section. In the submarine-fan column, the numbers in square brackets refer to proposed hierarchical classifications of depositional units that have not been extended yet to include specific types of bounding surface. The rankings are based on the work of Mutti and Normark (1987), and have been discussed by Miall (1989), who showed how some of the architectural-element concepts derived from fluvial sediments could be applied to coarse, deep-sea channel-fill clastic systems.

The examples of sedimentary processes shown in column three are not intended to comprise a complete list, but to provide a flavor of the variation in process with time scale. This list is amplified in the following discussion, and many examples of fluvial deposits classified into the various groups are discussed throughout this book. The groups pertain to both the depositional elements and to their bounding surfaces, where shown. Order-of-magnitude values for rates of sedimentation are indicated in column 4. The basis on which these values have been assigned is discussed in Sect. 3.4.

Group 1 deposits are those formed in a few seconds, such as the lamination developed by the burst-and-sweep process in traction currents (Leeder 1983), and the sediment layers accumulated by eolian grain fall and grain flow.

Group 2 deposits are those formed in periods of a few minutes to a few hours. Ripple trains are the typical depositional unit in many clastic environments. As shown by Southard and others (1980), the time taken for the migration of a waterlain ripple set by one wavelength typically varies from 20 min to 2 h, with shorter times (20–60 min) more common. Wind ripples may form and migrate in similar time periods.

Group 3 deposits are those that form in periods of a few hours to a day or two. Diurnal processes are the most important in this category of time scale, such as tidal cycles, daily variations in spring-melt runoff reflecting temperature variations, and changes in wind strength and direction (which commonly occur because of the different thermal properties of land and adjacent water masses). These processes typically develop distinctive sediment bundles or cyclic sequences on a small scale.

In fluvial environments, group 3 deposits consist of complete ripple sets and growth increments (foreset bundles) of megaripples (dunes). In eolian

Table 3.2. Hierarchies of architectural units in clastic deposits

Group	Time scale of process (years)	Examples of processes	Instantaneous sedimentation rate (m/ka)	Fluvial, deltaic Miall	Eolian Brookfield, Kocurek	Coastal, estuarine Allen	Shelf Dott and Bourgeois, Shurr	Submarine fan Mutti and Normark
1	10^{-6}	Burst-sweep cycle		Lamina	Grain flow grain fall	Lamina	Llamina	
2	10^{-5} -10^{-4}	Bedform migration	10^{5}	Ripple (microform) [1st-order surface]	Ripple	Ripple [E3 surface]	[3rd-order surface in HCS]	
3	10^{-3}	Diurnal tidal cycle	10^{5}	Diurnal dune incr., react. surf. [1st-order surface]	Daily cycle [3rd-order surface]	Tidal bundle [E2 surface]	[2nd-order surface in HCS]	
4	10^{-2} -10^{-1}	Neap-spring tidal cycle	10^{4}	Dune (mesoform) [2nd-order surface]	Dune [3rd-order surface]	Neap-spring bundle [E2], storm layer	HCS sequence [1st-order surface]	
5	10^{0} -10^{1}	Seasonal to 10-year flood	10^{2-3}	Macroform growth increment [3rd-order surface]	Reactivation [2nd, 3rd-order] surface, annual cycle	Sand wave, [E1], major storm layer	HCS sequence [1st-order surface]	
6	10^{2} -10^{3}	100-year flood	10^{2-3}	Macroform, e.g. point-bar levee, splay [4th-order surface]	Dune, draa [1st-, 2nd-order surfaces]	Sand wave field, washover fan	[facies package (V)]	macroform [5]
7	10^{3} -10^{4}	Long-term geomorphic processes	10^{0}–10^{1}	Channel, delta lobe [5th-order surface]	Draa, erg [1st-order, super surface]	Sand-ridge, barrier island, tidal channel	[Elongate lens (IV)]	Minor lobe, channel levee [4]
8	10^{4} -10^{5}	5th-order (Milankovitch) cycles	10^{-1}	Channel belt [6th-order surface]	Erg [super surface]	Sand-ridge field, c-u cycle	[Regional lentil (III)]	Major lobe [turb. stage: 3]
9	10^{5} -10^{6}	4th-order (Milankovitch) cycles	10^{-1}–10^{-2}	Depo. system, alluvial fan, major delta	Erg [super surface]	c-u cycle	[ss sheet (II)]	Depo. system [2]
10	10^{6} -10^{7}	3rd-order cycles	10^{-1}–10^{-2}	Basin-fill complex	Basin-fill complex	Coastal-plain complex	[Lithosome (I)]	Fan complex [1]
11	10^{7} -10^{8}	2nd-order cycle	10^{-1}–10^{-2}	Basin-fill complex				

Hierarchical subdivisions of other authors are given in square brackets; names of authors are at head of each column.

environments, wind ripples, depositing climbing translatent strata, form group 3 bundles (Hunter and Richmond 1988). In tidal environments, sand-wave cross-stratification may form bundles, reflecting tidal-velocity asymmetry (Allen 1980) or variations in the supply of sand fed to the foreset crest by superimposed megaripples (Dalrymple 1984). Various types of tidal ripples were described by Dalrymple and Makino (1989).

Group 4 deposits represent time periods of a few days to a few months. Many of the depositional events that occur over such a time frame were termed "dynamic events" by Jackson (1975). They are events that move disproportionately large volumes of detritus in relatively short time periods. Many marine storm events fall into this category. Hummocky cross-stratification (HCS) sequences are thought to form in a few days to a few weeks during major storms. This time span also includes the lunar neap-spring tidal cycle. Periodicity in tidal bundles was observed by Visser (1980) and in tidal ripples by Dalrymple and Makino (1989) and equated with the lunar cycle. In fluvial systems, complete trains of mesoforms, such as dunes, may be deposited during such time spans.

Group 5 deposits are those that represent time spans of 1 year to a few tens of years. Seasonal deposits, such as spring runoff events and glacial varves, are among the most important types of deposit in this category. The deposits of occasional major storms (shelf environment) and flash floods (fluvial environment), including those of strengths occurring only every few years (e.g., McKee et al. 1967), are also group 5 deposits. Annual changes in wind strength and direction lead to cyclic deposition of eolian dunes, as has been recognized in several ancient deposits (Hunter and Rubin 1983; Kocurek et al. 1991).

Growth increments on macroforms in subaerial and subaqueous environments are typical depositional products, and may be separated by large-scale reactivation surfaces showing low-angle truncations of underlying bedding (bounding surfaces are discussed later). Examples are shown in Fig. 3.5. Coleman (1969) showed that in the Brahmaputra River giant "sand waves" (possibly more accurately classified as macroforms) are active primarily during seasonal floods. Jackson's (1976) studies of the Wabash River showed that meander bends in that river migrated up to 2 km in 50 years, generating an equivalent width of a new point bar. The Ucayali River, a tributary of the upper Amazon in Peru, shows comparable migration rates of 40–60 m/year (Dumont and Fournier 1994). Iseya and Ikeda (1989) demonstrated that, in the ridge-and-swale topography of a point bar in a meandering river in Japan, the recurrence interval of the ridges is estimated to be 25 years. However, preservation of point-bar deposits depends on the direction and style of meander migration, which could include meander expansion, translation, and rotation. Net sedimentation-accumulation rates are highest during phases of meander expansion (Jackson 1976b). Studies of the Atchafalaya delta in Louisiana indicate that sedimentation during high-discharge events may add as much as 1000 km^2 of new land to the delta during the next 50 years (Roberts et al. 1980).

Many storm sequences containing hummocky cross-stratification, although developing in a few days or weeks, may also represent the action of violent storms or storm groups that occur only every few years (Dott and Bourgeois 1982). Separation of such deposits from storm deposits of group 4 is arbitrary in this classification, because there is probably a continuum of depositional variability between the sequences that represent individual storms during an annual storm season, and those that represent the "10-year" or "100-year" event. As noted later, work on hurricane deposits is beginning to suggest some criteria for discriminating storm magnitude in carbonate environments, but there are no such data relating to hummocky cross-stratification or other terrigenous clastic deposits.

Group 6, in this classification, represents deposits accumulating over time spans of hundreds to a few thousands of years. Particularly violent dynamic events are thought to occur infrequently in many environments, and are commonly termed for convenience (if not accuracy) the "100-year events". These would include violent hurricanes, catastrophic flash floods, and sediment gravity-flow events in many environments (in some settings, sediment gravity-flows are infrequent enough to be group 7 events). Many of the resulting depositional sequences are very similar to those of group 5, such as those described earlier, and could be distinguished only by such clues as the large volume or thickness of the deposit, or the scale of the resulting bedforms. For example, Wanless and others (1988) took advantage of the passage of a major hurricane across the Caicos Platform in the British West Indies to compare its effect on carbonate tidal-flat sediments with that of "normal" winter storms. Hurricanes, which are group 5 or 6 events in this area, deposit centimeter-scale-thick beds of peloidal grainstone, whereas winter storms (group 5) generate millimeter-thick laminae. Much more work of this type is needed to provide criteria for the recognition of event magnitude in the sedimentary record.

Complete macroforms are group 6 deposits (Figs. 3.5, 3.6). Examples include point bars, sand flats, and crevasse splays in fluvial systems, and similar elements in submarine channels; major segments of coastal barriers and spits, and large washover fans in

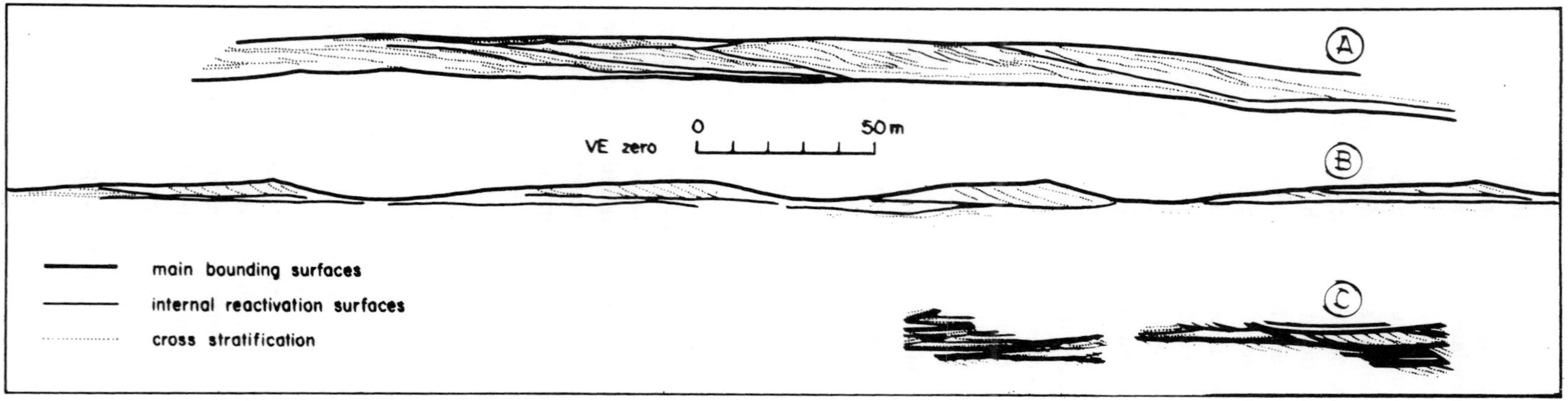

Fig. 3.5. Examples of depositional units that accumulate in a few hundred to a few thousand years (group 6). Internal reactivation surfaces subdivide the deposits into internally conformable assemblages. The reactivation surfaces may represent the result of erosion during seasonal or longer-term dynamic events, and the beds between them typically accumulate in periods of a few years to a few tens of years (group 5). **A** Fluvial or estuarine point bar, Torrivio Member, Gallup Sandstone, near Gallup, New Mexico (Miall 1991b). **B** Field of modern sand waves (dunes), Bay of Bourgneuf, France (Berné et al. 1991). **C** Triassic eolian deposits, Holy Cross Mountains, Poland (Gradzinski et al. 1979). Diagrams are drawn without vertical exaggeration

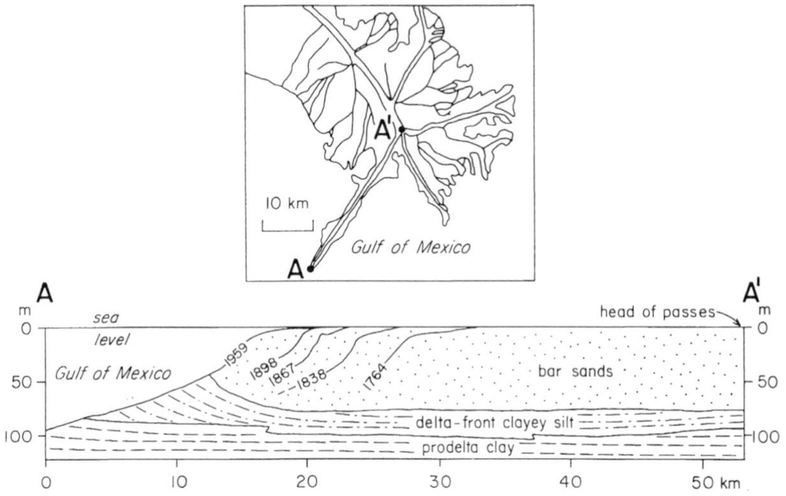

Fig. 3.6. Sedimentation rates (both vertical and lateral) are extremely rapid in such major deltas as the Mississippi. This is an example of a group 6 deposit, a bar-finger sand forming the core of a delta lobe (Southwest Pass). Such lobes are formed and abandoned in the Mississippi system in about 1000 years. *Time lines* show the historic development of this channel-fill and mouth-bar deposit, which is about 7 km wide, 70 m thick and at least 40 km in length. (Gould 1970)

coastal settings; and fields of sand waves in continental shelves.

A few quantitative estimates of sedimentation rates are available, including the following (these data are used in a discussion of vertical accumulation rates in a later section): Allen (1984) estimated that estuarine sand-wave fields take approximately 1200 years to form. Data given by Hayes and Kana (1976) show that coastal spits on the east coast of the United States grow laterally by average rates of about 200 m in 100 years, with local short-term growth rates of up to 90 m/year. Galveston Island, a regressive barrier system, has built seaward at an average rate of about 100 m per 100 years (Bernard et al. 1962). Gagliano and van Beek (1970) documented the growth of major subdeltas in the Mississippi system using old maps, and showed that deltaic growth could fill an interdistributary bay by progradation from a single crevasse in about 100 to 150 years. Diversion of major distributaries in this river system is capable of generating major subaqueous deposits in the space of less than 100 years. Roberts and others (1980) documented the effects of the diversion of flow from the Mississippi into the Atchafalaya River since the early part of this century. A major floodplain lake has been filled with sediment, and a subaerial delta appeared at the mouth of the river in 1972. Complete abandonment of the main course of the Mississippi River in favor of the Atchafalaya would probably already have occurred but for river-control emplacements constructed by the US Army Corps of Engineers (Sect. 10.3.1). Evidence from the Holocene record shows that construction and abandonment of major delta lobes in the Mississippi system take about 1000 years (Kolb and Van Lopik 1966; Frazier 1967). Individual major distributary mouths in this system advance about 3 km in 100 years, with the formation of levees and mouth-bar deposits (Gould 1970; Fig. 3.6). It is not yet known whether separate growth increments corresponding to a group 6 (or any other) time scale can be distinguished in these fluvial and coastal deposits on the basis of internal bounding surfaces.

Group 7 deposits include complete major elements of depositional systems, such as channels, major delta lobes and some delta complexes, draas or entire ergs, barrier islands, tidal channels, sand ridges on the continental shelf, and submarine-fan channels and compensation cycles. They are formed over periods of thousands to tens of thousands of years. Examples are illustrated in Fig. 3.7. Most Holocene depositional systems formed since the postglacial sea-level rise, if considered in their entirety, are group 7 deposits. Storm-generated sand ridges "take many thousands of years to build" (Brenner et al. 1985). The preserved deposits of the Galveston Island Barrier system represent about 3500 years of accumulation (Bernard et al. 1962). Sandy Neck Spit, Massachusetts, is 10 km long and dates back to about 3.3 ka (Hayes and Kana 1976). Tidal-inlet fills formed by lateral migration near Sapelo Island, Georgia, are at least 3.4 ka old (Hoyt and Henry 1967).

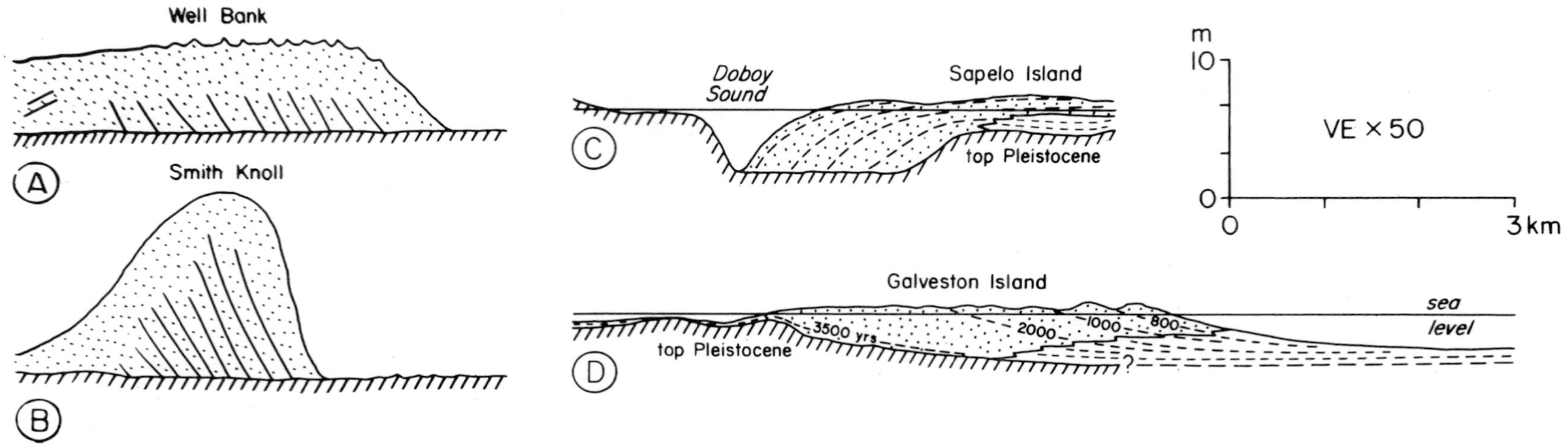

Fig. 3.7. Examples of group 7 deposits. **A, B** Sand ridges in the North Sea. Stratification within the ridges has an actual dip of about 3°, and represents master bedding surfaces or reactivation surfaces (see discussion in text; Houbolt 1968). **C** Channel-fill and spit deposit formed by lateral migration of a tidal channel. *Dashed lines* are time lines. Section is parallel to axis of barrier island forming the coast of Sapelo Island, Georgia (Hoyt and Henry 1967). **D** The Galveston Island barrier system, Texas. *Dashed lines* are time lines, with dates derived from ^{14}C analyses (Bernard et al. 1962). Time lines in **C** and **D** may be preserved in the rock record as distinct stratification surfaces comparable to the reactivation surfaces of Fig. 3.5

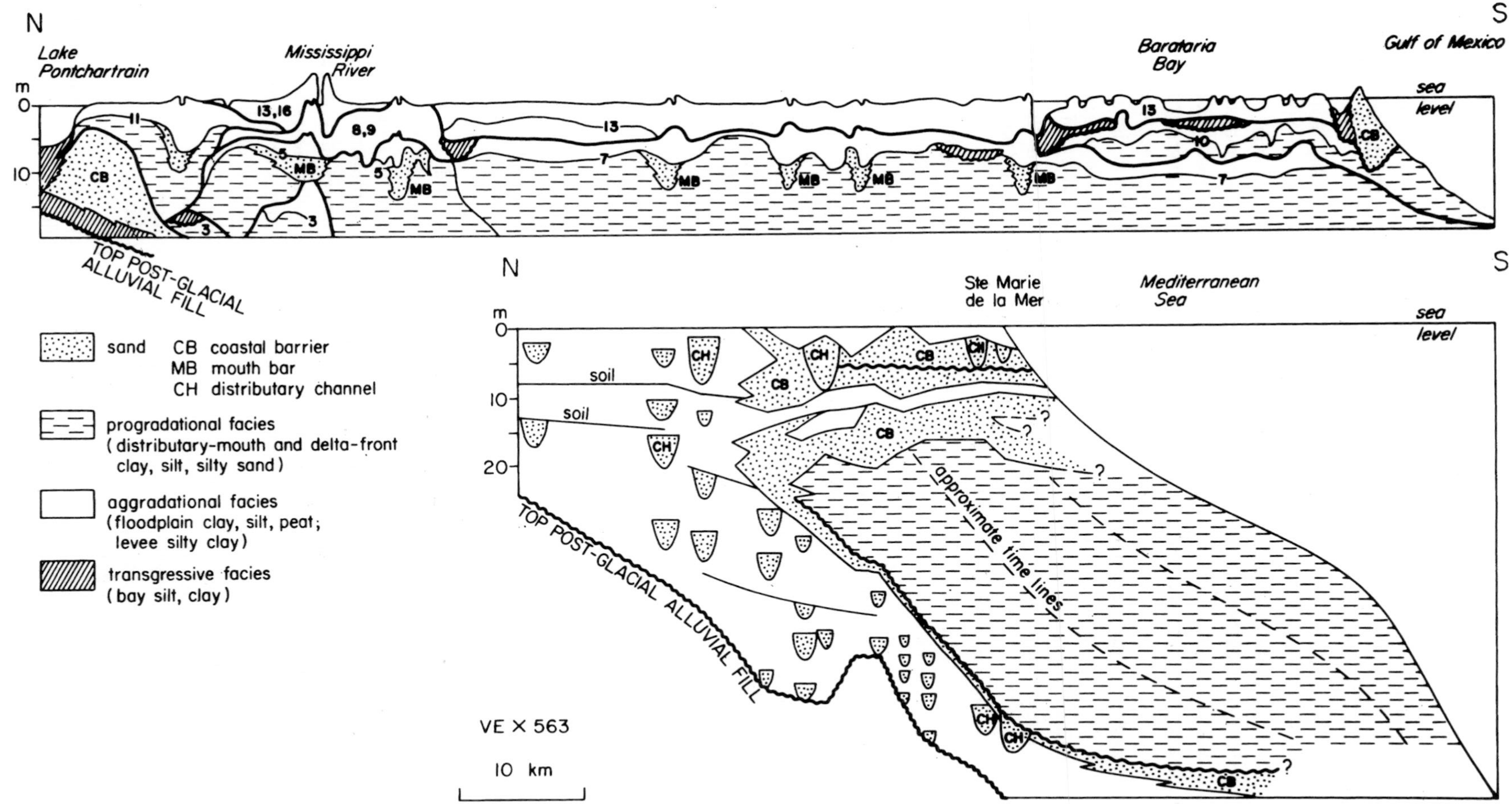

Fig. 3.8. Two examples of postglacial delta complexes built during the last 6 to 8 ka. *Above* North-south cross section through the Mississippi complex, showing the thin, overlapping lobes that developed on the continental shelf (*numbered 1–16*). Lobe 3, the oldest seen here, is about 4.6 ka old (Frazier 1967). *Below* The Rhône delta. The greater thickness of this delta probably reflects its location on the edge of the continental shelf. This delta did not build its deposit by lateral delta switching but by steady seaward growth of barrier complexes, behind which a fluvial and lagoonal delta-plain aggraded. (Oomkens 1970)

Turning to the alluvial environment, paleosols require several thousand years to develop, and may take at least tens of thousands of years to reach "maturity" (Leeder 1975; Retallack 1984; Wright 1990; see Sects. 9.2, 10.3.1 for additional discussion). Most river valleys aggraded with fluvial and estuarine deposits immediately following the postglacial sea-level rise. Following this, many rivers have had 6000 to 8000 years to build major delta complexes. The modern (postglacial) Mississippi delta complex is one example (Kolb and Van Lopik 1966; Frazier 1967). The Rhône delta is another. Both are illustrated in Fig. 3.8.

Fluvial deposits of groups 8 to 11 are described at length in Chap. 9. The sequence stratigraphy of these deposits is discussed in Chap. 13. The following are some brief notes on marine and nonmarine deposits.

Group 8 sedimentary bodies represent time periods of tens to hundreds of thousands of years. Examples include major fluvial channel-belt deposits, such as the "third-order" cycles of the Siwalik Group in Pakistan (as demonstrated by detailed studies of the magnetostratigraphy by Johnson et al. 1985). The nine eolian "complexes" bounded by supersurfaces described by Kocurek et al. (1991) each represent an average of about 700000 years. Other examples include such major depositional elements as delta complexes and alluvial fans, shelf sand-ridge fields (group III "regional lentils" of Shurr 1984), and major lobes of submarine fans (third-order "turbidite stages" of Mutti and Normark 1987). The distribution in time and space of such elements is commonly controlled by regional tectonism or by regional sea-level change which itself is commonly caused, in turn, by tectonism, astronomical (Milankovitch) forcing, such as glacioeustasy, or other eustatic causes.

Many stratigraphic sequences are of group 8, such as the "minor" cyclothems of Heckel (1986), the "punctuated aggradational cycles" of Goodwin and Anderson (1985), and the Cardium Sandstone cycles of Plint et al. (1986). These have been termed "fifth-order cycles" in earlier syntheses (Vail et al. 1977; Miall 1984a, 1990). The Cardium sequences may record tectonic thrust-loading events, as suggested by Swift and Rice (1984), who pointed out the common occurrence of repeated coarsening-upward cycles in shelf sandstone bodies of the Cretaceous Western Interior. This model has now been developed by many workers, as discussed in Sect. 13.4. Other fifth-order cycles (group 8 deposits of this classification) may be caused by glacioeustatic sea-level change. For example, Suter et al. (1987) described six sequences that have formed on the Louisiana continental shelf since about 150 ka, with an average duration of 25000 years. There is increasing evidence of astronomical forcing in the development of cycles of this duration – the so-called Milankovitch cycles (Fischer 1986). For example, Olsen (1990, 1994) postulated a Milankovitch cyclicity to explain variations in fluvial cycle thickness and calculated stream discharge in some Devonian deposits of East Greenland (Sects. 9.2, 12.13).

Major bounding surfaces within group 8 deposits may not be caused by autogenic mechanisms, such as channel migration, but may be regional in extent and allogenic in nature. Such are the ravinement surfaces and other types of diastem described by Nummedal and Swift (1987), and the sequence boundaries of Van Wagoner et al. (1990; see Chap. 13).

Group 9 deposits represent major depositional systems, accumulating over hundreds of thousands to a few millions of years. Some stratigraphic sequences, including many cyclothems (the "major" cycles of Heckel 1986), are of this group. These have been termed "fourth-order cycles" in earlier syntheses (Vail et al. 1977; Miall 1984a, 1990), in a numbering system that uses lower "order" numbers for cycles of increasing duration (the reverse of the "group" numbering used here).

Group 10 deposits are basin-fill complexes (depositional systems tracts), representing time spans of millions to tens of millions of years. They correspond to "third-order cycles" in earlier syntheses. There is ample evidence for such cycles in the stratigraphic record, including Cambrian grand cycles (Aitken 1966, 1978; Chow and James 1987), Carboniferous megacyclothems (Heckel 1986) and mesothems (Ramsbottom 1979), and the Cretaceous mesothems of the Western Interior (Weimer 1960; Kauffman 1969, 1984). Regional or global sea-level changes caused by regional tectonism, or by global changes in seafloor spreading rates, may be the main mechanism leading to the development of this type of cyclicity. Another promising idea regarding the origin of third-order cycles is that of plate-margin tilting as a result of in-plane stress transmitted across plate interiors from extensional or compressional plate margins (Cloetingh 1988). Miall (1990, 1991c) provided reviews of this topic, and the ideas are discussed further in Sects. 11.3.5 and 13.4.

Second- and first-order cycles, in the terminology of Vail et al. (1977) and Miall (1990), may be designated stratigraphic groups 11 and 12, if required.

3.3 Definition of Sediment Groups by Bounding Surfaces

For some clastic deposits, as noted earlier, attempts have been made to recognize the types of facies composition, vertical succession, geometry, scale, and bounding surfaces that characterize the deposits formed over particular time scales. Where this can be done it permits the deposits to be subdivided empirically into a hierarchy of architectural units. Characteristics of the bounding surfaces that enclose these units may also include features useful for distinguishing the depositional units themselves, such as the relationship of the surfaces to overlying and underlying strata (erosional, gradational), their shape (flat, irregular, concave- or convex-up), areal extent, and the nature of associated facies (e.g., they overlie mud drapes, or are followed by poorly sorted lag deposits). However, sets of field criteria are not yet available that can provide unambiguous interpretations of the time scale of the sedimentary processes for every field case. Many deposits, therefore, cannot readily be classified into the groups listed in Table 3.2, but could fall into any one of two or three of the groups. For example, diurnal, seasonal, and random meteorological processes may produce similar results in eolian (G. Kocurek, pers. comm., 1989) and fluvial strata, particularly if preservation of the products is incomplete. For these reasons, it is not yet possible (and may never be) to establish an all-encompassing classification of the architectural subdivisions and bounding surfaces of all terrigenous clastic deposits. A discussion of the progress to date in this direction is the subject of the remainder of this section, and the fluvial bounding-surface classification used in this book is described in Sect. 4.5.

Lamination and small-scale ripples develop rapidly beneath traction currents in a wide range of sedimentary environments. These are group 1 and 2 deposits, respectively. The superimposition of trains of ripples may generate surfaces of bedform climb, or may be represented by virtually flat bounding surfaces. These are set boundaries, in the terminology of McKee and Weir (1953). Those that occur in fluvial sediments were classified as first-order surfaces by Miall (1988a,b). This type of surface is one where little or no erosion is apparent, and the surface records the virtually continuous sedimentation of trains of similar bedforms. Changes in flow characteristics may lead to changes in the type of bedform, with resulting changes in cross-bed style. Bounding surfaces separating different assemblages of cross-bed structures are coset boundaries (McKee and Weir 1953), and were classified as second-order surfaces by Miall (1988a,b). Dynamic superimposition of bedforms, such as the migration of megaripples over sand waves (Jackson 1975), may also lead to the development of second-order surfaces.

In the tidal environment, the individual small-scale bedforms that make up class V and VI sand waves are group 2 structures of the present classification. The boundaries between them were termed E3 surfaces by Allen (1980).

In storm cycles containing hummocky cross-stratification, two types of internal bounding surfaces have been recognized by Dott and Bourgeois (1982). Individual laminae are separated by what they termed third-order surfaces, and are thought to represent lamination produced by individual wave oscillations or pulsations of wave trains. This process represents periods of seconds or minutes, and so the deposits are of group 1 or 2 in the present classification.

Diurnal changes in bedform migration may develop discrete bundles of cross-strata, or rhythms, commonly separated by minor reactivation surfaces. Tidal bundles are particularly distinctive examples of group 3 deposits. Boundaries between tidal bundles are E2 surfaces, in Allen's (1980) terminology, and are formed by erosion by the subordinate current in the tidal cycle, according to him. However, Dalrymple (1984) showed that, at least in some cases, the spacing between E2 surfaces is several times the net bedform-migration distance per tidal cycle. He interpreted E2 surfaces and variation in tidal foreset thickness between the surfaces as a response to the arrival at the crest of the sand wave of the peaks and troughs of superimposed megaripples. The equivalent structures in fluvial environments are simple reactivation surfaces, some of which probably reflect diurnal changes in stage and therefore separate units of group 3 rank. The variations in tidal-bundle thickness related to neap-spring changes that were described by Visser (1980) are also bounded by E2 surfaces. In this case, the surfaces are useful for defining sediment packages of group 4 rank.

Pulses within individual storm events may occur over periods of hours, days, or weeks, giving rise to bundles of hummocky bedforms separated by second-order bounding surfaces (Dott and Bourgeois classification), and representing group 3 deposits (present classification). Storm events characteristically develop sequences that rest on scoured surfaces, pass upward into hummocky cross- stratification, and are capped by plane lamination and ripples. This is the idealized storm sequence of Dott and Bourgeois (1982). The first-order surfaces (their terminology) that bound these sequences represent the completion of single storm events or storm seasons, and therefore may be used to define deposits of group 4 or 5 of the present work. At present, there are no criteria for the distinction between the storm sequences that might accumulate during individual storms within an annual storm season, and those that represent rarer, and perhaps more violent events, such as hurricanes.

In the eolian environment, dune cross-stratification may include interbedded grain flow, grain fall, and ripple cross-lamination. These can be separated in the field by their distinctive facies characteristics, but determining the temporal significance of each unit may be extremely difficult (G. Kocurek, pers. comm., 1988). In coastal environments, differential heating and cooling of the sea and the land lead to diurnal fluctuations in the strength and direction of sea breezes, with a resultant distinctive bundling of eolian wind-ripple cross-lamination (Hunter and Richmond 1988). Such bundles are thought to have low preservation potential, but yearly cycles have been tentatively identified in ancient deposits (Hunter and Rubin 1983). The eolian third-order surfaces of Brookfield (1977) are dune reactivation surfaces, but these could represent daily, seasonal, or longer term (e.g., 100-year wind storm) erosional events (G. Kocurek, pers. comm., 1988). The depositional units enclosed by these surfaces may therefore be of group 3, 4, or 5 of the present classification. Some of the eolian stratification surfaces shown in Fig. 3.5C may include third-order surfaces.

The architecture of estuarine and shelf sand waves (Allen 1980; Dalrymple 1984), and sand ridges (Houbolt 1968; Harris 1988), and that of downstream-accreted macroforms in fluvial and deltaic deposits (Miall 1988a,b; Yang and Nio 1989), is superficially similar (cf. Figs. 3.5B and 3.7A,B, although these drawings are at different scales). In each case, the deposit rests on a flat erosion surface and contains gently dipping (< 10°) accretion surfaces that commonly truncate underlying bedding at a low angle (the reactivation surfaces of Fig. 3.5). Bedding, including cross-bedding, commonly downlaps onto these accretion surfaces. The deposits typically are lens-shaped, with a convex-up upper surface. In the bounding-surface classification of Miall (1988a,b), the internal accretion surfaces in fluvial macroforms are classified as third-order surfaces. The C1 and C2 surfaces in the ebb-tide deltas described by Yang and Nio (1989, p. 187) are comparable. Houbolt (1968) termed those that occur in sand ridges "master bedding surfaces", and this term has also been applied to the large modern sand waves studied by Berné et al. (1988, 1991). Berné et al. (1988) implied an analogy between their master bedding surfaces and the E2 surfaces of Allen (1980). There may be a confusion in bedform terminology here, because what Berné et al. (1988) termed sand waves are large structures 3.5 to 7.5 m high, compared to an average height of 0.81 m for those described by Allen (1980) and Dalrymple (1984; his height data).

These estuarine and fluvial macroforms vary widely in physical scales and probably represent sedimentary processes acting over a wide range of time scales, although data relating to the migration and accumulation rates of sand ridges are sparse. Because of this, it is premature to attempt to develop a common architectural classification scheme for these deposits. Fluvial third-order erosion surfaces (e.g., the reactivation surfaces of Fig. 3.5A) are considered to be related to seasonal or longer term (e.g., "10-year") floods. They may represent the internal sedimentary breaks between individual ridges in point-bar ridge-and-swale topography. They therefore enclose packages of strata of group 5 rank. As noted above, the E2 surfaces in sand waves enclose packages of groups 3 and 4. Berné et al. (1988) suggested that the master bedding surfaces in their bedforms may represent longer-term phenomena, longer than neap-spring tidal cycles. Possibly, they are related to storm scour on a yearly or longer-term scale, in which case the sediment packages between the surfaces are group 5 deposits of the present classification. Berné et al. (1991) carried out seismic and coring investigations through fields of giant sand waves, and illustrated a threefold hierarchy of bounding surfaces. Their "second-order" surfaces (equivalent to master bedding surfaces, labeled D2 surfaces in core; their Figs. 10, 11) probably equate to E2 surfaces and are of group 5 rank. The internal structure and migration dynamics of sand ridges are very poorly known. Harris (1988) suggested that they consist of superimposed assemblages of sand waves, separated by master bedding surfaces. If this is the case, the master bedding surfaces may represent major erosional events, such as the "100-year event" of group 6, for example, the scour of rare violent storms.

The bounding surfaces that enclose entire fluvial macroforms, sand waves, and sand ridges typically have convex-up shapes. This reflects accretionary growth terminated by a shift in the depositional milieu, such as a channel-avulsion event. As noted above, however, similarity of form does not necessarily indicate similarity in origins. Fluvial point bars and downstream-accreted macroforms are group 6 deposits. They are bounded by fourth-order surfaces (Miall 1988a,b), although commonly, in the ancient record, the caps of these deposits are planed off during subsequent erosional events, and the top surface is not preserved. Fluvial macroforms are typically capped by the fifth-order surfaces that define the base of succeeding channel-fill units. Little is known about the bounding surfaces that enclose sand-wave and sand-ridge deposits in the ancient record.

In many outcrops, the most prominent bounding surfaces are those between group 7 elements, such as the fifth-order surfaces that define the base of fluvial channels (Miall 1988a,b), the "first-order" eolian surfaces of Brookfield (1977), and the "first-order" surfaces in the sand-wave deposits described by Berné et al. (1991). These commonly have planar geometry, except for the curvature at channel cutbanks (rarely preserved), and may represent hundreds to thousands of years of nondeposition, for example, between events of fluvial-channel or draa migration. There may be little to distinguish these, at the outcrop level, from the surfaces that define sediment bodies of groups 8 through 10 (e.g., eolian supersurfaces, sequence boundaries), and recognition and classification of such surfaces then depend on careful mapping and regional stratigraphic interpretation.

Lithostratigraphic and allostratigraphic subdivisions (formation and member boundaries) and sequence boundaries constitute a still larger scale of bounding surface. These include the sixth-order fluvial surfaces of Miall (1988a,b) and the eolian supersurfaces of Kocurek (1988). They are discussed further in Sects. 4.5 and 9.6.

3.4 Sedimentation Rate and Its Relation to Depositional Recurrence Interval

It is a common observation that measured or calculated sedimentation rates vary in inverse relation to the length of time over which the measurement is made (Sadler 1981, provided detailed documentation). In general, the sedimentation rate of the depositional units in each of the 11 groups described here differ markedly from those of the next group, commonly by an order of magnitude. The reason for this is that sedimentation is rarely continuous for more than a few weeks or months, at most, at any one location, and typically is interrupted by erosion or nondeposition, so that net accumulation is almost always substantially less than total sedimentation. Areas where sedimentation might be expected to be most nearly continuous include the deep oceans, where very slow pelagic settling occurs, and in the sea off the mouths of perennial rivers, especially below the influence of waves. In this section, an attempt is made to evaluate variations in sedimentation rate with respect to the ten sediment groups defined in this chapter.

An important consideration in evaluating sedimentation rates is the concept of recurrence interval. As noted by Jackson (1975), disproportionate volumes of sediment are commonly moved by infrequent dynamic events. The sedimentation (and accumulation) rate during the event may be extremely high, but must be discounted for geological purposes by factoring in the (typically) much longer periods of time when sedimentation is slow, or might even be negative, as a result of erosion. For example, a magnetostratigraphic study of a record of fluvial flash flood deposits in Argentina indicated that only one flood event every 50 to 500 years is preserved in the rock record there (Beer 1990). Spasmodic sedimentation also occurs in areas characterized by lateral accretion or progradation. A depositional element, such as a delta distributary mouth, or a fluvial point bar, builds rapidly in a horizontal direction, giving rise to temporarily rapid local sedimentation rates, but then the dispersal system switches elsewhere for an extended period of time, with resultant abandonment and, perhaps, erosion of the new deposit.

These points can be illustrated by a discussion of some of the sedimentation rates that have been measured for the various types and scales of deposits that develop in rivers, tidal estuaries, and deltas. Estimates of order-of-magnitude instantaneous sedimentation rates averaged over the time span of each of the ten groups are given in Table 3.2. An "instantaneous sedimentation rate" is one calculated for a short period of time and extrapolated to longer periods for purposes of comparison. Examples of the type of calculation that can be made for some of the groups are given later.

Among the smallest sedimentary structures are small-scale ripples (group 2 of this work). These typically migrate a distance equivalent to their own wavelength in 20 to 60 min (Southard et al. 1980). A

5-cm-high ripple that forms in 30 min is equivalent to an instantaneous sedimentation rate of 876000 m/ka. Clearly, this number bears no relationship to real long-term sedimentation rates, but it serves to emphasize the extremes of sedimentation rate, to compare with more geologically typical rates discussed later. Tidal sand waves (group 3) have similarly very high instantaneous rates. In the Bay of Fundy, Dalrymple (1984) demonstrated that in one tidal cycle sand waves migrate a distance about equivalent to their average height, which is 0.8 m. Bay of Fundy tides are semidiurnal, and so this migration is equivalent to a sedimentation rate of 584000 m/ka. Climbing-ripple stratification represents intervals of extremely rapid sedimentation. Data collected by Bristow (1993a) from large bars in the modern Brahmaputra River suggested that for very short periods as much as 3.75 m of sediment may accumulate in the space of 1 h, an instantaneous sedimentation rate of 33×10^6 m/ka!

The deposits formed by seasonal or more irregular runoff events in rivers (group 5) have extremely variable instantaneous sedimentation rates. The flood deposit in Bijou Creek, Colorado, described by McKee et al. (1967) was formed by the most violent flood in 30 years. It formed 1 to 4 m of sediment in about 12 h, an instantaneous sedimentation rate of 730000–2920000 m/ka. Assuming no erosion, and a repetition of such floods every 30 years, this translates into a rate of 33–133 m/ka averaged over a few hundred years. In fact, scour depths during the flood ranged from 1.5 to 3 m, and true net preservation of any one flood deposit over periods of hundreds or thousands of years may be negligible. Long-term rates measured over hundreds to thousands of years are likely to be an order of magnitude less, in the range of 10^{-1} m/ka.

The point bars (group 6) in the Wabash River are on average about 5 m thick. The active area of each bar is about 200 m wide. If the bars migrate at a maximum rate of 2 km in 50 years (Jackson 1976b) they would take 5 years to migrate one point-bar width, which is equivalent to an instantaneous sedimentation rate of 1000 m/ka. Comparable rates may be calculated from the migration of distributary mouth bars. A typical Mississippi distributary migrates 3 km in 100 years (Gould 1970; Fig. 3.6). The mouth-bar deposits, from the mouth of the channel to the toe of the distal bar, are about 4 km wide (in a dip direction) and about 45 m thick. This lateral migration is equivalent to an instantaneous vertical sedimentation rate of 340 m/ka. Data from modern Dutch tidal deposits summarized by Yang and Nio (1989) showed that ebb-tide deltas (group 6 deposits) accumulate at rates of 100 to 450 m/ka for periods of about 20 years, before abandonment occurs.

If we assume that a distributary will only build out across a given area of the delta front once during the migration of one major delta lobe, we can calculate the sedimentation rate of the mouth bar averaged over the life of the lobe (group 7). In the postglacial Mississippi delta, major lobes are formed and abandoned in about 1000 years (Kolb and Van Lopik 1966; Frazier 1967; Fig. 3.8 above), giving an average sedimentation rate for that period of 45 m/ka. The recurrence interval of delta lobes themselves depends on subsidence rates, sediment supply, and the configuration of the continental shelf. In the case of the Mississippi complex, the river is attempting to switch discharge (and delta construction) to the Atchafalaya River, where one of the earliest lobes developed about 6 to 8 ka before present (Sect. 10.3.1). On this scale deposition of a 45-m-thick mouth-bar deposit represents a sedimentation rate of 5.6 to 7.5 m/ka, although this calculation does not take into account the bay-fill and other facies interbedded with the mouth bar deposit. This compares with typical values for Holocene sedimentation rates of 6 to 12 m/ka that are commonly quoted for the Mississippi delta complex (e.g., Weimer 1970). Other deltas do not accumulate so rapidly. For example the Rhône delta has a thickness of only 50 m, accumulated since about 8.2 ka, indicating an average sedimentation rate of 0.6 m/ka (Oomkens 1970).

A few sedimentation rates can be calculated for tidal-inlet and barrier deposits. The Galveston Island barrier is 12 m thick and 3.5 ka old at its base (Bernard et al. 1962; Fig. 3.7D), indicating an average sedimentation rate of 3.4 m/ka. A tidal inlet at Fire Island, New York, has migrated 8 km in 115 years (Kumar and Sanders 1974). The depositional slope from spit crest to channel floor is about 500 m wide, suggesting that at any one point the entire tidal-inlet fill could form by lateral accretion in about 7 years. The sequence is 12 m thick, indicating an instantaneous sedimentation rate of 1714 m/ka. This migration rate is unusually rapid. Tidal inlets at Sapelo Island, Georgia, appear to have migrated only about 2.5 km since the postglacial sea-level rise (Hoyt and Henry 1967; Fig. 3.7C), indicating a sedimentation rate of 4.5 m/ka.

It is instructive to compare these rates with those obtained by Oomkens (1970) and McKee et al. (1983) as a result of their studies of modern sedimentation rates on the continental shelf off river mouths. A short-term rate of 4.4 cm/month (528 m/ka) was determined close to the mouth of the Yangtze River,

China, by McKee et al. (1983), who studied the decay of short-lived radionucleides in the uppermost 15 cm of recent deposits, representing about 100 days of accumulation. The uppermost 200 cm of section, representing about 100 years of accumulation, yielded a rate an order of magnitude lower, 5.4 cm/year (54 m/ka). Oomkens (1970) quoted sedimentation rates of 35 cm/year (350 m/ka) for the delta front of the modern Rhône River. The monthly rate at the Yangtze mouth and that determined for the Rhône compare with the migration rates calculated for Holocene point bars and distributary-mouth bars, whereas the yearly rate at the Yangtze Mouth is similar to that of a delta lobe of the Mississippi.

Of particular relevance to the subject of this book are the sedimentation rates of fluvial deposits at the groups 6 to 8 level, corresponding to deposits formed by long-term geomorphic processes, including those related to geomorphic thresholds, and the largest scale of autogenic depositional feature (as discussed in detail in Chap. 10). Rates of sedimentation on modern alluvial fans and fluvial floodplains have been measured using ^{14}C dates on plant material, and tephrochronology. Available data were summarized by Miall (1978d) and shown to encompass a wide range, from 0.08 to 50 m/ka. However, these measurements were not correlated to specific scales of architectural units, such as the depositional groups defined here. Bridge and Leeder (1979) also assembled data from modern fluvial systems in order to establish meaningful sedimentation rates for use in their computer simulation models. They used a range from 5 to 40 m/ka, with an average of 20 m/ka, for their models of channel-belt deposition based on autogenic avulsion processes (group 7).

Leeder (1975) developed a quantitative model for the estimation of floodplain accretion rates based on the ages of paleosols in Recent sediments, particularly those in the arid American southwest region. The maturity of the paleosol was assessed using various facies criteria (see Sect. 7.4.2), and this was related to the age of the deposit, using ^{14}C dating. It was suggested that rates of sedimentation in ancient alluvial deposits could be estimated using this model. However, Wright (1990) indicated that the ages of the original Recent American paleosols have been substantially revised, and he adduced a considerable body of additional data to indicate that the relationship between paleosol maturity and age varied by as much as an order of magnitude. He showed that Leeder's (1975) mature "stage 4" paleosols in the Quaternary record require as little as 3 ka or as much as 1 million years to develop. The reasons for these variations are discussed in Sect. 7.4.2. Kraus and Bown (1993) also employed paleosols in an estimation of sediment accumulation rates, but used a different technique that avoids problems of correlation to Recent equivalents. In their well-studied Eocene Willwood Formation succession in Wyoming, they recorded numerous paleosols, and were able to construct an artifical composite section composed entirely of paleosol units, by substituting channel and floodplain clastics with a paleosol by detailed lateral correlation. They were able to estimate the relative time required for each type of paleosol to form by this same process of lateral correlation. Thus, it was found that four vertically stacked stage 1 paleosols are stratigraphically, and therefore temporally, equivalent to a single stage 3 paleosol. Each paleosol could therefore be weighted according to its relative time of development, and a time-thickness plot developed. Given precise biostratigraphic and radiometric brackets on the top and base of the succession, they could then convert this graph to absolute accumulation rates, and correct this to a sediment accumulation curve by correcting for compaction. The results indicated a variation in sedimentation rate between 0.1 and 2 m/ka. An examination of unsteady sedimentation processes by Friend et al. (1989) suggested that in the Siwalik sediments of Pakistan individual beds of a few meters in thickness, corresponding to channel-fill increments (group 6) and complete channel-fills (group 7) were deposited at rates of up to 1 m/ka.

Badgley and Tauxe (1990) carried out a detailed study of a transition interval between normal and reversed magnetic polarity within the interfingering fluvial units previously described by Behrensmeyer and Tauxe (1982). This interval, estimated to last between 4 and 10 ka, varied in thickness from 0.5 to 2.7 m, indicating a fivefold variation in net-accumulation rate. Badgley and Tauxe (1990) were also able to demonstrate diachroneity of a major paleosol horizon and a channel-fill sandstone unit, both of which crossed this transitional polarity interval.

It is not immediately clear why the data of Bridge and Leeder (1979) should suggest sedimentation rates an order of magnitude greater than those of Kraus and Bown (1993) and Friend et al. (1989). In the case of Kraus and Bown (1993), the main control on absolute time was provided by radiometric dating of bracketing volcanic units in an ancient succession, and in the study of Friend et al. (1989) absolute time limits were derived by magnetostratigraphic correlation. This contrasts to the estimates from modern floodplains that comprise the Bridge and Leeder (1979) data base. It is possible that despite every effort to relate the calculations to specific beds the estimates from ancient units incorporate significant intervals of a missing section of a type that are not present in the modern successions used by Bridge and Leeder (1979).

Groups 8 and 9 of this book include the Quaternary shelf-margin sequences of Suter et al. (1987) and the minor and major cyclothems of Heckel (1986). The sequences described from the Gulf Coast by Suter et al. (1987) averaged 25000 years in duration and range in thickness from about 25 to 160 m, indicating average accumulation rates of 1–6.4 m/ka. Heckel (1986) documented the chronology of 55 cycles of Westphalian-Stephanian age in the US midcontinent. Estimates of the length of this time span range from 8 to 12 million years. The thickness of the sucession varies from 260 m in Iowa to 550 m in Kansas. These values indicate an average accumulation rate of between 0.02 and 0.07 m/ka. Many of the cyles contain substantial fluvial-deltaic sandstone units and, according to Ramsbottom (1979), who studied similar cyclothems in Europe, rates of lateral deltaic growth must have been about as rapid as that of the modern Mississippi; yet, the average sedimentation rate is two orders of magnitude less than that of the Holocene Mississipi delta complex and its Pleistocene shelf-margin precursors on the Louisiana Gulf Coast. Part of the explanation for this marked contrast is that the Carboniferous cyclothems that were the subject of Heckel's study are located in a cratonic region, where subsidence rates would be expected to be substantially lower than on the continental margin of the Gulf Coast. In addition, many of the cycles are separated by erosion surfaces. Rapid sedimentation rates of an order-of-magnitude more were measured by Johnson et al. (1985) and Friend et al. (1989) in their study of the Siwalik sediments of Pakistan. Detailed documentation of a nearly 2-km-thick section spanning more than 9 million years yielded constant long-term sedimentation rates that increase from 0.1 m/ka at the base to 0.3 m/ka at the top. Individual channel-fill cycles in this succession are estimated to have return periods of $1–4 \times 10^5$ years (group 8 deposits). Bentham et al. (1993) used magnetostratigraphic correlation to estimate accumulation rates of 0.17 to 0.57 m/ka over a 4 million year time period, for a braided fluvial system of Eocene age in Spain. Beer (1990) recorded a maximum accumulation rate of 0.9 m/ka in a Miocene sheetflood deposit in Argentina.

The long-term sedimentation rates of groups 8 to 10 depend largely on rates of generation of sedimentary-accommodation space. This depends on both basin subsidence, which is controlled by tectonic setting, and changes in base level, such as eustasy. Miall (1978) showed that nonmarine basins, in various tectonic settings, have sedimentation rates averaged over millions of years of 0.03 to 1.5 m/ka. This subject is reviewed at greater length by Miall (1990), and in Chap. 11 of this book.

3.5 Application of Scale Concepts to Basin Analysis and Petroleum Geology

The discussion in this chapter is intended to provide a framework for the analysis of clastic deposits at all scales, from the individual core sample, through outcrop-scale analysis, to the mapping of reservoir heterogeneities, depositional systems, stratigraphic sequences, and entire basin-fill complexes.

Research at each level of the hierarchy requires different techniques. Depositional units in the rock record up to group 7 and, in exceptional cases, group 8, may be studied using outcrop data, especially if the lateral-profiling technique described by Miall (1988a,b) is employed (Chap. 4). Mapping and classification of bounding surfaces between depositional elements are a key to such analyses. Detailed subsurface studies may be successful, if supported by detailed core analyses.

In the case of modern environments, side-scan sonar techniques coupled with high-resolution, shallow seismic profiling provide many critical new data regarding depositional patterns, especially in the study of estuarine and shelf bedforms and larger features of groups 6 to 8 (e.g., Berné et al. 1988, 1991; Harris 1988).

Units as small as group 6 may be detailed on exceptionally high-quality petroleum exploration seismic data, especially on three-dimensional seismic surveys (e.g., Brown 1985, 1991). Larger features, including stratigraphic sequences of groups 8 to 10 (the third- to fifth-order stratigraphic cycles of Vail et al. 1977; Miall 1984a, 1990), are now analyzed routinely on seismic records. Surface and subsurface lithostratigraphic correlation, with the aid of magnetostratigraphy, marker beds, and refined biostratigraphic indicators, is also employed (Miall 1984a, 1990). The application of these various methods to the study of fluvial stratigraphic architecture is described in Sect. 9.5, and examples of various scales of heterogeneity in fluvial reservoirs (Fig. 3.4) are described in Chaps. 14 and 15.

Chapter 4

Methods of Architectural-Element Analysis

4.1 Introduction

A wealth of information is contained in many outcrops of fluvial deposits, that is not captured by conventional methods of facies analysis, such as the measurement of vertical-profiles. In this chapter, the formal methods of field architectural-element analysis are described. Analysis and interpretation of the results are the subjects of later chapters.

Earlier work on the development of architectural concepts for fluvial deposits is discussed in Sect. 2.4. This chapter describes current techniques, including the writer's contributions (Miall 1985, 1988a,b), together with the modifications, improvements, and additional suggestions contributed by other researchers, notably those of Soegaard (1990, 1992), S.A. Smith (1990), and DeCelles et al. (1991), who worked mainly with conglomerates.

Methods of conventional facies analysis are not covered in this book. The reader is referred to Miall (1990, Chaps. 2, 4) for a detailed discussion of the methods of facies analysis, including the measurement and interpretation of vertical profiles.

4.2 Construction of Outcrop Profiles

Architectural methods are based on the two- and three-dimensional mapping of large outcrops, using outcrop profiles. These are essentially vertical geological maps, and are constructed in much the same way. The geologist begins with a base map, and then carries out traverses across the map, making observations and walking out key structures and contacts. The base map may be either a line-drawing of the outcrop, made with or without the aid of surveying equipment, or it may be a photomosaic. The latter is the easiest to construct and to use in the field.

In the simplest case, an entire outcrop may be encompassed in a single photographic frame, but the largest and most interesting outcrops tend to be too large to be viewed within one field of view, unless the photographer is required to move so far back from the outcrop that the scale of the photograph becomes too small to be useful. The obvious alternative is to construct a mosaic from overlapping frames. The geologist moves several tens or hundreds of meters back from the face, and carries out a traverse parallel to it, taking a series of overlapping photographs. Dipping beds may be viewed from an adjacent hilltop, or from a low-flying aircraft. It is important to stay the same distance from the face, so that the frames are all as nearly as possible the same scale. Ideally, this may be accomplished with the use of surveying equipment, but the high degree of precision obtainable in this way is not really necessary for sedimentological purposes. It is also useful for the line of sight towards the outcrop to be situated in the plane of the bedding, in order to minimize vertical distortions. For a road cut, which can be photographed from the other side of the road, these requirements may be a simple matter to satisfy. For many hillside exposures and river cliffs, it may be considerably more difficult to generate an adequate suite of photographs, with variations in the terrain and obscuring trees causing problems of position and distance. In practice, the writer has found that useful mosaics can almost always be constructed with a little effort, although perspective problems, and variations in the scale or orientation of different parts of the profile may remain problematic.

Variations in scale from one frame to the next can be accomodated by varying the enlargment size of the prints. This is particularly easily accomplished if each frame includes a scale, or some object whose dimensions are known. Perspective problems are less easily eliminated. They can be troublesome if the outcrop is not flat, but includes projecting and receding regions, because such features will not overlap exactly when adjacent frames are joined.

An almost inevitable element of vertical distortion will be present in most profiles, because the camera cannot view all beds along the plane of their bedding from a single viewpoint. For example, when the line of sight is slanted upward toward a cliff face, the upper edge of the frame has a smaller scale than the lower edge. Upper and lower edges cannot both be exactly overlapped, and some compromise match must be achieved. All these perspective problems can be reduced by moving further away from the outcrop and taking the pictures through a telephoto lens.

Use of advanced cameras with adjustable bodies can eliminate some of the regular, geometric distortions. Wizevich (1992a) provided a useful discussion of equipment and techniques.

An alternative or supplement to photographic methods is to construct a base map of the outcrop using surveying techniques. Kocurek et al. (1991) carried out a very detailed architectural study of eolian sandstones using a laser theodolite. This method is, of course, complex and time-consuming. It may be the only approach that can be used when the topography of an outcrop is very complex. The researcher should carefully assess, before beginning such a survey, how accurate the outcrop base map needs to be. Precision in the placement of bounding surfaces to within a few decimeters is claimed by Kocurek et al. (1991). While this is impressive, it may not be necessary. Much depends on the measurements that are to be made from the profile. Bearing in mind that most fluvial depositional elements have very irregular shapes, and change their thickness by several meters within a few tens of meters laterally, striving for a high degree of base-map precision may be largely a wasted effort. A simple photographic mosaic, with all its shape and scale distortions, is ideal for plotting data, and provides a powerful basis for visualizing and interpreting the geometry of the architectural elements. Vertical profiles do not need to be documented directly in the field, but can be measured from the completed profile if required, for example, to provide comparisons with subsurface records.

In most countries, development of field film can now be done very quickly. "One-hour" commercial photographic laboratories are available in many small towns in North America and Europe, especially in tourist areas, and even in many Third-World countries. Unless the services of a professional photographer are available it is recommended that the mosaic be made from color-print film, as this is the most popular type of film and the one which most commercial tourist laboratories are equipped to deal with. Instant cameras, such as Polaroid, may also be usable, although print quality does not tend to be as good. Commerical photographic laboratories are equipped to produce standard prints (such as those of 6×4 inches, in the United States) very quickly. Enlargements may take a few days. In any case, it is now possible to make a reconnaissance of an outcrop, and take the necessary photographs, and then return with the results to carry out ground measurements within a few hours or days.

The next step is to mount the photographic mosaic on a lightweight board suitable to be carried into the field (thin plywood or Bristol board), and to attach an overlay of tracing paper or drafting film. The geologist should then spend a considerable amount of time poring over the mosaic and drawing in every visible bedding plane. For a large outcrop this may take several hours of careful work. The first round of drawing may be done in the office, but it is essential that the placement of these lines on the overlay be checked against the actual outcrop. Commonly, a complex corner of an outcrop, when reduced to the two dimensions of a photograph, can yield confusing and conflicting bedding traces, which can only be resolved by going back to the point from which the picture was taken. For very large or inaccessible outcrops, Dueholm and Olsen (1993) and Olsen (1993) recommended the use of photogrammetric techniques to analyze the outcrop photographs. The construction of steroscopic overlaps between the photographs should be done to facilitate this. This technique will permit an analysis of the channel-scale lithosomes, but interpretation of the smaller-scale features, such as individual macroforms, cannot be carried out reliably without ground checking.

In the rocks, a bedding plane is a continuous surface, but it may not be equally visible everywhere on an outcrop face, because of variable weathering characteristics. For example, even major channel scours may be virtually invisible where the basal channel lag has been removed by erosion, and identical sandstones are amalgamated across a clean bounding surface. Inexperienced interpreters tend to avoid tracing bounding surfaces through such areas of faint bedding traces, leaving the lines "hanging". It is an essential part of the interpretive process to eventually complete and join up all these hanging lines. It may be necessary, however, to leave the completion of this process until lithofacies and paleocurrent observations are underway, so that where there is ambiguity, alternative correlations can be evaluated in the light of the alternative fluvial architectures they would create. Architectural-element analysis necessarily includes this type of ongoing interpretation, and the geologist needs to be clear about this as the work proceeds. Eventually, as discussed in later sections of this chapter, each resulting complete surface can be assigned an appropriate interpretation and rank, and can then be colored or coded accordingly on the final drawing.

Before leaving the outcrop, the geologist should be sure to record two critical items of observation: the orientation of the outcrop and its scale. Both may vary along the profile, in which case this should be measured and carefully recorded. Scale should be measured vertically to reduce possible distortions (very little of the face can be photographed from a point perpendicular to it in the horizontal plane). A surveying rod may be included in one or more of the photographs, or the size of a recognizable feature of the outcrop may be measured and marked on the overlay.

The final, drafted profile may need to be published as a foldout, because even the A4 and quarto-sized paper that is becoming fashionable for geological journals, is too small to contain a large profile illustration without severe reduction. I recommend publishing the photomosaic separately from the line drawing interpretation, preferably with the former positioned immediately above the latter. The photograph should be marked only with the necessary identifying information. This is analogous to the practice now adopted for the publication of seismic sections. An uninterpreted line is given together with the interpretation, so that the reader can make an independent assessment of the raw data. Alternatively, the overlay can be printed in place on the outcrop photograph, or it can be published on its own, without the photograph. Both methods have the disadvantage of the loss or obscuring of information.

4.3 Classification of Lithofacies

Observation and classification of lithofacies are now standard components of the facies-analysis methodology for studying sedimentary rocks. Good summaries of the principles and methods are contained in Reading (1986), Miall (1990, Chap. 4), and Walker and James (1992). Beds are classified on the basis of their primary depositional attributes, notably (in the case of fluvial clastic deposits) bedding, grain size, texture, and sedimentary structures. Biogenic structures and fossils may be important locally as additional descriptive attributes. Chemical sediments, such as pedogenic calcretes, coal, and evaporites, typically form minor components of most fluvial systems, but need their own careful characterization. As noted by Miall (1990, p. 150), the scale of an individual lithofacies unit depends on the level of detail incorporated in its definition. Facies may be defined very broadly to encompass mappable stratigraphic units, or they may be defined finely to accomodate the level of detail obtainable in the centimeter-by-centimeter logging typically carried out on core. For the purpose of architectural-element analysis, a relatively fine degree of descripion and subdivision is required.

It is good research practice to approach each new rock unit afresh, with the aim of making complete, unbiased observations of all important lithofacies attributes. However, sedimentological research has demonstrated that much of the apparent variability in sedimentary units disguises a limited range of basic lithofacies and biofacies types. The depositional processes which control the development of clastic fluvial lithofacies, such as traction-current transportation, with its accompanying fluid turbulence and its effects on beds of clastic grains, are common to all rivers and obey the same physical laws everywhere, with the production of similar suites of lithofacies. For example, hydrodynamic structures, such as ripples and cross-bedding, are formed by the migration of ripples and dunes. Considerable experimental work has shown that in the development of these bedforms there are consistent empirical relationships between bedform size and shape and a limited suite of physical parameters, of which the most important are the depth and velocity of the flow, and the grain-size of the sediment (e.g., Harms et al. 1982). Repetition of similar conditions leads to repetition of similar depositional products, which are then susceptible to a universal empirical classification in the field.

Miall (1977) reviewed braided-river deposits, and demonstrated a consistency in the lithofacies assemblages occurring in a wide range of modern and ancient sandy and gravelly sediments. However, many of the researchers whose work was examined in this review had erected their own local classification, which had obscured the similarities between the deposits and prevented the recognition of common depositional themes. Miall proposed a simple classification, making use of a two-letter code to facilitate quick field and laboratory identification and documentation. Use of this lithofacies scheme by a number of workers led to the recognition that the scheme could be applied to all fluvial deposits, not just those of braided type, but also demonstrated that it was incomplete. In a later paper, Miall (1978c) expanded the classification with the definition of a number of additional minor but significant lithofacies types. Subsequently, the classification has been used by dozens of researchers and has become a standard field methodology for the examination of fluvial deposits.

Not all professionals agree with the use of predetermined lithofacies schemes, such as that described here. Bridge (1993a) remains amongst the most sceptical of their value. He is particularly concerned that the use of such schemes by the inexperienced will lead to their uncritical application, with a possible loss of important detail where new observations might have been made of deposits and structures that differ from the "standard". This is an important warning that should be borne in mind by all researchers.

The scheme is presented in Table 4.1, and is used throughout this book. Modifications have been made to the scheme originally presented in Miall (1978c) in an attempt to eliminate some superfluous categories and to better discriminate between others. The capital letter in the facies code indicates dominant grain size (G = gravel, S = sand, F = fine-grained facies, including very fine sand, silt, and mud). The lowercase letter serves as a mnemonic for the characteristic texture or structure of the lithofacies (e.g., p = planar cross-bedding, ms = matrix-supported). The lithofacies are described in detail, with illustrations, in Chap. 5.

It is recommended that the researcher use Table 4.1 as a basis for field research, while remaining alive to the possibility that refinements are always possible, based on detailed observations of new units. For example, Leblanc Smith (1980) incorporated details on grain size into the coding scheme. Haszeldine (1983a,b), who did not, in fact, use the lithofacies scheme of Table 4.1, subdivided trough cross-bedding (lithofacies St) into two types based on trough thickeness, and was able to demonstrate the utility of this refinement in interpreting the accretionary development of large bar forms. A similar subdivision was used by Eberth and Miall (1991), who employed the codes Sti and Stii for large- and small-scale trough cross-bed sets, respectively.

Some of the lithofacies classes are gradational with others. For example, there may be no clear distinction in the field between lithofacies Sh and Sl; lithofacies Fl may contain minor coal streaks or carbonate nodules, leading to questions about a workable definition of lithofacies C and P for logging purposes; how thick should a sand bed be before it is separated out from Fl? Bed thickness cutoffs, and accessory percentage limits should be established at the outset of a project exercise in order to facilitate consistent logging practices.

The reader should be aware of the recent review of bedform classifications undertaken by the Bedforms and Bedding Structures Research Group of the Society for Sedimentary Geology (SEPM)(Ashley 1990). A synthesis of modern work on bedforms of all types, in fluvial, tidal, and other environments, resulted in a uniform approach to classification and a better understanding of the causes of morphological variation. This is discussed in Chap. 5.

4.4 Principles of Paleocurrent Analysis

Paleocurrent data are essential for the purpose of architectural-element analysis. For the typical, relatively flat, two-dimensional cliff outcrop, these data provide the essential third dimension. Orientation information on hydrodynamic sedimentary structures reflects the internal geometry of bar complexes, channels, and sand sheets. The dip and strike of bounding surfaces reveal the orientation of accretionary bar growth and meander migration (Fig. 4.1). Overall, the data will show the orientation of the outcrop relative to channel and bar trends, which is important to know when carrying out visual examination of the architecture in order to assess bar and channel evolution. For such purposes, even very limited data can be extremely useful.

The tradition in basin-analysis practice has been to collect very large numbers of readings without necessarily recording very precise data regarding the location of the measurements or, at best, keying such measurements to vertical profiles. While such methods provide good estimates of local and regional paleocurrent trends and of vertical changes in these trends, they do not help much in the elucidation of the internal architecture of the sand bodies. This approach also does not address the concerns regarding the architectural scale of depositional units (Chap. 3), and its effects on paleocurrent variability (Sect. 2.3.5; Fig. 2.10).

The method recommended here is to use the preliminary interpretations of the bounding surfaces drawn up prior to detailed field examination of the outcrop as a guide to the architectural subdivision of the rocks. Attempts should then be made to collect a suite of readings from each of the major elements

Table 4.1. Facies classification. (Modified from Miall 1978c)

Facies code	Facies	Sedimentary structures	Interpretation
Gmm	Matrix-supported, massive gravel	Weak grading	Plastic debris flow (high-strength, viscous)
Gmg	Matrix-supported gravel	Inverse to normal grading	Pseudoplastic debris flow (low strength, viscous)
Gci	Clast-supported gravel	Inverse grading	Clast-rich debris flow (high strength), or pseudoplastic debris flow (low strength)
Gcm	Clast-supported massive gravel	–	Pseudoplastic debris flow (inertial bedload, turbulent flow)
Gh	Clast-supported, crudely bedded gravel	Horizontal bedding, imbrication	Longitudinal bedforms, lag deposits, sieve deposits
Gt	Gravel, stratified	Trough cross-beds	Minor channel fills
Gp	Gravel, stratified	Planar cross-beds	Transverse bedforms, deltaic growths from older bar remnants
St	Sand, fine to very coarse, may be pebbly	Solitary or grouped trough cross-beds	Sinuous-crested and linguoid (3-D) dunes
Sp	Sand, fine to very coarse, may be pebbly	Solitary or grouped planar cross-beds	Transverse and linguoid bedforms (2-D dunes)
Sr	Sand, very fine to coarse	Ripple cross-lamination	Ripples (lower flow regime)
Sh	Sand, very fine to coarse, may be pebbly	Horizontal lamination parting or streaming lineation	Plane-bed flow (critical flow)
Sl	Sand, very fine to coarse, may be pebbly	Low-angle (< 15°) cross-beds	Scour fills, humpback or washed-out dunes, antidunes
Ss	Sand, fine to very coarse, may be pebbly	Broad, shallow scours	Scour fill
Sm	Sand, fine to coarse	Massive, or faint lamination	Sediment-gravity flow deposits
Fl	Sand, silt, mud	Fine lamination, very small ripples	Overbank, abandoned channel, or waning flood deposits
Fsm	Silt, mud	Massive	Backswamp or abandoned channel deposits
Fm	Mud, silt	Massive, desiccation cracks	Overbank, abandoned channel, or drape deposits
Fr	Mud, silt	Massive, roots, bioturbation	Root bed, incipient soil
C	Coal, carbonaceous mud	Plant, mud films	Vegetated swamp deposits
P	Paleosol carbonate (calcite, siderite)	Pedogenic features: nodules, filaments	Soil with chemical precipitation

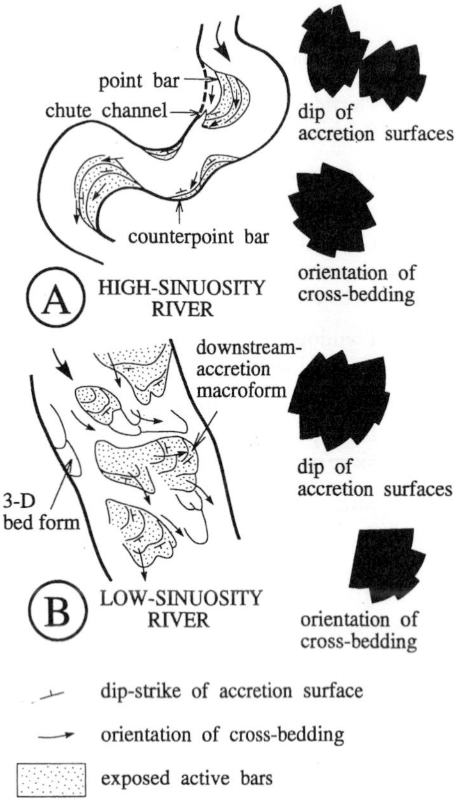

Fig. 4.1. Hypothetical examples of fluvial styles, indicating the range of orientations of dipping accretion surfaces and the variation in cross-bed orientations. **A** High-sinuosity river, such as a typical meandering river. Note that although there is a regional overlap of cross-bed and accretion-surface orientations, locally the two types of dip are oriented nearly perpendicular to each other, and the rose diagrams reflect this divergence. A counterpoint bar is shown. Such bars can be distinguished from the more common point bars by the fact that the accretion surfaces exhibit curvature in plan view that is concave in a downdip direction, contrasting with the convex curvature of point-bar surfaces. Commonly, such bars are characterized by finer grain sizes than other macroforms. **B** Low sinuosity river – a typical sandy-braided river. Such rivers may exhibit relatively high local channel sinuosity, in which case macroforms accreting downstream may also be oriented at a high angle to the regional trend. In the example shown here mean directions are skewed to the southeast, diverging from the overall channel orientation of south-southeast because of a major channel and bar complex (near the center of the reach) oriented in an easterly direction. (Miall 1994)

exposed in the face. It is essential to key each measurement to structure type and lithofacies, which can readily be done using the lithofacies code scheme discussed in Sect. 4.3. The usual rules about the statistical validity of small samples apply at the level of the individual architectural element. Even within a single bar or bar complex, indicated directions can vary widely, sometimes over a range of more than 180°. However, calculation of statistically significant mean and variance data for each element may be less valuable than the information provided on the evolution of the bar with time. For example, paleocurrent directions may show regular lateral changes across the bar, indicating a change in accretion directions as a result of shifts in meander position. Less than half a dozen readings may be adequate to document this, especially if the data are combined with information on the orientation of bounding surfaces. Readings from successive elements show how the entire river system evolves through time.

In order for such interpretations to be made, it is obvious that the location of each paleocurrent reading must be recorded very precisely. Each measurement should be recorded by a numbered point on the profile overlay. Normally, the location within the profile can readily be determined by comparing the shape of the outcrop and the location of vegetation around the observation point to the same details on the photograph. Sometimes, however, the monotony of the outcrop may make this difficult, or the geologist may be required by the steepness of the face to stand too close to the outcrop to be able to determine an exact location. In such cases, it is useful for records on the profile overlay to be made by an assistant, who locates him/herself some distance away from the face. Information can be conveyed to the assistant orally, if necessary with the aid of two-way radios.

Details on what structures to measure and what they indicate, methods to use for incompletely exposed structures, corrections for structural disturbance, and some references to statistical and computer methods, are contained in Miall (1990, Sect. 5.9).

Paleocurrent data can readily be displayed on the final profile that is prepared for publication. Individual readings should be shown by arrows, with the head of the arrow located at the measurement point, and the tail of the arrow oriented to indicate direction. Symbols added to the tail, such as varying numbers of ticks, or variations in the design of the arrow head, or some other device, may be used to indicate the type of structure indicated by the arrow. Orientation measurements on bounding surfaces should be clearly differentiated, for example, with differently sized arrows. The orientation of the arrow tail may be plotted in one of two ways to suit the needs of the particular study. The more conventional way is to define paleocurrent north as the top of the diagram, so that, for example, arrows pointing horizontally to the left indicate flow due west (azimuth 270°). In some cases, it renders the profile more easy to read if the paleocurrent data are oriented with respect to the outcrop face. Readings parallel to the face then appear as horizontal arrows pointing left or right, flow directly into the face is shown as a vertical arrow pointing upward, and so on. The second method has the advantage that the reader can more easily integrate the architecture of the face and the corresponding paleocurrent readings when assessing the profile visually. The disadvantage comes when it is necessary to compare profiles taken from outcrops that have different orientations. Comparisons between paleocurrent directions in different outcrops are then not as obvious visually.

Summary paleocurrent rose diagrams may usefully be added to the profile, if space permits. A rose for each element is a particularly useful addition. Summary statistics can be prepared as a table for inclusion in the text.

The use of paleocurrent analysis as a mapping tool is discussed in Sect. 9.5.7.

4.5 Classification of Bounding Surfaces

The principles of the hierarchical classification of depositional units are detailed in Chap. 3. In this section, it is shown how the ideas can be applied more specifically to fluvial deposits. A version of Table 3.2 that focuses on fluvial deposits is included here as Table 4.2, and incorporates a column summarizing the nine major types of bedding and bounding surface that characterize fluvial deposits. Most of these types are illustrated in Figs. 4.2 and 4.3. The method was developed initially for sandstone deposits, but further work by S.A. Smith (1990) and Soegaard (1990, 1992) has demonstrated the applicability of the classification to conglomerates as well. DeCelles et al. (1991) used the system, with some modifications, to study alluvial-fan deposits. As S.A. Smith (1990) pointed out, the textural and structural monotony of conglomerates may make recognition and correlation of bounding surfaces more difficult in these types of deposit.

An alternative classification of depositional units and bounding surfaces was erected by Bridge (1993a), but based on very similar principles to those discussed here. A comparison between the two nomenclature schemes is provided below, and a discussion of the differences between the schemes was given by Miall (1995a).

Bridge	Miall
microscale set (e.g., ripple)	1st-order unit
mesoscale set (e.g., dune)	1st-order unit
micro/mesoscale coset	2nd-order unit
macroscale inclined stratum	3rd-order unit (macroform increment)
macroscale inclined strata set	4th-order unit (macroform)
group of macroscale sets	5th-order unit (channel)
group of macroscale sets	6th-order unit (e.g., channel-belt)

A question of scale arises in the use of such classifications, because rivers and their deposits vary enormously in their size. However, Bristow and Best (1993) have pointed out that rivers show a considerable degree of self-similarity over the complete range of scales (see also Sect. 8.3), and the classification described here can be applied to rivers of all scales providing due attention is paid to river size. For example, Fig. 4.4 illustrates the range of depositional scales in the Mississippi, one of the world's largest rivers. It is instructive to compare this with the subdivisions of alluvial fans shown in Fig. 4.3. One of the challenges in the study of ancient rocks is to determine the scale of the rivers responsible for forming the deposits, based on architectural studies of the preserved record.

First- and second-order surfaces record boundaries within microform and mesoform deposits. The definition of first-order surfaces is unchanged from Allen (1983a). They represent cross-bed set bounding surfaces (Fig. 4.5). Little or no internal erosion is apparent at these boundaries, and they represent the virtually continuous sedimentation of trains of bedforms of similar type. Subtle modifications in attitude, with minor erosion, may be caused by reactivation following stage changes (Collinson 1970), or may be the result of changes in bedform orientation (Haszeldine 1983b). In core, these surfaces may not be very prominent, but the presence of a bounding surface can be recognized by truncation or wedge-out of cross-bed foresets.

Second-order surfaces are simple coset bounding surfaces, in the sense of McKee and Weir (1953). These indicate changes in flow conditions, or a change in flow direction, but no significant time break (Fig. 4.6). Lithofacies above and below the

Table 4.2. Hierarchy of depositional units in alluvial deposits. (Modified from from Miall 1991b)

Grp	Time scale of process (a)	Examples of processes	Instantaneous sedimentation rate (m/ka)	Fluvial, deltaic depositional units	Rank and characteristics of bounding surfaces
1	10^{-6}	Burst-sweep cycle		Lamina	0th-order, lamination surface
2	10^{-5} -10^{-4}	Bedform migration	10^{5}	Ripple (microform)	1st-order, set bounding surface
3	10^{-3}	Bedform migration	10^{5}	Diurnal dune increment, reactivation surface	1st-order, set bounding surface
4	10^{-2} -10^{-1}	Bedform migration	10^{4}	Dune (mesoform)	2nd-order, coset bounding surface
5	10^{0} -10^{1}	Seasonal events, 10-year flood	10^{2-3}	Macroform growth increment	3rd-order, dipping 5–20° in direction of accretion
6	10^{2} -10^{3}	100-year flood, channel and bar migration	10^{2-3}	Macroform, e.g., point bar, levee, splay immature paleosol	4th-order, convex-up macroform top, minor channel scour, flat surface bounding floodplain elements
7	10^{3} -10^{4}	Long-term geomorphic processes, e.g. channel avulsion	10^{0}–10^{1}	Channel, delta lobe, mature paleosol	5th-order, flat to concave-up channel base
8	10^{4} -10^{5}	5th-order (Milankovitch) cycles, response to fault pulse	10^{-1}	Channel belt, alluvial fan, minor sequence	6th-order, flat, regionally extensive, or base of incised valley
9	10^{5} -10^{6}	4th-order (Milankovitch) cycles, response to fault pulse	10^{-1}–10^{-2}	Major dep. system, fan tract, sequence	7th-order, sequence boundary; flat, regionally extensive, or base of incised valley
10	10^{6} -10^{7}	3rd-order cycles. Tectonic and eustatic processes	10^{-1}–10^{-2}	Basin-fill complex	8th-order, regional disconformity

surface are different, but the surface is usually not marked by significant bedding truncations or other evidence of erosion, except for the same kinds of minor modification which occur on first-order surfaces, as noted above. In core, these surfaces may be distinguished from first-order surfaces by a change in lithofacies. Decelles et al. (1991), who studied a Cenozoic alluvial-fan complex, defined the contacts enclosing individual sediment-gravity-flow deposits as first-order surfaces. However, by the definitions used here such contacts should be ranked as second-order, as they indicate a significant change in flow conditions.

Third- and fourth-order surfaces are defined when architectural reconstruction indicates the presence of macroforms, including lateral-accretion deposits (LA; see next section for definition of architectural elements and the element codes) and downstream-accreting macroforms (DA). Individual depositional units ("storeys" or "architectural elements") are bounded by surfaces of fourth-order or higher rank.

Third-order surfaces are cross-cutting erosion surfaces within macroforms that dip at a low angle (normally < 15°) and may truncate underlying cross-bedding at a low angle (Figs. 4.7–4.10). They may cut

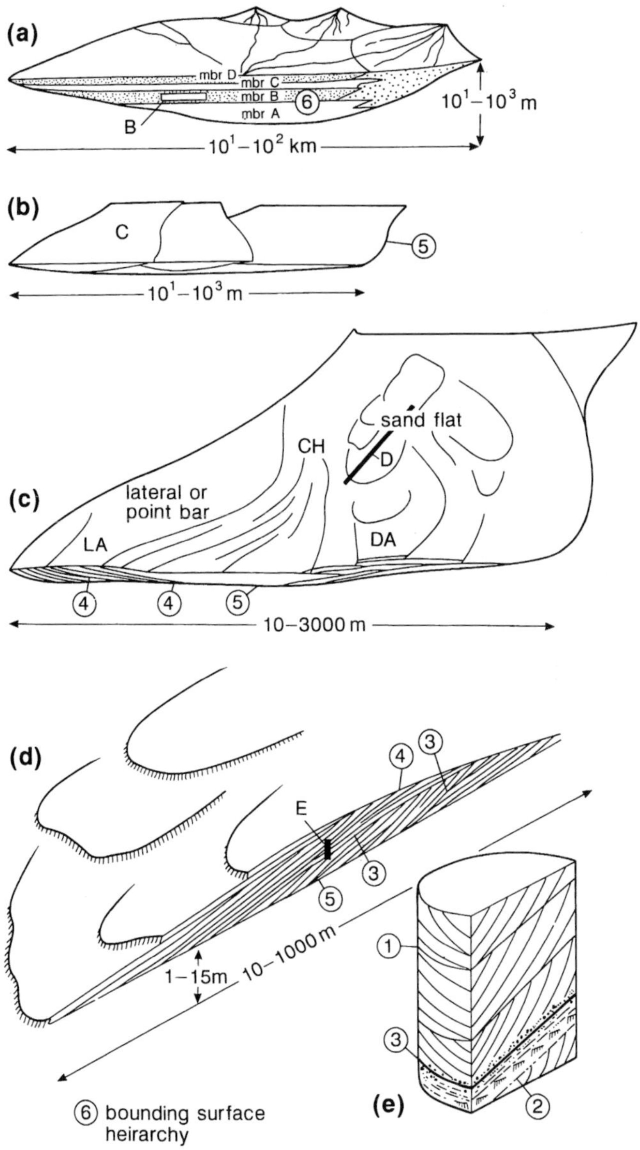

Fig. 4.2. The scales of depositional elements in a fluvial system, showing the bounding-surface hierarchy (Table 4.2). *Circled numbers* indicate the ranks of the bounding surfaces. In **c**, the two-letter codes are for architectural element types (Table 4.3). In **d**, the sand flat is shown as being built up by migrating "sand waves". Foreset terminations of these are shown at *top* of diagram, but the internal cross-bedding that results has been omitted for clarity

through more than one cross-bed set. They are commonly draped with mudstones and followed by an intraclast breccia. Facies assemblages above and below the surface are similar. Such surfaces are equivalent to the "epsilon" cross-bedding of Allen (1963a). Third-order surfaces may also develop at the top of minor bar or bedform sequences, and are draped by mudstone or siltstone, indicating the falling stage. Succeeding strata commonly contain a basal intraclast breccia composed of rip-up clasts of the draping fine-grained sediments. These characteristics are readily recognized in core. These surfaces indicate stage changes, but no significant change in sedimentary style or bedform orientation within the macroform. They indicate a form of large-scale "reactivation", to adapt Collinson's (1970) term. DeCelles et al. (1991) assigned a third-order rank to large-scale cross-strata in fanglomerates, which he also interpreted as accretion surfaces on macroforms.

Fourth-order surfaces represent the upper bounding surfaces of macroforms, in the original

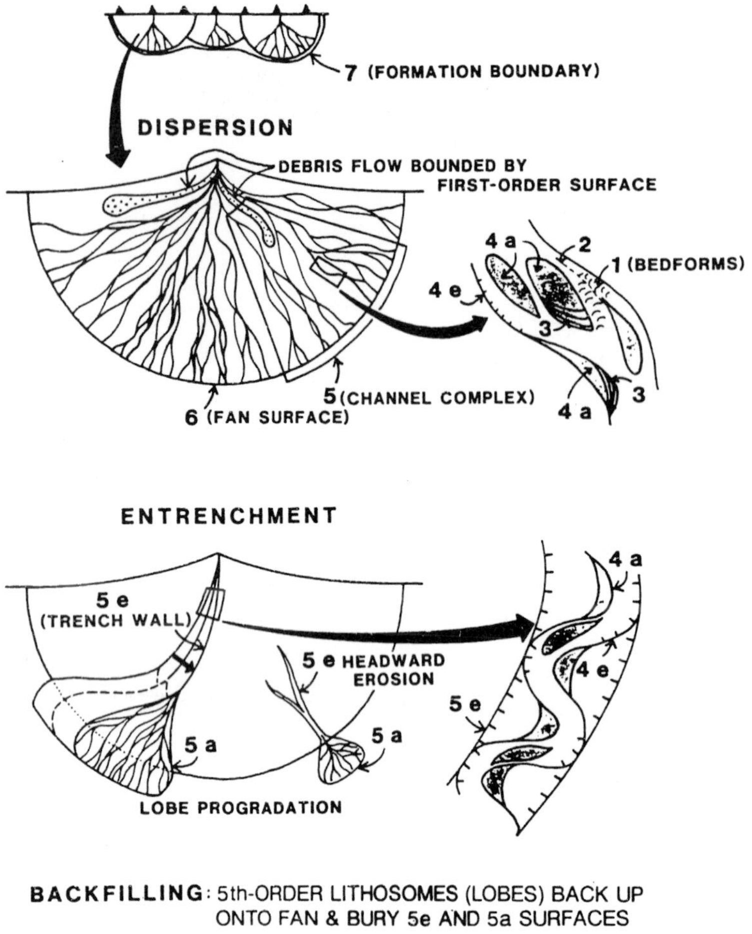

Fig. 4.3. The major geomorphological features of alluvial fans, showing the rank numbering of lithosomes and bounding surfaces used by DeCelles et al. (1991). The cycle of dispersion, entrenchment and backfilling is adapted from Schumm et al. (1987) and is discussed in Chap. 10

scheme of Miall (1988a,b). However, several researchers have used this rank in different ways, and there is potential for confusion if a consistent usage is not observed. In the braided-fluvial systems studied by the writer, fourth-order surfaces are typically flat to convex-upward (Figs. 4.8, 4.9). Underlying bedding surfaces and first- to third-order bounding surfaces are truncated at a low angle or may be locally parallel to the upper bounding surface, indicating that they are surfaces of lateral or downstream accretion. The form of the accretionary surface is commonly echoed by that of internal third-order surfaces within the underlying macroform element. Mud drapes on the element underlying the surface are common.

Macroforms are "group 6" deposits in the classification of Table 3.2. They represent sedimentary events with a duration or recurrence interval of 10^2–10^3 years. It is appropriate to define other types of fourth-order surface for other lithosomes of this rank. The basal scour surface of minor channels, such as chute channels, would be included in this category. Where major channels are present these would be bounded by higher order surfaces. Individual crevasse-splay lithosomes within the floodplain are bounded by fourth-order surfaces, and may contain separate growth increments bounded by third-order surfaces.

In the work of DeCelles et al. (1991) two types of fourth-order surface were recognized, erosional (4e) and accretionary (4a). Macroform bounding surfaces, analogous to those defined by Miall (1988a,b) are designated type 4a. Type 4e is defined for the "erosional bases of large lenticular lithosomes." DeCelles et al. (1991) interpreted these as channels, which is a potential source of confusion with the writer's scheme (but see below). DeCelles et al. (1991) stated that fourth-order surfaces are by far the most easily recognizable surfaces in his rocks, whereas Cowan (1991) found it difficult to distinguish surfaces of this type in his braided-fluvial sandstones, and concluded that accretionary upper

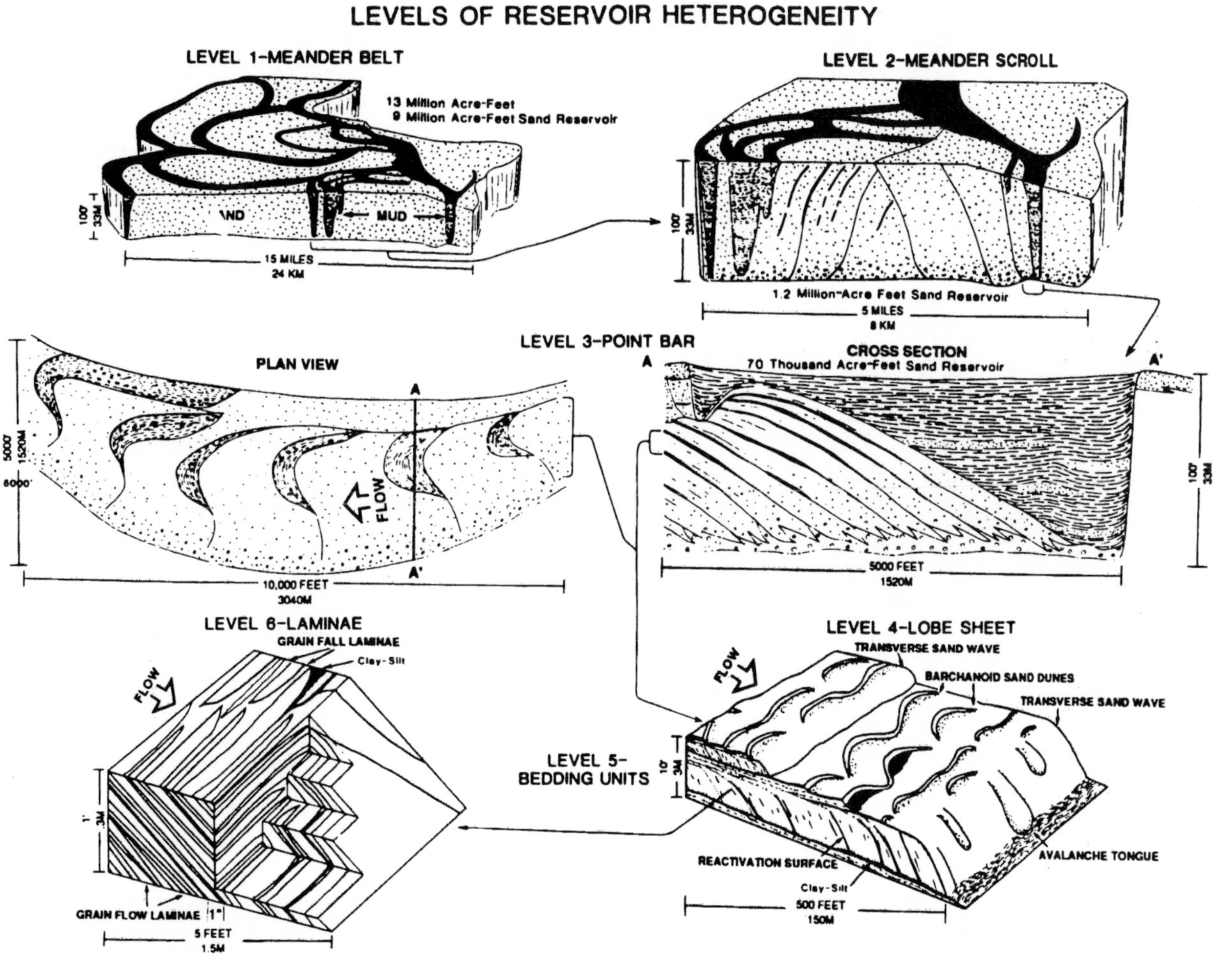

Fig. 4.4. Hierarchies of depositional units in the modern Mississippi River. Although this river and its deposits are at a much greater scale than most fluvial deposits encountered in the rock record, the same principles of hierarchical classification apply. Compare this figure with Fig. 4.3. *Level 1* Bounded at the base by a fifth-order surface. *Level 2* Individual point-bar increments are defined by several third-order surfaces. The entire point-bar deposit is enclosed by a fourth-order surface. *Level 3* A point-bar increment, bounded by a third-order surface. Bedding units of different type are in contact across second-order surfaces. *Level 4* Lobe sheets consist in part of stacked bedforms of comparable type, separated by first-order surfaces. (Jordan and Pryor 1992, reprinted by permission)

bounding surfaces of macroforms are rarely preserved in his particular fluvial system.

Third- and fourth-order surfaces may be recognized in the subsurface by their low depositional dips (Figs. 4.8–4.10). The dip should be apparent in core and may also be revealed by dipmeter data and such tools as the Schlumberger MicroScanner device (Sect. 9.5.8). However, third- and fourth-order surfaces may be very difficult to distinguish in individual wells. The low dips and draping breccias or mud lenses are similar. The best way to distinguish these surfaces from each other is if the lithofacies assemblages above and below the surface are different, indicating a change in element type. Correlation between cores would be possible only with very close well spacing, as the units defined by third-order surfaces are estimated to have dimensions of less than about 10 ha (Miall 1988a; Doyle and Sweet 1995).

Second-, third-, and fourth-order surfaces, in this classification, were all included in the second-order category of Allen (1983a; see Fig. 2.30). Third- and fourth-order surfaces correspond to the "minor surfaces" of Bridge and Diemer (1983; see Fig. 2.29).

Fifth-order surfaces are those bounding major sand sheets, such as broad channels and channel-fill complexes (Figs. 4.7–4.12). They are generally flat to slightly concave-upward, but may be marked by local cut-and-fill relief and by basal lag gravels. These are the third-order surfaces of Allen (1983a) and the

Fig. 4.5. Coset of planar cross-bedding (lithofacies Sp) showing first-order bounding surfaces. Cretaceous, Banks Island, Arctic Canada

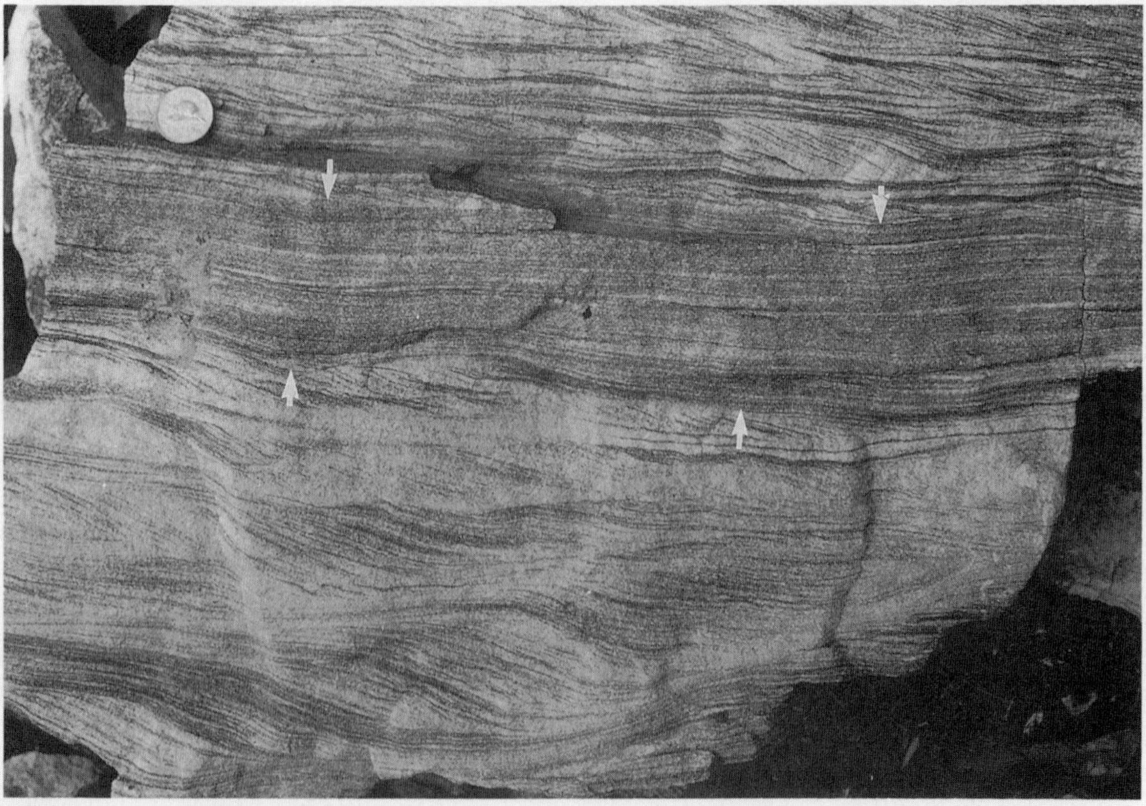

Fig. 4.6. Two second-order surfaces (*arrows*) enclose a bed of plane-laminated sandstone (lithofacies Sh), which shows gradational contacts above and below with ripple cross-lamination (Sr). Eocene, Axel Heiberg Island, Arctic Canada

Fig. 4.7. A third-order surface (*arrows*) within a homogeneous, sandstone-dominated lateral-accretion deposit that accreted to the left. The macroform rests on a channel scour (fifth-order) surface (*arrows*) above a siltstone unit. *Numbers* indicate ranks of bounding surfaces. Eocene Eureka Sound Group, Axel Heiberg Island, Arctic Canada

"major surfaces" of Bridge and Diemer (1983; see Figs. 2.29, 2.30). Many paleosol horizons may be correlated with specific channels, indicating contemporaneous development. These horizons within floodplain sequences may therefore be classified as fifth-order surfaces.

As first pointed out by Williams and Rust (1969), there is a hierarchy of channels in many multiple-channel rivers (Fig. 2.28). The major channel scour is defined as the fifth-order surface, and the minor channel fills within the complex are then bounded by fourth-order surfaces. In the work of DeCelles et al. (1991), fifth-order surfaces are considered to be "mosaics of fourth-order surfaces", and represent the basal bounding surfaces of fan trenches and lobes. They defined erosional (5e) and accretionary (5a) surfaces. Type 5e is comparable with the fifth-order surfaces bounding channel belts, as defined here.

In conglomerates, S.A. Smith (1990) noted that "abrupt lateral transitions between distinct macroform types occur in the abandoned reaches of some modern gravel-bed rivers. These gravel macroforms are juxtaposed as a result of channel migration and the accretion of gravel as discrete macroforms. This creates horizons of laterally continuous, but genetically distinct macroforms which are separated by high-angle contacts." Such high-angle contacts would include the fifth-order contacts between fragments of separate channel-fills, the fourth-order contacts separating discrete macroforms, and the third-order contacts developed by macroform reactivation. All might appear similar in outcrop. Very careful textural and clast-imbrication studies may assist in identifying the correct rank of each surface. Soegaard (1990, 1992) proposed slightly different terminology for the definition of the various ranks, in his study of a Pennsylvanian fanglomerate, but in practice the geometrical and genetic characteristics of the units he defined do not differ markedly from those discussed here.

Sixth-order surfaces define groups of channels or paleovalleys. Mappable stratigraphic units such as members or submembers are bounded by sixth-order surfaces (Fig. 4.12). Surfaces of sixth-order and higher rank were not defined by Allen (1983a). DeCelles et al. (1991) defined the major bounding surfaces enclosing entire alluvial fans as sixth-order

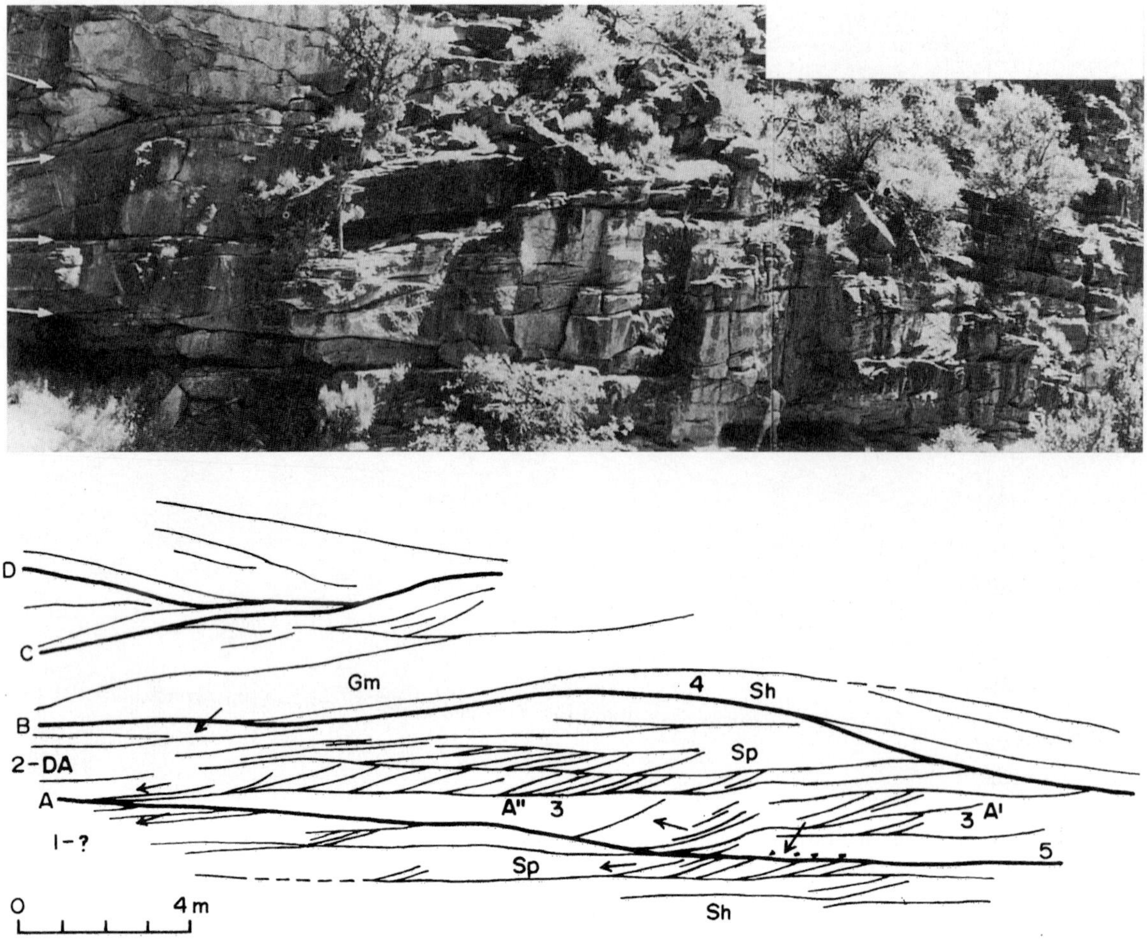

a

b

Fig. 4.8. A downstream-accreted macroform (element DA) bounded at the top by a convex-up fourth-order surface. One of several internal third-order surfaces is also indicated, and the macroform rests on a channel-floor fifth-order surface. Kayenta Formation (Jurassic), Colorado (see Miall 1988c)

surfaces. Where fan deposits are distinguished by different clast lithologies or textures, such surfaces may be prominent and easily mapped, but fan interfingering may generate ragged contacts corresponding to the surfaces of lower rank defining component channels and macroforms, and these may be difficult to trace.

Fourth-, fifth-, and sixth-order surfaces may appear very similar to third-order surfaces in core. They are best differentiated by careful stratigraphic correlation between closely spaced cores, an objective that is best achieved in the most intensively developed field, where well spacing may be a few hundred meters, or less. Considerable potential also exists to identify and map surfaces of this rank using 3-D seismic data. Brown (1991) provided several examples of channels mapped using seismic data (fifth-order surfaces), and it may well prove possible to map individual components of a channel fill if the interpreter knows what to look for.

Seventh-order surfaces enclose the major lithosomes representing discrete allogenic events. DeCelles et al. (1991) recognized that such a surface encloses the discrete, but laterally equivalent fan bodies assigned to the Beartooth Conglomerate stratigraphic unit, all of which represent the sedimentary response to a pulse of tectonic activity along the basin-margin fault. Many sequence boundaries also constitute seventh-order surfaces. Lithologic contrasts across surfaces of this type are likely to be significant. As discussed in Chaps. 11 and 13, distinguishing between the various allogenic mechanisms that can generate seventh-order lithosomes and bounding surfaces (such as tectonism and eustasy) is one of the current major research tasks in stratigraphic geology.

Eighth-order surfaces are regional disconformity surfaces that develop in response to continental-scale or global geologic events. The boundaries of third-order sequences are of this rank.

Fig. 4.9. A lateral-accretion deposit 4 m thick, resting on a major fifth-order surface and capped by a gradational fourth-order surface, above which is the floodplain sequence. Several internal third-order erosion surfaces are present. All surfaces are indicated with *arrows* showing rankings. Pennsylvanian, Alabama

Correct identification and correlation (where possible) of these first- to eighth-order bounding surfaces are clearly essential if the various scales of reservoir heterogeneity are to be identified and dealt with appropriately in the design of production models (Chap. 14). Even in excellent outcrop the correct classification of bounding surfaces is not always easy. Three useful rules may make the task easier:

1. A surface of any given order may be truncated by a surface of equal or higher order, but not by one of lower order. In this way, the rank of the surface always defines the major process that generated it.
2. Accretionary surfaces may be removed by erosion preceding the deposition of the next unit. In such cases, the top of a lithosome may be more logically defined on the basis of what follows it. For example, the top of a macroform is defined by a fourth-order surface, except where it has been cut into by a major channel, the base of which typically constitutes a fifth-order surface.
3. Minor surfaces may change rank laterally. For example, the upper fourth-order bounding surface of a macroform may merge into a second-order surface in the floor of an adjacent channel. Third-order surfaces may downlap onto channel-scour surfaces of sixth order or higher rank where these define sequence boundaries.

4.6 Classification of Architectural Elements

Early attempts at the classification of architectural elements are discussed in Chap. 2 (see Figs. 2.39, 2.40). The nomenclature proposed by Miall (1985), as revised in later work (Miall 1988a,b) is reviewed here. Recent work in which the method has been used is listed by Miall (1994, 1995a) and is briefly discussed by Fielding (1993b). A discussion of some of the problems with the architectural approach (Bridge 1993a; Miall 1995a) indicates the need for caution in the definition of architectural elements and the use of the technique in field research.

A brief note on the meaning of the terms architectural element and lithosome is in order. In this book, the term lithosome is used to refer to a discrete depositional unit bounded by a distinct bounding

Fig. 4.10. A macroform, probably of lateral-accretion type, with several internal third-order erosion surfaces indicating changes in accretion direction. These are indicated by *arrows*. Cretaceous, Gamtoos Basin, South Africa

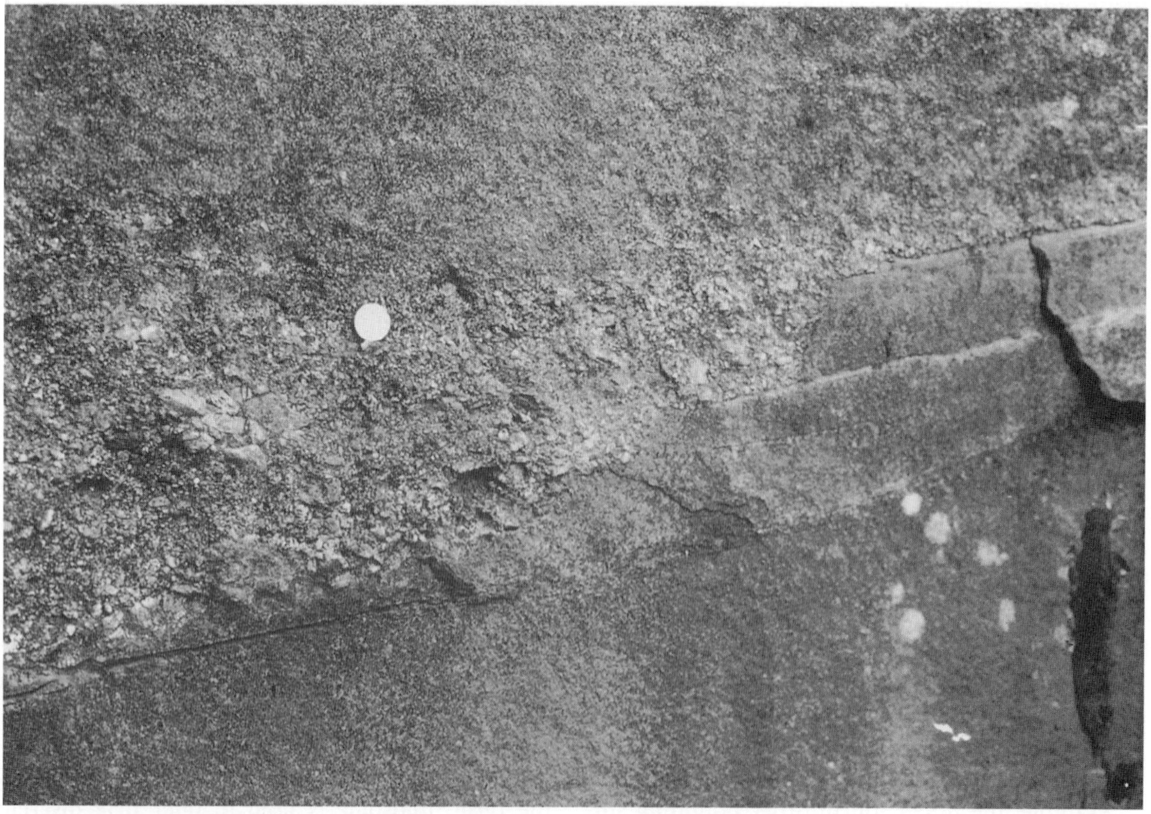

Fig. 4.11. Closeup of a fifth-order surface at the base of a major sandstone sheet. Note minor downcutting and intraclast breccia. There is little to distinguish this surface from others of third- to sixth-order in small outcrops or core. Permian, Poland

Fig. 4.12. Two major sixth-order surfaces (*arrows*) divide this unit into three submembers, which can be traced for about 200 km. The middle submember contains a scour hollow (element *HO*: see Sect. 6.9) about 200 m wide, floored by a concave-up fifth-order surface. Morrison Formation (Jurassic), New Mexico

surface, of any rank. The word is thus a useful, descriptive, nongenetic term.

When a river is viewed from the air, it can be seen to consist of various straight and curved channel reaches, and large areas of exposed gravel, sand, or mud, termed bars. In many rivers the shape and distribution of these various features might seem chaotic, but most such units have distinctive surface forms, the terms for which, such as point bar, side bar, sand flat, chute channel, crevasse splay, etc., constitute the familiar lexicography of fluvial geomorphology. Their development follows certain relatively predictable patterns (Bridge 1985, 1993b) that leave their record in the resulting deposits. The channels and bars are the component depositional elements of the river, and the sediments that comprise them are termed architectural elements. An architectural element may be defined as a component of a depositional system equivalent in size to, or smaller than a channel fill, and larger than an individual facies unit, characterized by a distinctive facies assemblage, internal geometry, external form. and (in some instances) vertical profile. The term is used here for units enclosed by bounding surfaces of third- to fifth-order rank.

Architectural elements are amenable to descriptive and genetic classification, as are their component lithofacies. Miall (1985) proposed the following components of a descriptive classification:

1. Nature of lower and upper bounding surface: erosional or gradational; planar, irregular, curved (concave-up or convex-up).
2. External geometry: sheet, lens, wedge, scoop, U-shaped fill.
3. Scale: thickness, lateral extent parallel and perpendicular to flow direction.
4. Lithology: lithofacies assemblage and vertical sequence.
5. Internal geometry: nature and disposition of internal bounding surfaces; relationship of bedding and first- to second-order surfaces to these surfaces (parallel, truncated, onlap, downlap)
6. Paleocurrent patterns: orientation of flow indicators relative to internal bounding surfaces and external form of element.

Miall (1985) suggested that there are eight basic architectural elements in fluvial deposits, based on a review of studies available up to the early 1980s (Fig. 4.13). Subsequent work has provided a number of elaborations of this original classification and, as Fielding (1993b) pointed out, researchers may need to erect their own classification to better reflect the characteristics of the fluvial unit being investigated. Thus, Cowan (1991) defined a completely new element, the scour hollow, based on his work in the Morrison Formation of New Mexico, and Eberth and Miall (1991) showed how detailed work on a specific unit, in this case the Cutler Formation of New Mexico, could yield a useful subdivision of some of the more important elements. Miall's (1985) original element classification, as modified by recent work, is given in Table 4.3. A classification of the architectural elements in the Cutler Group is given in Table 4.4. A standard classification of architectural elements in the overbank environment is proposed in Table 7.1 (details of these various element types are given in Chaps. 6 and 7).

Other element classifications have been used by other workers. For example, Lang (1993) subdivided element GB into four element types, and developed his own subdivisions of sandy channel-fill and floodplain deposits. Brierley (1991) developed the following classification based on surfaces studies of a modern river floodplain. Corresponding element codes of this book are given in the third column.

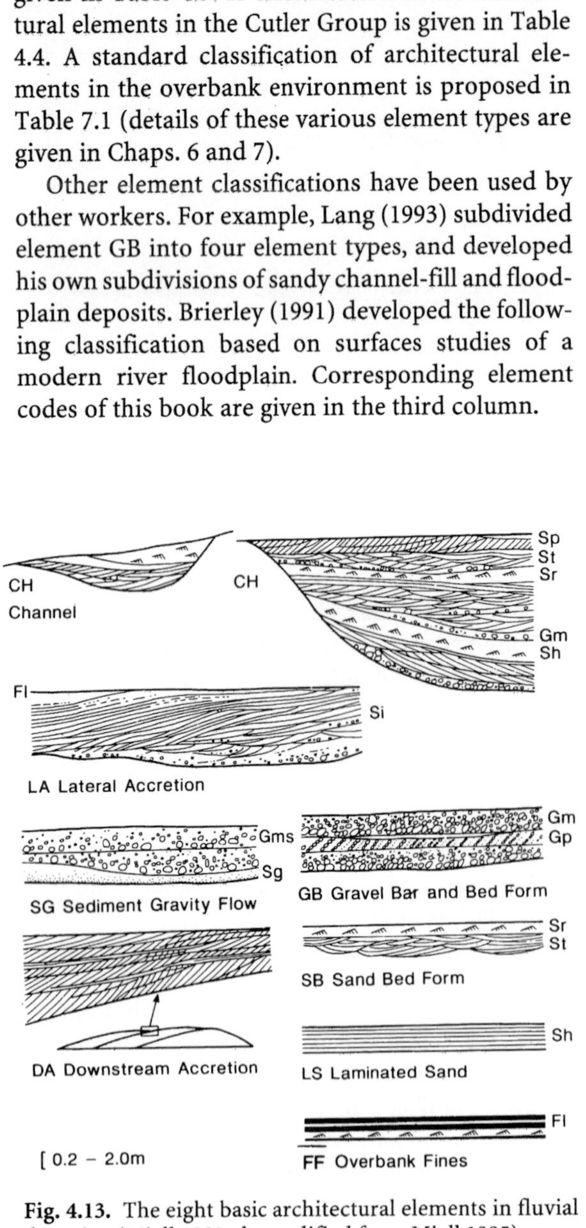

Fig. 4.13. The eight basic architectural elements in fluvial deposits. (Miall 1988a,b; modified from Miall 1985)

Element	Interpretation	Element code
Top stratum	floodplain deposits	FF
(A) sand wedge	levee	LV
(B) proximal sand sheet	marginal overbank sheet	CS
(C) distal sand sheet	distal overbank sheet	CS
Ridge	accretionary deposit associated with channel shifting	CH, LA, DA
Chute channel	cutoff channel	CH
Bar platform	coarse, in-channel deposits	GB, SB, etc.
Basal channel gravels	channel floor lag	GB

For the purpose of interpretive annotation of outcrop profiles and efficent note taking, each element type may be assigned a two- or three-letter symbol. Most of these are self-explanatory. The three-letter codes given in Table 4.4 are designed to permit subdivision of some of the elements. For example, channel-fill sandstone bodies (CH) can be readily subdivided on geometrical grounds into sheets (CHS) and ribbons (CHR). Crevasse channels (CR) are distinguished from the main channels by their small size and fine-grained fill. They include those filled primarily by sandstone (CRS) and those with a mixed, and largely fine-grained fill (CRM). Floodplain fines (FF) include two distinct assemblages, those containing sheets of ripple-marked sandstone, which were probably deposited relatively proximal to the channels (FFP), and those lacking this lithofacies, which represent distal flood-basin deposits (FFD).

In employing architectural-element classifications in the field, three considerations should be borne in mind:

1. Scale: The question of scale arises because some elements occur on several different scales within the same deposit, and the researcher needs to adjust his/her sense of scale to the problems to be addressed. Thus, many deposits could simply be subdivided into channels (CH) and floodplain fines (FF). This is not very useful. Most of the important detail is derived from a subdivision of these elements into the smaller channels and bars, different types of flood-basin deposits, and so on. At a small scale of observation, limited outcrop may lead to the classification of an assemblage of

Table 4.3. Architectural elements in fluvial deposits. (Modified from Miall 1985)

Element	Symbol	Principal facies assemblage	Geometry and relationships
Channels	CH	Any combination	Finger, lens or sheet; concave-up erosional base; scale and shape highly variable; internal concave-up 3rd-order erosion surfaces common
Gravel bars and bedforms	GB	Gm, Gp, Gt	Lens, blanket; usually tabular bodies; commonly interbedded with SB
Sandy bedforms	SB	St, Sp, Sh, Sl, Sr, Se, Ss	Lens, sheet, blanket, wedge, occurs as channel fills, crevasse splays, minor bars
Downstream-accretion macroform	DA	St, Sp, Sh, Sl, Sr, Se, Ss	Lens resting on flat or channeled base, with convex-up 3rd-order internal erosion surfaces and upper 4th-order bounding surface
Lateral-accretion macroform	LA	St, Sp, Sh, Sl, Se, Ss, less commonly Gm, Gt, Gp	Wedge, sheet, lobe; characterized by internal lateral-accretion 3rd-order surfaces
Scour hollows	HO	Gh, Gt, St, Sl	Scoop-shaped hollow with asymmetric fill
Sediment gravity flows	SG	Gmm, Gmg, Gci, Gcm	Lobe, sheet, typically interbedded with GB
Laminated sand sheet	LS	Sh, Sl; minor Sp, Sr	Sheet, blanket
Overbank fines	FF	Fm, Fl	Thin to thick blankets; commonly interbedded with SB; may fill abandoned channels

Facies classification from Miall (1978b).

Table 4.4. Architectural elements in the Cutler Formation, New Mexico. (Adapted from Eberth and Miall 1991)

Element	Symbol	Lithology	Geometry	Interpretation
Major ss. sheet	CHS	Gm, Se, St, Sh, Sl, rare Sp	w/d > 15, with 'wings', storeys 2–5 m thick, units < 1 km wide	Braided fluvial channel
Major ss. ribbon	CHR	Gm, Se, St, Sh, Sl, rare Sp	w/d< 15, with 'wings', storeys <5 m thick	Anastomosed fluvial channel
Minor ss sheet	CS	Se, Sh, Sl, rare St, Sr	Tabular, flat-based, < 1 m thick, wing of CHS sheet	Sheet splay
Minor ss lens	CRS	Se, St, Sl, Sh, Sr	w/d< 10, < 1 m thick	Minor crevasse channel
U-shaped mixed fill	CRM	Se, Sr, Fl, P	1–10 m thick 1–30 m wide	Deep, stable crevasse channel
Interlam. mixed sheets	FFP	Sr, Fl, P	Tabular, < 15 m thick	Proximal floodplain deposits
Siltstone-mudstone sheets	FFD	Fl, P	Tabular < 5 m thick	Distal floodplain deposits

cross-bedded sandstone or conglomerate lithofacies as element SB or GB, respectively, whereas better exposure of the internal geometry of the beds may lead to the recognition that the unit is actually part of a macroform, such as DA or LA.
2. Interbedding: Elements may be interbedded at the scale of the individual lithofacies set or coset. For example, alluvial fanglomerates, comprising stacked GB elements, commonly contain thin stringers of cross-bedded sandstone (element SB). Many sandstone macroforms are draped by thin mudstone lenses, representing abandonment or preservation of FF deposits. At what point is element subdivision cut off and generalized? This is a common problem in facies definition, and the solution depends only on practical criteria of data availability and descriptive convenience.
3. Intergradation: Elements tend to grade laterally into each other. For example, a mid-channel sand flat in a braided system may accrete laterally along its flanks and in a downstream direction at its downstream end. Therefore, a DA unit may pass laterally along strike into an LA unit, or the same transition may occur down depositional dip, indicating an evolution of macroform geometry with time (Fig. 6.24c).

It is not suggested that the classifications offered in Tables 4.3 and 4.4 should be rigidly followed in all future fluvial research. Some other classifications that have been published are noted above. However, it is claimed that the method of systematic observation and classification required to construct such descriptive frameworks should be an integral part of future research endeavours. Other researchers may wish to construct their own classifications and codes. However, it is incumbent on all researchers, where they modify an existing nomenclature for their own purposes, to explain and justify such revisions. Use of the same lithofacies and element codes and bounding-surface rank numbers in ways different from those proposed by earlier workers can lead to unnecessary confusion.

4.7 Classification of Channels and Larger Bodies

Lithosomes bounded by fifth-order surfaces, corresponding to individual channel fills and channel-fill complexes, may be classified using the proposals by Friend et al. (1979) and Friend (1983). According to these authors, there are essentially three types of fifth-order lithosomes: sheet sandstones, formed by nonchannelized flow, for example during ephemeral flash floods; fixed channels, where the channel is laterally stable; and mobile-channel belts, where a broad channel or channel complex occupies a fixed position for a lengthy period of time, and develops minor channels and bars within a broadly defined basal scour surface. Mobile channels may evolve by lateral migration or in-channel avulsion, or a combination of the two processes. Fixed channels give rise to ribbon sandstones, and Friend et al. (1979) proposed that these be defined as those linear sandstone bodies having a width/depth ratio of less than 15. Sheet sandstones and mobile-channel-belt sandstones have ratios greater than 15, and are commonly in excess of 100.

Sheet sandstones formed by sheet floods commonly have flat bounding surfaces showing little evidence of erosion. This type of sheet sandstone, and ribbon sandstones formed by aggradation of fixed channels, commonly have simple fills, in which lithosomes and bounding surfaces of third- and fourth-order are absent. By contrast, sheet sandstones formed as the fill of mobile channels typically have complex fills composed of the deposits of bar complexes and minor channels.

Channel stacking patterns depend on such factors as total sediment load, channel migration, and switching rates, and basin subsidence rate (Chap. 10). Friend (1983) suggested that where the sediment load is dominated by coarse, bed-load materials, overbank areas would be poorly developed and channel fills would be superimposed directly, with floodplain deposits thin or absent. By contrast, a significant suspension load may lead to significant deposition of floodplain successions, and the isolation of channel fills from each other. These variations led Friend (1983) to propose a sixfold classification of fifth-order lithosomes (Fig. 2.35), although no examples of stacked fixed channels had been identified at the time he proposed his classification (nor has such a depositional pattern been identified since, according to the writer's knowledge).

The description and classification of the channels and their component elements are part of the basis for the erection of local "summaries of the environment", or overall facies models for particular fluvial styles. This process, and the results, are the subject of Chap. 8. However, it is worth pointing out, at this stage, that such interpretations are by no means straightforward, and the researcher should be wary of falling into the simplistic traps set in earlier fluvial literature, for example, that lateral-accretion deposits indicate point bars in meandering channels. As noted in Chap. 8, LA deposits occur in a wide range of fluvial environments. Brierley and Hickin (1991) carried out a detailed element analysis of a modern gravel-sand river that showed braided, wandering, and meandering reaches. The three fluvial channel styles could not be distinguished from their deposits on the basis of either lithofacies assemblages or architectural-element types. In Chap. 8, it is shown that most element types occur in several different fluvial settings. In order to distinguish them and to arrive at reliable interpretations of channel style it is necessary to take into account all available data on element geometry and orientation. Detailed paleocurrent analysis is a very important part of this interpretive process.

Lithosomes of sixth-order and higher rank require classification and documentation using the principles of lithostratigraphy or allostratigraphy (Sect. 9.6). Increasingly, sequence concepts are now being used to classify and describe such deposits (Chap. 13). Large-scale channel belts and other discrete facies tracts may be mappable as individual stratigraphic units and assigned member or formation rank. For example, DeCelles et al. (1991) defined the surface enclosing individual fanglomerate bodies as sixth-order in rank, and designated the entire assemblage of contemporaneous conglomerate units as a formation-rank unit, enclosed by a seventh-order surface.

Deposits whose large-scale architecture is controlled by tectonic pulses or repeated base-level changes may be suited to a stratigraphic approach that employs allostratigraphic principles. As discussed in Chaps. 11 and 13, fluvial depositional systems tracts in the rock record are commonly bounded by major erosion surfaces (of sixth-order and higher rank), and may contain marine flooding surfaces. These surfaces are typically regional in extent, and indicate major allogenic controls on sedimentation. They provide the basis for subdividing stratigraphic successions using regional contemporaneous or near-contemporaneous erosional or flooding events. This is the formal basis for allostratigraphy, as defined in recent stratigraphic codes (see Miall 1990, Sect. 3.6; Sect. 9.6, this Vol.). Both regional erosion surfaces and flooding surfaces have been proposed as sequence boundaries by different authors (e.g., Galloway 1989a; Van Wagoner et al. 1990), that is, as surfaces used to define the beginning and end of sets of transgressive and retrogressive processes that are repeated in the rock record. Regardless of the terminology used, the genetic concepts that underlie sequence-stratigraphic principles are important aids for the regional stratigrapher (Chap. 13). Employing such principles can facilitate the understanding of regional stratigraphic processes more readily than the classic methods of lithostratigraphy, which classify similar facies together regardless of their age range, and separate dissimilar facies even where they are interbedded and clearly contemporaneous. Although basic geological mapping is more easily carried out using lithostratigraphic methods, this older approach is much less useful as a basis for interpretation. Formal allostratigraphic terminology is now being proposed for some such successions. For example, Plint (1990) subdivided a suite of shoreface deposits (that included some fluvial units formed during times of low base level) into a series of allomembers defined by regional disconformities and cycle boundaries. In another study, referred to at length in Chap. 13, Van Wagoner et al. (1990) attempted to adapt existing lithostratigraphic terminology to a new sequence framework. A unit originally called the Grassy Member was found to contain a major regional disconformity as a result of their sequence analysis, and was split into upper and lower parts, each of which comprised part of two separate stratigraphic sequences. The result is a clumsy stratigraphic nomenclature that needs a complete recasting in sequence terminology.

Research on nonmarine sequences is actively underway (Chap. 13), and a descriptive architectural classification of sequence geometries and bounding surfaces will likely be an eventual outcome. Terms derived from the Exxon work on sequence stratigraphy, such as low-stand and high-stand systems tracts (Posamentier and Vail 1988; Posamentier et al. 1988; Van Wagoner et al. 1990), are interpretive in nature, and only to be employed following thorough analysis, as discussed later. They cannot, in any case, be applied without qualifications to rocks that are entirely nonmarine.

4.8 Annotation of Outcrop Profiles

A completed profile is analogous to a geological map. It is basically a descriptive diagram of the exposed rocks. However, like a geological map, it may contain a degree of interpretation. On geological maps uncertain contacts and structures are indicated by dashed or dotted lines, and geological units that cannot be mapped separately may be shown unsubdivided. The same illustrative devices should be used on outcrop profiles. It is important to remember, however, that the classification of bounding surfaces and lithosomes, according to the systems described in this chapter, is an interpretive process. The classification is firmly grounded on observable attributes, such as the shape of the bounding surfaces, and the facies associations of the deposits immediately above and below, but a degree of interpretation is commonly required to produce a fully annotated profile.

It is recommended that the major bounding surfaces in each profile, of fifth-order rank and higher, be lettered in sequence from the base of the outcrop upwards, using capital letters. The architectural elements bounded by these surfaces are then numbered in the same order, using Arabic numerals. This system provides a basic reference framework for describing the profile.

Each profile should be lettered separately, using a local lettering and numbering sequence, unless profiles can be correlated laterally. If specific bounding surfaces can be traced between outcrops, it may be useful to use the same lettering and numbering sequence throughout, but it must be remembered that fluvial deposits tend to be characterized by rapid lateral changes in facies and architecture, and it cannot be assumed that surfaces and elements can be correlated across unexposed areas, unless there is a considerable amount of internal evidence to suggest such correlations.

Additional details can be added to the annotation, where this is deemed useful. Bounding surfaces of third- and fourth-order rank may be designated by primes or some other form of superscript added to the surface lettering. Architectural elements so subdivided may be designated by the letters A, B, C, etc. added to the element number. The element code may be added after this number, where a reliable interpretation can be arrived at. A complete profile annotation may then look something like this:

4-SB
D ———————— (5th order)
3C-DA
C^2 ———————— (4th order)
3B-DA
C^1 ———————— (4th order)
3A-LA
C ———————— (5th order)
2-LA
B ———————— (5th order)
1-SB
A ———————— (5th order)

Some sandstones sheets are built by a complex process of cutting and filling, so that only fragments of the bars and minor channel fills are preserved in the final deposit, many of which may not extend completely across the outcrop. Numbering of the elements and subelements must then be carried out

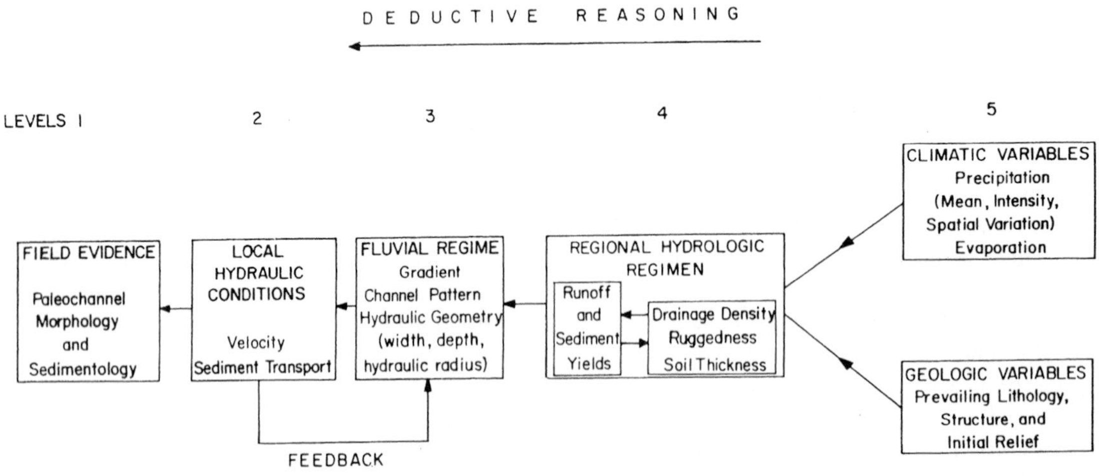

Fig. 4.14. Flow diagram showing the relationships between geological and climatic variables to hydraulic conditions and the resulting fluvial style and sedimentary record. The controls, in nature, act from *right to left*, as shown by the *boxes* connecting the *arrows*. Geologists attempt to reconstruct these parameters by inductive reasoning, moving from *left to right*, except for independent evidence relating to regional geological variables. (Baker 1978a)

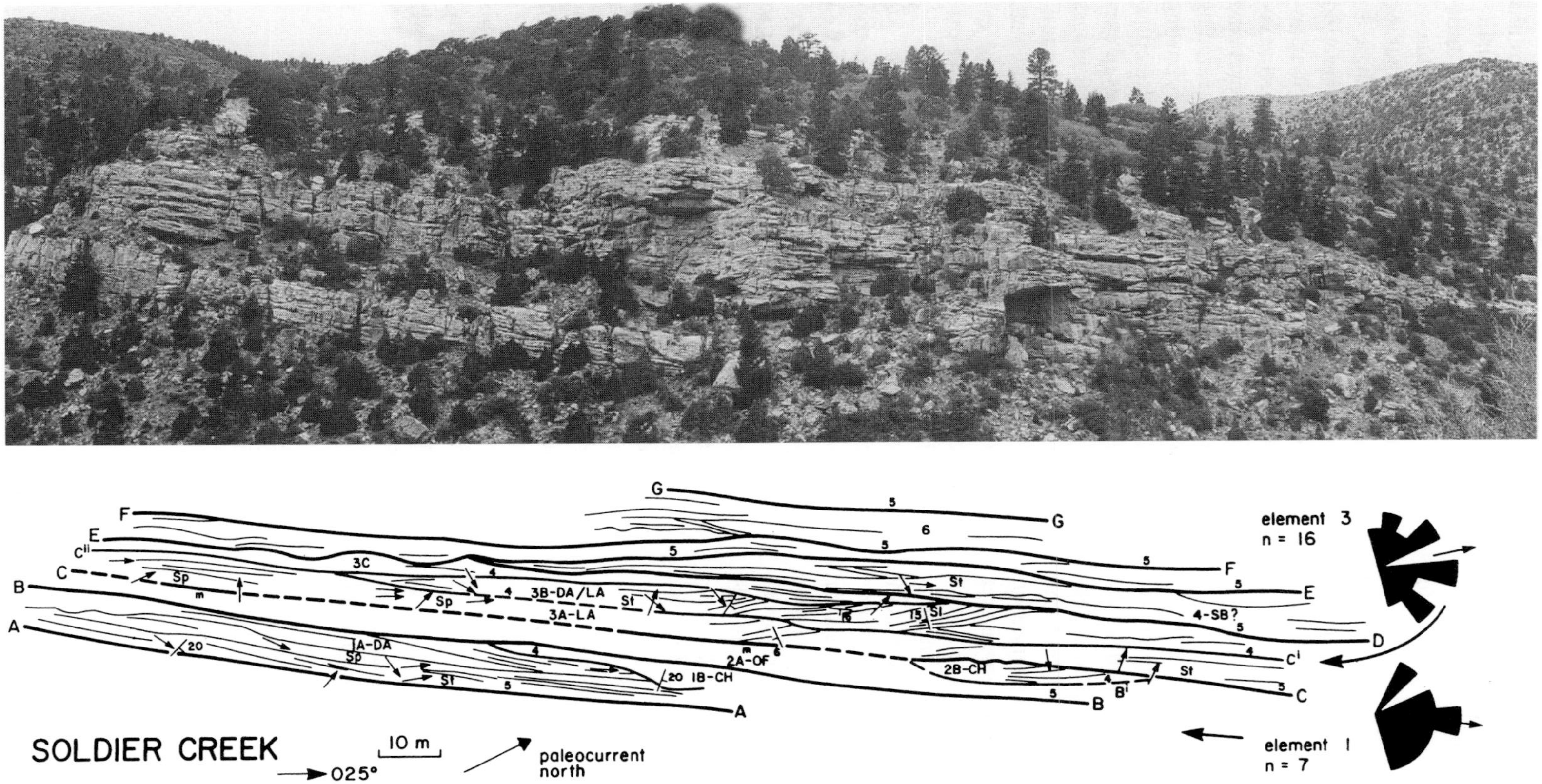

Fig. 4.15. Example of a completed, annotated photomosaic through a fluvial deposit. Castlegate Sandstone (Upper Cretaceous) at Soldier Creek, Utah. Major bounding surfaces are indicated by *capital letters*, architectural elements are indicated by *numbers with element code*, where known. Paleocurrent data are oriented such that flow parallel to the outcrop face is given by *horizontal arrows*. A detailed interpretation of one of the elements in this profile is provided in Fig. 6.32. (Miall 1994)

with care, and no single vertical profile across the outcrop will contain a complete sequence of surface and element notations. Profiles of this type can provide a graphic illustration of the lengthy process of channel migration and bar accretion that can occur within a single mobile channel before it is finally abandoned.

4.9 Summary of Methods

The field methods described in this chapter are designed to lead to detailed descriptions of stratigraphic architecture for purposes of resource exploration and production, and to facilitate the interpretation of sedimentary controls. Observation leads to interpretation by inductive reasoning. Baker (1978a), in an examination of the methods of fluvial geomorphology, illustrated the links between sedimentary controls and the real-world outcome by using a flow chart (Fig. 4.14). Architectural methods for detailed outcrop study, which form the basis for inductive reasoning, contain the following components, to be carried out in essentially the order given. References are made to the chapters where the relevant detail is provided. A typical example of a completed, annotated photomosaic is illustrated in Fig. 4.15.

1. Photograph the outcrop and construct a photomosaic.
2. Document the scale and orientation of the profile.
3. Enter all visible bounding surfaces on a profile overlay.
4. Check the correctness of the surface drawings in the field, and carry out preliminary element interpretation. This can be done by viewing the outcrop from the same place as the photographs were taken. It is particularly useful to distinguish the major lithosomes by defining fifth- and higher-order surfaces wherever possible.
5. Commence traverses of the outcrop, making detailed observations of lithofacies (Chap. 5) and paleocurrent directions. Mark observation points precisely on the profile overlay, if necessary with the assistance of a second worker located at the photo site (who has a better overview of the outcrop). Attempt to complete the tracing of bounding surfaces left "hanging" following the preliminary photointerpretation. Attempt to obtain a representative suite of readings of current indicators and dipping bounding surfaces for at least each fifth-order lithosome, and preferably for third- and fourth-order units as well.
6. Carry out element interpretation using the architectural details obtained from facies and orientation data (Chaps. 6, 7).
7. Annotate elements and bounding surfaces.
8. Correlate adjacent outcrops. If possible, correlate specific major lithosomes and bounding surfaces by walking them out or by visual tracing from a distance, but be particularly cautious of attempting to correlate units and surfaces lower than fifth-order in rank (Chap. 9).
9. Synthesize lithofacies, element, and paleocurrent data to arrive at interpretations of fluvial style and channel pattern (Chaps. 8, 9).
10. Where appropriate, develop a regional allostratigraphic and sequence framework from such regional correlations (Chap. 13).
11. Where required, measure the scale and geometry of lithosomes at each level of the hierarchy, for the purpose of documenting reservoir heterogeneities (Chap. 10).

Chapter 5

Lithofacies

5.1 Introduction

Most fluvial deposits are clastic. A simple tripartite subdivision into gravel, sand, and fine-grained (mud, silt, and very fine-grained sand) lithofacies is a useful first step in description and classification. Most beds may be classified readily into one or the other of these groups, reflecting a natural separation of processes and sorting of the sediment load. Some mixed lithofacies, such as pebbly sandstones, do occur but can normally be classified according to their dominant grain-size class.

This chapter is intended primarily as descriptive. Other texts, such as Middleton and Southard (1977), Collinson and Thompson (1982), and Allen (1984), deal with the mechanics of sediment transport and the theory of bedform generation in some depth, and it is not the purpose of the present book to duplicate this material. Some of the major recent advances in this area are touched on in Chap. 2. Recent work on bedform genesis and classification (especially Ashley 1990) is discussed in this chapter only where it provides a useful genetic underpinning for the lithofacies classification. The major lithofacies types are listed in Table 4.1. Most were initially proposed by Miall (1977, 1978c); references to other definitions are given where appropriate.

5.2 Gravel Facies

5.2.1 Depositional Processes in Gravel-Bed Rivers

5.2.1.1 Introduction

Gravel transport takes place under a wide spectrum of physical conditions and leads to a range of textural and structural variations in the ensuing deposits. The major controlling factors are the condition of the flow, either laminar or turbulent, and the sediment concentration. Specific depositional responses (lithofacies) are associated with particular processes, but individual flows may evolve from one set of conditions to another, so that the final deposit may reveal a range of textures and structures within one lithosome. Figure 5.1 illustrates a model that attempts to summarize the relationships between the various processes and their results.

The main differentiation is between two types of process: (1) flows that are turbulent and have low sediment concentrations, in which transport and deposition take place by traction and very limited suspension, and (2) flows in which the higher viscosities associated with high sediment concentrations dampen turbulence, leading to laminar flow and grain support by buoyancy. This second class of event is given the general term sediment-gravity flow.

Useful work on gravel sedimentation by traction currents and the resulting lithofacies has been carried out in modern rivers by Williams and Rust (1969), Rust (1972), N.D. Smith (1974), Boothroyd and Ashley (1975), Lewin (1976), Hein and Walker (1977), Allen (1983c), Southard et al. (1984), and Carling (1990). Ashmore (1991) carried out a very useful series of model experiments. Sediment-gravity flows are locally important, particularly in alluvial fan settings, and have been studied by Blissenbach (1954), Bluck (1967), Johnson (1970), Larsen and Steel (1978), Schultz (1984), Blair (1987), Maizels (1989, 1993), Hubert and Filipov (1989), and Blair and McPherson (1992, 1994).

5.2.1.2 Traction Currents, Fluid Flows

The processes of gravel transport and gravel bedform, channel and bar formation are less well understood than those of sand-bed rivers. The reasons for this are partly observational. It is much more difficult to observe gravel-bed rivers during active bedload transport because the high transport energies make in-place measurement and sampling hazardous, because direct observations may be prevented by the opaqueness of the water, which commonly

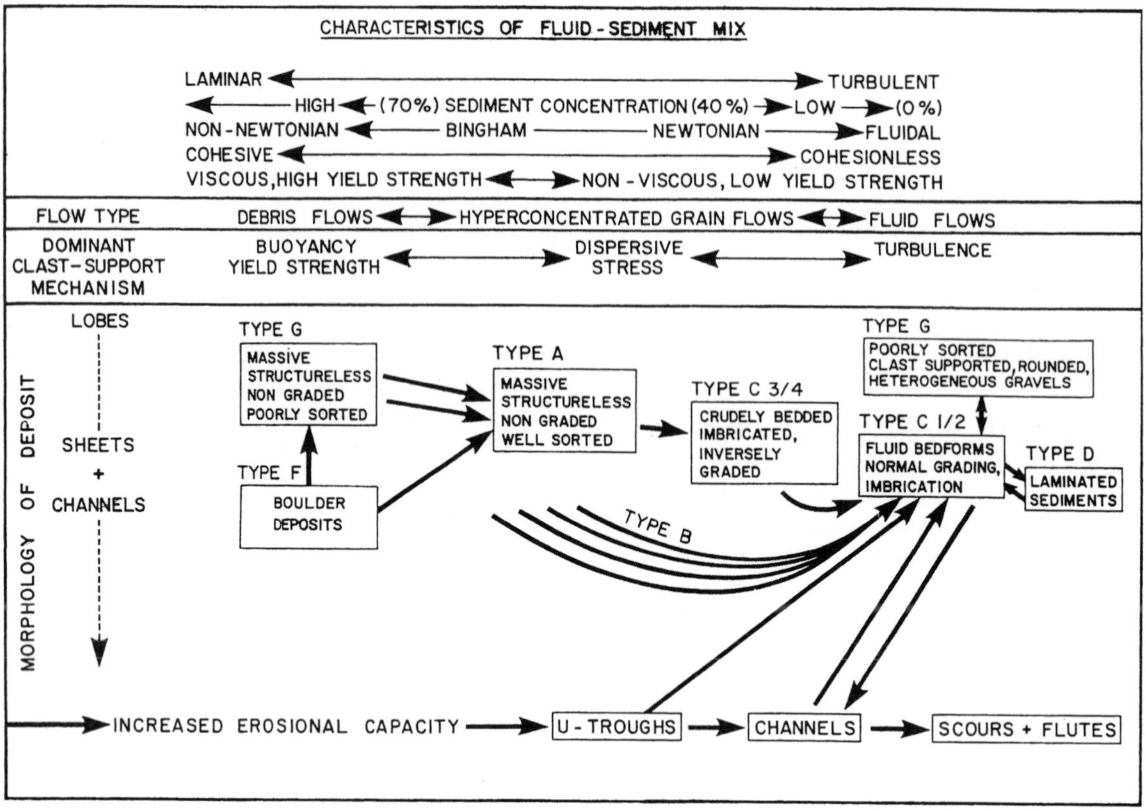

Fig. 5.1. Relationships between the physical characteristics of a fluid-sediment mixture, the resulting flow type, and the corresponding deposit. Note that an individual flow event may evolve through more than one of these stages (in the direction shown by the *arrows*), resulting in vertical and downstream gradations in texture and structure. (Maizels 1989)

bears a heavy suspended-sediment load, and because trenching of coarse gravels is difficult. It has been found that there are multiple processes affecting the morphology and stratification of gravel deposits, and that these interact and change very quickly, so that relating hydraulics to bedforms and stratification is a complex problem. Flume modeling of gravel-bed rivers has been slow to generate useful results, because of equipment limitations. However, it has now been demonstrated that modeling of gravel transport can be done in the flume by using sand, and by careful scaling down of all appropriate physical parameters.

Analysis of gravel-bed rivers and their deposits is carried out at two scales. At the sediment-water interface, variations in turbulence patterns, sediment transport rate, flow speed, and depth control bed textures and structures, and are reflected in the resulting stratification patterns and lithofacies characteristics. These processes, in turn, are part of the larger-scale evolutionary development of channels and bars that governs the development of architectural elements in the final deposit. In addition to changes in conditions imposed by external factors, such as diurnal and seasonal variations in discharge and temperature, deposition and erosion at the bar and bedform scale have their own continuous effects on water velocities, depths, and sediment loads immediately over, around, and downstream from each feature, resulting in corresponding localized changes in the conditions nearby. For example, deposition of a gravel lobe locally raises the water level, and can lead to small-scale flooding and channel diversion. Bank and bar erosion feeds slugs of bed load into the system, which may be deposited as a lobe in an area of shallowing or flow expansion a short distance downstream. In this way, momentary changes in the bed are propagated downstream and the river is maintained in a continually changing state of dynamic disequilibrium. Preservation of any given depositional unit is due to chance abandonment and burial. It is rare for complete macroforms to be preserved in a gravel deposit.

In this section, we consider the develoment of textures and structures and the resulting lithofacies variations. Gravel macroforms are described and interpreted in Chap. 6.

Variations in the texture and structure of gravel deposited from traction currents reflect the extreme unsteadiness of flow and transport rates in natural gravel-bed rivers. Gravel tends to be transported in repeated pulses as loosely defined 'slugs' or 'lobes'. This reflects turbulence patterns in the river, and also reflects the way in which gravel is released spasmodically into the river from bedform migration and cutbank erosion. Under conditions of high bed-shear stress only the coarsest clasts are deposited, leading to open-work, clast-supported gravels. Finer clasts, and sand, infiltrate the clast framework at lower flow velocities. Changes in flow velocity and shear stress can occur over very short periods, reflecting changes in river discharge and responses to channel migration and avulsion patterns immediately upstream. As a result, the texture of the deposits may show rapid vertical changes (Fig. 5.2).

Clasts deposited from traction currents commonly have a regular fabric, termed imbrication. This is an orientation of each clast such that the plane formed by the long and intermediate axis of the clast dips upstream, and is the most stable position for a clast to take up. It is a characteristic feature of gravel-bed rivers (Fig. 5.3) and is commonly preserved in clast-supported conglomerates, where it provides a useful paleocurrent indicator. Suspension of clasts during floods may generate a fabric with the long axis parallel to flow.

On a bed-by-bed scale the most important variation in texture and structure is that between horizontally bedded gravel showing crude stratification, and cross-bedded gravel. Hein and Walker (1977) provided the most convincing explanation for the relationship between these two major types of gravel lithofacies (Fig. 5.4). A slug of gravel, released from an eroded bank or bar upstream, is transported as a lag in an active channel. It comes to rest at a point of shallowing or flow expansion, where it constitutes a channel-floor lag a few clasts thick, in the form of a diffuse gravel sheet. The sheet accretes into a bedform by vertical and downstream growth. The balance between these two directions of growth seems to determine the final structure. Under condi-

Fig. 5.2. Exposure of alluvial gravel deposit, showing the variation in texture resulting from constant changes in discharge and sediment-transport patterns. Quaternary glaciofluvial deposits, near Banff, Alberta, Canada

Fig. 5.3. Imbrication in a modern gravel-bed river, Lake Hazen area, Arctic Canada. Flow was towards the *right*

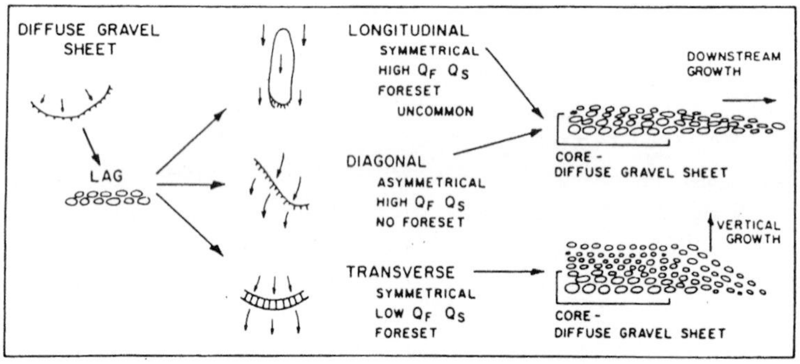

Fig. 5.4. Model for the development of the major types of bedform structure in gravels, depending on flow symmetry and fluid (Q_F) and sediment (Q_S) discharges. (Hein and Walker 1977)

tions of rapid gravel transport the sheet lengthens downstream faster than it aggrades, and the result is a low-relief sheet elongated downstream, with flat or very gently inclined stratification (Fig. 5.5). This structure was termed a longitudinal bar in some older literature. Hein and Walker (1977) suggested that where gravel transport is slower, vertical accretion takes place. An avalanche face develops, and a separation eddy may result, which serves to amplify the bedform-forming process. A transverse, cross-bedded bedform is generated (Fig. 5.6). In both cases, decreasing water depths over the growing bedform may lead to decreased shear stress and a corresponding upward reduction in clast size.

In some deposits gravel and sand are segregated within the same bedform. Topset beds are composed of well-sorted sand, commonly with parting lineation, and bottomset beds consist of cross-bedded gravel. Cross-bedding continues from the gravel up into the sand, where it undergoes a reduction in slope (Fig. 5.7). Form sets, where preserved, have a convex-up form and have been referred to as humpback dunes (Allen 1983c). Segregation occurs where the sediment mixture is bimodal, and where the

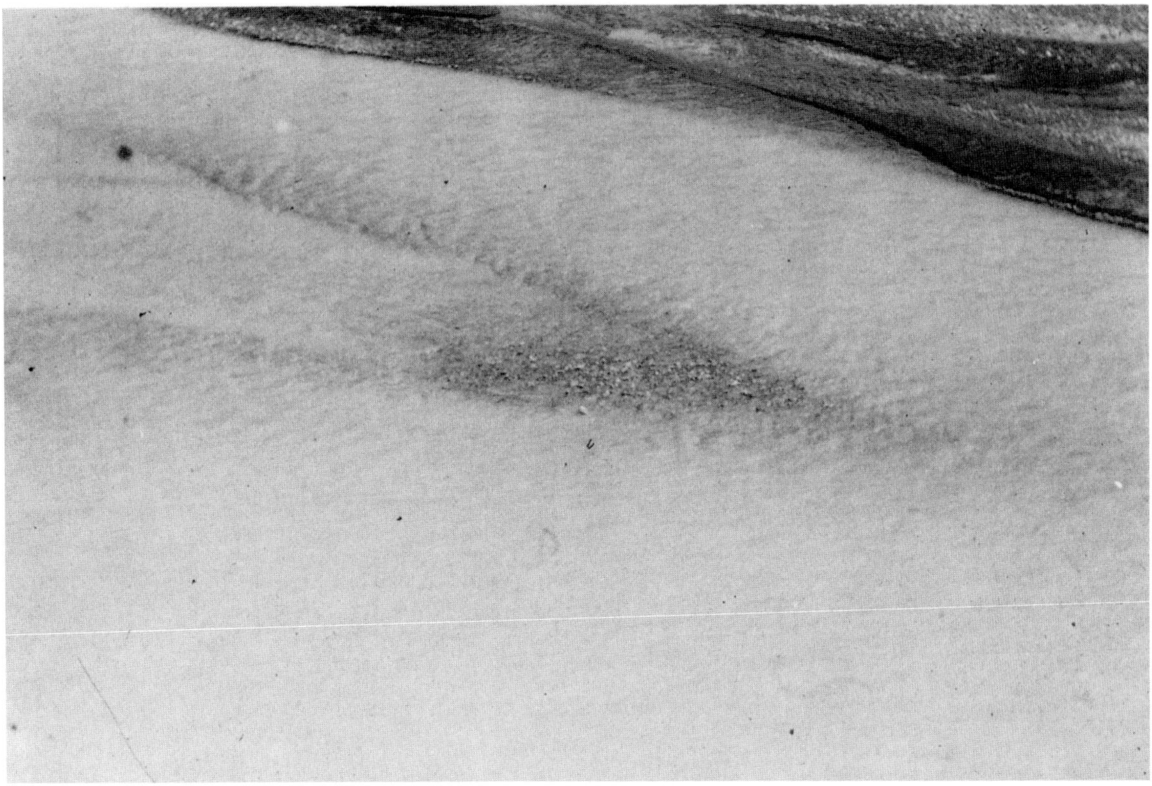

Fig. 5.5. Active growth of a longitudinal gravel bedform. Flow is from *bottom right to top left*. Field of view is approximately 20 m wide. Banff National Park, Canada. (Photo courtesy of D.G. Smith)

gravel clasts are well-rounded, facilitating sorting by rolling on the bed. Dunes develop, in which the topsets are under critical flow conditions, equivalent to the plane-bed condition of flow-regime theory (see Sect. 5.3.1). Gravel clasts are readily rolled over this smooth surface, and accumulate in the lee of the dune, as a forward-accreting apron. The process has been termed "gravel overpassing" (Allen 1983c).

Modeling experiments by Carling and Glaister (1987) and Carling (1990) have shown that the overpassing process can also explain cases where gravel cross-bed sets show a considerable internal variation in texture, as illustrated in Fig. 5.8. Gravel clasts show an upward decrease in grain-size, as a result of sorting on the avalanche slope. The matrix-filled and matrix-supported gravel forming the base of the sets results from the simultaneous deposition of gravel clasts that roll down the foreset slope, and sand settling from suspension in the separation eddy. This sand component is absent on the foreset slope, so that the gravel retains an open-work texture. At the top of the set infiltration occurs from the intermittent-suspension sand load beneath the separating flow, resulting in a matrix-filled, fine-grained gravel. This dynamic model dispenses with the need to invoke separate depositional processes for each textural variation.

Bluck (1979, 1980) showed that in some cases mesoforms are capped by coarse gravels, which may interfinger with finer gravel or pebbly sand foresets, resulting in small coarsening-upward sequences. Forbes (1983) referred to this as surface armoring. Such an arrangement may develop in several ways, such as the sweeping of gravel sheets across the bar tops at high stage, and development of sandy scour fills at the toe of the foreset during low-water stages (Massari 1983). Crowley (1983) showed that similar coarsening-upward textures occur in some large sandy bedforms, and are the product of changing water velocity and depth over the bar crest during active bar growth (element DA). Coarsening-upward, therefore, is probably a dynamic component of many large gravel elements.

Gravel sheets building into deeper water or areas of flow expansion, or those covered by gradually waning floods may develop lee-side separation eddies. This is accompanied by and encourages the growth of foresets, leading to the development of

Fig. 5.6. Development of a transverse gravel bedform. Cross-bedding formed where the bedform prograded toward the right into a pool of deeper water. This bedform has been abandoned by falling water. Height of bedform above water is about 1 m. Near Mito, Japan

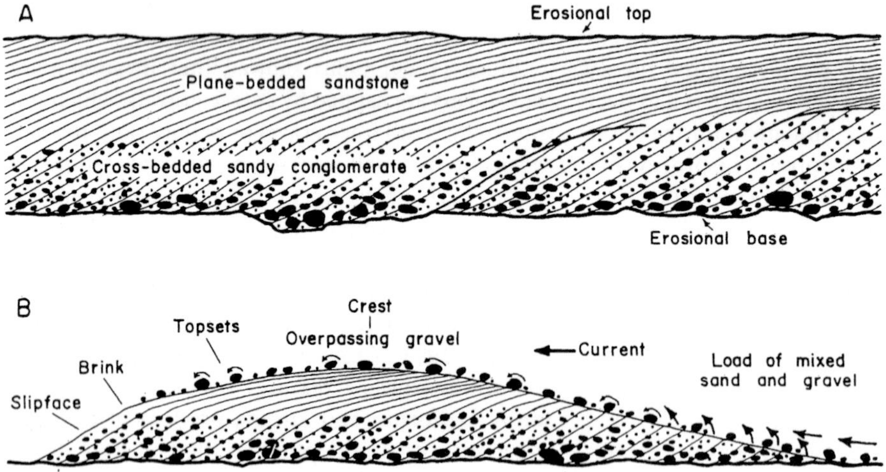

Fig. 5.7A,B. Typical form of stratification developed in segregated gravel-sand mixtures. **B** Form sets have been termed humpback dunes. (Allen 1983c)

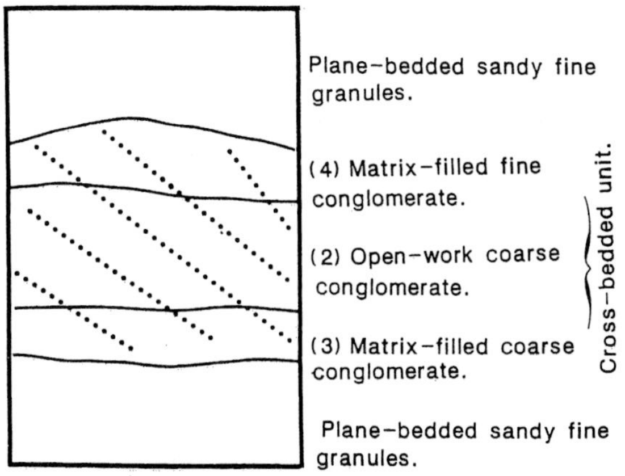

Fig. 5.8. Cross-bedded gravel set showing variation in textures. Foresets are indicated by *dotted lines*. Simultaneous deposition of gravels with different textures is explained by the overpassing process, as discussed in the text. (Carling 1990)

transverse bedforms (lithofacies Gp), as described by Hein and Walker (1977). Some examples of lithofacies Gt also represent migration of transverse bedforms with curved crests, while others represent the fill of minor channels. Where such channels debouch into pools they develop cross-bedded chute bars (lithofacies Gp) (Massari 1983; Ramos and Sopeña 1983). Large-scale Gp sets (> 2 m thick) suggest deposition in deep, confined channels (Kraus 1984; Middleton and Trujillo 1984). S.A. Smith (1990) recorded sets of cross-bedded gravel up to 3 m thick with foreset dips of 24–35°, indicating deposition of bedforms in water at least 3 m deep. In rare cases, lateral-accretion sets can be recognized. Such deposits are defined as part of the LA element because of their implication for relatively long-term lateral migration of a channel-bar complex, resulting in a distinctive architecture (Sect. 6.7).

Southard et al. (1984) reported a gravel transport process that they observed in shallow channels, which they described as the formation of chutes and lobes. The lobes are distinctive, coarse gravel accumulations, but their preservation potential appears to be low.

It has been observed that on the bed of many gravel rivers there is a distinctive arrangement of the large clasts into ribs or stripes perpendicular to flow. Koster (1978) demonstrated that these structures, conveniently termed "transverse ribs", form at hydraulic jumps between supercritical and subcritical flow. They also occur beneath kinematic standing waves that can develop during subcritical flow conditions. These structures have rarely been identified in the ancient record. Rust and Gostin (1981) described Holocene examples.

5.2.1.3 Sediment Gravity Flows

In flows with a high sediment concentration, grain transport is a result of buoyancy or matrix strength. At modest levels of sediment concentration the flow is cohesionless and may be internally turbulent. Crude lamination, with imbrication of clasts, may result. These flows are erosive, and may produce flutes and other forms of basal scour. At higher levels of sediment concentration (> 40%) the flow is said to be hyperconcentrated. Shear stress is transmitted through the flow by a dispersive pressure resulting from intergranular collisions. Coarser grains move to regions of lower shear, at the edge of the flow, resulting in inverse grading. Such flows commonly contain isolated large clasts that have been rafted on the top or edge of the flow. At still higher sediment concentrations the flow has pseudoplastic characteristics. The matrix is cohesive, and has a strength adequate to support large boulders. Little internal sorting takes place, although larger clasts gradually drop out, so that vertically and downstream there may be a crude clast grading. The upper part of the flow undergoes little internal shear, and may be transported as a semi-rigid 'plug'. Events of this type are termed debris flows, and are common in some alluvial-fan settings. Their high viscosity leads to lobate flow outlines, with convex-up margins. The size of clast that can be transported in debris flows depends on the yield strength of the flow, which is related to the viscosity and thickness of the flow. For this reason, the maximum clast size in a debris-flow deposit commonly varies with flow thickness.

Individual flows may evolve through the three stages of increasing concentration as a result of downstream loss of water, through infiltration into the substrate. Downstream reduction in stream slope results in a loss of momentum and transport energy. Eventually, the increase in internal friction resulting from a loss of momentum and of internal fluid lubrication leads to "freezing" of the flow. As this occurs internal shear may develop shear lamination and crude clast fabrics. The relationships between the various flow types and their lithofacies are shown in Fig. 5.9.

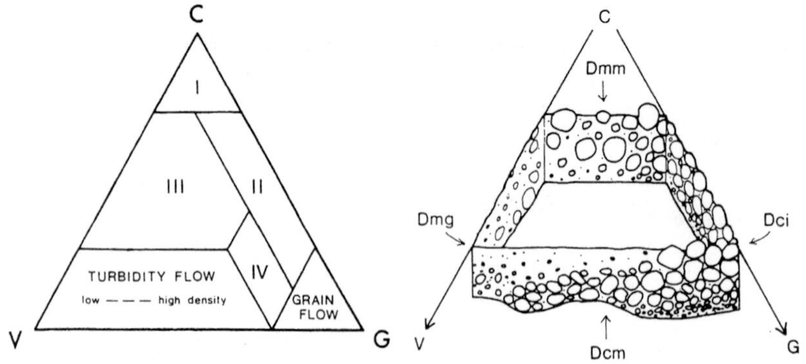

Fig. 5.9. *Left* Conceptual classification of debris flows. Poles represent cohesive-plastic behavior with increasing clay content (*C*), viscous-fluid behavior with increasing water content (*V*), and granular-collisional behavior with increasing clast content and shear rate (*G*). Flow types are: *I* plastic debris flow, *II* clast-rich debris flow, *III* pseudoplastic debris flow, and *IV* pseudoplastic debris flow with inertial bed load. *Right* The four main lithofacies classes and their intergradation. Facies codes are discussed in the text. (Schultz 1984)

5.2.2 Gravel Lithofacies

The seven major gravel lithofacies that occur in fluvial systems are differentiated initially on the basis of texture, and secondly, on the basis of their internal structure. It is useful to record the average grain size of the clasts, and any vertical changes in clast grain size. Such changes within an individual lithofacies unit reflect changes in shear stress over the bedform, while larger-scale changes, over several or many lithofacies units, indicate progressive changes in the channel system. A close relationship between bed thickness and maximum clast size is common in sediment-gravity-flow deposits (Bluck 1967). A convenient field method for rapidly recording maximum clast size is to measure the ten largest clasts present within a small area of exposure, and to take the mean of these readings.

The lithofacies codes have been modified from those presented by Miall (1978c) in order to incorporate additional observations on debris flows in an ancient fan deposit by Schultz (1984), the classification of which, in turn, was based on work on subaqueous debris flows by Eyles et al. (1983). A code scheme for poorly sorted, subaqueous, glacigenic diamicts was proposed by Eyles et al. (1983), using the capital letter D for diamict, followed by the lowercase letters m or c for matrix-supported or clast-supported, respectively. A second lowercase letter, m, i, or g, was added for massive, inverse, or normally graded textures. Schultz (1984) showed that the same scheme could be applied to subaerial debris flows in alluvial fans. This scheme has been adapted for use here, retaining the capital letter G to indicate the dominant grain size. Lithofacies Gms, as originally proposed by Rust (1978b), has been replaced by Gmm and Gmg. Lithofacies Gm of Miall (1978c) has been replaced by Gh to avoid confusion in the use of the lowercase letter m.

Lithofacies Gmm and Gmg: Matrix-Supported Gravel. These lithofacies were recognized by Rust (1978b), based on his work in some Carboniferous conglomerates. He used the code Gms (for matrix-supported gravel), which does not permit a distinction between massive and graded units. The most distinctive attribute of Gmm and Gmg is the absence of a clast framework. The clasts are poorly sorted and are supported by a poorly sorted matrix of sand, silt, and mud. The beds may be massive (Gmm: Fig. 5.10), or may show grading of clasts and/or matrix (Gmg: Fig. 5.11). Imbrication is normally absent, but tabular clasts may assume approximate horizontal orientations. Beds of this lithofacies have sharp but nonerosional relationships with underlying beds. They commonly have sharp lateral terminations. These characteristics reflect the formation of the lithofacies by the process of high-strength debris flows. Flows passively occupy preexisting alluvial topography – they will, for example, occupy channels and assume a channelized form. Flows are lobate in plan view and, because they have internal strength, they develop lobate, convex-up margins. This form is preserved when the flows stop forward movement as a result of the development of internal friction due to water loss.

Lithofacies Gci: Clast-Supported, Inverse-Graded Gravel. This lithofacies can occur in two ways, as a clast-rich, high-strength debris flow, or as a low-strength flow with an inertial bed load transported by laminar to turbulent flow.

Fig. 5.10. Debris-flow deposit; lithofacies Gmm. Note the poor sorting and matrix-supported framework. Carboniferous, Poland

Fig. 5.11. Debris-flow deposit; lithofacies Gmg. Clasts are logs and other plant debris, which show a crude upward size decrease and which also become somewhat more dispersed in the matrix. Tertiary, northern Kyushu, Japan

Fig. 5.12. Clast-supported gravel with crude horizontal stratification (lithofacies Gh), and minor interbedded sand lenses. Note upward decrease in clast size. Devonian, Prince of Wales Island, Arctic Canada

Lithofacies Gcm: Clast-Supported, Massive Gravel. This lithofacies represents low-strength, pseudoplastic debris flows, deposited from viscous, laminar or turbulent flows.

Lithofacies Gh: Clast-Supported, Horizontally Stratified Gravel. This lithofacies (Fig. 5.12; first defined as Gm in Miall 1978c) consists of clast-supported pebble and cobble gravel with crude horizontal stratification. Most deposits have a clast framework and an abundant sand matrix. Individual beds are typically a few decimeters thick, with multistory units reaching several meters in thickness. Bed contacts may be obscure because of the absence of well-defined bedding. Clasts are commonly imbricated.

Lithofacies Gt: Trough-Cross-Bedded Gravel. This lithofacies is distinguished by the presence of broad, shallow scoop-shaped bodies, typically 0.2–3 m deep and 1–12 m wide. These units cut into each other both laterally and vertically. The erosional base may be followed by a lag deposit of coarser grain size than the trough fill. Foreset dip is steepest in the smaller troughs, reaching a maximum of about 30°.

Lithofacies Gp: Planar-Cross-Bedded Gravel. Planar cross-bed sets in gravel reach 4 m in thickness, but are typically less than 1 m. Textural variation within these sets may be considerable, because of variations in sorting resulting from changing hydraulic conditions and gravel-clast overpassing (Fig. 5.13).

Fig. 5.13. Example of cross-bedded gravel deposit (lithofacies Gp), with variable textures, controlled by the overpassing process. Pleistocene, near Mito, Japan

5.3 Sand Facies

5.3.1 Sand Bedform Genesis and Classification

Sand lithofacies in fluvial systems result from the transport of sand by traction currents, as bed load and in intermittent suspension ("saltation"). The principles of fluid turbulence that control bedform generation are discussed briefly in Sect. 2.4.1. Empirical observation of bedforms in flumes and in a variety of modern environments, including rivers and estuaries, have led over the years to a proliferation of morphological terms for the various types that have served to confuse rather than elucidate their origins. However, a recent synthesis and review of this work have cut through much of the confusion and resulted in some useful simplifications (Ashley 1990).

The morphology of sand bedforms depends primarily on three parameters, sand grain size, flow depth, and flow velocity. Other parameters, such as fluid viscosity and temperature, exert a minor control over bedform morphology, but can be safely ignored for the study of most natural systems. For some years it has been known that the main types of bedform are stable in flowing systems only under limited ranges of these three main parameters, and that these stability ranges, for the most part, do not overlap significantly. Figure 5.14 illustrates the stability fields for the principal bedform types. Each of these classes gives rise to a characteristic lithofacies, as described in the next section.

One of the problems with the application of this type of classification to natural systems is that bedforms occurring in the dune field have a wide range of morphologies and sizes (height, wavelength), resulting in a corresponding variability in sedimentary structures. There has been considerable debate regarding the genetic significance of these variations, including whether they indicate a corresponding range of distinct depositional processes.

The major and most obvious subdivision of dunes is into those of two-dimensional (2-D) type and those of three-dimensional (3-D) type. Two-dimensonal forms tend to occur at lower flow speeds for a given grain size. They have simple, prismatic cross sections and give rise to planar-tabular cross-bedding, bounded by simple, planar first-order

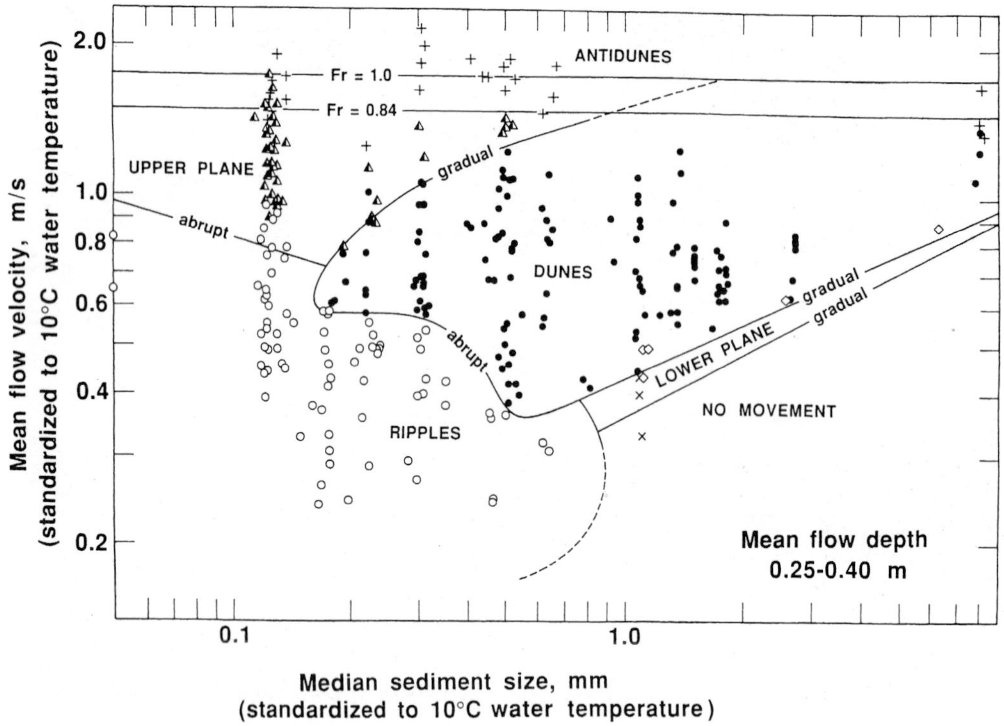

Fig. 5.14. Plot of mean flow velocity against median sediment size, showing stability fields of bed phases in sand. Different *symbols* indicate the various bed phases, as indicated by the names in the center of each field. Note the overlap between some of the fields. (Ashley 1990)

surfaces (Fig. 5.15). Three-dimensional dunes are characterized by curved lee faces and by deep, circular or oval scour pits at the foot of the lee slopes. The typical depositional product is trough cross-bedding. In most deposits the troughs are stacked into overlapping, mutually erosional sets bounded at the base by curved first-order surfaces (Fig. 5.16). An alternative term for dune is megaripple; 2-D forms have been referred to as transverse bars and sand waves; 3-D forms include linguoid and lobate bars. The use of the term "bar" for these features is now regarded as incorrect. It is retained for macroforms, which are long-term products of river systems (primarily the "group-6" deposits of Chap. 3).

Ashley (1990) argued that the apparent distinctions between the various morphological forms encompassed by these different terms are largely artifacts of our limited data. A review of available data from fluvial and tidal environments indicates that 3-D forms always occur at higher flow speeds than 2-D forms where other parameters, such as grain size and depth, are equal (Fig. 5.17). This differentiation relates to the growth of three-dimensional separation vortices in the lee of the dunes as shear stress increases within the turbulent outer layer. A review of height-spacing data for flow-transverse bedforms over a range of spacing from 0.01 to > 1000 m (Fig. 5.18) reveals a continuum, except for a single discontinuity at a spacing of 0.5 to 1 m. This corresponds to the break that has commonly been observed between ripples and the larger bedforms, the cause of which was discussed by Leeder (1983; see Fig. 2.22). The lack of any natural grouping in the larger forms is significant, and suggests that they comprise a single genetic population. Ashley (1990) reported a consensus view (of a discussion panel which examined this question) that the distinction between the various large-scale 3-D forms relates simply to the space, time, and volume of sediment available for their formation. The morphology responds to fluctuating water levels and velocity, and to the different turbulence patterns of unsteady and reversing flow. Dunes of different scales that are dynamically superimposed on each other indicate the development of nested boundary layers wherever space is sufficient. Large bedforms develop a boundary layer in which the smaller forms are stable. The smaller forms may remain active even where the large forms upon which they rest are inactive, but this simply reflects the limited availability of sedi-

Fig. 5.15. Two-dimensional dunes. *Above* Block diagram showing the typical surface form, and the resulting stratification (Harms et al. 1982); *Below* Modern example, Platte River, Nebraska. (Miall and Smith 1989)

ment, or inadequate time, for the large forms to migrate. In other words, no genetic distinction needs to be made between the various sizes of 3-D dunes (and the resulting cross-bedding) in a given fluvial system. However, it may well be useful to make a descriptive subdivision of these forms, in order to clarify the structure of the deposits and the mechanisms of channel and bar construction. Ashley (Table 6, 1990) suggested the terminology for such descriptions.

As noted in Sect. 2.4.1, experiments on the dune-upper flat bed transition by Saunderson and Lockett (1983) documented the existence of modified dune forms which preserve distinctive styles of stratification. Most characteristic are the humpback dunes that form by the sweeping of sand over a dune crest without the generation of a separation eddy (Fig. 5.19). The dune crest is not coincident with the top of the foreset, but is located behind it. The continued development of humpback forms leads to low-angle

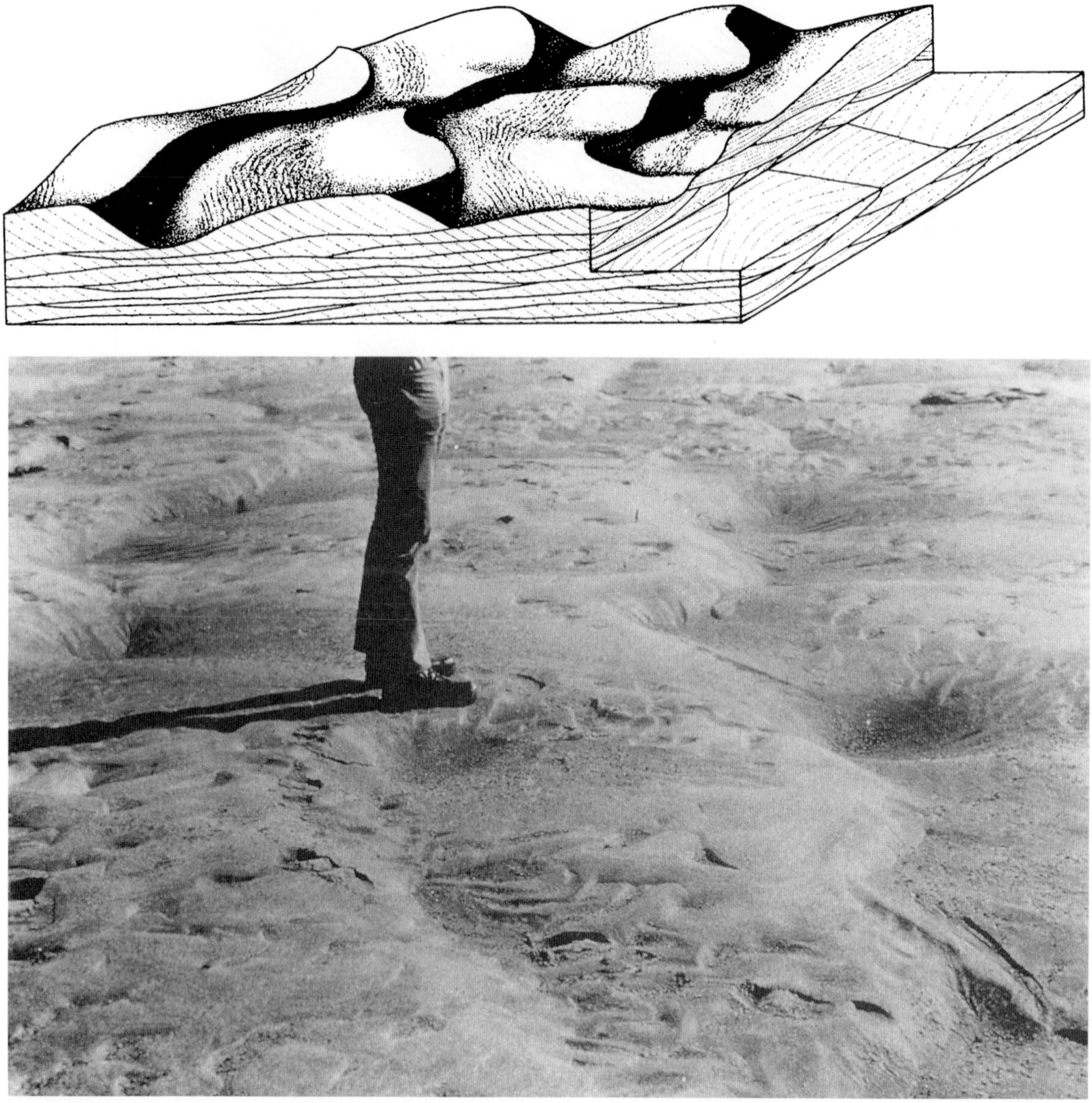

Fig. 5.16. Three-dimensional dunes. *Above* Block diagram showing the typical surface form, and the resulting stratification (Harms et al. 1982); *Below* Modern example. (Miall and Smith 1989)

(< 15°) sigmoidal cross-bedding. The crests of the humpbacks may be capped by plane bedding. The combination of flat bedding and low-angle cross-bedding is characteristic of the stratification that is preserved from this transitional phase.

Antidunes form under conditions of supercritical flow, but are rarely preserved in nature as they tend to be readily modified by the changing turbulence patterns that develop during decreasing flow speeds. Langford and Bracken (1987) described examples from modern stream deposits.

5.3.2 Sand Lithofacies

Sand lithofacies are classified on the basis of the dominant primary sedimentary structure. Detailed measurements of grain size are not required. Determination of modal grain size in the field, preferably with the aid of a hand-held grain-size comparator, is sufficient for the purposes of facies analysis. Qualifications should be observed wherever appropriate as to the degree of sorting, and the presence of accessory components such as intra- or extraclasts of pebble or larger size.

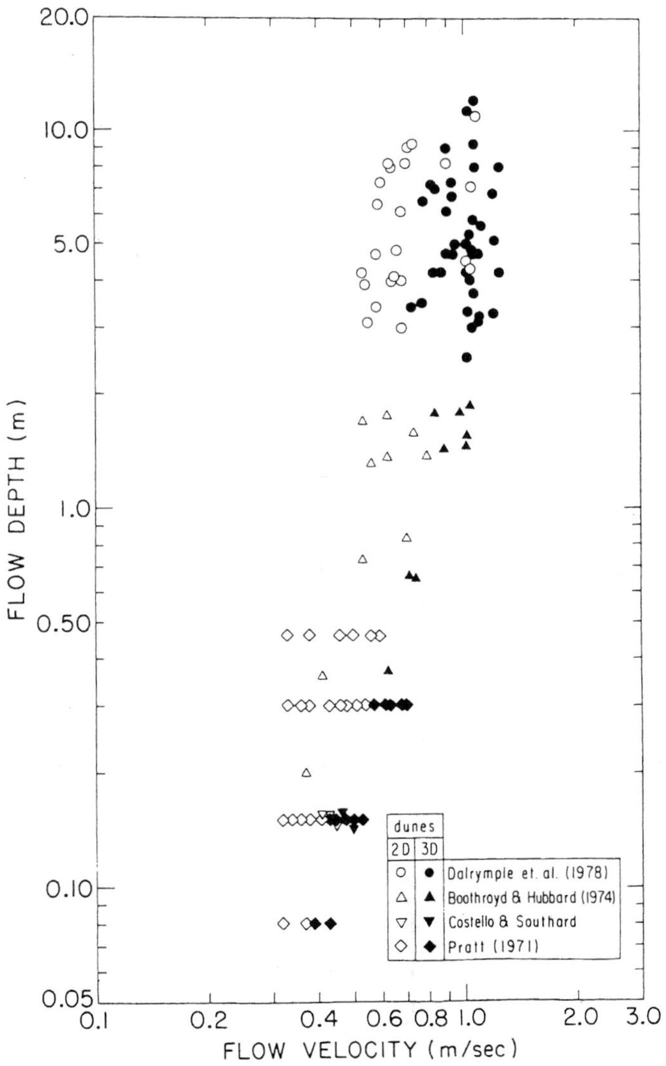

Fig. 5.17. Depth-velocity diagram, showing the stability ranges of 2-D and 3-D dunes in medium to coarse sand. (Compiled by Ashley 1990)

It is suggested that lithofacies Se, which was proposed by Rust (1978b) and incorporated in the scheme of Miall (1978c), can be abandoned as a separate category, as its recognition is now considered superfluous (see discussion below). No other changes from the original suite of seven sand lithofacies are proposed, although some additional refinements in definitions are now possible, and one code has been modified.

Lithofacies Sp: Planar-Cross-Bedded Sand. This lithofacies forms by the migration of 2-D dunes (Fig. 5.15). Sand is transported up the flank of the bedform by traction and intermittent suspension (commonly forming a carpet of small-scale current ripples), and deposited at the crest, where bed-shear stress drops at the point of flow separation. Cross-bedding is typically at or near the angle of repose (15–35°), with sharp, angular, upper and lower terminations, indicating avalanching of sand on foresets (Fig. 4.5). Upper and lower bounding surfaces are typically flat, with little evidence of scouring. Sand is typically sorted by the process of ripple migration up the stoss side of the dune, resulting in foresets in which the modal sand grain size may differ by several size classes.

The basic form of the cross-bedding is modified under different flow conditions (Fig. 5.20). At high flow speeds, approaching the transition to the plane-bed condition, separation eddies become smaller, and the foreset flattens out. Curved toesets may develop at this time. Minor sand transport by the backflow beneath the separation eddy may deposit small current ripples at the toe of the avalanche slope, with foresets oriented in the opposite direction. Falling water may lead to the abandonment and

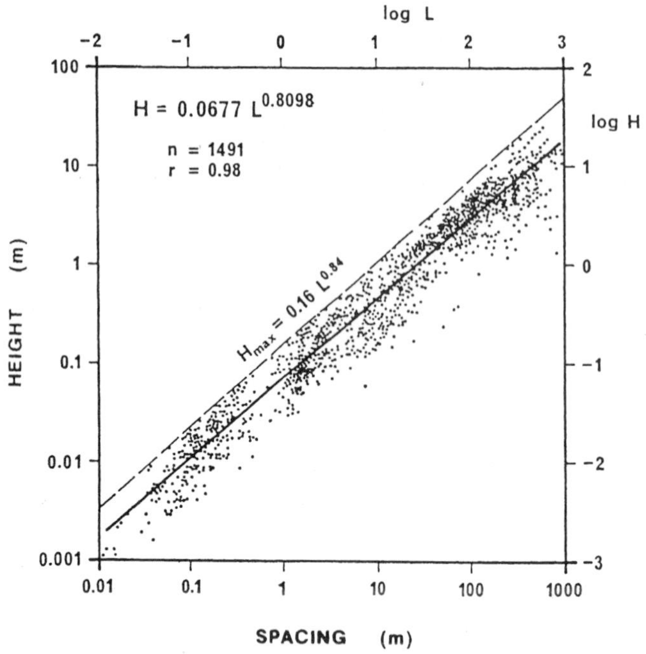

Fig. 5.18. Plot of height versus spacing of 1491 flow-transverse subaqueous bedforms. Note the discontinuity at 0.5 to 1 m spacing. (Ashley 1990; after Flemming 1988)

subaerial erosion of the dune, producing a curved upper dune surface. If this is covered again by rising water and renewed sand transport, the dune may be reactivated, so that the erosion surface is preserved as a cross-cutting surface within the dune, termed a reactivation surface (Collinson 1970).

Planar cross-bed sets form in sand of very fine to very coarse grain size. The lower limit of set thickness is 5 cm. This corresponds to the natural size break in the height of transverse bedforms determined by the size of separation eddies in the turbulent outer layer (Sect. 2.4.1). Set thicknesses of 0.5 to 1.5 m are typical of most fluvial sandstones. Sets several meters thick are abundant in the Hawkesbury Sandstone, near Sydney, Australia (Conaghan and Jones 1975; Rust and Jones 1987), reaching a maximum thickness of 7.5 m. Coleman (1969) recorded dunes 1.5 to 8 m in height in the modern Brahmaputra River of Bangladesh, and what he termed sand waves that had a height range of 8 to 15 m. However, some of the sand waves may be macroforms rather than mesoforms. An example of a solitary large-scale set is illustrated in Fig. 5.21. Cosets comprising thicknesses of several meters of sand are common. Individual sets may be traced for tens of meters parallel to bedding, although complete form sets are rarely preserved because of the continual channel shifting and scour occurring in most fluvial systems.

Lithofacies St: Trough-Cross-Bedded Sand. Troughs develop by the migration of 3-D dunes (Fig. 5.16). They occur in fine- to very coarse-grained sand. Pebbles may be present, and there is commonly a lag of poorly sorted sand with intraclasts of siltstone or mudstone at the base of the trough. Cross-stratification consists of curved sets, with an angle of dip that rarely reaches the angle of repose. The cross-stratification normally curves out at the base of the trough, which invariably shows an erosional relationship to underlying stratification. The curved foresets and the asymptotic downlap at the base are the features that distinguish trough cross-bedding from planar cross-bedding in small outcrops and in core.

Troughs occur as solitary or grouped forms. Solitary sets are commonly observable on bedding planes, where the distinctive basal scour surface and the curved trace of the infilling cross-bedding may be commonly traced for distances down paleoflow of up to 6 m (Fig. 5.22). Grouped sets (termed "festoon cross-bedding" in older literature) may make up thicknesses of several meters of sand (Fig. 5.23).

The lower limit of set thickness is 5 cm, as in the case of 2-D dunes. In most fluvial sandstones, trough sets are rarely larger than 1 m in thickness, but larger forms have been observed, indicating correspondingly greater water depths and dune heights, as in the case of the 2-D dunes discussed above.

In rare cases, dunes may be abandoned by rapidly falling water, and the dune scours subsequently

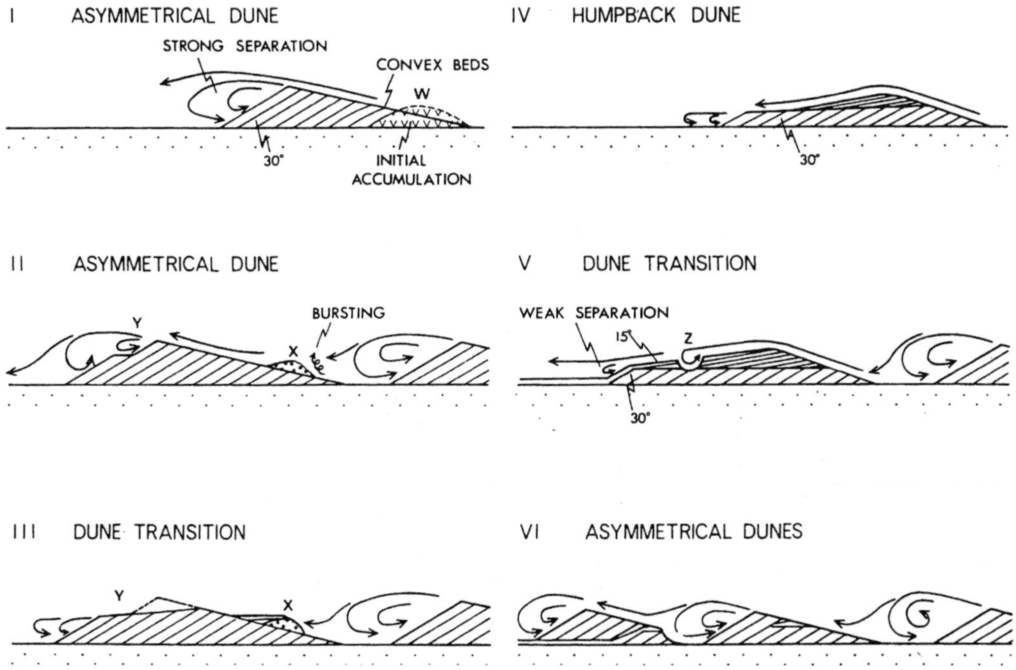

Fig. 5.19. Schematic diagram showing the transition from asymmetrical dune to humpback dune and back to asymmetrical dune. A mound of sand forms a symmetrical, convex feature (**I**, point *W*) with convex bedding, which then develops as an asymmetrical dune (**II**). Bursting takes place near the reattachment point (*X*). Erosion of the topset and progradation of low-angle sets from the reattachment point (**III**) leads to the development of a humpback dune (**IV**). These have small foreset slopes reflecting the small scale of the separation eddies. Draping of the humpback form leads to low-angle, sigmoidal cross-bedding (**V**). Continued vortex development (**Z**) may lead to dissection of the humpback form and regeneration of asymmetrical forms (**VI**). (Saunderson and Locket 1983)

filled with other bedforms, including lithofacies Sl and Sr. This is an example of bedform "lag", where the form of the structure indicates that it was not in equilbrium with flow conditions throughout its generation (Allen 1984).

Lithofacies Sr: Ripple Cross-Laminated Sand. A variety of asymmetric ripple types characterize this lithofacies. Sand grain size ranges from coarse to very fine, but fine- to medium-grained sand is most typical. A wide variety of internal structures may be generated from ripple migration, depending on flow velocity and the rate of sediment supply (Jopling and Walker 1968; Allen 1984). Migration of trains of ripples with a low rate of sedimentation from suspension leads to ripples that are mutually erosive (Fig. 5.24), termed "type-A" ripples by Jopling and Walker (1968). Where sediment is added from suspension during ripple migration, mutual erosion of the ripples is incomplete, leading to partial preservation of the stoss sides of the ripple, and "climbing" of the ripple train ("type-B" ripples; Fig. 5.25). The latter condition is common in fluvioglacial outwash, and in other settings where the river carries a large load of very fine sand and silt.

Ripples develop at low flow speeds (< 1 m/s), and are very sensitive to changes in flow conditions. Ripple trains with double crests may develop where an older ripple set is superimposed by a younger set. Ladderback ripples occur where flow directions change, as in a partially abandoned pool subject to fluctuating vortices entering from a main channel, or where temporary currents in very shallow water are driven by wind. Examples of these variations in ripple morphology are illustrated in Fig. 5.26.

Ripples are, by definition, less than 5 cm in height, many are less than 2 cm. Solitary ripple trains are common, and may be well exposed on bedding-plane surfaces. Cosets form thicknesses of decimeters to a few meters.

Lithofacies Sh: Horizontally Bedded Sand. This lithofacies occurs under two quite different conditions. The most important is that which represents the upper plane bed condition, at the transition from subcritical to supercritical flow (Fig. 5.14). This

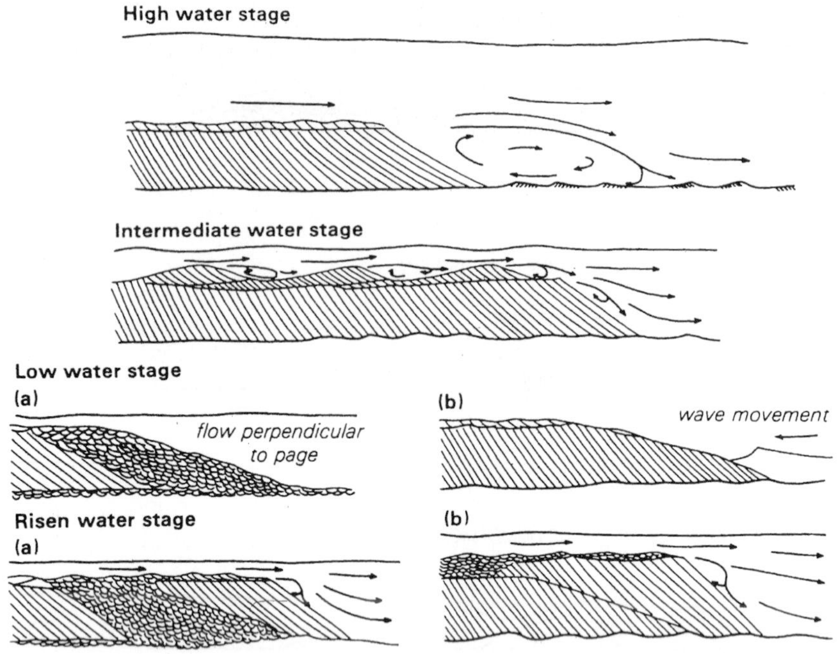

Fig. 5.20. Changes in morphology and internal structure associated with changes in stage over a 2-D dune. At high stage a strong separation eddy gives rise to asymptotic foresets and countercurrent ripples. The separation eddy is reduced in strength during falling water, resulting in angular foreset terminations. At low stage the dune may become inactive, with minor structures draped over it, commonly with different indicated flow directions. Erosion by minor waves and surface runoff as the dune emerges leads to a rounding of the dune crest, and this is then preserved as an internal reactivation surface when stage rises and the dune commences advancing again. (Collinson 1970)

Fig. 5.21. Solitary set of large-scale planar cross-bedding, approximately 4 m thick . Despite the size of this unit, it is a mesoform rather than a macroform, and is assigned to lithofacies Sp. Devonian, Scotland

Fig. 5.22. Solitary set of trough cross-bedding (lithofacies St) exposed on a bedding-plane surface. Proterozoic Matinenda Formation, near Elliott Lake, Ontario, Canada

Fig. 5.23. Coset of trough cross-bedding (lithofacies St) viewed in cross section. Tertiary Eureka Sound Group, Axel Heiberg Island, Arctic Canada

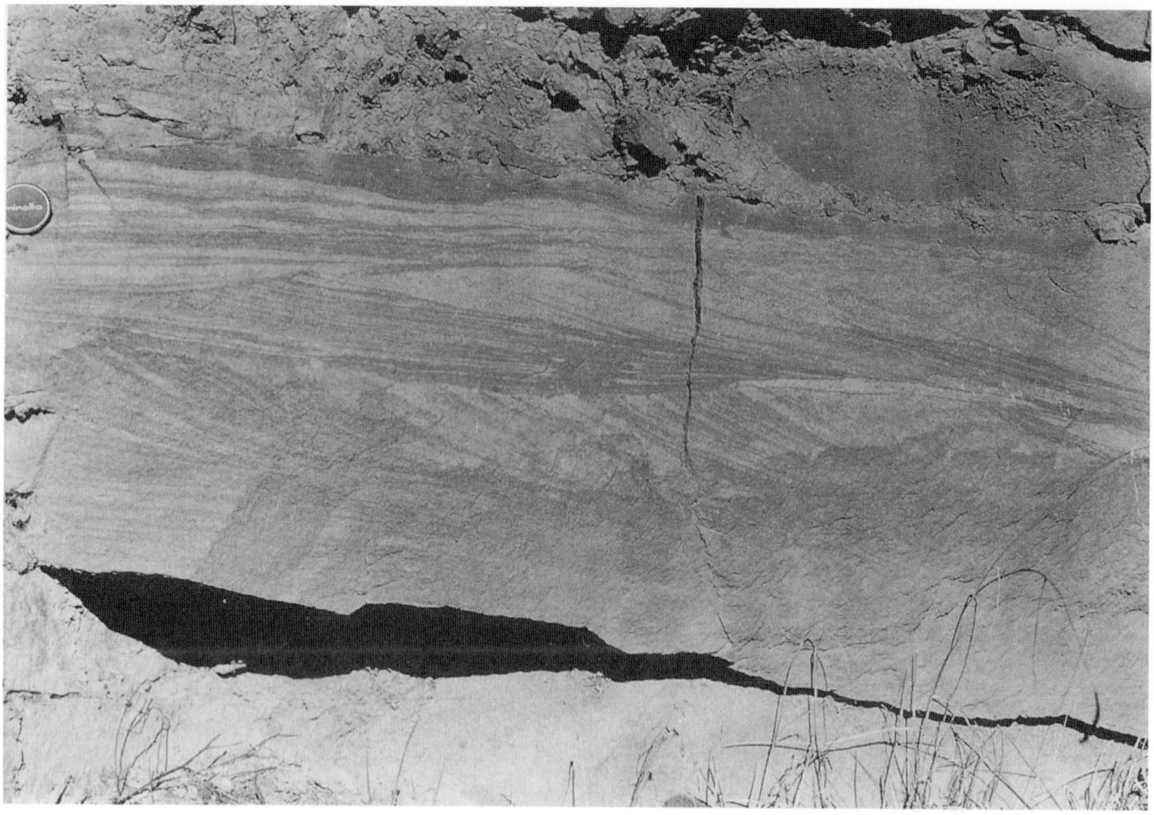

Fig. 5.24. Type-A ripple cross-lamination. Devonian Old Red Sandstone, Scotland

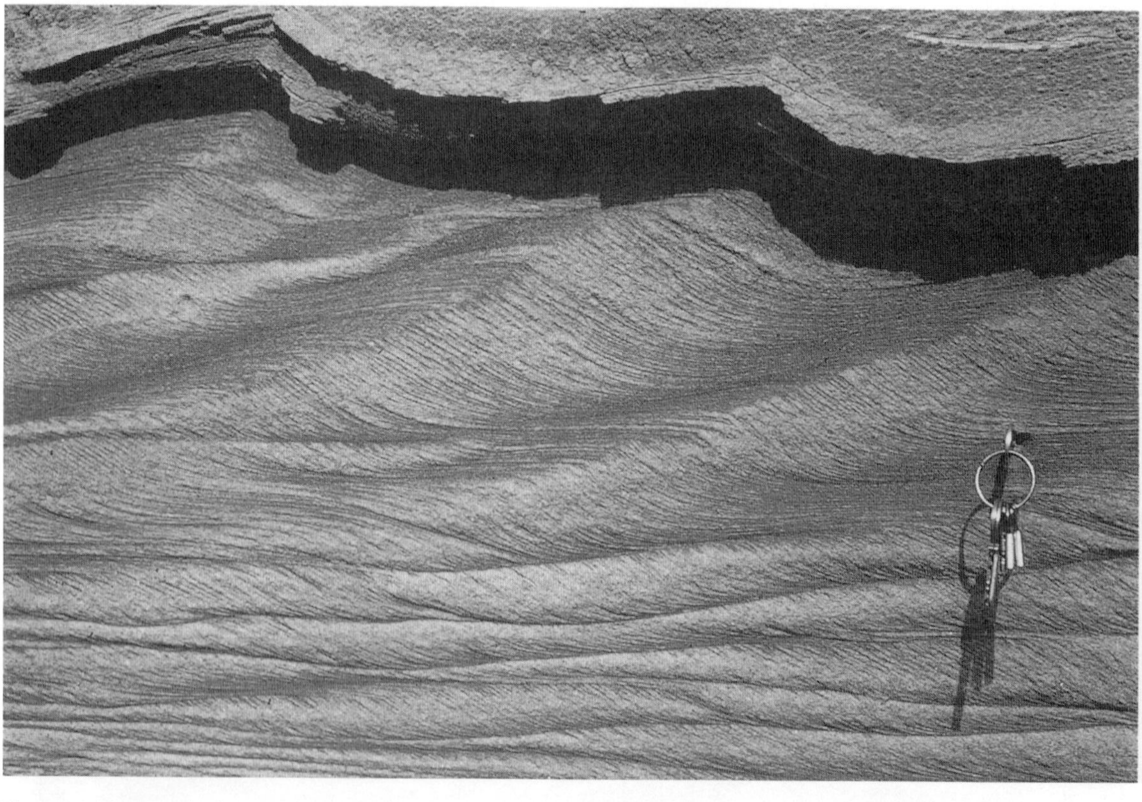

Fig. 5.25. Type-B ripple cross-lamination, Quaternary fluvioglacial deposits, near Ottawa, Canada

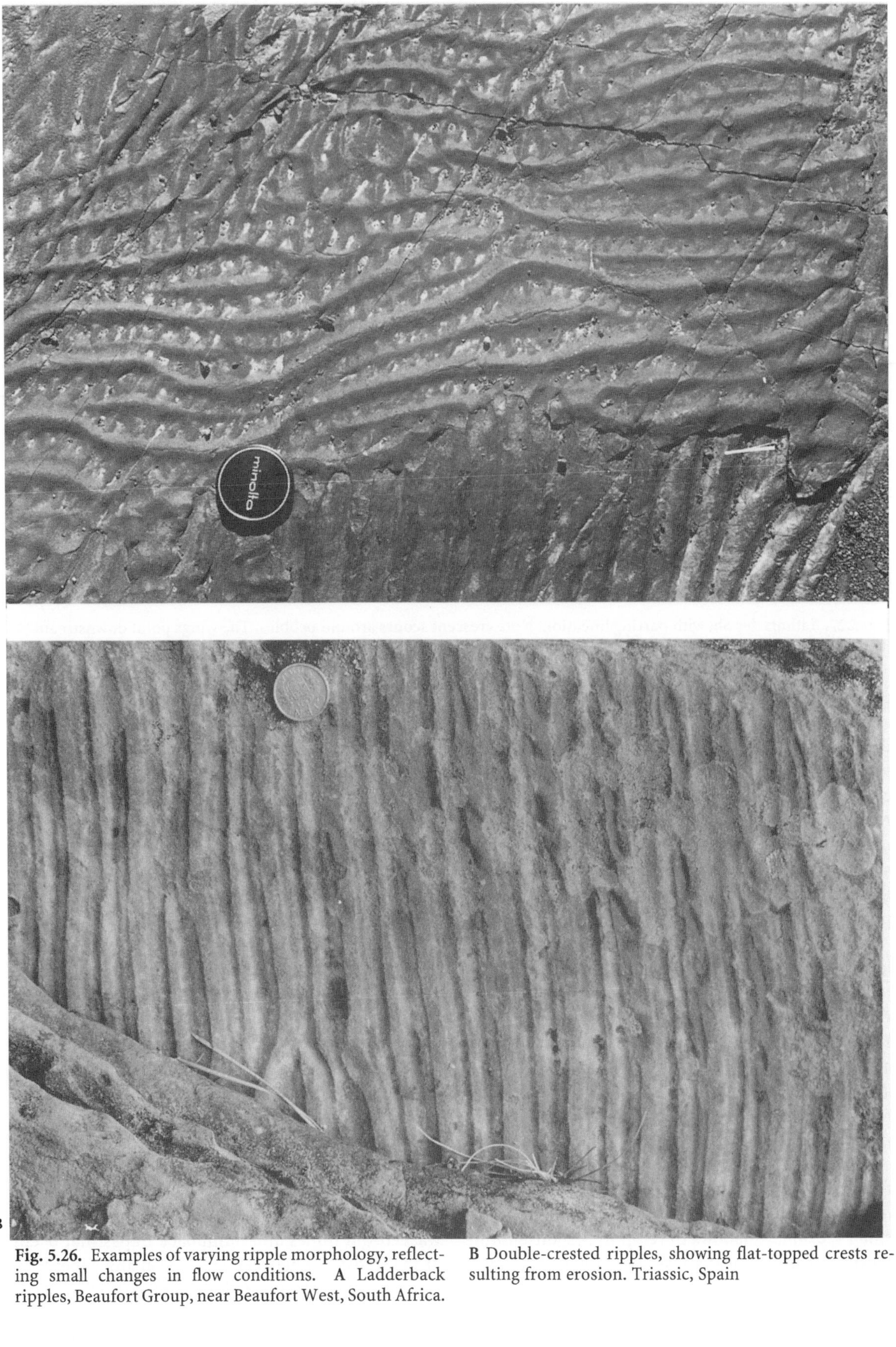

A

B

Fig. 5.26. Examples of varying ripple morphology, reflecting small changes in flow conditions. **A** Ladderback ripples, Beaufort Group, near Beaufort West, South Africa. **B** Double-crested ripples, showing flat-topped crests resulting from erosion. Triassic, Spain

Fig. 5.27. Lithofacies Sh, with parting lineation. Note crescent scours around pebbles. The wings point downstream (to the *left*). Ordovician, near Toronto, Canada

phase is most stable in very fine- to medium-grained sandstone at velocities of around 1 m/s and water depths of 0.25 to 0.5 m, but also occurs at lower velocities at shallower depths. Coarser grains, including pebbles, are rare, and are emplaced by being rolled along in the sand traction carpet. Units of Sh up to several meters thick may be deposited during single dynamic events, such as flash floods, when flow conditions may remain in the critical stage for periods of many hours.

Lithofacies Sh is distinguished by flat, parallel lamination, with parting lineation (otherwise termed streaming lineation or primary-current lineation) occurring on bedding planes (Fig. 5.27). This structure is generated by small longitudinal vortices at the base of the inner turbulent layer, and provides an excellent paleocurrent indicator. The orientation of the lineation is parallel to flow, but does not indicate which of the two opposing directions is the correct one. This information can be obtained by reference to interbedded cross-bedding structures, or by examining contained pebbles, around which small erosional shadows develop in response to the locally enhanced shear stress. The wings of the crescent scours point downstream (Fig. 5.27).

Plane beds also develop in coarse to very coarse sand at low flow speeds (~0.4 m/s), during initial sand bed movement, but this condition is rarely preserved in nature.

Lithofacies Sl: Low-Angle Cross-Bedded Sand. This lithofacies was proposed by Rust (1978b), following descriptions by Miall (1977) and Cant and Walker (1976, their facies G). Grain size and bedding characteristics of this lithofacies are similar to that of lithofacies Sh, with which it is commonly associated. The main distinguishing characteristic is the presence of low-angle cross-bedding, dipping at < 15°, and normally < 10° (Fig. 5.28). The cross-bedding is commonly asymptotic with reference to upper and lower set bounding surfaces. Parting lineation may be present, with an orientation parallel or oblique to the dip of the cross-bedding.

Some occurrences of lithofacies Sl represent plane beds deposited on initially dipping surfaces, such as scour hollows. In other cases, the lithofacies represents distinct bedform geometries. The washed-out dunes and humpback dunes that occur at the transition between subcritical and supercritical flow typically form this lithofacies (Fig. 5.19).

Fig. 5.28. Succession of sandstone beds deposited under near-critical flow conditions. Beds behind notebook are Sh showing gentle structural dip to the left. A low-angle cross-bed set of lithofacies Sl downlaps onto this unit. Jurassic Morrison Formation, near Gallup, New Mexico

Antidunes also give rise to a similar lithofacies, but are rarely preserved in the ancient record.

The distinction between plane beds on dipping surfaces and bedforms with low-angle internal stratification is best made with reference to the overall associations of the lithofacies. Interbedding of lithofacies Sh and Sl, with low-angle cross-bedding downlapping onto Sh bedding surfaces, indicates deposition close to the upper plane-bed condition, with the generation of a variety of dune-like structures. Where possible, the two types of Sl may be distinguished by additional lithofacies coding. For example, Sli may be used for inclined plane bedding, and Slb for low-angle cross-bedding representing bedform generation. Cowan (1991) used the codes Si and Sl for these contrasting cases.

Cotter and Graham (1991) pointed out some similarities between beds of Sl, which they observed in an ephemeral-fluvial sandstone, and hummocky cross-stratification, but concluded that the beds were formed under unidirectional, transitional to upper flow regime current conditions.

Lithofacies Se: Erosional Scours. This lithofacies was proposed by Rust (1978b), who compared the facies characteristics to lithofacies SS of Cant and Walker (1976). It comprises an erosional scour surface followed by up to 0.25 m of massive, coarse-grained sandstone containing intraclasts of mudstone and siltstone.

The original definition of this lithofacies was made by Cant and Walker (1976) in order to permit them to incorporate bounding surface information into a Markov chain analysis of their sandstone succession. However, this writer now considers that it is a superfluous category. Bounding surfaces may now be analyzed and described using a separate terminology, as described in Sect. 4.5. This must be carried out before Markov chain analysis is attempted and, as shown in Chap. 8, such analytical methods can now be seen to have a severely limited usage in fluvial deposits. The beds overlying the scour surface may be classified into any of the other lithofacies classes. Lithofacies Ss, as originally proposed by Miall (1977), is suitable in many cases.

Fig. 5.29. Lithofacies Ss forming overlapping scour-fill sets. Basal scour surfaces are indicated by *arrows.* Jurassic Morrison Formation, near Gallup, New Mexico

Lithofacies Ss: Scour-Fill Sand. Above fluvial erosion surfaces there commonly are a few decimeters to a meter of coarse- to very coarse-grained, poorly sorted sand, which may show poorly defined cross-bedding and contain abundant intraclasts and other lag materials (e.g., waterlogged plant debris) (Fig. 5.29). The bedding may drape over the irregular shape of the basal bounding surface. Sole markings, including flute marks and tool marks, are common on these basal bounding surfaces.

The characteristics of lithofacies Ss indicate rapid deposition of poorly sorted, coarse bed load. While the process is clear from the facies characteristics, the details of the environmental interpretation of the lithofacies depend on its facies associations. For example, what is the type of bounding surface on which it rests? Questions of this type relate to the architectural-element level of the analysis. For example, lithofacies Ss may comprise the basal fill of elements CH (channels) or HO (scour hollows). The same type of deposit may also occur at the base of macroforms, or draping the third-order surfaces that define their growth increments. A description of the geometry of the basal bounding surface should not, therefore, comprise part of the lithofacies definition, as originally proposal by Miall (1977), but should be part of the architectural description of elements and bounding surfaces.

Distinction between Ss and large, solitary St sets migrating along a channel floor may be difficult and even, perhaps, arbitrary. However, possible confusion about a lithofacies classification is less important than what the observations tell us about sedimentary processes during deposition of a unit of interest.

Lithofacies Sm: Massive Sand. Sandstone beds in outcrop may appear massive if weathering does not pick out the lamination. However, true massive sandstones do exist. They may show grading, or reveal very faint, patchy lamination. Such beds are the deposits of sediment gravity flows. A characteristic occurrence of this facies is in small channels resulting from bank collapse (Sect. 6.2). Massive texture may also be produced by postdepositional modification, for example by dewatering and bioturbation. Faint residual sedimentary structures may reveal such origins.

5.4 Fine-Grained Clastic Facies

Fine-grained clastic sediments are deposited primarily from the suspension load of rivers. This cannot take place within active channels, because the shear stress and turbulence are such that the load remains in suspension. Deposits of mud, silt, and very fine sand therefore indicate deposition in floodplain areas, in abandoned channels, and in abandoned areas of normally active channels, for example during seasonal low-water stages. The thickness of depositional units ranges from the few millimeters of mud drape formed in small abandoned channels, to thicknesses of tens to even hundreds of meters in the floodplains of major, suspension-load streams.

The original suite of two fine-grained lithofacies erected by Miall (1977) was modified by Rust (1978b), and McLean and Jerzykiewicz (1978) proposed two additional categories, one of which is now considered to be superfluous, as discussed below.

Lithofacies Fl, Fsm, and Fm represent a gradation in grain size and bedding characteristics from relatively coarse, proximal floodplain deposits to more distal deposits.

Lithofacies Fl: Laminated Sand, Silt and Mud. Interlamination of mud, silt, and very fine-grained sand is common in overbank areas, and represents deposition from suspension and from weak traction currents (Fig. 5.30). Very small-scale ripples may be present in the sand and silt beds. Undulating bedding, scattered bioturbation (including animal footprints), desiccation cracks, plant roots, coal streaks, and scattered pedogenic nodules may be present. Typical thicknesses of continuous Fl deposits range from a few centimeters to many meters, depending on sediment supply, fluvial style, and basin subsidence rates. Individual beds may be as thin as a few millimeters. Careful observation may indicate that units are organized into thin fining- or coarsening-upward successions.

Discrete sand beds thicker than a few centimeters may be separately logged as lithofacies Sr, St etc., and may represent more energetic floodplain traction currents, such as those that deposit crevasse splays.

A

B

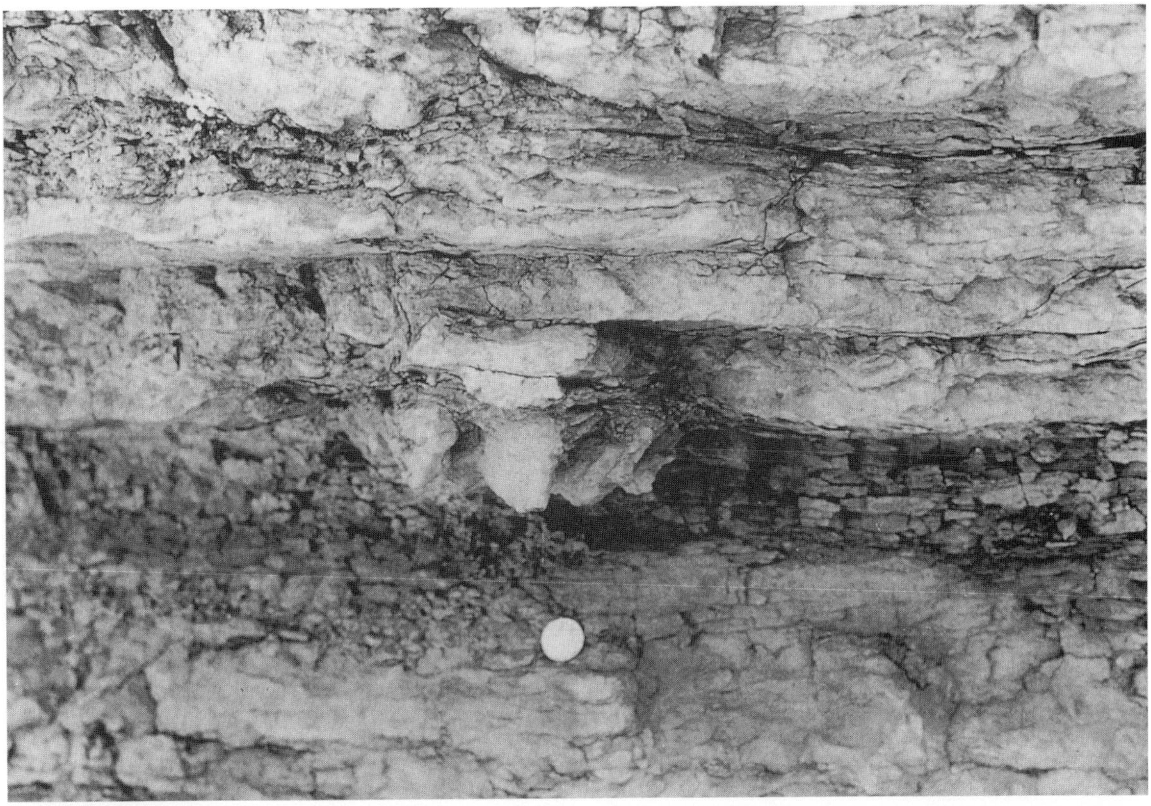

C

Fig. 5.30. Lithofacies Fl. **A** Channel sandstone overlying dark floodplain siltstones containing thin interbedded silty sandstone beds. Devonian Old Red Sandstone, Scotland. **B** Thinly laminated sandstone-shale succession with flow rolls resulting from syndepositional loading. Chinle Formation, Triassic, near Winslow, Arizona. **C** Interlaminated shales and thin sandstones. The cast of a dinosaur footprint filled with sandstone extends downward from an overhanging sandstone bed at *center* of picture. Triassic, Spain

Similarly, beds dominated by or consisting of more than a few centimeters of continuous coal, root beds, or pedogenic nodules should also be logged as separate lithofacies. The reader is referred to comments in Sect. 4.3 regarding the need to establish limits between lithofacies classes, where a possible overlap exists, for example between lithofacies Fl containing coal streaks, and separate beds of coal.

Lithofacies Fsm: Siltstone, Claystone. A lithofacies coded Fsc was proposed by McLean and Jerzykiewicz (1978), and appears as such in Miall (1978c). The lowercase letters stand for silt and clay, respectively. For consistency, this is relabeled here as Fsm, with the m standing for mud. The main feature of this lithofacies, which distinguishes it from Fl, is the absence of sand beds. In other features, such as bedding and presence of accessories, this lithofacies compares with Fl. It probably represents a floodplain deposit somewhat more distal relative to clastic sources such as nearby fluvial channels. In practice, the absence of thin sand beds in a heterogeneous, fine-grained succession may be difficult to observe, and the distinction between Fl and Fsm is then difficult or arbitrary. I have not used this lithofacies class in any of my field programs.

Lithofacies Fm: Massive Mud, Silt. Miall (1977) proposed this lithofacies for the mud drapes that commonly occur within gravelly and sandy braided sediments, where they represent the deposits from standing pools of water during low-stage channel abandonment. Thicknesses of a few millimeters to a few centimeters are typical. Carbonaceous streaks, plant rootlets, and desiccation cracks are common. The lower bounding surface of the bed may conform to the shape of any underlying bedforms, such as a train of ripples. Rust (1978b) suggested expanding this lithofacies to include any bed of massive mudstone. As such, it may represent the most distal floodplain facies, including deposition in floodplain ponds.

A

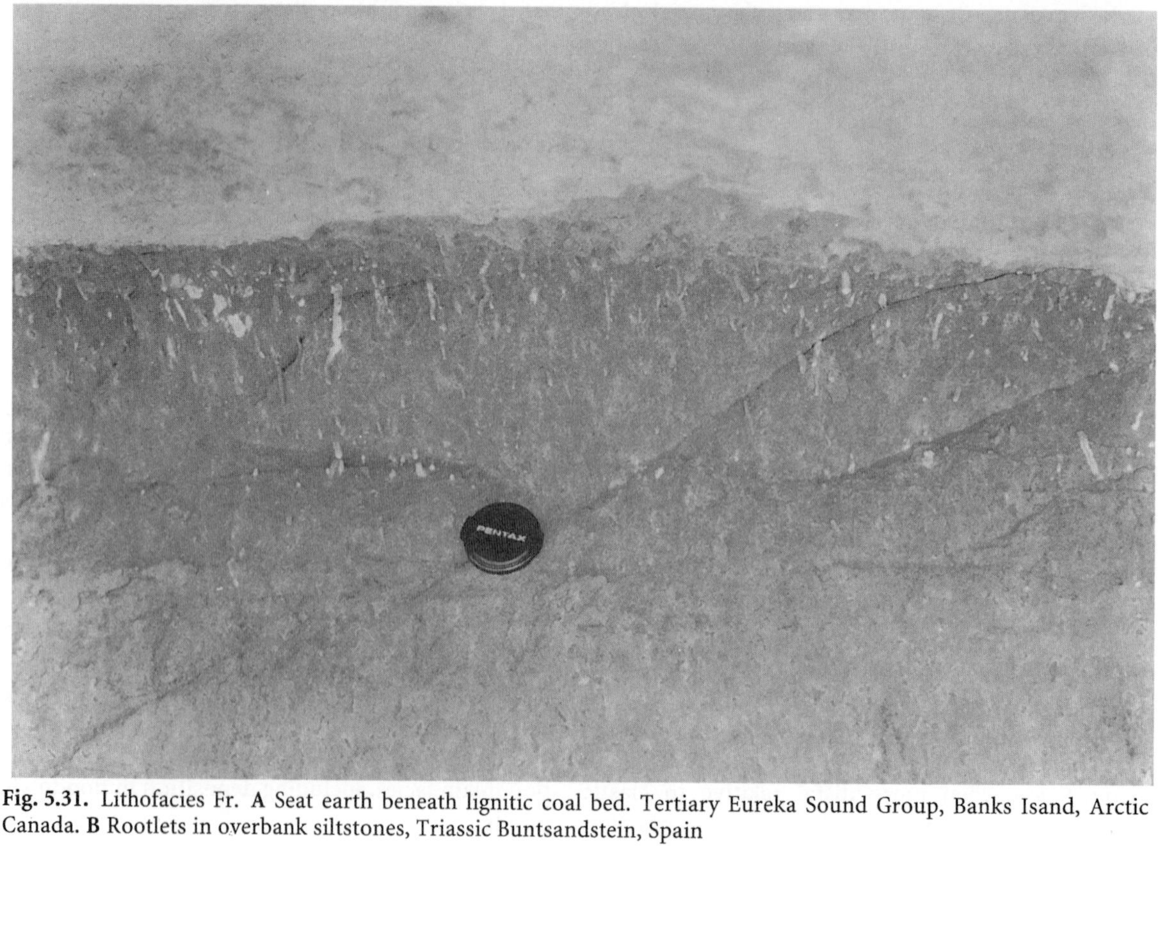

B

Fig. 5.31. Lithofacies Fr. **A** Seat earth beneath lignitic coal bed. Tertiary Eureka Sound Group, Banks Isand, Arctic Canada. **B** Rootlets in overbank siltstones, Triassic Buntsandstein, Spain

Lithofacies Fcf: Claystone with Freshwater Molluscs. This lithofacies class was proposed by McLean and Jerzykiewicz (1978). However, it is regarded here as superfluous. Lithologically, the beds are identical to Fm, and the molluscs constitute a lithologic accessory, which is insufficent to require definition of a different lithofacies class.

Lithofacies Fr: Root Bed. These are common in vegetated floodplains (Fig. 5.31). This lithofacies represents soil development in a humid climate. Carbonate nodules are typically absent, but silica cementation may have occurred as a result of leaching. The host lithology may be sand, silt, or mud, with any primary stratification, such as cross-bedding, obscured or destroyed by root emplacement and bioturbation, giving the beds a mottled appearance. The roots may retain their original carbon, or may be replaced by calcite or siderite.

Where associated with overlying coals (Fig. 5.31A), these beds have been termed seat earths or underclays. As an alternative to the clastic coding discussed here, seat earths may be discussed and classified under the heading of paleosols, as described in Chap. 7.

5.5 Nonclastic Facies

A limited range of chemical sediments occurs in fluvial deposits, mainly but not exclusively in the floodplain. The most important of these are the soils, which are best developed in semiarid climates. In humid, tropical settings coal may occur, while in arid climates minor evaporites may develop in inland sabkhas.

Lithofacies P: Pedogenic Carbonates. A wide variety of textures and structures develops where floodplains are exposed to surface weathering processes for extended periods (thousands of years). Rain infiltration leaches dissolvable ions downward, whereas evaporation and capillary groundwater flow during arid periods concentrate the same ions near the surface. The result is the gradual development of carbonate cements that coalesce into nodules and these, in turn, into more or less continuous carbonate substrates, commonly with a blocky fracturing pattern (Fig. 5.32). Modern calcitic soils are referred to as caliche or calcrete, and the terms have been adopted by sedimentologists for paleosols. Calcitic paleosols are most typical of arid to semiarid, oxidizing climates, whereas siderite nodules occur in waterlogged, reducing settings, and are commonly associated with coals.

The stratigraphy and facies of paleosols constitutes a specialized study that is touched on further in Sect. 7.4.2, where additional petrographic and facies classifications are provided.

Paleosols constitute useful local marker beds, and can therefore yield valuable information on basin stratigraphy (e.g., Allen and Williams 1982; Behrensmeyer 1987; Eberth and Miall 1991). Paleosol maturity can be deduced from color, soil-layer differentiation, and nodule morphology, providing data on floodplain exposure times and clastic influxes (Allen 1974b; Leeder 1975; Bown and Kraus 1987). These deposits are therefore of considerable value in the study of large-scale basin architecture, a subject that is considered at length in Chaps. 9 and 10.

Soil texture and geochemistry are sophisticated subjects, yielding much information on local biochemical processes, local climate, etc. (Johnson 1977; Bown and Kraus 1981, 1987; Retallack 1981; see also books referenced in Sect. 7.4.2). However, much of this is beyond the purpose and scope of this book, and is not discussed in detail.

Lithofacies C: Coal. Coal is typically associated with deltaic and fluvial floodplain environments (Fig. 5.33). McCabe (1984) has suggested that the presence of thick coals, as opposed to carbonaceous mudstones, indicates the presence of raised peat swamps undergoing rapid plant accumulation under humid-tropical conditions.

Other Facies. Carbonates occur as tufas, and as bedded deposits, with stromatolites, oolites and oncolites, in freshwater ponds and in some fluvial channels (Ordóñez and García del Cura 1983; Eberth and Miall 1991). Algae form encrustations and replacements wherever suitable substrate material is present, such as dead plants or shells, and clasts. Oolites and oncolites in the form of grains are moved by fluvial transport (Fig. 5.34). Eberth and Miall (1991) reported lacustrine marls interbedded with fine-grained floodplain clastic units. They consisted of laminated micrite, microspar, siltstone, and ferric oxide.

A

B

Fig. 5.32. Lithofacies P. **A** Isolated caliche nodules (behind and below hammer) in overbank deposits below a channel sandstone. Devonian Old Red Sandstone, Scotland. **B** Mature caliche. Triassic, near Siguenza, Spain

Fig. 5.33. Exposure of coal in a fluvial setting. Tertiary Eureka Sound Group, Ellesmere Island, Arctic Canada

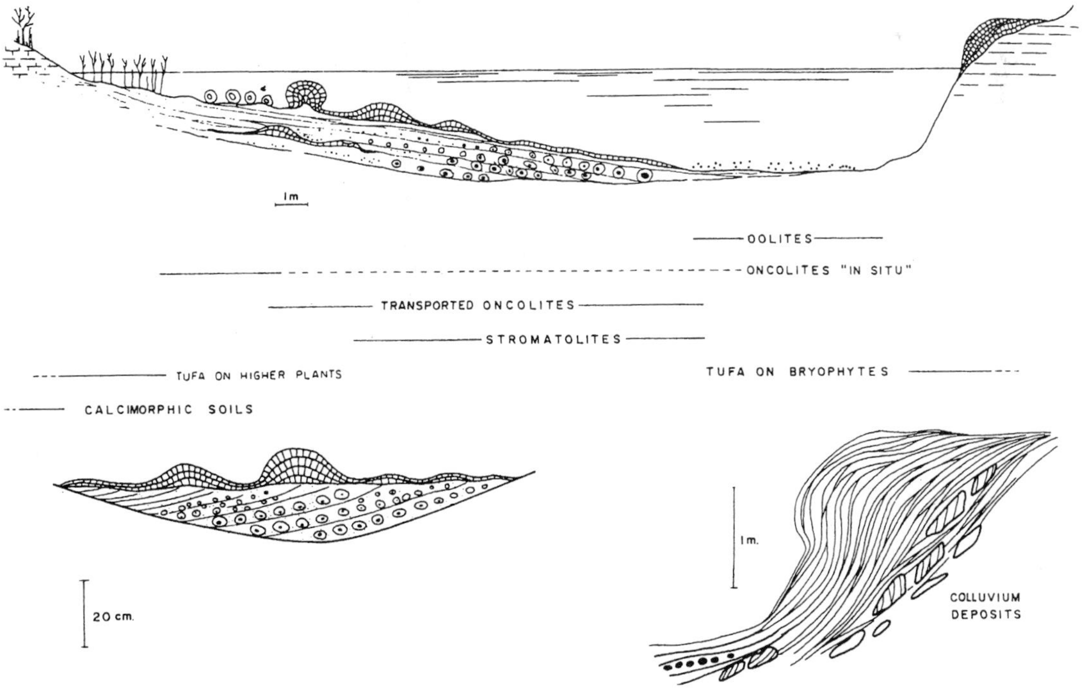

Fig. 5.34. Idealized sedimentological model of fluvial carbonate formation in the Tertiary Madrid Basin. (Ordóñez and García del Cura 1983)

5.6 Associated Facies

Fluvial deposits are commonly intimately interbedded with the deposits of other environments, notably lakes, eolian deserts, estuaries, glaciated environments, and volcanic environments. It is not the purpose of this book to describe such environments or their deposits in detail. The following are some brief notes on these deposits, and some additional reference is made to some of the related environments in Sect. 12.9, which is a discussion of the climatic information that can be gleaned from such deposits.

In arid fluvial settings, windblown sediment is common, consisting of individual ventifacts, isolated eolian dunes, or discrete laminae or beds of eolian dust (loess) or sand. Good descriptions of interbedded fluvial and eolian strata were provided by Langford (1989) and Langford and Chan (1989), to which additional reference is made, with a brief description of other examples, in Sect. 12.9.4.

Fluvial floodplains commonly contain ponds, and the distinction between ponds and lakes may become somewhat arbitrary. Many fluvial systems also drain into and out of lakes. Fluvial and lake sediments are therefore intimately interbedded in many settings. Fluvial crevasse splays may be difficult to distinguish from lacustrine deltas. Associated lake sediments may include oil shales, palustrine limestones (Sect. 12.9.5), and shales containing lacustrine faunas, such as pelecypods, ostracods, or fish (e.g., Gersib and McCabe 1981).

Estuarine environments may contain fluvial and tidal deposits in close association (Zaitlin et al. 1994). Tidal influence may be indicated by the presence of typical tidal sedimentary structures, such as tidal rhythmites, flaser bedding, and inclined heterolithic stratification (Thomas et al. 1987; Figs. 2.38, 13.18) in otherwise typical fluvial successions (e.g., Shanley et al. 1992). The fluvial-tidal relationship is a common one where coastlines undergo repeated changes in base level, and has therefore received considerable attention from sequence stratigraphers (Sect. 13.3.3).

In glacial environments, fluvial sediments occur on outwash plains (sandurs) in front of continental glaciers and ice sheets. There is, therefore, a common association with till and moraine deposits, and the deposits of glacial lakes (Eyles et al. 1983; Miall 1983b; Eyles and Eyles 1992).

Fluvial sedimentation is also an important component of the volcanic environment. The rapid production of abundant volcanic debris and the violent rainfall that commonly accompanies volcanic eruptions typically leads to catastrophic debris-flow activity (e.g., Vessell and Davies 1981; Lajoie and Stix 1992). The sedimentary apron around volcanos may be interbedded with lava flows.

Chapter 6

Architectural Elements Formed Within Channels

6.1 Introduction

A basic set of eight elements was recognized by Miall (1985), and has been updated, as noted below. The purpose of this chapter is to provide a brief summary description and key illustrations of the main element types. Further descriptions are provided in the case-studies that illustrate the discussion of fluvial style in Chap. 8.

Most of these elements are macroforms which, in the definition of Jackson (1975), are the product of the cumulative effect of sedimentation over periods of tens to thousands of years. They include major channels and bars (such as point bars, side bars, sand flats, and islands), flood sheets, and sediment-gravity-flow lobes. It is the plan view of these macroform elements that generates the familiar channel styles, so commonly illustrated by low-level aerial photographs of modern rivers.

Macroforms are "group 6" deposits, in the classification of Table 3.2. In principle, they are bounded by fourth-order surfaces (Table 4.2). However, where an element rests on the floor of a channel its base is defined by the base of the channel, which is a surface of fifth-order rank or higher (e.g., sequence boundary of seventh order). In addition, elements are commonly truncated by overlying channels, in which case their upper bounding surface is the fifth-order surface at the base of that channel. Fourth-order surfaces are, for this reason, commonly not preserved. Where they are present, they may commonly be recognized by their convex-up shape which, in part, parallels stratification in the underlying beds, and indicates that they are surfaces of accretion. Large channel-fill sheets may consist of stacked macroforms separated by fourth-order surfaces. All the elements may contain internal growth increments, defined by third-order surfaces. Typically, these show evidence of cut-and-fill erosional relief. They may rest on mud drapes capping the underlying element, and they may be followed by poorly sorted beds forming the base of the next element, such as gravel lags.

Bristow and Best (1993, p. 2) summarized the complexities in channel-fill sedimentation that may occur during changing stage:

"Where bars exist for periods of time in excess of a single flood event they will experience a complex history of erosional and depositional modifications related to changes in stage. At higher flow stages when the largest volumes of sediment are tranported, the channels are often scoured, bars may be reduced in height or in some cases completely eroded. However, during falling stage maximum deposition occurs as discharge and flow competence are reduced. Channel beds aggrade, the high stage bedforms may be modified and new bars may be formed or enlarged as sediment is deposited. As discharge continues to fall, bars may become emergent and dissected by low stage channels. Additionally, the nature of the falling limb recession (rate and length of recession) will be important not only in the reworking of higher stage sediments, but also in the deposition and spatial distribution of the finer grained sediments (silts and clays) ..."

Modifications in the original element scheme of Miall (1985) are as follows: The proposed code for midchannel macroforms that accrete downstream was changed from FM (foreset macroform) to DA (downstream accretion) by Miall (1988a) because the old code seemed to lead to confusion, and because DA is comparable to LA, in terms of the element code and the element it defines. A new in-channel element, HO, is introduced here, bringing the total to nine. These are listed in Chap. 4 (Table 4.3). Figure 4.13 is a summary illustration of the eight in-channel gravel and sand elements that are the focus of this chapter. Examples of a typical range of macroforms and other architectural elements in a sheet-braided fluvial deposit are illustrated in Fig. 6.1.

6.2 Channels (Element CH)

Most coarse (gravel, sand) deposits in fluvial systems are deposited in channels, therefore designation of a channel element in an ancient unit is not a particularly significant step forward. On outcrop profiles, use of the CH coding is useful only where the chan-

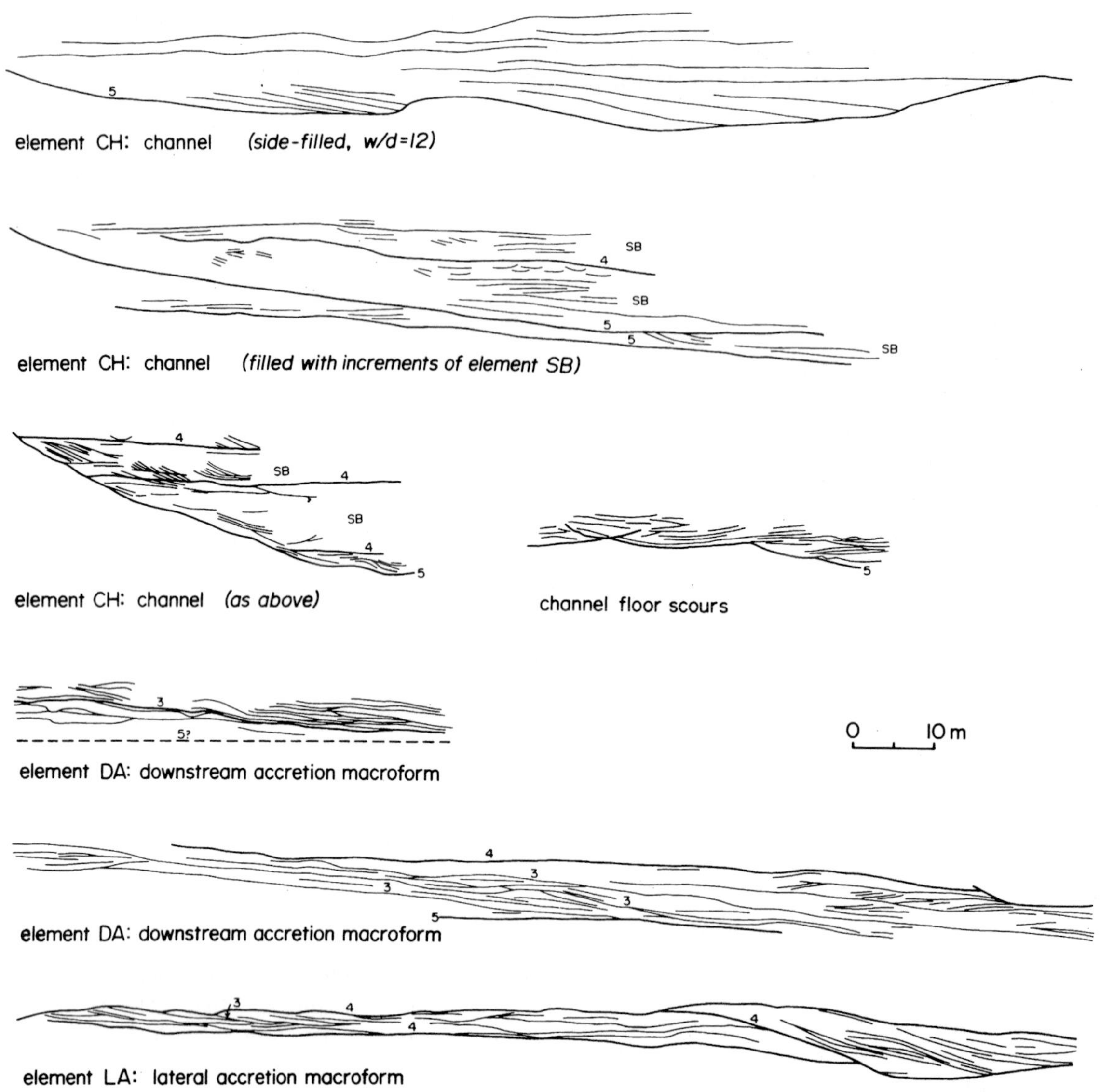

Fig. 6.1. Examples of a typical range of architectural elements in a sheet-braided sandstone, the Westwater Canyon Member of the Morrison Formation, San Juan Basin, New Mexico. The drawings are at the same horizontal and vertical scale. Bounding surfaces are *numbered* according to rank. *Unnumbered surfaces* are of first- and second-order. Individual lithofacies consist mainly of St, Sp, Sh, and Sl. (Miall 1988a)

nel-fill is simple, and cannot be further broken down into components such as DA or LA. This section describes and illustrates some of the architectural characteristics of channel bounding surfaces and simple aggradational fills.

Major channels are bounded by fifth-order surfaces. However, there may be a hierarchy of channels (and bars) in a fluvial system, with smaller channels within the larger, major channel (Fig. 6.2). Examples of this were described by Williams and Rust (1969) and Bristow (1987). Where components of a channel hierarchy can be mapped in the ancient record, the main basal bounding surface may be labeled as a major fifth-order surface, with component fifth-order surfaces indicated by letters with superscripts, using the annotation system described in Sect. 4.8. Minor channels include partially to completely abandoned channels, chute channels cutting across point bars, other channels crossing sandflats that are generated during falling water, channels generated by slumping and mass-flow processes at the cutbank of the main channel, and crevasse channels that feed crevasse splays on the floodplain. These are bounded by fourth-order surfaces.

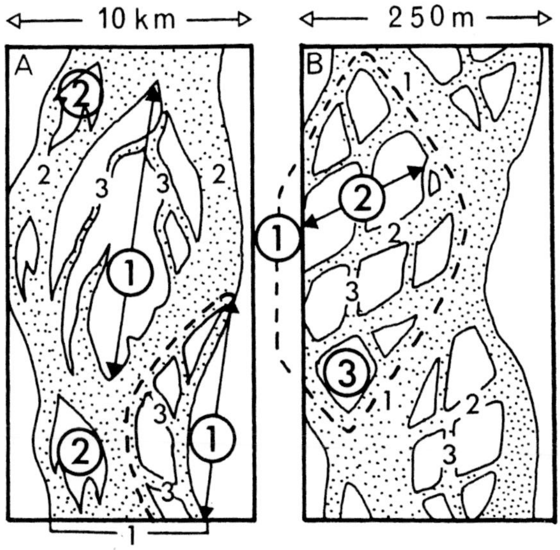

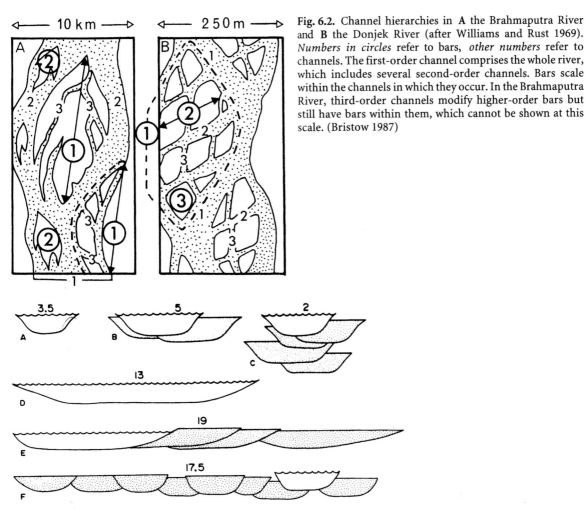

Fig. 6.2. Channel hierarchies in **A** the Brahmaputra River and **B** the Donjek River (after Williams and Rust 1969). *Numbers in circles* refer to bars, *other numbers* refer to channels. The first-order channel comprises the whole river, which includes several second-order channels. Bars scale within the channels in which they occur. In the Brahmaputra River, third-order channels modify higher-order bars but still have bars within them, which cannot be shown at this scale. (Bristow 1987)

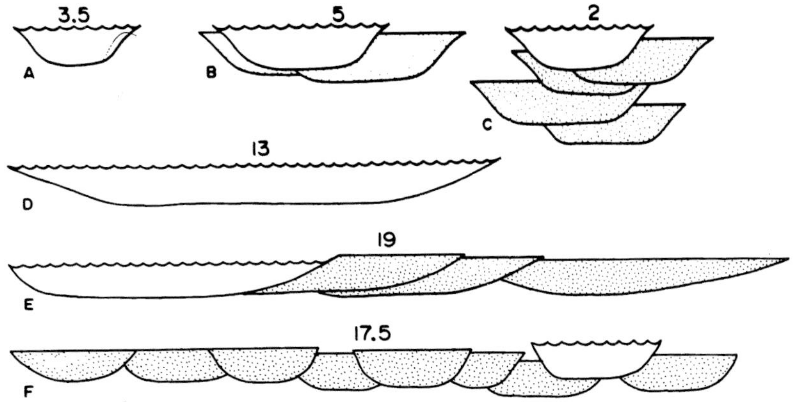

Fig. 6.3. Diagram to show the lack of relationship between the geometry of an individual active channel and the geometry of resulting channel-fill sand bodies. *Numbers* above each channel are the width/depth ratios of the sand bodies. **A,D** Simple channels; **B, E, F** broad channel-fill complexes formed by lateral channel migration or switching with little contemporaneous subsidence; **C** stacked channel complex formed by vertical aggradation. (Miall 1985)

Channel geometry is conveniently defined by depth, width/depth ratio, and sinuosity. Sand body thickness is commonly employed as a guide to original channel depth, based on reasoning explained in Sect. 10.4.1. Methods of analysis of channels were discussed by Rust (1978a) and Friend (1983), as summarized in Chap. 2 (Figs. 2.33, 2.34, 2.35). The resulting classification is further discussed in Chap. 4. Studies of the relationships between width, depth, and fluvial style, and their investigation in the subsurface, are discussed in Sect. 10.4. There may not be a simple relationship between sand body width and channel dimensions. As shown in Fig. 6.3, channels can aggrade both vertically and laterally, leading to sand bodies having very different width/depth ratios than the channel which deposited them. Various channel types are illustrated in Figs. 6.4–6.12, to which the following discussion is keyed.

Channels have concave-up erosional bases (Figs. 6.5–6.12). The top of the channel may be erosional (Fig. 6.5) or gradational (Fig. 6.6). Channels commonly have multistory fills, with each story bounded by an erosion surface (Figs. 6.4, 6.7). Channel margins become gentler in slope with increasing channel width. Cutbank slopes in excess of 45°, possibly even vertical or undercut, are not uncommon bordering narrow channels, especially those developed by rapid erosion in arid regions ("wadi"; Figs. 6.11, 6.12). Sheet-like channels may have practically imperceptible channel margins, sloping at a few degrees or less (Fig. 6.4). These variations reflect bank stability. Channels cut into mud-dominated fines,

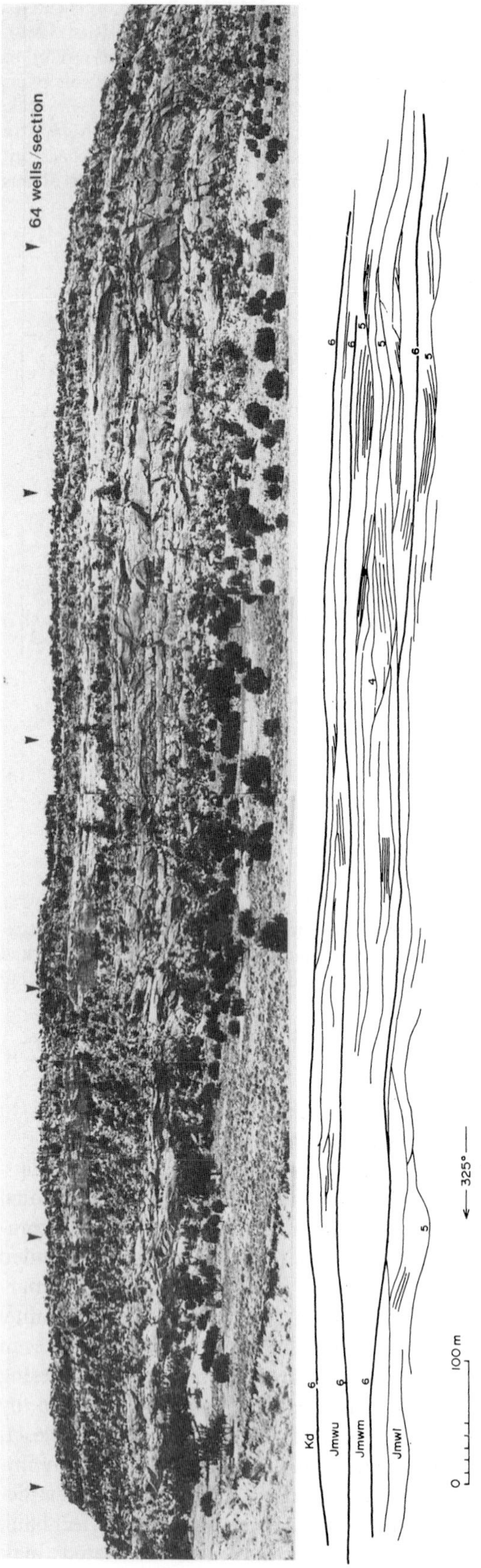

Fig. 6.4. Example of braided sheet sandstone, Westwater Canyon Member, Morrison Formation, near Fort Defiance, Arizona. Note the absence of clearly defined channel margins. *Numbers* indicate rank of bounding surfaces. *Kd* Dakota Sandstone; *Jmw* Westwater Canyon Member; *u* upper; *m* middle; *l* lower. (Miall 1988a)

Fig. 6.5. Ribbon sandstone body 8 m thick, deposited in a fixed channel of anastomosed type.The fill consists of simple superimposed cross-bed sets. The channel is cut into thick floodplain fines containing pedogenic carbonate units. Cutler Group (Upper Paleozoic), El Cobre Canyon, New Mexico (see Eberth and Miall 1991)

particularly where the banks are stabilized by a dense root network, offer considerable resistance to erosion (D.G. Smith 1976) and tend to be steep (Figs. 6.5, 6.10). Peat is particularly difficult to erode (McCabe 1984), and may result in steep channel margins developed by sudden bank failure (Fig. 7.16). Cemented paleosols, such as calcretes, also offer considerable resistance to erosion and may result in steep channel margins (Allen and Williams 1982; see also discussion of this point by Gibling and Rust 1990). By contrast, those channels which cut into unconsolidated sand and gravel are easily eroded and may retreat rapidly, giving rise to lower channel-margin slopes, or stepped margins with steep cutbank sections alternating with flat terraces formed by bar complexes and partly filled minor channels. Where the sediment load of a river is dominated by sand or gravel, a braid plain may develop, with almost unconfined, sheet-like channels (Fig. 6.4). The channel cross section geometry is therefore not necessarily an inherent property of a river with a particular slope, discharge, or sediment load, but at least partly reflects the nature of the preexisting sediment into which the channel is cut (Carson 1984a–c; Church and Rood 1983; Crowley 1983). This is one reason why attempts to define fluvial facies models on channel geometry have not always been successful.

Recognition of the channel-fill element in a fluvial deposit depends on the ability to define the sloping channel margins. Ribbon sandstones, being narrow, are commonly completely exposed (e.g., Fig. 6.5) but, apart from this, cutbanks are rarely preserved in outcrop (examples are illustrated in Figs. 6.6–6.8, 6.10). Examples of complete, or nearly-complete major channels exposed at the surface were described by Okolo (1983) and Hopkins (1985). Channel reconstruction is commonly attempted by correlation of closely spaced outcrop or subsurface sections (e.g., Campbell 1976; Hopkins et al. 1982; Putnam 1982a,b) but, because of the presence in most deposits of a hierarchy of channels of different scales, such correlation may be difficult or impossible (e.g., Wightman et al. 1981; see discussion in Sect. 9.5.3 of Mannville Sandstone in Lloydminster area, Alberta). As shown in Sect. 10.3.2, the well-

Fig. 6.6. Cutbank of channel in sand-dominated system. The channel is about 5 m deep and is overlain by a thick floodplain succession. Eureka Sound Group (Tertiary), Bylot Island, Arctic Canada

known channel reconstruction of the Morrison Formation in New Mexico by Campbell (1976), which was constructed by this process of outcrop mapping, has been shown to be unreliable. Large channel-fill complexes may be better termed paleovalleys, and contain the accumulated deposits of many of the other types of element described later in this chapter. They are bounded by sixth-order bedding contacts. Good examples up to 8 km across and 90 m deep were described by Blakey and Gubitosa (1984). Many small to medium-sized petroleum fields are contained in fluvial paleovalley fills and paleochannels (e.g., Mannville Sandstone, Alberta; Muddy sandstone, Wyoming; "J" and "D" sandstone, Colorado; Red Fork Sandstone, Oklahoma; see Chaps. 14, 15).

Where the channel is of broad mobile or sheet type, defining the channel margins may be difficult or impossible. Large channels filled by continually shifting minor channels (the familiar braided pattern) may contain evidence of several or many temporary channel margins (Fig. 6.4), and the overall channel-fill geometry then means little in terms of conventional channel classifications (Schumm 1963; see Fig. 6.3). Attempts to determine the channel width and depth for the purpose of paleohydraulic reconstruction are likely to result in large errors.

If the channel margins cannot be defined, field analysis is likely to result in a classification of the fill in terms of one or more of the other architectural elements. For example, the fills of ephemeral channels on arid braid plains, particularly on lake margins, are typically sheet-like and may consist mainly of elements SB: sandy bedforms, and LS: laminated sand sheets.

Channels filled by simple vertical aggradation commonly show fining-upward successions, reflecting one of two processes, progressive abandonment as a result of upstream avulsion, or the plugging action of a few dynamic events (e.g., flash floods). Typical cycles include:

GB $\rightarrow$ DA $\rightarrow$ SB $\rightarrow$ OF
LS $\rightarrow$ SB $\rightarrow$ OF

The thickness of such cycles cannot exceed the depth of the channel, and is likely to be much less where the dynamic event strips away earlier deposits before depositing their sediment load. As shown by Godin (1991) cycles in channel fills may represent flood events, or the accretion of individual macro-

Fig. 6.7. Rare preservation of a cutbank in multistory sandstone sheets formed by flash-flood sedimentation in an arid environment. The channel fill consists mainly of lithofacies Sh with numerous low-angle internal third-order erosion surfaces. Kayenta Formation (Jurassic), Arizona

Fig. 6.8. Close-up of cutbank illustrated in Fig. 6.7. Note blocks of siltstone on the cutbank that have been eroded from the siltstone bed at the *upper left*. Kayenta Formation (Jurassic), Arizona

Fig. 6.9. Abandoned channel filled with mudstone, within a delta-plain succession. Note sandy levee deposits high on the flanks of the channel. Carboniferous, Kentucky

Fig. 6.10. Channel cutbank that became a slide plane, along which the channel fill slumped down to the right. Note the tilted sandstone block at *right*. Carboniferous, Kentucky

Fig. 6.11. The person in the center is standing within the fill of a narrow, steep-sided channel eroded in the Canyon de Chelly Sandstone, and filled with Shinarump Formation (Triassic). Edges of channel are shown by *arrows*. This is a good example of a fossil "box canyon", a steep-sided wadi, typical of erosional valleys in arid regions. Canyon de Chelly National Monument, Arizona

forms, and so caution should be taken in the extrapolation from cycle thicknesses to channel dimensions.

Channels, particularly in high-sinuosity streams, may be abandoned by chute or neck cutoff, in which case they will be filled by fine-grained deposits [element CH(FF); see Table 7.1] showing a channelized, concave base (Fig. 6.9). Hopkins (1985) described three well-exposed channels that contain concave-up fills deposited mainly by vertical aggradation during progressive or sudden abandonment. Okolo (1983) described channels filled obliquely from the side during gradual abandonment. Components of these channel fills may be interpreted as alternate bars (Sect. 6.7).

Minor crevasse-, chute-, and bar-top channels, bounded by fourth-order contacts, typically contain assemblages of lithofacies Ss, Se, and Sl, with gravel lags and thin units of flow-regime bedforms (element SB: sandy bedforms; lithofacies St, Sp, Sh, Sr), showing no particular cyclic order.

A distinctive type of minor channel containing massive sandstones (lithofacies Sm) was described by Jones and Rust (1983), Rust and Jones (1987), Turner and Munro (1987), and Wizevich (1992b). Some deposits occur as sheets, bounded by irregular channel bases with linear scours; other deposits occur in narrow channels with steep margins. Both types originate as channel-bank failures, with slumped bank materials evolving from landslip masses (e.g., Fig. 6.10) into sediment-gravity flows. The orientation of the channels and their basal scours is typically at a high angle to the main channel trend.

6.3 Gravel Bars and Bedforms (Element GB)

Lithofacies Gh, Gp, and Gt define three main kinds of mesoforms. The formation of these is described in Sect. 5.2.1.2. In preserved gravel deposits, the various mesoforms and constituent lithofacies are commonly complexly interbedded (Fig. 5.2), and it is therefore important, in a description of the GB element, to understand how varying water and sediment discharge may lead to rapid lateral and temporal changes in lithofacies and architecture.

Mesoform and macroform development in gravel-bed rivers is strongly affected by the nature of gravel transport. Gomez et al. (1989) and Hoey

Fig. 6.12. Close-up of the vertical margin of the wadi fill shown in Fig. 6.11

(1992) showed that transport is not regular at any point in a braided system, but occurs as a series of pulses or sediment "slugs" at different spatial and temporal scales. Stochastic variations in entrainment are responsible for some pulses, and bar erosion, channel avulsion, and tributary supply may feed gravel into the system at irregular rates. In natural systems, these sediment pulses are several hours in duration, and have a significant effect on channel and bar development, as summarized below.

The simplest deposits are the thin, diffuse gravel sheets of Hein and Walker (1977), which are a few clasts thick, have diffuse, lobate margins, and move only during peak flow (lithofacies Gh). During episodes of high water and sediment discharge, these sheets grow upward and downstream by the addition of clasts to form horizontally stratified gravel sheets (Gh), the structure termed longitudinal bars in some older literature (Rust 1972). These mesoforms reach about 1 m in height, and may show either an upward increase or decrease in clast size, depending on their mode of accretion. Clast accumulation in place tends to result in upward fining, as the deposit builds to shallower water levels. However, these mesoforms tend to fine downstream, and they may also migrate downstream. In such cases, the coarser topmost deposits migrate over the finer deposits that form the base of the mesoform (Gustavson 1978). Processes that lead to gravel-sand segregation and to upward fining and coarsening within the gravel deposits are discussed in Sect. 5.2.1.2. Gustavson (1978) and Massari (1983) noted the close relationship between the three main gravel lithofacies and confirmed the general model developed by Hein and Walker (1977).

These gravel mesoforms are amalgamated into macroforms (bars) in several ways. Our present understanding of gravel macroforms follows from the classic work of Leopold and Wolman (1957), who modeled the formation of midchannel bars in a flume. Observations in modern rivers by Hickin (1969), Bluck (1971), Lewin (1976), Gustavson (1978), Forbes (1983), Ferguson and Werritty (1983), Hooke (1986), and Carson (1986) were synthesized by Ashmore (1991), who carried out a series of flume runs to demonstrate the major processes. Ferguson (1993) also provided a useful review of braiding processes in gravel-bed rivers.

Bar types include midchannel and bank-attached types. The latter are termed point, alternate, or lateral bars depending on their shape and position. S.A. Smith (1987) described a minor type of midchannel bar, termed a counterpoint bar, that develops on the outside of meander bends as a result of turbulence and shoaling near the bend entrance (fine-grained equivalents, termed concave benches, were described earlier by Nanson and Page 1983). Each bar type may evolve into the other, as a result of flow diversions resulting from meander migration and channel avulsion.

Large-scale turbulence leads to the development of alternate bars in initially straight channels, as a straight thalweg develops a meandering trace (Hickin 1969; Lewin 1976). These evolve into fully developed point bars (element LA), the development of which is discussed briefly in Sect. 6.7.

Ashmore (1991) recognized three main methods of midchannel gravel bar construction. These are illustrated in Figs. 6.13–6.15 using diagrams from his flume runs. The first method is a process whereby point and alternate bars are separated from the bank by the development of a chute channel (Fig. 6.13). This occurs when sediment is eroded rapidly from a

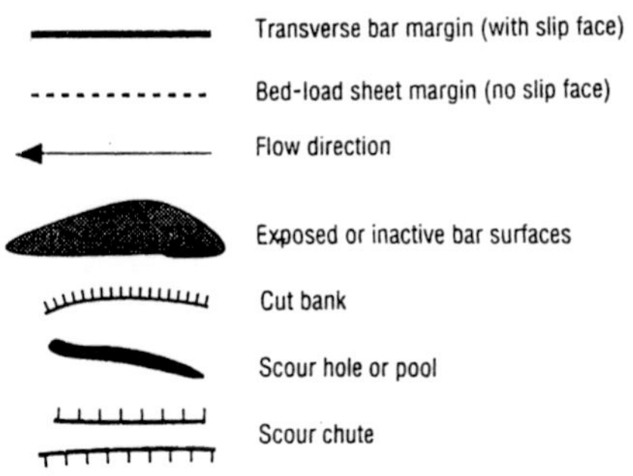

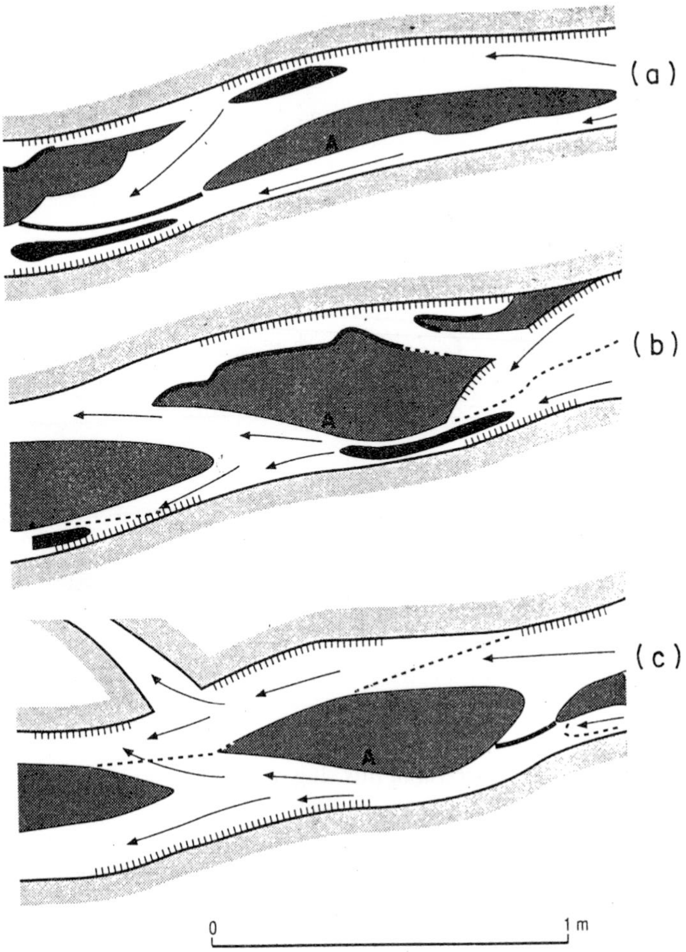

Fig. 6.13. Evolution of midchannel bar in a gravel-bed river by process of chute cutoff, as modeled in a flume. An initial point bar *A* in diagram **a** is separated from the bank by the development of a scour pool, as shown in **b**. A new point bar develops on the opposite bank, but is removed by erosion as the new midchannel bar enlarges and stabilizes (**c**). (Ashmore 1991)

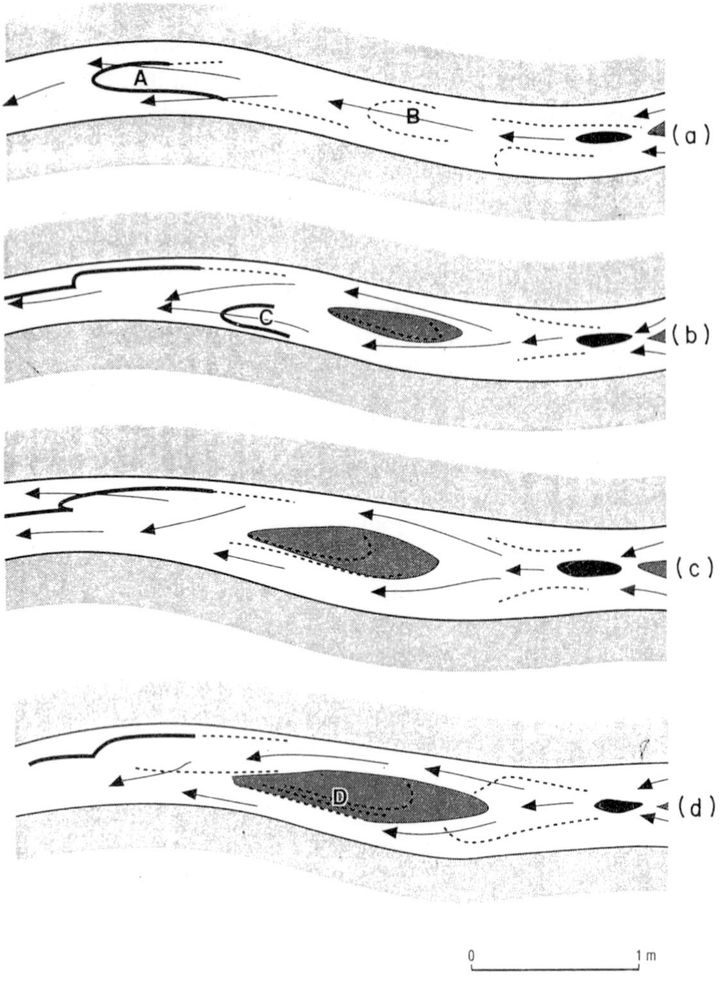

Fig. 6.14. The central-bar braiding mechanism in a gravel-bed river. **A**, **B**, and **C** are successive bed-load sheets. Eventually, one of these stalls (the *stippled area* in **b**), and becomes the locus of gravel accumulation. Drawn from flume model by Ashmore (1991). *Symbols* as in Fig. 6.13

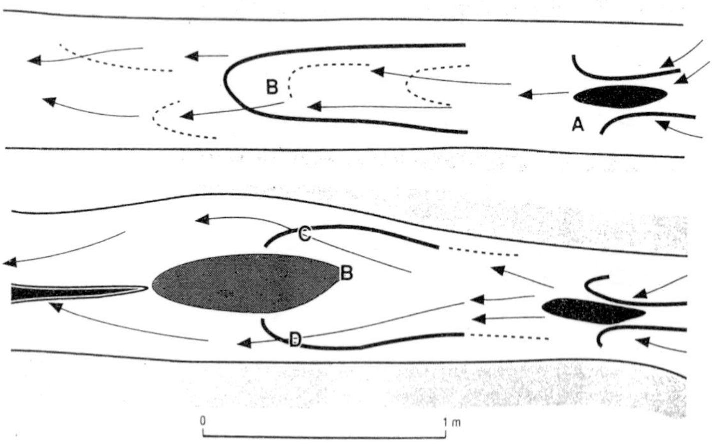

Fig. 6.15. Development of midchannel bar in gravel-bed river by the migration and coalescence of transverse dunes. From flume experiment by Ashmore (1991). *Symbols* as in Fig. 6.13

bar upstream. The arrival of this slug of sediment at the head of the point bar causes additional flow to be directed over the bar. The second process, central bar initiation, is one that was first modeled by Leopold and Wolman (1957), in an experiment that has been widely quoted in the fluvial literature. The process involves the migration of thin bed-load sheets that become stalled in midchannel positions because of slight changes in shear stress (Fig. 6.14), for example at places where the channel widens or shallows. Under different conditions of bed-shear stress transverse bedforms with slip faces develop.

The migration and convergence of these bedforms also lead to midchannel bar formation, the third of the main processes of bar development (Fig. 6.15).

Leddy et al. (1993) also modeled braid bar development in the laboratory, and proposed three slightly diferent mechanisms of bar development. "Choking avulsion" occurs when a sediment lobe migrates through an active channel into an area of flow expansion, where deceleration and flow divergence occur. Part of the flow is diverted into a different course. Typically, this will involve diversion across an older bar deposit, and a remnant of the sediment lobe is left as the nucleus of a new midchannel bar. "Constriction avulsion" occurs when a sediment lobe migrates into an area of channel narrowing, as at a channel junction. Blocking of the channel leads to overbank flooding, crevassing, and diversion of part of the flow into a different course, leaving the sediment lobe as a new bar nucleus. "Apex avulsion" occurs as a result of the lateral and downstream growth of macroforms, accompanied by corresponding cutbank erosion on the opposite side of the channel. Eventually, this erosion may lead to breakthrough into lower areas in the channel complex, and partial flow diversion.

Germanoski and Schumm (1993) modeled the effects of adding and removing sediment load from experimental braided streams, and documented the subtle but significant changes in fluvial style that result. In the case of gravel-bed rivers, addition of sediment load leads to increased braid-bar development and aggradation, whereas removal of load leads to amalgamation of braid bars and to channel incision. Changes in sediment load could be very local effects resulting from changes in the materials introduced into the channel by bank erosion, or longer-term, regional changes brought about by tectonic or climatic causes. Base-level changes will also modify sediment load, as discussed in Sect. 11.2.2.

The different conditions that lead to the formation of gravel sheets and slip-face bedforms have been discussed earlier (Sect. 5.2.1.2), drawing on the work of Hein and Walker (1977). In actual streams, all these processes may be expected to occur simultaneously, or in succession, as the organization of the channel and bar complex changes, and conditions of water and sediment discharge vary. Unfortunately, there is virtually no information on the internal architecture of the different types of bar deposits formed by these three processes. Lithofacies types are as described in Chap. 5, and it is anticipated that, internally, the bars would consist of lenses and wedges of gravel organized between third- and fourth-order bounding surfaces. However, at present, we can only speculate on the geometry and relative orientation of the surfaces and current structures. Some comments follow on what is known about ancient GB deposits. An example is illustrated in Fig. 6.16.

Element GB typically forms multistory sheets tens to hundreds of meters thick. Flat or irregular erosion surfaces between mesoforms are common. Steeply dipping channel margins are rarely seen, partly because they tend to be minor parts of a gravelly fluvial landscape. Actively migrating channels may undercut older bar gravels, producing cutbanks 1–2 m high, but when filled with later bar gravels of similar composition and texture, the cutbanks may be difficult to identify (Fig. 6.17).

S.A. Smith (1990) defined three styles of GB accretion in the Triassic Budleigh Salterton Pebble Beds of southern England. Couplets of Gh and St separated by fourth-order surfaces represent deposition in topographically segregated parts of a deep braided channel (Fig. 6.18a). Units of Gp alternating laterally with Gh and sandstone lithofacies (Fig. 6.18b) represent the variations in gravel sheet growth described by Hein and Walker (1977), that S.A. Smith (1990) postulated took place in a relatively deep channel, as evidenced by the height of some of the Gp sets (3 m). Superimposed Gt wedges bounded by fourth-order surfaces represent scour fills (Fig. 6.18c).

Element GB may be interbedded with minor to predominant sheets or lenses of element SG: sediment gravity flows. Element SB typically comprises 5–10% of even the coarsest gravel succession, and represents slack-water deposits, such as abandoned channel fills (minor CH element, where identifiable), bar-edge sand wedges, and microdeltas (Rust 1972; Miall 1977), or the deposits of topographically elevated parts of a deep gravel river (Fig. 6.18a), such as parts of the Donjek River, Yukon (Williams and Rust 1969; Rust 1972; S.A. Smith 1990). These element associations are discussed further and illustrated in Chap. 8. The development of fourth- and fifth-order lithosomes in gravelly alluvial fans is discussed in Sect. 10.2.

It is now known that channel confluences are sites of significant scour in gravel-bed rivers (Ashmore 1993; Bridge 1993b; Ferguson 1993; Siegenthaler and Huggenberger 1993), and that large scoop-shaped deposits may occur there. As Ashmore (1993, p. 130) stated, "They are one of the few places in gravel braided stream where flow is sufficiently deep to

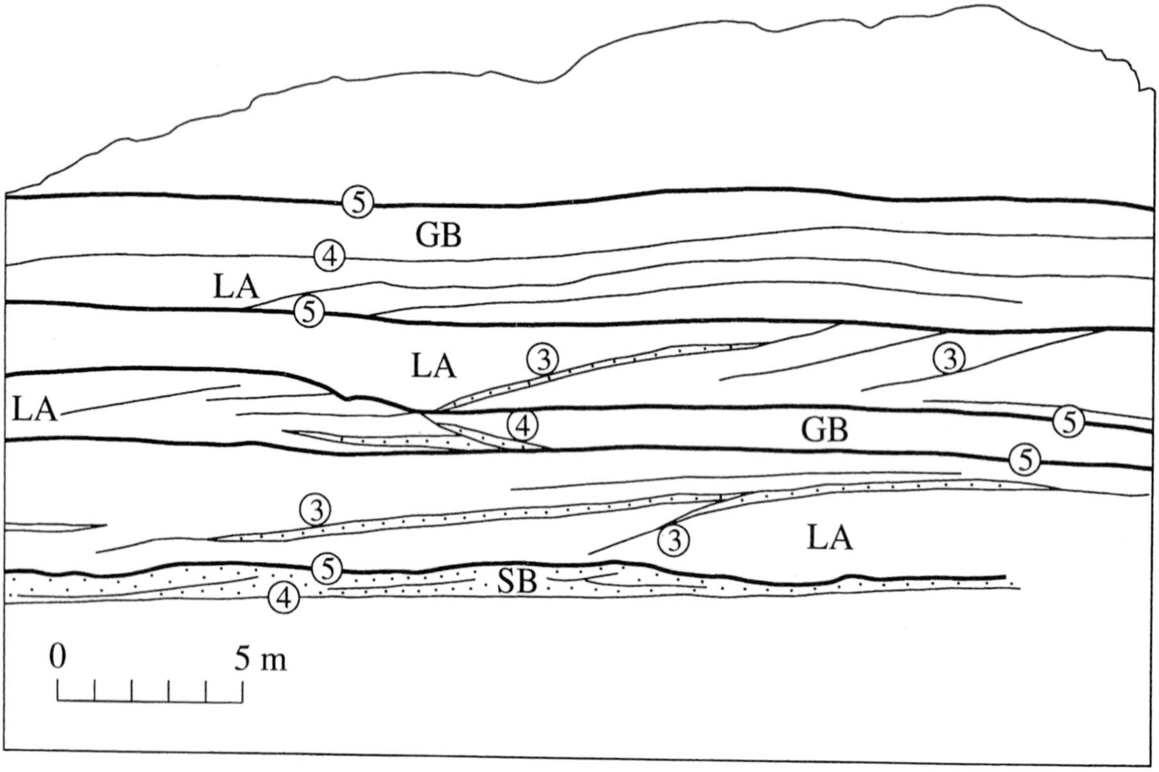

Fig. 6.16. Examples of LA, GB, and other elements in gravels. Buntsandstein (Permian-Triassic), central Spain. Interpretation is adapted from Ramos and Sopeña (1983)

Fig. 6.17. Beds of Gh draping the floors and margins of minor channels. At least three channel-floor bounding surfaces, of fourth- or fifth-order rank, are present in this outcrop (*arrows*). Enon Conglomerate, Cretaceous, near Oudtshoorn, South Africa

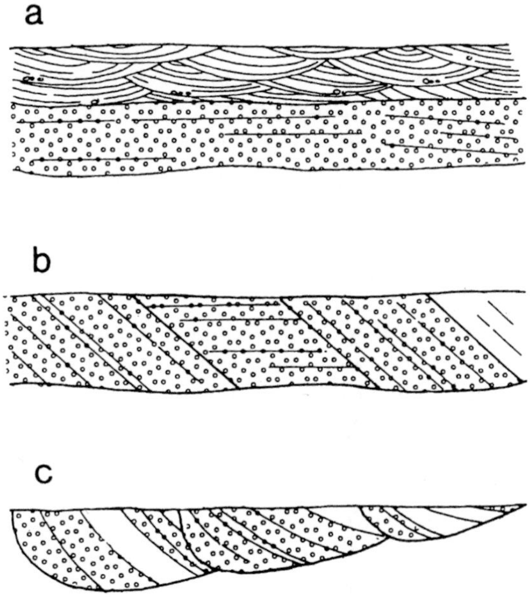

Fig. 6.18. Three styles of element GB in the Budleigh Salterton Pebble Beds (Triassic), England. **a** Couplets of trough cross-beded sandstone (lithofacies St, element SB) and horizontally bedded conglomerate (lithofacies Gh, element GB). **b** Planar cross-bedded conglomerate (Gp) interbedded with horizontally bedded conglomerate and sandstone (Gh, St). **c** Large-scale trough cross-bedded conglomerate wedges (Gt) (S.A. Smith 1990)

allow the development of avalanche faces at unit bar margins." A distinctive element termed the "hollow" (HO) is the result. This is described in Sect. 6.9.

6.4 Sediment-Gravity-Flow Deposits (Element SG)

This element occurs as narrow, elongate lobes or multistory sheets, and is typically intimately interbedded with element GB or SB (Fig. 6.19). Lithofacies Gmm, Gmg, Gci, and Gcm are predominant. The element forms by debris flows and related mechanisms, as described in Sect. 5.2.1.3. Individual beds average 0.5–3 m in thickness, rarely exceeding 3 m. Flow units may be lobate in plan view, with widths of up to about 20 m, and downstream lengths of several kilometers (Hooke 1967; Wasson 1977a,b; Vessell and Davies 1981; Nemec and Muszynski 1982; Blair and McPherson 1992). Amalgamated flows with total thicknesses of several meters are common. A single, exceptionally extensive SG unit, triggered by a catastrophic rock fall, was described by Bürgisser (1984).

Flow units typically have irregular, nonerosive bases. Flow events passively occupy existing ero-

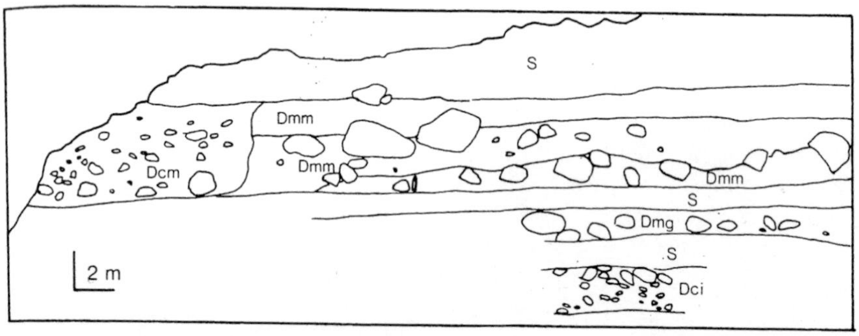

Fig. 6.19. Typical outcrop of element SG, showing various sediment-gravity-flow deposits (substitute G for D for appropriate facies codes used in this book) and interbedded sandstone lenses (S). Cutler Formation, Colorado. (Schultz 1984)

sional channels or the irregular topography formed by earlier sediment-gravity-flow and sheet-flood events. Internally, they may show a wide range of textures and fabrics. Disorganized textures are typical of rigid plugs that are rafted at the center of some debris flows (Bull 1977). Grading and inverse grading are common. Nemec and Muszynski (1982) described an upward transition in some flow types (their facies C) from graded to low-angle cross-stratified gravels, which they interpreted as a transition from debris-flow to traction-transport mechanisms. Buck (1983) described a sand-dominated diamictite facies interpreted as mud-flow deposits. Shultz (1984) proposed a fourfold classification of sediment gravity-flow deposits based on matrix content, packing characteristics, and grading. His lithofacies scheme has been adapted for use in this book (Sect. 5.2.1.3; Table 4.1).

6.5 Sandy Bedforms (Element SB)

The familiar flow-regime bedforms that form in sand-dominated river systems have been described by many writers (e.g., Allen 1968, 1984; Southard 1971; Harms et al. 1975, 1982; Miall 1977; Collinson and Thompson 1982; Ashley 1990). Dunes (3-D dunes of Ashley 1990; lithofacies St), sand waves, and transverse bedforms (2-D dunes of Ashley 1990; lithofacies Sp), linguoid bedforms (large 3-D dunes, lithofacies Sp in small outcrops), upper flow-regime plane beds (Sh), washed-out and humpback dunes (Sl), and ripple marks (Sr) occur in a wide variety of fluvial settings and show a range of assemblages and vertical sequences.

In some cases, large exposures show that short sequences of bedforms, resting on dipping bounding surfaces, are interbedded with each other over wide areas below major (fourth-order) convex-up bedding contacts, indicating that they were dynamically related and formed simultaneously. This type of architecture is the key diagnostic characteristic of elements LA and DA (lateral and downstream accretion units), described in the next sections. Where these architectural features can be conclusively ruled out, the deposits probably represent fields or trains of individual bedforms that accumulated predominantly by vertical aggradation. In some cases, evidence of both vertical and lateral accretion may be present in a given element. Gibling and Rust (1990) proposed the calculation of an "aggradation index" to quantify the relative importance of the two different styles of accumulation, but not enough measurements of this have been made to determine where numerical cutoffs should be placed between dominantly aggradational (e.g., SB) and accretionary (LA, DA) geometries.

Vertical stacking of different bedform types indicates long- or short-term changes in flow regime. Short-term changes occur during stage changes (flash floods, seasonal fluctuations; group 5 deposits of Table 4.2). Longer term changes reflect aggradation and reduction in water depth over periods of several to many years (groups 6, 7 deposits of Table 4.2). Both can result in similar lithofacies assemblages and successions (which is one of the problems with vertical profile analysis; see Godin 1991), requiring examination of the architecture and overall context of the deposits in order to arrive at correct interpretations. Such deposits contain first-, second-, and third-order contacts. A brief discussion of some typical examples of the SB element follows.

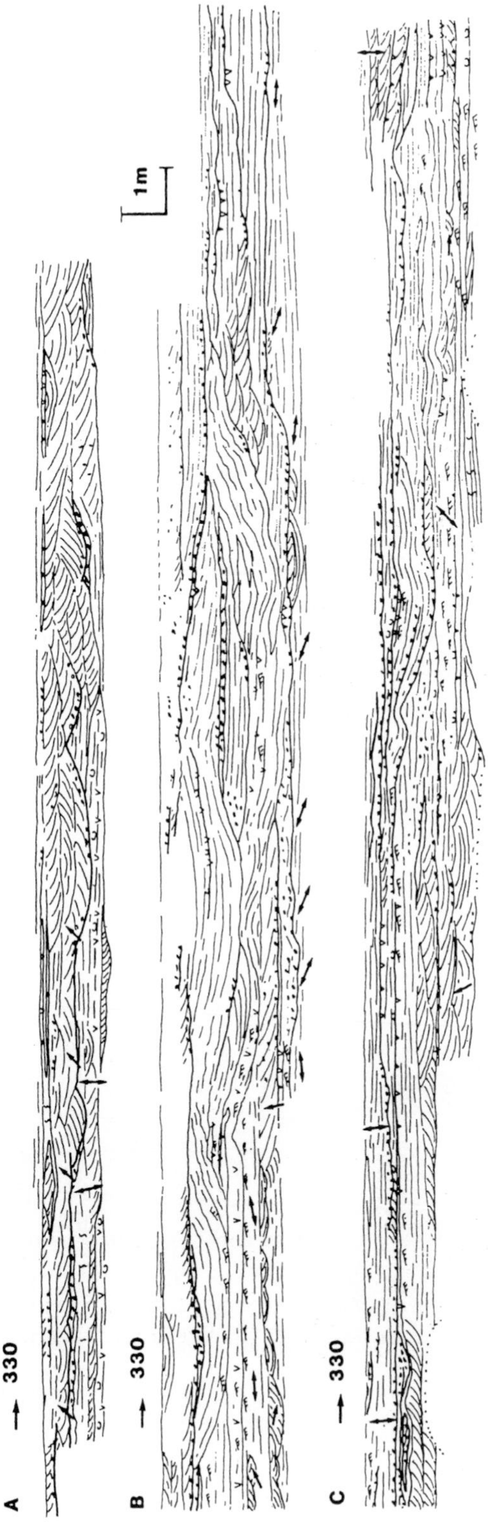

Fig. 6.20A–C. Lateral profiles through sheet sandstones of element SB. Association 2, Battery Point Formation (Lower Devonian), Gaspé, Québec, Canada. (Lawrence and Williams 1987)

Channel-Floor Dune Fields. Fields of 3-D dunes (lithofacies St), characteristically occupy the deeper portions of active channels wherever the bed load is predominantly sand (e.g., Fig. 6.20). Lenses or lobes of trough cross-bedded sand a few meters thick and tens to hundreds of meters wide may result. They may be cut by broad, shallow scours and erosion surfaces of second and third order, indicating stage fluctuations (Harms et al. 1963; Harms and Fahnestock 1965; McGowen and Garner 1970; Jackson 1976a; Cant and Walker 1978; Nijman and Puigdefabregas 1978; Plint 1983; Buck 1983; Stear 1983; Tyler and Ethridge 1983; Lawrence and Williams 1987). In the shallower parts of channels, including the tops and flanks of macroform elements (such as point bars and sand flats), transverse bedforms and sand waves (2-D dunes) are common. These generate sheets of planar-tabular cross-bedding (lithofacies Sp) (N.D. Smith 1970, 1971, 1972; Levey 1978; Jackson 1976a; Nijman and Puigdefabregas 1978; Cant and Walker 1978; Blodgett and Stanley 1980; D.G. Smith 1983).

Shallow Channel-Fill Assemblages. Miall (1977) defined a distinctive lithofacies assemblage dominated by Sp and showing little or no internal cyclicity (Fig. 5.20). This was named the Platte type of braided river deposit, after the Platte River, Nebraska. This assemblage represents the migration of fields of linguoid or transverse bedforms, many of which are capped or draped by Sr or Fl during falling water (N.D. Smith 1970, 1971, 1972; Blodgett and Stanley 1980). The "cross-bedded simple bars" of Allen (1983a) are similar. Many ancient examples of this assemblage occur, for example, in the Cretaceous and Lower Cenozoic deposits of Arctic Canada (Miall 1976, 1984b) and the Nubian Sandstone (Jurassic-Cretaceous) of North Africa (Harms et al. 1982; Bhattacharyya and Lorenz 1983).

Crowley (1983) suggested that in at least some cases these bedforms form part of much larger macroform structures 200–400 m in length and 0.7–1.5 m high. The architecture of these structures is discussed in the next section. One of their diagnostic characteristics may be an upward-coarsening

Fig. 6.21. Sandstone sheet (element SB) resting on recessive coal bed, at base of outcrop. The sandstone unit contains gently dipping accretion surfaces at *left*, indicating that it developed in part by lateral progradation, but dominantly by vertical aggradation. It is interpreted as a crevasse-splay deposit in an upper delta-plain setting. Pocahontas Basin, Carboniferous, Kentucky. (Horne and Ferm 1976)

succession generated during high-stage flow conditions. Not all linguoid bedforms fields seem to be interpretable in terms of Crowley's (1983) macroform hypothesis – the distinctive meandering main channel of Crowley (1983; his Fig. 10A) is not present in much of the Platte River (Blodgett and Stanley 1980; their Figs. 2, 3), and even where the macroforms are present the complete sequence may not be preserved. Miall's (1977) original Platte model may therefore still have its uses, but clearly the search for the macroform architecture should be pursued wherever possible.

Bar-Top Assemblages. Many workers have described the characteristic small-scale cross-bedding that occurs in shallow areas of active channels, particularly on bar tops. Various types of ripple cross-lamination (lithofacies Sr) are the result. These small-scale structures are typically deposited during falling water and, where preserved, their capping of larger bedforms or bars produces local fining-upward sequences. Such sequences are almost ubiquitous in fluvial environments and their occurrence has little diagnostic value in terms of fluvial style.

Crevasse Channels and Crevasse-Splay Deposits. These are typically composed of element SB. Channels should be identifiable by the concave-up channel floor (a fourth-order bounding surface), and proximity to the main channel. Splays are sheet-like bodies tens to hundreds of meters across and typically 1–2 m thick that thin and pass laterally into element FF: overbank fines. Both upward-coarsening and upward-fining successions may be present, indicating progradation or gradual abandonment, respectively. Plant litter and vertebrate bones are common (Collinson 1978; Ethridge et al. 1981; Gersib and McCabe 1981; D.G. Smith 1983; Bridge 1984; Eberth and Miall 1991). Where architectural description of these deposits is possible, they may be assigned their own element codes, as described in Sect. 7.2. Figure 6.21 illustrates an example of a broad, sheet-like splay deposit in which local lateral accretion can be observed.

Distal Braid-Plain Sandstone Sheets. On distal braid plains, such as those bordering playa lakes, fluvial deposits may be entirely composed of element SB. Sheets of sand develop in broad, virtually unconfined chan-

Fig. 6.22. Superimposed sheets of element SB, bounded by fifth-order surfaces, and constituting successive flood cycles in a distal braid-plain setting. Lithofacies consist mainly of St, Sh, and Fl. Peel Sound Formation (Lower Devonian), Somerset Island, Arctic Canada. (Miall and Gibling 1978)

nels (Fig. 6.22). Aggradation and progressive abandonment of these channels occur slowly or during single flood events. In either case, fining-upward cycles are commonly the typical result. Williams (1971) and Hardie et al. (1978) described modern examples, and Miall and Gibling (1978) documented an ancient example. In the latter case, cycles are mostly between 1 and 3 m in thickness, and show an upward transition from a scoured base through Sh, Sp, or St, to Sr and Fl, or directly to Fl omitting Sr. Similar deposits characterize the arid "terminal fan" deposits of northern India (Parkash et al. 1983) and elsewhere, a depositional style that is gaining increasing recognition in the ancient record (Sect. 8.3).

Other Occurrences Of Element SB. In the pre-Devonian, the lack of vegetation is thought to have resulted in a predominance of weakly channelized bed load streams (Schumm 1968a; Cotter 1978). The architecture and composition of the resulting fluvial deposits were probably in many cases similar to the distal braid-plain sand sheets described here. Long (1978) discussed some Proterozoic examples.

Particularly vigorous flood events in ephemeral channels may produce a distinctive type of lithofacies assemblage and sand body geometry, described below under the heading of element LS.

A few workers have documented the ocurrence of giant cross-bedding, indicating the former existence of deep channels which migrated down very large bedforms. Such bedforms are mesoforms. They are internally of simple structure, with steep foreset dips (e.g., Fig. 5.21), and are not to be confused with the more complex macroforms described in the next section. Giant bedforms are particularly common in the giant sandy rivers draining the Himalayas, where sand waves up to 15 m high have been recorded (Coleman 1969; Singh and Kumar 1974). However, as noted below, at least some of these sand waves are probably macroforms. Possible ancient analogs were described by Conaghan and Jones (1975). McCabe (1977) and Jones and McCabe (1980) analyzed cross-bed sets up to 40 m thick and 1 km wide that they interpreted as the deposits of large prograding alternate bars in a major delta distributary (Fig. 2.27). Reactivation surfaces (third-order surfaces) were interpreted in terms of fluctuating water depth. If, as in

Fig. 6.23. Example of a midchannel sand flat in a modern sandy-braided river. These compound bars develop by lateral and downstream accretion, and contain internal bounding surfaces of first- to third-order dipping in the direction of accretion. (Photograph courtesy of D.G. Smith)

this case, the deposits can be interpreted as alternate bars, they should be assigned to a category of macroform rather than to element SB (see Sect. 6.7).

6.6 Downstream-Accretion Macroforms (Element DA)

Downstream-accretion and lateral-accretion deposits are the principal products of accretion within the bar complexes of major sand-bed channels. Upstream accretion also may occur locally (e.g., Bristow 1987, 1993). The bars are scaled to the size of the containing channel, and their height is a rough guide to minimum channel depth. Three-dimensional architectural analysis is essential for complete description of these deposits because, unlike elements GB, SG and SB, they contain significant internal three-dimensional geometrical complexities. Large compound bar forms have been described from many modern rivers (Fig. 6.23), including the side bars of the Tana (Collinson 1970), the sand flats of the South Saskatchewan (Cant and Walker 1978), and the sand waves of the Brahmaputra, as reinterpreted by Bristow (1987, 1993). The bars are characteristically 1–15 m high and 10–1000 m long. The maximum recorded height and length refer to the bars of the Brahmaputra. It is only in recent years that such large-scale deposits have been described in ancient deposits. Research by the author indicates that DA and LA elements are very common in braided sheet

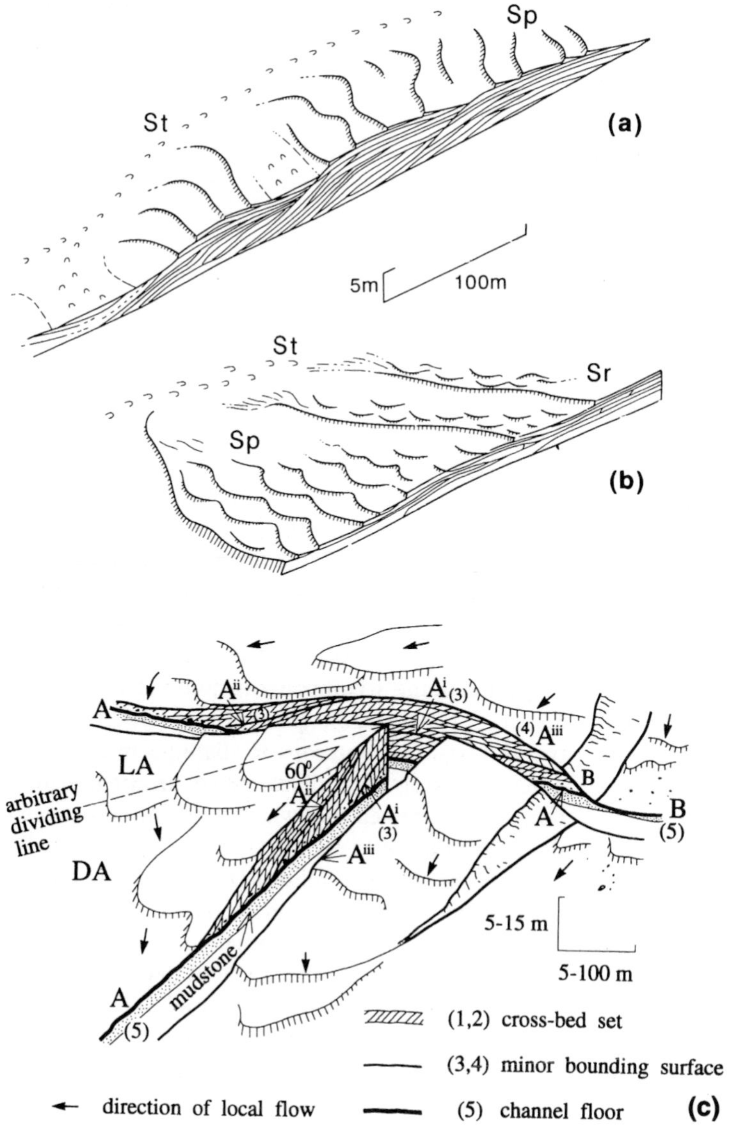

Fig. 6.24. Generalized models of downstream-accretion (DA) architectural elements. **a** Loosely based on Allen (1983a) and Kirk (1983); **b** loosely based on Cant and Walker (1978) and Haszeldine (1983a,b). **a** and **b** from Miall (1985). **c** Model developed from a synthesis of macroforms in the Castlegate Sandstone, Utah, showing the system of annotation of bounding surfaces used in this book. Note gradation between LA and DA architecture within the same element (Miall 1994). Scales are approximate. Internal geometry varies considerably depending on channel depth, grain size, discharge amount and variability

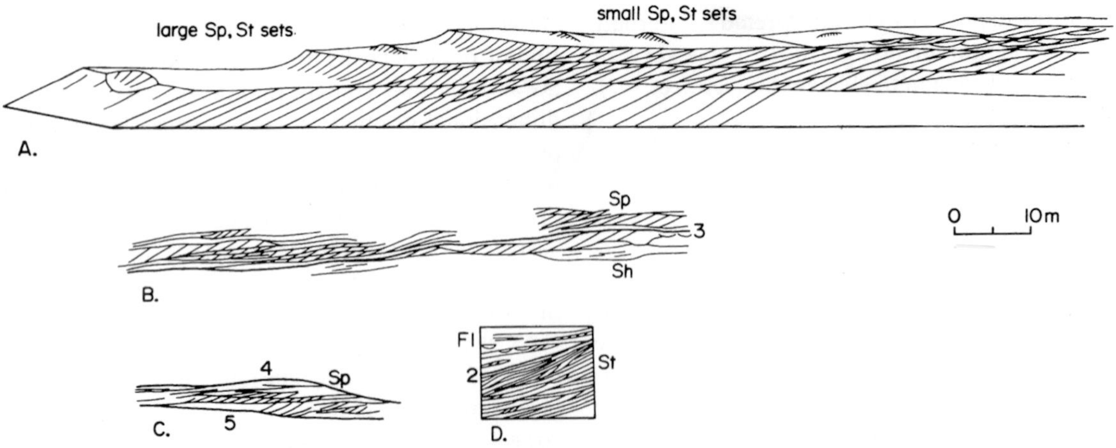

Fig. 6.25. Ancient examples of DA elements showing the distinctive, internal, downstream-dipping bounding surfaces, down which trains of medium-scale bedforms migrated. *Letters* are lithofacies codes (Chap. 5), *numerals* are rank of bounding surfaces (Chap. 4). **A** Carboniferous example, England (Haszeldine 1983a,b); **B** Poison Canyon Sandstone, Morrison Formation, New Mexico (Miall 1988a); **C** Example from Kayenta Formation, Colorado (Miall 1988c); **D** part of a "sand wave" in the modern Brahmaputra River (Coleman 1969). (Miall 1988a)

sandstones (e.g., Figs. 4.8, 4.15), a fact that appeared to be missed by those focusing on vertical-profile studies.

The essential characteristic of a downstream-accretion element are that it consists of several (possibly many) cosets of downstream-oriented flow-regime bedforms dynamically related to each other by a hierarchy of internal downstream-dipping bounding surfaces (Figs. 6.24, 6.25). These reveal the former existence of an active, nonperiodic, possibly irregularly shaped bar form comparable in height and width to the channel in which it formed. The bars contain second- and third-order surfaces that generally dip gently (< 10°) downstream (Banks 1973; Allen 1983a; Haszeldine 1983a,b; Kirk 1983; Miall 1988a,c, 1992a, 1994), although oblique or gentle upstream dips, around and over low-relief bar cores (sand shoal of Allen 1983a) may also be present. Between these surfaces are sets or cosets of Sp, St, Sh, Sl, or Sr. The Sh and Sl laminae are organized parallel or subparallel to the internal bounding surfaces. Detailed paleocurrent studies show that the flow-regime bedforms advance generally down the slopes defined by the second- and third-order surfaces (Haszeldine 1983a,b) or oblique to the surfaces draping the bar cores (Allen 1983a; Miall 1994). These data reveal a picture of fields of bedforms driving across, around and down the bar forms (Fig. 6.24c).

Most sandstone elements are accumulated by both vertical aggradation and lateral accretion. In DA and LA elements, the evidence of accretion is obvious, and forms the main basis for the definition of the element, but in some cases the accretionary geometry may be very subtle and difficult to define in small outcrops. Figure 6.21 illustrates a crevasse-splay sandstone sheet that contains some evidence of lateral progradation, yet is classified here in the SB element because of the dominance of vertical aggradation. For more precise work, use of Gibling and Rust's (1990) "aggradation index" may be useful to assist in discriminating betwen the two styles of accumulation.

It has been shown that midchannel macroforms may accrete downstream at their downstream ends and laterally along their flanks (Bristow 1987, 1993). Upstream accretion on the upstream flank has also been recorded (Bristow 1987, 1993). Preserved macroforms may therefore grade laterally from DA to LA architecture (Allen 1983a; Miall 1993, 1994), as suggested in the accompanying model diagram (Fig. 6.24c). Care must be taken in the interpretation of two-dimensional outcrops to distinguish these geometries, using all available paleocurrent information. It is suggested that where a gradation between the LA and DA end members can be documented a cutoff be employed as follows (Fig. 6.24c). Where the orientation of the accretion surface and that of the cross-bedding within the same element are within about 60° of each other, it indicates that the element grew by accretion in a direction parallel or oblique to local flow, and the element is designated a DA unit, even if local flow is oriented at a high angle to the regional trend. Where the orientations of the accre-

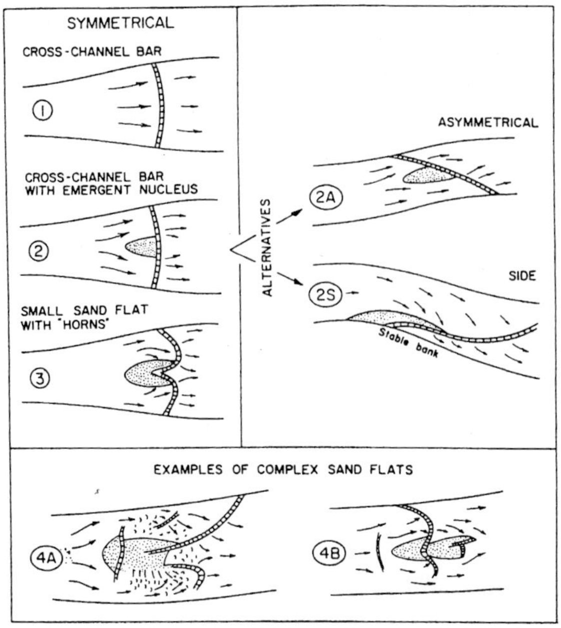

Fig. 6.26. Development of sand flats, based on the surface-process studies of Cant and Walker (1978)

tionary surfaces and the cross-bedding are more nearly perpendicular (> 60° difference), the element is designated an LA unit. Where adequate data cannot be gleaned from the outcrop, the macroforms may be designated DA/LA.

Cant and Walker (1978) referred to the large, midchannel macroforms in the South Saskatchewan River as sand flats. They showed that these evolve from large, simple (2-D), flow-transverse bedforms, termed cross-channel "bars" by Allen (1983a) and Cant and Walker (1978) Elevated parts of these bedforms (which may be emergent at low water) become the nuclei of new sand flats, which anchor part of the bar in the middle of the channel. Sediment is added to the cross-channel bedform by the migration of fields of dunes and ripples. These may move more slowly over the crest of the bar, which may become anchored completely if an emergent nucleus is present. The opposite end of the crest line, in deeper water, continues to advance more rapidly, so that the entire bedform swings around oblique to the channel direction (Cant and Walker 1978, Fig. 9; Allen 1983a, Fig. 19). The macroforms accrete sediment partly by this process of bedform capture on the upstream or flanks, and partly by rapid burial and preservation of superimposed bedforms on the advancing downstream face (Fig. 6.26).

Crowley (1983) described a Platte-type macroform consisting of suites of large, linguoid (3-D) dunes arranged in an en echelon pattern. Internally, they are composed of a single, large-scale Sp set resting on an apron of fines and draped by coarser-grained St or Sr sets. The upward coarsening reflects varying shear stress in relation to increasing water depth from top to bottom of the advancing foresets (Fig. 6.27). Crowley (1983) suggested that these macroforms are comparable to, and grade into, the alternate bars of moderately sinuous rivers.

Many of the variations in composition and geometry between described macroforms probably reflect fluctuations in stage and local changes in sediment supply (Germanoski and Schumm 1993). Many of the second- and third-order surfaces have the character of reactivation surfaces (Collinson 1970). The "sand flat" macroforms of Cant and Walker (1978) are cut by numerous erosional channels during falling water. Kirk (1983) described a distinctive low-stage lithofacies assemblage draping the macroform, distinguished from the body of the structure by divergent paleocurrents that reflect falling-water surface runoff and bar-top channel orientation.

Bridge (1993b) developed simple models for the development of midchannel macroforms such as DA units, based on his dynamic models of flow in river bends, and observations of modern rivers. Figure 6.28 provides a general architectural model, and Fig. 6.29 suggests how patterns of deposition and erosion change during changes in flow stage.

Wizevich (1992b) developed a model for a typical DA unit in his Carboniferous deposits (Fig. 6.30). Resting on a fourth-order surface is a thin unit of pebbly sandstone with poorly defined cross-beds (lithofacies Ss). This is followed by a large Sp set, above which are various assemblages of smaller-scale St and Sp. The element is capped by small scoops or channels filled with asymmetric Sp sets. The large Sp set at the base of this succession compares with the similar set in Haszeldine's (1983a,b) model (Fig. 6.25A), and probably corresponds to the cross-channel "bar" that initiates DA development in Cant and Walker's (1978) model.

Descriptions of macroforms in modern rivers typically suffer from the lack of three-dimensional control. Thus, Cant and Walker's (1978, Fig. 14) sand flat model predicts a simple tabular sheet of Sp cosets, and Crowley's (1983, Fig. 10) macroform model similarly predicts simple superimposed sets of Sp and Sr. Missing from these descriptions are any indications of dipping second- or third-order internal bounding surfaces. Crowley's (1983) model shows reactivation surfaces that could be third-order surfaces, but their distinctive characteristics

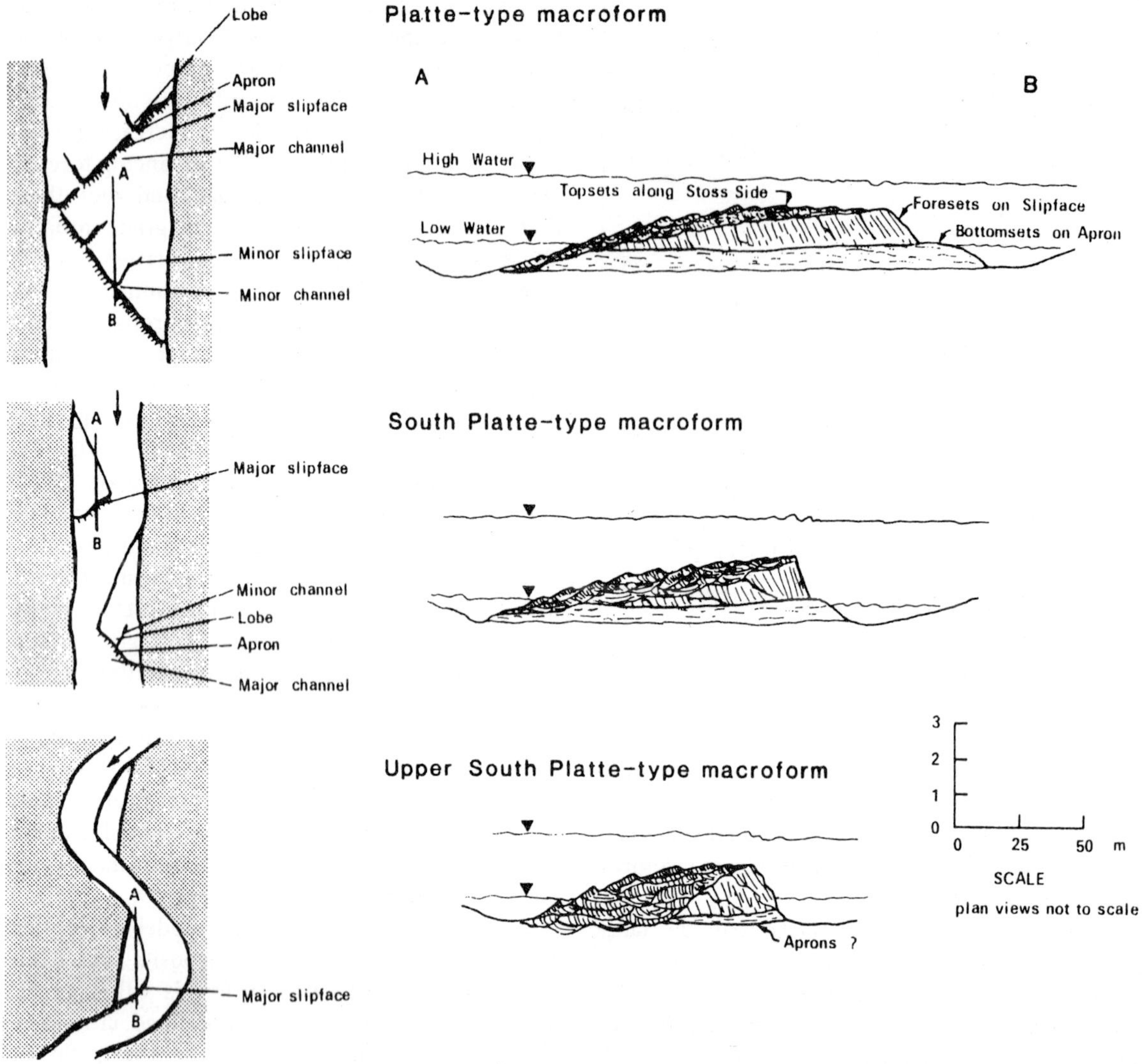

Fig. 6.27. The macroform models of Crowley (1983). The Platte-type macroform is comparable to the downstream-accretion element described here. The South Platte-type macroform is comparable to the alternate bars described in Sect. 6.7

(notably their low depositional dip) are not emphasized, and they could be mistaken for bedform reactivation surfaces. Bristow (1993) provided a three-dimensional facies model for the bars of the Brahmaputra River, based on natural cutbank outcrops along the modern channel, but the model only relates to the upper few meters of these very large bar forms (Fig. 6.31).

In the ancient record, detailed architectural studies rarely reveal simple "textbook" examples of macroforms, such as those shown in the model diagrams in this section. Bar accretion and channel scour can lead to changes in channel orientation, resulting in complex accretionary geometries, commonly with superimposed minor channels scoured out during stage changes. Miall (1994) documented a suite of LA and DA units in the Upper Cretaceous Castlegate Formation of Utah, which illustrate some of the resulting complexities. An example is illustrated in Fig. 6.32.

In some cross sectional configurations certain types of lateral-acretion deposit could be confused with downstream-accretion elements, as noted in the next section.

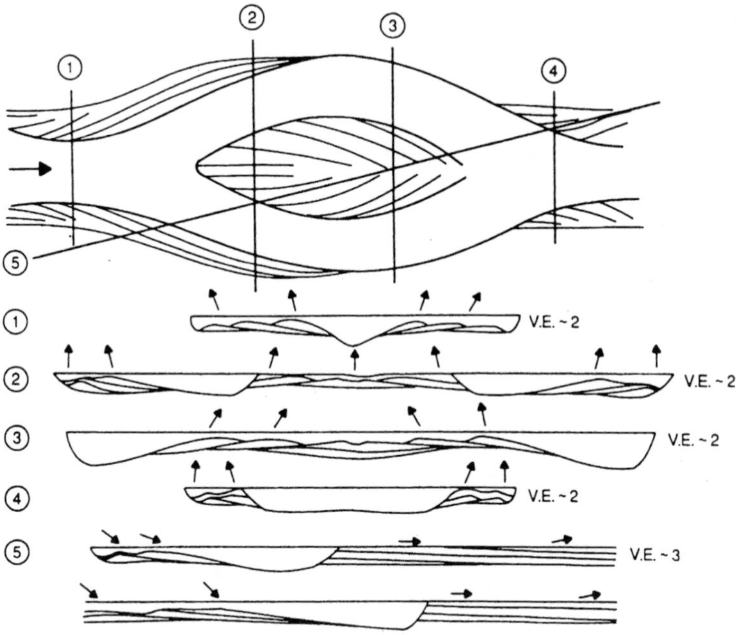

Fig. 6.28. Theoretical depositional model for a simple braided channel pattern (*above*) in which bars and channels show oblique to downstream migration, and showing locations of *cross sections 1–5* (*below*). Lines in plan and cross sections indicate configuration of first- to third-order bounding surfaces. *Arrows* indicate orientation of bedforms during deposition of uppermost units in the cross sections. (Bridge 1993b)

6.7 Lateral-Accretion Deposits (Element LA)

Where the main flow of the channel is directed away from the bank, as on the inside of a meander, surface flow impinges on the outer bank, leading to a "setup" (raised water level as a result of pressure against the bank), and active cutbank erosion (Fig. 6.33). The flow turns downward, developing a helical overturn pattern. The return flow, at depth, passes obliquely up the bed of the inner bank. Because of the reduced shear stress associated with this current, significant sedimentation takes place, and the bank accretes laterally at a high angle to the principal flow direction (Fig. 2.20). The helical flow pattern decays as the flow emerges from the bend, and is replaced by a helical overturn in the opposite direction as the flow impinges on the cutbank of the next bend downstream. Sediment removed from the cutbank is incorporated in the overall sediment load of the river. Large slump blocks may accumulate in the deepest part of the channel as a lag deposit, whereas material broken down into individual grains is incorporated into the bed load and suspension load and is swept downstream, much of it becoming deposited in bedforms and bars.

Depending on the grain size of the sediment load, and the meander geometry, transverse flow across the accreting point bar surface may take several forms. Large or small gravel or sand bedforms migrate subparallel to the strike of the accretion surface, with larger mesoforms ("transverse bars") being driven inward, up onto the upper point-bar surface, until they are stranded by a loss of depth and shear strength (Jackson 1976b). Flow is then diverted around them to the lower part of the point bar, and the mesoform becomes part of the point-bar deposit, enclosing a swale against earlier bar sediments. On tight bends separation zones may be set up, with complex vortex flow causing the development of banks and benches on the point-bar surface (Nanson 1980; Fig. 2.37).

Meanders evolve in several different ways, including cross-channel and down-channel migration, rotation or expansion (Willis 1989; Fig. 6.34). Sediment is added to the inner bank at a rate comparable to that at which it is removed by erosion along the outer bend of the meander. The result is the development, by lateral growth, of a compound, bank-attached macroform, termed a point bar. A distinctive architectural element results, characterized by large-scale, gently dipping third-order bounding surfaces that correspond to the successive increments of lateral growth (Figs. 4.7, 4.9). These surfaces have traditionally been termed epsilon cross-bedding, following the classification of bedforms by Allen (1963a). They usually show offlapped upper terminations, followed by fine-grained facies of the FF element. Their lower terminations downlap onto the channel floor. The height or thickness of the element approximates the bankfull depth of the channel, and can be used to estimate channel scale, as discussed in Sect. 10.4.

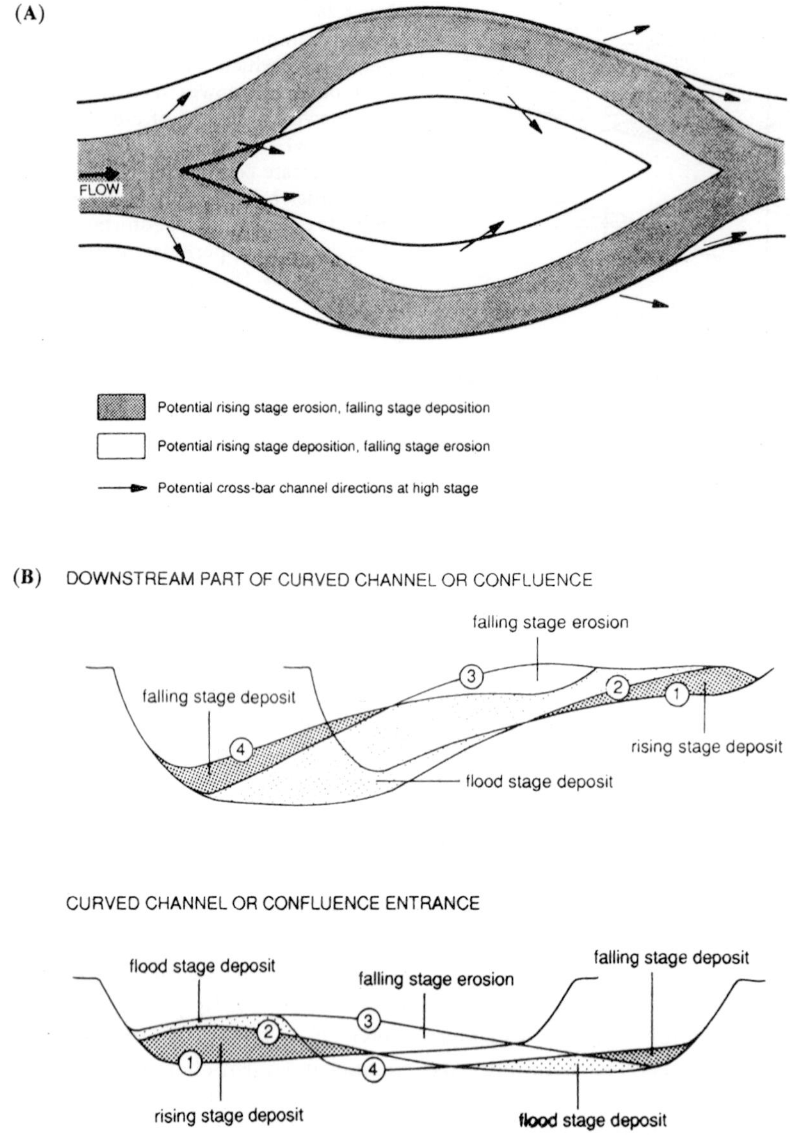

Fig. 6.29. Theoretical model of erosion and deposition during flow stage in a simple braided channel. Cross sections show *1* low stage geometry, *2* flood stage geometry, *3* flood stage geometry following cutbank erosion and bar deposition, *4* subsequent low stage geometry. (Bridge 1993b)

Fig. 6.30. a Typical succession of lithofacies within a DA unit in the Carboniferous Lee Formation, central Appalachian basin. **b** Model of the development of the DA macroform. *Numbers 3 and 4 within arrows* indicate locations and ranks of bounding surfaces. (Wizevich 1992b)

Fig. 6.31. Model for bar-top sedimentation in the Brahmaputra River, based on observations of natural bank exposures in the modern river. Upstream accretion occurs in the upper portion of the bar, lateral accretion in the middle portion, and downstream accretion in the lower portion. Accretion surfaces all dip at very low angles, which may be imperceptible in small outcrops. Significant sedimentation takes place during falling stage, when the bar top is dissected by minor channels and large amounts of suspended sediments are deposited on bar margins. (Bristow 1993a)

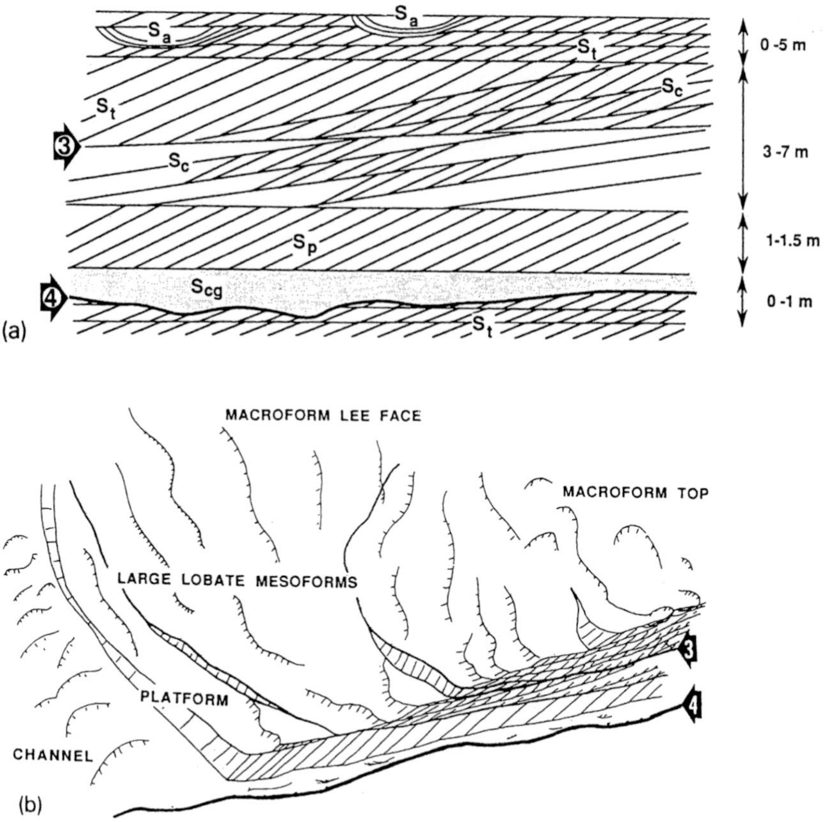

Fig. 6.30

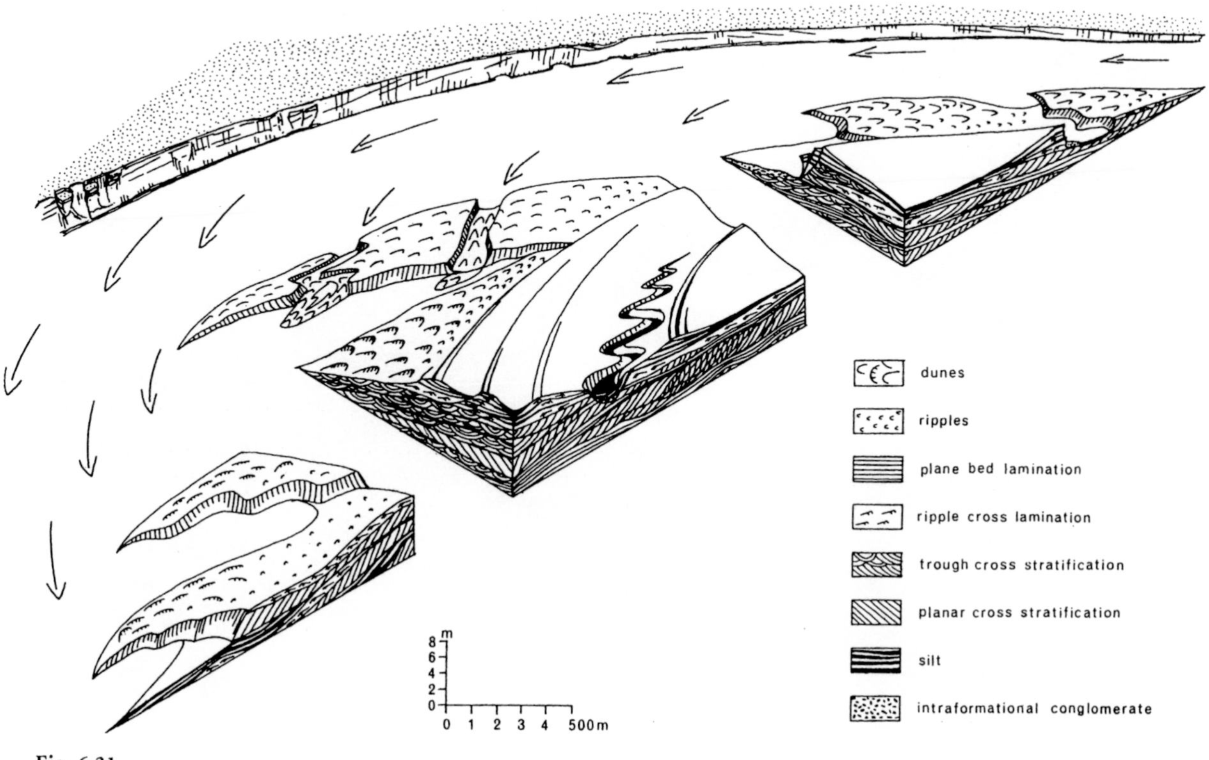

Fig. 6.31

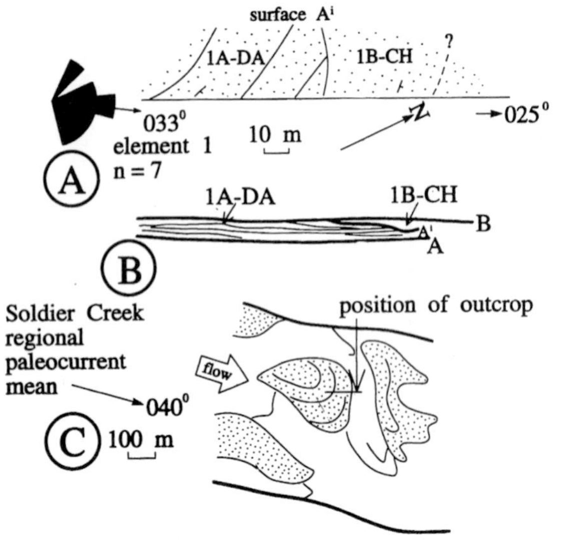

Fig. 6.32. Use of orientation data to reconstruct bar architectures in an ancient braided stream deposit, Castlegate Sandstone (Upper Cretaceous), Utah. Observations of accretion surfaces in the outcrop profile (**B**) are used to construct a map of the top surface of the element (**A**). These data, combined with paleocurrent information, are combined in a reconstruction of bar geometries (**C**). In this case, element 1A is interpreted as a DA unit, crosscut at its downstream end by a minor channel (Miall 1994). The entire profile from which this example was selected is shown in Fig. 4.15

Crowley (1983) suggested that LA deposits of high-sinuosity channels are dynamically comparable to the DA deposits of lower sinuosity channels and that both reflect the long-term behavior of large-scale vortices affecting the entire turbulent boundary layer (Fig. 6.27). Carson (1986) showed that in many gravel-bed rivers much of the point-bar development takes place as a result of flow expansion where the flow enters the bend. LA elements may also occur as midchannel bars in multiple-channel rivers, such as the Brahmaputra (Bristow 1987; Fig. 6.36).

The internal geometry and lithofacies composition of LA elements is highly variable, and depends on channel geometry and sediment load, but the presence of lateral accretion sets is the common theme (Fig. 6.35). The width averages two-thirds of the channel width (Sect. 10.4.1), at least in single-channel rivers, so that the dip of the lateral-accretion surface varies according to the width/depth ratio (Leeder 1973). Dips of up to about 25° have been recorded in some fine-grained point bars (e.g., Miall 1979a), in rivers corresponding to model 7 of Miall (1985; see Chap. 8). In wide channels, particularly where the sediment load is gravelly, the accretion surface is covered by bedforms and by minor bars

Fig. 6.33. Examples of river meanders, with point bars forming on the insides of the bends. Channel is about 20 m wide. Milk River, Alberta, Canada. (Photo courtesy of D.G. Smith)

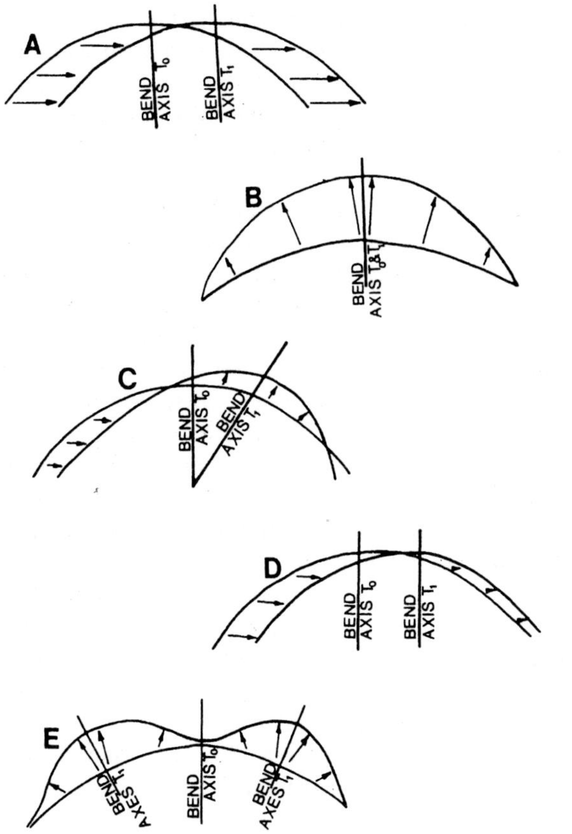

Fig. 6.34. Models of meander migration. Arrows show migration paths. **A** Translation; **B** expansion; **C** rotation; **D** wavelength variation; **E** complex evolution. (Willis 1989)

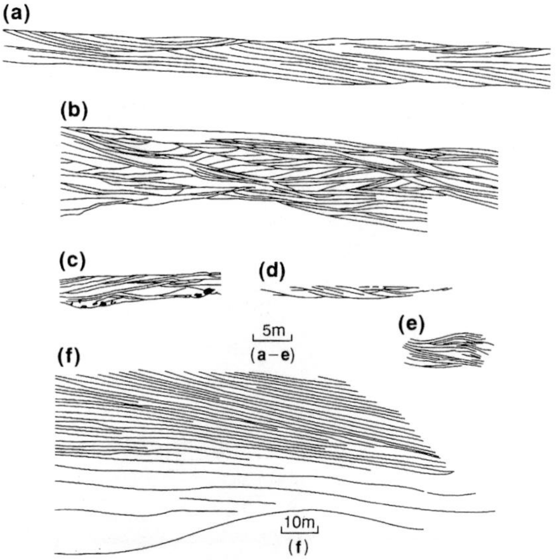

Fig. 6.35. Examples of lateral accretion elements. No vertical exaggeration. Fluvial-style model numbers of Miall (1985) are indicated. **a** Conglomerate point bar (lithofacies Gm), with chute channels (lithofacies Gt), model 4 (Ori 1979); **b** element composed of medium-grained sandstone, with abundant internal planar-tabular cross-bedding (lithofacies Sp), model 6 (Beutner et al. 1967); **c** fine- to very-coarse sandstone and pebbly sandstone with cobble to boulder conglomerae lag. Abundant internal cross-bedding (lithofacies Sp, St, Sh, and Sl), model 5 (Allen 1983a); **d** small sandy point bar with abundant dune and ripple cross-bedding (lithofacies St, Sr), model 6 (Puigdefábregas 1973); **e** point bar composed mainly of fine sandstone and siltstone (lithofacies Sl) with minor medium- to coarse-grained, cross-bedded sandstone (lithofacies St) at base, model 7 (Nanson 1980); **f** giant point bar with thick, fine-grained trough cross-bedded sandstone at base (lithofacies St) passing up into accretionary sets of alternating fine sandstone and argillaceous siltstone showing evidence of tidal bundling (lithofacies Se), model 7 (Mossop and Flach 1983). (Diagram from Miall 1985)

and channels, and this obscures the simple geometry of the LA element (e.g., Schwartz 1978). Lateral-accretion surfaces may be difficult to identify in such deposits.

Lithofacies assemblages within lateral-accretion deposits vary markedly, depending on the caliber of the sediment load, and on discharge variability. LA elements can be classified into four groups, according to grain size and facies. Earlier classifications by Jackson (1981) and Smith (1987) have been incorporated into the present subdivision. Gradations between the four groups are to be expected even within a single fluvial deposit, as a result of variations in the energy of discharge events, meander evolution, etc. (some of these variations are discussed in Chap. 8), plus longer-term changes in tectonic and climatic control. Examples of descriptions of these groups are as follows, with the fluvial-style model numbers of Miall (1985) indicated:

Gravel Rivers (Model 4). Gravel-dominated deposits are relatively rare, and in most gravelly fluvial deposits the LA element is subordinate to element GB. Examples are illustrated in Figs. 6.16, 6.35a, and 6.37. Modern examples are e.g.: Bluck (1971), Lewin (1976), Gustavson (1978), Forbes (1983), Hooke 1986; ancient examples: Ori (1979, 1982), Arche (1983), Ramos and Sopeña (1983), Cavazza (1989).

Gravel-Sand Rivers (Model 5). Deposits consisting of sand or pebbly sand contain a wide variety of lithofacies reflecting vigorous bedform and bar progradation and chute development. Bedding within this type of LA element is complex and may obscure the underlying laterally accreted geometry. Indicated flow directions in these cross-bedded deposits are parallel or subparallel to the strike of the accretion surfaces. Modern examples are: Bernard and Major (1963),

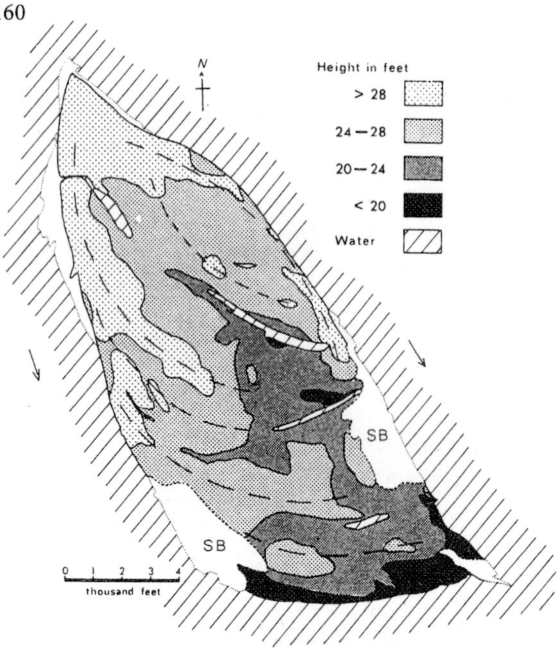

Fig. 6.36. Midchannel macroform, Brahmaputra River. *Arrow at right* shows flow direction. The bar is elongated downstream and is enlarging by accretion on the west and south sides. Surface traces of the growth are shown by the *dashed lines*, and active scroll bars (SB) are also indicated. This macroform is in part an LA unit (in its central part and on its west side), and in part a DA unit (along the east flank and at the downstream end). (Bristow 1987)

McGowen and Garner (1970), Bridge and Jarvis (1976), Jackson (1976b), Levey (1978), Campbell and Hendry (1987); ancient examples: Nijman and Puigdefábregas (1978), Allen (1983a), Campbell and Hendry (1987), Díaz-Molina (1993).

Sand Rivers (Model 6). Figure 6.38 illustrates a modern example and Figs. 4.7, 4.9, 4.10, 6.35b,d and 6.39 are ancient examples. For modern rivers, see: Sundborg (1956), Davies (1966); for ancient examples: Beutner et al. (1967), Puigdefábregas (1973), Shelton and Noble (1974), Nami and Leeder (1978), Puigdefábregas and Van Vliet (1978), Ethridge et al. (1981), Hobday et al. (1981), Plint (1983), Stear (1983), Mossop and Flach (1983), Bridge and Diemer (1983), Link (1984), Diemer and Belt (1991), Muñoz et al. (1992), Willis (1993a), Alexander and Gawthorpe (1993).

Sand-Silt-Mud Rivers (Model 7). The simplest LA elements are those composed of fine sand, silt, and mud. Secondary bedforms on the accretionary surfaces are rare and small in scale, and the accretion surfaces are steeply dipping and more readily identified in outcrop. Benches may be generated by horizontal sepa-

Fig. 6.37. *GB* and *LA* elements in Pleistocene gravels. Bounding surfaces are also indicated by *arrows with rank numbers* Near Mito, Japan. Note person at *lower right* for scale

Fig. 6.38. View across a modern point-bar surface, looking upstream. A field of 2-D dunes that was migrating toward the viewer has been abandoned by falling water. The point bar is accreting toward the left. Note that the crest lines of these dunes are oriented down the dip of the point-bar accretion surface. (Photo courtesy of M.O. Hayes and R. Levey)

ration eddies at sharp bends (Figs. 2.37, 6.40). For modern examples, see: Nanson (1980), Jackson (1981); for ancient examples: Miall (1979a), Stewart (1983), D.G. Smith (1987), Wood (1989).

LA deposits do not retain a constant geometry or composition around any given meander bend. As a result, the classic fining-upward profile (Allen 1964, 1965a,b) may not be present. In gravelly rivers, Bluck (1971), Lewin (1976), and Bridge and Jarvis (1976) showed that the coarsest part of the point bar is located at the upstream end of the bar (bar head) and may migrate downstream over sandy bar tail deposits. Wood (1989) interpreted facies variations in some ancient point-bar deposits on this basis. Jackson (1976b) found that in the Wabash River (sand and pebbly sand) the helical flow patterns responsible for the fining-upward point bar profile tend to develop only in the downstream part of a meander bend (Fig. 6.41). Carson (1986) made a similar observation on gravel-bed rivers in New Zealand. Nanson and Page (1983) showed that within tight meanders flow separation may occur at the downstream end of a point bar. Eddy currents there form significant deposits of fine sand, silt, and mud in "concave bench complexes" (Figs. 2.37, 6.40). In some muddy rivers, it has been found that accretionary development of mud, silt, and fine sand may take place on both sides of the channel. The deposits on the inside of the meander bend are referred to as an "inner accretionary bank." This element type is common in some of the muddy, ephemeral rivers of interior Australia (Taylor and Woodyer 1978; M.R. Gibling, pers. comm. 1995). A detailed study of the ridge-and-swale topography that develops in the uppermost levels of point bars was provided by Gibling and Rust (1993). They described variations in some exhumed Carboniferous examples, and related the variations to slight differences in the style of scroll-bar accretion.

Variations in the style of meander evolution (Fig. 6.34) lead to subtle but important differences in the internal architecture of the resulting LA deposit. These variations were described by Willis (1989), who used a computer model to simulate various meander configurations. Figure 6.42 illustrates his cross sections through a meander that evolves by

Fig. 6.39. View of a point bar in the Chinle Formation (Triassic), central Arizona. View is downstream, along the strike of the bar, the accretion surfaces of which dip gently to the right. The accretion surfaces are covered in trough cross-bed sets (lithofacies St), the axes of which are oriented away from the viewer

down-valley translation. The channel orientation at each stage of point-bar growth (arrows in each cross section) should be indicated by paleocurrent measurements. These cross sections are therefore useful as templates for the interpretation of randomly oriented outcrops through point-bar deposits. Note that obliquely downstream-oriented cross sections (Fig. 6.42C) reveal an architecture similar to that of downstream-accretion deposits, indicating the need for caution in the interpretation of individual, isolated outcrops. The application of these ideas to actual ancient examples is discussed by Willis (1993a). Detailed studies of some ancient lateral-accretion deposit, revealing complexities in internal accretionary geometry, are described by Díaz-Molina (1993) and Miall (1994).

As noted in Sect. 6.6, Crowley (1983) has described a gradation between the midchannel macroforms of low-sinuosity, braided channels (element DA), and the laterally accreted bars of high-sinuosity rivers (element LA). An intermediate form is the alternate bar, which develops on the insides of the meandering thalweg in rivers of low to moderate sinuosity (Fig. 6.27). This element type is common in straight deltaic distributary channels, such as the modern Mahakam delta, Indonesia (Verdier et al. 1980), and the Carboniferous fluvial-deltaic deposits of northern England (McCabe 1975, 1977; Okolo 1983). It has also been observed in fluvial settings, such as in straight reaches of modern anastomosed rivers (D.G. Smith 1983), and the Carboniferous sandstone deposits of the Appalachian basin (element type DA2 of Wizevich 1992b). The modern gravel bars described by Lewin (1976) are comparable. The bars consist of macroforms developed by accretion oblique to the channel trend (Fig. 6.43). They may consist of simple, large-scale cross-bed sets, in which the cross-bed foresets are the accretion surfaces (McCabe 1975, 1977). In some cases minor, bedforms develop by secondary flow oblique to the slip face, as in the DA2 elements of Wizevich (1992b). McCabe (1977) described large-scale internal reactivation surfaces (third-order, in the classification of this book) that indicate major stage changes during the generation of the bar. It is a matter of convenience how alternate bar elements are classified and coded. McCabe (1977) treated them as large-scale bedforms. Okolo (1983) described them as channel

Fig. 6.40. LA element composed of very fine-grained sand and silt. Note the steep accretionary dip, and the flat benches formed by the development of horizontal separation eddies at the downstream ends of tight bends (Nanson and Page 1983; see Chap. 8). Eureka Sound Formation (Tertiary), Ellesmere Island, Arctic Canada. (Miall 1979a)

fills. Wizevich (1992b) designated them as DA2 – a distinctive type of downstream-accretion element.

6.8 Laminated Sand Sheets (Element LS)

Sheets of laminated sand (lithofacies Sh, Sl) with minor Sp, St, or Sr commonly dominate in some ancient rock sequences, and have been interpreted as the product of flash floods depositing sand under upper flow-regime plane bed conditions (Miall 1977, 1984b; Tunbridge 1981, 1984; Sneh 1983). The flood deposits of Bijou Creek, Colorado (McKee et al. 1967), are invariably quoted as a close modern analog; they provided the basis for the Bijou Creek model of Miall (1977). Ephemeral streams of the Lake Eyre Basin also contain local accumulations of this assemblage (Williams 1971).

The characteristic architecture of this element has been best described by Tunbridge (1981), Sneh (1983), and Stear (1985). Individual sand sheets are 0.4–2.5 m thick, and rest on flat to slightly scoured erosion surfaces. They may be capped, gradationally, by Sp, St, or Sr, indicating waning flow conditions at the end of a flood event. Individual sheets may be traced laterally for more than 100 m. At the edges they thin and split into thinner units dominated by finer-grained sands and silts of lithofacies Sr. These beds probably represent the margins of individual flood sheets. Channel cutbanks are rare to absent. Stacked sequences may reach tens of meters in thickness.

6.9 Hollows (Element HO)

Research by Cowan (1991) in the Westwater Canyon Member of the Morrison Formation has indicated that some architectural units previously classified as channels require a quite different interpretation, and justify the erection of a new type of architectural element. Although the features to be described here

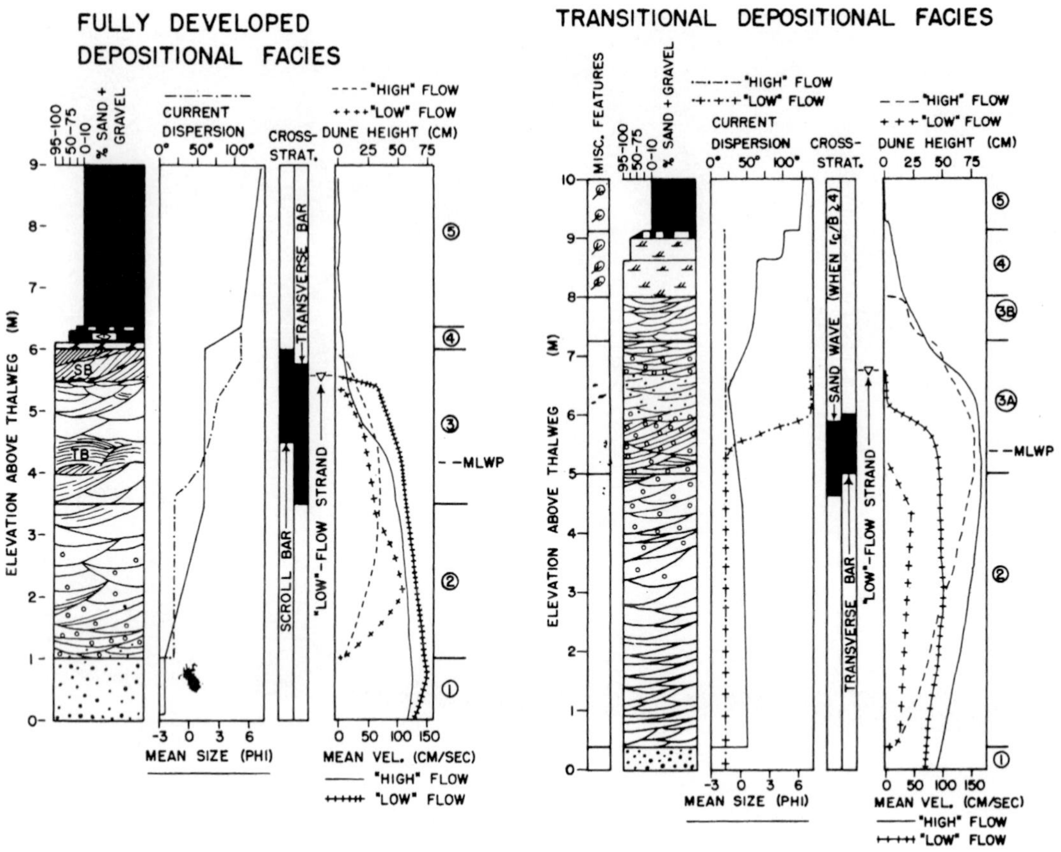

Fig. 6.41. Point bar successions in a gravel-sand river, the Wabash River (model 5 of Miall 1985). *Numbers in circles at right* are subfacies: *1* channel lag, *2* lower point bar, *3* upper point bar, *4* levee, *5* overbank. *Transitional* model applies to upper part of meander bend where helical flow is not well developed. *Fully developed* model applies to lower part of bend, where helical flow is established. (Jackson 1976b)

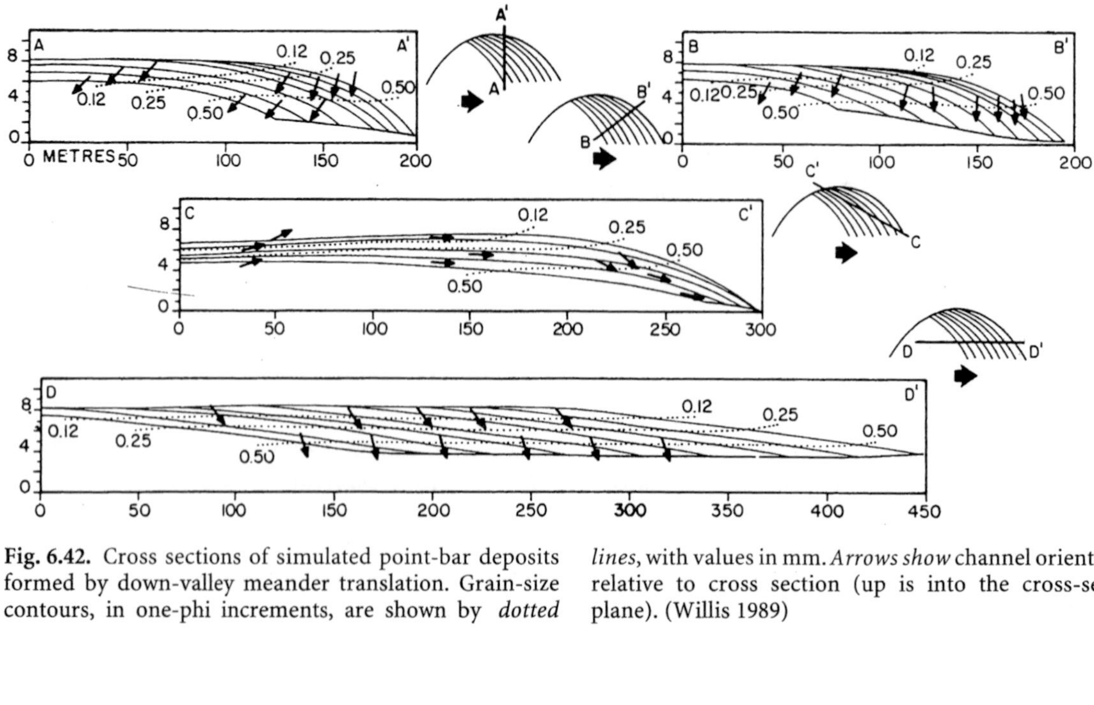

Fig. 6.42. Cross sections of simulated point-bar deposits formed by down-valley meander translation. Grain-size contours, in one-phi increments, are shown by *dotted lines*, with values in mm. *Arrows show* channel orientation, relative to cross section (up is into the cross-section plane). (Willis 1989)

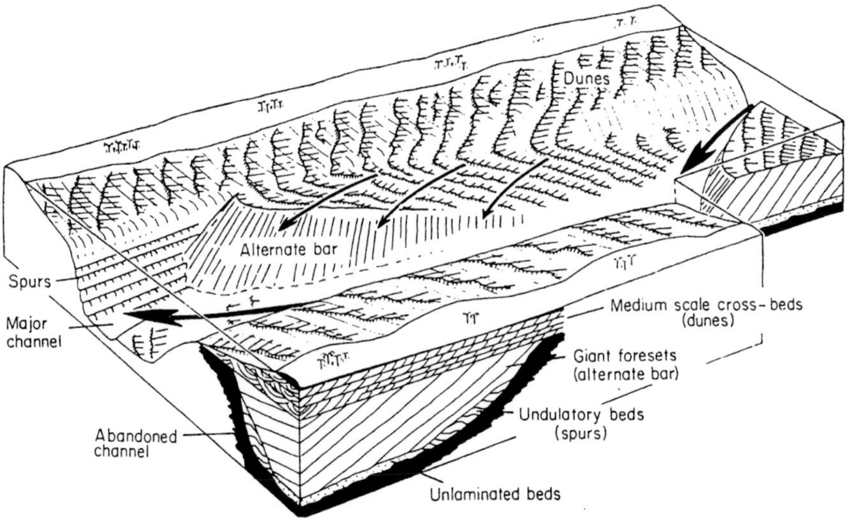

Fig. 6.43. Model for the development of a straight distributary channel with alternate bars, the latter comprising large-scale, simple cross-bed sets up to 40 m thick. (McCabe 1975, 1977)

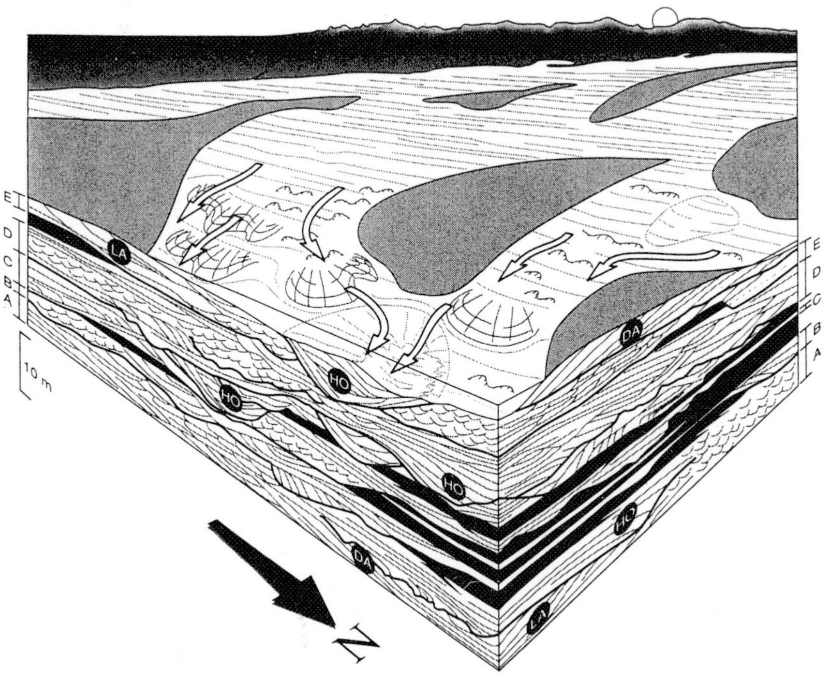

Fig. 6.44. Architectural diagram for the Westwater Canyon Member of the Morrison Formation, northern New Mexico, showing the predominance of element HO (Cowan 1991). A typical outcrop example of HO is illustrated in Fig. 4.12

are common in the Westwater Canyon Member, ongoing research is indicating that they are present in other gravel-braided and sand-braided systems. In the Westwater Canyon Member element HO resembles the "smaller channels" of Campbell (1976). They are up to about 20 m deep and 250 m wide, and are bounded at the base by curved, concave-up fourth-order surfaces (Figs. 4.12, 6.44). The dip on these basal surfaces reaches 26°, but is normally less. They are not cylindrical in shape, as are channels, but are scoop-shaped. The fill typically consists of lithofacies Sh and Sl, commonly with the dip of the Sl bedding planes parallel to the lower bounding surface, indicating that deposition took place on a

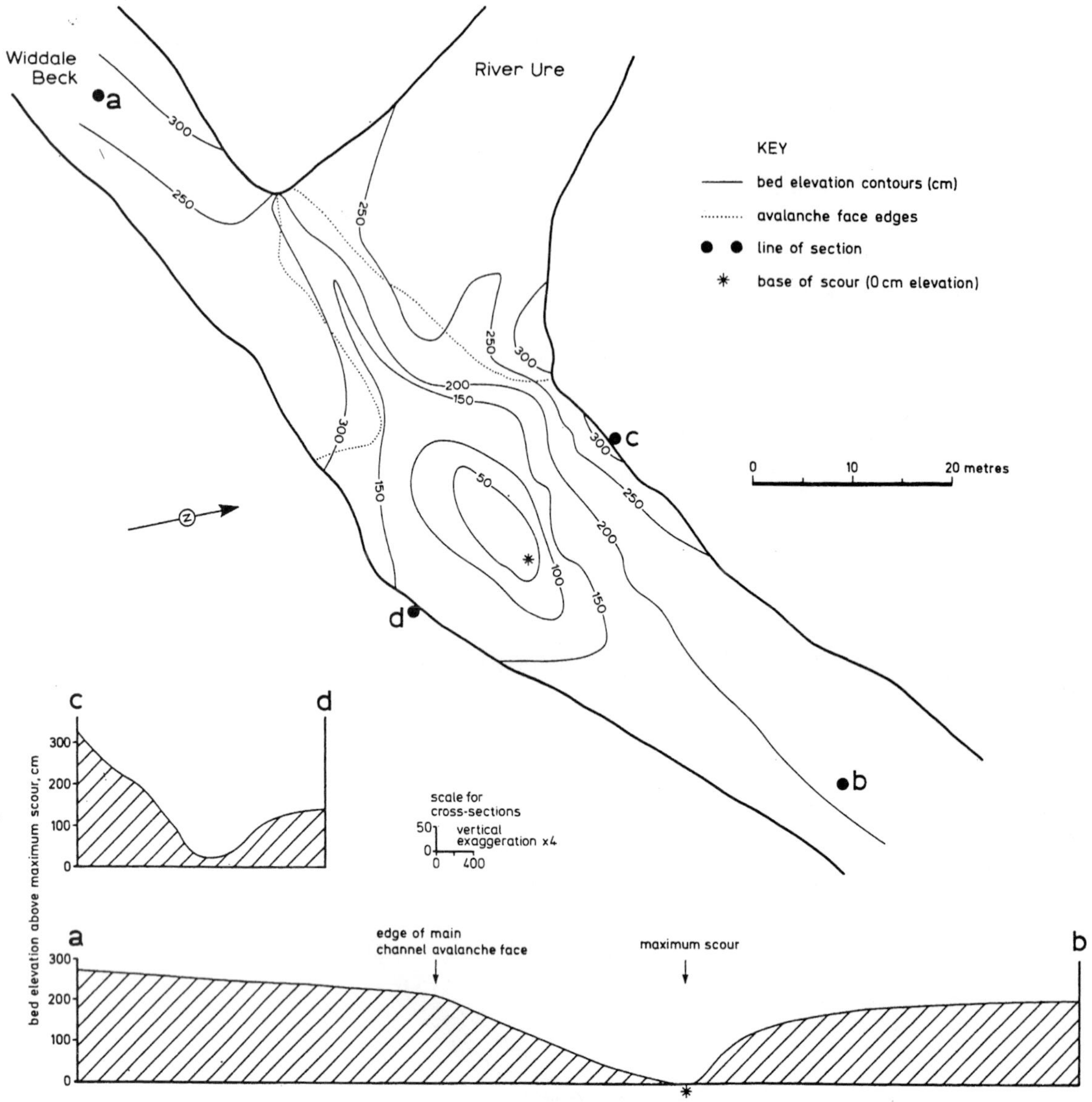

Fig. 6.45. Confluence of two channels, River Ure, Yorkshire, England. Note the scour, up to 2.5 m deep and 40 m long with avalanche faces at the mouth of each channel. (Best 1987)

sloping surface. The hollows fill by lateral or vertical accretion or combinations of both. They may be symmetrical or asymmetrical in cross section, and the orientation of dipping Sl sets may be perpendicular or oblique to the axis of the hollow. This axis, itself, is commonly oriented at a high angle to regional paleoflow. Siegenthaler and Huggenberger (1993) reported very similar scour-fill structures in Pleistocene gravel deposits in Switzerland. The scours are up to 100 m wide and 6 m deep.

These hollows may be interpreted in several ways. They are not channels. Channel cutbanks in sand are rarely as steep as observed in these hollows. Furthermore, the fact that the hollows are most likely trough-shaped in three dimensions suggests an origin other than by elongated channelization. They may be ancient analogs of large, elongate flute-like scours formed at channel bases, as documented by Coleman (1969) from the Brahmaputra River. The physical processes involved in the formation of these flute-like scours are not yet known. Similar but much smaller-scale structures have been documented from an ephemeral stream deposit by Olsen (1989), and were interpreted to have been formed by the erosional action of spiral vortices during a sheet flooding event.

Cowan (1991) proposed that these hollows formed by a process of deep scouring at points of channel convergence. Scour depths of up to six times the mean channel depth have been documented in rivers and laboratory flumes by Mosley (1976), Mosley and Schumm (1976), Best and Brayshaw (1985), Best (1987, 1988), Best et al. (1989), and Salter (1993). Bristow et al. (1993) proposed a facies model for the scour and fill processes at channel confluences, and Ashmore (1993) modeled their development in gravel-bed rivers. These studies suggest that a mechanism of deep scouring may be a significant process at channel junctions (Fig. 6.45) and below mid-channel macroforms in braided streams, where stream junctions and bar forms abound (Best 1987, p. 34; Best et al. 1989). Salter (1993) documented scour where flow is deflected against banks or islands. Best (1987) showed that avalanche faces can develop on the upstream end of these scours (Fig. 6.45). This allows the scours to be filled laterally, obliquely, or vertically by an avalanche deposit in a short period of time during channel switching or a flood event. The scours are, therefore, envisaged to form as clusters or isolated features, depending on the density and spacing of the channels within a braided channel belt.

Cant (1976, p. 125) interpreted deep scours, reaching nearly three times the mean braid-channel depth, to have formed upstream of a large emergent bar. This process is analogous to scouring on the upstream margin of an obstacle clast in a flow (cf. Best and Brayshaw 1985), with the emergent bar acting as an obstacle within the channel.

Because these scours result in the deposition of sediment below mean channel depth, they have a high preservation potential. In the deposits studied by Cowan (1991), they are common, whereas other types of macroforms, such as DA and LA units, are preserved only in fragmentary form.

Chapter 7

Architectural Elements of the Overbank Environment

7.1 Introduction

In contrast to the considerable amount of research devoted to the study of channel-fill sediments, overbank deposits have received much less attention, as has been noted by several researchers (e.g., Farrell 1987; Kraus 1987; Kraus and Bown 1988). The main exception to this is a recent focus on paleosols, stimulated by their use as lithostratigraphic and time markers (Sects. 7.4.2, 9.5.1, 10.4.3) and as climatic indicators (Sect. 12.9.2). In the original architectural-element classification (Miall 1985), overbank deposits were assigned to a single element, designated OF. This is clearly inadequate as many researchers have pointed out. In fact, deposits formed outside the main fluvial channels can be classified into three broad classes.

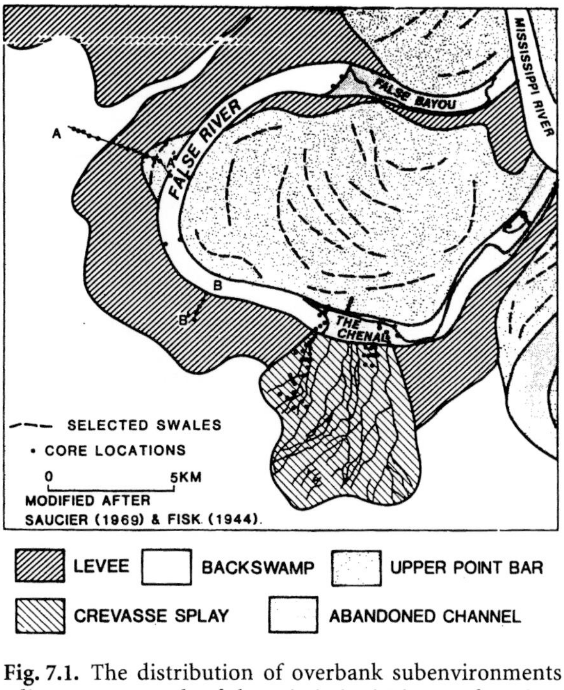

Fig. 7.1. The distribution of overbank subenvironments adjacent to a reach of the Mississippi River, False River cutoff, Louisiana. (Farrell 1987)

1. The relatively coarse deposits formed by overbank flow of channel bed load, constituting levee, crevasse-channel and crevasse-splay deposits.
2. Fine-grained deposits formed in low-energy environments, including ephemeral sheet floods and more permanent floodplain ponds.
3. Biochemical sediments formed by pedogenesis, evaporation, or organic activity.

Geomorphologists recognize a relatively simple subdivision of floodplain landforms and processes. For example, Brierley (1991) erected the following classification:

Top stratum (floodplain deposits), consisting of:

(A) sand wedge (levee)
(B) proximal sand sheet (marginal overbank sheet)
(C) distal sand sheet (distal overbank sheet).

His classification of in-channel deposits is given in Sect. 4.6.

Nanson and Croke (1992) listed six processes of floodplain sedimentation. Four of these, lateral point-bar accretion, braid-channel accretion, counterpoint accretion, and oblique accretion refer to channel aggradation, although the latter two processes may deposit fine-grained sediment that could be confused with floodplain deposits in small outcrops. These processes and their deposits are described in Chap. 6. The other processes of floodplain construction are overbank vertical accretion and abandoned-channel accretion.

Figure 7.1 illustrates the overbank elements developed adjacent to a reach of the Mississippi River, and Fig. 7.2 is a cross section showing the distribution and architecture of a coal-bearing fluvial unit in which overbank deposits constitute about 55% of the total rock volume. Figures 7.3 and 7.4 illustrate an element classification and depositional model developed for the Miocene Lower Freshwater Molasse of Switzerland. Another example of an element classification of a fluvial unit that includes significant thicknesses of overbank deposits is given in Table 4.4. Standardized element classification and coding for

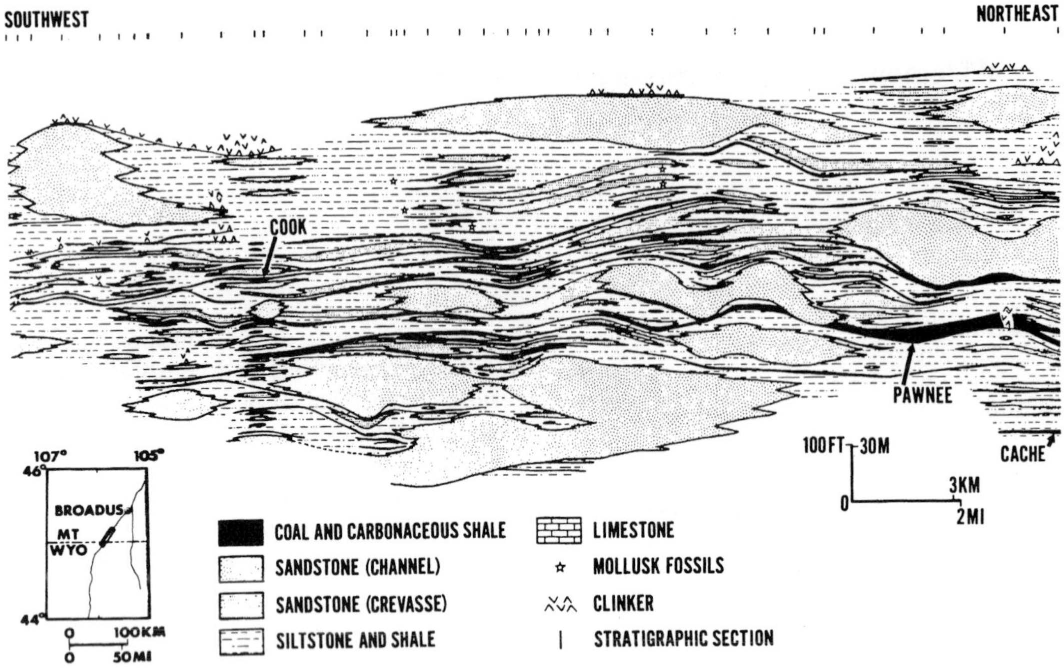

Fig. 7.2. Stratigraphic cross section showing the architecture of a fluvial depositional system made up of isolated and multistory channel sandstones and thick overbank deposits. The latter include overbank fines, crevasse sandstones, and coal. Paleocene, Wyoming-Montana. (Flores 1981)

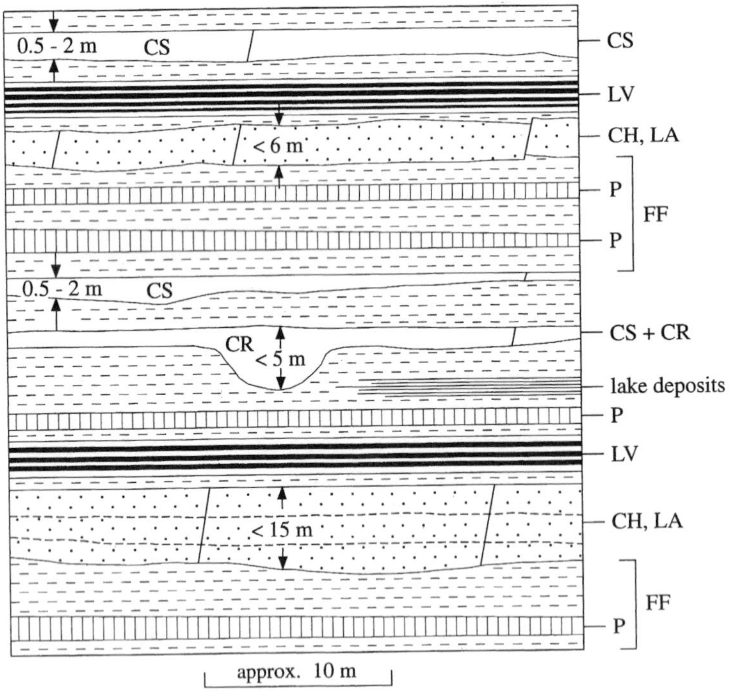

Fig. 7.3. Stratigraphic model of the architectural elements within the Lower Freshwater Molasse (Miocene) of Switzerland, showing the thickness of the major sandstone units. Element codes (in *capital letters*) are those given in Tables 4.3 and 7.1. Oblique lines are faults. (Modified from Platt and Keller 1992)

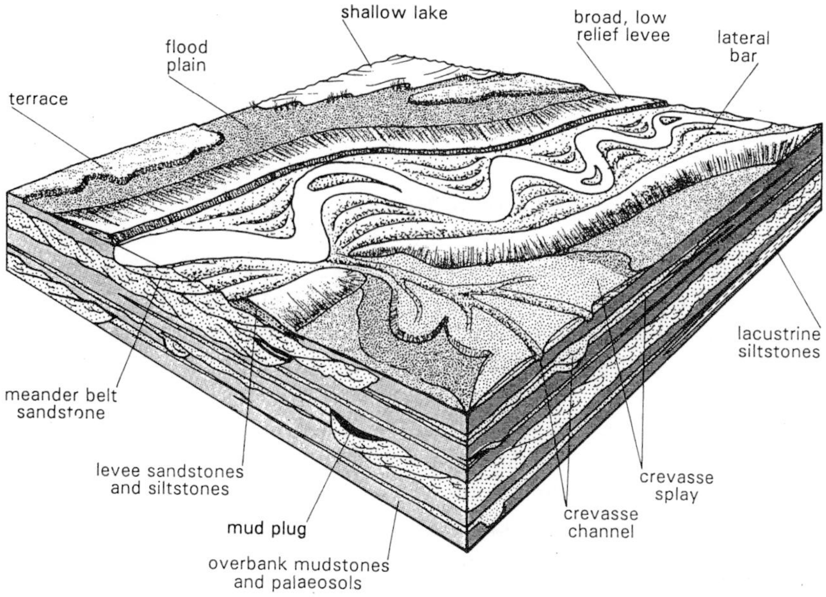

Fig. 7.4. Summary facies model for the Lower Freshwater Molasse. (Platt and Keller 1992)

Table 7.1. Clastic architectural elements of the overbank environment

Element	Symbol	Lithology	Geometry	Interpretation
Levee	LV	Fl	Wedge up to 10 m thick, 3 km wide	Overbank flooding
Crevasse channel	CR	St, Sr, Ss	Ribbon up to a few hundred m wide, 5 m deep, 10 km long	Break in main channel margin
Crevasse splay	CS	St, Sr, Fl	Lens up to 10 × 10 km across, 2–6 m thick	Delta-like progradation from crevasse channel into floodplain
Floodplain fines	FF	Fsm, Fl, Fm, Fr	Sheet, may be many km in lateral dimensions, 10s of m thick	Deposits of overbank sheet flow, floodplain ponds and swamps
Abandoned channel	CH(FF)	Fsm, Fl, Fm, Fr	Ribbon comparable in scale to active channel	Product of chute or neck cutoff

clastic deposits are suggested in Table 7.1 and are discussed below. This should avoid any confusion between different authors For example, Miall (1985) used the code LS for laminated sand sheets formed in ephemeral channels, whereas Platt and Keller (1992) employed this code for levee sandstones.

Overbank deposits are particularly important components of the deposits of rivers bearing a fine-grained sediment load, including fine-grained meandering rivers (Sect. 8.2.9) and anastomosed rivers (Sect. 8.2.10). These are the fluvial styles that are characterized by broad floodplains. Several maps and photographs of rivers and their deposits that fall into these two classes of fluvial style are provided in Chap. 8. These include a profile and interpretation of an outcrop that exposes sheet-like bodies of crevasse-splay sandstone in a Tertiary unit in the Canadian Arctic Islands (Figs. 8.44, 8.45), maps of two large splays in the modern Magdalena River, Colombia (Fig. 8.52), and cross sections through two splay deposits in the Cumberland Marshes, Saskatchewan (Fig. 8.53). Lithofacies are described and illustrated in Chap. 5 (Figs. 5.30–5.34).

Gravel-bed and sand-bed braided rivers have channel systems that tend to occupy most of the river floodplain, so that overbank environments are small in area and their deposits constitute a minor component of the fluvial assemblage (e.g., Reinfelds and Nanson 1993). A significant exception to this generalization was described by Bentham et al. (1993). The Eocene Escanilla Formation of the Spanish Pyrenees consists of more than 40% by volume of fine-grained bioturbated, mottled, and pedogenically modified fine siltstones and silty mudstones. Bentham et al.

(1993) attribute the preservation of such a large volume of overbank material to rapid subsidence and sedimentation and frequent overbank flooding.

7.2 Levee and Crevasse Deposits

The lithofacies and architecture of these deposits have been described by several workers studying modern fluvial environments, including Coleman (1969: Brahmaputra River), D.G. Smith (1986: Magdalena River), Farrell (1987: Mississippi River), N.D. Smith et al. (1989: Cumberland Marshes, Saskatchewan) and Brierley (1991: Squamish River).

Good descriptions of examples in the ancient record have been given by Gersib and McCabe (1981: Carboniferous, Nova Scotia), Allen and Williams (1982: Devonian, southwest Wales), Flores and Hanley (1984: Paleocene, Wyoming), Rust et al. (1984: Carboniferous, Nova Scotia), Bridge (1984: Devonian, Ireland), Fielding (1986: Carboniferous, England), Eberth and Miall (1991: late Paleozoic, New Mexico), and Platt and Keller (1992: Miocene, Switzerland). In addition, there is a considerable body of literature dealing with coal deposits that is beyond the scope of this book (e.g., see Rahmani and Flores 1984; Diessel 1992).

7.2.1 Levee Deposits (Element LV)

Adjacent to the lower Mississippi River natural levees are up to 3 km wide and 9 m high (Farrell 1987; Fig. 7.1). Similar widths were encountered in the Magdalena Valley by D.G. Smith (1986). Along the Brahmaputra River of Bangladesh, levees extend 700 m from the channel margins and are built up of overlappping lenses up to 1 km in width, parallel to the channel margin (Coleman 1969). Adjacent to smaller rivers, such as reaches of the Saskatchewan in the Cumberland Marshes, levees are less than 100 m wide (N.D. Smith et al. 1989). Depositional dips away from the channel margins of up to 10° were observed in some Eocene levee deposits by Bown and Kraus (1987), but dips of 2–4° are more common in these beds.

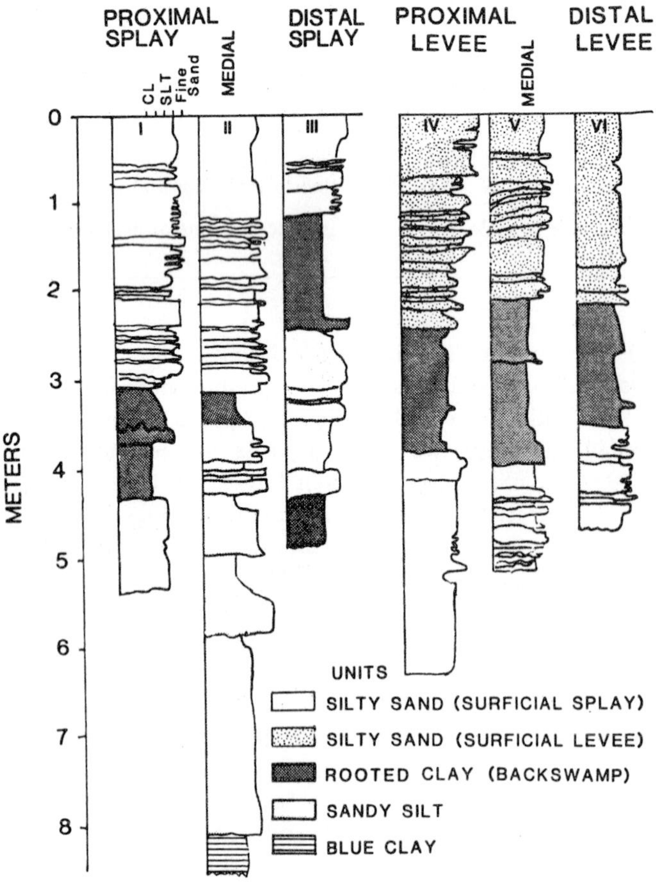

Fig. 7.5. Stratigraphic successions through overbank deposits, modern floodplain of the Mississippi River. (Farrell 1987)

Levees typically consist of rhythmically bedded units of silty, ripple-laminated sand a few decimeters thick (lithofacies Fl; Fig. 7.5). Bioturbation or root development commonly obscures lamination. Abundant vertebrate remains may be present (Bown and Kraus 1987). Each sedimentary rhythm represents a flood event. The levee element may reach 10 m in thickness, and constitutes a tapering wedge of deposit, thinning and fining away from the channel margin. Paleocurrent measurements made on contained ripple cross-lamination indicate flow perpendicular to the channel or obliquely downstream, away from the channel margin (e.g., Coleman 1969).

7.2.2 Crevasse-Channel Deposits (Element CR)

The scale of crevasse channels depends on the scale of the river. Those associated with the lower Mississippi River are up to 5 m deep (Farrell 1987). Fielding (1986) recorded crevasse channel fills up to 6 m thick. Even in smaller fluvial systems such as the Cumberland Marshes of Saskatchewan crevasse channels reach 4.5 m depth (N.D. Smith et al. 1989). Rhee et al. (1993) reported small channels less than 1 m in depth in a coarse-grained terminal fan deposit. Crevasse channels range from a few tens of meters to a few hundred meters in width. These channels incise levee and other backswamp deposits, and typically form small, delta-like distributary systems, becoming shallower away from the main channel (Fig. 8.52). In the case of the Magdalena River, Colombia, the network of crevasse channels extends up to 10 km away from the channel margin (Fig. 8.52), but this seems to be exceptional. Those flanking the Mississippi River are up to 5 km long (Fig. 7.1), and those flanking the Brahmaputra River rarely exceed 500 m in length (Coleman 1969). The bounding

Fig. 7.6. Overbank deposits of the Moenkopi Formation (Triassic), Arizona. Resistant units are distal crevasse splay deposits. *light-colored units* are caliche. *Dark, recessive units* are mudstones and siltstones of the overbank environment. Note the channel-shaped unit near the base, filled with fine-grained deposits. This may be the fill of an abandoned channel: element FF(CH), or a distal crevasse channel (element CR). Outcrop is about 10 m high

Fig. 7.7. Floodplain deposits of the Chinle Formation (Triassic), near Ghost Ranch, New Mexico. Fine-grained deposits are floodplain fines (element FF). Sandstones are interpreted as crevasse splays (element CS). Massive, irregular-weathering unit at *top* is the Jurassic Entrada Formation, an eolian unit. Outcrop is about 10 m high

surfaces of crevasse channels are classified as fourth-order surfaces (Table 4.2).

Crevasse channel-fills are ribbon-like bodies typically consisting mainly of fine- to medium-grained sandstone with trough cross-bedding and ripple cross-lamination (lithofacies St, Sr). Internal scours (lithofacies Ss) are common. The coarsest units are equivalent in grain size to the deposits of the main channel, but in average finer-grained facies are to be expected, as these channels represent a "stripping" of the top of the flow in the main channel as a result of overbank flow and incision.

Coarse lags containing fossil remains, such as abraded bones, are a common component of crevasse channels (e.g., Bown and Kraus 1987). Eberth and Miall (1991) described a type of crevasse channel filled with fine-grained deposits which they termed "U-shaped mixed-fill units." These contained thin-bedded, fine-grained facies (lithofacies Fl), commonly with a pebbly-sandstone lag at the base, and abundant disarticulated vertebrate remains. A similar deposit is shown in Fig. 7.6.

Crevasse channels are particularly important in anastomosing systems, because it is by the process of crevassing that the river compensates for the aggradation or plugging of individual channels. Flow from crevasse channels may eventually rejoin the main channel downstream, or be diverted into the course of a paleochannel, and may, in this way bring about a significant diversion of flow (D.G. Smith 1983; N.D. Smith et al. 1989; Richards et al. 1993). Such avulsive processes are discussed further in Sect. 10.3.

7.2.3 Crevasse-Splay Deposits (Element CS)

Crevasse splays are the delta-like deposits that form adjacent to the margins of main channels (Figs. 7.1, 7.2, 7.4, 8.52). In fact, they may be virtually indistinguishable from lacustrine deltas. Typical examples of crevasse splays in outcrop are illustrated in Figs. 7.6 to 7.9 and 8.44. The sediment is introduced into the backswamp environment by crevasse channels, and is deposited as a result of flow expansion and

Fig. 7.8. Overbank deposits of a fluvial delta plain, showing laterally extensive, fine-grained deposits (several units of element FF) and two crevasse-splay lenses containing internal, gently dipping accretion surfaces. Carboniferous, Kentucky

loss of flow power as discharge leaves the confines of the channels and spreads out as sheet floods. These deposits are particularly important components of the anastomosed fluvial environment, where the formation of crevasse channels and splays forms an intermediate step in the shifting of main channels into new positions on the floodplain (N.D. Smith et al. 1989; see Sects. 8.2.10, 10.3.3).

Crevasse splays form lens-shaped bodies up to 10 km long and 5 km wide. They are typically 2–6 m thick. They are cut by their feeder crevasse channels, and interfinger at their margins with fine-grained floodplain deposits. They overlap, and may pass imperceptibly into levees along the flanks of the main channels. The bounding surfaces of splay deposits are classified as fourth-order surfaces (Table 4.2). Bown and Kraus (1987) reported cumulative levee-crevasse splay successions 15–20 m thick and extending for up to 10 km away from channel margins.

The deposits of crevasse splays typically consist of fine- to medium-grained sandstone with abundant hydrodynamic sedimentary structures, plant roots, and bioturbation. Trough cross-bedding and ripple cross-lamination are common. Interbedded laminae of siltstone and mudstone constituting lithofacies Fl are also common. The assemblages are characterized by thin bedding and abundant surfaces of non-deposition and small-scale erosion (third-order surfaces), reflecting the origin of the splays by periodic or irregular sheet flooding. Grain size decreases away from the main channel toward the fringes of the splay. Internally, splay deposits may exhibit low-angle accretion surfaces recording growth by lateral progradation (Figs. 6.21, 7.8). The same process commonly leads to upward coarsening through the splay deposit (Fig. 7.10). The top of the splay may show upward fining as abandonment takes place. This upper part of the succession may show abundant bioturbation.

N.D. Smith et al. (1989) examined the evolution of crevasse channels and splays in the Cumberland Marshes anastomosed fluvial system of Saskatchewan. They subdivided splays into three classes (Fig. 7.11), and demonstrated an evolutionary development from the first to the third class, as channels lengthen and deepen, and flow becomes concen-

Fig. 7.9. A crevasse-splay sandstone overlying a coal seam. Carboniferous, Kentucky

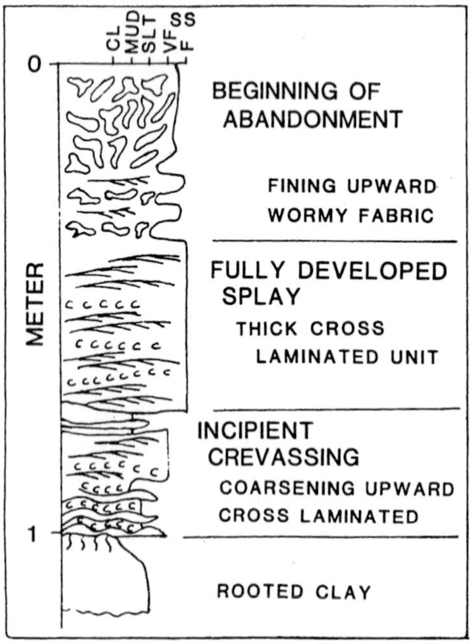

Fig. 7.10. Generalized stratigraphic section through a crevasse-splay deposit, based on coring of the splays of the lower Mississippi River. (Farrell 1987)

trated in a few of the channels. The third stage is that of a few larger channels isolated in floodplain fines, and is analogous to the channel pattern of the anastomosed fluvial style.

7.3 Fine-Grained Clastic Deposits

This category of deposits includes sheet-like units formed by settling from sheet floods, deposits formed in floodplain ponds, and the fill of abandoned channels. Stratification is tabular, and on floodplain surfaces individual units extend for hundreds of meters, or even for distances of kilometers (e.g., Willis and Behrensmeyer 1994). In some fluvial settings, particularly anastomosed rivers, and in any floodplains distant from the meander belt, these deposits may comprise the bulk of the total fluvial depositional thickness.

Lithologically these deposits are rather monotonous, although important variations in depositional processes lead to subtle variations in color, texture,

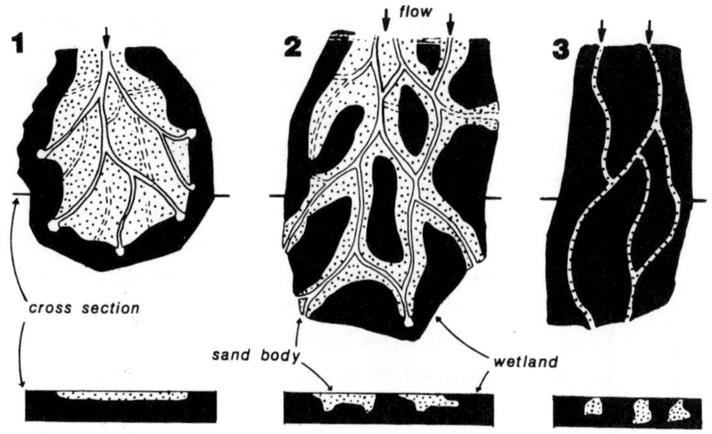

Fig. 7.11. The three types of crevasse splay in the Cumberland Marshes anastomosed fluvial system, Saskatchewan. Splays initially form type-1 systems, following an avulsive event, and progressively evolve into type 3. (N.D. Smith et al. 1989)

and accessory components. For example, in her work on the Mississippi River floodplain, Farrell (1987) distinguished between blue clays deposited in poorly drained swamps, and rooted clays deposited from suspension following overbank flooding events, and subsequently colonized by vegetation. This second facies is interpreted as the deposit of a well-drained swamp, possibly a levee.

In general, the presence of lamination indicates deposition from suspension or from very low-energy underflows. Such stratification is commonly destroyed by the presence of organisms, the activity of which leads to bioturbation, producing trace fossils and, ultimately, a churned and mottled texture in which stratification may be rare or absent altogether. Load structures may represent footprints of land vertebrates. Pedogenic processes and colonization by plants also disturb lamination. The common occurrence of invertebrate fossils suggests that the deposits did not dry out during deposition, and may indicate the presence of permanently saturated swamps or ponds.

In areas distal to channel margins, floodplain fines are commonly intercalated with distal splay deposits (element CS). Good examples of this are illustrated in Figs. 7.6 to 7.8. It is a matter of convenience whether sand units a few decimeters or less in thickness are given a separate element classification, or are grouped with the floodplain fines.

7.3.1 Floodplain Fines (Element FF)

This element consists of sheet-like units many hundreds of meters or even several kilometers in lateral extent (e.g., Willis and Behrensmeyer 1994). There may be considerable vertical lithologic variability, reflecting the fact that the depositional surface is flat and readily susceptible to small changes in depositional processes. Sedimentation may take place in separate increments representing individual flood events, or by continual slow settling of fine-grained sediment from suspension in permanent swamps or ponds (Sect. 10.3.1). Where seasonal or longer-term drying-out of the floodplain takes place, desiccation and pedogenic processes may be important, and this element may be interbedded with paleosols.

A typical profile through floodplain deposits is illustrated in Fig. 7.12. Architectural diagrams showing the scale of typical floodplain deposits are shown in Fig. 7.13. Typical outcrops of this element are seen in Figs. 7.6–7.8.

7.3.2 Abandoned Channel Fills (Element FF(CH))

Variously termed oxbows, bayous (Mississippi), or billabongs (Australia), abandoned channels are common components of many fluvial styles, particularly the finer-grained systems, including sandy meandering, fine-grained meandering, and anastomosed systems. An example in the Mississipi valley is illustrated in Fig. 7.1. A cross section through an ancient example is illustrated in Fig. 8.33. An outcrop photograph of an ancient example on a delta plain is illustrated in Fig. 6.9. A fluvial example is shown in Fig. 7.14.

Commonly, these channels remain as lakes or ponds for a considerable length of time. The mouth of the channel, at the cutoff, may gradually become

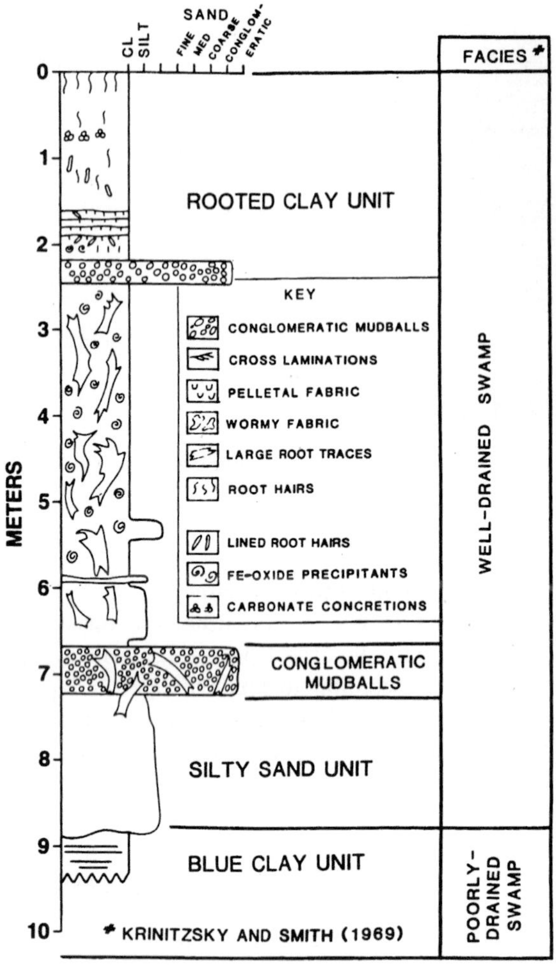

Fig. 7.12. Typical vertical profile through a backswamp deposit. From the modern Mississippi floodplain, Louisiana. (Farrell 1987)

silted up, leaving the abandoned channel to become progressively more and more undisturbed by turbulent eddies entering from the main channel. The waters of this elongate pond may therefore become gradually more static, and it slowly fills with clays deposited from suspension. Fine-grained lithofacies are therefore typical. Peat and other organic remains may also be common.

A variant of the common abandoned meander model is where the sediment load in main channels is so large that the mouths of tributary streams are blocked, causing the tributaries to back up as small lakes. Two examples of this, in very different environments, were described by Baker (1978a) and Kuenzi et al. (1979). In the first case, Baker (1978a) shows that the sediment load in the main tributaries of the Amazon system is very high, as these rivers have their headwaters in Andean regions underlain by loosely consolidated Quaternary glacial deposits. Tributaries rising within the Amazon lowlands carry much smaller sediment and water loads, and are dammed at their junctions with the main tributaries to form lakes. The sediment entering these lakes from the lowland tributaries forms small deltas, and the remainder of the lake may be expected to slowly fill with mud. In the second case, Kuenzi et al. (1979) provided an interesting study of fluvial-deltaic sedimentation on the Guatemalan coastal plain (Fig. 7.15). Fluvial styles here seem unusual, but may, in fact, be typical of nonmarine forearc (or backarc) regions characterized by active intermediate volcanism. Explosive volcanic eruptions coupled with very high rainfall lead to the catastrophic transportation and sedimentation of large quantities of coarse volcaniclastic debris. This may be confined to the main trunk rivers, and the result is that tributaries are dammed at their junction with the main river and back up to form lakes. These then fill with muds and small Gilbertian deltas. A possible ancient example of this volcaniclastic fluvial style with tributary lakes was described by Ballance (1988).

The clay plugs that commonly fill abandoned channels may be of considerable importance in subdividing channel-sand reservoir bodies into compartments, or providing seals on the updip flanks of channel sands. Several examples of this are described in Chaps. 14 and 15.

7.4 Biochemical Sediments

There is a considerable body of literature dealing with the sedimentary geochemistry of these rocks, which is beyond the scope of this chapter. The main interest in fluvial biochemical sediments from the point of view of this book is as architectural components of a stratigraphic assemblage (this section), as indicators of original depositional environments (this section and Chap. 8), as stratigraphic marker beds (Chap. 9), and as indicators of changes in autogenic and allogenic sedimentary controls (Chaps. 10, 11, 12). Coals and paleosols have a particular significance in the study of nonmarine sequence stratigraphy because of what they tell us about the balance between subsidence and sedimentation. In many cases, their widespread extent is an indication of regional changes in sedimentation patterns, and has led to their designation as markers of significant stratigraphic surfaces, such as sequence boundaries (Chap. 13).

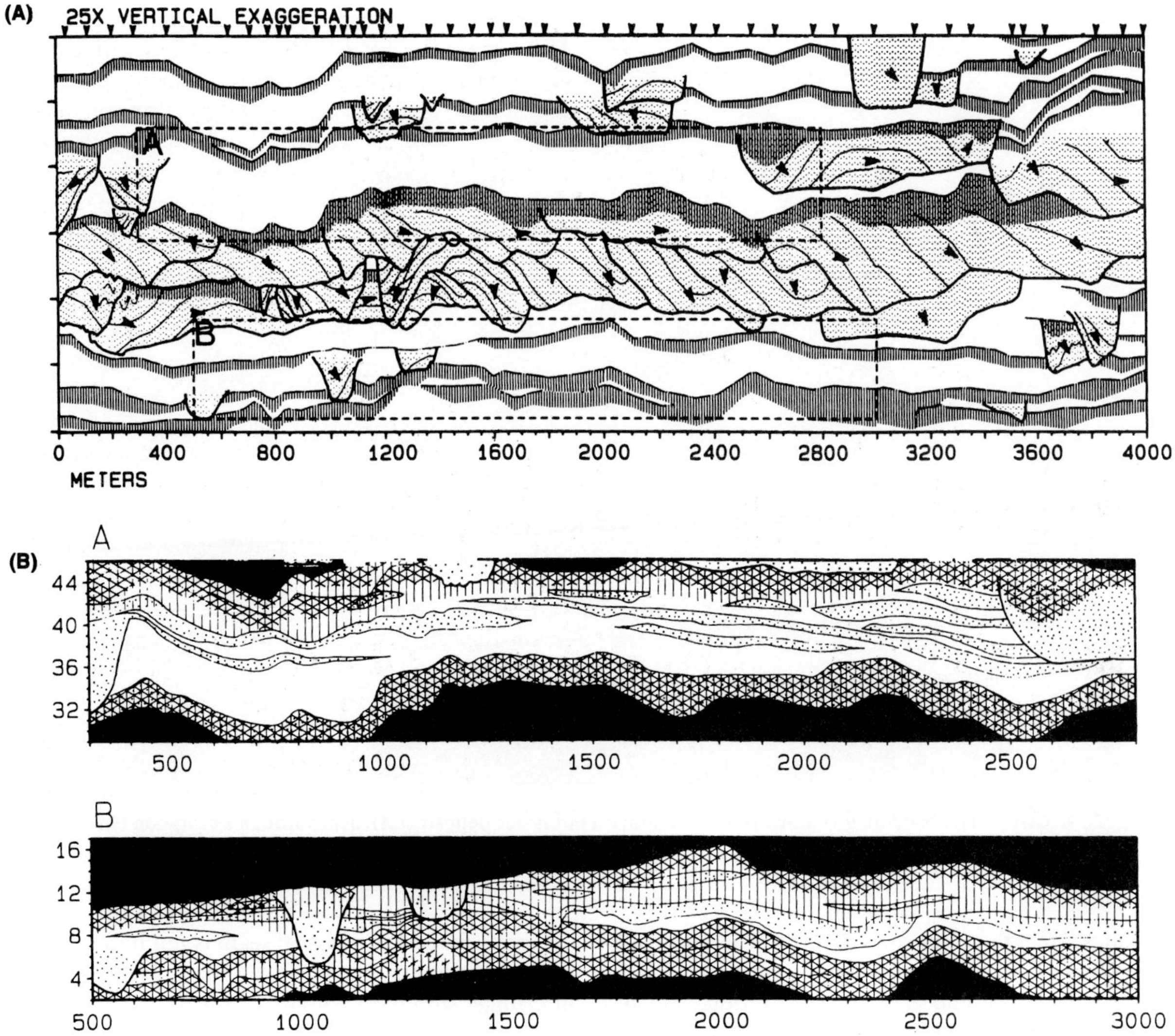

Fig. 7.13. **A** Architecture of major channel and overbank bodies in the Chinji Formation (Siwalik Group), near Chinji, Pakistan. *Stippled units* are major channel fills, showing accretionary geometries and direction of paleoflow (down is north, which is directly into the outrop plane). Paleosols are shown by *vertical ruling*, fine-grained deposits by *blank areas*. **B** Detail of *areas A and B* shown at enlarged scale. *Stipple* Coarse-grained units (crevasse splays); *vertical ruling* disruption, e.g., by bioturbation; *cross-hatching* carbonate-leached horizons of well-developed paleosols; *blank areas* undisturbed lamination. (Willis and Behrensmeyer 1994)

Paleosols may be subdivided into various layers and have been classified into many different types, with appropriate codings for field description. There seems to be no particular need to assign element codings to coals or evaporites.

7.4.1 Coal

Coal occurs in a wide range of depositional environments (Diessel 1992), but is most commonly associated with fluvial and delta-plain settings. As McCabe (1984) noted, modern depositional models for coal evolved from the work of Fisk (1960), who emphasized the importance of peats in the overbank regions of the modern Mississippi river and delta, and mapped the association between peat beds, crevasse, levee, and channel deposits. Most of these modern peats are high in clastic ("ash") content, and would not constitute true coals when lithified. McCabe (1984) demonstrated that in most clastic settings the influx of detritus from channels and crevasse splays during overbank flooding dilutes the organic content and leads to the formation of carbonaceous

Fig. 7.14. Clay-filled abandoned channel [CH(*FF*)] above a point-bar deposit (*LA*), Scarborough Formation (Jurassic), Yorkshire

shales rather than true coals. Low-ash coals probably require raised swamps to form. These develop in areas where rainfall exceeds evaporation, and organic growth is rapid (McCabe 1984; Moore 1987). Peat undergoes considerable compaction during the formation of coal, as a result of water loss, etc. Ryer and Langer (1980) compiled data showing peat:coal thickness ratios ranging between 1.4:1 and 30:1, with a median of 7:1. A further discussion of coal development, and its climatic significance, is given in Sect. 12.9.1.

Coals may also form by allochthonous processes. Degraded peat mats may be washed into lakes, and accumulate as sapropels. Cannel coals form from the accumulation of windblown spores.

Coal seams are typically interbedded with fine-grained overbank sediments of element FF (Figs. 7.2, 8.44, 8.45). They may also overlie or underlie crevasse-splay deposits (Figs. 6.21, 7.9) and fluvial channel-fill deposits, including point-bar deposits (Fig. 4.9). Figure 7.16 illustrates a case where a peat bog was incised by a fluvial channel, and rapid undercutting of the peat by lateral erosion formed a steep step in the base of the resulting channel sandstone.

The architecture of a coal seam is not necessarily that of a simple sheet. Commonly, clastic facies interfinger with the seam forming "splits". Examples are shown in Figs. 7.2 and 7.17. On a large scale, splits are caused by differential subsidence of the depositional basin (Diessel 1992). However, on a smaller scale (the examples illustrated here), the causes are autogenic, and are related to channel migration and avulsion and to the growth and abandonment of crevasse splays. Detailed mapping of coal seams in mines may reveal the pattern of channels and crevasses, where the coal has been "washed out" by fluvial channel erosion. An example of such a map is illustrated in Fig. 7.18. The ribbons are several kilometers wide, and consist of a main fluvial channel plus amalgamated levee and crevasse-splay deposits.

The coal seam itself may be subjected to detailed stratigraphic and facies analysis. Hacquebard and Donaldson (1969) carried out a detailed analysis of seams in the Sydney and Pictou coal fields in Nova Scotia. Based on maceral content they recognized three broad types of coal-swamp depositional environments that reflect original water-table levels:

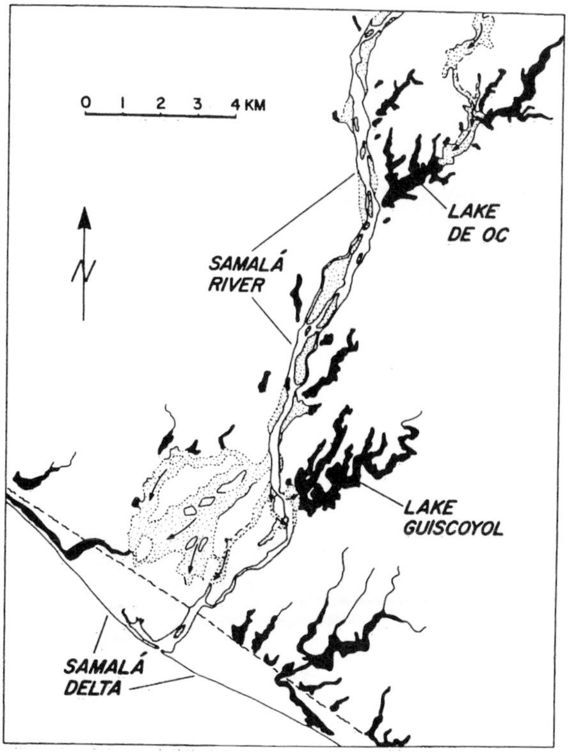

Fig. 7.15. Formation of lakes on a fluvial floodplain by damming of the mouths of tributaries. Guatemala coast. (Kuenzi et al. 1979)

1. Forest moor: well-drained swamp, characterized by the presence of trees, and yielding vitrinite- and fusinite-rich deposits.
2. Reed moor: poorly drained swamp, resulting in spore-rich coal.
3. Open moor: subaquatic, with sporinite, cannel and boghead coals and carbonaceous shales.

The distribution of these environments within one seam is shown in Fig. 7.19. This cross section illustrates the Harbour Seam, which is estimated to have taken 2800 years to accumulate. The seam is characterized by forest-moor deposits, with occasional flooding producing the other coal types. An increase in reed moor intervals in the upper half of the seam indicates a gradual rise in the water table. Figure 7.20 is a detailed section through this seam, showing details of coal petrography and palynology. Note the dominance of *Lycospora* in the lower half of the seam. This spore type was derived from the tree *Lepidodendron*, and confirms the forested nature of the swamp at that time. *Punctatosporites*, a herbaceous plant, is characteristic of reed-moor coals, and occurs near the top of the seam.

Diessel (1992) used maceral-derived ratios for investigating vegetation variations and mire type.

Fig. 7.16. A coal seam cut by a channel sandstone in a fluvial delta-plain setting. Note the steep cutbank, probably formed by the sudden breakup and undercutting of a peat mat by channel erosion. Carboniferous, Kentucky

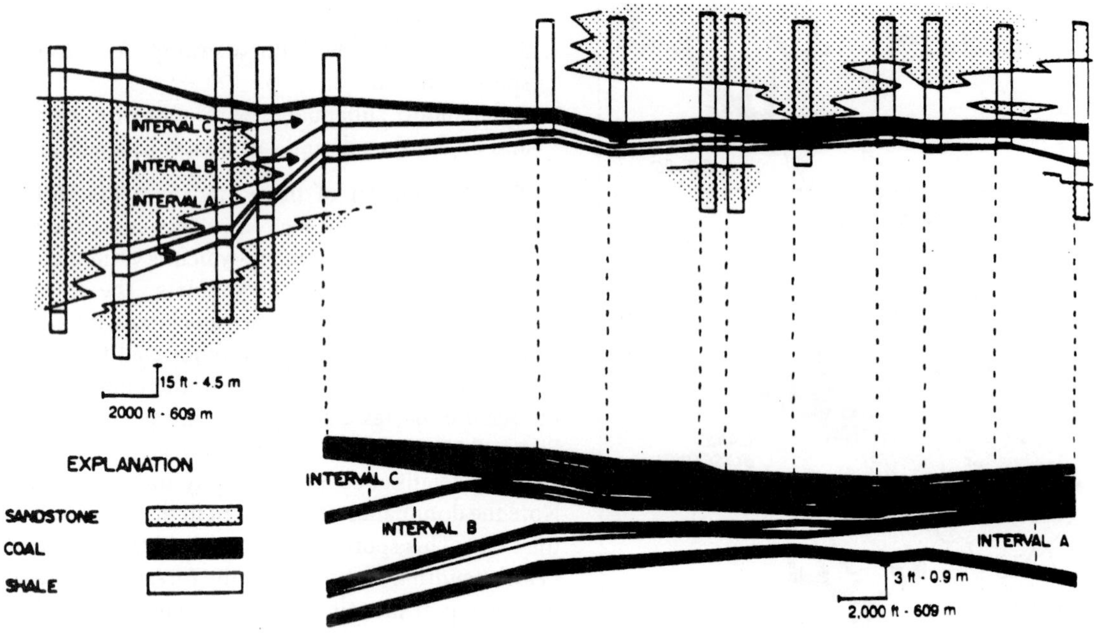

Fig. 7.17. Examples of splits in a coal seam. The Stockton seam, east Virginia. *Lower diagram* shows the splits at an enlarged vertical scale. (Ferm and Staub 1984)

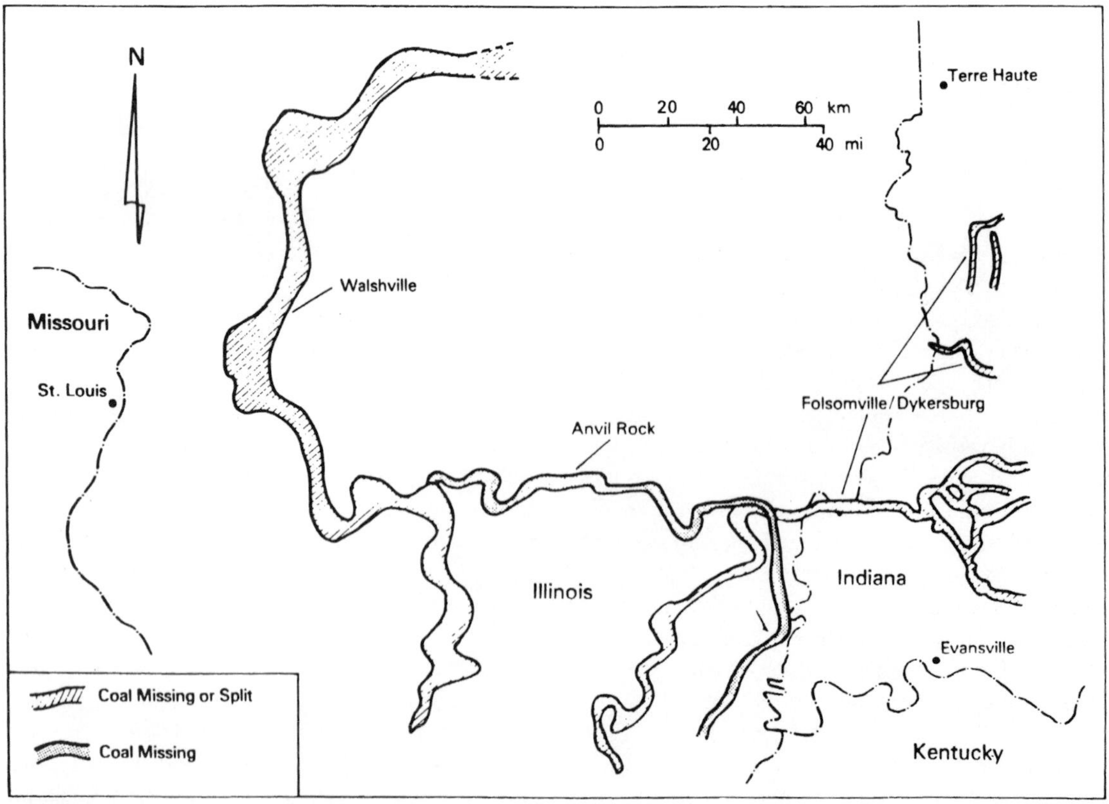

Fig. 7.18. Map of three major sandstone ribbons in Carboniferous coals of Illinois and Indiana. (Compiled by McCabe 1984)

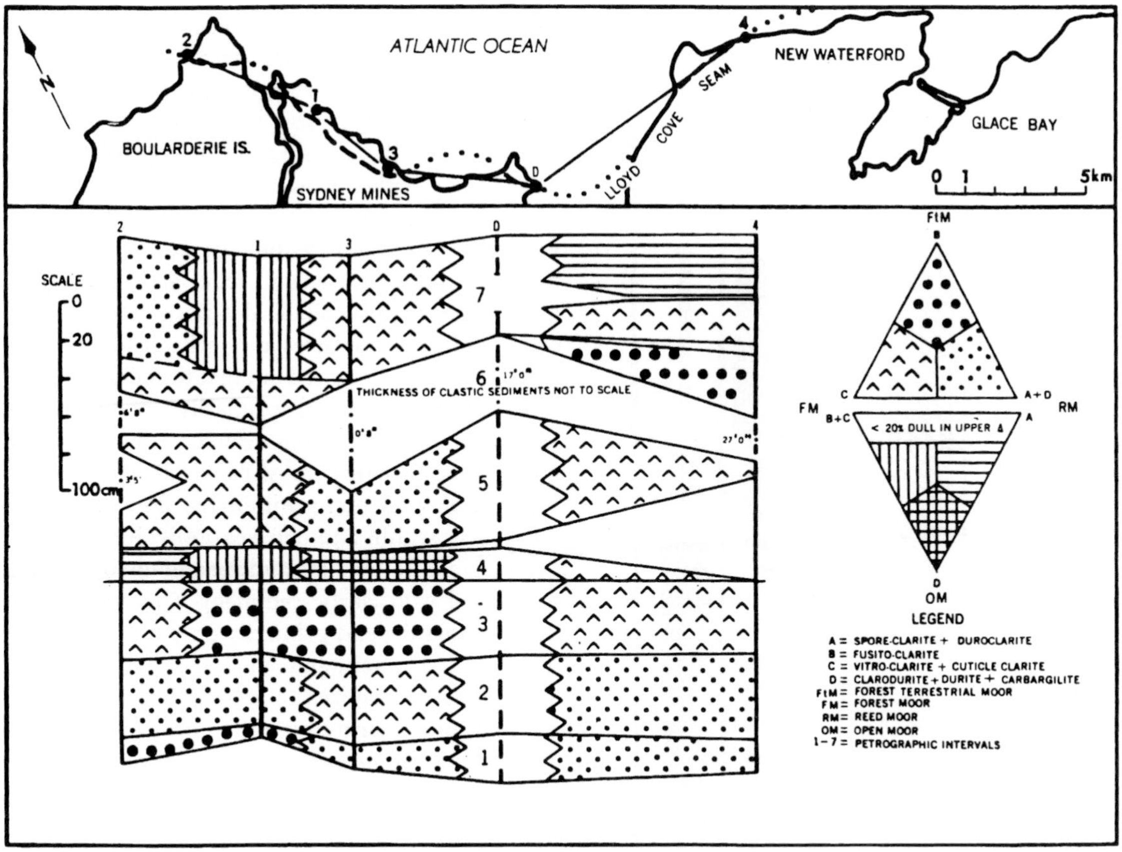

Fig. 7.19. Stratigraphy of a coal seam, the Lloyd Cove seam, Sydney, Nova Scotia. See text for discussion. (Hacquebard and Donaldson 1969)

7.4.2 Paleosols

Fossil soils have received a considerable amount of attention from sedimentologists, geochemists, and Quaternary geologists because of the information they yield regarding climates and rates of sedimentation, plus information on the evolutionary patterns of alluvial floodplains. Several research syntheses on these topics have now been published (Wright 1986; Reinhardt and Sigleo 1988; Martini and Chesworth 1992). (Additional publications dealing with modern soils and their geomorphology are referenced in these texts.) In this section, some basic descriptive data regarding paleosols are provided. Their use as stratigraphic markers is discussed in Chap. 9, and the information they can provide on autogenic and allogenic sedimentary processes is considered in Chaps. 10 to 13.

Some paleosol types are so distinctive that they have been given specific geologic names. These include cornstone, a nodular calcrete, and ganister, a silicified sandy soil horizon (Retallack 1988).

The three main field features of paleosols are root traces, soil horizons, and soil structures (Retallack 1988). The original carbonaceous root material may be preserved, but commonly this is replaced by or encrusted with carbonate or siliceous material. Movement of soluble material by groundwaters and evaporation (translocation) develops complex crystalline textures. Commonly, there are networks of irregular, mineralized planes within the soil, termed *cutans*. These may consist of clay skins, formed by leaching (eluviation). Cutans may also consist of ferruginized planes, manganese encrustations, or cracks filled with clastic material (clastic dikes). Sheets of calcite, barite, or gypsum may also occur. Aggregates of soil material are termed *peds*. They are bounded by cutans or voids. Compaction of peds and voids may develop slickensided surfaces. Mineral concentrations, termed *glaebules*, are common in soils. They are typically nodular or concretionary, consisting of calcareous, ferruginous, or sideritic material. Clay glaebules also occur. Diffuse mineral aggregation leads to mottling.

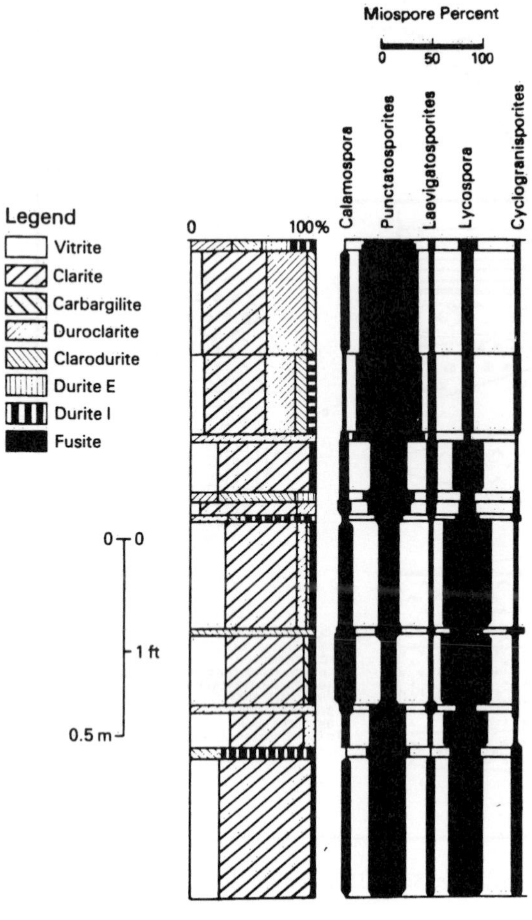

Fig. 7.20. Palynology and petrography of a typical vertical profile through the Harbour Seam, Nova Scotia. (McCabe 1984; modified from Hacquebard and Donaldson 1969)

Variations in soil components, color, and layering relate to climate, source terranes, and rates of subsidence and sedimentation (which control the rate of dilution of the soil by clastic material) (Fig. 7.21). In addition, a very important control from the point of view of the stratigrapher is the length of time over which the soil matures before burial. The lateral variability (stratigraphy) of soils within an alluvial basin reflects differences in relief and drainage. A soil unit showing such lateral variations is termed a *catena* (Milne 1935; Bates and Jackson 1987). Bown and Kraus (1987) pointed out that the concept of the *catena* was introduced primarily to encompass variations in modern soils related to their position on hillslopes, whereas most paleosols of interest to sedimentologists occur within the floodplain, and differences within and between individual soils depend on maturation time and the amount of dilution of the soil with clastic sediment. They found that these factors depend primarily on proximity to active fluvial channels, and suggested that a different terminology is required to accomodate such observational and interpretive differences. They introduced the term *pedofacies*, which they defined to encompass the

"laterally contiguous bodies of sedimentary rock that differ in their contained laterally contiguous paleosols as a result of their distance (during formation) from areas of high sediment accumulation" (Bown and Kraus 1987, p. 599).

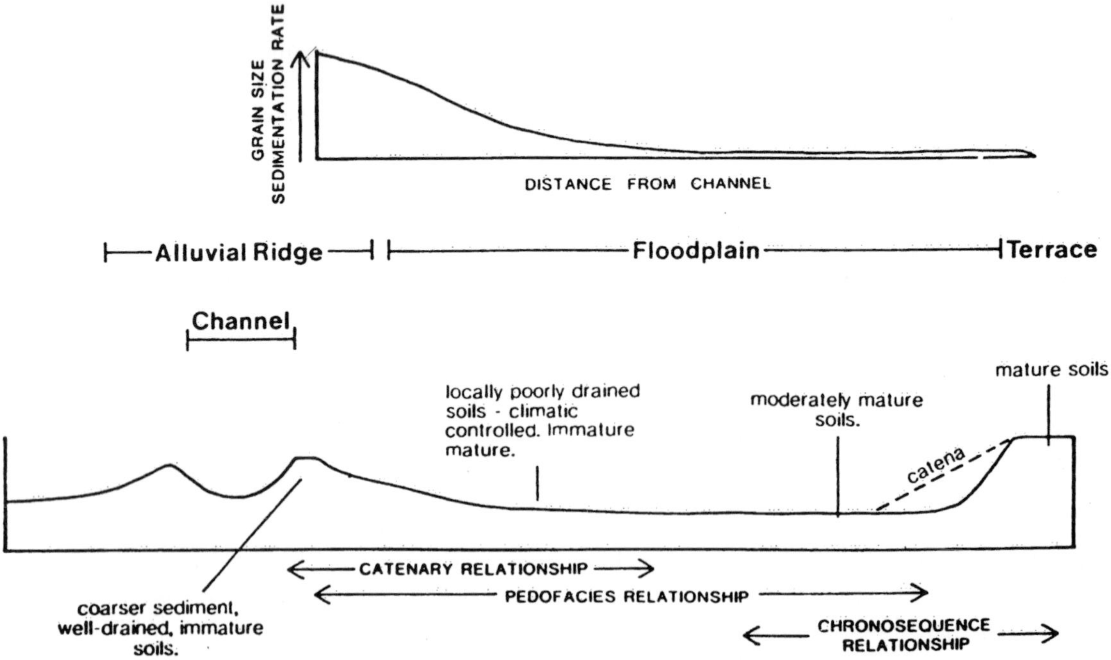

Fig. 7.21. Soil landscape relationships on a floodplain. (Wright 1992)

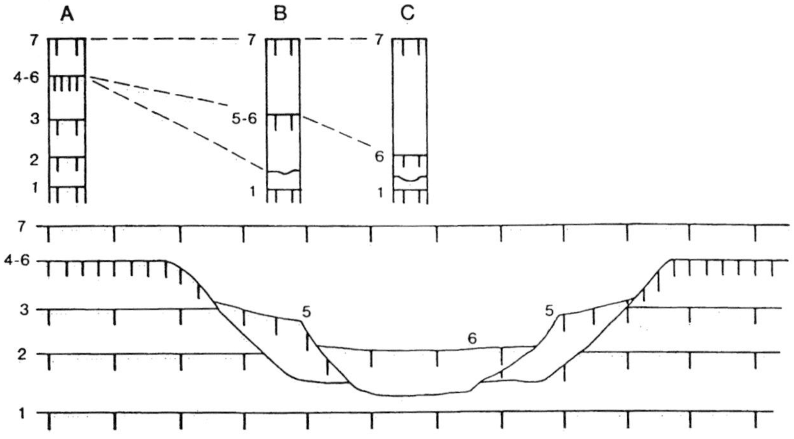

Fig. 7.22. The complex of soil stratigraphies that can be caused by a single phase of valley incision and filling, such as characterizes the base-level cycles associated with tectonism and sea-level change (Chap. 11). *1-7* Geomorphic surfaces with corresponding paleosols; *A-C* stratigraphic sections. (Wright 1992)

Table 7.2. Descriptive codes for classification of paleosol horizons. (After Retallack 1988)

Code	Description
O	Surface accumulation of organic materials (peat, lignite, coal), overlying clayey or sandy part of soil
A	Usually has roots and a mixture of organic and mineral matter; forms the surface of those paleosols lacking an O horizon
E	Underlies an O or A horizon and appears bleached because it is lighter colored, less organic, less sesquioxidic, or less clayey than underlying material
B	Underlies an A or E horizon and appears enriched in some material compared to both underlying and overlying horizons (because it is darker colored, more organic, more sesquioxidic, or more clayey) or more weathered than other horizons
K	Subsurface so impregnated with carbonate that it forms a massive layer
C	Subsurface horizon, slightly more weathered than fresh bedrock; lacks properties of other horizons, but shows mild mineral oxidation, limited accumulation of silica carbonates, soluble salts, or moderate gleying
R	Consolidated or unweathered bedrock

Adapted from Retallack (1988).

Internal erosion caused by channel shifting and base-level changes can develop erosional terraces within alluvial plains, leading to variations in maturity of individual soils, lateral erosion of soils, and superimposition of younger upon older soil horizons (Wright 1992; Fig. 7.22). These complexities all require a rigorous field observational methodology, which Retallack (1988) set out to provide. Some of the basic descriptive techniques and classifications summarized by Retallack (1988) and Bown and Kraus (1987) are provided here, but the reader is referred to these papers for a more complete discussion of field methods, and to the other books and papers referred to here for additional detail.

Paleosols typically consist of several distinct layers that can be defined on the basis of their color, texture, or mineralogy. Retallack (1988) provided a descriptive classification, which is given in Table 7.2. He recommended that the application of the classification be based in the first place on careful fieldwork, which should include detailing the thickness and vertical relationships of the various layers, as well as the taking of samples for subsequent laboratory analysis. Retallack (1988) also provided a code system for gradational layering and for subordinate descriptive characteristics, but the reader is referred to his original work for these details. Many different types of soil have been recognized, based on agricultural and other work on modern soils. Descriptions of these types, adapted for the study of paleosols, are provided in Table 7.3. The use of these names requires considerable experience. More amenable to field use by the nonspecialist clastic sedimentologist are the paleosol classifications that reflect simple maturation processes. Two alternative schemes are provided here. That in Table 7.4 is a generalized classification provided by Retallack (1988). The alternative scheme of Table 7.5 summarizes the

Table 7.3. Paleosol types. (Adapted by Retallack 1988 from US Department of Agriculture classification)

Type	Features
Vertisol	Abundant swelling clay (mainly smectite) to a presumed uncompacted depth of 1 m or more to a bedrock contact, together with hummock and swale structure, especially prominent slickensides or clastic dikes
Entisol	No horizons diagnostic of other types; very weak development
Inceptisol	No horizons diagnostic of other types; weak development
Aridisol	Light coloration, thin calcareous layer close to surface of profile, or evidence of pedogenic gypsum or other evaporite minerals
Mollisol	Organic (but not carbonaceous or coaly), well-structured (usually granular) surface (A) horizon, usually with evidence of copious biological activity (root traces, burrows) and with subsurface horizons often enriched in carbonate or clay
Histosol	Surface organic (O) horizon of carbonaceous shale, peat, lignite, or coal, precompaction thickness of at least 40 cm
Alfisol	Thick, well-differentiated (A, B, C horizons) profile, with subsurface B horizon appreciably enriched in clay and often red sesquioxides, or dark with humus, and also high cation content (Ca^{2+}, Mg^{2+}, Na^{+}, K^{+})
Ultisol	Similar to Alfisol but with sparse cation concentration
Spodosol	Thick, well-differentiated (A, B, C horizons) profile, with sandy subsurface B horizon cemented with iron or aluminum oxyhydrates or organic matter; little or no clay or carbonate
Oxisol	Thick, well-differentiated to uniform profile, clayey texture, subsurface horizons highly oxidized and red, and almost entirely depleted of weatherable minerals

Table 7.4. Stages of paleosol development. (Retallack 1988)

Stage	Features
Very weakly developed	Little evidence of soil development apart from root traces; abundant primary textures of parent material
Weakly developed	Surface rooted zone (A horizon), incipient subsurface clayey, calcareous, sesquioxidic, or humic, or surface organic horizons
Moderately developed	Surface rooted zone and subsurface clayey, sesquioxidic, humic or calcareous or surface organic horizons
Strongly developed	Thick, red, clayey, or humic subsurface (B) horizons, or surface organic horizons (coal or lignite), or especially well-developed soil structure, or calcic horizon
Very strongly developed	Unusually thick subsurface (B) horizon, or surface organic horizons (coal or lignite) or calcic horizon; mostly developed at major geologic unconformities

Table 7.5. Stages of paleosol development. (Bown and Kraus 1987)

Stage	Description
1	No horizon formation. Yellow or orange mottling, occasional development of yellow or orange beds with reddish streaks; color contacts diffuse. Carbonate accumulation initiated. *Entisol*
2	First incipient horizon formation. A horizon thin. B horizon red, sometimes darker red at top. Color contacts diffuse. Significant clay and plasma translocation initiated. Carbonate accumulation rapid. *Alfisol*
3	Marked horizon formation. Thick A horizon. B horizon purplish at top, dark and/or light red in middle, yellow or orange at base. Incipient, impermanent spodic zones sometimes present at top of B horizon. Color contacts diffuse. Significant plasma translocation, much concentrated as glaebules. Clay translocation near peak, carbonate translocation at peak. *Alfisol-Spodosol*
4	Profound horizonation. A horizon differentiation initiated, with beginning of clay leaching. B horizon thick. Aluminum and iron translocation at peak. Decrease in color mottling. Colour contacts more distinct. Well-defined spodic horizons. *Spodosol*
5	Profound horizon formation. Thick leached A horizon. A and B horizons much thickened relative to stage 4. Mottling rare. Color contacts sharp. Thick spodic horizons. *Spodosol*

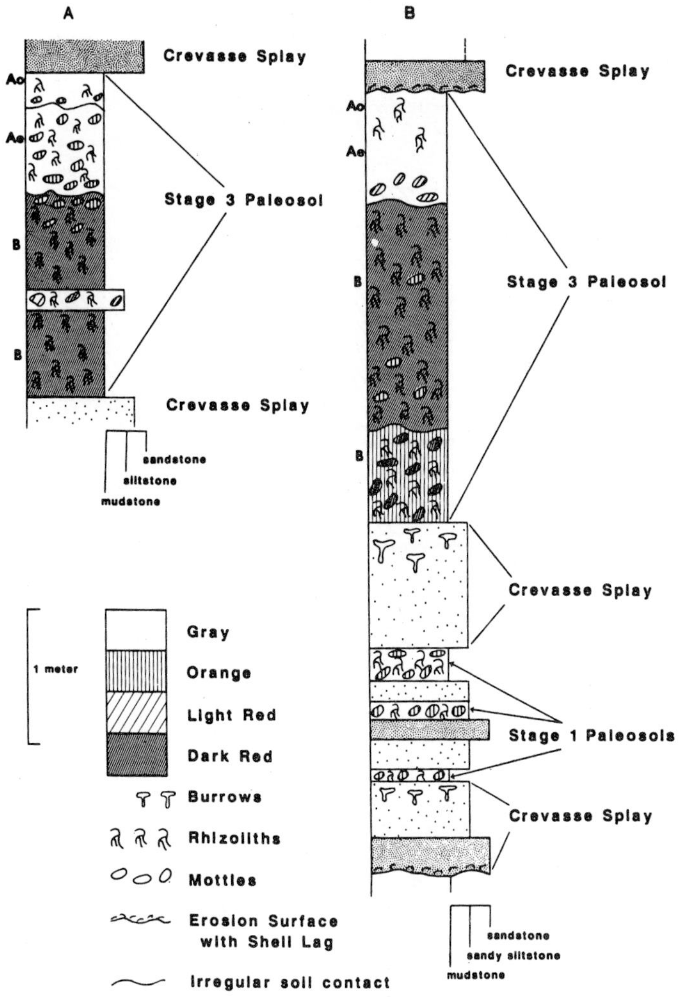

Fig. 7.23. Stratigraphic sections of two pedofacies successions. *Letters on left* indicate soil horizons. Willwood Formation (Eocene), Wyoming. (Kraus 1987)

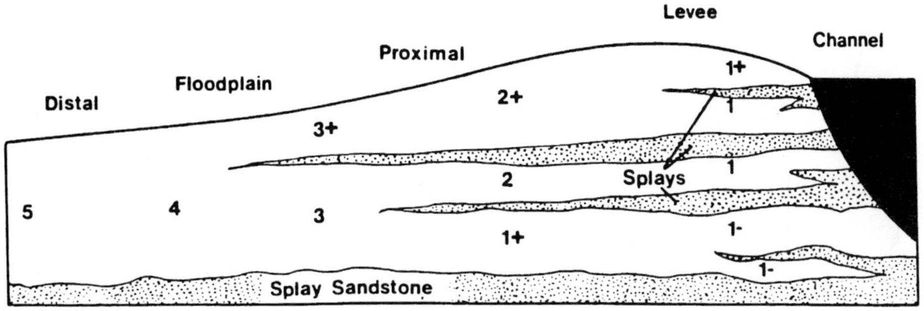

Fig. 7.24. The pedofacies model of Bown and Kraus (1987) and Kraus (1987, 1992). *Numbers* refer to the paleosol stages of Table 7.5

paleosol types identified in the Eocene Willwood Formation of Wyoming, which formed the basis for the pedofacies concepts of Bown and Kraus (1987).

Figure 7.23 shows two stratigraphic sections through two simple pedofacies successions, illustrating the graphic plotting methods of Bown and Kraus (1987) and Kraus (1987). More complex graphic techniques are described by Retallack (1988). The pedofacies model, relating paleosol "stage" to distance from the alluvial channel, is shown in Fig. 7.24. Examples of detailed profiles through the five types of paleosol stage are shown in Fig. 7.25. Caution is required in the application of such models, because it has been shown that other

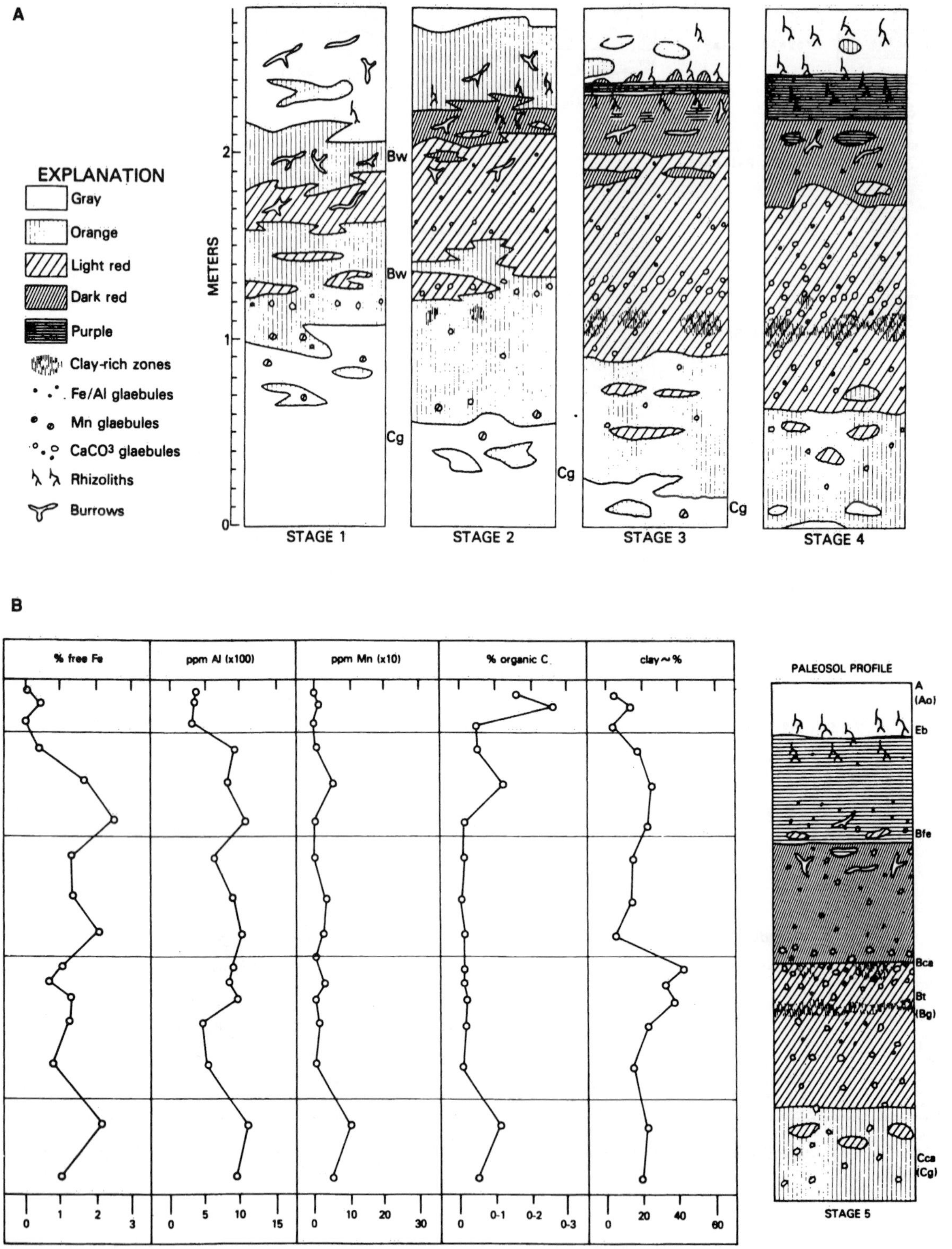

Fig. 7.25. A The five basic paleosol stages identified in the Willwood Formation of Wyoming. **A**, **B**, and **C** are horizon types. Subtypes are as follows: *Ao* epipedon; *Bca* calcrete; *Bg* gleyed horizon; *Bt* clay enrichment; *Bfe* iron enrichment; *Cca* calcrete; *Cg* gleyed; *Eb* eluviated zone in A horizon. **B** Distribution of important soil components in a stage-5 paleosol. (Bown and Kraus 1987)

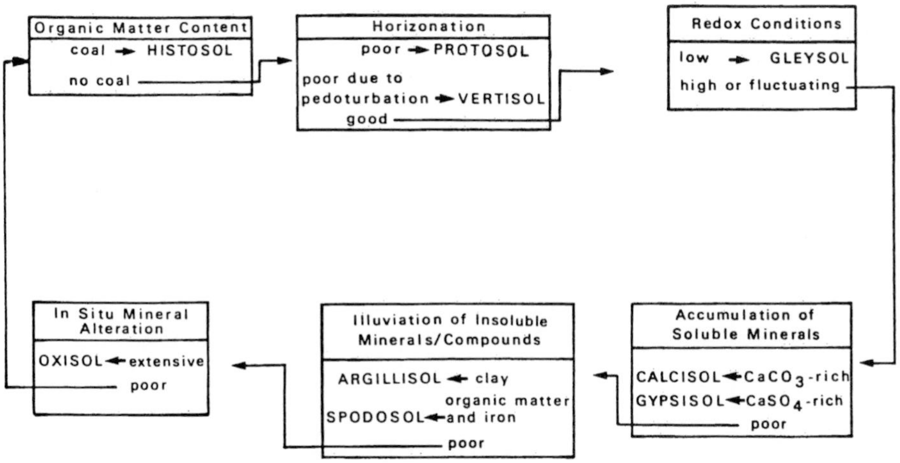

Fig. 7.26. Main Criteria used in the paleosol classification of Mack et al. (1993), presented as a flowchart showing how individual paleosols may be identified

Fig. 7.27. General view of Death Valley, California, showing the pattern of ephemeral stream channels and gypsum-encrusted playas on the valley floor, and the marginal alluvial fans. (Photo courtesy of S. Prosser)

Fig. 7.28. The lowest point in Death Valley, showing a pool of hypersaline water and part of the playa flats occupying the valley center

factors may affect paleosol maturity and pedofacies. For example, Atkinson (1986) showed that variations in paleosols in an Eocene succession in Spain could be related to regional changes in fluvial style and sedimentation rate which, in turn, reflected tectonically driven subsidence rates. Theriault and Desrochers (1993), who studied a Carboniferous succession in the Canadian Arctic, related areal changes in calcrete type in part to variations in substrate lithology. Wright (1990) demonstrated that variations in biogenic activity and the supply of the calcium ion markedly affected rates of pedogenesis.

One of the problems with most paleosol classifications currently in use is that they are based on classifications developed for agricultural purposes, which originally included features such as moisture content that are not relevant for geological purposes. A simpler classification of paleosols, emphasizing preservable criteria, was developed by Mack et al. (1993). Their flowchart for identifying paleosol type is illustrated in Fig. 7.26. In a later paper, Mack and James (1994) discussed the uses of this classification in studies of paleoclimate (Sect. 12.9.2).

The implications of pedofacies studies for interpretation of autogenic channel processes and floodplain evolution are discussed in Chaps. 9 and 10. The calculation of sedimentation rates using paleosol maturity and stratigraphy is discussed in Sect. 3.4. The presence of widespread paleosols may indicate low rates of generation of accommodation space in a sedimentary basin, and such an interval may correlate with a sequence boundary, as discussed in Sect. 13.3.1.

7.4.3 Evaporites

Evaporite deposits, including bedded gypsum, anhydrite, halite and other facies, and evaporite crystals developed within clastic facies, are a characteristic deposit of ephemeral lakes, or playas, occurring at the distal fringes of sheetflood fluvial units, or at the center of basins with inland (centripetal) drainage systems. Their sedimentology is described well by Hardie et al. (1978) and will not be repeated here.

Figures 7.27 and 7.28 illustrate the playa ephemeral fluvial systems of Death Valley, California.

Chapter 8

Fluvial Styles and Facies Models

8.1 Controls on Channel Style

The development of end-member facies models for a few distinctive fluvial styles has obscured the continuum that exists between fluvial channel patterns. The fourfold classification of channel styles discussed in Sect. 2.3.6.5 (Fig. 2.18) has served sedimentologists well, but should be replaced by a more sophisticated approach to channel classification. This needs to be based on an understanding of the controls that determine channel style.

Considerable theoretical and experimental work has been carried out to explore the causes of meandering and braiding. The natural meandering of channels, in particular, has received much attention. It is now known that meandering occurs in all fluid systems as a result of turbulence, internal shear, and bank and bed friction. Straight channels, including canals and channels in flumes, commonly develop meanders because of these controls. Air and ocean currents meander despite the absence of banks to impose a frictional shear. Threads of water flowing down smooth surfaces of glacial ice and tilted glass plates in a laboratory also develop meanders. Details of the mathematical and hydraulic models that have been developed to quantify the meandering process in rivers are beyond the scope of this book. We are concerned here with what can be determined about channel style from the geological record, because this provides information about the architecture of the beds, and about the basin controls that were operating during deposition.

Observations and measurements of modern rivers, and experiments with flumes, have shown that the channnel pattern in alluvial rivers (those flowing in their own sediment) is primarily dependent on discharge, sediment load, and slope. Lane (1957) and Leopold and Wolman (1957) demonstrated a natural transition between braided and meandering styles, dependent upon channel slope and discharge (Fig. 8.1). Thus, for a given slope, a river changes from meandering to braided as discharge is increased. Schumm (1968a) proposed the following general equations:

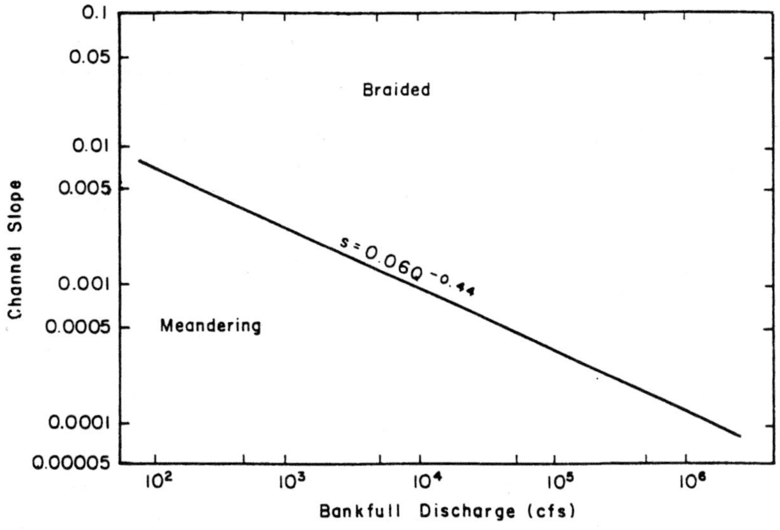

Fig. 8.1. The relationship between channel slope, bankfull discharge, and channel style. (Leopold and Wolman 1957; Schumm 1985b; Reproduced, with permission, from the Annual Review of Earth and Planetary Sciences, v. 13, © 1985, by Annual Reviews Inc.)

$$Q_w \propto \frac{w,d,l}{s}$$

$$Q_s \propto \frac{w,l,s}{d,P}$$

where Q_w=discharge, Q_s=sediment load, w=width, d=depth, l=meander wavelength, s=slope, and P= sinuosity. For a few sets of rivers in specific climatic ranges, some of these relationships have been quantified, as summarized by Ethridge and Schumm (1978). However, the data are inadequate to permit generalizations that can be applied to all geological conditions, a fact that considerably reduces the usefulness of quantitative geomorphology for geological reconstructions. Additional details on the relationship between discharge and slope and the braided-meandering threshold are given by Bridge (1993b).

One of the primary prerequisites for braiding is a high coarse-grained sediment load (bed load). Leopold and Wolman (1957, p. 50) stated:

"Braiding is developed by sorting as the stream leaves behind those sizes of the load which it is incompetent to handle ... if the stream is competent to move all sizes comprising the load but is unable to move the total quantity provided to it, then aggradation may take place without braiding."

In addition to the relationship between discharge and channel style shown in Fig. 8.1, Miall (1977) and Rust (1978a) noted that braided rivers are commonly characterized by high discharge variability; for example, rivers in alpine and arctic areas, with highly seasonal discharge variations, tend to be braided, as are ephemeral rivers in arid regions. Miall (1977, p. 7) argued from these and other observations that discharge variability is a primary control on the braiding pattern:

"In rivers of highly variable discharge competency will be similarly variable, and there will be long periods of time throughout which the river will be unable to move at least the coarsest part of its bed-load. The incidence of bar initiation, flow diversion and the creation of new channels (braiding) will thus be high."

On the scale of the alluvial basin the nature of the vegetation cover has an important effect on discharge characteristics, sediment load, and fluvial style. In vegetated areas presently, runoff following major rainfalls is rarely catastrophic because the precipitation is absorbed by soil and plants and released slowly. Similarly, sediment is stabilized by roots, and sediment yields are relatively low. Vegetated banks are highly resistant to erosion, as demonstrated in an experiment by D.G. Smith (1976). This limits the supply of sediment to the river, and also reduces the bank widening that commonly accompanies braiding. In vegetated areas, therefore, because of the effects on discharge variability, bank stability and sediment supply, braiding is inhibited. Baker (1978a) quoted examples in tropical Africa and South America where dense bank vegetation inhibits braiding, even where abundant coarse sediment is present in the banks. Stanistreet et al. (1993) described the channels of the Okavango fan in Botswana, which are low-sinuosity to meandering rivers, in which banks are stabilized by peat levees. The changes in river behavior when forests in upland catchment areas are destroyed by fire or deforestation are marked and well known. Runoff becomes more flashy, the sediment load increases, and a tendency may develop for debris flows to occur.

Rivers are sensitive to longitudinal changes in the controlling variables and to temporal changes in these variables. Carson (1984a,b,c) showed from his studies of rivers in the Canterbury Plains area of New Zealand that the transition from meandering to braided takes place at higher slopes as the caliber of the sediment load is increased. He also showed that the development of the braided pattern is very sensitive to the local rate of supply of bed load to the river. A large bed load, supplied, for example, from easily erodible banks, leads to channel shoaling and local flooding, bank incision, and avulsion. N.D. Smith and D.G. Smith (1984) described the William River in northern Saskatchewan, a river that changes downstream from a single-channel to a multiple-channel, braided pattern where a large sand bed load is introduced by eolian processes. Stollhofen and Stanistreet (1994) demonstrated a change in fluvial style from meandering to braided when volcanoes erupted in a fluvial plain and provided a large supply of pyroclastic debris. Friend and Sinha (1993) recorded much local variablity in sinuosity and braiding character in the large rivers of India. They confirmed the conclusions of Carson (1984a,b,c), and noted that the nature of the local alluvial substrate, and the sediment input from tributaries, will also contribute to continuing downstream adjustment and consequent change in the configuration of the river. Important experimental work that confirms these trends has been reported by Ashmore (1991), Leddy et al. (1993), and Germanoski and Schumm (1993), as noted in Sects. 6.3 and 11.2.2.

Temporal changes may be exemplified by several studies of rivers in the United States. Nadler and

Schumm (1981) and Schumm (1985b) described examples of two rivers, the South Platte and Arkansas, that evolved from braided to meandering patterns within historic times as a result of damming and irrigation that resulted in lower discharge variability, lessening of flood peaks, and stabilizing of banks by the growth of vegetation. Conversely, Schumm and Lichty (1963) described the case of the Cimarron River, Kansas, that changed from a suspension-load, meandering morphology to a broad, shallow, bed-load, braided morphology during a single major flood in 1914. Channel widening continued until 1942, during a period of below-normal precipitation, which inhibited vegetation growth.

Changes in regional slope may to some extent be accomodated in the river system by changes in channel pattern (Schumm 1993; Wescott 1993). As noted by Wescott (1993), channel slope S_c, valley slope S_v and sinuosity P are related:

$$P = \frac{S_v}{S_c}$$

Given constant discharge and sediment load the river will tend to maintain a constant channel slope. Therefore, an increase in regional slope (S_v) as a result of tectonic activity will tend to be compensated by an increase in sinuosity. This ability of the fluvial system to respond to regional changes in the major external controlling variables must be taken into account in interpretations of the fluvial response to tectonic activity and changes in base level. Some of the early models of sequence stratigraphy did not recognize this, and are overly simplistic (Miall 1991c), as discussed in Sect. 11.2.2.

Turning to long-term temporal changes, it has been suggested that the evolution of vegetation has had a major effect on fluvial styles through geologic time, because of the implications for bank stability and erodibility, and the rate of supply of sediment into river systems. Schumm (1968a) developed a suite of hypothetical curves for the relationship between precipitation and sediment yield through time (Fig. 8.2), in which he developed this idea. Prior to the Devonian, there was little or no land vegetation, and the land surface probably appeared much as arid areas do today, even where rainfall was high. Runoff would have been flashy and sediment yield large (Fig. 8.2, curve 1). From the Devonian to the end of the Paleozoic, vegetation was probably confined to nearshore and coastal plain areas, so that bank stabilization would have begun, although discharge and sediment-yield characteristics, controlled mainly by processes in the headwaters, would have changed little (Fig. 8.2, curve 2). Schumm (1968a) noted that primitive flowering plants appeared by Permian time, but would not likely have occurred outside tropical rain forests. Conifers were abundant by the Jurassic and modern deciduous forests appeared by the Mid-Cretaceous. The Devonian-Cretaceous period therefore marked a period of increasing stabilization of land surfaces by vegetation, but it was probably not until the early Cenozoic that interfluve and upland areas were colonized by plants capable of surviving severe weather and climatic fluctuations (Fig. 8.2, curve 3). Grasses appeared in the Miocene, and since that time runoff and sediment yields would have been much as they are today (Fig. 8.2, curve 4).

Developing these ideas further, Schumm (1968a) deduced that bed-load streams would have been predominant in early geologic time. Cotter (1978) car-

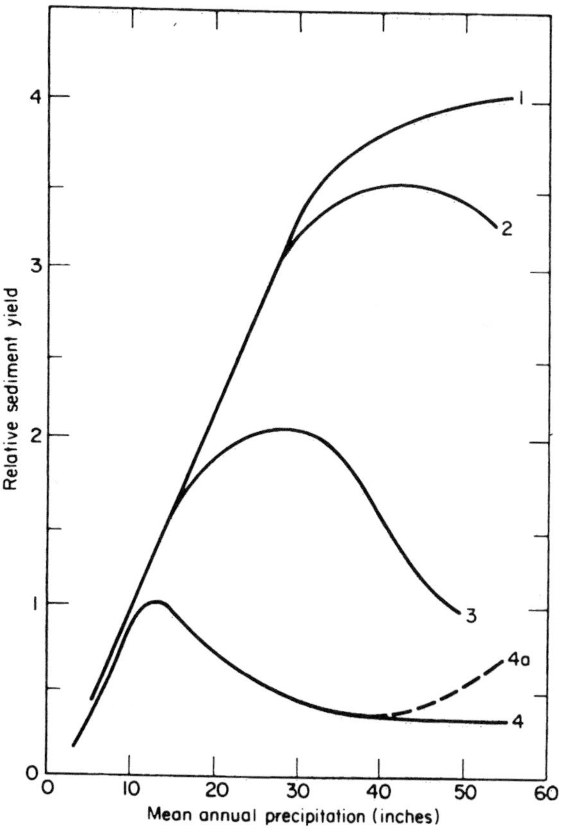

Fig. 8.2. Hypothetical suite of curves illustrating the relationship between precipitation and sediment yield during geologic time. *1* Before the appearance of land vegetation; *2* after the appearance of primitive vegetation; *3* after the appearance of flowering plants and conifers; *4* after the appearance of grasses; *4a* increase in sediment yield for tropical monsoonal climates. (Schumm 1968a)

ried out a test of this proposal, using the Paleozoic fluvial sedimentary record of the Appalachian basin as his data base. He was able, from a limited number of case studies, to confirm that pre-Devonian fluvial deposits in this basin were formed in braided fluvial environments, whereas Devonian and Carboniferous deposits recorded the appearance of meandering fluvial styles. Long (1978) confirmed that most Proterozoic fluvial deposits were formed in braided environments.

Bridge (1985, 1993b) disagreed with the idea that discharge variability is a primary control on braiding, pointing out that "most channel patterns can be formed in laboratory flumes at constant discharge, and many rivers with a given discharge regime show downstream changes in pattern." However, downstream changes in channel pattern may be caused by variations in bank erodibility and consequent local changes in bed load, as suggested by the work of Carson (1984a,b,c). Carson's work, together with that of Schumm (1969, 1981, 1985b) on river metamorphosis, suggests that it is average peak discharge and the rare violent flood that may determine channel patterns. Thus, it may not be discharge variability, as such, that determines channel style, but the fact that the channel-forming high-discharge events that lead to rapid bank erosion and macroform evolution have the highest stream power in rivers with the most variable discharge.

A useful idealized model for the evolution of different channel styles was developed by Bridge (1985), based on much experimental and theoretical work (Fig. 8.3). It has been known since the early 1960s that flow around a bend leads to a pressure setup at the cutbank, and a helical overturn pattern (Sect. 6.7; Figs. 2.15, 2.20). Flow velocity and shear stress increase transversely across the convex side of the bend, toward the thalweg, which, downstream, progressively shifts to the outer side of the bend at the point of maximum curvature. The transverse variation in velocity leads to an obliquity in the advance of bedforms, as shown in Fig. 8.4. Essen-

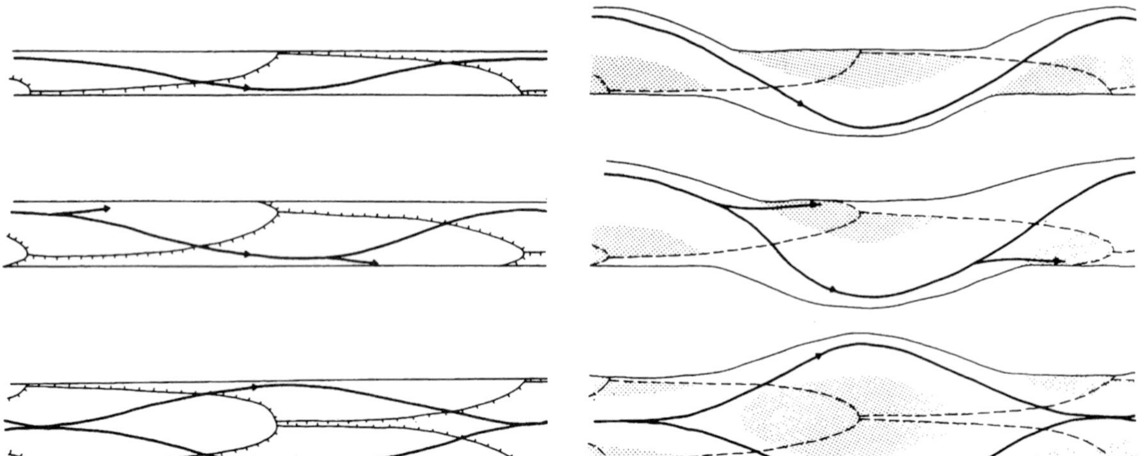

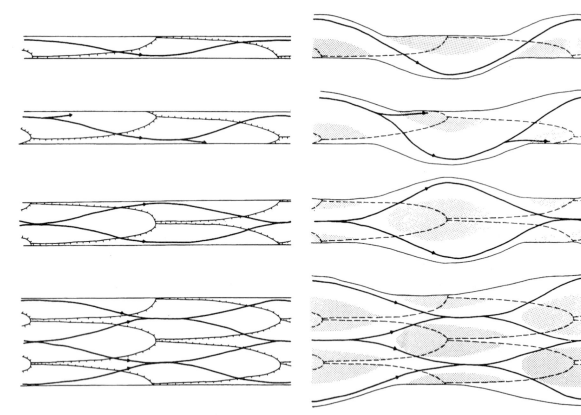

Fig. 8.3. Idealized model for the early evolution of different channel patterns (*right*) from straight channels of different widths (*left*). Heavy curved lines show location of thalweg, *lines with ticks* show original position of alternate bars, and these are shown as remnants by *dashed lines* in diagrams at *right*. *Stippled areas* are topographic highs (macroforms). Sinuous thalwegs lead to the development of rows of alternate bars. Narrow channels with single rows of bars evolve into wandering, and then into meandering channels with point bars. Wider channels develop mid-channel bars and become braided. (Bridge 1985)

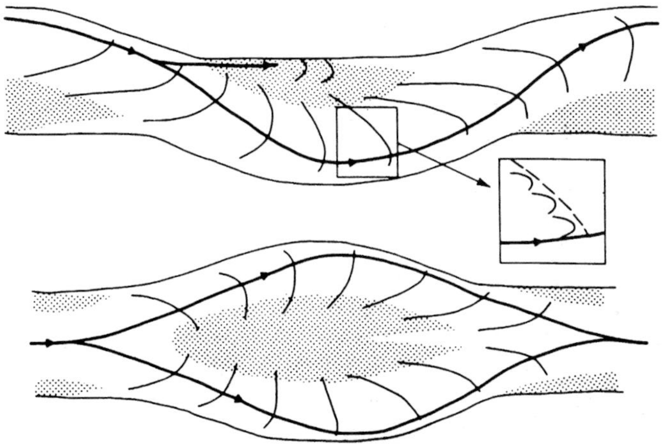

Fig. 8.4. Idealized patterns of flow velocity and bedform orientation in fluvial channels. *Symbols* as in Fig. 8.3, with the addition of *short curved lines* transverse to flow that show schematically the orientation of dune crest lines. This model mimics the development of scroll "bars" on point bars (*above*) and cross-channel "bars" in sandy braided systems (*below*). (Bridge 1985)

tially, the same model may be applied to the development of midchannel braid bars, such as in the sandflat model of Cant and Walker (1978; Fig. 6.26). The basic similarities in flow patterns, bedform configurations, and bar development at channel bends in all rivers (Fig. 8.4) indicate that studies of the geological results of these processes in small outcrops cannot hope to lead to useful conclusions about fluvial style, a point now made by many writers (e.g., Miall 1980; Bridge 1985; see also Sect. 2.4.2). For example, the identification of DA or LA units and the description of their internal architecture cannot, by itself, indicate channel sinuosity or channel multiplicity. Bridge's conceptual models are developed further in a later paper (Bridge 1993b), in which detailed speculations are provided regarding the lithofacies successions and architectural construction to be expected in a set of basic evolutionary cases for simple channel-macroform configurations. Development of a complete model of a given ancient river depends on a thorough architectural analysis of the sediments, incorporating local and regional stratigraphic, facies, architectural-element, and paleocurrent data (Chap. 4).

The unifying models of Bridge (1985, 1993b) provide a useful counterweight to the increasing complexities introduced by the facies-models approach to fluvial sedimentology. To a considerable extent we are prisoners of our terminology. The different names that have evolved for each fluvial style tend to emphasize their differences rather than their similarities. The flume runs of Ashmore (1991) demonstrated the arbitrary nature of some of the terminology. As shown in Fig. 6.13 gravel-bed "braided" rivers can evolve from "meandering" rivers by the development of simple chute channel across the inside of a point bar. The outer part of the bar, once isolated, evolves into a typical midchannel macroform. As noted by Bridge (1985), "channel bends with chute channels are braided channels."

The relationships between sediment load and channel form were explored by Schumm (1963, and reproduced in much of his later work, e.g., 1981, 1985b) in a widely used channel classification (Table 8.1, Fig. 8.5). This classification summarizes many of the most important trends, for example, that rivers transporting a coarse sediment load are more likely to be of low-sinuosity, multiple-channel type, and that they are more unstable and more prone to channel avulsion than are rivers with a fine-grained, suspended load. However, the classification is simplified, as shown in Fig. 8.6. Straight channels can occur where the sediment load is dominantly of bed-load, mixed-load, or suspended load type. Braided channels are of bed-load or mixed-load type.

Rust (1978a) showed that two simple parameters could be used to define channel style for most rivers. These are channel sinuosity, P, and a braiding parameter. The latter expresses the number of bars or islands per meander wavelength of the river. Rust (1978a) adapted this second parameter from the "braiding index" of Brice (1964). He pointed out that simply counting bars and islands is highly dependent on the stage of the river. During full flood, most topographically elevated areas of the river may be under water, resulting in a braiding index of near zero. He suggested that braids be defined by mapping the channels that surround them at low water. With these two measures Rust (1978a) erected a fourfold classification (Table 8.2) that corresponds to the subdivision referred to at the beginning of this section (Fig. 2.18). Friend and Sinha (1993) proposed a modified braiding index, and suggested natural ranges for braiding and sinuosity, based on measurements in Indian rivers.

Table 8.1. Classification of alluvial channels by sediment load. (Schumm 1963, 1985b)

Type of channel	Bed load (% of total load)	Single-channel systems	Multiple-channel systems
Suspended load	< 3	Suspended-load channel W/D ratio < 10, sinuosity > 2.0, gradient relatively gentle	Anastomosing system
Mixed load	3–11	Mixed-load channel W/D ratio 10–40, sinuosity < 2.0, gradient moderate, may be braided	Delta distributaries Alluvial plain distributaries
Bed load	> 11	Bed-load channel W/D ratio > 40, sinuosity < 1.3, gradient relatively steep, may be braided	Alluvial fan distributaries

W/D=width/depth ratio.

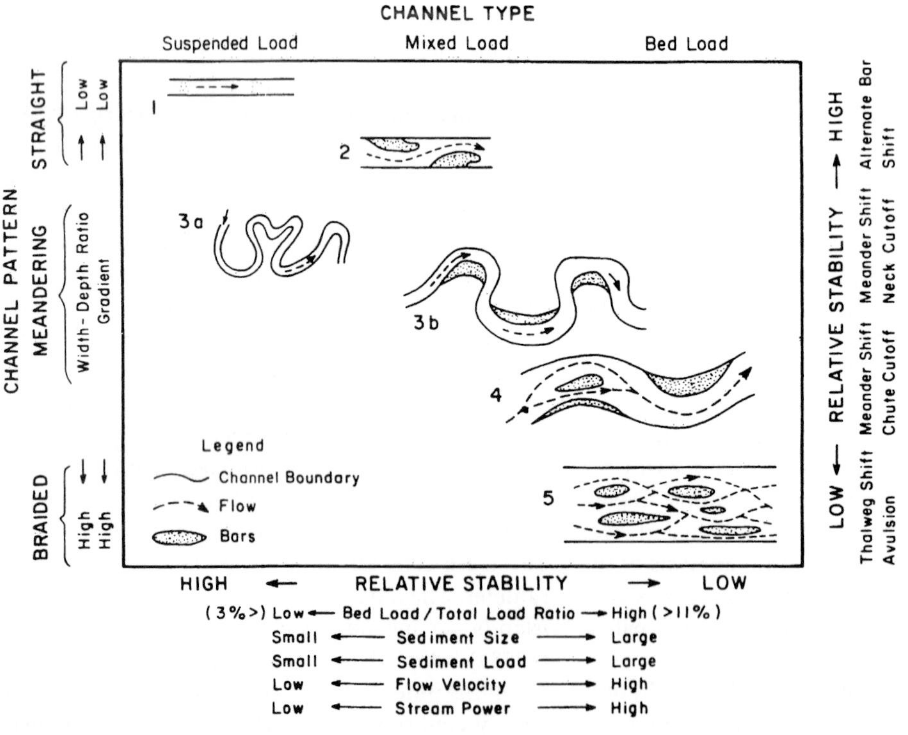

Fig. 8.5. A simple channel classification based on sediment load. Variations in some other variables are shown. (Schumm 1981, 1985b; Reproduced, with permission, from the Annual Review of Earth and Planetary Sciences, v. 13, © 1985, Annual Reviews Inc.)

The cutoffs used in Table 8.2 (braiding parameter=1, sinuosity=1.5) were suggested by Rust (1978a) based on a review of modern rivers and some earlier literature discussion to which he made reference. It is suggested here that they need revision. In particular, a type of gravel-bed river termed "wandering" has now been recognized that is intermediate between braided and meandering. Church (1983) used the Bella Coola River, British Columbia, as his type example. Study of his maps indicates that the river has a braiding parameter of 2.8. A wandering reach of the Squamish River, British Columbia, has a braiding parameter of 2.5 (Brierley and Hickin 1991). The braided rivers illustrated by Rust (1978a) all have braiding parameters greater than 8. It is suggested here that a range of 1.0–3.0 be used to define the "wandering" category. Similarly, it is suggested that a range of P=1.2–1.5 be used to define the

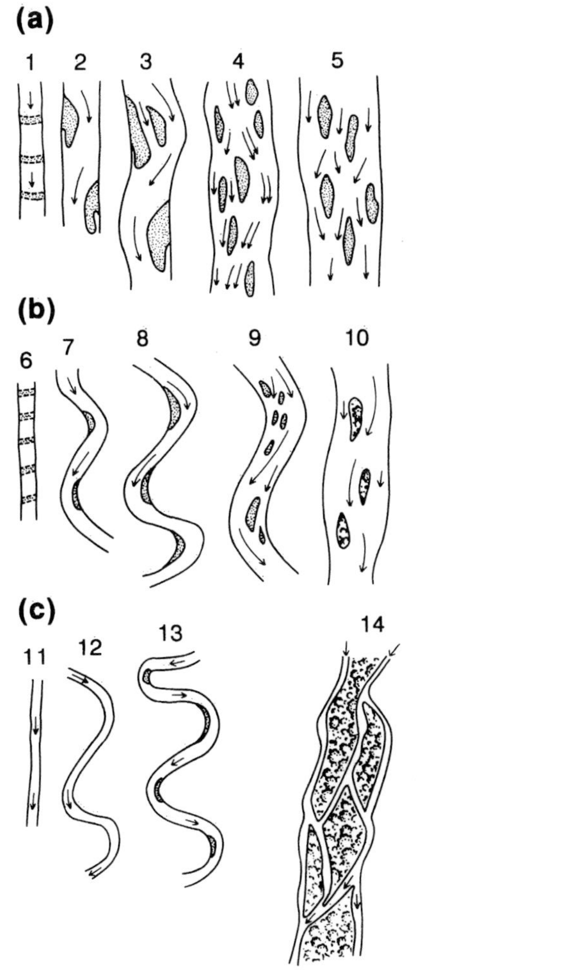

Fig. 8.6. The range of alluvial channel patterns. **a** bed load channels; **b** mixed-load channels; **c** suspended-load channels. (Schumm 1981)

intermediate sinuosity range. Lowest sinuosities are those of "straight" rivers. Schumm and Khan (1972) showed that this condition develops at very low slopes (Fig. 8.7), such as near the mouths of some delta distributaries.

For a variety of reasons, the geometric and hydraulic characters of a river may not precisely reflect local conditions. For example, local bedrock controls may slow down erosion and set back the response of a river to climate change. Many rivers are "underfit", that is, their discharge is too small to be correlated with existing channel characteristics (Dury 1964). For example, many rivers have yet to adjust to the reduced discharge that accompanied the postglacial climate changes of the Holocene. In the case of many of the world's very large rivers, the sediment load and discharge may partially reflect the climate, relief, and sediment-yield characteristics of a source area many hundreds of kilometers distant (Potter 1978). For example, the Ganges, Brahmaputra, and Mekong, all of which are partly braided, have headwaters located in the Himalayan Mountains. Rivers in the Amazon Basin show widely varying fluvial style, reflecting the location of their catchment areas. Those heading in the high Amazon are characterized by strongly seasonal discharge and derive sediment from relict Pleistocene fluvioglacial deposits. They are of moderate sinuosity and partially braided. Rivers sourced within the lowland areas carry much smaller sediment loads and are highly sinuous streams (Baker 1978a). Climatic controls on channel patterns and fluvial style are discussed at greater length in Chap. 12.

Several channel patterns illustrated in Fig. 8.6 are not amongst the classic suite of fluvial styles (the four that became most well known in the 1970s; Table 8.2), but are receiving greater attention as sedimentological techniques improve. For example, the "wandering" style (pattern 4 in Fig. 8.5, pattern 3 in Fig. 8.6) is intermediate between braided and meandering (Church 1983). Such rivers have a single, relatively stable, dominant channel of intermediate to high sinuosity, but also contain short braided reaches with bars and islands. This style is shown by some gravel-bed rivers, as discussed later in this chapter. The causes of the wandering style are not clear, although Brierley and Hickin (1991) described a downstream change within a single river from braided to wandering to meandering as slope and sediment grain-size decrease. It may, therefore, simply represent a transitional condition.

Pattern 2 in Fig. 8.6 shows a straight channel with a sinuous thalweg, with alternate bars developing on the insides of the meanders. This pattern occurs in delta distributaries of low slope. Crowley (1983) also suggested that it develops as an intermediate form between low-sinuosity braided and meandering systems in some sand-bed rivers. The major depositional forms are alternate bars, the characteristics of which are discussed in Sect. 6.7. The overall fluvial style is described later in this chapter.

Pattern 14 in Fig. 8.6 is that of the anastomosed river, as described in the definitive work of D.G. Smith (1973, 1983) and D.G. Smith and N.D. Smith (1980). These rivers are characterized by a network of relatively stable, interconnected channels of low to high sinuosity. They commonly show low slopes and high aggradation rates. D.G. Smith's work suggests that the anastomosed pattern is favored by high rates of aggradation. This can occur where the river flows across a basin that is undergoing rapid subsi-

Table 8.2. Classification of alluvial channels by geometric characteristics. (Rust 1978a)

	Single-channel (braiding parameter < 1)	Multichannel (braiding parameter > 1)
Low sinuosity ($P < 1.5$)	Straight	Braided
High sinuosity ($P > 1.5$)	Meandering	Anastomosing

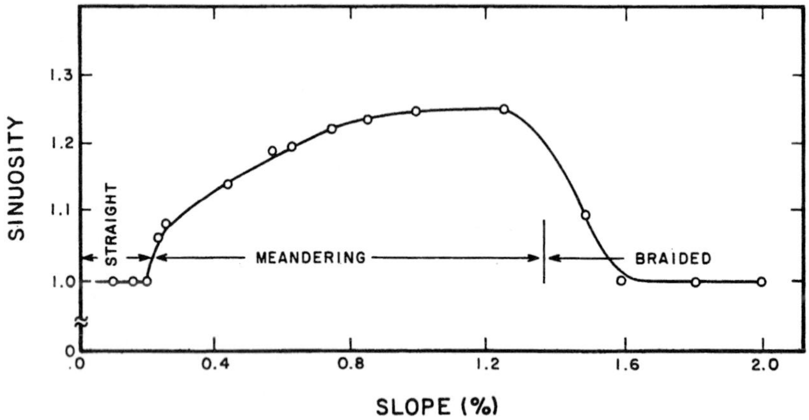

Fig. 8.7. Relationship between slope and sinuosity at constant discharge, based on flume experiments. (Schumm and Khan 1972; Schumm 1985b; Reproduced, with permission, from the Annual Review of Earth and Planetary Sciences, v. 13, © 1985, by Annual Reviews Inc.)

dence, but with a downstream "damming" effect, caused by local bedrock control.

As pointed out by Bristow and Best (1993) studies of channel deposits and dynamics have ranged across at least five orders of magnitude of channel scale, from laboratory models to the largest rivers, such as the Brahmaputra and Mississippi. Channels show a considerable degree of self-similarity across this range of scales, which is why laboratory models are a useful way to explore fluvial processes. Ashley (1990) noted a continuum in the sizes of flow-transverse bedforms, and suggested that bedforms also adjust their scale and geometry to the space and sediment available. Scaled flume models are therefore a useful way to explore fluvial processes, but much work remains to be done to clarify scale considerations in ancient deposits. The same architectural principles can be applied at all field scales, as demonstrated in Chap. 4 (cf. Figs. 4.2, 4.3, and 4.4). Part of the task of architectural analysis of an ancient deposit is to determine the scale of the channels, bars and islands that formed the deposit under consideration. This can be approached by documenting such attributes as the thickness and lateral extent of channel and bar deposits, following the quantitative approaches described in Sect. 10.4.

8.2 Facies Models

Most of the various channel styles shown in Fig. 8.6 have now received considerable attention by sedimentologists, with data collected from both modern and ancient examples. In Miall's (1985) review, fluvial styles were discussed with reference to 12 characteristic examples, which were given model numbers. Additional work has demonstrated the need to recognize a few more categories, for example, in order to distinguish some useful intermediate categories (e.g., the "wandering" style), and in order to clarify some differences in depositional behavior that reflect climatic controls. In this chapter, 16 fluvial styles are described.

As stated in the earlier review (Miall 1985, p. 289):

"It must be emphasized that this section does NOT represent an attempt to provide a comprehensive suite of fluvial models, to replace or add to those already in existence. The purpose is solely to illustrate some of the variablity in style that is possible in natural environments, much of it barely appreciated by sedimentologists. .... Every conceivable gradation between any two of the models illustrated here is to be expected ... the writer accepts no responsibility for any uncritical pigeonholing or force-fitting of field examples into any of these models."

Table 8.3. Architectural characteristics of some common fluvial styles. (After Miall 1985)

Style name	No.[a]	Sinuosity[b]	Braiding parameter[c]	Sediment type	Characteristic elements[d]	Modern examples and interpreted ancient examples
Part 1: Gravel-dominated rivers						
Gravel braided with sediment-gravity flows	1	Low	High	Gravel, minor sand	SG (GB, SB)	Hooke (1967), Bull (1972), Wasson (1977), Nemec and Muszynski (1982), Schultz (1984), Blair and McPherson (1992)
Shallow gravel braided "Scott type"	2	Low	High	Gravel, minor sand	GB (SB)	Boothroyd and Ashley (1975), Boothroyd and Nummedal (1978), Miall and Gibling (1978), Vos and Tankard (1981), Ramos and Sopeña (1983), Dawson and Bryant (1987), Muñoz et al. (1992)
Deep gravel braided "Donjek type"	3	Low to int.	Int. to high	Gravel, minor sand, fines	GB, SB, DA (FF)	Williams and Rust (1969), Rust (1972, 1978b), Steel (1974), Minter (1978), Massari (1983), Miall (1984b), Morison and Hein (1987), S.A. Smith (1990), Reinfelds and Nanson (1993)
Gravel wandering	–	Int. to high	Int.	Gravel, minor sand, fines	GB, DA, LA (SB, FF)	Church (1983), Ferguson and Werrity (1983), Desloges and Church (1987), Billi et al. (1987), S.A. Smith (1990), Brierley and Hickin (1991)
Gravel meandering	4	High	Low to int.	Gravel, minor sand, fines	GB, LA, FF (SB)	Bluck (1971), Lewin (1976), Gustavson (1978), Schwartz (1978), Ori (1979, 1982), Arche (1983), Ramos and Sopeña (1983), Forbes (1983), Hooke (1986)
Part 2: Sand-dominated, high-sinuosity rivers						
Gravel-sand meandering (the "coarse-grained meandering" model)	5	Int. to high	Low to int.	Sand, pebbly	SB, LA, FF (GB)	Bernard and Major (1963), McGowen and Garner (1970), Bridge and Jarvis (1976), Jackson (1976a,b,d), Levey (1978), Nijman and Puigdefábregas (1978), Crowley (1983), Allen (1983a), Campbell and Hendry (1987), Todd and Went (1991)
Sandy meandering (the "classic" meandering model)	6	High	Low	Sand, minor fines	LA, SB, FF LV, CR, CS FF(CH)	Sundborg (1956), Allen (1963b, 1970b), Davies (1966), Beutner et al. (1967), Puigdefábregas (1973), Nami and Leeder (1978), Puigdefábregas and Van Vliet (1978), Ethridge et al. (1981), Hobday et al. (1981), Plint (1983), Bridge and Diemer (1983), Link (1984), Farrell (1987), Alexander (1992), Jordan and Pryor (1992), Diemer and Belt (1991), Muñoz et al. (1992), Alexander and Gawthorpe (1993)
Ephemeral sandy-meandering		High	Low	Sand, minor fines	LA, SB (FF)	Stear (1983), Shepherd (1987), Lawrence and Williams (1987)
Fine-grained meandering	7	High	Low	Fine sand, silt, mud	LA, SB, FF LV, CR, CS FF(CH)	Taylor and Woodyer (1978), Miall (1979a), Nanson (1980), Jackson (1981), Stewart (1983), Mossop and Flach (1983), D.G. Smith (1987), Wood (1989)

Table 8.3 (*Contd.*)

Style name	No.[a]	Sinuosity[b]	Braiding parameter[c]	Sediment type	Characteristic elements[d]	Modern examples and interpreted ancient examples
Anastomosed	8	Low to high	High	Sand, fines	SB, CH (LA) FF, LV, CR, CS	Rust (1981), D.G. Smith (1983, 1986), Rust and Legun (1983), Rust et al. (1984), Flores and Hanley (1984), N.D. Smith et al. (1989), Shuster and Steidtmann (1987), Eberth and Miall (1991), Kirschbaum and McCabe (1992)
Part 3: Sand-dominated, low-sinuosity rivers						
Low sinuosity braided-meandering with alternate bars	–	Low	Low	Sand	DA-LA, SB, FF	McCabe (1977), Galloway (1981), Crowley (1983), Okolo (1983), Bridge et al. (1986), Olsen (1988), Wizevich (1992b)
Shallow perennial braided "Platte type"	9	Low to int.	High	Sand	SB (FF)	Miall (1976, 1984b), Blodgett and Stanley (1980), Crowley (1983), Allen (1983a), Smith and Smith (1984), Røe and Hermansen (1993)
Deep perennial braided "S. Saskatchewan type"	10	Low to int.	Int. to high	Sand, minor fines	DA, LA, SB (FF)	Cant and Walker (1978), Kirk (1983), Haszeldine (1983a,b), Allen (1983a), Ramos et al. (1986), Lawrence and Williams (1987), Miall (1988c, 1994), Bristow (1987, 1988, 1993a,b), Wizevich (1992b), Luttrell (1993) Willis (1993a,b)
High-energy, sand-bed braided	–	Low to int.	Int. to high	Sand, minor fines	DA, SB, HO, (FF)	Miall (1988a), Cowan (1991)
Sheetflood distal braided	11	Low	High	Sand, minor fines	SB (FF)	Williams (1971), Miall and Gibling (1978), Parkash et al. (1983), Sneh (1983), Lawrence and Williams (1987), Olsen (1989), Cotter and Graham (1991), Muñoz et al. (1992)
Flashy, ephemeral sheetflood "Bijou Creek type"	12	Low	High	Sand minor fines	LS (FF)	McKee et al. (1967), Miall and Gibling (1978), Rust (1978b), Tunbridge (1981, 1984), Sneh (1983), Miall (1984b), Lawrence and Williams (1987), Langford and Bracken (1987), Mertz and Hubert (1990), Cotter and Graham (1991), Bromley (1991a)

[a] Model numbers from Miall (1985).
[b] Suggested sinuosity (P) values: low < 1.2, intermediate 1.2–1.5, high > 1.5.
[c] Suggested braiding parameter values: low < 1, intermediate 1–3, high > 3.
[d] Elements shown in brackets are minor components. Element CH is shown only where it forms a significant proportion of the assemblage, and cannot be subdivided into component elements.

A summary of the models and the data base from which they have been built is given in Table 8.3.

Nanson and Croke (1992) provided a review of fluvial floodplains from a geomorphological perspective (although they quoted much of the sedimentological literature), and developed a classification that is very similar to that used here. Their focus was on surface processes, and their emphasis was on floodplain development rather than channel sedimentation, but the convergence of ideas between sedimentology and geomorphology that seems to be evolving is reassuring. Their classification is given in Table 8.4, omitting all the descriptive detail, which would largely repeat what is presented in the rest of this chapter. Class A in this classification deals mainly with upland streams, which do not deposit large, permanent alluvial accumulations. This class of river is therefore of minor importance to sedimen-

Table 8.4. A classification of floodplains. (Nanson and Croke 1992)

Class A: High-energy, noncohesive floodplains
Disequilibrium floodplains which erode in response to extreme events, typically located in steep headwater areas where channel migration is prevented by valley confinement

- A1 Confined, coarse-textured floodplains
- A2 Confined, vertical-accretion floodplains
- A3 Unconfined, vertical-accretion sandy floodplains
- A4 Cut-and-fill floodplains

Class B: Medium-energy, noncohesive floodplains
Equilibrium floodplains formed by regular flow events in relatively unconfined valleys

- B1 Braided river floodplains
- B2 Wandering gravel-bed river floodplains
- B3 Meandering river, lateral-migration floodplains
  - B3a Lateral-migration, nonscrolled floodplains
  - B3b Lateral-migration, scrolled floodplains
  - B3c Lateral-migration/backswamp floodplains
  - B3d Lateral migration, counterpoint floodplains

Class C: Low-energy, cohesive floodplains
Floodplains formed by regular flow events along laterally stable, single-thread or anastomosing low-gradient channels

- C1 Laterally stable, single-channel floodplains
- C2 Anastomosing river floodplains
  - C2a Anastomosing river, organic-rich floodplains
  - C2b Anastomosing river, inorganic floodplains

tologists. Classes B and C encompass rivers that develop broad floodplains underlain by significant alluvial accumulations, and are therefore the rivers of interest to sedimentologists. The subdivision of these two classes corresponds closely to the classification used to subdivide fluvial styles in the following sections of this chapter. The subdivisions of class B3, the floodplains of meandering rivers, reflect the style of floodplain sedimentation, which in part reflects variations in sediment grain size. These subclasses are therefore similar to the subclasses of meandering river defined in Table 8.3 (part 2). Other floodplain detail is discussed in Chap. 7.

Sixteen examples of fluvial architectural style are reviewed in the following sections. Each of these is a "facies model" in the sense used by Walker (1984) and Reading (1986). That is, it is a summary of a particular environment, in which local details have been distilled away, leaving the "pure essence" of the environment, its facies, and its architecture (see Miall 1990; Chap. 4, for a complete discussion of facies analysis methods). The resulting summary acts as a norm for purposes of comparison, and as a framework and guide for future observation. As Dott and Bourgeois (1983) remarked, fluvial facies models have "multiplied like rabbits", and this undoubtedly makes facies studies more difficult. However, it also makes them more realistic, as use can now be made of ever more refined observational details to reconstruct local variations in fluvial style, and the subtleties of their dependency on slight changes in depositional controls, including tectonism, climate, base-level change, source-area geology, vegetation, the effects of human interference, and so on. Walker (1990) suggested that the retreat from a limited suite of end-member facies models to a multiplicity of models based on shifting assemblages of architectural elements would lead to sedimentological anarchy. However, as attempts have been made to demonstrate in earlier chapters of this book, facies and element analysis serve to simplify and codify the study of complex real-life fluvial assemblages.

Earlier workers expressed concern that the accretionary geometries of large bars would be difficult to identify in the geological record. They were concerned mainly with the study of point bars, but the same concerns apply to all macroforms. Thus, Collinson (1986) suggested that the scour and-fill associated with the transport of coarse bed load, and the presence of numerous bedforms on the point-bar surface, would obscure the presence of lateral-accretion surfaces and render point bars difficult to identify. It has also been argued that because the dip of these surfaces is dependent on the width/depth ratio of the river, the low accretionary dips would make it difficult to recognize accretionary geometries in large, wide rivers (Leeder 1973). As the discussion in the subsequent sections should make clear, recent work has resulted in numerous documented examples of macroforms in a wide range of fluvial settings. Nevertheless, the concerns quoted above are real. The best architectural methods may not be able to identify accretionary geometries where exposures are inadequate, for example, where an outcrop is oriented parallel to the strike of gently dipping accretion surfaces. It remains very difficult to document macroforms in the subsurface.

It is important to note that almost all the developments that have taken place in the documentation of the two- and three-dimensional architecture of fluvial systems have relied on studies of the ancient record, and several of the fluvial styles described in this section are based almost entirely on studies of the ancient. There is a pressing need for research into the internal architecture of modern river deposits, preferably using high-resolution seismic or ground-penetrating radar methods, such as are described in Sect. 9.5. Only work of this type can provide the undisputed link between surface processes and the preserved deposits that is necessary for the reliable definition of process-response models. What is reported here is therefore work in progress toward that goal.

One important test of the architectural method in a modern river was carried out by Brierley and Hickin (1991). They documented the facies sequences and architectural elements in bank exposures of a modern gravel-sand river in British Columbia that undergoes a downstream style change from braided to wandering to meandering. Facies successions and architectural cross sections were similar for all three fluvial styles, and the authors concluded that these kinds of data would not be able to provide predictions of fluvial style in other comparable fluvial settings. It is possible to question their results because the river is undergoing rapid evolution, and the surface channel style may not be an accurate representation of the fluvial style at the time the sampled channel and floodplain assemblages were deposited. Nevertheless, their conclusion might seem disturbing. In fact, their results bear out the point this writer has been trying to make, that it is only by carrying out a detailed architectural documentation that the researcher can hope to arrive at precise and accurate interpretations. In the case of Brierley and Hickin's (1991) study, their work did not include paleocurrent analysis, which would provide vital data for the interpretation of channel sinuosity and the range of channel and bar orientations – two of the key differences between meandering, wandering, and braided gravel-bed rivers. As noted elsewhere in this book, fluvial style should not be interpreted from single exposures of one channel-fill succession, the method tested by Brierley and Hickin (1991), but should be based on three-dimensional data from several successions. Such a data base would provide a more quantitative measure of the relative importance of the various facies and the types of channel and bar present in the river.

Several example of intermediate fluvial styles are referred to in the succeeding sections. The "wandering gravel-bed" river and the "low-sinuosity braided-meandering river with alternate bars" are by definition intermediate fluvial styles. Reference is also made to transitions between the "high-energy sand-bed braided" and the "flashy, ephemeral sheetflood" styles, and between the latter and the "sheetflood distal braided style". Differences between some of these fluvial styles are small, and it could be argued that no purpose is served by proposing so many "models". The point in doing so is simply to recognize the real-life variability of natural systems and to encourage flexibility in their interpretation. Many actual examples of fluvial successions will display characteristics of more than one of the models described here. This is not "sedimentological anarchy", as suggested by Walker (1990), but a recognition of the complex reality of fluvial systems. Brierley (1993) refers to element analysis as a "constructivist" approach to facies studies. Observation and measurement of two- and three-dimensional outcrops using the architectural-element approach provide a data base from which measurements and calculations can be made concerning channel dimensions, flow characteristics, etc. In this way, each river and each fluvial deposit are treated as an individual, possibly unique entity. End-member force-fitting, which is what tends to occur when few models are available, is therefore avoided.

An excellent example of an ancient fluvial system that does not fit any of the "standard" models (including those described below), and yet contains many features tantalizingly similar to several of them, is the Escanilla Formation (Eocene) of the Spanish Pyrenees. A regional architectural reconstruction is provided by Bentham et al. (1993), and additional details of the gravel and sand bodies are described by Dreyer et al. (1993). This unit consists of poorly interconnected sheet sandstones and conglomerates, separated by fine-grained sediments constituting more than 40% by volume of the total

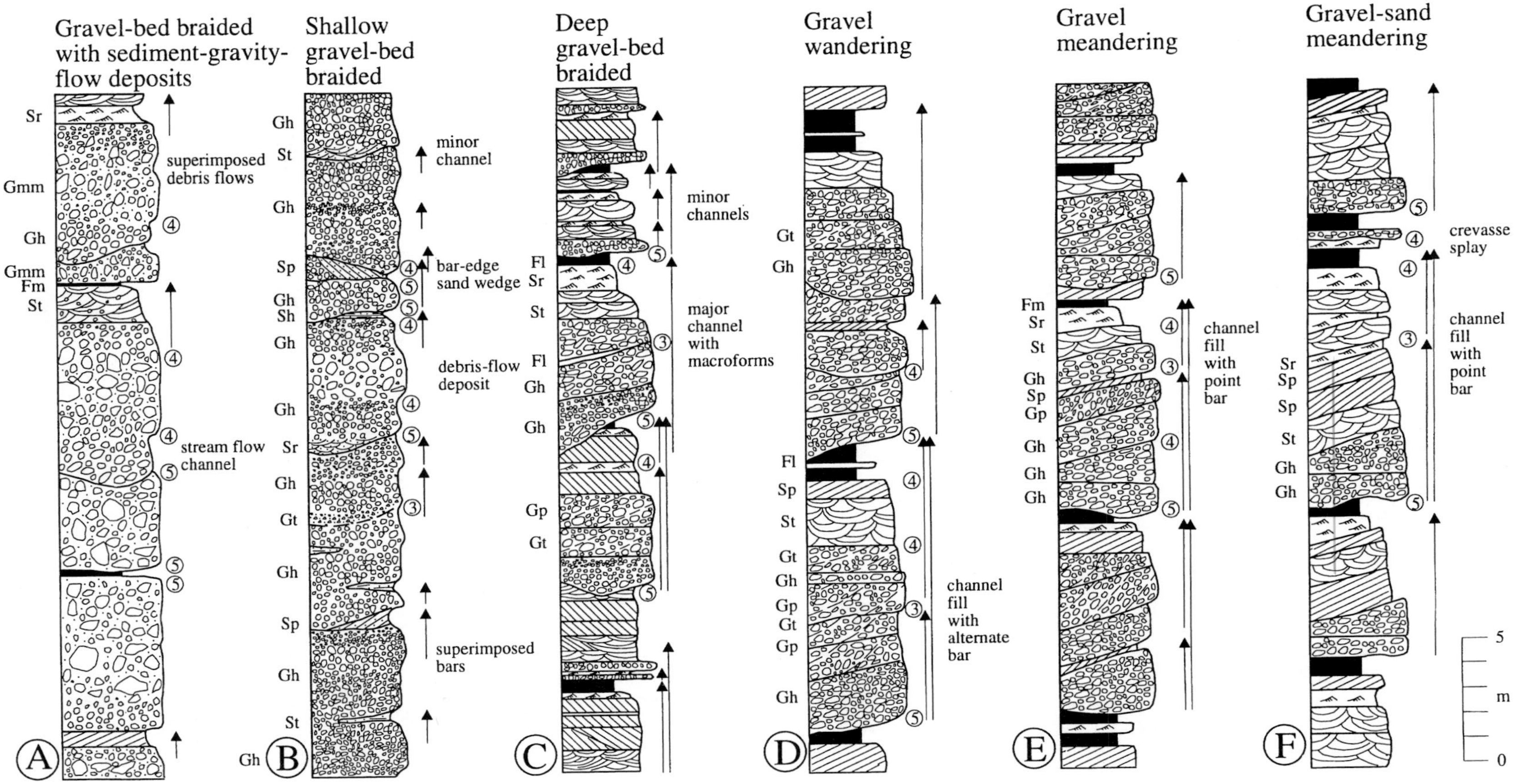

Fig. 8.8A–P. Typical fluvial lithofacies assemblages and vertical profiles for the 16 fluvial styles discussed in Chap. 8. *Vertical arrows* indicate cyclic successions of various types, showing direction of fining and bed thinning. *Numbers in circles* indicate rank of bounding surfaces. Lithofacies codes are given *at left* of column. (Adapted from Miall 1977, 1978b)

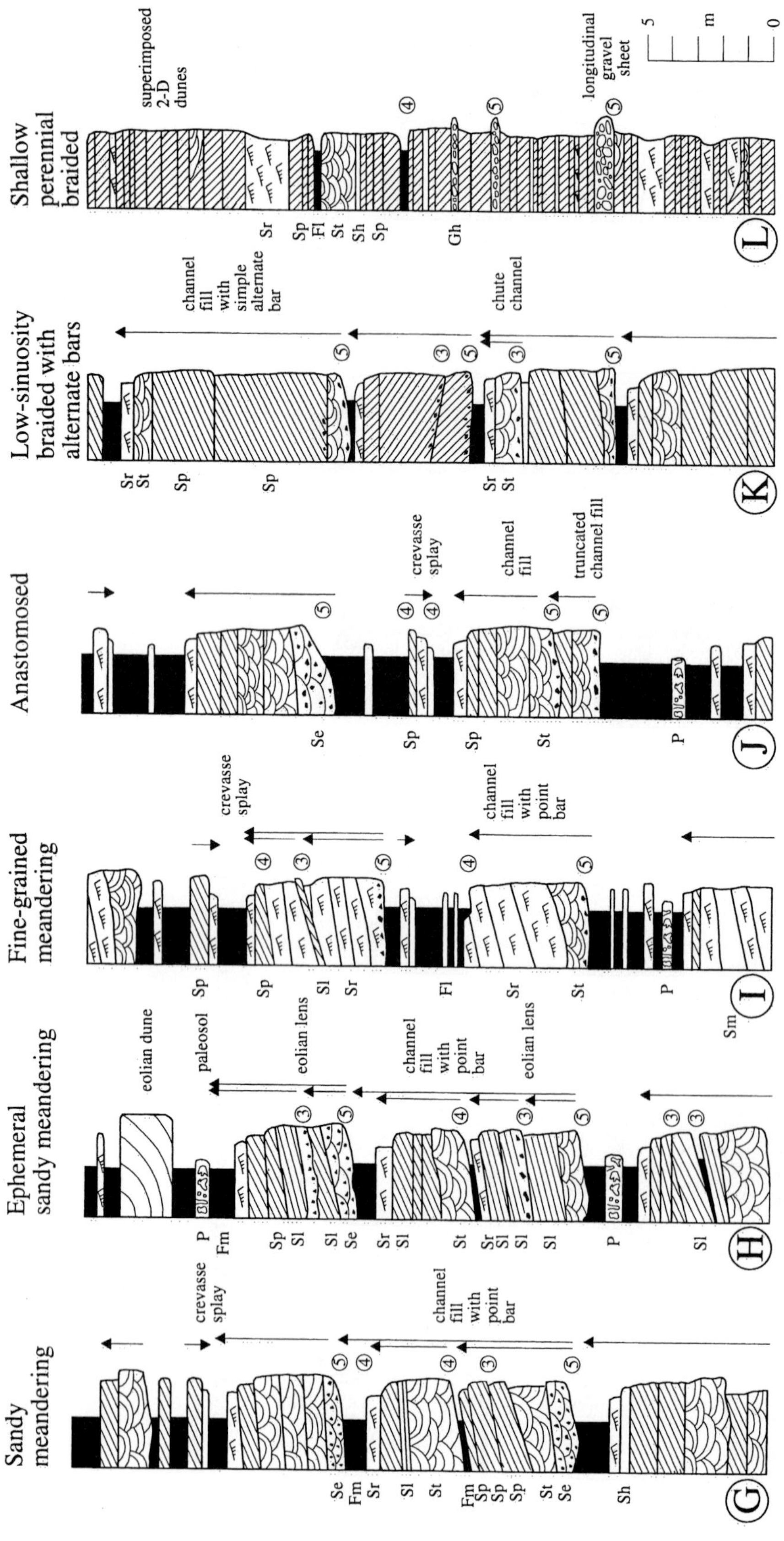
Sandy meandering
crevasse splay
channel fill with point bar
Se Fm Sr Sl St Fm Sp Sp Sp St Se Sh
G
Ephemeral sandy meandering
eolian dune
paleosol
eolian lens
channel fill with point bar
eolian lens
P Fm Sp Sl Sl Se Sr Sl St Sr Sl Sl Sl P Sl
H
Fine-grained meandering
crevasse splay
channel fill with point bar
Sp Sp Sl Sr Fl Sr St P Sm
I
Anastomosed
crevasse splay
channel fill
truncated channel fill
Se Sp Sp St P
J
Low-sinuosity braided with alternate bars
channel fill with simple alternate bar
chute channel
Sr St Sp Sp Sr St
K
Shallow perennial braided
superimposed 2-D dunes
longitudinal gravel sheet
Sr Sp Fl St Sh Sp Gh
L
5
m
0

Fig. 8.8G–L

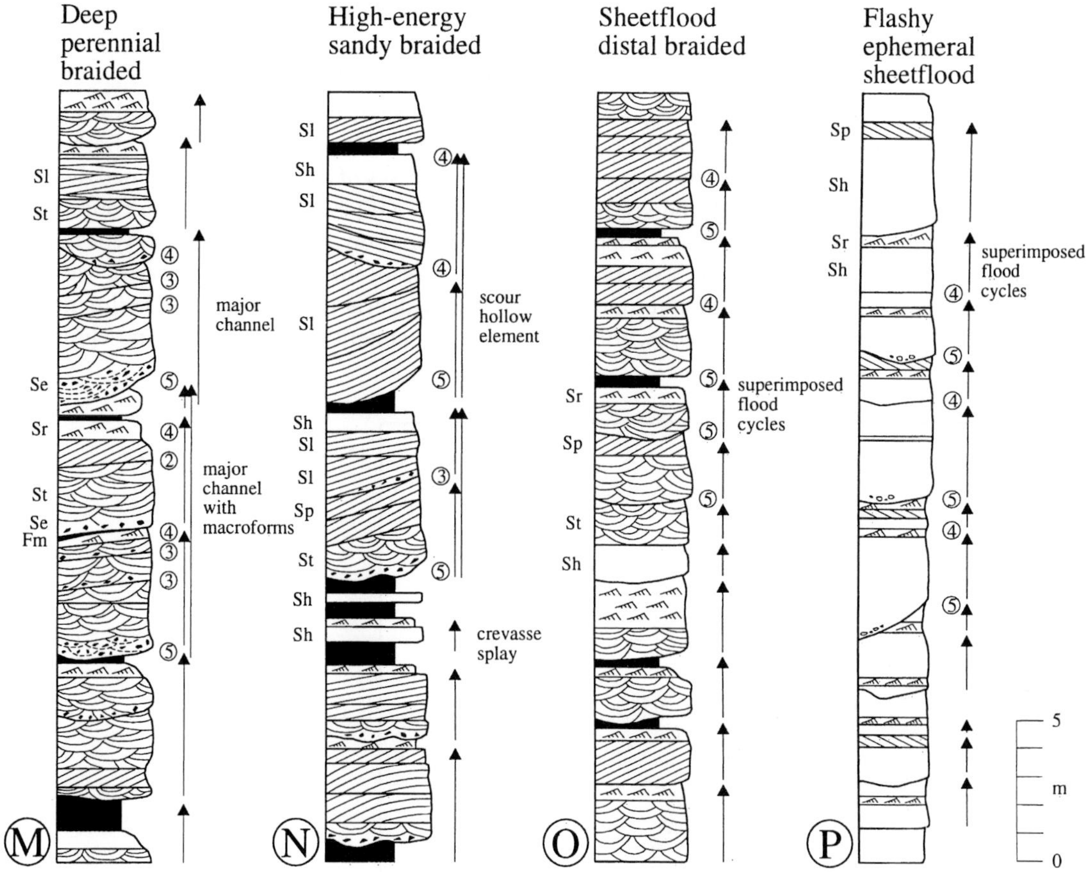

Fig. 8.8M–P

formation. Bentham et al. (1993) pointed out the superficial similarity of this broad architectural style to that of anastomosing rivers, but they concluded that the geometry and internal features of the coarse members indicated a braided style. Many of these consist of simple conglomerate and sandstone sheets (elements GB, SB), comparable to those forming in shallow, gravel-bed braided rivers (Sect. 8.2.2) and shallow, perennial, sand-bed braided rivers (Sect. 8.2.12). Ribbon bodies, and sheets with lateral accretion bedding, indicating sinuous channel reaches (compare with the sand-bed meandering river of Sect. 8.2.7), are also present. This unusual mix of coarse-member architectures is contained within thick floodplain deposits, itself an unusual feature of a dominantly braided system. Bentham et al. (1993) attributed this last feature to rapid sedimentation and frequent channel avulsion in an unconfined basin.

Some characteristic fluvial assemblages are illustrated in Fig. 8.8. Some of these profiles have been adapted from the original model descriptions of Miall (1977, 1978b). Others are new. Each represents an "ideal" profile, and should therefore not be relied upon as a template for interpretation. The local variability in each deposit, and the similarities between many of the vertical sections given in Fig. 8.8, should serve to emphasize the need for two- and three-dimensional architectural analysis in sedimentological studies.

The descriptive sections that comprise the remainder of this chapter follow the organization of Table 8.3. Gravel-bed rivers, the first to be described, are characteristic of proximal, basin-margin settings, including alluvial fans. Sand-bed rivers are typical of alluvial plains, including more distal settings. They are conveniently subdivided into those of high- and low-sinuosity.

8.2.1 Gravel-Bed Braided River with Sediment-Gravity-Flow Deposits

Many basin margins are characterized by wedges of gravel deposits. Commonly, these are deposited by distributary fluvial systems that form the cone-shaped deposits of alluvial fans. Two fluvial styles are typical, one the subject of this section, in which sediment-gravity-flow deposits (element SG), related deposits of high-energy stream flow, and their winnowed derivatives make up the successions, and one (described in Sect. 8.2.2) in which sediment-gravity flows are rare or absent, with sedimentation dominated by element GB (the role of depositional processes and their deposits in the classification of alluvial fans is discussed in Sect. 8.3). In both cases, minor thicknesses of sand and fine-grained deposits (elements SB, OF) represent low-water and overbank sedimentation, respectively. Both types of fluvial style may occur in the same basin, with the frequency of debris flows and other sediment-gravity flows dependent on the geology and climate of

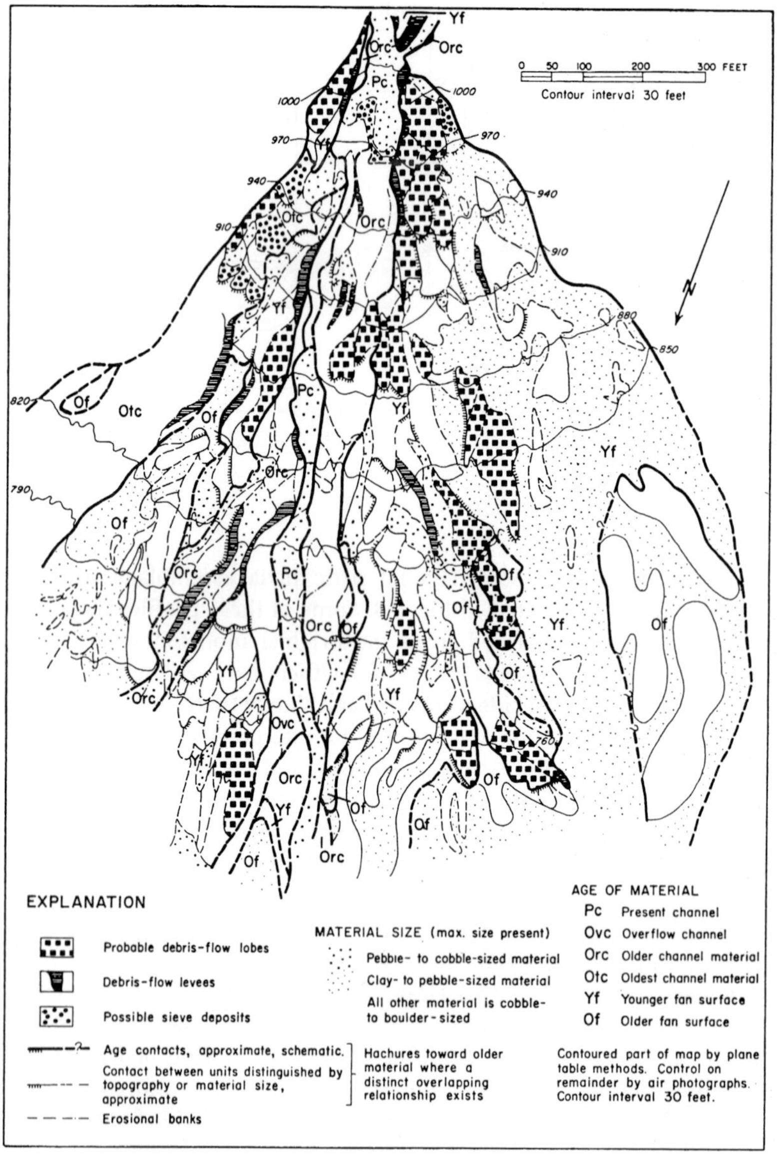

Fig. 8.9. Surface of the Trollheim fan, Death Valley, California, showing the distribution of the major types of fluvial deposit, and the position of the modern fan channel (Hooke 1967). This fan is dominated by sediment-gravity-flow processes (Blair and McPherson 1992), but other fans where this type of flow is important may also be strongly influenced by traction currents and their deposits

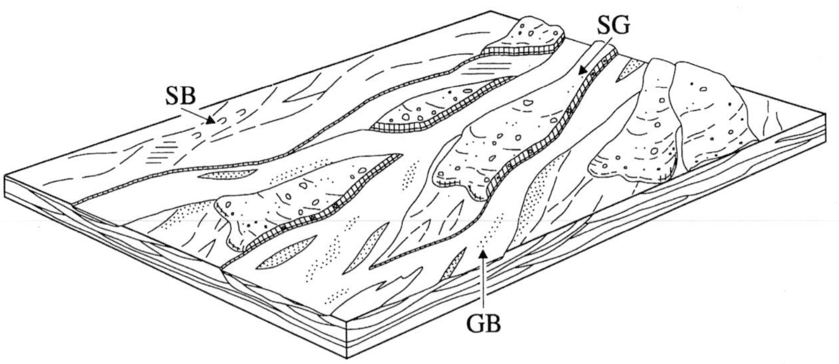

Fig. 8.10. Model of gravel-bed braided river showing dissected lobes of sediment-gravity-flow deposits (element *SG*). Model 1 of Miall (1985)

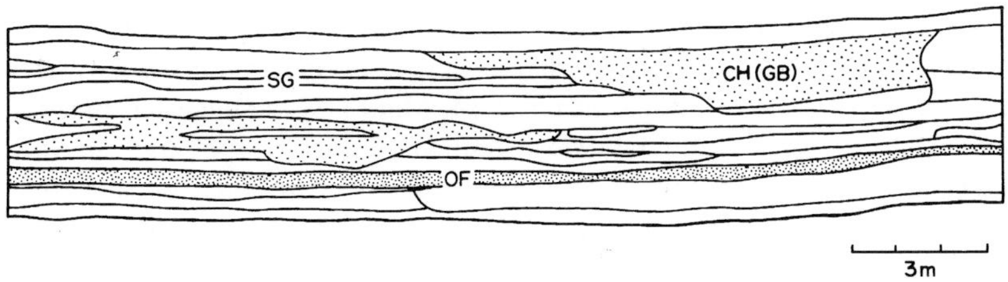

Fig. 8.11. Cross section through postglacial fan deposits, Tasmania, showing interbedding of elements *GB* and *SG*. (Wasson 1977)

the source area. Even adjacent fans may show different fluvial styles because of differences in the source-area geology, or because of differences in forest cover, leading to different weathering rates, and consequent differences in the rate at which loose sediment is supplied to the fluvial system, so that one fan may be formed entirely by traction-current sedimentation, with the next fan significantly affected by sediment-gravity flows (Hooke 1967). Debris flows, and other types of sediment-gravity flow, are particularly common in volcanic areas (e.g., Nemec and Muszynski 1982; G.A. Smith 1987), but the distinctive characteristics of volcaniclastic deposition are not addressed in this book.

Figure 8.9 illustrates the modern Trollheim fan in Death Valley, California, which Miall (1978b) suggested be used as the type example of this style of sedimentation. On the surface of the Trollheim fan there are numerous remnants of debris flow deposits. These are initially lobate in shape, immediately after deposition, but are subject to incision by subsequent steam-flow processes. The edge of each flow tends to be abrupt. They have convex-up margins, somewhat like viscous lava flows. Characteristic lithofacies are Gmm, Gmg, Gci, and Gcm. Other deposits in the Trollheim fan consist of element GB. Figure 8.10 illustrates a fluvial model for this type of channel, and Fig. 8.11 is a cross section through a typical ancient deposit. Note the tabular shape of the SG units, their nonerosional bases, and their curved, convex-up margins. Units composed of element GB typically rest on irregular or channelized, erosional bases.

Blair and McPherson (1992) reexamined the Trollheim fan, the basis for the debris flow-dominated fan model of Miall (1978c) and Collinson (1986). They concluded from the field evidence that the Trollheim fan was formed exclusively from the products of debris flows, with postflow winnowing and reworking processes being responsible for the modification of some units prior to burial and preservation. They suggested that the vertical profile model of Miall (1978c) is in error in including traction-current-deposits, including thicknesses of Gh and various sandy facies, because they did not find these facies in the deposits of the Trollheim fan. However, other descriptions of fan deposits by Bull (1972), Wasson (1977), and Schultz (1984) clearly indicated that these other facies do occur interbedded with debris-flow deposits in modern and ancient alluvial fans. Therefore, the facies model may be incorrect only in the choice of the Trollheim fan as the type example. Figure 8.8A illustrates a vertical profile containing these accessory facies.

Rust and Koster (1984) suggested that the presence of debris-flow deposits should be used as a primary criterion in the recognition of the alluvial fan environment, a proposal favored by Blair and McPherson (1994). However, the work of Hooke (1967), quoted above, on the controls of sediment-gravity-flow processes, suggests that this criterion could produce misleading results. Also, it would exclude other types of depositional systems that have traditionally been included in the fan category. Stanistreet and McCarthy (1993) presented a more inclusive classification of alluvial fans, as discussed further in Sect. 8.3.

Large-scale autocyclic processes that control the development of fourth- and fifth-order lithosomes in alluvial fans are discussed in Sect. 10.2.

8.2.2 Shallow, Gravel-Bed Braided River

Proximal gravel-bed rivers and braid deltas, in which sediment-gravity flows are rare to absent, consist of a shifting network of unstable, low-sinuosity channels in which a variety of gravel bedforms is deposited (Fig. 8.12; Sect. 5.2). Channel depths on the order of 1 m are typical. Channel margins are rarely identifiable in outcrop (Fig. 6.17). Element GB predominates (Sect. 6.3), and consists of tabular bodies with numerous minor internal erosion surfaces, and varying assemblages of gravel traction-current deposits (predominantly lithofacies Gh, Gp, Gt). Channels may be abandoned at low stage, in which case thin lenses and wedges of sand may be deposited (Fig. 8.13), comprising element SB. Thin, sandy, bar-top sheets may be formed, and small delta-like wedges of sand are formed at bar margins by surface runoff. Typically, element SB comprises about 5% of most fluvial successions formed in this type of river (Fig. 8.8B).

An architectural model for this type of river is shown in Fig. 8.14. Miall (1977) suggested that this fluvial style be named the "Scott-type" after the Scott fluvioglacial outwash river in Alaska (Boothroyd and Ashley 1975; Boothroyd and Nummedal 1978). Figures 8.15 and 8.16 are outcrop examples of the resulting deposits. Figure 8.15 is most typical of this fluvial style, showing the characteristic thick, multistory conglomerate deposit formed in gravel-domi-

Fig. 8.12. View of a typical gravel-bed braided river. Athabasca River, Alberta, shown at low-water stage in early spring, prior to main spring runoff

Fig. 8.13. An abandoned channel in a gravel-bed braided river, Athabasca River, Alberta. (Photo courtesy of T. Jerzykiewicz)

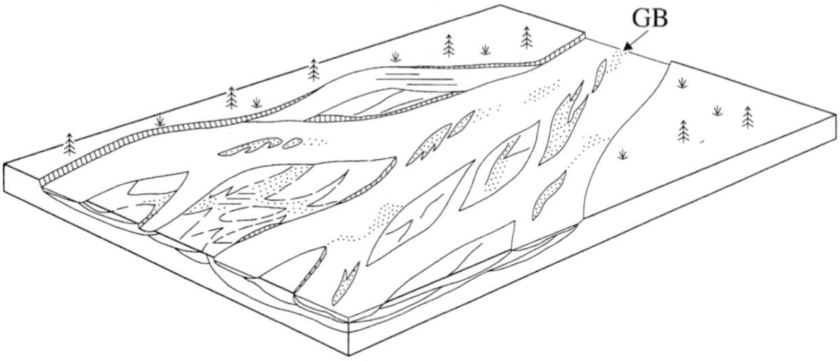

Fig. 8.14. Model of gravel-bed river dominated by traction-current deposits (element *GB*). Model 2 of Miall (1985)

nated alluvial-fan distributaries. In this case, the deposits were probably formed in an arid tropical environment (Miall and Gibling 1978), but comparable deposits are formed in fluvioglacial settings (e.g., Fig. 5.2; see Chap. 12 for a discussion of the climatic significance of the various gravel assemblages). Figure 8.16 is an example in which unusual thicknesses of overbank deposits have been preserved. It seems likely that subsidence was rapid in this foreland basin setting, leading to high rates of deposition and preservation of overbank deposits.

8.2.3 Deep, Gravel-Bed Braided River

In some braided rivers, several distinct topographic levels may be recognized, including major and minor channels, bar surfaces, and the floodplain, spaced over a vertical range of several meters. These levels may represent stages of progressive downcutting by the river, and are most recognizable in areas undergoing active degradation. The Donjek River, Yukon Territory, Canada, is an example of this type (Fig. 8.17A), and was the first to be de-

Fig. 8.15. Typical outcrop of gravel-bed river dominated by element GB. Thin lenses and wedges of SB represent the fill of minor abandoned channels. Peel Sound Formation (Devonian), Somerset Island, Arctic Canada. Scale is 1.5 m

scribed in this way (Williams and Rust 1969, p. 650–652). The floor of the main channel is here at least 3 m below the level of the floodplain. Figure 8.17B illustrates another example, the Rangitata River of South Island. Miall (1977) proposed the name "Donjek type" as a general name for this type of river and its deposits. In rivers undergoing degradation, the topographic levels appear as terraces within the river valley, but similar topographic differentiation occurs in some rivers on open basin plains, for example, in parts of the Scott and Yana outwash "fans" of southern Alaska (Boothroyd and Ashley 1975; Fig. 3), and in some sandur plains (Church and Gilbert 1975, p. 62), where the differentiation is constructional. The lithofacies succession that occurs in this type of river is comparable to those in other deep, gravel-bed rivers as discussed in the sections on those rivers (Sects. 8.2.4, 8.2.5).

Four levels may be recognized in the Donjek River:

Level 1: level of main channels and principal sediment dispersal route. Little or no vegetation, bars exposed only during low stage.

Level 2: level active during flood stages, with few active channels at other times, sparse vegetation cover.

Level 3: little continuous water movement, low-energy flow during flood stage, moderate vegetation cover.

Level 4: dry islands with dense vegetation.

The lowest level is that of the active channel, and is similar in all respects to the gravel-bed rivers of Sect. 8.2.2. Higher levels are active only during flood stages, and characteristically accumulate deposits of element SB. A floodplain may or may not form a significant part of the system, depending on valley width and channel stability. Lateral migration of channels, as for example by distributary shifting on alluvial fans, is accompanied by deposition on the insides of meander bends and the aggradation of abandoned channel reaches, resulting in the superimposition of deposits formed at successively higher topographic levels, and the generation of fining-upward successions (Williams and Rust 1969; Rust 1972). These will be at least as thick as the depth of the channel, but may be greater if they are developed

Fig. 8.16. Outcrop of Molasse Rouge, Provence, France. Stratigraphic top is to the *right*. Lenses of conglomerate represent channel fills formed in gravel-bed river. These lenses are interbedded with thick sandstone-siltstone-mudstone deposits representing rapid floodplain sedimentation

by distributary shifting on a rapidly subsiding fan. As shown by Reinfelds and Nanson (1993), the upper topographic levels of this type of river represent what is normally classified as the floodplain of the river. They may have a fair preservation potential, and accumulate significant, if thin, deposits of fine-grained sediment (Sect. 10.3.2).

Figure 8.18 is an architectural model of this type of river. Element GB is predominant, but large-scale gravel bars forming macroform elements LA or DA may also be present. An increasing predominance of the element LA indicates a gradation into one of the higher sinuosity fluvial styles described in the subsequent sections. Several types of cycle may be present, cycles meters to tens of meters thick representing distributary migration and fan evolution (Sect. 10.2), channel-fill cycles up to a few meters thick (Fig. 8.8C), and those a few decimeters thick representing bar progradation or flood events. Origins of cycles are discussed further in Chap. 10. Figure 8.19 illustrates a conglomerate-sandstone succession interpreted to have formed in such a river. The deposits may show upward fining, as in the stratigraphic section measured at this and an adjacent cliff face (Fig. 8.20). Another example is illustrated in Fig. 8.21. Conglomerate occurs in sheets 1.5–4 m thick, with erosive bases showing up to 2 m of relief (element GB). Tops are esentially flat, but are cut by the scours at the base of the succeeding trough cross-stratified sandstone (element SB). S.A. Smith (1990, p. 212) interpreted the sandstone-conglomerate contact as a fourth-order bounding surface, indicating successive stages of deposition within a single channel.

8.2.4 Gravel-Bed, Wandering River

The definition of this class of river follows from work in the gravel-bed braided rivers of British Columbia, Canada (Church 1983; Desloges and Church 1987; Brierley and Hickin 1991), but comparable rivers have now been studied elsewhere (e.g., Ferguson and Werrity 1983), and it is generally realized that this

A

Fig. 8.17. A Vertical air photo of part of the Donjek River, Yukon Territory, showing the four topographic levels (*1–4*) defined by Williams and Rust (1969). Part of photo A15728-89, National Airphoto Library, Canada. Widest reach of the river is 3 km across. B Oblique aerial view of the Rangitata River, South Island, New Zealand, a typical deep, gravel-bed braided river. Flow is toward the viewer. At least three topographic levels can be defined, based on the density of vegetation. Note alluvial fans entering the river from both sides. (Photograph courtesy of Lloyd Homer, Institute of Geological and Nuclear Sciences Limited, Lower Hutt, New Zealand)

type of river occupies an intermediate class between low-sinuosity, multiple channel rivers - the classic braided river - and high-sinuosity, single channel rivers - the classic meandering river (Fig. 8.22). Sinuosity is intermediate on average (in the range 1.2–1.5), although nearly straight and highly sinuous stretches may occur, and the braiding parameter is also intermediate (in the range 1–3). In places, these rivers have a single channel, in others two or more. Large, bank-attached point-bars or side bars are common. A model for this class of river is shown in Fig. 8.23.

The lithofacies assemblage is similar to that of deep, gravel-bed-braided (Sect. 8.2.3), and gravel-meandering (Sect. 8.2.5) rivers. Channel-fill deposits are dominated by lithofacies Gh, Gp and Gt, with sand lithofacies (St, Sp, Sr, Sh) forming the upper part of bar deposits. Fining-upward successions are typical (Fig. 8.8D). At the architectural-element level, the style is dominated by lateral-accretion deposits (element LA). Slough deposits are floored by gravel and may be filled with massive silt and clay or by cross-bedded gravel or sandstone.

The similarities between the three classes of major gravel-bed rivers (those illustrated in Fig. 8.22), in terms of lithofacies and vertical profile, indicate the need for careful architectural work to distinguish them. Orientation data, as obtained from paleocurrent analysis of cross-bedding and clast imbrication, are critical. An example of such a study was reported by Billi et al. (1987). They showed that gravel bars accreted downstream and developed a model of al-

Fig. 8.17B

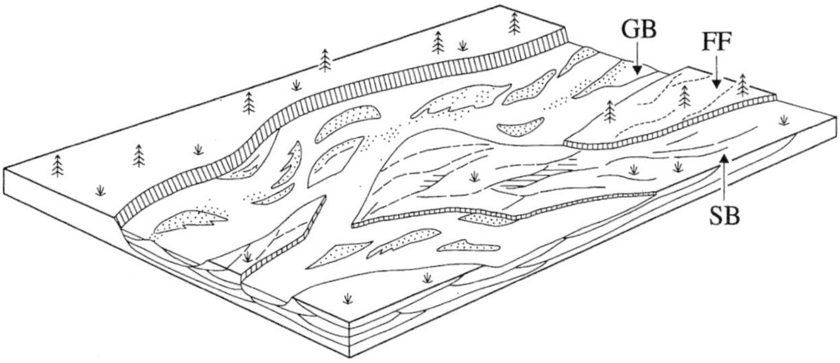

Fig. 8.18. Model of deep, gravel-bed braided river with well-defined topographic levels. Model 3 of Miall (1985)

ternate-bar growth. Where paleocurrent data can be obtained from successive bar deposits, it may be possible to determine sinuosity from the change in orientation of the bars, using the sinuosity equation of Miall (1976).

8.2.5 Gravel-Bed, Meandering River

As shown by Jackson (1978), high-sinuosity rivers range from steep, torrential, gravel-bed rivers, to sluggish, suspended-load streams. Gravel-bed me-

Fig. 8.19. Example of a conglomerate-sandstone sucession formed in a major braided system with topographic levels, probably an alluvial-fan complex. Peel Sound Formation (Devonian), Prince of Wales Island, Arctic

andering rivers have not received much attention by sedimentologists, but increasing use of architectural methods to study conglomerate deposits is showing that they are relatively common in the stratigraphic record (Ori 1979; Arche 1983; Ramos and Sopeña 1983). One of the first studies of this environment was by Bluck (1971), who examined the modern Endrick River, Scotland.

There is typically one main, active channel, with scattered bars and islands, and occasional subsidiary channels (Figs. 8.22, 8.24). New channels are commonly initiated as chute channels, as modeled in the flume experiments by Ashmore (1991; see Fig. 6.13). Sedimentation occurs on large, flat-topped point bars and side-bar complexes. These commonly show a downstream decrease in grain size, with gravel sheets, lobes or dunes at the head, and sand dunes at the tail. Lateral accretion deposits (element LA) are characteristic (Fig. 6.16). Information about the floodplain of this class of rivers is sparse, and is is not known whether crevassing and crevasse splays are common. Arche (1983) recorded the presence of fine-grained deposits in abandoned channels, and thin overbank silt-mud successions, in his study of Quaternary river terraces in Spain. Forbes (1983) showed the modern floodplain of the Babbage River, Yukon, to be composed entirely of fine-grained sediments.

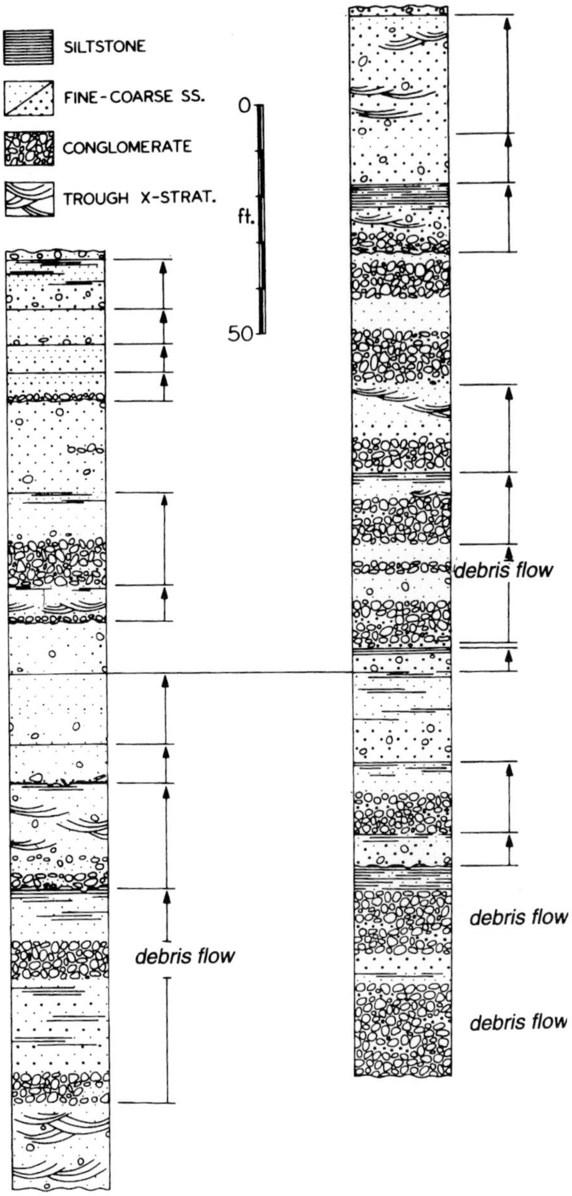

Fig. 8.20. Stratigraphic successions formed in a major braided alluvial-fan complex. One of the outcrops is illustrated in Fig. 8.19. (Miall 1970b, 1973)

An architectural model for this class of river is shown in Fig. 8.25. A characteristic vertical profile is illustrated in Fig. 8.8E, and examples of ancient deposits are illustrated in Figs. 6.37 and 8.26. Note the presence of well-defined lateral-accretion deposits in these exposures. Other examples are illustrated in Figs. 6.16 and 6.35A.

8.2.6 Gravel-Sand Meandering River

Rivers of this type, in the coastal plain of Texas and Louisiana (e.g., Brazos and Amite rivers), were amongst the first to be studied by sedimentologists in the 1950s and 1960s (Fig. 2.16; Bernard et al. 1962; Bernard and Major 1963; McGowen and Garner 1970), and were used in the construction of the classic point-bar model (Fig. 2.20). Other modern rivers of this type include stretches of the Wabash River, Indiana-Illinois (Jackson 1976a,b,d), part of the upper South Platte River, Colorado (Crowley 1983; Fig. 8.27), the upper Congaree River, South Carolina (Levey 1978), and the South Saskatchewan River of east-central Saskatchewan (Campbell and Hendry 1987). They are examples of the so-called coarse-grained meandering streams. They are bed load streams, in which the channel and macroform sediments consist of sand and pebbly sand, or sand with a lag of gravel. Abandoned channels and meander scars are common on the floodplain, and are preserved as clay-silt plugs.

Several excellent studies of hydraulics and sedimentation have been carried out on these rivers. Jackson (1976d) and Levey (1978) demonstrated the development of the spiral or helical flow pattern at a meander bend (Fig. 8.28). This pattern is best developed in bends of intermediate curvature. During high-stage conditions, the channel floor and bar surfaces are covered with numerous bedforms. A fathometer trace along one meander wavelength of

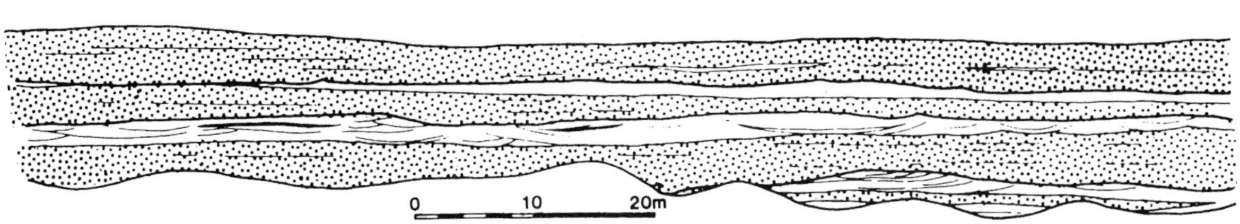

Fig. 8.21. Couplets of horizontally bedded conglomerate (lithofacies Gh) and trough cross-bedded sandstone (St). Thin mudstone lenses are present in places above the St sets and in some dune troughs (*thick black lines*). Flow is into the face. Budleigh Salterton Pebble Beds (Lower Triassic), southwest England. (S.A. Smith 1990)

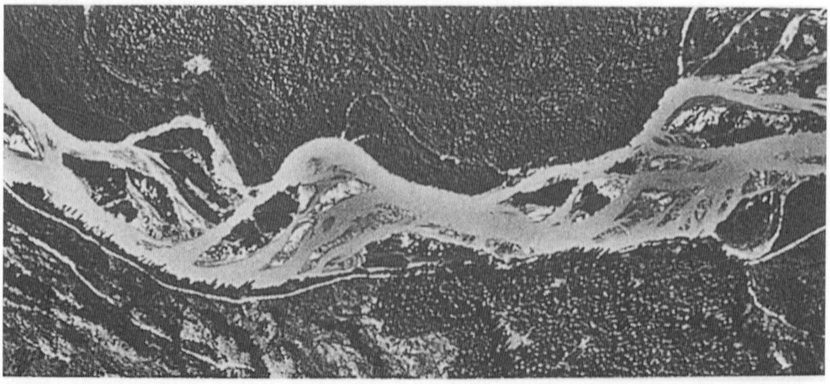

A. BRAIDED REACH: Note three to four active channels and extensive bar platform areas.

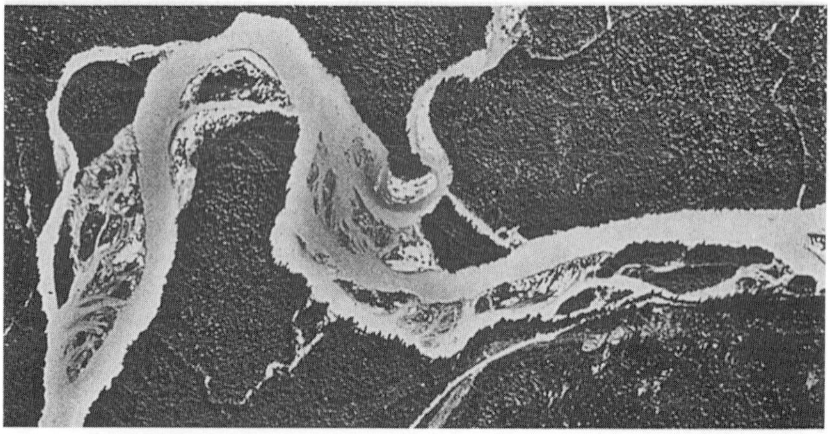

B. WANDERING GRAVEL-BED REACH: Note one to two active channels, sinuous outline, and less bar platform area than up-valley.

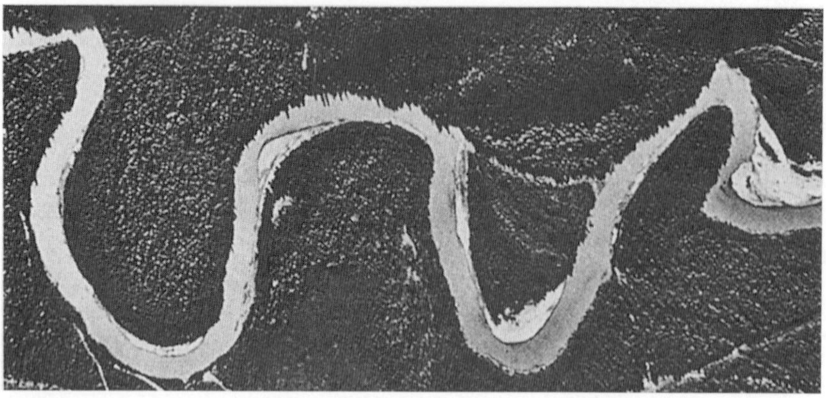

C. MEANDERING REACH: Note single channel, sinuous outline, and restriction of bar platform areas to the insides of bends.

Fig. 8.22A–C. The transition from braided, through wandering, to meandering styles in the Squamish River, British Columbia, Canada. (Brierley and Hickin 1991)

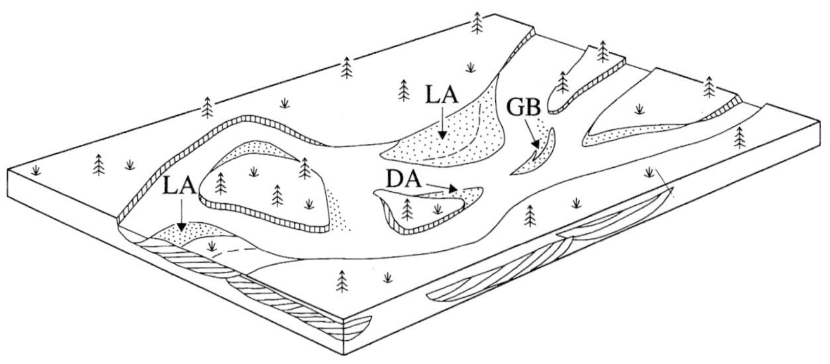

Fig. 8.23. Architectural model for wandering, gravel-bed river

the upper Congaree River showed three types of bedform (Fig. 8.29). Large, straight-crested dunes up to 2 m high, termed "transverse bars" by Levey (1978), occur on point-bar surfaces. In the relatively straight reach between the bends, a population of smaller, straight-crested dunes ("sandwaves") averaging 85 cm in height are the predominant bedform. Superimposed on these, and in dynamic equilibrium, are fields of small, straight to sinuous-crested dunes ("megaripples") averaging 18 cm in height. Small-scale ripples may be superimposed on these. Chute channels with large chute "bars" occur in the downstream parts of point bars. Figure 6.38 illustrates a field of two-dimensional dunes migrating across the upper surface of one of the point bars in the upper Congaree River.

A useful study of a river terrace by Campbell and Hendry (1987) and a well-exposed Eocene deposit by Nijman and Puigdefábregas (1978) provide many facies and architectural details for the documentation of this type of river. These and other examples were used to construct the architectural model shown in Fig. 8.30 and the vertical profile in Fig. 8.8F (see reference list in Table 8.3).

In the South Saskatchewan example, the lower part of the point bar consists of gravel sheets, typically dipping across the channel at accretionary angles of about 5° or less (Fig. 8.31). Lithofacies Gh is predominant. The upstream part of the bar in this type of river tends to be coarser than the downstream end, which may be composed largely of sand (Fig. 8.32). The top of the gravel bar may be a distinct fourth-order bounding surface, but in places (as in the center of the cross section shown in Fig. 8.31) the gravel interfingers with the overlying facies, which consists of cross-bedded sand and pebbly sand (lithofacies St). This facies develops as lobes that migrate across the gravel bar platform, filling chute channels and sloughs, migrating up against the inner accretionary bank of the point bar, and occupying the downstream tail of the point bar (Fig. 8.32). The upper point-bar facies consists of interbedded sand, silt and clay, including layers of cross-laminated and parallel-laminated sand (lithofacies Sr, Sh), and fine-grained units (Fl). These beds have accretionary dips of up to 12°. They are interbedded with the units of St, and typically downlap onto the surface of the gravel bar. Their uppermost surface forms the active inner accretionary bank of the point bar. The difference in accretionary dip between these three facies is noteworthy, It has also been observed in the exhumed Eocene point bar studied by Nijman and Puigdefábregas (1978), a model of which is shown in Fig. 8.33. The clay plug deposited after the abandonment of this channel is well preserved, where it rests on the cutbank.

These examples show the upward fining of the point bar that was noted in some of the classic early research (Bernard et al. 1962; Bernard and Major 1963). More recent work has shown that this profile is produced only in the median portion of a meander bend. The upstream part of the bar (bar head) may show upward coarsening because of the presence of gravel lobes that migrate out of the channel and across finer bar deposits (e.g., Bluck 1971). Jackson (1976b) termed this the "transitional" zone, because the helical flow pattern is not fully developed at the entrance to each meander bend. His profile shows the coarsest part of the section near the middle of the bar profile (Fig. 6.41). The bar tail may consist entirely of one facies, and show no grain-size trends (Jackson 1976b).

8.2.7 Sand-Bed Meandering River

This fluvial style was amongst the first to be examined by sedimentologists, and provided us with the

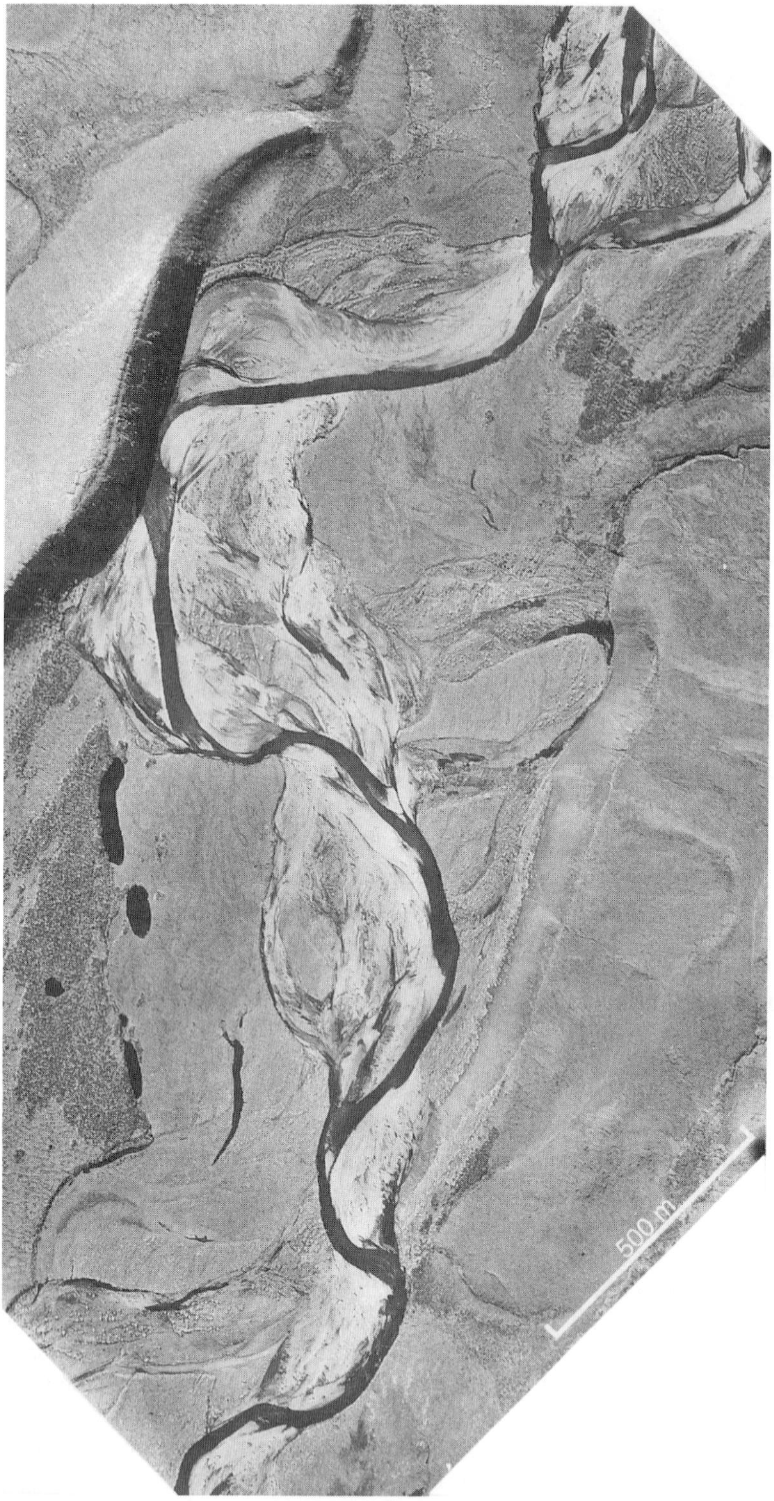

Fig. 8.24. Typical gravel-bed meandering river. Babbage River, Yukon, Canada. Part of photo A21924-199, National Airphoto Library, Canada (see McDonald and Lewis 1973)

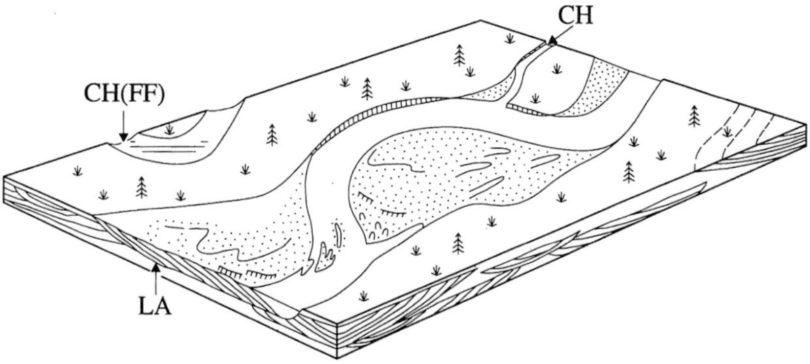

Fig. 8.25. Architectural model for gravel-bed meandering river. Model 4 of Miall (1985)

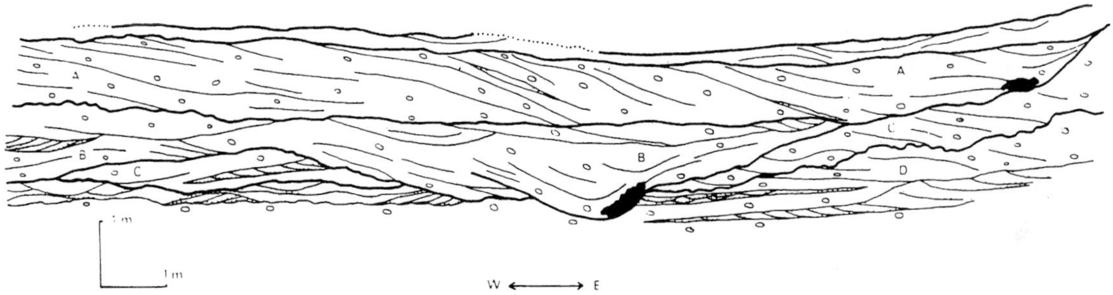

Fig. 8.26. Example of Quaternary gravel-dominated succession deposited in meandering river. Jarama River Terrace, central Spain. (Arche 1983)

classic sandy-meandering model of Allen (1963b, 1970b), as illustrated in Fig. 2.15. It is one of the most-studied fluvial styles in the ancient record, as can be seen from the list of selected references in Table 8.3, although studies of modern rivers of this type are few. The modern Mississippi River is a good example (Fisk 1944, 1947, 1960; Farrell 1987; Jordan and Pryor 1992). Channel and bar sediments are dominantly sand, although some intraformational conglomerates may be present as channel-floor lag deposits, where they accumulate as a result of cutbank erosion and caving. The sedimentology of this type of river is similar to that of the gravel-sand meandering river, differing only in the details of the lithofacies of the channel and point-bar deposits. Crevasse-channel and crevasse-splay deposits may be more common, as suggested by descriptions of this type of river by Farrell (1987), Plint (1983), and Muñoz et al. (1992)(see Chap. 7 for examples from the Mississippi River).

Figure 8.34 is an architectural model for this type of river, Fig. 8.8G is a typical vertical profile, and Figs. 4.9, 6.35B,D, 8.35, and 8.36 are outcrop photographs of ancient examples. Architectural details of a Triassic coastal-plain system are seen in Figs. 8.37 and 8.38. Evaporite pseudomorphs and calcrete horizons are present, and the unit is interbedded with the deposits of an inland sabkha, suggesting a relatively dry environment, although the presence of abundant plant and vertebrate remains indicates that the climate was not truly arid. Accretionary bundles within the lateral-accretion deposits commonly show cyclic character, indicating that they were formed by floods or seasonal high-discharge events (Fig. 8.37). Some sandstone sheets are made up of thin, overlapping channel-fill bundles or accretionary wedges separated by scours, indicating considerable channel instability (Fig. 8.38). Crevassing took place onto well-developed floodplains, preserving small abandoned channels filled with alternating thin mudstone and sandstone beds (Figs. 8.38, 8.39).

Another example of a crevasse-splay deposit is illustrated in Fig. 6.21. In this case the environment was humid, as indicated by the underlying coal.

8.2.8 Ephemeral, Sand-Bed Meandering River

This fluvial style (Figs. 8.8H, 8.40) differs in detail from that of the previous model, based mainly on subtle lithofacies features that reflect climatic differences. As discussed in Sect. 12.6, the determination

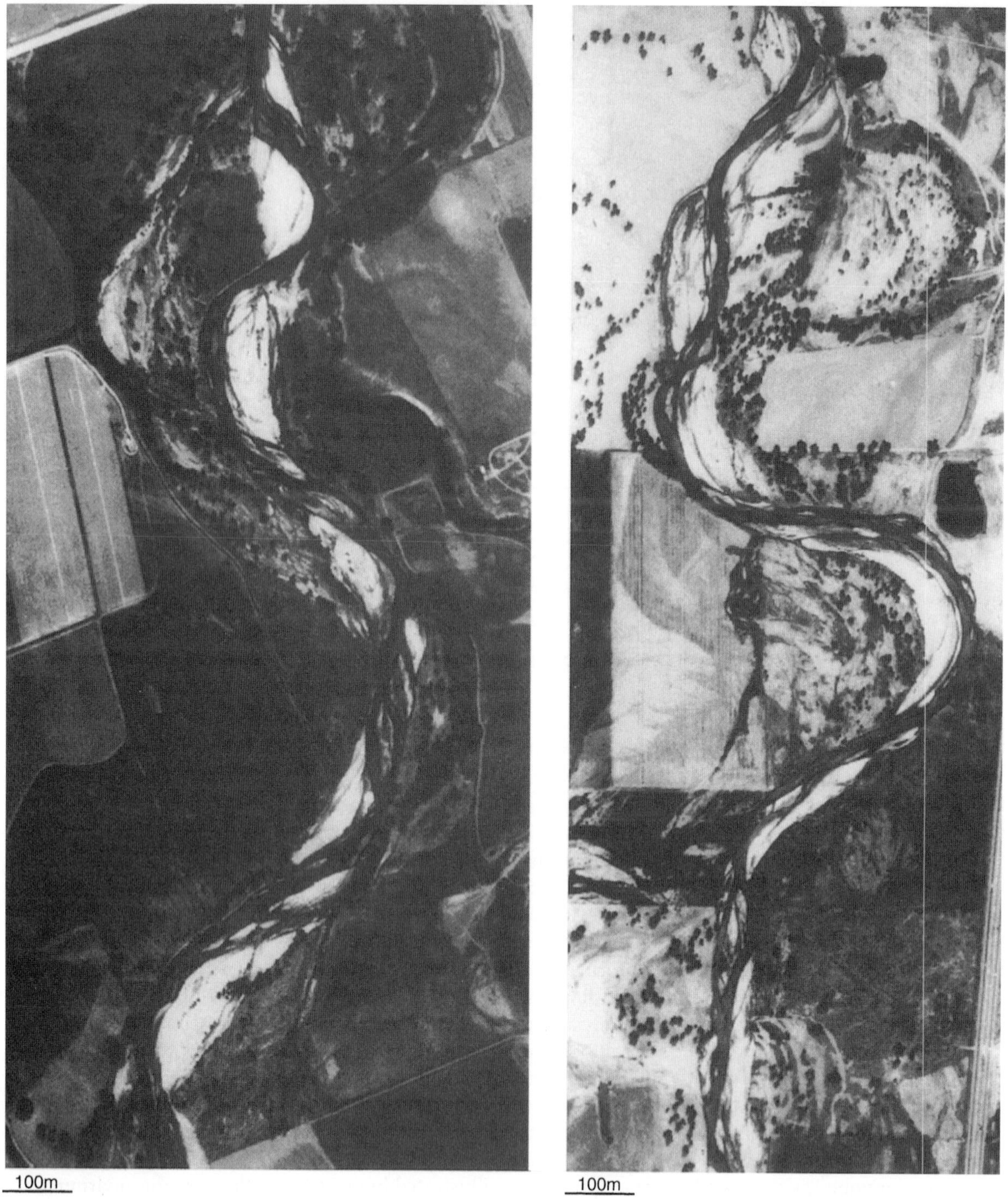

Fig. 8.27. Examples of a coarse-grained meandering stream. Upper South Platte River, near Denver, Colorado. US Soil Conservation Service photograph (Crowley 1983). The present style of this river, as with many rivers in Europe and North America, has radically changed from its natural state as a result of flood control works, artificial channel diversions, and the extraction of water for agricultural and other needs

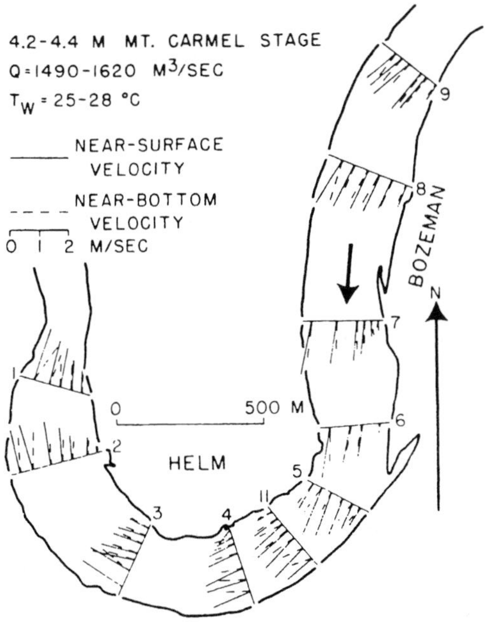

Fig. 8.28. Velocity patterns around two successive meander bends of the Wabash River. The development of the spiral flow pattern is demonstrated by the divergence in surface and bottom flow vectors. Note the shifting of the position of maximum surface velocity across the channel, and the fact that this does not impinge against the outer bank until half way around the Helm bend. (Jackson 1976d)

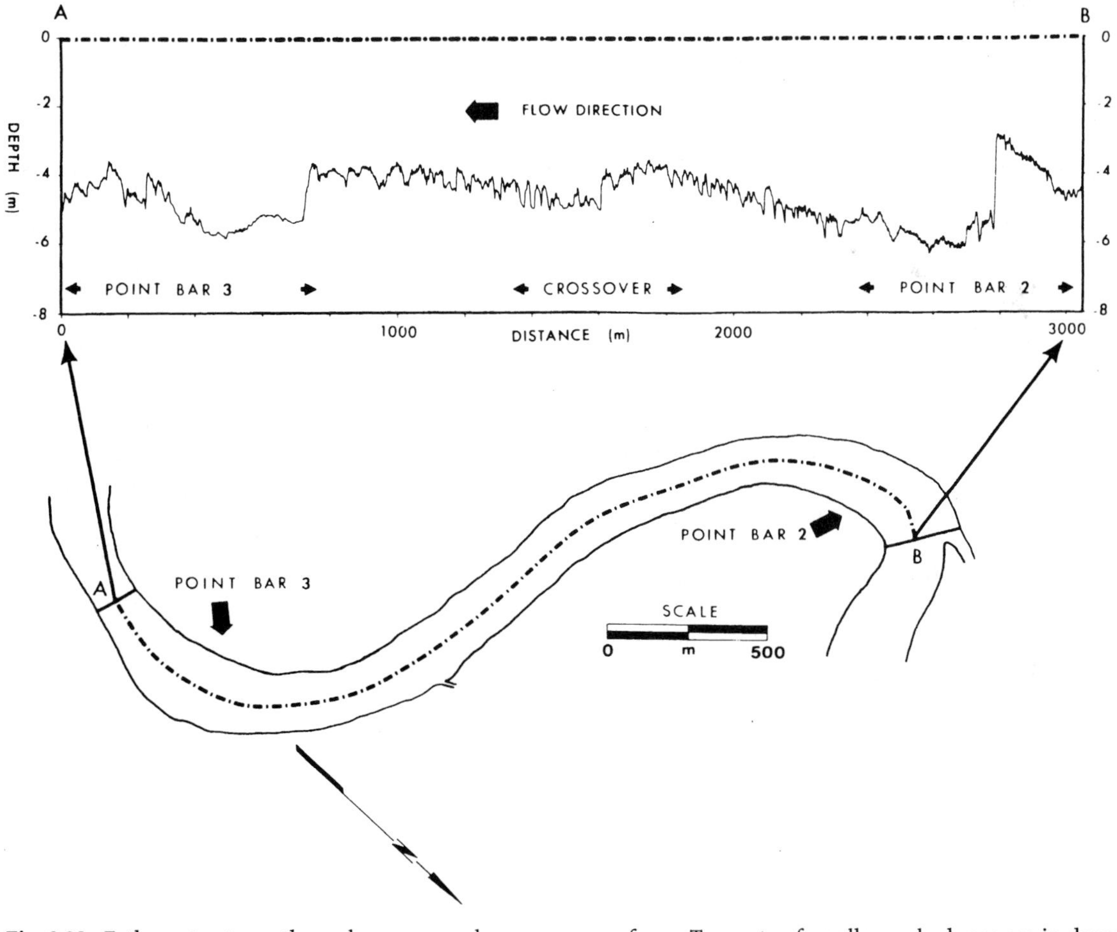

Fig. 8.29. Fathometer trace through one meander wavelength of the upper Congaree River during bankfull discharge. Large dunes up to 2 m high occur on point-bar surfaces. Two sets of smaller-scale dunes are in dynamic equilibrium in the relatively straight stretch betwen the bends. (Levey 1978)

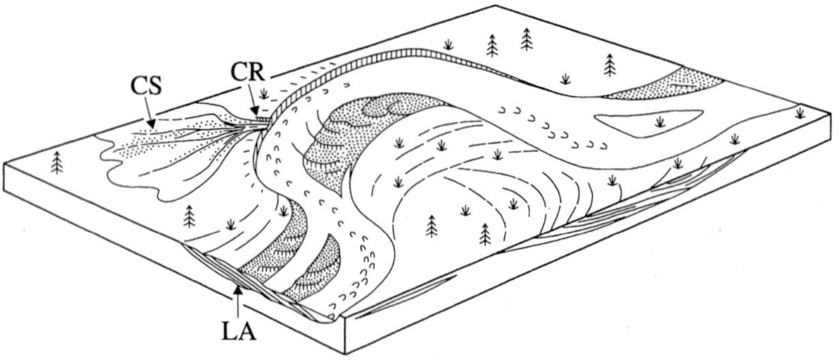

Fig. 8.30. Architectural model for gravel-sand meandering river. Model 5 of Miall (1985)

of climatic signatures from the facies record of sand-bed rivers is a difficult undertaking, especially since most of the discussion is based on interpretations of the ancient record, not from studies of modern examples. To what extent, for example, can rivers characterized by seasonal discharge, such as those which are interpreted to have deposited the succession illustrated in Figs. 8.37 and 8.38, be distinguished from deposits formed in arid environments characterized by ephemeral flow? This section is included here to emphasize the need for careful facies analysis.

The key indicator of ephemeral flow is that evidence of aridity and flashy discharge is abundant. The deposits contain numerous internal scour surfaces, and lithofacies Sh and Sl are abundant, indicating the common occurrence of high-energy, shallow flow near the transition from subcritical to supercritical. Mud drapes occur on second- and third-order surfaces, and these may be associated with desiccation cracks, mud curls, and various types of evaporite nodules and crystals. Wedges and laminae of eolian sand may also occur along major bounding surfaces, as in the modern examples described by Shepherd (1987). These can be distinguished by their distinctive texture, presence of eolian ripples, etc.

The architecture of this fluvial style is characterized by sandstone ribbons and sheets, the latter including lateral-accretion deposits. These sand bodies may not be very different from those formed in perennial sand-bed rivers. A brief description of a modern example in an ephemeral river was given by Shepard (1987), and ancient examples were described by Lawrence and Williams (1987; upper part of their association 2) and Stear (1983). Figures 8.41 and 8.42 illustrate Stear's examples, from the Beaufort Group of the Karoo Basin, South Africa. Stear (1983) estimated that the channels which deposited these sandstone bodies were 3–5 m deep and up to 100 m wide. Sections C-G in Fig. 8.42 illustrate complex sandstone bodies, in which lateral accretion is only one of several mechanisms that appear to have occurred. In places, this has preserved overlapping, laterally accreted wedges separated by mud drapes and third-order surfaces, as in sections C, F, and G, the left side of section D, and the right side of section E. Note the excellent preservation of scroll-bar morphology on the top of the point bar, and the shale-filled abandoned channel, in section E.

Floodplain deposits in this fluvial environment include thin sandstone sheets and mudstones. The sandstone beds consist of laterally coalesced lenses deposited from an anastomosing network of ephemeral floodplain channels. As noted by Stear (1983), internal scour and fill structures and siltstone drapes between accretionary units indicate multiple episodes of erosion and sedimentation. Irregular, undulating top surfaces to the sandstone sheets indicate erosion by rivulets crossing the floodplain. Well-developed levees are lacking in these Beaufort Group examples, but broad wedges of alternating sandstone and siltstone flank some channel sandstone bodies, and are interpreted as low-relief levee deposits.

8.2.9 Fine-Grained Meandering River

Figure 8.43 illustrates a highly sinuous, suspended-load stream. D.G. Smith (1987) reviewed this fluvial style and provided several well-documented examples. The overall geometry is similar to that of the sand-bed meandering streams, but differs in detail because of the finer-grained sediment load (fine sand, silt, mud). Point-bar accretion surfaces dip steeply (up to 25°) and have a simple geometry, typically planar or with banks and benches (Figs.

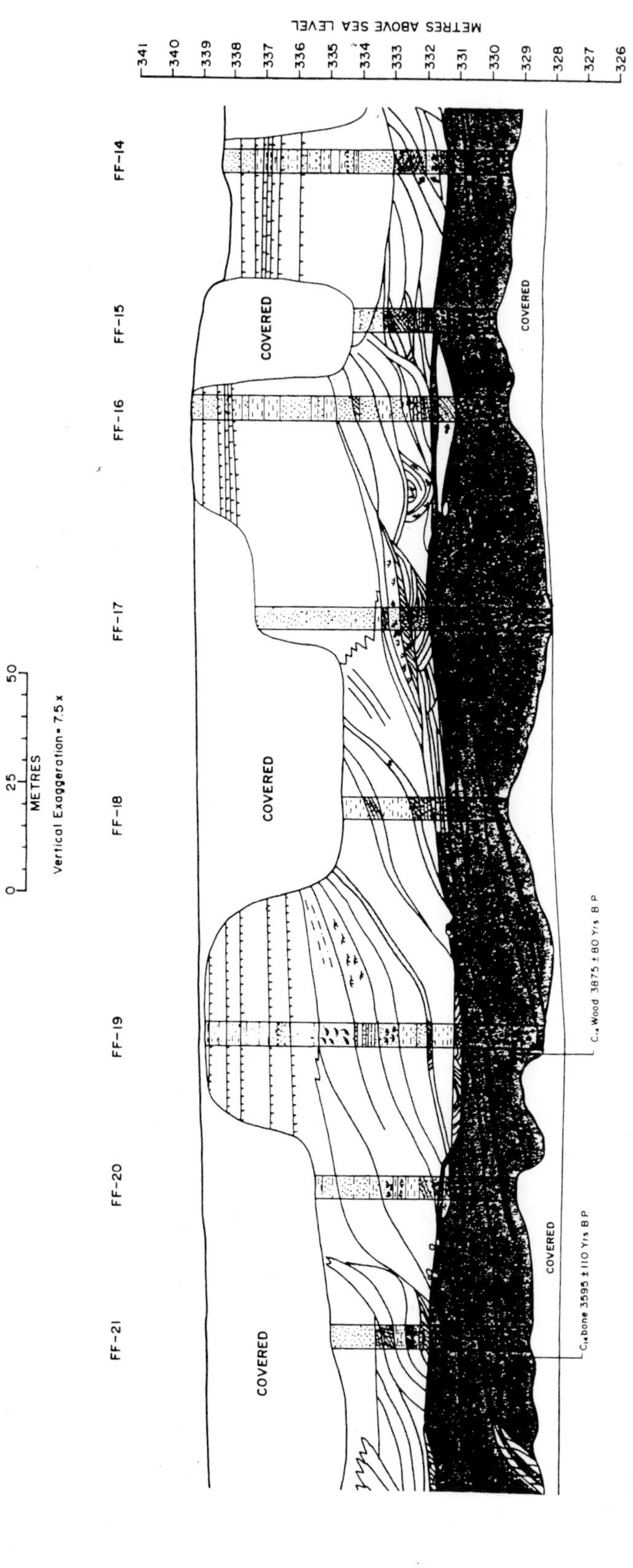

Fig. 8.31. Cross section through river terrace adjacent to the South Saskatchewan River, east-central Saskatchewan, Canada (*dark shading*): gravel (lithofacies Gh); *medium shading* cross-bedded sand and pebbly sand (St), *light shading* interbedded sand, silt, and clay (Sr, Sh, Fl). (Campbell and Hendry 1987)

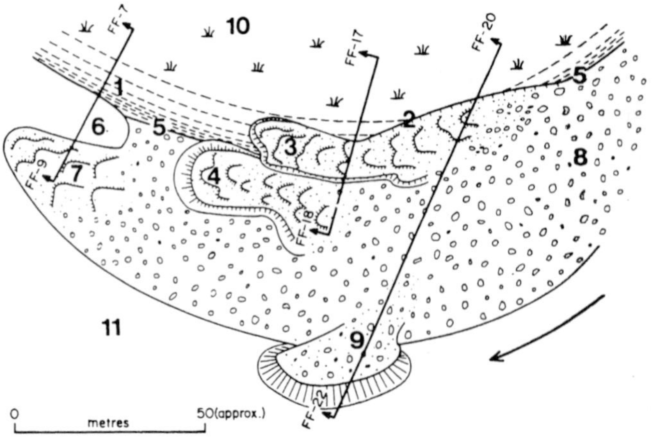

Fig. 8.32. Reconstruction of meander bend in the Saskatchewan River, based on data shown in Fig. 8.31, and other sections. *1* Inner accretionary bank (IAB); *2* scarp eroded in IAB; *3* sand sheet with 3-D dunes lapping onto IAB; *4* sand sheet on gravel platform or filling chute channel; *5* gravel platform buried by encroaching IAB; *6* slough (former chute channel) floored by gravel, and or mud; *7* bar-tail sand; *8* bar-head gravel; *9* microdelta formed by late-stage runoff; *10* floodplain; *11* channel. (Campbell and Hendry 1987)

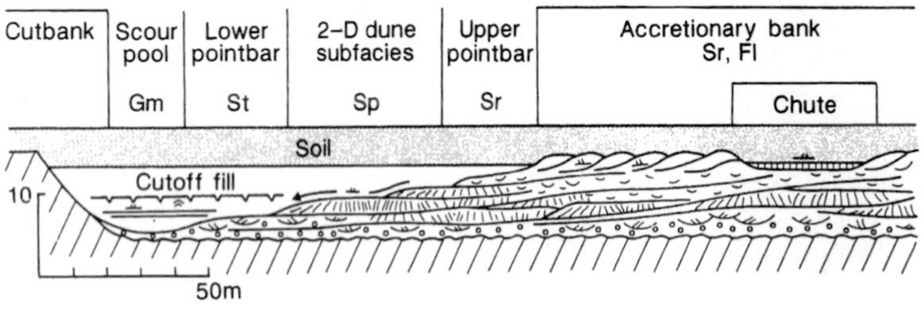

Fig. 8.33. Cross section model of point bar in the Castisent Sandstone, Spain. (Nijman and Puigdefábregas 1978)

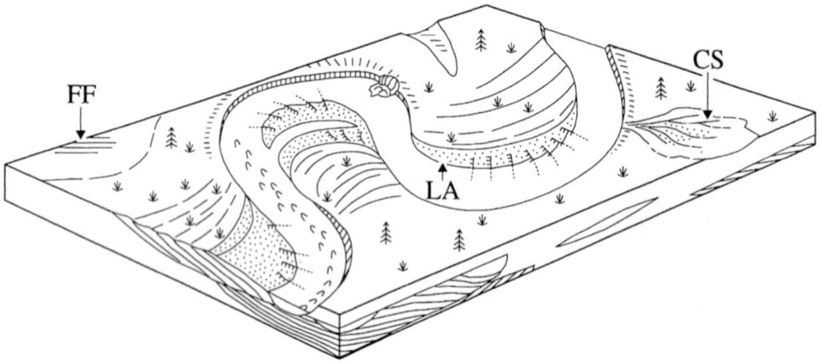

Fig. 8.34. Architectural model for a sand-bed meandering stream. Model 6 of Miall (1985)

6.40, 8.44, 8.45) indicating downstream flow separation and development of incipient scroll bars (Fig. 2.37; Nanson 1980). Ripple marks are typically the most abundant flow-regime bedform present (lithofacies Sr). Gravel lags and cross-bedded medium to coarse sands may occur at the base of the point bar (typically lithofacies St). Upward fining of grain size and thinning of bedding may occur (Figs. 8.8I, 8.46; Stewart 1983). Thick, fine-grained floodplain deposits may be present, including thin sandstone splays and coal seams (Figs. 8.44, 8.45).

As noted in Sect. 2.4.4.1, this fluvial style is characteristic of low-energy estuarine environments, in which the rivers may be subject to tidal influence (D. G. Smith 1987; Thomas et al. 1987; Zaitlin et al. 1994). Decimeter- or centimeter-scale sand-mud couplets may form on the point bar, and have been termed inclined heterolithic stratification (IHS; Fig. 2.38). An example of a unit containing IHS is the Athabasca Oil Sands of Alberta, Canada (Mossop and Flach 1983; Flach and Mossop 1985), which contains point-bar successions up to 25 m thick (Fig. 8.47). Typically, these consist of a lower 5–10 m thick succession of trough cross-bedded sandstone, overlain by up to 15 m of low-angle, cross-bedded, heterolithic deposits (IHS), consisting of decimeter-scale couplets of sand and mud dipping at an average angle of 12° (Fig. 8.48). Bioturbation of the upper part of the succession is common.

Other, very well-exposed examples of this fluvial style were described by Wood (1989). He demonstrated variations in lithofacies within the point-bar

Fig. 8.35. Point bar in lower delta plain deposits. Carboniferous, Pocahontas Basin, Kentucky

Fig. 8.36. Close-up of bar seen in Fig. 8.35, showing slump blocks of mudstone at the base of the bar, resulting from bank caving

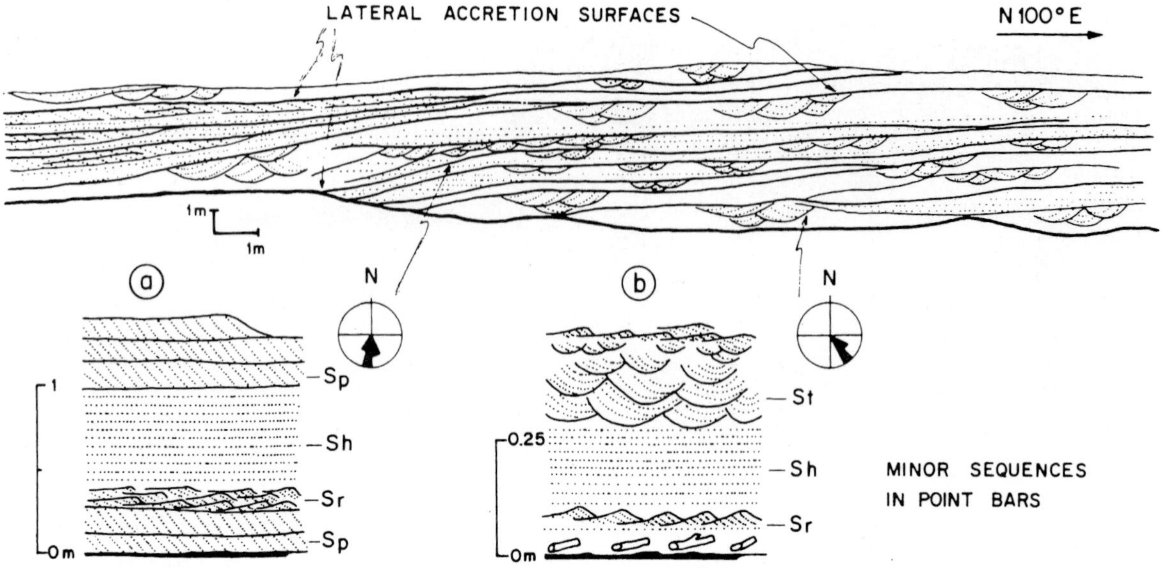

Fig. 8.37. A lateral-accretion deposit in the Buntsandstein (Triassic) of central Spain, showing details of cyclic accretionary bundles. (Muñoz et al. 1992)

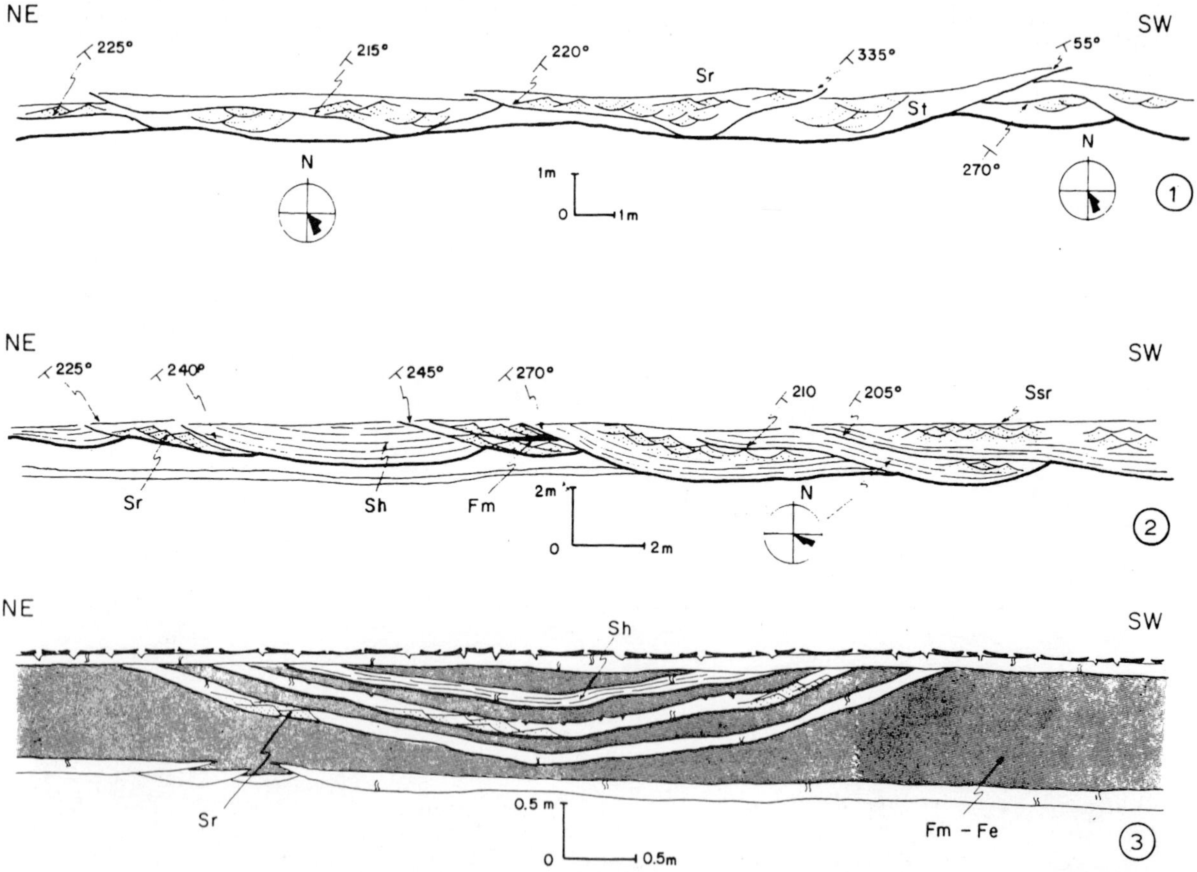

Fig. 8.38. **1** and **2** Complex sandstone sheets resulting from repeated shifting of small channels. **3** Fill of small abandoned crevasse channel. Buntsandstein (Triassic), central Spain. (Muñoz et al. 1992)

Fig. 8.39. Outcrop photograph of the floodplain deposits illustrated in Fig. 8.38, showing small crevasse channel

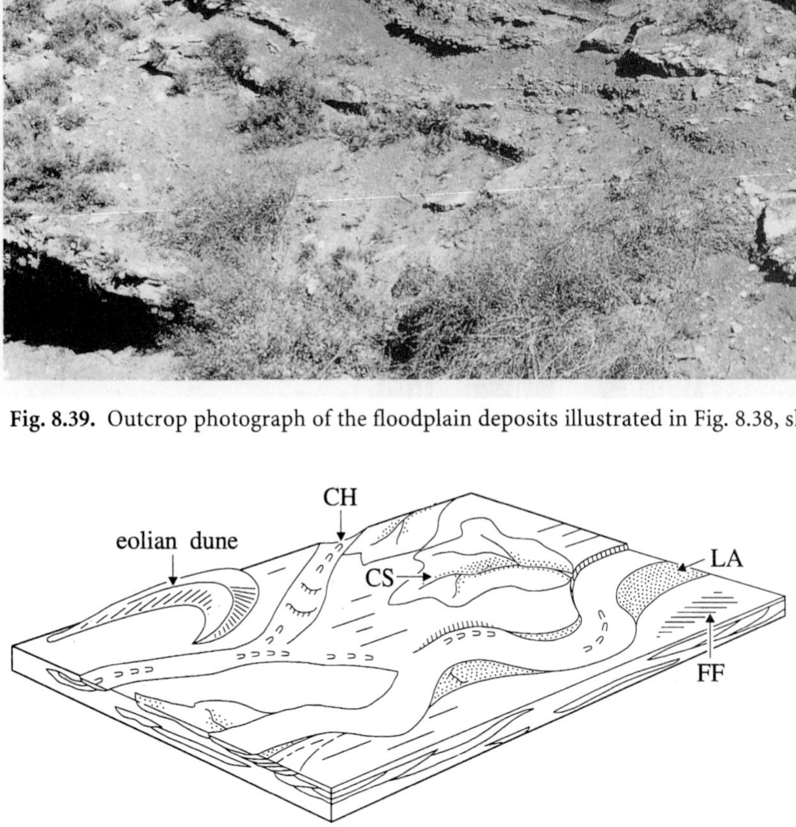

Fig. 8.40. Architectural model of ephemeral, sand-bed meandering river

Fig. 8.41. Outcrop of ribbon and sheet sandstone bodies interbedded with floodplain fines, Beaufort Group. Karoo Basin, South Africa. Many of the sheets sandstones contain evidence of lateral accretion, indicating deposition within high-sinuosity channels (Fig. 8.42)

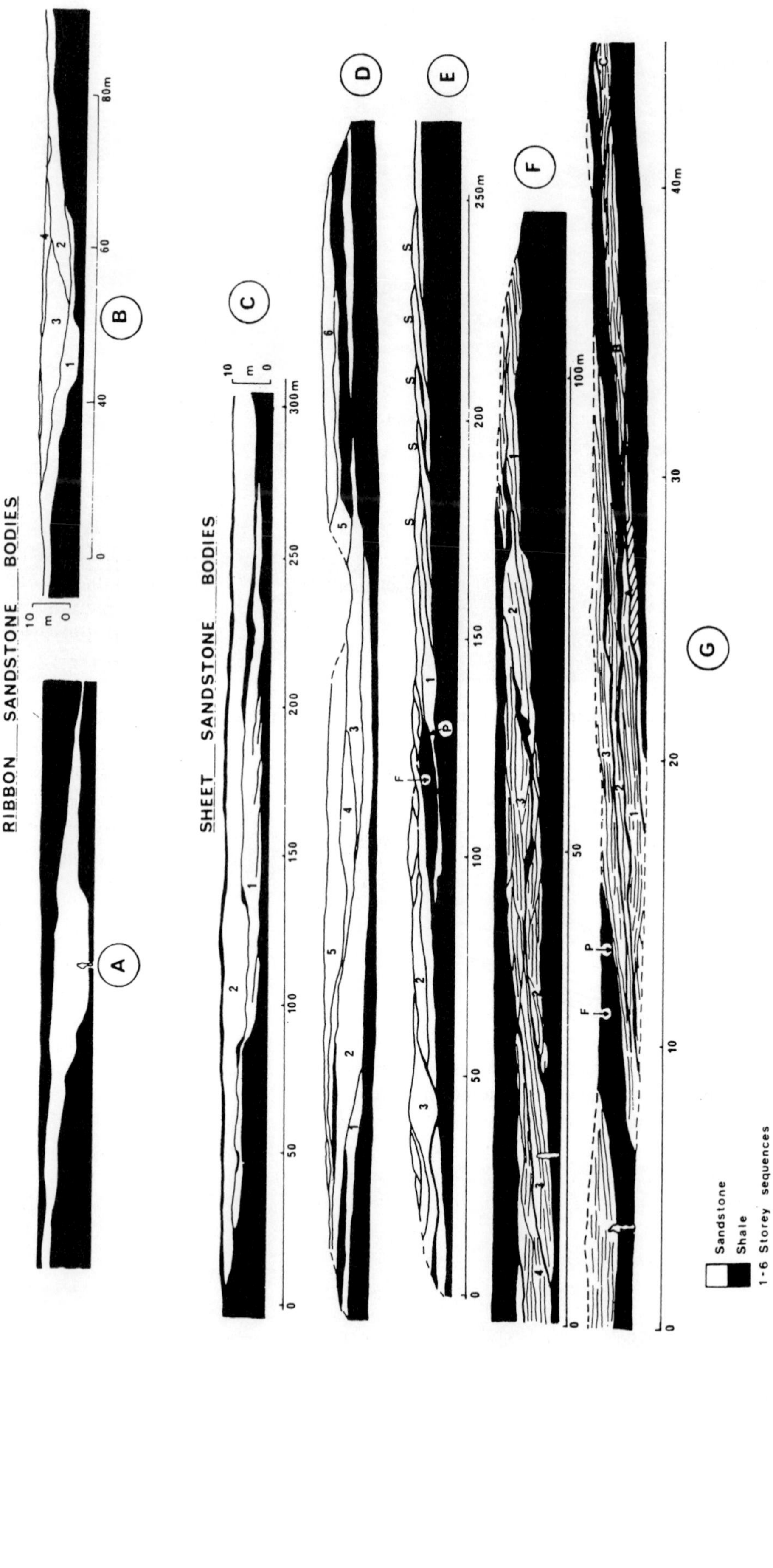

Fig. 8.42A–G. Ribbon and sheet sandstone bodies in the Beaufort Group, Karoo Basin, South Africa. Note the variable scale. *S* Scroll bar; *P* point-bar surface; *F* shale fill of abandoned channel. (Stear 1983)

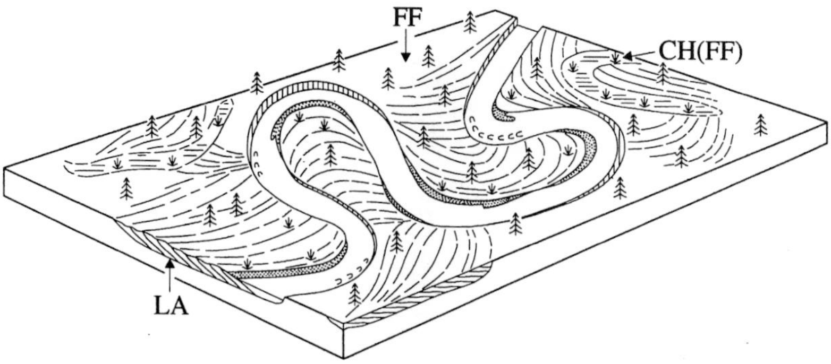

Fig. 8.43. Architectural model of a fine-grained meandering stream. Model 7 of Miall (1985)

deposits from sand-dominated, trough cross-bedded units to muddy IHS successions, and suggested that these variations reflected varying transport energies within a meander bend (sand-dominated in the upper part of the bend, mud dominated downstream), much as have been reported from coarse-grained meander bends. Wood argued that the processes of meander evolution, including the changes in local sediment load and slope caused by neck and chute cutoffs, would affect the grain-size and facies assemblage of the point-bar, accounting for lateral variations in point-bar style within the same macroform. Wood (1989) also reported the presence of ribbon sandstones within a dominantly sheet-sandstone succession, and suggested that these represented rapid cut-and-fill of channels, followed by avulsive abandonment before significant lateral migration could occur.

8.2.10 Anastomosed River

The distinctiveness of the anastomosed fluvial style was established by D.G. Smith and N.D. Smith (1980) and D.G. Smith (1983), based on work on small modern rivers in the Rocky Mountains of Canada (Sect. 2.4.4.3). Most of the examples of this type of river that have been described in the literature have low gradients, and low stream power. Lateral channel migration is minimal, and the floodplains therefore lack scroll bars and oxbow lakes (Fig. 8.49). Channels tend to be isolated – bounded in most places by floodplain deposits. As these are normally fine-grained, channel banks are typically cohesive and steep-sided. Channel evolution takes the form of crevassing, and the development of stable crevasse channels, which may rejoin a main channel downstream and divert the flow from the channel from which they branched. These characteristics of anastomosed rivers apply in both humid and arid climates, in both settings anastomosed rivers are to be found. Floodplains are characterized by muds which may contain pond deposits, coals, calcretes, or evaporites, depending on climate (Sect. 12.7). Sandy crevasse-channel and splay deposits are also common. Table 4.4 lists the architectural elements identified in a typical ancient anastomosed fluvial deposit, the Cutler Group of New Mexico. Figure 8.8J is a typical vertical profile.

Channel bodies are typically straight to sinuous ribbon sandstones (e.g., Fig. 6.5). Friend (1983) suggested that a width/depth ratio of < 15 is typical of ribbon sandstones, but the new data and review provided by Nadon (1994) indicate that a W/D ratio of < 30 is a more appropriate definition. Commonly, the channel sandstones consist of simple, cross-bedded channel elements, as in most of the channels described by Eberth and Miall (1991). Figures 8.50 and 8.51 illustrate typical examples from that study. Note the presence of minor internal erosion surfaces in these sandstone bodies, These are third-order surfaces that divide the sandstone bodies into depositional increments. In body B (Fig. 8.51), the inclination of the surfaces suggests local lateral accretion and minor lateral channel migration. Kirschbaum and McCabe (1992) demonstrated that anastomosed channels may fill by lateral or vertical accretion or by concentric accretion, in which bedding in the channel-fill parallels the channel floor. Nadon (1994) demonstrated that in his examples the channel sandstones aggraded vertically at the same time as the levees, so that channel and levee facies are interbedded.

Crevasse channels and splays typically form complex sheet-sandstone bodies cut by numerous internal erosion surfaces. Figure 8.52 illustrates two large,

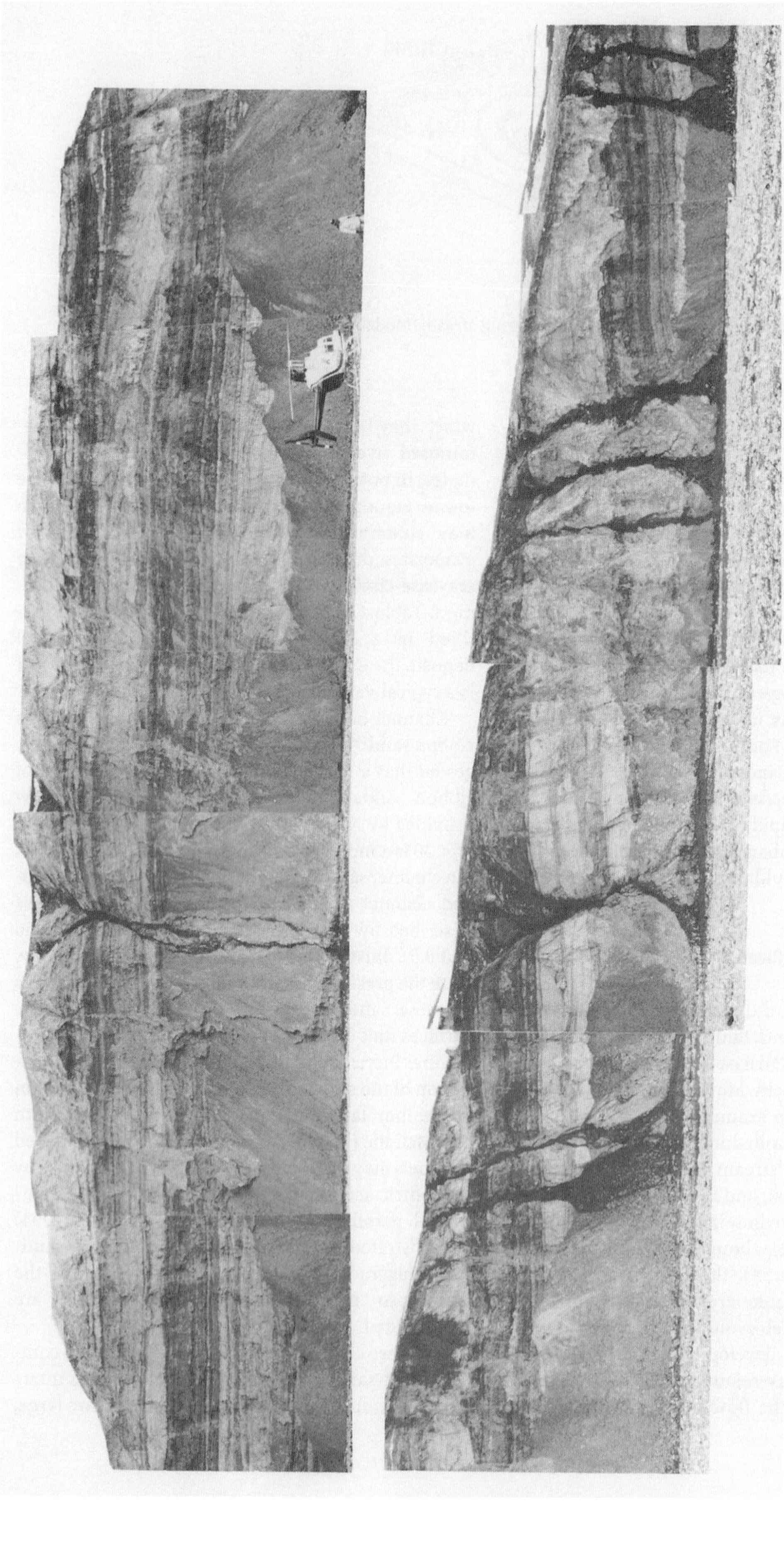

Fig. 8.44. Outcrop of fluvial succession deposited by fine-grained meandering stream. Figure 6.40 is a close-up of part of this outcrop, and Fig. 8.45 provides an interpretation of the entire outcrop. Tertiary, Ellesemere Island, Arctic Canada

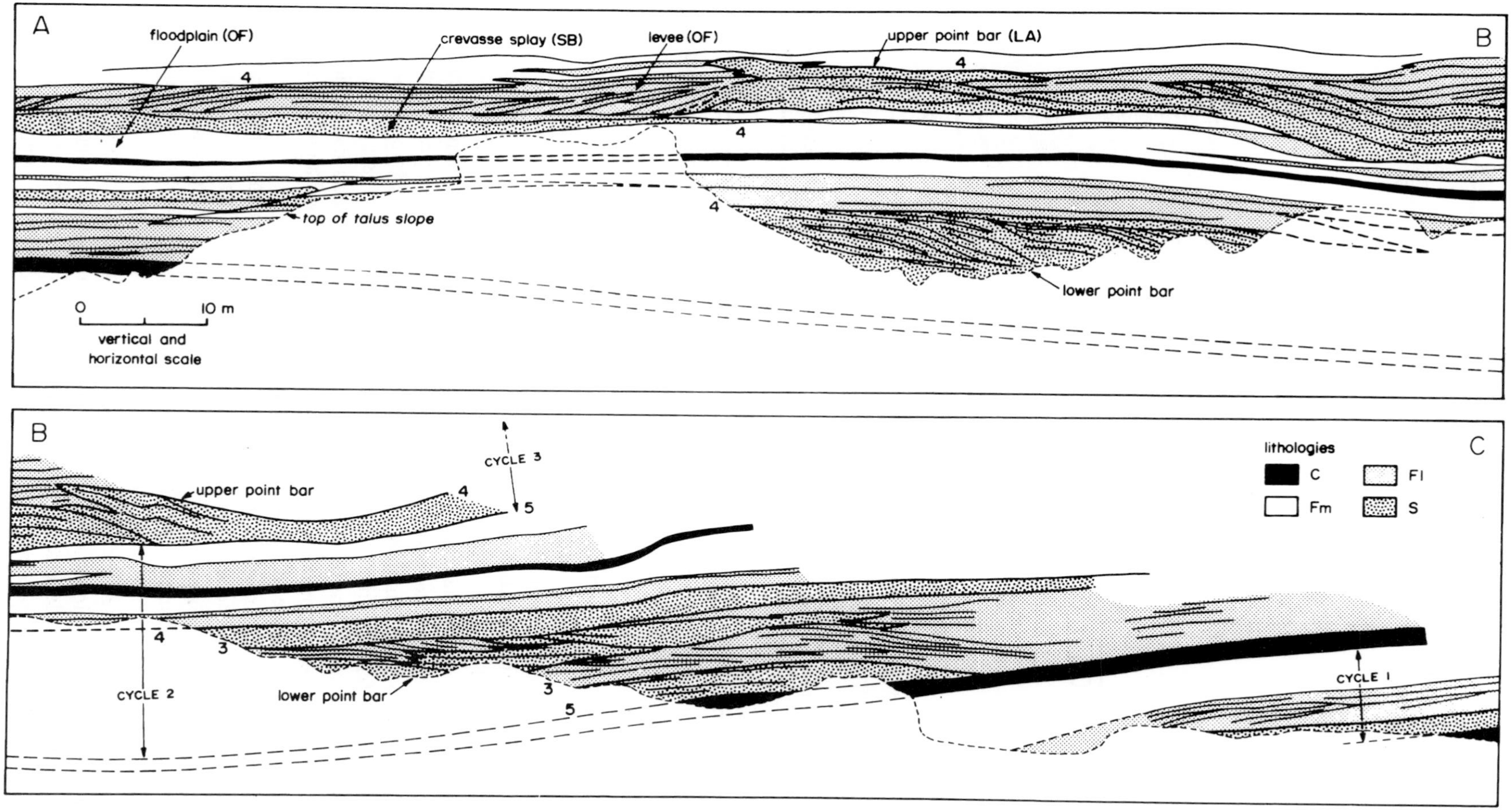

Fig. 8.45A–C. Interpretation of large outcrop showing fluvial architecture formed by a fine-grained meandering stream system. Figures 6.40 and 8.44 illustrate parts of this outcrop. Tertiary, Ellesmere Island, Arctic Canada. (Miall 1979)

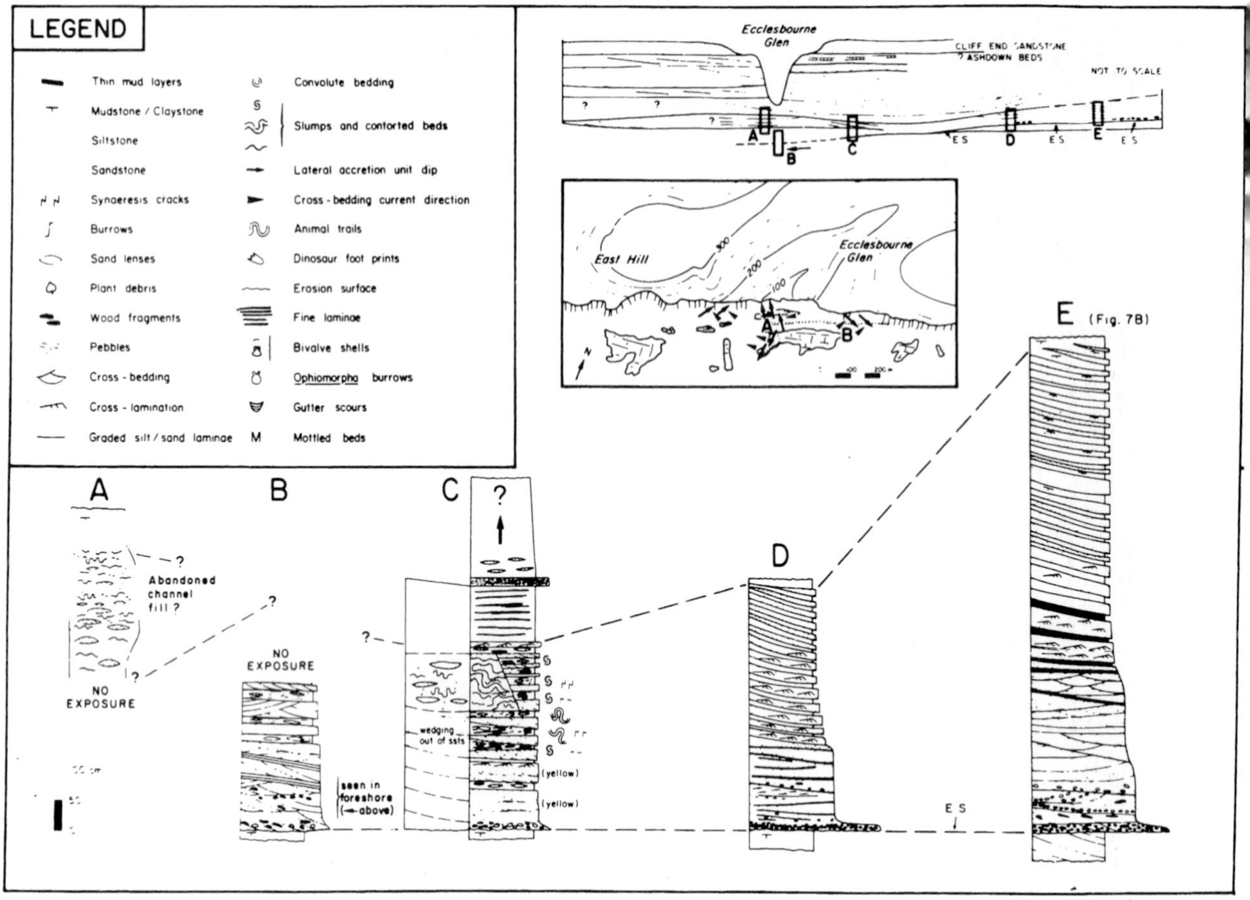

Fig. 8.46A–E. Architectural cross section and detailed vertical profiles through the deposit of a fine-grained meandering river system. Hastings Beds (Cretaceous), SE England. (Stewart 1983)

modern splays, and Fig. 8.53 provides cross sections through two small splays. Note the large difference in scales between the rivers illustrated in these diagrams. Anastomosed channels of the Magdalena River are up to 0.5 km wide and 10 m deep, with splays extending up to 10 km away from the channel. Splay channels are 2–10 m deep. The main channels of the Saskatchewan River are up to about 5 m deep and 100–200 m wide, with splays extending only 1–2 km into the floodplain. Where interpretations of an ancient deposit are made from limited exposure, it may be difficult to assess the overall scale of the system. Thus, the main channels of the modern Saskatchewan system and the ribbon sandstone bodies described by Eberth and Miall (1991; Fig. 8.51) are comparable in scale to the crevasse channels of the modern Magdalena splay systems.

The larger-scale architecture of an inferred anastomosed system is illustrated in Fig. 8.54, in which ribbon sandstones can be seen to comprise less than 50% of the total rock volume. They show generally moderate to poor lateral and vertical interconnectedness, and are separated by flood-basin mudstones containing interbedded coal seams.

In some large rivers, such as the Brahmaputra, a form of anastomosis develops, in which normally braided channels flow around relatively stable islands. The islands develop by the process of amalgamation of braid bars that build up to floodplain level (Coleman 1969; Bristow 1987; Thorne et al. 1993). Switching between channels and the development of new channels occurs by channel aggradation and overbank flooding.

Modern anastomosed channels that had been identified in the deserts of the Lake Eyre Basin, Australia (Rust 1981), are now regarded as a distinct fluvial style (Nanson and Croke 1992; Nadon 1994). The channels are incised into the floodplain and are not in equilibrium with it, in the sense that channel and floodplain deposits aggrade together. Sand in the channels is derived by erosion of an underlying sand bed, and the floodplains are underlain largely

Fig. 8.47. Outcrop of Athabasca Oil Sands, near Fort McMurray, Alberta, Canada, illustrating very large-scale point bars of the fine-grained-meandering fluvial style. (Photo courtesy of G. Mossop)

by relict mud "braids" composed of sand-sized mud aggregates (Nanson et al. 1986; Rust and Nanson 1989).

8.2.11 Low-Sinuosity River, with Alternate Bars

Just as the wandering gravel-bed river (Sect. 8.2.4) is regarded as an intermediate form between braided and meandering patterns, this fluvial style represents sand-bed rivers that are intermediate between braided and meandering rivers. The style is illustrated in Figs. 8.8K, 8.55, and 8.56. It contains characteristics of both of the end members of which it is an intermediate form, and great care is needed to distinguish this fluvial style.

In some respects, the fluvial style is that of a braided river with only one active channel. Channel sinuosity is low, and there may be both midchannel and bank-attached bar forms. The main in-channel architectural element is the alternate bar. This element is a macroform comparable to a large braid bar but deposited by accretion of a bank-attached bar. The alternate bar is therefore analagous to the lateral-accretion elements of point bars. However, there are significant differences to point bars. Firstly, the direction of accretion is obliquely downstream rather than perpendicular to the channel margin, as in high-sinuosity rivers. Secondly, the accretion commonly takes the form of a single, large-scale, high-angle cross-bedded unit, rather than an assemblage of smaller sets advancing down first- to third-order surfaces. Examples of large macroforms developed in this way were described by McCabe (1977; Figs. 2.27, 6.43). Crowley (1983) recognized a continuum of geometries from the braid bar to the point bar in the Platte River of Colorado and Nebraska, and designated the intermediate form the South Platte-type macroform (Fig. 8.56). Wizevich (1992b) distinguished alternate bars as a special type of downstream-accretion element, to which he gave the field code DA2 (Fig. 8.57).

Other modern and ancient low-sinuosity fluvial systems have been described by Okolo (1983), Bridge et al. (1986), and Olsen (1988). Bridge et al. (1986) described a process whereby bank-attached bars

Fig. 8.48. Close-up of inclined heterolithic stratification in the Athabasca Oil Sands of Alberta, showing the decimeter-scale interbedding of sand and mud and the abundant bioturbation. (Photo courtesy of G. Mossop)

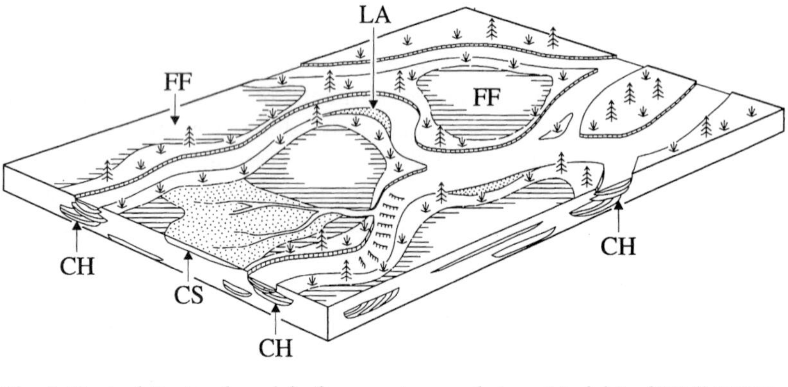

Fig. 8.49. Architectural model of an anastomosed river. Model 8 of Miall (1985)

evolve into midchannel forms by the development of chute channels (Fig. 8.58). The resulting macroform may evolve in several ways. Figure 8.59 illustrates one model, in which the chute channel fills with sediment and the bar subsequently accretes laterally into the larger channel, as flow is concentrated there and the incipient bend translates downstream.

Little information is available on the nature of floodplain deposits of this type of river, but it is anticipated that they would be similar to those of meandering rivers.

8.2.12 Shallow, Perennial, Sand-Bed Braided River

This is the classic "Platte-type" braided river, the model for which was proposed by Miall (1977), based mainly on work by N.D. Smith (1970, 1971, 1972) on the Platte River of Colorado and Nebraska. Later work on this river by Blodgett and Stanley (1980) and Crowley (1983) is also of importance (Sect. 2.4.4.2). The William River of northern Saskatchewan is another example of a modern river of this type (N.D. Smith and D.G. Smith 1984). An architectural model

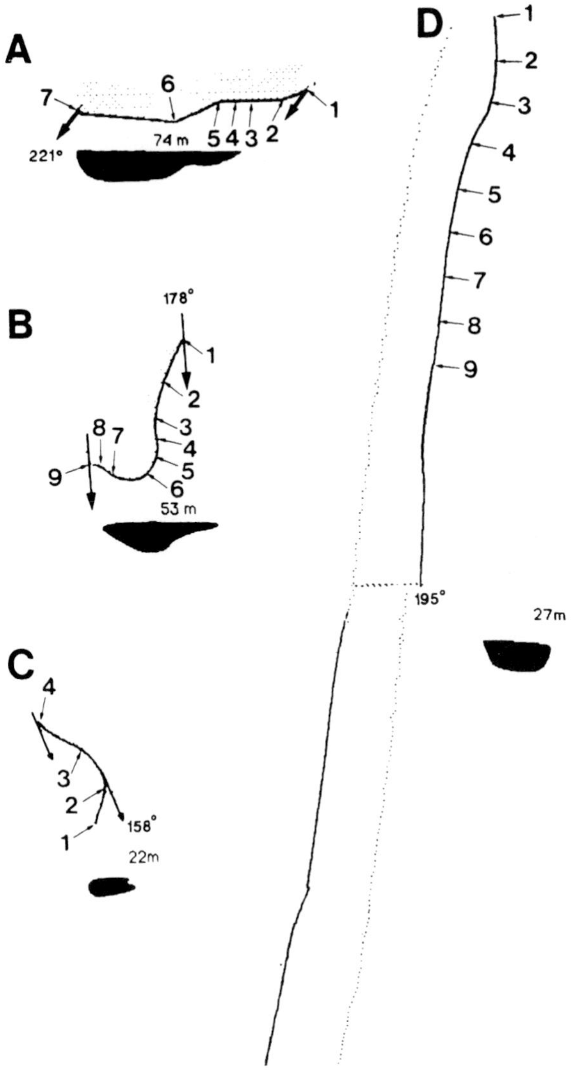

Fig. 8.50A–D. Outcrop patterns of four typical ribbon sandstones in the anastomosed Cutler Group, New Mexico. *Heavy arrows* indicate paleocurrent directions, *small arrows with numerals* indicate positions of vertical sections illustrated in Fig. 8.51. (Eberth and Miall 1991)

is illustrated in Fig. 8.60, and aerial photographs of the Platte River are shown in Fig. 8.61. Figure 8.8L reproduces the vertical-profile model of Miall (1977).

The fluvial style is characterized by the presence of fields of large, flat-topped three-dimensional dunes, termed linguoid bars in the older literature (see Ashley 1990), which are active during high stage. Internally, these are constructed of planar cross-bed sets. Smaller 3-D dunes building trough cross-bed sets may occupy deeper channels. These rivers are braided only during low discharge stages, when the tops of the dunes become exposed. At times of high discharge there may be a single, very broad, but shallow channel occupying almost the full width of the floodplain. As a result, fine-grained floodplain deposits constitute a minor part of the lithofacies assemblage. The Platte River itself no longer shows these extremes of stage because of human interference, including the presence of dams and the withdrawl of water for agricultural purposes.

The main architectural component is the presence of extensive sheets of in-channel, lower flow-regime dunes constituting element SB. The deposits are typically dominated by either lithofacies Sp or St, but not both together, because these lithofacies represent dune types formed under different hydraulic conditions. Extensive references to examples are given in Sect. 6.5, and a typical example is illustrated in Fig. 4.5. Simple macroforms may be present, such as the cross-bedded or plane-bedded simple bars of Allen (1983a), and the Platte-type macroforms of Crowley (1983), but available information on these deposits indicates that the lithofacies sets are typically not organized into higher-order lithosomes, and bounding surfaces of fourth-order are uncommon. The architecture of the deposits tends to be that of simple, tabular sandstone sheets. Cyclic successions of autogenic origin, such as are common in deeper rivers of braided and meandering type, are rare in Platte-type river deposits.

8.2.13 Deep, Perennial, Sand-Bed Braided River

In braided rivers larger and deeper than the Platte, the channel topography shows greater differentiation between channel floor, bar, and bar-top environments, and the corresponding depositional facies are similarly more complex and varied. Most of the constructional activity takes place within large macroforms that have variously been termed compound bars (Miall 1977), sand flats (Cant and Walker 1978), or sand shoals (Allen 1983a). Miall (1978c) suggested that the South Saskatchewan River of southern Saskatchewan provided a good type example of this type of river, based on the work of Cant and Walker (1976, 1978). An architectural model is illustrated in Fig. 8.62.

The most characteristic depositional feature of this type of river is the presence of macroforms that contain internal accretionary geometries, quite different from the tabular bedding style typical of the shallow braided rivers described in the previous section (Figs. 8.8M, 8.63). The macroforms are bounded by fourth-order surfaces (which may or may not be preserved, depending on the amount of scour that occurs between depositional events), and commonly

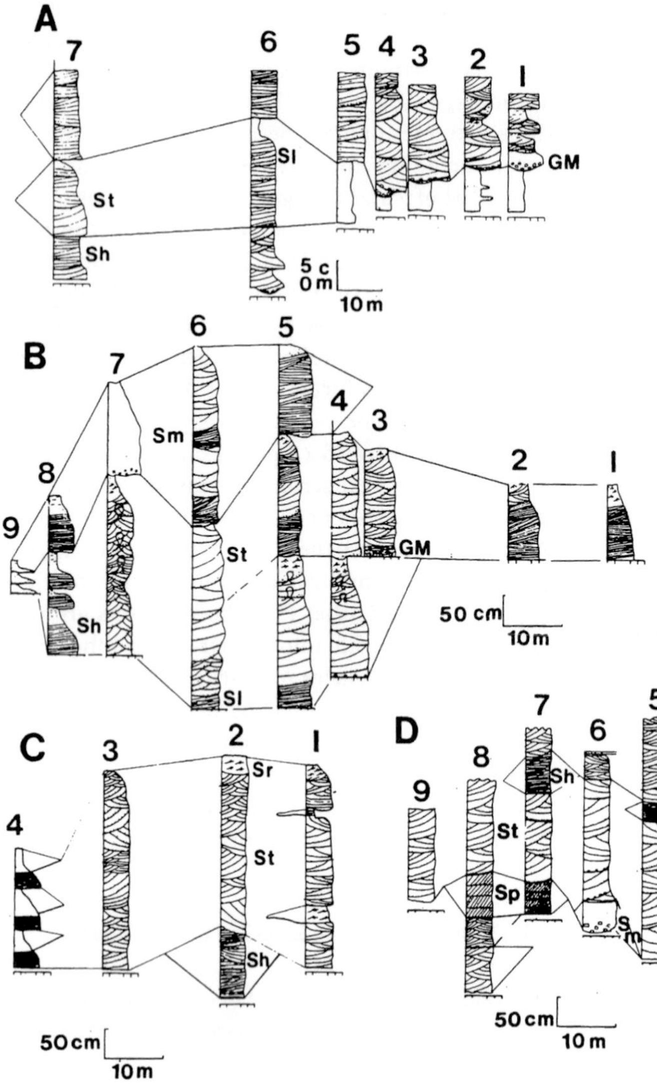

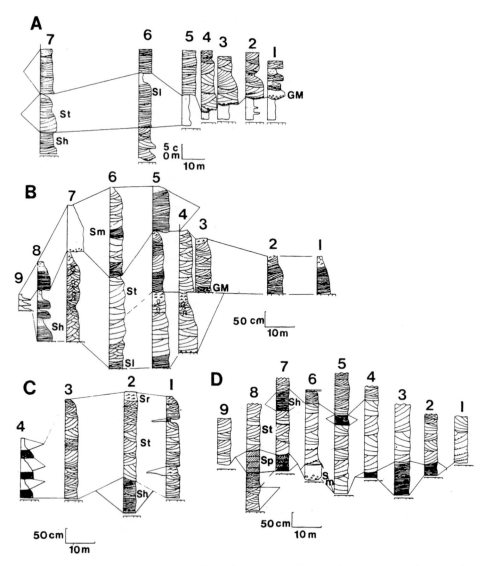

Fig. 8.51A–D. Vertical profiles of the channel sandstones illustrated in Fig. 8.50. (Eberth and Miall 1991)

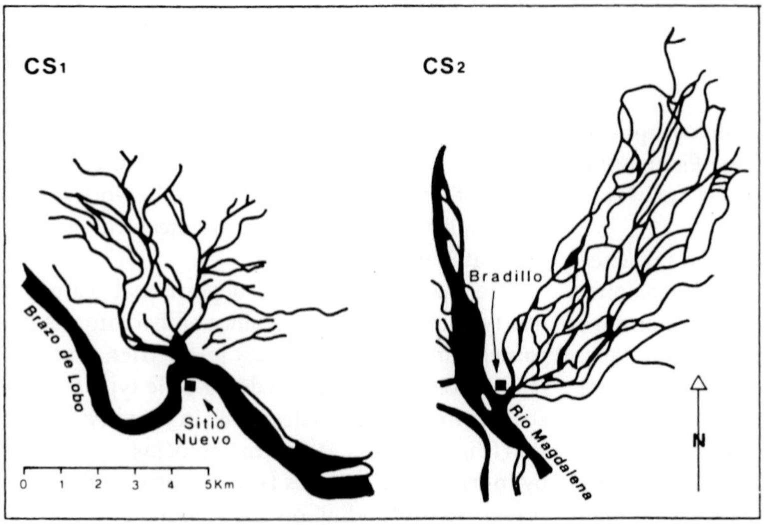

Fig. 8.52. Maps of two large splays of the modern Magdalena River, Colombia. (Smith 1986)

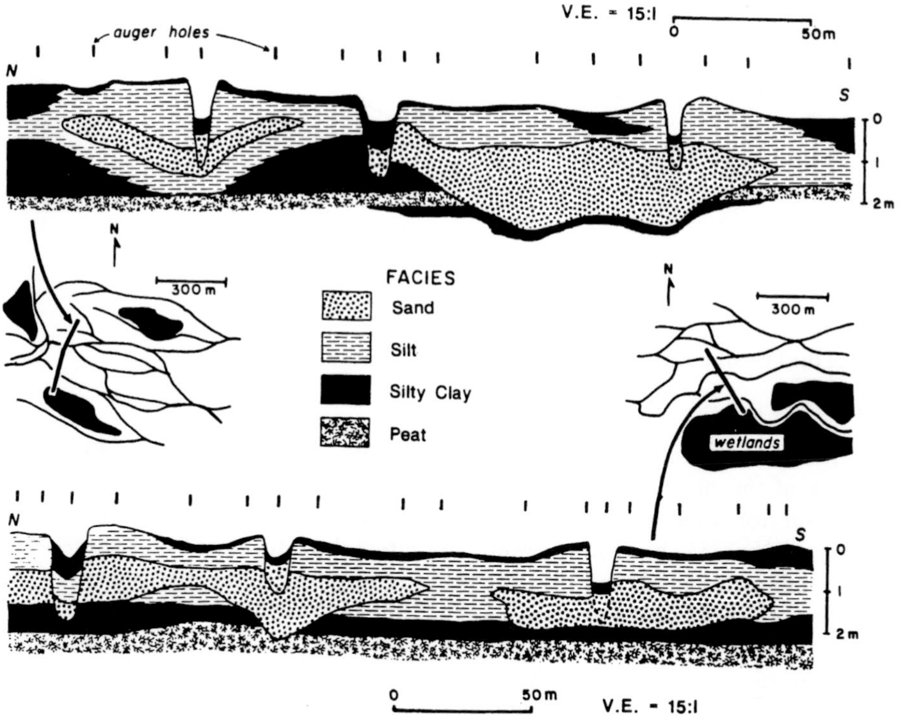

Fig. 8.53. Cross sections through two abandoned splays in the Cumberland Marshes, Saskatchewan. (N.D. Smith et al. 1989)

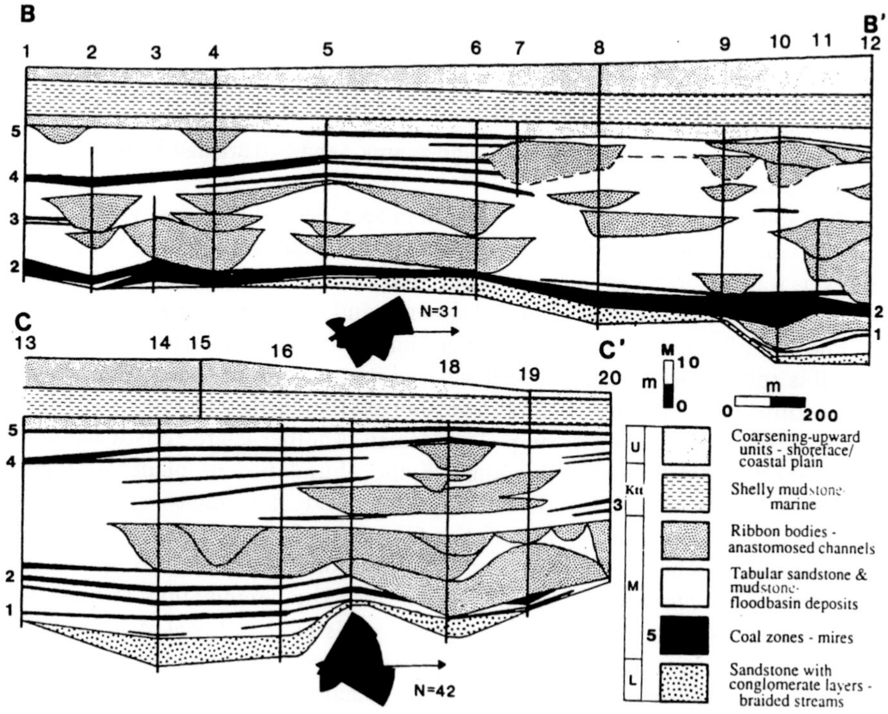

Fig. 8.54. Cross sections through an interpreted ancient anastomosed fluvial deposit, Dakota Sandstone, southern Utah. The anastomosed interval is that characterized by ribbon sandstones with floodbasin mudstones and coals (*unit M in the legend at right*). Note the relatively low paleocurrent dispersion. (Kirschbaum and McCabe 1992)

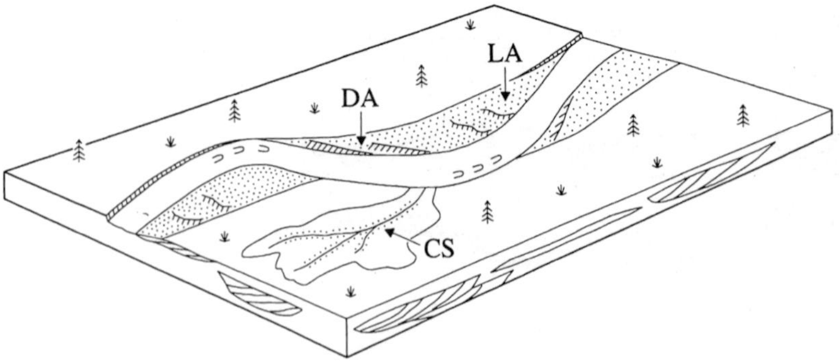

Fig. 8.55. Architectural model of the low-sinuosity river, with alternate bars

consist of successive growth increments separated by third-order surfaces. Many of the macroforms show an upward decrease in grain size or the scale of sedimentary structures (Haszeldine 1983a,b), and fine-grained bar-top facies, including thin muds and carbonaceous deposits, may be preserved (Cant and Walker 1978). Therefore, vertical profiles through the macroforms commonly constitute fining-upward successions. The macroforms may be classified as downstream-accretion elements (DA) or lateral-accretion elements (LA), or may contain both geometries in different parts of the same bar complex. Their sedimentology is described in detail in Sects. 6.6 and 6.7.

Early descriptions of rivers of South Saskatchewan type, by Cant and Walker (1976, 1978) did not discuss the three-dimensional geometry of the macroforms because little deep trenching was carried out as part of that study. Subsequent work in ancient deposits (references in Table 8.3) has shown that accretionary geometries are common in ancient braided river deposits. However, it needs to be emphasized that this work, and the distinction that it suggests between the "shallow" and "deep" sandy braided systems described in this and the preceding section, are based entirely on architectural interpretations of the ancient record. Data derived from small-scale trenches in modern sandy braided rivers reveal very similar lithofacies assemblages. There is a need for architectural studies of modern sandy braided deposits, perhaps using ground-penetrating radar, comparable to the studies of modern point bars reported by Gawthope et al. (1993; see Sect. 9.5.5).

Sandy braided rivers vary enormously in scale. The reaches of the South Saskatchewan River described by Cant and Walker (1978) are about 600 m wide, with individual channels ranging from 70–200 m in width. They average 3 m in depth, with deeper scour pools. The Brahmaputra River, above its confluence with the Ganga, in Bangladesh, averages 10 km in width, with a maximum depth of 45 m. Even the smallest-scale individual channels are up to hundreds of meters wide. As shown by Bristow (1987, 1993b), the internal arrangement of channels and bars and the architecture of the macroforms of this river are not unlike those of smaller braided rivers, although greatly scaled up. A particular challenge in the analysis of this type of river is therefore to determine the scale of the river and its constituent channels and macroforms from the complex hierarchy of preserved lithosomes. An ancient example of this type of river, comparable in scale to the large modern rivers of the Indian foredeep, was described in detail by Willis (1993a,b). Bars in this ancient system are up to 3 km in length.

8.2.14 High-Energy, Sand-Bed Braided River

As discussed in Sect. 6.9, a distinctive type of scoop-shaped architectural element, termed a hollow (element code HO) is common in certain types of braided systems, and may be rare or absent in others. The recognition of this element type in the ancient record is based on the work of Cowan (1991) in the Westwater Canyon Member of the Morrison Formation, New Mexico. Another characteristic feature of this unit is the abundance of plane lamination (lithofacies Sh) and low-angle cross-bedding (lithofacies Sl), which Cowan (1991) interpreted as the result of common transitional- to upper-flow-regime conditions. The combination of these features suggests a distinctive fluvial style, in which sedimentation takes place during high-energy, possibly shallow, discharge events. These events lead to considerable scour of the channel floor, and the planing off of the tops of macroforms such as DA and LA units, so that these may not even be recognizable in the resulting deposits. Scour hollows are formed at channel confluences, at tributary junctions, and below macroforms where flow converges. The rapid

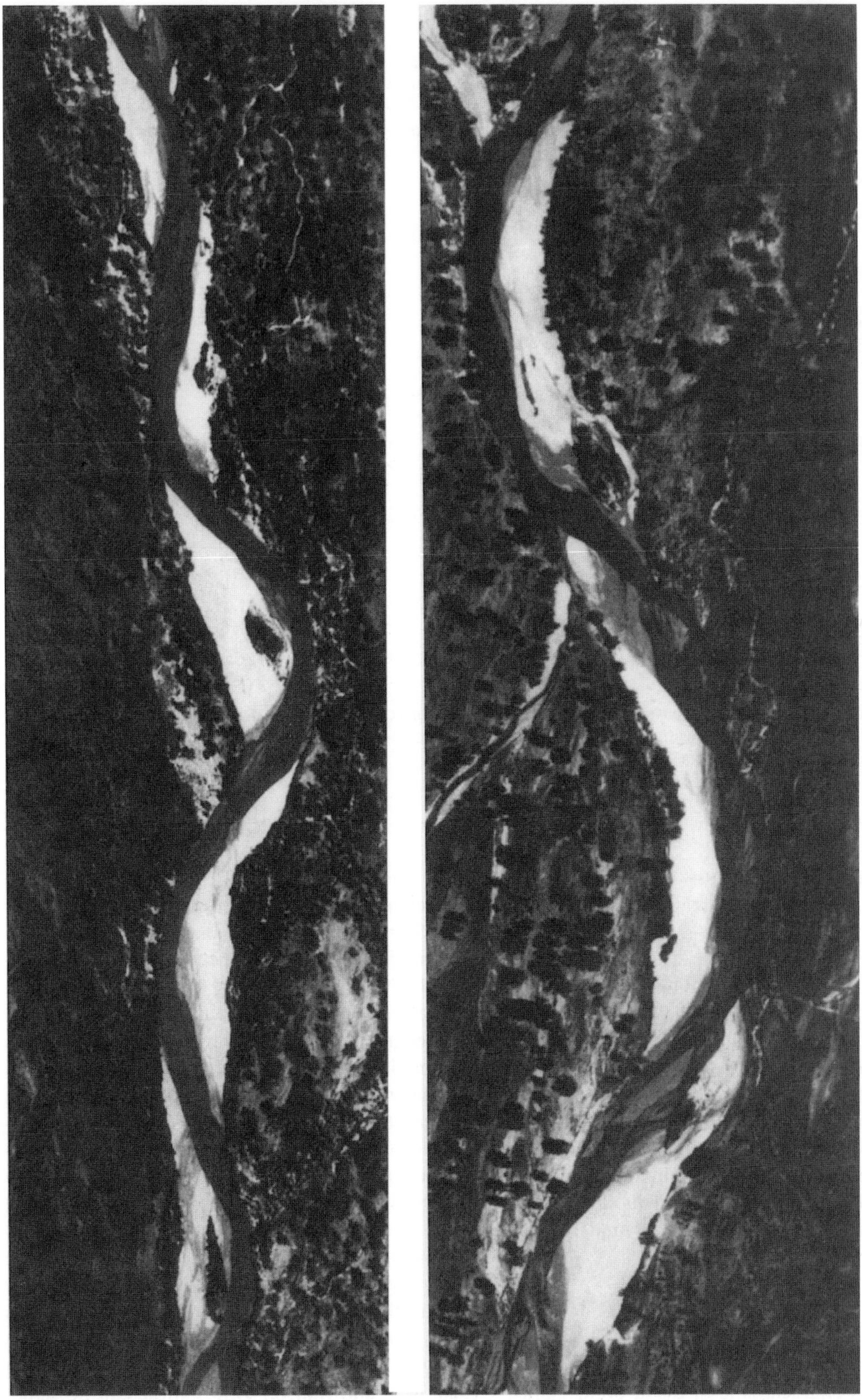

Fig. 8.56. Aerial photographs of the South Platte River, Colorado, showing reaches that are organized into low-sinuosity channels with alternate bars. Photo from the US Soil Conservation Service. (Crowley 1983)

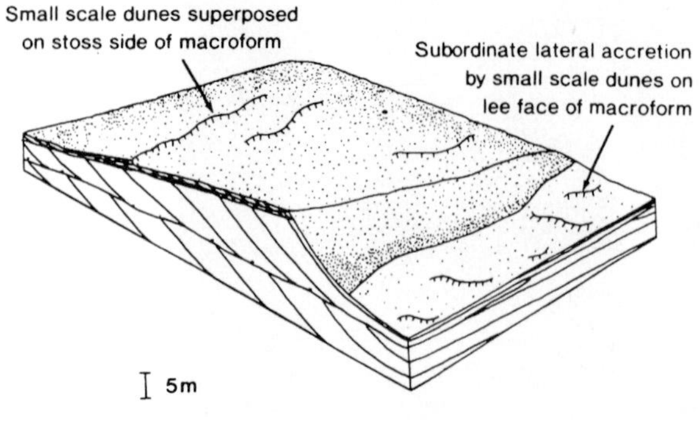

Fig. 8.57. Block diagram model of an alternate bar in the Pennsylvanian Lee Formation, central Appalachian Basin. The main slip face is oriented in a downstream to obliquely-downstream direction. (Wizevich 1992b)

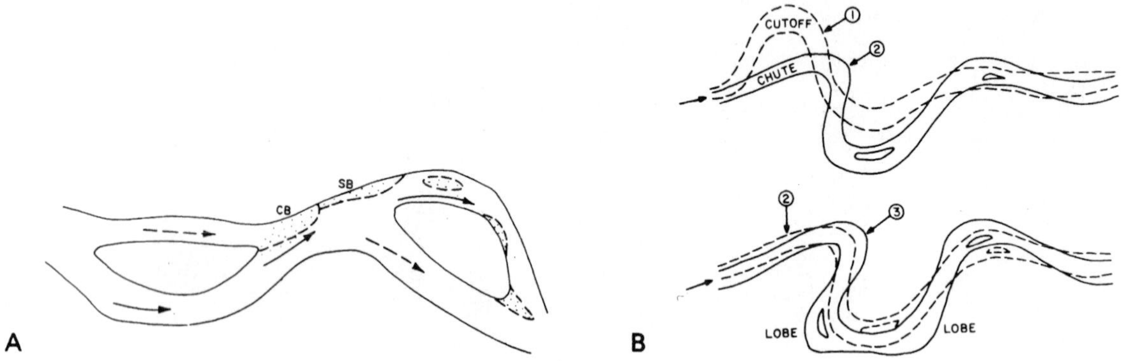

Fig. 8.58. Evolution of channels and bars in low-sinuosity, single-channel rivers, based on the Calamus River, Nebraska. **A** Initiation of chute cutoffs (*dashed arrows*). *CB* Chute bar; *SB* scroll bar. **B** Changes in channel geometry following chute cutoff. successive stages in channel position are shown by the numerals *1*, *2*, and *3*. (Bridge et al. 1986)

flow conditions are the cause of the distinctive lithofacies assemblage. An architectural model of this fluvial style is shown in Fig. 6.44, and a typical vertical profile is given in Fig. 8.8N.

Although the HO element has been observed in a few other stratigraphic units (Miall 1993; Stephens 1994) it does not constitute a major component of the architectural assemblage of these units, and it remains for future work to demonstrate whether the Westwater Canyon Member of the Morrison Formation is anomalous. It is suggested that a gradation into the flashy, ephemeral fluvial style (Sect. 8.2.16) may occur in some cases. This fluvial style has not yet been documented in any modern river system.

8.2.15 Distal, Sheetflood, Sand-Bed River

This model is characteristic of distal braid plains, particularly in arid regions where ephemeral runoff forms a network of shallow, interlacing, possibly poorly defined channels. The sedimentology of such rivers was well described by Williams (1971). An architectural model is given in Fig. 8.64. Terminal fans (Sect. 8.3) may consist partly or largely of the deposits formed in this fluvial style (Parkash et al. 1983; Kelly and Olsen 1993).

The sediments are dominated by sheets, lenses, and wedges of element SB, consisting of a wide range of sandy lithofacies. Flood cycles up to a few meters thick may be preserved (e.g., Miall and Gibling 1978; Muñoz et al. 1992; Figs. 6.22, 8.8O, 8.65). Fine-grained overbank deposits are likely to be rare to absent, as the rivers, when active, tend to overtop their banks and form broad sheets transporting bed load.

A variation on this fluvial style has been identified in the arid basins of interior Australia, where it now appears that thick mud beds are not overbank deposits but were formed as accumulations of sand-sized, pedogenic mud aggregates (Nanson et al. 1986; Rust and Nanson 1989). These can be transported as sand grains and may form hydrodynamic sedimentary structures, but the distinctiveness of these may be lost as a result of diagenesis. Micro-

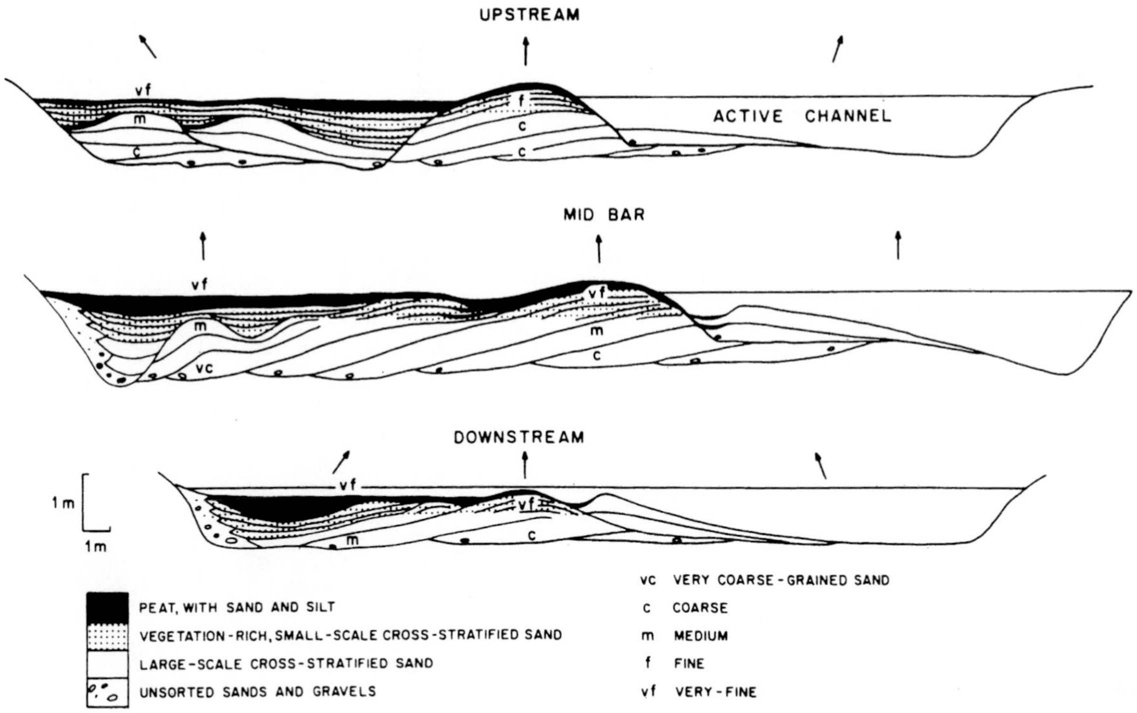

Fig. 8.59. Generalized model of a midchannel macroform in a low sinuosity river. Three sections are shown, oriented perpendicular to channel trend. At the *center*, are obliquely downstream-oriented accretion deposits of a midchannel macroform consisting of cross-stratified sand and gravel. Fine-grained, vegetation-rich deposits fill a chute channel at *left*. The main active channel is shown at *right*. Late-stage lateral accretion deposits are shown on the midchannel flanks of this channel, reflecting downstream translation of the adjacent meander bend. (Bridge et al. 1986)

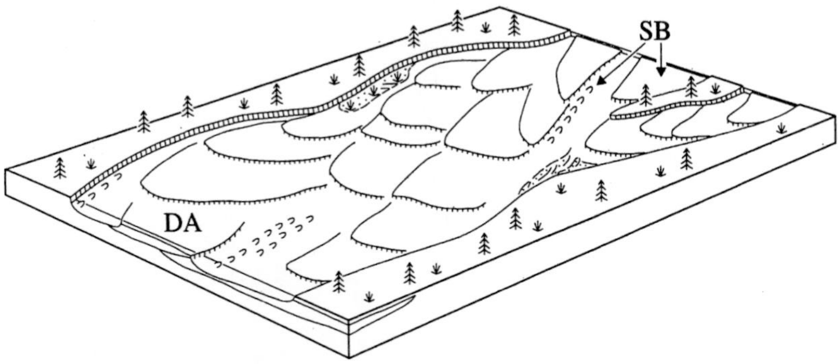

Fig. 8.60. Architectural model for the shallow, perennial, sand-bed braided river. Model 9 of Miall (1985)

scope examination of the mudstones may, however, reveal relict "clotted" textures and other pedogenic features. Ancient examples of this fluvial style have been described by Rust and Nanson (1989), Ekes (1993), and Talbot et al. (1994).

A particularly common association in hot, arid regions is that between eolian dunes, interdune pond deposits, and ephemeral-fluvial flash-flood deposits. Fluvial deposits typically have the character described in this section. Ancient examples include the interbedding of the fluvial Kayenta and eolian Navajo formations in Arizona (Middleton and Blakey 1983; Herries 1993), the Silurian Tumblagooda Sandstone in Western Australia (Trewin 1993a), Triassic rift-fill deposits in the Irish Sea area (G. Cowan 1993), and the Permian Rotliegende Formation of the North Sea Basin (Glennie 1972, 1983; George and Berry 1993).

Fig. 8.61. Aerial views of the Platte River, Nebraska. Photos from US Soil Conservation Service. (Crowley 1983)

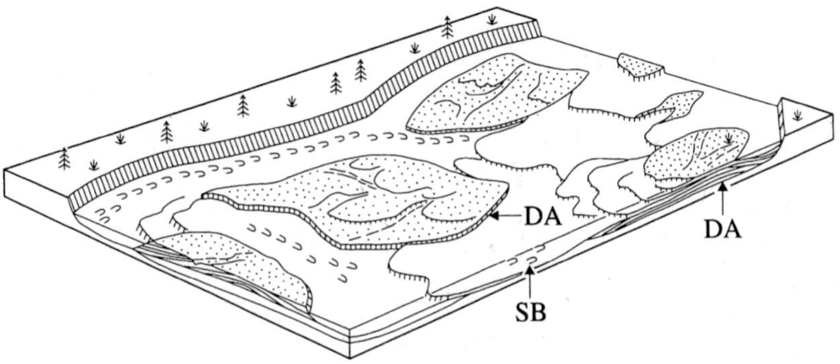

Fig. 8.62. Architectural model for the deep, perennial, sand-bed braided river. Model 10 of Miall (1985)

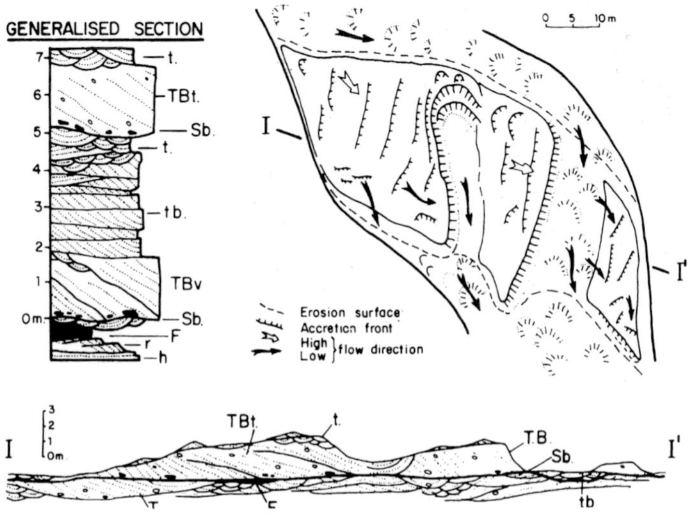

Fig. 8.63. Vertical profile through a macroform in a deep, perennial, sandy-braided river, showing architectural reconstructions of the cross-section profile and plan view of the bar. (Ramos et al. 1986)

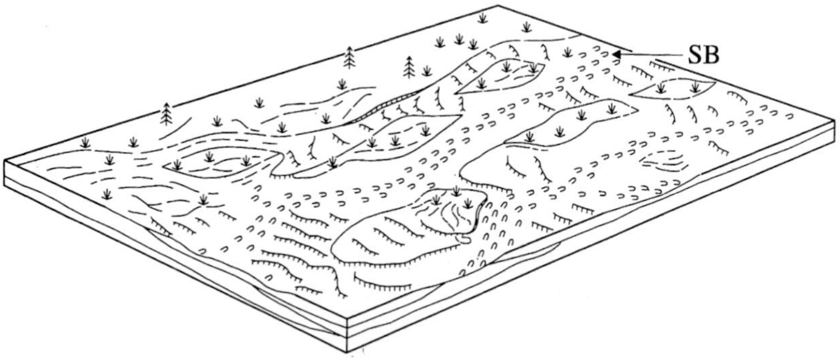

Fig. 8.64. Architectural model of the distal, sheetflood, sand-bed river. Model 11 of Miall (1985)

8.2.16 Flashy, Ephemeral, Sheetflood, Sand-Bed River

This fluvial style was distinguished from the distal, sheetflood braided model by Miall (1985) based on a recognition of differences in the lithofacies assemblages. Miall (1977, 1978c) had proposed a vertical-profile facies model named the "Bijou Creek-type", based on a comparison with flash-flood deposits in the creek of that name in Colorado, which had been described by McKee et al. (1967). The difference between this and the previous model is that in this model high-velocity, flashy discharge leads to the preferential deposition of transitional to upper flow-regime beds, dominantly plane-laminated sand (lithofacies Sh) and low-angle cross-bedded sand (Sl). Together, successions of these lithofacies con-

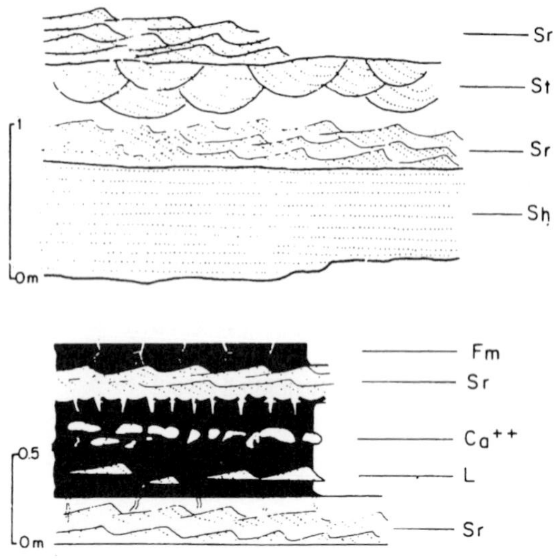

Fig. 8.65. Typical vertical profiles of the "second stage" of Upper Buntsandstein (Triassic) sedimentation, showing flood cycles. (Muñoz et al. 1992)

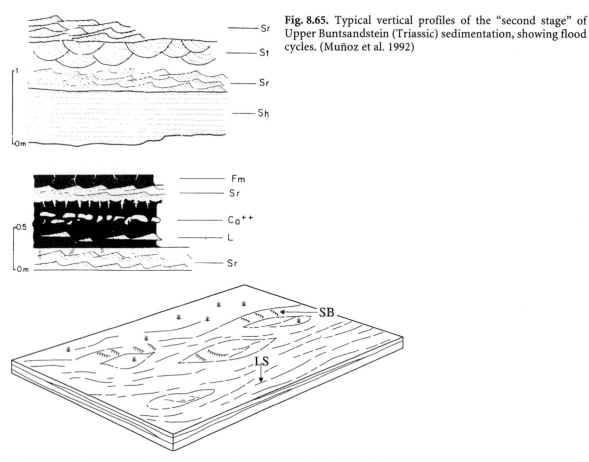

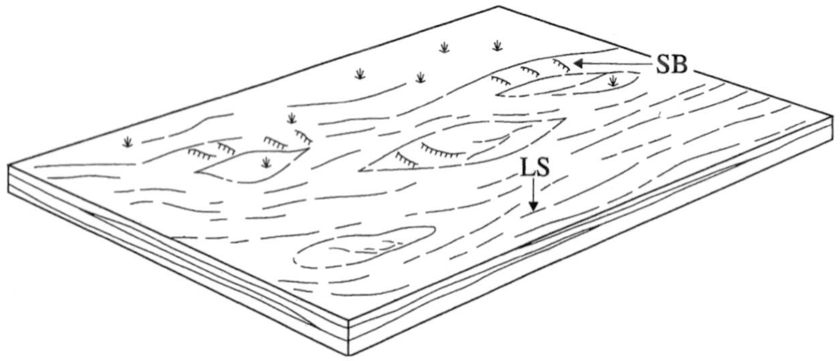

Fig. 8.66. Architectural model of the flashy, ephemeral, sheetflood, sand-bed river. Model 12 of Miall (1985)

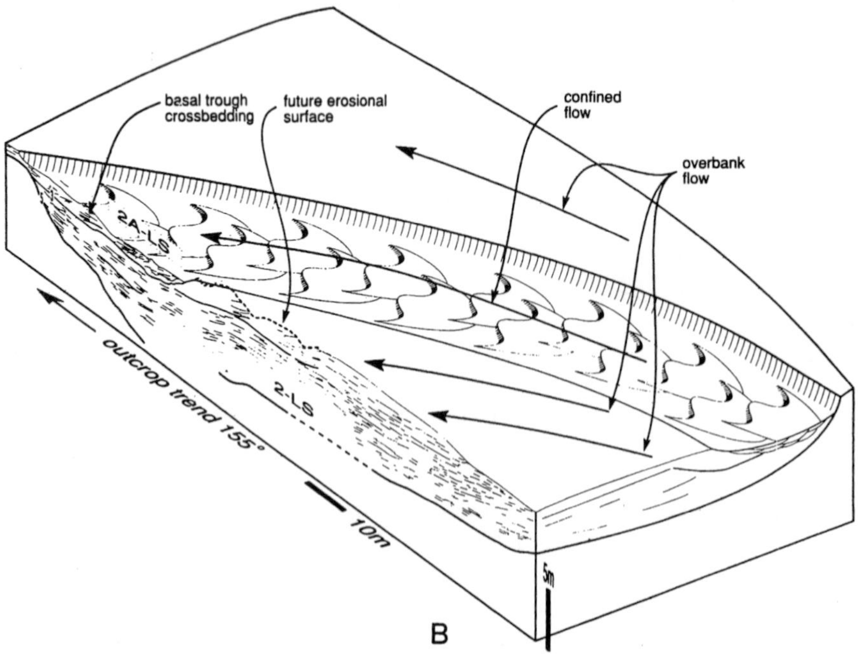

Fig. 8.67. Simplified lateral profile and interpreted block-diagram facies model for part of the Kayenta Formation (Jurassic) of the Mesa Creek area, western Colorado. (Bromley 1991a)

stitute sandstone sheets of the element LS (Figs. 8.8P, 8.66). Cotter and Graham (1991), in their studies of ephemeral-fluvial sandstones of Devonian age in Ireland, described examples of lithofacies Sl which they likened to hummocky cross-stratification, and concluded that the facies represented deposition from transitional to upper-plane bedforms, similar to those described by Cowan (1991). Channels are poorly defined or absent altogether. Flood cycles may be recognizable, showing cut-and-fill relief along their basal bounding surfaces. Overbank fines are rare to absent. Antidunes are common in the modern stream deposits described by Langford and Bracken (1987), but do not appear to have a very high preservation potential. Evaporite crusts may occur on the upper surface of sand beds, and disruption of bedding by evaporite crystallization is common (Mertz and Hubert 1990).

The succession of events in a single flood cycle was described by Bromley (1991a), based on his studies of the Jurassic Kayenta Formation of western Colorado. He mapped a broad channel with gently dipping margins that appears to have focused flow in the initial stages of the flood (Fig. 8.67). At the beginning of the flood event, a phase of channel-floor erosion and dune sedimentation occurred. During peak discharge, flow extended in a shallow sheet over a broad area, beyond the confines of the original, poorly defined channel, and sheets of Sh were deposited. During the waning flood stage, incision of these sheets occurred, and pockets of two- and three-dimensional dunes (lithofacies Sp, St) were deposited.

8.3 Alluvial Fans and Other Fluvial Distributary Systems

Alluvial fans are a form of fluvial depositional system distinguished on the basis of geomorphic character rather than by a characteristic fluvial style. As discussed here, the term is in broad, general use amongst sedimentologists, but there is considerable confusion and disagreement as to the range of types of fluvial depositional systems that should be included under the heading of alluvial fans.

An alluvial fan is defined in the Glossary of Geology (Bates and Jackson 1987, p. 17) as:

"A low, outspread, relatively flat to gently sloping mass of loose rock material, shaped like an open fan or a segment of a cone, deposited by a stream (esp. in a semiarid region) at the place where it issues from a narrow mountain valley upon a plain or broad valley, or where a tributary stream is near or at its junction with the main stream, or wherever a constriction in a valley abruptly ceases or the gradient of the stream suddenly decreases; it is steepest near the mouth of the valley where its apex points upstream, and it slopes gently and convexly outward with gradually decresing gradient."

Miall (1990, p. 345) suggested that:

"the term alluvial fan may be used for any entirely nonmarine and nonlacustrine fluvial system whose channel network is distributary rather than contributary. Fans that prograde directly into a standing body of water are termed fan deltas or coastal alluvial fans. ... Alluvial fans and fan deltas are a sedimentary response to flow expansion at a basin margin."

The term "fan" places natural emphasis on the surface geomorphology of this depositional system. The radiating distributaries and the cone-shaped architecture of the deposit are the two features that would be the key elements of most geologists' definition of the term alluvial fan. However, there is considerable controversy amongst sedimentologists as to the scope of the definition beyond this initial point. Miall's (1990) definition (which reflects widespread usage) implies that the term has no unique facies sense. As noted by Miall (1992, p. 134), there are small, steep, gravel-dominated fans, within many of which debris-slow processes are important; there are sandy systems deposited by braided streams; there are giant modern fans, such as the Kosi of India, which grades from boulder conglomerate near the mountains to fine sand-silt-mud 140 km downslope at its distal end; and there are fine-grained, arid systems, termed "terminal fans" deposited by ephemeral flows that percolate and dissipate into the substrate at their distal fringes. Geomorphologically, these are all alluvial fans, and all fall under the definitions of this term provided by Bates and Jackson (1987) and Miall (1990, 1992). A recent review by Stanistreet and McCarthy (1993) has followed this inclusive approach. They described a new class of large, low-energy fan based on the example of the modern Okavango fan of Botswana, and erected a simple classification of alluvial fans, including all the types mentioned above. This is described below. They also discussed comparisons with submarine fans, a class of depositional system about which there have also been problems of definition.

Contrasting with this approach is that of McPherson and Blair (1993) and Blair and McPherson (1994). These authors accepted many of the standard components of the geomorphic definition of fans, noted above, including their conical, radiating form, their steep depositional slopes, and their association with high-relief source areas. They pointed out a significant contrast between fans and most other fluvial systems. The latter typically exhibit concave-up transverse profiles, the channel being the lowest point, whereas fan systems are characterized by convex-up transverse profiles, which lead to overbank flow and sheet flooding of bed load. However, these authors argue that there is a distinct break in the range of longitudinal slopes exhibited by the various types of fluvial distributary system. They recommended limiting the term alluvial fan to steep systems, which have a range of slopes between 1.5° and 25° (slope: 0.026–0.466). These fans are characterized by sediment-gravity-flow deposits, especially debris flows, by hyperconcentrated flows and sheetfloods. Traction-current deposits are rare to absent. According to the same authors, fluvial depositional systems having slopes in the range 0.4–1.5° (0.007–0.026) are rare, with most meandering and braided fluvial systems, including the radial, distributary systems that many authors would classify as fans, exhibiting slopes less than 0.4°. This provides a natural break, which they proposed to emphasize in their classification. They recommended excluding depositional systems on slopes of < 1.5° (< 0.026) from the category of alluvial fans, mainly on the basis of an historical analysis of the evolution of the term fan. They pointed out that the original work of such researchers as Drew (1873) and Surell (1841, 1870), which helped to establish the term "fan", dealt with small, steep, coarse-grained fans (Sect. 2.2.2). Heward's (1978a) important review of fan depositional cycles also focuses on coarse-grained fans, although he did not discriminate or subdivide fans on the basis of sediment transport mechanisms. Blair and McPherson (1994) suggested that the inclusion of fan-shaped glacial outwash dispersal systems under the heading of the term alluvial fan serves no useful purpose.

Following this approach, fine-grained, fan-shaped deposits, such as those of the Kosi River, terminal fans, and those not associated with mountainous uplifts, would not be included under the heading of alluvial fans. They would be termed distributary fluvial systems or braid deltas. While this proposal is clear, and may be applied with some consistency, at least to modern environments where slopes can be measured, it leaves many depositional systems and deposits that have traditionally been termed fans out of the classification. This may be of little importance in the case of sandy systems, such as the Kosi fan, even though the pattern of channel wandering and avulsion documented on this fan (Holmes 1965; Gole and Chitale 1966; Wells and Dorr 1987; Singh et al. 1993) has been cited as a "textbook" example of one style of channel evolution within fans (e.g., Collinson 1986). Also, the term "terminal fan" (Mukerji 1976; Friend 1978; Parkash et al. 1983) is now widely used (e.g., Kelly and Olsen 1993), but most terminal fans are not alluvial fans according to the definition of Blair and McPherson (1994).

Continuing detailed work on modern and Recent fans suggests that the proposals of Blair and McPherson (1994) may need modification. Nemec and Postma (1993) provided a detailed description of some Quaternary alluvial fans in Crete that do not fit their definitions. A suite of fan deposits there has preserved depositional dips of up to 13°, yet the deposits consist largely of clast-supported streamflow gravels forming gravel sheets (element GB).

Use of the restricted definition of Blair and McPherson (1994) will lead to the recognition of fans in the ancient record on the basis of their facies characteristics alone, the deposits being coarse-grained debris-flow and related facies banked against an uplifted source area. Such classification already is implied by the use of such terms as "fanglomerate" for ancient conglomerates, even where evidence of the geomorphology of the original depositional system is not available. Such a designation will likely be suggested without evidence of the distributary nature of the fluvial system, a feature that is difficult to prove in the rock record without detailed paleocurrent evidence. Whether the use of the Blair and McPherson (1994) definition in this way will prove to be useful and will serve to reduce controversy and confusion remains to be learned from experience. In the meantime, the more inclusive classification of Stanistreet and McCarthy (1993) is provided below.

In this classification (Fig. 8.68), it is recognized that various fluvial processes may occur in different proportions, depending on climate, the nature of the source terrane, etc. This led Stanistreet and McCarthy (1993) to propose a triangular classification that illustrates the varying importance of the three main processes, sediment-gravity flows, braiding, and meandering (Fig. 8.69). The first class of fan in their classification is the debris-flow-dominated fan, the only type accepted by Blair and McPherson (1994) as a "true" fan. These fans are small, typically less than 10 km in radial length (Fig. 8.70), and steep (slope > 0.1). They are particularly common in arid regions, where sediment-gravity flow processes are more common (but are not exclusively found in this climatic setting, as detailed in Sect. 12.5). Blair and McPherson (1994) recognized two varieties. The first type are those constructed mainly of sediment gravity-flow deposits, together with colluvial slides, bedrock slides, rock avalanches, and rockfalls. Sandy interbeds are not common. Secondary processes, including overbank sheet flow, wind erosion, weathering, bioturbation, soil development, and case hardening are common. The second variety is built primarily of sheet flow, and hyperconcentrated flow deposits. Sandy overbank deposits are common, and a distal sand fringe may also be present.

The other two types are the braided fluvial fan (e.g., tributary fans seen in Fig. 8.17B), and what Stanistreet and McCarthy (1993) termed the *losimean* fan, an abbreviation of "low sinuosity-meandering". The latter type is exemplified by the Okavango fan of Botswana, the largest subaerial fan system yet described, at 150 km in axial length. Depositional processes in these braided and meandering systems are those of the respective fluvial styles, described in Sect. 8.2. As discussed by Stanistreet and McCarthy (1993), such "fluvial" fans as the Kosi and the Okavango, are difficult to recognize in the ancient record, particularly in the subsurface where the absence of paleocurrent data would

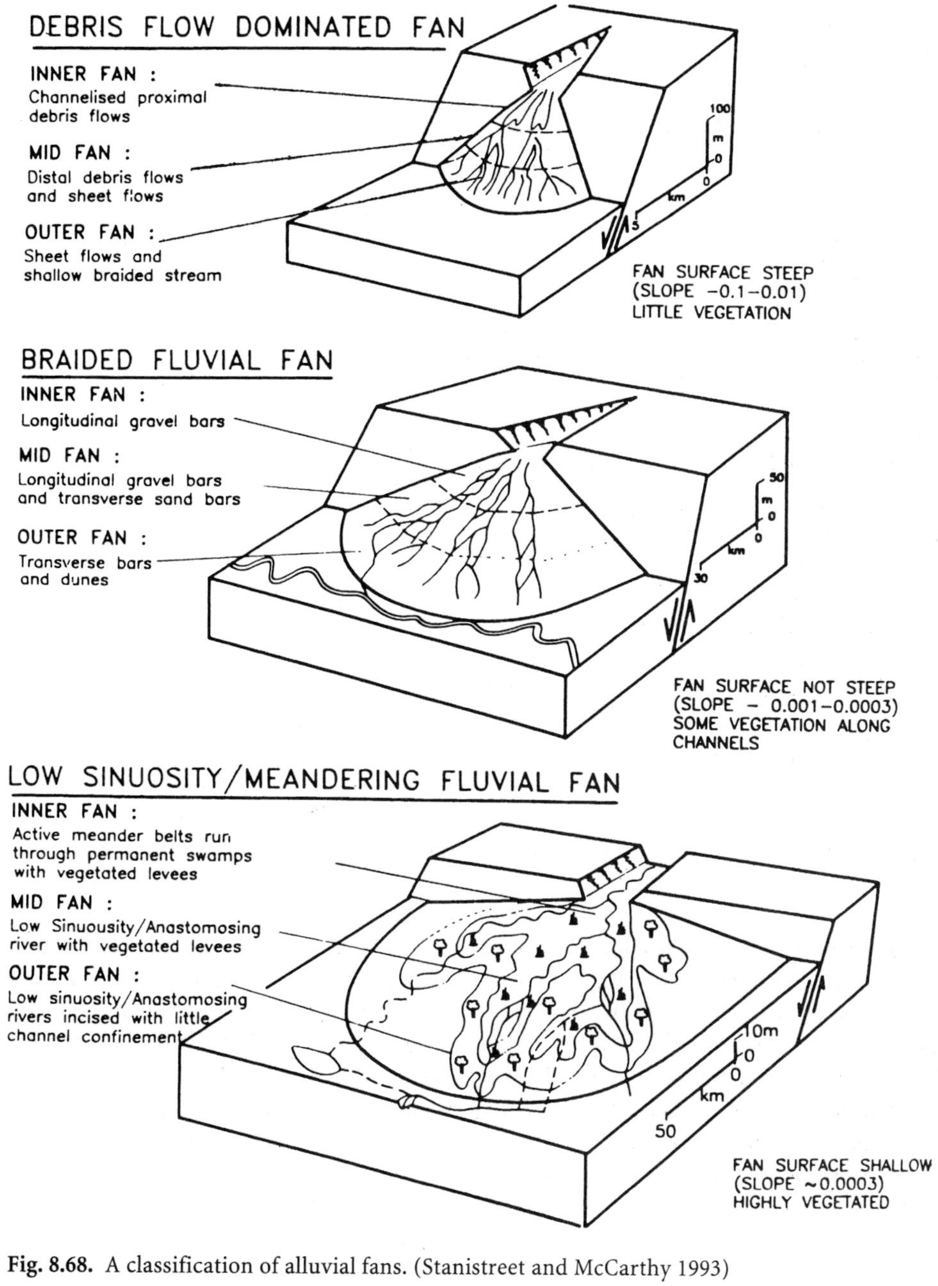

Fig. 8.68. A classification of alluvial fans. (Stanistreet and McCarthy 1993)

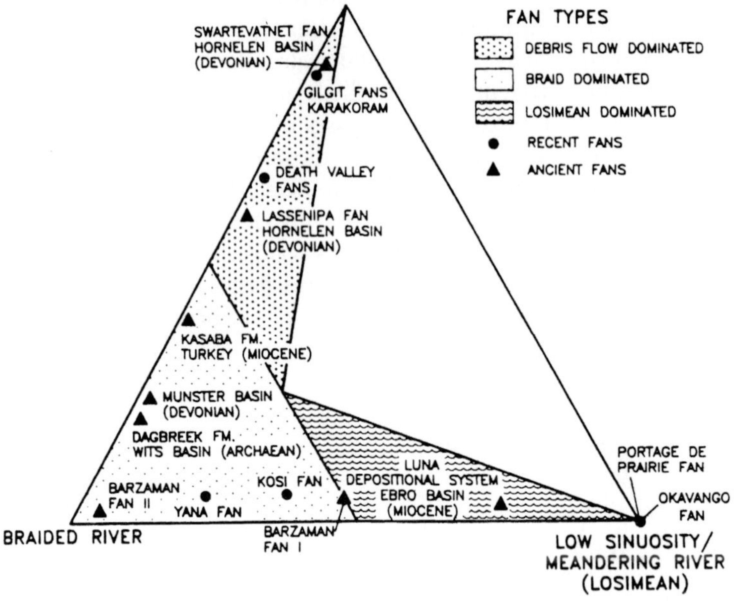

Fig. 8.69. A triangular classification of alluvial fans, showing the subdivision into the three main depositional styles. (Stanistreet and McCarthy 1993)

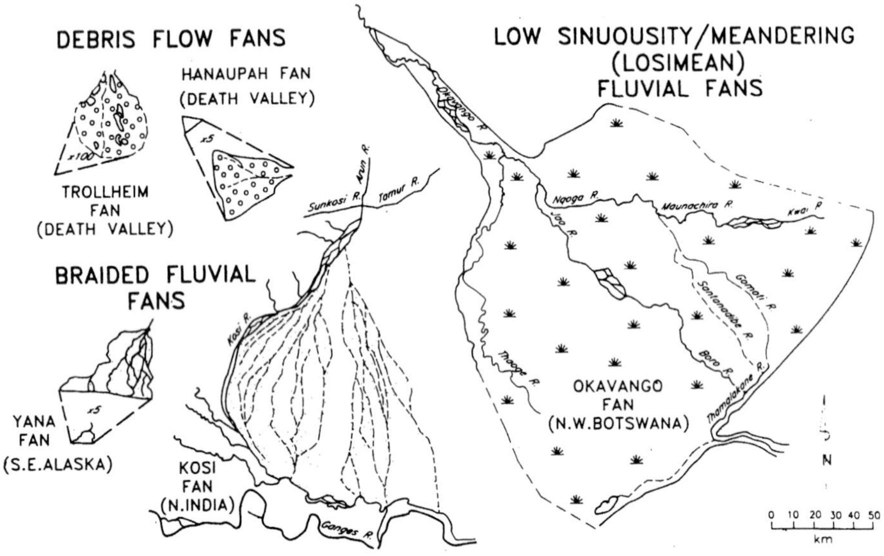

Fig. 8.70. A comparison of the sizes of modern alluvial fan systems. (Stanistreet and McCarthy 1993)

make the radial dispersal pattern almost impossible to prove. Two examples of sandy systems from the ancient record were described by Nichols (1987) and Hirst (1991). A gravelly system was described by Miall (1970a). In the first two cases, paleocurrent data were used to reconstruct radial dispersal patterns. In the gravel example, clast counts were interpreted in terms of radial dispersal within a system of laterally coalescing alluvial fans.

Some ancient fluvial deposits are marked not only by a downstream decrease in grain size, but by a decrease in interpreted channel scale, and a distal transition into fine-grained flood basin or lacustrine deposits. Friend (1978) was the first to discuss the sedimentological implications of these characteristics. He suggested the terminal fans of India as analogs for these ancient systems (Mukerji 1976), and a modest literature has now developed in which

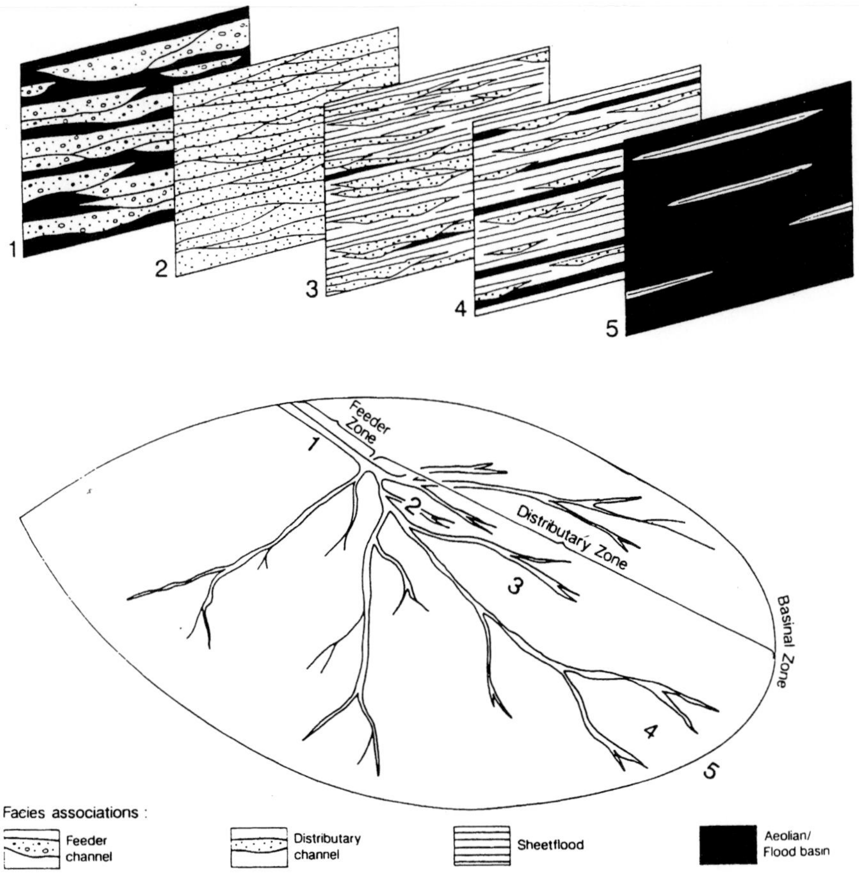

Fig. 8.71. A facies model for terminal fans. *1* Feeder zone, possibly with more than one channel; *2–4* correspond to proximal, medial, and distal fan zones; *5* floodbasin. (Kelly and Olsen 1993)

his embryonic ideas have been applied to sedimentological interpretations (Parkash et al. 1983; Abdullatif 1989). A useful summary, with additional examples, was provided by Kelly and Olsen (1993), who developed the general sedimentological model illustrated in Fig. 8.71. They concluded that individual fans are unlikely to exceed 100 km in radius. Fluvial styles within terminal fans are characterized by broad, shallow channels in the proximal and feeder zones (typically the shallow, perennial type of Sect. 8.2.12), grading distally into sheet-flood styles with poorly defined channels (the distal, sheet flood, and ephemeral styles of Sects. 8.2.15 and 8.2.16).

DeCelles et al. (1991) pointed out the hierarchical nature of alluvial fan deposits, and applied Miall's (1988a) bounding-surface hierarchy to an analysis of some Paleocene conglomerates in the Rocky Mountains. These sediments are characterized by stream-flow facies deposited mainly under lower flow-regime, traction-current conditions, and would be classified as the deposits of distributary fluvial systems or braid deltas by Blair and McPherson (1994). The classification of DeCelles et al. (1991) is described and illustrated in Sect. 4.5 (see Fig. 4.3), and is discussed in Sect. 10.2.

Other syntheses of alluvial fan sedimentology are given in the books edited by Nilsen (1985) and Rachocki and Church (1990). Alluvial fans and braid deltas are very sensitive indicators of basin-margin tectonism (Heward 1978a), a point discussed at length in Chaps. 10 and 11.

Chapter 9

The Stratigraphic Architecture of Fluvial Depositional Systems

9.1 Introduction

In the preceding chapters we have seen how to describe and interpret the various lithofacies components of a fluvial depositional system (Chap. 5) and how these components may combine to form channel fills and floodplain blankets (Chaps. 6, 7). Studies of lithofacies and channel geometry then permit the recognition of fluvial style, of which 16 typical examples are described (Chap. 8). These earlier chapters describe units comprising groups 1–7 of Miall's (Miall 1991b) hierarchical classification (Table 3.1). The next step is to reconstruct fluvial systems on a basin scale, that is, at the level of groups 8–11. How many separate fluvial systems were there in a given basin? Which way did they flow, and how did they interact? What is the nature of vertical and lateral transitions into other nonmarine facies and into marine deposits?

Table 9.1 provides a listing of examples of large-scale fluvial units and depositional processes organized in accordance with the hierarchical classification of Chap. 3. Many terms, such as cyclothem and megasequence, have been assigned a variety of different meanings. Also, terms such as first-, second-order, etc., have been employed in many conflicting ways for processes and products, with the numbering arranged in order from both large- to small-scale features and long- to short-term processes, and the reverse. The result has been that the use of "order" classifications has become very confused. It is hoped that the listing in Table 9.1 of examples of processes and depositional products according to the time scale they represent (recurrence interval) and the use of the nongenetic "group" classification will help to clarify terminology and provide insights into processes. Many more examples could be cited in this table. A representative selection of ancient examples has been given, in which constraints on timing and rates of sedimentation are available to enable the units to be classified into the appropriate group.

The construction of a basin-scale stratigraphy is a question of correlation and mapping, and it is the practical details of this procedure that are addressed later in this chapter. General discussions of stratigraphic correlation have been provided in several recent textbooks (e.g., Miall 1990, Chap. 3). Conventional stratigraphic methods are concerned with correlation at the formation or member level and are not the main interest here. The focus of this chapter is on methods whereby the stacking of individual large-scale elements of a fluvial succession, such as the meander belts and floodplains and alluvial fan lobes can be reconstructed using techniques of outcrop and subsurface mapping. On the larger scale, fluvial units combine into sequences of various types. Unless specific depositional elements can be traced, and their scale, geometry, and lateral relations determined, it is not possible to address the larger questions regarding the action of autogenic and allogenic sedimentary controls. Several of the examples of the various scales of stratigraphic unit discussed in Sects. 9.2–9.4 are used to illustrate mapping methods in Sect. 9.5.

9.2 Channel Belts

The subject of this section is the deposits of groups 7 and 8 (Table 9.1), corresponding to depositional units formed over time periods of 10^3–10^5 years. Over this time scale channels and channel belts are formed, and alluvial fans may undergo a progradational-trenching cycle. Deposits of group 8 compare in duration to the high-frequency (fifth-order) Milankovitch cycles of Vail et al. (1977) and Miall (1990).

Examples from studies of the Siwalik Group of Pakistan are illustrated in Figs. 2.32 and 9.1 (see also Fig. 7.13). The Siwalik sections were constructed from the measurement and physical correlation of numerous outcrops in areas of excellent exposure. The isochron and time scales were derived from

Table 9.1. Terminology and examples, large-scale fluvial stratigraphic architecture

Group[a]	Time scale years	Terminology and examples	References
6	10^2	Architectural elements:	Miall (1985)
	-10^3	e.g., crevasse splay, macroform.	
		Fan channels	DeCelles et al. (1991)
		Simple pedofacies sequences	Kraus (1987)
7	10^3	Channels.	
	-10^4	Autogenic channel-belt models	Allen (1974b), Bridge and Leeder (1979), Bridge and Mackey (1993)
		Sand-belt architecture models	Galloway (1981), Hirst (1991), Cowan (1991)
		Avulsion events (buff system, Siwaliks)	Behrensmeyer and Tauxe (1982), Behrensmeyer (1987)
		Fan progradation-trenching cycle (fifth-order lithosome)	DeCelles et al. (1991)
		Fan sequences.	Heward (1978a)
		Compound pedofacies sequences	Kraus (1987)
		Typical coal seams	Nemec (1988)
		Units I to VIII, Old Red Sandstone	Allen and Williams (1982)
8	10^4	Channel belt, alluvial fan.	
	-10^5	Long-term geomorphic thresholds	Schumm (1977)
		(third-order geomorphic cycles)	Schumm (1977)
		Fifth-order sequence: Milankovitch cycle:	Vail et al. (1977), Miall (1990)
		Return period of autogenic channel-fill Siwalik cycles	Johnson et al. (1985),
		Channel-belt migration (buff/blue-grey syst.)	Behrensmeyer and Tauxe (1982)
		Fan megasequences.	Heward (1978a)
		Major coarsening-upward cycles.	Gloppen and Steel (1981)
		Sixth-order lithosomes	DeCelles et al. (1991)
		Pedofacies megasequences	Kraus (1987)
		Devonian cycles, E. Greenland	Olsen (1990)
		Minor cycles	Heckel (1986)
		Incised valley-fills, Louisiana shelf	Suter et al. (1987)
		Channel-belt alloformations, Louisiana	Autin (1992)
9	10^5	Major depositional system, fan tract.	
	-10^6	Long-term geomorphic thresholds	Schumm (1977)
		(third-order geomorphic cycles)	Schumm (1977)
		Fourth-order sequence: Milankovitch cycle:	Vail et al. (1977), Miall (1990)
		Depositional-system axis	Galloway (1981)
		Seventh-order lithosomes	DeCelles et al. (1991)
		Pedofacies megasequences	Kraus (1987)
		Catenas of tectonic origin	Atkinson (1986)
		Cyclothems	Klein and Willard (1989)
		Minor cycles	Holdsworth and Collinson (1988)
		Major cycles	Heckel (1986)
		Castlegate Sandstone sequences, Utah	Van Wagoner et al. (1990, 1991), Miall (1993)
		Medium-order cycles, Tr-Jur, North Sea	Nystuen et al. (1989)
10	10^6	Basin-fill complexes	
	-10^7	Second-order geomorphic cycles	Schumm (1977)
		Tectonic cyclothems	Blair and Bilodeau (1988)
		Third-order stratigraphic sequences	Vail et al. (1977)
		Mesothems	Ramsbottom (1979)
		Major cycles	Holdsworth and Collinson (1988)
		Ancient examples:	
		Third-order Brazeau-Paskapoo cycles	McClean and Jerzykiewicz (1978)
		Tertiary formations, Gulf Coast	Galloway (1981)
		Karoo basin cycles, S. Africa	Visser and Dukas (1979), Turner (1983)

Table 9.1 (*Contd.*)

Group[a]	Time scale year	Terminology and examples	References
		c-u cycles of L. and M. Siwaliks	Johnson et al. (1985)
		Formations in U. Cret clastic wedge of Utah	Molenaar and Rice (1988)
		High-order cycles/megasequences of Triassic-Jurassic, North Sea	Nystuen et al. (1989)
			Steel and Ryseth (1990)
		Megacycles, Cutler Gp. New Mexico	Eberth and Miall (1991)
		Red Molasse (Olig), Digne, France	Crumeyrolle et al. (1991)
		Sequences of Straight Cliffs Fm, Utah	Shanley and McCabe (1993)
		Chinle Sandstone members, Utah	Dubiel (1991)
		Cañizar Sandstone sheets, Spain	López-Gómez and Arche (1993)
11	10^7 -10^8	Basin-fill complexes	
		First-order geomorphic cycles	Schumm (1977)
		Second-order stratigraphic sequences	Vail et al. (1977)
		Sloss-type sequences	Sloss (1963)

[a] As defined by Miall (1991b): see Table 4.2; c-u=coarsening-upward

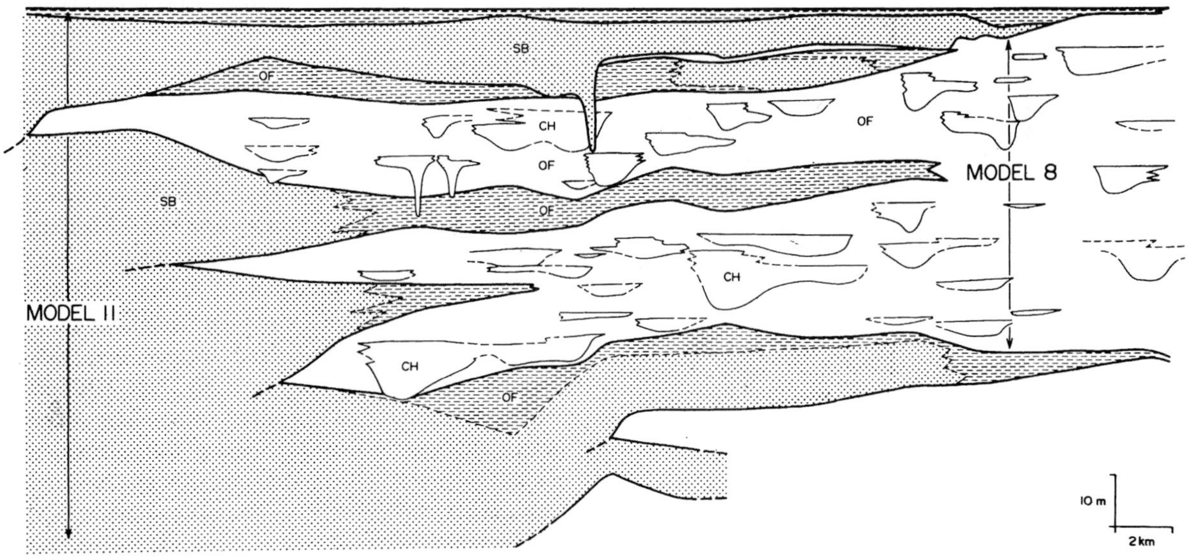

Fig. 9.1. Interbedding of contrasting fluvial styles. The depositis of fluvial model 11 of Miall (1985) are the "blue-gray" deposits of Behrensmeyer and Tauxe and were deposited by a trunk-braided fluvial system. Model 8 represents the "buff" deposits laid down by an anastomosing tributary system

magnetostratigraphic studies. The deposits consist of major channel sandstone bodies 10–30 m thick and extending laterally for up to tens of kilometers. They include both ribbons and sheets (Behrensmeyer 1987). Ribbon sandstones are particularly well documented in section 4 of Fig. 2.32 Overbank deposits include clay, marl, pedogenic carbonate, sand, and intraformational conglomerate, representing levee, crevasse-channel, crevasse-splay, and minor-channel environments. Several major abandoned channels more than 1 km wide are indicated in Fig. 2.32 by cut-and-fill erosion surfaces with 10 m or more of relief, followed predominantly by fine-grained deposits.

Behrensmeyer and Tauxe (1982) demonstrated that these deposits represented two quite different fluvial systems that interfingered with each other over a time period of about 8×10^4 years (Fig. 9.2). A sandy braided fluvial system formed sheet-like channel bodies. Behrensmeyer and Tauxe (1982) referred to this as the "blue-grey system" (Fig. 9.1). Tributaries, consisting of narrower, possibly anastomosing channels, formed mainly ribbon sand bodies, constituting the "buff system" of Behrensmeyer and Tauxe (1982). The interaction of these two fluvial systems, and the autogenic processes that controlled them, are discussed in Chap. 10.

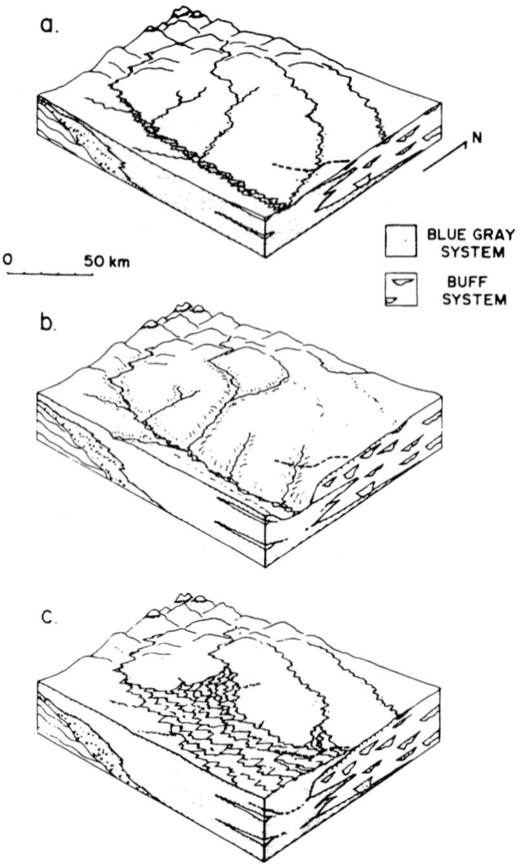

Fig. 9.2. Reconstruction of the two interacting fluvial systems that constructed the Siwalik sediments in a part of Pakistan. The outcrops illustrated in Fig. 9.1 are shown by the *heavy dashed line* near the right-hand edge of each block diagram. (Behrensmeyer and Tauxe 1982)

Several detailed studies of the fluvial stratigraphy of coal-bearing units have been reported by R.M. Flores and his colleagues (e.g., Flores 1981, 1983b, 1984; Flores and Hanley 1984; Warwick and Flores 1987). An example is illustrated in Fig. 7.2. In each case, coals provide major marker beds, permitting lateral correlation of stratigraphic sections for distances of up to tens of kilometers. As discussed in Chap. 13, coals have been used in some studies for the definition of stratigraphic sequences.

Theoretical sand-belt models have been constructed by many workers for the purpose of exploring the lateral spacing and vertical stacking pattern of channel belts. A basic classification was offered by Friend (1983) and is illustrated in Fig. 2.35. These models, the most recent of which employ computer simulation procedures, are discussed in Chap. 10. In this section we illustrate and briefly discuss some practical examples based on studies of the ancient record.

Galloway (1981) described Cenozoic fluvial systems of the Gulf Coast of Texas from a combination of outcrop and subsurface data. The distribution of sand bodies in the subsurface is shown schematically in Fig. 9.3. Interpretations of one of these, the Gueydan system, are shown in Figs. 9.4 and 9.5. This system has the characteristics of the "low-sinuosity river with alternate bars," the style described in Sect. 8.2.11. According to Galloway (1981), channel sand bodies are locally stacked vertically along "depositional axes" (as described in Sect. 9.3). Channels and associated splay deposits form sand belts up to several kilometers wide. Several other fluvial styles occur in this Cenozoic succession and are described by Galloway (1981).

Continuous lateral profiling of large outcrops in the Ebro Basin, Spain, produced detailed maps of sand-body distribution of an Oligocene-Miocene fluvial system by Hirst (1991). He classified the sand bodies into ribbons, sheets, and amalgamated complexes, following the definitions of Friend et al. (1979) and Friend (1983). Figure 9.6 shows examples of the various sandstone types, and Fig. 9.7 is a small example of part of one of his lateral profiles.

The pedofacies concepts of Bown and Kraus (1987) and Kraus (1987) may be useful for gaining information on channel architecture in areas of limited exposure. As shown in Fig. 7.24, pedogenic maturity is related to distance from a main channel, which governs the exposure time of the paleosol, and its dilution with clastic detritus. Kraus (1987) suggested that vertical successions of pedofacies record lateral movements of major channels. Such ordered vertical successions she termed "compound pedofacies sequences." For example, Fig. 9.8 shows a stratigraphic section in which paleosols are interbedded with levee and crevasse-splay deposits. The paleosols show an upward increase in maturity from stages 1 to 3 (0–19.5 m interval of section), but the last paleosol then returns to stage 2. Kraus's (1987) interpretation of this succession is shown in Fig. 2.31. The section begins with the channel and flanking levee deposits formed when the channel was in its first position and ends with the channel in position VI, having returned to approximately its original position on the floodplain. Guccione (1993) suggested that, in much the same way, the grain size of overbank clastic deposits could be used as an indicator of distance from the source channel. As noted in Sect. 7.4.2, however, other factors, including substrate type and differential subsidence rates, may

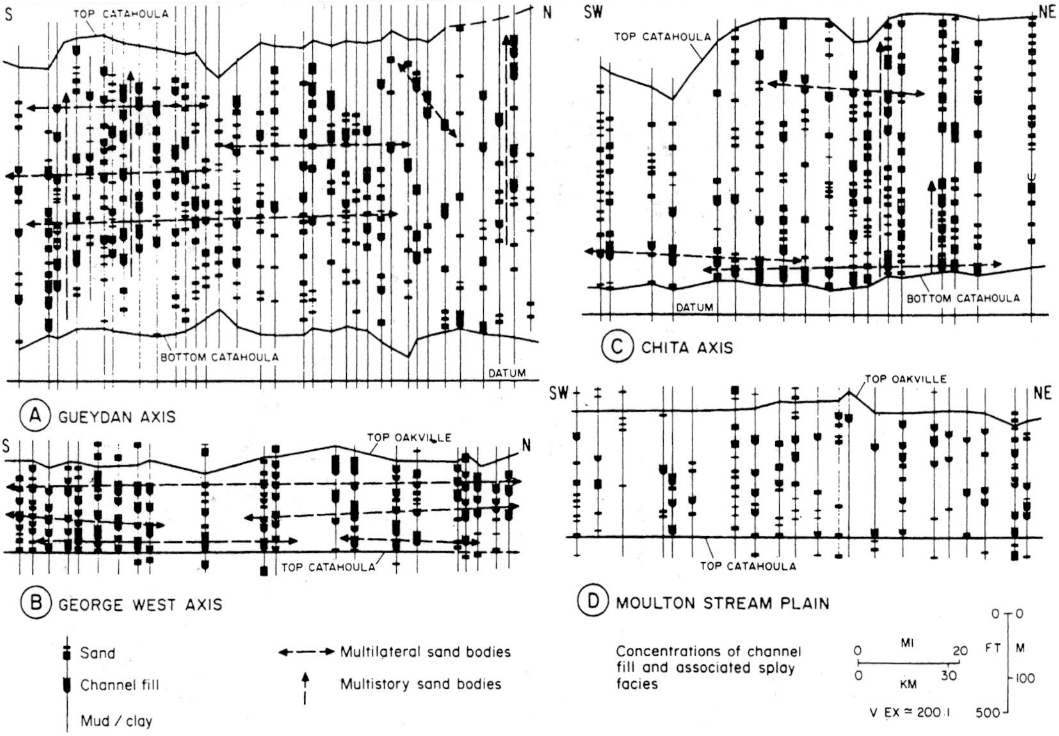

Fig. 9.3. Representative strike-parallel cross sections of major coastal-plain fluvial deposits, Cenozoic of the Gulf Coast, Texas. The fluvial style of the Gueydan system, shown in diagram A, is illustrated in Figs. 9.4 and 9.5. The locations of these cross sections are shown in Fig. 9.14. (Galloway 1981)

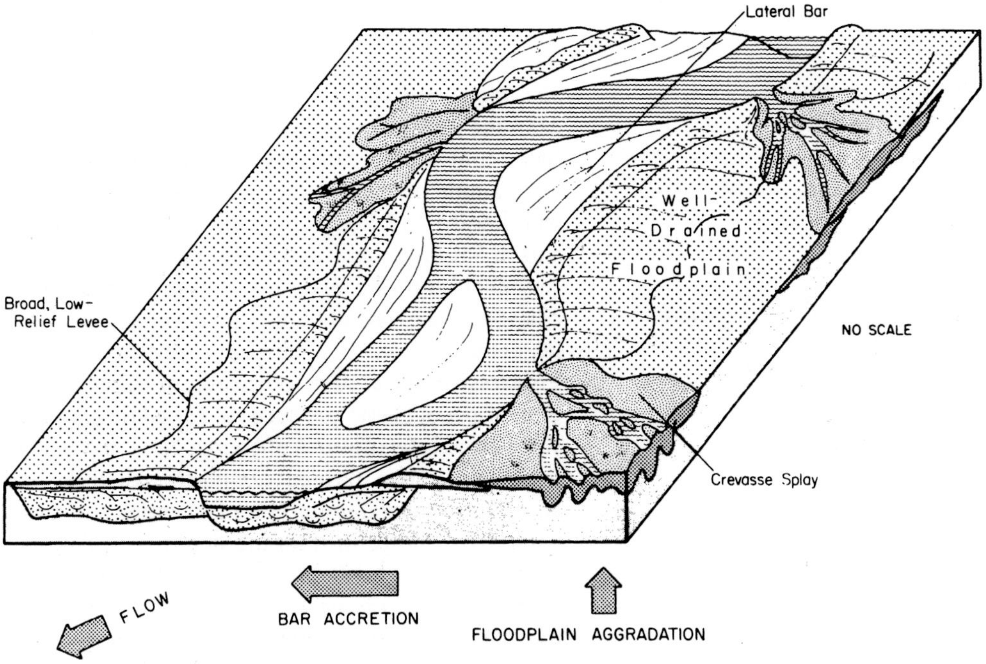

Fig. 9.4. Schematic reconstruction of the Gueydan fluvial style, Texas. (Galloway 1981)

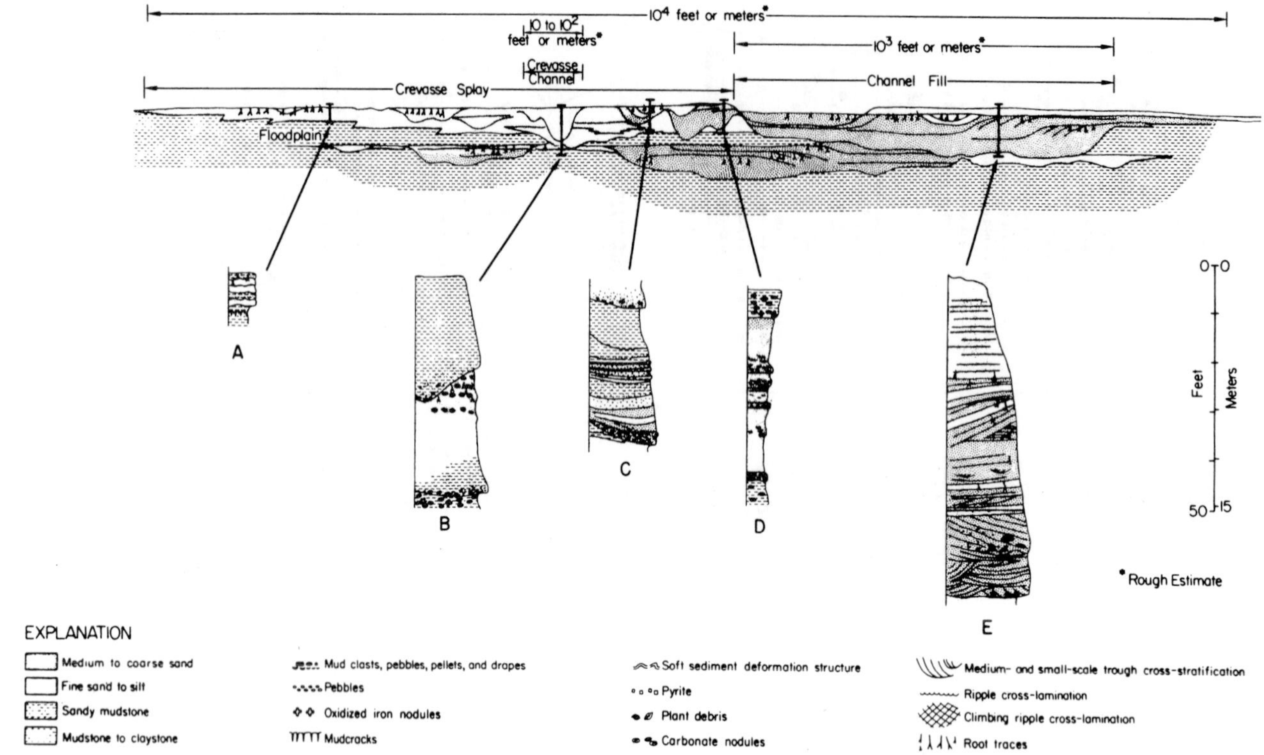

Fig. 9.5. Schematic facies architecture of the Gueydan fluvial system, Texas, illustrating the lateral relationships and scales of the various channel and overbank elements. Measured sections illustrate *A* distal crevasse-splay, *B* proximal, mudstone-filled crevasse channel, *C* mixed-fill crevasse channel, *D* sand-silt-filled channel, *E* main fluvial channel fill. (Galloway 1981)

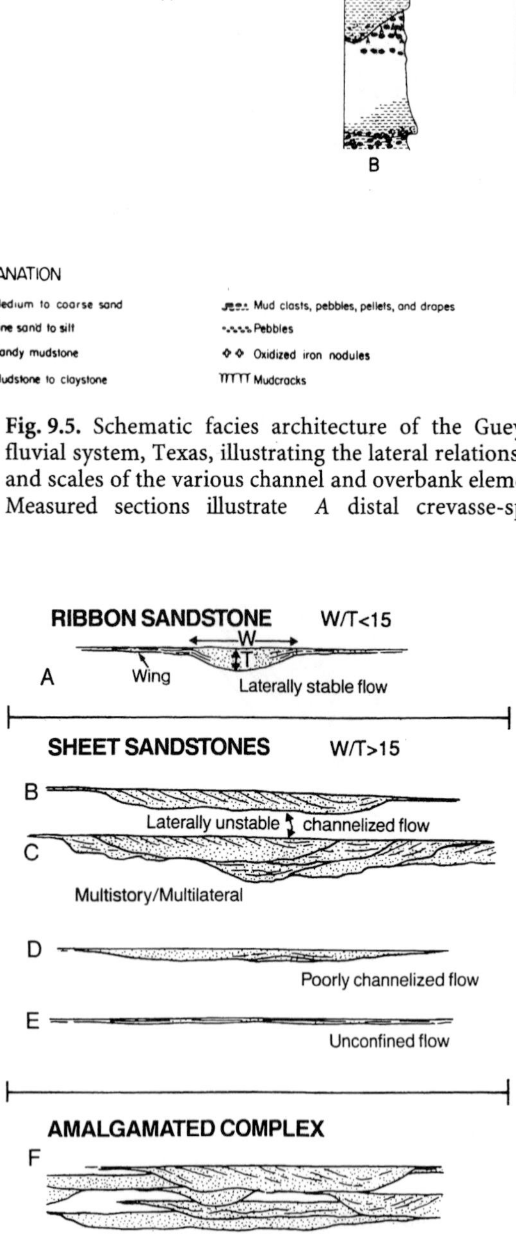

Fig. 9.6. The range of sand-body geometries in the Huesca fluvial system (Oligocene-Miocene), Ebro Basin, Spain. "Wings" are levee deposits. (Hirst 1991)

also affect soil character, and so it is necessary to examine and interpret paleosols within an overall basin context. Marriott and Wright (1993) provided an example of such a study with their work on the well-known Old Red Sandstone (Devonian) paleosols of South Wales. They showed how variations in sedimentation rate, erosional episodes, and overall changes in the style of floodplain development could be deduced from the details of paleosol profiles. Progressive burial of a soil profile without disruption of pedogenesis produces a *cumulate* paleosol, short-lived interruptions of pedogenesis may lead to overlapping soils, termed *composite* paleosols, and so on. Spectacular outcrops of channel and overbank architecture in Miocene deposits in Pakistan enabled Willis and Behrensmeyer (1994) to offer a more thorough discussion of the possible styles of meander-belt development. These deposits, together with a description of the nature of avulsive processes and their control on meander-belt architecture, are discussed in Sect. 10.3.1.

Conglomeratic alluvial fan deposits commonly display coarsening- or fining-upward successions

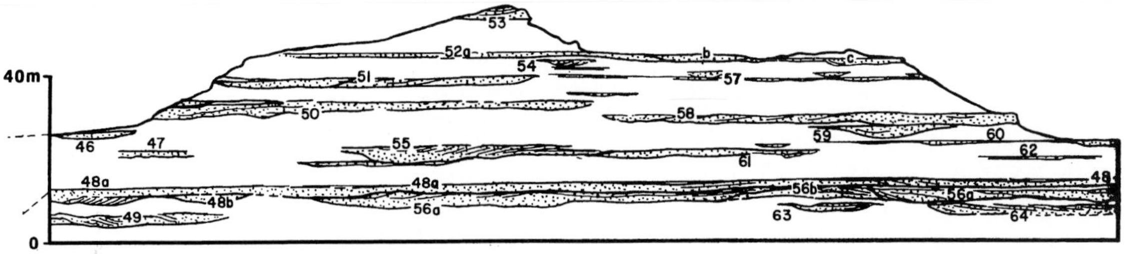

Fig. 9.7. A small example of the profiles constructed from outcrop data by Hirst (1991). The drawing is at true scale (without vertical exaggeration), and the complete profile extends for more than 1 km. Sand bodies are numbered in order of deposition within the entire profile

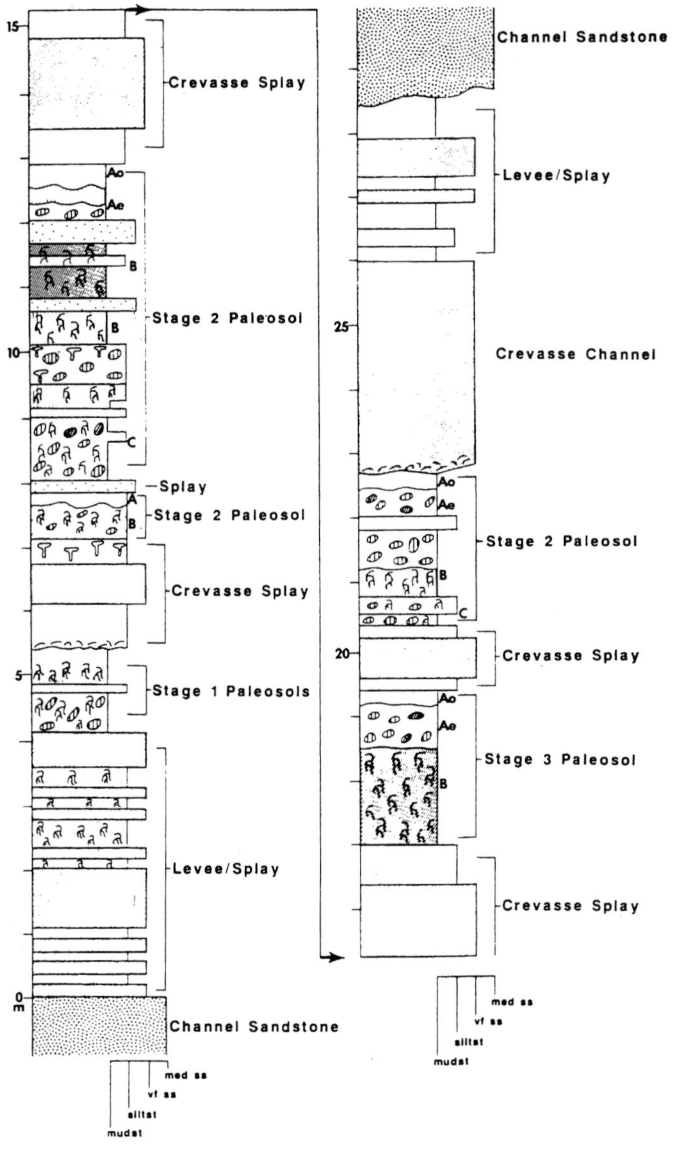

Fig. 9.8. Stratigraphic section of the Willwood Formation (Eocene), Wyoming, showing a compound pedofacies sequence. (Kraus 1987)

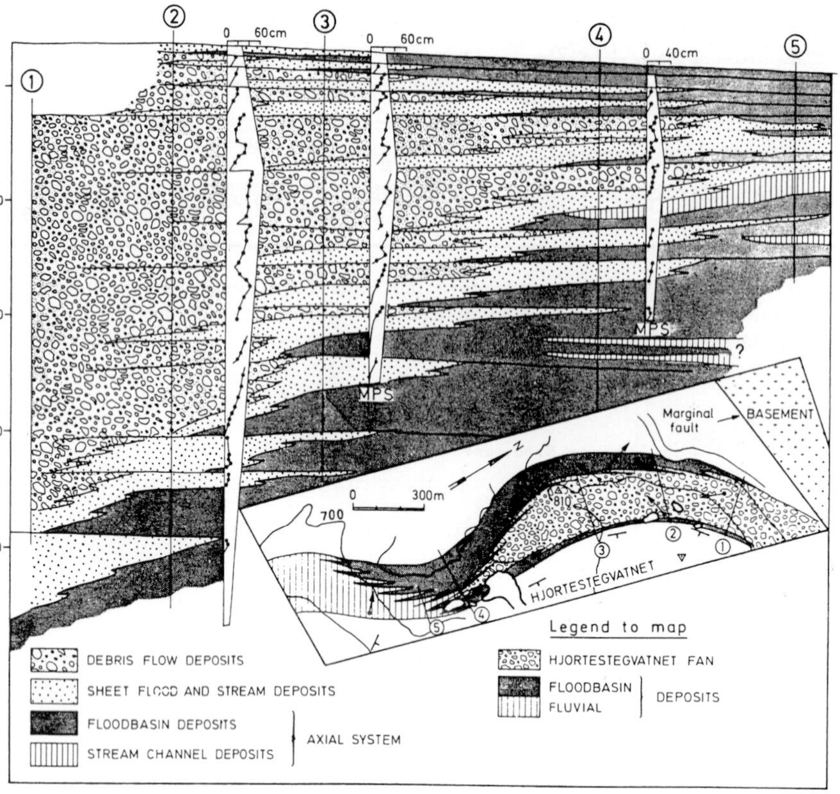

Fig. 9.9. Cross section of the Hjortestegvatnet fan (Devonian), Norway, showing coarsening- and fining-upward on at least two scales. *Vertical graphs with dots* show maximum particle size in centimeters. (Gloppen and Steel 1981)

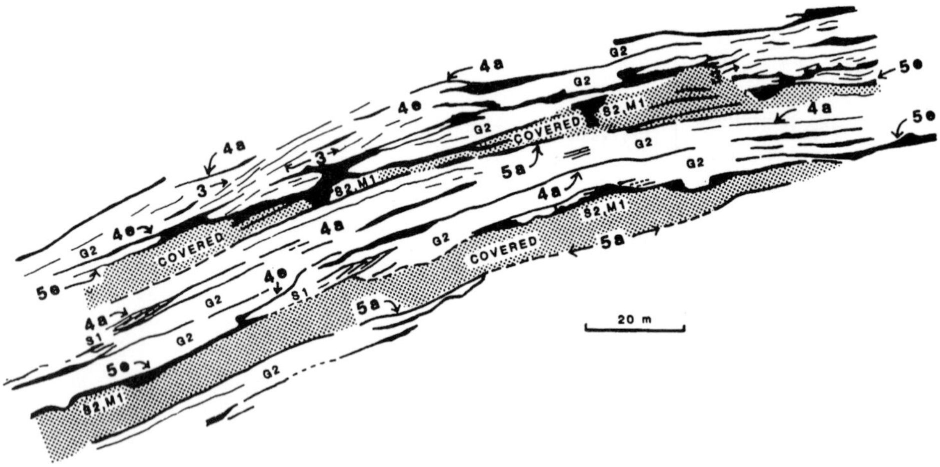

Fig. 9.10. Drawing and interpretation of an outcrop of the Beartooth Conglomerate (Paleocene), Wyoming. *Numbers* are bounding surfaces, the classification of which is given in Fig. 4.2. *Stippled areas* are poorly exposed siltstones. (DeCelles et al. 1991)

(cycles), and these have been much studied for the information they yield regarding autogenic and allogenic processes (Heward 1978a; Gloppen and Steel 1981; DeCelles et al. 1991). Examples of profiles from two fan complexes are shown in Figs. 9.9 and 9.10. Sequences of 15- to 65-m thickness showing either coarsening- or fining-upward trends are common in these examples. They are bounded by fifth-order surfaces in the classification by DeCelles et al. (1991; Fig. 4.3), and may extend laterally for hundreds of meters. The examples illustrated in Fig. 9.9 are upward coarsening in type. Those in Fig. 9.10 are upward fining. They are bounded at the base by erosion surfaces with up to several meters of relief. The top of each sequence may be marked by a thick siltstone unit several meters thick, which may contain pedogenic features.

Thicker successions, termed megasequences by Heward (1978a), are also common in alluvial fan deposits. Figure 9.9 shows that the entire succession of the Hjortestegvatnet fan in Norway constitutes a megasequence 250 m thick, which coarsens upwards through the first 200 m of the section and then fines upward. Heward (1978a) illustrated several examples of fan profiles showing sequences and megasequences of several types and thicknesses (Fig. 9.11). Individual fans constitute "group 8" deposits in the hierarchical classification of this book. In the classification of DeCelles et al. (1991) entire fan bodies are bounded by sixth-order surfaces (Fig. 4.3) but, as DeCelles et al. (1991) noted, such surfaces may be difficult to map in practice, because fans coalesce laterally, with complex interfingering relationships developing as individual distributaries on adjacent fans meander and shift by avulsion. Compositional differences between adjacent fans may develop because of different source-area geologies, and this may permit individual fans to be mapped on the basis of sandstone petrography or clast composition, a mapping technique that has been in use for many years (e.g., Miall 1970a).

Coastal fluvial systems are strongly influenced by base-level change, a topic that forms part of the basis for the study of sequence stratigraphy. This subject is considered in detail in Sect. 11.2.2 and Chap. 13, but a few descriptive points are noted here for the sake of completeness. Among the most detailed studies of coastal stratigraphy are those undertaken on the Gulf Coast of Texas and Louisiana. Late Cenozoic deposits contain the evidence of numerous glacioeustatic changes in sea level, and fluvial systems contain much evidence of this. For

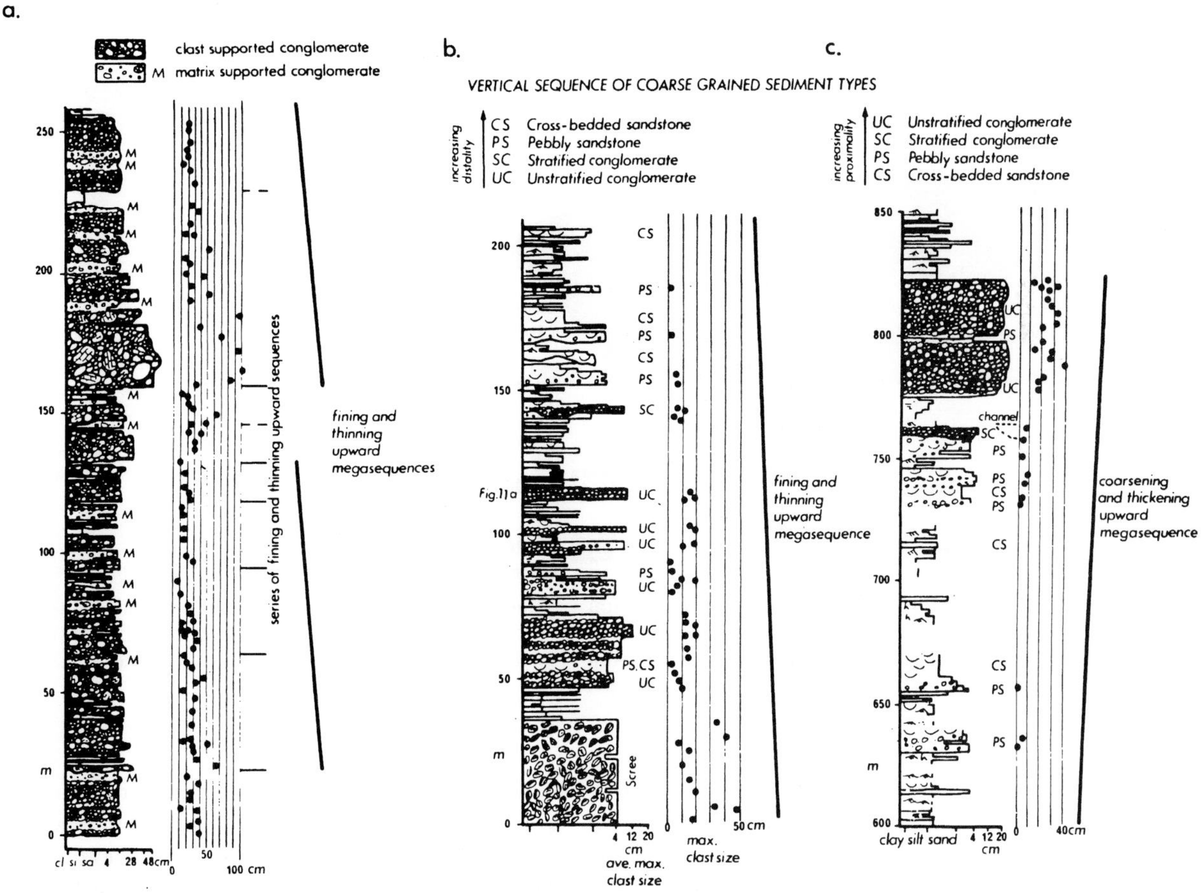

Fig. 9.11. Examples of fan sequences and megasequences in the Carboniferous alluvial-fan succession of northern Spain. (Heward 1978a)

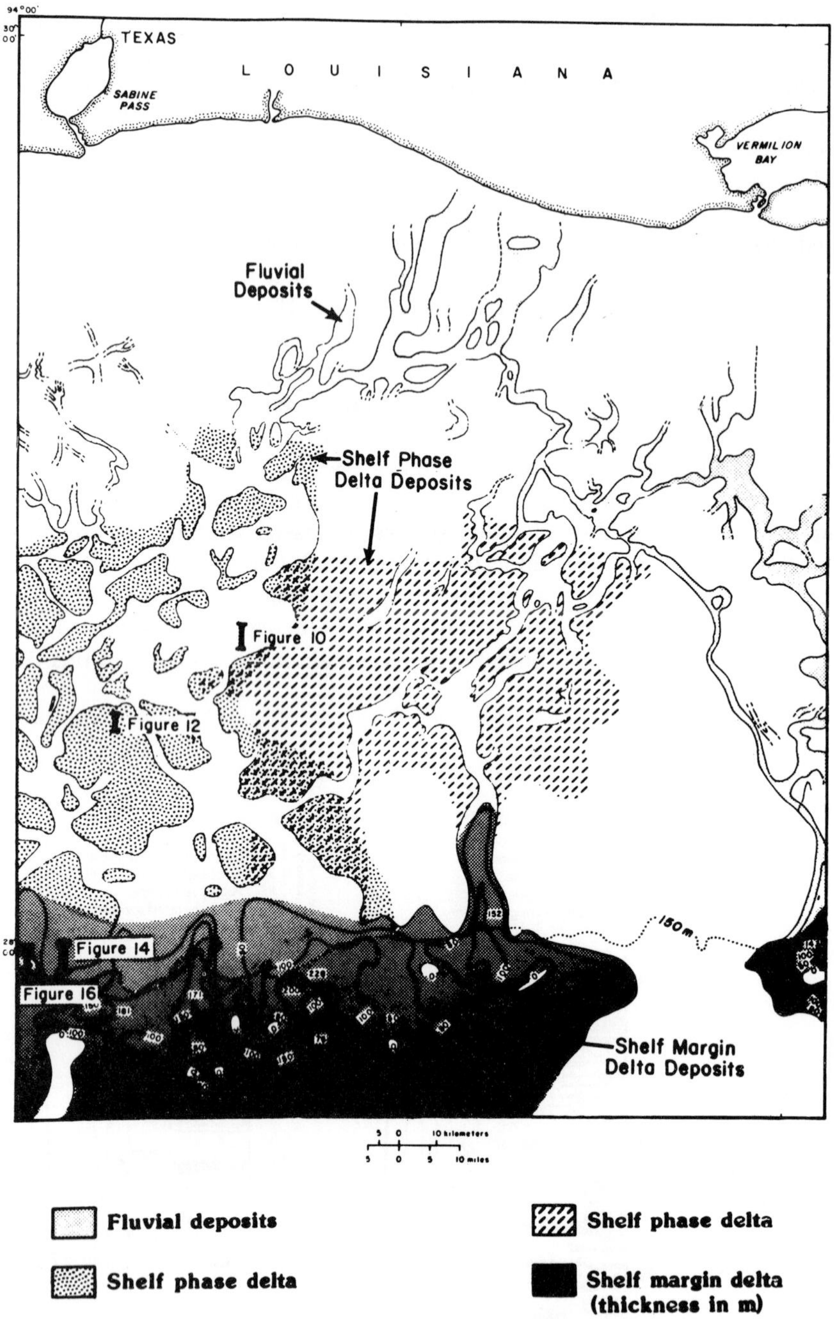

Fig. 9.12. Depositional systems on part of the continental shelf of Louisiana that developed during the Pleistocene at a time of low sea level. Note the fluvial systems occupying incised valleys. The sea-level cycle during which this paleogeography developed was one of many fifth-order glacioeustatic (Milankovitch) cycles that affected continental margins during the late Cenozoic. (Suter et al. 1987)

example, during low stands of sea level, fluvial systems become incised into the continental shelf and extend their courses to the shelf margin. During the subsequent sea-level rise, these broad incised valleys are then filled with fluvial or estuarine deposits (depending on the balance between sediment supply and rate of base-level rise). An example of the depositional systems that result is illustrated in Fig. 9.12. Commonly, the scale of such incised valleys is little different from that of channel systems generated by autogenic processes, and correct interpretations require very careful mapping.

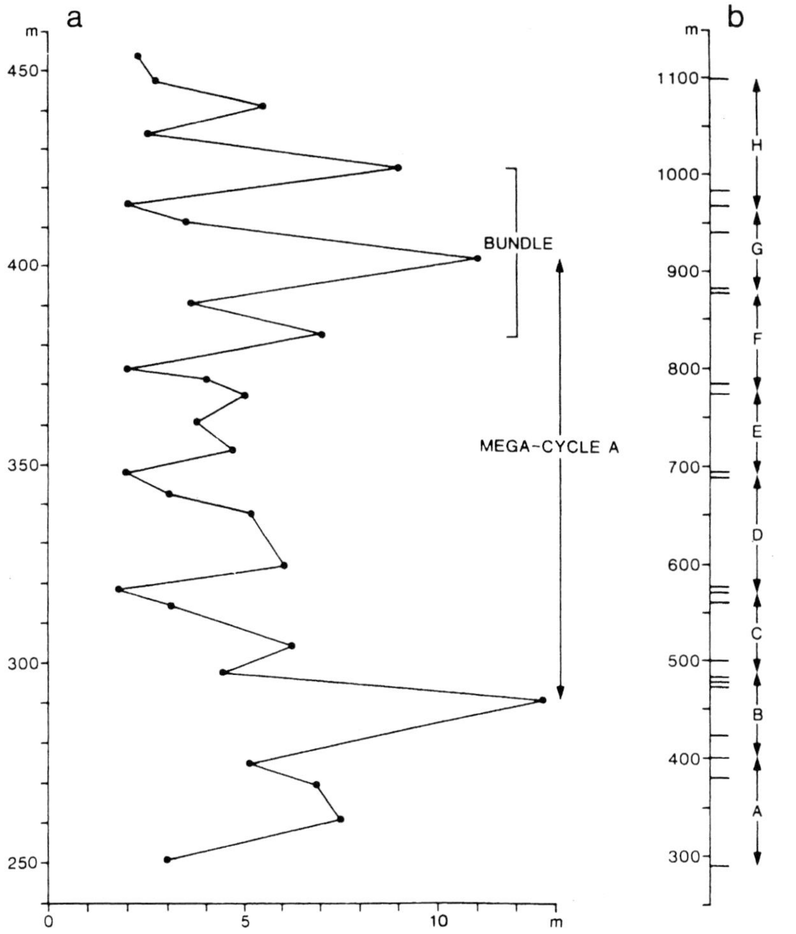

Fig. 9.13. Variations in sand-body thickness in a Devonian fluvial section in East Greenland. (H. Olsen 1990)

Another possible example of fifth-order cyclicity preserved in nonmarine deposits is that given by H. Olsen (1990). He described a Devonian fluvial section in East Greenland containing prominent fining-upward cycles that show systematic variations in thickness. Variations in channel sand-body thickness are shown in Fig. 9.13. The average thickness of the fluvial fining-upward cycles is 20 m, with a range of 13–30 m. Channel sandstones vary from 2 to 13 m in thickness. Bundles of 3–6 cycles can be defined, as can thicker trends that trace out megacycles. Megacycles range from 79 to 135 m in thickness, with an average of 101 m. H. Olsen (1990) suggested that these cyclical changes in thickness are the result of discharge variations brought about by orbital forcing, a point discussed further in Sect. 12.12.

9.3 Depositional Systems

Group 9 deposits are those representing hundreds of thousands of years of deposition. The generation and abandonment of entire depositional systems or alluvial fan tracts may be accomplished during such time intervals. Long-term (fourth-order) Milankovitch cycles are also of this time duration. Deposits of this scale are typically regionally extensive and are "mappable", in the sense that they can be defined on conventional geological maps on scales of 1:50000 or larger. Many such units have been therefore assigned lithostratigraphic formation or member names in the older literature. They are bounded by seventh-order surfaces (Tables 4.2, 9.1).

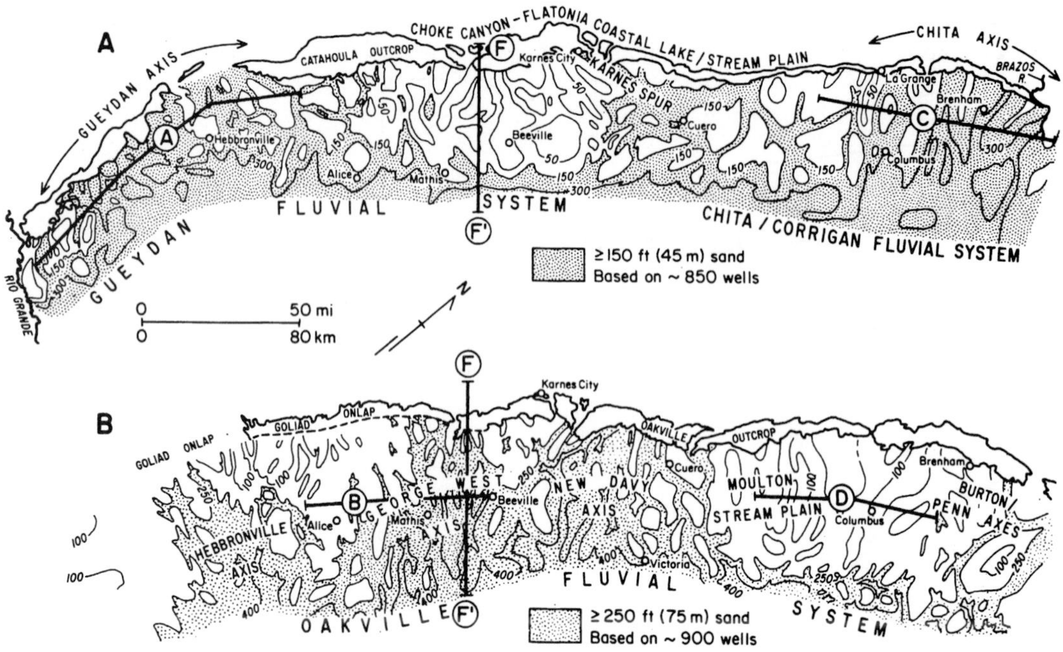

Fig. 9.14A,B. Simplified total-net-sand isolith maps of two Cenozoic fluvial systems on the Gulf Coast of Texas. **A** Catahoula Formation, **B** Oakville Formation. Areas containing more than 75 m of total sand are indicated by *stippling* and define broad, sand-rich belts or aprons that correspond to channel-belt axes. These are separated from each other by sand-poor zones corresponding to interfluve areas, crossed by small tributary streams and also occupied by floodplain lakes. (Galloway 1981)

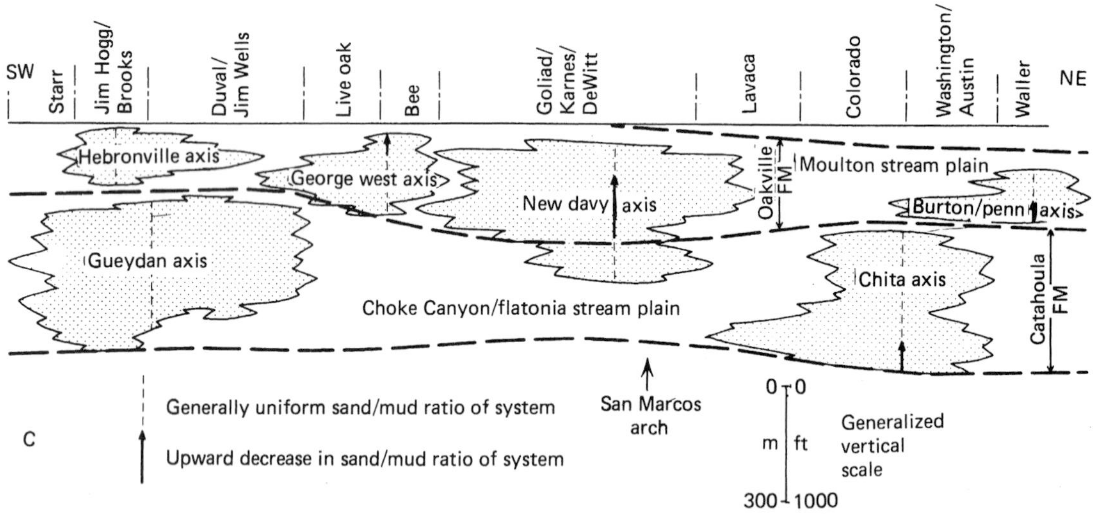

Fig. 9.15. Generalized strike-parallel cross section through the Oligocene-Miocene succession of Texas, showing the principal depositional elements and fluvial "axes". Note the long-term vertical persistence of these axes. (Galloway 1981)

Galloway (1981) showed from detailed subsurface work that large-scale depositional systems may occupy constant positions within a basin margin for extended periods of time (Figs. 9.14, 9.15). His fluvial "axes" within the Cenozoic Gulf Coast fluvial succession correspond to group 9 deposits.

Kraus (1987, 1992) noted that over vertical intervals in the order of hundreds of meters of section, average pedofacies maturity may undergo long-term changes. For example, in the northern Bighorn Basin of Wyoming, compound pedofacies sequences near the base of the Willwood Formation are characterized by stages 1–3, whereas at the top they are typically stages 3, 4, and rarely 5. She termed such long-term changes "pedofacies megasequences" and attributed them to changes in deposition rates. In the Willwood example it seems likely that the basin subsidence rate decreased during accumulation of this formation, permitting individual paleosols to mature for longer periods before being covered by overbank flooding. Atkinson (1986) described variations in paleosol character displayed laterally on a regional scale and interpreted the variations in terms of differences in drainage conditions and fluvial topography related to differential subsidence within a foreland basin. This is an example of a paleosol catena controlled by tectonism. Examples from the Recent were described from the Gangetic plains of India by Srivastava et al. (1994), who attributed local changes in fluvial style and paleosol character to neotectonic movements of the basin.

The classic cyclothems of North American geology are fourth-order stratigraphic sequences and correspond to group 9 of the classification in this

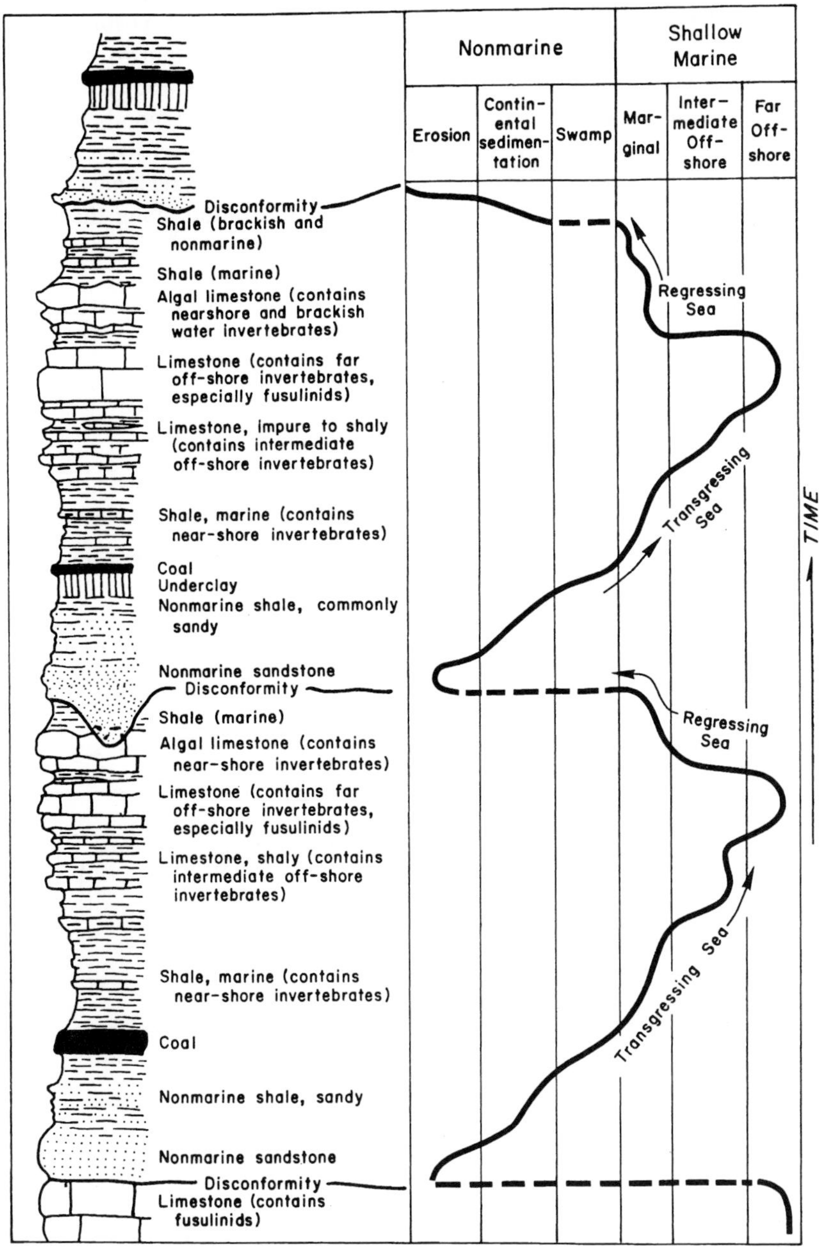

Fig. 9.16. Two typical Illinois-type Carboniferous cyclothems showing interpretation in terms of transgression and regression. (Crowell 1978, based on Moore 1964)

book (Table 9.1). They may contain a significant fluvial component, representing incised valley fills deposited following the period of maximum regression. Two typical Illinois-type cyclothems are illustrated in Fig. 9.16, and their origins are discussed further in Chap. 13. Similar regressive fluvial-deltaic sandstones, termed "grits" in the older literature, form regressive units in comparable British cyclothems (Ramsbottom 1979; Holdsworth and Collinson 1988). Note the confusion in terminology for these cycles: In Heckel's (1986) work on the Pennsylvanian rocks of the Midcontinent of the United States these cycles constitute the larger scale of cyclicity, and they are termed "major cycles". The study by Holdsworth and Collinson (1988) of similar cyclicity in rocks of broadly comparable age in Britain refers to cycles of similar 10^5-year duration as "minor cycles", compared with the major cycles, or "mesothems" of Ramsbottom (1979).

Another example of this type of cyclic sedimentation on a 10^5-year scale is that of the Dunvegan delta complex of Alberta (Bhattacharya 1991; Fig. 9.17). The Dunvegan Formation consists of seven subdivisions, classified as allomembers, each of which is estimated to represent about 200 ka. Fluvial channel fills occupy valleys incised into the top of several of the allomembers.

Examples of fourth-order cycles that include more extensive fluvial deposits are those that constitute the Castlegate Sandstone and Upper Desert Member of the Blackhawk Formation of Utah (Van Wagoner et al. 1990, 1991; Miall 1993; Yoshida et al., in press; Fig. 9.18). Each of the fourth-order sequences represents several hundred thousand years of geologic time. The sequence corresponding to the lower Castlegate Sandstone is an extensive braided-fluvial sandstone sheet resting on a regional erosion surface. It can be traced downdip along the Book

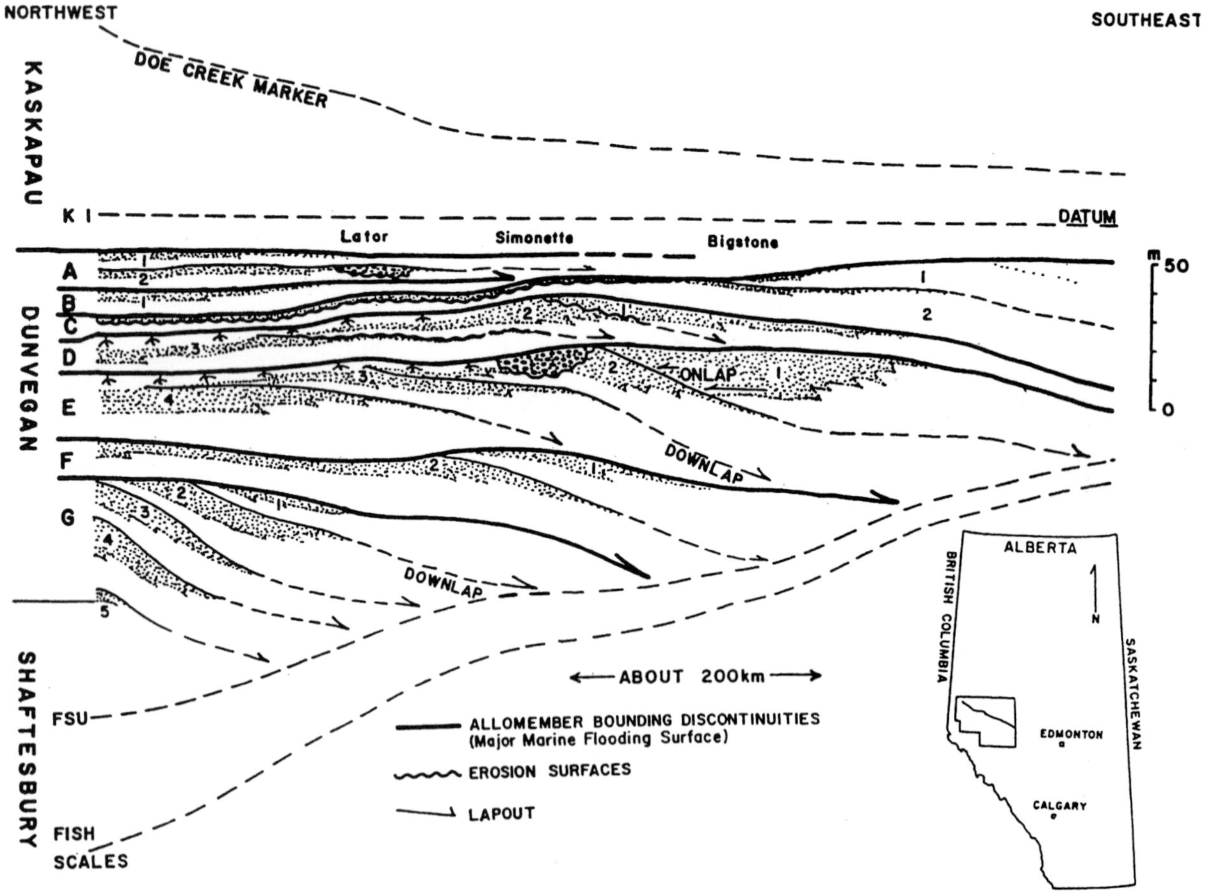

Fig. 9.17. Schematic regional dip-oriented cross section of the Dunvegan Formation, west-central Alberta. *Heavy lines* are regional flooding surfaces that subdivide the formation into seven allomembers. Within each allomember separate offlapping shingles can be mapped, as shown by the *numbers. Root symbol*, nonmarine facies; *light stipple*, marine sandstone; *heavy stipple*, channel fills; *blank*, marine shale. (Bhattacharya 1991)

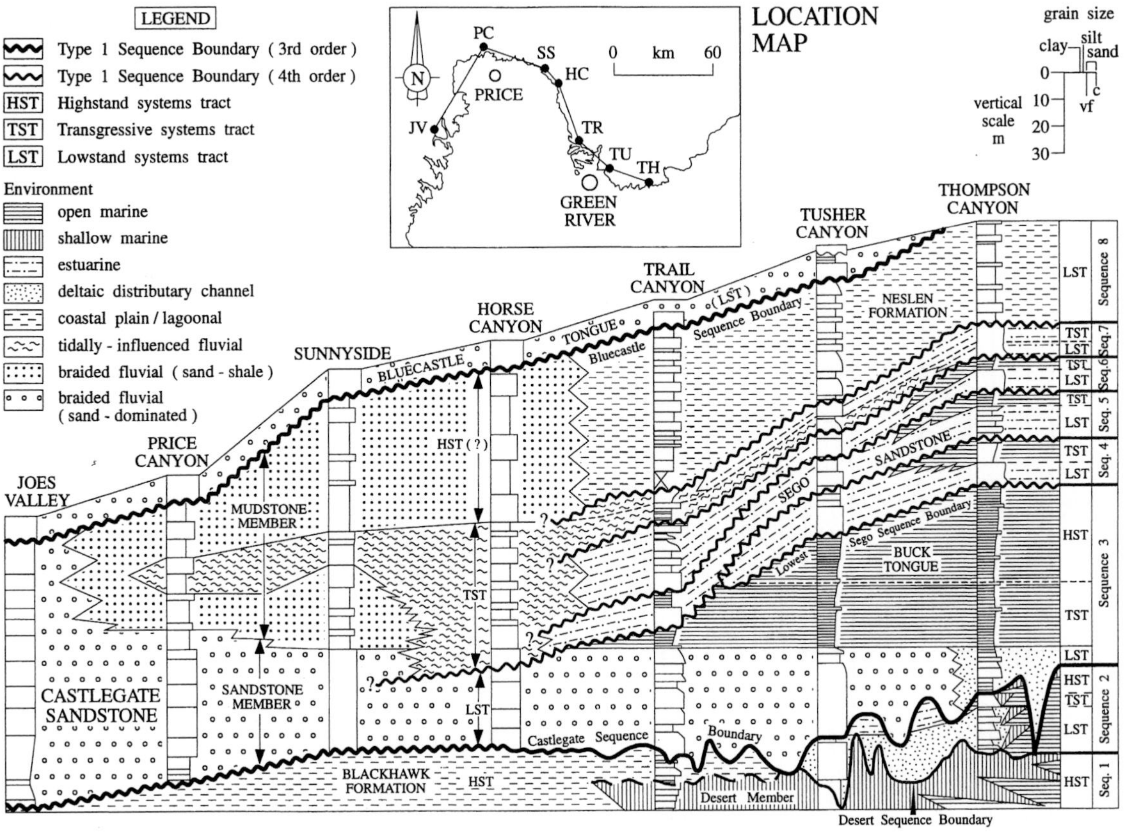

Fig. 9.18. Sequence stratigraphy of the Castlegate Sandstone and equivalent deposits, Book Cliffs, Utah. In proximal areas (*left-hand sections*) the succession consists of a single third-order sequence spanning about 5 million years. This passes down dip into a series of fourth-order sequences consisting of fluvial and estuarine sandstones interbedded with shales. (Yoshida et al., in press)

Cliffs for some 140 km and passes eastward into deltaic and marine deposits. It is assumed to extend laterally along strike (north-south) for a considerable distance. As discussed in Chap. 13, the sediments are interpreted as having been deposited during or immediately following a low stand in regional base level.

Some other important examples of nonmarine fourth-order cycles are those which form part of complex cyclic sequences in the North Sea Basin ("medium-order" cycles of Nystuen et al. 1989); these are discussed in the next section.

9.4 Basin-fill Complexes

Groups 10 and 11 correspond to sedimentary successions having durations of a few million to a few hundred million years. This includes third- and second-order stratigraphic sequences, in the classification of Vail et al. (1977) and Miall (1990). Regional tectonism and eustatic sea-level variations are the major generating causes of these sequences, as discussed in Chap. 11.

Ramsbottom (1979) argued that cyclothems in the British Upper Carboniferous succession could be grouped into larger-scale packages, which he termed mesothems, corresponding to group 10 deposits in the present terminology. Part of the evidence for this form of cyclicity is that regressions are more extensive at the close of mesothems than at the termination of individual component cyclothems, leading to thicker and more areally extensive fluvial-deltaic successions, and the more widespread establishment of fully nonmarine conditions, with deposition of major coals. The major sandstones that terminate each cycle are called "grits" in the British stratigraphic literature. A chronostratigraphic section through Namurian mesothems in part of northern

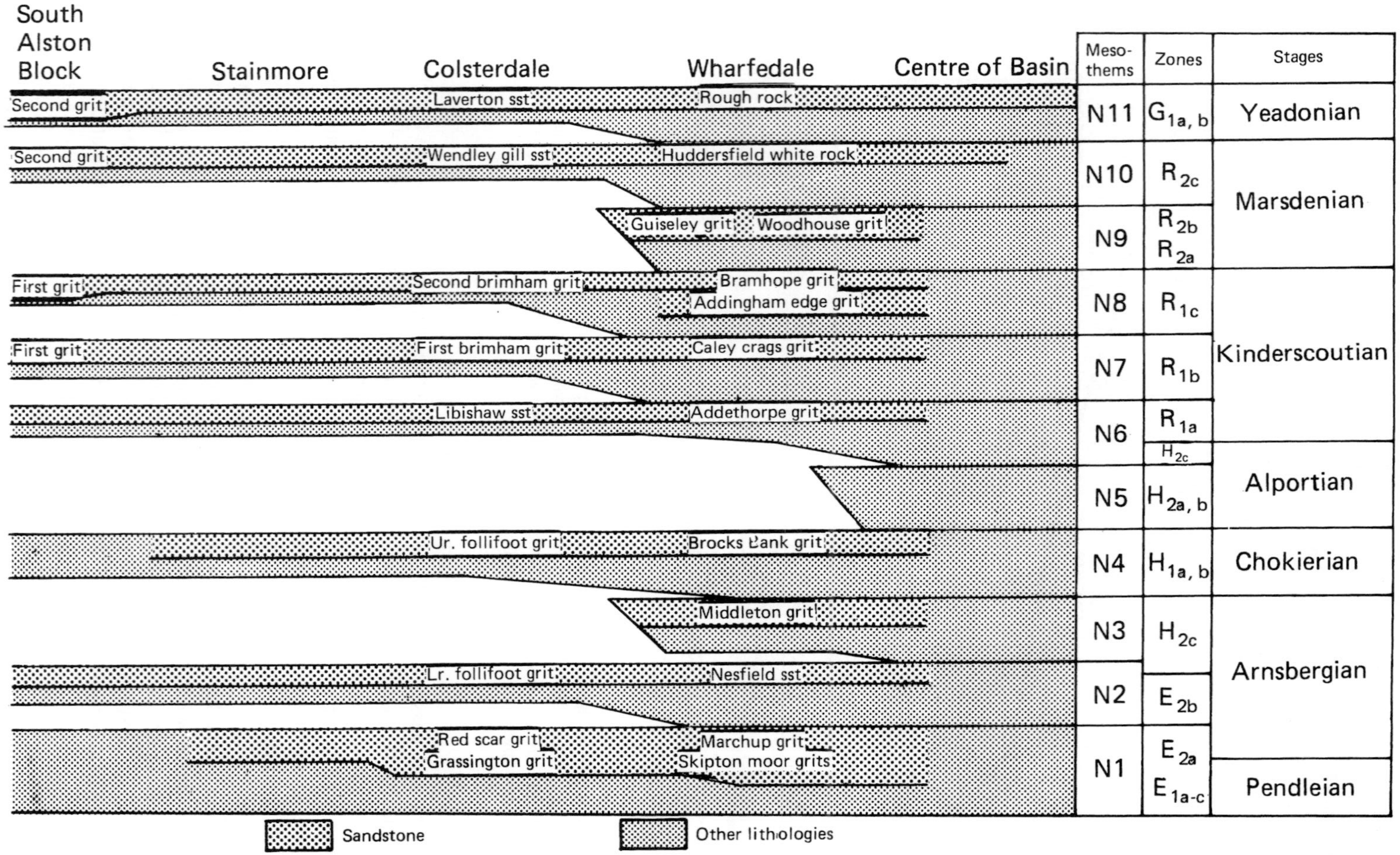

Fig. 9.19. Chronostratigraphic section through Namurian mesothems in part of northern Britain showing the repeated occurrence of regressive sandstones ("grits") of fluvial deltaic origin. (Ramsbottom 1979)

England is shown in Fig. 9.19. The debate regarding the reliability of the mesothem concept (Holdsworth and Collinson 1988; Leeder 1988) illustrates the difficulty in reconstructing complex litho- and biostratigraphic patterns in tectonically active areas and in unraveling the various causes of cyclicity (Chaps. 11, 13). Leeder (1988) noted that new biostratigraphic data and revised sedimentological interpretations that emphasize autogenic deltaic progradation cast doubt on the reality of Ramsbottom's mesothems.

Large-scale, regionally extensive, sand-dominated, fining- and coarsening-upward cycles, typically 100 m or more in thickness, are common in nonmarine deposits. References to several examples are given in Table 9.1. Commonly, such cycles are areally very extensive. For example, Fig. 9.20 illustrates the extent of six such cycles in the Triassic Molteno Formation of the Karoo Basin in South Africa. One of these cycles extends for more than 400 km down depositional dip into the basin interior. A comparable subsurface example, from a different basin, is illustrated in Fig. 9.21. This diagram shows that the "high-order" cycles (those corresponding to group 10 of this book) form the largest of three scales of cyclicity, including allogenic "medium-order" cycles and fully autogenic "low order" cycles. Steel and Ryseth (1990) extended this type of analysis throughout much of the northern North Sea basin, on the basis of seismic evidence (Fig. 9.22) and additional subsurface well correlation (Fig. 9.23). They were able to relate the origins of the sequences to the tectonic history of the area, as discussed further in Chap. 13. Biostratigraphic evidence available to

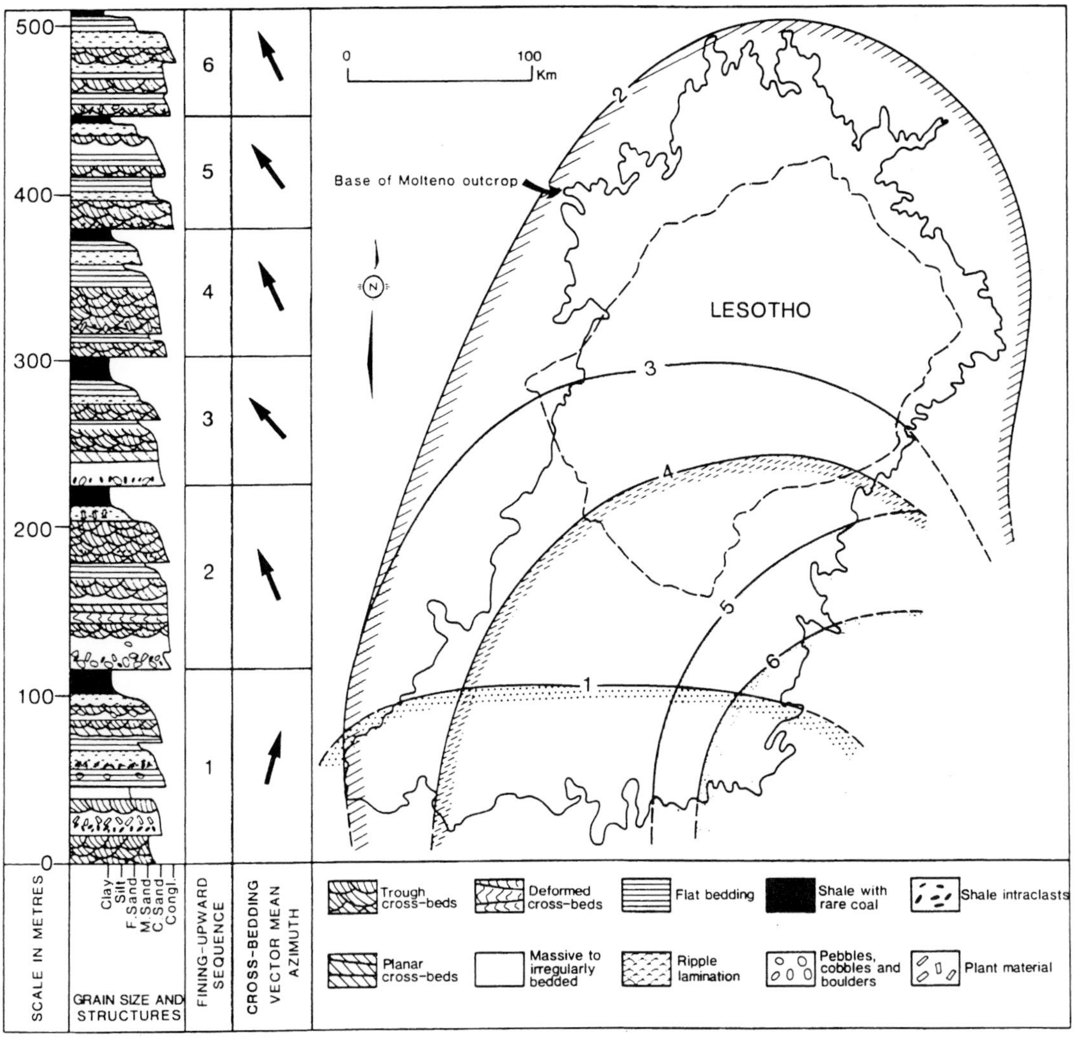

Fig. 9.20. Six cycles in the Molteno Formation (Triassic) of the Karoo Basin in South Africa. (Turner 1983)

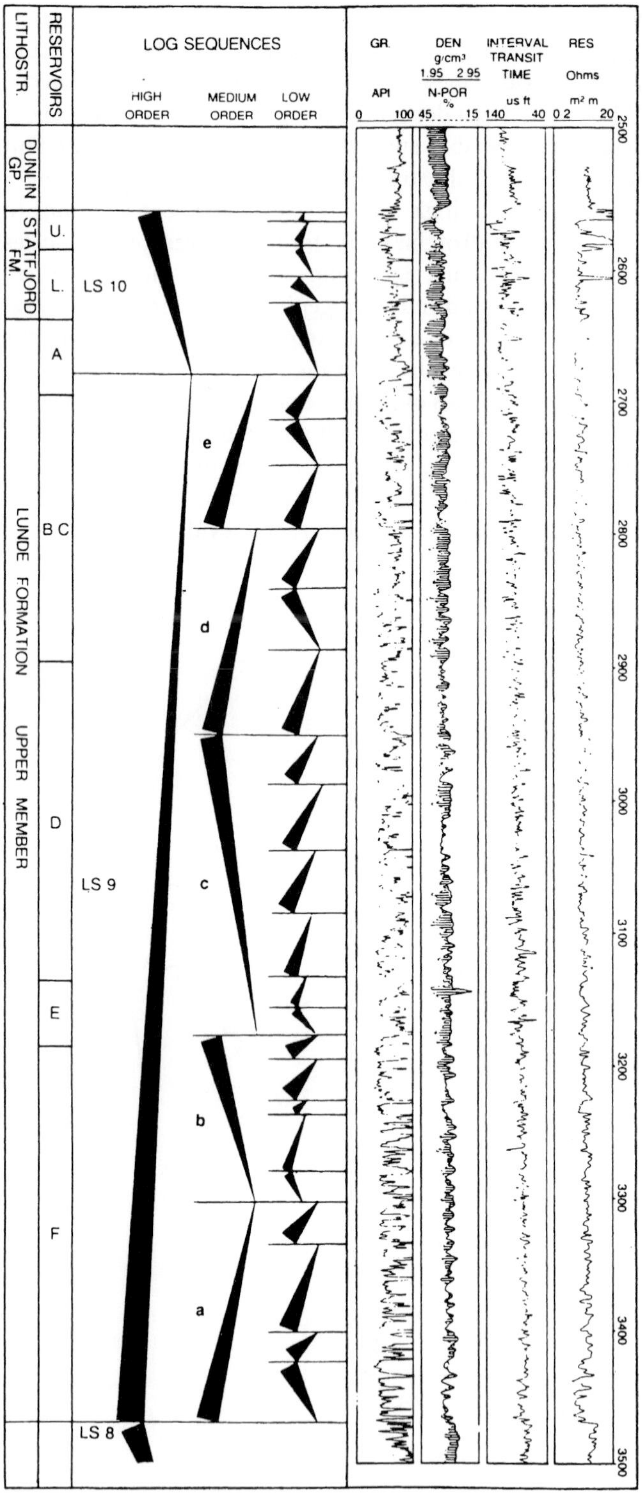

Fig. 9.21. Example of a subsurface wireline-log record through Triassic nonmarine deposits in the Snorre field area of the northern North Sea, showing three categories of nonmarine sequence. The high-order sequences correspond to third-order cycles of Vail et al. (1977) and are allogenic in origin. (Nysteun et al. 1989)

Nystuen et al. (1989) indicated that their high-order cycles have durations of 4–12 million years. Steel and Ryseth (1990) suggested a duration of 10–15 million years for their megasequences.

Nested sequences on two scales occur in the Castlegate Sandstone of Utah (Fig. 9.18), as noted above. The main Castlegate sequence in proximal areas had an estimated duration of about 5 million

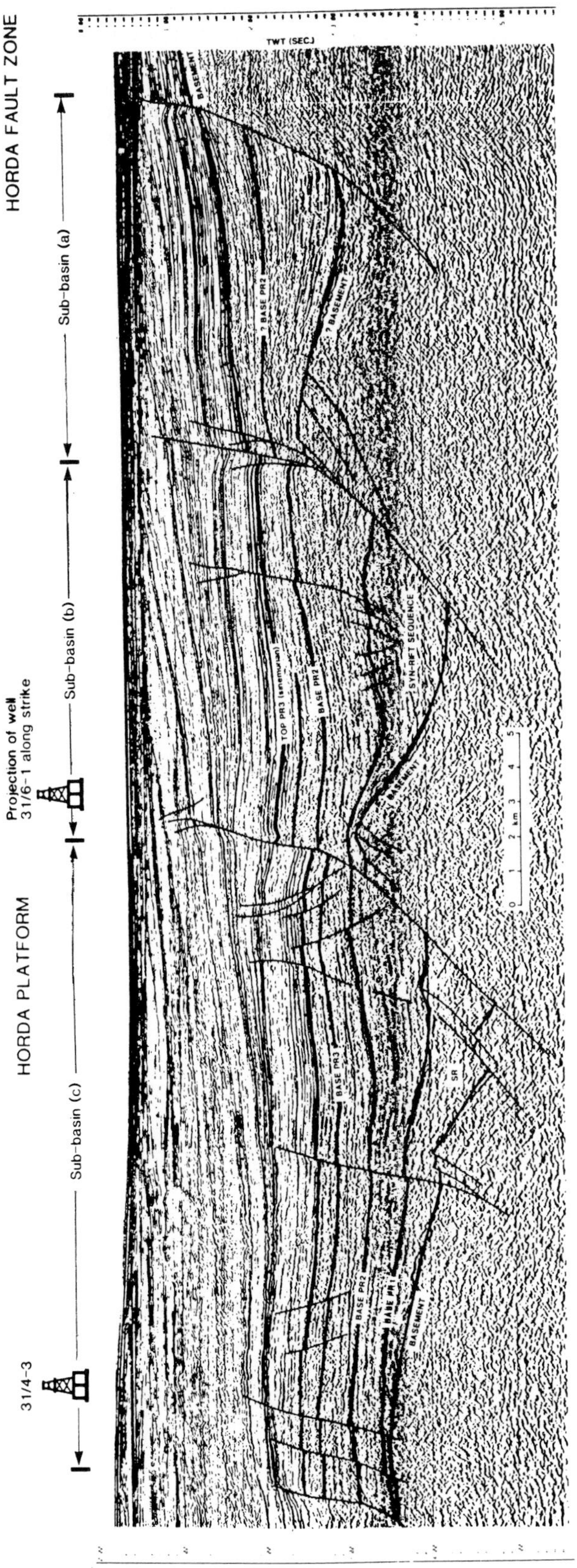

Fig. 9.22. Seismic cross section through the Permian-Lower Jurassic succession in the Horda Platform area of the northern North Sea Basin, showing the subdivision of the Triassic succession into megasequences *PR1*, *PR2*, and *PR3*. (Steel and Ryseth 1990)

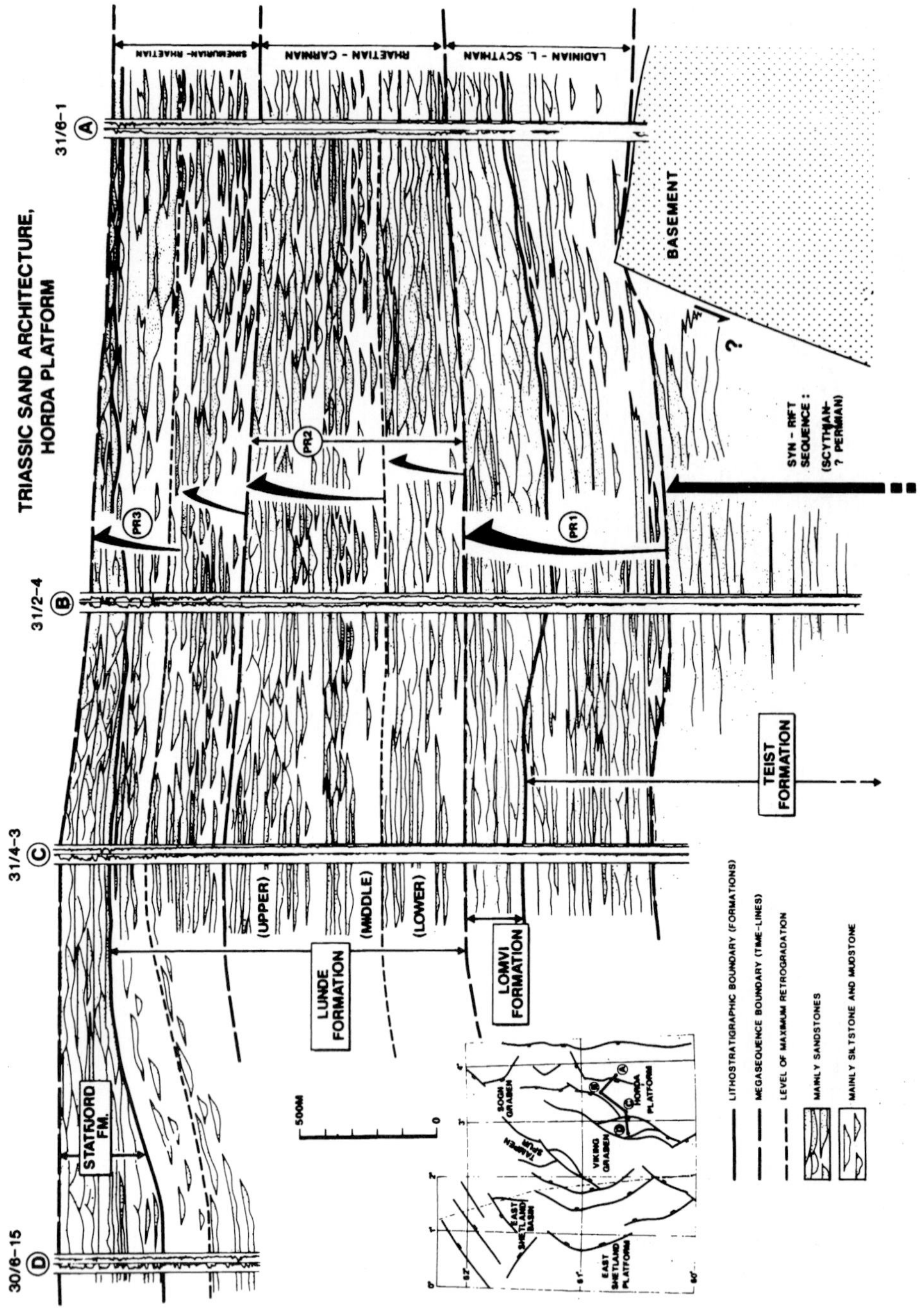

Fig. 9.23. Correlation of nonmarine Triassic megasequences on the Horda Platform, North Sea basin. (Steel and Ryseth 1990)

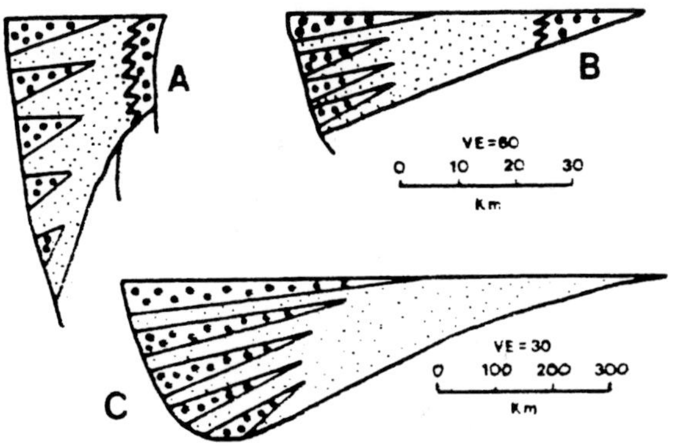

Fig. 9.24. Schematic cross sections perpendicular to active margin of A pull-apart, B rift, and C foreland basins, showing tectonic cyclothems. *Large-dot pattern*, alluvial-fan and braidplain deposits; *small-dot pattern*, fluvial, lacustrine, and marine deposits. (Blair and Bilodeau 1988)

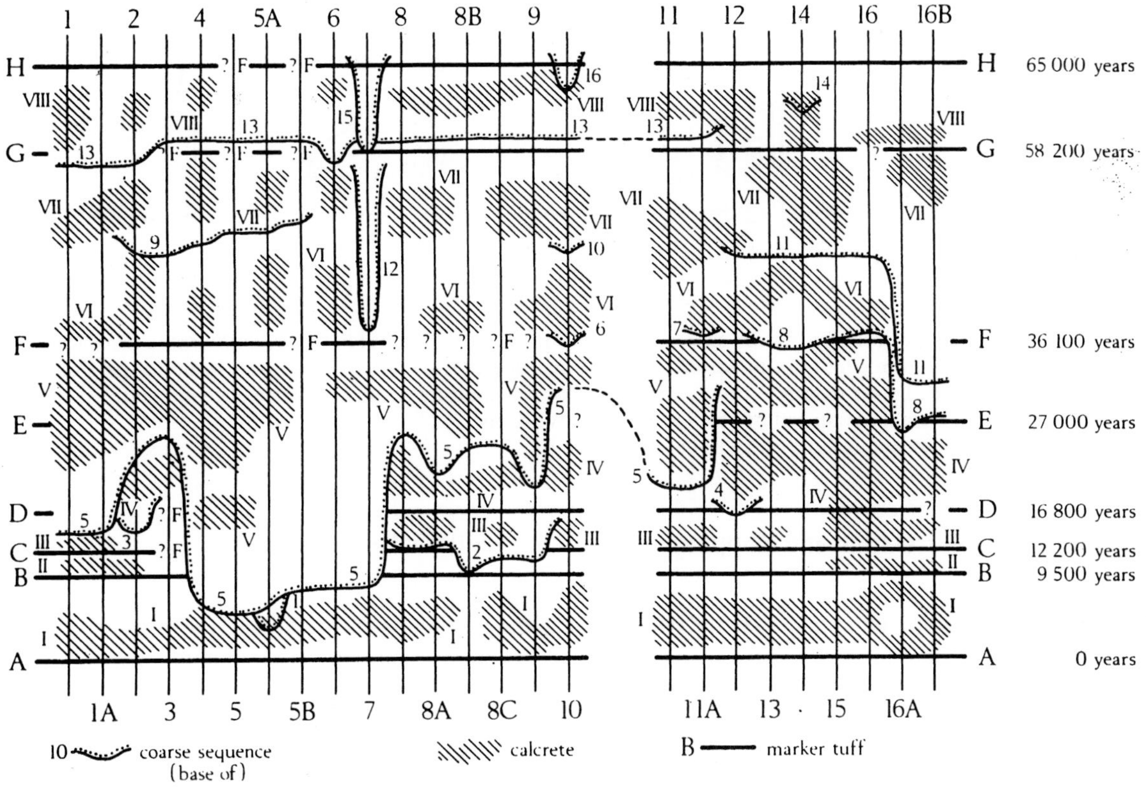

Fig. 9.25. Schematic summary of an alluvial suite containing up to seven marker tuff beds, showing the use of these beds to provide a detailed subdivision of the succession. *Horizontal scale* is arbitrary, but includes vertical profiles mapped over a lateral distance of about 35 km. *Vertical scale* is proportional to time. Thickness of the succession varies between about 13 and 27 m. Old Red Sandstone (Lower Devonian), southwest Wales. (Allen and Williams 1982)

years. These sequences are thought to reflect the independent operation of tectonic mechanisms on two time scales (Yoshida et al., in press; see Chap. 13).

Blair and Bilodeau (1988) reviewed a wide range of literature on sedimentation and tectonics and pointed out that large-scale cyclic successions of tectonic origin are common in the geological record, in all types of tectonic setting (Fig. 9.24). They proposed the term "tectonic cyclothem" for these successions. Their tabulation shows cycles of 120 m to more than 2 km in thickness and representing 2.5–15 million years of sedimentation.

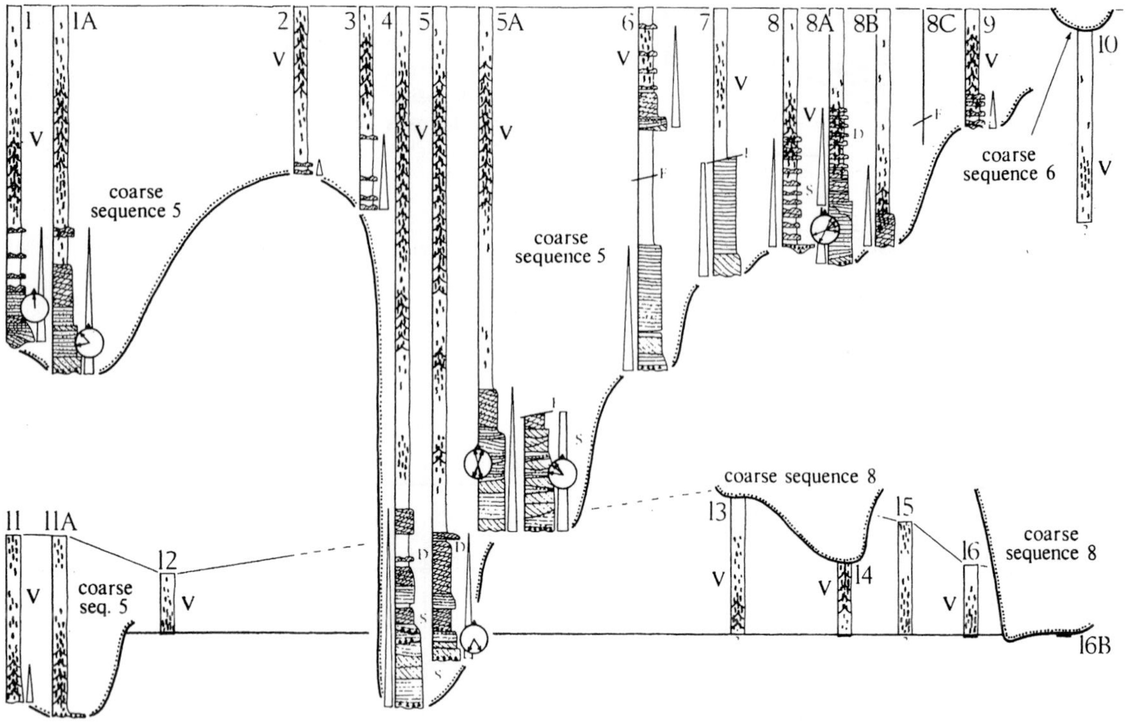

Fig. 9.26. Details of unit V of the succession shown in Fig. 9.25 (section continues from profile 10 in *upper right* to profile 11 in *lower left*). Lower datum is tuff E, which is truncated by a major channel to the left of profile 12. Left (west) of profile 10 the section is hung on an upper datum corresponding to tuff F. *Circles with arrows* show variable paleocurrent patterns. *Dark spots* in upper part of each profile indicate nodular calcrete development. (Allen and Williams 1982)

9.5 Methods of Correlation and Mapping

Fluvial deposits are difficult to map in detail because of the frequency of lateral facies changes and the lack of distinctiveness of individual beds in successions consisting of repeated similar channel and overbank units.

Basin mapping methods in general are discussed in Miall (1990, Chap. 5), and most of this detail is not repeated here. The focus in this section is on specific techniques that have been found to be useful in fluvial settings. The discussion is organized to flow from the more generalized and widely applicable methods (Sects. 9.5.1–9.5.4) to the more specialized methods that are suitable for use only in certain field situations (Sects. 9.5.5–9.5.9). Both outcrop-based and subsurface methods are discussed.

Mapping may be based on the tracing of unconformity-bounded units (sequences) in the surface or subsurface, as illustrated in Figs. 9.18, 9.21, and 9.23. The recognition of such sequences is an interpretive procedure and is discussed in Chap. 13.

9.5.1 The Use of Marker Beds

Channel-fill deposits are not laterally extensive relative to many other types of sedimentary deposit. This limits their usefulness for mapping purposes. Superimposed channel units may be lithologically very similar, so that, in the absence of complete exposure or some other data, lateral correlation of partial outcrops of channels is not likely to be very reliable. Detailed architectural examination of laterally extensive outcrops, including "walking out" of individual beds or tracing them on photomosaics, may provide enough data for detailed mapping (e.g., Fig. 9.7), but such situations are rare. However, a few other types of fluvial facies may be extensive enough to be useful for regional mapping purposes. Notable among these are coal seams and paleosols. In addition, nonmarine environments are the ideal location for the preservation of tephras, and these may extend for tens or hundreds of kilometers, providing ideal marker horizons.

Allen and Williams (1982) documented a succession of Lower Devonian beds in southwest Wales consisting of channel sandstones and interbedded floodplain units with calcretes and including eight airfall tuffs. They were able to map the tuffs over a lateral distance of more than 30 km, and this enabled them to subdivide the succession into eight units ranging from about 2 to 13 m in thickness. Each unit represents an average of about 8000 years of sedimentation, so that very detailed architectural reconstructions are possible from these rocks. Figure 9.25 shows the architectural summary, and the details of one of the depositional units is given in Fig. 9.26. In this interval a major channel up to 10 m deep incises the alluvial plain, and is only partially filled with typical channel-fill sand lithofacies. This led Allen and Williams (1982) to suggest that the channel in fact represents an incised valley formed during an interval of base-level lowering, and that the channels that subsequently filled it with sediment were on a smaller scale. Note the mature pedogenic calcretes that blanketed the area prior to the deposition of tuff F.

Individual coals and paleosols (discussed in Sects. 7.4.1 and 7.4.2) may extend for tens of kilometers across a floodplain and are therefore among the most laterally extensive and mappable units within fluvial systems. For example, Fig. 7.2 illustrates a coal-bearing alluvial succession in Wyoming and Montana; Fig. 9.26 illustrates a thick and areally extensive calcrete unit. Nemec (1988) argued that because coal seams are laterally persistent and can be correlated over wide areas, they can be used to examine the details of differential subsidence within a basin. He developed a form of graphic correlation between coal seams to be used for this purpose and demonstrated the technique using detailed correlation profiles from the Carboniferous South Wales Basin (from Woodland and Evans 1964).

As discussed in detail in Chap. 13, base-level changes may lead to the generation of widespread units reflecting regional increases or decreases in accommodation space. Coal seams and paleosols are commonly deposited in response to such base-level changes. Estuarine and lacustrine flooding during base-level rise can generate distinctive tidal sand bodies and widespread shales, the study of which can aid in the definition and interpretation of non-marine sequences.

9.5.2 Wireline Logs

The recognition of characteristic log "shapes" and distinctive vertical-profile character may be a useful tool for correlation purposes and an aid for interpreting fluvial style. The technique is almost as old as modern sedimentology itself (e.g., Nanz 1954), as noted in Chap. 2. The technique depends on the distinctiveness of the vertical profile through a sandstone body and on our ability to interpret this profile in terms of depositional processes and environment. The well-known bell-shaped gamma ray or spontaneous-potential log response yielded by a typical fluvial fining-upward cycle is a classic example. However, such interpretations are simplistic and may be quite incorrect. Thus, gamma ray logs record the presence of natural background radioactivity, which typically is highest in clay minerals because of the concentration in these minerals of naturally radioactive isotopes of potassium and thorium. The log response is therefore interpreted as an indicator

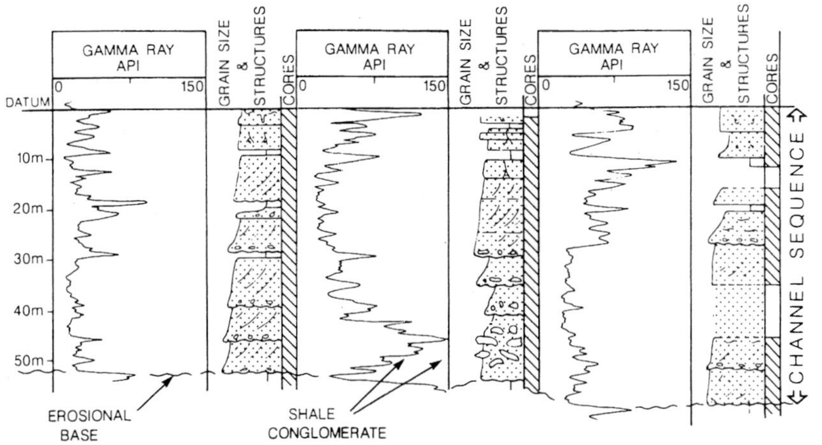

Fig. 9.27. Channel sandstone succession as seen in three adjacent wells, and corresponding gamma-ray logs. Note the interval of high gamma-ray values in the *center well*, reflecting the presence of a shale-clast conglomerate. (Rider 1990)

of "clayeyness", which is normally inversely proportional to grain size. However, the presence of clay-rich clasts in an otherwise clean, coarse sandstone may distort the reading (Fig. 9.27), and feldspar-rich sandstones (e.g., "granite wash") also yield high gamma-ray readings because of their high potassium content. Other problems with interpretation are addressed by Rider (1990).

A quite different type of problem is that, as discussed in Chap. 8, fining-upward cycles in fluvial deposits are not amenable to unique interpretations. Channel fill and abandonment, point bar growth, and certain tectonic processes can all yield fining-upward cycles and corresponding bell-shaped log patterns. In the first instance, attention should be paid to the scale of the cycle. Cycles thicker than about 20 m are unlikely to be the product of within-channel (autogenic) processes and are more probably related to allogenic causes, such as sequence development or tectonic pulses.

Wireline logs may reveal the presence of regionally mappable marker beds, distinguished by a recognizable type of log deflection or because they subdivide the vertical profile into intervals of distinctive log character. An example of such a profile containing marker beds is shown in Fig. 9.28. Such markers may serve to divide a thick nonmarine succession into thinner, more readily mappable horizons or "operational units". These markers typically reflect allogenic controls, such as changes in base level (stratigraphic sequences) or regional changes in paleoslope brought about by tectonic tilting. Their

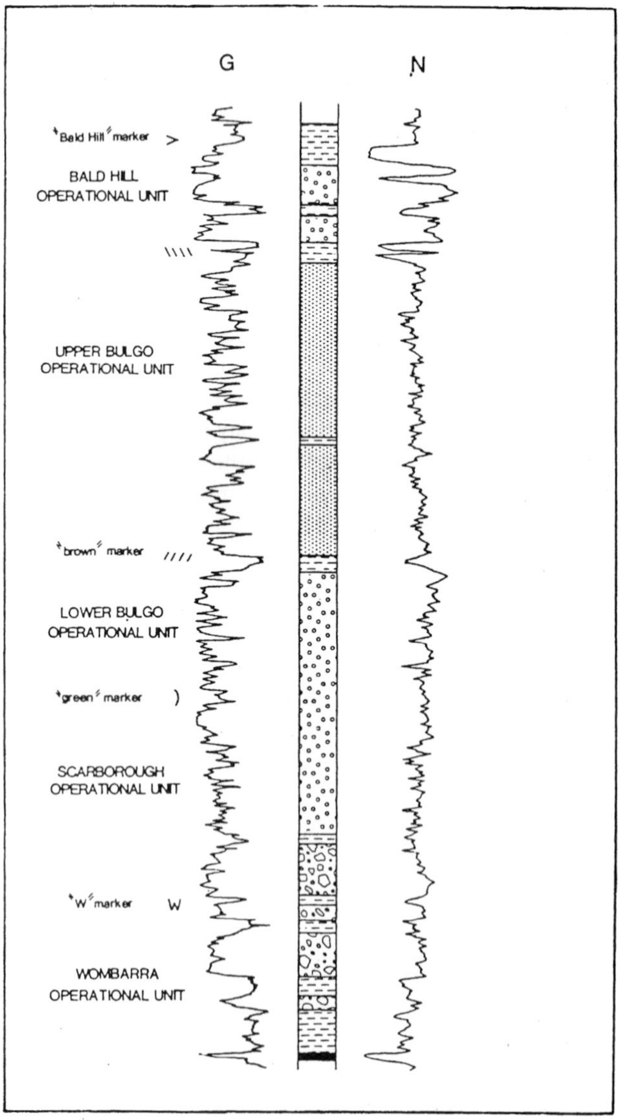

Fig. 9.28. Operational subdivision of a nonmarine succession using wireline log character. Note the presence of various "marker" beds. *G*, Gamma ray log; *N*, neutron log. Narrabeen Group (Permian), Sydney Basin, Australia. (Hamilton and Galloway 1989)

origins are discussed in Chap. 11. Hanneman et al. (1994) discussed the subsurface mapping of paleosols and their use for defining sequence boundaries. Calcic paleosols within clastic fluvial successions are characterized by their high density and high interval velocities, which makes them readily identifiable on density and sonic logs and in seismic records (Sect. 9.5.4).

The use of sequence-stratigraphic concepts can considerably aid in regional mapping, even within entirely nonmarine successions. For example, the Statfjord and underlying units in the northern North Sea can be subdivided into shale-rich and sandstone-rich intervals, the development of which reflects the balance between changing accommodation space and sediment supply, as described in detail in Chap. 13. Mapping of these variations and correlation of major internal erosion surfaces interpreted to be sequence boundaries forms the basis for field-wide and regional correlation (Steel and Ryseth 1990; MacDonald and Halland 1993). Exploration in the mixed marine-nonmarine Latrobe Group of the Gippsland Basin is now based largely on the sequence-stratigraphic model of Rahmanian et al. (1990). Melvin (1993) discussed the influence of tectonism on basin development in the Endicott field area of Alaska.

Ryseth (1989) reported a subsurface study of the Middle Jurassic Ness Formation in the northern North Sea Basin. He was able to subdivide this 100-m-thick succession into its main architectural components and to correlate them between four exploration wells over a distance of about 10 km using a combination of core and wireline-log information. Coals and paleosols defined the tops and bases of a series of autogenic cycles which could be traced between wells.

Visher et al. (1971) used log character and shape to discriminate various fluvial and deltaic environments in the Bartlesville Sandstone of Oklahoma. Wightman et al. (1987) reported on a similar type of exercise applied to the Mannville Group of the

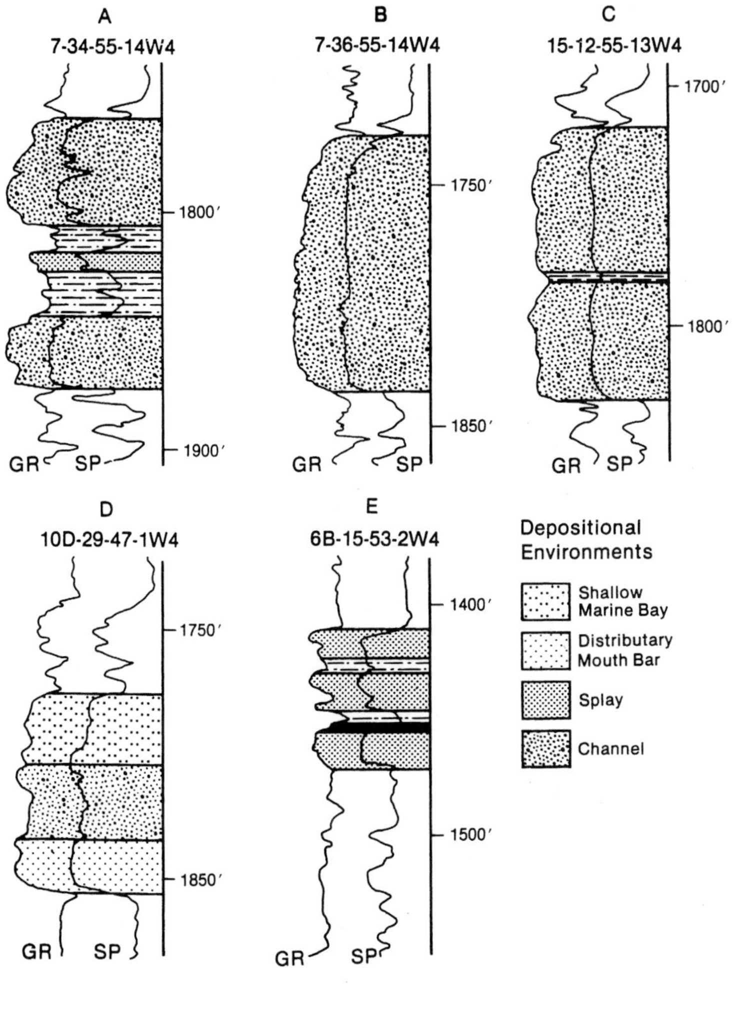

Fig. 9.29. Typical log character of sandstones in the Upper Manville Sandstone, Lloydminster area, Alberta, showing interpreted environments. (Wightman et al. 1987)

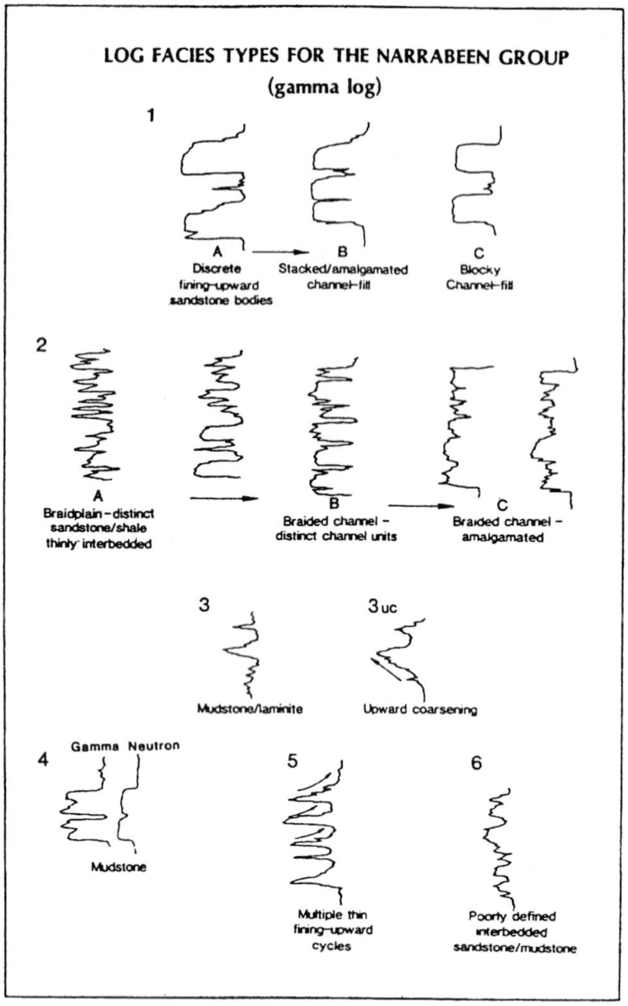

Fig. 9.30. Log facies types for the Permian Narrabeen Group, Sydney Basin, Australia. (Hamilton and Galloway 1989)

Lloydminster area, Alberta (Fig. 9.29). Hamilton and Galloway (1989) discriminated various fluvial styles in a Permian fluvial unit in Australia using log shape (Fig. 9.30). Golin and Smyth (1986) used the technique for a Jurassic unit in Surat basin, and Jirik (1990) used the same technique to recognize fluvial subenvironments in a study of the Frio Sandstone in Texas. Davies et al. (1991, 1993) suggested that vertical-profile character could be used to discriminate meandering from straight (including braided) channel styles in a study of the Travis Peak Formation (Lower Cretaceous) of east Texas, although this type of approach may not always be reliable, as argued by Miall (1980).

The most effective use of log shape as a mapping tool is in conjunction with other lithofacies mapping techniques, such as those discussed in the next section. Thus Fig. 9.31 shows the distribution of log "facies" within an informal subsurface stratigraphic unit that had been defined on the basis of log character (see Fig. 9.28). A sand distribution map of the same unit is given in Fig. 9.32. The two maps provide a good insight into the paleogeography of the unit.

9.5.3 Lithofacies Mapping

Conventional methods of lithofacies mapping are described in detail in standard textbooks (e.g., Miall 1990, Sect. 5.3). Such techniques as mapping of net sand content, or sand/shale ratio, or net pay are standard procedures for the delineation of reservoirs and for use in subsurface trend prediction. However, these techniques are of limited usefulness for exploration in fluvial environments, because of the extreme facies variation in these rocks, the thinness of the sandstone bodies, and their lack of directional predictability. In most exploration situations well spacing is considerably greater than the average

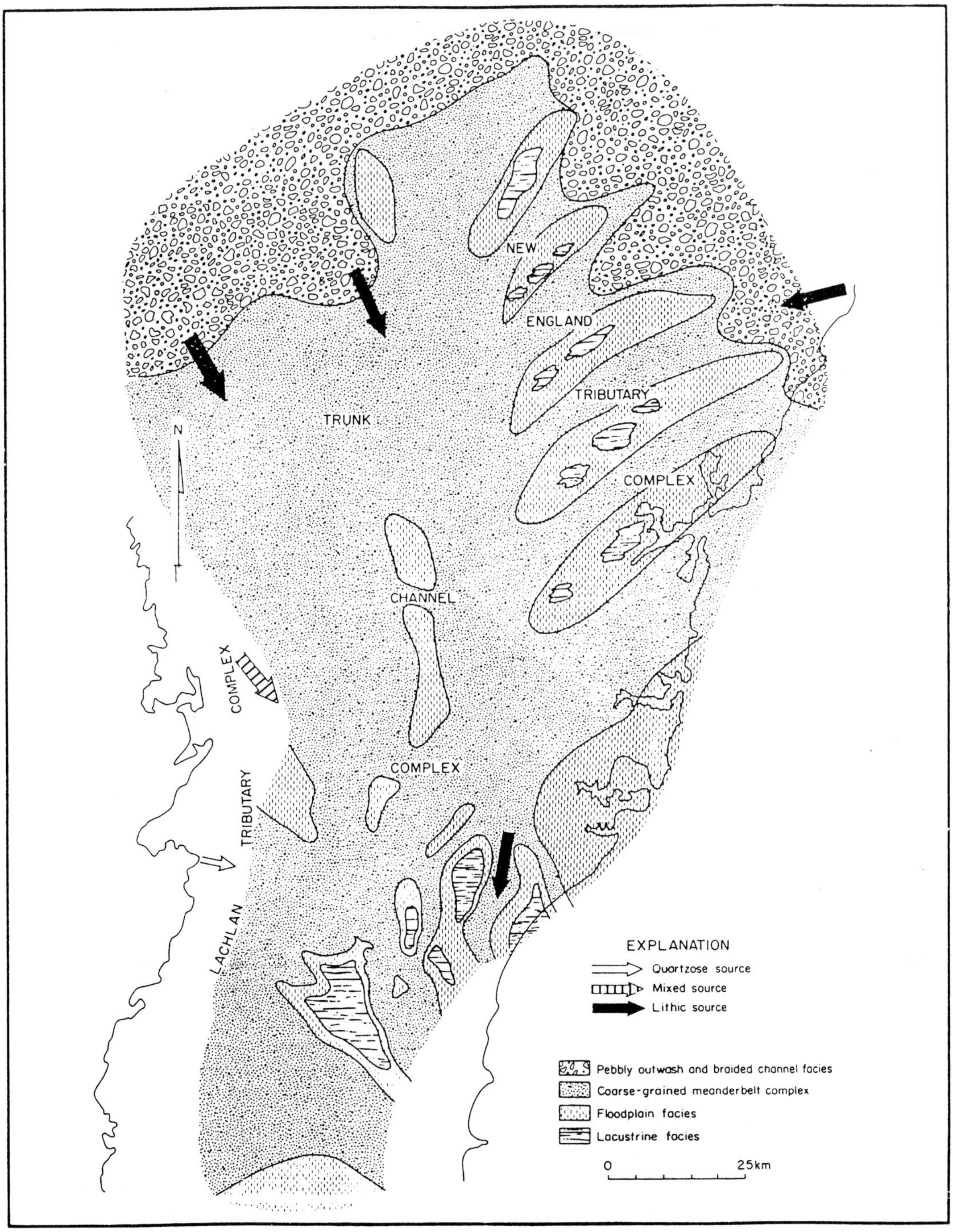

Fig. 9.31. Distribution of characteristic log facies types in the Wombarra Operational Unit, Sydney Basin. *Arrows* indicate sediment sources interpreted from petrographic data. The combination of data types provides an excellent paleogeographic map. (Hamilton and Galloway 1989)

dimensions of the typical sandstone body (except, perhaps, for the length of major paleovalleys). Narrow channels and small point bars may therefore be missed, except where well spacing is unusually dense, as in some enhanced recovery projects. It is quite possible for the same data to lead to different mapped contour patterns, reflecting the different paleogeographic models of the authors. For example, compare the "fan" model for the Westwater Canyon Member of the Morrison Formation of New

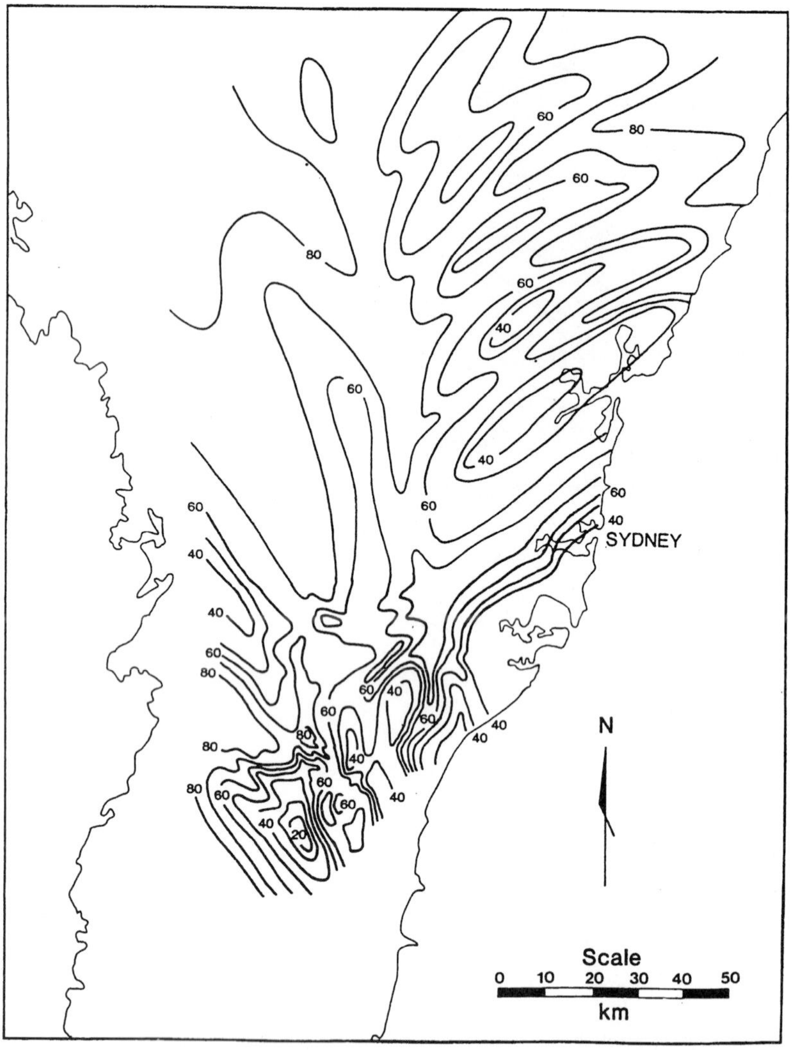

Fig. 9.32. Lithofacies map of the Wombarra Operational Unit, Sydney Basin, showing sandstone thickness expressed as a percent of total unit thickness. (Hamilton and Galloway 1989)

Mexico (Galloway 1980) with the channel model of Saucier (1976).

Mapping of units defined by the sixth- and higher-order surfaces according to this book can be based on conventional methods of subsurface wireline log correlation, provided well spacing is adequate (many examples of this are described and illustrated in the sections on petroleum geology). In most mature basins, with a well spacing of one to a dozen wells per township (wells 1.5–5 km apart), such correlation should be possible down to the "sequence" or submember level, although local problems may arise as a result of internal erosion and the resulting amalgamation of sandstone sheet units. An example is illustrated in Fig. 9.23. Computer modeling of lithologic units at this scale, for reservoir engineering purposes, is normally carried out by subdividing the producing formation into "sheets" or "flow units" or "petrophysical zones" that extend across the entire field (e.g., Wadman et al. 1979; Jin et al. 1985; Yinan et al. 1987; Lawton et al. 1987; Struijk and Green 1991; Melvin 1993). Principles of sequence stratigraphy, incorporating ideas regarding eustatic and tectonic control of fluvial depositional systems, may be introduced at this stage to facilitate stratigraphic interpretation (e.g., Melvin 1993). Production engineering and computer modeling for development purposes are normally based on this type of stratigraphic subdivision of the reservoir units (Stanley et al. 1990; Melvin 1993; Martin 1993).

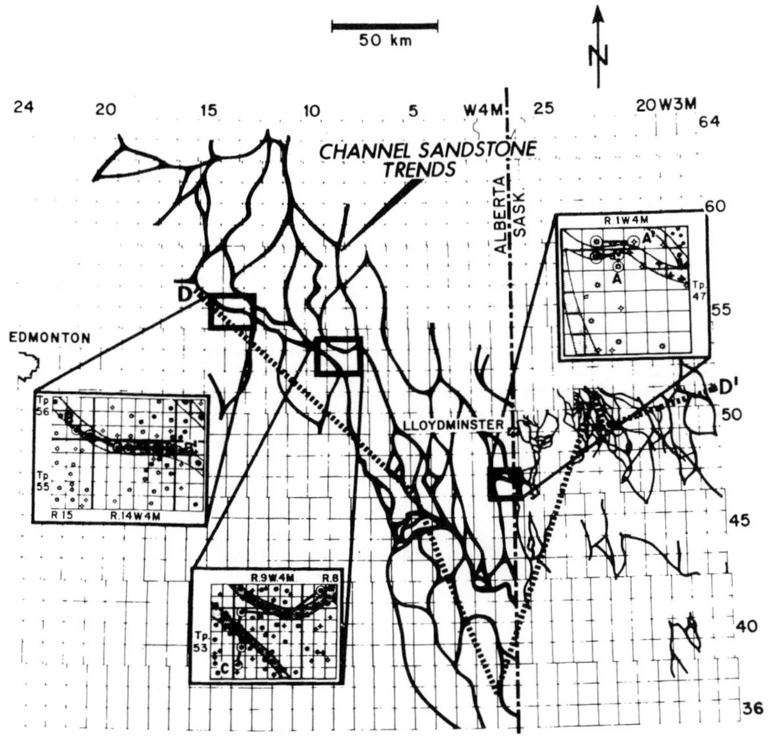

Fig. 9.33. Pattern of ribbon sandstone bodies in the Upper Mannville Sandstone, as proposed by Putnam (1983). *Insets* show some of the details of data-point distribution

Mapping of ribbon sandstone bodies in the subsurface can cause particularly difficult problems, because the width of the sandstone bodies is usually less than well spacing, except in densly drilled fields. In addition, it may be difficult on wireline logs to discriminate between channel sandstones and the various overbank sandy facies, including levees and splays, which may occupy a much wider belt flanking the channel than the channel itself. An example of the problems in interpretation that can arise is the pattern of anastomosed ribbon sandstone bodies that were reported in the subsurface Mannville Formation of Alberta and Saskatchewan by Putnam (Smith and Putnam 1980; Putnam 1982a,b, 1983; Fig. 9.33). His analyses were challenged by Wightman et al. (1981) on several grounds. They pointed out that Putnam used an arbitrary classification of petrophysical log character to define channel sand bodies, that some of the bodies correlated between adjacent wells are of different thicknesses, and that they are at slightly different stratigraphic levels, suggesting that the channel patterns may be an amalgamation of different channels that were of different ages, and not all active at one time. The channels in Putnam's reconstructions are also much larger than any described from modern anastomosed systems. In a later paper Wightman et al. (1987) presented detailed sedimentological evidence in support of an alternative interpretation. They showed from detailed core studies of lithofacies, trace fossils, and vertical sequences that many of the sand bodies are of deltaic origin, including mouth-bar and splay sandstones, and others are marine sheet sands (Fig. 9.29). The anastomosed fluvial model was decisively rejected in favor of a more elaborate deltaic model (a deltaic setting does not, by itself, exclude the possibility of an anastomosed fluvial style. As discussed in Sect. 13.3.3, this fluvial style is commonly developed during times of rising base level).

Figures 9.34 and 9.35 show differences in stratigraphic interpretation that illustrate the difficulties in defining narrow channel sand bodies. The interpretation by Putnam (1982b) shows the sandstone as a continuous, ribbon-like body (Fig. 9.34). However, Wightman et al. (1987) demonstrated, with pressure data, that the sandstone in well 7-33 is not in fluid connection with that at 7-34 and 7-35. There are, therefore, at least two separate sandstone bodies (Fig. 9.35). Note also the interpreted presence of a lower channel sandstone in well 7-34 in Fig. 9.35, a unit that is incorporated into the main sand body in Putnam's intepretation.

Mapping and reservoir modeling of ribbon sandstone bodies is best attempted when a large volume of subsurface data is available from a well-developed field (well spacing of 32–54 ha; 8–4 wells per one-

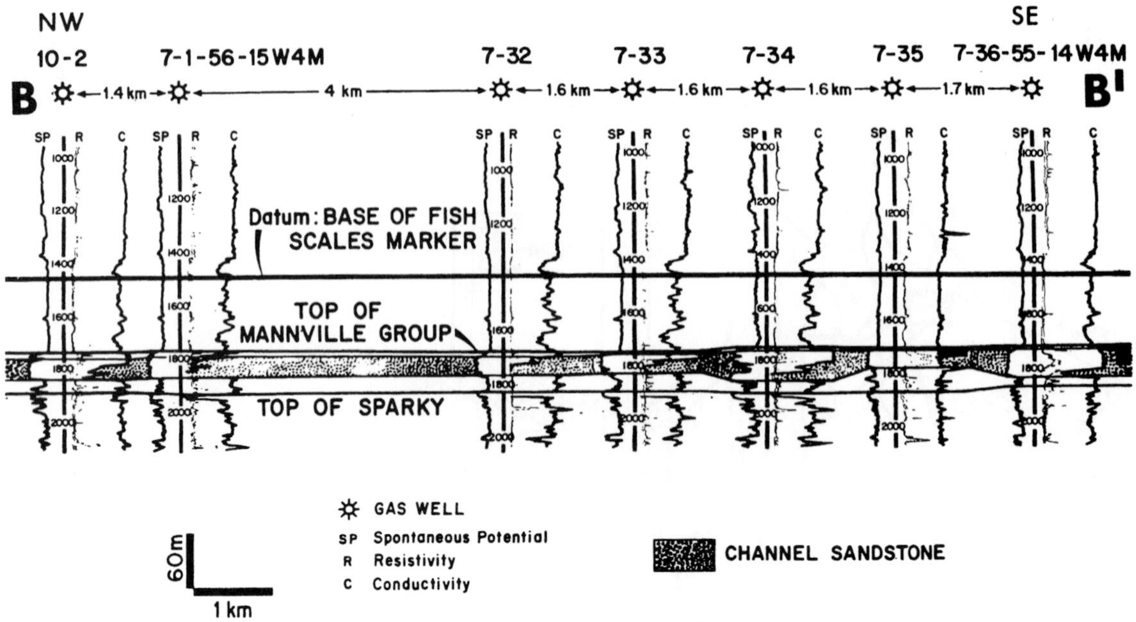

Fig. 9.34. Stratigraphic interpretation of sandstone bodies in the Upper Mannville, near Lloydminster. (Putnam 1982b)

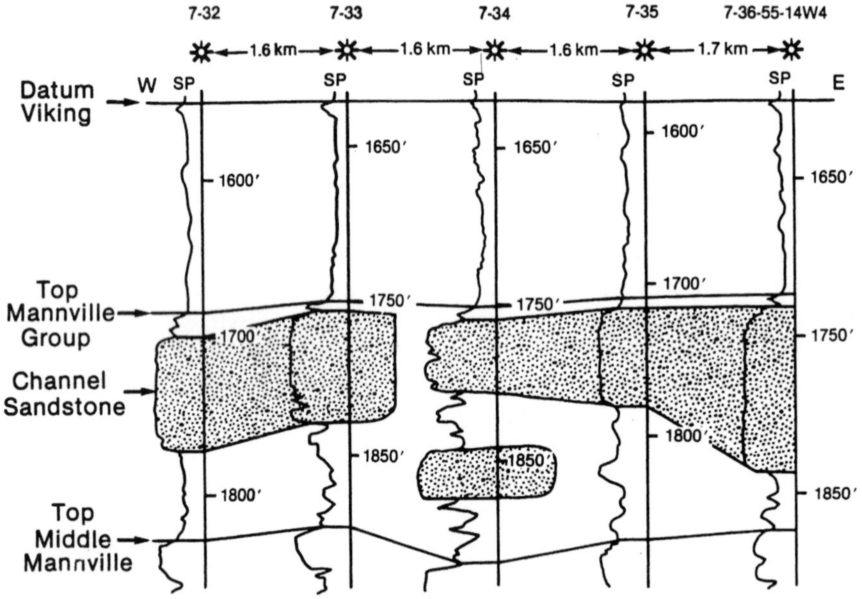

Fig. 9.35. Stratigraphic interpretation along same line of cross section as Fig. 9.34, showing alternative interpretation. (Wightman et al. 1987)

mile-square section). Yinan et al. (1987) noted the difficulties in producing from "straight" and "confined channel" (anastomosed?) sandstone bodies because of the discontinuous lateral geometry and the need for a particularly dense well spacing. Robinson (1981) discussed the problem of mapping paleovalleys from a statistical point of view (probability of intersection depending on well spacing). He demonstrated that only where well spacing was equal to or less than the average valley or channel width could a reliable map be constructed.

Conventional lithofacies mapping of fluvial stratigraphic units, for example, sand/shale ratios, net sand isopachs, etc. (e.g., Tyler and Ethridge 1983; Hamilton and Galloway 1989), may not be a very effective technique for reservoir development pur-

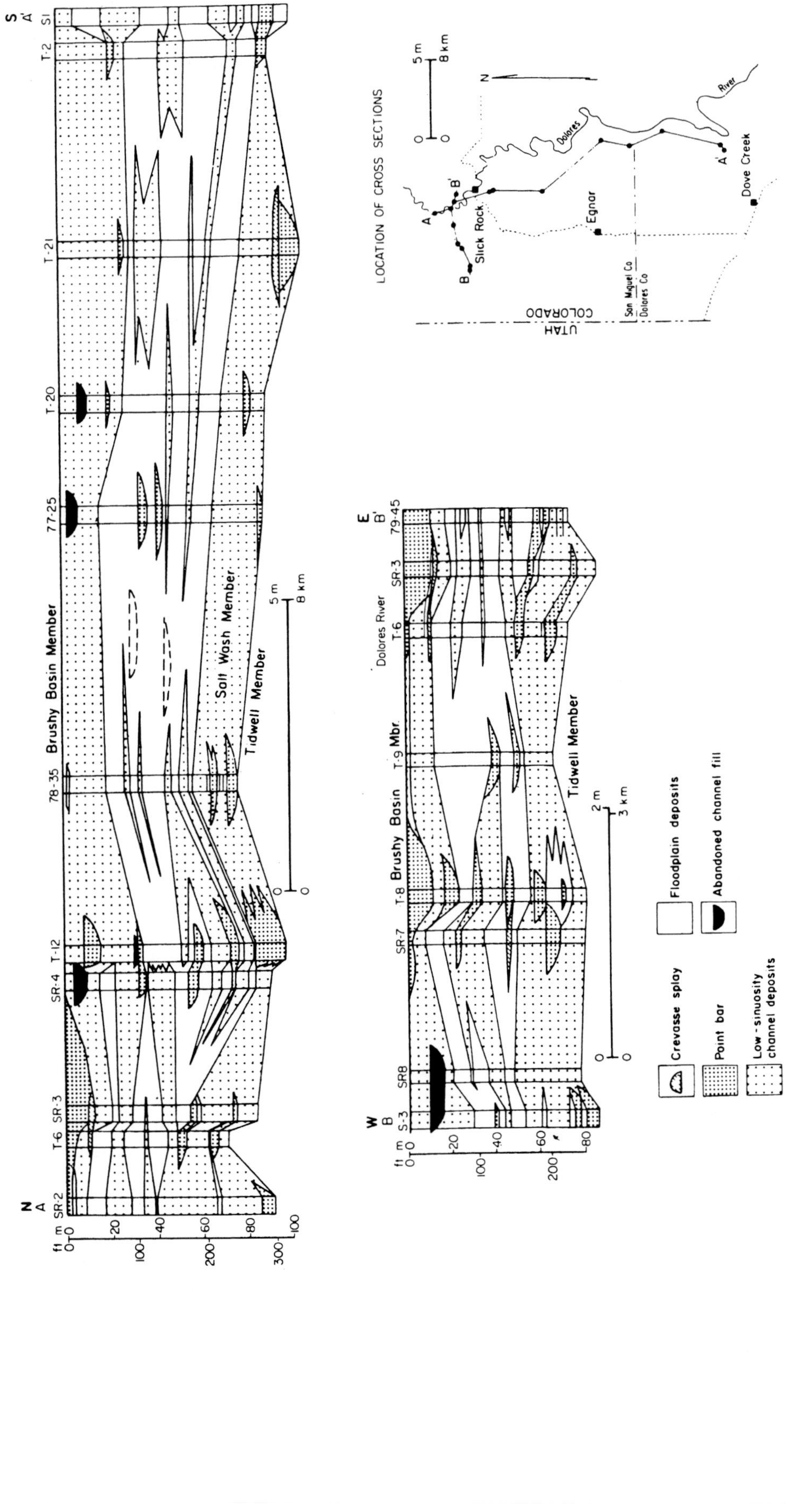

Fig. 9.36. Mapping of sand bodies in the subsurface. Strike- (*A–A'*) and dip- (*B–B'*) oriented cross sections through the Salt Wash Member of the Morrison Formation (Jurassic), westernmost Colorado. (Tyler and Ethridge 1983)

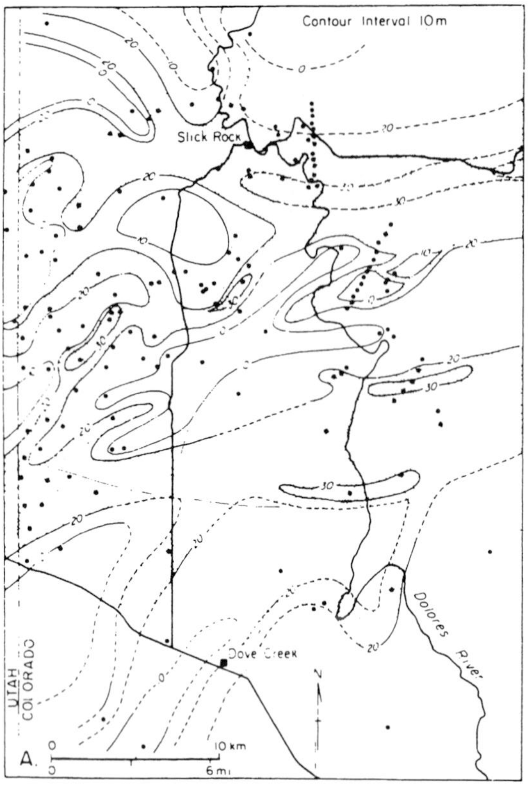

Fig. 9.37. Isopach map of the upper Salt Wash Member, Colorado. Note linear trends of isopach "thicks" which indicate the general position of channel belts. (Tyler and Ethridge 1983)

poses. The construction of such maps requires the definition of a mappable stratigraphic unit, but in fluvial deposits there are few mappable markers, and so such units tend to be much thicker than the channel belts, with the result that the maps show the average position of several or many channels with different position and trend. Figure 9.32 is a "sand" map of one of the Permian operational units in the subsurface of the Sydney Basin, Australia. Figures 9.36 and 9.37 provide cross sections and an isopach map of the Salt Wash Member of the Morrison Formation in Colorado. These diagrams illustrate this point. The isopach maps provide only the most generalized information regarding channel position and could not be relied upon as a subsurface mapping tool for locating actual channel bodies. In the case of the Salt Wash Member, channel belts are 2–10 km wide (Fig. 9.37), whereas individual channels are only a few hundred meters wide (Tyler and Ethridge 1983).

However, where well density is sufficient, detailed lithostratigraphic subdivision and channel mapping may be a very effective exploration tool. Based on experience with the Westwater Canyon Member (Miall 1988a), macroform bar units, bounded by fourth-order surfaces, and the subunits within them that are defined by third-order surfaces would require well spacing of 32 ha (8 wells/section) or less if their geometry is to be reliably reconstructed from wireline log and core data (Fig. 9.38, bottom). In the Westwater Canyon Member, even with a well spacing of 4 ha (64 wells/section) many of the smaller individual fifth-order units (e.g., Miall 1988a, Fig. 4) would not be correlatable. In larger river systems the macroforms are correspondingly more extensive, and a wider well spacing might be adequate. For example, a point bar 6 km in diameter described by Busch (1974) was mapped using a 16-ha (16 wells/section) spacing (Fig. 9.39). In their study of the Gypsy Sandstone of Oklahoma, Doyle and Sweet (1995) determined that channel deposits bounded by fifth-order surfaces were the smallest scale of unit that could be reliably correlated with a 100-m well spacing. Other examples of subsurface channel and point-bar mapping are described in Chap. 15.

The necessary data for detailed lithofacies mapping should become available from pilot studies carried out at the commencement of enhanced recovery projects. Engineering grid-block modeling at this scale may be a useful intermediate step between models of the large flow units and those constructed from core-test data.

Mapping of a point bar provides information on the scale of the river, such as its meander wavelength. This may become a useful prospecting tool, as scattered well data may be used to predict meander position by the sketching in of possible meander positions to accommodate the available data (e.g., sandstone penetration corresponding to bar, shale possibly interpreted as abandoned channel fill). This technique has, to the writer's knowledge (consulting contacts), been used as an aid in the exploration of Tuscaloosa prospects in the Gulf Coast area and Permian sands in Texas and may be in widespread use in other mature basins. As discussed in Sect. 10.4.1, use of surface information to model the scale of channel sandstones may also be a helpful approach.

One of the most detailed published examples of subsurface mapping that has been carried out in fluvial sandstones is illustrated in Fig. 9.40. Here gamma-ray logs from very closely spaced wells have been used to construct detailed cross sections across two sandstone bodies. They show that minor internal erosion surfaces can be detected and mapped in the subsurface, given adequate data. These may be

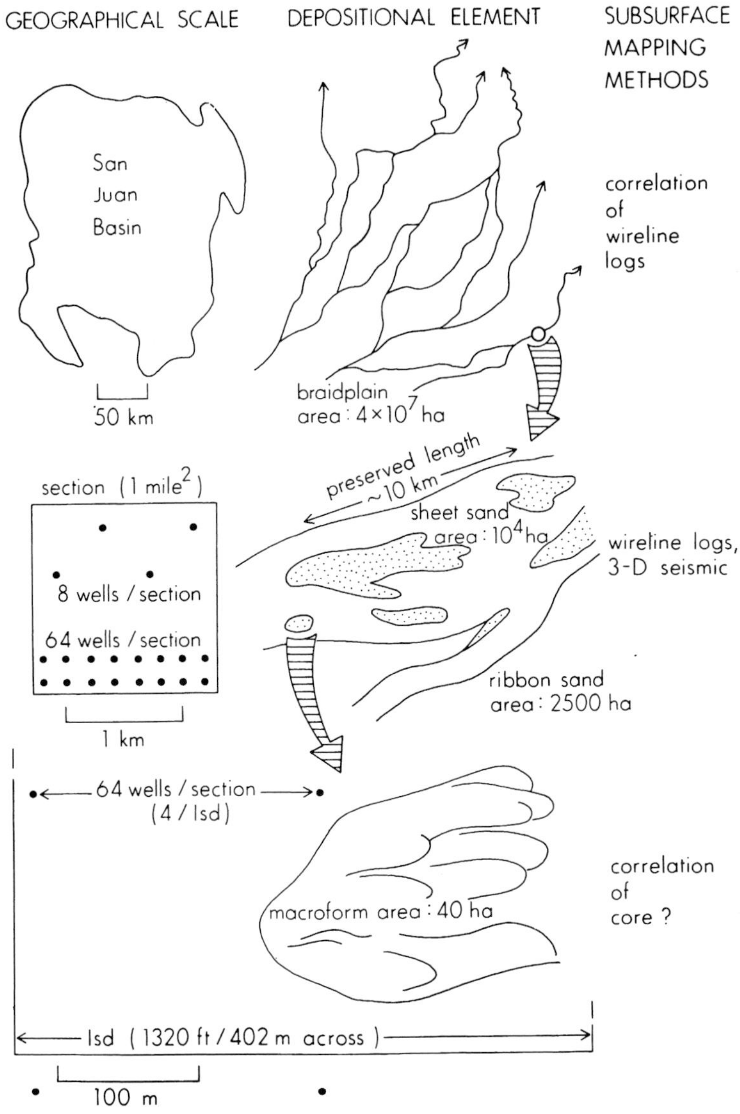

Fig. 9.38. Depositional systems and geomorphic elements compared at three scales. *Top*: The San Juan Basin (Colorado-New Mexico) and a speculative braidplain system (Westwater Canyon Member), shown side by side at the same scale; *middle*: sheet and ribbon sandstones in the Westwater Canyon Member (*right*) shown at the same scale as a 1-square-mile section with two different densities of well spacing (*left*); *bottom*: enlargement of a section, shown at the same scale as a small macroform. *lsd*, Legal subdivision (Canada). (Miall 1988a)

fifth-order surfaces, recording successive positions of a channel as it migrated laterally, or they may be third-order surfaces, indicating erosional breaks in the filling of the channel by lateral or downstream accretion.

9.5.4 Seismic Methods

Many of the major structural features typically associated with fluvial deposits are readily detectable on reflection-seismic records. For example, as shown in Chap. 11, thick clastic wedges are likely to be associated with rift faults or fold-thrust belts, and regional unconformities may be crossed by incised paleovalleys filled with fluvial deposits. These large-scale features are readily mapped using seismic methods. For example, Fig. 9.41 illustrates the essential seismic-facies characteristics of a fluvial-deltaic clastic wedge prograding from a fault-bounded basin margin. The mountains (*a*) may be flanked by an alluvial-fan zone (*b*), which would likely yield reflections of variable amplitude and poor continuity, because of the poorly stratified and variable lithofacies of most alluvial-fan deposits. Braidplain deposits, downstream from the fan (*c*), may yield good to discontinuous reflections, depending on the degree of channelization and preservation of floodplain deposits. Shale beds in sandy braided stream deposits are commonly areally extensive (e.g., Morrison Formation: Miall 1988a), especially where they have developed in response to regional tectonic adjustments

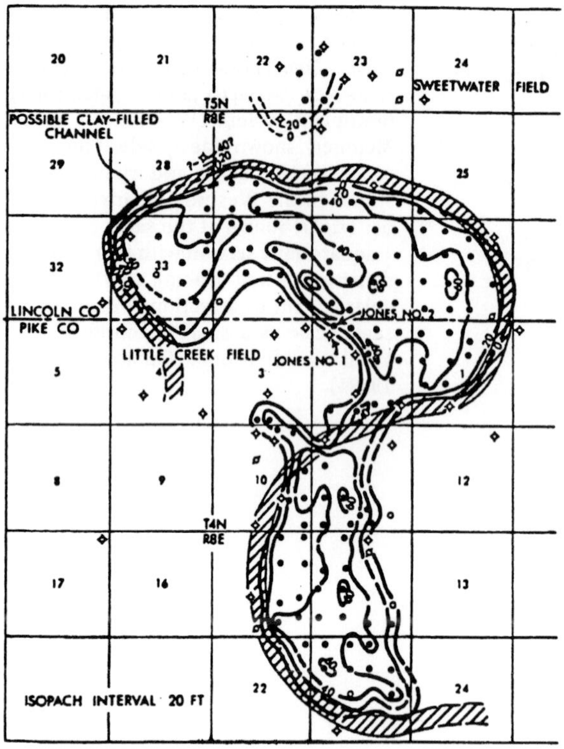

Fig. 9.39. Isopach of producing sandstone, Little Creek field, Mississippi. (Busch 1974)

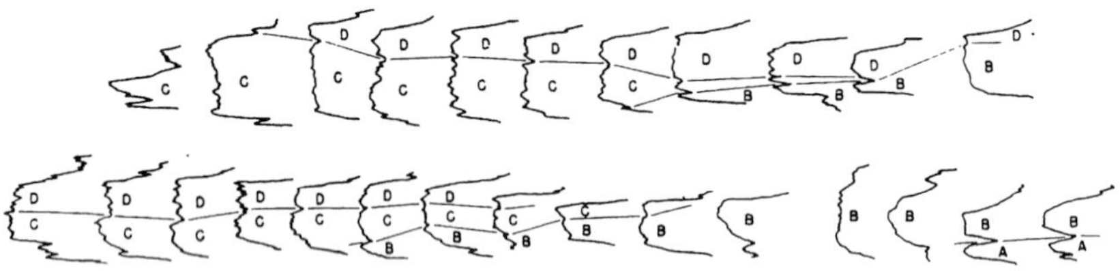

Fig. 9.40. Two gamma-ray log cross sections through the producing sandstone in the Bronson-Xenia field, Kansas. Note presence of internal, dipping bounding surfaces that subdivide the sandstone into four discrete lenses. (Walton et al. 1986)

or base-level changes, and may be thick enough to yield good reflections. As in the alluvial-fan zone, channels marked by lithologic contrasts between the channel fill and the channel banks may give rise to seismic scatter and to the development of short, confused reflections, or to diffractions. Close to the coastline (spot *d* in Fig. 9.41) seismic facies may change as more floodplain deposits are incorporated into the succession, giving rise to a greater and more areally extensive acoustic contrast between sandstone bodies and overbank materials. Reflections may therefore be more continuous and of higher amplitude. The fluvial wedge may pass laterally into a delta, distinguished by clinoform relections (*e*). Where the clastic wedge is part of a sequence controlled by base-level change or tectonism it may be sandwiched between other facies, as marked by a regional unconformity. For example, there may be widespread transgressive deposits formed on a "maximum flooding surface" (*f*). Such facies, and the unconformity with which they are associated, are commonly readily traceable on seismic records (Chap. 11). An example is illustrated below.

Meadows and Beach (1993) demonstrated how seismic data could be used to distinguish between fluvial-dominated and eolian-dominated sandstone successions in a Triassic rift fill, and they documented the distribution of these broad facies variations relative to active faults in the basin. Brown and Fisher (1977) described the typical seismic-facies characteristics of alluvial and delta-plain facies in Brazilian offshore basins. Reflections are typically "horizontal, parallel, rarely divergent, layered to locally reflection-free; locally, erosional channels may

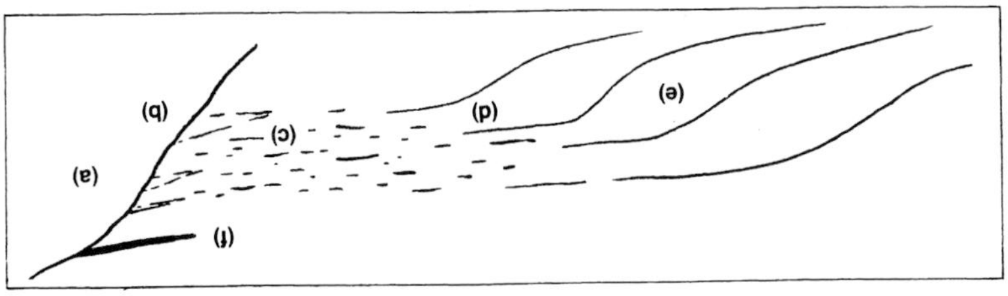

Fig. 9.41. Simplified seismic facies characteristics of a fluvial-deltaic clastic wedge prograding from a fault-bounded basin margin (Anstey 1980). See text for explanation

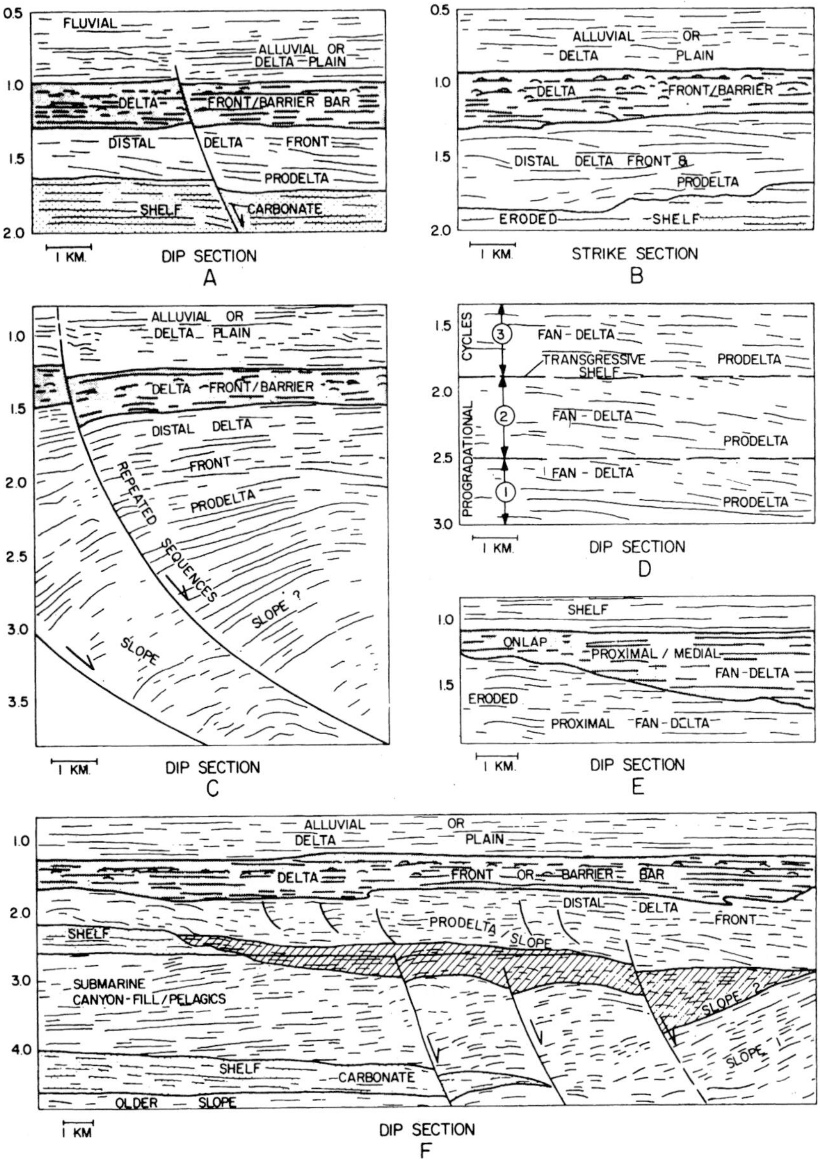

Fig. 9.42. Deltaic and associated seismic facies patterns, redrawn from original seismic sections through Brazilian offshore basins. (Brown and Fisher 1977)

be inferred. In strike sections, the reflections are weak, parallel-layered to subtle-mounded, chaotic-to-drape patterns. Continuity of reflections ranges from excellent to fair in dip sections ..., but continuity is poor to fair in strike sections ...; amplitude is variable (high in continuous reflections and poor in chaotic zones), and spacing is very regular in zones of high-continuity reflections but irregular in the remainder of the unit. The reflections collectively define a tabular external geometry ..." (Fig. 9.42).

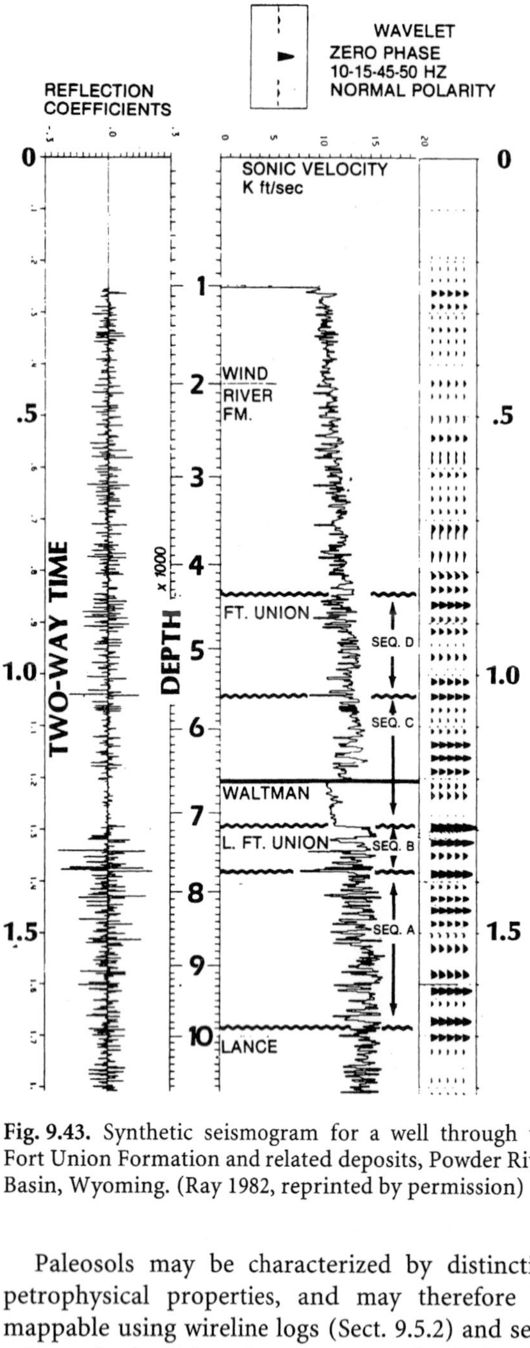

Fig. 9.43. Synthetic seismogram for a well through the Fort Union Formation and related deposits, Powder River Basin, Wyoming. (Ray 1982, reprinted by permission)

Paleosols may be characterized by distinctive petrophysical properties, and may therefore be mappable using wireline logs (Sect. 9.5.2) and seismic methods. Paleosols are commonly distinctive stratigraphic markers (Sect. 9.2) and in some cases are used to define sequence boundaries (Sect. 13.3.1). They can therefore be of considerable stratigraphic importance. Hanneman et al. (1994) mapped calcic paleosols in some Miocene beds in Wyoming using seismic methods and wireline logs and demonstrated that their petrophysical visibility was caused by their high densities and interval velocities.

The Paleocene Fort Union Formation of the Powder River Basin, Wyoming, provides an example of a seismic reflection survey of a fluvial-lacustrine deposit (Ray 1982). A synthetic seismogram is shown in Fig. 9.43 and Fig. 9.44 is a seismic cross section. Five unconformities have been recognized in this succession, at 810, 1000, 1295, 1360, and 1670 ms (depths on the right-hand side of the section, Fig. 9.44). These unconformities separate distinct stratigraphic sequences. A particularly instructive example of the lateral changes in seismic facies discussed by Anstey (1980) is shown by the sequence between the 1000- and 1295-ms unconformities (sequence C). Note the east-dipping clinoform structure of this sequence. Note also the change in reflection character from low-amplitude, low-frequency, low-continuity reflectors in the upper part of the section, particularly in the west, to a region of high-continuity, high-amplitude, high-frequency reflectors in the middle of the unit (the clinoform structure is particularly clear here), to a zone of low- to moderate-continuity, low-frequency, low-amplitude reflectors in the lower part of the sequence. These changes are interpreted as lateral facies changes from discontinuous fluvial sandstones and shales in the west, to interbedded sandstones and shales of the lake-margin delta clinoforms, to uniform lacustrine shales in the east.

The ideas in the preceding paragraphs, including those of Anstey (1980), Brown and Fisher (1977), and Ray (1982), relate to the recognition of large-scale fluvial depositional systems within other types of stratigraphic units. Much more difficult is the mapping of individual sandstone bodies within a fluvial system. As pointed out by Neidell and Beard (1985), careful attention to processing and presentation of seismic data can lead to considerable refinements in our ability to map subtle stratigraphic objectives. A knowledge of the structure of the propagating waveform and the use of color to display amplitude variations are two vital aspects of this work. Fulthorpe (1991) demonstrated that seismic resolution of stratigraphic features is increased in areas of high sediment supply, because the deposits are thicker and therefore more readily resolvable. Also, deposits that develop by lateral progradation are much more readily subdivisible into their component units than those which accumulate by vertical aggradation, because of the greater scale of depositional units in the horizontal dimension. Cartwright et al. (1993) pointed out that the improved resolving power available from modern seismic acquisition and processing techniques, relative to those available when seismic stratigraphy was first introduced in the 1970s, has led to an increase in the demonstrable complexity of stratigraphic units and to a need for

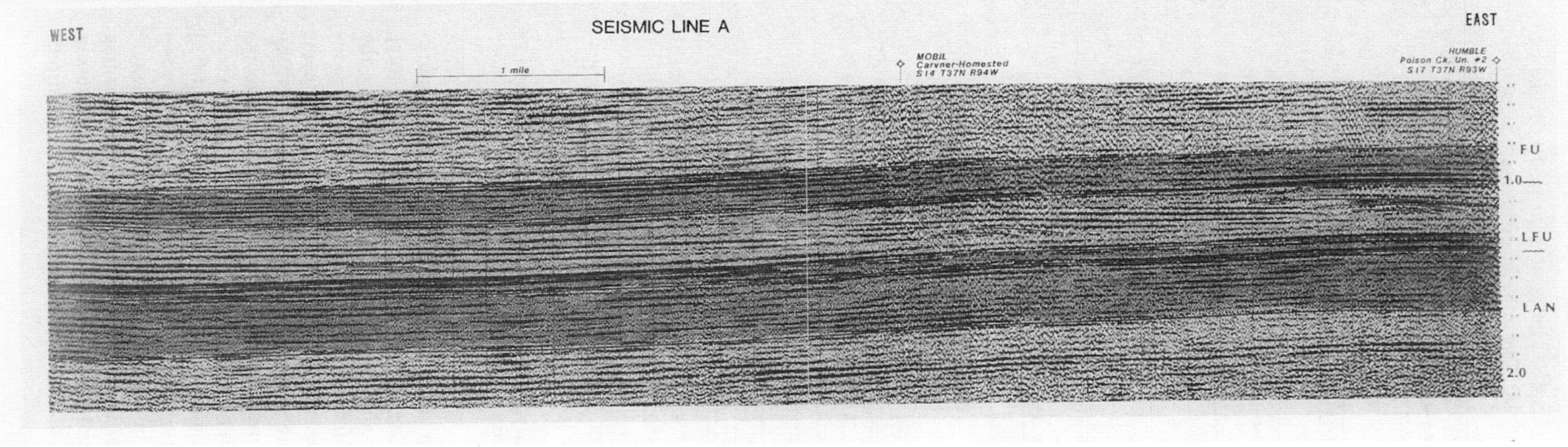

Fig. 9.44. A typical seismic line through the Fort Union Formation, Powder River Basin, Wyoming. The Lower Fort Union (LFU) shows a gradation from low-continuity, low-amplitude, low-frequency fluvial strata at *left*, through high-frequency, east-dipping clinoform shoreline deposits, to lacustrine deposits at *lower right* of unit. (Ray 1982, reprinted by permission)

A
B
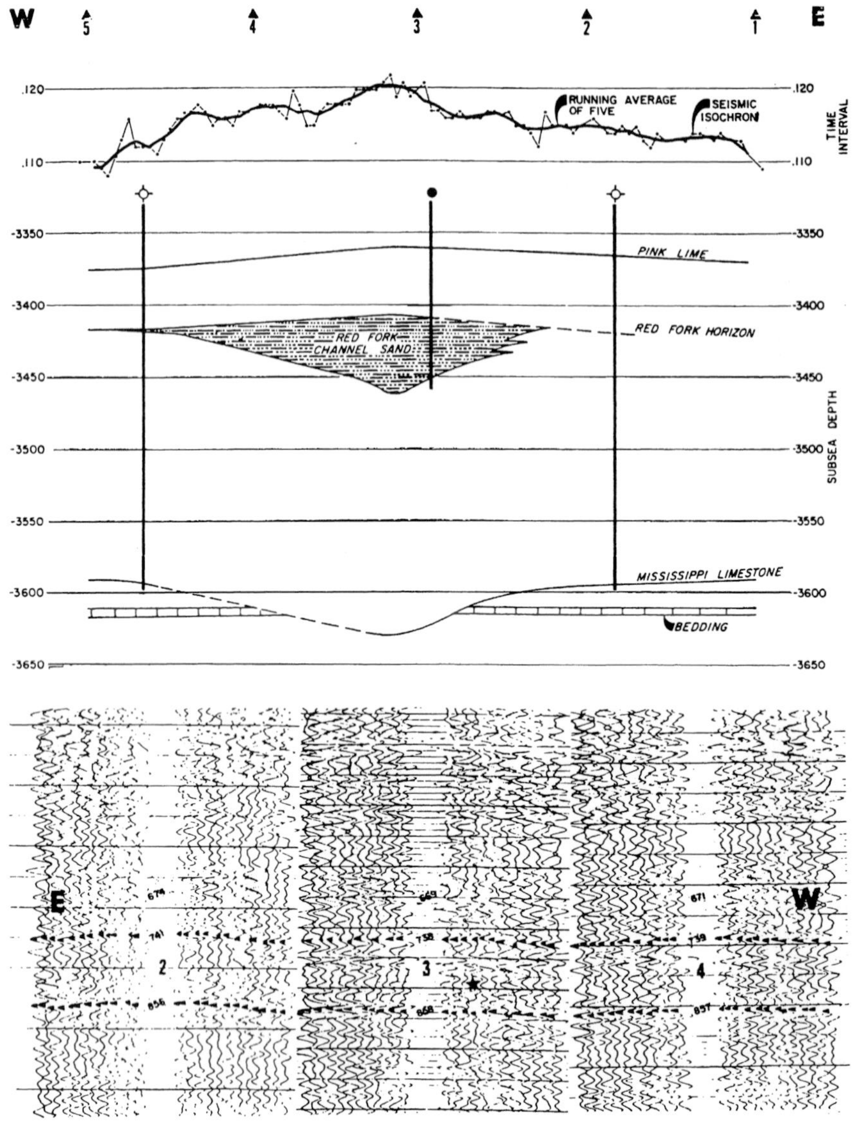

Fig. 9.45. A Cross section made from seismic records across a paleovalley filled with Red Fork sandstone, South Ceres field, Oklahoma. **B** Reflection picks which determined the channel. (Lyons and Dobrin 1972, reprinted by permission)

considerably more sophistication in the interpretation of seismic data. This is of particular importance in the study of stratigraphic sequences, as discussed in Chap. 13.

Fluvial deposits are commonly characterized by lenticular sandstone bodies, including paleovalleys, channels, and bars, but these may be difficult to detect on reflection-seismic records, because of their small size and the poor acoustic contrasts between the sandstone body and its host strata, such as a channel fill and the sediments into which it is cut. The general comments of Tipper (1989) regarding the nature of facies "mosaics" apply in particular to fluvial settings. Many fluvial sandstone units are too thin to be picked up on most seismic data. Dobrin (1977) reported the difficulty in mapping the three major productive fluvial sandstone lenses in the Candeias field, Recôncavo Basin, Brazil, using good-quality seismic data. The sandstone bodies themselves were not visible on seismic sections, and the structure was confused by a velocity pull-up generated by an adjacent, massive sandstone unit.

Sandstone bodies and channels can be identified and mapped using seismic data, but the best results tend to be achieved only when a considerable amount of local information is available regarding the stratigraphy, the seismic character of the units in question, and the likely associations, orientations,

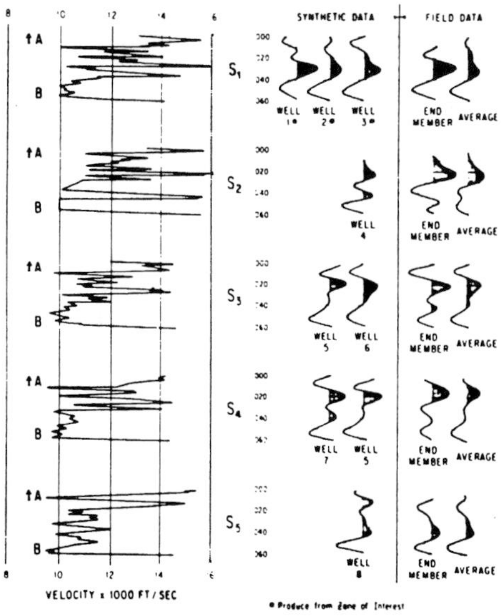

Fig. 9.46. Velocity logs and synthetic and recorded waveforms for five types of stratigraphic section; Morrow Sandstone, Oklahoma. (Dobrin 1977, reproduced from work by Waters and Rice 1975, reprinted by permission)

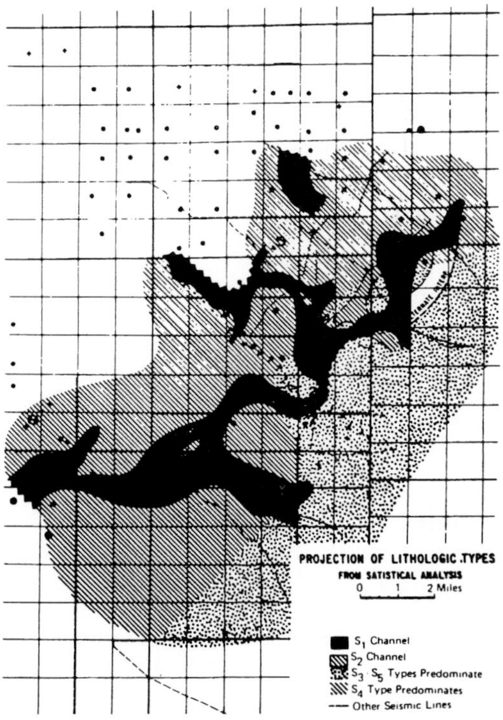

Fig. 9.47. Mapped distribution of types of channel body, based on statistical analysis of waveforms recorded along seismic lines. (Dobrin 1977, reproduced from work by Waters and Rice, reprinted by permission)

and configurations of the bodies to be mapped. In other words, seismic data cannot readily be used as a prospecting tool for fluvial sandstone bodies in frontier areas, but they may prove invaluable for extending fields or finding infill fields in well-known areas. Success depends on using the information obtained from existing fields to carry out very refined seismic modeling and interpretation.

Paleovalleys, where a sandstone is cut into a limestone, may yield good reflections, as may a sandstone that is porous and gas filled. These are ideal situations (Anstey 1980). In some cases the sandstone of the valley fill may resist compaction more than the host strata, in which case a slight structural anomaly may appear in the records. Figure 9.45 illustrates an early success story in the mapping of a valley fill in which the shale became compacted around the sandstone, leading to a slight drape structure.

The Morrow Sandstone (Pennsylvanian) of Kansas and Oklahoma has been extensively studied using seismic methods (Clement 1977). This is a transgressive marine unit that fills paleovalleys eroded by fluvial processes during a preceding regression. The sandstone bodies are therefore similar to fluvial paleovalley fills and other types of channel, and the details of the techniques by which they have been mapped are instructive.

Dobrin (1977) reported on a study of the channel-fill sandstones in Oklahoma using seismic modeling, followed by the application of pattern-recognition techniques in an attempt to map distinctive types of wavelets in the subsurface. Synthetic and actual mapped data are compared in Fig. 9.46, and Fig. 9.47 is a map showing the interpreted distribution of several types of channel in part of Oklahoma.

A modern and much more sophisticated study was reported by Clark (1987). Figure 9.48 is an idealized geological and velocity model for a Morrow channel in an area of southern Kansas, and Fig. 9.49 is the synthetic seismic response generated using this information. The channel is constructed to be 18 m deep on the left, thinning to 6 m on the right. It is filled with shale, except for 3 m of sandstone at the top. The channel is at 200 ms in Fig. 9.49. It is an obvious anomaly, caused by the slow velocity of the channel fill in contrast to that of the overlying and underlying rocks. Note also the sagging of the overlying Marmaton anomaly at 155 ms, as a result of differential compaction, with the channel fill compacting more than the host strata. The apparent faulting in the underlying Arbuckle reflector, at 260 ms, may represent only apparent breaks caused by

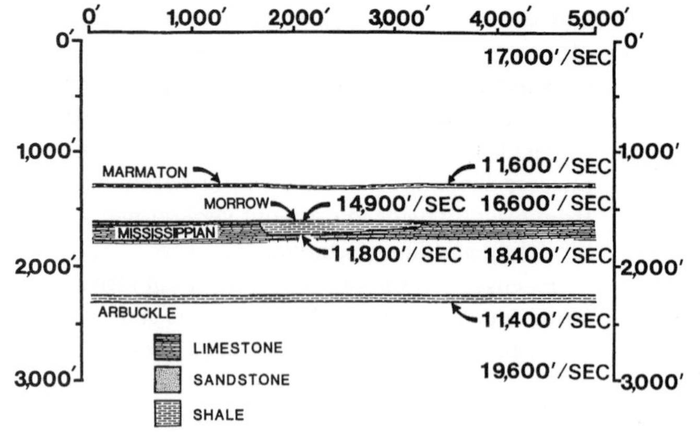

Fig. 9.48. Idealized geology and velocity model for a Morrow channel, southern Kansas. (Clark 1987, reprinted by permission)

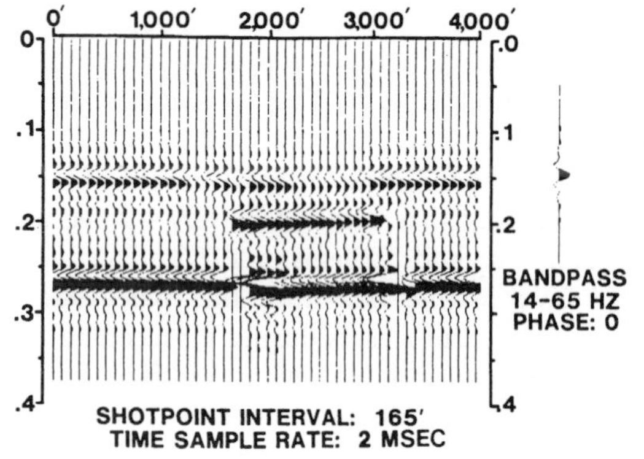

Fig. 9.49. Synthetic seismic cross section generated from the data shown in Fig. 9.48. (Clark 1987, reprinted by permission)

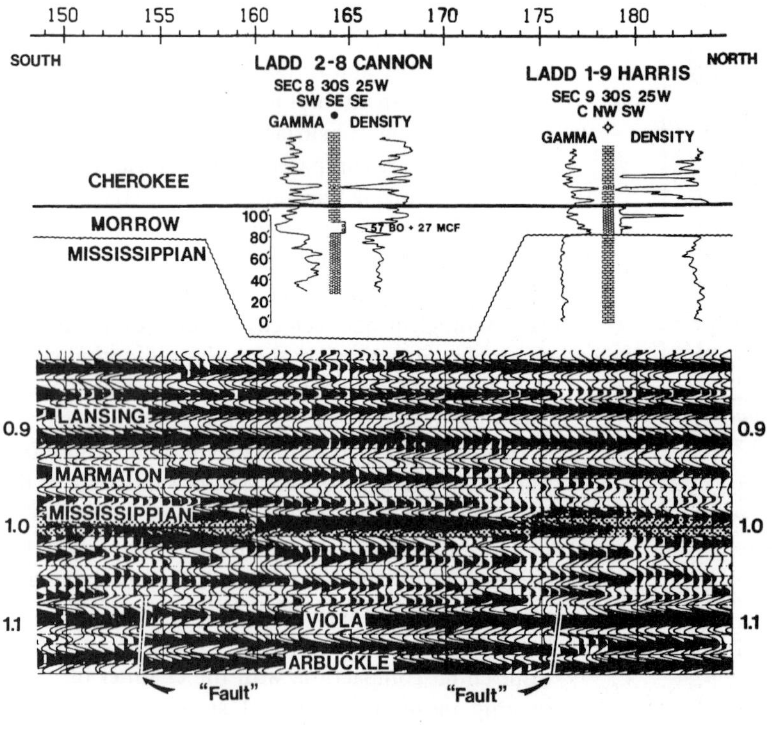

Fig. 9.50. Unmigrated seismic line across a Morrow channel, southern Kansas. Note three wells projected onto the line. The channel is the strong positive reflector immediately above the Mississipian limestone (*stippled*) at 1 s, between the 160 and 173 shot points. (Clark 1987, reprinted by permission)

ray deflection from the edge of the Morrow channel. All these anomalies may be used to locate actual channels. An actual sesimic line through this area is reproduced as Fig. 9.50. Note the "sagging" of the Marmaton, the "faults" in the Arbuckle, and the strong positive reflection from the channel itself.

Other examples of channel and paleovalley mapping include those by Wood and Hopkins (1989), who illustrated a seismic section across a Mannville paleovalley fill in the Little Bow area of Alberta. Here, the differential compaction of the valley fill relative to the host strata provided the key for the identification and mapping of the channels. Sonnenberg (1987) mapped paleovalleys in the "D" sandstone of the Denver Basin using techniques similar to those described above for the Morrow channels.

Many of the stratigraphic traps discussed later in this book, especially the CB-type fields, are characterized by structural drape over a laterally limited sandstone ribbon or lens (Chaps. 14, 15). Potentially, these small structural anomalies can be mapped using refined seismic techniques, once the acoustic properties of the rocks in question have been thoroughly investigated by regional seismic mapping and the use of well logs for velocity control.

Brown (1985, 1991) and Weber (1993) discussed the applications of three-dimensional seismic data to the mapping of stratigraphic objectives. Channels that yield very subtle reflections in vertical cross

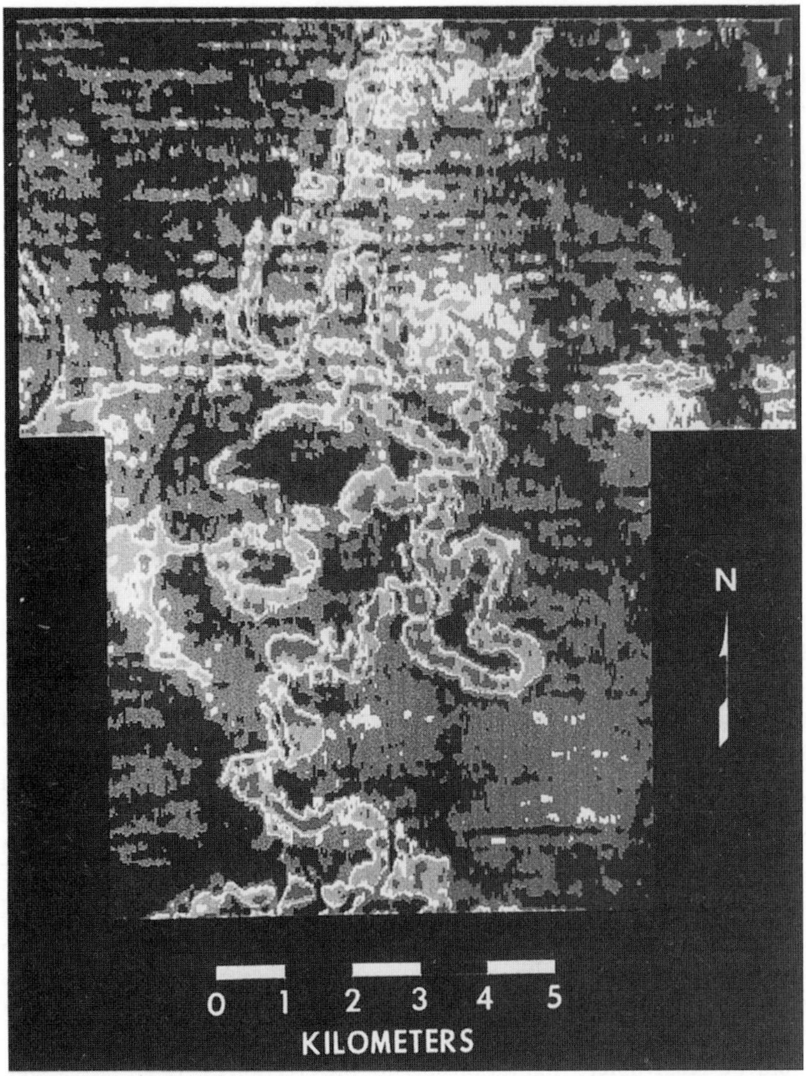

Fig. 9.51. Horizontal 3-D seismic section showing meandering channels in Miocene-Pliocene fluvial-deltaic deposits, Gulf of Thailand. The original illustration was in color. (Brown 1991, reprinted by permission)

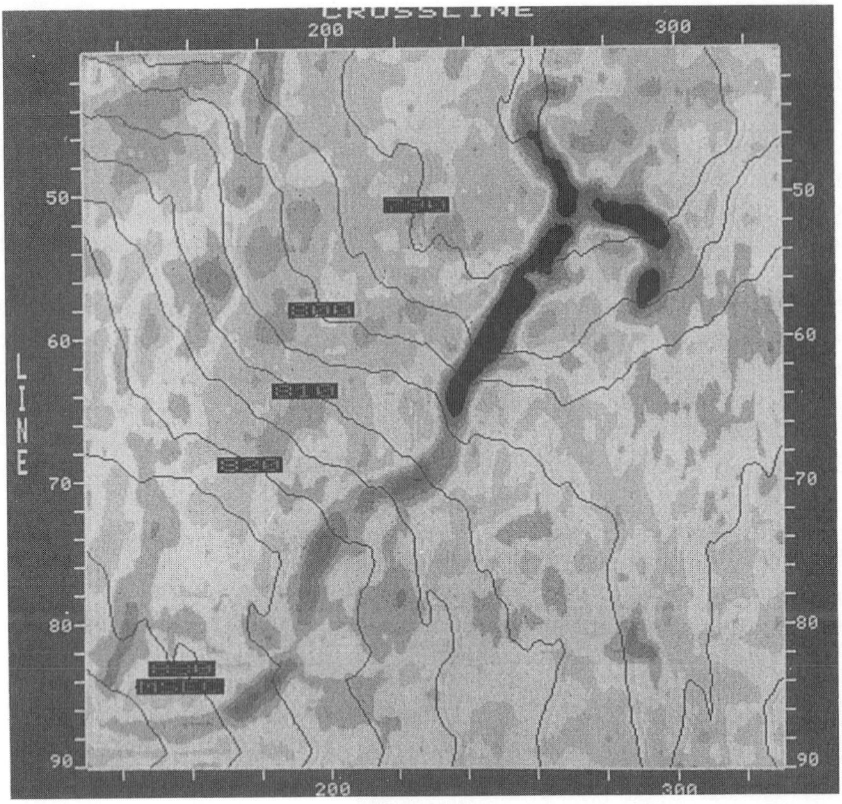

Fig. 9.52. Horizontal slice section, part of a 3-D seismic section showing bifurcating deltaic distributaries, Cenozoic, Gulf of Mexico. Structural contours, in milliseconds, are superimposed. The original illustration was in color. (Brown 1991, reprinted by permission)

section may be quite spectacular when viewed in horizontal sections. Figures 9.51 and 9.52 illustrate two examples of the mapping of channels. The use of 3-D seismic method holds great promise for the detailed mapping of channels and bar deposits in PV- and CB-type fields, and even for the better definition of reservoir heterogeneity in SH-type fields (terminology is defined in Chap. 14). Increasing refinement in acquisition and processing techniques will lead to the definition of ever-smaller features, although limitations in resolution are imposed by the nature of the seismic wave, which has a spherical wave front and generates reflections over a zone termed the "Fresnel zone" that may be several to many meters across. A. Nur (personal communication, 1993) noted that in some circumstances, such as in the shallow tar-sand studies being carried out in northern Alberta, 3-D seismic resolution may be able to resolve features as small as 4 m across. At depths of 2 km objects 20 m wide should be resolvable, which should permit the mapping of individual minor channels and bars.

9.5.5 Ground-Penetrating Radar

Ground-penetrating radar (GPR) is a relatively new technique that has found increasing application for the delineation of the shallow subsurface in a wide range of geological applications (Davis and Annan 1986, 1989; Moorman et al. 1991). The technique involves the transmission of high-frequency (10–100 MHz) electromagnetic pulses and the recording of reflected signals, which are processed much as are seismic signals. Reflections occur at the interface of beds with contrasting electrical properties, and the depth of penetration depends on the attenuation of the signal. Greatest penetration and lowest reflectivity are yielded by unconsolidated sands and gravels and dry sandstones. Attenuation increases with saturation, and with decreasing grain size, reaching the highest values in the case of wet muds. Characteristic reflection strengths and configurations permit the erection of "radar facies" and "radar sequences" in much the same way that seismic facies have been delineated (Huggenberger

1993; Gawthorpe et al. 1993; Alexander et al. 1994). Core or outcrop data are desirable to "ground-truth" the radar facies.

Several applications have demonstrated the ability of the tool to map architectural details in fluvial deposits down to depths of as much as 30 m, with resolution as high as 10 cm, although where the section consists of saturated fine-grained deposits, as in the case of modern floodplains, penetration may be only a few meters. Both modern (Moorman et al. 1991; Huggenberger 1993; Gawthorpe et al. 1993) and ancient (Stephens 1994) fluvial deposits have been examined.

There are two main potential applications of the technique.

1. For the the evaluation of sand-body architecture as an aid to the study of reservoir heterogeneity. A particularly powerful approach is to examine an ancient unit in outcrop, using two-dimensional lateral profiling techniques (Chap. 4), and to run a GPR survey over the top of the outcrop, in order to calibrate the radar data and to extend the architectural analysis back into the third dimension. Pratt and Miall (1993) reported a preliminary study of a Silurian carbonate shoal deposit carried out this way. Gawthorpe et al. (1993) reconstructed lateral-accretion bedding in a modern point-bar deposit. Stephens (1994) examined an ancient fluvial system and was able to document a history of complex braid-bar development.
2. The evaluation of the flow of groundwater and toxic wastes through shallow aquifers, including ancient sediments and modern surficial deposits (Huggenberger 1993; Knoll et al. 1994).

A single example of a radar line and its interpretation are illustrated in Figs. 9.53 and 9.54. This survey was carried out by Stephens (1994) over the top of a large outcrop of the Kayenta Formation, a Lower Jurassic ephemeral fluvial system in southwest Colorado. Most of the reflections are interpreted to represent bounding surfaces, probably of third-order rank, that record the lateral to oblique accretion of DA units in a broad sandy-braided river channel.

9.5.6 Magnetostratigraphy

The record of magnetic reversals preserved in sedimentary rocks has, in recent years, become a powerful tool for intrabasin and global correlation. Carefully sampled stratigraphic sections can lead to the erection of a local chronostratigraphy (examples are shown in Figs. 9.55 and 9.56), and with the assistance of biostratigraphic data or information from radiometric dating of interbedded tuffs, the local record may be correlated with the global time scale (Fig. 9.57). The procedure is summarized in Miall (1990, Sect. 3.7.5).

Not all fluvial deposits lend themselves to paleomagnetic study. The quality of the results decreases sharply with increasing age, because of diagenetic complications, and, because the magnetic signature is best obtained from fine-grained units, reliable results depend on the presence of such facies at least every few meters through the succession. The work on the Siwalik Group, which is the Late Cenozoic fill of the Himalayan foredeep basin in Pakistan (Behrensmeyer and Tauxe 1982; Behrensmeyer 1987; Johnson et al. 1985; Burbank and Raynolds 1988; Burbank et al. 1986; Mulder and Burbank 1993) is of the highest quality and has been widely quoted because of the insights the work has provided on stratigraphic architecture (this chapter) and tectonic controls (Chap. 11). The data permit the subdivision and regional correlation of fluvial successions down to the group 8 or 7 level, as discussed in Sect. 9.2. Under ideal conditions, this is comparable to the subdivisions that can be erected based on interbedded tuffs, coals, and paleosols, as described in Sect. 9.5.1.

9.5.7 Paleocurrent Analysis

Methods of outcrop paleocurrent measurement and documentation are discussed in Sect. 4.4, in which it is shown how orientation data relating to cross-bedding and macroform accretion directions may be used to reconstruct channel and bar configurations (Fig. 4.1). Elsewhere, the concept of the hierarchy of depositional units is introduced, and it is shown how the directional properties vary at each level of the hierarchy (Fig. 2.10). Reconstruction of macroforms using paleocurrent and facies data is discussed in Chap. 6 (e.g., see Fig. 6.24c).

Paleocurrent analysis may be used as a supplementary mapping tool to investigate the following kinds of information:

1. Changes in channel and bar orientation and directional variability through a stratigraphic unit, as one of several indicators of vertical or lateral changes in fluvial style
2. Reconstruction of tributary or distributary patterns, e.g., determination of radial alluvial-fan

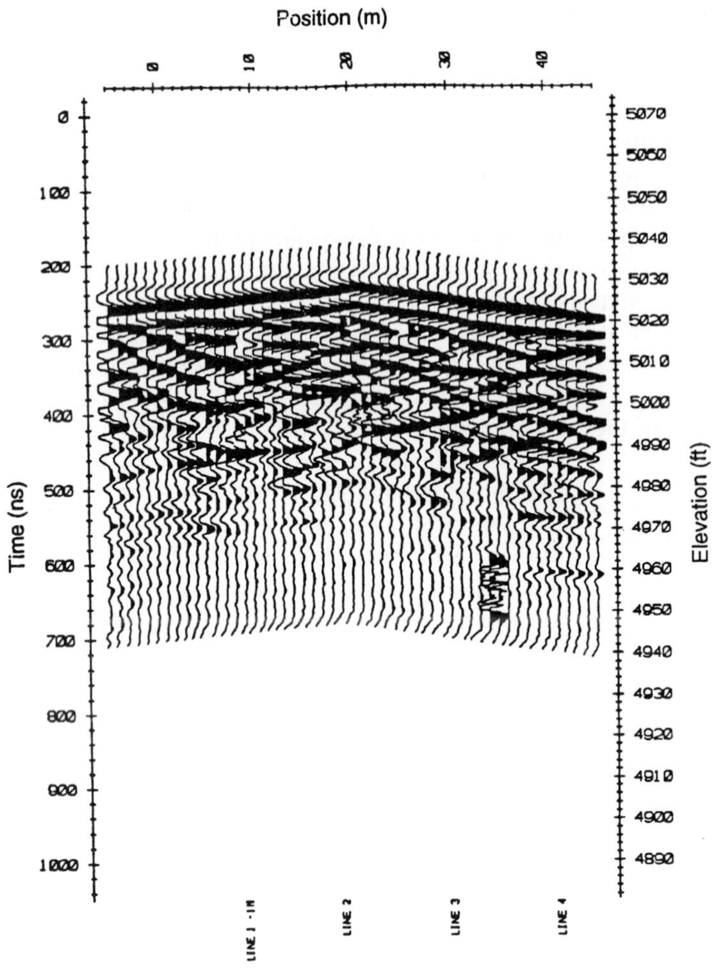

A

Fig. 9.53A,B. A radar line (A) and its interpretation (B), run across an outcrop of the Kayenta Formation (Lower Jurassic), Dolores River, southwest Colorado. (Stephens 1994)

drainage and location of entry point of drainage into a basin
3. Reconstruction of local and regional paleoflow patterns
4. Vertical changes in flow direction through a stratigraphic section as indicators of interacting fluvial systems (e.g., trunk and tributary drainages), or vertical changes in orientation of a system, in response to paleogeographic changes in a basin

An example of the use of detailed paleocurrent information to reconstruct channel and bar orientation is illustrated in Figs. 9.58 and 9.59, taken from the author's study of the Castlegate Sandstone in Utah (Miall 1993, 1994). At a location named Tusher Canyon-B, the outcrop is subdivisible into a series of elements separated by fifth-order bounding surfaces. The exposure extends around a bend above a dry creek bed. Figure 9.58A and B illustrate the paleocurrent data for the entire outcrop and for one of the elements, together with readings of the dip and strike of accretion surfaces. Element 2 contains an internal bounding surface, Ei, which separates increments showing opposite accretion directions (in the left-hand half of the outcrop). It is therefore interpreted as a fourth-order surface. Accretion surfaces at the left-hand end of the outcrop in element 2A dip NNW and NNE (Fig. 9.58A), while the mean cross-bed azimuth from the same beds is to the NNE (011°). These data indicate that element 2A is a DA unit, accreting obliquely downstream at about 70° to the outcrop mean. This suggests that the flow within and adjacent to the bar was oriented at a relatively high angle to the mean valley orientation, as indicated in Fig. 9.58C. Another example of this type of analysis is given in Fig. 6.32.

A repetition of this type of analysis through a vertical section can provide insights into how a river system evolved through time. Figure 9.59 illustrates three successive bar deposits superimposed on each other at the Tusher Canyon locality. There is no way

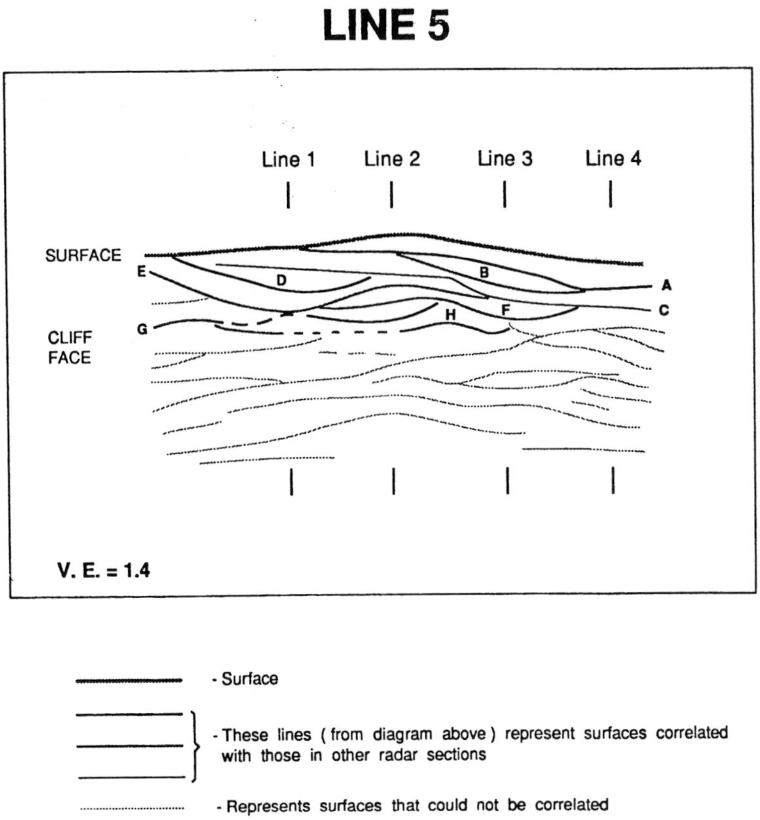

Fig. 9.53B

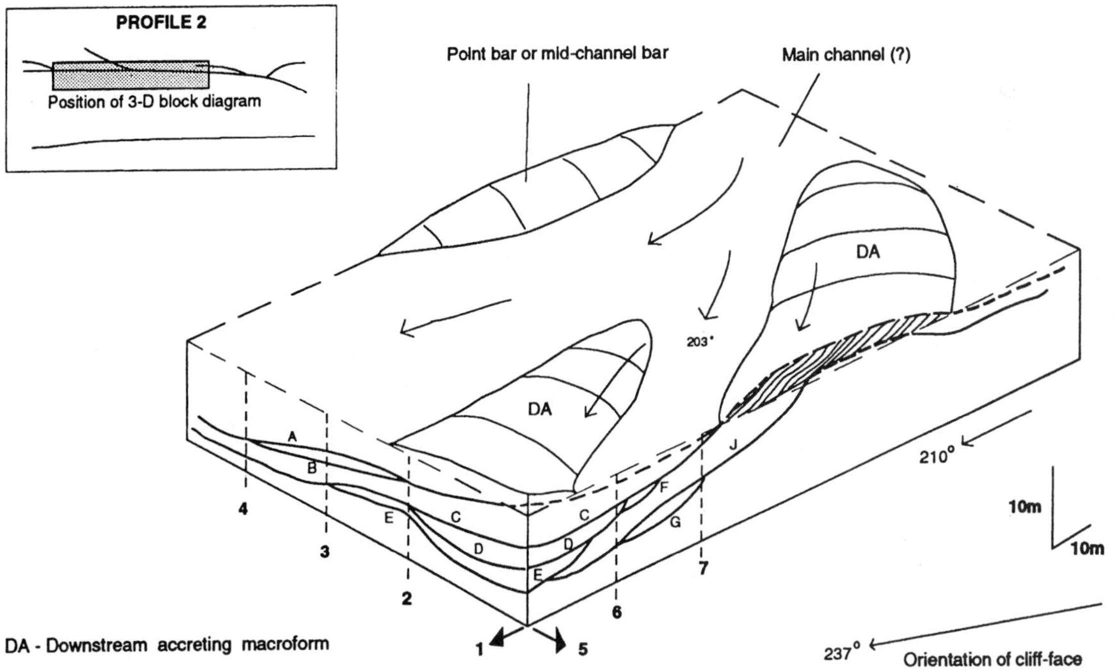

Fig. 9.54. Block-diagram reconstruction of a fluvial system, based on the correlation of seven radar lines run through the Kayenta Formation, SW Colorado. Line 5 (the front of the block diagram) is illustrated in Fig. 9.53B. (Stephens 1994)

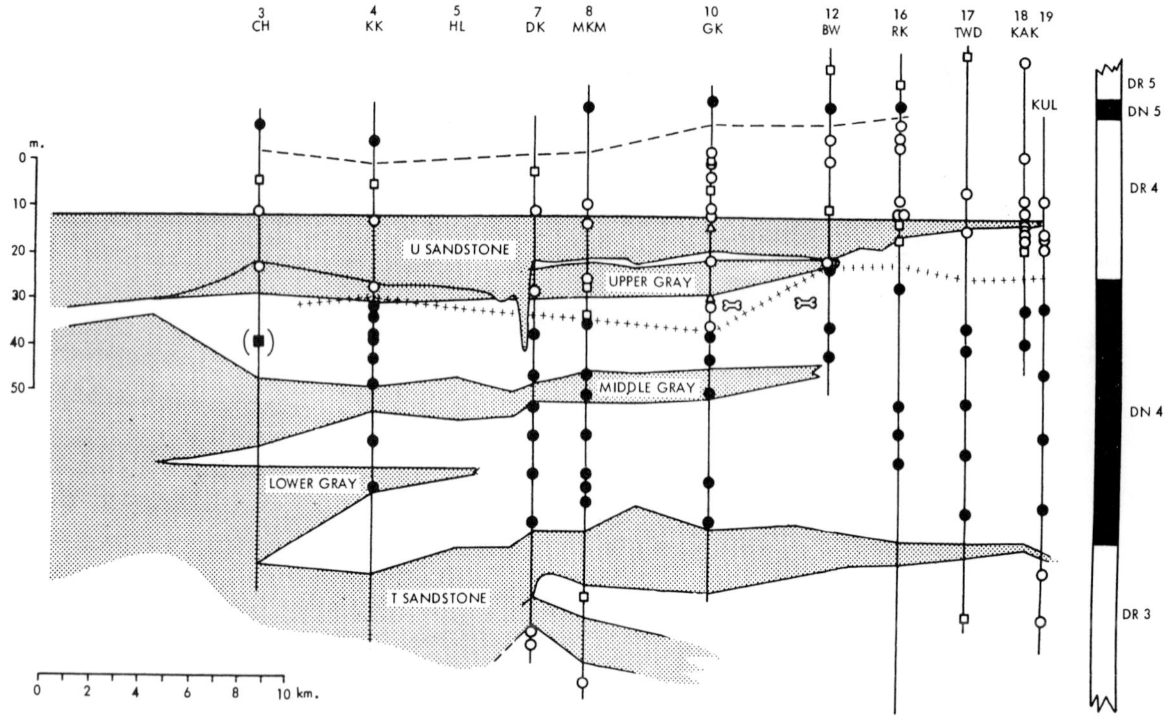

Fig. 9.55. Magnetic sampling of a fluvial succession, the Miocene-Pliocene Siwalik Group of Pakistan. *Black sample points*, normal polarity; *open points*, reversed polarity. This is the field data base from which the fluvial reconstruction shown in Fig. 9.1 was derived. (Behrensmeyer and Tauxe 1982)

to determine whether the three bars followed each other closely in time, or whether they were separated by long or short intervals of time in which other bars were formed and destroyed. However, this type of analysis does provide some insights into the variability of river behavior at a given site and may provide information on subtle vertical changes in fluvial style brought about by allogenic mechanisms, such as the response of the river to a base-level change or a regional tectonic tilt. These processes are discussed at length in Chap. 11.

In an earlier analysis of vertical changes in paleocurrent patterns it was suggested that the vertical changes in mean paleocurrent orientation could be used to calculate river sinuosity (Miall 1976). It was proposed that the angular change in mean orientation, as measured by a moving average technique, could be converted to sinuosity using an empirical equation adapted from the work of Langbein and Leopold (1966). A slightly different approach was taken by Le Roux (1992, 1994). These techniques now seems simplistic, because paleocurrent directions record the orientation of bedform and bar deposits, not the channel itself, and may not correspond closely to channel orientation at any one location. Also, the problem of section missing at major bounding surfaces (as noted above) is insurmountable. Another problem is that flood-stage deposits may be preferentially preserved, and these are likely to preserve lower directional variance than other deposits in the system because of the tendency for sinuosity to decrease at high discharge (Bridge 1993).

The use of paleocurrent data to reconstruct the details of a fluvial depositional system is a well-established practice (e.g., see Miall 1990, Sect. 5.9). Integrated basin-analysis methods couple paleocurrent data with data on facies and grain-size trends, and perhaps information on detrital sediment composition and possible sediment sources, the combination providing a powerful mapping technique. A few selected examples of the application of this integrated methodology to the study of alluvial basins include Friend and Moody-Stuart (1972), Friend et al. (1976), and Miall (1979a,b). Lawton (1985, 1986a,b) used paleocurrent and petrographic data to explore the history of tilting of the regional paleoslope and basement uplift in the Rocky Mountain Foreland Basin of Utah during the Late Cretaceous and Tertiary (Sect. 11.3.3). Miall and Gibling (1978), Nichols (1987), and Hirst (1991) reconstructed radial paleocurrent patterns along a basin

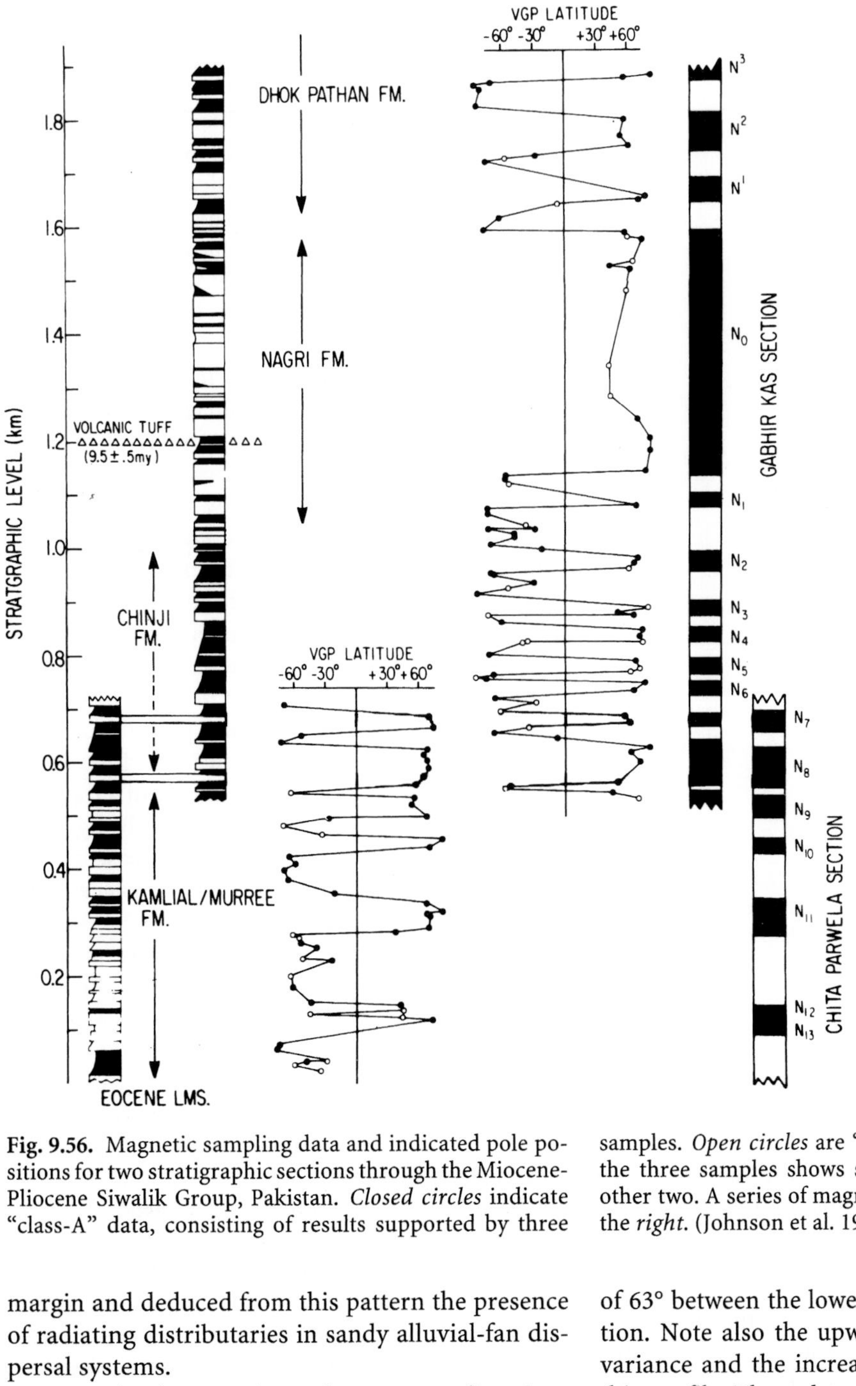

Fig. 9.56. Magnetic sampling data and indicated pole positions for two stratigraphic sections through the Miocene-Pliocene Siwalik Group, Pakistan. *Closed circles* indicate "class-A" data, consisting of results supported by three samples. *Open circles* are "class-B" data, in which one of the three samples shows statistical divergence from the other two. A series of magnetic chrons (*N*) is identified at the *right*. (Johnson et al. 1985)

margin and deduced from this pattern the presence of radiating distributaries in sandy alluvial-fan dispersal systems.

Vertical changes in paleocurrent directions through a stratigraphic section, coupled with changes in detrital composition, may be a simple means to reveal tectonic movements in a basin (e.g., Miall 1979a). In some cases, however, the changes in bedform character and orientation may be very subtle. Figure 9.60 is a plot of bedform and paleocurrent data obtained from a vertical profile through a sandy braided stream deposit near a faulted basin margin. Note the change in mean current direction of 63° between the lower and upper part of the section. Note also the upward decrease in directional variance and the increase in set thickness through this profile. These data are interpreted as the result of the rejuvenation of a tributary stream flowing into the basin from the southeast following uplift along the basin-margin fault. The northwestward-flowing stream may have been deeper and more vigorous, accounting for the larger bedforms and the lower channel sinuosity. Information on bounding surfaces and architectural elements is not available for this locality.

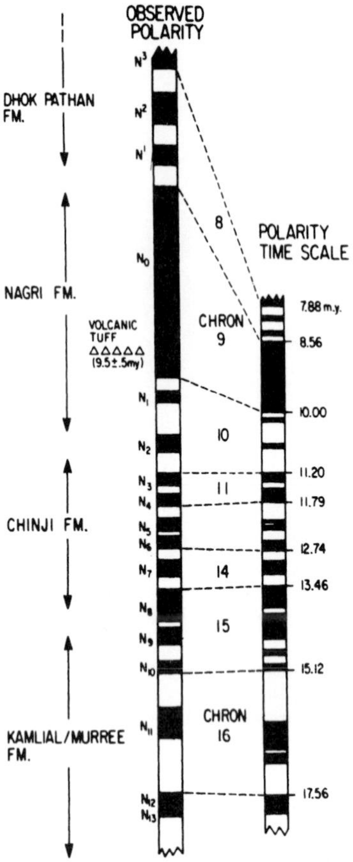

Fig. 9.57. Correlation of the magnetic chrons of Fig. 9.56 with a standard magnetic time scale. (Johnson et al. 1985)

9.5.8 The Dipmeter

Logging companies, such as Schlumberger, promote the use of the dipmeter and related tools, such as the Formation MicroScanner and the Fullbore Formation MicroImager, as tools for the detection and mapping of sedimentary dips in the subsurface (Serra 1989). There are three main potential applications in the study of fluvial sandstones:

1. The mapping of dipping fourth- and fifth-order surfaces corresponding to bar-top and channel-floor surfaces, and the drape associated with them, providing information on the shape and orientation of these features
2. The mapping of internal, second- and third-order erosion surfaces, that would facilitate the mapping of macroforms, such as point bars
3. The mapping of cross-bed orientations for the paleocurrent information they yield

These three applications are arranged in order of decreasing scale and utility, in terms of their practicality for studying fluvial systems. Very few published examples of the successful application of these tools are available, which either may be a reflection of the industry confidentiality surrounding an invaluable technique, or may (more likely, in this author's experience) indicate that successful applications are sparse. Descriptions of the technique and processing routines are given by Schlumberger (1970) and Vincent et al. (1977, 1979) and have been summarized by Miall (1990, Sects. 5.4.3, 5.9.6).

Figure 9.61 illustrates the principal of the mapping of drape over bars or within channels. The only example known to this writer of an application of this technique to a practical case study is that of Muwais and Smith (1990). They found that in the large, tidally influenced fluvial channel fills of the Athabasca Oil Sands of Alberta, the main surfaces recorded by the dipmeter were those defined by the fill of vertically accreting channels and the lateral-accretion surfaces of large point bars. Cross-bedding is rare in these deposits, except near the base of the channels, and was not normally picked up by the dipmeter. Figure 9.62 illustrates the interpretive principles, and Fig. 9.63 is an example of a dipmeter log interpreted in terms of three vertically aggraded channels. The interpretation in this case is confirmed by outcrop studies of large surface mines nearby.

Potentially, third- and fourth-order surfaces, such as those mapped in Fig. 9.40, should be recognizable in cores and dipmeter logs (especially using the Schlumberger Formation MicroScanner) by their gentle depositional dip and their association with shale drapes or lag deposits, but in practice their identification is very difficult (author's consulting experience). However, there is considerable potential here for the detailed study of sand-body anatomy in field development situations, where abundant core and log data should be obtainable. Figure 9.64 illustrates a typical point bar and the type of dipmeter pattern that might be expected to result from a well drilled through it. Figure 9.65 illustrates an example from the Athabasca Oil Sands, where the accretionary dips are not confused by the presence of dipping cross-bed surfaces.

The potential for mapping cross-bedding (Figs. 9.66, 9.67) would seem to be high and, indeed, this technique has been successfully applied to the study of large-scale eolian cross-bedding, in which the structure of the dunes is simple and their orientation relatively consistent (see Miall 1990, p. 327 for discussion). However, in fluvial deposits there are many

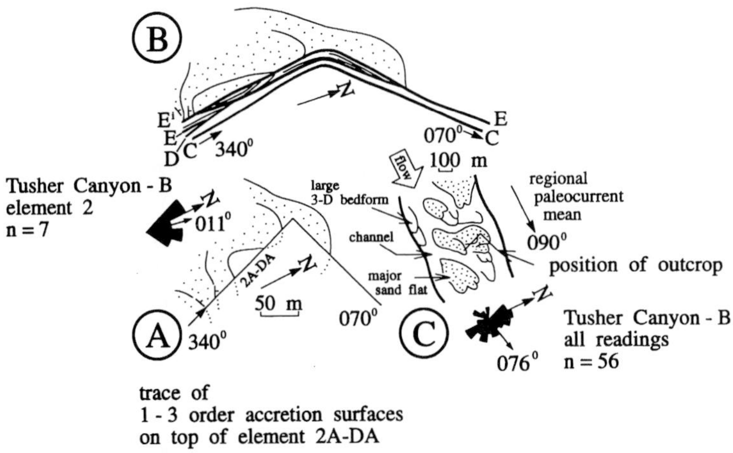

Fig. 9.58. Example of the interpretation of an architectural element using orientation data from cross-bedding and accretion surfaces. *A*, Orientation of Tusher Canyon-B section, Castlegate Sandstone, Utah, showing measured accretionary dip directions and interpreted accretion surfaces on the top of *element 2A-DA*, and current rose diagram obtained from measurement of cross-bed orientations in *element 2*. *B*, Perspective view of element 2A-DA. *C*, Interpretation of braided river that deposited the Tusher Canyon section. Note the position of the *Tusher Canyon-B* section. The river is oriented according to the regional paleocurrent mean. The rose diagram for this outcrop is also shown. Note the slight divergence between the outcrop mean and the regional mean, which may reflect the dominance of channels and macroforms oriented oblique to the main channel trend at the Tusher Canyon outcrop. Note also the large divergence between the element-2 paleocurrent mean and the outcrop mean, reflecting the locally high divergence of accretion directions from the mean channel trend. Accretion in element 2A is interpreted as DA rather than as LA because of the within-element similarity between indicated paleoflow and accretionary dip directions. (Miall 1993)

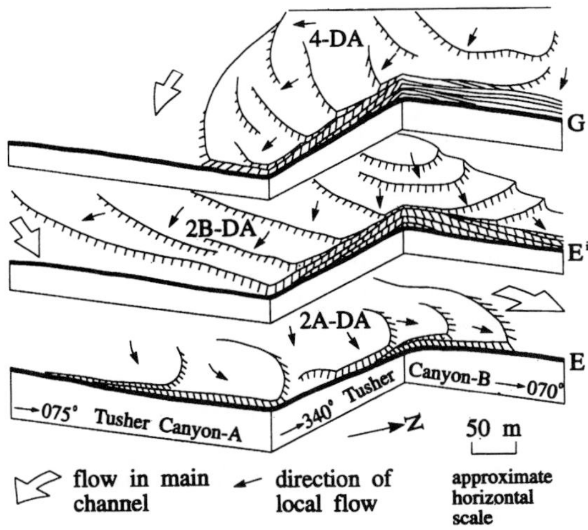

Fig. 9.59. Superimposition (stacking) of macroforms in a multistory sand body, showing three elements at Tusher Canyon. Each panel shows the two adjacent outcrops, Tusher Canyon-A and -B, straightened out to facilitate viewing. Each element is shown in the process of formation. The reconstruction of the lowest element, *2A-DA*, is shown in Fig. 9.58

complications. The scale of the cross-bedding, at a few tens of centimeters or less, approaches the limit of resolution of the dipmeter tool (although it is well within the range of modern imaging tools). In addition, there are many types of surface in cross-bedded units that can produce confusion. For example, trough cross-beds have curved dips, and many cross-bed sets are deformed by slumping or overturning, or contain reactivation surfaces. Cameron et al. (1993) provided a useful discussion of the problems and the filtering techniques that can be used to sharpen the results.

Williams and Soek (1993) tackled the problem of dip complexity by progressively filtering out low dips and testing the variability of the indicated orientations at each stage. Their data were collected from outcrop measurements in order to simulate subsurface dipmeter records, the purpose being to facilitate interpretation by selecting a stratigraphic unit already well known from surface studies. One of their stratigraphic sections with dip records is shown in Fig. 9.68, and Fig. 9.69 shows the results of filtering. Note that as the range of dip magnitudes is progressively narrowed in favor of readings approaching 35° (typically the highest angle of repose for loose sand), the percentage of readings falling within the known channel orientation (north to northeast) increases. A similar result was obtained by Cameron et al. (1993) in their outcrop simulation

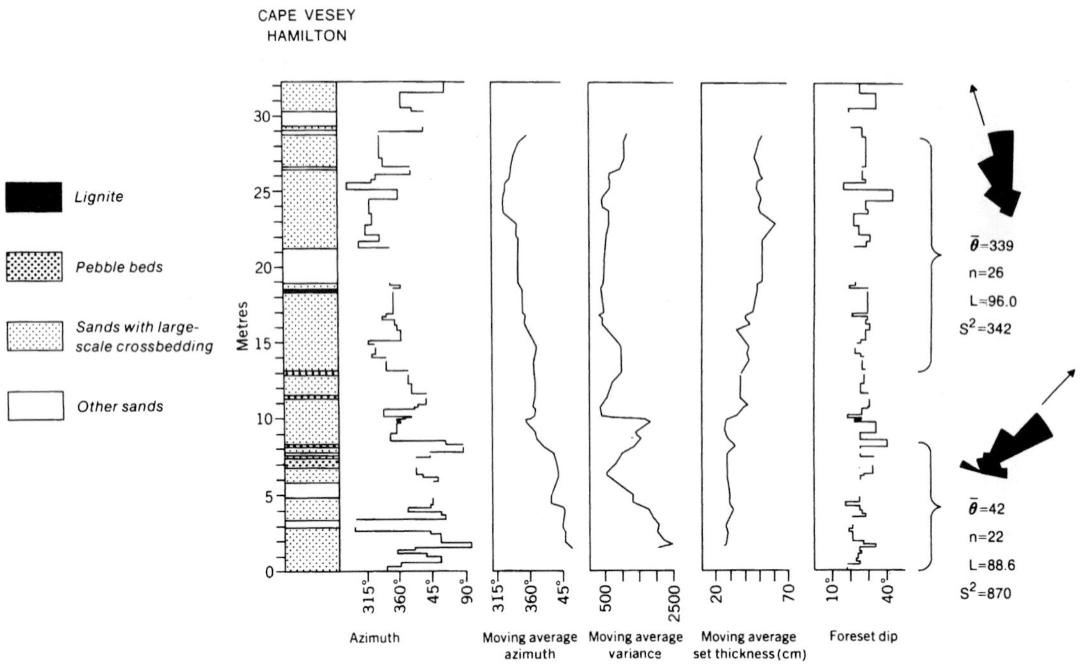

Fig. 9.60. Paleocurrent data and bedform scale measured over a vertical profile through a sandy braided stream deposit, the Isachsen Formation, Banks Island, Arctic Canada. Moving averages are based on a moving set of ten readings. (Miall 1976)

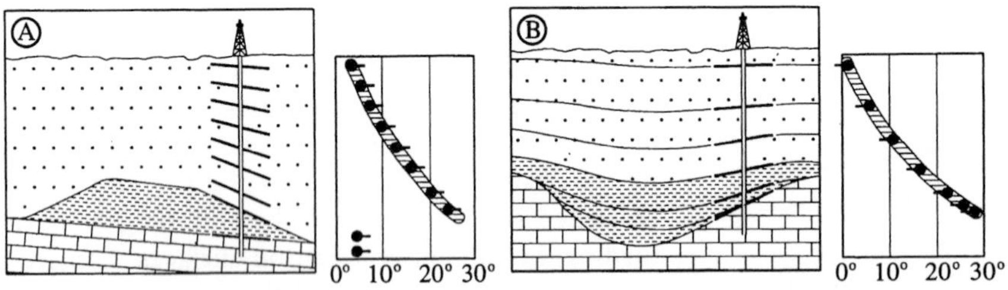

Fig. 9.61. Dipmeter patterns associated with bars and channels. Dip magnitudes are indicated by the position of the head of the "*tadpoles*", and azimuth by the orientation of the tails, with north being toward the *top*. (Schlumberger dipmeter manual 1986; see also Miall 1990 for general interpretive principles)

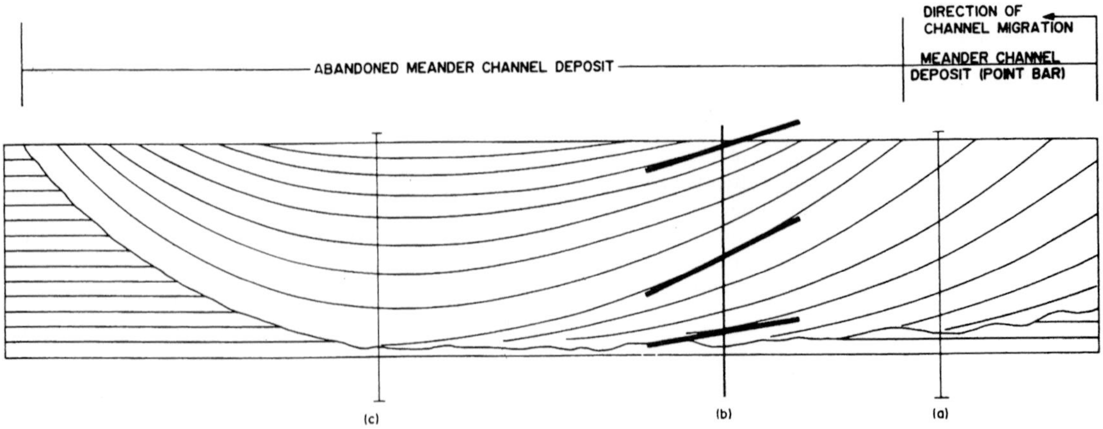

Fig. 9.62. Schematic illustration showing how vertical profiles from dipmeter logs may be interpreted in terms of the large-scale features of a channel fill. Upward increase in dip (*profile a*) may indicate lateral accretion, whereas upward decrease in dip (*profile c*) may indicate vertical aggradation. (Muwais and Smith 1990)

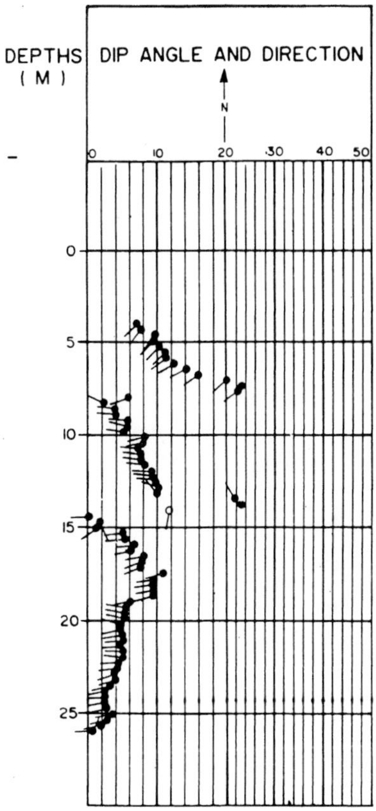

Fig. 9.63. Dipmeter log interpreted as the record of three stacked, vertically accreted channels, in the intervals 26–14, 14–8, and 8–4 m. (Muwais and Smith 1990)

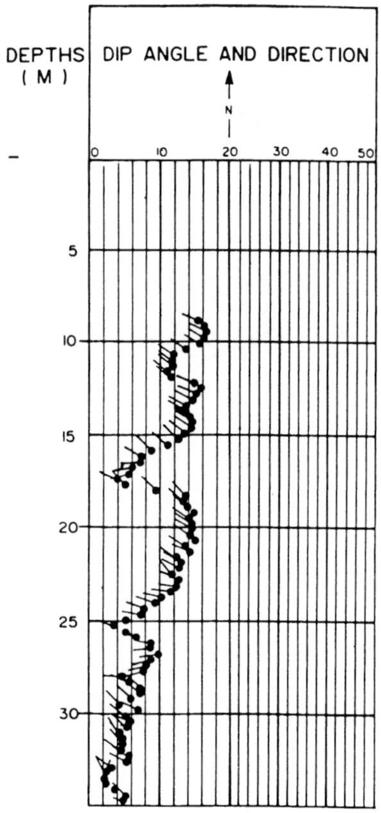

Fig. 9.65. Dipmeter log interpreted as the record of three stacked lateral-accretion deposits, in the intervals 35–25, 25–17, and 17–8 m. (Muwais and Smith 1990)

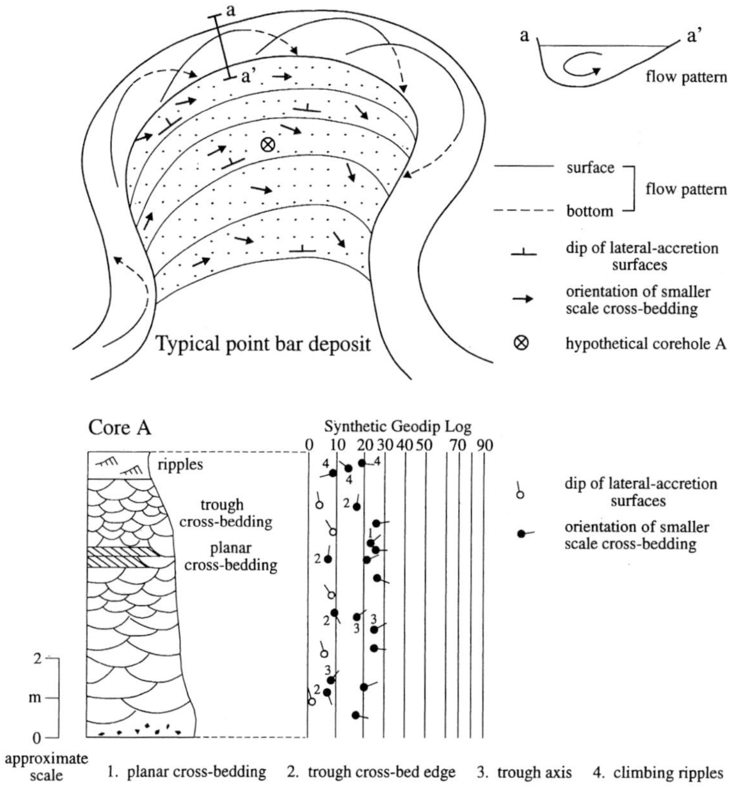

Fig. 9.64. Typical point bar and the dipping surfaces associated with it. Note the distinctive low dip and orientation of the third-order surfaces (epsilon cross-bedding). A speculative dipmeter log is also shown

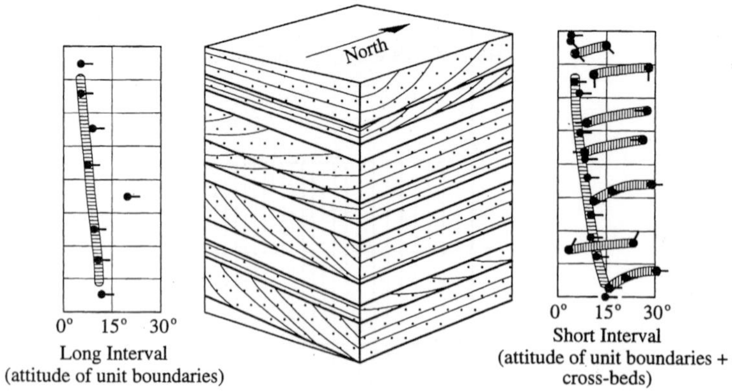

Fig. 9.66. Detection of cross-bedding using the dipmeter and its dependency on the sensitivity and sample interval of the tool. (Schlumberger dipmeter manual 1986)

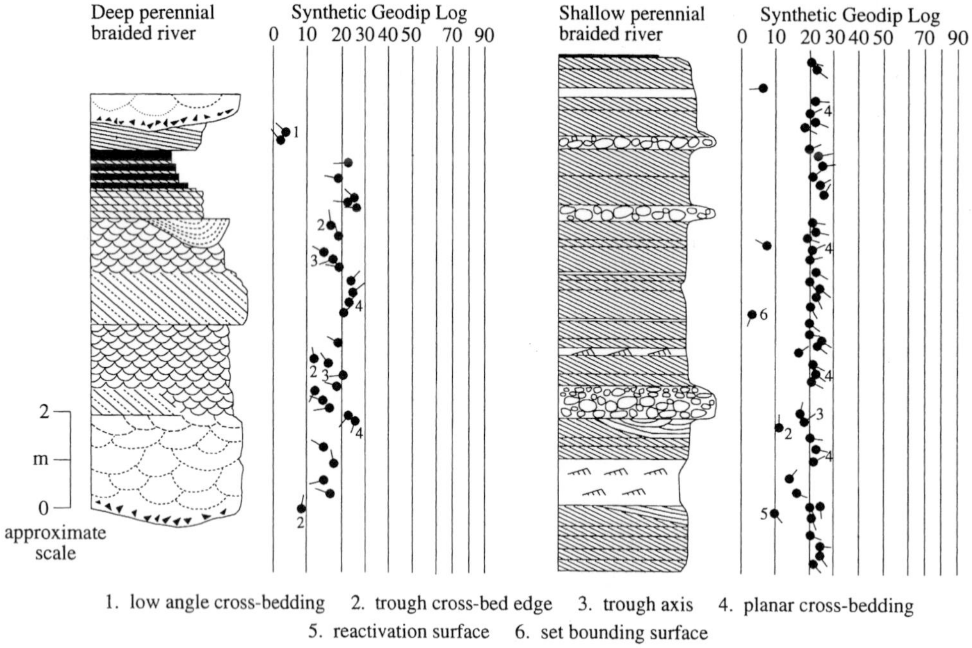

Fig. 9.67. Two types of braided river profile (see Chap. 8 for a discussion of these) and the dipmeter patterns that might be expected from them

of a dipmeter survey run through deltaic distributary channels. The conclusion of both studies is that actual subsurface dipmeter studies are likely to provide useful paleocurrent data, even when obtained from single exploration holes.

Dueck and Paauwe (1994) provided an excellent case study of the gradual development of small-scale – but nonetheless productive – exploration prospects in channel sandstones based mainly on the careful use of paleocurrent data obtained from borehole imaging techniques. They were able to plot cross-bed foreset surfaces as little as 5 cm apart within a borehole succession, allowing for the demonstration of variable dip angle from base to top of individual sets. By carefully considering the "rank" of the cross-bed sets within the overall hierarchy of directional indicators (Fig. 2.10), they were able to make predictions about channel trend from these readings, predictions that revealed the presence of a previously unsuspected meander bend. Several new producing oil wells were drilled using these data.

9.5.9 Surveillance Geology

The term surveillance geology is used to describe the work done by production geologists to monitor the pressure and fluid composition of producing wells. Pressure data can be used to determine the connectedness of specific sandstone bodies, and fluid com-

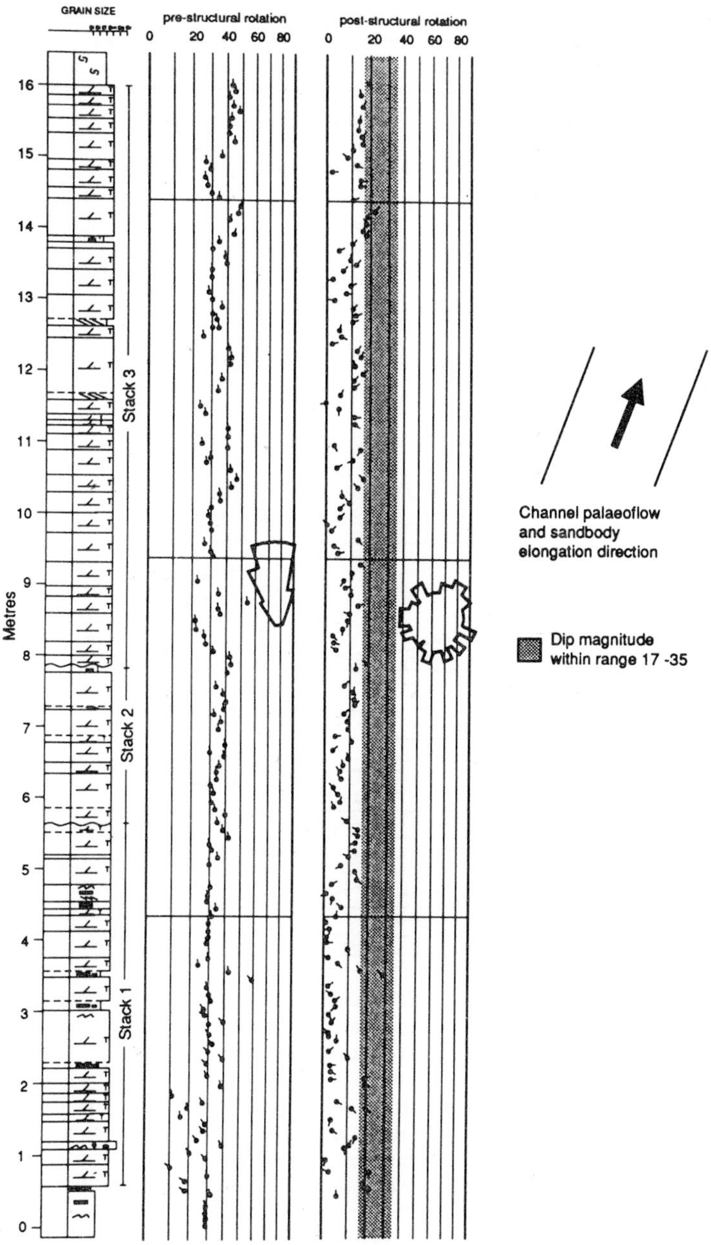

Fig. 9.68. Example of sedimentary log and dip profile through stacked channel sands, Carboniferous, western Ireland. Note that rotation of readings was carried out to remove regional structural tilt. (Williams and Soek 1993)

position, especially water saturation, can be used to track the movement of the oil front through a reservoir body as it is swept by injected water or steam. Thakur (1991) and Martin (1993) provided reviews of this specialized subject. Lorenz et al. (1991) described a useful case study.

Pressure-depth plots can be used to test reservoir body connectedness (e.g., Lorenz et al. 1991). Although this information is of importance in its own right, as an aid to proper planning of a production program, the information is of use in the lithostratigraphic mapping of the reservoir and therefore provides data which can be used to refine the depositional model. Figure 9.70 illustrates a pressure-depth plot constructed for the Mannville sandstones of part of Alberta.

The pattern of the flow of oil and water depends on the porosity-permeability architecture of the reservoir, and careful attention to the details of the fluid movement through surveillance methods can lead to continual refinements in the architectural model of the reservoir. Figure 9.71 illustrates patterns of wa-

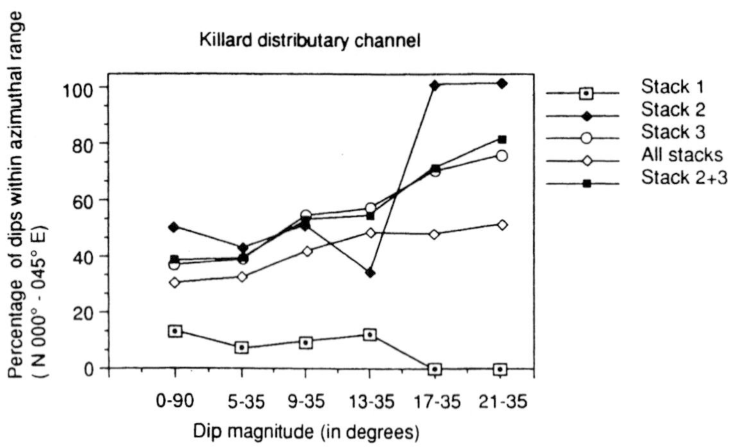

Fig. 9.69. Filtering of dip readings from the Carboniferous of western Ireland to progressively narrow the dip range in favor of steep dips expected to represent cross-bedding. Each curve shows data from a different stack of fluvial channels. Note that, with one exception, the percentage of readings within the expected range (channel orientation measured from outcrop) increases, as dips are limited to high readings. (Williams and Soek 1993)

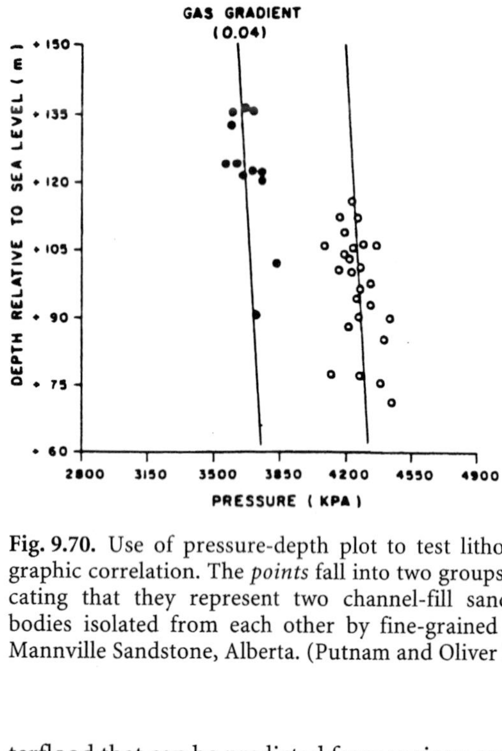

Fig. 9.70. Use of pressure-depth plot to test lithostratigraphic correlation. The *points* fall into two groups, indicating that they represent two channel-fill sandstone bodies isolated from each other by fine-grained units. Mannville Sandstone, Alberta. (Putnam and Oliver 1980)

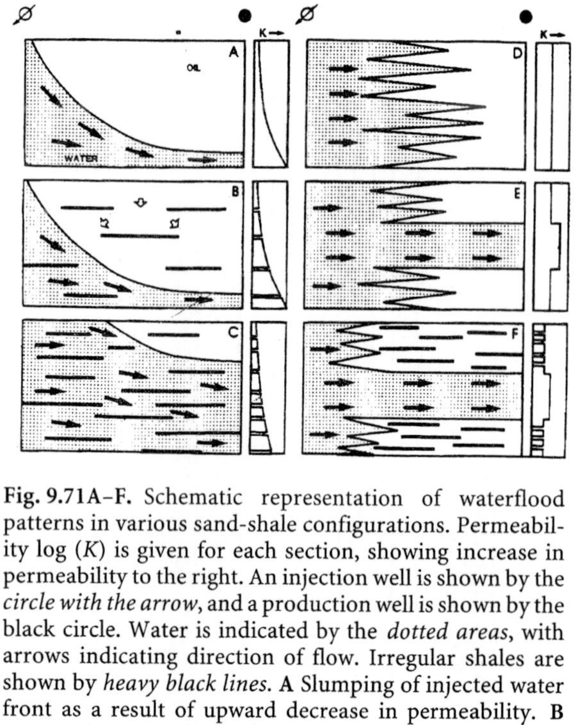

Fig. 9.71A–F. Schematic representation of waterflood patterns in various sand-shale configurations. Permeability log (*K*) is given for each section, showing increase in permeability to the right. An injection well is shown by the *circle with the arrow*, and a production well is shown by the black circle. Water is indicated by the *dotted areas*, with arrows indicating direction of flow. Irregular shales are shown by *heavy black lines*. **A** Slumping of injected water front as a result of upward decrease in permeability. **B** Gravity drainage of oil around shale beds (*open arrows*). **C** High sweep efficiency reflecting presence of shales to guide the flow. **D** Viscous fingering of waterfront in homogeneous strata. **E,F** Viscous fingering, plus channeling along a high-permeability zone. (Hopkins et al. 1991, reprinted by permission)

terflood that can be predicted from various configurations of an interbedded sandstone-shale unit. The presence of high-permeability zones in a reservoir may lead to the development of "fingers" or "channels" in the flood front (Fig. 9.71D,E,F) and to early breakthrough of water to a production well. "Slumping" of the water front may also occur where there is an upward decrease in permeability (Fig. 9.71A,B,C). This can lead to isolation of pockets of oil, which may be partly recovered by gravity drainage down to the main flood pathway (Fig. 9.71B). Hopkins et al. (1991) used monthly fluid production data to test the applicability of these models to a water-injection production situation in the Upper Mannville Sandstone of Alberta. The reservoir is an estuarine sand body within an incised valley-fill succession. Figure 9.72A shows a longitudinal cross section through the field, and Fig. 9.72B is an interpretation of flow patterns through the field. Slumping and channeling are interpreted to be occurring in this field, as indicated by water-production patterns in the northwest end of the field.

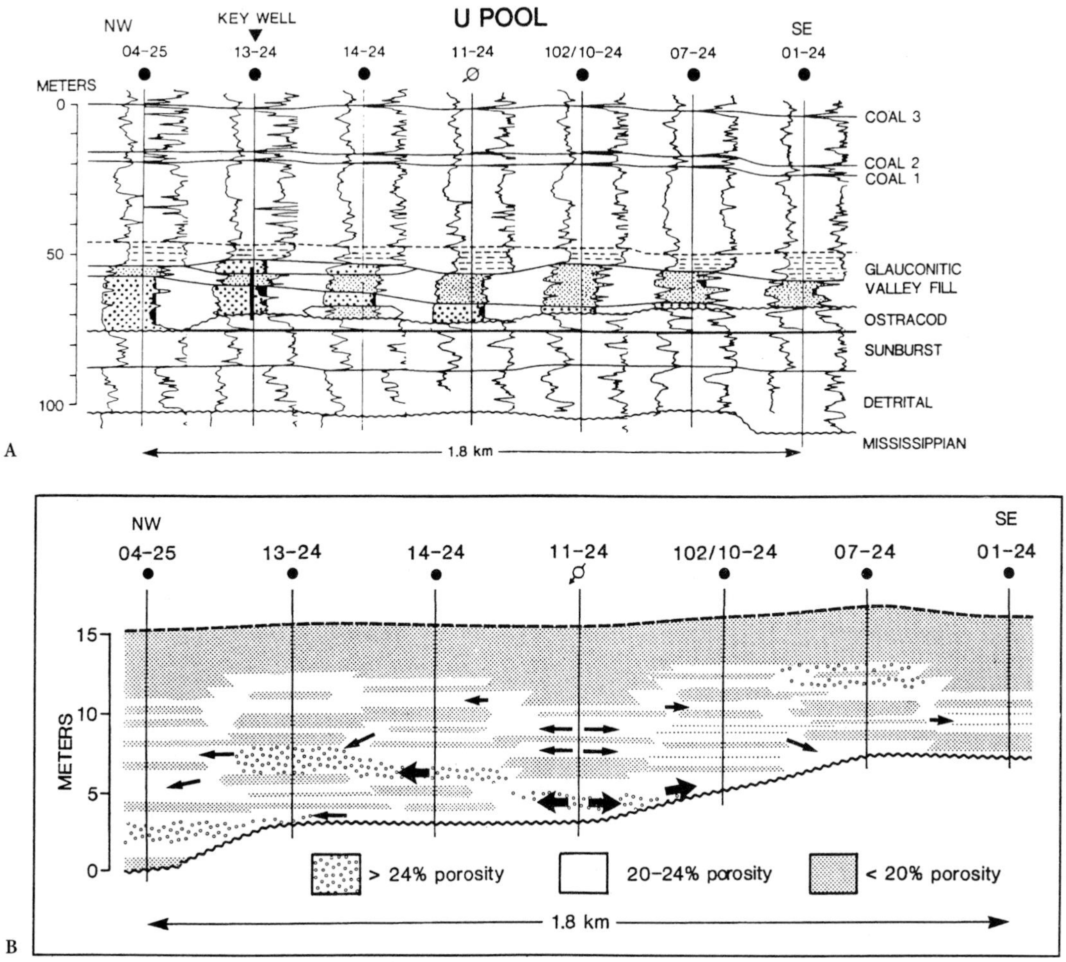

Fig. 9.72A,B. U Pool, The Little Bow field, Alberta. **A** Stratigraphic cross section. Porosity values higher than 21% are shaded black on porosity logs. *Large dots*, sandstone; *small dots*, sandstone-shale; *horizontal dashed lines*, shale. **B** Porosity distribution and inferred fluid flow pattern. (Hopkins et al. 1991, reprinted by permission)

Martin (1993) described various engineering techniques that are used to evaluate production in braided fluvial reservoirs. Pressure tests can be used to assess the size and orientation of permeable flow pathways. Pulse tests are used to determine connectedness between wells. One well is produced, and the degree of pressure drop at an adjacent well is monitored. Lorenz et al. (1991) described comparable procedures as part of their detailed case study of the Mesa Verde Group of Colorado. They also used injected tracers to test reservoir heterogeneity.

Robertson (1991) described the use of 3-D seismic reflection data to monitor fluid flow through a reservoir. Using identical data acquisition and processing routines, repeated surveys are run over intervals of several months. If fluid composition can be established by appropriate processing, for example, by using bright-spot techniques to illustrate the presence of gas, then the movement of the fluid can be charted.

Karlsen and Larter (1989) demonstrated that the petroleum in a given field is not homogeneous, but varies in terms of gas/oil ratio and oil type. These variations occur because the diffusion of petroleum into and through a reservoir is a slow process, requiring millions of years for complete homogenization. Detailed study of the petroleum itself may therefore reveal reservoir heterogeneities that are impeding the homogenization process.

Aquifer studies and problems of toxic waste dispersal through alluvial deposits may also constitute a form of surveillance geology, in which studies of fluid movement throw light on facies architecture models and thereby provide data for improving the flow models. An example is given by Blair et al.

(1991), who examined the flow of pentachlorphenol in groundwaters around a chemical plant.

Information on reservoir architecture and fluid movement patterns is used in production simulation models, as discussed in Chaps. 14 and 15. Field checking of these models against measured data on pressures and fluid contents is a process called history matching. Examples are described in Sneider et al. (1991) and Martin (1993).

9.6 Stratigraphic Nomenclature

The methods of lithostratigraphy have served sedimentologists and basin analysts well in the establishment of the broad stratigraphic and paleogeographic subdivisions of a basin fill. These are standard methods and are described in several textbooks (e.g., Miall 1990, Sect. 3.4). However, the advent of sequence stratigraphy and a renewed emphasis on the importance of breaks in sedimentation have created the opportunity for a much more refined approach to stratigraphic documentation. The current North American Code of Stratigraphic Nomenclature (North American Commission on Stratigraphic Nomenclature 1983) includes proposals for a new class of stratigraphic units and a new method of defining and mapping them. This is the method of allostratigraphy. "An allostratigraphic unit is a mappable stratiform body of sedimentary rock that is defined and identified on the basis of its bounding discontinuities." (NACSN 1983, p. 865) A hierarchy of units, including the allogroup, alloformation, and allomember, was proposed, and rules were established for defining and naming these various types of units. Allostratigraphic methods enable the erection of a sequence framework that avoids the cumbersome and inappropriate rules of lithostratigraphy, whereby lateral changes in facies within a unit of comparable age require a change in name. An example is illustrated in Fig. 9.73, in which it can be seen that by using an allostratigraphic approach to the subdivision of a fluvial-lacustrine assemblage the natural subdivision of the succession into four sequences, bounded by breaks in sedimentation, can readily be formalized in a stratigraphic nomenclature.

It will be apparent that the ideas behind the bounding-surface hierarchy used throughout this book may readily be adapted to the methods of allostratigraphy. In fact, the three systems of stratigraphic subdivision, allostratigraphy, the bounding-surface hierarchy, and sequence stratigraphy, are in

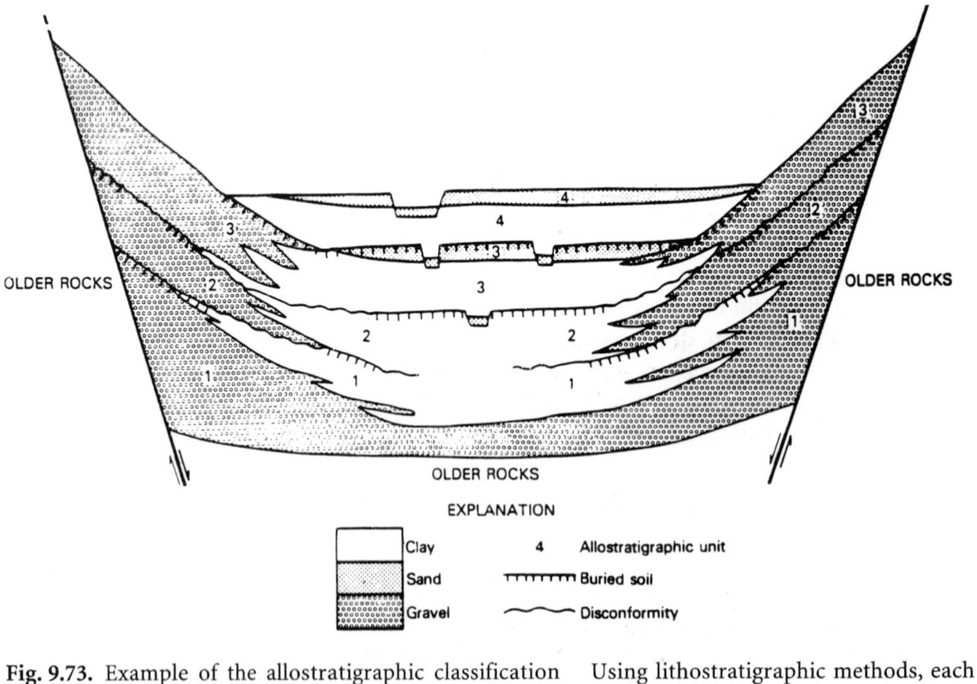

Fig. 9.73. Example of the allostratigraphic classification of a fluvial-lacustrine assemblage in a graben. *Numbers 1–4* correspond to alloformations which are defined by breaks in sedimentation and cut across facies boundaries. Using lithostratigraphic methods, each gravel, sand, and clay unit would typically be given a separate name. (NACSN 1983, reprinted by permission)

reality closely related, differing mainly in emphasis. Allostratigraphy has been proposed as a general system of descriptive, nongenetic, regional stratigraphic subdivision based on the recognition of breaks in sedimentation. The bounding-surface hierarchy proposed by Miall (1988a,b) and used throughout this book focused initially on smaller-scale features, such as the individual components of depositional systems but has now been expanded to include larger-scale features. For example, the numbered alloformations of Fig. 9.73 correspond to group 9 depositional systems of Tables 3.2 and 9.1 and are bounded by sixth- or seventh-order surfaces in Miall's (1988a) scheme. The subdivisions of each of the alloformations into allomembers, indicated by the clay, sand, and gravel areas in Fig. 9.73, represent the group 8 subdivisions of the depositional system into alluvial fans, channel belts, floodplains, etc., bounded by fifth-order surfaces. The scheme of DeCelles et al. (1991), illustrated in Fig. 4.3, should be compared with this. There are some differences, but the same principles of hierarchical subdivision

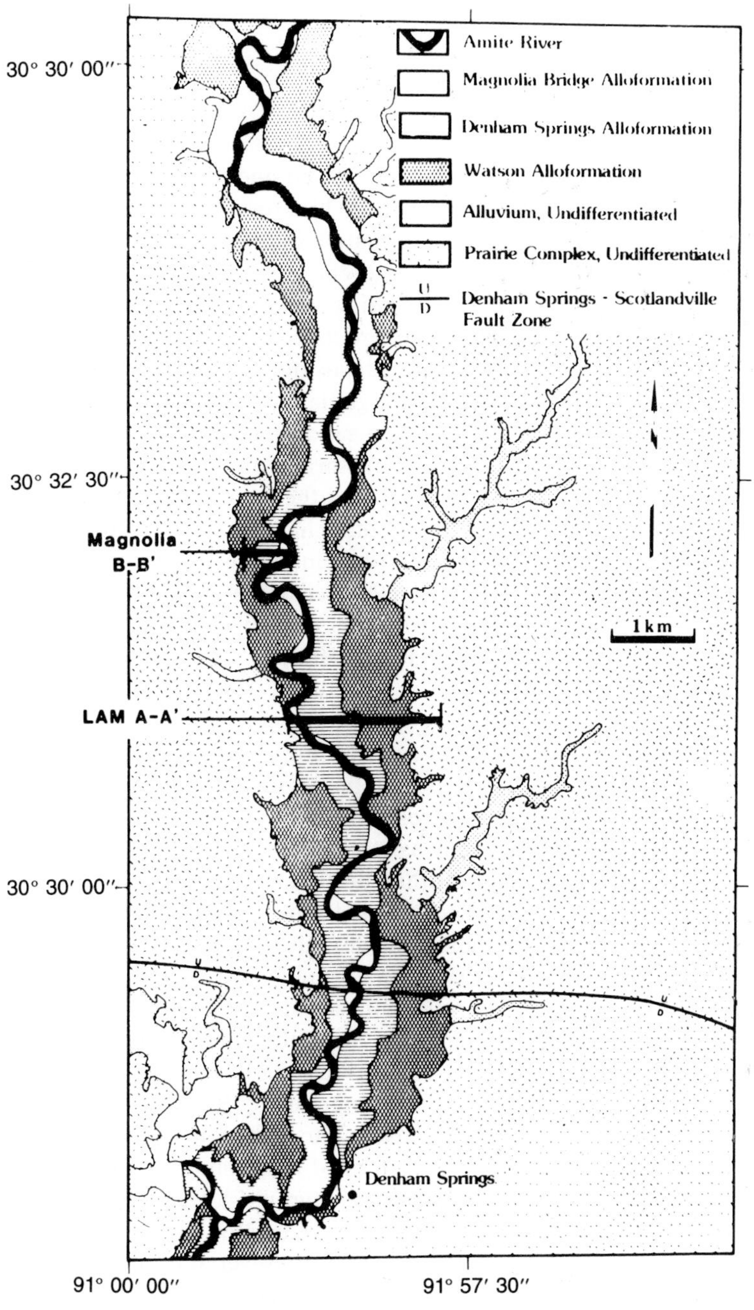

Fig. 9.74. Distribution of alloformations in the Recent sediments of the Amite River valley, Louisiana. (Autin 1992)

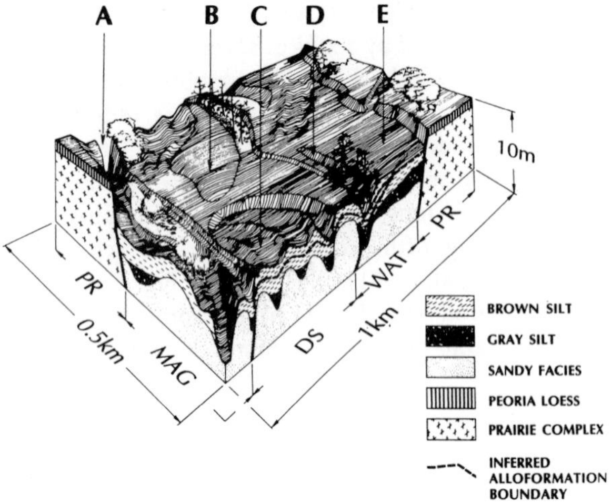

Fig. 9.75. Conceptual block diagram of the relationships between the various alloformations shown in map view in Fig. 9.74. The oldest unit is the Prairie Complex (*PR*). Successively younger alloformations, *WAT*, *DS*, and *MAG*, are incised into each other, with steep erosional lateral contacts and diffuse horizontal margins on the floodplain, where fine-grained units are in contact across boundaries that have been disturbed by pedogenesis. *A*, Modern Amite River; *B*, point bar topography of DS alloformation; *C*, DS ridge and swales topography; *D*, WAT abandoned channels; *E*, WAT floodplain flats. (Autin 1992)

apply. Groups 1–7, bounded by surfaces of first- to fifth-order rank, represent the smaller-scale stratigraphic features, not shown in Fig. 9.73. Sequence stratigraphy focuses on the larger-scale features of this subdivision and interprets them in a genetic context, as discussed in Chap. 11.

Several groups of workers are now explicitly employing an allostratigraphic methodology. Walker and James (1992) recommended the general use of allostratigraphic methods in stratigraphic and facies studies. A few examples of the application of the technique to fluvial and deltaic deposits are now available. For example, Bhattacharya (1991) subdivided the deltaic Dunvegan Formation of west-central Alberta into a series of allomembers using mappable flooding surfaces (Fig. 9.17). Autin (1992) subdivided the postglacial terraces and associated sediments in a Holocene fluvial floodplain succession in Louisiana into alloformations (Figs. 9.74, 9.75). Each represents a separate channel-belt complex that formed in response to changes in climate and sea level. They are incised into each other and partly overlap, and are therefore comparable to successive, superimposed incised valley fills of the type described in Chap. 13.

In other cases, data may be available for allostratigraphic subdivision, but the terminology has not yet been employed. For example, the detailed subdivision of the Old Red Sandstone using tephras by Allen and Williams (1982) led to the erection of a detailed stratigraphy based on regional channel-belt erosion surfaces (Figs. 9.25, 9.26). The eight units defined by Allen and Williams (1982) are in effect allomembers. López-Gómez and Arche (1993) mapped seven sixth-order surfaces in a 170-m-thick Triassic sandy-braided system in Spain. They traced the surfaces for more than 60 km and suggested that they are probably, in fact, basin-wide in extent. They subdivided the formation into six sandstone sheets. López-Gómez and Arche (1993) suggested that the seven "main bounding surfaces" are regional erosion surfaces that developed as a result of reorganization of fluvial depositional systems following pulses of basin-margin faulting and gentle regional tectonic tilting. The sandstone sheets could therefore be defined as alloformations.

No rules for relating the various systems of nomenclature to each other have yet evolved. However, the examples described in this section indicate that alloformations are defined by surfaces that would be classified as sixth-order or higher in Miall's (1988a) scheme, and that allomembers are bounded by surfaces of fifth- or sixth order. The examples quoted here reveal a range of time scales for alloformations and allomembers, reflecting the wide variation in the rates of change imposed on alluvial systems by allogenic controls. For this reason, it is not possible to definitively assign allostratigraphic units to specific groups within the hierarchy of this book (Tables 3.2, 9.1). The six sandstone sheets comprising the Cañizar Sandstone described by López-Gómez and Arche (1993) span 10 million years, indicating an average duration for each of 1.7 million years (group 9). It is suggested here that they correspond to alloformations. The schematic alloformations of the NACSN (1983) proposal (Fig. 9.73) are equivalent to group 9 deposits. The eight units of Allen and Williams (1982), termed allomembers above, span an average of 8000 years (group 7). The Recent floodplain alloformations of Autin (1992) have time spans of a few thousand years. However, it is questionable whether Autin's (1992) units would be assigned alloformation rank if they comprised part of an ancient nonmarine succession. A comparison with the eight units of Allen and Williams (1982) is suggested here, with a change in rank to allomembers. The entire postglacial valley-fill complex would then rank as an alloformation.

A proposal for a correlation between allostratigraphic units, the bounding-surface hierarchy, and sequence concepts is presented in Chap. 13.

A recent study of a stratigraphic succession in Wyoming was carried out for the specific purpose of comparing the use of lithostratigraphic, allostratigraphic, and sequence nomenclature for the subdivision and interpretation of an interfingering marine-nonmarine clastic succcession. The study was by Martinsen et al. (1993), who discussed the stratigraphy of the Hanna Basin, an Upper Cretaceous foreland basin. A system of lithostratigraphic nomenclature was available from earlier work and provided a complete framework of named units for descriptive purposes. Allostratigraphic concepts enabled the complex of laterally interfingering units to be related to each other and provided the basis for a simpler system of named units, but uncertainties remained regarding the definition of boundaries, because in places the units are conformable, and no discernible lithologic changes could be discovered to enable placement of the boundaries to be completed. To then compare this allostratigraphic framework with one constructed using sequence concepts is perhaps inappropriate, because allostratigraphy is intended to be purely descriptive, whereas sequence stratigraphy is genetic and can lead to very specific predictions regarding stratigraphic composition and geometry (Chap. 13). Sequence concepts can be used to construct a complete stratigraphic framework as long as some of the basic elements of the sequence model can be observed, although in this case difficulties arose because the rapid subsidence and the particular tectonic style of the Hanna Basin made comparisons with standard sequence models difficult.

Chapter 10

Fluvial Depositional Systems and Autogenic Sedimentary Controls

10.1 Introduction

Beerbower (1964) was the first to clearly distinguish between within-basin and extrabasinal sedimentary controls. The first, which he termed autocyclicity, refers to the redistribution of sediment within a depositional system as a result of processes inherent to the system, such as channel migration and bar development. As discussed in Sect. 2.3.6.2, during the 1950s and 1960s it was realized that fining-upward cycles in fluvial systems could be explained by such processes. The term autogenic processes is now preferred, because they are not necessarily strictly cyclic in nature. Much of the sedimentological research in the 1960s and 1970s was concerned with the recognition and description of autogenic processes, as discussed in Chap. 2. The purpose of the present chapter is to review the present state of our understanding of the subject. Extrabasinal, or allogenic (formerly allocyclic), processes, are discussed in Chap. 11.

The most important autogenic processes are those which lead to changes in the position of channels. Channels may change in position and shape by gradual lateral movement, or they may change position entirely to another part of the floodplain by the process termed avulsion.

10.2 The Evolution of Distributary Fluvial Systems

Distributary systems include alluvial fans and deltas. They are characterized by networks of channels radiating from a single trunk channel. There may be one or more active channels, together with several or many dry channels that are occupied only during floods or as a result of the autogenic processes described in this section. The current controversy regarding the definition of the term "alluvial fan" and the range of distributary fluvial systems that should be included under this heading is discussed in Sect. 8.3. This section is concerned primarily with what Blair and McPherson (1994) term braid deltas.

There is a very large literature on alluvial fans and other distributary systems, dealing with such data as depositional processes, downstream changes in slope, sediment grain size, sorting, facies, facies associations, and clast shape (Denny 1967; Bull 1972, 1977; Heward 1978a; Nilsen 1985; Rachocki and Church 1990; Fraser and DeCelles 1992; Blair and McPherson 1994). A significant body of field data is presented in these publications, together with much useful theoretical discussion. A summary of this would require a book of its own, and the discussion below is limited to a presentation of the more important concepts which control large-scale fan architecture and vertical profile. The main theoretical framework for interpretation has been built from studies of modern and ancient gravelly braided systems, supplemented by experimental work on small laboratory models. However, it can be shown that the same principles apply to all distributary systems, such as terminal "fans" and the giant rivers of the Himalayan foredeep.

Basin-margin gravel rivers are highly susceptible to allogenic influences, especially tectonism. Much research has been published on the sequences and megasequences in the resulting deposits generated by pulses of basin-margin faulting (e.g., Steel et al. 1977; Steel and Aasheim 1978; Heward 1978a; Blair and Bilodeau 1988). These processes act over a physical scale and time scale corresponding to sediment groups 8 and above of this book (Table 9.1) and are discussed in Chap. 11. However, autogenic processes are of equal importance. They concern primarily depositional processes and products of group 7 and are the subject of this section.

There are two main autogenic processes in fluvial distributary systems, the avulsive shifting of distributary channels and a cycle of trenching and backfilling. The experimental work of Schumm et al. (1987), and the observations of DeCelles et al. (1991) and Fraser and DeCelles (1992) suggest that these processes are intimately related. As shown in Fig. 4.3, conglomeratic distributary systems cycle through two main depositional styles, a cycle of entrenchment and a cycle of trench backfilling and dispersion of flow. Avulsive shifting of distributaries may be associated with the establishment of new depositional lobes during the entrenchment phase, or with the more minor shifting of distributaries that can occur during the dispersion phase. Events upstream in the drainage basin may also affect the development of the deposits, as the tributary network grows and sediment is stored and periodically flushed from the hinterland out into the basin (Fraser and DeCelles 1992). In detail, over periods of 10^3–10^5 years, the resulting stratigraphy is controlled by the action of various geomorphic thresholds, as described by Schumm (1977, 1979).

Schumm et al. (1987; Table 9.1) listed numerous mechanisms that lead to fan-head entrenchment, including climate change and reduction in source area-basin relief as a result of normal erosion. However, their experimental work on small, artifical fans

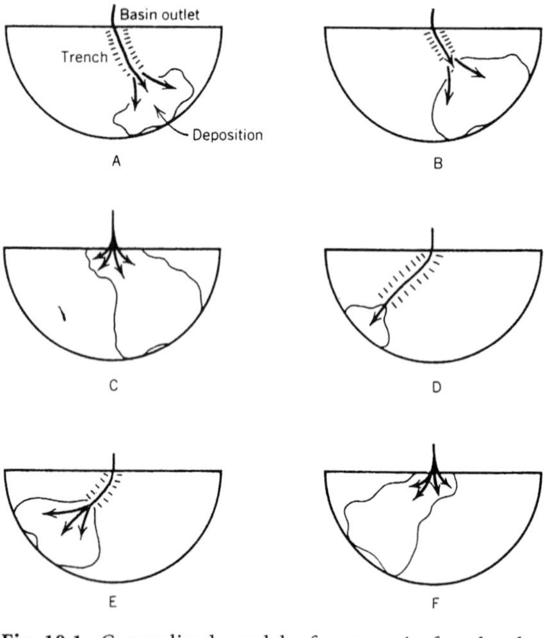

Fig. 10.1. Generalized model of autogenic fan development, based on experimental work on small, artificial fans. *A* Apex incision results from source-area erosion and leads to deposition near the toe; *B* the channel backfills and the locus of sedimentation (the intersection point of channel slope with the fan surface) moves up to the midfan region; *C* channel is obliterated by backfilling, and flow is dispersed from near the apex in all directions; *D* oversteepening of the apex occurs again, and incision along a steep flank of the fan leads to renewed entrenchment; *E*, *F* depositional backfilling commences in a new part of the fan. (Schumm et al. 1987)

indicated that a natural, autogenic cycle of entrenchment occurs, and DeCelles et al. (1991) confirmed the existence of this cycle in their architectural mapping of a Paleocene gravel complex in Wyoming and Montana. Figure 10.1 is the schematic and generalized model of the entrenchment-backfilling-dispersion cycle developed from experimental work on small laboratory models (from which Fig. 4.3 was developed by DeCelles et al. 1991), and Fig. 10.2 is an architectural model of a distributary complex that develops by this process, based on observations of ancient deposits.

The cycle begins with an interval of rapid aggradation on the upper part of the river system as a result of the lateral flow expansion and abruptly decreased gradient that occurs at the mouth of the feeder canyon. "Eventually the local gradient on the upper fan becomes steep enough that a trench forms in order to re-establish grade between the floor of the feeder canyon and local base level ... Sediment then begins to bypass the upper fan, through the trench, and accumulate in the form of a depositional lobe on the lower part of the fan, below the intersection point ... This part of the cycle is referred to as entrenchment" (DeCelles et al. 1991, p. 583; Fig. 10.1A). Trenching may also be initiated by a major flood event (Denny 1967). A new drainage system is thereby established, a new sediment lobe forms, and the trench begins to backfill with sediment (Fig. 10.1B). The trench is eventually obliterated, and sediment is then spread out in a fairly uniform manner across the fan surface – the stage termed "dispersion" by Schumm et al. (1987) (Fig. 10.1C). Oversteepening of gradients on the upper part of the system replicates the first stage, and a new cycle is initiated, with trenching occurring in a different part of the distributary complex showing a lower topographic profile (Fig. 10.1D–F).

Other processes may also lead to diversion of the main distributary. Denny (1967) and Schumm (1979) described stream capture by secondary streams occupying topograpically lower profiles, such as streams developed on the basin floor that cut back toward the apex of the distributary system by headward erosion.

DeCelles et al. (1991) recognized a category of fifth-order lithosomes and two types of fifth-order bounding surface that they interpreted as the product of the cycle of entrenchment and backfilling (Figs. 4.3, 10.2). Fifth-order surfaces, in this model, may be erosional or accretionary. Erosion surfaces are produced by the cutting and lateral migration of entrenched channels. These may include trench

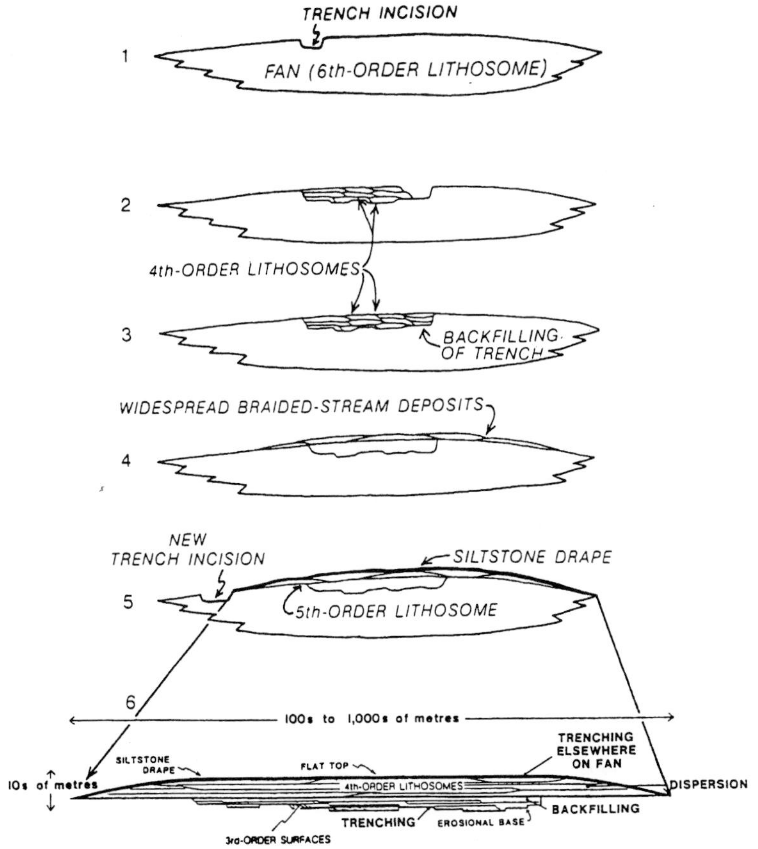

Fig. 10.2. Architectural cross section through the upper part of a fan, transverse to fan axis, showing the development of a fifth-order lithosome (see also Figs. 4.3 and 10.1). *1* Trenching; *2* lateral migration and filling of trench; *3* backfilling of trench; *4* dispersion of flow; *5* abandonment of active part of fan, with development of pedogenic modifications and initiation of renewed trenching elsewhere; *6* cross section through resulting fifth-order lithosome, showing architectural features and types of bounding surface. (DeCelles et al. 1991)

cutbanks several meters deep, although such surfaces may be difficult to detect within continuous conglomerate successions (see Fig. 6.17 for possible examples). Accretionary surfaces are developed by lobe construction during the backfilling stage. These surfaces are hundreds to thousands of meters in lateral extent in the conglomeratic deposits studied by DeCelles et al. (1991).

Fifth-order lithosomes formed in each cycle may be separated from the next overlying lithosome by thin siltstone intervals showing evidence of pedogenesis. These likely form during the interval of entrenchment that initiates a new cycle. Wind erosion and the formation of desert varnish may also characterize these temporarily abandoned areas of the fan surface (Denny 1967), a process particularly important on debris-flow dominated fans (Blair and McPherson 1994).

Some differences in architecture may occur between systems dominated by traction-current processes and those with significant sediment-gravity-flow (debris-flow, mud flow) sedimentation (the "true" fans of Blair and McPherson 1994). Trenches on fans that are formed predominantly by traction currents tend to migrate laterally, whereas on fans with significant sediment-gravity-flow deposition the trenches formed by traction-current incision are simply plugged by flow deposits during the backfilling stage (Schumm et al. 1987).

Much attention has been paid to the nature of the vertical profile through fan deposits. In a thorough discussion of this topic, Heward (1978a) stated:

"Despite the apparent individual characteristics of modern fan flood events, vertical sequences of progressively changing grain size, bed thickness and depositional processes characterise many ancient alluvial fan deposits ... Such sequences are generally interpreted to represent the gradual progradation, avulsion, abandonment, or lateral migration of depositional elements; rather than fractionation of sediment within a depositional element, or progressive changes in the sediment supplied" (Heward 1978a, p. 676).

The term "sequence" was used by Heward (1978a) to refer to autogenic fan successions. They are typically tens of meters in thickness and are classified as group 7 deposits (Table 9.1). Examples are illustrated in Figs. 10.3 and 10.4. Variations in upward fining and coarsening character reflect the details of

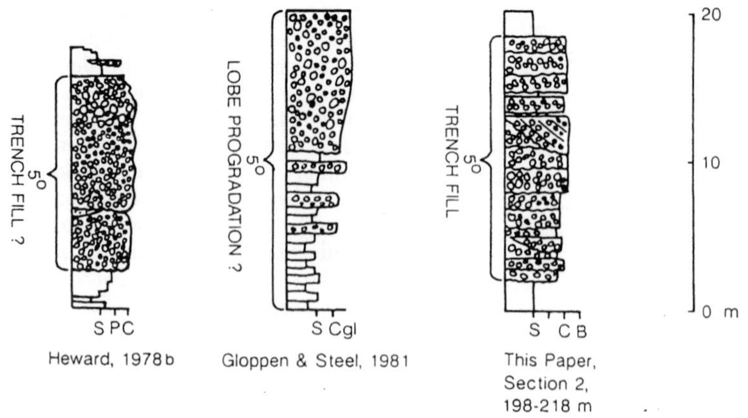

Fig. 10.3. Vertical profiles through three ancient conglomeratic fan deposits, showing different types of autogenic succession. The *left-hand example* is a Carboniferous succession in Spain, from Heward (1978b), and is interpreted to have formed from a single debris-flow event. *The middle profile*, from Gloppen and Steel (1981), is from Devonian basin in Norway. The *right-hand profile* is from the Beartooth Conglomerate (Paleocene) of Wyoming-Montana. (DeCelles et al. 1991)

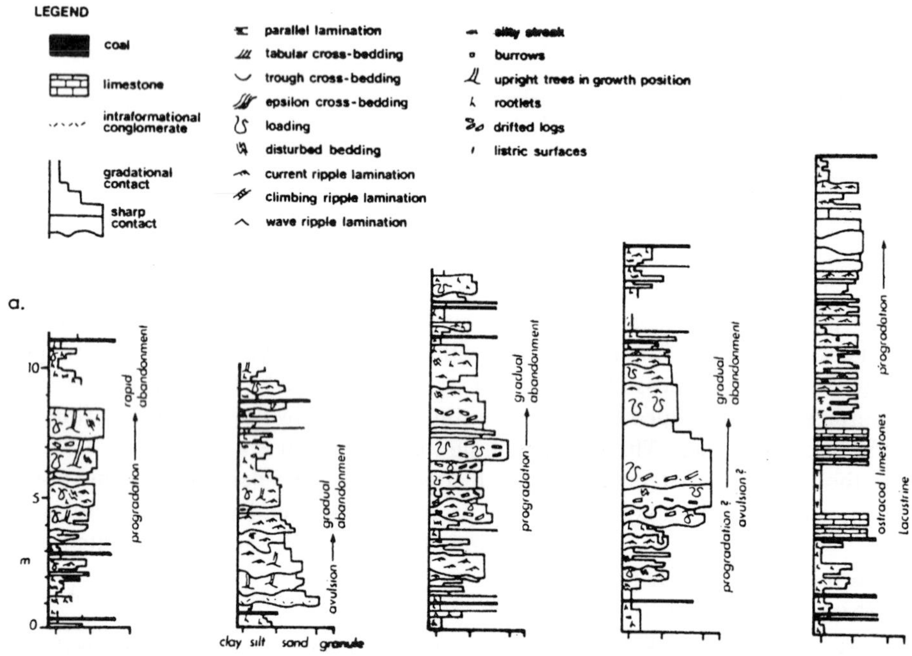

Fig. 10.4. Examples of distal, sandstone-dominated, sheetflood successions in ancient fan deposits, Carboniferous, Spain, showing variations in upward coarsening and fining character. (Heward 1978)

channel and lobe initiation (Fig. 10.5). Upward fining above a basal erosion surface indicates erosional trench or distributary initiation by avulsion, or emplacement by lateral migration, followed by gradual filling and abandonment. Upward coarsening indicates progradation. DeCelles et al. (1991) found that their fifth-order lithosomes typically did not show upward coarsening or fining (e.g., right-hand profile, Fig. 10.3), and attributed this to a predominance in their field area of sharp-based trench-fill successions. Progradational units are thought to be more likely to predominate in distal parts of fans, especially where the fans interfinger with basinal deposits, such as lacustrine facies or the deposits of a trunk river, as in the rightmost profile of Fig. 10.4.

The concepts developed in the preceding paragraphs can be applied, with some modifications, to all nonmarine distributary systems, including termi-

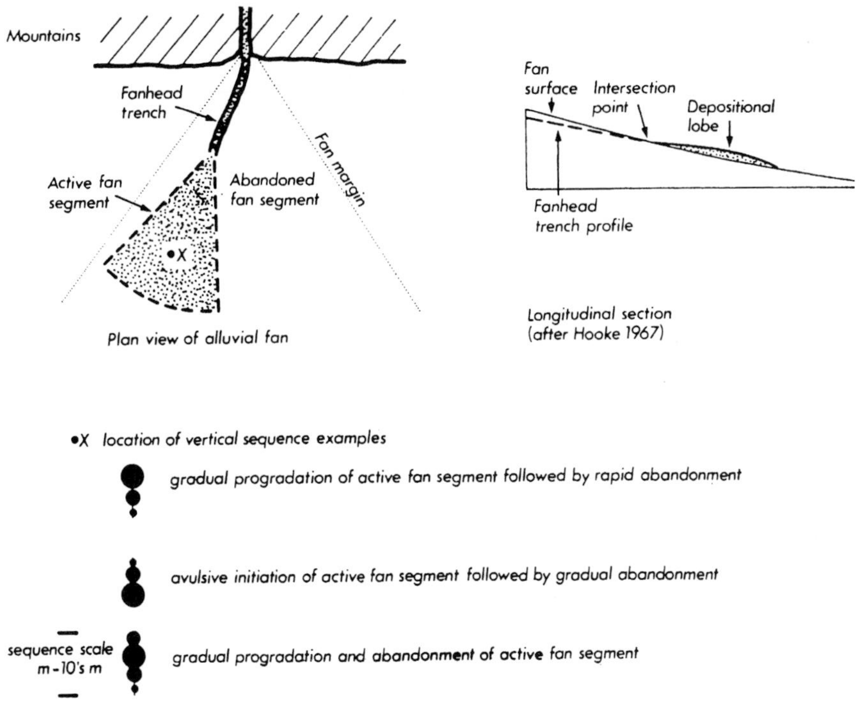

Fig. 10.5. Fan profile and model of possible vertical successions resulting from channel and lobe switching. (Heward 1978)

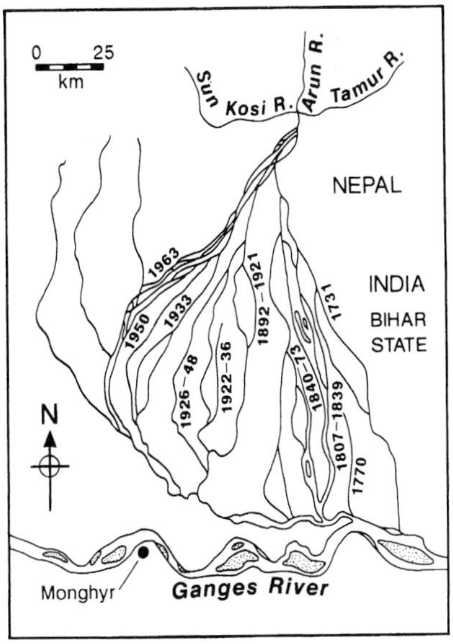

Fig. 10.6. Courses of the Kosi River, India, between 1731 and 1963. (Schumm 1977, after Gole and Chitale 1966)

nal fans and the large sandy systems discussed in Sect. 8.3. In particular, the process of channel or lobe aggradation followed by avulsion is common to all distributary systems, although the details differ.

In the case of the large, sandy Okavango system of Botswana anastomosis takes place as channels aggrade. Switching is gradual, because channel banks are stabilized by vegetation and resist erosion; but avulsive switching of channels is nevertheless an important part of the evolutionary process (McCarthy et al. 1992). It is not known what vertical successions develop in this river. The Oligocene-Miocene Huesca fluvial system of Spain (Figs. 9.6, 9.7) was compared to the Okavango fan by Hirst (1991), and avulsive mechanisms that operated there were also attributed to channel aggradation, perhaps with the additional trigger of tectonic tilting at the fan apex.

In the case of the Kosi River of India, many studies have reported the avulsive migration of the main distributary channel across this sediment surface (Holmes 1965; Gole and Chitale 1966; Schumm 1977; Gohain and Parkash 1990; Wells and Dorr 1987). The main channel has migrated westward a distance of 110 km between 1736 and 1964 (Fig. 10.6). The cause of this migration was examined thoroughly by Wells and Dorr (1987), who were able to demonstrate that avulsive events were not correlated in time with major floods or with tilting episodes that occurred during earthquakes. They confirmed the conclusions of earlier workers that the channel shifting is an entirely autogenic process. Gole and Chitale (1966) demonstrated that channels shift to occupy

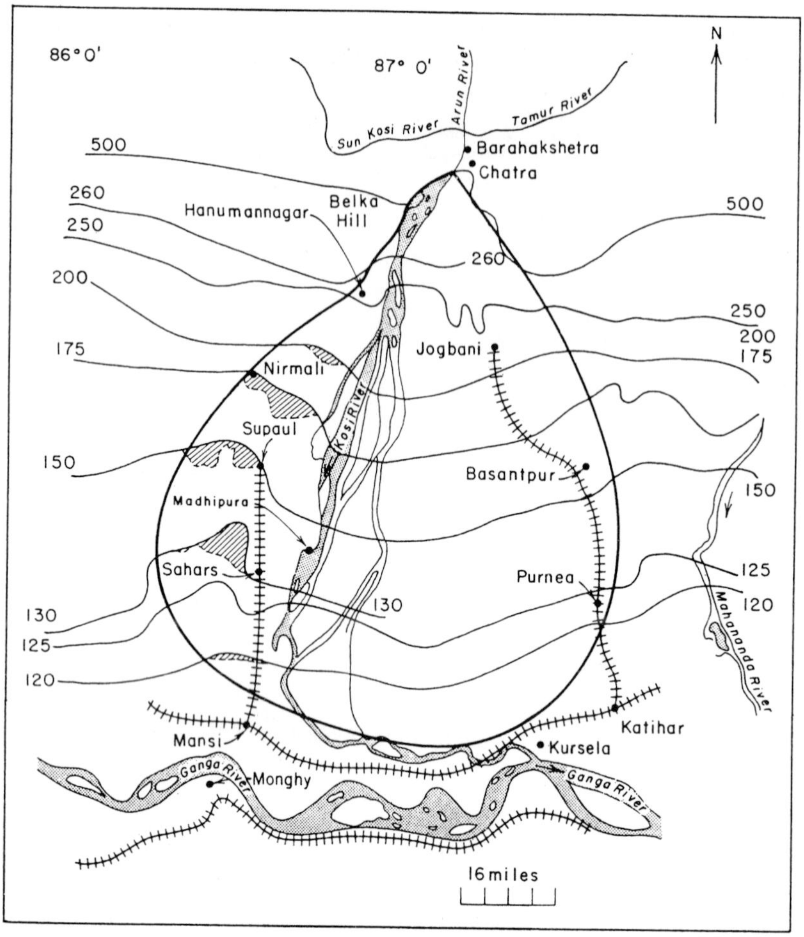

Fig. 10.7. The Kosi fan in 1936. *Solid contour lines* show configuration of the fan at that time. *Dashed lines* are contours re-surveyed following an avulsion event in 1957, with the *shaded areas* indicating deposition. (Schumm 1977, after Gole and Chitale 1966)

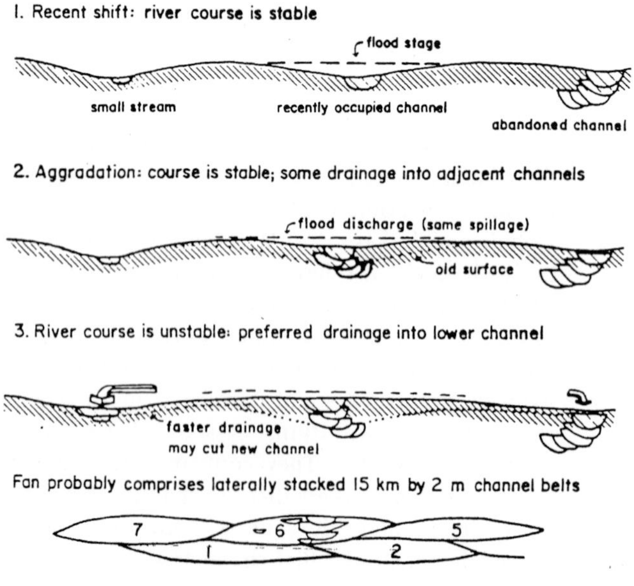

Fig. 10.8. Model for the avulsive shifting of the main channel of the Kosi River. (Wells and Dorr 1987)

low-lying areas on the depositional surface (Fig. 10.7). The triggering process seems to be instability caused by channel aggradation, with the main channel moving to occupy low-lying areas traversed by old, former channels, or tributaries of the Kosi system (Fig. 10.8). Wells and Dorr (1987) noted several other examples of similar lateral shifting of distributaries on modern distributary systems, but it is unknown whether the very regular one-way (westward) sweeping of the Kosi distributaries is typical.

Bluck (1979, 1980) proposed a model for the regular, autogenic, pendulum-like sweeping of an alluvial fan by its major distributary, based on some observations of modern fans and Devonian fan deposits. Viseras and Fernández (1994) described a Pliocene succession in Spain which they suggest fits this model. The mechanism for the model, suggested by Bluck (1979, p. 201), involves the preferential building of bars on one side of the distributary, with bank erosion and channel widening occurring preferentially on the opposite bank. There is some stratigraphic evidence for this, as reported by Bluck (1980) and Viseras and Fernández (1994), in the form of bar accretion sets dipping preferentially in one direction at a given stratigraphic level. However, the cause of this preferential lateral migration of the channel remains unclear. As discussed in Sects. 6.6 and 8.1, bar growth and cutbank erosion are related to meander development both in braided and meandering systems, and this, in turn, is related to the turbulent properties of flowing water. Therefore, it seems unlikely that lateral bar accretion and cutbank erosion would take place on the same sides of a river for more than one meander length. It seems more likely that a process of channel aggradation and avulsive reoccupation of low-lying areas is the main cause of lateral channel movement in fans, as demonstrated (above) for large fans such as the Kosi. A possible tectonic mechanism may also be sought for the lateral migration of channels and the preferential preservation of accretion sets dipping in one direction, and just such a mechanism is described in Sect. 11.2.1.1.

10.3 Avulsion in Fluvial Systems and Its Effect on Alluvial Stratigraphy

Fluvial style is governed mainly by a range of extrabasinal controls, as discussed in Sect. 8.1. However, given a normally graded fluvial system flowing in a steadily subsiding basin under conditions of dynamic equilibrium, normal autogenic processes will lead to the gradual or sudden and episodic shifting (avulsion) of the channel, with consequences for the ongoing development of the alluvial stratigraphy (Schumm 1977; Richards et al. 1993).

There are two kinds of avulsion, that which occurs within the active channel system, and that which displaces flow into a new channel position. The simplest example of the first case is where the natural migration of a river meander results in one reach of the meander bend cutting into another reach, resulting in a neck cutoff. This process has been known and understood for a very long time. For example, it was described and illustrated in Lyell's (1830) great textbook (Fig. 2.2). Chute cutoff occurs where a flood channel develops across the top of a point bar (Fig. 8.58). The complex of channels, cutoffs, crevasse splays and levees created by this process may become slightly elevated from the main floodplain, and is termed an alluvial ridge.

The avulsive shifting of braided channels is an integral part of the braiding process. Several autogenic mechanisms occur. They are discussed in Sect. 6.3, where the emphasis is on the development of gravel bars and bedforms. Avulsion is also the main process whereby channels in anastomosing systems evolve, as described in Sects. 7.2 and 8.2.10.

Maps of meander migration in sinuous rivers may be produced by observing the relict topography of abandoned channels and of the scroll bars on point-bar surfaces. Such maps began appearing in textbooks in the early part of this century, as exemplified by those in Chamberlin and Salisbury (1909; Figs. 2.7 and 2.8 of this book). Maps of this type, or schematic versions of them, are popular as generalized illustrations of sedimentary processes (e.g., see the cover of Walker and James 1992). They are easily made from aerial photographs, in which patterns of ancient meanders are commonly quite spectacular. Current work on meander evolution and migration and its stratigraphic results is discussed in Sect. 6.7. Cutoffs in modern rivers were reviewed by Lewis and Lewin (1983). Within-channel avulsion in gravel-bed braided rivers was modeled by Ashmore (1991) and Leddy et al. (1993) and is discussed in Sect. 6.3.

The second, more stratigraphically significant, form of avulsive channel shifting is where the entire river changes position, abandoning its original course and flowing in a different part of the valley. This process is most important in the case of meandering rivers, because these normally occupy only a small part of their total valley width, so that when a shift occurs it has a profound effect on the resulting stratigraphic architecture. However, avulsion is important in the case of other fluvial styles as well, as discussed later in this section.

On coastal plains avulsion may lead to the abandonment of a delta lobe and to the initiation of a new locus of deposition laterally displaced by tens of kilometers. Maintenance of a new channel in such cases is commonly facilitated by the fact that the diversion results in the occupation of a shorter, steeper route to the sea. This avulsion process is analogous in some ways to that which occurs on alluvial fans.

10.3.1 The Development of Meander Belts

Much work on avulsion in meandering systems has been carried out by geomorphologists and civil engineers, because of the importance of the process from the point of view of navigation and flood control. The review by Schumm (1977, pp. 297–305) is particularly useful in this regard.

There are two types of avulsive process in meandering rivers: firstly, high discharge, or aggradation of the channel by a large sediment load, may lead to the channel overtopping its banks, causing crevassing and the diversion of flow into the floodplain. Under certain conditions, this can evolve into long-term channel diversion. Secondly, avulsion can occur when normal, gradual, meander migration results in the channel encountering a pre-existing, older channel at a lower elevation, resulting in the reoccupation of that channel. Such earlier channels may have developed as a result of the first of these two avulsive process.

Documented cases of channel avulsion in modern rivers are fairly limited. In some river valleys the existence of older channel belts has been reconstructed from archeological or historical data in areas of early human civilization; for example, Holmes (1965) referred to the lateral shifting of the Euphrates River in Mesopotamia (Iraq) by a distance of some 16 km, around the position of the ancient city of Ur. MacKay (1945; quoted by Schumm 1977) documented the rise and fall of ancient cities, including those in Mesopotamia, whose prosperity depended on the proximity to a navigable river. Russell (1954) noted the existence of "old" and present courses of the Meander River in Turkey, and summarized historical evidence relating to the change in position of the meander belt. According to Russell (1954) this river was mentioned in the work of Herodotus and Strabo, and its name became used as a general term for highly sinuous rivers because of its characteristic winding pattern. Anderson (1961) documented shifts in the Rufiji River in Tanzania. Schumm (1977) illustrated old and new courses of the Murrumbidgee River, Australia, using aerial photographs, and discussed the climatic changes that accompanied this avulsive shifting. Holmes (1968) described the shifting of the modern Indus River.

By far the best-known history of channel avulsion is that of the Mississippi River, which was investigated in detail by Fisk (1952). Numerous changes in the course of this river have been documented, and many of these triggered the growth of new delta lobes along the Gulf Coast (Fig. 10.9). Fourteen separate delta lobes are known to have developed in the past 7500 years, since the last deglaciation (Frazier 1967). All of these required a major channel avulsion to occur. Other avulsive events which occurred far up the Mississippi valley, such as at Cairo and Thebes Gap (Fisk 1944), are probably not associated with deltaic diversion. According to Fisk (1952), the Mississippi has followed 20 distinct courses during the past 2000 years, meaning a change in course somewhere in the lower alluvial valley on average every 100 years. Not all workers agree with this reconstruction. For example, Saucier (1974) recognized only five main meander belts (one shift every 400 years). Leeder (1978) concluded that in major rivers such as the Mississippi "realistic mean avulsion periodicity" is in the order of one to two events per 1000 years. In the hierarchical classification of this book this would assign meander-belt avulsion to group 6 or 7 (Table 9.1).

The case of the Atchafalya River, a distributary branching off the Mississippi (Fig. 10.9), may be taken as a type example of channel avulsion in a meandering system (details from Fisk 1952; summary in Schumm 1977). The Atchafalaya is an old, former course of the Mississippi, known from historical records to have been an active distributary in the early 1500s (Shlemon 1975). It was intersected by the modern river as a result of channel migration in the late 1940s. Its course from this point to the Gulf is 225 km long. The modern channel is much longer, at 515 km, partly because of the lengthening that has occurred at the mouth by delta construction. By 1951 the Atchafalaya was carrying 21% of the flow of the Mississippi, and by 1956 it was the third largest of all rivers flowing into the sea from the United States. Diversion of the flow was gradual initially, because of the originally small size of the channel, but natural widening and deepening occurred, facilitated by the steeper course along the Atchafalaya relative to the main Mississippi, and it was estimated that once the flow reached 40% of the total, complete diversion would be accomplished rapidly. Large new deltaic deposits developed very rapidly in the Gulf of Mexico at the mouth of the river (Shlemon 1975). In 1963, engineering control works were completed which maintain the diversion at no more than 30% of Mississippi flow, and in this way New Orleans, one of the great port cities of the United States, has been at least temporarily saved from the fate that befell Ur and other ancient cities when their navigable waterways were abandoned.

The development of crevasses in rivers such as the Mississippi is not thought to lead to avulsion in the normal course of events (Fisk 1952; Schumm 1977). Crevasse splays may form (Sect. 7.2.3), but unless there is a gradient away from the main channel there is no force to drive the diversion further, and the crevasse channel will simply become plugged with sediment and abandoned. Only where the channel floor is aggraded by bedload sedimentation is there likely to be a significant elevation difference between the channel and its floodplain, providing a force to drive flow diversion. However, this is an important cause of diversion in some braided river systems, as discussed in the next section.

What are the stratigraphic effects of major avulsive diversion in a meandering system? Most obviously, they include abandonment of the original meander belt and initiation of a new one. In the case of large rivers, such as the Mississippi, this can result in the displacement of the locus of sand-body sedimentation by several to many tens of kilometers. The steeper course of the new channel may result in straightening or scouring of the main channel above the point of diversion, although the geological results of this, in the form of unusually deep incised channels and changes in facies architecture, may be destroyed by continued evolution of the river system.

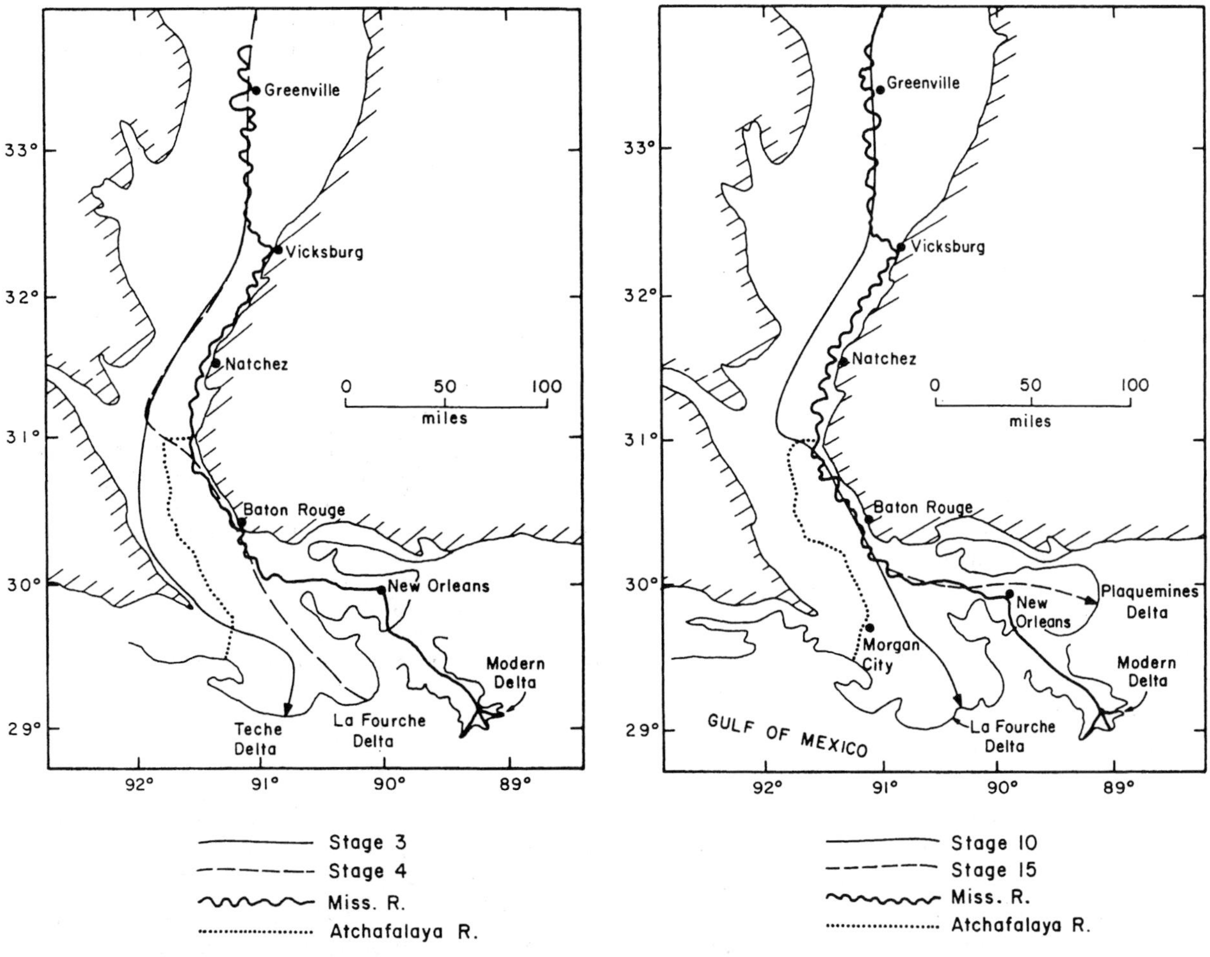

Fig. 10.9. Former courses of the Mississippi River, according to Fisk (1944), as redrawn by Schumm (1977)

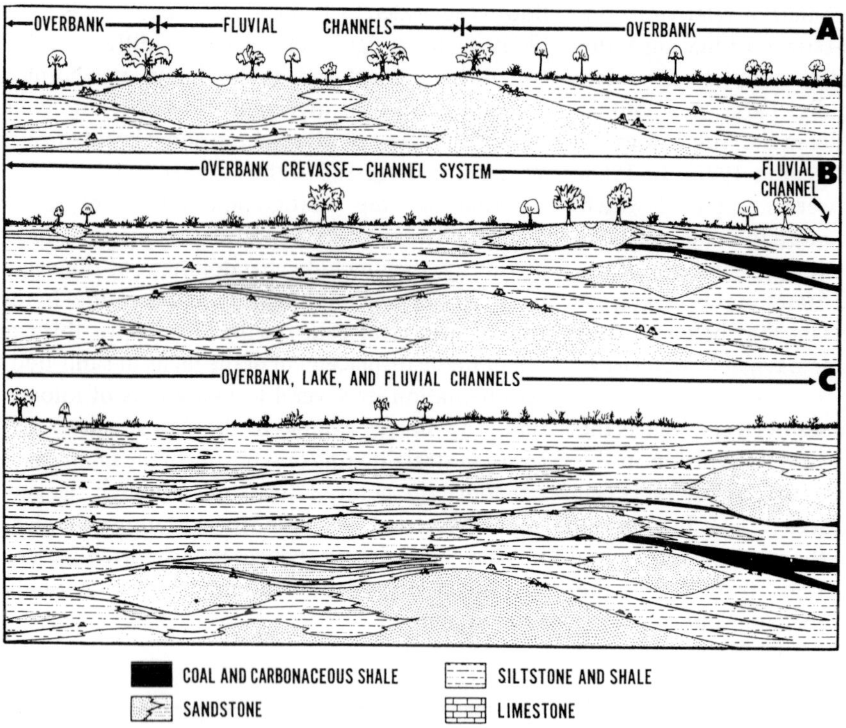

Fig. 10.10. Lithogenetic model for the autogenic evolution of a meander-belt complex, showing three stages in the development of the stratigraphic succession illustrated in Fig. 7.2. (Flores 1981)

An ancient example of channel avulsion interpreted to be by autogenic processes was described from the Himalayan foredeep by Behrensmeyer and Tauxe (1982) and is illustrated in Fig. 9.1. These authors suggested that the pattern of channel bodies in the "buff lithofacies" may represent avulsion along two channel belts spaced 10–15 km apart, comparable to the small drainage systems occupying the Himalayan front at the present day. Alternatively the observed architecture could represent the activity of a single river. In the first case, given the detailed magnetostratigraphic control available to date these rocks, it can be shown that avulsion would have had to occur on one or another of the rivers at a maximum spacing of once about every 6000 years. If only one river was present it would have had to undergo an avulsive shift at a minimum rate of once every 2000 years. This is therefore a "group 7" process.

Another excellent example of the alluvial architecture of a meander-belt complex is illustrated in Fig. 7.2. The interpretation of this composite cross section is illustrated in Fig. 10.10 and may serve as a good example of the kinds of stratigraphic complexity generated by autogenic crevassing and channel avulsion. The deposits have been interpreted as follows:

"The upper cross-section (A) portrays the buildup of a major channel complex prior to abandonment. The broad lateral extent of this channel system reflects high sinuosity or meandering of the stream. As the channels filled, adjoining peat swamps spread into submergent abandoned channel ridges, forming thick, laterally extensive peat coals. The abandoned channel system, which was maintained as a high topographic area, influenced the autocyclic shift of a new channel into the adjoining topographically low area to the right (northeast). The middle cross-section (B) shows that the area formerly occupied by the oldest channel complex was transformed into a floodplain that probably supported poorly drained backswamps. The backswamps were relatively stable and long-lived as evident by deposition of thick peat coals. Occasional intrusions by crevasse-splay and overbank detritus account for the splitting of the peat coals and testify to the last activity in [the] adjoining channel before being inundated by peat swamp. The detritus probably came from a major channel that was formed at this time to the right (northeast) of the oldest channel complex. The small channel sandstones associated with the crevasse-splay sandstones are deposits of active feeder channels. In the floodbasin to the right of the major channel, a very thick peat accumulated, indicating development in a more stable and long-lived backswamp. The absence of thick crevasse-splay deposits within the very thick, laterally ex-

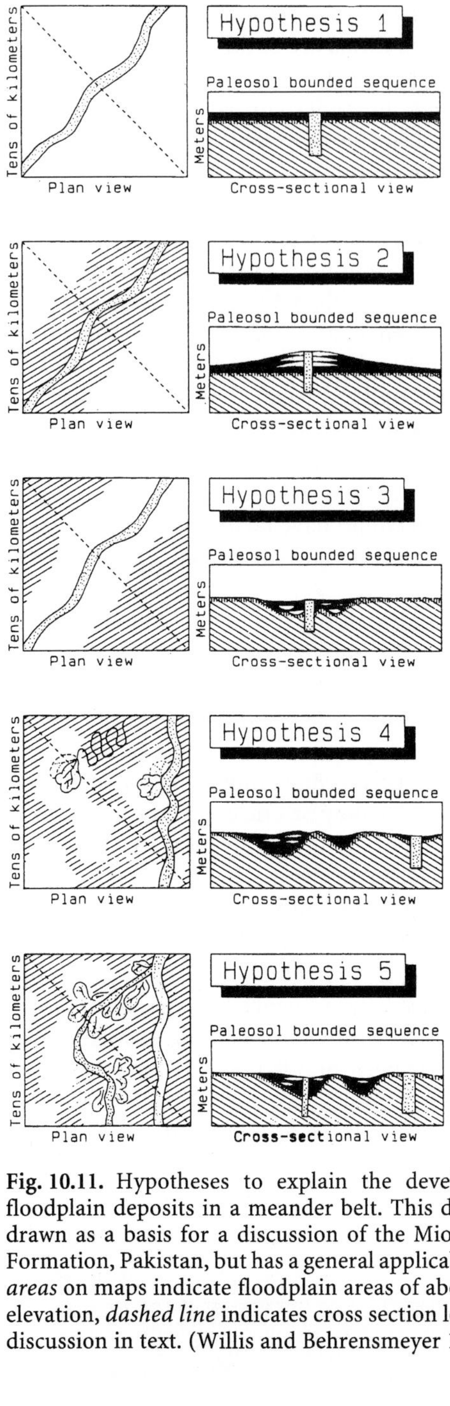

Fig. 10.11. Hypotheses to explain the development of floodplain deposits in a meander belt. This diagram was drawn as a basis for a discussion of the Miocene Chinji Formation, Pakistan, but has a general applicability. *Ruled areas* on maps indicate floodplain areas of above-average elevation, *dashed line* indicates cross section location. See discussion in text. (Willis and Behrensmeyer 1994)

tensive peat coal suggests that the northeast side of the channel system probably was not breached by flood waters, thus allowing uninterrupted peat deposition in an isolated, broad floodplain (e.g. marginal to abandoned meander channels propagated by avulsion). The lower cross-section (C) shows abandonment of the channels at the right (northeast) accompanied by autocyclic shifting of a major channel to the left (southwest). Correspondingly, the central floodbasin remained an interchannel low; however, it became a well-drained interfluvial lake or pond. This condition may have been in response to differential compaction of earlier formed interchannel and channel sediments, allowing subsidence of the floodbasin, which ponded some open water." (Flores 1981)

A contrast with the humid meandering system described above is the arid system documented by Stear (1983). These Permian deposits of the Karoo Basin, South Africa, exemplify an "ephemeral sand-bed meandering river" fluvial style (Sect. 8.2.8; Figs. 8.40–8.42). Isolated sandstone sheets and ribbons were deposited by channels that were free to meander and avulse on a broad floodplain. These deposits are comparable in facies to those of terminal fans, but Stear (1983) presented no evidence of downstream fining and thinning, or of radial paleocurrent patterns.

Willis and Behrensmeyer (1994) examined autogenic processes in a meander belt from a different perspective, that of the overbank deposits. Excellent outcrops of the Miocene Chinji Formation (Siwalik Group) in Pakistan (Fig. 7.13) provided details on the large-scale architecture of these deposits, and provided the basis for a discussion of the possible mechanisms for meander-belt evolution, with particular reference to the development of the floodplain. They proposed five hypotheses (Fig. 10.11), most of which have been applied in different settings to specific fluvial units by other workers. The first hypothesis predicts that the entire floodplain aggrades episodically, producing uniform, basin-wide floodplain units. There do not appear to be any documented examples of this case in the fluvial literature. The second hypothesis relates the thickness and grain size of floodplain units and paleosol type to proximity to an "alluvial ridge". This hypothesis has commonly been invoked and is the basis for the "pedofacies" concept described in Sect. 7.4.2. Hypothesis 3 is based on alternation of degradation and aggradation, as in the cycles of channel incision and filling that occur during changes in base level (Sect. 11.2.2, Chap. 13). Hypothesis 4 proposes that overbank sedimentation is controlled by deposition in topographically low areas of the floodplain, which shift with time as a result of channel-belt aggradation and differential compaction and subsidence. This hypothesis has been invoked for the modern Mississippi and Kosi rivers. The final hypothesis is similar to no. 4 but links floodplain aggradation in low-lying areas to the avulsive shifting of major channels. The specific flooplain units studied by Willis and Behrensmeyer (1994), which stimulated this discussion (see Fig. 7.13), appear to have evolved in the manner described by hypotheses 4 or 5.

10.3.2 Avulsion in Braided Fluvial Systems

Braided rivers tend to occupy most of their alluvial valley, leaving little room for floodplains, and therefore only minor areas for the channel to avulse into. However, avulsive shifting is a significant sedimentary control, and large- and small-scale effects on stratigraphic architecture can result, as discussed in this section.

By far the most well-known examples of the large-scale avulsive shifting of braided rivers are those of the Kosi River in India, and the Yellow River in China. The Kosi River is referred to elsewhere in this book as an alluvial fan (e.g., Sect. 8.3), and the avulsion processes there are described in Sect. 10.2. However, it may serve as the type example of a particular avulsion mechanism, that which is triggered by aggradational elevation of the channel on the floodplain surface. This process seems to be particularly important in braided systems, which are characterized by large bedloads and by rapid aggradation of channels and bars. Other types of distributive systems, such as terminal fans, also show this type of autogenic avulsion process (Sect. 10.2). Richards et al. (1993) noted that other fans in the Indo-Gangetic plains also show avulsive behavior similar to that of the Kosi. Several other examples of avulsive channel and bar changes in modern rivers are discussed in Best and Bristow (1993).

Where the river is in the form of a distributive system, as in the case of "inland deltas" such as the Kosi River or terminal fans in arid regions, the channel and adjacent floodplains are elevated with respect to the regional surface as a result of aggradation. Processes such as those which occur in the Kosi River system are therefore likely to be characteristic. The spectacular and catastrophic diversions of the Yellow River in China are of this type (Holmes 1965; Li and Finlayson 1993).

The Eocene Escanilla Formation of the Spanish Pyrenees appears to represent an exception to the generalization that in braided systems channels occupy most of the alluvial valley (Dreyer et al. 1993; Bentham et al. 1993; see Sect. 8.2). Channel-fill architecture indicates a braided fluvial style. Fine-grained deposits total more than 40% of the succession, by volume. Bentham et al. (1993) attributed this fluvial style to rapid subsidence and frequent avulsion within an unconfined basin. There are indications that the depositional system was distributary in nature, and there may therefore have been a tendency for the depositional surface to have been convex-up, which would have facilitated avulsion, as in the various alluvial fans described earlier in this chapter.

The alluvial architecture of the Old Red Sandstone (Lower Devonian) as exposed in coastal sections in southwest Wales, is discussed and illustrated in Sect. 9.5.1 (Figs. 9.25, 9.26) and serves as a good example of tightly documented regional autogenic control of a largely braided fluvial system. Although base-level change may be one contributing factor in the development of this alluvial suite, autogenic switching of the main channel belts on average about every 8000 years was also suggested by Allen and Williams (1982) as a major sedimentary control. The depositional environment is interpreted as that of a coastal-plain tidally influenced fluvial and mudflat complex, in which the channels were of low sinuosity and may have been braided. Earlier work on these beds by Allen (1974b), based on a much less extensive data base of vertical profiles, had led to specula-

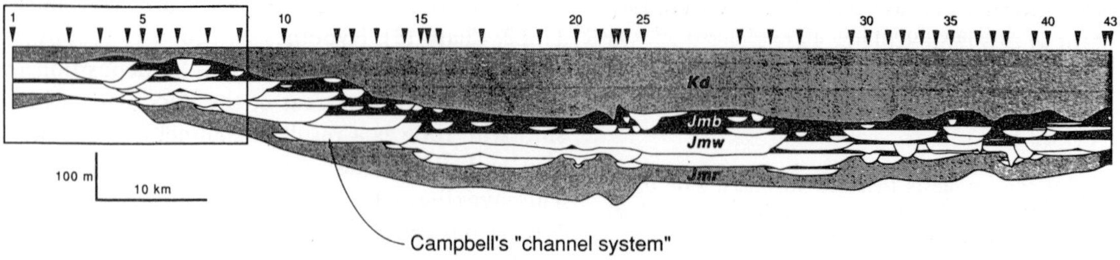

Fig. 10.12. Cross section of the Morrison Formation, northern New Mexico, redrawn from the work of Campbell (1976). Units prefixed with "J" are members of the Morrison Formation: *Jmr* Recapture Member; *Jmw* Westwater Canyon Member; *Jmb* Brushy Basin Member; *Kd* Dakota Formation. The diagram was constructed by reconnaissance correlation of channels within the Weswater Canyon Member, and the top of the diagram does not correspond to the top of the Dakota Formation. *rectangle at left* shows the area re-examined in detail by Cowan (1991)

tions by him regarding alluvial architecture and proposals for the nature of the architectural response to various combinations of autogenic and allogenic controls. The eight models that he illustrated in that paper formed the conceptual basis for all the later work in this area (Sect. 10.4), including the modern work on the architecture of fluvial deposits in a sequence-stratigraphic context (Chap. 13).

A well-known example of an early study of alluvial architecture is that by Campbell (1976), who described the predominantly sheet-sandstone bodies of the Westwater Canyon Member of the Morrison Formation in northern New Mexico. Some authors have interpreted this unit as the deposit of a giant sandy alluvial fan (e.g., Galloway 1980). Campbell's (1976) reconstruction is illustrated in Fig. 10.12. The apparent westward migration of the channels with time caught the attention of many sedimentologists, and the deposit has been compared to the Kosi fan of India (Miall 1981a; Collinson 1986), which shows a similar westward migration (Fig. 10.6). However, a recent detailed remapping of this area by Cowan (1991) demolished this reconstruction and proposed a quite different fluvial style for the beds (Fig. 10.13). Cowan (1991) was able to show that Campbell's (1976) mapping of individual sandstone sheets was based on outcrop color differentiation. He was able to demonstrate that the different colors of the sandstone bodies were formed by diagenetic leaching, which was only partly related to depositional architecture. The curved channel margins that make Campbell's reconstruction so appealing are seen to be mainly inferred. Cowan's (1991) new map of the extensive cliff outcrops shows that the Westwater Canyon Member in fact consists of laterally extensive sandstone sheets that commonly have indistinct, feathered margins and contain considerable internal complexity, resulting from the migration of bar complexes in a braided fluvial environment. Internal channel scour was particularly important in this river system. As discussed in Sect. 8.2.14, these deposits are now regarded as an example of a distinctive fluvial style, that of the "high-energy, sand-bed braided river."

On a smaller, intrachannel scale, the stacking of channels and macroforms in braided systems is largely the product of autogenic avulsion of minor and major channels triggered by channel and bar aggradation. For example, Fig. 10.14 illustrates stacked sandstone sheets in the type section of the Castlegate Sandstone (Upper Cretaceous), near Price, Utah. Most sheets are composed of laterally and vertically amalgamated macroforms that were built by lateral to downstream accretion. As far as can be determined, both the stacking of macroforms and, at the next level, the stacking of sandstone sheets represent autogenic migration and switching (avulsion) of major and minor channels on the braidplain. Through the limited vertical thickness of this outcrop there are no discernible systematic changes in channel or bar style, or in orientation, that would point to some external control on sedimentation. Autogenic stacking of macroforms was explored in more detail in another outcrop of this unit, at Tusher Canyon, near Green River, Utah, where detailed paleocurrent analysis of cross-bedding and accretion-surface orientations permitted the reconstruction of three of the bar units in a vertical succession. This example is illustrated in Sect. 9.5.7 (Fig. 9.59).

The example of the Castlegate Sandstone illustrates the existence of autogenic processes on two scales, the intrachannel switching of bars and their resulting macroform deposits, and the avulsive switching of the channels themselves, typically as a result of aggradation. Campbell (1976) used a comparable subdivision of sandstone bodies in his architectural reconstruction of the Westwater Canyon Member in New Mexico, the larger-scale features of which have now been shown to be in error, as discussed above. His reference to sandstone sheets as the deposits of "channel systems" composed of stacked and amalgamated channel deposits has, however, been confirmed by Godin (1991), although in modified, reinterpreted form. Wizevich (1993) performed a similar type of cycle analysis on the Lee Formation (Pennsylvanian) of Kentucky.

An example of Godin's (1991) detailed vertical profiles through the Westwater Canyon Member is illustrated in Fig. 10.15. Small-scale fining-upward cycles 1–6 m in thickness are amalgamated into four major cycles over the total 41.9 m thickness of the entire section. This section constitutes a complete profile through a mappable middle submember of the Westwater Canyon Member. Godin (1991) defined three types of autogenic cycle, based on analysis of this and other profiles in the Gallup area of New Mexico (Fig. 10.16). The small-scale cycles, bounded above and below by surfaces of third order or higher rank, he interpreted as the fining-upward successions formed by macroform accretion or scour fill. Channel-scale cycles range from 1 to 10 m in thickness and are composed of stacked macroform cycles. They are bounded by surfaces of fifth order or higher rank. These cycles are not distinguished in Fig. 10.15. Sheet-scale cycles are 4–16 m in thickness (heavy

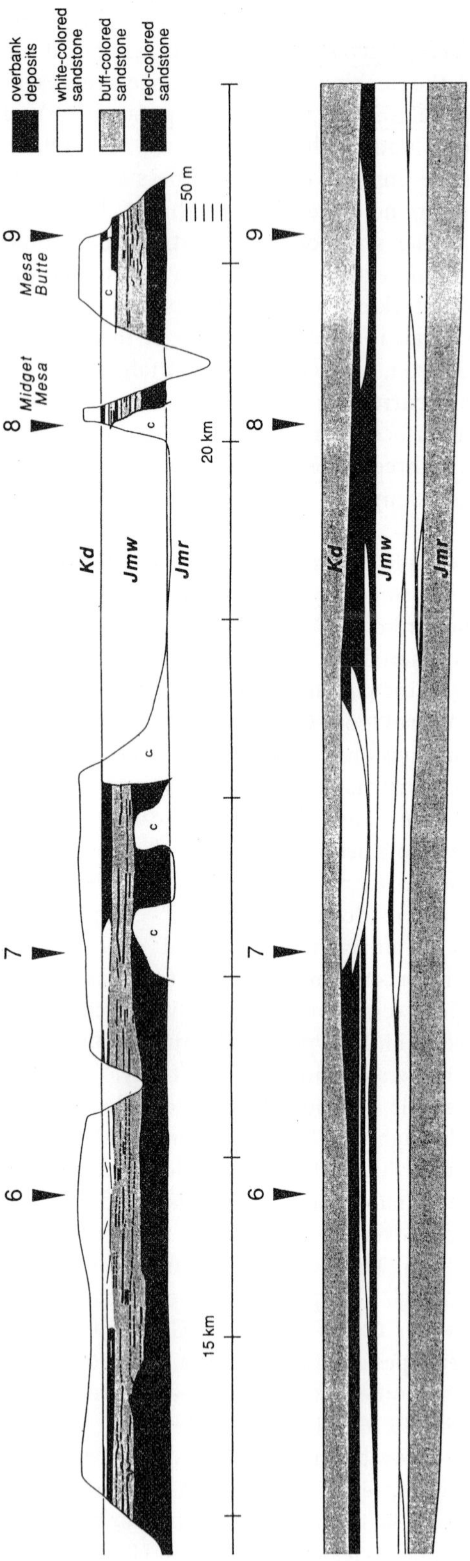

Fig. 10.13. Comparison of part of Campbell's (1976) reconstruction of the area between his sections 6 and 9, *below* (see Fig. 10.12 for location), and Cowan's detailed cliff map of the same exposures, *above* (Cowan 1991). Legend applies to Cowan's reinterpretation. Westwater Canyon sandstone bodies were not differentiated by color in Campbell's reconstruction. *Kd*= Dakota Sandstone, *Jmw*= Westwater Canyon Member, *Jmr*= Recapture Member

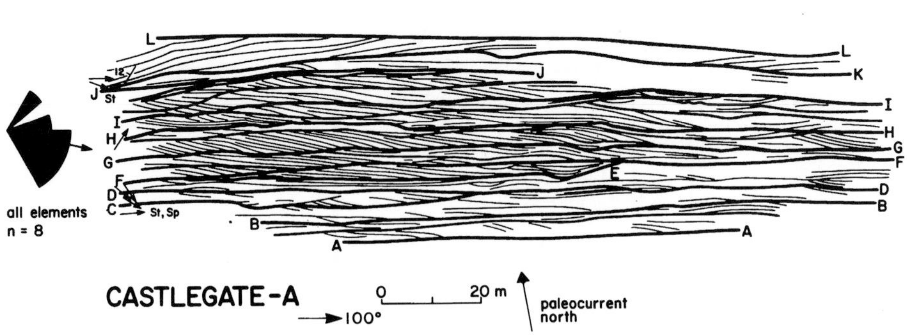

Fig. 10.14. Architectural analysis of part of the type section of the Castlegate Sandstone (Upper Cretaceous), near Price, Utah, showing subdivision of the unit into a series of stacked sandstone sheets, each of which is composed mainly of laterally to downstream-accreted macroforms. *A–L* Major (fifth-order) bounding surfaces. (Miall 1993)

arrows in Fig. 10.15) and may be hundreds of meters to several kilometers in lateral extent. Each type of cycle may be terminated at the top by fine-grained deposits representing bar-top deposits, channel-abandonment deposits, and floodplain deposits formed following channel-belt avulsion events. However, these fine-grained deposits commonly are absent because of scour accompanying the initiation of the overlying deposit. For example, the top of the first sheet-scale cycle in Fig. 10.15 is indicated not by fine-grained deposits but by a prominent intraclast lag breccia at the base of the next cycle. Amalgamated, multistory sand bodies showing little internal cyclicity are commonly the result, and distinguishing the various types of cycle from vertical profile data alone may be impossible.

Godin (1991) pointed out that because of the presence of several superimposed scales of cyclicity within the same sand body, the use of simple one-dimensional Markov-chain techniques for the analysis of cyclicity is not to be recommended. This technique was popular in the 1970s (e.g., Miall 1973; McDonnell 1978), but the advent of two- and three-dimensional architectural methods of analysis has revealed its inadequacy as a method of investigating the complexity of depositional processes in fluvial deposits.

A useful study of floodplain development in gravel-bed braided rivers was reported by Reinfelds and Nanson (1993). They mapped modern floodplains in the upper reaches of the braided Waimakariri River in New Zealand and showed that significant areas within the river valley are occupied by floodplain deposits (Fig. 10.17). These floodplain areas correspond in origin to the upper topographic levels of the Donjek River, Yukon, the type example of a gravel-bed braided river (Sect. 8.2.3).

Some stretches of the Waimakariri River valley are unoccupied by active channels for significant periods of time, for two main reasons. Mid-channel floodplain areas may represent large bar complexes that have accreted to the point that channel diversion around them conserves more energy than does their elimination by erosion (e.g., area A in Fig. 10.17). In other cases, floodplains develop in protected areas (e.g., area E in Fig. 10.17), although long-term migration of the channel complex may lead to periodic flushing out of any of these floodplain areas (Fig. 10.18). Aggradation of channels may lead to reoccupation of former channels on a floodplain surface, and this may contribute to the erosion of a long-term floodplain area.

The stratigraphy of floodplains in gravel-bed braided rivers consists of amalgamated units of channel-bed gravel deposits interbedded with thin units of floodplain fines, such as lithofacies Fl or Fm. The fine-grained units are a few decimeters thick, ranging up to a few meters thick where they fill abandoned channels. The lateral extent of these fine units is unlikely to be greater than a few hundred meters.

10.3.3 Avulsion in Anastomosed Fluvial Systems

The cause of anastomosis is generally agreed to be the low gradient of the river. Smith and Smith (1980) cited the backwater effect caused by the damming of a river by temporary obstructions, such as ice or log jams, or by more permanent damming, for example, by the progradation of a tributary alluvial fan across a confined valley. Immediately above such impediments the slope is reduced and crevassing and avulsion are frequent events. A rise in base level, or the

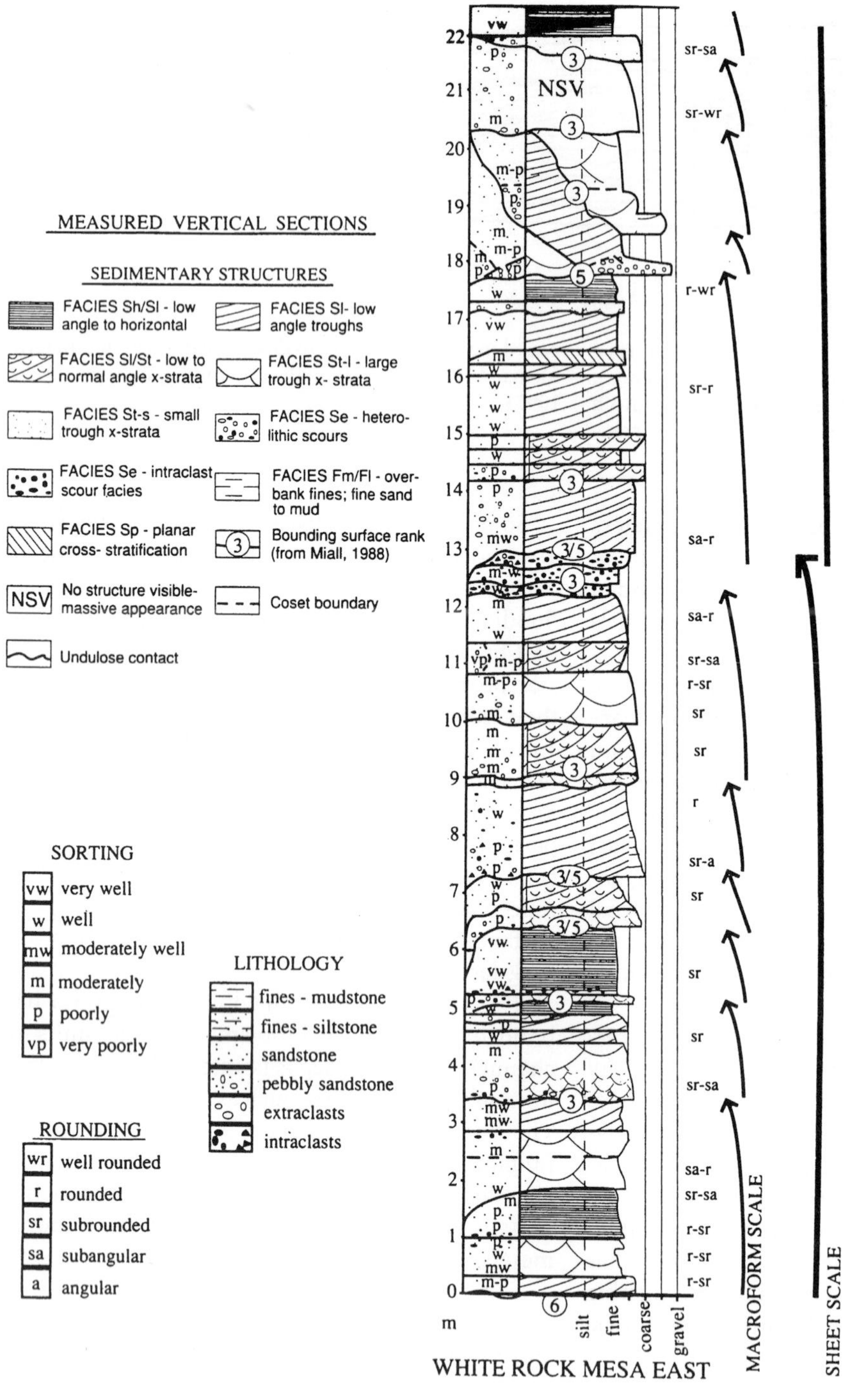

Fig. 10.15. Detailed vertical profile through the Westwater Canyon Member of the Morrison Formation (Upper Jurassic) at White Rock Mesa, near Gallup, New Mexico, showing the stacking of fining-upward cycles on at least three scales. (Godin 1991)

tectonic rejuvenation of a structural high crossed by the river, will also have the same effect of reducing slope (D.G. Smith 1983, 1986; Törnqvist et al. 1993). The last two controls are, of course, allogenic, not autogenic. Törnqvist (1994) demonstrated that in the Rhine-Meuse system of the Netherlands, the avulsion frequency has been about 1 ka during the past 10 ka, and that the number of active distributaries was greatest during the period 6–4 ka B.P., at a time when the postglacial sea-level rise was at its

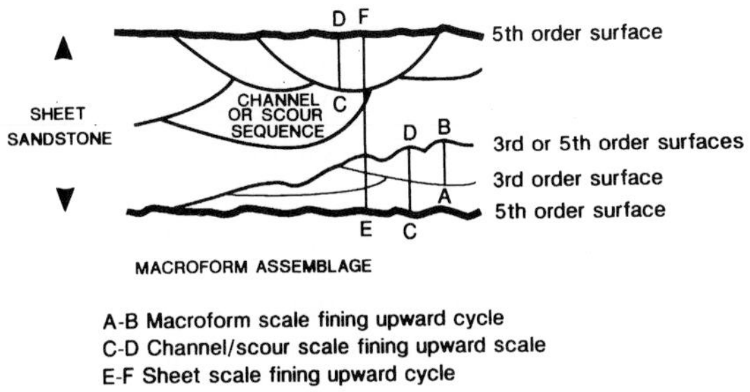

Fig. 10.16. Model of cyclicity in the Westwater Canyon Member of the Morrison Formation, based on vertical profiles, such as the one illustrated in Fig. 10.15. (Godin 1991)

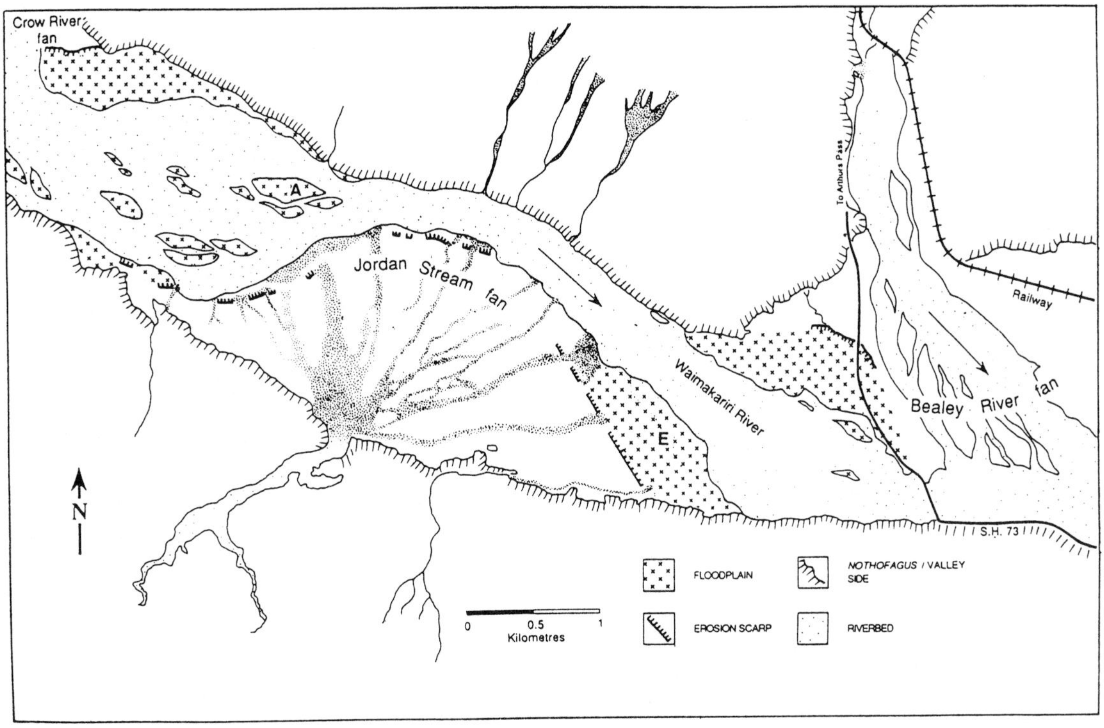

Fig. 10.17. Distribution of channels and floodplain areas within part of the Waimakariri River, New Zealand. Extensive areas of floodplain occur in unprotected midchannel location (*A*) and, more commonly, in areas protected by bedrock, bends in the valley, or projecting alluvial-fan deposits (*E*). (Reinfelds and Nanson 1993)

most rapid and the anastomosing fluvial style was dominant.

The Cumberland Marshes of east-central Saskatchewan have served as an excellent field location for the study of avulsion in anastomosed fluvial systems (Smith et al. 1989). This area, nearly 100 km across, consists of a complex of shallow lakes and anastomosed channels (Fig. 10.19). The Old Channel of the Saskatchewan River was dammed by ice in 1873, causing a diversion to the north, where a "New Channel" developed. Most of the other channels in this area are now largely inactive, with the exception of Carrot River, in the south. The New Channel has continued to evolve by processes of autogenic crevasse formation and avulsion, as described in Sects. 7.2.2 and 7.2.3. An example of the type of stratigraphy generated by avulsion in an anastomosed system is illustrated in Fig. 8.54.

10.4 Quantitative Studies of Alluvial Architecture

Information regarding the architecture of alluvial deposits is vital in the petroleum industry for the purpose of estimating reservoir volumes and petroleum productivity. There are three questions to be asked:

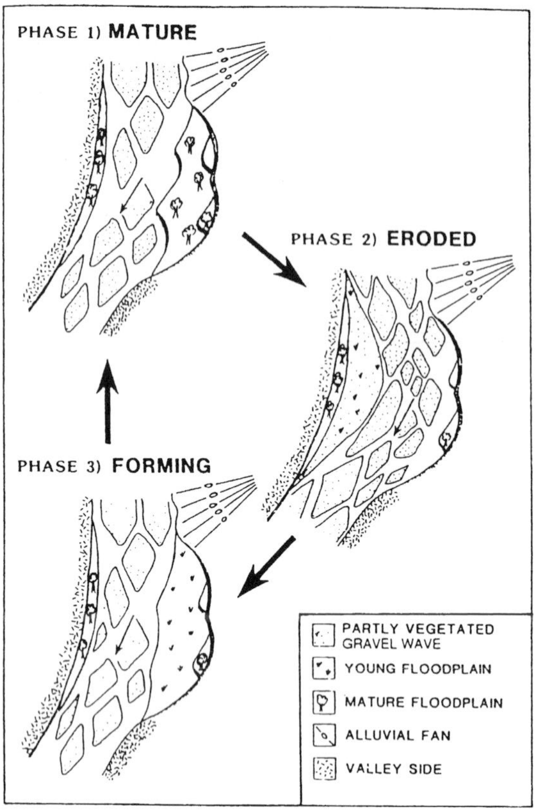

Fig. 10.18. Model for the development of floodplains in braided rivers. (Reinfelds and Nanson 1993)

1. What are the lateral dimensions of the sand body?
2. Does it contain internal heterogeneities, such as individual point bars, shale drapes, and abandoned channels, and if so what is their scale and disposition?
3. What is the connectedness of the sand bodies?

In this section we examine the various empirical studies and statistical and computer-based simulation models that have been developed to address these questions. We focus here on alluvial suites built by autogenic processes, those discussed in the first part of this chapter. The effects of tectonic and other allogenic modifications are discussed in Chap. 11.

10.4.1 The Dimensions of Fluvial Sand Bodies

Channel dimensions are set by external (allogenic) controls, including discharge and overall sediment supply, and by internal (autogenic) mechanisms that govern the geometry of the channel, including its width/depth ratio and sinuosity (Sect. 8.1). Channel-scale turbulence patterns and local bed and bank materials (within-basin autogenic controls) are largely responsible for determining channel form at

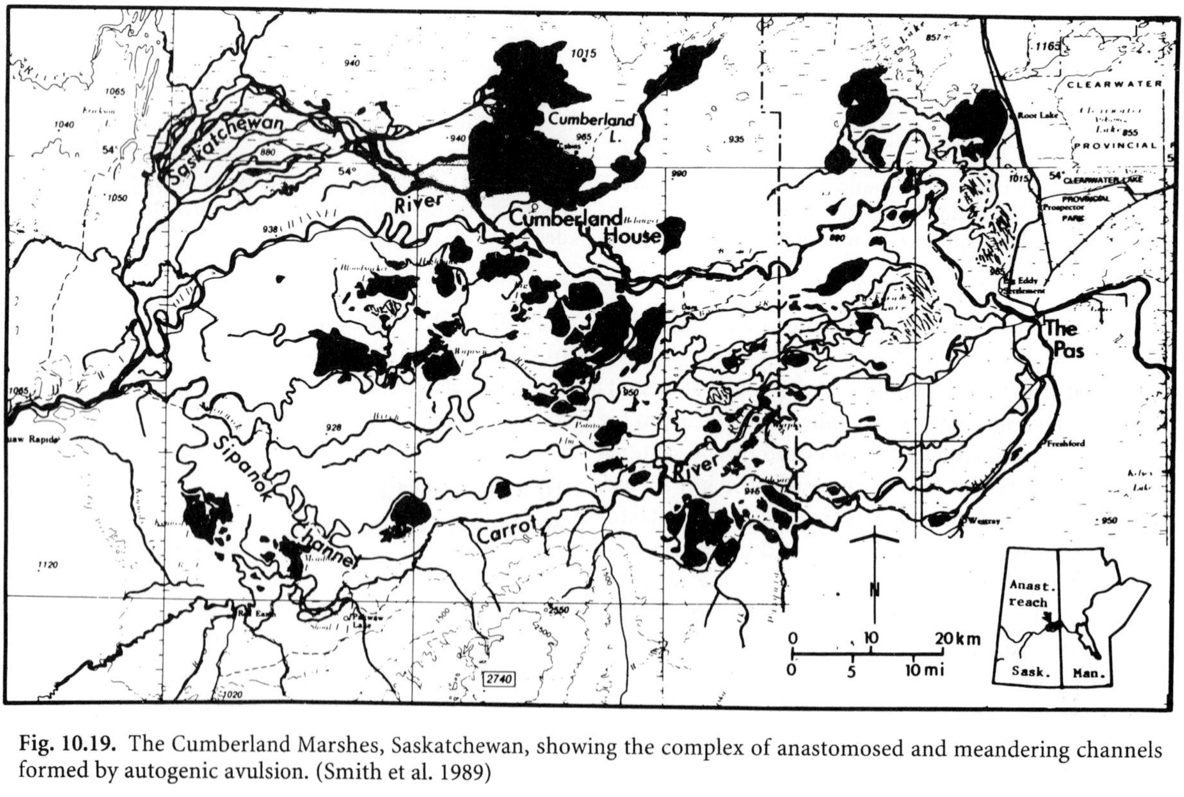

Fig. 10.19. The Cumberland Marshes, Saskatchewan, showing the complex of anastomosed and meandering channels formed by autogenic avulsion. (Smith et al. 1989)

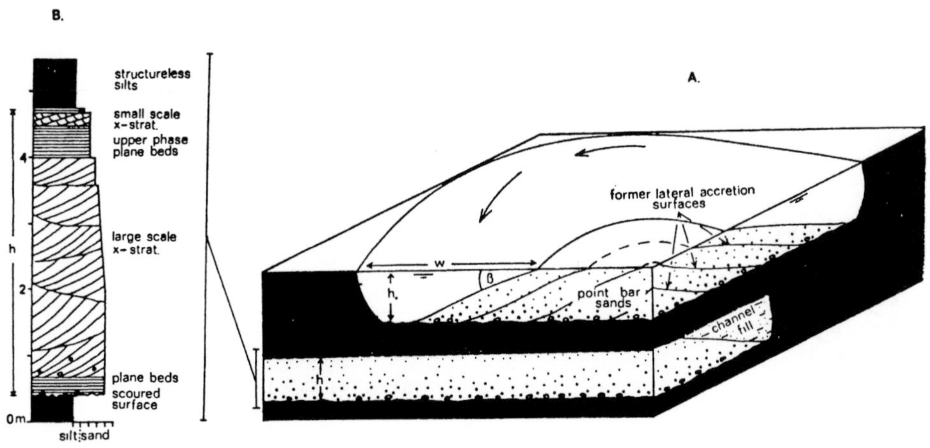

Fig. 10.20. Relationship of fining-upward cycle to depth of channel and thickness of point bar. (Leeder 1973)

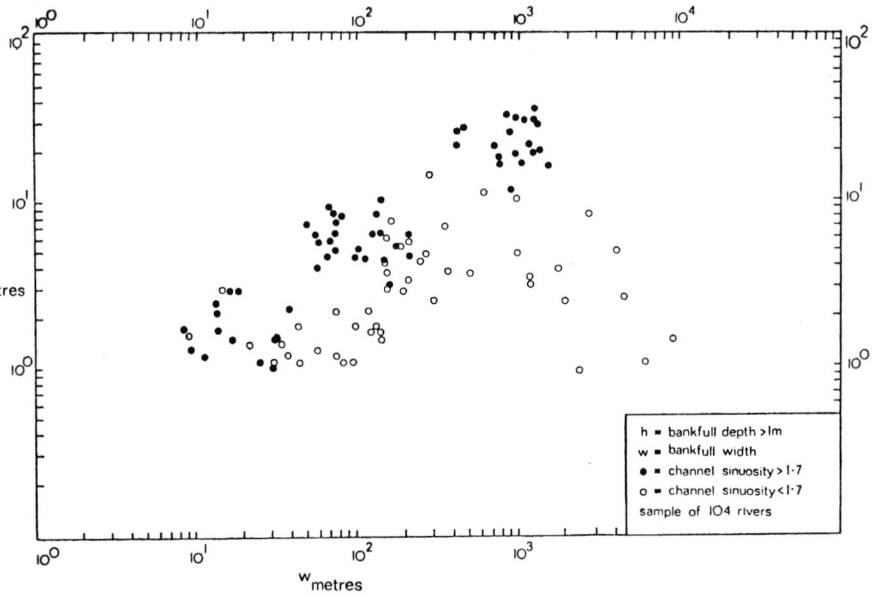

Fig. 10.21. Plot of 104 measurements of bankfull depth h against width w, for modern rivers. (Compiled from various sources by Leeder 1973)

a given location. Reconstruction of the local channel form is the subject of this section.

Vertical profile data from outcrops and drill holes provide information on sand-body thickness and internal lithofacies composition but yield no direct information regarding the lateral dimensions of the sand body. Where the outcrops are large and continuous it may be possible to map entire bodies, as described in an example below. However, this is rarely possible from subsurface data. Well spacing generally is inadequate for accurate sand-body mapping, except in cases of tight well spacing during enhanced recovery projects (see discussion of Mannville Sandstone in the Lloydminster area, Alberta, in Sect. 9.5.3 for an example of the difficulties encountered in such work).

Much work has been carried out to investigate the internal relationships between the various dimensions of fluvial channels, meander bends, and their major deposits. Measurements have been collected and tabulated for such data as channel width and depth, meander radius, point-bar dip angle, and the relationships between these data and sediment grain size and discharge. Two useful papers are those by Leeder (1973) and Ethridge and Schumm (1978). They showed that for typical meandering rivers

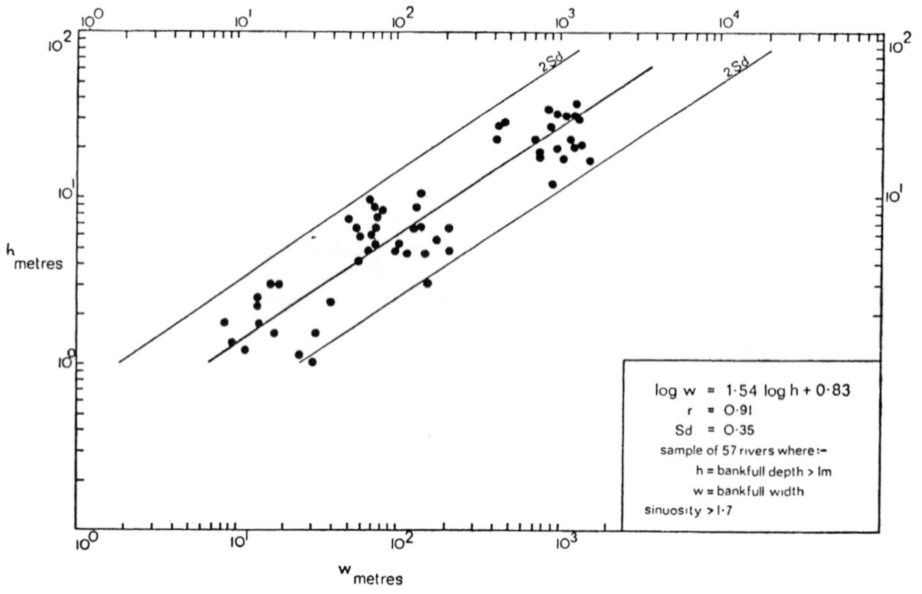

Fig. 10.22. Plot of 57 measurements of bankfull depth h against width w, for modern meandering rivers (sinuosity> 1.7). (Compiled from various sources by Leeder 1973)

some useful statistical relationships can be developed for predictive purposes. The thickness of a fining-upward cycle in the deposits of such rivers is approximately equal to the bankfull depth of the river (Fig. 10.20). The width of a point bar, as measured by the horizontal extent of a single accretion surface from top of bar to base of channel, is approximately two thirds of the former channel bankfull width (Leeder 1973). Unfortunately, accretion surfaces are rarely fully exposed in outcrop and are extremely difficult to reconstruct from subsurface data. As noted in Sect. 9.5.8, attempts may be made to use dipmeter data for this purpose, but results to date have not been promising.

Leeder (1973) compiled data to demonstrate the relationship between channel depth and width (Fig. 10.21) and showed that the width/depth ratio is greater for channels of low sinuosity, as expected. For meandering rivers (those of sinuosity > 1.7), Leeder (1973) was able to demonstrate a reasonably close statistical relationship between width and depth (Fig. 10.22). The relationship is:

$$\log w = 1.54 \log h + 0.83$$
$$\text{or } w = 6.8h^{1.54}$$

$$r = 0.91,\ Sd = 0.35 \text{ log units}$$

where w=bankfull width and h=bankfull depth. The value of this equation is that, using the 95% confidence limits, upper values of channel width can be estimated from cycle thickness. The scatter of the data reflects the fact that cycles may be incompletely preserved, and that cycle thickness will not necessarily be the same in every location around a meander bend. It should be noted that no comparable statistical relationship has been demonstrated for modern low-sinuosity (< 1.7) rivers, which are mainly those in the braided category.

If point-bar dip angle β can be measured in the field or obtained by identification of dipping second- or third-order surfaces in a core or dipmeter log, another method of estimating channel width is available (Leeder 1973). Width and depth are systematically related to this angle, as shown in Fig. 10.23. The relationship is:

$$w = 1.5\ h/\tan \beta$$

Point bars are discussed in detail in Sect. 6.7. Several studies of fluvial point bars in the subsurface are reported in Chaps. 14 and 15 and reveal the regular geometry of these bars. See, for example, the discussions of the Fall River Sandstone in Wyoming, the Tuscaloosa Sandstone in Mississippi, and the Cherokee Group in Missouri.

Leopold et al. (1964) demonstrated that channel dimensions are related to the scale of meander bends, through the equation:

$$Lm = 10.9w^{1.01}$$

where Lm=meander wavelength. Using this equation some estimate of the radius of a point bar can be

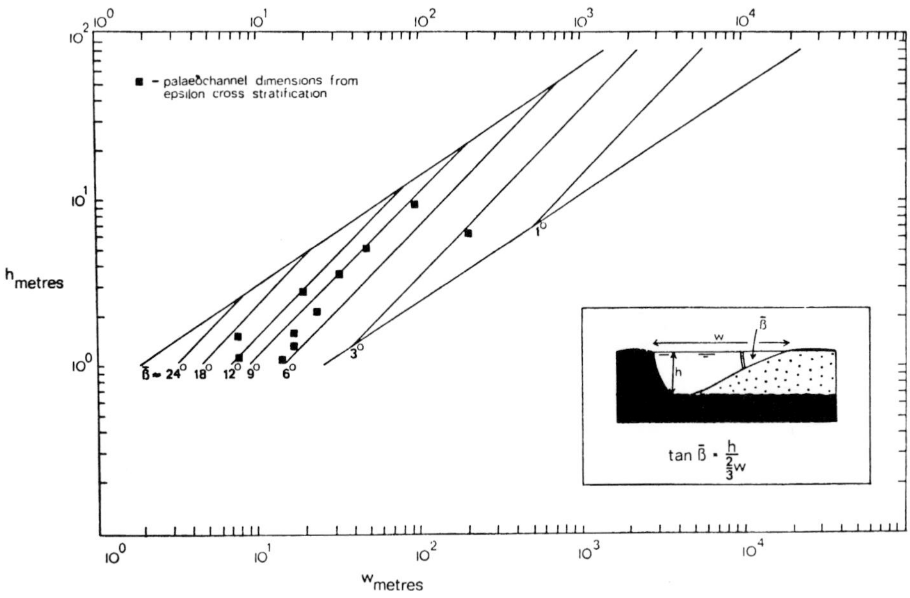

Fig. 10.23. Relationship between bankfull width and depth and point-bar dip angle, β. Note decrease of dip angle with increasing river size. (Leeder 1973)

obtained. This information may of considerable potential usefulness in attempting to determine the best location for step-out wells, when prospecting in point-bar sandstones. The user should be aware, however, of the built-in potential error in these estimates. Precise predictions cannot be made, and the technique is best suited for determining maximum possible dimensions of the various parameters of interest. Thus, there may not be a simple relationship between sand-body width and channel dimensions. As shown in Fig. 6.3, channels can aggrade both vertically and laterally, leading to sand bodies having very different width/depth ratios than the channel which deposited them. The sand body classifications dependent on geometry, such as that proposed by Friend (1983), should therefore be used with caution. An example of the variability in sand body geometries is illustrated in Fig. 6.4. Seventeen units bounded by fifth-order surfaces can be defined. These range in thickness from 5 to 43 m, with a mean of 16.3 m. The full width of many of these units is not exposed. They range from a minimum of 160 m to an exposed maximum of 860 m. The average exposed width is 379 m. Even if the true mean width is double this, the width/depth ratio is 46, which is low for a sandy-braided system.

Lorenz et al. (1985) suggested that, given average sedimentation rates and channel residence times, the width of a typical meandering-river sand body would be approximately equivalent to the amplitude of the meanders (the distance between two successive meander cutbanks measured perpendicular to the channel trend). This relationship would not apply where sedimentation rates are slow and the river has an extended period of time to comb back and forth across its own deposits. Lorenz et al. (1985) developed the following relationship between channel width and meander-belt amplitude (Wm), using data from Leopold and Wolman (1960) and Carlston (1965):

$$Wm = 7.44\, w^{1.01}$$

Assuming average conditions, this equation can be used to estimate the expected width of a sand body. This should be a useful technique for subsurface petroleum exploration, where information is needed on potential reservoir dimensions.

Collinson (1978) assembled data to explore the direct relationship between channel depth and meander-belt width. He was not able to be certain of eliminating some nonmeandering sheet sandstones from his data set, and the resulting equation shows some scatter:

$$Wm = 64.6 h^{1.54}$$

Lorenz et al. (1985) reported a test of his method, where both outcrop and core information were available to permit a cross-check between the various measurements and estimates. They measured cycle thickness in cores (Fig. 10.24), and plotted a

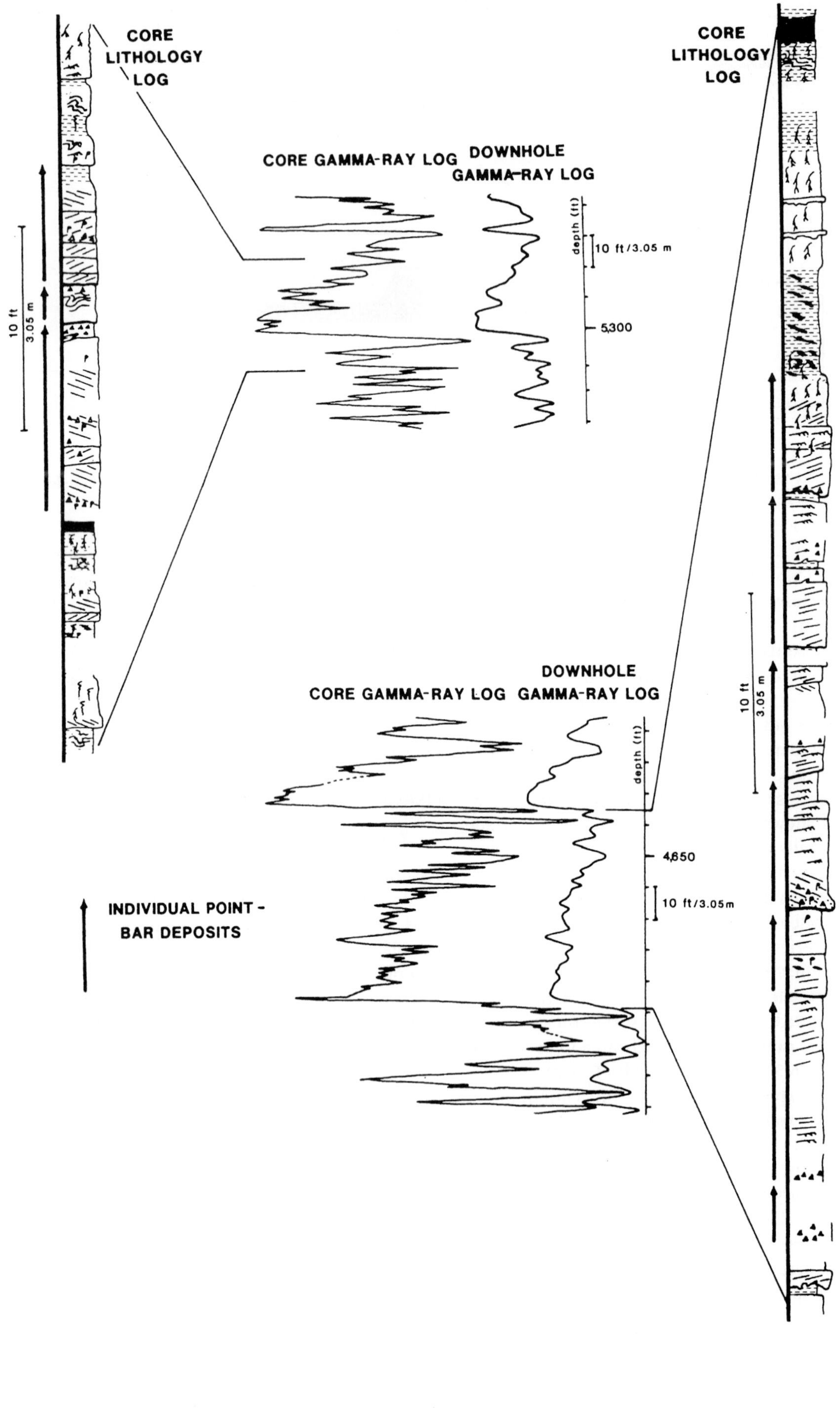
CORE
LITHOLOGY
LOG
CORE GAMMA-RAY LOG
DOWNHOLE
GAMMA-RAY LOG
depth (ft)
10 ft/3.05 m
5,300
10 ft
3.05 m
CORE
LITHOLOGY
LOG
CORE GAMMA-RAY LOG
DOWNHOLE
GAMMA-RAY LOG
depth (ft)
4,650
10 ft/3.05m
10 ft
3.05 m
INDIVIDUAL POINT-
BAR DEPOSITS

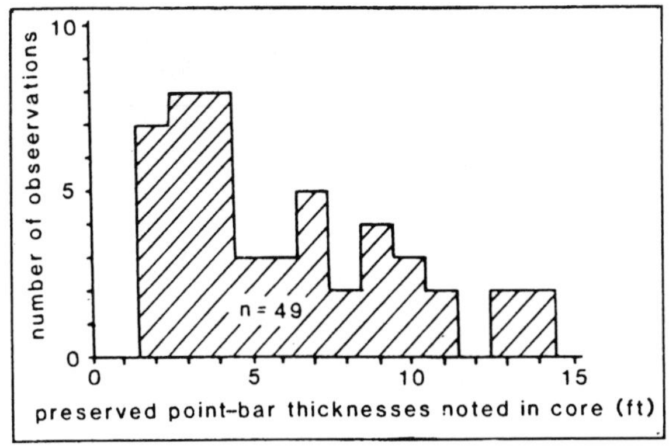

Fig. 10.25. Histogram of interpreted point-bar thicknesses in cores through the Mesa Verde Group, Colorado. (Lorenz et al. 1985, reprinted by permission)

histogram of the results (Fig. 10.25). Rare, complete fining-upward cycles (Fig. 10.24) provided a maximum thickness for the cycles, whereas many of the cycles had been truncated by internal erosion. The thickest cycle, at 4.3 m, was assumed to represent the high end of a normal distribution, with the lower end of the thickness distribution, at 2.7 m, suggested by the preservation of nontruncated cycles of that thickness. Using an estimated "typical" range of cycle thicknesses of 3–4 m, Lorenz et al. (1985) then used the equations given above to estimate channel width and meander-belt amplitude (=reservoir sandstone width). They arrived at a range of 350–520 m for predicted meander-belt width.

Nearby outcrops of the same stratigraphic unit enabled Lorenz et al. (1985) to test their estimates. Using air photographs, they measured the widths of meander-belt sandstone lenses, making allowances for the angle of the exposure and the orientation of the bodies relative to the outcrop face. This procedure yielded a range in width values of 60–2130 m, with a mean of 316 m, a standard deviation of 275 m, and a median of 250 m. In some cases point bars with lateral accretion surfaces were exposed. Lorenz et al. (1985) measured the width of these, multiplied these values by 1.5, and used the resulting estimated channel widths to determine meander-belt widths. The resulting estimated range of values is 430–650 m. These three methods of determining reservoir width yielded reasonably consistent results, and Lorenz et al. (1985) demonstrated their applicability to a few other cases described in the literature.

Davies et al. (1993) used the estimation methods described here to predict the scales of sand bodies deposited by high- and low-sinuosity rivers in a subsurface example in Texas, but no independent data from outcrop architectural studies or closely spaced wells were available to test the estimates, and it is risky to base predictions of fluvial style and sand body geomtry on a few examples of vertical profiles in wireline logs or core, as in this study. Thus, sheet sandstones are commonly the product of braided fluvial systems, as might be expected from the large width/depth ratio of such rivers, but can also be deposited by high-sinuosity rivers. For example, the width of the braided Brahmaputra River valley reaches 14 km, with an active channel up to 3 km wide (Coleman 1969). However, Bristow (1987, 1993a) showed that lateral accretion of sand flats is one of the dominant modes of bar construction in this river. Sheets consisting mainly of lateral accretion deposits (element LA), with occasional incised minor channels (element CH) are locally common and indicate an abundance of sinuous channel reaches. This is but one example to demonstrate that the presence of lateral-accretion deposits is not a reliable indicator of a meandering style and does not automatically permit the application of the equations discussed above. Stear (1983) described examples of sandstone sheets with LA deposits more than 250 m wide in the Beaufort Group of South Africa (Figs. 8.41, 8.42) that were deposited in an ephemeral meandering system. In the Dakota Sandstone of New Mexico and Arizona, similar sheets exceed 1 km in width. R.M.H. Smith (1987) described an exhumed sandstone sheet at least 6 km across, consisting of the preserved point bar and associated facies formed within four successive meander bends of a major high-sinuosity stream in the Permian of South Africa. The width of the meander belt in this example is approximately 3 km.

Fielding and Crane (1987) assembled data from numerous studies, including those described above, and plotted a graph in which five distinct styles of sand body geometry could be evaluated (Fig. 10.26). Their case 1A represents streams with the lowest width/depth ratio and is limited to incised, straight, nonmeandering rivers. Case 1B represents the upper bounding line for meandering channel deposits. Case 2A is the best-fit line for all the data and represents a geometric mean. This relationship could be

Fig. 10.24. Determination of fining-upward cycle thickness from core and wireline log data. Mesa Verde Group (Cretaceous), Colorado. (Lorenz et al. 1985, reprinted by permission)

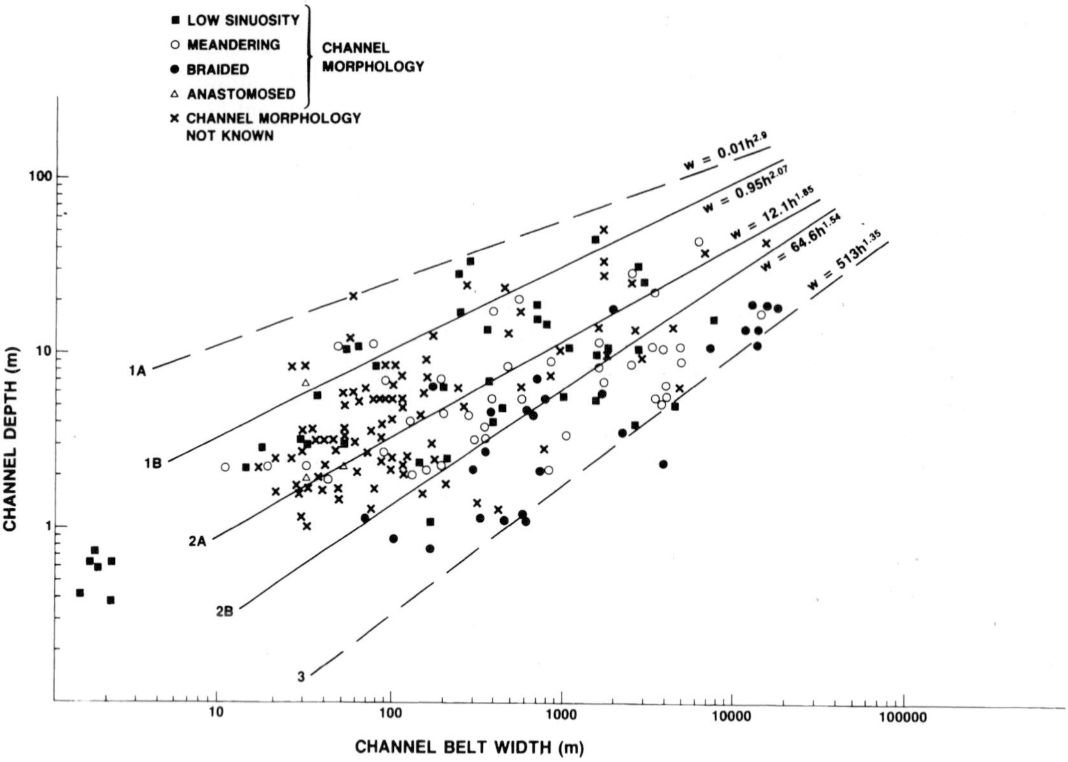

Fig. 10.26. Plot of channel depth, h, versus sand body width, *Wm*, for various types of modern and ancient fluvial systems, using 45 published sources. (Fielding and Crane 1987)

used where a variety of channel types are suspected to coexist, or where the formative channel type is not known. Case 2B is Collinson's (1978) relationship for typical, "fully developed" meandering streams. Case 3 is the lower relationship for typical, "fully developed" meandering streams. Case 3 is the lower bounding line for all the data and represents sand bodies deposited by laterally unrestricted rivers, such as braided rivers.

In recent years a great deal more field work has been carried out on the architecture of exposed fluvial sandstone bodies in order to improve the data base for subsurface modeling and prediction. A Norwegian program, led by T. Dreyer, was initiated following the discovery of large petroleum reserves in fluvial reservoirs in the Norwegian sector of the North Sea (Dreyer 1990, 1993a,b; Dreyer et al. 1993). Dreyer and his colleagues examined several fluvial and deltaic units using the kinds of architectural techniques outlined in Chap. 4. They reconstructed fluvial styles, examined the control of climate and tectonism, and compiled data on sandbody geometry at several scales of the architectural hierarchy. Another Norwegian group examined the exposed fluvial crevasse-splay deposits of the well-known Ravenscar Group of Yorkshire (Mjøs et al. 1993). Examples of the type of data collected by Dreyer (1993b) are illustrated in Fig. 10.27. These data were measured in an ephemeral system, in which sheet and ribbon sandstones were deposited on an arid fluvial plain and frontal splay sheets were formed at the mouths of incised channels. Other quantitative studies of sand body dimensions were reported by Cuevas Gozalo and Martinius (1993).

10.4.2 Estimating Probabilities of Sand Body Penetration and Interconnectedness in the Subsurface

Figures 10.28 and 10.29 illustrate the problem to be addressed by exploration geologists attempting to determine the location of sand bodies in the subsurface. The success rate of exploration wells depends on the width and spacing of the sand bodies. The degree of interconnectedness of the sand bodies will affect their petroleum productivity and determine the strategies to be used for field development. The first important parameter, that of the scale of the sand bodies, has been addressed in the previous section. Here, we review briefly the work that has been done to explore the distribution of sand bodies within a basin-fill.

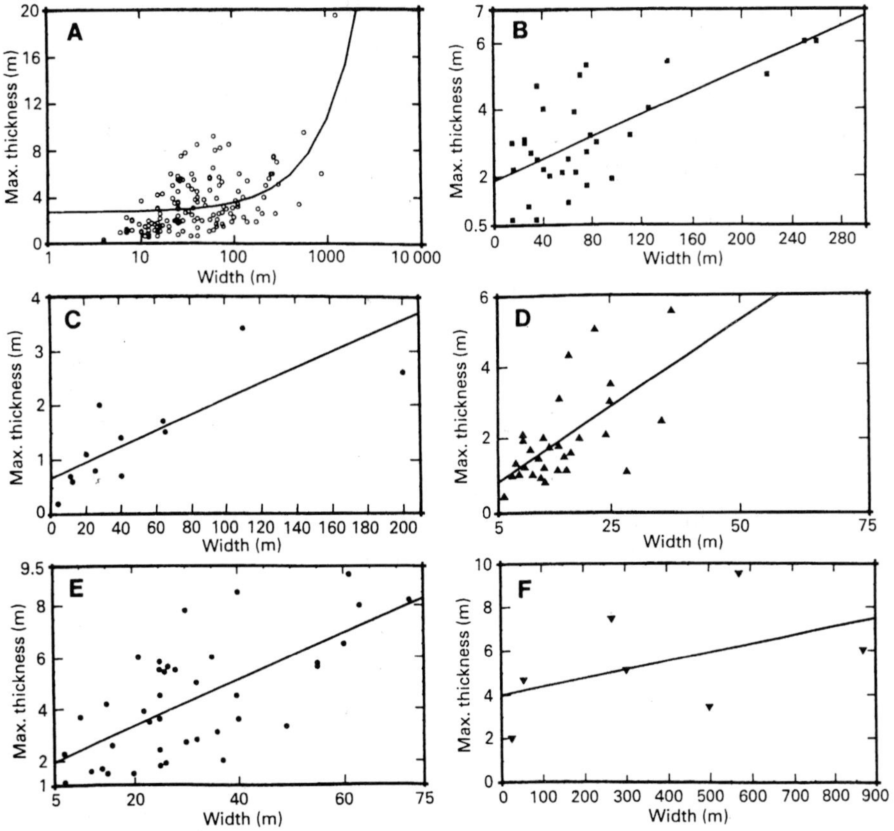

Fig. 10.27A–F. Plots of thickness versus width for various types of sand body in the ephemeral Esplugafreda fluvial system (Paleocene), northern Spain. **A** All coarse-member sand bodies; **B** channel-to-sheet transitional bodies; **C** frontal-splay sheetflood sand bodies; **D** single-story ribbon bodies; **E** multistory ribbon bodies; **F** multilateral channel bodies. (Dreyer 1993b)

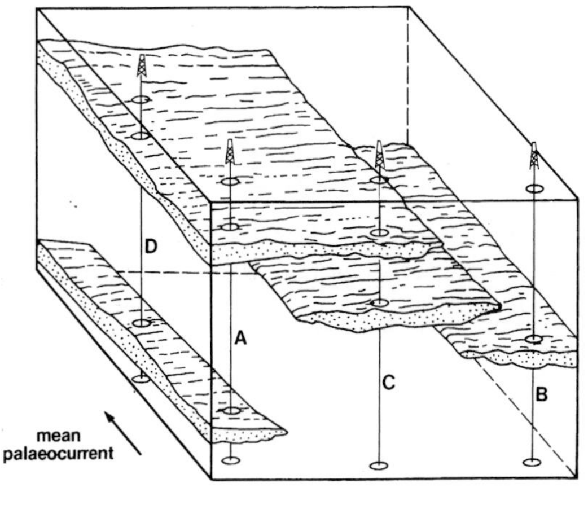

Fig. 10.28. The importance of well spacing in effective exploration for lenticular sand bodies. Where only wells A and B have been drilled a sand body of intermediate depth will not be discovered. (Fielding and Crane 1987)

Leeder (1978), Allen (1978), and Bridge and Leeder (1979) developed numerical models which explored the size and spacing of meander-belt sand bodies within a fluvial basin fill, depending on such factors as subsidence and sedimentation rates and the nature of the channel-shifting patterns, specifically, steady meandering versus sudden switches in position (avulsion). Given input data from preliminary studies within a basin, it may be possible to employ these models to predict sand body distribution and well-penetration probabilities.

Fielding and Crane (1987) developed a statistical model for prediction of well-penetration success rates. They proposed that first the channel sand bodies be identified in each well log by analysis of lithofacies characteristics and wireline-log response. The use of wireline logs is discussed in Sect. 9.5.2. An analysis of fluvial style must also be carried out so that the appropriate case from Fig. 10.26 can be selected. The thickness of each identified channel sand body is then multiplied by its predicted width, using the case selected from Fig. 10.26. In this way the cross-sectional area of all channel sand bodies in the well can be calculated and totaled. Given that the wells already drilled in an area have penetrated a

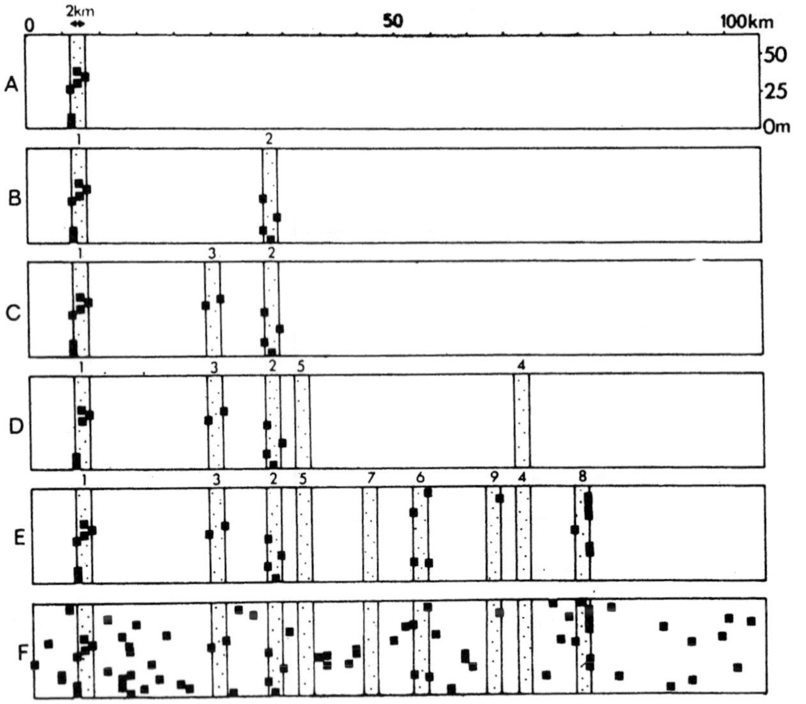

Fig. 10.29. A–E Results of successively denser sample spacing in the exploration of a fluvial section 100 km across. Black squares indicate location of channel bodies. *Stippled areas* represent areas of outcrop or well sections. The location of all channel bodies is given in F. Wells *4*, *5* and *7* fail to find any sandbodies because of the random spacing of these units, but such failure means little in terms of the overall distribution of the sandstone bodies. (Leeder 1978)

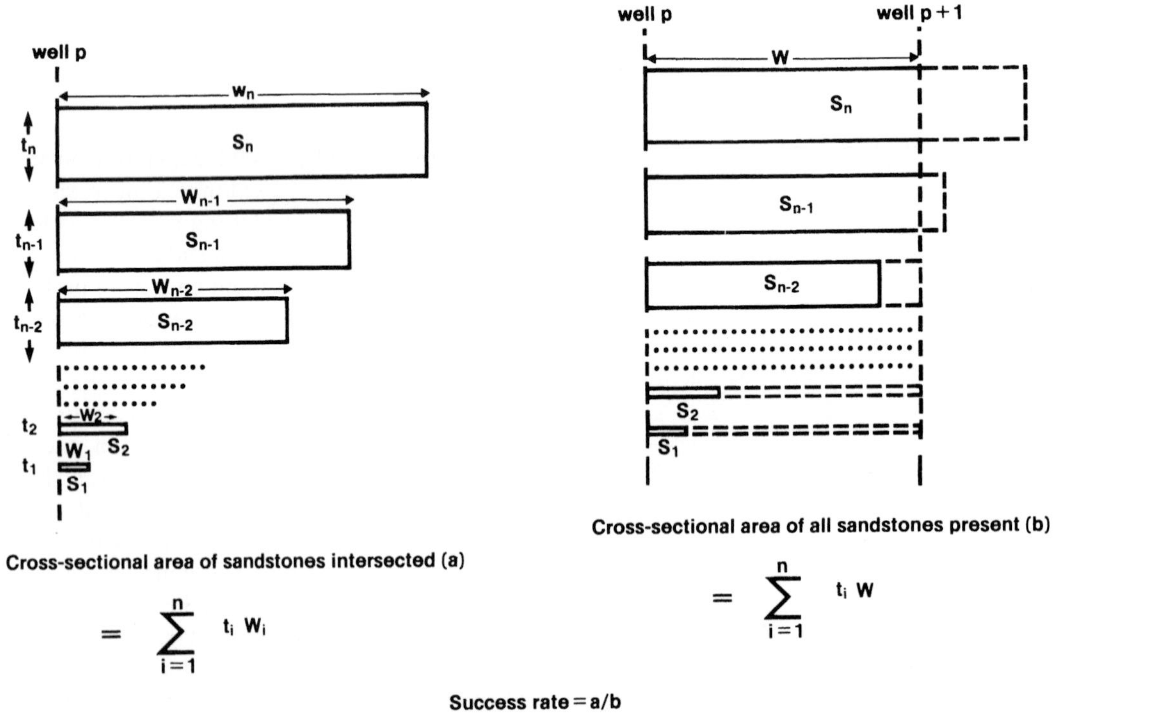

Fig. 10.30. Statistical description of sandstone beds. In average well *p* there are *n* sand bodies having thicknesses t_n and widths w_n; the latter values are estimated from Fig. 10.26. In well *p+1* the cross-sectional area of all sandstones present can be predicted based on well spacing. The well "success rate" is the proportion of the total cross-sectional area that is penetrated by the development wells. (Fielding and Crane 1987)

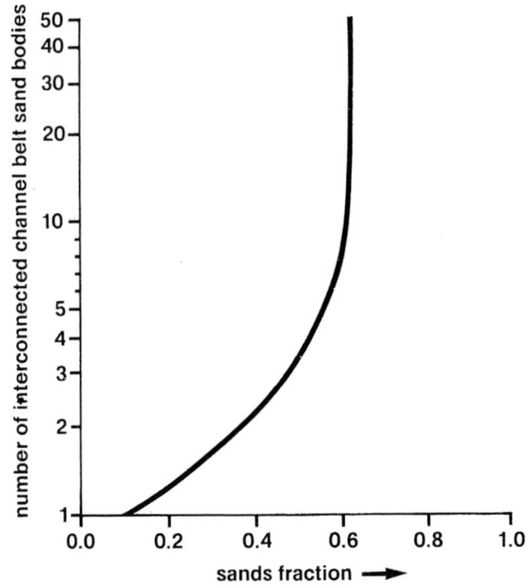

Fig. 10.31. Sand body interconnectedness as a function of the proportion of sand in the section. (Fielding and Crane 1987)

representative succession, then any hypothetical well drilled through the interval of interest will penetrate a set of sandstone beds equivalent to those found in the test well. A "success rate" can be defined, which is the proportion of the total channel-belt sandstone volume in a given interval that will be penetrated by planned development wells a distance W apart (Fig. 10.30).

Theoretical simulation studies (Fig. 10.31) suggest that channel bodies are effectively in isolation if they comprise less than about 50% of the succession (Allen 1978; Fig. 10.31), although interconnectedness may be higher locally where sand body distribution is not entirely random but is constrained by local tectonic factors, or by such factors as bank stability (e.g., presence of clusters of resistant shale plugs filling abandoned channels, which resist lateral channel migration).

Estimates of sand-body interconnectedness are one of the important outputs of the computer simulation models discussed in the next section.

As noted by Bridge and Mackey (1993b), great care must be employed in attempts to use outcrop data as a basis for statistical estimates or modeling of a subsurface fluvial unit. The exercise will be misleading if the fluvial systems in outcrop and the subsurface represent different fluvial styles, have different channel scales, and were deposited in different tectonic and climatic regimes. Unfortunately, many of the observational details that are used to interpret these autogenic and allogenic controlling parameters are very difficult, if not impossible to make from reflection-seismic or wireline data, and even core data provide very limited information on three-dimensional sand body architecture. Bridge and Mackey (1993b) also pointed out the inherent variability in fluvial systems and the consequent statistical error in the approaches summarized in this and the preceding sections. Most of the data sets from which the various equations have been derived represent limited suites of measurements collected in quite specific fluvial settings. The methods may therefore be used to suggest limits, or ranges, for sand body dimensions and relationships but should not be relied upon to provide precise estimates.

10.4.3 Alluvial Stratigraphy Models

The processes by which fluvial channel and floodplain deposits are stacked to form stratigraphic units have been the subject of several specialized studies over the past two decades. As noted by Friend (1983), Allen's (1965) speculative diagrams (Fig. 2.17) were very influential, because they represented the first attempt to interpret the details of alluvial stratigraphy in terms of autogenic and allogenic sedimentary controls. Allen (1974b) returned to this subject in an evaluation of the architecture of the Old Red Sandstone in South Wales, in which he drew attention to the significance of pedogenic carbonate units as indicators of periods of nonsedimentation in interfluve areas and as marker beds that could be used for correlation within suites of laterally impersistent channel and floodplain deposits. Based on qualitative, deductive reasoning, Allen (1974b) developed five distinct models, with additional variations bringing the total to eight. These represent essentially abstract models in which an interpretation of the Old Red Sandstone was an almost incidential byproduct of the analysis. Two of the models are reproduced here because of their historical interest (Fig. 10.32). The first represents a purely autogenic model, and one which has now been produced by computer simulation, as discussed below. The second represents, in effect, the first model of fluvial sequence stratigraphy, as discussed in Chap. 13.

Attempts to quantify these models were made by Allen (1978) and Leeder (1978). Both authors make reference to the need to quantify sand body geometry and architecture as an aid to the study of hydrocarbon reservoirs.

The generation of alluvial stratigraphy models by computer simulation was pioneered by Bridge and Leeder (1979). This is a much-quoted paper that was

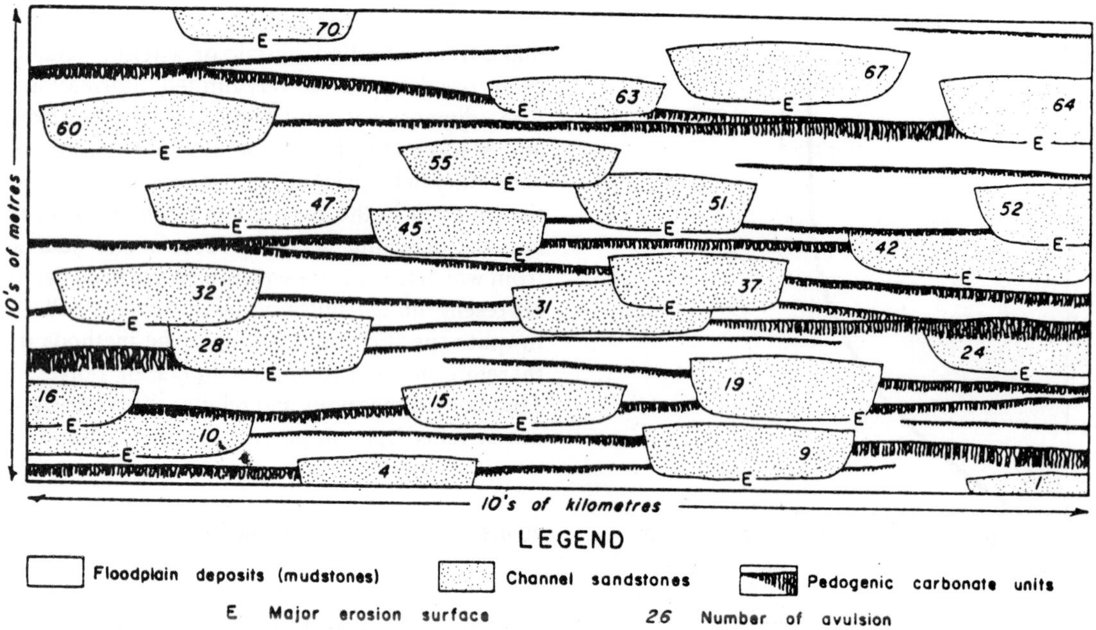

A

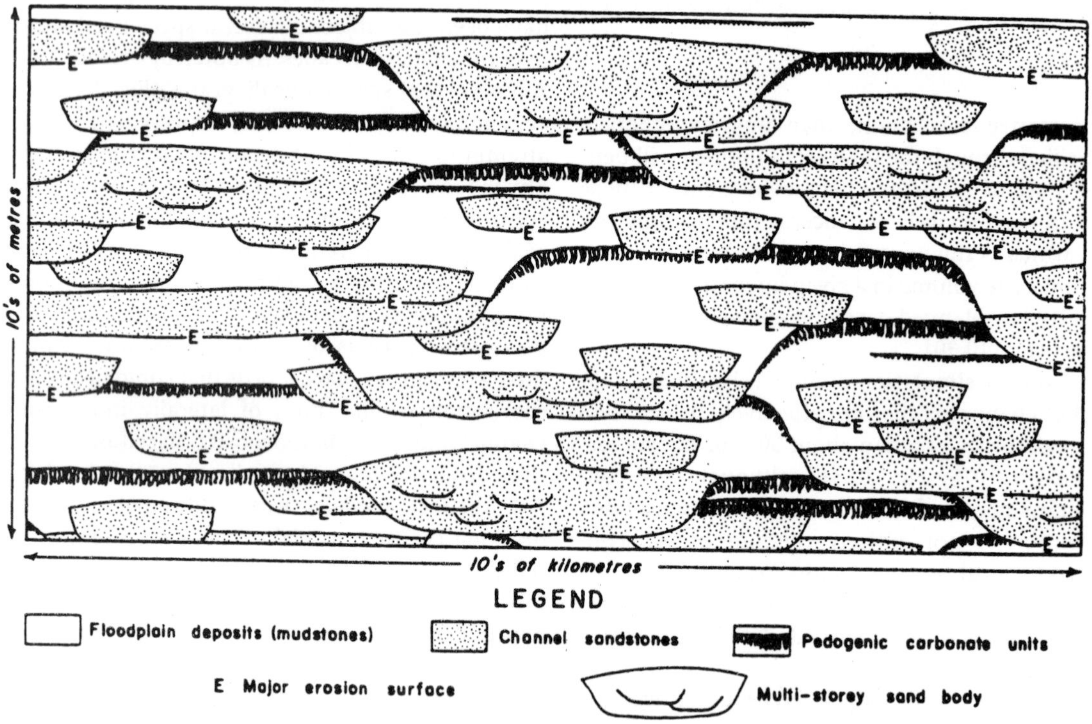

B

Fig. 10.32A,B. Two of Allen's (1974b) qualitative models of alluvial stratigraphy. **A** Model 3, developed by autogenic avulsion of the channel belt. **B** Model 5B, developed by a combination of autogenic channel avulsion and a cycle of dissection and aggradation, in which the base level changes over a vertical interval greater than channel depth

built on an analysis of the mechanics of channel-belt and floodplain construction and a systematic documentation of the rates and scales of various autogenic sedimentary processes. The main purpose of the model is to build cross sections of alluvial stratigraphy in an orientation perpendicular to paleoslope. The model is quantitative, in that magnitudes, rates, and scales are stated, and many of the parameters are designed as variable input for successive computer runs to permit an exploration of the complex interaction between dependent and independent variables. Channel-belt sand bodies are

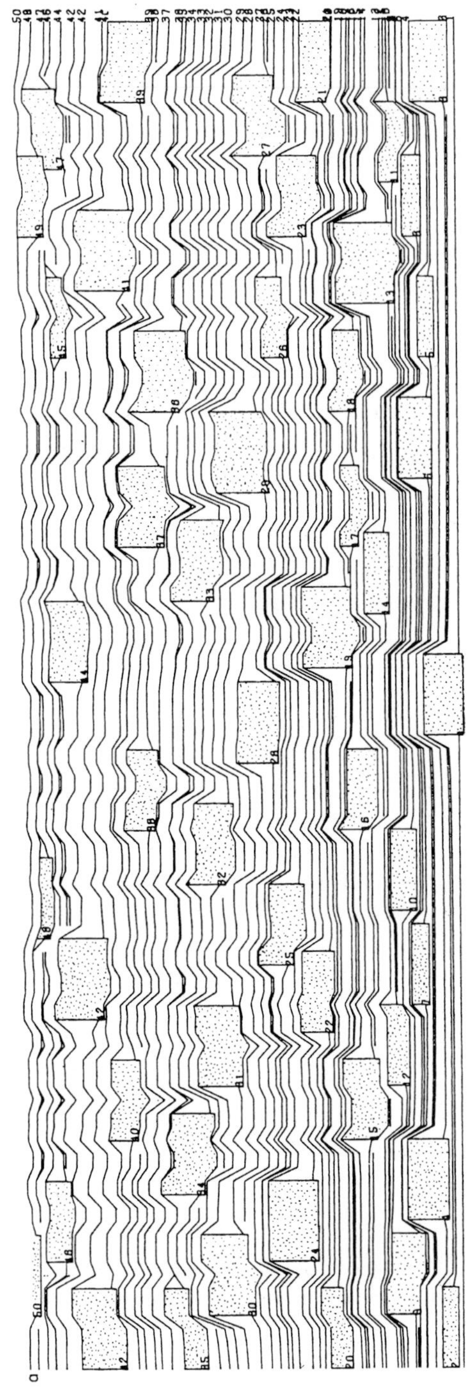

Fig. 10.33. Example of a simulated cross section of alluvial stratigraphy constructed using the numerical model of Bridge and Leeder (1979). The cross section is 10 km wide. Channel-belt aggradation rate is 20 m/ka. Channel width is 600 m and depth is 3 m

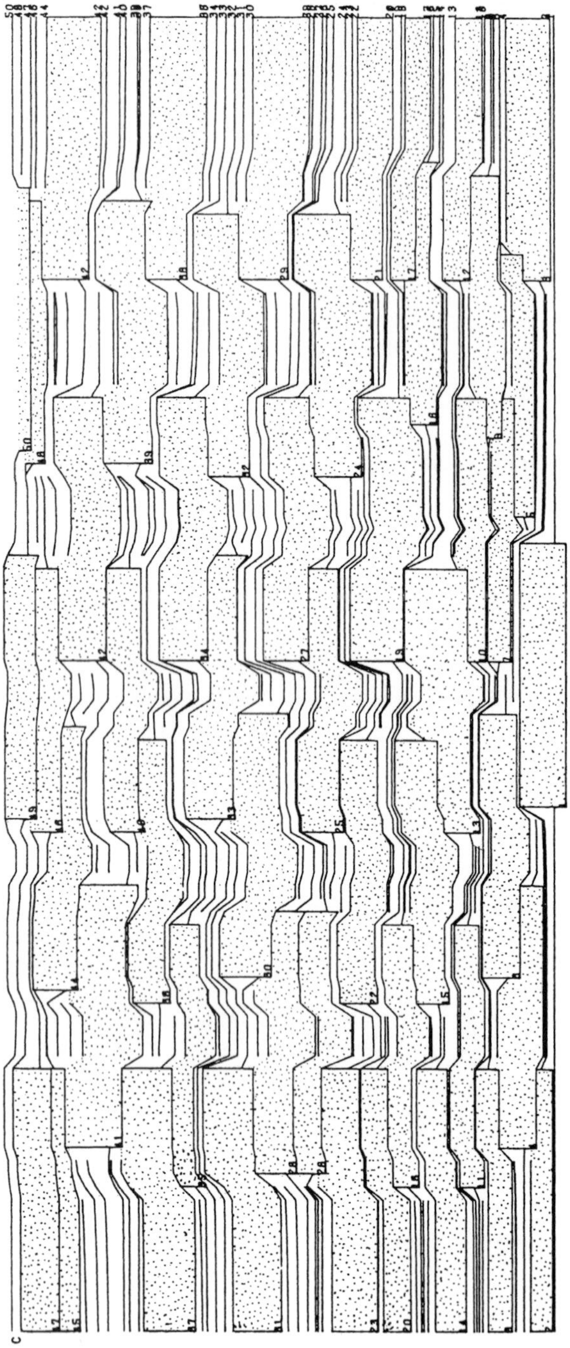

Fig. 10.34. Another model run from the Bridge and Leeder (1979) experiments. In this case channel units are set at 2 km wide and 7 m deep

assumed to be constructed by any appropriate fluvial style, following the hypothesis that the two major styles, meandering and braided [the "mobile channels" of Friend (1983)], both construct complexes of bars and minor channels within relatively stable meander belts. This is certainly the case for meandering systems, in which lateral migration is constrained by the alluvial topography of levees and the resistance met by channels when they erode laterally into the fine-grained deposits of abandoned channels and the floodplain. The assumption of a confined channel belt may not be so appropriate for braided systems.

The alluvial stratigraphy models discussed here are autogenic, in that the external, allogenic variables that control the long-term behavior of the river system are set at constant background values. Tectonism and sea-level change are not assumed to affect the system, so that average accumulation rates and regional paleoslope remain constant throughout a model run. Climate change and its effect on discharge, fluvial style, and sediment supply are also not considered. Bridge and Leeder (1979) were among the first to consider the quantitative effects of tectonism on fluvial aggradation, by building in the ability in the computer model to tilt the depositional surface in a direction perpendicular to paleoslope. This aspect of the model is noted in Sect. 11.2.1.

The most important controlling factor in autogenic models of alluvial stratigraphy is the determination of the rate and style of avulsion. As noted earlier in this chapter, little is known about the detailed avulsion history of most rivers, and so some generalizing assumptions have to be made. Leeder (1978) and Bridge and Leeder (1979) assumed in their models that avulsion is not dependent on the existing position of the channel within the model, but occurs sufficiently far upstream from the location of the model that the channel is free to move to the current lowest position on the floodplain. Avulsion frequency was determined by reference to the limited information on stratigraphic and historical data (Sect. 10.3), and the recurrence interval was assumed to follow a distribution function similar to that of earthquakes, which in any location of tectonic stress become more likely to occur as time passes. Mean avulsion periods ranging from 111 to 1780 years were used by Bridge and Leeder (1979).

Given an avulsion event, the channel is moved to the lowest point on the floodplain and allowed to accumulate a channel sand body at a preset accumulation rate, ranging from 5 to 40 m/ka. Floodplain accretion is accomplished using a simple equation that reduces the aggradation rate across the width of the model cross section in proportion to the distance from the channel belt. This reflects the fact that floodplain sediments are derived primarily from overbank flooding. Compaction of the accumulated deposits is accomplished, layer by layer, using preset compaction factors. This process generates relief on the floodplain surface because floodplain fines compact less than channel sands, and the least compaction occurs where channel sands are stacked vertically. The model therefore builds in a feedback effect, in that the site of the next lowest position for channel occupancy will be determined by the outcome of the preceeding sedimentation-compaction history. Two examples of Bridge and Leeder's (1979) model are illustrated here (Figs. 10.33, 10.34). The most realistic of these is the first, in which sand body distribution appears to be relatively random after the initial ten avulsion events. The second model shows a much too regular stacking of the channel bodies, an effect that occurs when channel width and depth are set too large relative to aggradation rate. The effects of the stacking and compaction of numerous preceding events are downplayed in this model, and successive channel positions are determined primarily by the position of the two or three immediately preceding channels. The result is a distribution of channel bodies not unlike that of bricks in a wall. Nevertheless, the Bridge and Leeder (1979) model has been much used, particularly by petroleum geologists, because of the insights it yields into channel stacking patterns and interconnectedness, factors of considerable importance in the understanding of reservoir predictability and fluid migration behavior. The model has recently been expanded and updated by Bridge and Mackey (1993a), but the improvements are at a level of refinement that does not need to be addressed here.

The application of these model concepts to the interpretation of real stratigraphic data can be a complex problem. Willis (1993b,c) provided extremely detailed architectural descriptions of the Siwalik Group in a segment of the Himalayan foreland basin in Pakistan, including quantitative studies of a 2-km-thick section keyed to the magnetostratigraphic time scale. Variations in sedimentation rates, sand body thickness and sand body density cannot readily be interpreted using simple autogenic models because of lack of knowledge of the number of rivers in the basin and a lack of detailed information on local tectonism.

Quite different approachs to modeling have been taken by other researchers. Several modeling

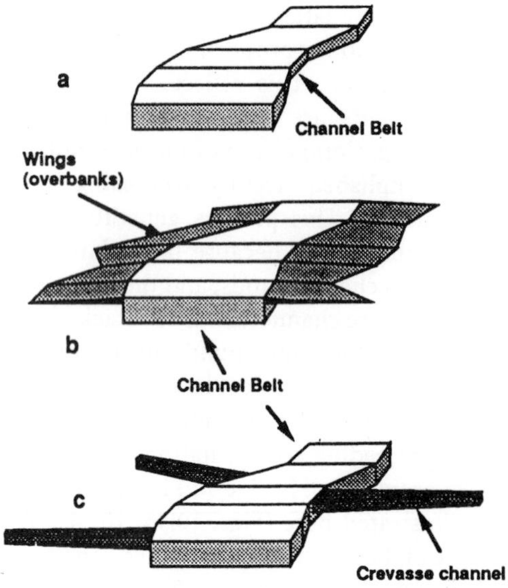

Fig. 10.35. Simulated channel belts, built by a stochastic computer model. Each belt is composed of rows of "boxes", with the size and shape of the boxes and the sinuosity implied by the nature of the contacts between them all determined by stochastic sampling from built-in sampling distributions. Crevasses and levees may be specified. (Hirst et al. 1993)

programs have been developed by petroleum companies to generate realistic sand body architectures for partially known petroleum reservoir intervals, in order to aid in the determination of essential reservoir data, such as total sandstone volume and sand body interconnectedness (e.g., Hirst et al. 1993; several papers in Buller et al. 1990). The nature of the depositional process is not the primary concern in these studies, and many of the models are simply designed to simulate sand body geometries using a stochastic approach, rather than building a stratigraphy layer by layer using a mainly deterministic model based on actual depositional processes, which is the approach used by Bridge and Leeder (1979). The model of Hirst et al. (1993) generates three-dimensional hexahedra corresponding to channel reaches and links these in chains by stochastic processes to form channel belts (Fig. 10.35). The sizes and shapes of the channel "boxes", the sinuosities of the simulated reaches, the presence or absence of channel bifurcations and of levee and crevasse-splay wedges are all determined by sampling from distributions that are supplied as input. Channel belts are positioned in a simulated three-dimensional space in order to agree with sand body sizes, dimensions, and positions, as determined from well penetrations. Channel belts are added until the observed sandstone/nonsand-stone ratio (the net/gross ratio) is achieved. The availability of a reliable data base for the various parameters of the sand body architecture is clearly essential for the running of this program. Some of the statistical relationships referred to in the preceding sections can be used, or the modeler may wish to generate a unique data base built on available know-ledge of the specific reservoir interval under investigation.

Chapter 11

Tectonic Control of Fluvial Sedimentation

11.1 Introduction

The major external, or allogenic, controls on fluvial sedimentation are tectonics and climate. This chapter considers tectonic controls; climatic controls are discussed in Chap. 12. Base-level changes, driven either by tectonism or eustasy, are of great importance in controlling fluvial style and sedimentation rates in the lower reaches of rivers, through the control base level exerts on regional slope and sedimentary accommodation space. The role of eustasy in controlling stratigraphic processes remains controversial and is described in Chap. 13, where the main theoretical discussion of sequence stratigraphy is presented. In this chapter changes in base level are considered as relative changes, and their effects on stratigraphic processes in the fluvial environment are considered without entering into the tectonism-versus-eustasy debate.

There is a vast literature on the tectonic control of sedimentation (Sect. 11.2). In this chapter it is considered under the following subheadings. In Sect. 11.2.1, the local effects of intrabasin fault and fold movement are considered, including the effects of rejuvenating fluvial systems and the diversion of drainages. Relative changes in base level are discussed in Sect. 11.2.2. Regional, basin-scale effects are discussed in Sect. 11.3, including the differing responses of depositional systems to the major types of tectonism. Particular attention is paid to the styles of nonmarine sedimentation that occur in extensional basins and foreland basins. Recent quantitative models for sediment dispersal are evaluated. Nonmarine sedimentation is then discussed within the context of plate tectonics (Sect. 11.4), and, finally, some generalizations regarding the tectonic control of regional facies and dispersal patterns are given (Sect. 11.5).

11.2 Tectonic Control of Alluvial Stratigraphy

It has been realized since ancient times that the deposition of coarse-grained sediment in rivers is related to the presence of elevated source areas. As noted in Sect. 2.2.1, a clear link between the grain size of gravels and proximity to a mountainous source area was suggested by Playfair (1802). It has been a standard interpretation, at least since the work of Barrell (1917), that beds of coarse clastic deposits close to a basin margin indicate tectonic rejuvenation of that margin.

The links between tectonic movement and coarse-clastic sedimentation have recently been reexamined by Jordan et al. (1988), Heller et al. (1988), and Blair and Bilodeau (1988). These researchers have challenged the conventional idea that wedges of coarse sediment are syntectonic in origin, presenting several lines of argument to suggest that in many cases the most widespread coarse units form after tectonism has ceased. The actual link between sedimentation and tectonics is not questioned; it is not doubted that rejuvenated relief is required to generate and transport coarse debris in large quantities. What is questioned is the detailed timing of the various events, relating to the speed of basinal and depositional response to uplift and subsidence. This topic is examined in Sect. 11.3. Here the discussion is limited to a presentation of the immediate evidence for tectonic influence in nonmarine basins – the effects of changes in slope on grain size and dispersal patterns. There are two main kinds of effects to be described: the effects of fault pulses at basin margins, and the growth of faults and folds within basins. Many data on these topics from the 1960s and 1970s were summarized by Miall (1978d) and Heward (1978a), and numerous case studies have been published since these papers appeared. A research program initiated by M.R. Leeder in the 1980s set out to investigate the quantitative relationships between geomorphic, sedimentary, and tectonic processes, and some of this work is referred to in this chapter (see also his summary article: Leeder 1993).

It should not be forgotten that sediments are simultaneously affected by both autogenic and allogenic processes, and the allogenic processes described here are overprinted on the autogenic variations in stratigraphy, such as those generated by geomorphic thresholds and lateral channel migration and avulsion, as described in Chap. 10. Recent work also suggests that changes in fluvial facies, channel density, channel style, and other features that have been attributed to tectonic mechanisms in earlier studies may in fact be driven by climate change. For example, in Sect. 12.12.2 it is shown how changes from aggradation to degradational regimes in Texas river terraces were caused by climate changes associated with the late Cenozoic glaciation and not by tectonism or base-level change. G.A. Smith (1994) documented a complex record of changes in fluvial style in a late Cenozoic extensional basin in Arizona at a time of tectonic quiescence, and was able to suggest climatic causes because of the availability of a proxy climatic record in the form of oxygen and carbon isotopic signatures.

11.2.1 The Effects of Syndepositional Fault and Fold Movements

11.2.1.1 The Effects of Basin-Margin Faulting

Extensional basins are commonly bounded by normal faults. These may be steep faults that penetrate to at least mid-crustal levels, or listric faults that may flatten out onto décollements at some shallower level. Graben basins are bounded on both sides by faults, whereas half-graben basins are underlain by tilt blocks and have a bounding fault on one side (Fig. 11.1). Tectonic slopes are produced by a combination of footwall uplift and hanging-wall subsidence (Leeder and Gawthorpe 1987; Prosser 1993). The footwall slope normally is steep, and typically is blanketed by coarse debris in the form of alluvial fans. The source area for these fans is small, because it is limited by the size of the area comprising the footwall scarp, unless repeated uplift occurs over a very long time period and vigorous downcutting results in the tapping of source areas behind the footwall scarp. The hanging-wall slope is gentler but may tap a much larger sediment source area, resulting in the formation of larger depositional systems depositing finer-grained sediment than those blanketing the footwall scarp. Subsidence of the basin is typically asymmetric, with rotation occurring about a fulcrum point (Fig. 11.1). In the case of basins bounded and underlain by listric faults, rotation against the fault results in the fulcrum point migrating up the hanging-wall slope, generating stratigraphic onlap onto the basement.

The stratigraphy of the clastic units derived from the marginal fault scarps depends on the balance between uplift and erosion rates on the basin margins and on that between subsidence and sediment-accumulation rates within the basin. Typically, the marginal deposits form lenticular clastic wedges, with the lateral extent of the wedge and vertical changes in grain size through it recording changes in the balance between these processes (Miall 1978d; Heward 1978a). These are large-scale, long-term controls on sedimentation, relative to the autogenic processes of avulsion and fan-trench evolution described in Sect. 10.2. They represent processes occurring over time spans of 10^4–10^6 years, corresponding to groups 8 and 9 of Table 9.1. Two styles of fan evolution may occur, fan segmentation and fan-head trenching (Bull 1964; Hooke 1967). Segmented fans are those in which the radial profile shows two or more straight-line segments steepening toward the apex (Fig. 11.2A). Typically, each segment repre-

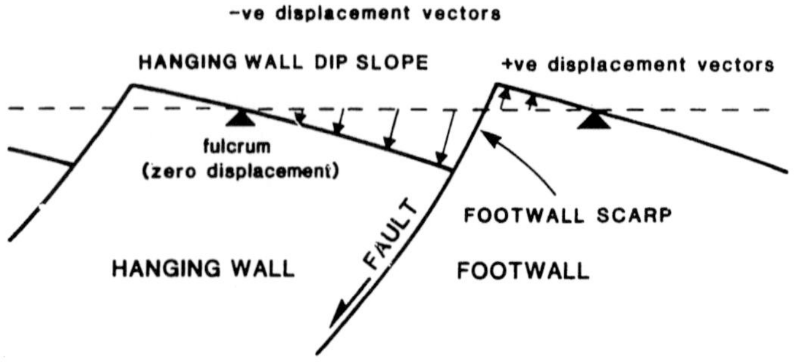

Fig. 11.1. Nomenclature for tectonic slope associated with a simple tilt-block, half-graben basin configuration. (Leeder and Gawthorpe 1987)

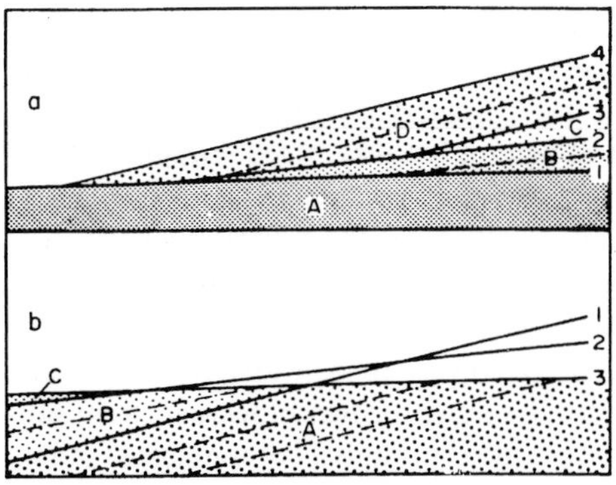

Fig. 11.2. Two alluvial fans in cross section, showing response to changes in the balance between tectonism and sediment transfer. *Numbers 1–4* are successive fan profiles, *A–D* are successive wedges of fan sediment. *a* Uplift (to the right) more rapid than basin subsidence and sedimentation, causing fan segmentation, upward coarsening of the debris, and basinward progradation; *b* fan-head trenching, caused by relative decrease in the rate of tectonic rejuvenation. Older fan segments are left as terrace remnants (*unornamented areas* of diagram), and fan sediments become finer-grained with time. (Miall 1978d)

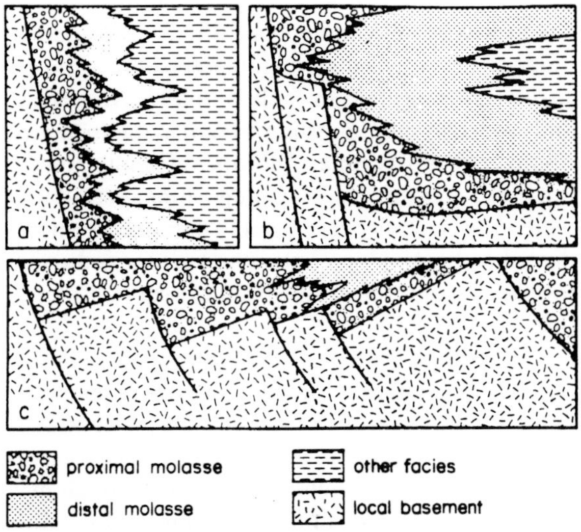

Fig. 11.3. Examples of alluvial-fan wedges banked against basin margins. *a* Simple faulted margin with continuous variations in width and grain-size of clastic wedge, reflecting both autogenic and allogenic processes; *b,c* retrogradation of fan deposits caused by backstepping of basin margin on successive faults. (Miall 1978d)

sents the sedimentary response to a pulse of uplift in the source area, resulting in steepening of the depositional slope and coarsening of the transported debris. Upward coarsening and lateral progradation of the fan deposits is the result. Between periods of tectonic rejuvenation the rate of source-area erosion may exceed the rate of isostatic rebound. Progressive downcutting results in channel incision, leaving earlier fan segments as terraces and depositing a new cone of sediment as an offlapping wedge, with a gentler depositional slope, at the distal end of the fan (Fig. 11.2B). Significant fan-head trenches may be formed in this way, although, as noted in Sect. 10.2, they can also be formed by autogenic processes.

Progradation and retrogradation of the fan deposits, with corresponding vertical changes in grain size, are a characteristic feature of basin-margin alluvial-fan deposits and have been described by numerous researchers (see references in Miall 1978d; Heward 1978a; Leeder and Gawthorpe 1987; Blair and Bilodeau 1988; Prosser 1993). Some examples of typical stratigraphic configurations are illustrated in Fig. 11.3 (see also Fig. 9.11). Heward (1978a) termed the resulting clastic wedges "megasequences", to distinguish them from the "sequences" of autogenic origin (the term sequence has now, of course, assumed a different meaning). Prosser (1993) recommended the erection of "tectonic system tracts," based on the three-dimensional architecture and composition of tectonically controlled stratigraphic units. As described in Sect. 9.2, allogenic cycles are commonly as much as 250 m thick. Clast size and bed thickness commonly increase upward through most or all of the section, attesting to the increasing competence and capacity of the fan streams.

The details of the tectonic geomorphology of modern faulted tilt blocks and sedimentary basins were considered in detail by Leeder et al. (1991) and Leeder and Jackson (1993), based on the study of tectonically active extensional basins in Greece and the Basin and Range area of the United States. The evolution of fault systems, the evolving geomorphology of the uplifted fault scarps and drainage basins, and the interaction between adjacent basins were considered in detail, and it was pointed out that these modern systems can be used as analogues for the interpretation of economically important areas of rift-basin tectonics, such as the structural traps in Mesozoic fluvial deposits in the northern North Sea Basin (Leeder and Gawthorpe 1987; Prosser 1993; see also Chap. 15).

The basins that develop along strike-slip faults have faulted margins similar to those of rift basins. However, the gradual lateral displacement of the marginal clastic belts relative to their feeder streams introduces some characteristic features. Nonmarine strike-slip basins in California were discussed by Crowell (1974a), and Crowell and Link (1982), and many examples of their structure and stratigraphy were given by Ballance and Reading (1980) and Biddle and Christie-Blick (1985). A general discussion of this class of basin was provided by Miall (1990, Sect. 9.3.3). A recent study by Mastalerz and Wojewoda (1993) focused on the details of the effects of strike-slip displacement on a modern alluvial fan prograding across the basin-bounding strike-slip fault. The deposits contain small syndepositional grabens and shear zones and show a complex pattern of laterally offlapping depositional lobes, formed as older fan deposits are structurally displaced along the basin margin (Fig. 11.4).

Syntectonic to post-tectonic fanglomerate wedges also characterize compressional basins bounded by thrust faults. Well-described examples occur worldwide, for example in the Rocky Mountain foreland basin (Lawton 1986a,b; DeCelles et al. 1987, 1991), in the foreland basins flanking the Spanish Pyrenees (Anadón et al. 1986; Puigdefábregas et al. 1986; Burbank et al. 1992), the Andes (Jordan et al. 1988), and the Himalayas (Burbank et al. 1986; Burbank and Raynolds 1988), in the Canadian Arctic (Miall 1991d), the modern Persian Gulf (Baltzer and Purser 1990), the Po basin, Italy (Ori 1993), Papua-New Guinea (Crook 1989), and in many other locations (other references in Blair and Bilodeau 1988; Miall 1995b). However, In contrast to extensional basins, where the downthrown block passively subsides, perhaps with a rotational component of movement, in compressional basins the downthrown block may undergo significant folding and uplift.

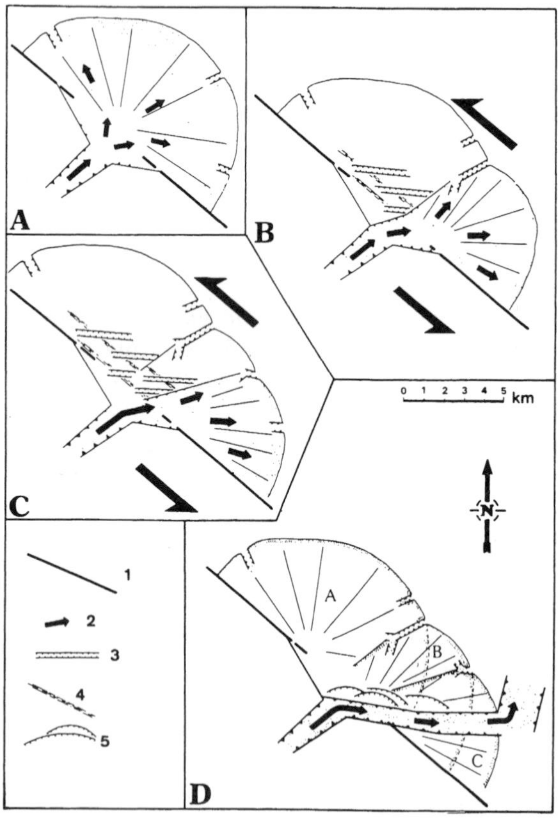

Fig. 11.4A–D. Evolution of an alluvial fan during left-lateral strike-slip displacement along the basin-margin bounding fault. *1* Main basin-margin fault; *2* sediment dispersal; *3* graben; *4* shear zones; *5* listric faults. (Mastalerz and Wojewoda 1993)

One of the most important effects of this local tectonism is the development in many foreland basins of a distinctive class of structures termed syndepositional or intraformational unconformities (Fig. 11.5). These were first described in detail using examples from the Ebro basin, Spain (Riba 1976; Anadón et al. 1986; Puigdefábregas et al. 1986) but have also been observed in many other foreland basins, including the Indo-Gangetic basin (Johnson and Vondra 1972), the Venetian Alps (Massari et al. 1993), and the Rocky Mountain foreland basin (Spieker 1946; DeCelles et al. 1991). In present day basins Recent alluvium may be tilted and incised by modern streams. Fanglomerate deposited against a rising (growing) fold or fault is uplifted, deformed, and eroded by continued movement of the structure. Younger deposits onlap the unconformity toward the basin margin. These syndepositional structures tend to be laterally restricted. The unconformities described by Anadón et al. (1986) typically die into the basin within 500 m and extend along strike for as little as 5–10 km.

The deepest part of a fault-controlled sedimentary basin is typically located close to the bounding fault. In the case of extensional basins it will be immediately adjacent to the fault (Fig. 11.1), whereas in the case of compressional basins the trough may be located a short distance out from the bounding fault, corresponding to the axis of the syndepositional fold generated in the footwall. The position of this axial basin and its transverse and longitudinal slopes will vary depending on the progress of deformation on the basin-margin structures, and this ongoing deformation has an important effect on the evolution of river systems flowing through the basin. Commonly, the axial trough is occupied by a river flowing parallel to the basin margin. This will migrate laterally or undergo avulsive switches in position in response to pulses of fault movement. Alexander and Leeder (1987) and Alexander et al.

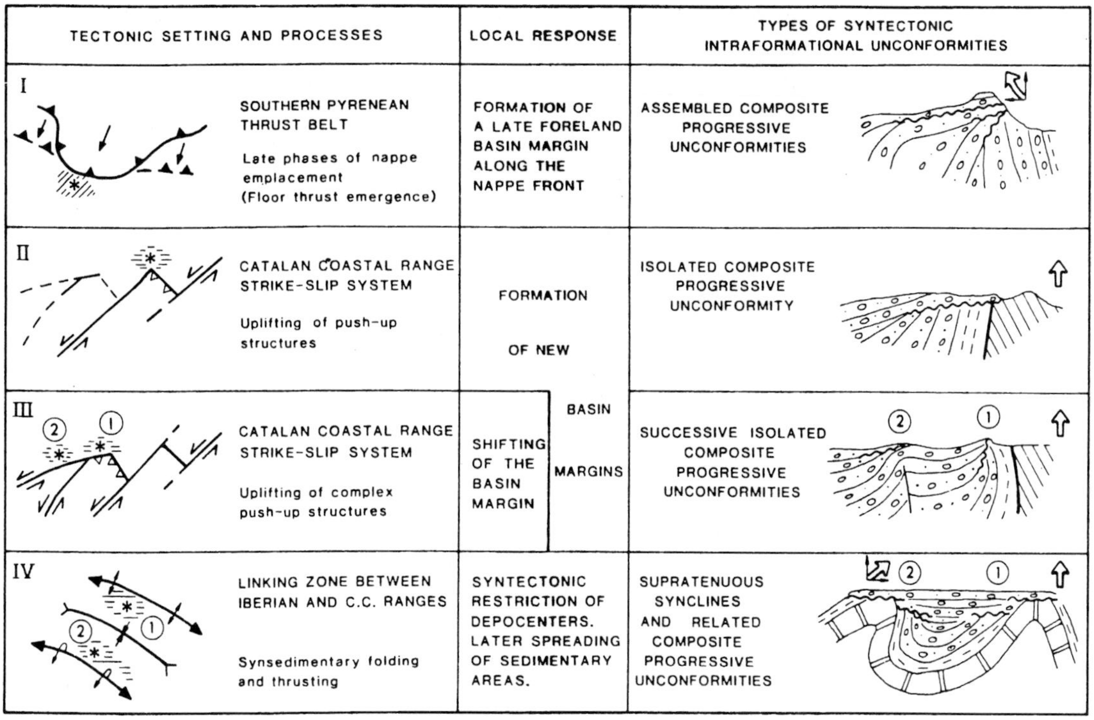
TECTONIC SETTING AND PROCESSES
LOCAL RESPONSE
TYPES OF SYNTECTONIC INTRAFORMATIONAL UNCONFORMITIES
I
SOUTHERN PYRENEAN THRUST BELT
Late phases of nappe emplacement (Floor thrust emergence)
FORMATION OF A LATE FORELAND BASIN MARGIN ALONG THE NAPPE FRONT
ASSEMBLED COMPOSITE PROGRESSIVE UNCONFORMITIES
II
CATALAN COASTAL RANGE STRIKE-SLIP SYSTEM
Uplifting of push-up structures
FORMATION OF NEW BASIN MARGINS
ISOLATED COMPOSITE PROGRESSIVE UNCONFORMITY
III
CATALAN COASTAL RANGE STRIKE-SLIP SYSTEM
Uplifting of complex push-up structures
SHIFTING OF THE BASIN MARGIN
SUCCESSIVE ISOLATED COMPOSITE PROGRESSIVE UNCONFORMITIES
IV
LINKING ZONE BETWEEN IBERIAN AND C.C. RANGES
Synsedimentary folding and thrusting
SYNTECTONIC RESTRICTION OF DEPOCENTERS. LATER SPREADING OF SEDIMENTARY AREAS.
SUPRATENUOUS SYNCLINES AND RELATED COMPOSITE PROGRESSIVE UNCONFORMITIES

Fig. 11.5. A classification of syndepositional (intraformational) unconformities in the Ebro Basin, Spain, showing the range of tectonic processes and resulting structural styles. (Anadón et al. 1986)

(1994) documented this effect in the case of the Mississippi River, where it crosses the New Madrid earthquake zone, and in the case of the South Fork Madison River in Montana, which has been affected by tilting of a half graben in a direction tranverse to the river (Fig. 11.6). They pointed out that the tectonic slope generated by individual earthquake episodes may be comparable in magnitude to the longitudinal slope of the river, so that significant immediate effects on drainage may be expected. Todd and Went (1991) and Olsen and Larsen (1993) described possible ancient examples of the lateral migration of rivers in response to tectonic tilting. The effect has also been modeled by Bridge and Leeder (1979). As discussed in Sect. 10.4.3, their computer model of avulsion selects the lowest position on the flooplain for each successive channel position. In the case of a lateral tectonic tilt the model automatically tends to shift channels downslope, much as has been actually observed by Alexander and Leeder (1987). However, in real basins the alluvial architecture may not always follow this pattern. Atkinson (1986) documented a case in the Tremp-Graus Basin, Spain, where channel densities are highest in areas of most rapid subsidence, the opposite of the effect predicted by the Bridge and Leeder (1979) model. Atkinson (1986) attributed this contrast to the fact that the simulation studies model the evolution of single rivers, whereas in real basins (such as the one he described), more than one river may be present. High channel densities in the Spanish example are attributed to the activity of several distributaries of a fan system. Other local complications are described by Alexander et al. (1994).

In addition to the shifts in meander pattern illustrated in Fig. 11.6, Alexander and Leeder (1987) and Alexander et al. (1994) pointed to other significant effects on the alluvial architecture that could be expected to be caused by tilting. For example, soils are likely to be more mature on the updip flank of the tilted basin, because such areas receive less clastic influx to dilute pedogenic processes and change the position of the water table.

It was noted by Alexander and Leeder (1987) and Leeder and Alexander (1987) that changes in river courses in general and meander patterns in particular could be predicted from a knowledge of the tectonic slopes developed by fault movement, and they developed this idea to explain the changes in meander pattern of the Mississippi River in the New Madrid area in response to the earthquakes of 1811–1812. Idealized river flow lines are drawn

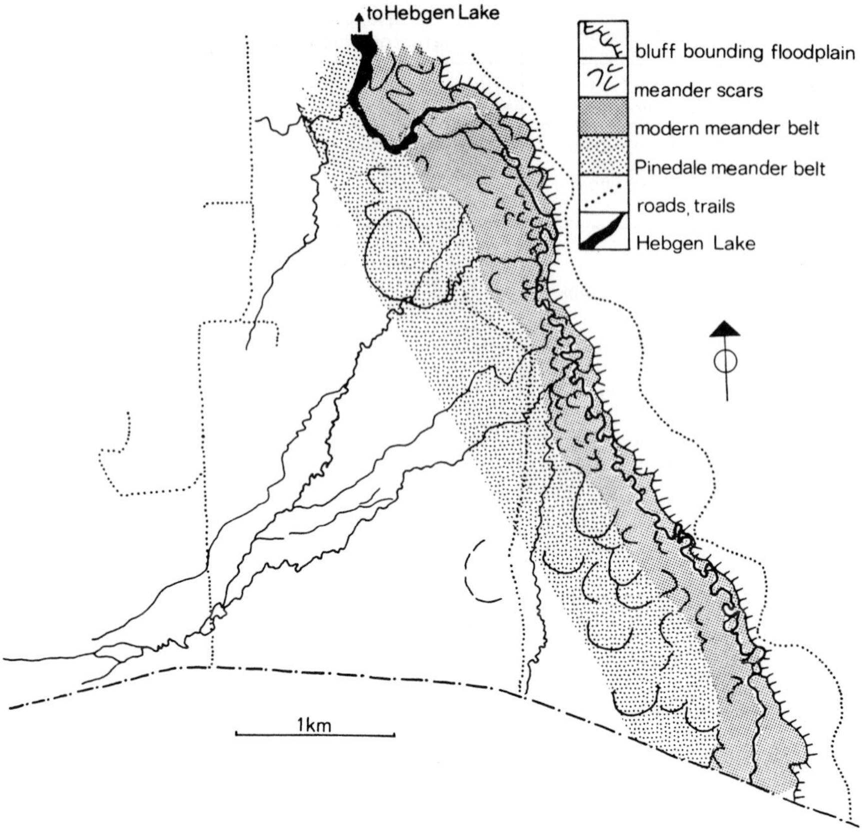

Fig. 11.6. The meander belt of the South Fork Madison River in SW Montana, showing the asymmetry of meander patterns and relict floodplain deposits. The river has gradually migrated northeast in response to a tectonic tilting in that direction, and the river now flows against the basin-margin fault. Note the presence of an abandoned Pleistocene meander belt (*Pinedale belt*) on the updip flank of the modern meander belt, and the preferential preservation of meander scars that are convex to the southwest, as the river migrated laterally away from them and eroded the meanders facing down structural dip. (Alexander and Leeder 1987)

perpendicular to the contours of the tectonic slope, a process that can be simulated by computer (Weston and Alexander 1993). Alexander (1986) combined information on syntectonic fault movement and paleochannel trends derived from outcrop paleocurrent data to project the distribution of sand bodies in the subsurface behind cliff exposures of a Middle Jurassic sand body on the Yorkshire coast, England.

11.2.1.2 The Effects of Faulting and Folding Within Basins

Rivers are sensitive to external forces, such as tectonic subsidence, uplift, and tilting. In tectonically active areas the drainage patterns may be directly related to the patterns of subsidence and uplift generated by growing structures. Baltzer and Purser (1990) showed that this is the case adjacent to the Zagros fold-thrust belt, which forms the border to the Mesopotamian foreland basin, and Räsänen et al. (1987) discussed the postglacial adjustment of the Amazon system in the Andean foreland basin of Brazil to tectonic movements. Changes in the magnitude of the basement slope lead to changes in river style (Schumm 1993; Wescott 1993), and changes in the orientation of the slope cause changes in flow direction, including reversals in flow, lateral shifts in meander belts, and river capture. Numerous neotectonic observations have been made of these features. In particular, the growth of folds within basins, such as the anticlines that develop over blind thrust faults, can be mapped on the basis of changes in the drainage system at the surface. For example, in Iraq, ancient drainage canals have been deformed by the growth of anticlines beneath them (Lees 1955). In central Asia, growing anticlines are bringing about diversions in drainage patterns (Trifonov 1978). Numerous changes in river patterns in the Hungarian Plains were attributed to tectonic tilting by Mike (1975). Several studies have examined the changes in fluvial morphology in response to modern tectonic movements, including changes in longitudinal profile, terrace patterns, channel sinuosity, meander shape and orientation, bar character, and lateral channel migration (Seeber and Gornitz 1983; Burnett and Schumm 1983; Ouchi 1985; Leeder and Alexander 1987; Räsänen et al. 1987; Marple and Talwani 1993). Modifications in fluvial style in the plains of the Ganges River in India and changes in local sedimentation rate that affected soil development were attributed to neotectonic tilting by Srivastava et al. (1994).

Blind structures may eventually penetrate the land surface and evolve into topographic barriers and sediment sources. Such patterns of development are particularly common in foreland basins (Morley 1986), where they may accompany the development of satellite or piggyback basins (Sect. 11.4.6).

Ouchi (1985) reported on a series of experiments to simulate the effects of anticlinal and synclinal folding across a river's longitudinal profile and

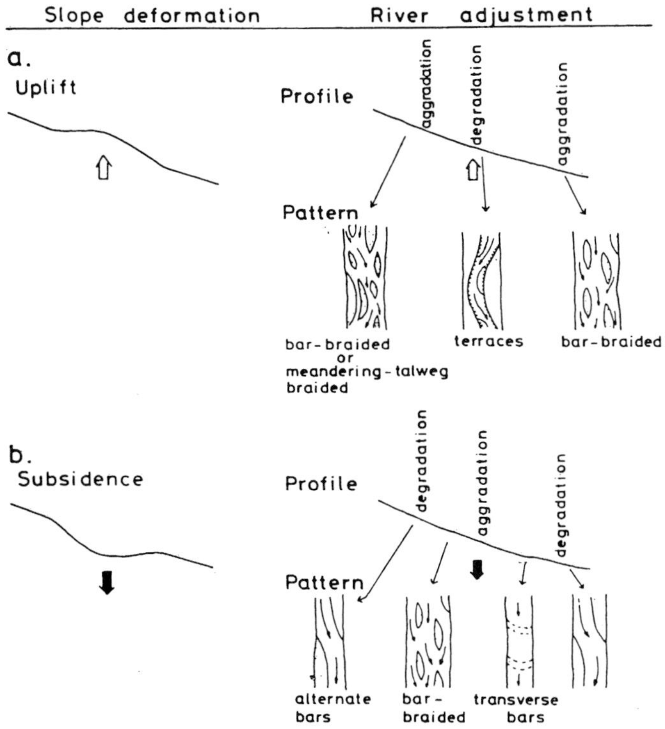

Fig. 11.7. Adjustment of a braided river to (*a*) anticlinal uplift and (*b*) synclinal subsidence across it. (Based on experimental work by Ouchi 1985)

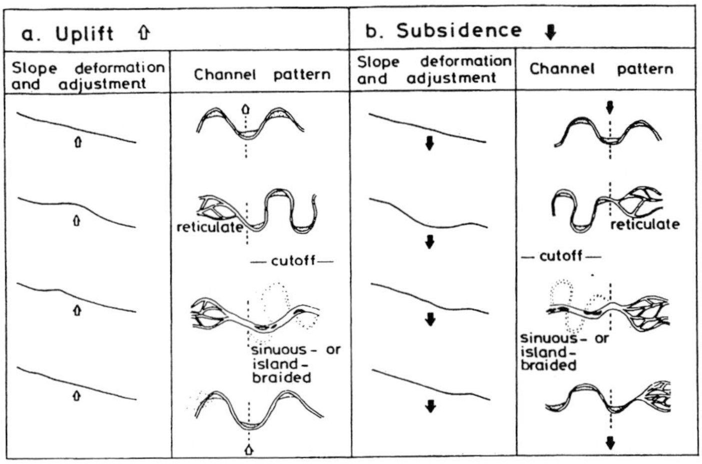

Fig. 11.8. Adjustment of a mixed-load meandering river to (*a*) anticlinal uplift and (*b*) synclinal subsidence across it. Four time steps are shown, in order from top to bottom. (Based on experimental work by Ouchi 1985)

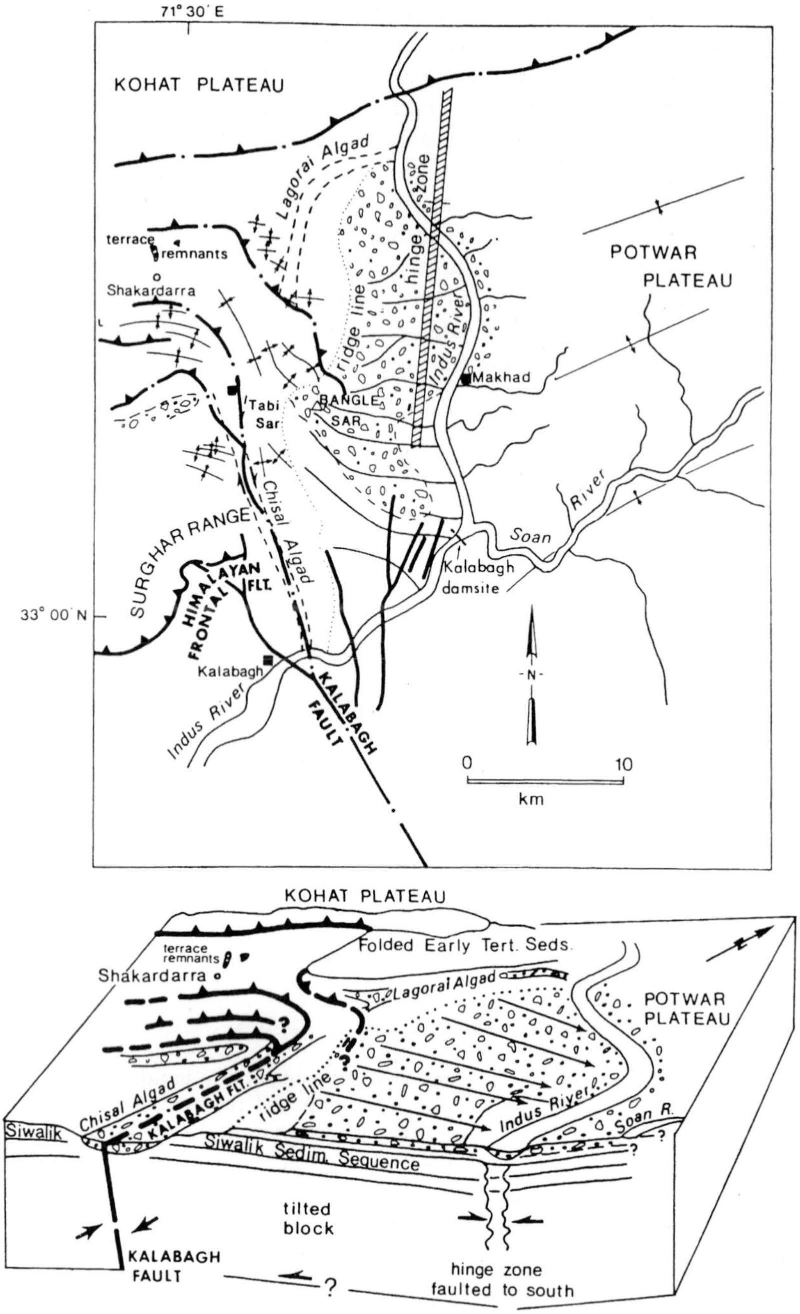

Fig. 11.9. Map and block diagram of part of the Himalayan frontal fault system, Kohat Plateau, Pakistan, showing the eastward migration of the Indus River in response to tilting and uplift of the alluvial plain. Transpressional uplift along the Kalabagh Fault is the main cause of this movement. Terrace remants in the northwest corner of this area represent former Indus gravels now elevated to as much as 630 m above their former position. (McDougall 1989)

described some actual examples of these changes based on detailed mapping at various locations in the United States, where cratonic warping at rates of up to 1 cm/year has been recorded. Burnett and Schumm (1983) described additional examples from the United States. Some of the major conclusions of this work are summarized in Figs. 11.7 and 11.8. In summary, changes in slope imposed on the longitudinal profile of a river are compensated initially by changes in sinuosity, so that the along-channel slope remains constant. Thus, an increase in slope leads to an increase in sinuosity, and vice versa. There are limits to the ability of rivers to change in this way. At high sinuosities the incidence of chute and neck cutoffs increases (reducing the sinuosity again), and bank erosion and incision ensue. Reduction in slope may lead to meandering rivers becoming flooded and developing anastomosed patterns, as was predicted by Smith and Smith (1980). At synclinal and anticlinal axes, respectively, aggradation and degradation occur. These ideas are of considerable importance in connection with the effects of base-level change on rivers and are discussed further in Sect. 11.2.2.

Recent applications of these ideas include the work of Marple and Talwani (1993), who employed all these indicators in a study of neotectonic deformation of the coastal plain of South Carolina. Upstream from a region of uplift, rivers are anastomosed and swampy. In addition, some rivers flow along one side of their floodplains, suggesting lateral migration due to tectonic tilting. Mather (1993) documented the uplift of a small fluvial basin in Spain, which was undergoing active alluvial-fan sedimentation in the Pliocene and now is characterized by incised streams, with the courses of the fan feeder channels truncated by river capture.

These observations based on neotectonic and experimental studies and observations of the Recent may be applied to analyses of the more ancient record, where examples abound of the effects of syndepositional, within-basin tectonic deformation. Some of the most spectacular examples occur in the Himalayan foreland basin of India and Pakistan, where fluvial deposits have been accumulating continuously since the mid-Cenozoic, and major trunk rivers such as the Indus, Ganges, and Brahmaputra have occupied the axis of the basin for millions of years. These rivers show many changes in channel patterns (e.g., Holmes 1968; Wells and Dorr 1987), at least some of which are related to fault movement and block tilting within the sedimentary basin (Morgan and McIntire 1959; McDougall 1989; Srivastava et al. 1994). For example, the Indus River has been diverted eastward a lateral distance of 10 km along the western side of the Potwar Plateau by uplift on a major transpressional fault during the past 100–500 ka. The former course of the river can be traced across uplifted fault blocks, where terrace remants are now as much as 630 m above their former elevation. Lateral migration of the Indus laid down a sheet of gravel which is now being drained by small consequent streams flowing across the newly tilted block (Fig. 11.9).

Earlier changes in the course of the Indus River in the Potwar plateau area of Pakistan were reconstructed by Mulder and Burbank (1993), based on an extensive synthesis of facies, petrographic, and paleocurrent data within a stratigraphic framework that has been rigorously dated using magnetostratigraphy. Several studies have shown that the Indus River flowed eastward into the Ganges system prior to about 5 Ma (Fig. 11.10). Its deposits are characterized by a distinctive white sandstone in the form of channel bodies bearing a characteristic variety of hornblende in the heavy-mineral suite. These beds are interbedded with a variegated assemblage of brown sandstones showing much greater lithologic variablity and generally smaller-scale cross-bed structures and thinner channel units. The brown units are interpreted as the deposits of smaller streams draining uplifts of the Himalayas and the Salt Ranges to the south of the Potwar Plateau. The structural emergence of the Salt Ranges can be determined on the basis of the appearance of brown sandstone units draining the emerging uplift and accompanying shifts in paleocurrent patterns and, later, by the appearance of a distinctive granite clast type known only from Salt Range sources. Initially, the uplift of the Salt Ranges shifted the Indus River system northward, and continued uplift then deflected it westward to near its present course (Fig. 11.10). Minor drainages of the Jhelum system captured the runoff from the emerging structures.

This picture of tectonic control of fluvial paleogeography, and the sedimentological evidence from which it can be reconstructed, is typical of many basins marginal to fold-thrust belts, particularly "piggyback" or "satellite" basins, as described in Sect. 11.3.

The development of coal is sensitive to tectonic setting. Given the appropriate climate for peat formation, the thickness and quality of a peat unit depend on subsidence rate and clastic influx. Titheridge (1993) described the effects of faulting and consequent local basinal tilting on the distribution and thickness of coals. He showed that the slightly increased subsidence rate on the downthrown block of a half-graben leads to an increase in seam thickness and also to a greater frequency of seam splitting. The difference between true low-ash coal and carbonaceous shale depends on the clastic influx into the coal swamp. McCabe (1984) demonstrated that low-ash coal may reflect the development of raised peat islands, which deflect drainage channels and their contained clastic influx away

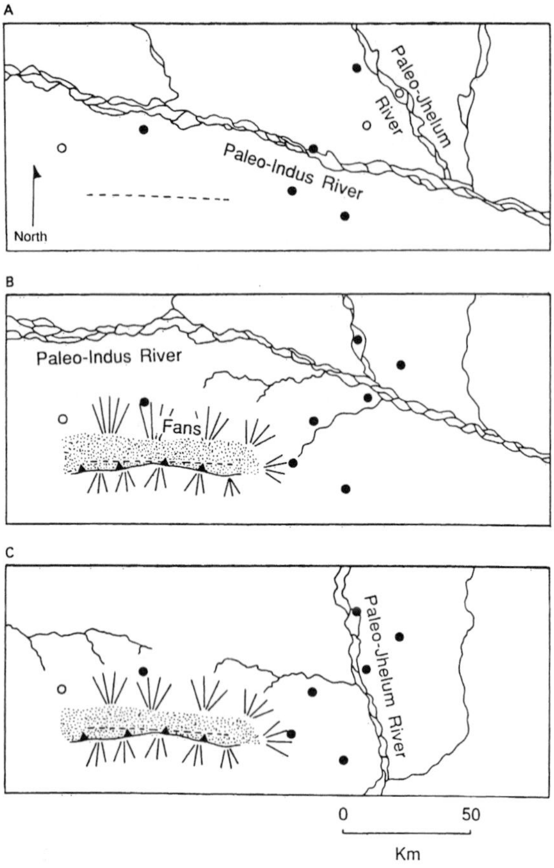

Fig. 11.10A–C. Paleogeographic reconstruction of the Potwar Plateau area, Pakistan, for **A** 6.0 Ma; **B** 5.2 Ma; and **C** 4.8 Ma. *Dashed line* marks position of basement normal fault used as a reference frame. *Dotted area* shows position of upthrust Salt Ranges. (Mulder and Burbank 1993)

from the peat swamp. Wise et al. (1991) and Belt (1993) argued that, on a local scale, basinal tectonics may also be a factor, with the growth of anticlinal structures over blind faults acting to dam or divert clastic drainage. They attempted to explain rapid (0.5 million years time scale) vertical changes between clastic and coal-bearing intervals in proximal parts of the Appalachian foreland basin by this mechanism. Belt (1993) suggested that differential uplift and subsidence during the evolution of the Laramide basins in the western United States could explain drainage-pattern changes in the Williston Basin and the development of coal swamps there on a similarly rapid time scale.

Because the effects of tectonism on fluvial style and paleogeography can be very subtle, it may be possible to read into the geological record more than can be conclusively demonstrated. Peterson's (1984) study of tectonic control of cratonic sedimentation in Utah and Arizona may be a case in point. Peterson suggested that very gentle structures mapped in a Jurassic fluvial system were active during sedimentation and controlled sediment thicknesses, paleocurrent patterns, river sinuosities, and the location of interfluvial lacustrine deposits. Heller et al. (1993) suggested that these structures are part of a suite of similar structures in the Western Interior Basin developed by intraplate stress (see below). Most of the supposed stratigraphic and facies relationships are not obvious to this writer, except that one major "basin", the Henry Basin (which is actually crossed by several of the minor anticlines), is clearly located at a confluence between two major fluvial systems and shows a thicker section and higher channel sinuosities (based on paleocurrent variability) than adjacent areas.

11.2.2 Base-Level Changes

There are several alternative definitions of base level. Schumm (1993, p. 279) stated: "Base level is the imaginary horizontal level or surface to which sub-aerial erosion proceeds. It is sea level." In their review of nonmarine sedimentation in relation to sea-level change, Shanley and McCabe (1994) pointed out that sea level is not always the critical surface in determining the level to which rivers tend to erode. In inland basins, lake level, or the level of a tectonic barrier, may be the practical base level. In eolian basins the surface of deflation, related to the level of the groundwater table, may be the local base level for fluvial processes. Shanley and McCabe (1994) proposed the adoption of the concept of the stratigraphic base level as encompassing all these varied controls on fluvial sedimentation.

In coastal plains, base level may change as a result of tectonic movements of the basin or as a result of eustatic changes of sea level. Changes in sea level and their effects on stratigraphy have, of course, come to form the basis for the subject of sequence stratigraphy and are therefore of paramount importance to sedimentary geologists (Chap. 13). However, the nature of the fluvial response to base level change is a subject that has been bedeviled by controversy and misunderstanding since the nineteenth century (see Schumm 1993 for a brief review). Much that has been written recently about the effects of base-level change is the work of sequence stratigraphers, or of geomorphologists explaining their science in response to developments in sequence stratigraphy. This section therefore involves a consideration of some of the older sequence literature, in order that a clearer understanding of the present state of understanding can be reached.

Over geological lengths of time, rivers develop a dynamic equilibrium in the form of a graded longitudinal profile (some of the historical steps in the emergence of this idea are summarized in Sect. 2.3.3). Butcher (1990, p. 376) stated:

"An equilibrium profile, or graded stream, is one in which slope, velocity, depth, width, roughness, pattern, and channel morphology delicately and mutually adjust to provide the power and efficiency necessary to tranport the load supplied from the drainage basin without aggradation or degradation of the channels (Leopold and Bull 1979). Rivers that are out of equilibrium will aggrade or incise in an attempt to achieve a graded profile which ... is a function of tectonic, isostatic, climatic variation ... and base-level controls."

Isolating the effects of base-level changes is difficult, because in many naturally occurring cases changes in base level are linked to other effects. For example, the major base-level changes that affected the world's coasts during the late Cenozoic are glacioeustatic in origin, and the effects on the rivers and their deposits of actual changes in base level are difficult to separate from changes in discharge and sediment load that occurred at the same time (Butcher 1990; Schumm 1993; Leckie 1994). Experiments by Begin et al. (1980) and Begin (1981) in modeling base-level changes in a flume have been useful in clarifying geomorphic concepts. Recent work by Blum (1994) and Blum and Price (1994) has demonstrated that climate change may be much more important than base-level change in affecting the balance between aggradation and degradation, because of the effects of climate change on discharge and sediment yield. This topic is dicussed at greater length in Sect. 12.12.2.

Posamentier et al. (1988, p. 135) stated:

"The fundamental principle which governs sedimentation in fluvial environments (assuming constant sediment supply) is that a stream will aggrade if its equilibrium profile shifts basinward or upward and incise if its equilibrium profile shifts downward ... subaerial accommodation occurs by basinward shifts of the stream equilibrium profile."

This statement is quite correct, but there has been debate regarding the causes and controls on the shift in fluvial profiles (Miall 1991c; Schumm 1993). The book in which the statement appears (Society of Economic Paleontologists and Mineralogists *Special Publication 42)* is still regarded as an authoritative source, containing two landmark papers by Posamentier et al. (1988) and Posamentier and Vail (1988). For this reason, it is important that the earlier models be described briefly and the problems with them mentioned. As noted below, H. Posamentier and his co-workers do not now use the part of the earlier work that is now described.

The basis for the model of Posamentier et al. (1988) and Posamentier and Vail (1988) is illustrated in Fig. 11.11, which is a variation on a diagram published by Lane (1955), as shown in Fig. 11.12. The shift in point of grade from O to F is thought to occur during the time of maximum rate of sea-level fall, when the equilibrium point in the basin (the point where vertical movements due to eustasy and tectonism are in balance) moves seaward across the continental shelf. Two observations of modern rivers which Posamentier et al. (1988) described in support of this model included the effect of the lateral migration of a modern river meander and the filling of a reservoir. In the case of the meander migration, the junction of the river with a tributary shifted across the river valley as the meander shifted, and this created accommodation space within the tributary, to which the tributary responded by aggrading. This effect was noticed because the aggradation resulted in deposition taking place over railroad tracks that run along the banks of the main river valley (Posamentier et al. 1988, Fig. 12, p. 133). In the second case, a river aggraded in its lower course as it built a coastal deltaic prism across a reservoir (their Fig. 13, p. 134).

However, there are three problems with the model of Posamentier et al. (1988):

1. No account was taken of the rate of change. How fast is the shift of the point of grade from O to F, relative to the ability of the stream to adjust? The family of curves AF, BF, CF, and DF is only one set of possibilities and represents a situation where the river is responding only to changes at the mouth by a backfilling process, ignoring upstream influences. Another possibility is the cre-

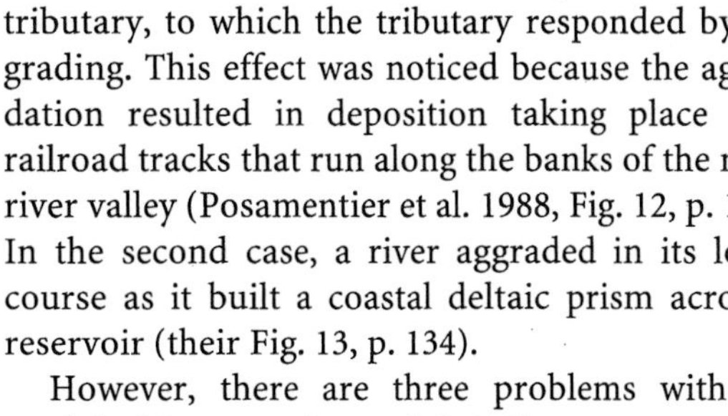

Fig. 11.11. Fluvial aggradation due to shift of stream equilibrium profile from *O* (original point of grade), with initial curve *AO*, to *F* (final point of grade). According to Posamentier et al. (1988) the river aggrades in response to this shift, developing in turn profiles *BF*, *CF*, and *DF* (letters added to original illustration of Posamentier et al. 1988, Fig. 15). (Diagram from Miall 1991c)

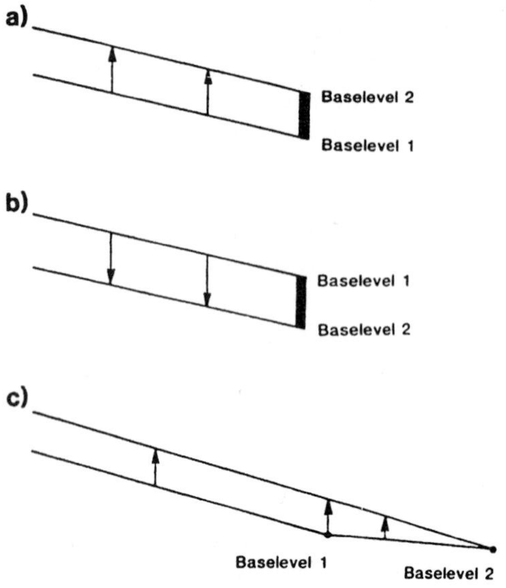

Fig. 11.12. Movement of stream profile in response to changes in base level. Note the increase in sedimentary accommodation space (upward movement of equilibrium profile) generated by a horizontal shift in the mouth of the stream in diagram (c). (Schumm 1993, after Lane 1955)

ation of a series of curves parallel to and above the original curve AO, as the point of grade shifts seaward slowly enough for the river to grade itself throughout. Much would depend on the rate of sediment supply.

2. It is not made clear how far upstream these profiles are intended to represent. The shift of the upper part of the profile from B to D ignores upstream controls. Projection of these profiles upward takes them back to the stream sources. The diagram seems to suggest that the response of a base-level fall is for the mountainous source area to rise upward and/or move forward (seaward) whereas, of course, it may be undergoing active degradation. There is, in any case, no reason why a change in base level should be linked to changes in the source area. Given slow tectonism (as on the passive margin assumed for these models by Posamentier and his co-workers), the upper parts of the later curves (C, D) should drop below B, not climb above it, leading to an overall decrease in slope.
3. The shift from O to F is drawn as a *horizontal* shift, whereas the whole point of the discussion is *vertical* changes in base level. This error follows from the use of the two particular examples by Posamentier et al. (1988), quoted above. In neither of these examples was there a vertical change in base level (the initial filling of the reservoir in their Fig. 13 is not at issue here).

In another diagram, Posamentier and Vail (1988, Fig. 8) showed fluvial coastal-plain progradation and coastal onlap occurring during the late-highstand phase of the eustatic sea-level cycle, during the time of maximum rate of fall (times T4-T6). They showed the equilibrium point migrating seaward, the bayline and fluvial profile shifting seaward, and the coastal plain expanding landward as a result of onlap. Progradation is shown by the conventional rising zig-zag facies contact in the deposited sediments. Landward of the equilibrium point is a zone of relative sea-level fall (this is what the equilibrium point means, as explained in the definition of this term by Posamentier et al. 1988, p. 116). These diagrams are not consistent with some of the text in the two companion papers in this volume, which discuss the tendency for fluvial incision during late-highstand, relative to sea-level fall. Figures 8 and 15 of Posamentier and Vail (1988) clearly indicate significant late-highstand fluvial deposition, coastal progradation, and coastal onlap during a time of relative sea-level fall (as do several figures in Jervey 1988), yet Posamentier and James (1993, p. 15) and Weimer and Posamentier (1993, p. 10) do not accept that this is what had been described in the earlier studies. Careful examination of their Fig. 8 indicates that the late-highstand fluvial aggradation and coastal onlap indicated in this diagram have been created in two ways. First, deposition has been allowed beneath the "tectonic hinge", which cannot occur if the hinge is the stable point about which the continental margin undergoes rotational subsidence. Second, most of the accommodation has been created by rotational subsidence at a rate comparable to the eustatic fall in sea level. The equilibrium profile is not shifted horizontally, but is extended obliquely downward in a seaward direction.

In retrospect, it is difficult to ascertain the basis for these models, such as that reproduced in Fig. 11.11. Many of the architectural ideas of Posamentier et al. (1988) are based on the computer simulations of Jervey (1988), which have led to schematic models of continental-margin configuration that are difficult to interpret because they lack scales, and therefore cannot show realistic slopes. Rates of subsidence and sea-level change are shown as having the same magnitude, which is commonly not the case. The late-highstand fluvial onlap model seems to have developed from an attempt by Vail and his co-workers to reinterpret their "sawtooth" coastal-

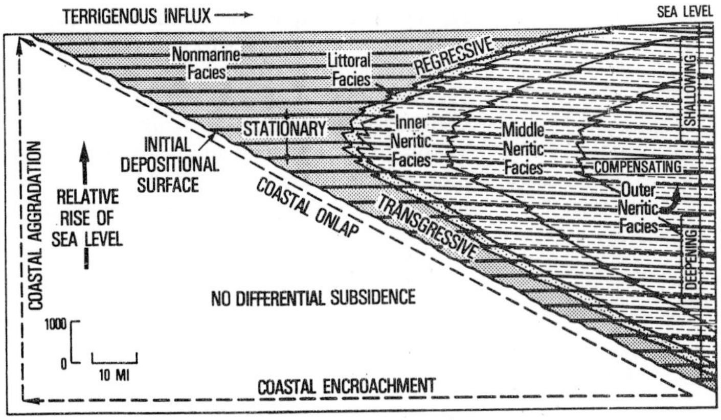

Fig. 11.13. Shifting of facies boundaries during eustatic rise and fall. Note the regression during sea-level fall, and the seaward extension of the coastal plain. (Vail et al. 1977, reprinted by permission)

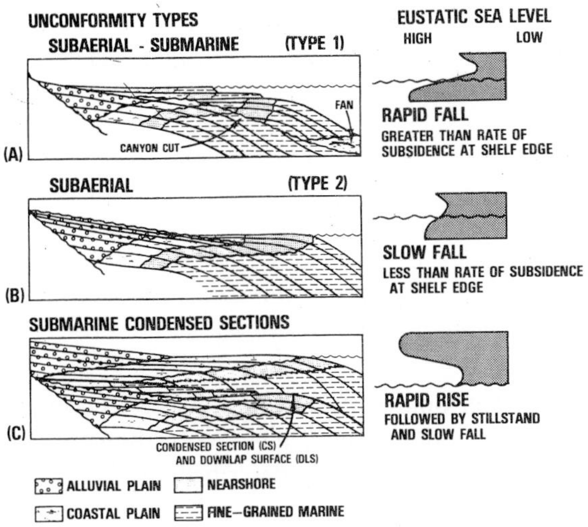

Fig. 11.14. Definition of type 1 and type 2 unconformities. (Vail et al. 1984, reprinted by permission)

onlap curves in order to arrive at symmetrical curves for eustatic sea-level change. The original sawtooth shape aroused considerable controversy (e.g., Pitman 1978), and their early interpretive diagrams contain the seed of a possible reinterpretation that is more fully explored in the papers by Vail and Todd (1981) and Vail et al. (1984). Figure 11.13 is reproduced from Vail et al. (1977) and shows their early ideas regarding regression of the shoreline during sea-level fall. Vail and Todd (1981, pp. 221–223, Fig. 2) incorporated these ideas into their methods for deriving sea-level change from measurements of coastal onlap and showed the same change from transgression to regression of the coastal zone towards the end of a cycle of eustatic rise and fall. Figure 11.14 is the definition diagram for type-1 and type-2 unconformities (from Vail et al. 1984) and shows what this writer regards as an impossible situation: the stratigraphic effects of sea-level rise and sea-level fall at the same time, in the same stratigraphic units. Onlap is shown at the left in each panel, where the stratigraphic units terminate against a steeply dipping basement surface (this is not a faulted contact). At the same time facies belts are shown migrating seaward, culminating in the development of an unconformity, indicating regression and relative sea-level fall. It seems possible that the observations of small-scale systems by Posamentier et al. (1988) referred to above suggested the possibility for this late-highstand regression, and the early model of Vail et al. (1977) (Fig. 11.13) was ready at hand. In Posamentier and Vail (1988), Fig. 8 shows a small segment of a "sawtooth" coastal onlap curve in which late-highstand alluvial onlap is clearly indicated as the reason for the sharp tip of the sawtooth.

As discussed by Miall (1991c), a drop in sea level would take the shoreline out and down across the continental shelf. The slope of this additional stream course may be comparable to that of the lower reaches of the original curve BO, but this would depend on local, *actual* continental margin configuration (Fig. 11.15). A drop in base level would therefore not necessarily *shift* the graded curve, but might simply *extend* it. It should not be forgotten that a newly exposed shelf might add new drainage area to a river, thus adding its discharge and tending to lower the slope.

Butcher (1990) summarized some long-known ideas regarding the response of river profiles to base-level change, and reminded geologists of the concept of the nickpoint (Fig. 11.16). In most cases, a fall in sea level leads to incision and to the headward migration of a nickpoint. The steepened lower reaches of the river may change in fluvial style as the river cuts down to its new equilibrium position, and remnants of the earlier course of the river are left as

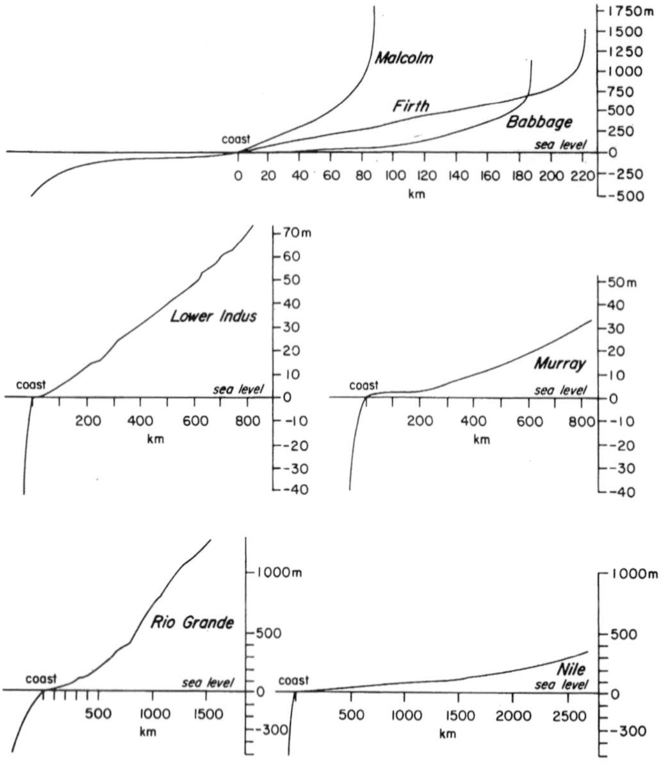

Fig. 11.15. Examples of continental margin profiles, showing the shapes of slopes that would be added to river profiles if sea level fell, exposing additional areas of the continental shelf. Profiles of Babbage and Firth rivers, Yukon, from McDonald and Lewis (1973); profiles of other rivers from Leopold et al. (1964). Profiles of shelves from *Times Atlas* and other sources. (Miall 1991c)

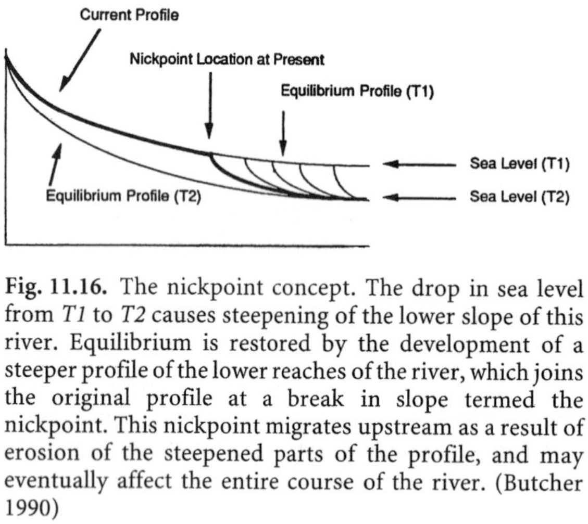

Fig. 11.16. The nickpoint concept. The drop in sea level from *T1* to *T2* causes steepening of the lower slope of this river. Equilibrium is restored by the development of a steeper profile of the lower reaches of the river, which joins the original profile at a break in slope termed the nickpoint. This nickpoint migrates upstream as a result of erosion of the steepened parts of the profile, and may eventually affect the entire course of the river. (Butcher 1990)

terraces on the valley margins. The response of the river to imposed changes in slope in terms of changes in fluvial style is complex, as discussed below. Rivers that have undergone such changes recently in response to glacioeustatic changes in sea level reveal complex cut-and-fill stratigraphies reflecting the integrated response to changes in base level and climate (Sect. 12.12.2).

Few workers have considered the actual slopes of continental margins and rivers. Pitman (1986) and Pitman and Golovchenko (1988) discussed a hypothetical case which might, in fact, correspond to sea-level change in the way described in the Exxon models:

"Let us consider a coastal region in which a large graded fluvial system enters the ocean at the coastline and in which the littoral zone has been transgressing slowly for a long period of time. In this case it is likely that the graded slope of the shelf will be less than the gradient of the lower course of the fluvial system that is being progressively transgressed. In that case, if there is a regression exposing the shelf, subaerial deposition will take place. Because the exposed shelf is at a gentler slope than the lower course of the fluvial system, the rivers crossing the exposed shelf must slow down and hence will deposit a part of their load. These will be entirely nonmarine coastal deposits. This coastal onlap will continue during regression."

Their illustration of this process is shown in Fig. 11.17. As Butcher (1990) pointed out, the scenario envisaged by Pitman and Golovchenko (1988) is less likely to occur than one where the lower course of the river has a lower slope than the continental shelf. This is especially likely to be the case where rivers have had an extended time period to build a coastal plain – the very situation most likely to be of interest to geologists because of the high preservation potential of such accumulations on slowly subsiding continental margins. In order to test these points, Miall

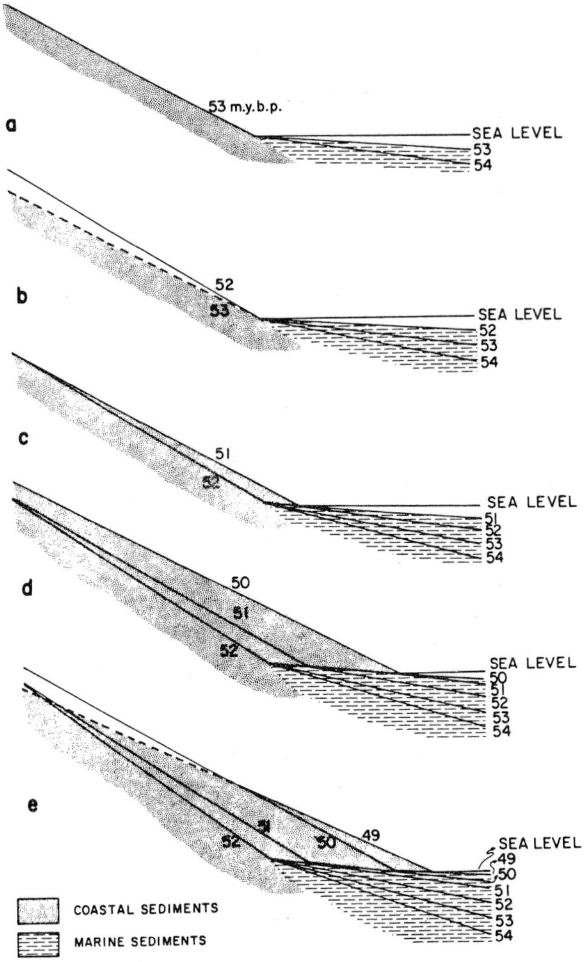

Fig. 11.17. Model of a fall in sea level from time 53 Ma (*a*), to 49 Ma (*e*). If the slope of the coastal plain is steeper than the slope of the shelf, regression and onlap will occur. (Pitman and Golovchenko 1988)

Table 11.1. Slopes of rivers and shelves. (Miall 1991c)

Location	Distance averaged (km)[a]	Slope
Rivers:		
Mississippi	650	0.000014
Indus	500	0.00008
Nile	1000	0.0001
Guadalupe (Texas)	?	0.00025
Rio Grande	500	0.0004
Babbage (Yukon)	100	0.0008
Firth (Yukon)	100	0.0036
Shelves:		
Guadalupe (Texas)	8	0.00023
Off Mississippi R	60	0.0003
South Carolina	75	0.00067
N. Slope, Alaska	65	0.0009
Murray	190	0.0011
N. Slope, Yukon	115	0.0017
Rio Grande	80	0.0026
Nile	20	0.005
Indus	40	0.005
Mississippi delta	20	0.0095

[a] Distance from coast to nearest available contour on shelf, or length of lowermost, flattest portion of river profile, measured from mouth.

(1991c) compiled a table of actual slopes of some modern rivers and shelves (Table 11.1). The results of this review are illustrated in Fig. 11.15. Contrary to the assumption of Pitman and Golovchenko (1988), many modern shelves have steeper slopes than the lower courses of the rivers flowing into them. This is particularly the case where the river has built a substantial delta out across the shelf (compare values for Mississippi and Nile) and also occurs where a submarine canyon heads near the river mouth (Indus). A fall in sea level would therefore result in steepening and a change in the fluvial style of the river. Incision of the lower course of the river would likely follow. Leopold et al. (1964, p. 476) predicted that, in general, glacioeustatic fall in sea level would lead to degradation at the mouth of the river. The only situation where a fall in sea level may lead to the development of accommodation space is where a relatively small, steep river flows out into a coastal plain flanked by a broad shelf. For example, if the three Yukon rivers (Babbage, Firth, Malcolm) were to be extended by sea-level fall across a shelf similar to that of the Gulf of Mexico between the Mississippi and the Guadalupe rivers, the newly extended lower courses of these rivers would be at a slightly lower slope, and a minor amount of aggradation might take place. In their present relationship with the Yukon continental shelf, the Malcolm may aggrade during a sea-level fall, the Babbage may undergo incision, and the Firth would likely not undergo significant change in grade. Such small rivers are of little geological significance. Wood et al. (1993) carried out laboratory experiments to study the influence of varying shelf angle on fluvial and deltaic deposition during rises and falls of base level and produced results very similar to those described here.

The numerical and graphical models of Nummedal et al. (1993) suggested that during deltaic progradation the longitudinal profile of the river will be shifted upward as the river mouth moves seaward, thereby creating accommodation space. They suggested that for this reason the trajectory of the river mouth has only an indirect relationship to the continental shelf. But under a condition of falling sea level a delta will prograde seaward only a limited distance before autogenic switching leads to the development of a shorter, steeper route to the sea. And however far the river mouth moves horizontally in response to progradation, it must also move vertically in response to base-level change. Whatever the slope of the continental shelf, a fall in sea level forces the river mouth to drop. The slope of the shelf is therefore relevant to questions of accommodation space, as discussed above.

There is therefore no *a priori* reason why a fall in sea level should be accompanied by creation of accommodation space and consequent fluvial aggradation and coastal onlap. The extension of a thin fluvial sheet seaward over shoreline deposits seems likely, but thick aggradation and coastal onlap does not seem probable. For example, note that in the thick clastic-wedge cycles described by Embry (1990a, 1991) there is no wedge of onlapping fluvial strata between the subaerial unconformity (sequence boundary) and the overlying transgressive deposits. This supports my objections to the Posamentier and Vail (1988) model. Additional support has been provided by the numerical models of Rivenaes (1992).

With regard to horizontal shifts in the mouth of a river (for example, as it "tracks" a prograding delta: Nummedal et al. 1993), the condition shown in Fig. 11.11, Leopold et al. (1964, pp. 260–266) and Leopold and Bull (1979) presented data showing that a rise in base level or horizontal extension of a profile, as in the filling of and progradation across a reservoir, has a very minor effect on the overall river profile. Aggradation takes place for a few hundred meters above the reservoir. Above this the longitudinal profile is characterized by a transitional reach, which curves upstream into the remainder of the river, whose slope and longitudinal profile remain unaffected. Schumm (1993) suggested that the rivers observed by Leopold et al. (1964) and Leopold and Bull (1979) were too small and the observation time too short for these to be useful examples from a geological point of view. The major rivers draining into the Gulf of Mexico show significant incision relating to glacioeustatic falls in base level; for example, the Mississippi valley is incised for 370 km upstream from its mouth, as far as Baton Rouge (Schumm 1993), with aggradation having taken place during the subsequent rise. However, the precise nature of the relationship between base-level change and changes in equilibrium profile in these rivers is confused by the nature of complex response referred to at the beginning of this section (and was also rendered more complex by climate change, as discussed in Sect. 12.12.2).

The shift in the entire profile suggested by Posamentier et al. (1988) has not been observed in any of the rivers described by Leopold et al. (1964), Leopold and Bull (1979), and Schumm (1993). Even if such a shift could occur, it would create little accommodation space. Given a slope of 0.001, typical of moderate to large rivers such as the Nile (Table 11.1), simple trigonometry indicates that moving the mouth of the river horizontally a distance of 100 km would create only 10 m of accommodation space, which is entirely inadequate to explain the large thicknesses of fluvial sediments present in some stratigraphic sequences (e.g., see Galloway 1989b). The Yellow River of China illustrates this point (Qian 1990, as quoted by Schumm 1993). The river aggraded significantly between the late seventeenth century and 1855 as a result of the growth of its delta. During this period the lower reach of the river was extended nearly 200 km by deltaic progradation, and the channel aggraded its bed to 10 m above the level of the North China Plain. However, this effect occurred only because of the confinement of the channel between artificial levees, which were breached by a major flood in 1855. Without the levees the increase in accommodation space would have resulted in channel and delta-lobe switching, probably without significant aggradation.

Recent work by Leckie (1994) on the rivers of the Canterbury Plains, New Zealand, demonstrated that transgression is not necessarily accompanied by the immediate creation of accommodation space. Because of very high wave energies along the east coast of South Island, New Zealand, much of the coastal plain is undergoing erosional shoreface retreat, and the rivers are incising their valleys near their mouths as they grade down to the retreating shoreline. The process is analogous to the development of the ravinement surface that develops during transgression of a shallow shelf.

Another assertion by Posamentier et al. (1988, p. 135) that should be commented on is the prediction of an "overall upward-fining trend" in grain size in intermediate positions of the stream profile (e.g., a short distance downstream from A in Fig. 11.11). This remark is likely to lead to many simplistic deductions made from observed grain-size trends in well logs. The argument is that in the new accommodation space slopes are initially lower, leading to a dumping of coarse sediment, followed by remobilization of this sediment as slopes increase again.

The basic hydraulic flaw in this argument is that the deposited sediment is assumed to become finer as slopes (and thus velocity and shear stress) increase (from slope AF to DF). Of course, the exact opposite would happen. An increase in slope results in increased competency and increased grain size. Sediments deposited on the fluvial plain represent temporary sediment storage, with geological preservation typically occurring because of abandonment of a deposit before it can be remobilized (e.g., lateral accretion on a point bar, channel avulsion). Therefore, if the pattern of Fig. 11.11 were actually to occur, an upward-coarsening profile would result.

In conclusion, the model of Posamentier et al. (1988) and Posamentier and Vail (1988), suggesting that significant fluvial sedimentation, coastal-plain progradation, and coastal onlap occur during times of rapidly falling sea level (the late-highstand phase, when eustatic fall outpaces subsidence, and an actual relative fall occurs: times T4–T6 in Posamentier and Vail 1988, Fig. 8), seems incorrect. It seems much more likely that the main coastal-plain fluvial accumulations occur either during relative rises in sea level or during periods of tectonism and source-area uplift (as described earlier in this chapter). These conclusions were supported by the numerical modeling experiments of Rivenaes (1992), but they need to be evaluated in each field case, in light of local variations in subsidence, sediment supply, and marine energy, as demonstrated by the work of Leckie (1994). The rise in sea level following the last ice age is a good example of a transgressive situation. Most river valleys were initially drowned and then became filled with estuarine and fluvial facies. Suter et al. (1987) showed that this happened several times on the Louisiana shelf during the late Quaternary sea-level changes (Chap. 13). Yet because of high sediment supply the major river on the Gulf Coast, the Mississippi, has prograded significantly.

A recent text on sequence stratigraphy by H.W. Posamentier and G.P. Allen (manuscript in preparation) contains many improvements on the 1988 synthesis. The following lengthy quote provides the current state of ideas by the acknowledged leading experts in the field of sequence stratigraphy:

"During the early lowstand phase, relative sea level is gradually falling and the highstand shelf or coastal plain is subjected to subaerial exposure and fluvial incision. Sediments pass through this area with no net deposition. Consequently, in the absence of tectonic tilting, within the incising river channels there is no net accumulation of fluvial sediment during this phase. Locally, however, fluvial deposits can be preserved in the form of stranded fluvial terraces. ... When relative sea level stops falling, fluvial incision ceases. At this time, examination of fluvial systems would reveal a minimum fluvial deposit concentrated along the axes of incised valleys. ... After relative sea-level fall ends, and stillstand and slow rise ensues, rivers stabilize and then begin to aggrade as coastal plains enlarge by normal (in contrast with "forced") shoreline regression. The total amount of fluvial aggradation during this late lowstand phase results from the combination of two mechanisms: 1) tectonically-induced aggradation by a landward tilting of the fluvial profiles, and 2) aggradation provoked by the seaward and upward migration of the fluvial equilibrium profile during normal shoreline regression under the influence of slow relative sea-level rise or stillstand." (H.W. Posamentier and G.P. Allen, manuscript in preparation, p. 51)

The landward tilt referred to in this excerpt applies in the case of foreland basins in which subsidence is more rapid in the proximal part of the basin, but would not apply in the case of extensional continental margins. "Early lowstand" corresponds to the late highstand of Posamentier and Vail (1988).

Experimental work and geomorphological observations on the effects of slope changes on rivers have

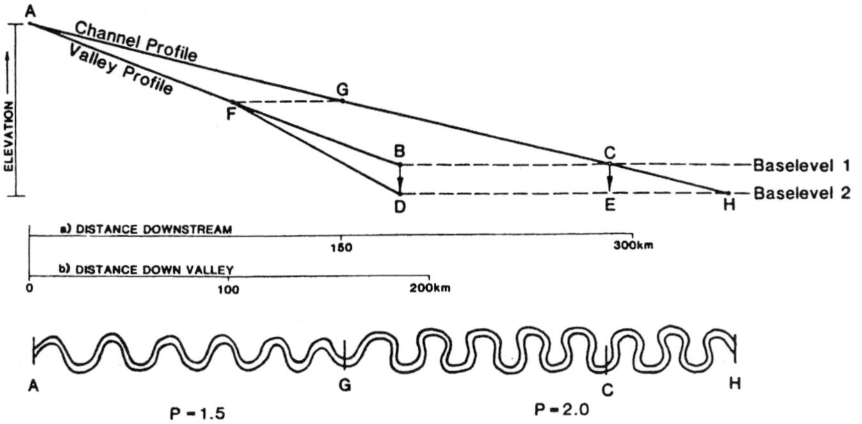

Fig. 11.18. Effect of base-level fall on channel lengths and patterns. See text for discussion. (Schumm 1993)

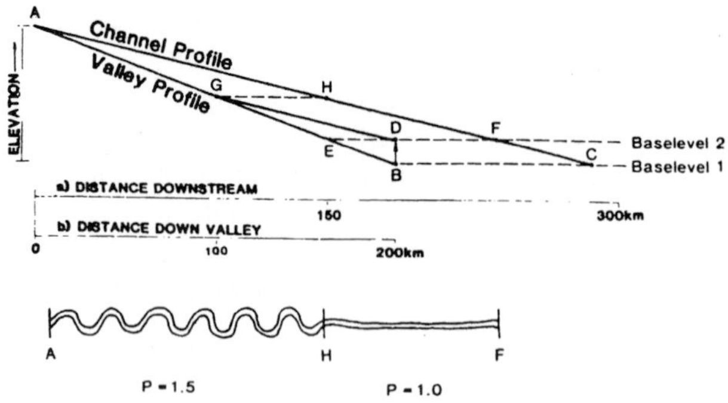

Fig. 11.19. Effect of base-level rise on channel length and patterns. See text for discussion. (Schumm 1993)

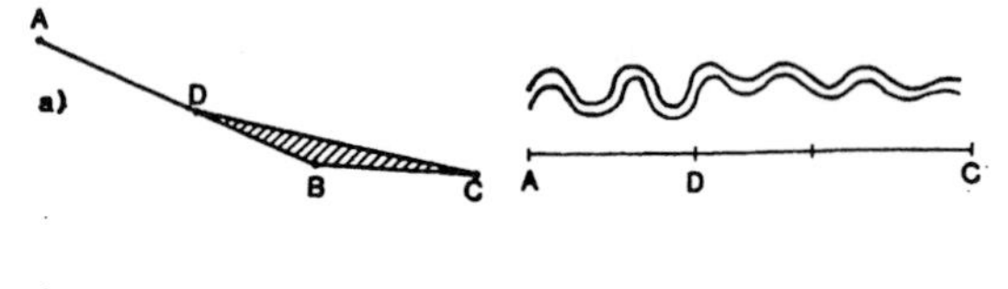

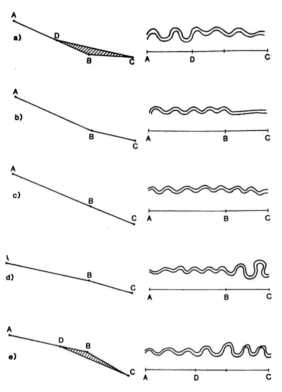

Fig. 11.20. Summary of effects of changes in slope in the lower course of rivers, for example, as a result of base-level lowering across continental shelves of different inclinations. See text for discussion. (Schumm 1993)

long been available, as described in Sects. 8.1 and 11.2.1.2 (e.g., Ouchi 1985; see Figs. 11.7, 11.8), and need to be fully integrated into the study of sequence stratigraphy. The publication of Posamentier et al. (1988) stimulated a revival of interest in this area of research, and Schumm (1993) and Wescott (1993) have now clarified this subject. Using geometrical constructions, Schumm (1993) illustrated a well-known geomorphic principle, i.e., that rivers are capable of making significant adjustment to changes in slope without incision or aggradation, simply by changes in sinuosity. Figures 11.18–11.20 demonstrate his arguments. It is important to distinguish between valley slope and channel slope. The latter is less, by an amount directly proportional to sinuosity (see Sect. 8.1). In Figs. 11.18 and 11.19, valley slope is shown by AB and channel slope by AC. Consider the effect of an arbitary base-level fall at a point - for example, by a change in dam height in a sediment-filled reservoir (a drop from B to D in Fig. 11.18). The channel profile can lengthen from C to H, as the stream incises downward to the new profile FD. An increase in channel length over the same valley distance without a change in channel slope or valley length means an increase in sinuosity. A rise in base level results in aggradation over the area FBD and a reduction in sinuosity (Fig. 11.19). These changes would be gradual. Point F in Figs. 11.18 and 11.19 represents the nickpoint, which retreats upstream. Changes in sinuosity are also accompanied by changes in channel depth and bed roughness, and by changes in the sediment supply. For example, in experimental studies of base-level lowering, the channel becomes shallower but wider, and valley incision feeds additional sediment into the river, which must be compensated by a steeper gradient.

The arbitrary conditions presented in Figs. 11.18 and 11.19 are given greater geological relevance in Fig. 11.20, where the effects of base-level lowering exposing continental shelves of different slope are summarized. In examples (a) and (b) the newly exposed shelf has a gentler slope than the stream channel. In (b) the change in slope can be accommodated by a reduction in sinuosity without aggradation, but in (a) aggradation must also take place to compensate for a much gentler slope (area DBC). In (c) the river extends its course without changes in fluvial style. In (d) the river can accommodate to a steeper course by an increase in sinuosity, and in (e) the natural limit on the ability of a river to increase its sinuosity because of chute and neck cutoffs leads to incision and to the erosion of area DBC.

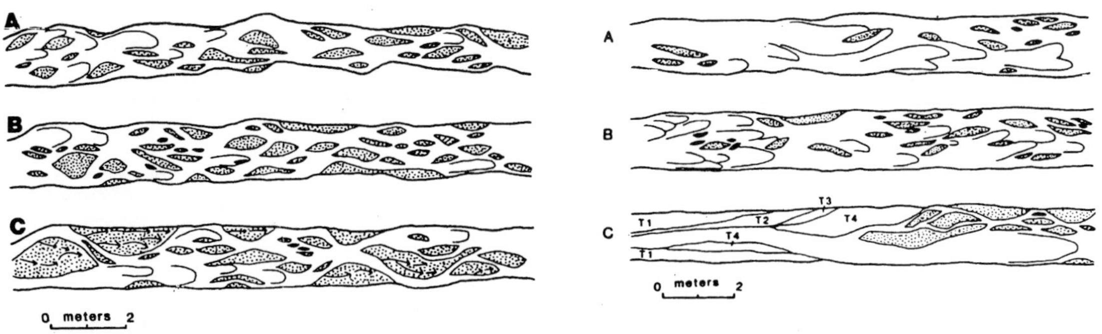

Fig. 11.21. Maps of flume experiments showing *A* equilibrium, *B* aggrading, and *C* degrading channels; gravel-bed rivers at *left* and sand-bed rivers at *right*. In degrading sand-bed channels incision leads to terrace development upstream (numbered in order from *T1* to *T4*) and to a wave of aggradation downstream. Eventually, a return to dynamic equilibrium would reduce the size of the aggraded braid bars in this downstream area. (Germanoski and Schumm 1993)

Experimental studies suggest that the responses described above may apply only to sinuous rivers (Schumm 1993). In the case of braided rivers, base-level lowering causes channel instability and incision. Incision itself has the apparently contradictory effect of increasing valley deposition in the lower reaches because of the increased local sediment supply. Deepening and widening is eventually compensated by significant valley-floor aggradation. Germanoski and Schumm (1993) reported on a series of experiments designed specifically to test the response of gravel-bed and sand-bed braided rivers to changes in sediment supply. Increases in sediment supply cause aggradation, resulting in an increase in the number of braid bars, whereas a reduction in the sediment supply leads to channel incision and the coalescence of braid bars into fewer, larger bars (Fig. 11.21). Changes in sediment supply and the changes in braided style resulting from them could be expected to occur as a result of changes in base level. Thus, incision resulting from a reduction in accommodation space increases the downstream sediment supply as a result of increased bank erosion and leads to downstream aggradation. Aggradation that occurs to fill increased accommodation space resulting from a rise in base level reduces the incidence of bank erosion, leading to a reduction in sediment supply. The experiments of Germanoski and Schumm (1993) suggest that the change in the dynamic equilibrium in favor of aggradation would therefore be balanced to some extent by the tendency for individual channels to incise and rework the bed materials. As they noted, however, their experiments were not scaled to actual full-size rivers, and it would be premature to translate their results directly to interpretations of large-scale natural systems. Section 12.12.2 deals with the response of rivers to changes in sediment supply resulting from climate change.

The changes in sinuosity illustrated in Fig. 11.20 carry implications for other changes in fluvial response, as described by Schumm (1993) and Wescott (1993). Reduction in sinuosity near the river mouth (Fig. 11.20a,b) is comparable to a mountain stream entering an inland basin. The formation of a fan delta may be the result. An increase in sinuosity (Fig. 11.20d,e) may be accompanied by development of other characteristics of high-sinuosity streams, such as levee development, with crevassing, cutoffs, and avulsion. Newly exposed areas of continental shelf that emerge as a result of base-level fall develop their own drainage networks that interact with the previously existing streams. They will be smaller in drainage area and consequently develop smaller channels. A variety of fluvial styles may therefore be present along continental margins. Other implications for sequence stratigraphy are discussed in Chap. 13.

Numerous studies have been carried out recently on the effects of tectonism and changing base level on alluvial fans and fan deltas (e.g., Muto 1987, 1988; Gawthorpe and Colella 1990; Bardaji et al. 1990; Fernández et al. 1993). The facies and architecture of fan deltas constitute a specialized topic that can now be illustrated by numerous case studies (Nemec and Steel 1988; Colella and Prior 1990), but the details of this are considered beyond the scope of this book. Muto (1987, 1988) discussed the implications of lowering base level across a slope that is greater or less than the slope of the fan and suggested that the fan would entrench or aggrade, respectively, based on arguments similar to those of Schumm (1993), as shown in Fig. 11.20. A complete cycle of base-level

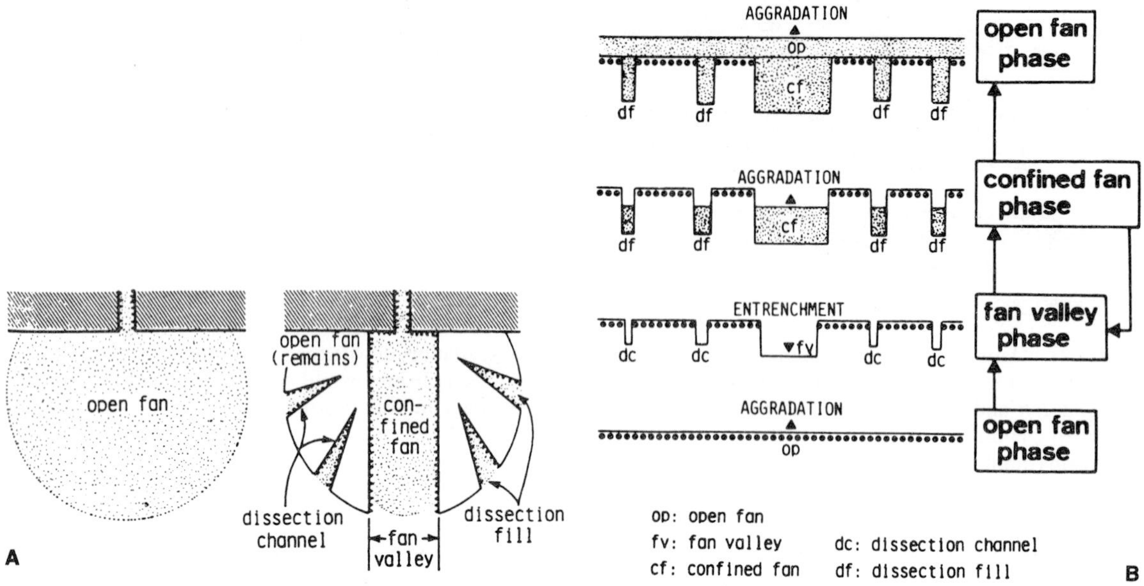

Fig. 11.22. A Terminology for alluvial fan and fan-delta morphology; B phases in the development of fan morphology in response to changes in base level. See text for explanation. (Muto 1988)

change generates a three-stage change in overall fan architecture comparable to the autocyclic entrenchment-backfilling-dispersal cycle described in Sect. 10.2 (Fig. 11.22). Entrenchment causes the development of confined fan valleys and dissection of channels. These may fill during a rise of base level, leading eventually to the spreading out of fan deposits across the entire fan surface. Muto (1988) showed how the superimposition of base-level changes on basins of varying subsidence rates would change the overall architecture of the succession, a topic discussed in Chap. 13. In a study of interfingering fan-delta and lacustrine deposits, Fernández et al. (1993) described a type of cycle similar to that of Muto, in which the entrenchment phase is represented by ribbon-shaped gravel bodies filling narrow, incised channels, and the open-fan phase by sheet-shaped gravel bodies and by interfingering with fine-grained lacustrine deposits.

11.3 Tectonic Control of Basin Style and Basin-Scale Fluvial Patterns

11.3.1 Big Rivers

Most of the very large rivers of the world have a long geological history and follow very long established courses through structural depressions, even though their courses may be modified or even reversed by changes in plate-tectonic configuration. For example, a river has occupied the course of the Mississippi River since at least the Late Paleozoic, in the position of an ancient Precambrian rift (Potter 1978). The Amazon River may also follow the course of ancient rifts, amplified and deepened during the Mesozoic opening of the Atlantic Ocean. Other rivers draining into the Gulf of Mexico were in similar positions throughout most of the Tertiary, as shown by the position of modern river courses immediately above thick Tertiary fluvial-deltaic lobes entering the basin from the same position on the continental margin (Fisher and McGowen 1967; Galloway 1981; Fig. 11.23).

Some of these large rivers cross major plate-tectonic boundaries, with the result that in their lower courses their size and discharge characteristics may bear no relationship to the local tectonic and climatic control, while their sediment load may reflect the mineralogy of far distant sources. Good examples are the Nile and the Amazon. The headwaters of the Nile are in the tropical grasslands and rain forests of central Africa, and discharge is strongly seasonal where the river flows through the arid deserts of Egypt. The Amazon headwaters are located in the Alpine snowfields of the Andean mountains, contrasting with the rain-forest basin along its lower course. The rivers draining the Alpine-Zagros-Himalayan foldbelt show similar contrasts between source area and basinal climate.

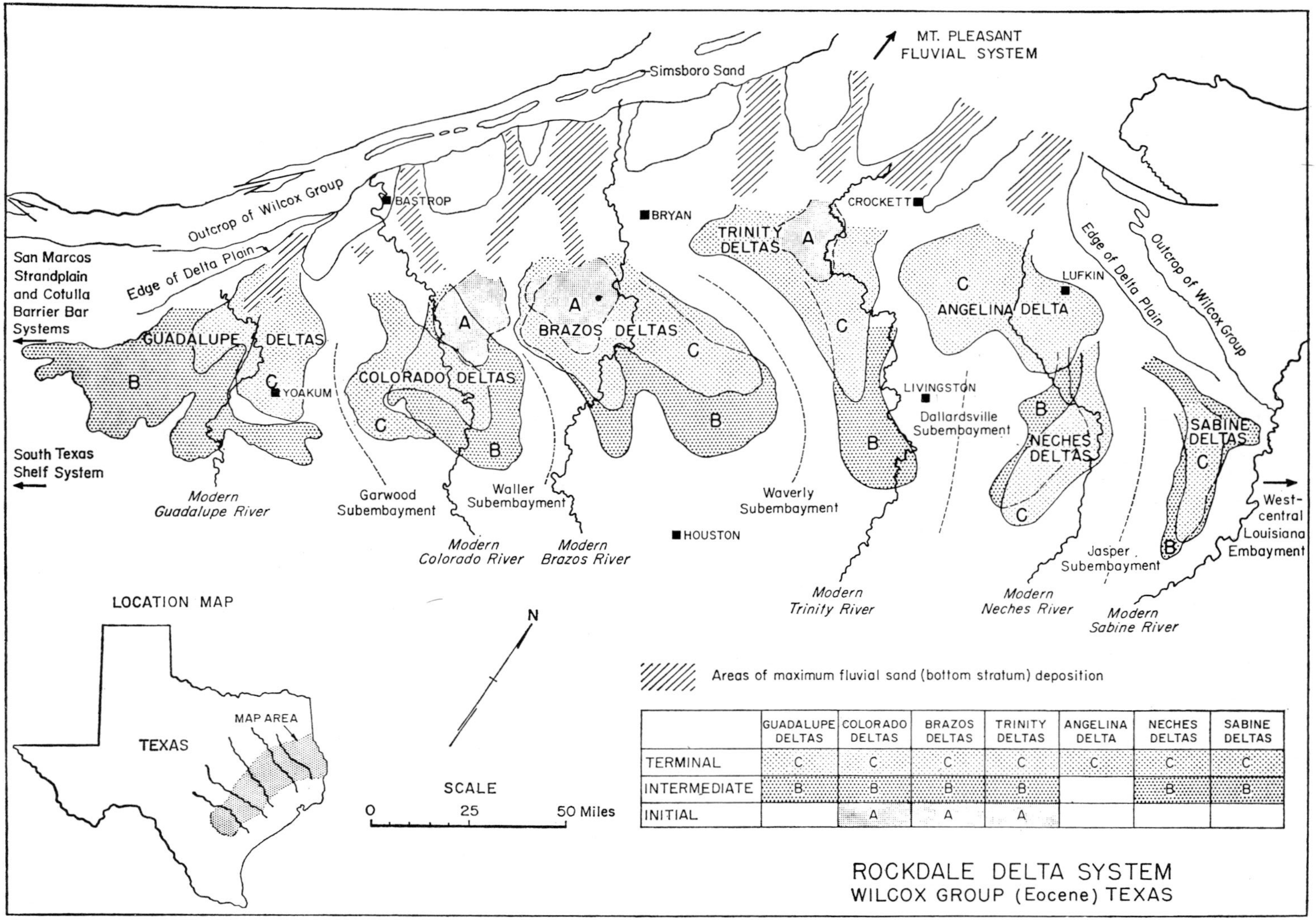

	GUADALUPE DELTAS	COLORADO DELTAS	BRAZOS DELTAS	TRINITY DELTAS	ANGELINA DELTA	NECHES DELTAS	SABINE DELTAS
TERMINAL	C	C	C	C	C	C	C
INTERMEDIATE	B	B	B	B		B	B
INITIAL		A	A	A			

Fig. 11.23. The longevity of fluvial depositional systems flanking the Gulf of Mexico. Simplified isopachs outline delta systems of the Eocene Wilcox Group. Note their coincidence in space with the course of modern rivers. (Fisher and McGowen 1967)

Most major rivers enter the sea at divergent plate margins (Mississipi, Rio Grande, Mackenzie, Murray, Amazon, Yellow, Plate, Congo) or where the axes of foreland basins terminate at the margins of continental plates (Po, Tigris-Euphrates, Ganges, Indus). Foldbelts that are marginal to continents, such as the Appalachians and Cordilleran belts of North America, have continental divides close to the continental margin, so that the rivers draining their seaward flanks are short. A major exception is the Columbia River, which crosses the coastal mountains of the western United States by a geologically complex route. On the landward flank of such foldbelts are the vast continental interiors, which by their very size contribute to the development of major rivers.

11.3.2 Axial and Transverse Drainage

A particular characteristic of the world's major rivers is that most are axial or longitudinal rivers, that is, their course parallels regional structural grain (Kuenen 1957; Potter 1978; Miall 1981a,b). The grain may be that of major rift systems, as in the case of the Mississippi, Rio Grande, Amazon, Niger, lower Rhône, Rhine, and upper Nile (where it flows through part of the East African Rift System), or it may be the axis of foreland and forearc basins, such as those occupied by the Mackenzie, Po, Tigris, Euphrates, Indus, Ganges, lower Brahmaputra, and Irrawaddy. Others flow along the structural grain of foldbelts until near their mouths, such as the Mekong. A few major rivers, such as the Columbia, are transverse systems; that is, they are consequent streams draining directly from an uplifted source area, or antecedent streams that have maintained their course as mountain systems rose in their path. It has been suggested that many of the major transverse rivers of the Himalayan region are of the antecendent type (Holmes 1965, p. 595). Transverse and axial systems normally have very different channel scales and styles and may carry very different sediment loads, resulting in the interfingering of contrasting facies assemblages with distinct sandstone colors, compositions, and sedimentary structures. Typically, transverse systems are much smaller and carry coarser sediment loads. They may, for example, be alluvial fans. Sinha and Friend (1994) carried out a detailed study of this point with reference to the large rivers of the Indo-Gangetic plains of northern India. Examples of the contrast-

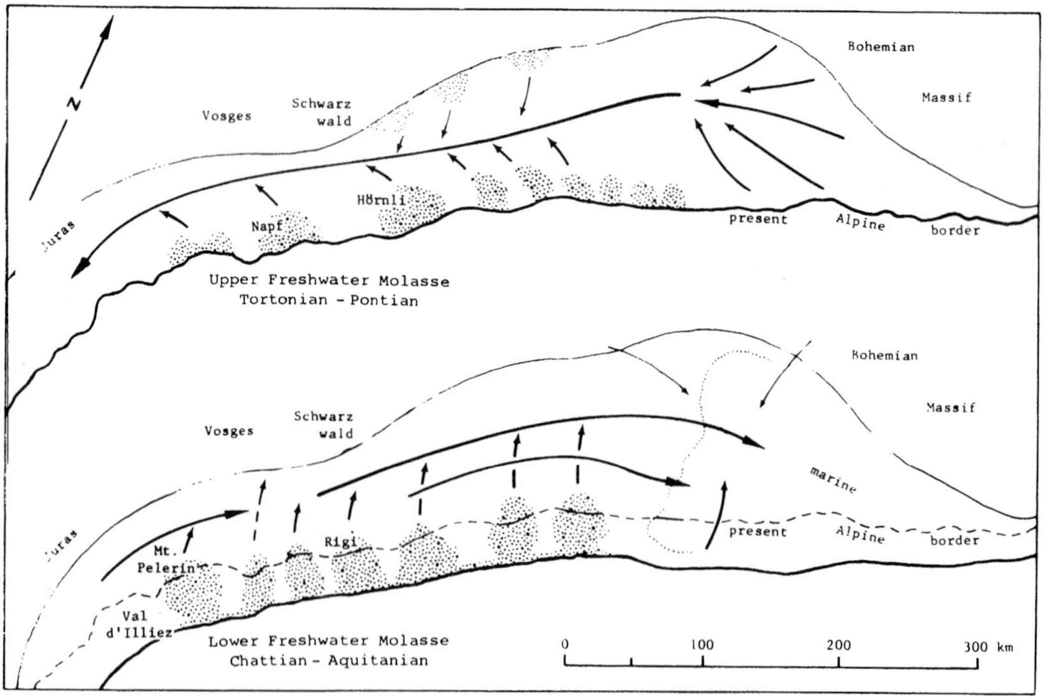

Fig. 11.24. Paleogeography of the Swiss Molasse basin during deposition of the Lower and Upper Freshwater Molasse. Transverse alluvial fans fed axial drainage, which reversed in direction between these depositional phases. (Van Houten 1981)

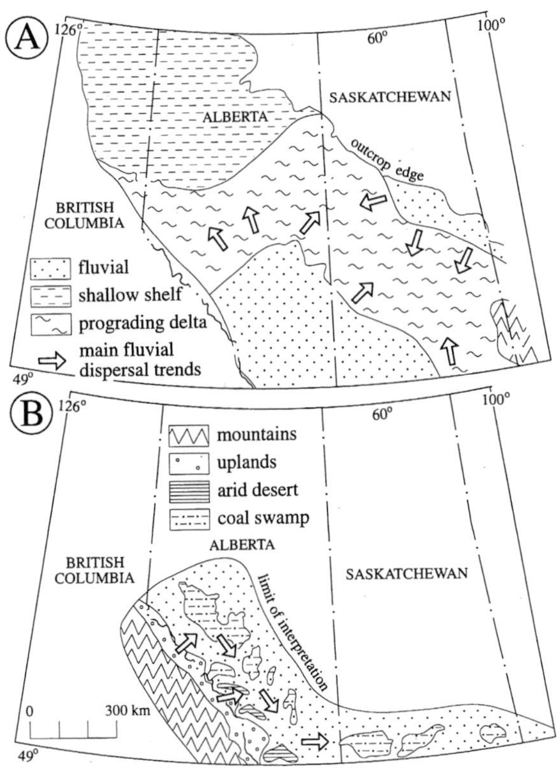

Fig. 11.25A,B. Regional drainage of the Alberta Basin at two stages in its development: **A** during the Albian (D.G. Smith, in Mossop and Shetsen 1994); **B** during the earliest Paleocene. (Dawson et al., in Mossop and Shetsen 1994)

ing styles of axial and transverse river systems are described in several places in this book (e.g., see discussion of Figs. 9.1, 11.10).

Axial drainages are particularly sensitive to the larger scale of plate-tectonic changes in a region. For example, as shown in Sect. 11.2.1.2, the Indus River, which formerly flowed longitudinally southeastward for the full length of the Himalayan foredeep (Burbank and Beck 1991), was diverted by growth of the Salt Range and uplift of the piggyback basin behind it and now flows southward. Van Houten (1981) demonstrated a reversal in axial drainage along the Swiss Molasse basin during the Oligocene, as a result of variations in rates of uplift and sediment supply along the Alpine Orogen (Fig. 11.24). A similar reversal in drainage occurred in the Alberta basin during the Cretaceous (Fig. 11.25), for similar reasons.

The relationships between drainage patterns, basin subsidence, and sediment supply are further discussed in Sect. 11.3.6. Regional paleogeographic models that incorporate the transverse/longitudinal subdivision of drainage patterns are discussed in Sect. 11.5.

11.3.3 Regional Tectonic Control Revealed by Basin Analysis

Various regional paleogeographic indicators can be referred to in the evaluation of tectonic influences. This has become important in recent years because of the debate regarding the tectonic-versus-eustatic origin for stratigraphic sequences. For example, Embry (1990a) studied the Mesozoic stratigraphic record of Sverdrup Basin, in Arctic Canada, a succession up to 9 km thick which he subdivided into 30 third-order stratigraphic sequences. Based on this work, Embry (1990a) suggested the following lines of evidence which may be used to indicate a significant tectonic control of the basin-fill architecture:

> "(1) [T]he sediment source area often varies greatly from one sequence to the next; (2) the sedimentary regime of the basin commonly changed drastically and abruptly across a sequence boundary; (3) faults terminate at sequence boundaries; (4) significant changes in subsidence and uplift patterns within the basin occurred across sequence boundaries; and, (5) there were significant differences in the magnitude and the extent of some of the subaerial unconformities recognized on the slowly subsiding margins of the Sverdrup Basin and time equivalent ones recognized by Vail et al. (1977, 1984) in areas of high subsidence."

Although applied here to the Sverdrup Basin, all of these criteria have a general applicability. They are particularly useful in cases such as the Sverdrup Basin, which does not flank any contemporaneous foldbelt that could have served as the sediment source. The sediments filling the Sverdrup Basin are now thought to have been derived by intraplate uplift of the cratonic interior of North America (Embry 1990a, 1991).

Lawton's (1986a,b) work on the foreland-basin clastic wedge of Utah is a good illustration of the use of basin-analysis techniques to demonstrate tectonic control. Lawson used paleocurrent and petrographic evidence to indicate shifting sediment sources and changes in regional paleoslope during deposition of the Mesa Verde Group and associated units in Utah (Fig. 11.26). He was able to demonstrate the unroofing of intrabasin uplifts from which early basin-fill sediments were cannibalized, and he also documented changes in dispersal patterns related to basin tilting. Cowan (1993) used petrofacies data to demonstrate shifting sediment sources in the Permo-Triassic Sydney Basin, Australia, resulting

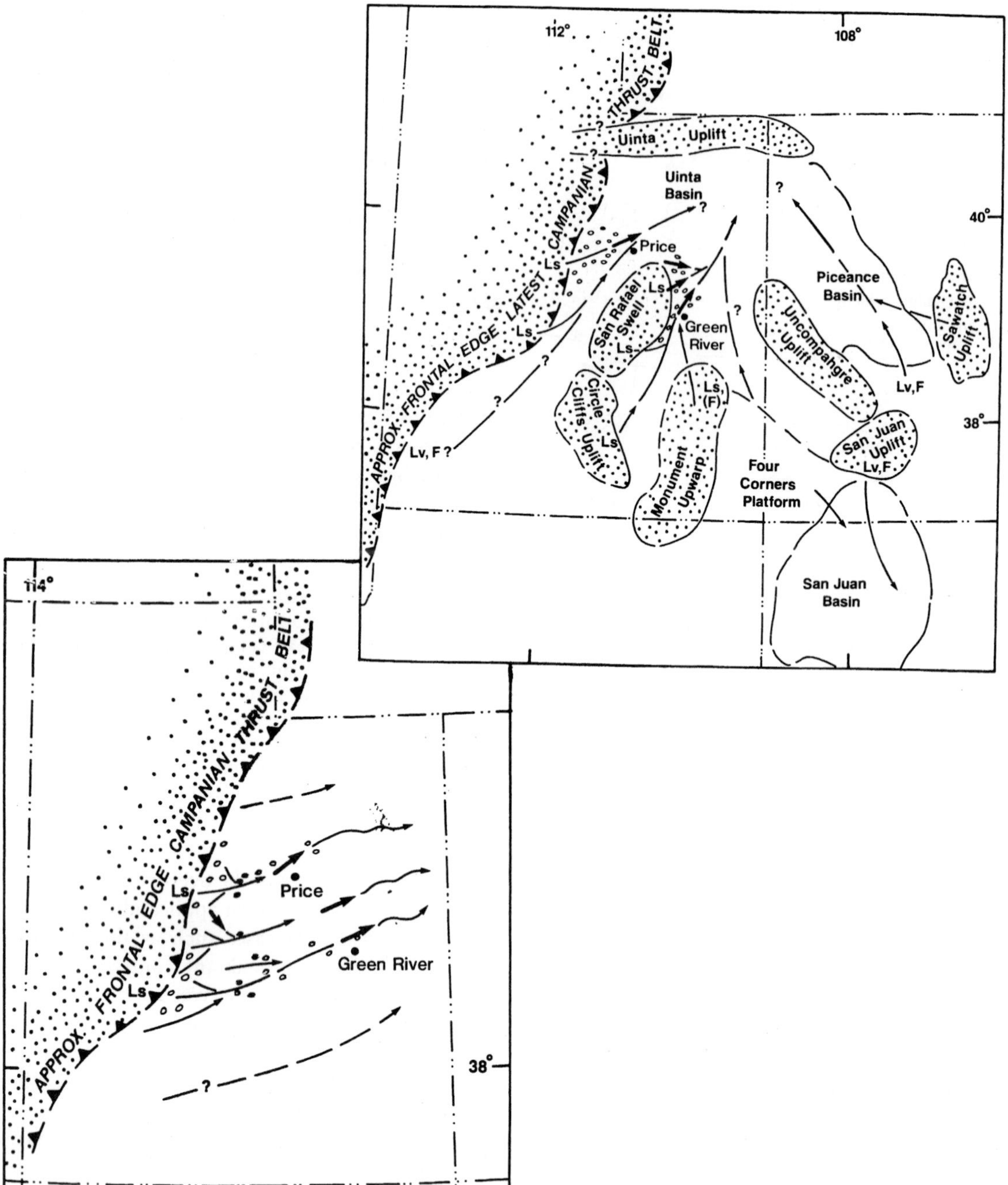

Fig. 11.26. Evolution of fluvial dispersal systems in Utah from the mid-Campanian (Castlegate Formation, *left*) to the late Campanian (Price River, Tuscher Formations, *right*), based on analysis of source terranes and paleocurrents (*heavy arrows*). The mid-Campanian pattern is typical of transverse-oblique drainage from a foreland fold-thrust belt. The late Campanian pattern indicates the development of basement uplifts within the basin and major reorganization of drainage patterns. Stratigraphic sequences formed in the basin clearly were strongly influenced by tectonism. (Lawton 1986b, reprinted by permission)

from differential uplift of the marginal fold-thrust belt and of the craton on the opposite side of the basin.

11.3.4 Tectonism and Sediment Supply

Sediment supply is primarily driven by tectonism, but in geologically simple areas sediment-supply considerations are likely to be subordinate to basin subsidence and eustasy as major controls of basin architecture. Such is the case where the basin is fed directly from the adjacent margins, and source-area uplift is related to basin subsidence - for example, through the yoking of subsidence to peripheral upwarps or the fold-thrust belts marginal to foreland basins. It is where the basin is supplied by long-distance fluvial transportation that complications are likely to arise.

Many sedimentary basins were filled by river systems that have long since disappeared from the continental surface, and it may take considerable geological imagination to reconstruct their possible past position. For example, McMillan (1973) envisaged a Tertiary river system draining from the continental interior of North America into Hudson Bay. Potter (1978) discussed several similar examples. As noted in Sect. 11.3.1.1, major river systems may cross major tectonic boundaries, feeding sediment of a petrographic type unrelated to the receiving basin into the basin at a rate unconnected in any way with the subsidence history of the basin itself. The modern Amazon river is a good example. It derives from the Andean Mountains, flows across and between and is fed from several Precambrian shields, and debouches onto a major extensional continental margin. This is an important point with respect to questions of sequence stratigraphy, because it means that large sediment supplies delivered to a shoreline may overwhelm the stratigraphic effects of variations in sea level. A region undergoing a relative or eustatic rise in sea level may still experience a major stratigraphic regression if large delta complexes are being built by major sediment-laden rivers.

A significant illustration of this point is the Cenozoic stratigraphic history of the Gulf of Mexico. This

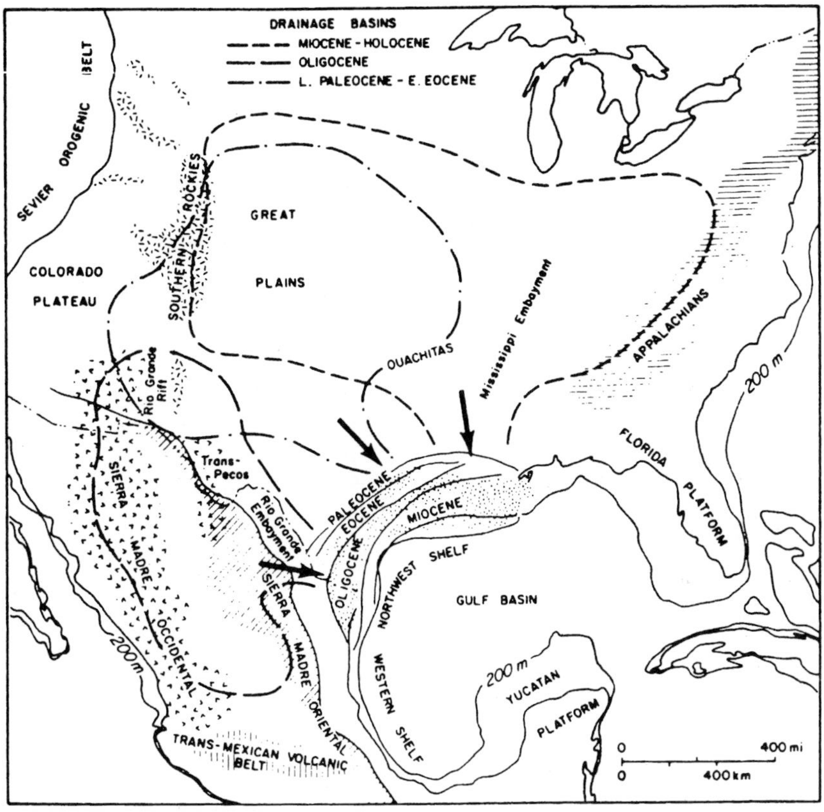

Fig. 11.27. The major drainage basins which fed delta complexes on the Gulf Coast during the Cenozoic. *Arrows* indicate the position of three long-lasting "embayments" through which rivers entered the coastal region. (Winker 1982; Galloway 1989b, reprinted by permission)

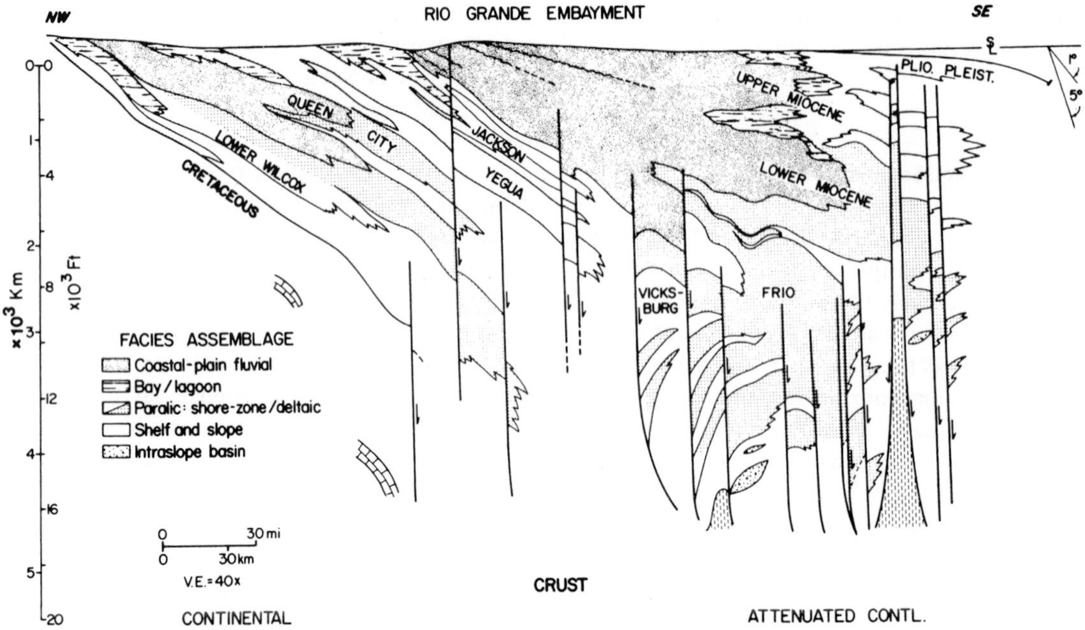

Fig. 11.28. Generalized dip-oriented stratigraphic cross section through the Rio Grande depocenter, in the northwest Gulf Coast (location shown in Fig. 11.27), indicating principal Cenozoic clastic wedges. Many of these thicken southward across contemporary growth faults. (Galloway 1989b, reprinted by permission)

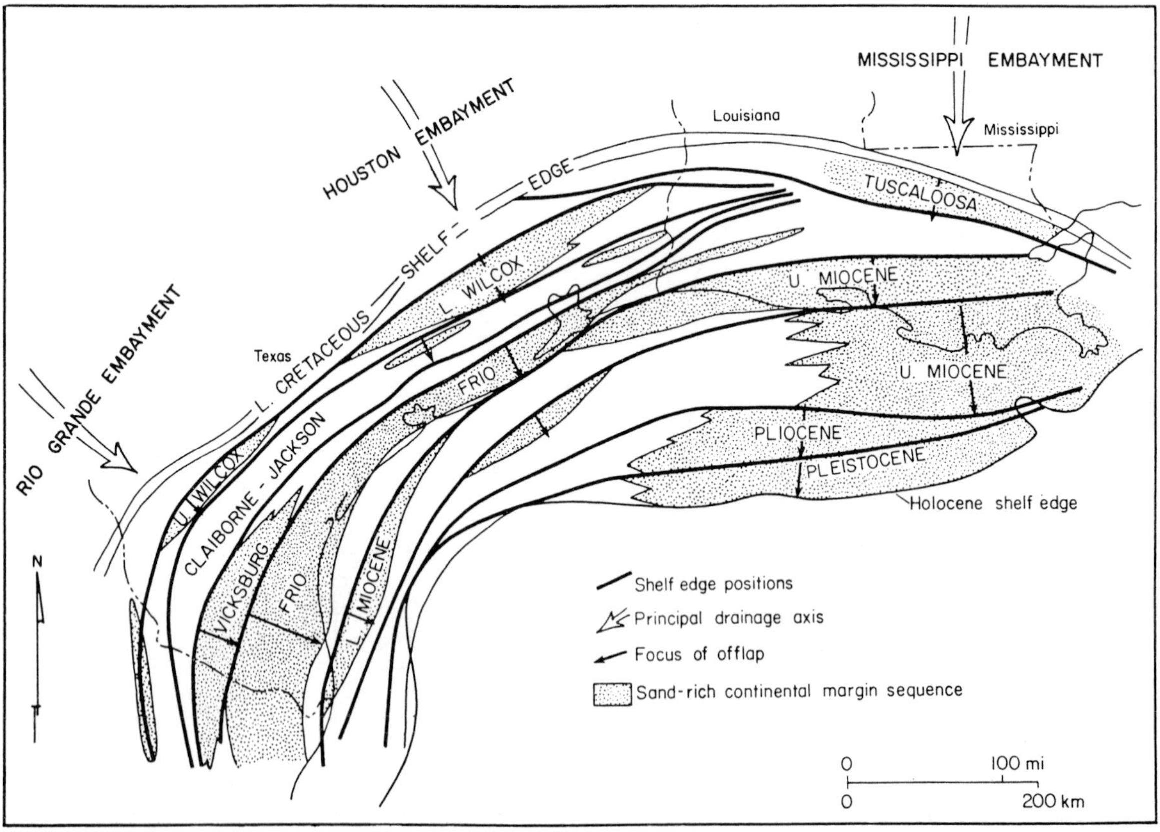

Fig. 11.29. Progradation of the Gulf Coast continental margin by the development of clastic wedges during the Cenozoic. Locations of the three principal embayments (structural sags) through which rivers entered the coastal plain are also shown. (Winker 1982; Galloway 1989b, reprinted by permission)

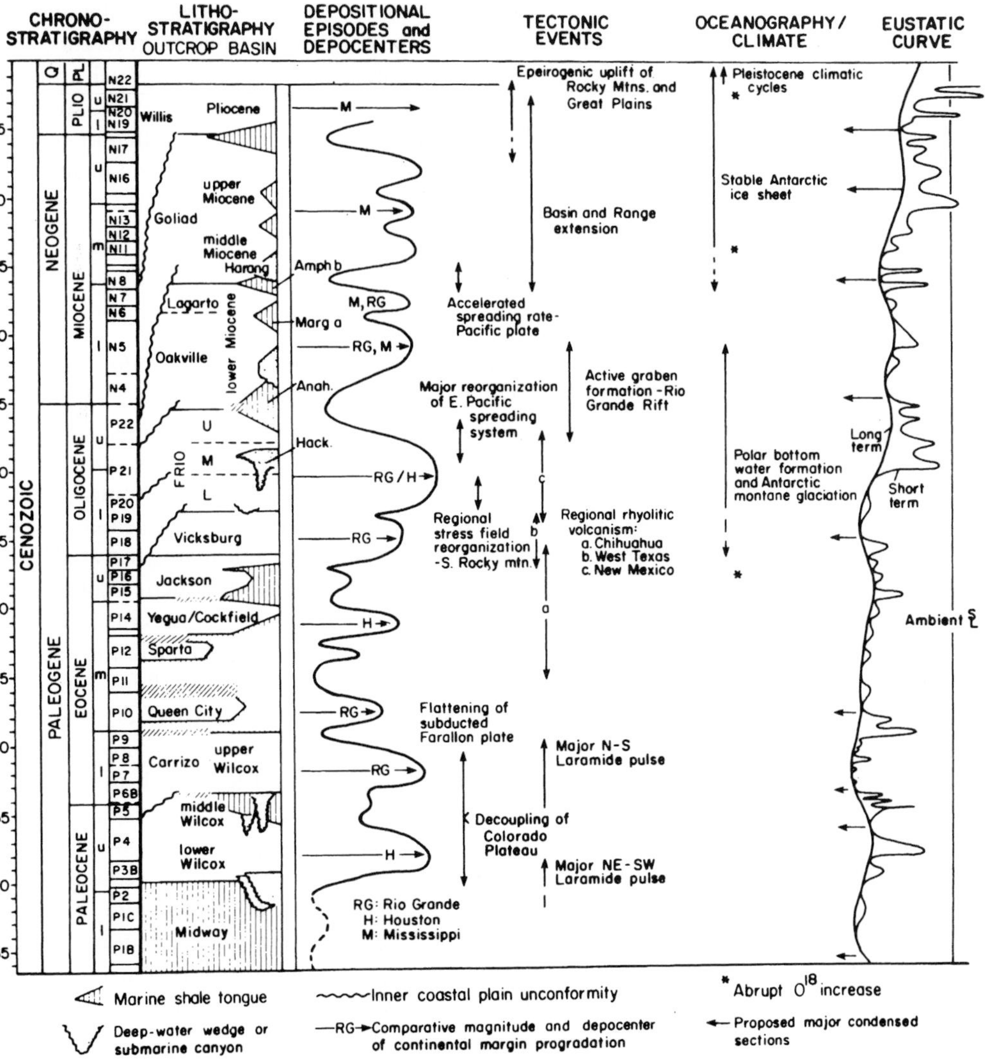

Fig. 11.30. Age of the major clastic wedges along the Gulf Coast, compared with the age range of tectonic events in the North American Interior. The global cycle chart of Haq et al. (1987, 1988) is shown for comparative purposes at the *right*. (Galloway 1989b, reprinted by permission)

continental margin is fed with sediment by rivers that have occupied essentially the same position since the early Tertiary (Fisher and McGowen 1967; Fig. 11.23). The rivers feed into the Gulf Coast from huge drainage basins occupying large areas of the North American interior (Fig. 11.27). Progradation has extended the continental margin of the Gulf by up to 350 km. This has taken place episodically in both time and space, developing a series of major clastic wedges some hundreds of meters in thickness (Figs. 11.28, 11.29). According to the Exxon sequence-stratigraphy models, these clastic wedges would be interpreted as highstand deposits (Chap. 13), but Galloway (1989b) demonstrated that their age distribution shows few correlations with the global cycle chart (Fig. 11.30). The major changes along strike of the thickness of these clastic wedges (Fig. 11.29) are also evidence against a control by passive sea-level change. Highly suggestive are the correlations with the tectonic events of the North American Interior, for example, the timing of the Lower and Upper Wilcox Group wedges relative to the timing of the Laramide orogenic pulses along the Cordillera. It seems likely that sediment supply, driven by source-area tectonism, is the major control on the location, timing, and thickness of the Gulf Coast clastic wedges. A secondary control is the nature of local tectonism on the continental margin itself, including growth faulting, evaporite diapirism, and gravity sliding.

In arc-related basins, volcanic control of the sediment supply may overprint the effects of sea-level change. For example, Winsemann and Seyfried (1991) stated:

"Sediment supply and tectonic activity overprinted the eustatic effects and enhanced or lessened them. If large supplies of clastics or uplift overcame the eustatic effects, deep marine sands were also deposited during highstand of sea level, whereas under conditions of low sediment input, thin-bedded turbidites were deposited even during lowstands of sea level."

A fluvial example of the volcanic control of sediment supply was given by Kuenzi et al. (1979). Fluvial styles in this area seem unusual but may, in fact, be typical of nonmarine arc-related environments characterized by active intermediate volcanism. Explosive eruptions, coupled with very high rainfall, lead to the catastrophic transportation and sedimentation of huge quantities of coarse volcaniclastic debris. This may be confined to the trunk rivers, and the result is that the tributaries are dammed at their junction with the main river and back up to form lakes (Fig. 7.15). These then fill with mud and small Gilbertian deltas. Lobate, river-dominated deltas form at the coast when sediment supply is high, but between major eruptions these may be reworked by waves into narrow, arcuate deltas. An ancient example of this fluvial style was described by Ballance (1988).

In more general, regional terms, sediment supply in volcanic terrains is markedly influenced by the uplift and subsidence generated by the thermal effects of volcanic activity. A good example is described by Dorsey and Burns (1994). Magmatic intrusion generates regional uplift, resulting in steepening of regional slopes, and the sediment supply is vastly increased by accompanying extrusive activity. Cessation of magmatic activity leads to thermal relaxation, subsidence, and transgression, with a reduction of sediment supply and the drowing of paleovalleys.

Other examples of the tectonic control of major sedimentary units are provided by the basins within and adjacent to the Alpine and Himalayan orogens. Sediments shed by the rising mountains drain into foreland basins, remnant ocean basins, strike-slip basins, and other internal basins. But the sediment supply is controlled entirely by uplift and by the tectonic control of dispersal routes, as described in several places earlier in this chapter. Brookfield (1992) reported on the beginnings of a project to investigate the shifting of dispersal routes through basins and fault valleys within the Himalayan orogen of central and southeast Asia. Some of the major rivers in the area (Tsangpo, Salween, Mekong) are known to have entirely switched to different basins during the evolution of the orogen. Much work remains to be done to relate the details of the stratigraphy in these various basins to the different controls of tectonic subsidence, tectonic control of sediment supply, and eustatic sea-level changes.

11.3.5 Intraplate Stress

"Ridge-push" and "slab-pull" are informal terms that have been used for some time in the plate-tectonics literature to refer to the horizontal (in-plane) forces associated with, respectively, the horizontal compressional effects resulting from the injection of magma at a spreading center and the tensional effects on an oceanic plate generated by the downward movement into a subduction zone, under gravity, of a cold slab of oceanic crust. In-plane (horizontal) stress is a central theme of the paper by Molnar and Taponnier (1975) describing a model of Himalayan collision, in which it was demonstrated that most of the Cenozoic structural geology of west China, extending for 3000 km north of the Indus suture to the edge of the Siberian craton, including that of many hinterland basins filled with nonmarine sediments (see Sect. 11.4), could be explained as the product of deformation resulting from intraplate stresses transmitted into the continental interior from the India-Asia collision zone. It came to be recognized that tectonic plates are not rigid and can store and transmit horizontal forces many thousands of kilometers from zones of plate-margin stress. The presence of residual stress fields in continental interiors has long been known from such evidence as the development of active joints ("break-outs") in exploration holes (summary in Cloetingh 1988). Cloetingh (1988, p. 206) suggested that

"the observed modern stress orientations show a remarkably consistent pattern [in northwest Europe], especially considering the heterogeneity in lithospheric structure in this area. These stress-orientation data indicate a propagation of stresses away from the Alpine collision front over large distances in the platform region."

Cloetingh et al. (1985) were the first to recognize the significance of these ideas for their possible effects on sedimentary basins. They argued

"that variations in regional stress fields acting within inhomogeneous lithospheric plates are capable of producing vertical movements of the Earth's surface or the apparent sea-level changes ... of a magnitude equal to those deduced from the stratigraphic record."

The important contribution which Cloetingh et al. (1985) made was to demonstrate by numerical modeling that horizontal stresses modify the effects of existing, known, vertical stresses on sedimentary basins (thermal and flexural subsidence, sediment loading), enlarging or reducing the amplitude of the resulting flexural deformation. They demonstrated that a horizontal stress of 1–2 kbar, well within the range of calculated and observed stresses resulting from plate motions, may result in a local uplift or subsidence of up to 100 m, at a rate of up to 0.1 m/ka. Compressional stresses generate uplift of the flanks of a sedimentary basin and increased subsidence at the center. Extensional stresses have the reverse effect. The magnitude of the effect varies with the flexural age (rigidity) of the crust, as well as with the magnitude of the stress itself.

It now seems likely that intraplate stresses may be transmitted for thousands of kilometers across oceanic and continental plate interiors, causing enhanced or subdued uplift and subsidence of cratonic and orogenic features. Regional tilts generated by this effect may have important consequences for sea level and sediment supply. For example, Kominz and Bond (1991) attributed a widespread mid-Paleozoic sea-level rise in cratonic North America to intraplate stresses associated with the assembly of the Pangea supercontinent and depression of the North-American continent. Embry (1990a, 1991) attributed the transgressive-regressive cyclicity of Mesozoic deposits in Sverdrup Basin to intraplate stresses associated with the breakup of Pangea, including changes in regional sediment dispersal patterns brought about by regional tilting. Heller et al. (1993) suggested that subtle warping of the Rocky Mountain foreland region during the Late Jurassic to Middle Cretaceous resulted from changing intraplate stress fields throughout this time period. Peper et al. (1992) suggested that intraplate stresses generated by flexural loads in response to episodic shortening on fold-thrust belts could generate tectonic cyclicity on a scale as short as 10^4 years, a proposal developed by Yoshida et al. (in press) in their study of the Castlegate Sandstone of Utah (see Sect. 13.4.4).

Cloetingh and Kooi (1990) reported one attempt to model paleostress in which they superimposed short-term intraplate stress on the long-term flexural subsidence of the continental margin of the US Atlantic margins, thereby generating a curve of estimated changes in the paleostress field (Fig. 11.31). Miall (1991c) explored the implications of this curve for fluvial sediment supply along the continental margin, making use of a study of Appalachian denudation rates during the Mesozoic and Cenozoic (Poag and Sevon 1989). Changes in the stress regime of the US Atlantic margin occurred because of the continual change in the plate kinematics of the Atlantic region as the ocean opened and the spread-

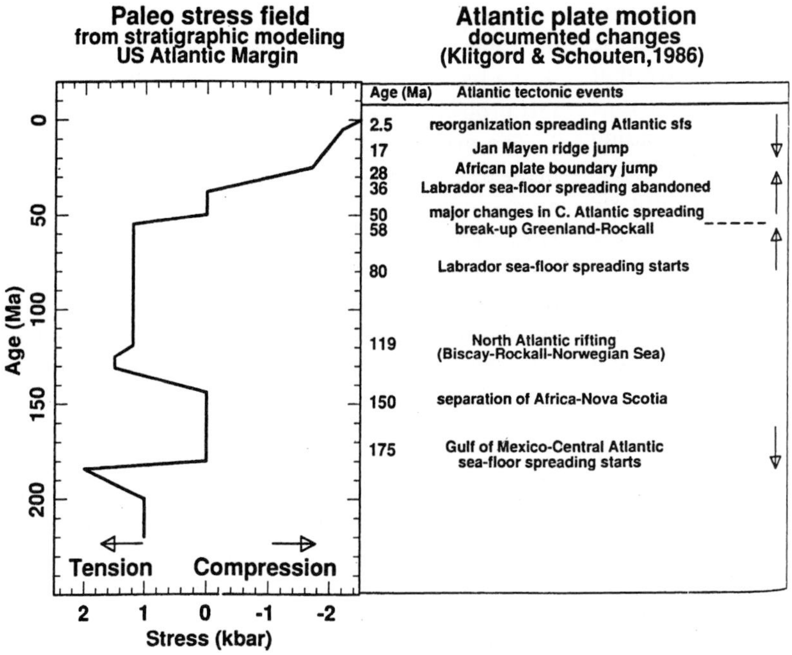

Fig. 11.31. Paleostress curve inferred from a stratigraphic model of the US Atlantic margins. Timing of tectonic events in the region is shown at right. (Cloetingh and Kooi 1990)

ing center extended and underwent various changes in configuration. Important events, such as the initiation of spreading in the Labrador Sea, the extension of spreading between Greenland and Europe, and various jumps in ridge position, brought about significant changes in plate trajectories, which would have been accompanied by changes in the direction and intensity of intraplate stresses.

Detailed studies of sea-floor spreading patterns in the north and central Atlantic Ocean by Srivastava and Tapscott (1986) and Klitgord and Schouten (1986) have yielded a detailed history of plate motions. Their reconstruction of a continuous succession of matched plate margins as the Atlantic Ocean opened indicates that the rotation poles of the North American plate relative to Europe and Africa changed at intervals of about 2–16 million years. This episodicity is comparable in magnitude to the duration of third-order stratigraphic cycles, a fact that is highly suggestive. Furthermore, some of the changes in rotation pole can be correlated in time with changes in the position of depocenters in the Mesozoic-Cenozoic stratigraphic record of the US Atlantic continental margin. Poag and Sevon (1989) compiled detailed isopach maps of the post-rift sedimentary record, and interpreted these in terms of the shifting pattern of denudation in and sediment transport from the Appalachian and Adirondack Mountains. It seems likely that many of the major existing rivers that presently carry sediment into this region have been in existence since the Mesozoic. Poag and Sevon (1989) were able to show many changes with time in the relative importance of these various sediment sources, and it is suggestive that several of these changes occurred at times of change in Atlantic plate kinematics. The sea-floor spreading record reveals at least 14 changes in plate configuration in the central Atlantic since the Mid-Jurassic (Klitgord and Schouten 1986). Seven of these changes, at 2.5, 10, 17, 50, 59, 67, and 150 Ma, correspond to times when the major sediment dispersal routes from the Appalachian Mountains to the continental shelf underwent a major shift (data from Poag and Sevon 1989).

Tectonism is diachronous and, in the case of the intraplate stresses discussed by Cloetingh (1988), the effects vary across the plate. Adjacent plates may be expected to have dissimilar tectonic histories, except where major extensional or collisional events affect adjoining plates. However, no such tectonic controls would be expected to be global in scope.

11.3.6 Quantitative Models of Sediment Supply, Transfer, and Accumulation

The purpose of this chapter is to address the question of allogenic controls on sedimentation. How do factors of tectonism (uplift, subsidence) and climate change affect sediment supply, sediment caliber, sediment transfer, and sediment accumulation? What does the large-scale architecture of the nonmarine stratigraphic record tell us about the working of these controls, and is it possible to relate the rates and magnitudes of the controlling processes to any quantitative, measurable attributes of the sedimentary product? As stated by Paola (1988, p. 231),

"The transition of basin analysis from a descriptive to an analytical science requires the development of well-founded theoretical models for basin development, including both the formation of basins and their infilling with sediments."

Models for basin formation, based on analysis of the thermal and mechanical behavior of the crust, have now reached a sophisticated level of development (see summaries by Miall 1990; Allen and Allen 1990). Models of basin filling have been slower to develop. Much work has been carried out on the physics of sediment transport (e.g., Middleton and Southard 1977), but the quantitative relationships that have been developed from such work relate only to instantaneous sedimentary conditions and cannot be used for the analysis of long-term (geological) sediment transfer and accumulation. The alluvial stratigraphy models summarized in Sect. 10.4.3 deal only with autogenic sedimentary processes. It is only in the past few years that successful approaches to this problem have begun to evolve.

The major developments in this area have been made by C. Paola, P.L. Heller, and their colleagues in a series of papers (Paola 1988, 1990; Paola et al. 1992; Heller and Paola 1992). This work addresses the question of mass balance of gravel in nonmarine basins – the transfer, sorting, and accumulation of geologically significant gravel deposits over geologically meaningful time periods. There are several independent factors to consider. Among the major controls are the rate of input of sediment and water, the rate of sediment transport and basin subsidence, the grain-size distribution and composition of the gravel, and its downstream reduction in grain size through abrasion and selective sorting. As noted by Paola (1988, p. 233):

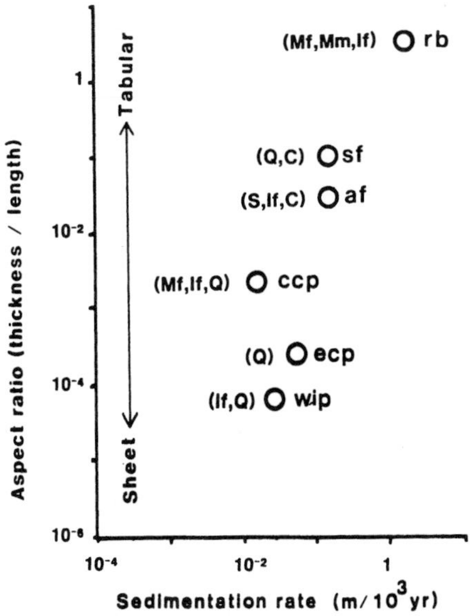

Fig. 11.32. Gravel body cross-sectional shape as a function of sedimentation rate. Dominant clast types are shown in *parentheses*: *Mf* felsic metamorphic; *Mm* mafic metamorphic; *If* felsic igneous; *C* carbonate; *Q* quartz and chert; *S* other sdimentary rocks. Basins are *rb* Ridge basin, California (Uper Miocene); *sf* Sevier foreland, Utah (Albian-Santonian); *af* Alpine foreland, Switzerland (Chattian-Aquitanian); *ccp* California coastal plain (Upper Eocene); *ecp* central Atlantic coastal plain, USA (Pliocene); *wip* Western Interior alluvial plain, USA (Pleistocene). (Diagram and data sources from Paola 1988)

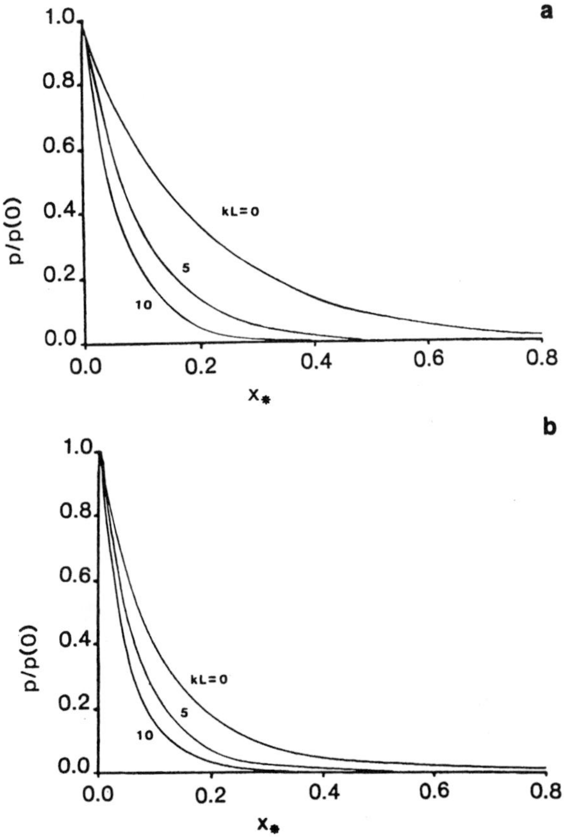

Fig. 11.33. Deposited gravel fraction *p* relative to its initial value *p(0)*, plotted relative to nondimensional transport distance across the basin *X∗*. In each case three different values of *kL* are shown. This is a nondimensional measure of the effectiveness of abrasion relative to selective deposition in causing downstream fining. (Paola 1988)

"Subsidence induces deposition, deposition causes downstream fining by selectively removing the coarsest clasts from the flow; and thus areas of rapid subsidence are areas of rapid downstream fining. In general, basins that subside slowly near their source areas transport gravels relatively long distances compared with basins that subside rapidly near their source areas."

Paola (1988) showed by theoretical argument, and with data collected from various ancient gravel deposits, that slow subsidence favored sheet-like gravel bodies composed predominantly of resistant, quartzose clasts (Fig. 11.32). Less resistant clasts have higher preservation rates when they are buried more rapidly. Figure 11.33 illustrates calculated gravel accumulation rates in basins with two different subsidence configurations and three different values for clast durability. The data are shown in nondimensional form. The first case (Fig. 11.33a) is for a basin showing uniform subsidence, the second case (Fig. 11.33b) for a basin undergoing asymmetric subsidence, more rapid at the upstream end and zero at the downstream end. The results confirm that the rate of downstream fining is much greater in the case of basins which subside asymmetrically, and that in the other case, symmetric subsidence, deposition of gravel is more sensitive to abrasion rate. These basins are more efficient at compositional fractionation.

In a later paper, Paola et al. (1992) developed a fully quantitative numerical model for predicting the distribution of gravel across an asymmetrically subsiding basin, in which four major parameters could each be independently varied in turn, while keeping the others fixed (Fig. 11.34). In the first case (Fig. 11.34A), slow variation of sediment flux in a sinusoidal manner leads to the development of a prograding pattern accompanied by increased sedimentation rate (increased spacing between the isochrons). Paola et al. (1992) termed this a "flux-driven" style of progradation. The second case (Fig. 11.34B) shows that sediment is trapped in more proximal positions when subsidence rate is increased. This is the so-called subsidence-driven

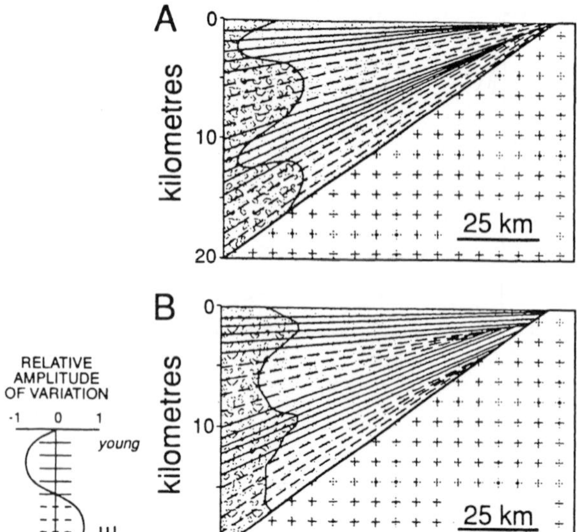

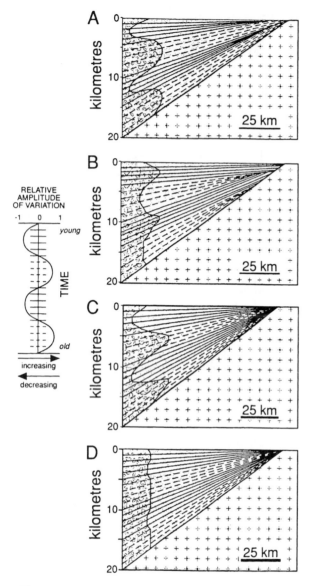

Fig. 11.34. Cross sections of a hypothetical basin showing the results of slow variations in four parameters: *A* sediment flux; *B* subsidence rate; *C* gravel fraction; and *D* diffusivity. Direction of transport is from *left* to *right*. Thin *dashed lines* are isochrons drawn every 1 million years. (Paola et al. 1992)

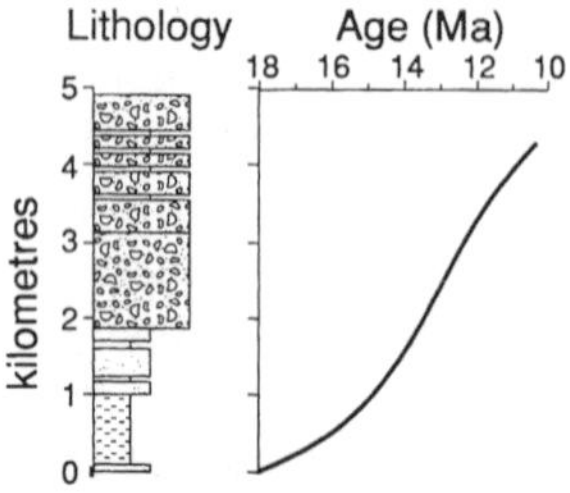

Fig. 11.35. Generalized stratigraphic section at Las Juntas, Argentina, showing calculated sedimentation accumulation rates, corrected for compaction. (Heller and Paola 1992)

model, in which progradation accompanies a diminished subsidence rate. The third case, termed the "distribution-driven" model (Fig. 11.34C), shows the progradation that takes place when the gravel proportion is varied while total sediment flux and subsidence rate are held constant. The fourth case (Fig. 11.34D) shows that varying the amount of water available for transportation makes little difference to the gravel distribution in the basin. Paola et al. (1992) developed a second set of models in which these same changes in parameters were imposed at much more rapid rates, producing some interesting variations on the results illustrated here.

Paola et al. (1992) also explored the relationship between basin scale and the rate of response to the various sedimentary controls. Each basin is characterized by a factor termed the "basin equilibrium time", which is defined as the square of basin length divided by the water flux, or diffusivity. The equilibrium time is the time scale over which the basin is able to fully respond to periodic or cyclic changes in controlling parameters (subsidence, sediment flux, etc.). In basins with long equilibrium times progradation is controlled primarily by changes in diffusivity or sediment supply. Where equilibrium times are relatively short, changes in subsidence rate appear to be the major control.

In a companion paper, Heller and Paola (1992) collected some basic data from several actual gravel-filled basins to explore the application of the models. An accurate chronostratigraphic framework is a prerequisite for this kind of testing and is not, in fact, available for many basins. Two examples of lithology-age plots are illustrated in Figs. 11.35 and 11.36. With the aid of the numerical modeling approach summarized above, some useful insights can be gained regarding the relationship between sedimentation and tectonics in these basins. The first plot, from Argentina, shows sedimentation rate increasing upsection, accompanying a coarsening upward into gravel. This suggests a flux-driven situation. In the second case, from Pakistan, sedimentation rates decrease upsection, accompanying an upward coarsening into gravel. This suggests an example of subsidence-driven progradation.

The importance of these models is the light they throw on the temporal relationship between tectonics and sedimentation. It has commonly been assumed that wedges of coarse sediment prograding from a basin margin are "syntectonic", and in many earlier studies the dating of such wedges has been used to infer the timing of major orogenic episodes (see summary and references in Miall 1981b; Rust and Koster 1984). It can now be seen that this inter-

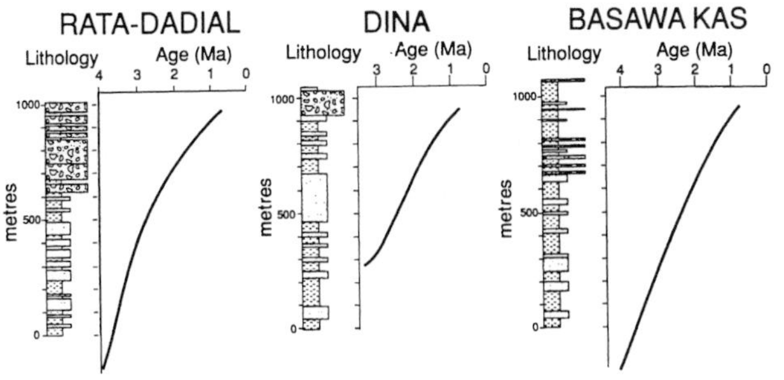

Fig. 11.36. Lithology and age of Siwalik Group, NW Himalayan foreland basin, Pakistan. The three sections are arranged in order from proximal (left) to distal (*right*) over a total distance of 40 km. (Heller and Paola 1992)

pretation is simplistic. In fact, the relationship between tectonism and sedimentation is complex, and depends on the balance between a range of controls, including sediment supply and sediment type. More recent studies have demonstrated that tectonism does not necessarily coincide with the progradation of wedges of coarse sediment but may precede such progradation by a significant period of time (Blair and Bilodeau 1988; Jordan et al. 1988; Heller et al. 1988). Tectonism, in the form of basin-margin fault movement, may lead to increased rates of subsidence, particularly at the edge of the basin. Foreland basins, in particular, are developed as a result of crustal loading by overriding thrust sheets and are characterized by syntectonic subsidence of the proximal regions of the basin. The immediate basinal response is likely to be marine or lacustrine incursions, with ponding of coarse debris against the basin margin. Tectonism eventually leads to increased basin-margin relief and hence to a rejuvenatation of the supply of coarse sediment, but this process take time, and it has been suggested that clastic-wedge progradation is largely a post-tectonic phenomenon (Blair and Bilodeau 1988; Heller et al. 1988). Increased rates of subsidence accompanying tectonism are more likely to be recorded by increased rates of sedimentation of fine-grained deposits, particularly shallow-marine and lacustrine sediments, at the basin margins, and at this time river systems may actually flow *towards* the foldbelt (Burbank et al. 1988). Heller and Paola (1992) referred to this model of coarse sedimentation as *antitectonic.* Only in the most fault-proximal regions is tectonic activity likely to be recorded by the rapid development of a clastic wedge.

The differences between the syntectonic and antitectonic models are explained using the quantitative models of Paola et al. (1992) and Heller and Paola (1992). The so-called flux-driven model of sedimentation (Fig. 11.34A) represents the traditional syntectonic explanation for conglomerate development and is illustrated by the example from Argentina (Fig. 11.35). Subsidence-driven sedimentation (Fig. 11.34B) is exemplified by the Siwalik sediments of Pakistan (Fig. 11.36), showing that coarse clastic progradation occurred at a time of decreasing sedimentation rate, following the end of thrust-induced flexural loading of the basin.

There has been considerable debate regarding these models, and much additional testing is underway. For example, widespread Early Cretaceous gravels in the Western Interior Basin, particularly in Utah, Colorado and Wyoming, have long been interpreted as syntectonic, related to thrust movements in the Sevier Orogen. Heller and Paola (1989) demonstrated that the gravels have a stratigraphic distribution inconsistent with that which would have been expected if they simply filled the flexural "moat" developed in front of the Sevier fold-thrust belt. They interpreted the gravels as the product of a regional thermal uplift, distinct from the uplift associated with Sevier tectonism. Burbank et al. (1988, 1989) defended the "syntectonic" model of gravel progradation in the case of the Siwalik sediments of Pakistan, a model disputed by Heller et al. (1989). The details of the dispute range beyond the scope of this book, involving discussions of the importance of local structural controls, such as folding of the proximal foreland leading to uplift and sediment bypass, the importance of flexural rigidity of the underlying crust in controlling the flexural behavior during thrust-sheet loading, etc. It was also pointed out (Burbank et al. 1989) that the models of Heller et al. (1988) and Blair and Bilodeau (1988) are two-dimensional and do not acknowledge the importance of *axial* transport of detritus into sedimentary basins, a very important factor in the case of the Himalayan foreland (see further discussion of this basin in Sect. 11.4.6). Gordon and Heller (1993) studied late Cenozoic sedimentation in Pine Valley, Nevada, an extensional basin in the Basin and Range province of the western United States. They found that the distance of progradation of alluvial gravels within the basin is inversely proportional to the contemporaneous subsidence rate. Jo et al. (1994) argued that local structural controls and autogenic mechanisms introduce complications that may have confused the simple picture painted by Gordon and Heller (1993), but, in a reply to this discussion, the authors maintained that these are essentially small-scale details which do not obscure the broad patterns predicted by the basin-scale model (Heller and Gordon 1994). Lang (1993) attempted to apply the new basin models to a Devonian alluvial complex in northeast Australia and suggested that examples of both syntectonic and antitectonic progradation are present. In a study of a modern syntectonic fan system in Argentina, Damanti (1993) suggested that thrust-proximal fan systems fed by large sediment source areas (large drainage basins for individual fans) are able to respond rapidly to thrust-fault activity by renewed basinward progradation, especially where there is adequate rainfall to maintain high rates of sediment transport. G.A. Smith (1994) warned of the ability of rapid climate change to generate stratigraphic effects (such as changes in grain size and channel style) very similar to those traditionally thought to be controlled by tectonism. It is clear that a variety of local factors influence the relationship between tectonics and sedimentation, and that in each case the nature of this relationship needs to be determined by detailed local studies.

11.4 Plate-Tectonic Setting of Alluvial Basins

The purpose of this section is to illustrate the broad stratigraphic architecture of alluvial depositional systems and their relationship to the structural style and tectonic processes dependent on plate-tectonic setting.

11.4.1 Basin Classification

Basin-forming mechanisms and the classification of sedimentary basins according to plate-tectonic principles are subjects that have been discussed by numerous authors (Dickinson 1974; Bally and Snelson 1980; Bally 1984; Klein 1987; Ingersoll 1988; Miall 1990; Allen and Allen 1990; Busby and Ingersoll 1995), and this discussion will not be repeated here. Table 11.2 presents the basin classification used in this book. A simple numerical code has been added to the classification, and this is used in Table 11.3, which lists some examples of modern depositional systems and ancient stratigraphic examples.

Most of the discussion in this section deals with primary plate-tectonic relationships, that is, basins where the alluvial stratigraphy and architecture are directly related to first-order contemporaneous tectonic controls. However, in many cases contemporaneous tectonism, including long-distance transmission of intraplate stress, reactivates ancient structures, generating topographies and regional paleoslopes in positions and with orientations that would not be predicted from simple plate-tectonic models. Many of the most interesting details of local basin architecture cannot be understood unless the effects of such reactivation are incorporated into paleogeographic reconstructions.

The thickest successions of alluvial deposits occur in rift basins (including failed rifts and aulacogens), forearc basins, retroarc and peripheral foreland basins, and those associated with plate-collisional hinterland deformation, particularly foreland and

Table 11.2. Classification of the types of sedimentary basins discussed in this chapter. (Miall 1990)

1 Divergent-margin basins
- 1.1 Rift basins
  - 1.1.1 Rifted-arch basins
  - 1.1.2 Rim basins
  - 1.1.3 Sag basins
  - 1.1.4 Half-graben
- 1.2 Continental-margin basins
  - 1.2.1 Red Sea type (youthful)
  - 1.2.2 Atlantic type (mature)
- 1.3 Aulacogens and failed rifts
- 1.4 Oceanic islands, seamounts, plateaus

2 Convergent-margin basins
- 2.1 Trenches and subduction complexes
- 2.2 Forearc basins
- 2.3 Interarc and backarc basins
- 2.4 Retroarc (foreland) basins

3 Transform- and transcurrent-fault basins
- Basin setting:
  - Plate boundary transform fault
  - Divergent-margin transform fault
  - Convergent-margin transcurrent fault
  - Suture-zone transcurrent fault
- Basin type:
  - 3.1 Basins in braided fault systems
  - 3.2 Fault termination basins
  - 3.3 Pull-apart basins in *en echelon* fault systems
  - 3.4 Transrotational basins

4 Basins developed during continental collision and suturing
- 4.1 Peripheral (foreland or foredeep) basins (on underriding plate)
- 4.2 Intrasuture embayment basins (remnant ocean basins)
- 4.3 Hinterland foreland, strike-slip, and graben basins (on overriding plate)

5 Cratonic basins

Table 11.3. Examples of modern and ancient fluvial deposits classified according to tectonic setting

Basin type[a]	Modern examples[b]	Ancient examples
1.1	East African rifts Rio Grande rift, New Mexico	Sadlerochit Fm, Alaska (Permian-Tr.) Newark Gp., N. America (Triassic) Bjarni Ss, Labrador (Cretaceous)
1.2.1	Fan deltas, Red Sea margins	
1.2.2	Gulf Coast	Woodbine-Tuscaloosa, Gulf Coast (Cret.) Frio Formation, Gulf Coast (Oligocene)
1.3	Benue Trough	Statfjord-Brent Ss, North Sea (Tr.-Jur.) Latrobe Gp., Gippsland Basin (Tert.) Natal Embayment, S. Africa (Paleozoic)
1.4	Sandurs, Iceland	
2.2	Pacific coast, Chile Pacific coast, Guatemala	Old Red Ss (Dev), Midland Valley, Scotland Kenai Gp, Cook Inlet (Tertiary) Western Trough, Burma (Tertiary) San Joaquin Basin, California (Olig.-Q.) Luzon Central Valley, Philippines (Plio.-Q.)
2.3	Cagayan Basin, Luzon Rifts in Aegean Sea	Ordovician fluvial-deltaic systems, N. Wales Carboniferous basins of N. England Kayenta Formation, Colorado Plateau (Jur.) (?) Songliao Basin, China (Cretaceous) Tertiary ss of W Japan Basin-and-Range, USA (Cenozoic)
2.4	Upper Amazon basin Bermejo Basin, Argentina Papuan basin, Papua-New Guinea	Hawkesbury Ss (Triassic), Sydney Basin Hutton Ss, Eromanga Basin (Jurassic) Rocky Mt. basin, Utah-Colorado (Cret.)
3.1	N. Anatolian basins, Turkey	San Andreas system (U. Tertiary)
3.3	Salton Trough, California Dead Sea	Ridge basin, California (Tertiary)
4.1	Indus-Ganges Trough Western plains of Taiwan	Catskill Delta, New York (Devonian) Cherokee Gp., Arkoma Basin (Penn.) Swiss Molasse Basin (Tertiary) Siwalik Gp, India-Pakistan (Tertiary) Aikawa, Ashigara basins, Japan (Tertiary)
4.3	Baikal Rift, Russia Tarim Basin, China Lake Hazar, Turkey	Midland Valley, Scotland (Devonian) Cutler Gp, Ancestral Rockies (Permo-Penn.) Alberta Basin (Cret.-Tertiary) Pattani Trough, offshore Thailand (Tertiary)
5	Lake Eyre Lake Chad	Late Paleozoic, Illinois Basin

[a] Numerical classification as in Table 11.2.
[b] "Modern" in the sense that the current plate-tectonic classification of the basin is as shown. However, some of these basins have had a very lengthy geological history, in some cases extending back to the Cretaceous.

strike-slip basins (Fig. 11.37). Fluvial deposits do not occur in some types of basin, such as oceanic trenches, subduction complexes, and remnant ocean basins, because these basins are floored by oceanic crust, and therefore are unlikely to include any areas of nonmarine deposition. Nonmarine sedimentation is also very limited on oceanic seamounts and plateaus for the same reason.

11.4.2 Extensional Basins

Crustal extension occurs in several distinct tectonic settings. The stretching and rifting of major continental plates initiates the development of rift basins (basin type 1.1 in Table 11.3) and failed rifts (basin type 1.3) and is followed by the development of broad continental-margin basins as a result of

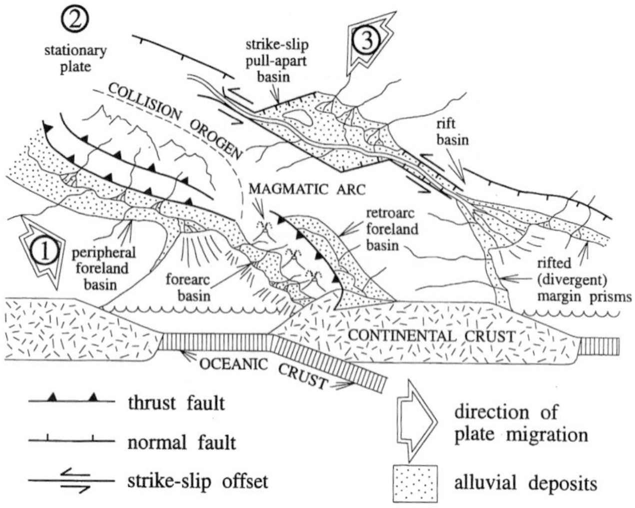

Fig. 11.37. Tectonic setting of fluvial deposits. Note the distinction between *transverse* drainage, which is oriented perpendicular to structural grain, and axial, or *longitudinal* drainage, which typically consists of trunk streams flowing along the basin axis. (Miall 1992)

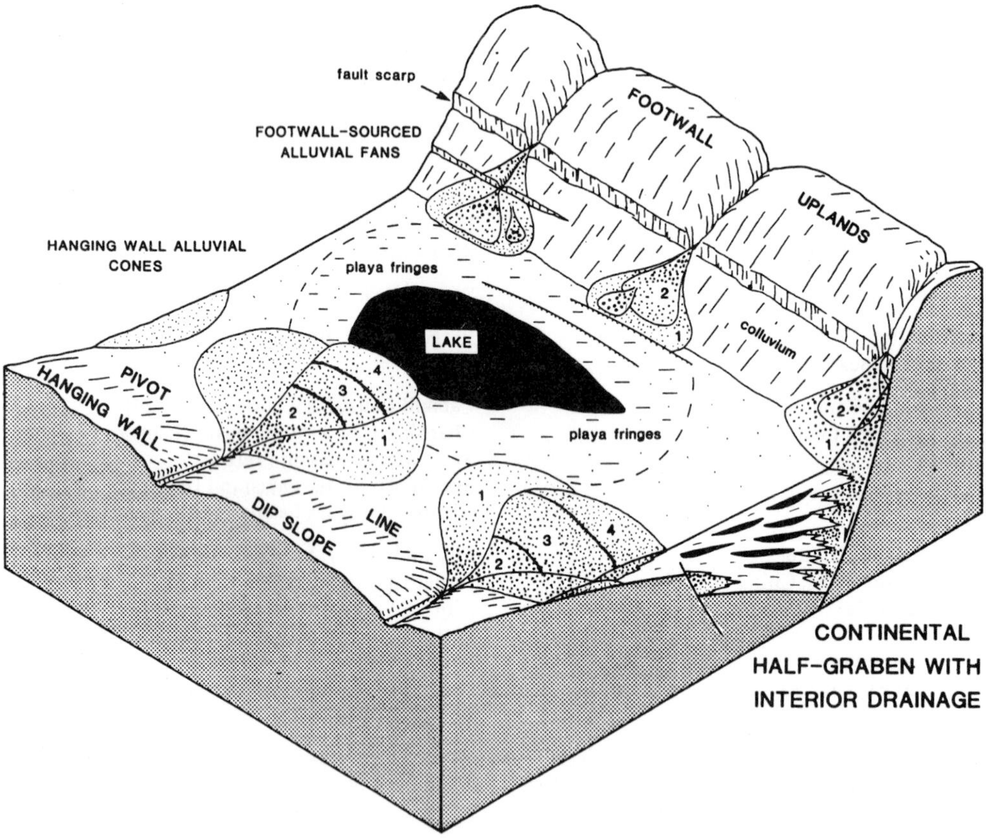

Fig. 11.38. Model for continental half-graben rift basin with interior drainage. *Numbers* indicate successive fan lobes. (Leeder and Gawthorpe 1987)

thermally driven flexural subsidence (basin type 1.2). Rift-type basins also occur in areas of continental back-arc spreading (basin type 2.3), in transtensional continental transform settings (basin type 3), and in areas of major plate collision, where lateral translation of continental blocks along strike-slip faults ("escape tectonics") develops pull-apart and other, more complex, basin types (basin type 4.3). The structural and stratigraphic style of rift basins that develop in these various plate-tectonic settings are similar.

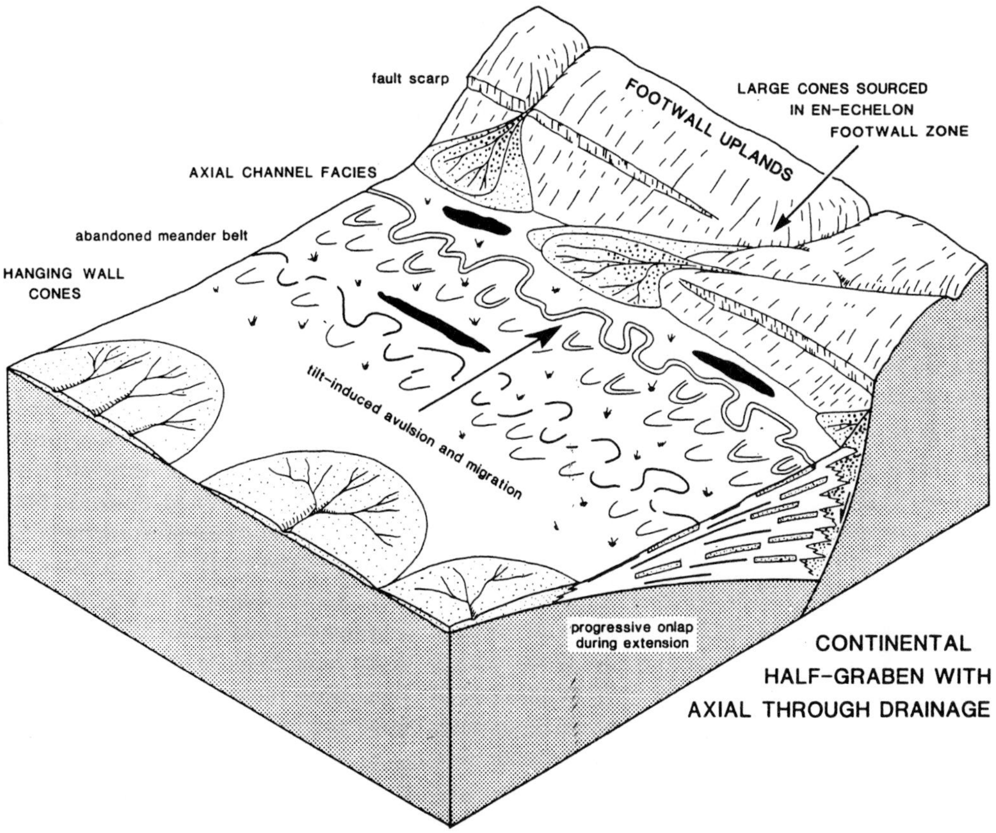

Fig. 11.39. Model for continental half-graben rift basin with axial through drainage. (Leeder and Gawthorpe 1987)

11.4.2.1 Rift Basins

The broad stratigraphic architecture of rift basins and its relationship to basin tectonics have been studied in detail by Leeder and Gawthorpe (1987) and Leeder and Jackson (1993). Numerous modern and ancient examples have been identified in various tectonic settings, some of which are noted in Table 11.3.

Two useful generalized models for rift basins have been proposed by Leeder and Gawthorpe (1987), based on research in the Basin and Range province of the United States and in backarc rifts of southern Greece. The first model (Fig. 11.38) is based mainly on studies of the Basin and Range, where this basin style is characteristic. Low-gradient hanging-wall dip slopes are the site for broad alluvial cones, while small, coarse alluvial fans prograde from the steeper slope of the footwall scarp. Lakes characteristically occupy the basin center; they may be perennial lakes or ephemeral playas, depending on climate. In the latter case, evaporite deposits may be important (Rosen 1994). A related model, illustrated in Fig. 11.39, is for a half-graben basin with axial through drainage. Flanking transverse alluvial fans may be similar in both models, the main difference between this model and the first being the environments and facies developed at the basin center. Axial rivers may be major regional trunk rivers with large channels and contrasting facies characteristics to the flanking fans. The position of the axial river within the basin is determined partly by the pattern of subsidence and partly by the scale of transverse fans, which may be large enough to displace the axial river laterally. Differential subsidence within the basin may cause the meander belt to migrate laterally, in the manner documented by Alexander and Leeder (1987; see Fig. 11.6) and Alexander et al. (1994). Gradual subsidence along the marginal fault may lead to rotation of the hanging wall and onlap by the floodplain deposits marginal to the axial river system.

Among the best-known ancient rift systems is that filled by the Triassic-Jurassic Newark Supergroup of eastern North America (Fig. 11.40). These rifts represent the initial rift phase that preceded the development of the modern Atlantic Ocean. They consist of a series of half-grabens, many now buried beneath the Jurassic and younger sedimentary cover of the Atlantic continental margin. Many of the basins are linked by continuous border faults with the same sense of dip that represent reactivated Paleozoic compressional faults, and in this way they differ from the East African rift system, which consists of segments of opposite-facing border faults (P.E. Olsen 1990). Newark basins show two different depositional patterns: one is fluvial dominated, consisting of basin-wide channel systems, commonly including axial trunk rivers; the other is lacustrine dominated, in which fluvial deposits comprise basin-margin alluvial fans and small deltas.

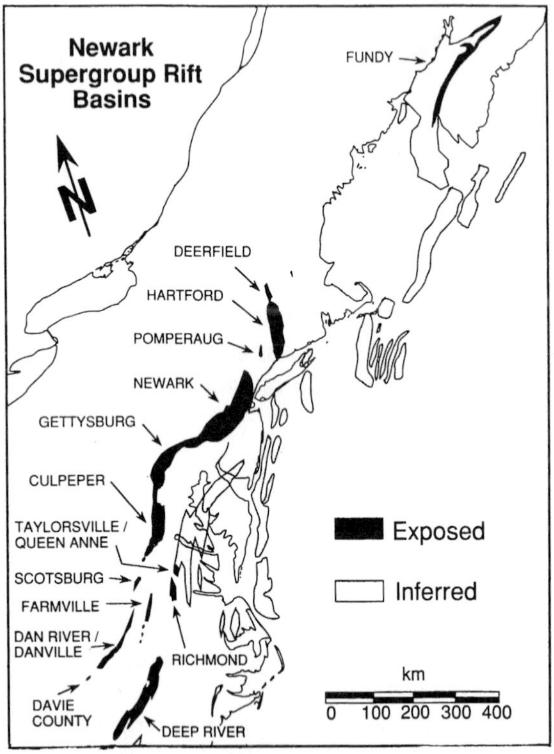

Fig. 11.40. Distribution of Triassic-Jurassic Newark basins in eastern North America. (P.E. Olsen 1990, reprinted by permission)

The Newark rifts are part of an immense system of rift basins that developed between North America, Europe, and Africa during the initial phase of extensional tectonics that led to the breakup of Gondwana (Ziegler 1988). Numerous individual case studies of nonmarine rift sedimentation have been carried out in eastern Canada and the United States, Spain, Britain, and the North Sea Basin.

Major rift systems developed in Africa in the Mesozoic-Cenozoic during the breakup of Gondwana. In some of these transtensional tectonism occurred, leading to strike-slip components of fault offset (Fairhead 1986; Genik 1993). Sedimentary patterns in the East African rift system were documented by Frostick et al. (1986), and several studies carried out during petroleum exploration work have now documented the subsurface geology of these rifts and other Mesozoic-Cenozoic rift systems in Central and West Africa (e.g., Genik 1993; Winn et al. 1993). Figure 11.41 illustrates the typical structure of African rifts, based on an example in Niger. Neocomian to Albian sediments in this and adjacent basins constitute a "phase-I synrift tectono-stratigraphic sequence" consisting largely of terrigenous, nonmarine clastics, fining upward from a basal fluvial feldspathic sandstone through a middle fluvial deltaic sandstone-shale facies, into an upper lacustrine shale facies. The phase-II synrift sequence (Cenomanian-Maastrichtian) is largely marine in the Western African basins but may consist of 1.5–3 km of fluvial deposits in the Central African rifts, organized into fining-upward channel-overbank cycles with interbedded lacustrine shales. In the West African basins phase-III rift deposits consist of

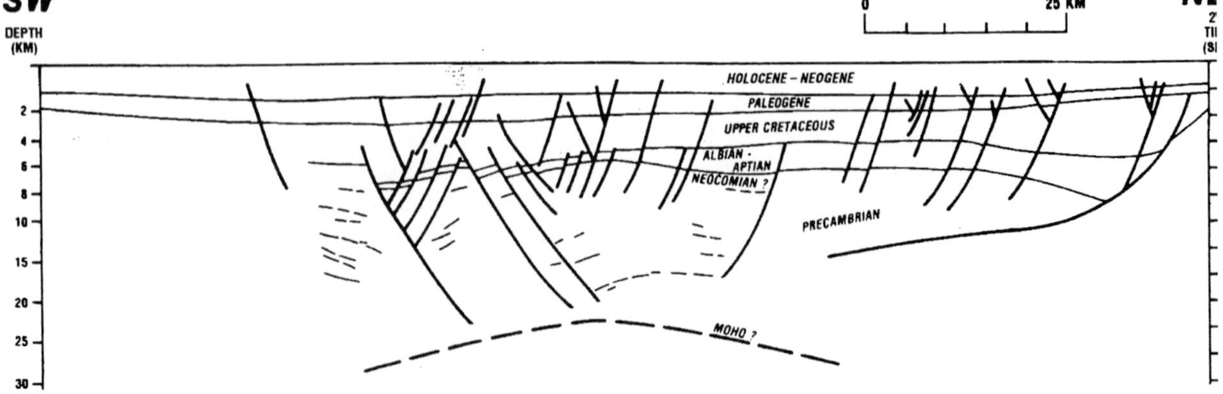

Fig. 11.41. Structural cross section through the Termit Basin, part of the West African Rift System in Niger, drawn from reflection-seismic data. (Genik 1993, reprinted by permission)

up to 2 km of fluvial sandstones overlain by marine or lacustrine deposits. Miocene-Holocene sediments constitute a post-rift assemblage and consist of a few tens of meters of fluvial deposits along the basin margins, with much thicker lacustrine assemblages in the center. All the fluvial deposits in these basins appear to have accumulated within systems of internal drainage (Genik 1993).

11.4.2.2 Continental-margin Basins

Extensional continental margins that develop once oceanic crust has evolved along the rift axis are commonly not characterized by significant thicknesses of nonmarine deposits, except where major rivers cross the continental margin. As noted in Sect. 11.3.1, many of the world's major rivers enter the sea at divergent plate margins, and their fluvial deltaic systems form part of the continental-margin basin stratigraphy. Away from such fluvial axes, however, flexural subsidence leads to the development of gently dipping continental shelves which are transgressed by the sea, with alluvial coastal plains typically occupying only the fringes of the basin. As discussed in Chap. 13, alluvial deposits on these shelf areas may become significant only during times of low sea level, and their stratigraphy may yield much useful information regarding local sea-level history.

The Jurassic-Tertiary section flanking the Gulf Coast of the United States is probably the thickest and most extensive suite of alluvial deposits on a divergent continental margin (Fisher and McGowen 1967; Galloway 1981; Worrall and Snelson 1989). Figure 11.42 is a cross section through the Gulf of Mexico, Fig. 11.43 shows part of the Cenozoic section in more detail (see also Figs. 11.23, 11.28), and Fig. 11.29 shows the position of major Cenozoic depocenters. This map shows that the coastal plain has prograded by up to 350 km since the Cretaceous. In places the clastic section is at least 15 km thick. Much of this consists of marine deposits, ranging from deltaic and barrier-lagoon deposits to deep-marine submarine fan successions. However, updip equivalents include significant fluvial units. Nonmarine stratigraphic units include parts of the Denkman Sandstone (Upper Jurassic), Travis Peak Formation (Early Cretaceous), Woodbine-Tuscaloosa Formation (Upper Cretaceous), the Wilcox Group (Paleocene-Eocene), the Frio Formation (Oligocene), and the Catahoula and Oakville formations (Oligocene-Miocene). One of the most interesting features of this continental margin is the extreme long-term persistence of many of the sediment transport pathways. As noted by Potter (1978), a river has occupied the course of the Mississippi since the Late Paleozoic. All the major rivers draining across the Texas coast have remained in essentially the same position since at least the Eocene, as demonstrated by Fisher and McGowen (1967), who showed that each is underlain by a major Wilcox deltaic depocenter. Galloway demonstrated the vertical persistence of many of the Cenozoic fluvial depositional axes (Fig. 9.15). Many additional details of these fluvial units are discussed in Chap. 9.

The long-term persistence of fluvial axes on the Gulf Coast contrasts with the wide variation in sedimentation rates documented by Galloway (1989b) and Galloway and Williams (1991). Sedimentation was most rapid in the deltaic to shallow-marine environment of the continental margin, where the generation of accommodation space by subsidence was most rapid. There, values during the Cenozoic varied by more than an order of magnitude, from 0.03 m/ka to more than 1.4 m/ka (Galloway and Williams 1991; the rates were calculated for each chronostrati-

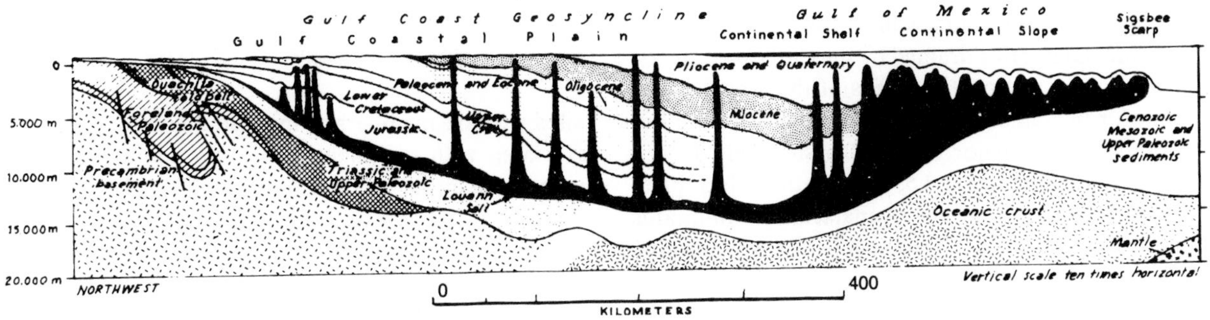

Fig. 11.42. Structural cross-section through the northern margin of the Gulf of Mexico. showing the immense thickness of Jurassic to Recent terrigenous clastic sediments. Reprinted by permission from King PB: The evolution of North America, rev. edn. Copyright © 1977 by Princeton University Press

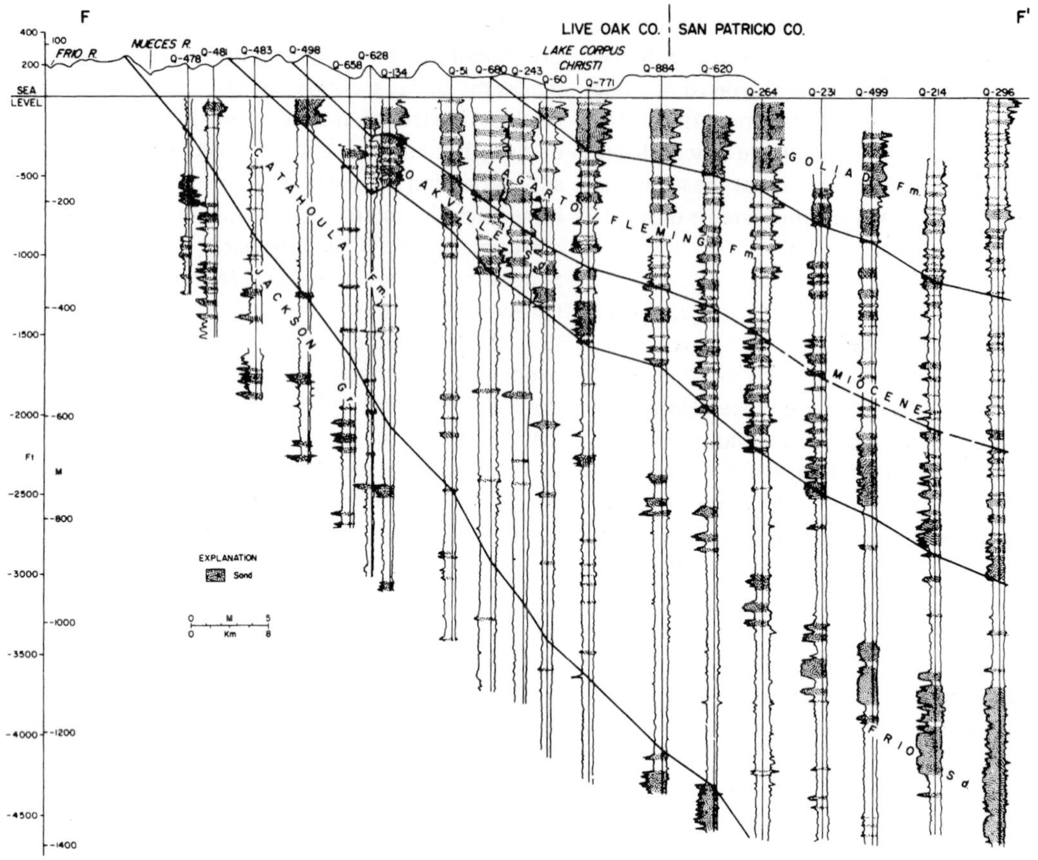

Fig. 11.43. Typical dip-oriented cross section through the Upper Tertiary clastic succession of south Texas. The section crosses an interfluvial, sand-poor area of the Catahoula and a major sand-rich fluvial axis of the Oakville Formation. (Galloway 1981)

graphic unit, of 1–2 million years duration. Therefore, the sedimentation rates should be compared with the group 9 and 10 rates and processes discussed in Chap. 3). Winker (1982) and Galloway (1989b) argued that the Cretaceous-Cenozoic Gulf Coast stratigraphy, with its alternation of sand-rich clastic wedges and shale-rich intervals, is largely a reflection of the large variations in sediment supply brought about by changes in tectonic regime in the continental interior drained by the Gulf Coast rivers. These rivers rise on the flanks of the Appalachian and Cordilleran orogens and, in the case of the Rio Grande, drain a major, active rift system (Fig. 11.27). As discussed in Sect. 11.3.5, changes in the intraplate stress regime along the Appalachian Mountains are probably largely responsible for geomorphic adjustments that controlled changes in sediment supply to the Atlantic continental margin, and the rivers draining from the opposite flank of the mountains into the interior, and hence into the Mississipppi system, would have been similarly affected. Winker (1982), Galloway (1989b), and Galloway and Williams (1991) argued that Laramide tectonics along the Cordilleran orogen were largely responsible for controlling the sediment supply to the Missouri and other rivers draining across the Great Plains into the Gulf of Mexico (Fig. 11.29). This is discussed further in Sect. 11.3.5.

11.4.2.3 Failed Rifts and Aulacogens

Failed rifts are members of the initial family of rifts associated with continental breakup, along which much of the initial crustal extension takes place, but which cease active extensional activity when complete continental separation takes place elsewhere and the generation of oceanic crust commences. Failed rifts undergo the initial stages of normal faulting and crustal extension, but remain floored by continental crust. The term is normally limited in its application to major rifts striking inland at high angles to the continental margins, which for periods of a few millions or tens of millions of years act as the major loci of crustal extension. Aulacogens are failed rifts that have subsequently been affected by closure of the ocean with which they are tectonically associated. Plate convergence and collision commonly leads to compressional tectonics trending parallel to the basin axis. Both the initial rifting and the subsequent plate convergence and collision may be associated with some oblique or transcurrent plate motion, giving rise to a component of strike-slip on the bounding faults.

These basins commonly constitute important petroleum provinces because of the combination of good reservoir characteristics, good petroleum source rocks, appropriate maturation levels, and excellent structural trapping configurations. Examples of failed rifts containing important fluvial units are the Viking Graben of the North Sea and its flanking East Shetland Basin, Gippsland Basin off southeast Australia, Benue Trough, Nigeria, and Kutei Basin, Borneo (Kalimantan).

The North Sea is an intracratonic basin which developed by a complex and lengthy series of extensional movements beginning in the late Paleozoic (Ziegler 1988). The Upper Permian to Jurassic rocks of the North Sea were developed during a phase of rift faulting and thermal subsidence related to the incipient opening of the Atlantic Ocean (Fisher 1984; Ziegler 1988; Steel and Ryseth 1990). The Viking-Central Graben was the major structure to develop at this time (Figs. 9.22, 9.23, 11.44). Regional subsidence that accompanied rifting activated regional drainages, leading to the development of major fluvial units, the Teist, Lomvi, Lunde, and Statfjord Formations (Upper Permian-Sinemurian; Steel and

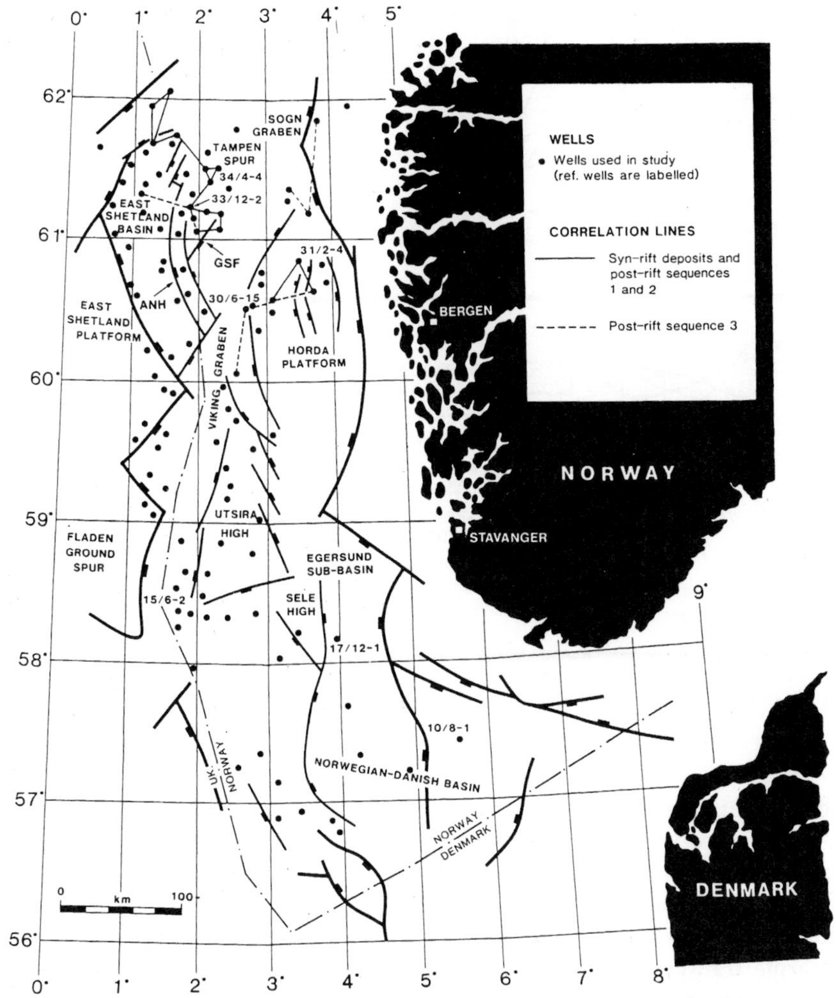

Fig. 11.44. Location of Viking Graben and related structures in the North Sea Basin. (Steel and Ryseth 1990)

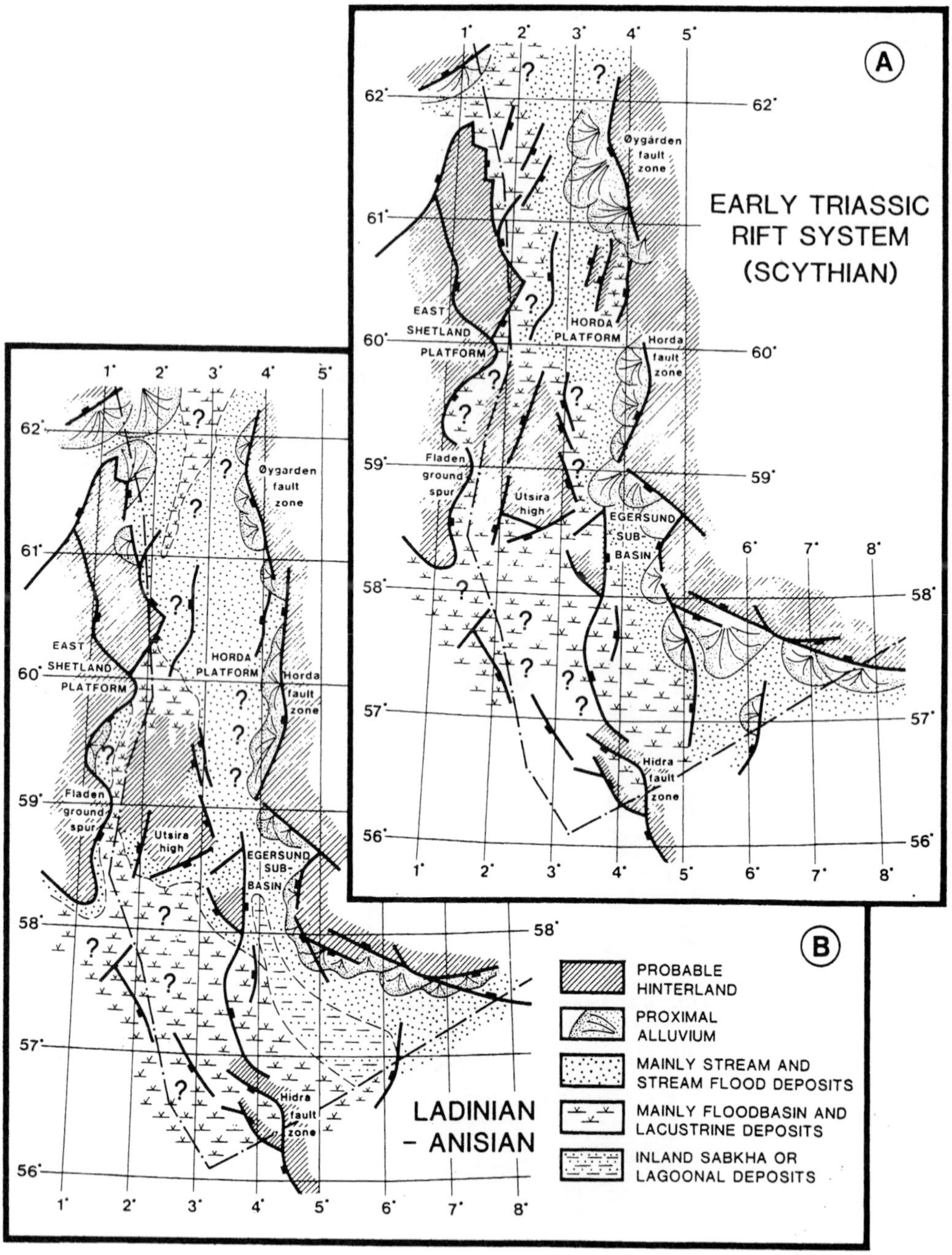

Fig. 11.45. Tentative paleogeographic maps for post-rift megasequence 1 in the northern North Sea Basin. (Steel and Ryseth 1990)

Ryseth 1990). These can be subdivided into a series of megasequences (Fig. 9.23). The first of these is correlated with the rift fault activity, while the three subsequent post-rift sequences are attributed to post-rift thermal subsidence (Steel and Ryseth 1990). With continued subsidence, marine transgression occurred and the largely deltaic Brent Group (Aalenian-Bajocian) was deposited (Brown 1984). Faulting and folding occurred in mid to Late Jurassic time, and the fault blocks were tilted, uplifted, and eroded (Fig. 9.22). Subsequent Late Jurassic (Kimmeridgian) and Cretaceous transgressions covered the eroded fault blocks with shales, including the Kimmeridge Shale, a prolific source rock. The shales also provide the seals for the Viking Graben oil fields.

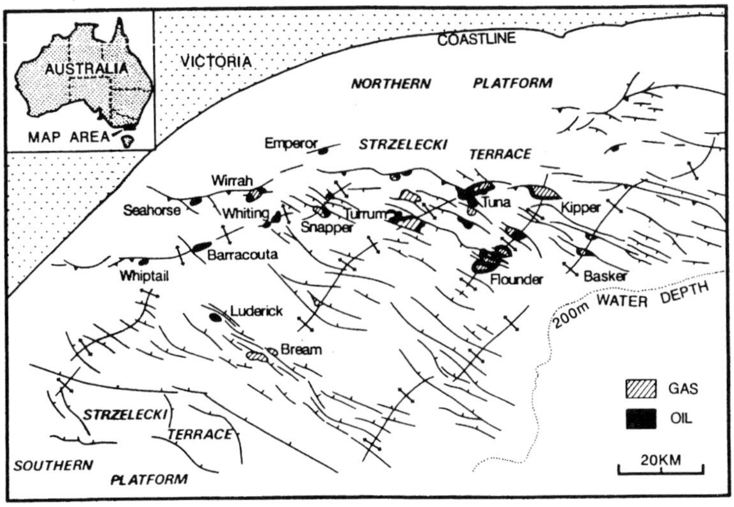

Fig. 11.46. Major faults, anticlines, and oil and gas fields in the Gippsland Basin, off the southeast coast of Australia. (Rahmanian et al. 1990)

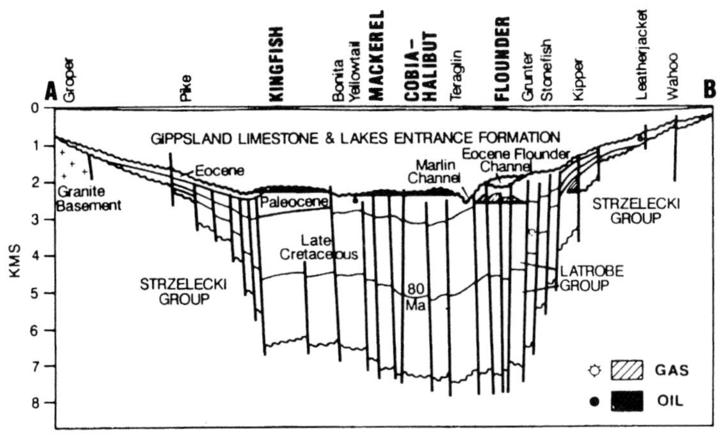

Fig. 11.47. Regional cross section through Gippsland Basin. The section is oriented SW-NE, approximately parallel to the coast of Victoria. The location is given by the position of Flounder field in Fig. 11.46. (Rahmanian et al. 1990)

Speculative paleogeographic maps for the lower and upper part of post-rift megasequence 1 are shown in Fig. 11.45, indicating a complex of fluvial and lacustrine environments. It is not clear whether these basins drained internally, or whether there was a through-going axial drainage. Transgression spread a tongue of shallow marine deposits southward along the Viking Graben in the Sinemurian (Steel and Ryseth 1990), and this may have been the location for trunk drainage for part or all of the Triassic, with the river system flowing northward toward the location of the present continental margin.

Gippsland Basin is located offshore on the continental shelf of southeastern Australia (Fig. 11.46). It is similar in structure and origin to the Viking-Central Graben of the North Sea, having been interpreted as a failed rift, an arm of the Tasman Sea Basin (Veevers 1984; Rahmanian et al. 1990). The basin commenced faulting and subsidence in the early Cretaceous. By Oligocene time, subsidence had expanded beyond the original confines of the basin, suggesting that it had begun the flexural subsidence phase of development. The Gippsland Basin is oriented NW-SE. A typical cross section is shown in Fig. 11.47. The basin was filled during the Late Cretaceous-Eocene with 5800 m of nonmarine sediment assigned to the Latrobe Group (Fig. 11.48). Sediment was shed into the basin from the north and consists predominantly of coarse-grained, well-sorted sandstone, with minor fine-grained sandstone, siltstone, shale, and coal. The basin was uplifted and deeply incised by fluvial channels late in the Eocene (Fig. 11.48). Marine mudstones of Oligocene age transgressed this erosion surface, and they provide an important regional seal.

Kutei Basin in Borneo (Kalimantan) is a failed rift associated with the short-lived sea-floor spreading of Makassar Strait (Hamilton 1979; Hutchison 1989). The basin contains 16 km of Tertiary fluvial-deltaic

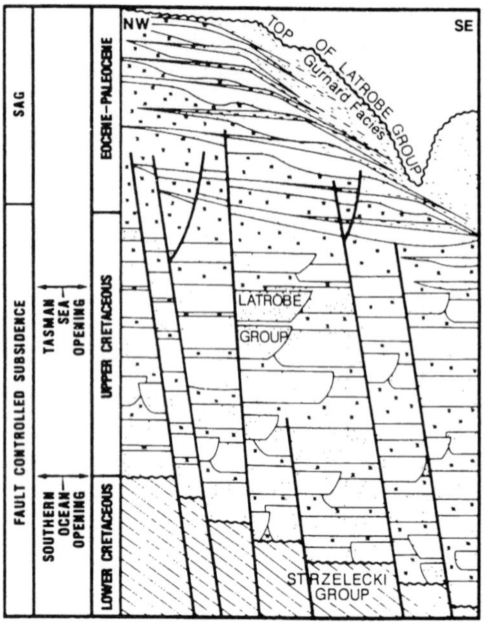

Fig. 11.48. Schematic stratigraphic diagram of the Gippsland Basin, showing major tectonic events. (Rahmanian et al. 1990)

clastics deposited by the Mahakam River and delta, and these rocks constitute a significant petroleum province (Verdier et al. 1980).

11.4.3 Convergent-margin Basins

11.4.3.1 Forearc Basins

Many forearc basins are floored by oceanic crust and are filled with marine sediments, typically turbidites and related facies (Dickinson and Seely 1979). Nonmarine deposits occur where an arc extends onto continental crust, and here sediment thicknesses may be considerable. In such settings the basin typically is floored by accreted continental rocks and is fed by active tectonic highlands of the magmatic arc. The subduction complex, on the other side of the basin, may become uplifted and is a significant sediment source late in basin development. Early basin-fill sediments typically are marine, with nonmarine deposits forming the youngest part of the basin fill. Considerable thicknesses of nonmarine sediment may be present. In Cook Inlet, Alaska, the largely nonmarine Kenai Group is more than 7.6 km thick.

Modern examples of nonmarine sedimentation in forearc basins include the Pacific coast of Guatemala (Kuenzi et al. 1979) and the Osorno Basin, Chile (González 1990). Forearc basins with a dominantly marine fill, but including late-stage nonmarine sediments, include parts of the Great Valley and related basins of the western United States (Nilsen 1987; Bartow 1987; Figs. 11.49, 11.50), Luzon Central Valley, Philippines (Bachman et al. 1983), Xigaze basin, Tibet (Einsele et al. 1994), and the East Coast Depression of North Island, New Zealand (van der Lingen 1982). Forearc basins with great thicknesses of nonmarine deposits include the Western Trough of Burma (Rodolfo 1975; Mitchell and McKerrow 1975; Hutchison 1989), Cook Inlet, Alaska (Kirschner and Lyon 1973; Figs. 11.51, 11.52), the Rosario Embayment, Baja California (Fulford and Busby 1993), and the Old Red Sandstone (Devonian) of the Midland Valley, Scotland (Mitchell and McKerrow 1975; Coward 1990).

Forearc basins are susceptible to tectonic modification at late stages during plate collsision. For example, the Western Trough of Burma and the Midland Valley of Scotland were both characterized by late-stage strike-slip faulting accompanying plate collision events. The forearac basin that developed over the Luzon arc prior to the collision with the Chinese mainland in the Pliocene evolved into a collisional basin that filled primarily with marine sediment-gravity-flow deposits until it was uplifted about 0.5 million years ago, at which point fan and braided stream deposition began (Lundberg and Dorsey 1988).

11.4.3.2 Backarc Basins

These basins develop by crustal extension and thermal subsidence in the backarc region. Commonly, they show a stratigraphic evolution that is the inverse of forearc basins. Nonmarine sedimentation is most typical of early stages of basin development, before crustal extension has led to significant subsidence, with basins deepening through time. Rift basins in two backarc regions, the modern Aegean Sea and Cenozoic rifts of the Basin and Range, in the western United States (Fig. 11.49), formed the basis for the work on tectonic geomorphology reported by Leeder and Gawthorpe (1987), Leeder and Jackson (1993), and Leeder et al. (1991). This work led to proposals for generalized paleogeographic models that are discussed in Sect. 11.4.2.1 (Figs. 11.38, 11.39).

Two good examples of ancient nonmarine sedimentation in backarc settings are the mid- to Upper Carboniferous sediments of northern Britain and the Jurassic Kayenta Formation of the Colorado Plateau region. During the Namurian-Stephanian period northern Britain underwent crustal extension behind the north-dipping Variscan subduction zone (Leeder 1982; Besly and Kelling 1988; Coward 1990), resulting in the development of a series of graben and half-graben basins. Initially, in the Dinantian, these filled with carbonates and basin-margin fan deposits. Later, during the mid to Late Carboniferous post-rift phase, regional subsidence led to the development of a broad basin occupying much of northern Britain, and this filled with thick clastic deposits derived from the hinterland, to the north (Besly and Kelling 1988; Fraser and Gawthorpe 1990; Leeder and Hardman 1990). These include important coal-bearing deltaic deposits (the British "Coal Measures"), delta-top fluvial systems, and prodelta turbidites (Figs. 11.53, 11.54).

The Lower Jurassic Kayenta Formation of the Colorado Plateau area developed in the backarc region between the Sierra Nevada arc and the Ancestral Rocky Mountains of the continental interior (Middleton and Blakey 1983; Luttrell 1993). Again, the hinterland constituted a major sediment source. Braided fluvial systems drained southwestward from this source area into an arid basin center, the Utah-Idaho Trough, which filled partly with eolian sands (Fig. 11.55). Ordovician fluvial-deltaic systems in North Wales constitute a third ancient example of an ancient partly nonmarine backarc system (Orton 1988).

There are many Cenozoic to Recent examples of backarc basins in Indonesia and the Philippines. For example, Cagayan Basin on Luzon, in the Philippines (Mathisen and Vondra 1983). In this basin, 900 m of nonmarine volcaniclastic sediment accumulated during the Plio-Pleistocene. The basin deepened and widened during sedimentation as a result of increasing relief of the arc and extensional faulting in the basin. Sedimentary styles are similar to those occurring in the nonmarine aprons of some forearc basins, such as in Guatemala. Other examples include the Meervlakte, Sepik, and Ramu basins of northern New Guinea, which contain thick Miocene to Recent fluvial, peat-bearing successions (Hamilton 1979). The Pannonian Basin of the Carpathian region in southeast Europe is another example of a backarc basin floored by continental crust, although Burchfiel and Royden (1982) and Royden et al. (1983) showed that microplate collision and strike-slip faulting complicated the history of basin development.

A particularly well-studied example of a backarc basin on continental crust is the Basin and Range province of the western United States and northern Mexico (Fig. 11.49). This area consists of a series of tilted fault blocks and half-graben basins filled with up to 3 km of nonmarine upper Cenozoic sediment. Fault-generated relief between uplifted blocks and the basement floor of the basins is 2–5 km, and the entire area has been elevated by 2–3 km (Stewart 1978) and extended to at least double its original width (Hamilton 1987). The province is characterized by an anomalously thin crust and high heat flow. It contains abundant early Cenozoic calc-alkaline and silicic volcanics, whereas younger volcanics are bimodal, including the voluminous Columbia Plateau basalts, of Miocene age. Rifting began locally in the Eocene, but the main phase of active extension and basin subsidence did not commence until the mid to late Miocene.

Songliao Basin, China, may be interpreted as a backarc basin. Klimetz (1983) interpreted the basin as a backarc rift, but Burke and Sengör (1986) suggested that the basin may be the product of "escape tectonics" during convergence between Asia and a North China block (basin type 4.3).

11.4.3.3 Retroarc (Foreland) Basins

Foreland basins occur in three plate-tectonic settings (Miall 1995b). One class of basins, termed retroarc basins by Dickinson (1974), occurs behind (on the continental side of) compressional arcs. Hamilton (1985) attributed crustal thickening and subsidence in compressional arcs "to the gravitational rise and spread of batholithic plutons." Crustal shortening, in the form of fold-thrust development, also occurs in compressional arcs (Jordan 1995). The second and third classes of foreland basin result from collision and suturing, including arc-arc, arc-continent, and continent-continent collision, and are described in Sect. 11.4.5. Some sedimentary and tectonic features common to all classes of foreland basin are described in Sect. 11.4.6.

Examples of retroarc basins with significant nonmarine fills occur throughout the length of the Cordilleran-Andean ranges of the east Pacific continental margin. They include the Sevier Basin of Utah and Idaho (Jordan 1981; Cross 1986) and many Andean basins (Jordan 1995), such as the Llanos Orientales and Putumayo basins of Columbia (Salazar 1990; Stabler 1990), the Marañón and Beni basins of Peru (Dumont and Fournier 1994), the

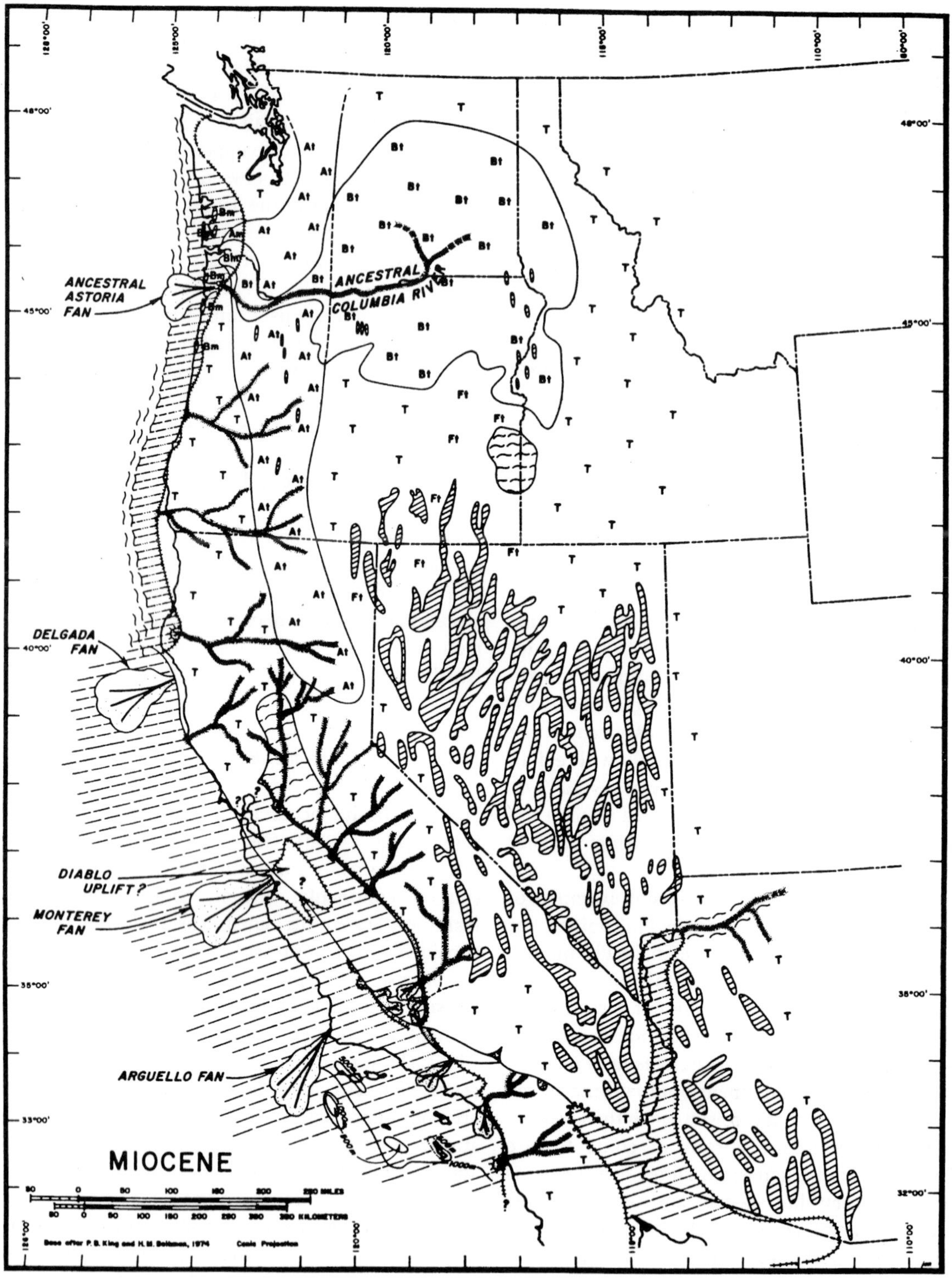

a

Fig. 11.49. Miocene paleogeography of the western United States, showing nonmarine infill of Great Valley and westward fluvial drainage from andesitic arc across the forearc toward the Pacific Ocean. Nonmarine basins of the Basin and Range province are also shown. (Cole and Armentrout 1979)

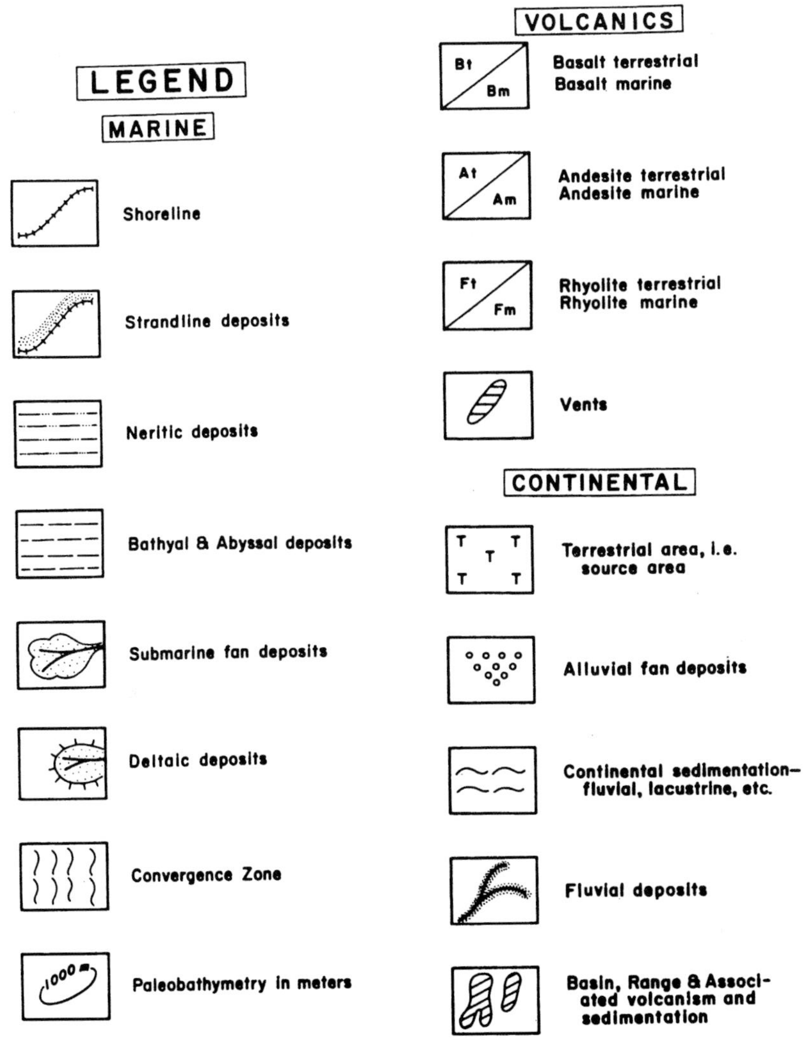

Fig. 11.49b

Cuyo and San Jorge basins of Argentina (Stabler 1990), and the Magellanes Basin of southern South America (Biddle et al. 1986). The Bermejo Valley of northwest Argentina is a small but well-studied basin (Johnson et al. 1986; Jordan et al. 1988; Stabler 1990; Jordan 1995).

The Sevier Foreland Basin spanned the Jurassic to Early Tertiary, but the main phase of rapid subsidence occurred during the Albian-Santonian, during a phase of rapid convergence between North America and the Farallon Plate (Dickinson 1981a; Jordan 1981; Fouch et al. 1983; Cross 1986). Weimer (1970) showed that the Cretaceous section underwent a crude upward coarsening and a facies transition from marine mudstone and limestone to nonmarine sandstone and conglomerate. The same facies transition occurs in reverse toward the east, away from the mountainous source. Syndepositional synclines developed progressively further from the source, in part as a response to sediment loading in shifting depocenters. Jordan (1981) showed that the isopach pattern of mid-Cretaceous strata could be explained by a model of flexural loading by thrust sheets. Campanian-Maastrichtian isopachs, however, do not correspond to a typical foreland-basin pattern, and Cross (1986) related them to a change in subduction patterns and subcrustal thermal events, the details of which are beyond the scope of this book.

The Sevier Basin was disrupted by the basement-involved tectonics of the Laramide Orogeny in the Late Cretaceous-Paleogene, and became segmented

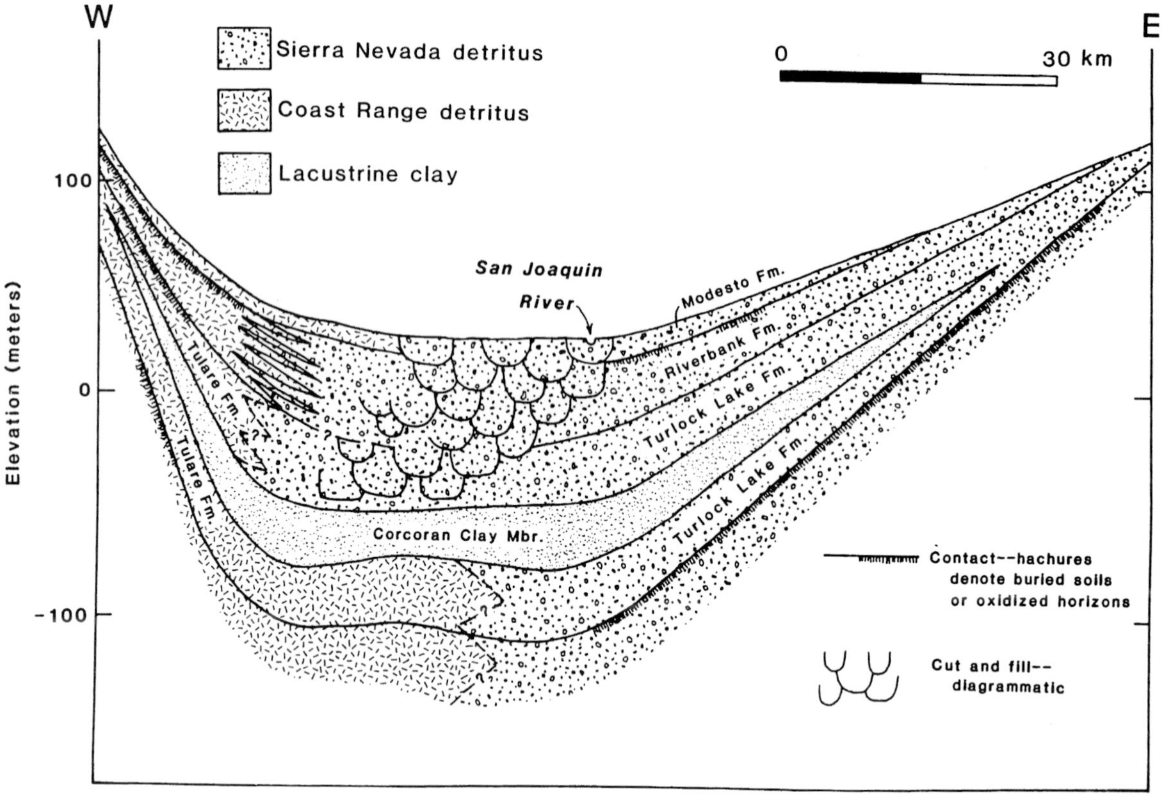

Fig. 11.50. Diagrammatic cross section through the northern San Joaquin valley, showing stratigraphic relationships of Quaternary alluvial deposits. Differences in tectonic history for the Coast Range subduction complex and the Sierra Nevada arc and differences in the composition of detritus dervied from these two source areas make correlation of units across the valley difficult. (Bartow 1987)

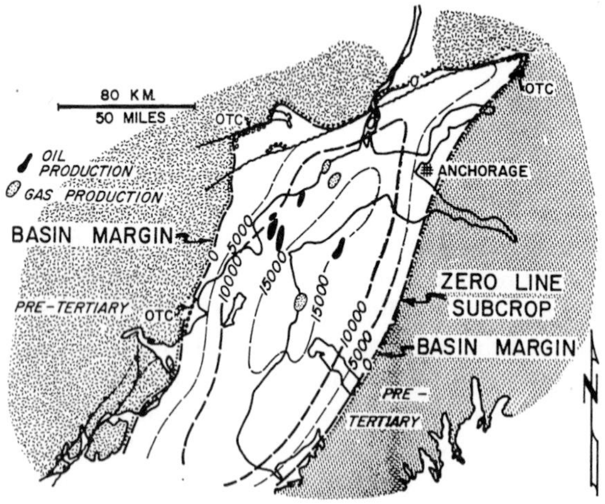

Fig. 11.51. Generalized isopach map of Kenai Group (Oligocene-Pliocene), Cook Inlet, Alaska. (Kirschner and Lyon 1973, reprinted by permission)

into a series of smaller basins separated by fault-bounded uplifts (Dickinson et al. 1988). Ingersoll (1988) classified these as foreland intermontane basins. Jordan (1995) refers to them as broken foreland basins. The basins are entirely nonmarine, containing thick synorogenic fluvial deposits, commonly preserving proximal fanglomerates, and lacustrine successions. Drainage is mainly internal, although Dickinson et al. (1988) summarized evidence for spillage between basins and possible drainage eastward into the continental interior and westward across the Sierra Nevada magmatic arc to the Pacific Ocean. There are many paleogeographic similarities between these basins and the hinterland basins of the Himalayan orogen (Sect. 11.5.5.2).

Two studies of basins adjacent to the southern Andes serve as examples of retroarc foreland basins developed in the absence of collisional events. The Neogene Bermejo Basin of northwest Argentina contains a crudely coarsening upward succession of fluvial siltstone, sandstone, and conglomerate. Johnson et al. (1986) and Jordan et al. (1988) discussed the uses of various facies and stratigraphic criteria in an attempt to relate sedimentation to tectonic events and to develop an accurate picture of the subsidence history. Biddle et al. (1986) provided a stratigraphic synthesis of the Magallanes Basin of southern

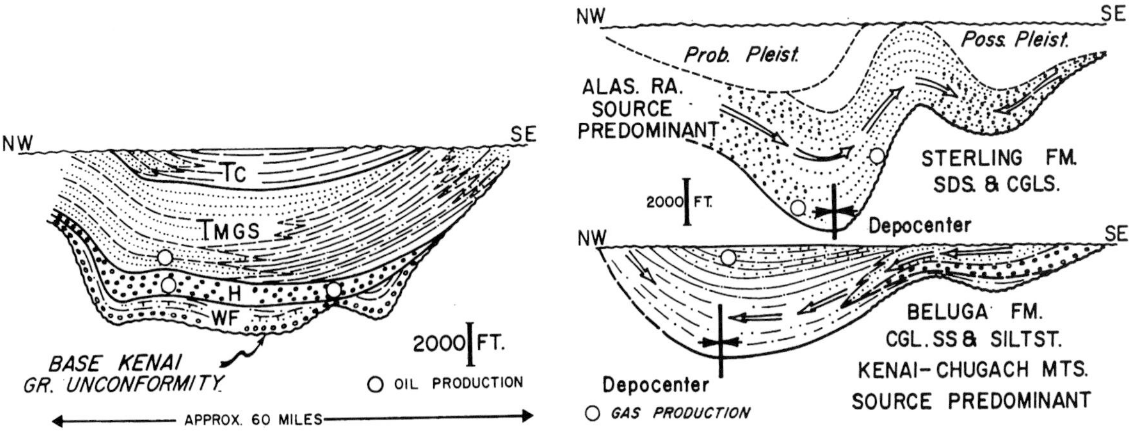

Fig. 11.52. Schematic cross sections through the Lower Kenai Group (*left*) and Upper Kenai Group (*right*). *WF* West Foreland Formation; *H* Hemlock Formation; *T* Tyonek Formation. (Kirschner and Lyon 1973, reprinted by permission)

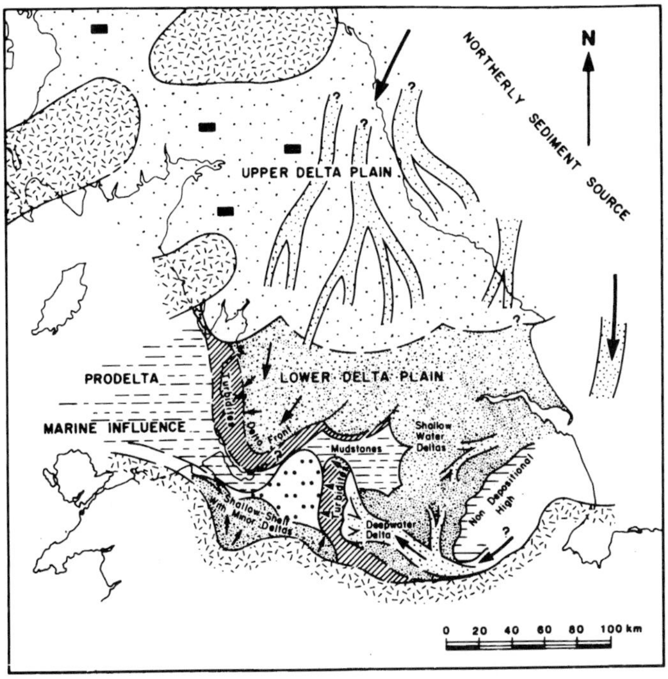

Fig. 11.53. Paleofacies map for the Kinderscout-Crawshaw delta system, central and northern England. The landmass to the south is the London-Brabant massif, a microcontinent located to the north of the Variscan subduction zone. (Fraser and Gawthorpe 1990)

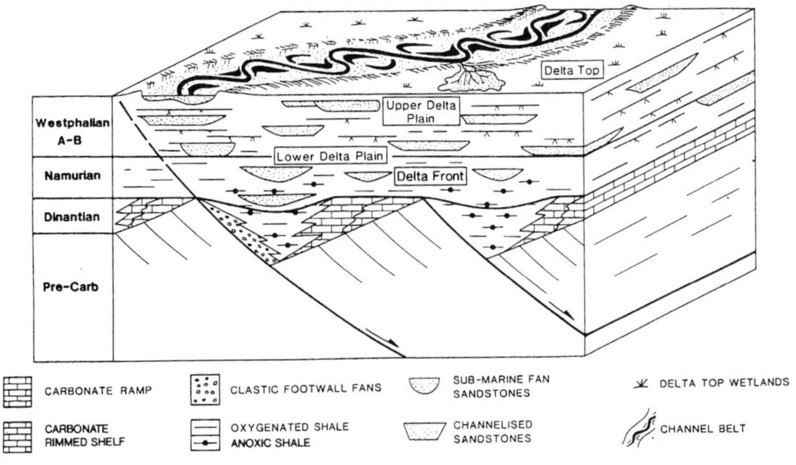

Fig. 11.54. Schematic block diagram for Carboniferous extensional basins of the southern North Sea. (Leeder and Hardman 1990)

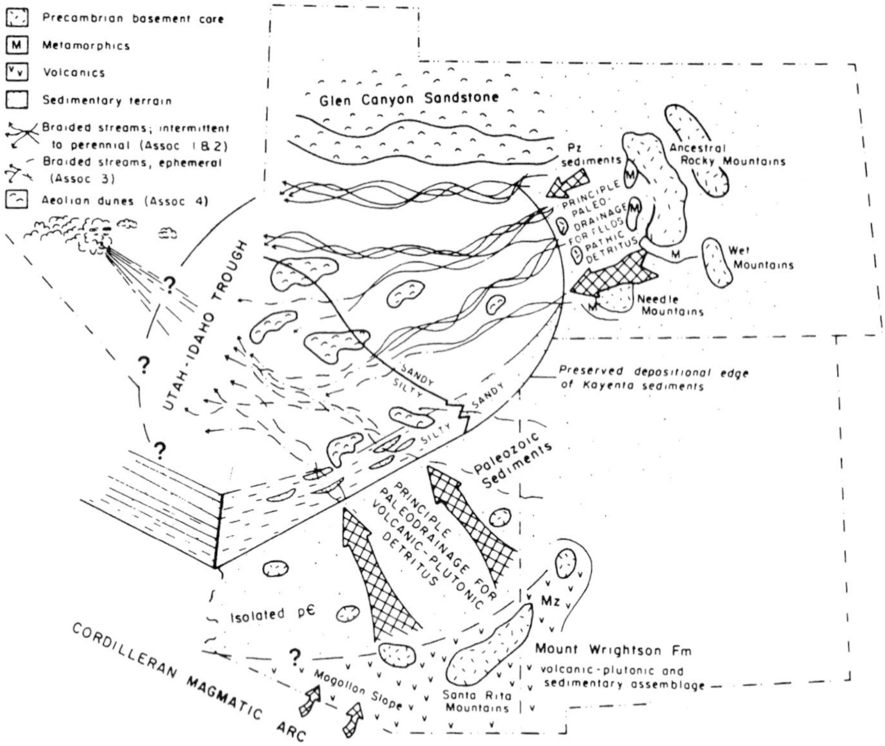

Fig. 11.55. Paleogeographic synthesis of the Kayenta Formation (Lower Jurassic), southwest United States. The main outcrop area of the formation is in western Colorado, southern Utah, and northeastern Arizona. (Luttrell 1993)

Argentina and Chile. This basin developed during the mid-Cretaceous and Cenozoic above a collapsed marginal basin during a major phase of Andean igneous activity and crustal shortening. The basin fill is primarily marine shale. Coarse sediments derived from the Andean uplift are restricted entirely to the Andean margin of the basin.

Another series of retroarc basins occurs behind the Sunda arc in Sumatra, Java, and the Sunda Shelf (Ben-Avraham and Emery 1973). The foreland basins of Sumatra and Java include the Madura, East Java, Tjeribon, West Java, and South Sumatra basins (Hamilton 1979; Hutchison 1989). Sediment thicknesses reach maxima of about 3–6 km in these basins. The Sunda and part of the West Java Basin are graben basins bounded by north-south faults. These faults were active during the Cenozoic, but their trend suggests basement control. The West Java-Tjeribon and Madura basins are separated by a basement high. Sedimentation there began with marine transgression during the Oligocene. Bathyal conditions persisted until the later Neogene in the east, but thick deltaic clastics accumulated in the West Java Basin in the Oligocene. A widespread reef limestone developed in the Miocene. Regression occurred everywhere during the Neogene, and these basins are now largely continental. Basement rocks contributed sediment to the basins at first, but contemporaneous volcanics became the main sediment source during the younger Neogene. Throughout this period the magmatic arc gradually migrated northward. Syndepositional folding commenced first in the south and is probably continuing, resulting in deformation that decreases in intensity northward. Hamilton (1979) attributed the folding to lateral pressures of magmatic intrusion. The sedimentary basins probably owe their origin to downbowing of the crust beneath the weight of volcanic and plutonic rocks in the magmatic belt. A fold-thrust belt has not yet developed here. The geology of the Sumatra foreland is similar. Basement control by north to north-northwest trending structures had an important influence on sediment thicknesses and facies changes. Miocene strata include quartzose deltaic sandstones derived from the Malay Peninsula and Sunda Shelf to the north. Drainage in these basins is typically transverse.

11.4.4 Basins Formed Along Strike-Slip Faults

As shown by Miall (1990, Table 9.6), strike-slip faults occur in a variety of plate-tectonic settings. Of concern here are those which constitute intracontinental plate-boundary transform faults and the boundaries of divergent plate margins. In both settings, basins with significant nonmarine fills occur. Plate boundaries undergoing oblique convergence may also undergo strike-slip displacement, with the offset typically localized along the axis of the magmatic arc. The Sunda Arc of Sumatra offers an example of this, and the forearc and backarc basins of this area show the influence of wrench tectonics, as described by Hamilton (1979) and Hutchison (1989). Another important class of strike-slip-fault-bounded basin is that which occurs as a result of escape tectonics in the hinterland regions adjacent to continental sutures. These basins are discussed in Sect. 11.4.5.2.

Following the work of Luyendyk and Hornafius (1987), Ingersoll (1988) proposed the definition of a class of transrotational basins developed by block rotation around vertical axes within strike-slip fault zones. Such rotation causes thrust faulting where blocks overlap and basins where blocks are rotated apart. Luyendyk and Hornafius (1987) demonstrated that many Neogene basins in southern California are of this type.

It is important to realize that strike-slip motion may, in fact, be a superficial representation of deep crustal processes that show quite different structural style. For example, the construction of the Carpathian Mountains during the Neogene involved differential movement of nappes along flat thrust faults. These nappes, which moved at different times, are separated in places by steep faults, which show a history of strike-slip movement where the nappes were displaced relative to each other. The Vienna Basin is a pull-apart basin that was created in this way (Royden 1985). Conversely, it has now been shown that the Hornelen Basin of the Norwegian Caledonides, long cited as a classic example of a strike-slip basin (e.g., see Steel and Gloppen 1980) is soled by a low-angle extensional detachment surface. This surface is scoop shaped. Where the edges of the scoop reach the surface, extensional movement shows a strike-slip motion relative to adjacent undisturbed strata (Hossack 1984; Seranne and Seguret 1987). May et al. (1993) have demonstrated that the San Gabriel fault, bounding the Ridge Basin, is also a low-angle detachment fault.

11.4.4.1 Basins Associated with Intracontinental Transform Faults

The San Andreas fault system in California and the Alpine fault of New Zealand are the best-known intracontinental transform faults, and much of our knowledge of fault tectonics and basin style is derived from studies in these areas.

Ridge Basin, near Los Angeles, is a well-exposed example of a basin developed by supracrustal loading in a transpressional setting, and excellent studies by Crowell (1974b), Link and Osborne (1978), Crowell and Link (1982), and Link (1984) have turned it into a classic of its kind. The basin was active during Miocene and Pliocene time, when the San Gabriel Fault was the master fault (it is now inactive). Salton Trough, at the head of the Gulf of California, is an example of a pull-apart basin. It was initiated during the Miocene and is still active, so it has not been uplifted and dissected for our examination as has the Ridge Basin. Volcanism, heat flow data, and tectonism in the Salton Trough suggest that it is an incipient spreading center (Crowell 1974a) and may eventually open fully to expose new oceanic crust, as in the Gulf of California. This would accord with what appears to be a pattern of oblique spreading in the area (a "leaky transform"). Based on several Californian basins, Crowell (1974b) proposed a general model of pull-apart structure and stratigraphy as shown in Fig. 11.56.

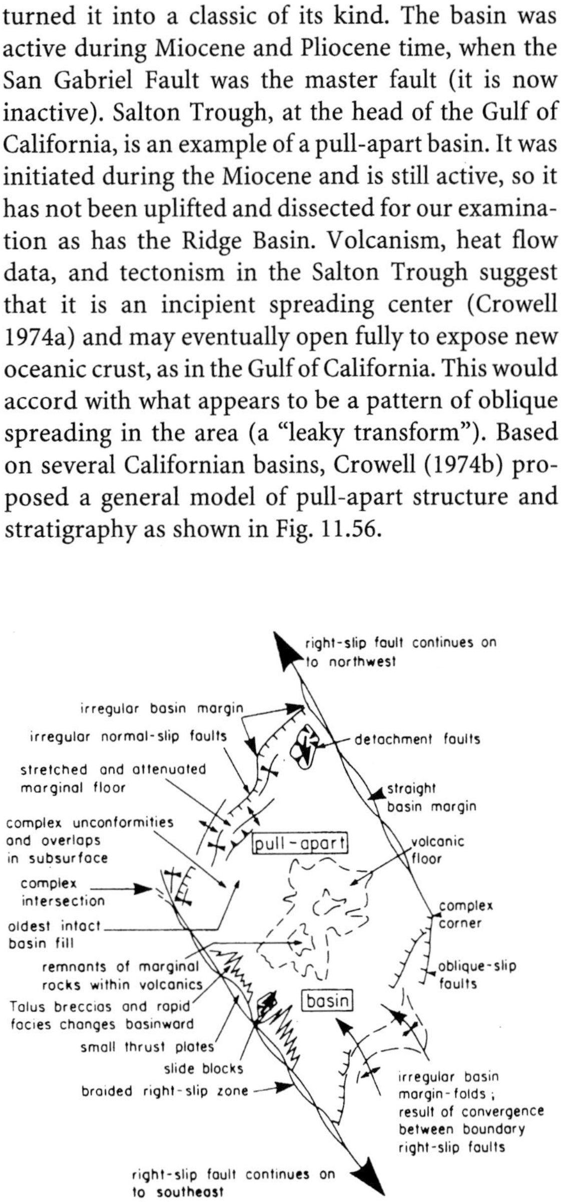

Fig. 11.56. A general model for pull-apart basins. (Crowell 1974b)

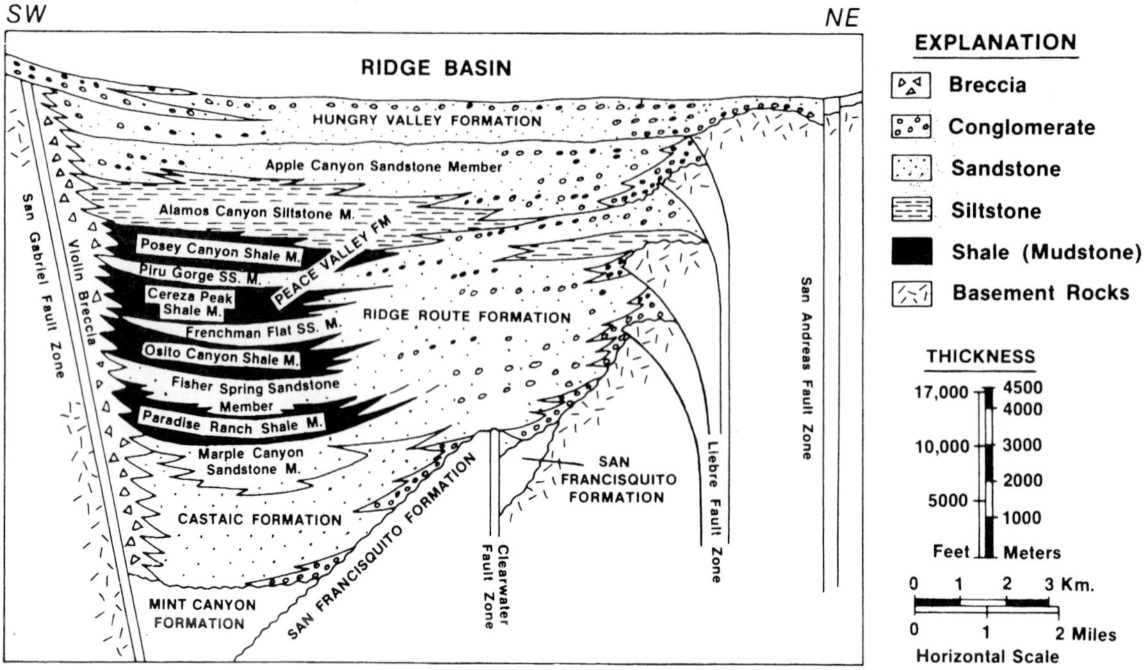

Fig. 11.57. Stratigraphic cross section through Ridge Basin. (Crowell and Link 1982)

Ridge Basin contains more than 12 km of marine, lacustrine, and fluvial sediments that show numerous rapid local facies changes (Fig. 11.57). The San Gabriel Fault bounded the basin to the southwest. It was the main active fault of the San Andreas transform system during sedimentation and is flanked by the Violin Breccia, an alluvial fan wedge 11 km thick. Much of this great thickness is only apparent and reflects offlapping of the basin-fill to the southeast as the basin subsided and enlarged in that direction. The Liebre Fault Zone, on the northeast side, shows the typical splayed or flower pattern of strike-slip fault zones (Fig. 11.57). Uplift on this fault was intermittent and localized, and generated local, short-lived basement sediment sources. The basin center is filled with lacustrine mudstones, fluvial-deltaic sandstones, and minor conglomerates deposited by longitudinal and transverse depositional systems flowing from the north and northwest.

Nilsen and McLaughlin (1985) compared three nonmarine basins bounded by strike-slip faults: Ridge Basin, Little Sulphur Creek Basin in northern California, and Hornelen Basin in Norway. Despite the differences in scale, the basins show remarkably similar paleogeographic patterns (Fig. 11.58). All three are characterized by rapid lateral facies changes and rapid subsidence.

11.4.4.2 Basins Associated with Divergent Plate Boundaries

The mid-Atlantic spreading center includes long stretches of ridge with few transform offsets, and zones of short spreading segments with long transform offsets (Wilson and Williams 1979). During the early stages of continental separation, in the Jurassic-Cretaceous, the transform offsets functioned as intracontinental plate boundaries, and important nonmarine basins developed. Transform offsets commonly continued into the continent as transfer faults, along which significant dip slip and strike slip occurred. Rift and transtensional basins with nonmarine sedimentation were the result. Examples include the following:

1. Central Tertiary basin, Spitzbergen, now bounded to the west by the Hornsund Fault and initially formed as a transtensional basin in the zone separating Svalbard from Greenland. The Early Paleocene Firkanten Formation includes a basal fluvial-deltaic, coal-bearing member, followed by marine deposits (Steel et al. 1981, 1985).
2. The rifts of Central and West Africa developed by dip-slip rift faulting and transtensional strike-slip faulting and include major nonmarine successions. The Benue Trough, Nigeria, is controlled

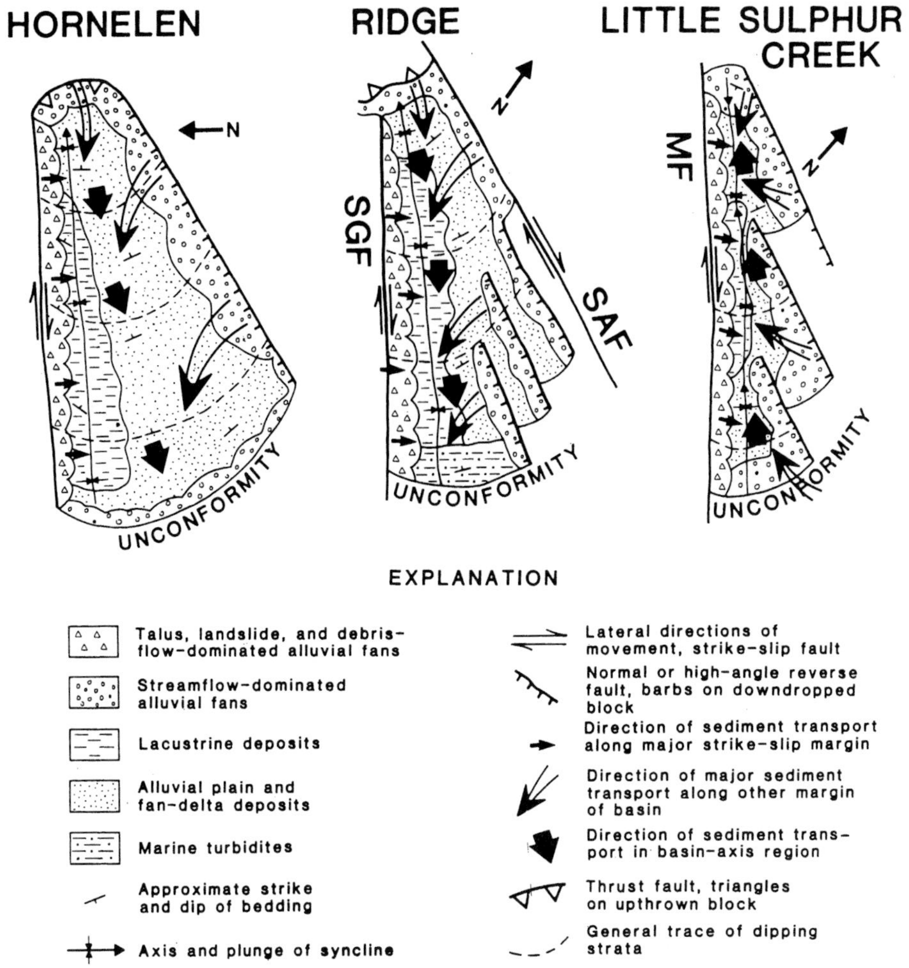

Fig. 11.58. Comparison of three strike-slip-fault-bounded basins. The three basins are drawn at different scales. Hornelen Basin, Norway, is 70 km long, Ridge Basin, California, is 40 km long, and Little Sulphur Creek Basin, California, is 13 km long. *SGS* San Gabriel fault; *SAF* San Andreas fault; *MF* Maacama fault. (Nilsen and McLaughlin 1985)

structurally by landward extensions of the Romanche, Chain, and Charcot fracture zones. Nonmarine sediments range from Albian to Cenozoic age. For example, the Benin Formation (Eocene-Miocene) is a succession of continental sandstones more than 2 km thick (Benkhelil 1982; Fairhead 1986; Shannon and Naylor 1989; Genik 1993). Additional discussion of these rifts is provided in Sect. 11.4.2.1.

3. Basins of southeast South African continental margin, of which the Algoa Basin is the largest. Paleozoic structural lineaments oriented at an acute angle to the continental margin were reactivated during the Late Jurassic by shear accompanying the transform separation of Africa from the Falkland Plateau. A series of transtensional rift basins developed, filled by coarse alluvial-fan conglomerates and basin-center braidplain deposits (Tankard et al. 1982; Fig. 11.59).

11.4.5 Basins Related to Plate Collision

Nonmarine deposits occur in two classes of collisional basin, peripheral foreland basins and hinterland basins, as defined below. A third class of basin, remnant ocean basins, are floored by oceanic crust and therefore do not accumulate nonmarine sediments.

There are two distinct types of foreland basin associated with plate collision (Miall 1995b). First, collisional retroarc basins are hinterland basins, located on the overriding plate, behind the arc. At certain stages of its development, the Alberta basin is

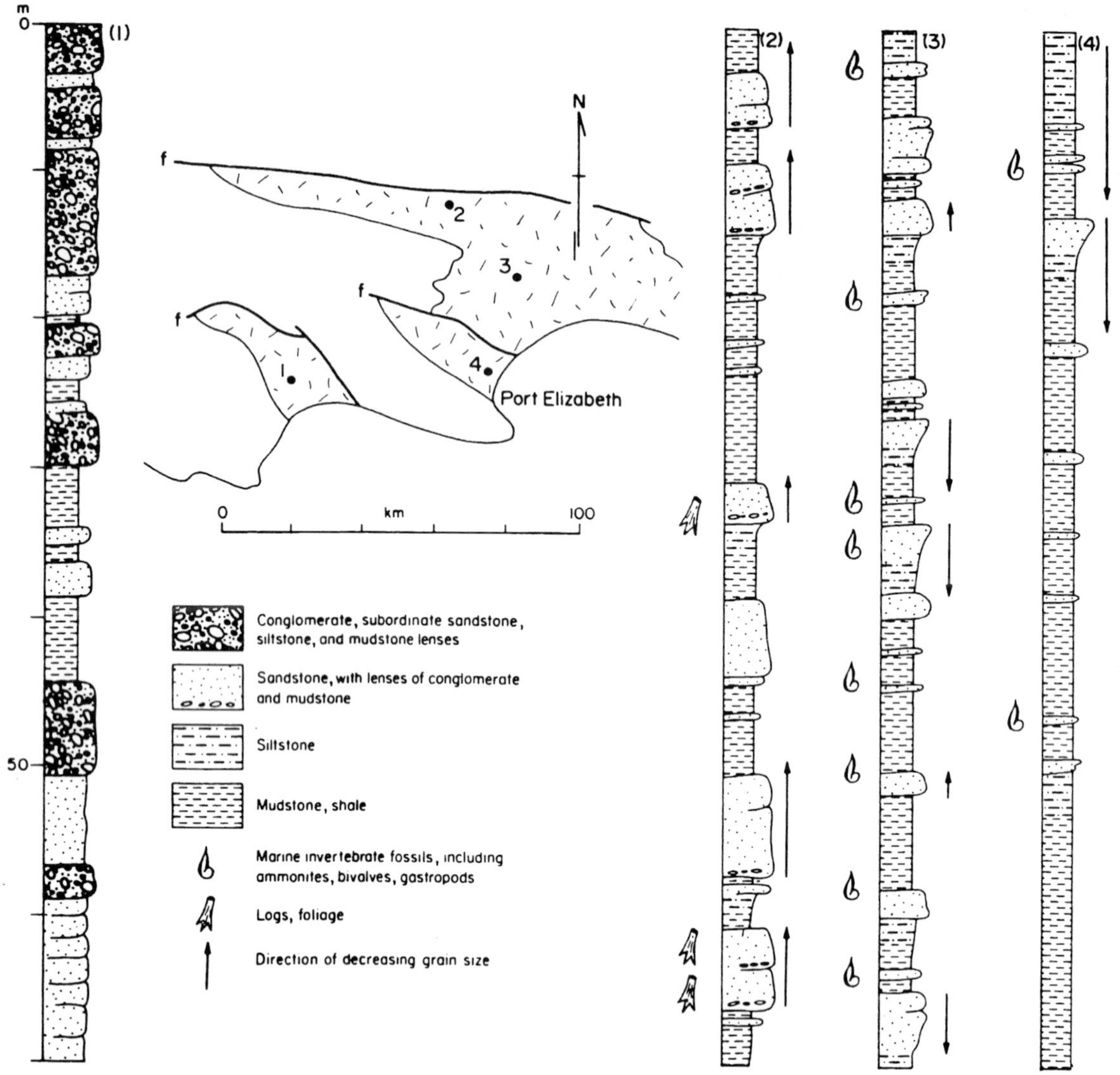

Fig. 11.59. Stratigraphic sections in the Gamtoos Basin, a rift basin developed in the Late Jurassic during the strike-slip offset between southern Africa and the Falkland Plateau. *Section 1* is through the Enon Conglomerate, a basin-margin fanglomerate succession; *sections 2–4* are through finer-grained basin-center deposits. (Tankard et al. 1982)

a good example, and some of the interior basins of western China, located behind (north of) the collision mountains of Tibet, are also of this type. These are described in Sect. 11.4.5.2. The second type of collision-related foreland basin was termed the peripheral basin by Dickinson (1974) and is described in the next section. Peripheral basins are initiated by subduction and loading of the subducting plate in a forearc setting. They overlie, adjoin, and may be overriden by the subduction complex of the arc and overlap the marginal sediment wedge of the subducting plate.

Distinguishing the two classes of collision-related foreland basin from each other and from retroarc basins (Sect. 11.4.3.3) in the ancient record may be difficult because most orogens undergo several phases of accretion, changes in subduction polarity and changes in convergence angle, leading to such complications as strike-slip displacement of a basin and its source area(s) and the superposition of basins controlled by different plate-tectonic mechanisms. Some features common to all types of foreland basin are described in Sect. 11.4.6.

11.4.5.1 Peripheral Foreland Basins

The western Pacific offers several examples of peripheral basins formed by arc-arc collision. For example, a series of small peripheral basins developed in central Honshu, Japan, as a result of the collision of the Izu-Bonin arc with Honshu during the late Cenozoic (Ito and Masuda 1986). Fan deltas and submarine fans are the dominant types of depositional system. Arc-continent collisions may lead to the development of larger peripheral basins. For example, during the late Cenozoic, the Luzon arc collided with the Chinese mainland, creating the island of Taiwan, the western part of which constitutes a fold-thrust belt and associated peripheral basin (Covey 1986; Teng 1990).

Many of the world's largest and best-known foreland basins (Himalayan foredeep, Appalachian basin, Swiss Flysch-Molasse basin, Adriatic foreland, and Persian Gulf) resulted from continent-

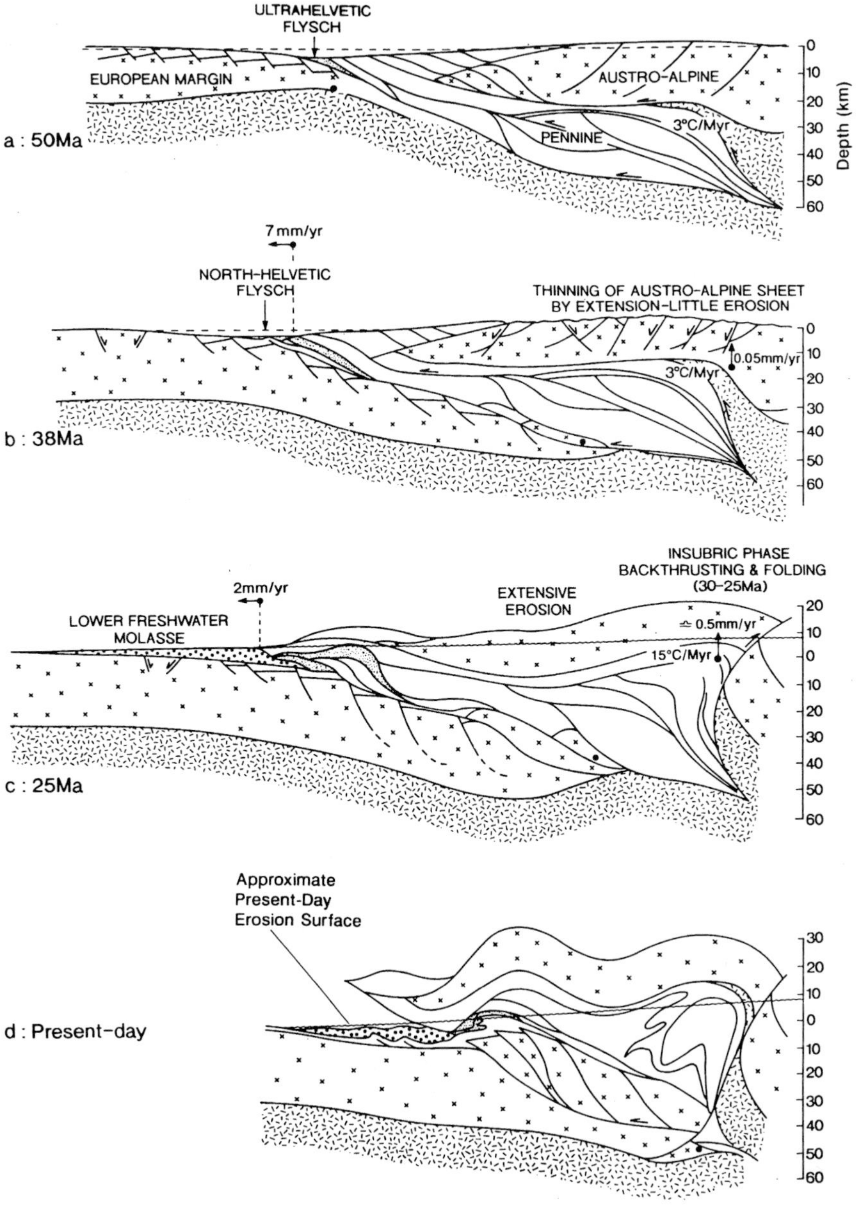

Fig. 11.60. The structural evolution of the Alpine thrust wedge and the North Alpine foreland basin, showing the shallowing of the basin as the accretionary wedge climbed the ramp onto the continental margin. The molasse was deposited in the mid to late Oligocene, during the "continental wedge" phase of development. (Sinclair and Allen 1992)

continent collision. Typically, these basins form above, and rest on, the extensional continental margin of the downgoing plate that developed during the preceding cycle of continental rifting and sea-floor spreading. For example, the Himalayan foredeep basin is formed above the Paleozoic through early Cenozoic Tethyan continental margin of the Indian subcontinent (Gansser 1964). Development of the peripheral basin may involve several distinct but diachronous phases, as first the arc on the overriding plate undergoes compression, and later the main mass of continental crust reaches the subduction zone. Typically, the basins become shallower with time, as the colliding arc climbs the ramp formed by the subducting continent, and the basin fills with sediment (Cant and Stockmal 1989; Stockmal et al. 1986, 1992; Stockmal and Beaumont 1987; Sinclair and Allen 1992). Nonmarine deposits, the typical "molasse" (Van Houten 1981), constitute the terminal basin fill (Figs. 11.24, 11.60).

The Indus-Ganges trough of India and Pakistan is an excellent modern example of a peripheral foredeep basin. It is underlain by the Siwalik Group, a molasse-type deposit of Miocene-Pliocene age accumulated in conditions apparently identical to those of the present day (Johnson and Vondra 1972; Parkash et al. 1983). The Siwaliks have been much studied in recent years for the information they yield on rates and styles of sedimentary and tectonic processes, information that can be obtained from the rocks because of the excellent chronostratigraphic control provided by magnetostratigraphy (e.g., see Sects. 11.2.1, 11.3.2, 11.3.6).

The early "collisional facies", the marine Murree Formation (Cretaceous-Eocene) passes gradationally up into the Siwaliks, an entirely nonmarine unit locally more than 6 km thick (Burbank et al. 1986; Burbank and Raynolds 1988; Burbank and Beck 1991; Burbank 1992: Fig. 11.61). Depositional environments were similar to those existing at the present day, namely, giant alluvial fans draining transversely from the rising mountains into longitudinal trunk rivers, such as the Indus, Ganges, and Brahmaputra. Indeed, the major north-south rivers are regarded as type examples of antecedent drainage (Holmes 1965; Seeber and Gornitz 1983). They have maintained their course, perhaps since the Miocene, while cutting through ranges that are now higher than their source. The Siwaliks have been divided into three units. The Lower Siwalik comprises a coarsening-upward megacycle consisting of sandstone-mudstone alternations passing up into a predominantly sandy sequence. The Middle Siwalik consists mainly of medium to coarse sandstones with interbedded mudstones, and it grades up into a largely conglomeratic Upper Siwalik, the Middle and Upper Siwaliks together comprising a second megacycle. Fining-upward cycles 1–35 m thick occur throughout the sequence. The three subdivisions of the Siwaliks are time transgressive, representing southward and southeastward progradation of fluvial depositional systems into and down the axis of

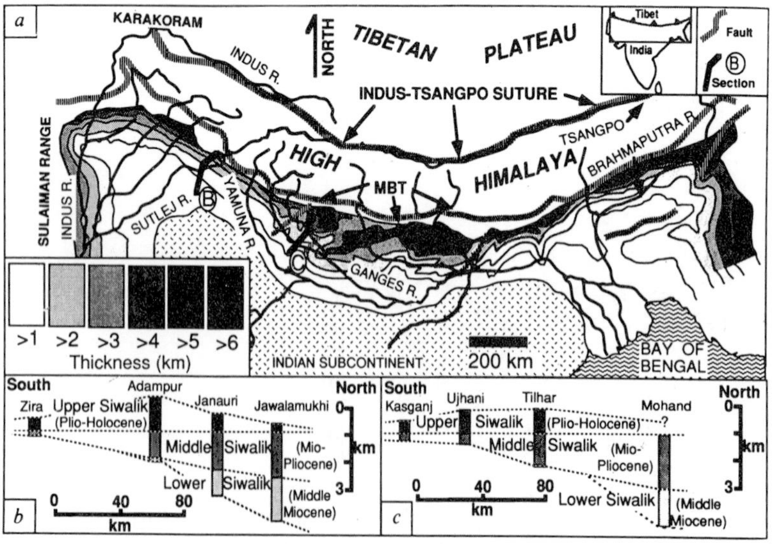

Fig. 11.61. The Himalayan peripheral foreland basin, showing major drainages, and isopach of Miocene-Recent deposits. Major faults are indicated by *vertically hatched lines*. *MBT*, main boundary thrust. (Burbank 1992)

the trough. The Siwaliks are folded and faulted near the Main Boundary Fault but pass out into flat-lying beds overlain conformably by modern alluvium. The alluvium itself is tilted near the mountain front, and Landsat images indicate active folds in parts of the trough. Petrographic studies of light and heavy minerals and of the conglomerate clasts document the progressive and rapid unroofing of Tethyan rocks that were metamorphosed deep in the collision zone (Cerveny et al. 1988).

The Swiss molasse basin was initiated as a peripheral basin, extending from France to Austria. Its strongly deformed basin fill is exposed in the western Alps of Switzerland and France (Van Houten 1981; Homewood et al. 1986; Crumeyrolle et al. 1991; Sinclair and Allen 1992; Fig. 11.60). The molasse is subdivisible into several marine and freshwater units reflecting the regional tectonic history. In the Digne Basin of France more than 2 km of molasse deposits are preserved. In Switzerland the basin fill reaches at least 5 km. The nonmarine part of the Molasse is dominated by thick conglomerate units, termed the Nägelfluh. As noted in Sect. 11.3.2, axial drainage developed during two distinct phases of basin evolution, separated by marine transgression. The axial rivers showed opposite drainage directions during the two phases of basin development, reflecting variations in basin subsidence patterns (Fig. 11.24).

The Adriatic foredeep of northern Italy and the Adriatic Sea developed as a result of collision between the Apennines (a spur of Africa) and the Alps, beginning in the Oligocene and not yet completed (Ricci Lucchi 1986). The basin is asymmetric in cross section, with the deepest part of the basin located adjacent to the Apennine thrusts, in the south. The basin has been suplied with terrigenous sediment from the northeast, northwest, and southwest sides, with the largest nonmarine depositional system constituting the axial Po River drainage system (Fig. 11.62). Quaternary nonmarine deposits in this basin are locally more than 400 m thick (Ori 1993). They consist of three main depositional systems (Fig. 11.63): (a) the axial meander belt of the Po River, up to 25 km wide; (b) transverse drainage from the Alps to the north, including large gravel alluvial fans and sandy braided and meandering tributary rivers; and (c) relatively small fans and tributary rivers forming an alluvial apron flanking the Apennines to the south. These deposits contain intraformational unconformities, attesting to their syndepositional origins.

A final example of a peripheral basin containing a significant nonmarine sedimentary fill is the Mesopotamian plain. Baltzer and Purser (1990) classified rivers in this basin into four types (Fig. 11.64): Type 1 – Consequent streams drain active anticlinal slopes of the Zagros fold-thrust belt and develop small alluvial fans having low preservation potential. Type 2 – Axial rivers flow along the synclines and drain into the Gulf where structural trends are oriented obliquely to the present coastline, indicating axial plunge to the east. Deltas form at the mouths of these rivers. Type 3 – A few major antecendent rivers

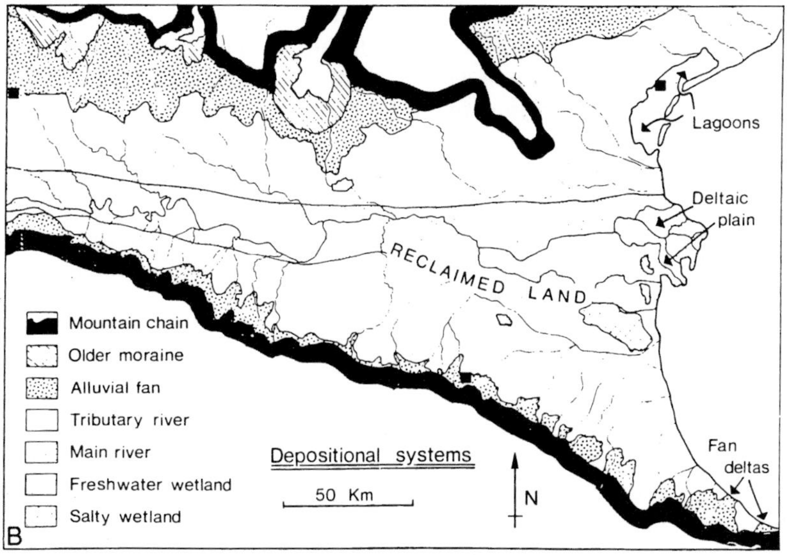

Fig. 11.62. Depositional systems of the modern Po River, Italy. (Ori 1993)

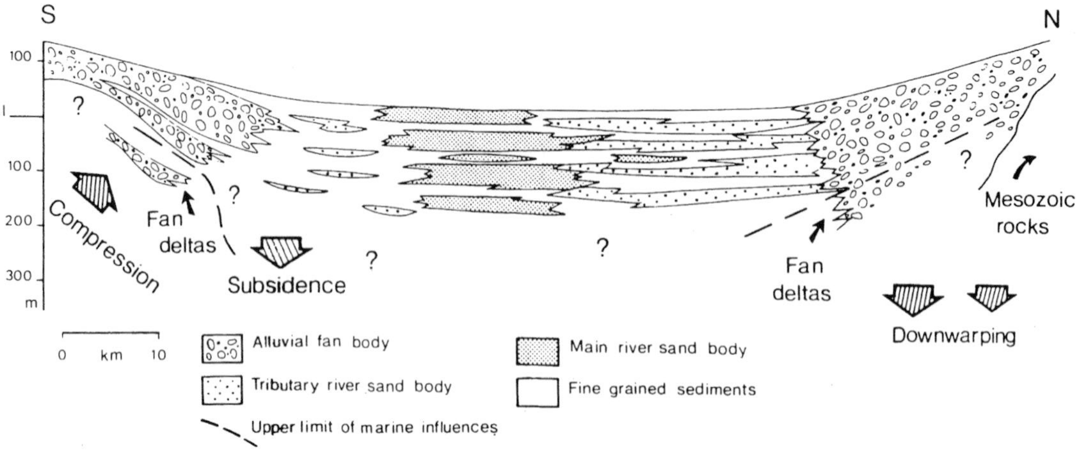

Fig. 11.63. Schematic stratigraphic architecture of the Po Basin deposits. (Ori 1993)

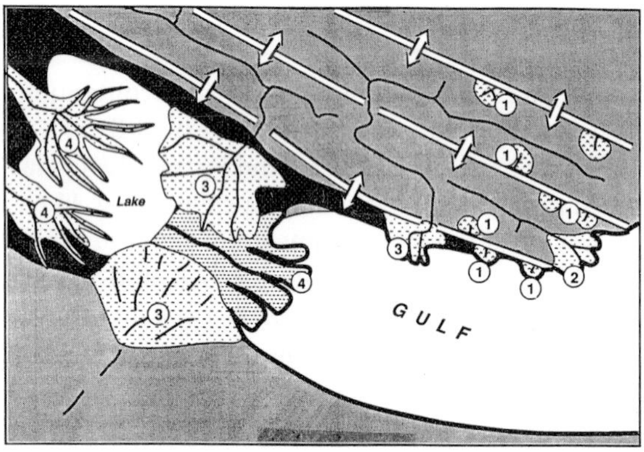

Fig. 11.64. The four main types of river in the Persian Gulf foreland basin. (Baltzer and Purser 1990)

cut across the fold-thrust belt, linking several elongate drainage basins occupied by the first two types of river. Type 4 – The Tigris and Euphrates rivers occupy the main structural depression of the foreland basin and form the Mesopotamian Plain and the large deltas at the head of the Gulf. Baltzer and Purser (1990) described the modern depositional systems of these rivers, but little or no information is available on the stratigraphy of the deposits formed by them.

11.4.5.2 Hinterland Basins

In large orogens, crustal shortening may be accommodated by sideways squeezing of major crustal blocks, with the development of strike-slip faults in the foreland or the hinterland or both. This has been termed escape tectonics (Burke and Sengör 1986). Strike-slip faulting is particularly characteristic of late-stage adjustments to plate collision. Both transpressional foreland-type basins and transtensional pull-apart basins are common in such settings. Retroarc-type foreland basins may also be present. Because these basins are located mainly within continental plates, they are dominated by nonmarine depositional systems, and because they are surrounded by tectonically active mountain belts, subsidence and sedimentation tend to be rapid. The type example of a region dominated by escape tectonics is southeast Asia, the late Cenozoic geology of which is dominated by the effects of the collision between India and Asia (Tapponnier et al. 1986). Brookfield (1993) discussed the complex sequence of tectonic and stratigraphic events that has occurred in the Himalayan orogen in the India-China-Afghanistan area since the collision of India with Asia in the Eocene. The tectonics and sedimentary basins of this region are illustrated in Figs. 11.65 and 11.66. Collisional retroarc basins include Tarim and Jiuquan basins, while important transtensional pull-apart basins include the Baikal and Shansi rift systems and major basins in the South China Sea and the Gulf of Thailand. Nonmarine sediments predominate in all these basins.

Most orogens contain the evidence of late-stage strike-slip adjustments, with the presence of large to small, mainly nonmarine, intermontane sedimentary basins. Examples include the Cordilleras of Canada (Long 1981; Eisbacher 1985; Ridgway and DeCelles 1993), the late Cenozoic and Recent development of Turkey (Hempton and Dunne 1984; Sengör et al. 1985; Dewey et al. 1986), the Devonian "Old Red continent" formed by the

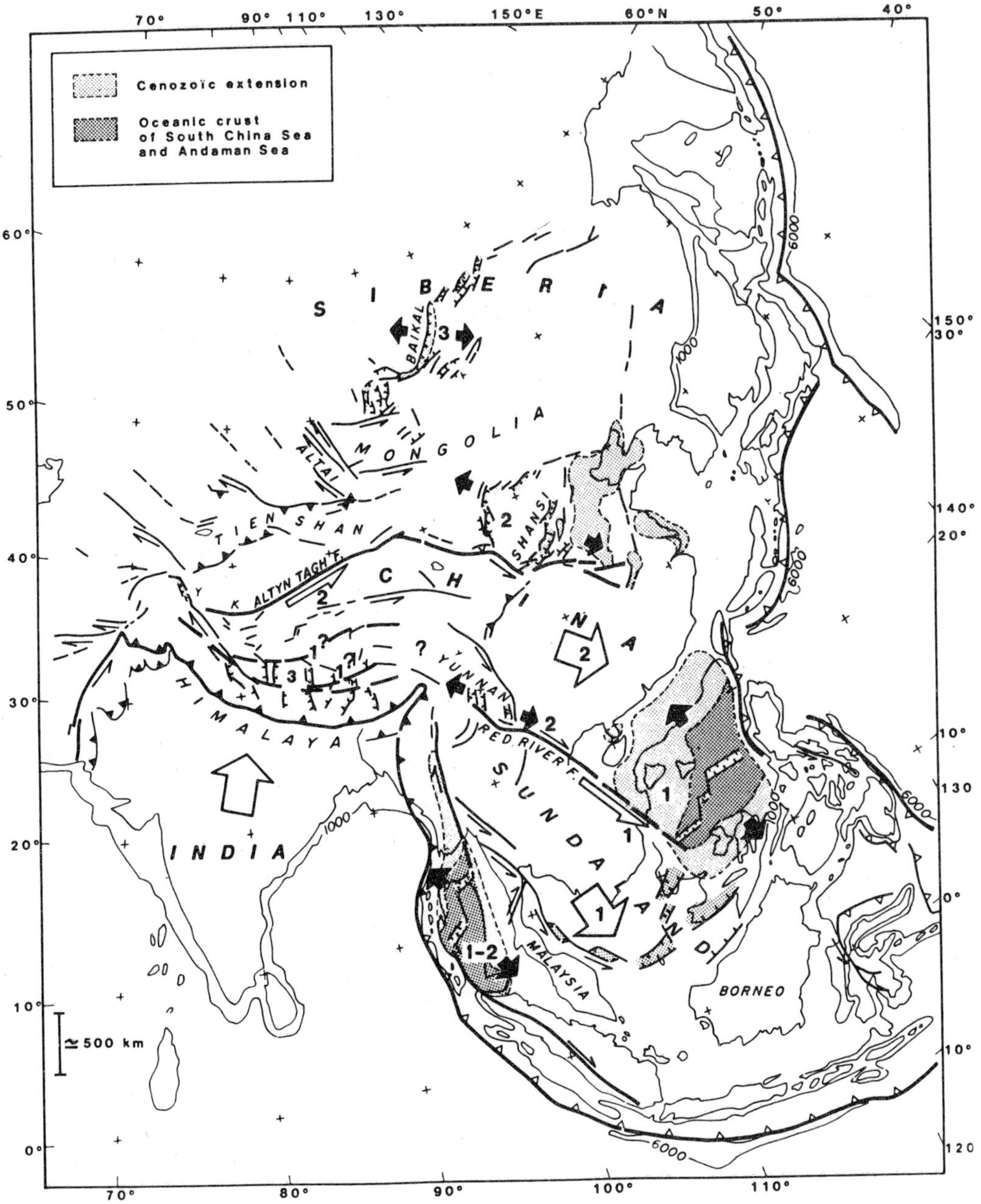

Fig. 11.65. Generalized tectonic map of east Asia, showing the effects of the India-Asia collision. *Large open arrows* represent major block movements since the Eocene. *Smaller solid arrows* indicate direction of extensional movements. *Numbers* refer to extensional phases: *1*, 50–17 Ma; *2*, reprinted by permission, 17 Ma to present; *3*, most recent, including probable future movements. (Tapponnier et al. 1986, reprinted by permission)

amalgamation of North America and Europe (Dewey 1982; Friend 1985; McClay et al. 1986; Harris and Fettes 1988; Coward 1990), the Devonian-Carboniferous (Acadian-Alleghanian) orogen of eastern Canada (Hamblin and Rust 1989; Gibling et al. 1992), and the late Paleozoic Ancestral Rockies and related structures formed by the oblique collision between North and South America (Kluth and Coney 1981; Dewey 1982), which provided the basins and sediment sources (e.g., Paradox Basin, Uncompahgre Uplift) for the Cutler Group of New Mexico and Colorado (see Eberth and Miall 1991). Stanistreet (1993) compared the history of the Precambrian Witwatersrand Basin of South Africa with the modern evolution of the Maracaibo Basin of Venezuela and attributed both to escape tectonics in response

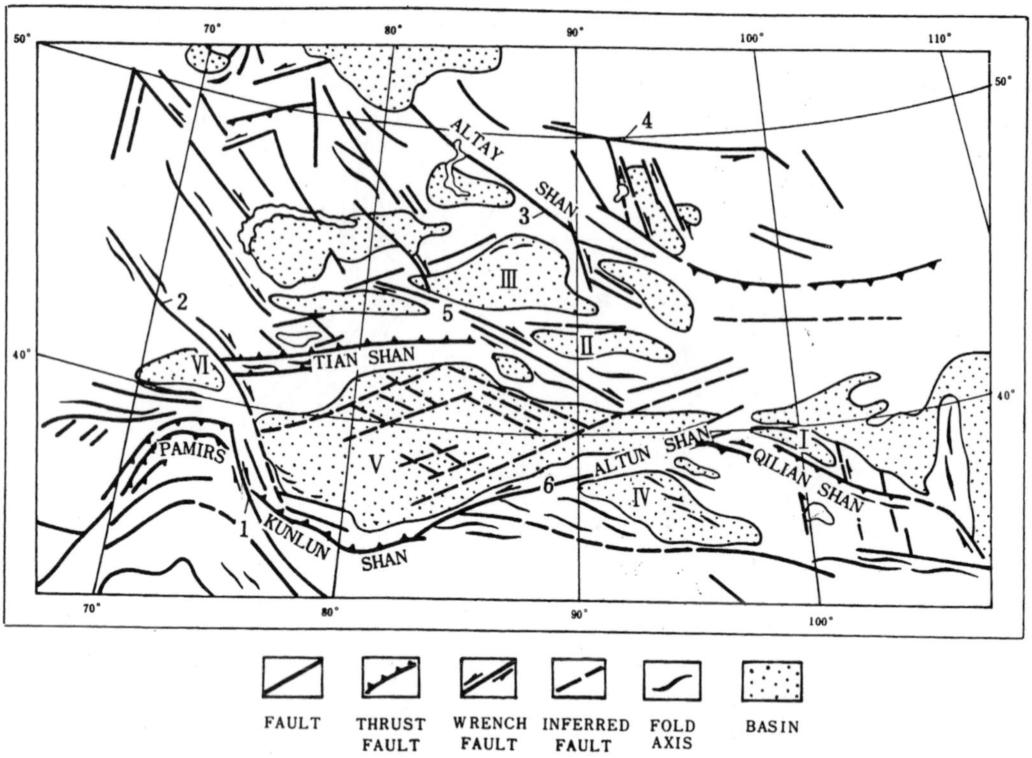

Fig. 11.66. Basins and major structures in west China and adjacent areas of Russia, Afghanistan, and the Indian subcontinent. Many of the basins are bounded by fold-thrust belts and have the character of foreland basins. Basins: *I* Jiuquan; *II* Turpan; *III* Junggar; *IV* Qaidam; *V* Tarim; *VI* Fergana. Faults: *1* Karakoram; *2* Talaso-Fergana; *3* Ertix; *4* Changajn; *5* Borohoro; *6* Altun Shan. (Liu 1986)

to lateral compression within colliding continental blocks.

Complex orogens undergo changes in convergence vectors and multiple terrane-accretion events during progressive crustal shortening, and this may lead to extremely complex stratigraphic and structural histories in hinterland basins. Tectonic styles may change during basin development, and the result may be a hybrid basin that is difficult to classify by plate-tectonic origin. The progressive (diachronous) collision of irregular margins and the reactivation of inherited basement structures may lead to repeated pulses of loading and subsidence. The simple rheological responses predicted by the geophysical models are then overprinted by the local responses to inherited structural anisotropies. Relative sea level and sediment source relief are affected by the resulting tectonism on both local and regional scales, with the possibility for developing very complex patterns of intersecting clastic wedges (e.g., Ricci Lucchi 1986; Hsü et al. 1990; Cluzel et al. 1990; Brookfield 1993). Within large orogens, such as the Mediterranean-Alpine region and the Himalayan orogen of west China, basins may become isolated from the world ocean system within surrounding mountain ranges as a result of regional collisional events (e.g., Tarim, Jiuquan, Qaidam basins: Fig. 11.66), leading to centripetal drainage, changes from marine to fresh water, lowering of sea level, anoxic events, and changes in climate (Sect. 12.12), possibly including basin evaporation, causing salinity crises (Ricci Lucchi 1986). Paleoflow patterns may undergo major changes, even reversals, as a response to basin tilting, diachronous closure, or the influence of reactivated basement structures.

11.4.6 Structural and Stratigraphic Patterns Common to Foreland Basins

The crustal shortening, thickening, and loading that leads to the development of foreland basins causes an expulsion and cratonward migration of pore fluids. Earlier-formed faults become locked, and subsequent faults tend to step outward further into the basin (Dahlstrom 1970). Syndepositional tectonism is common, with the development of growth folds, intraformational unconformities (Anadón et al. 1986; Sect. 11.2.1.1), and cannibalism of older sediments by younger alluvial systems. A wide variety of basin-fill patterns can develop, depending on the configuration of the continental hinterland. Transverse drainage systems may develop during times of high sea level, when the axis of the basin is occupied by the sea. During times of low sea level the basin axis, between the fold-thrust belt and the forebulge, may be occupied by a major longitudinal trunk river. The Alberta Basin (which is a hinterland foreland basin) showed both these conditions during the Cretaceous and Cenozoic.

Axial drainage is particularly important in basins that develop by oblique closure, with the migration of the suture longitudinally and the transportation of detritus away from the locus of maximum compression and toward any remnant marine basin. Modern examples of this include (a) the Adriatic foredeep, undergoing filling by the Po river system; (b) the Persian Gulf, filling from the northwest end by the Tigris and Euphrates systems; (c) the Himalayan foredeep; and (d) Huon Gulf and Markham Valley in Papua-New Guinea, which are the deep-water and nonmarine components, respectively, of a small peripheral basin currently developing by diachronous closure of a small ocean basin to the east (Crook 1989; Brierley et al. 1993).

As shown in Sect. 11.3.6, the timing relationship between tectonism, subsidence, and nonmarine sedimentation in foreland basins is variable, depending on the rigidity of the underlying crust and the magnitude and grain-size distribution of the sediment supply (including that introduced by axial drainage). Two contrasting models represent end-member conditions, the conventional "syntectonic" model of gravel progradation occurring contemporaneously with tectonism, and the "antitectonic" model, in which gravel progradation postdates tectonism during isostatic uplift and unroofing of the fold-thrust belt. The latter model is illustrated in Fig. 11.67. Climatic factors may also be important, including the amount of rainfall, which controls the rate of sediment dispersal. Uplift of the fold-thrust belt increases orographic rainfall, which increases erosion and sediment yield, a process discussed in Sect. 12.11.

Drainage patterns within foreland basins have been discussed by several authors (e.g., Eisbacher et al. 1974; Graham et al. 1975; Sengör 1976; Dewey 1977; Miall 1978d, 1981b; Van Houten 1981; Burbank 1992). The subdivision into transverse and axial (longitudinal) flow is discussed in Sect. 11.3.2. In basins receiving a large sediment supply from the fold-thrust belt, the transverse drainage may extend across the entire basin. Such basins are said to be overfilled (Sinclair and Allen 1992; Jordan 1995). In contrast, axial drainage may take place along the deepest part of the basin. It collects and funnels the

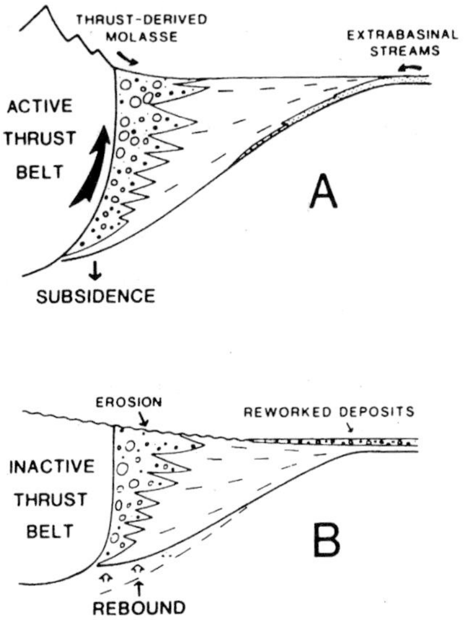

Fig. 11.67. The alternative, "antitectonic" model for foreland-basin development. *A* During active thrust loading gravel deposition is limited to the most proximal parts of the basin. *B* Gravel is transported far into the basin during postorogenic tectonic rebound, when the proximal part of the basin may undergo erosion. (Heller et al. 1988)

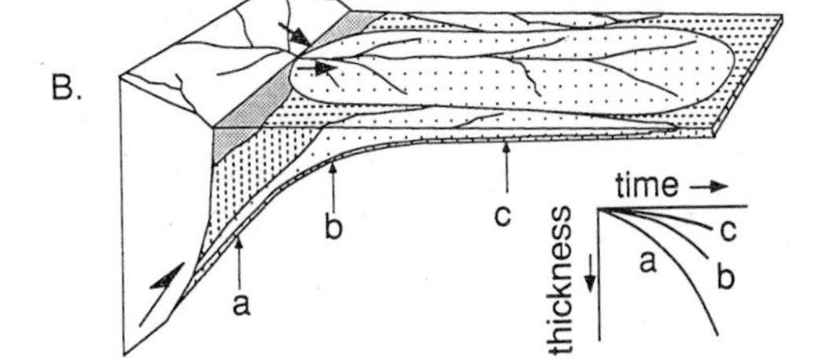

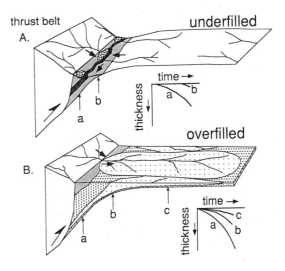

Fig. 11.68. Overfilled and underfilled basins. In underfilled basins the peripheral bulge is a sediment source, whereas in overfilled basins the bulge is covered by sediment. (Jordan 1995)

sediment entering the basin along transverse drainage systems and therefore tends to be characterized by larger channels. Such basins are commonly underfilled, in the terminology of Jordan (1995). These differences are illustrated in Fig. 11.68. Examination of these differences in individual cases may throw important light on basin development. For example, Burbank (1992) demonstrated that during the Pliocene the Himalayan foreland basin of India underwent a transition from an underfilled condition, with an axial ancestral Ganges river system flowing in a relatively proximal position, to an overfilled condition, dominated by transverse drainage of giant fan systems, such as the Kosi, and with the axial drainage having migrated laterally to a very distal position within the basin. He related this change to post-tectonic uplift of the Himalayan ranges and the proximal basin and to the increased sediment supply that resulted from this unroofing, the initiation of the Indian monsoon, and from the late Cenozoic Alpine glaciation. His data for India therefore support the "antitectonic" model for foreland basin development (Sect. 11.3.6), in constrast to that part of the basin located in Pakistan, which Burbank et al. (1988, 1989) argued developed according to the traditional "syntectonic" model of sedimentation and tectonics.

Eisbacher et al. (1974) demonstrated that transverse drainage derives mainly from structural salients along fold-thrust belts, such as the central, most uplifted part of thrust plates, where the belt tends to have a convex curvature facing the basin, whereas many major longitudinal streams enter the basin at reentrants, such as at the termination of thrust plates or at terrane sutures. They cited the modern Himalayan foredeep as an example. Major alluvial fans, such as that of the Kosi River, represent transverse drainage from structural salients, whereas the Brahmaputra River, one of the major axial drainage channels in the basin, enters at the major syntaxis that marks the end of the Himalayan ranges, in Assam. Many transverse drainages are thought to be antecedent rivers; that is, they predate the development of the mountain ranges through which they cut. Examples in the Himalayas are cited by Holmes (1965) and Seeber and Gornitz (1983), and in the Zagros Ranges by Baltzer and Purser (1990).

Stratigraphic successions in foreland basins are strongly cyclic. Regional tectonism has long been known to be a primary control, and in recent years there has been much debate regarding the importance of eustasy as a major stratigraphic control (e.g., Posamentier et al. 1988; Miall 1991c; Posamentier and Weimer 1993). The relationship between thrust-sheet loading and the development of a sedimentary moat is well established (e.g., Bally et al. 1966) and has been modeled quantitatively by several workers (e.g., Beaumont 1981; Jordan 1981; Jordan and Flemings 1990). Recently, attempts have been made to examine the correlation between specific stratigraphic sequences, such as major molasse pulses, and major regional tectonic events such as terrane collisions (Cant and Stockmal 1989; Stockmal et al. 1992) and thrust movement episodes (Jordan et al. 1988; Burbank and Raynolds 1988). The nature of the theoretical relationship between tectonism, subsidence, and sedimentation is examined in detail in Sect. 11.3.6. Tectonic cyclicity is on several times scales. Of most concern here are the so-called third-, fourth-, and fifth-order sequences, which have indicated durations of 1–10, 0.2–0.5, and 0.01–0.2 million years, respectively. Longer-term sequences (> 10 million years) are related to global plate-tectonic controls and are beyond the concern of this chapter (see Sect. 13.4 and review in Miall 1990). The relative importance of tectonism and eustasy in generating these sequences is by no means clear, either in general global terms or in the case of the specific sequences in foreland basins. The following paragraphs provide brief discussions of the nonmarine cyclicity in several foreland basins around the world.

The Oligocene-Miocene molasse in the Digne area of southeast France is discussed briefly in Sect. 11.4.5.1. Crumeyrolle et al. (1991) suggested that the overall molasse succession is a long-term (second-order?) trangressive-regressive cycle induced by foreland-basin tectonism. The succession is punctuated by unconformities and is subdivisible into third-order sequences, which they suggested may be related to eustatic events and can be correlated with the Haq et al. (1987, 1988) global cycle chart. However, recent debate (Miall 1992b) indicates that this type of correlation is speculative and should be viewed with caution.

The South Pyrenean foreland basin of northeast Spain contains a mixed carbonate-siliciclastic succession of Paleocene-Eocene age 3 km thick that has been subdivided into nearly 20 third-order sequences. Contrasting analyses of different, but overlapping portions of this succession were provided by Puigdefábregas et al. (1986) and Luterbacher et al. (1991). In the first of these papers, the authors attributed the development of the sequences to tectonism, the gradual southward overthrusting of the fold-

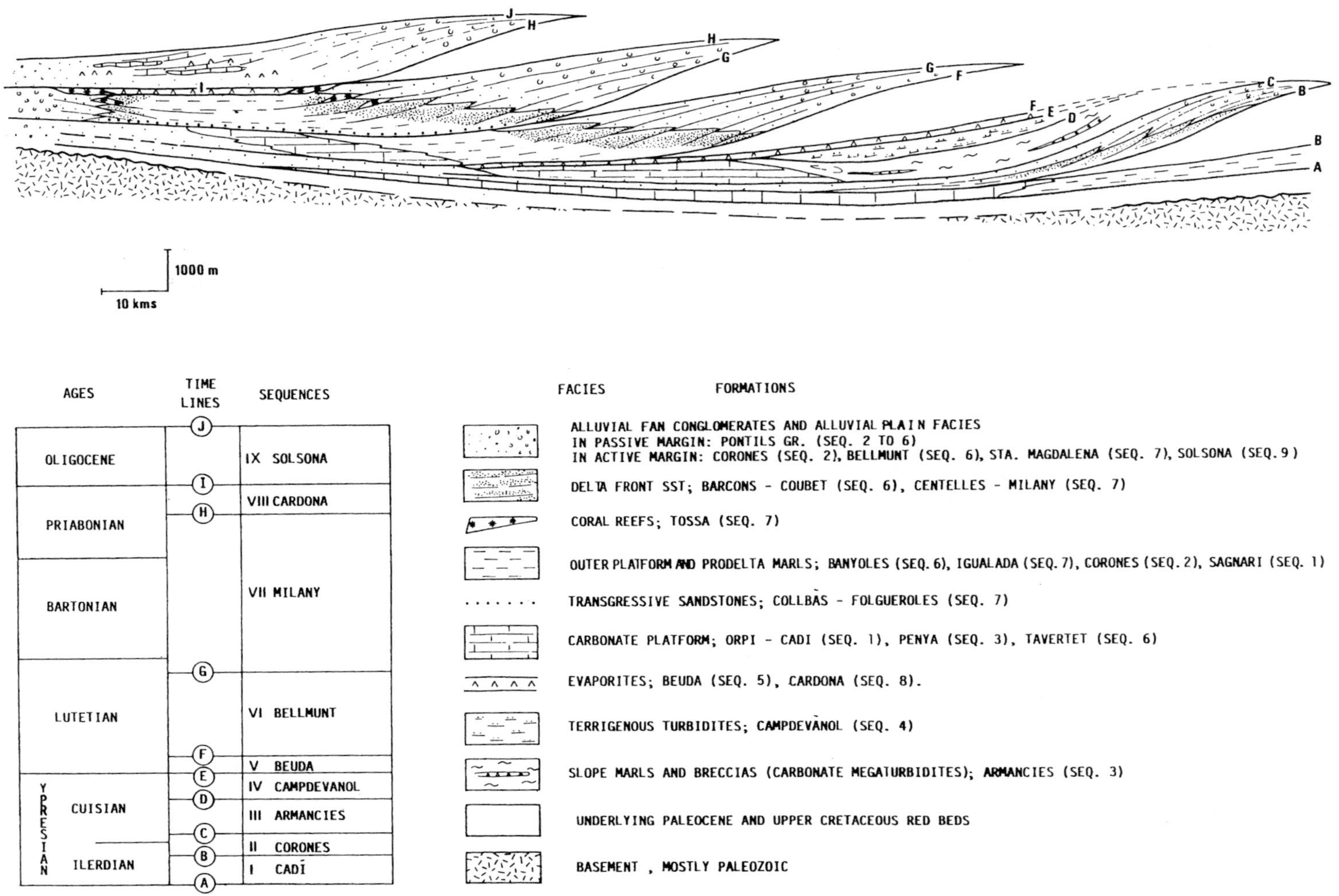

Fig. 11.69. Idealized cross section of the Eocene-Lower Oligocene sequence stratigraphy of the South Pyrenean basin. (Puigdefábregas et al. 1986)

thrust belt leading to a migration and offlapping of depocenters (Fig. 11.69). However, in the second paper, it is claimed that correlation of the sequences with the Haq et al. (1987, 1988) chart can be carried out, and the authors interpreted eustasy as the main driving mechanism.

An idealized cross section of the Eocene-Oligocene section is shown in Fig. 11.69. The overall structure of the sequences is that of a coarse, nonmarine clastic wedge prograding southward from the thrust front into basin-center marine marls. Shallow-water carbonate deposits are formed on the distal ramp and forebulge of the foreland basin, away from the influence of the fold-thrust-belt clastic source. Carbonates are found in the transgressive stage of each sequence. They may be associated with evaporites formed during the subsequent regressive stage. Careful dating of these sequences by Burbank et al. (1992) using magnetostratigraphy indicated that "progradation was controlled primarily by subsidence rather than sediment supply or base-level changes".

Ten major third-order trangressive-regressive cycles have been recognized in the Western Interior basin of the United States (Weimer 1960, 1986; Kauffman 1984). Most of these are shown in Fig. 11.70. High sea level was characterized by the development of thick and areally extensive mudstone units (e.g., Mancos Shale), and by fine-grained limestone (including chalk) in areas distant from sediment sources (e.g., Niobrara Formation). Lower sea level gave rise to extensive clastic wedges, in which nonmarine sandstone and conglomerate passed basinward into shoreline and shelf sand bodies. The Indianola and Mesa Verde groups of Utah and Colorado are good examples of major regressive stratigraphic packages which contain numerous minor transgressive-regressive cycles nested within them (Weimer 1960; Fouch et al. 1983; Lawton 1986a,b; Swift et al. 1987; Van Wagoner et al. 1990, 1991; Yoshida et al., in press; Fig. 11.71).

Some of the third-order cycles may be global in origin and related to eustatic changes in sea level. As discussed by Weimer (1986), there is a fair degree of correlation between the sea-level curve for the western United States and that for northern Europe, although correlations are bedeviled by the difficulties in reconciling various chronostratigraphic time scales. Arguments regarding the validity of this type of correlation are beyond the scope of this book.

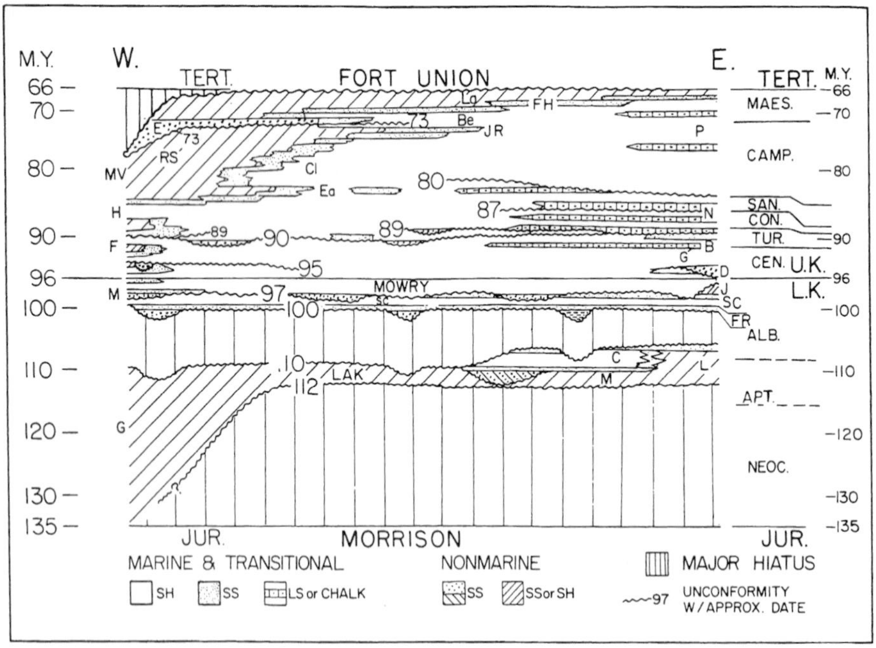

Fig. 11.70. Diagrammatic west-east cross section through the Western Interior Seaway of the Rocky Mountains, showing stratigraphic positions and approximate dates of major transgressive units and interregional unconformities. Formations or groups to the west are: *G* Gannett; *SC* Skull Creek; *M* Mowry; *F* Frontier; *H* Hilliard; *MV* Mesa Verde; *RS* Rock Springs; *E* Ericson; *Ea* Eagle; *Cl* Claggett; *JR* Judith River; *Be* Bearpaw; *FH* Fox Hills; *La* Lance. To the east formations are: *L* Lytle; *LAK* Lakota; *FR* Fall River; *SC* Skull Creek, *J* and *D* sands of Denver basin; *G* Greenhorn; *B* Benton; *N* Niobrara; *P* Pierre; *M* and *C* McMurray and Clearwater of Canada. (Weimer 1986, reprinted by permission)

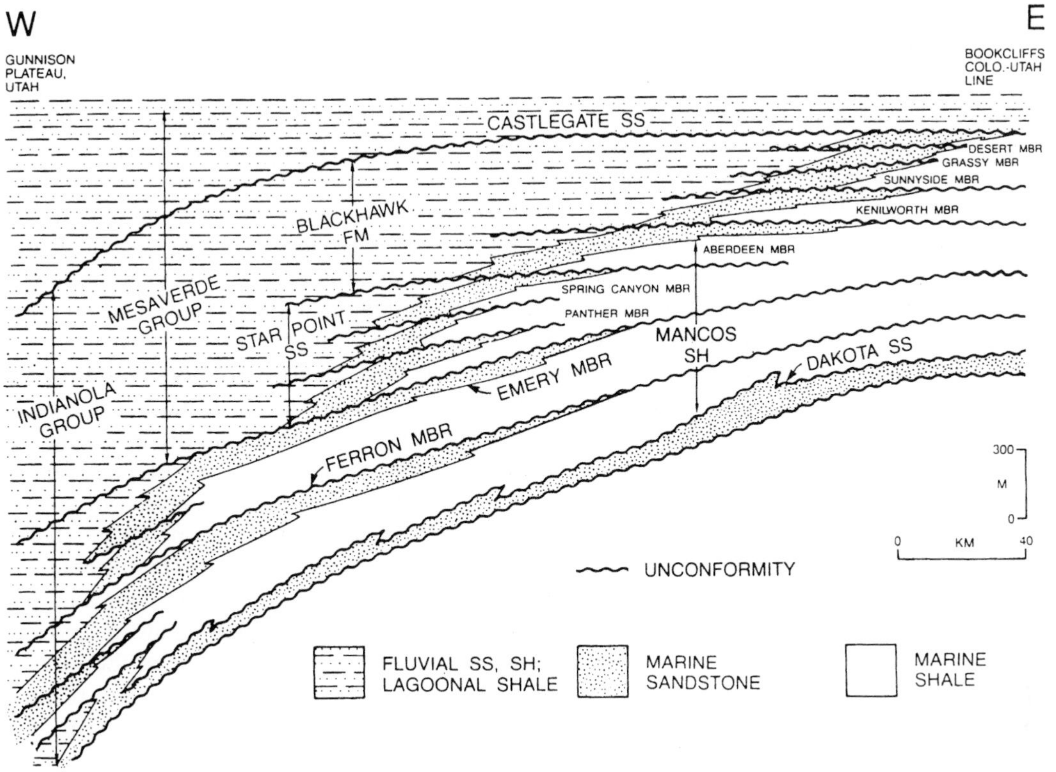

Fig. 11.71. Regional stratigraphic framework of Upper Cretaceous strata of east-central Utah. (Swift et al. 1987)

Kauffman (1984) claimed that there is a consistent correlation among transgression, thrusting (in Wyoming and Utah), and volcanism in the Western Interior, but his data are not convincing to this writer. However, there is no doubt that some of the transgressions were contemporaneous with thrust-faulting episodes within the Sevier orogen of Utah, possibly including episodes in the Albian, Santonian, and Maastrichtian (compare Kauffman 1984; Lawton 1986a,b; and Weimer 1986). These transgressions were regional in scope, and cannot therefore be attributed to flexural loading by individual thrust plates. However, where thrusting is part of regional compression, possibly related to terrane accretion or variations in subduction parameters, the mechanical drag effect described by Mitrovica et al. (1989) may be invoked as a cause of regional basin subsidence and transgression.

Other thrusting events within the Sevier orogen are not clearly correlated with regional changes in sea level in the basin but are correlated with the development of major clastic wedges that prograded across the basin margins. Lawton (1986a,b) indicated a link between thrusting and clastic-wedge formation in Utah between the mid Albian and the late Campanian, although it is not possible to provide the tight correlation between individual tectonic and stratigraphic events that is now available for parts of the Himalayan foredeep.

Are individual clastic tongues within major wedges tectonic or eustatic in origin? Figure 11.71 illustrates the broad stratigraphic architecture of part of the Upper Cretaceous clastic wedge of Utah. There is no doubting an overriding tectonic control of sedimentation for many of these successions. Paleocurrent and petrographic evidence indicates shifting sediment sources and changes in regional paleoslope during deposition of the Mesa Verde Group and associated units in Utah. Lawton (1986a,b) documented unroofing of intrabasin uplifts, from which early basin-fill sediments were cannibalized, and changes in dispersal patterns related to basin tilting. The volume of sediment within the wedge is too great to have been controlled by passive changes in sea level (see also Galloway 1989a,b). But the question remains whether tectonically induced sediment input was modulated by sea-level control, leading to fourth- and higher-order cyclicity along the fringes of the clastic wedge (such as in the Book Cliffs of Utah: see Swift et al. 1987; Van Wagoner et al. 1990, 1991; Yoshida et al., in press). As discussed in Sect. 11.3.5, intraplate stresses related to episodic thrust-sheet loading may be responsible for stratigraphic cyclicity on time scales between 10^4 and 10^6 years.

McLean and Jerzykiewicz (1978) examined the uppermost of the major nonmarine clastic wedges in the Alberta Basin, the Brazeau-Paskapoo formations, and showed that this succession, up to 3600 m thick, encompasses at least four crudely upward-fining cycles, ranging from 400 to 1600 m in thickness. Age control on these cycles is poor, but deposition of the entire wedge spanned about 20 million years, and so each cycle represents an average of about 5 million years. In the Karoo Basin of southern Africa, which is also a foreland basin, the Beaufort Group comprises three basin-wide upward-fining cycles ranging from 160 to 500 m in thickness (Visser and Dukas 1979), whereas the overlying Molteno Formation consists of six upward-fining cycles that reach maximum thicknesses of 140 m (Turner 1983). Tectonic control of these foreland basin cycles in Alberta and southern Africa is indicated by sedimentological evidence for shifting source terranes (Rahmani and Lerbekmo 1975; Visser and Dukas 1979).

Tight correlation between individual tectonic and stratigraphic events is now available for some foreland basins, such as parts of the Himalayan foredeep. There, Burbank and Raynolds (1984, 1988) used magnetostratigraphic techniques to date the fluvial deposits of the Himalayan foredeep of Pakistan. The technique permits a precision of dating to within the nearest 10^5 years. These results show that, in the Himalayan foredeep, thrusting and uplift episodes were very rapid and spasmodic, and that they did not occur in sequence into the basin, as the classic model (e.g., Dahlstrom 1970) would predict (Fig. 11.72). In one case, Burbank and Raynolds (1988) were able to demonstrate the uplift and removal of 3 km of sediment over the crest of an anticline within the basin over a period of 200000 years, an average uplift and erosion rate of 1.5 cm/year. In most basins, the precision of dating obtained by these authors cannot be attempted, and this therefore suggests that caution should be used in assessing published reconstructions of the rates of convergence, crustal shortening, and subsidence, except where these are offered as long-term (> 1 million year) averages.

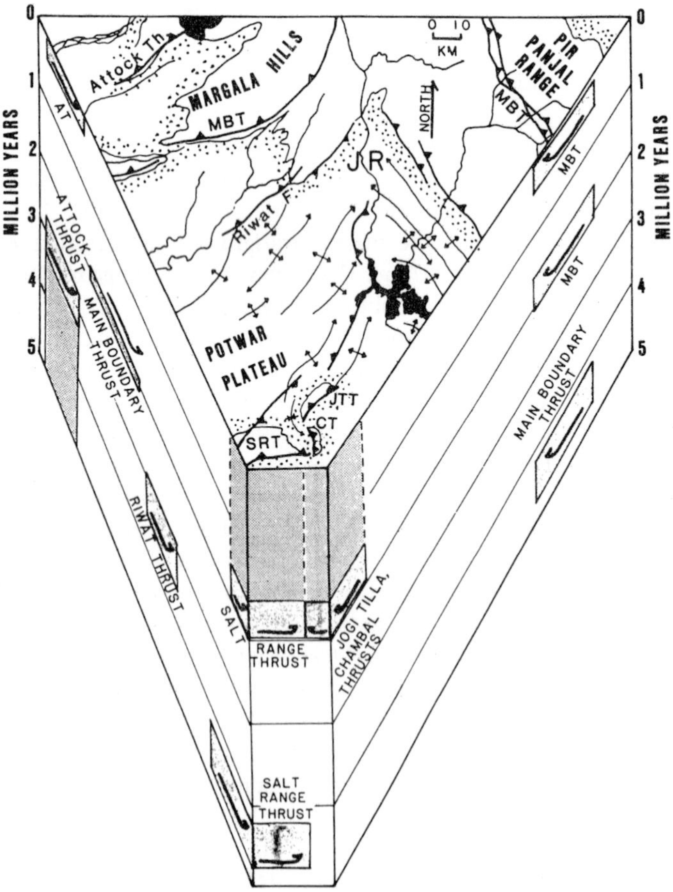

Fig. 11.72. Chronology of fault motions within part of the Himalayan foredeep of Pakistan. *Shaded boxes* indicate the time and space domain over which each thrust is interpreted to have been active. (Burbank and Raynolds 1988)

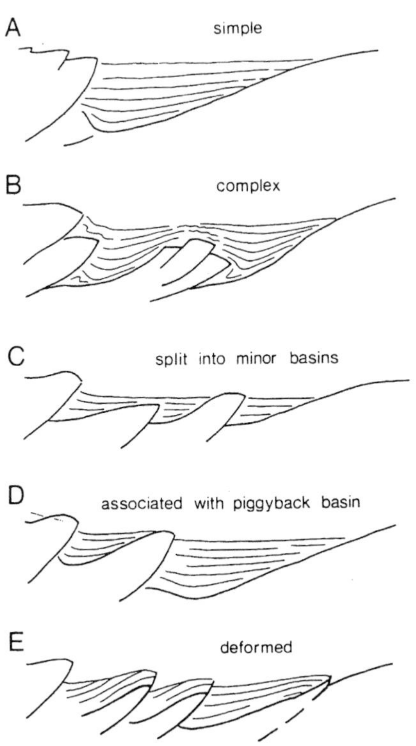

Fig. 11.73. Variations in the relationship between the fold-thrust belt and the foreland basin, based on seismic profiles. Minor basins and piggyback basins are varieties of satellite basins. (Ricci Lucchi 1986)

In the simplest case, a foreland basin is an asymmetric moat, deepest adjacent to the fold-thrust belt and shallowing toward the continental interior (Fig. 11.73A). However, as a result of the complexities in thrust kinematics noted above, the proximal edge of the basin may evolve into several separate sub-basins, isolated or semi-isolated during sedimentation by faults growing within the basin (Fig. 11.73B-E). Ori and Friend (1984) introduced the term piggyback basin for basins carried on the backs of thrust plates (Fig. 11.73D), but Ricci Lucchi (1986) suggested that the specific condition described by Ori and Friend is only one of several possibilities, which he preferred to include in a wider category of minor or satellite basins (Fig. 11.73C,D).

Minor satellite or piggyback basins develop as fault ramps that cut the foreland basin and are uplifted, isolating part of the foreland and acting as sediment barriers (Figs. 11.73, 11.74). Ponding of sediment takes place within the minor basins, with diversion of flow along the basin axes toward reentrants in the main thrust front. Such reentrants commonly form at terminations or overlaps of thrust plates. Careful stratigraphic work on the fill of these satellite basins may assist in unraveling the thrust kinematics of the fold-thrust belt (e.g., Ori and Friend 1984; Johnson et al. 1986; Ricci Lucchi 1986; Bentham et al. 1992). Continued crustal shortening may culminate in the uplift of these minor basins, so that only remnants may be preserved within the eroded foldbelt, as in the Swiss Alps (Homewood et al. 1986; Pfiffner 1986) and the Appalachian orogen (Lash 1990). Several recent studies of modern thrust-fault-influenced alluvial-fan environments provide excellent analogs for the interpretation of the details of sedimentary styles and the response of depositional systems to tectonic tilts and changing sediment source areas (Jolley et al. 1990; Allen et al. 1991; Räsänen et al. 1992; Dumont 1993; Damanti 1993; Fig. 11.74).

11.4.7 Sedimentary Basins and Allochthonous Terranes

The collision of major continental plates involves crustal thickening and shortening and late-stage strike-slip displacement (escape tectonics), as exemplified by the India-Asia collision, discussed in the preceding section. The basins that form and are filled during this process are commonly termed intermontane or successor basins. They may preserve much internal evidence of the collision that can be analyzed to determine its style and timing. Such basin analysis may be particularly useful in the case of orogens composed of numerous small plates, tectonic flakes, or slivers of uncertain affinities, so-called suspect terranes, for which evidence bearing on the amount and timing of relative motions may be limited. However, compression and lateral displacement of the terranes in the final stages of orogenic assembly may disrupt and displace the successor basins (Fig. 11.75), making their analysis very difficult. Basins associated with terrane accretion are not necessarily fluvial, but many are so, particularly during the final phases of collision, because collision invariably leads to uplift.

The most useful contribution fluvial basin analysis can make to the unraveling of a history of accretionary tectonics is to throw light on sediment transport directions and detrital provenances (Saleeby 1983; Schermer et al. 1984; Kleinspehn 1988; Howell 1989; Bluck 1991). The timing of the suturing between terranes is most readily determined in one of three ways: (a) by the dating of intrusive igneous bodies that cross-cut the suture, (b) by dating overlap assemblages of sediments that

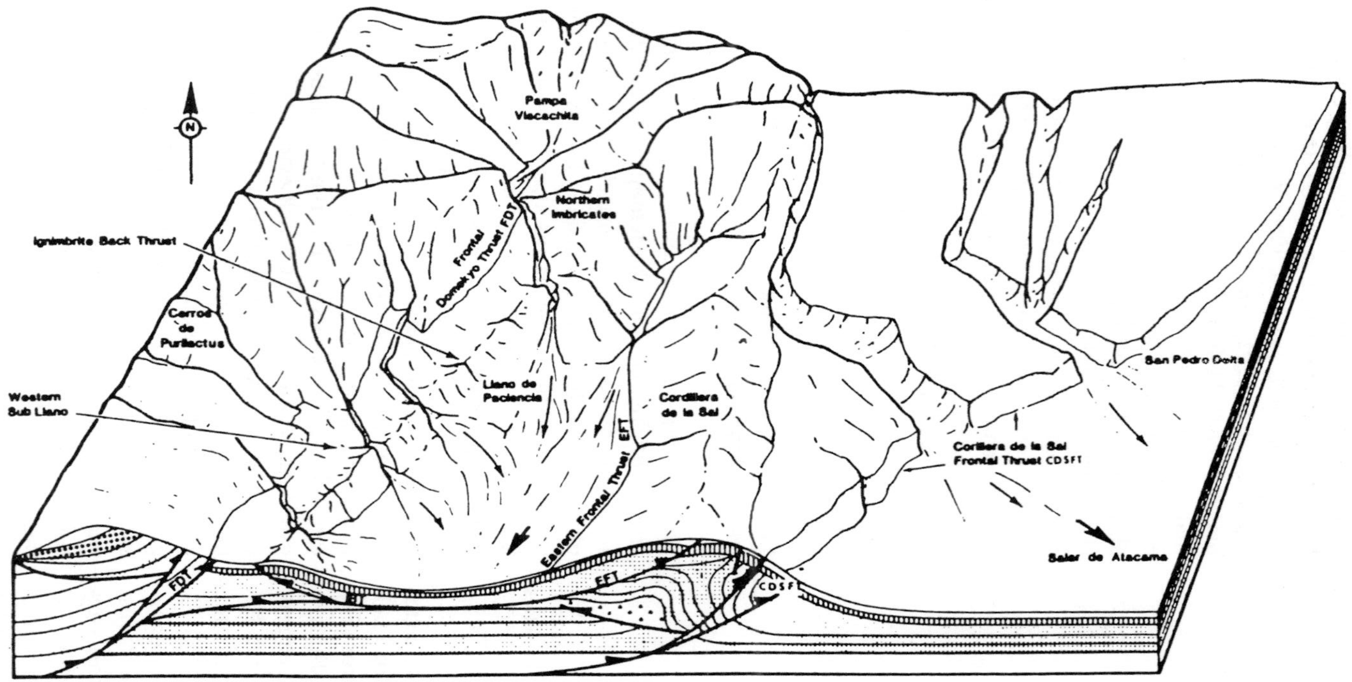

Fig. 11.74. Relationship between paleogeography and structure in an active thrust-faulted basin margin. The basin at lower left is a satellite basin. (Jolley et al. 1990)

are deposited across the suture, or (c) by identifying detritus in a basin-fill that can only have been derived from the adjacent terrane. In the case of the third criterion, provenance studies may also be used to demonstrate post-suture displacement, if the detritus does not match the petrology of the source from which paleogeographic analysis indicates the detritus should have come. Examples of both these

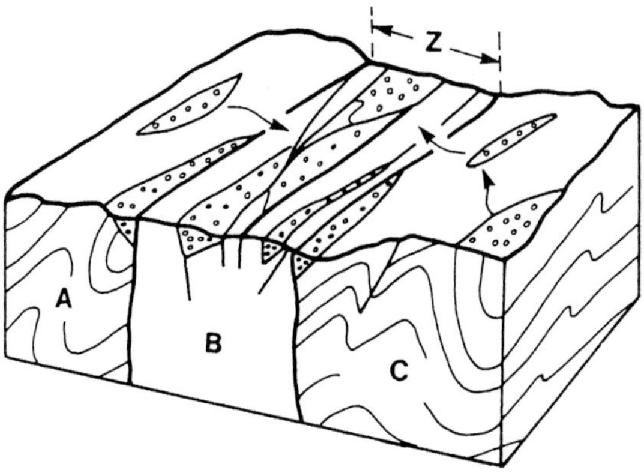

Fig. 11.75. The development of successor basins at the margins of terranes. A and C are terranes which amalgamated along the complex suture zone *Z*. *Arrows* indicate typical dispersal of sediment following terrane amalgamation. (Bluck 1991)

cases have been cited by Schermer et al. (1984) and Bluck (1991). Schermer et al. (1984) stated:

"An example of an overlap assemblage is the Late Cretaceous Gravina-Nutzotin belt of southern Alaska, which links the Wrangellia and Alexander terranes; an example of a provenancial linkage is provided by the Middle to Upper Jurassic strata of the Bowser Basin of British Columbia, which are deposited on the Stikine terrane and contain debris from the Cache Creek terrane."

These relationships are both indicated in the terrane-amalgamation diagram compiled by Saleeby (1983; Fig. 11.76). Discussing the Devonian sedimentary fill of the Midland Valley of Scotland, Bluck (1991) stated that sediments banked against the fault which forms the southeast margin of this basin, the Southern Uplands Fault, "contain abundant igneous and metamorphic clasts up to boulder size, but in the direction from which they were transported lies the Southern Uplands, a sequence of greywackes and shales. In this instance a source block has been laterally displaced or overthrust by the greywacke sequence."

An example of the complex problems in plate kinematics that terrane amalgamation histories can set up was described by Kleinspehn (1985). She examined the Tyaughton-Methow Basin in southwestern British Columbia, a basin formed adjacent

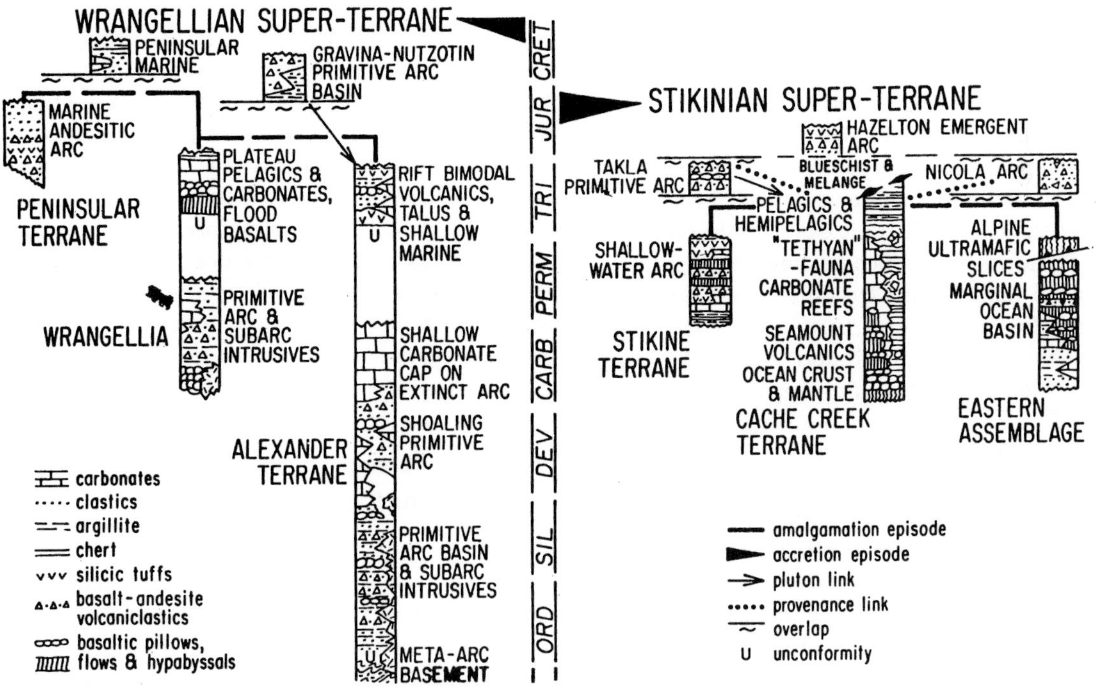

Fig. 11.76. Amalgamation history of the major terranes of British Columbia and southern Alaska. Note the use of provenance linkages and overlap assemblages to document amalgamation histories. (Saleeby 1983, reproduced with permission from the Annu Rev Earth Planet Sci vol 11, © 1983, by Annual Review Inc.)

to and partly overlapping six Cordilleran terranes. Paleocurrent and provenance studies provide partial constraints on the reconstruction of terrane amalgamation and post-suture displacement on a series of terrane-bounding faults. Kleinspehn (1985) described four main hypotheses of relative terrane motion.

11.4.8 Cratonic Basins

The origin of sedimentary basins and uplifts within stable areas such as the Canadian Shield and adjacent stable platform is controversial (Miall 1990, Sect. 9.3.6.1). Most hypotheses involve the reactivation of ancient lineaments or zones of weakness to provide loci of subsidence and uplift, although the mechanisms for generating vertical movements within stable cratons are not clear. Many cratonic tectonic elements may be related to plate-margin processes, and intraplate stress is increasingly being invoked to explain the initiation or reactivation of major intracratonic structures at distances far removed from the main locus of extension or compression (e.g., Cloetingh et al. 1985; Sanford et al. 1985; Miall 1986; Heller et al. 1993; Sect. 11.3.5). Embry (1990a, 1991) suggested that during the Mesozoic intraplate stress led to tilting on a continental scale and provided the cratonic sediment sources and paleoslope for the delivery of large volumes of terrigenous detritus to Sverdrup Basin, on the northern margin of the North American craton. Miall (1986) suggested that intraplate stress transmitted from Caledonian compression at the suture between North America and Europe generated thrust fault-bounded uplift of the Boothia Uplift, in the Canadian Arctic Platform, from which was shed the alluvial fan and fluvial wedge of the Devonian Peel Sound formation (Miall and Gibling 1978). Dott et al. (1986) demonstrated that fluvial and eolian processes formed widespread but thin clastic sheets in the Wisconsin area, flanking the southern margin of the Canadian Shield, during two periods of transgression following low stands of sea level, in the Late Cambrian and again in the Middle Ordovician. Isopachs of the Tertiary Eyre Formation of the Great Artesian Basin, Australia, indicate rejuvenation of Paleozoic structures; basins became deeper and upwarps were uplifted (Wopfner et al. 1974)

Descriptions of the regional geology of large platform areas such as interior Australia (Veevers 1984; Morgan 1993) and cratonic Canada (Stott and Aitken 1993) indicate the long-term persistence of ancient basins and upwarps, with the widespread distribution of thin nonmarine sheets at times when much thicker clastic wedges are being deposited in marginal basins. Veevers (1984, p. 161) stated:

"The Cainozoic record of central Australia is a continuation back to 60 Ma of the present pattern of local sags, such as Lake Amadeus, within highlands flanked by broad plains. Today, little sediment is shed by the highlands, and the wholly carbonate province on the south, in the Great Australian Bight, is traceable in the Eucla Basin back to the Eocene. What little sediment escapes central Australia today, via the coherent stream system on the east and southeast – the Finke River and others to the south are the only streams that maintain flow out of the region – is concentrated in the sump of Lake Eyre and the dunefields of the Simpson Desert, and this pattern (but not facies) of deposition persists to the beginning of the Cainozoic in the form of the Eyre Formation (Wopfner et al. 1974). The aeolian sand dunes of the Great Sandy Desert of the Canning Basin and of the area to the north between 18° and 22° S, in which only a few tens of meters of sediment have accumulated, are the only other sediment derived from central Australia."

Sloss (1979) has argued for the existence of vertical crustal movements within the craton – true epeirogeny – but mechanisms to explain it have been lacking until recently. Gurnis (1990, 1992) proposed that mantle convection beneath the crust results in continental-scale differential heat distribution and consequent heating and cooling of different parts of the overlying earth's surface, resulting in broad regional uplifts and downwarps because of the effects on crustal densities. These processes maintain what is called dynamic topography. The effects of crustal heating cause uplift. This occurs along the flanks of new continental rift systems (e.g., parts of present-day East Africa) and above mantle plumes. Subsidence takes place over cooling areas of the earth's crust, such as areas of aging oceanic crust distant from spreading centers, and over regions of mantle downwelling. The stratigraphic effects of this process have yet to be evaluated in detail.

11.5 Basic Paleogeographic Models for Nonmarine Basins

On a basin scale, the architecture of the alluvial fill can be classified into a series of distinctive styles, governed by the tectonic control of basin shape and paleoslope (Miall 1981b). A basin-fill classification is given in Table 11.4, in which depositional systems are simplified into nine basic types: 1. lacustrine, 2. alluvial fan/fan delta, 3. low-sinuosity fluvial, 4.

high-sinuosity fluvial, 5. river-dominated delta, 6. wave-dominated delta, 7. tide-dominated delta, 8. nondeltaic coast (beach and barrier systems), and 9. estuarine.

For the purpose of this analysis, a twofold subdivision of fluvial environments into rivers of high and low sinuosity is adequate. This serves, in most cases, to distinguish proximal from distal alluvial settings, although some rivers of exceptionally high discharge and sediment load (e.g., Amazon, Ganges, Brahmaputra) do not show significant increases in sinuosity near their mouths. The threefold delta classification is that of Galloway (1975).

Table 11.4. Alluvial basin-fill patterns[a]

Model	Proximal	Medial	Distal
1	T fan	T braid plain	Lake margin/nondeltaic coast
2	T fan-delta	–	Lake margin/sea coast
3	T fan/river	T river	River-dominated delta
4	T fan/river	T river	River-dominated delta with barrier lagoon
5	T fan/river	T river	Wave-dominated delta
6	T fan/river	L river	Lake margin/estuary
7	T fan/river	L river	River-dominated delta
8	T fan/river	L river	Tide-dominated delta
9	T fan/river	L river	Wave-dominated delta

T, Transverse; *L*, longitudinal.
[a] From Miall (1981b).

An alluvial drainage system can be divided into four parts: 1. headwaters in upland source areas; 2. proximal environments, immediately below the fall line along the margins of sedimentary basins; 3. medial environments, including axial fluvial systems; and 4. distal environments, where the river interacts with some terminating environment, such as a lake, playa system, eolian dune system, delta front, or tidal flat. The headwaters zone is insignificant from the point of view of basin analysis, as it is rarely preserved. As discussed in Sect. 11.3.2, rivers are classified as either transverse, in which case they flow directly from the uplifted source across structural grain (e.g., the bounding faults), or longitudinal (axial), flow being parallel to strike, along the basin axis. These generalizations may hold only locally, for foldbelts show syntaxes, island arcs are

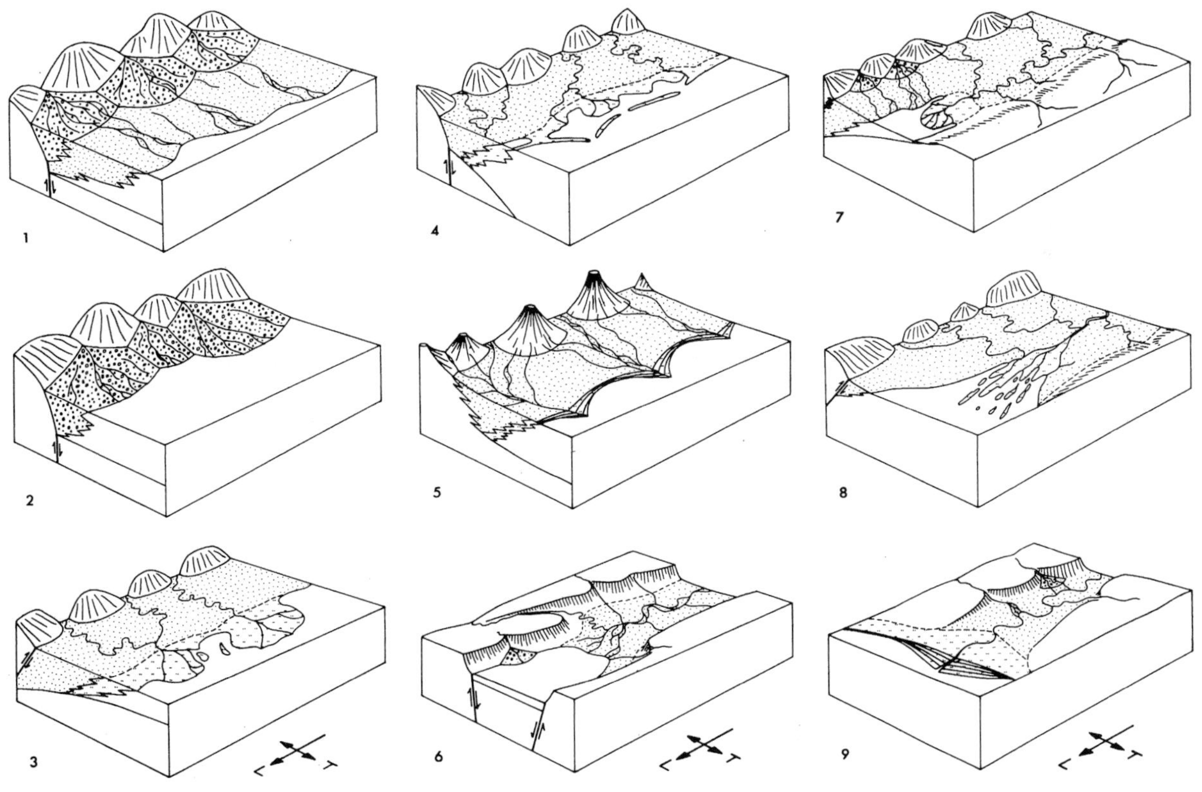

Fig. 11.77. The nine basin-fill patterns discussed in this section. See Table 11.4. *T* Transverse; *L* longitudinal. Scales are variable. Structural controls are given as examples only. Most basin-fill patterns occur in several tectonic settings. (Miall 1981b)

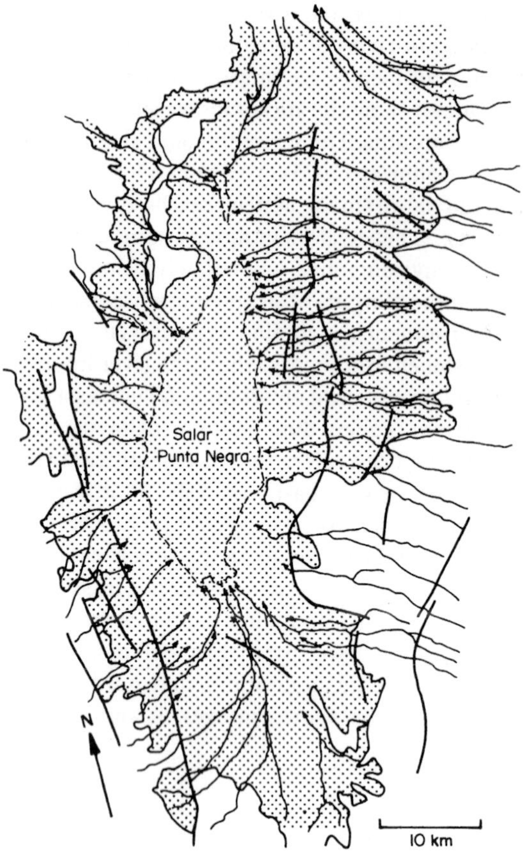

Fig. 11.78. The Salar de Punta Negra, Chile, illustrating basin-fill models 1 and 6. (Miall 1981b, after van Overmeeren and Staal 1976)

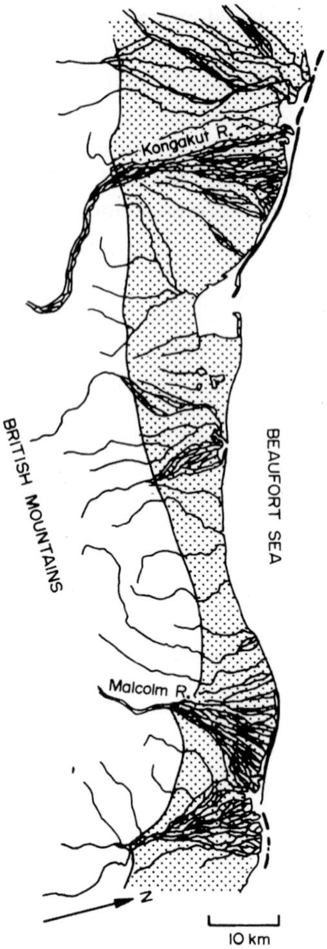

Fig. 11.79. Part of the North Slope, Alaska and Yukon, illustrating basin-fill model 2. (Miall 1981b)

arcuate, and plate margins may be irregular. However, the generalizations are a useful aid to classification of basin paleogeography.

The distinction between transverse and longitudinal rivers is an important one sedimentologically. Commonly, transverse rivers provide the tributaries to a longitudinal trunk river, and the two may be characterized by quite different fluvial styles and by different paleocurrent means and variances. Some longitudinal rivers originate as major transverse rivers which emerge from structural reentrants in the mountainous source area and curve downstream into a longitudinal orientation (Eisbacher et al. 1974). The modern Brahmaputra is an example. Figure 9.1 illustrates an ancient example of an alluvial basin filled by both transverse and longitudinal depositional systems of contrasting fluvial style. The major sediment packages deposited by each system are separated by a bounding surface of sixth order, according to the classification used in this book.

Using the simple ninefold environmental classification provided above, the proximal-medial-distal subdivision of fluvial systems, and the transverse/longitudinal criterion, it is possible to define nine major basin-fill models (Table 11.4; Fig. 11.77). This classification is not intended to be all-inclusive, nature being highly variable. However, it suggests that these nine models are the most likely to be found in practice because of the constraints imposed on river and coastline configurations by plate geometry. A major complication is that of "big rivers" (Potter 1978), which flow for hundreds or thousands of kilometers, perhaps through or across several tectonic zones or terranes, before debouching into the sea. Alluvial systems that are predominantly erosional in nature are not of concern here.

The first model, that of a transverse fan and braidplain system, is a common one in relatively small drainage basins, where river length is measurable in tens of kilometers. Runoff may be ephemeral. There is no single trunk river, and water and sediment discharge at base level is inadequate to develop major deltas. Rivers may terminate as a network of shallow distributaries or as a zone of sheetflooding on a lake margin, playa, or tidal flat. Intermontane lake basins (basin type 4.3 in Table 11.2) are commonly bordered by this type of alluvial pattern (e.g., parts of the Salar de Punta Negra, Chile: Fig. 11.78). The pattern is also common in rift basins (basin type 1.1), such as Death Valley, California, and the East African Rift System. Some of the Triassic rift basins of eastern North America also fit this pattern (Hubert et al. 1976; P.E. Olsen 1990), as does the Tertiary of Uinta Basin, Utah (basin type 2.4). Ballance (1980) described an example in a modern wrench-fault basin. Miall and Gibling (1978) described a Devonian example of such a pattern flanking the fault-bounded Boothia Uplift in Arctic Canada. Alluvial fans along the flanks of the mountains drained into a sandy braidplain and an arid tidal flat.

Model 2 is a variant of the first model, in which the alluvial fans drain directly into a standing body of water as fan deltas or braid-deltas (Fig. 11.79). McPherson et al. (1987) and Nemec and Steel (1988) provided discussions and classifications of these types of depositional system. Fan deltas typically are small, gravel-dominated systems, a few kilometers across. This pattern occurs on coastlines flanking small ocean basins, such as the Red Sea, where fan deltas prograde into the ocean.

Models 3, 4, and 5 are systems where water and sediment discharge are adequate to construct deltas at base level. They may build major alluvial-deltaic plains with river lengths measurable in hundreds of kilometers. The three models are distinguished on the basis of delta type. None of the world's major rivers correspond to any of these three models, because most of these follow structures, and are therefore longitudinal. The typical pattern is of a broad coastal plain in front of a foldbelt mountain range, a volcanic arc, or a downwarping continental margin. The coastal plain is crossed by numerous parallel,

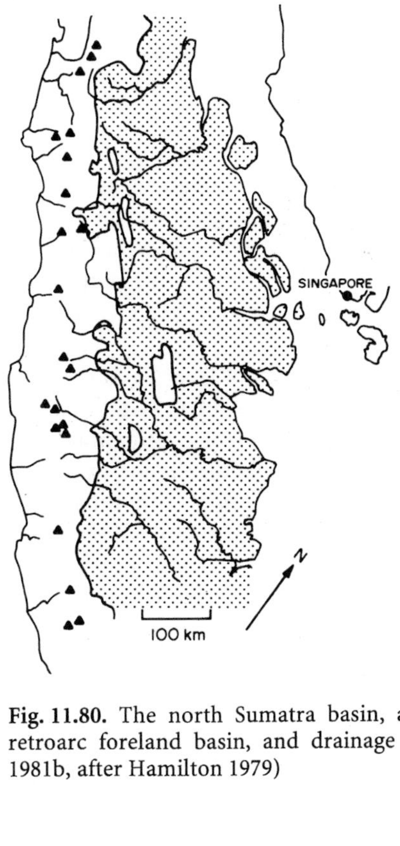

Fig. 11.80. The north Sumatra basin, an example of a retroarc foreland basin, and drainage model 3. (Miall 1981b, after Hamilton 1979)

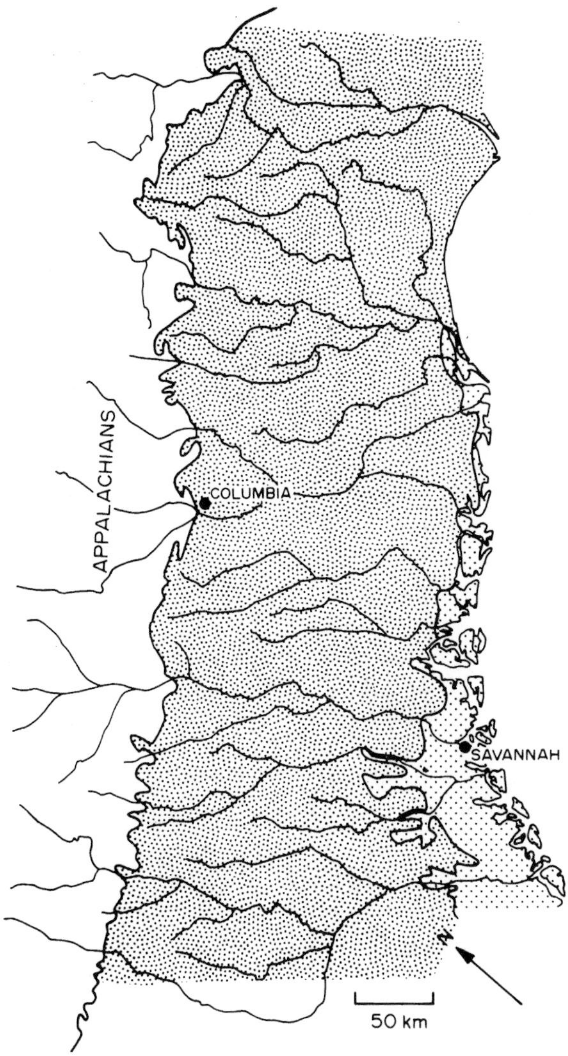

Fig. 11.81. Part of the coastal plain of Georgia and North and South Carolina. An example of basin-fill model 4. *Open stippling* indicates areas of active sedimentation; *closed stippling* indicates older deposits formed in the same tectonic environment. (Miall 1981b)

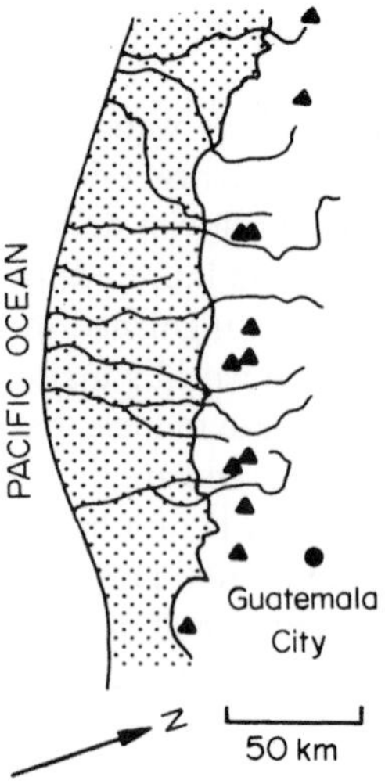

Fig. 11.82. Part of the Guatemala coast, an example of basin-fill model 5. (Miall 1981b)

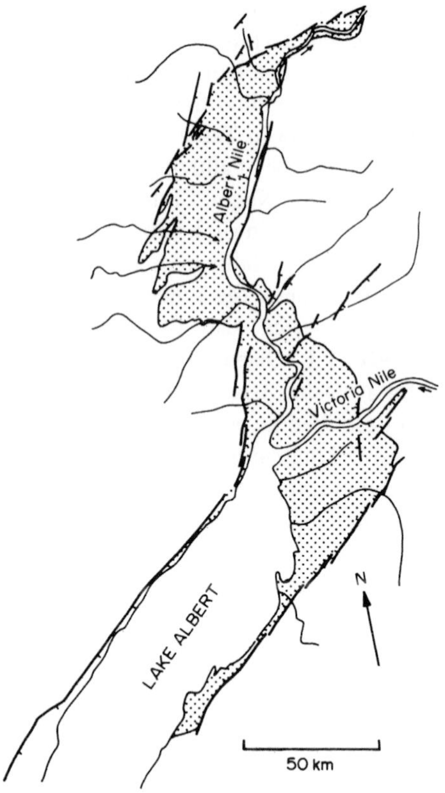

Fig. 11.84. Part of the East African Rift System, Uganda, an example of basin-fill model 6. (Miall 1981b)

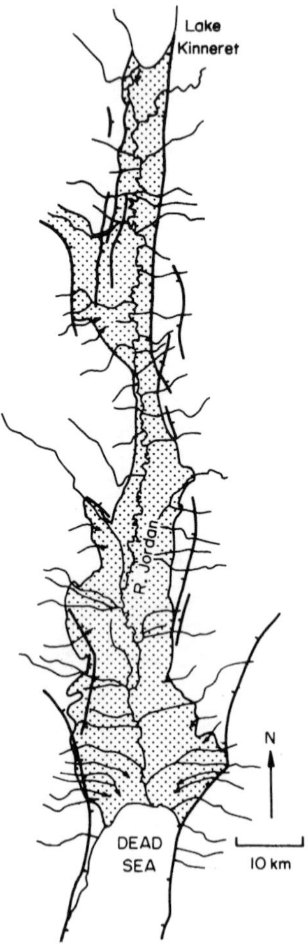

Fig. 11.83. Jordan Valley and northern Dead Sea Basin, an example of basin-fill model 6. (Miall 1981b)

transverse rivers, commonly flanked by peat swamps. River termination patterns are controlled by coastal processes.

Model 3 is well developed on the north coast of Sumatra (basin type 2.4; Fig. 11.80) and New Guinea. The petroliferous Cherokee Group of Oklahoma and Kansas (basin type 4.2, but with sediment source from the continental side) displays this pattern, as does the Permian Beaufort Group, South Africa (basin type 4.1; Hobday 1978).

The Atlantic coastal plain of Georgia and North and South Carolina (Fig. 11.81) and the Gulf Coast of Texas are Atlantic-type divergent continental-margin basins (basin type 1.2.2), and are good examples of model 4. These are mesotidal and microtidal coasts, respectively, and are characterized by major barrier island-lagoon systems into which the coastal plain rivers drain. The Carboniferous coal-bearing fluvial, deltaic, and barrier-island deposits of Kentucky and West Virginia are a good ancient example (peripheral foreland basin, basin type 4.1; Horne et al. 1978).

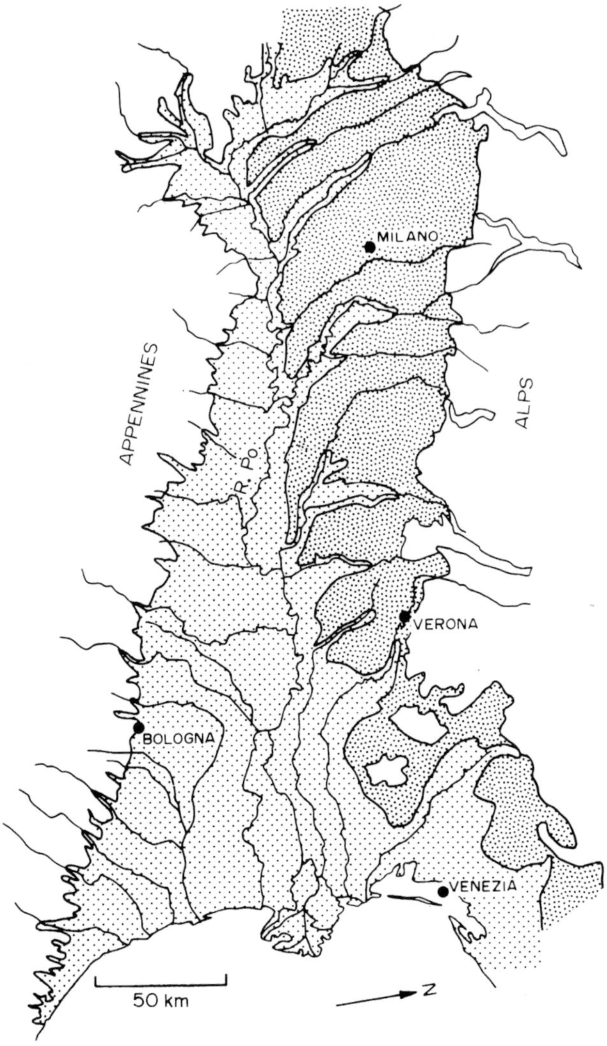

Fig. 11.85. The Po Basin, Italy, an example of basin-fill model 7. (Miall 1981b)

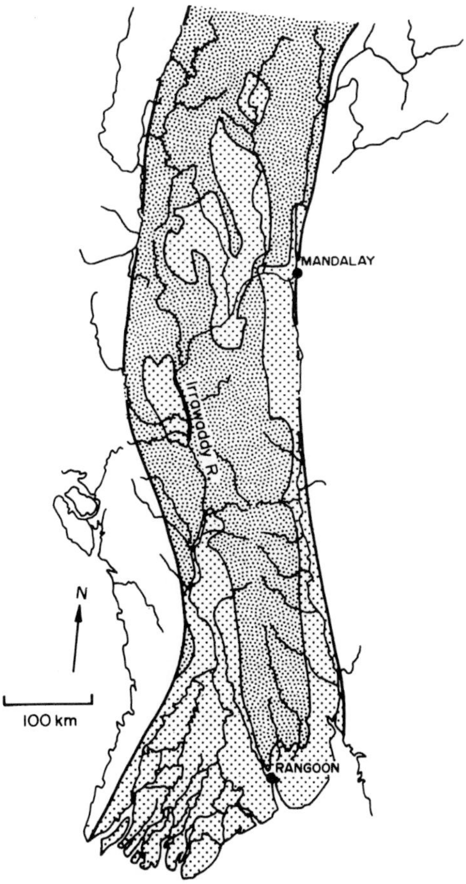

Fig. 11.86. The Central Lowlands of Myanmar (Burma), an example of basin-fill model 8. (Miall 1981b)

Model 5 occurs mainly in areas facing open oceans, which are characterized by high wave energy. Deltaic deposits are reworked into superimposed beach ridges flanking the river mouth, which may be deflected downcurrent into a spit. Modern examples of such alluvial systems include the Pacific coast of Guatemala (basin type 2.2; Fig. 11.82) and the Sao Francisco delta of Brazil (basin type 1.2.2). Ancient examples have been described from the Gulf Coast (Fisher 1969).

Models 6, 7, 8, and 9 are combinations of longitudinal trunk rivers with transverse tributaries. The latter are commonly alluvial fans at the basin margins. Most major rivers and large inland basins show one or other of these four patterns. Trunk river length may be thousands of kilometers. The models are distinguished mainly on the basis of river termination patterns. In the case of smaller rivers with low sediment discharge, or those emptying into a rapidly subsiding basin, there may be inadequate sediment load to form a delta. This is model 6. Rivers may be ephemeral, terminating in a sheet flood zone along a playa lake margin (Fig. 11.78), or they form an estuary with tidal flats. Modern examples of the latter include the Thames (U.K.), the Amazon, the Rhine, the Susitna (Cook Inlet), and the Colorado at the Gulf of California. Many intermontane basins in California and Nevada (backarc rifts, basin type 2.3) show longitudinal rivers terminating in a playa or permanent lake. Some strike-slip basins (e.g., Jordan Valley, basin type 3.3; Fig. 11.83) and rift systems (e.g., East African rifts, basin type 1.1; Fig. 11.84) also show this pattern. Ancient nonmarine examples include the Devonian fluvial sediments of Spitsbergen, which flow into a playa (basin type 4.3; Friend and

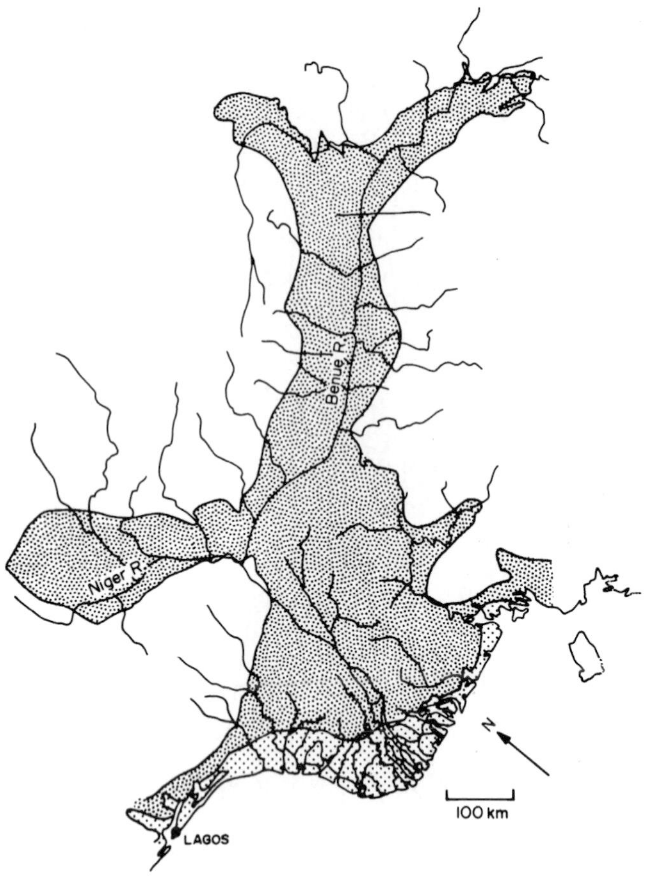

Fig. 11.87. Benue Trough, Nigeria, an example of basin-fill models 8 and 9. (Miall 1981b)

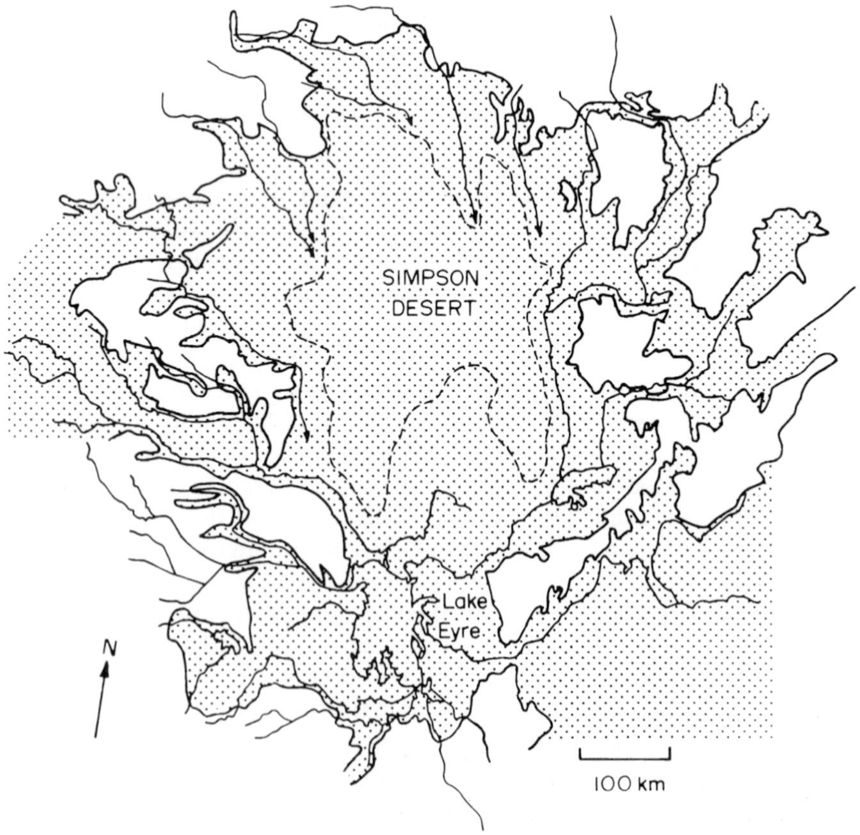

Fig. 11.88. Lake Eyre basin, Australia, an example of centripetal drainage in a cratonic basin. (Miall 1981b)

Moody Stuart 1972), and the Permo-Triassic of northwest Scotland (Steel and Wilson 1975). Ancient marine examples are few and far between. Possibly Cook Inlet, in the Tertiary, may have consisted of a paleo-Susitna River flowing into an estuary, as it does at the present day.

Model 7 is that of a transverse and longitudinal river system which terminates in a river-dominated delta. This includes several of the world's major present-day alluvial basins, such as those of the lower Danube, Po (Ori 1993; Fig. 11.85), Tigris-Euphrates (Baltzer and Purser 1990), Indus, and Mississippi. The Mississippi is an example of a river flowing along the axis of a graben or embayment striking at a high angle to the continental margin. Some of the best ancient examples include the Tertiary precursors of the modern rivers listed above (e.g., Gulf Coast, basin type 1.2.2; Fisher and McGowen 1967). Other examples include the Alpine Lower Freshwater Molasse (basin type 4.1; Van Houten 1981), and the Middle-Upper Devonian Melville Island Group of Arctic Canada (basin type 4.3; Embry and Klovan 1976).

Tide-dominated deltas are most likely to occur at the head of a bay or gulf at the terminus of a longitudinal river (model 8). Good examples are the Ganges, Irrawaddy (Fig. 11.86), and Niger (Fig. 11.87) fluvial-deltaic systems (basin types 4.1, 2.2, and 1.3, respectively). Hobday and Von Brunn (1979) interpreted the Table Mountain Sandstone of eastern South Africa in terms of a longitudinal river flowing southwest to a tide-dominated delta system located near the southern end of the continent, prior to the breakup of Gondwana (basin type 1.3). Much of the Cretaceous-Tertiary molasse of the Alberta basin (basin type 4.3) was deposited with a model 8 geometry. For example, the Cadomin conglomerate was a unit deposited by transverse alluvial fans draining the ancestral Rocky Mountains. These fed a trunk river flowing along strike, which probably flowed into an interior epeiric sea through a tide-influenced estuary or delta system (McLean 1977; McLean and Wall 1981).

Model 9 is that of a transverse-longitudinal river terminating in a wave-dominated delta. The best examples of this type are the modern Rhône and Nile, both of which flow along grabens that trend virtually perpendicular to straight coastlines. They are therefore open to the effect of waves oriented around a 180° arc. Mutch (1968) described a Carboniferous example in New England.

A tenth model is not listed in Table 11.4 because it does not fit the transverse-longitudinal drainage pattern discussed here. This is the centripetal drainage pattern shown by some cratonic basins, such as Lake Eyre basin (Fig. 11.88).

Chapter 12

What Does Fluvial Lithofacies Reveal About Climate?

12.1 Introduction

Integrated analyses of past climates are now becoming commonplace, as attempts are being made to produce complete reconstructions of ancient environments incorporating numerical global climate models linked to the details of local plate-tectonic position and paleogeography. Evidence from the stratigraphic record, including the distribution of certain types of lithology, plus fossil evidence and isotopic data are combined in such analyses. This chapter is concerned with the nonmarine lithostratigraphic record and the information it can yield regarding climate.

Several climatic indicators of chemical or biochemical origin are present in the nonmarine sedimentary record, such as coal, evaporite, bauxite, laterite, and paleosols, all of which yield limited information on some of the main attributes of climate, particularly rainfall and temperature (Frakes 1979; Parrish and Barron 1986). Some of the clearest climatic signals available in nonmarine sediments are the wind patterns that can be deduced from giant fossil ergs (e.g., Parrish and Peterson 1988), the chemical rhythms in lacustrine sediments and the climatically driven changes in lake levels recorded in lacustrine-fluvial interbedding (Katz 1990), redox and water-table information from paleosols (Wright 1986; Retallack 1981, 1988), and the various indicators of continental glaciation, such as tills, glaciolacustrine dropstones, and striated pavements (Hambrey and Harland 1981). It was once thought that continental redbeds were a clear indicator of oxidizing and probably arid depositional environments (see Pettijohn 1957; Van Houten 1973), but Walker (1967) clearly demonstrated that the red color may be produced by diagenesis, requiring that color be interpreted with caution.

Eolian, lacustrine, and continental glacial sediments, and the chemical sediments in fluvial floodplain deposits may therefore yield some evidence of climate. What about clastic fluvial deposits – the coarse fluvial channel and bar deposits and the detrital overbank deposits? Are there any characteristics of the assemblage, vertical succession, and associated sedimentary structures of fluvial gravels, sands, and muds that can be interpreted in terms of local climate? Many studies of ancient fluvial deposits include an interpretation of the contemporary climate, but in most cases the interpretation is based on the stratigraphic context, on paleosols, or on associated faunas and floras. Other studies infer climate from regional paleogeography and make predictions of temperature, rainfall and seasonality from knowledge of comparable geographic configurations on the modern earth [e.g., the discussion of monsoonal conditions in Pangea by Dubiel et al. (1991) and Parrish (1993)]. The purpose of this chapter is to focus on the lithofacies of clastic deposits in fluvial systems to determine whether they contain any useful independent information regarding climate. This chapter specifically does not focus on paleosols or the paleoecology of nonmarine systems, which have received ample coverage elsewhere.

Although the history of global climates during the Phanerozoic is in general fairly well known (e.g., Frakes 1979; Parrish and Barron 1986; Parrish 1993), many local details remain to be resolved, and the kinds of field criteria described in this chapter may therefore be useful, especially where the more obvious evidence from lacustrine, eolian, glacial, or other related facies is absent. In addition, there is increasing interest in climate change, particularly changes driven by orbital forcing. The recognition of Milankovitch rhythms in the nonmarine sedimentary record is a subject barely in its infancy, and again the kinds of observational approach reviewed here may provide useful tools for this type of study (Sect. 12.12).

The main body of this chapter (Sects. 12.5–12.7) provides a systematic discussion of the lithofacies evidence that can be obtained from detailed local studies of conglomerates, sandstones, and fine-grained facies, focusing on the major climatically sensitive attributes for each sediment type. The evidence may be complex, however, because sedimentation at any one location in a basin may be affected by both local and regional factors. For example, the channel and floodplain processes of tributary rivers may reflect local microclimates, which may vary across a basin (e.g., rain-shadow effects), whereas a trunk river may bear the discharge and sediment-load signatures of a distant source area that may be characterized by a quite different climate (see also Sect. 11.3.1). These and other complexities are introduced below and are discussed in Sect. 12.10.

12.2 Climatic Variables

What is climate, and how might it be recorded in terrigenous detrital sediments? The two most important attributes of climate are rainfall and temperature. Wind, its strength, direction, and consistency, are important climatic attributes in the study of eolian deposits, but not of other nonmarine sediments. The production, dispersal, and deposition of terrigenous detrital sediment depends on water, and therefore on rainfall, its surface runoff in fluvial channels, and in overbank flow. Both the amount of rainfall and its distribution with time (e.g., seasonal, monsoonal, flashy) are climatic factors that have an important influence on sedimentation. Temperature is important as a control on the physical behavior of water – freezing, thawing, and evaporation. Temperature also controls the rate of various chemical reactions that affect syndepositional diagenesis. Rainfall and temperature together control the amount and type of vegetation, which influences depositional styles in channels and overbank areas and, through biochemical processes, the type of weathering and erosion processes in fluvial source areas, thereby affecting sediment yield.

Climatologists have classified global climates in many different ways, recognizing up to as many as 14 different climate types that reflect the ranges of variability and seasonality of temperature and rainfall on the present-day earth (Huggett 1991; Barry and Chorley 1992), but most of these subtleties are only crudely represented in the lithofacies record. Perlmutter and Matthews (1989) used the following temperature categories in their discussion of Milankovitch forcing: tropical, temperate, and polar, and the following rainfall categories: very humid, humid, subhumid, dry, arid. Mack and James (1994) proposed a simple climatic differentiation of the earth related to latitudinal position and discussed how paleosol distribution is related to this classification. Their four broad climatic zones are wet equatorial, dry subtropical, moist midlatitude, and polar. Complications in this zonation are caused by the regional distribution of land and water masses, and by topographic effects.

From the point of view of the preservability of different bedform styles, Jones (1977) recognized five basic types of fluvial discharge pattern:

1. Perennial rivers with relatively steady discharge and a single maximum peak annual flood; common in temperate climates, e.g., Rhine
2. Perennial rivers with double or many maximum discharge events; common in tropical and temperate regions with seasonal variations in rainfall, e.g., Platte
3. Perennial rivers with a pronounced low-discharge period and a long flood period; common in tropical areas with monsoons, e.g., Brahmaputra
4. Perennial rivers with pronounced low-discharge period and short flood period; common in high-latitude rivers and those draining alpine mountains, where there is a short spring flood period and a long period of slowly tapering summer discharge, e.g. Tana, Norway; sandur rivers in Iceland and Alaska
5. Ephemeral rivers with flashy discharge; characteristic of arid regions

These categories cannot yet be resolved in terms of the facies styles described in Chap. 8. Given the obvious relationships between discharge regime, fluvial style, and climate discussed in these paragraphs, it is perhaps surprising that climate is not singled out as a significant controlling parameter in the major geomorphic classification of fluvial floodplains proposed by Nanson and Croke (1992), although several variables affected by climate are discussed, such as the style of flooding and the influence of vegetation.

In this chapter the following broad climatic zonations are used, based on features observable from the lithofacies record:

- Hot-arid
- Hot-humid
- Tropical to semitropical with wet-dry seasonality
- Temperate-humid
- Boreal-paraglacial

There are several complications involved in determining climate from the fluvial sedimentary record, as follows:

1. Large variations in fluvial discharge produce recognizable effects in the facies record, such as the interbedding of very coarse and very fine facies, but the time scale of variation may not be obvious and the climatic implications therefore unclear. Seasonal effects, such as those arising from the freeze-thaw cycle, monsoonal variations, and those that occur in arid environments characterized by occasional violent flash floods, months or even years apart, all generate the same wide variation in discharge and consequent wide variation in facies, but the climatic interpretation of the resulting facies may be quite ambiguous (considerations of time scale are discussed at length in Chap. 3).
2. The climate of the sediment source area and that of the depositional basin may not be the same. For example, rivers flowing southward from the Alpine-Himalayan mountain chain are all strongly seasonal, reflecting the alpine climate of their mountainous source areas, but they enter very different climates in the plains to the south. European rivers, such as the Rhone and Po, have coastal plains in the subtropical Mediterranean region. The Tigris and Euphrates enter the highly arid Mesopotamian basin. The fluvial plains of northern India are located in a tropical climate where seasonal monsoonal rainfall strongly influences discharge characteristics (see also Sect. 11.3.1 for other characteristics of "big rivers").
3. Related to point 2 above is the fact that flow hydraulics of a river, and therefore the facies and architecture of the bedload deposits, are determined largely by the climate of the source area, whereas floodplains and their finer-grained deposits are influenced primarily by the climate of the depositional basin.
4. Topographic effects may generate local climates that complicate the depositional record. For example, mountain ranges rising in the path of prevailing humid winds may generate high orographic rainfalls on their upwind slopes and rain shadows downwind. The topographic diversion of air masses is an important general effect. Very high plateaus, such as Tibet, are also a significant source of heat, which has its own effect on air temperatures and circulation. The possible wider linkages between tectonics and climate are discussed in Sect. 12.11.
5. Even within a fluvial basin, levels of water saturation (the water table) and redox conditions may vary from channel to floodplain and between the proximal, possibly more elevated basin-margin rivers and the topographically lower basin-center channel systems.
6. Climate is commonly recorded in the sedimentary record by vegetation, by the fossil remains of vegetation itself, and by the effects vegetation has on sediment erodibility, sediment yield, channel style, and so on. However, the styles of vegetation that characterize the earth today have evolved with time and were very different in the geological past. The development of large land plants in the Devonian, of plants that could survive seasonal climate changes in the Mesozoic, and of grasses in the Miocene all brought about changes in fluvial hydrology and channel style (Schumm 1968a). Modern analogues therefore have a limited applicability to the distant past, pre-Devonian landscapes probably having functioned sedimentologically much as do arid regions of the present day, even where rainfall was abundant (Fig. 8.2).

Correct identification of these various effects and separation of them from the sometimes very similar effects of tectonic control require meticulous observation and very careful deductive reasoning.

12.3 Distinguishing Tectonic from Climatic Control

The most obvious effects of tectonic controls are the changes that tectonism induces in sedimentary accommodation space and in the quantity and caliber of the sediment load. Upward coarsening and fining of alluvial successions, in the form of cycles, sequences, and megasequences, have long been interpreted in tectonic terms (Steel et al. 1977; Heward 1978a), although the nature of the relationship between tectonic movement and the sedimentary response is currently undergoing debate, as discussed in Sect. 11.3.6. However, similar sedimentary effects could be generated by climate change. Langbein and Schumm (1958) showed that in the United States sediment yield is related to rainfall, varying from a minimum in arid climates, to a maximum in subhumid climates, to a low in humid climates when the effectiveness of vegetation in binding soil and retarding erosion reaches a maximum. However, later work has shown that this is not a simple relationship that applies to all regions of the earth. For example, increased seasonality of the climate also increases sediment yield, because the vegetation cover is limited by the constraints of the dry season, so that runoff is more flashy and therefore more competent in the wet season (Walling and Webb 1983). This effect is particularly marked in monsoonal climates. Cooling is induced by uplift, as well as by climate change, and the sedimentological effects are the same in both cases – changes in fauna and flora, and typically an increase in sediment caliber and quantity as a result of a reduction in the density of vegetation cover. What sedimentological criteria can be used to distinguish climatic from tectonic controls in the generation of these kinds of changes in detrital caliber and volume?

In their study of the fan gravels flanking the Dead Sea rift, Frostick and Reid (1989) noted a sharp interbedding of coarse and fine deposits, with no upward coarsening or fining. This they interpreted to indicate an absence of progradation or retrogradation, which are the typical alluvial responses to tectonic adjustments (Heward 1978a). They suggested that the coarse units indicate periods of increased rainfall.

De Boer et al. (1991) studied a complex interfingering of marine and nonmarine deposits in a foreland basin adjacent to the Pyrenees, in Spain. They demonstrated that in the proximal part of the basin a cyclic alternation of fine and coarse sedimentation persists vertically through about a dozen cycles, totaling over 100 m of beds. In the most proximal areas the cycles are entirely nonmarine, and in more distal areas entirely marine, with the facies boundary (shoreline) remaining fixed vertically. These are therefore not transgressive-regressive cycles. The fixing of the facies boundary is attributed to the influence of growth anticlines which stabilized areas of uplift and subsidence, but the absence of transgression and regression suggests that there were no vertical movements of the basin or of sea level during this time period. It therefore seems likely that climatic control of rainfall and sediment yield is the explanation for the observed cyclicity.

An important debate regarding the distincton between global climate change and large-scale uplift in the Cenozoic is discussed in Sect. 12.11.

Table 12.1. The occurrence of debris flows in gravel deposits

Environment	Debris-flow deposits common to dominant	Debris-flow deposits rare to absent
Arid fans	Beaty (1963); Bull (1963); Hooke (1967); Schultz (1984); Flint (1985); Hubert and Filipov (1989); Harvey (1990); Blair and McPherson (1992); Nemec and Postma (1993)[b]	Frostick and Reid (1989); Jolley et al. (1990)[a]; Smith (1990); Harvey (1990); Maizels (1990); Hartley (1993)[a]
Temperate-humid fans	Winder (1965); Pierson (1980); Kochel and Johnson (1984); Wells and Harvey (1987); Kochel (1990); Ono (1990)	Ori (1982); Blair (1987)[a]; Kochel (1990)[a]; DeCelles et al. (1991); Crews and Ethridge (1993); Ridgway and DeCelles (1993)
Tropical-humid fans	Heward (1978b); Iwaniw (1984); Ahmad et al. (1993); Brierley et al. (1993)	Rust (1984); Darby et al. (1990); Wescott (1990); Evans (1991)
Upland periglacial-humid fans	Church and Ryder (1972); Wasson (1977); Derbyshire and Owen (1990)	Nemec and Postma (1993)[c]
Outwash "fans" (sandurs)	Maizels (1993)	Le Blanc Smith and Eriksson (1979); Boothroyd and Nummedal (1978)

[a] Deposits of hyperconcentrated flows are important.
[b] Stage 1 fans.
[c] Stage 2 fans.

12.4 Review of Climatic Criteria

The following sections review the major facies attributes of clastic fluvial lithofacies that convey some climatic information. For each lithofacies class the focus is on the most climatically sensitive criteria, and these are emphasized in the subheadings. The data are summarized in accompanying tables and figures, as follows: Table 12.1 classifies some major recent descriptions of modern and ancient gravel deposits according to climatic setting and according to the predominance of mass-flow or stream-flow deposits. Tables 12.2 and 12.3 summarize the descriptive criteria of gravel- and sand-dominated fluvial systems, respectively, subdividing them into various climatic classes and highlighting the attributes that are climatically sensitive. Figure 12.1 illustrates two contrasting styles of gravel-dominated fluvial deposit, Fig. 12.2 does the same for sand-dominated deposits, and Fig. 12.3 illustrates a range of climatic criteria, mostly those preserved in the overbank environment.

12.5 Conglomerates: The Significance of Texture and Petrology

12.5.1 Mass-flow Versus Traction-current Processes

The textures of gravels reveal a gradation in sediment transport mechanisms from mass-flow to traction-current processes. Such characteristics as a matrix-supported framework and the presence of grading are typical of mass flow, including debris flows, mudflows, and hyperconcentrated flows, whereas a clast-supported framework, stratification, and cross-bedding are typical of traction-current deposits (Sect. 5.2). Several factors are important in determining the type of flow that occurs in gravel-bed fluvial systems, including the magnitude and

Table 12.2. Field criteria for the interpretation of climate from clastic fluvial lithofacies: gravel-bed alluvial-fan and braided-channel systems (*climatically distinctive criteria are indicated by italics*)

Hot-arid systems
- Typical clastic lithofacies: *Mass-flow deposits, including Gmm, Gmg, Gci, Gcm*
- Typical architecture: *Sheets, lenses and lobes of SG*
- Major bounding surfaces: *Fan lobe deflation surfaces (fifth- and sixth-order) overlying winnowed gravels*
- Autogenic cyclicity: *Debris-flow events, may show winnowed tops*
- Petrology: *Survival of chemically unstable lithologies, such as carbonates*
- Other structures and associated lithologies: *Silcrete, calcrete, dreikanters, eolian lenses, evaporitic cements. Lenses of plant hash may be present*

Hot-humid systems
- Typical clastic lithofacies: Stream-flow gravels, typically Gh, Gt, Gp. Mass-flow deposits may be present
- Typical architecture: Channel-form CH, sheets and lenses of GB, SB
- Major bounding surfaces: Channel-form fourth- and fifth-order cycles, fan-lobe sixth-order surfaces
- Autogenic cyclicity: Macroform and channel-fill cycles, progradation-abandonment cycles of fan lobes
- Petrology: *Chemical destruction of unstable minerals and lithologies, ghost clasts with diagenetic clay matrix*
- Other structures and associated lithologies: *Bauxite, laterite, coal*

Temperate-humid systems
- Typical clastic lithofacies: Stream-flow gravels, typically Gh, Gt, Gp. Mass-flow dpeosits may be present
- Typical architecture: Channel-form CH, sheets and lenses of GB, SB
- Major bounding surfaces: Channel-form fourth- and fifth-order cycles, fan-lobe sixth-order surfaces
- Autogenic cyclicity: Macroform and channel-fill cycles, progradation-abandonment cycles of fan lobes
- Petrology: No distinctive features
- Other structures and associated lithologies: Coal may be present

Boreal-paraglacial systems
- Typical clastic lithofacies: Stream-flow gravels, typically Gh, Gt, Gp. *Mass flow deposits generated by jökulhlaups, including Gmm, Gmg, Gci, Gcm.* Minor traction-current sand lithofacies
- Typical architecture: Channel-form CH, sheets and lenses of GB, SB, lobes of SG
- Major bounding surfaces: Channel-form fourth- and fifth-order cycles, fan-lobe sixth-order surfaces
- Autogenic cyclicity: Macroform and channel-fill cycles, progradation-abandonment cycles of fan lobes
- Petrology: *Survival of chemically unstable lithologies, such as carbonates. Composition-shape inversion (weak clasts fractured by frost, resistant clasts retain roundness)*
- Other structures and associated lithologies: *Till, ice-wedge casts, ice-push stone clusters, ice melt depressions, lacustrine deposits with dropstones*

Table 12.3. Field criteria for the interpretation of climate from clastic fluvial lithofacies: sand-bed fluvial systems, i.e., meander-belts, braid plains, terminal fans (*climatically distinctive criteria are indicated by italics*)

Hot-arid systems

Braid plains with playas, terminal fans, flash floods common

- Typical clastic lithofacies: St, Sh, Sl, Sr. *Large Sp sets rare to absent*, Fl, Fm
- Typical architecture: *Proximal: steep-sided wadis*, medial: broad channels with macroforms (elements LA, DA) and *scour hollows (element HO), distal: unchannelized sheet bodies interbedded with playa mudstones*
- Major bounding surfaces: Channels bounded by fifth-order surfaces, *macroforms with abundant second- and third-order surfaces defining flood-related accretionary bundles; distal fans: Flat fourth- and fifth-order surfaces*
- Autogenic cyclicity: *Flood cycles within macroforms and sheet floods indicating rapid deceleration, commonly from upper flow regime, plane-bed condition*
- Petrology: *Detrital carbonate sand common*
- Other structures and associated lithologies: *Intraclast siltstone breccias, desiccation cracks, mud curls, silcrete, calcrete, evaporite lenses, nodules or crystals, dreikanters, eolian lenses*

Hot-humid systems

Meandering and anastomosing styles most typical; floodplains with peat bogs typical

- Typical clastic lithofacies: Most sand and fine-grained lithofacies present
- Typical architecture: Meander-belts consisting of sand-bed channels of low to high sinuosity, with *steep, root-bounded banks, braiding rare*
- Major bounding surfaces: Channel-form fifth-order surfaces, macroform-bounding fourth-order surfaces, accretionary second- and third-order surfaces
- Autogenic cyclicity: Macroform and channel-fill cycles
- Petrology: *Chemical destruction of unstable minerals and lithologies, ghost clasts with diagenetic clay matrix*
- Other structures and associated lithologies: *Peat swamps common, bauxites and laterites in elevated interfluves*

Tropical to semitropical systems with wet-dry seasonality

Most fluvial styles; perennial but with seasonally dry floodplains

- Typical clastic lithofacies: Most sand and fine-grained lithofacies present. Large Sp sets common, calcrete breccias
- Typical architecture: Meander-belts with channels of varying style
- Major bounding surfaces: Channel-form fifth-order surfaces, macroform-bounding fourth-order surfaces, accretionary second- and third-order surfaces
- Autogenic cyclicity: Macroform and channel-fill cycles
- Petrology: No distinctive climatic signatures
- Other structures and associated lithologies: *Common association between peat swamps or drab soils flanking channels with calcretes, and evidence of desiccation on upland interfluves*

Temperate systems

Most fluvial styles. Some seasonal discharge variations

- Typical clastic lithofacies: Most sand and fine-grained lithofacies present. Large Sp sets common
- Typical architecture: Meander-belts with channels of varying style
- Major bounding surfaces: Channel-form fifth-order surfaces, macroform-bounding fourth-order surfaces, accretionary second- and third-order surfaces
- Autogenic cyclicity: Macroform and channel-fill cycles
- Petrology: No distinctive climatic signatures
- Other structures and associated lithologies: Peat swamps or drab soils may be present

Boreal-paraglacial systems

Braided outwash plains, sandurs, fans

- Typical clastic lithofacies: Most sand and fine-grained lithofacies present
- Typical architecture: Complex interfingering of channel lenses and fan lobes
- Major bounding surfaces: Channel-form fifth-order surfaces, macroform-bounding 4th-order surfaces, accretionary second- and third-order surfaces
- Autogenic cyclicity: Macroform and channel-fill f-u cycles, *c-u cycles filling ice melt depressions*, fan lobe progradation and abandonment cycles
- Petrology: *Survival of chemically unstable detritus, till balls*
- Other structures and associated lithologies: peat swamps or drab soils may occupy inter-fan lows. *Interbedded till or glaciolacustrine muds and sands with dropstones, ice-wedge casts, ice-push stone clusters, ice-melt depressions*

f-u, Fining-upward, *c-u*, coarsening upward

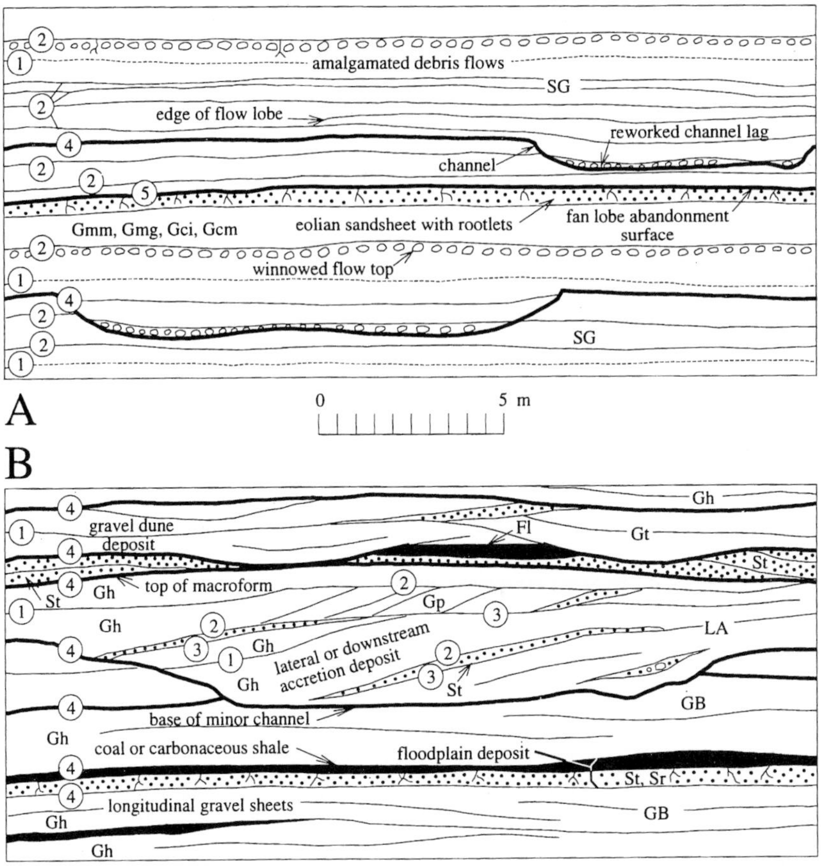

Fig. 12.1A,B. Architectural diagrams of conglomerates. *Numbers in circles* are bounding surfaces. **A** Typical appearance of debris-flow-dominated alluvial-fan deposit, such as are formed in hot, arid climatic settings. Debris-flow deposits and their derivatives – winnowed flow tops and channel-floor lags – constitute virtually the entire deposit. The lateral or distal fringes of fan lobes may interfinger with eolian sand sheets. (Based on Wasson 1977; Hubert and Filipov 1989; Blair and McPherson 1992). **B** Gravel assemblage deposited by stream-flow processes, which are the dominant gravel transport mode in most humid climatic settings. (Based mainly on Ramos and Sopeña 1983; Smith 1990; DeCelles et al. 1991)

variability of runoff and the quantity and type of debris available in the catchment area.

The presence or absence of debris flows, hyperconcentrated flows, and sheetfloods has been proposed by McPherson and Blair (1993) and Blair and McPherson (1994) as the major criterion to be used in the definition of alluvial fans, a proposal opposed by Stanistreet and McCarthy (1993) and Galloway and Hobday (1995), who preferred an all-embracing definition that would include the large, sand-dominated systems deposited by braided and meandering rivers, such as the Kosi and the Okavango (Sect. 8.3). The controversy has focused attention on the nature of fluvial processes and has thereby broadened our understanding of the autogenic and allogenic (tectonic, climatic) controls that govern the development of basin-margin alluvial wedges. Climatic factors are commonly taken to be among the more important controls on gravel transport processes. Blair and McPherson (1994) demonstrated that the hydraulic conditions for debris flows and stream flows are quite different, and the two types of deposit will therefore tend not to be associated in the same depositional system. They prefer the terms braid delta and floodplain delta for dispersal systems built by stream-flow (lower flow-regime traction-current) processes.

Because of the importance of American work in the deserts of the arid southwest US on the nature of mass-flow processes in particular, and alluvial-fan processes in general (see below), there is a tendency to associate mass flow with the catastrophic runoff that occurs in arid and semiarid regions. The absence or sparsity of vegetation in arid regions is an important factor, because vegetation binds the sediment and acts as a buffer to rapid runoff during all

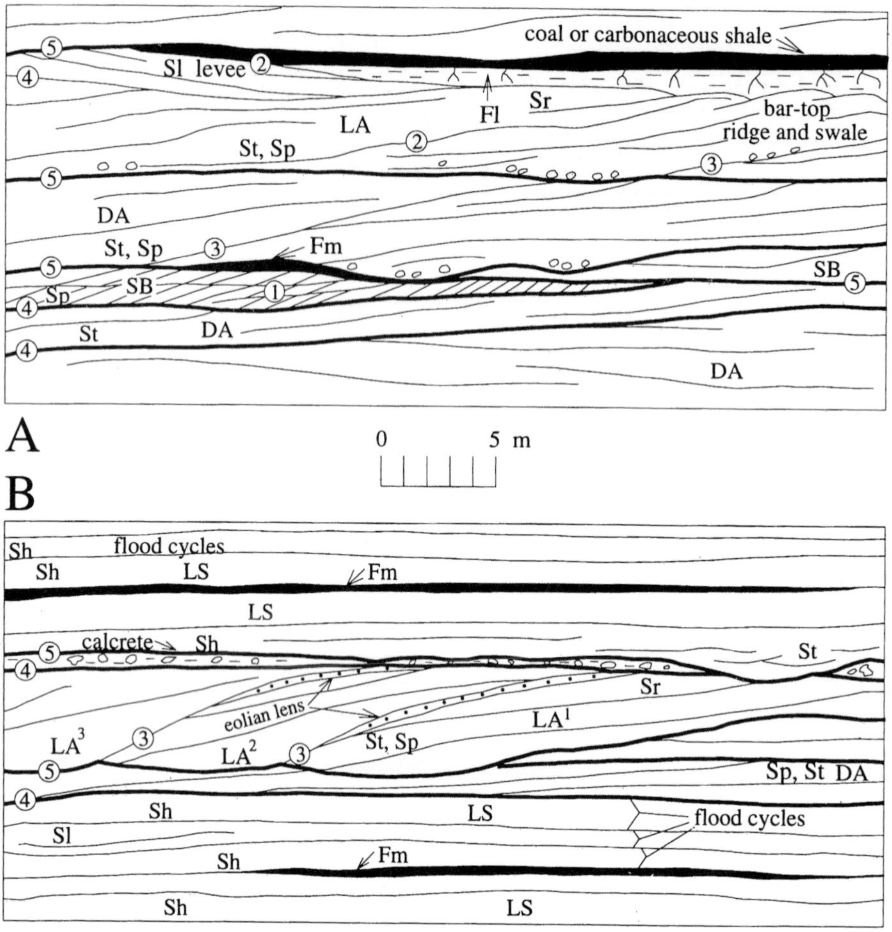

Fig. 12.2A,B. Architectural diagrams of sandstones. **A** Typical mobile-channel-belt deposit formed under perennial flow conditions, with large macroforms showing few internal sedimentary breaks (third-order surfaces), well-developed bar-top ridge-and-swale features, levees and swamp deposits. (Based on many examples, including Wizevich 1992b, 1993; and Miall 1993, 1994). **B** Medial to distal terminal-fan deposits formed under arid conditions, showing macroforms with accretionary bundles and unchannelized sand sheets, both architectural characteristics of rivers with flashy discharge. (Based on Tunbridge 1981, 1984; Stear 1983; Shepherd 1987; Bromley 1991; Muñoz et al. 1992; and Kelly and Olsen 1993)

but the most violent rainstorms. However, it is important to take account of the significant number of observations of mass-flow processes occurring in other climatic settings, and other studies of arid environments where mass-flow deposits are of limited importance. Some of the more important recent studies are listed in Table 12.1. Interpretations of climate in each of the ancient studies is based on the accumulated evidence from sedimentary features and contained fauna and flora, and in most cases are taken from the conclusions of the authors. A much more lengthy tabulation is provided by Blair and McPherson (1994, Table 1), but it is limited to distinguishing desert from nondesert settings for debris flows. The fluvial styles under discussion here are described in Sects. 8.2.1–8.2.5.

12.5.1.1 Arid Climates.

Based on the pioneering work in the desert areas of the southwestern United States by Chawner (1935), Sharp and Nobles (1953), Blissenbach (1954), Beaty (1963), Bull (1963, 1972), Lustig (1965), Denny (1967), and Hooke (1967), it has long been known that alluvial fans in modern arid and semiarid regions are characterized by debris flows (Fig. 12.1A). More recent work in this area, including detailed textural and facies studies, has been reported by Hubert and Filipov (1989), Beaty (1990), and Blair and McPherson (1992, 1994). Several of the older papers quoted here include eyewitness accounts of debris-flow events, which occur every few decades. Much of the time the fan distributaries are dry, with

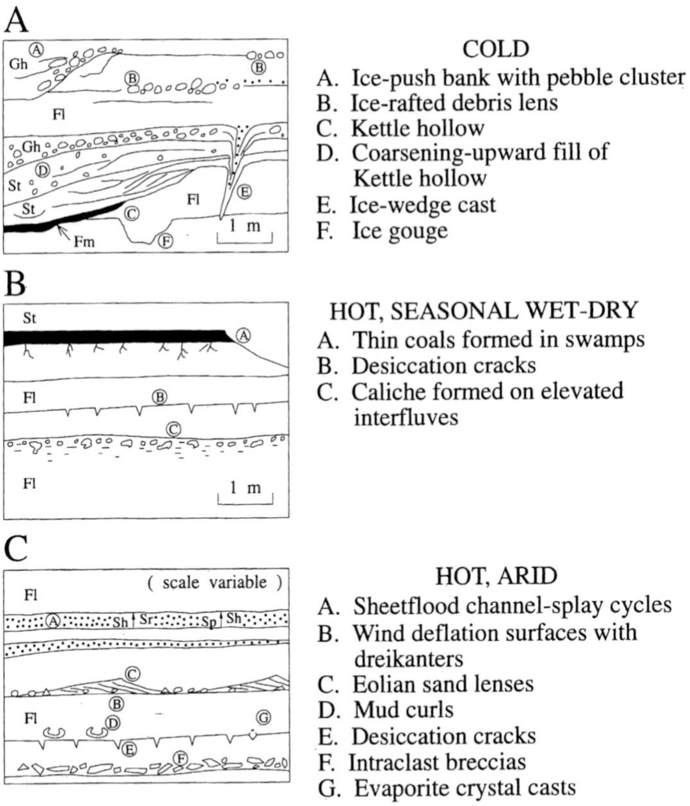

Fig. 12.3A–C. Some climatic lithofacies indicators of the overbank environment. **A** Cold climates with seasonal ice. (Based on Collinson 1971b; Bryant 1983; Martini et al. 1993). **B** Hot, seasonal wet-dry climates. (Based on Hallam 1984; Jerzykiewicz and Sweet 1988). **C** Hot, arid climates. (Based on Tunbridge 1981, 1984; Stear 1983, 1985; Muñoz et al. 1992)

surface runoff occurring during spring melt and following cloudbursts. However, it takes a considerable length of time for weathering to generate enough debris for mass flow to be triggered – an example of a significant "geomorphic threshold" that must be passed before a particular type of process can occur (Schumm 1979). Blair and McPherson (1992) emphasized that debris-flow deposits and the product of their reworking constitute the entire accumulated sedimentary record of the Trollheim fan in California (Fig. 8.9), contrary to the interpreted sedimentary model of Miall (1978c) and others. Traction current deposits, such as stratified gravels and cross-bedded sands, are absent.

The importance of sediment-gravity flows in the fan deposits of other arid areas is quite variable. Modern fans in the arid intermontane valleys of southeast Spain were subdivided by Harvey (1990) into three types, based on their facies composition. Debris flows are more common in fans draining small, steep drainage basins and are absent in some of the fans fed by larger drainage basins. The modern fans draining the flanks of the Dead Sea rift valley, a very arid area, do not include significant mass-flow deposits, according to Frostick and Reid (1989). Quaternary fan (braid-delta) deposits in the deserts of Oman also do not contain evidence of mass-flow processes (Maizels 1990). Likewise, some of the Recent and Cretaceous-Tertiary fan (braid-delta) deposits in arid intermontane basins in the Chilean Andes do not contain debris-flow deposits. Jolley et al. (1990) reported that ephemeral stream-flow deposits are dominant in their study of modern syntectonic fan deposits, and concluded that they were formed by high-concentration turbulent tractional flows. They stated that "[t]he absence of mass flow deposits is inspite of favourable depositional slopes, a persistently arid climate, the coarse-grained nature of the source material and the availability of mud in the [source] sediments."

The facies composition of ancient arid-climate fan deposits is similarly variable. In an ancient arid-fan system described by Hartley (1993), hyperconcentrated flood gravels and sheetflood traction-current gravels characterize proximal areas. By contrast, the Tertiary deposits described by Flint (1985) include significant debris-flow deposits. A Triassic conglomerate deposit in SW Britain is interpreted as a fan complex by S.A. Smith (1990), who stated:

"A semi-arid climate during the deposition of at least part of the Budleigh Salterton Pebble Beds is suggested by the evidence for large variations of discharge (e.g. drapes of the scour troughs of sand dunes and drapes of the lower portions of gravel macroforms). Also, rare aeolian sandstones have been found in the upper part of the sequence, where they occur as thin lenses capping bar tops."

The conglomerates consist entirely of stratified gravels, commonly with cross-bedding, organized into large in-channel macroforms, and were deposited by traction-currents in gravel-bed braided rivers (Figs. 6.18, 8.21). The term braid-delta would be more suitable for these deposits, following the recommendations of Blair and McPherson (1994).

12.5.1.2 Temperate-Humid Climates

Several studies have documented mass-flow processes in modern temperate-humid climates. Debris flows are particularly important in mountainous regions of southwestern Japan (Ono 1990) and New Zealand (Pierson 1980), where in many cases they contribute to the reworking and redeposition of glacial fan deposits. Kochel and Johnson (1984) described active fans in Virginia. The fan deposits consist of poorly sorted, mud-supported gravels with inverse grading and poor stratification, with a bed thickness varying from 50 cm to 5 m. The main depositional process is interpreted as debris flows and debris avalanches triggered by intense rainfall. From 20 to 70% of the fan surface is affected at any one time. The recurrence interval of flood events is estimated at 3–6 ka, based on radiocarbon dating. Colluvial processes are also very important. This study was expanded to other parts of the Appalachians by Kochel (1990), who confirmed the widespread distribution of debris-flow-dominated fans but also documented the local occurrence of fans formed by stream-flow and hyperconcentrated flow processes, where lithologic controls on source sediment and hydrologic controls on runoff prevented the generation of debris flows. Wells and Harvey (1987) described the results of a catastrophic cloudburst in northern England. In a temperate-humid environment such rare catastrophic floods can generate erosional and depositional effects that may overshadow the products of "normal" weather. In this case a single flood generated violent stream flows and debris flows. In this particular study, which examined numerous fans, the debris-flow deposits were found to constitute a greater part of the accumulated sediment on the fans draining smaller, steeper drainage areas. The description of a small modern fan by Blair (1987) indicates an absence of debris-flow deposits, although he reported evidence of hyperconcentrated flows.

It is perhaps significant that several recent descriptions of ancient alluvial-fan/braid-delta deposits from interpreted temperate-humid climatic regimes indicate that stream-flow deposits are dominant, and that mass-flow deposits are of minor importance. This depositional style is illustrated in Fig. 12.1B. In the study by Crews and Ethridge (1993), the dominance of braided stream-flow traction processes was demonstrated, with evidence of lateral channel migration. Debris flows were determined to be rare to absent. Likewise, DeCelles et al. (1991) reported mass-flow deposits in only one of numerous localities. In their study, Ridgway and DeCelles (1993) drew comparisons with the modern gravel sandurs of southern Alaska (Boothroyd and Nummedal 1978). Similar clast-supported gravels deposited by stream-flow processes are the dominant facies in their rocks. Debris flows are rare. One possible inference that could be drawn from the studies noted here is that while mass-flow processes may be very spectacular in modern humid-temperate settings, they may have low preservation potential and are therefore of limited geological significance. However, this ancient/modern contrast is based on only a few reports and should not be regarded as definitive.

12.5.1.3 Tropical-Humid Climates

Deposits containing significant mass-flow deposits and examples with a lack of such deposits have both been described in case studies of the modern and ancient record.

Modern fan/braid-delta deposits in Papua-New Guinea were described by Brierley et al. (1993). The area has a monsoonal, savannah-type climate, with a significant dry season. In addition to stream-flow deposits, debris flows and hyperconcentrated flows are present, as indicated by gravel textures. The mass flow deposits are limited to proximal reaches of the fans, where they comprise up to about 30% of the total. However, their preservation potential is estimated to be low, because of continued incision of the deposits, and this is aggravated by syndepositional tectonic uplift. The modern Yallahs braid delta of Jamaica lacks debris-flow deposits (Wescott 1990).

Several ancient conglomerate deposits have been interpreted to have been formed in a tropical-humid setting. The Devonian Malbaie Formation of Quebec contains no mass-flow deposits (Rust 1984), whereas in a study of Carboniferous fan and paleovalley deposits in Spain, Heward (1978b) reported substantial accumulations of debris-flow deposits. The paleovalleys are filled with debris-flow conglomerates. The deposits of fan channels and the overbank areas of fans consist of interbedded debris-flow and stream-flow deposits, with debris flows more important in proximal reaches. Rust (1984) reported an unusual abundance of cross-bedded gravels (lithofacies Gp) in the Malbaie Formation, and attributed this to the presence of deep braided channels and perennial discharge in a tropical, humid setting. In another study in Spain, Iwaniw (1984) reported that debris-flow deposits are important, whereas in a study of a Paleogene fan (braid-delta) complex in Washington state it was found that debris-flow deposits are of minor importance and can be related to volcanic influence (Evans 1991). Late Cenozoic deposits described by Darby et al. (1990) contain no evidence of mass-flow processes.

12.5.1.4 Boreal-Paraglacial Climates

It is not suprising that mass-flow deposits have been found to be a dominant facies in ice-margin settings, where abundant glacial debris is available to be transported by large volumes of glacial meltwater. Reports of Late Cenozoic deposits from various Alpine environments around the world are consistent on this point (Church and Ryder 1972; Wasson 1977; Derbyshire and Owen 1990). An interesting exception is the complex of Quaternary fan deposits in Crete constituting "stage 2" of Nemec and Postma (1993). These deposits are interpreted as periglacial in origin, yet were deposited almost entirely by stream-flow traction-current processes.

Glacial outwash deposits provided the first model for humid fans (Boothroyd and Ashley 1975; Miall 1977; Sect. 8.2.2), although some have objected to the use of the term "fan" for such deposits (McPherson and Blair 1993; Blair and McPherson 1994). The term "sandur" is to be preferred. The deposits are normally characterized by stream-flow traction-current deposits (Boothroyd and Ashley 1975; Boothroyd and Nummedal 1978; Leblanc Smith and Erikkson 1979; Miall 1983b; Ashley et al. 1985), although mass-flow processes are important in the event of the catastrophic floods – *jökulhlaups* – that characterize many outwash plains. These floods are triggered by the failure of ice dams or by other sudden glacial movements that release large volumes of meltwater. In Iceland jökulhlaups are also triggered by volcanic eruptions under ice caps. These floods may do little more than erode and rework pre-existing fluvioglacial sands and gravels, moraine, or till. Rounded till balls are common in some outwash deposits (Ashley et al. 1985). However, jökulhlaups may trigger mass flows, and in Iceland such deposits consititute a significant proportion of the outwash succession (Maizels 1993). Another product of catastrophic floods may be the deposition of trains of giant bedforms, as in the floods that affected the Columbia River valley during the draining of Lake Missoula (Baker 1978b).

12.5.1.5 Summary and Conclusions

To summarize, it is clear that while mass-flow deposits, particularly debris-flow conglomerates, and associated upper flow-regime stream-flow deposits are very common in arid environments (Trollheim fan being the type example), they are not always generated in this climatic setting and can occur under other climatic regimes, ranging from paraglacial to tropical (Table 12.1). The association with an arid climate is a reflection of the requirement for abundant loose debris, mobilized by rapid runoff, uninhibited by a vegetation cover. However, catastrophic floods can occur in all environments. They are triggered by cyclonic storms in temperate and tropical settings and by jökulhlaups in glacial settings. Climates characterized by seasonal rainfalls and seasonal dry periods are candidates for flash floods because of the lower density of vegetation in such areas than in humid settings. Floods need not be very frequent to impose a dramatic change on the depositional system that takes a long time to eradicate under "normal" weather conditions.

Many of the studies reported here indicate a considerable variation in the frequency and importance of mass-flow events and their deposits between one fan and another within the same study area. The steepness of the relief in the drainage basin and the nature of the bedrock are important controls. Mass flows are favored by steep slopes and by bedrock that weathers to produce large volumes of fine-grained sediment, especially clay. Therefore, within the same sedimentary basin, under constant tectonic and climatic conditions, the composition of the basin-margin conglomerate wedge may vary along strike because of controls imposed by the character of the fan drainage basins.

Conglomerates dominated by mass-flow deposits (Fig. 12.1A) may be somewhat more likely to have been deposited under arid climatic conditions, the Trollheim fan being the extreme example (Blair and McPherson 1992), but other criteria should be sought to test a climatic interpretation. Paraglacial alluvial fans may also contain significant debris-flow deposits. In most other humid environments lower-flow-regime stream-flow conglomerates with interbedded cross-bedded sandstones are dominant (Fig. 12.1B). As discussed in succeeding sections, the sandstones, fine-grained sediments, and biochemical sediments that commonly are interbedded with fans and braid deltas contain many other criteria, none of which are definitive, but which, in combination, should provide reasonably certain indications of broad climatic characteristics (Fig. 12.3, Tables 12.2, 12.3).

12.5.2 The Influence of Climate on Texture and Composition of Gravels

Climatic variables affect not only the production and transportation of clastic debris, but also the nature of the debris itself (Basu 1985; Johnsson and Basu 1993). Styles of weathering and erosion are governed largely by levels of water saturation, by temperature, and by biochemical processes. Mechanical weathering is favored in cold and arid climates; chemical weathering is more important in humid environments, especially in tropical settings, where organic activity is at its most intense. These variations have measureable effects on the petrology and texture of gravel and sand in fluvial systems. Detrital sandstone composition (including the detrital matrix of conglomerates) may contain much climatic information. This is a specialized area of study that is beyond the scope of this book, but some key points are noted in this section.

Several workers have commented on the abundance of carbonate sedimentary clasts and sand-sized detritus in deposits interpreted to have been deposited in cold or arid climates, the inference being that in humid climates chemical weathering would have dissolved such material before significant transportation and accumulation could occur (Miall 1970a; Besly 1988). Some arid braidplain and playa-margin deposits consist largely of detrital mud and sand fragments composed almost entirely of carbonate, including hematite-stained calcareous granules derived from reworking of soil nodules (Wells 1983), or detrital limestone and dolomite grains eroded from older sedimentary sources (Miall et al. 1978). Stanley and Wayne (1972) found that in the Pleistocene deposits of Nebraska, transport of mechanically weak clast types (limestone, anorthosite cobbles, sandstone, hornblende schist, and mica schist) was favored during glacial periods, when chemical weathering was at a minimum and transport rates were rapid.

Darby et al. (1990) set out to determine the ways in which gravels deposited in humid tropical environments might differ from those deposited in other climatic settings, but they were unable to identify any definitive facies criteria. However, they found that deep tropical weathering of clasts produced much in-place matrix, and clasts tended to be reduced to ghosts. Evans (1991) attributed gravel roundness to spheroidal weathering prior to erosion and transportation. Iwaniw (1984) interpreted well-rounded clasts in his deposits as the product of extreme weathering in a humid tropical climate. Kochel and Johnson (1984) found that during the long periods between depositional events on modern fans in Virginia unstable minerals tend to weather, producing clay films around the weathered clasts and stable minerals. By contrast, in another study of tropical alluvial fans, Brierley et al. (1993) found no evidence of deep weathering of clasts and no thick weathered zones, and suggested that this reflects the reduced importance of deep chemical weathering in the seasonally dry, savannah-type climate of Papua-New Guinea.

Periglacial and boreal environments are characterized by several distinctive petrological and textural features. As noted above, the chemically more easily weathered clast types survive transportation in greater abundance in dry or cold climates, where rates of chemical weathering are lower (Stanley and Wayne 1972). A composition-shape inversion may develop in channel deposits, whereby resistant clasts of plutonic and igneous rock retain their rounded shapes, whereas softer sedimentary clasts are subject to frost breakage and therefore tend to be angular (Martini et al. 1993). Clay balls, including till balls, are common in channel-lag deposits and in the lenses of gravel that are commonly pushed by seasonal ice onto floodplains. Such heterolithic clasts are commonly generated by the severe erosion that occurs during jökulhlaups (Ashley et al. 1985; Martini et al. 1993). Casshyap and Tewari (1982) described large disrupted beds of diamictite in Late Paleozoic fluvioglacial gravels in India, which they interpreted as ice-rafted blocks of frozen morainic material. Collinson (1971b) described distinctive structures formed in the Tana River, Norway, by the action of seasonal river ice. They included ice gouge marks up to 1 m wide and several tens of meters long, eroded in the gravel and sand of the channel floor and also on finer overbank sediments. A more subtle effect included flattening of current-formed clast imbrication by pressure from stranded ice masses, and scour holes arounded stranded ice blocks. Kettle holes are depressions in river bed formed by the melt of buried ice masses. Some coarsening-upward gravel-sand cycles in Quaternary outwash (Costello and Walker 1972) have been interpreted as the result of channel diversion into these depressions, which deepen as melting takes place (Rust 1975). Some of these criteria are illustrated in Fig. 12.3.

12.6 Sandstones: The Significance of Sand Body Architecture and Sedimentary Structures

12.6.1 Fluvial Style in Sand-bed Rivers

Channel style in sand-bed rivers is controlled by several factors, of which the amount and variability of discharge, the nature of the sediment load, bank composition, and bank stability, are among the more important (Sect. 8.1). Some of these variables are strongly affected by climate. Thus, humidity is the main control on vegetation, which has an important effect on discharge characteristics, sediment load, and fluvial style. In upland vegetated areas at the present day, runoff following major rainfalls is rarely catastrophic because the precipitation is absorbed by soil and plants and released slowly. Also, sediment is stabilized by roots, and sediment yields are relatively low. Therefore, as noted in the previous section, mass-flow processes are less common in humid climatic settings. Vegetated banks are highly resistant to erosion, as demonstrated in an experiment by Smith (1976). This limits the supply of sediment to the river, and also reduces the bank widening that is necessary for braiding to develop. In vegetated areas, therefore, because of the effects on discharge variability, bank stability, and sediment supply, braiding is inhibited. Baker (1978a) quoted examples in tropical Africa and South America where dense bank vegetation inhibits braiding, even where abundant coarse sediment is present in the banks. Stanistreet et al. (1993) described the channels of the Okavango fan in Botswana, which are low-sinuosity to meandering bedload rivers, in which banks are stabilized by peat levees. The contrasts in river behavior when forests in upland, catchment areas are destroyed by fire or deforestation are marked and well known. Runoff becomes more flashy, the sediment load increases, and a tendency may develop for debris flows to occur. The erosion of steep-sided arroyos or box-canyons is typical of flash-flood processes along basin margins and may be an indicator of this type of sedimentation where proximal fluvial deposits are preserved.

Temporal changes related to climatic causes may be exemplified by several studies of rivers in the United States. Nadler and Schumm (1981) and Schumm (1985b) described examples of two rivers, the South Platte and Arkansas, that evolved from braided to meandering patterns within historic times as a result of damming and irrigation that resulted in lower discharge variability, lessening of flood peaks, and stabilizing of banks by the growth of vegetation. A climatic reduction in seasonality would produce the same effect. Conversely, Schumm and Lichty (1963) described the case of the Cimarron River, Kansas, that changed from a suspension-load, meandering morphology to a broad, shallow, bed-load, braided morphology during a single major flood in 1914. Channel widening continued until 1942, during a period of below-normal precipitation, which inhibited vegetation growth. However, local factors unrelated to climate may produce the same effects on channel style, as described in Sect. 8.1.

There may be major proximal-distal differences in fluvial style between rivers in humid regions and those in arid regions. In humid climates flow is perennial, and channel scale increases downstream as discharge is added by tributaries, whereas in arid climates there may be a loss of discharge downstream as a result of infiltration and evaporation, with a consequent reduction in channel scale. Rivers may be perennial close to an upland, possibly alpine source, but become ephemeral downstream toward an arid basin center. For example, many of the present-day rivers draining into the vast inland basins of western China show this characteristic. Sedimentologically, these differences may be recorded in the scale of preserved bedforms and macroforms, channel-fill cycles and flood cycles.

The terminal-fan model encapsulates many of the features of arid sandy systems, including the downstream decrease of channel scale, absence of channel incision, and the lobate, distributary nature of the river system (Fig. 8.71). this has now been well documented in several studies, as described in Sect. 3. By contrast, the downstream increase in channel scale in perennial systems seems to be taken for granted and has not been specifically tested in many studies. In a study of a Cretaceous braidplain system in Utah, Miall (1993) demonstrated a downstream increase in sand body thickness over a distance of 140 km and attributed this to a downstream deepening of fluvial channels as tributary streams merged into fewer, larger rivers.

12.6.2 Sand Body Architecture

Given a channel form determined by discharge characteristics and bank materials, the main climatic control on sand body architecture is the control exerted by discharge variability on lithofacies assemblages and bounding surface characteristics (Fig. 12.2). It is to be expected that lithofacies heterogeneity will increase with increasing seasonal or longer-term variability in discharge, and that the frequency of minor internal bounding surfaces (second- and third-order surfaces) will likewise increase. As noted in Sect. 12.2, Jones (1977) classified river hydrography into five basic patterns, reflecting climatic and topographic controls on discharge variability. These variations are undoubtedly reflected in lithofacies and architectural differences within fluvial deposits, but the differences are expected to be subtle and not amenable to simple interpretation. As noted earlier, the sedimentological differences between high-stage and low-stage sedimentation may provide no clear information on the time gap (diurnal, seasonal, flood-related, or longer term) between these stages. Macroforms and channel-fill assemblages typically display upward fining, but this can be generated both by accretionary shallowing at the depositional site and by waning discharge. The gradual filling of a channel or accretion of a macroform under perennial flow conditions generates a fining-upward succession, but flood cycles also generate upward fining, because most sedimentation during floods takes place during the waning stages, following the erosion that occurs during peak runoff. The two causes of upward fining were not at first clearly distinguished from each other in sedimentological analyses. When first proposed, the South Saskatchewan model for sandy braided alluvium was defined largely on the basis of its cyclicity (Miall 1978c), but the studies of such cyclicity that were cited include both types. One of these studies described a portion of the Battery Point Formation (Devonian) of Quebec (Cant and Walker 1976), deposits which have now been interpreted as the product of channel migration and macroform aggradation in low-sinuosity rivers (association 3 of Lawrence and Williams 1987). Another example was the Devonian distal braidplain deposit in the Canadian Arctic described by Miall and Gibling (1978). These are now interpreted as flood cycles formed in an arid environment (Miall 1985). The original South Saskatchewan model of Miall (1978c) therefore includes at least two distinct fluvial styles (see Sects. 8.2.13 and 8.2.15).

Luttrell (1993) suggested that "it takes a medium to large-sized river with sufficient depth and width before long-term sand flats will develop. ... Only in the large perennial to intermittent rivers do channel-macroform assemblages dominate the alluvial architecture. Under sustained discharge conditions well-developed macroforms show a vertical decrease in both the scale of bedding and grain size." In her study of the Kayenta Formation (Jurassic) of the Colorado Plateau area she interpreted proximal sediments as the deposits of perennial streams largely on the basis of the abundance of large macroforms and channel scours indicating water depths of up to at least 8 m. In their study of Triassic sediments in New Mexico, DeLuca and Erikkson (1989) also proposed distinguishing deposits formed under perennial flow conditions from those formed by ephemeral flow on the basis of the occurrence of lateral-accretion deposits and a comparison with the Platte- and South-Saskatchewan-type profiles of Miall (1978c), which were models based partly on modern perennial rivers (Sects. 8.2.12, 8.2.13).

Other studies of the Kayenta Formation indicate that the deposits of the upper flow-regime plane-bed condition, lithofacies Sh, are important in the proximal regions of the depositional system, in southwestern Colorado (Miall 1988b; Bromley 1991a,b). As discussed later in this chapter (and in Sect. 8.2.16), the presence of sand bodies several meters thick composed predominantly of this lithofacies is normally an indicator of flash-flood sedimentation, a style of discharge that occurs most typically in ephemeral rivers. The Triassic deposits described by DeLuca and Erikkson (1989) also contain an ephemeral stream facies, and their descriptions of the deposits seem to confirm aridity throughout the succession. It is suggested that neither Luttrell (1993) nor DeLuca and Erikkson (1989) have made a convincing case for perennial flow in their studies.

In fact, it is quite possible for ephemeral rivers to develop deep channels with large macroforms that accrete only spasmodically, during flash floods (Fig. 12.2B). Shepherd (1987) described examples in the modern Rio Puerco, New Mexico. The point bars are up to 5 m thick and contain interbedded eolian lenses. Turning to the ancient record of ephemeral-stream sedimentation, Stear (1983) described ribbon and composite sheet sandstone bodies in the Lower Beaufort Group (Permian) of the Karoo Basin, south Africa, that were deposited in channels up to 5 m deep and 100 m wide. Many of the sandstone bodies consist of accretionary bundles of plane laminated sandstone (lithofacies Sh) and trough cross-bedded sandstone (St) separated by numerous internal scour surfaces. Lateral-accretion surfaces within many of the sheet sandstones indicate the development of point bars. Well-developed fining-upward cycles are also present (Turner 1978). Various lines of evidence, including numerous scour surfaces, intraclast breccias, and desiccation features, indicate an ephemeral, flash-flood style of sedimentation. In a later paper, Stear (1985) was able to compare these deposits with the products of a recent flash flood in the arid Karoo Basin. Muñoz et al. (1992) described accretionary bundles within Triassic lateral-accretion deposits in Spain. They commonly show a cyclic character, indicating that they were formed by floods or seasonal high-discharge events. The Westwater Canyon Member of the Morrison Formation (Jurassic) in New Mexico consists largely of sandstone sheets several meters thick organized into macroforms (Cowan 1991) and containing well-developed fining-upward cycles of several types (Godin 1991). This unit has also been interpreted as having been deposited by flash-flood sedimentation in moderate-sized channels (Cowan 1991) under arid climatic conditions (Parrish 1993). In their study of a Devonian unit in Quebec, Lawrence and Williams (1987) described lateral-accretion deposits which they compared with the ephemeral Rio Puerco examples described by Shepherd (1987). These fluvial styles are described further in Sects. 8.2.8 and 8.2.14. In their study of terminal fans, Kelly and Olsen (1993) erected a "feeder zone" lithofacies assemblage for the deposits of the main channel in the apical regions of the fan. This channel may be ephemeral, or perennial with widely fluctuating discharge. The lithofacies are characterized by channel sand bodies up to several meters thick (their examples indicate channel depths up to 6 m) and may contain the evidence of braid-bar complexes (Sect. 8.3).

The presence of channels several meters deep or cross-bedded assemblages organized into macroforms with evidence of lateral or downstream accretion (architectural elements LA and DA) should not, therefore, be interpreted as necessarily indicating perennial flow conditions. The presence of numerous internal scour surfaces indicates frequent discharge fluctuations. Macroforms consisting of accretionary bundles a few decimeters thick, separated by third-order erosion surfaces, as described by Stear (1983), Shepherd (1987), Langford and Chan (1989), and Muñoz et al. (1992), may be indicative of ephemeral or strongly seasonal conditions (Fig. 12.2B).

Another possible architectural indicator of high-discharge sedimentation, such as is characteristic of flash floods, is the scour hollow (architectural element HO). Deep scours form at channel confluences, as observed in flumes and in modern rivers, and a facies model for the deposits of these scours has been proposed by Bristow et al. (1993). This element is described and illustrated in Sect. 6.9.

Numerous studies have confirmed the predominance of unchannelized sheet flow in the medial to distal parts of arid braidplains, based on observations of modern floods (McKee et al. 1967; Williams 1971) and ancient deposits (Miall et al. 1978; Tunbridge 1981, 1984; Smoot 1983; Mertz and Hubert 1990; Hirst 1991; Dreyer 1993b; Kelly and Olsen 1993). The proportion of interbedded fine-grained deposits increases distally, depending upon the amount of fines in the sediment load. The presence of abundant sheet sandstone deposits decimeters to a few meters thick, showing very little evidence of channelization and no macroforms, but characterized by fining-upward cycles a few meters thick or less (commonly only a few decimeters) and interbedded with playa mudflats with abundant evidence of desiccation (mudcracks, evaporite nodules), is now regarded as a definitive indication of ephemeral, flash-flood sedimentation (Sects. 8.2.15, 8.2.16). These observations have now been incorporated into the terminal fan model of Kelly and Olsen (1993; Sect. 8.3). This style of sedimentation is illustrated in Fig. 12.2B.

Feeder streams along the margins of modern arid basins commonly occupy narrow, steep-sided canyons. In the arid southwestern United States they are termed box canyons or arroyos. The distinctive shape is probably generated by erosion during sudden floods in unvegetated areas (Bendix 1992). Such channels are rarely preserved (an example is illustrated in Figs. 6.11 and 6.12), but arroyo erosion may deliver large quantitites of unconsolidated or loosely consolidated sediment, including pedogenic nodules, into the dispersal system (Marriott and Wright 1993).

12.6.3 Bedforms and Cycles

Jones (1977) suggested that the style of preservation of bedforms in fluvial deposits should reflect the discharge regime of the river in which they were formed. He classified discharge regimes into five broad types, as noted earlier in this chapter. Critical parameters are the ratio of maximum to minimum discharge and the rate at which discharge changes. Bedforms have a characteristic "lag" or "relaxation time", which is the time required for them to respond to changes in flow conditions. This time will be much longer for large forms because of the volume of sand to be moved and, in fact, very large dunes in major rivers, such as the Brahmaputra, are unable to undergo continual adjustment to changed flow conditions and tend simply to be abandoned during the falling stage. Bedforms are rarely in perfect equilibrium with flow because of constantly changing conditions, but in general, the type of bedform assemblage and the various styles of bedform modification should provide clues as to the discharge regime. Jones (1977) proposed several models for field testing, but his work has not been systematically followed up by subsequent researchers. Following are some points regarding the generation and preservation of features sensitive to discharge regime.

Is evidence of the discharge regime likely to be preserved in the sedimentary record? Rising-stage changes are modified by peak flows and have low preservation potential. Falling-stage changes are more preservable, taking into account the effects of bedform lag. Jones (1977) suggested that "sand waves" (also termed "linguoid bars") are most likely to preserve differences in discharge regime because of their relatively long lag time. These correspond to what we would now term medium- to large-scale two-dimensional dunes (Ashley 1990; see Sect. 5.3.1). Most three-dimensional dunes are small enough that they are able to respond rapidly to changes in discharge, and differences between high and low stage are not as likely to be preserved.

Rivers characterized by short, sharp floods, such as the Tana (Norway), have rapidly falling discharge, during which the larger bedforms may be abandoned. They may become emergent, or may be draped by smaller forms deposited by low-stage runoff flowing in small channels eroded between the larger forms. Flow around emergent bedforms may erode part of some avalanche faces and deposit cross-bed sets at a high angle to the dip of the major foreset. The final abandonment surface becomes a surface of reactivation during the next flow cycle. The contrast between large, abandoned forms and the small, low-stage modifications may survive into the sedimentary record as evidence of this distinctive discharge regime.

In cases of less rapid fall in discharge, a discharge style displayed by such large rivers as the Brahmaputra and the Mississippi, dune scale may have time to become adjusted to reduced flow conditions, leading to the superimposition of smaller forms on the large dunes. Small eddies in the lee of the superimposed forms are likely to erode the crest of the major form, generating convex-up erosion surfaces that are buried by avalanching as the superimposed forms continue to migrate. In perennial rivers with slow or minor changes in discharge, such as the Platte and Loup rivers, complete reworking of the bedforms can occur in any channel that remains active. Channel migration and avulsion events therefore result in the preservation of dunes displaying a wide range of scales formed under varying discharge conditions, with no clear preserved signal of the discharge regime.

Application of these ideas to ancient fluvial deposits has not, to this writer's knowledge, been attempted. However, the concepts would appear to be worth testing in conjunction with the other climatic criteria discussed in this chapter.

In channels deep enough for macroforms to be developed, accretion and gradual shallowing of the channel at any given depositional site lead to a progressive lowering of flow velocity and shear stress, resulting in a decrease in grain size and a predictable change in the style of bedform generation. This has long been known to occur on point-bar surfaces and has formed the basis for the depositional model of the fining-upward cycle, and for computer models of meandering-river sedimentation under perennial flow conditions (Allen 1970b; Bridge 1975; Sect. 6.7). Are there patterns in the style of macroform preservation that could be interpreted in terms of discharge regime?

The original descriptions of deep, perennial, sand-bed braided rivers, such as the Brahmaputra (Coleman 1969) and the South Saskatchewan (Cant and Walker 1978), did not recognize the three-dimensional nature of the large sand flats, and it is now known that they develop by lateral, oblique, downstream, and even upstream accretion (Bristow 1987; Sect. 6.6). Cosets of downstream-climbing bedforms are common (Banks 1973) and were first observed in ancient deposits by Allen (1983a) and Haszeldine (1983a,b). Upward fining and an upward decrease in the scale of cross-bedding have been observed in some of these deposits. Much work remains to be done to document the details of the architecture and vertical profile of these macroforms under different flow conditions, and at present there are inadequate data for Jones' (1977) ideas regarding discharge regimes to be tested.

Luttrell (1993) suggested that ephemeral streams are characterized by a paucity of bedforms formed at intermediate flow strengths, notably two-dimensional dunes (lithofacies Sp). She suggested that "the association of high-energy structures grading vertically and laterally into waning-flow structures reflects ephemeral flood events." It is a matter of observation that fluvial systems containing abundant planar-tabular cross-bedding (Sp) are predominantly shallow perennial streams, or at least show steady summer discharges, the modern Platte and William rivers being excellent examples (Smith 1970; Blodgett and Stanley 1980; Smith and Smith 1984; Sect. 8.2.12). Planar-tabular cross-bedding also forms a significant component of many deep, perennial, sand-bed braided rivers, such as the modern South Saskatchewan, where fields of three-dimensional dunes commonly occupy the deeper channels and large, flat-topped two-dimensional dunes develop in shallower reaches (Cant and Walker 1978; Sect. 8.2.13). This river was the focus of a useful discussion of channel aggradation and the paleocurrent characteristics of different bedform types by Cant and Walker (1978). Flume observations indicate that at any given flow depth two-dimensional dunes form at lower flow velocities than three-dimensional forms (Sect. 5.3.1).

Flood cycles in ephemeral streams commonly consist of sets of St overlain by Sh or Sr, reflecting high-energy sheet flow followed by a rapid decline in discharge. Planar-tabular cross-bed sets (Sp) may be present, but the large flat-topped bedforms are not as likely to form as in perennial rivers. Reactivation surfaces within bedforms are rare, most forms being deposited rapidly in a single event and then completely abandoned. Many variations on this succession have been described following work in various modern and ancient ephemeral systems (McKee et al. 1967; Miall and Gibling 1978; Williams 1971; Tunbridge 1981, 1984; Sneh 1983; Glennie 1987; Luttrell 1993; Dreyer 1993b; Sects. 8.2.15, 8.2.16).

Godin (1991) documented three scales of autogenic cyclicity in the Westwater Canyon member of the Morrison Formation, in northern New Mexico, a unit probably formed under high-energy ephemeral flow. As discussed earlier, this unit contains the distinctive "hollow" architectural element formed by scour at channel confluences (Cowan 1991). The three types of cycle are related to macroform accretion, channel filling, and overall sand body aggradation (Sect. 10.3.2). A common lithofacies association is that between horizontally laminated sandstone (Sh) and low-angle cross-bedded sandstone (Sl). The low-angle cross-bedding is commonly asymptotic with reference to upper and lower set bounding surfaces. Parting lineation may be present, with an orientation parallel or oblique to the dip of the cross-bedding. Some occurrences of lithofacies Sl represent plane beds deposited on initially dipping surfaces, such as scour hollows. In other cases the lithofacies represents distinct bedform geometries. The washed-out dunes and humpback dunes that occur at the transition between subcritical and supercritical flow typically form this lithofacies (Saunderson and Lockett 1983).

Observations of modern floods by Frostick and Reid (1977) and Stear (1985) indicated that they occur in pulses. The passage of storms across different parts of a catchment area triggers waves of high-energy flow from different tributary systems which arrive in the depositional area at different times. This may be recorded in the deposits in the form of changes in the thicknesses of flood laminae (Frostick and Reid 1977) or as heterogeneous lithologies (Stear 1985).

12.7 Overbank Fines: The Significance of Bedding and Minor Sedimentary Structures

Overbank areas are the main sites for the accumulation of various climatically sensitive chemical and biochemical deposits, which are discussed in a later section. However, the detrital sediments can also yield some clues as to discharge regime, which depends on climate (Fig. 12.3).

Interfluve areas flanking the large channels of perennial, seasonal, or ephemeral rivers are areas that are only spasmodically affected by high-discharge events. In tropical-humid areas the rivers may be flanked by raised peat bogs that keep out overbank flows (McCabe 1984) or by swamps that absorb them (Stanistreet and McCarthy 1993). In temperate-humid and arid settings rare flows may spread over the floodplains, and normal overbank processes lead to the generation of crevasse channels and crevasse splays (Sect. 7.2), the geometry and composition of which convey little useful information regarding climate. Lenses of gravel may be deposited in areas close to the channel by rare flood events (Costa 1974) or by ice push (Martini et al. 1993).

The most common detrital sediments on floodplains in all climatic settings are fine silts and muds. These are normally finely laminated, but commonly the lamination is disturbed by pedogenesis, by roots, or by invertebrate bioturbation. The style of lamination in these fine-grained clastics normally reveals nothing about climate, although Broadhurst (1988) developed an argument suggesting that laminated siltstone-claystone couplets in a Carboniferous succession in Britain are seasonal in origin. He recorded some 45 couplets in 9 m of section (average couplet thickness 20 cm). The siltstone-claystone units are interpreted as lacustrine or overbank deposits, and interbedded sandstones as distal crevasse-splays. A bivalve colony shows evidence of being buried by the deposits and not replenished. Given the 20-year average life span of modern bivalves, the persistence of the colony through eight or nine of the overlying couplets suggests that the couplets may be seasonal. The sedimentation rate calculated from this evidence is 30 cm/year, which Broadhurst (1988) regarded as high, but in fact it is in agreement with the average rate for seasonal deposits (group 5 of the depositional hierarchy, Chap. 3). These deposits correlate with and are interbedded with coals, which indicates permanent saturation, and which appears to conflict with the seasonality suggested by the couplets. Broadhurst (1988) suggested that the coals were probably formed in permanent swamps, while the rivers were affected by seasonal discharge fluctuations.

In many arid systems fine-grained sediments deposited on distal interfluve mudflats are interbedded with lacustrine or playa margin sediments (Figs. 7.27, 7.28). Layers of very fine grained sandstone and siltstone may display a millimeter-scale lamination and commonly show graded bedding, indicating rapid deceleration of flow into very shallow standing bodies of water (Smoot 1983; Flint 1985; Mertz and Hubert 1990; Hartley 1993). Evaporite crystallization shortly after deposition may disrupt bedding (Mertz and Hubert 1990). Couplets of interbedded clastic and chemical sediments may reflect seasonal variations in precipitation, evaporation, and temperature, as in the siltstone-carbonate cycles in parts of the Chinle Formation (Triassic) of the Colorado Plateau (Dubiel et al. 1991). Clastic-chemical cycles also develop in lake-margin settings under the influence of longer-term climatic rhythms, as discussed in a later section.

Several types of minor sedimentary structure carry useful climatic signatures (Fig. 12.3). Desiccation features are common in hot, arid fluvial systems, including large mudcracks and mud curls on bedding planes (Miall et al. 1978; Tunbridge 1981, 1984; Rust and Legun 1983; Smoot 1983; Flint 1985; Stear 1985; Glennie 1987; Mertz and Hubert 1990; Hartley 1993). Cracks may be up to at least a meter deep and a decimeter or more in width. Even larger cracks, up to 1.5 m deep, have been recorded in evaporitic eolian sandstones (Kocurek and Hunter 1986). They are similar to ice-wedge casts formed in frigid climates (Bryant 1983), and both processes may form polygonal crack networks on bedding-plane surfaces. Supplementary evidence might be needed to differentiate such features. Rip-up clasts and intraclast breccias are very common features of hot, arid environments (even in high intertidal settings) because of the rapid lithification of freshly deposited muds caused by desiccation (Smoot 1983; Stear 1985).

Parts of the arid Lake Eyre drainage basin of Australia are characterized by an unusual combination of fluvial styles. A series of seasonally active anastomosed channels containing permanent pools of water cross a broad system of shallow braided channels which appeared to Rust (1981) to be relict channels dating from an earlier, wetter, climatic phase. This change in fluvial style has, in fact, long been regarded as a classic example of "river metamorphosis" in response to climate change (Schumm 1969). However, detailed studies by Nanson et al. (1986) and Rust and Nanson (1989) showed that the braided channels and intervening bars are composed of mud aggregates forming sand-sized particles that retain their coherence during fluvial transport and function as sand grains. The high "sand" load causes the rivers to develop the braided pattern during floods. The clay aggregates are interpreted as peds formed by volume changes of the floodplain mud in response to wetting and drying cycles in the arid to semiarid climate. Earlier work had suggested that the presence of salts is required for aggregate formation, but Nanson et al. (1986) disputed this, pointing out that the evaporite concentration is low. Flume experiments indicated that mud aggregates can form flow-regime type bedforms, and it was suggested that mud beds in the ancient record should be reexamined, as some may be mud braids, not floodplain deposits (Rust and Nanson 1989). In the modern Lake Eyre basin the anastomosed channels are deeper channels which have cut through lower levels in the muds, where the mud has greater cohesion, and into an underlying sand bed. Ancient examples of pedogenic mud aggregates have now been described by Rust and Nanson (1989), Ekes (1993), and Talbot et al. (1994).

12.8 The Significance of Color

The significance of color in nonmarine sediments has been the subject of much debate (Van Houten 1973; Walker 1967; Turner 1980) that will not be repeated here. Drab colors (gray, green, brown) are generated largely by ferrous oxides or carbonates (e.g., siderite) in reducing environments, whereas red, yellow and orange are created by ferric oxides, notably hematite, in oxidizing environments.

Studies in hot arid environments and in humid tropical regions show that red beds can be generated by diagenetic processes following deposition (Walker 1967, 1974). Yellow and brown iron oxides, in initial stages of hematite production, are present above the water table in the modern alluvium of the Colorado River on the arid shores of the Gulf of California, and well below the water table in alluvium in Puerto Rico, under hot, humid conditions. Therefore, red beds are not necessarily an indication of arid desert climates, as has long been thought. Turner (1980) summarized mineralogical, diagenetic, and grain-size data relating to post-depositional generation of red-bed coloration, drawing on the work of Walker. He distinguished between (a) a desert-evaporite red-bed association in which red beds are associated with eolian sands, desert fluvial sediments, and evaporites formed in playas and inland sabkhas, and (b) a moist-climate red-bed association in which red beds are interbedded and interfinger with coal-bearing strata.

The oxidation state is commonly set during early diagenesis rather than at the time of deposition, and so caution must be used in making environmental interpretations from color. However, many red units contain evidence of an arid or semiarid climate, such as pedogenic carbonates and facies associations indicating ephemeral deposition, whereas drab units are typically associated with coals or carbonaceous shales, indicating saturated environments of deposition and a humid climate. The preservation of organic matter is a function of organic productivity, burial rate, the depth of the water table, and the rate of groundwater movement.

The main attribute controlling color differences is the level of the water table. An oxidizing environment is favored by a low water table, such as occurs in arid climates and those with seasonal dry periods. Such a setting favors pedogenic carbonate development (Retallack 1981, 1986). A reducing environment occurs in saturated environments and is necessary for the preservation of plant matter. These differences are commonly preserved following deposition. In fact, different facies associations with the expected contrasting colors may be interbedded within the same basin fill, indicating variations in water table across the basin during or shortly after deposition (e.g., Besly 1988). The existence of these complex color patterns also indicates that, in such cases at least, colors are set during early diagenesis and not subsequently modified.

Although colors reflect levels of water saturation, the climatic implications of this may not be easily determined, especially where saturation levels vary across a basin. This is discussed further in a later section.

12.9 Associated Clastic, Chemical, and Biochemical Sediments

Many other nonmarine deposits constitute much clearer climatic indicators than do fluvial lithofacies. Lakes are sensitive indicators of climate because of the delicate balance that commonly exists between climate and the water level and chemistry of lakes (Katz 1990). Large-scale eolian depositional systems may indicate aridity, and large ancient ergs preserve evidence of persistent wind patterns that may be related to the patterns of global atmospheric circulation (Parrish and Peterson 1988). Chemical and biochemical sediments that occur in interfluve environments, particularly paleosols, coal, and evaporites, also contain useful signatures of the climate at the time of deposition. It is not the purpose of this chapter to review these climatic indicators at length, most of which have been extensively reviewed elsewhere. However, some brief comments are included here in order to provide the basis for an integrated interpretation of climate from an ancient alluvial suite.

12.9.1 Coal

McCabe (1984) stated:

> "The necessary conditions for peat accumulation can ... be regarded as a balance between plant production and organic decay. Both are a function of climate. Plant production is fastest in hot humid areas and is slowest in dry or cold areas. Organic decay is principally a function of

temperature, being slower in cooler climates. ... today the vast majority of peat is accumulating in cool climates – mostly between 50° and 70° N. Most tropical rain forests are not sites of peat accumulation because the organic matter rapidly deteriorates on the forest floor. The major area of peat formation in the tropics today occurs in parts of Southeast Asia where the rainfall is over 2 m year^{-1} and where substantial precipitation occurs throughout the year. Because evaporation is also a function of temperature, it is wrong to assume that coal is an indicator of high rainfall; much of the vast peat swamps of Siberia and Canada receive less than 50 cm of precipitation per year. ... Correlation of palaeolatitides with areas of coal deposition ... shows that coal has been deposited at all latitudes from the equator to polar regions, with the majority forming in mid-latitudes. ... Coal itself is not a good palaeoclimatic indicator, although characteristics of a coal may be related to palaeoclimate."

McCabe and Parrish (1992) reviewed the global distribution of coal in Cretaceous strata. They noted that coals require nonsaline wetlands for formation. Continuous wetness is required to permit growth and the replenishment of the nutrient supply for plants. Wetness is also a necessary condition for the anoxic environment required for preservation. They stated:

"Climate directly influences plant productivity and preservation of plant matter and therefore determines the amount of organic accumulation. High water tables, necessary for the accumulation of peat, tend to occur in climates where precipitation exceeds evaporation, especially where rainfall is evenly distributed throughout the year (Ziegler and others 1987). Seasonal climates, by contrast, often lead to such large annual fluctuations in the water table that organic matter accumulated during the wet season decays during the dry season, and no net accumulation of organic matter takes place (Ziegler and others 1987)."

McCabe and Parrish (1992) pointed out that an excess of precipitation over evaporation does not necessarily imply high rainfall. Such a condition may develop in cool climates with modest rainfall.

Very little vegetation cover may suffice to generate coaly layers in a fluvial deposit. Hubert and Filipov (1989) described the debris-flow-dominated sediments of Owens Valley, California, an arid area with scattered scrub vegetation. Many of their sections contain layers of plant debris, which would become coaly streaks following compaction and diagenesis. Therefore, climatic interpretations should not be attempted from the presence of coal unless the accumulations are substantial.

12.9.2 Paleosols

Pedogenesis takes place on any interfluve surface not receiving frequent additions of sediment from overbank flow. Soil types depend on substrate conditions, including chemistry and water saturation, and undergo continual modification with time. Two major concepts are useful in understanding the variability of paleosols. A "catena" is a soil unit showing lateral variations reflecting differences in relief and drainage, for example, the differences between the immature soils developing on a well-drained levee undergoing periodic addition of increments of sediment, the soils of a saturated backswamp, and the mature soils of an adjacent, well-drained, upland region (Wright 1992). "Maturity" refers to the state of completion of the pedogenesis process, and therefore is related to the age of the soil. Bown and Kraus (1987) pointed out that the concept of the catena was introduced primarily to encompass variations in modern soils related to their position on hillslopes, whereas most paleosols of interest to sedimentologists occur within the floodplain, and differences within and between individual soils depend on maturation time and the amount of dilution of the soil with clastic sediment. They found that these factors depend primarily on proximity to active fluvial channels, and they suggested that a different terminology is required to accommodate such observational and interpretive differences. They introduced the term "pedofacies", which they defined to encompass the "laterally contiguous bodies of sedimentary rock that differ in their contained laterally contiguous paleosols as a result of their distance (during formation) from areas of high sediment accumulation." These ideas are discussed further in Sect. 7.4.2.

The dependency of soil type on substrate and groundwater chemistry, on water saturation, and on the rate of groundwater flow through the soil introduces climatic factors into soil genesis, although care must be taken to distinguish depositional from diagenetic conditions. Thus ferric nodules and concretions typically indicate well-drained, oxidized soils; siderite nodules indicate neutral to alkaline, waterlogged soils; calcareous nodules indicate well-drained alkaline soils (Retallack 1988). Retallack (1986) reached several generalized conclusions based on his studies of the Upper Eocene and Oligocene succession of Badlands National Park, South Dakota. Several features reflect rainfall and water-table factors. For example, calcic soils characterize dry subtropical climates, where evapotranspiration exceeds precipitation for most of the year, and the depth to the top of the calcic horizon is related to mean annual rainfall, becoming deeper in wetter climates. Clay production in soils increases with humidity. The type of clay also varies with climate, with kaolinite indicating humid climates and illite more common in more arid climates. This reflects the leaching of soluble cations, which decreases with aridity. Quartz/feldspar ratios in soils also indicate humidity, with feldspar being destroyed by acidic weathering in humid climates. Indicators of temperature are harder to find. In general, soil becomes redder in warmer climates, but the amount of rainfall and the age of the soil complicate this relationship.

However, the climatic signature may be very difficult to interpret. The difficulties discussed above with regard to the interpretation of color apply equally to paleosols. Wright (1990) pointed out, with regard to the calcretes that develop in semiarid areas, that variations in exposure time (determined by subsidence and channel return time) and local variations in rainfall and calcium content all lead to markedly different rates of soil development.

Given these difficulties, some generalizations may be made, using the soil-type terms and descriptions adapted by Retallack (1988) from the U.S. Department of Agriculture classification. Aridisols develop in arid climates, and are characterized by light coloration and a thin calcareous layer close to the surface of the profile, and may include pedogenic gypsum or other evaporite minerals. Vertisols contain abundant swelling clays (smectite) to depths of a meter or more and may display slickensides or clastic dykes. This type of soil develops in climatic settings with seasonal variations from saturated to dry. Oxisols are a type of soil developed in humid tropical settings and consist of a thick, well-differentiated to uniform profile with a clayey texture, red, highly oxidized subsurface horizons, and an absence of easily weathered detrital minerals. Many other soil types are listed by Retallack (1988).

Mack et al. (1993) developed a paleosol classification better adapted to the study of ancient rocks than those developed for agricultural purposes, and Mack and James (1994) discussed the paleoclimatic significance of this classification.

12.9.3 Evaporites

Evaporite minerals are deposited within fluvial substrates by the evaporation of groundwaters and on the surface of playa mudflats during the evaporation of sheet floods. Thin beds, laminae, nodules, and individual crystals (or crystal casts) of evaporite, particularly gypsum and halite, are very common in the deposits of arid fluvial systems, especially in the distal regions where the braidplain or terminal fan merges imperceptibly into a playa lake or arid tidal flat (Miall et al. 1978; Smoot 1983; Glennie 1987; Mertz and Hubert 1990; Dreyer 1993b). Mertz and Hubert (1990) reported a gradual increase in humidity during deposition of the Triassic-Jurassic Blomidon Formation of Nova Scotia, based in part on the upward change in intergranular cements from halite and gypsum to calcite.

12.9.4 Eolian Interbeds

Many fluvial deposits formed in arid settings contain interbedded layers or lenses of eolian sand. Large-scale interbedding may occur at the margins of ergs, or where these are crossed by ephemeral river systems, but minor eolian units can occur in any fluvial system where aridity and wind strength are sufficient for the eolian reworking of fluvially transported detritus. Glennie (1970) published a classic study of desert sedimentary environments that contains much descriptive information on fluvial-eolian interactions, and a detailed local study was presented by Langford (1989).

Langford and Chan (1989) recognized four major facies in mixed eolian-fluvial systems, based on their study of the Cutler-Ceda Mesa Sandstones (Permian) of Utah. 1. Large-scale cross-stratified eolian sandstone; 2. the deposits of wet interdune areas, consisting of interbedded lenses of sandstone and limestones. The sandstones are structureless or reworked into wave ripples. Gypsum crystal casts and bioturbation are common; 3. Fluvial channel sandstones. Common lithofacies are St and Sh, which may be organized into accretionary macroforms (elements DA and LA), consisting of bundles of trough cross-bedded sandstone separated by third-order surfaces; 4. overbank interdune sandstones and shales in thin sheets and lenses up to 2 m thick. Mudcracks and bioturbation are common. Sandstones consist of lithofacies Sr, Sh, and Sm. Small channel forms up to 2 m thick may be present, formed from minor overbank flows. Isolated eolian dune lenses up to 1.5 m thick may also be present, formed by dune migration over desiccated interfluve surfaces between floods. Langford and Chan (1989) traced some fluvial flooding surfaces over areas of > 400 km^2, indicating extensive flooding episodes. Bioturbation and bleaching indicate incipient pedogenesis, but true calcretes are rare. Dreikanters (ventifacts) may be present on eolian deflation surfaces, but they require lengthy exposure to the wind to develop and therefore may not be common in areas dominated by fluvial sedimentation.

Bromley (1991a,b) mapped an eolian rippled-siltstone unit interbedded with braided sandstones in the Kayenta Formation (Jurassic) of western Colorado. Similar mixed facies were described by Luttrell (1993), based on her regional study of this unit.

Minor eolian lenses interbedded with fluvial conglomerates or sandstones have been recorded in several studies. For example, in his study of the Triassic Budleigh Salterton Pebble Beds of SW England, Smith (1990) noted sand lenses with distinct texture and petrography. Unlike the fluvial sands in the system, they contain no mica and are well sorted and free of pebbles. Shepherd (1987) documented eolian lenses within the modern lateral-accretion deposits of Rio Puerco, New Mexico, distinguishing them from fluvial units on the basis of textural analyses. Clemmensen and Tirsgaard (1990) recognized a distinctive type of bounding surface, which they termed "sand-drift surfaces", separating eolian from fluvial deposits.

Flint (1985) noted an absence of eolian sands in his study of the arid mid-Tertiary Pacencia Group of Chile. He attributed this to armoring of the sediment surface by evaporite crystallization, which prevented eolian deflation.

12.9.5 Palustrine Limestones

Freshwater limestones containing a pedogenic overprint are present in some alluvial successions, and indicate the presence of temporary ponds in the overbank environment (Platt and Wright 1992). They occur in a variety of climatic setttings, the nature of which may be indicated by associated features, such as evaporite nodules, desiccation features, karst solution, or thin coals.

12.10 Contrasting Climatic Indicators

In many recent regional studies sedimentological analysis has suggested that different parts of a sedimentary basin were characterized by different saturation levels, suggesting variations in the balance between rainfall, runoff, infiltration, and evaporation. In other cases it can be argued that the basin and its fluvial source area were characterized by different climates. Such differences are commonly observed in present-day sedimentary basins, but their presence requires subtlety and care in the interpretation of climatic indicators.

A good example of apparently conflicting climatic indicators occurs in the Late Carboniferous to Early Permian coal-bearing succession of central England and was highlighted by Besly (1988). He documented a diachronous interbedding between an alluvial coal-bearing succession, an alluvial-plain red-bed association, and an alluvial-fan red-bed association. The transition from gray to mottled to red between these assocations indicates improved drainage of the floodplain. Additional details on the paleosol catena that indicates a lateral transition from dry (red) to wet (gray) beds, corresponding to fan progradation, was provided by Besly and Fielding (1989). According to Besly (1988), "the transition from the deposition of coal-bearing sediments to that of red beds resulted from an improvement in drainage conditions at the site of deposition. This was initially achieved by the progradation of a topographically elevated, well-drained floodplain area over poorly drained alluvial backswamps. Repeated progradation led to a progressive drop in groundwater table levels, possibly reinforced by seasonal fluctuations ... which eventually led to the accumulation of a continuous red-bed sequence." Besly (1988) commented that "while the Late Carboniferous red-bed sequences do allow climatic reconstruction, they do not in all cases indicate aridity." He noted that the fairly rapid onset of more arid conditions during the Westphalian D and Stephanian corresponds to the acme of coal formation in basins within the mountain belt to the southeast. This contrast suggests the growth of a rain shadow in the lee of the Variscan mountains.

Blakey and Gubitosa (1984) found what appeared to be conflicting evidence of climate in the Chinle Formation (Triassic) of the Colorado Plateau. An abundant and diverse fossil fauna and flora suggests a humid climate, whereas the presence of caliches and other signs of overbank aridity suggest a dry climate. Their suggested explanation is that the rivers were perennial, fed by a humid source area, but that the depositional basin itself was more arid, with at least a seasonal dry period dominating conditions on the floodplains. These suggestions have recently been confirmed and amplified by Dubiel et al. (1991). The juxtaposition of humid upland source areas and arid basins is a common one, especially where the upland region generates orographic rainfall, leading to a permanent rain shadow downwind. The San Joaquin valley in California and the San Luis valley of Colorado are good modern examples of such contrasting climatic regimes.

DeLuca and Eriksson (1989) distinguished two contrasting fluvial assemblages in the Chinle Formation of New Mexico. They suggested that proximal, basin-margin rivers forming a tranverse drainage system were ephemeral, whereas the main basin-center rivers were perennial. Their evidence for perennial flow is not strong, as discussed earlier in this chapter, consisting as it does mainly of comparisons with the mainly perennial rivers that formed the basis for the Platte and South Saskatchewan fluvial models, with which their rocks are compared. The evidence for ephemeral flow and arid conditions is much stronger, consisting of many of the sedimentological attributes discussed in this chapter. However, the explanation DeLuca and Eriksson (1989) offered for the contrasting fluvial styles is not unreasonable. They suggested that in the basin center the water table was closer to the surface, in part because of groundwater recharge which kept the rivers flowing during seasonal dry periods. Hobday et al. (1981) noted a similar climatic differentiation in present-day south Texas, where "semiarid, caliche-soil, chaparal conditions commonly prevail on uplands while nearby contemporaneous riverbottoms sustain a humid subtropical assemblage of bald cypress and willows, all owing to different microclimate as dictated by water table conditions and associated soil moisture regimes."

Luttrell (1993) suggested a downstream change from perennial to intermittent in the rivers that deposited the Kayenta fluvial system of the Colorado Plateau. Again, the evidence for perennial flow seems weak, based mainly on the scale of the channels and macroforms in the proximal part of the basin (a criterion discussed earlier in this chapter). However, the downstream change from perennial to ephemeral is one that characterizes many arid basins in modern deserts, for example, the inland basins of western China, where snow-fed mountain streams enter hot, lowland deserts, such as the Taklamakam Desert of Tarim Basin.

Miocene fluvial and lacustrine deposits of the Madrid Basin, Spain, reveal numerous differences between northern and eastern margins of the basin (Alonso Zarza et al. 1992). In part, these differences reflect differences in sediment sources and tectonic activity, but it also appears that there were subtle climatic differences between the two areas that may, in part, have been caused by topographic effects. The northern margin may been under the influence of a relatively dry climate and is characterized by alluvial fans with abundant debris flows and immature calcrete paleosols. The eastern margin may have been in a more humid setting and was the site for the accumulation of sandy braided stream deposits with generally more mature paleosols that show pedofacies relationships to the channels.

12.11 The Interrelationship Between Tectonics and Climate

The importance of the orographic effect on the control of local climates has long been known. For example, it is common knowledge that the temperate rain forests of the west coast of North America depend in large measure on the adiabatic rainfall generated from westerly winds as they rise over the Cordilleran mountains. The presence of a significant rain shadow downwind, in such areas as the upper Fraser Valley (Kamloops) and the Okanogan Valley of British Columbia, is equally well known. Another example is the Dead Sea Rift (Manspeizer 1985). The orographic effect of the uptilted rift shoulder west of the rift is the generation of high rainfall; some of this flows west in perennial streams draining into the Mediterranean, but it also feeds ephemeral streams draining into the Dead Sea rift. The difference in annual rainfall between the rift crest and the rift valley is substantial, at > 800 mm/year on the rift shoulder, in the vicinity of Jerusalem, and < 100 mm/year in the Dead Sea area.

Knowledge of this type of effect is readily applied to regional paleogeographic and paleoclimatic interpretations. For example, Besly's (1988) suggestion of the rainshadow effects of the Variscan Mountains (Late Paleozoic) in northwest Europe was discussed earlier in this chapter. Miall (1984b) invoked regional tectonically-related climatic variations as a possible control on the distribution of fluvial lithofacies assemblages in the Paleogene fluvial deposits of Arctic Canada. He suggested that the unusual abundance of coals (and of fossil vertebrates) in central Ellesmere Island could be an indication of a locally humid climate related to adiabiatic rainfall generated by the effects of regional uplifts to the east and north of the depositional basin, whereas lithofacies in Eclipse Trough, in the area of northwest Baffin Island, indicate a predominance of ephemeral rivers and a drier climate, possibly related to the passage of westerly winds over the broad interior plains of the Arctic Platform.

Evidence from fluvial lithofacies, coupled with floral, isotopic, and other evidence, indicates a gradual global cooling during the Cenozoic. Major regional uplift of such areas as western North America and Tibet has also been documented. This has led to a lengthy and complex debate regarding the role of uplift in generating the cooling trend. This debate is largely beyond the scope of this chapter, but because discussions of fluvial lithofacies assemblages and petrology have formed part of the evidence, some relevant parts of the controversy are briefly summarized here.

Ruddiman and Kutzbach (1990) summarized extensive paleobotanical evidence for widespread global cooling during the Cenozoic and developed numerical climate models to analyze global climates with and without the effects of large-scale plateau uplift. They pointed out that areas such as southern China, Burma, and Thailand, which did not undergo significant uplift during the Cenozoic, show no botanical evidence of cooling during the Cenozoic, which would seem to confirm the tectonic causes of cooling. However, they did not point out that these areas are located in equatorial regions, which are not necessarily strongly affected by global climate change. Their field evidence and modeling data support the concept of plateau uplift as a cause for global cooling, but the data are inadequate to verify this as a cause (or at least a trigger) of the Late Cenozoic glaciation. Regional uplift is also cited as a cause of increased aridification because of rain-shadow effects and because of changes to global circulation patterns.

Molnar and England (1990) stated: "Late Cenozoic uplift has been inferred for mountain ranges throughout the world, yet globally synchronous changes in plate motions, if they have occurred, have been small. Climate change, on the other hand, has been a global phenomenon." Indicators of cooling climate in sediments include increased grain size, increased detrital volume and sedimentation rates reflecting increased erosion rates (Stanley and Wayne 1972; Stanley 1976), changes in fauna and flora, and changes in oxygen and carbon isotope ratios. These changes could mean regional or global climate change, or uplift without climate change. Rapid uplift in the past 2 million years has been interpreted for the Himalayas and elsewhere, but Molnar and England (1990) suggested that this may be incorrect. Plate tectonic reconstructions of India-Asia convergence suggest that the Tibetan Plateau would have risen gradually during the past 50 million years. This would have set off a long-term slow cooling, amplified by feedback effects, that was then enhanced in the Late Cenozoic by Milankovitch processes, leading to glaciation.

Hoffman and Grotzinger (1993) speculated that the climatic belt in which a rising orogen develops actually influences the tectonic style of the orogen and the architecture of the adjacent foreland basin. Areas of high precipitation, as in monsoonal belts, are characterized by rapid erosional unroofing, leading to rapid uplift, deep erosion, and the development of a foreland basin overfilled with nonmarine sediments (but of low preservation potential). The Grenville orogen (1.1 Ga) of eastern North America may be cited as a possible example. This contrasts with orogens such as the Alleghenian, which may have developed in arid climates. Erosional unroofing would not compensate uplift, leading to preservation of the fold-thrust belt and an underfilled foreland basin containing deep-water sediments. Sinclair and Allen (1992) pointed out the feedback effects of elevating a fold-thrust belt, resulting in greater orographic rainfall, increased rates of erosional unroofing, and consequently enhanced uplift rates.

In a recent lengthy review, Eyles (1993) demonstrated convincingly the importance of tectonic uplift in generating the climatic cooling necessary for the triggering of widespread continental glaciation throughout geologic time. Many of the major Phanerozoic and Proterozoic glacial episodes are correlated with times of regional continental extension and rifting. However, glaciofluvial deposits represent a very small fraction of the glacially generated clastic successions in the ancient record.

12.12 Orbital Forcing

12.12.1 Sedimentary Evidence of Orbital Forcing

Rhythmicity in glaciation is now attributed to orbital forcing, the so-called Milankovitch effects, named after the Serbian mathematician who first developed the quantitative relationships between the orbital behavior of the earth and its effects on incoming solar radiation. There is now a very considerable literature on the theory and sedimentological implications of orbital forcing, most of which is beyond the scope of this book (e.g., Berger et al. 1984; Fischer et al. 1990; de Boer and Smith 1994a). This section makes only very brief reference to the theory and focuses on the few examples of studies in the nonmarine record where some evidence of the influence of orbital forcing may be demonstrated, or at least postulated.

There are several separate components of orbital variation. The present orbital behavior of the earth includes the following cyclic changes (Imbrie 1985):

1. Variations in orbital eccentricity (the shape of the earth's orbit around the sun); several "wobbles", which have periods of 413, 100, and 54 ka.
2. Changes of up to 3° in the obliquity of the ecliptic, with a period of 41 ka.
3. Precession of the equinoxes. The earth's orbit rotates like a spinning top, with a period of 23 ka. This affects the timing of the perihelion (the position of closest approach of the earth to the sun on an elliptical orbit), which changes with a period of 19 ka.

Each of these components is capable of causing significant climatic fluctuations, given an adequate degree of global sensitivity to climate forcing. For example, when obliquity is low (rotation axis nearly normal to the ecliptic), more energy is delivered to the equator and less to the poles, giving rise to a steeper latitudinal temperature gradient and lower seasonality. Variations in precession alter the structure of the seasonal cycle by moving the perihelion point along the orbit. This changes the earth-sun distance at every season, thus changing the intensity of insolation at each season. "For a given latitude and season typical departures from modern values are on the order of ±5%" (Imbrie 1985). Because the forcing effects have different periods they go in and out of phase. One of the major contributions of Milankovitch was to demonstrate these phase relationships on the basis of laborious time-series calculations. These can now, of course, be readily carried out by computer. The success of modern stratigraphic work has been to demonstrate the existence of curves of temperature change and other variables in the Cenozoic record that can be correlated directly with the orbital variations. For this purpose, sophisticated time-series spectral analysis is performed on various measured parameters, such as oxygen-isotope content or cycle thickness. This approach has led to the development of a special type of quantitative analysis termed cyclostratigraphy (Fischer et al. 1990). Many examples of sedimentary cycles generated by orbitally forced sea-level change or climate change have now been published. The record is particularly clear in successions controlled by Cenozoic glacioeustasy and in certain shallow-marine platform carbonates and lacustrine rhythms.

In the case of nonmarine deposits, the effects of orbital forcing are thought to be much more subtle and difficult to prove. Forcing may be felt through the climatic control of rainfall and evaporation, upon which depends the rate of erosion, the rate of production of detritus and its type, and the rate of sediment transport. de Boer and Smith (1994b) suggested that orbital forcing in nonmarine settings may be particularly important in the 30–40° latitude belt, where the effects on precipitation and evaporation are thought to be the most pronounced. However, the effects are difficult to demonstrate, particularly where tectonism and sea-level change may have introduced complications into the stratigraphic record.

H. Olsen (1990, 1994) described a Devonian fluvial section in East Greenland containing prominent fining-upward cycles that show systematic variations in thickness. Variations in channel sand body thickness are shown in Fig. 12.4. The average thickness of the fluvial fining-upward cycles is 20 m, with a range of 13–30 m. Channel sandstones vary from 2 to 13 m in thickness. Bundles of 3–6 cycles can be defined, as can thicker trends that trace out megacycles. Megacycles range from 79 to 135 m, with an average of 101 m in thickness. H. Olsen (1990, 1994) calculated disharge values for the channels using a suite of geomorphic equations based on channel dimensions. The results are shown in Fig. 12.5 and demonstrate the same cyclic variability. H. Olsen (1990, 1994) argued that the simplest explanation for the rhythmic variations is changes in discharge brought about by orbital forcing of climate changes, and he attempted to fit the pattern of thickness and discharge variations to patterns of orbital variation.

However, Algeo and Wilkinson (1988) offered a warning:

"Despite an often-claimed correspondence between cycle and Milankovitch orbital periods, factors independent of orbital modulation that affect cycle thickness and sedimentation rate may be responsible for such coincidence. For example, nearly all common processes of sediment transport and dispersal give rise to ordered depositional lithofacies sequences that span a relatively narrow range of thicknesses ... Further, long-term sediment accumulation rates are generally limited by long-term subsidence rates, which converge to a narrow range of values for very different sedimentary and tectonic environments (Sadler 1981). In essence, the spectra of real-world cycle thicknesses and subsidence rates are relatively limited, and this in turn constrains the range of commonly determined cycle periods. For many cyclic sequences, calculation of a Milankovitch-range period may be a virtual certainty, regardless of the actual generic mechanism of cycle formation."

Algeo and Wilkinson (1988) studied cyclicity in more than 200 stratigraphic units, determining periodicities from cycle thickness, sedimentation

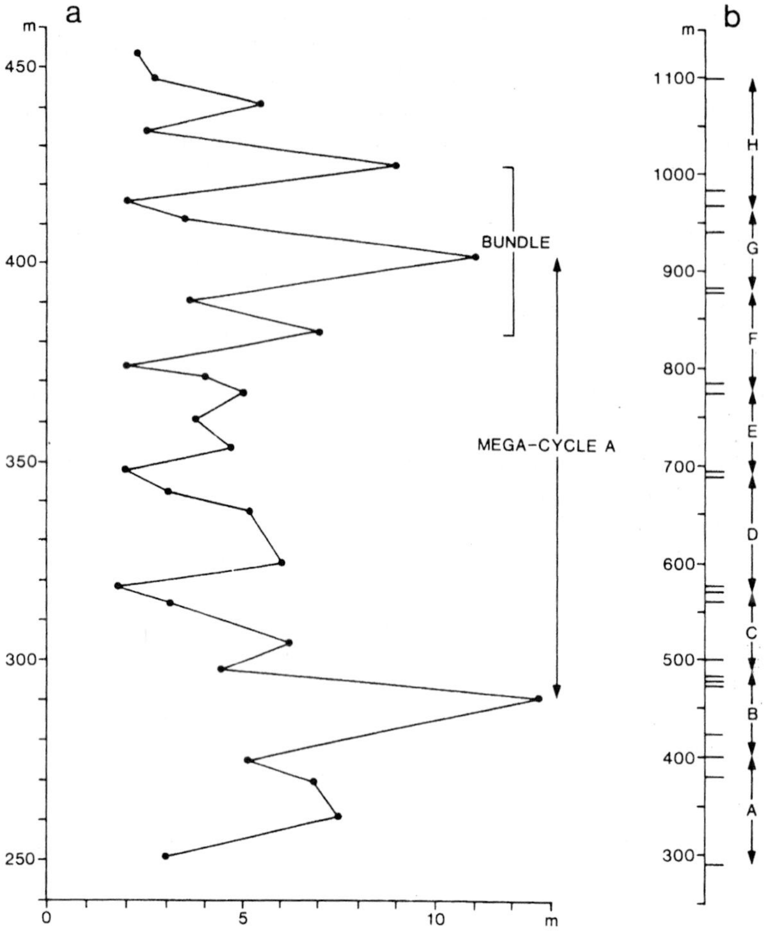

Fig. 12.4. Variations in sand body thickness in a Devonian fluvial section in East Greenland. (H. Olsen 1990)

rate, depositional environment, and age range. They found that calculated periodicities are randomly distributed relative to all the Milankovitch periods except the 413-ka eccentricity cycle. They concluded that "if, in fact, a calculated average period within the broad range of Milankovitch periodicities is not a sufficient test of orbital modulation of sedimentary cycles, demonstration of such control becomes significantly more difficult than hitherto appreciated." Demonstration of a hierarchy of periodicities may be one approach. For example, where there are two orders of cycles with recurrence ratios of 5:1, a precession (20 ka) and eccentricity (100 ka) combination may be indicated. Fischer (1986) discussed examples of this, especially from lacustrine environments, but much work remains to be done to conclusively demonstrate Milankovitch effects in the nonmarine record.

A different type of argument than cycle thickness was used by de Boer et al. (1991) in their postulation of climatic control of sedimentation. They demonstrated that a cyclic alternation of fine and coarse sedimentation persists vertically through about a dozen cycles totaling over 100 m of beds in a proximal foreland basin succession (Fig. 12.6). In proximal areas the cycles are entirely nonmarine and in more distal areas entirely marine, with the facies boundary (shoreline) remaining fixed vertically. These are therefore not transgressive-regressive cycles. The fixing of the facies boundary is attributed to the influence of growth anticlines which stabilized areas of uplift and subsidence, but the absence of transgression and regression suggests that there were no vertical movements of the basin or of sea level during this time period. It therefore seems likely that climatic control of rainfall and sediment yield is the most likely explanation for the observed cyclicity. Calculations of cycle duration suggested an approximate correlation with Milankovitch periodicity.

The Late Paleozoic cyclothems of NW Europe and North America have long been attributed to glacio-

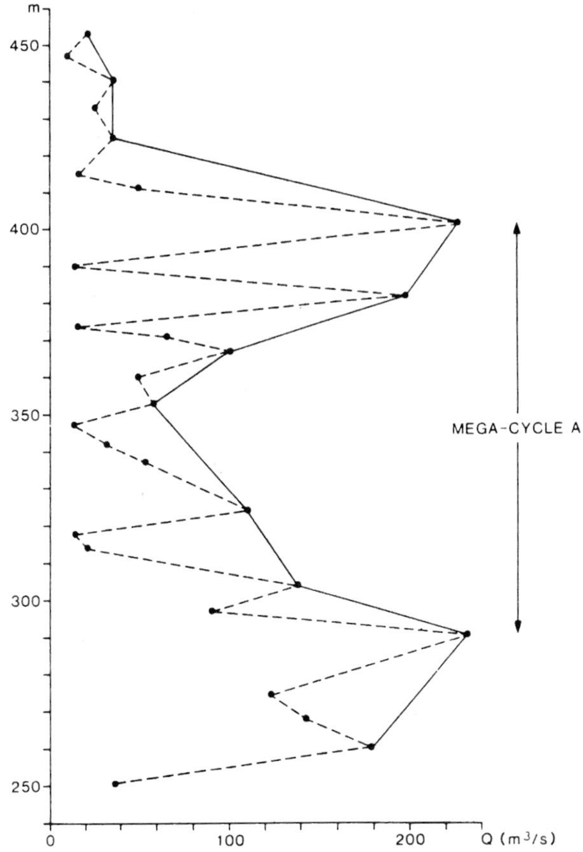

Fig. 12.5. Variations in calculated discharge for sandstone bodies in a Devonian section in East Greenland. *Dashed line* connects individual values for each sand body; it defines a form of short-term cyclicity. Maximum values for each cycle are connected by *solid line* and show a longer-term cyclicity. (H. Olsen 1990)

eustatic mechanisms generated by the Gondwanan glaciation (see review in Miall 1990, Chap. 8), but only recently have studies demonstrated climate change during the formation of the cyclothems themselves. In Nova Scotia, calcareous paleosols occur at cyclothem sequence boundaries, formed at times of low stands of sea level, indicating seasonally dry climates during glacial episodes in the southern hemisphere. This contrasts with the humid (interglacial) climates indicated by coal-bearing deposits of the highstand systems tracts (Tandon and Gibling 1994; see also Sect. 13.3.1). Yang and Nio (1993) demonstrated a similar oscillation between arid conditions during lowstand and more humid conditions during highstand in the eolian-fluvial cycles of the Rotliegend Sandstone of the southern North Sea Basin.

12.12.2 Fluvial Response to the Late Cenozoic Glaciations

The rapid recent developments in sequence stratigraphy, and the consequent focusing of interest on the response of depositional systems to base-level change, has renewed interest in the fluvial deposits formed during the late Cenozoic glaciation, including coastal-plain successions and inland terraces. These contain the record of substantial sea-level changes. However, they also contain the record of major climate changes. In coastal fluvial deposits the effects of base-level change and climate change are

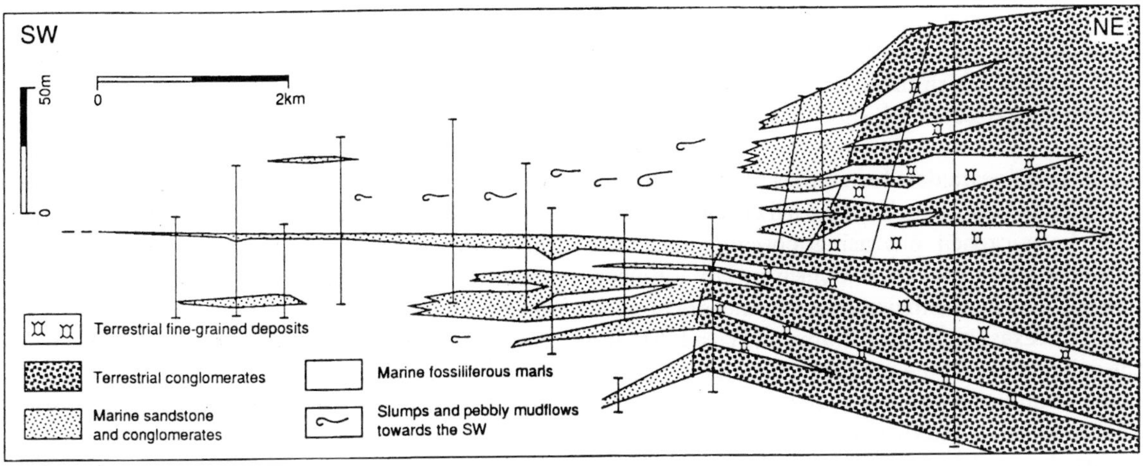

Fig. 12.6. Cross section through an Eocene cyclic succession in the Tremp-Graus foreland basin, Spain. Note the vertical persistence of the marine-nonmarine facies boundary. (de Boer et al. 1991)

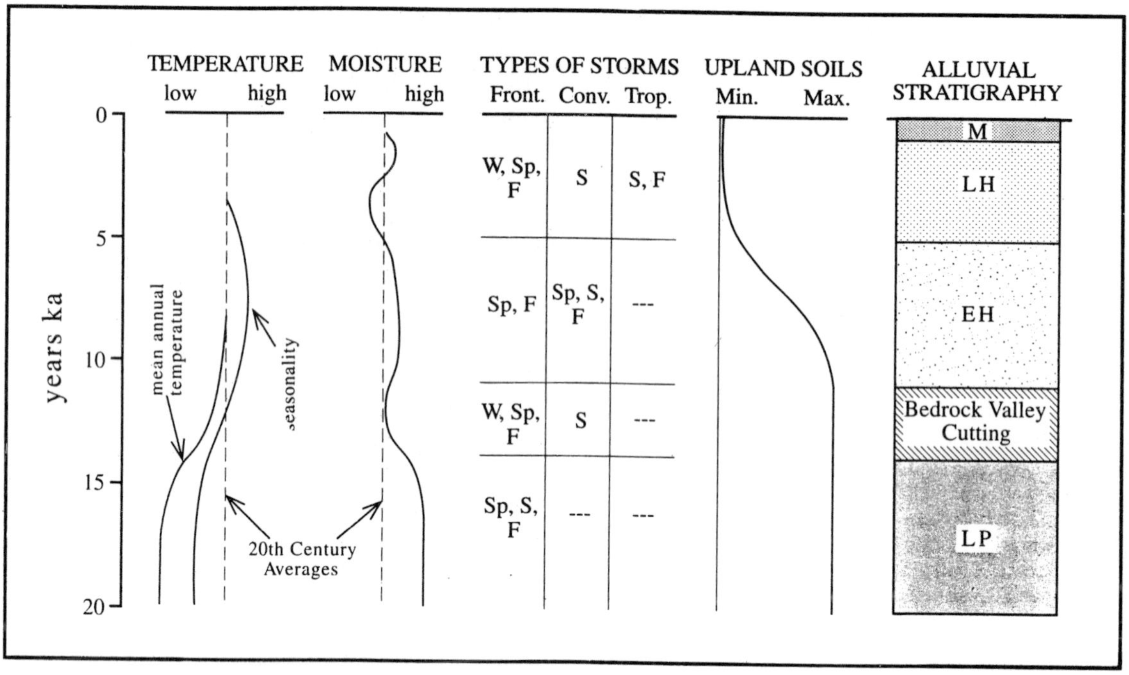

Fig. 12.7. Late Pleistocene-Holocene climatic and environmental changes on the Edwards Plateau, Texas, correlated with the alluvial stratigraphy of the upper Colorado drainage of the same area. *Front*, mid-latitude cyclones; *Conv*, convectional storms; *Trop*, tropical storms. Seasons are given by *W* winter; *Sp* spring; *S* summer; *F* fall. *Soil curve* represents initial thick red clay formation during the late Pleistocene and its gradual removal during the Holocene. *LP* Late Pleistocene; *EH* early Holocene; *LH* late Holocene; *M* modern alluvial deposits. (Blum 1994, reprinted by permission)

superimposed, and they are complex in nature. The major effects of climate change on river systems are to alter the discharge regime and, coupled with changes in vegetation cover, to alter the sediment yield. Because of lags in the response to climatic forcing and various geomorphic threshold effects, these changes do not occur in phase with the base-level changes, which themselves may be affected by isostatic responses to ice-sheet loading and unloading.

In recent years detailed studies of terrace stratigraphies and the deposits of buried channel systems have been carried out using borehole data and chronostratigraphic correlations based on radiocarbon dating. The two major areas of study are the Gulf Coast of Texas (Blum 1994; Blum and Price 1994) and the valley of the Rhine-Meuse system in the Netherlands (Vandenberghe 1993; Törnqvist 1993, 1994; Törnqvist et al. 1993). Vandenberghe et al. (1994) reported a comparative study of the Maas River in the Netherlands and the Warta River in Poland. These studies are beginning to unravel the complex and out-of-phase responses of river systems to Milankovitch-band external forcing.

Figure 12.7 shows the correlation between the alluvial stratigraphy of the Upper Colorado drainage, Texas, and independently derived climatic changes. Figure 12.8 shows a model of fluvial processes in relationship to glacially controlled changes in climate and vegetation, based on the Dutch work. Both these studies deal with periglacial regions, where climate change was pronounced but the areas were not directly affected by glaciation. It is interesting to note that a major episode of valley incision occurred in Texas not during the time of glacio-eustatic sea-level lowstand but at the beginning of the postglacial sea-level rise, which commenced at about 15 ka. The Dutch work explains why this may have been the case. Vandenberghe (1993) and Vandenberghe et al. (1994) demonstrated that a major period of incision occurs during the transition from cold to warm phases because runoff increases while sediment yield remains low. Vegetation is quickly able to stabilize river banks, reducing sediment delivery, while evapotranspiration remains low, so that the runoff is high. Fluvial styles in aggrading valleys tend to change from braided during glacial phases to meandering during inter-

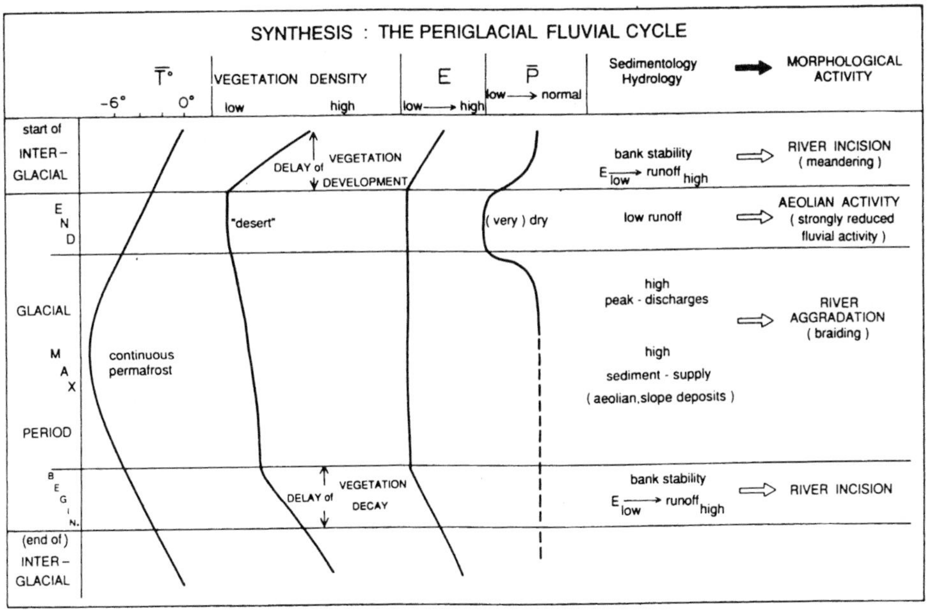

Fig. 12.8. A model showing the relationship between temperature, vegetation density, evapotranspiration (E), precipitation (P), and sedimentary processes in river systems during glacial and interglacial phases. (Vandenberghe 1993)

glacials (Vandenberghe et al. 1994). With increasing warmth, and consequently increasing vegetation density, rivers of anastomosed or meandering style tend to develop, the former particularly in coastal areas where the rate of generation of new accommodation space is high during the period of rapidly rising base level (Törnqvist 1993; Törnqvist et al. 1993; Fig. 13.16). Vandenberghe (1993) also demonstrated that valley incision tends to occur during the transition from warm to cold phases. Reduced evapotranspiration consequent upon the cooling temperatures occurs while the vegetation cover is still substantial. Therefore, runoff increases, while sediment yield remains low. With the reduction in vegetation cover as the cold phase becomes established, sediment deliveries increase and fluvial aggradation is reestablished.

Several other studies of the effects of early Holocene climate change have also revealed important sedimentary effects (Bull 1991). As noted by G.A. Smith (1994), "decrease in effective precipitation initiated time-transgressive changes in plant communities and diminished hill-slope vegetation densities that increased both sediment yield and runoff. The impact of this climate change was so profound that it controlled alluvial-fan sedimentation processes regardless of regional variations in tectonic activity." Abrahams and Chadwick (1994) documented multiple levels of alluvial fan development on the flanks of the coastal mountains of Oman and related these to climatically driven changes in runoff and sediment load. They found no evidence of tectonic control of this stratigraphy.

It is apparent that fluvial processes inland and those along the coast are likely to be completely out of phase during the climatic and base-level changes accompanying glaciation. Within a few tens of kilometers of the sea, valley incision occurs at times of base-level lowstand, during cold phases, but the surface may be modified and deepened during the subsequent transgression until it finally becomes buried. Inland, major erosional bounding surfaces correlate to times of climatic transition, from cold to warm and from warm to cold, that is to say, during times of rising and falling sea level, respectively. Minor climatic fluctuations lead to alternations of aggradation (terrace development) and valley incision, as has now been documented for several Gulf Coast rivers by Blum (1994) and Blum and Price (1994), as summarized in Fig. 13.11.

Although providing a fascinating explanation for fluvial processes during the Late Cenozoic, this body of ideas must be used with caution for the interpretation of more ancient fluvial successions, because of changes in vegetation type. Whether the same model can be applied to the interpretation of Late Paleozoic cyclothems, which were deposited before the advent of upland vegetation, remains to be examined.

12.12.3 Conclusions

The question remains whether high-frequency climatic forcing, on the 10^5-year time scale, will be detectable in the pre-Pleistocene stratigraphic record. To detect change on such a frequency requires stratigraphic mapping on at least as refined a time scale. This scale of periodicity corresponds to the cyclicity of many major geomorphic processes, such as the return period of major channels in large-scale fluvial systems (sediment groups 8 and 9 of Table 9.1), and is at or below the resolving power of most chronostratigraphic correlation methods, including magnetostratigraphy.

12.13 Discussion

This review has demonstrated that in very few fluvial settings are lithofacies criteria sufficiently distinctive that they can be relied upon, in the absence of supplementary evidence, to indicate climatic setting. Probably the most climatically distinctive lithofacies assemblage is that of the sandy terminal fan (ephemeral braid delta of Blair and McPherson 1992), which is characteristic of arid basin settings. The downstream change from channelized to sheet flow and the decrease in sand body thickness and the scale of hydrodynamic sedimentary structures are particularly characteristic. The architecture of the distal deposits, consisting of sandstone sheets composed of flood cycles and interbedded with fine-grained deposits showing evidence of evaporation, is very distinctive (Fig. 12.2B). Most other lithofacies and architectural features of fluvial deposits have been shown to be nondiagnostic of climate. Mass-flow deposits and other upper flow-regime fluvial sediments, while very characteristic of alluvial fans in hot-arid climates (Fig. 12.1A), are not confined to this climatic setting (Table 12.1), although it is possible that their abundance, and therefore their preservation potential, may be higher in this type of setting. A quantitative testing of this possibility has yet to be attempted.

However, a consideration of every aspect of a fluvial assemblage, including features of the overbank environment such as chemical and biochemical sediments, soils, minor sedimentary structures, and bed color, may yield useful clues as to climate. Conflicting criteria may indicate that the climate of the basin center is different from that of its margin, or that at least the water table is at a different level. Climates may also be different on either side of a major mountain range. The major criteria are summarized in Tables 12.2 and 12.3, and Figs. 12.1–12.3 are schematic diagrams of lithofacies assemblages in which climatically sensitive criteria are identified.

As noted earlier in this chapter, the evolution of land vegetation has probably had a profound effect on fluvial styles. The lack of roots to bind bank sediments in pre-Devonian time would have led to a tendency for rapid bank erosion and channel widening, with the supply of abundant bedload to the system. This would have favored braided over other fluvial styles, regardless of the discharge regime. Long (1978) considered Schumm's (1968a) speculations on this matter, in his review of Proterozoic fluvial sedimentation. He confirmed that rivers with an interpreted meandering style are indeed rare in the Precambrian sedimentary record, but much work has been carried out on Precambrian sedimentation since the late 1970s, and this subject should be reviewed again. In the meantime, the climatic ambiguity of fluvial lithofacies, and the absence of coal and of much other relevant fossil material means pre-Devonian climates are very difficult to interpret from the fluvial record.

It may be possible to make quite subtle climatic interpretations based on a summation of all the available evidence. For example, Hallam (1984) defined an "intermediate" type of climate, which shows evidence of both aridity and humidity. The sediments contain "thin seams of lignite or coal, modest in thickness and localised in distribution, and minor evaporites, such as scattered thin gypsum lenses and seams, [these] could both form in intermediate, savannah-type environments where wet and dry seasons are of roughly equal importance." The coexistence of trees and evaporites indicates a savannah or Mediterranean-type climate with alternating wet and dry seasons.

In a general discussion of the distribution of various climatic indicators, especially coal, Ziegler et al. (1987) provided this interpretation of the Nubian Sandstone of North Africa, based on many sources: Cross-bedding type and paleocurrent variance are interpreted in terms of braided or low-sinuosity streams. Flood activity is indicated by repetition of channel-based scour surfaces and their lag deposits. Prevalence of soil horizons and common preservation of flora suggest well-vegetated alluvial plains, but the common tree *Dadoxylon* was probably restricted to channel margins and swamps. Floral diversity is not high, and the ubiquitous braided stream deposits would seem to rule out the possibility of forest vegetation and especially the lowland rain forest. "We propose a savannah environment with its typical seasonal rainfall pattern for the Nubian Sandstone and its equivalents in NE Africa and Arabia."

Dubiel et al. (1991) were able to trace a lateral gradation from seasonally dry floodplains to permanently waterlogged swamps and ponds. In the Monitor Butte Member of the Chinle Formation purple-mottled sandstones contain large, irregular mottles of dark purple, lavender, yellow, and white that reflect varying concentrations of iron-bearing minerals. These mottles reflect variations in the water table, giving rise to alternating oxidizing and reducing conditions, with redistribution of iron and the formation of gleyed paleosols. The purple-mottled units can be traced laterally into gray and purple siltstones with plant debris, and then into gray to black, thinly laminated organic-rich shales with fish debris deposited in permanent ponds, marshes, and lakes. A seasonal climate is suggested, with the ponds occupying permanently saturated areas in topographically low parts of the basin.

Jerzykiewicz and Sweet (1988) distinguished three climatic regimes in the Upper Cretaceous-Paleocene sediments of the Foothills of Alberta. A humid fluvial facies association consists of organic-rich to coal-bearing floodplain deposits associated with meandering channels of high sinuosity. A semi-arid fluvial facies association includes varicolored floodplain mudstones with mature caliche paleosol horizons. The alluvial plain was drained by broad mobile channels. Periodic heavy rainfall modified the floodplain into a braidplain with numerous ephemeral streams eroding the caliche soil. An intermediate fluvial facies association can also be distinguished, consisting of gray or greenish mudstone, barren of coal and devoid of caliche, associated with a low-sinuosity meander-belt facies. They demonstrated the stratigraphic and geographic shifting of these climatic belts across Alberta through the Late Cretaceous and early Tertiary.

An example of a climatic interpretation of a pre-Devonian fluvial system is that by Trewin (1993b). He described sand sheets of mixed eolian-fluvial origin. Discharge is interpreted as flashy on the basis of a dominance of trough cross-bedded sandstone, the great lateral extent of thin units, the extreme rarity of channel features, strongly unimodal current directions, and planar upper bounding surfaces. These deposits contain abundant trace fossils. Red grain coatings indicate oxidizing conditions. Evidence of desiccation is very rare, but in the absence of significant land vegetation an arid climate is not necessarily implied by this evidence. A high water table is suggested by the trace fossils, which indicate invertebrate activity. Trewin (1993b) stated: "Before the advent of land plants sandy alluvial outwash areas were associated with a wider range of climate than is seen at the present day. Flashy discharge, and the lack of sediment binding resulted in lateral expansion of streams in times of flood to produce laterally extensive thin sandstone dominated by trough cross-bedding in medium- to coarse-grained sands." Eolian reworking took place between floods or during periods of low discharge.

In conclusion, it is clear that climatic interpretations can be made from clastic fluvial lithofacies, but this is best carried out in a regional context, with due regard for the tectonic setting and incorporating data from all interbedded nonmarine facies, especially chemical and biochemical sediments. Many features of clastic fluvial lithofacies, and the architectural elements and successions of which they are a part, are indicative of a discharge regime, the relationship of which to climate may be indirect or complex. In many cases climatic and tectonic signatures may be ambiguously similar. Nonetheless, a careful attention to the question of local climate may yield valuable information and insights to assist in regional basin studies and paleogeographic reconstruction.

Chapter 13

Sequence Stratigraphy

13.1 Introduction

The currently accepted definition of "sequence" is that proposed by Vail et al. (1977, p. 53). It is a "stratigraphic unit composed of a relatively conformable succession of genetically related strata ... bounded at its top and base by unconformities or their correlative conformities". However, the term sequence has had a checkered history. Sloss (1963) defined stratigraphic sequences as "rock-stratigraphic units of higher rank than group, megagroup, or supergroup, traceable over major areas of a continent and bounded by unconformities of interregional scope." This useage has not received universal acceptance. In some of the literature "sequence" has been used as a synonym for "succession", and it has been employed for cyclic or unconformity-bounded units of varying dimensions and for time spans ranging from thousands of years to hundreds of millions of years.

Since the ground-breaking publication by Vail et al. (1977), sequence stratigraphy has evolved into one of the central, theoretical underpinnings of the science of stratigraphy. It comprises a set of predictions that can be made about the facies and architecture of sediments formed in a given range of environments under specified allogenic controls. It is the predictability of sequences that give them their value for regional correlation and for one of the main practical applications of such correlation – petroleum exploration. However, modern work has demonstrated that sequences are generated by a variety of causes, including local- to continental-scale tectonism, eustatic sea-level change, and climate change (see reviews by Miall 1990, Chap. 8; Plint et al. 1992). More than one generating mechanism may be active at any one time, and it is not uncommon for sequences of several types and scales to be superimposed on each other within the same succession. Interpreting generating processes from the rock record requires the construction of a detailed stratigraphic framework. Questions about the way in which various allogenic controls actually work has stimulated much numerical modeling of basin evolution, aided in recent years by graphical simulation using computers.

Most recent work has focused on the sequence architecture of shallow marine deposits, because one of the major preoccupations of sequence stratigraphers has been sea-level change, and facies characteristics in the lower coastal-plain to shallow-shelf environment are particularly sensitive to changes in sea level. Much less is known about the sequence stratigraphy of nonmarine deposits, although this is now rapidly changing, as summarized in the recent reviews by Shanley and McCabe (1994) and Posamentier and Allen (in press).

Sequence-stratigraphic analyses of fluvial deposits commenced, in effect, with the work of Allen (1974b). As noted in Sects. 2.3.7 and 10.4.3 (Fig. 10.32), Allen developed a series of models of alluvial stratigraphy based on simple deductive reasoning. His model 5B is an interpretation of the stratigraphic architecture that might develop in an alluvial system characterized by a cycle of dissection and aggradation, controlled by base-level changes over a vertical interval greater than channel depth. This is precisely the set of conditions under which nonmarine stratigraphic sequences develop, according to modern work. Building on this work and on that of Bridge and Leeder (1979), Ross (1990) made some simple predictions regarding variations in channel density in response to variations in aggradation rate, but he did not formalize a sequence model. The current state of knowledge of the effects of base-level change on fluvial systems and of the control tectonic movements exert on rivers is discussed at length in Chap. 11 (see in particular Sect. 11.2.2). It is beyond the scope of this book to enter into the current controversy regarding the relative importance of regional tectonism versus eustatic sea-level change in the generation of base-level change.

The first modern sequence models, those of Vail et al. (1977), were based essentially on the architecture of divergent or passive continental margins, although the limitations imposed on the models by this fact were not clearly acknowledged by Vail and his colleagues until recently (Posamentier and Weimer 1993; Posamentier and Allen 1993). In the meantime, working on the Mesa Verde Group of the Western Interior foreland basin of Utah, Swift et al. (1987) proposed a quite different type of sequence model. In foreland basins the relationship between subsidence and sediment supply is reversed from that prevailing in extensional continental-margin basins. In the latter, sediment enters the basin near the tectonic hinge and is transported toward the more rapidly subsiding side of the basin. Sequences in foreland basins are "clastic wedges", as defined by Sloss (1962). "They are sediment prisms poured into the deeper side of a linear half-basin undergoing subsidence along its *landward* margin. Hence it is the proximal (landward) end of the wedge that is most completely preserved" (Swift et al. 1987, p. 449). Accommodation space is generated more slowly in the deeper part of the basin because of slower subsidence there, and so stratigraphic units tend to thin and pinch out toward the center of the foreland basin, whereas many units thicken toward the basin center in the case of extensional continental margins. The major architectural differences between the two types of basin setting are illustrated in Fig. 13.1.

Nonmarine deposits are much better preserved in foreland basins than in extensional basins because of the relationship between subsidence and sediment supply, and many of the recent studies of nonmarine sequences have been carried out in deposits formed in this tectonic setting (Aubry 1989; Hanneman and Wideman 1991; Holbrook and Dunbar 1992; Shanley and McCabe 1991, 1993; Martinesen et al. 1993; Miall 1993; Aitken and Flint 1995; Olsen et al., in press; Yoshida et al., in press). These studies effectively answered the question that had been raised by Posamentier and Weimer (1993) regarding the applicability of sequence concepts to nonmarine deposits. Posamentier and Allen (1993) provided a useful discussion of the particular characteristics of sequences formed in foreland basins and made several predictions regarding the architecture of nonmarine deposits in such basins that are discussed later in this chapter.

The purpose of the present chapter is to review recent work on sequence models. Does the response of fluvial systems to allogenic controls, including source-area uplift (tectonic control) and base-level change (tectonic or eustatic control), lead to predictable facies successions and architectures? To turn this question around, what can be deduced regarding allogenic controls from the detailed documentation of such facies and architectures? The research that has taken place since the late 1970s on the effects of base-level change on fluvial systems is discussed at length in Sect. 11.2.2. The most important such research, because of the attention it received, was that by Posamentier et al. (1988), which was based on extensive computer modeling and observations on small modern rivers. As argued by Miall (1991c), this research was seriously flawed. In Sect. 11.2.2 the work of later researchers, particularly that of Schumm (1993) and Wescott (1993), is reviewed, and it is demonstrated how rivers respond to base-level change by incision or aggradation and by changes in sinuosity. The most recent work by Posamentier and Allen (in press) is touched on in this section. The effects of tectonism on the middle and upper reaches of river courses are also discussed at length in Chap. 11. The purpose of the present chapter is to review current attempts to put these data and ideas into a sequence framework, that is to say, a predictive model.

Modern sequence models for fluvial deposits have been proposed by Shanley and McCabe (1991, 1993, 1994), Wright and Marriott (1993), and Posamentier and Allen (in press). Very few modern sequence studies of fluvial deposits have been published. Those by Legarreta and Gulisano (1989), Shanley and McCabe (1991, 1993), Van Wagoner et al. (1990, 1991), Hanneman and Wideman (1991), Miall (1993), Gibling and Bird (1994), Gibling and Wightman (1994), Olsen et al. (1995), and Yoshida et al. (in press) are the most important. This chapter builds mainly on the work of these researchers.

It is important to note that, unless otherwise stated, no particular time connotation is placed upon the sequence models discussed in this chapter. The examples of alluvial stratigraphy described in Sects. 9.2 to 9.4 (summarized in Table 9.1) include sequences of third to fifth order (groups 8 to 10, in the classification of this book: Tables 3.2, 4.2, 9.1). These range in duration from a few thousand years to a few million years. These sequences all display a certain degree of "self-similarity", indicating that the dynamic balance between the various allogenic controls can be achieved and maintained over a wide range of time scales. The various temporal scales of nonmarine sequences, and their causes, are reviewed in Sect. 13.4.

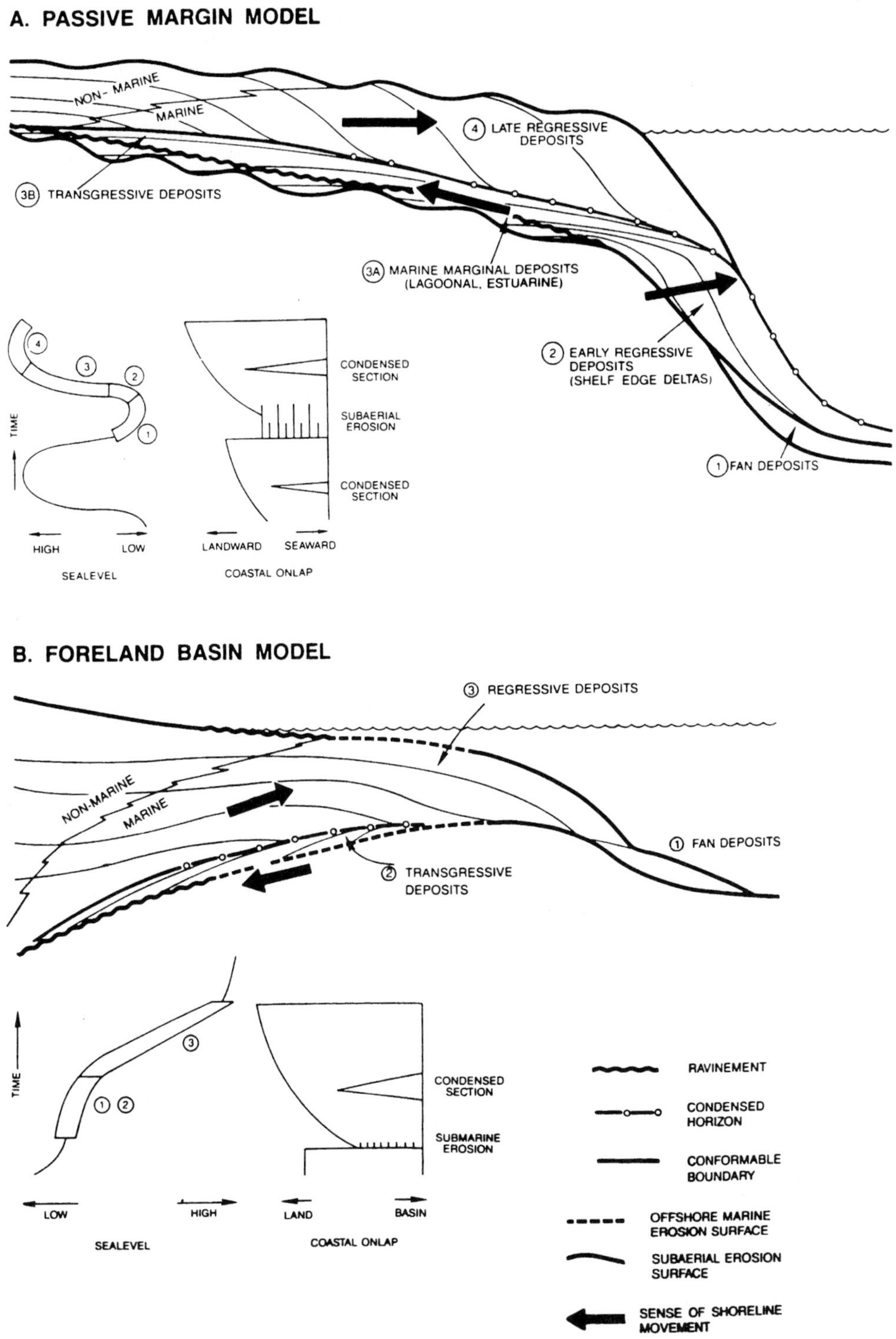

Fig. 13.1A,B. Comparison of sequence architecture on extensional continental margins (A) and in foreland basins (B). Note the much thicker record of nonmarine deposits in foreland basins. These deposits also have a higher preservation potential than the highstand deposits of extensional continental margins, because of the slow subsidence rates on the proximal margins of such basins and the tendency for nonmarine HST deposits to be eroded during subsequent base-level falls. (Swift et al. 1987)

13.2 Accommodation Space

The concept of accommodation space is fundamental in sequence stratigraphy. Jervey (1988, p. 47) defined it as "the space made available for potential sediment accumulation [where] in order for sediments to be preserved, there must be space available below base level (the level above which erosion will occur)." The word "accommodation" implies space; therefore, the use of the word "space" in the term "accommodation space" is, strictly speaking, superfluous.

Jervey was modeling marine sequences, for which base level corresponds to sea level; the sequence architecture of marine deposits is governed mainly by the response of depositional systems to sea-level change. However, this definition does not apply in the case of the fluvial environments because fluvial deposits, by their very nature, are deposited above sea level. Schumm (1993, p. 279) defined base level as "the imaginary horizontal or level surface to which sub-aerial erosion proceeds. It is sea level." Shanley and McCabe (1994) noted some important additional local "base levels" that may be significant in nonmarine settings, such as the deflation surface, defined by the water table, in eolian environments. So how does the concept of accommodation apply in the case of nonmarine deposits?

Base level is the level to which rivers grade their course, as discussed in Sect. 11.2.2. Rivers adjust their slope, sinuosity, and other hydraulic characteristics to the power and efficiency necessary to transport the water and sediment supplied from the drainage basin. Given equilibrium conditions, a graded profile will develop, which flattens out to the horizontal at the precise point at which base level is reached, at the mouth of the river. Posamentier et al. (1988, p. 135) stated:

"The fundamental principle which governs sedimentation in fluvial environments (assuming constant sediment supply) is that a stream will aggrade if its equilibrium profile shifts basinward or upward and incise if its equilibrium profile shifts downward ... subaerial accommodation occurs by basinward shifts of the stream equilibrium profile."

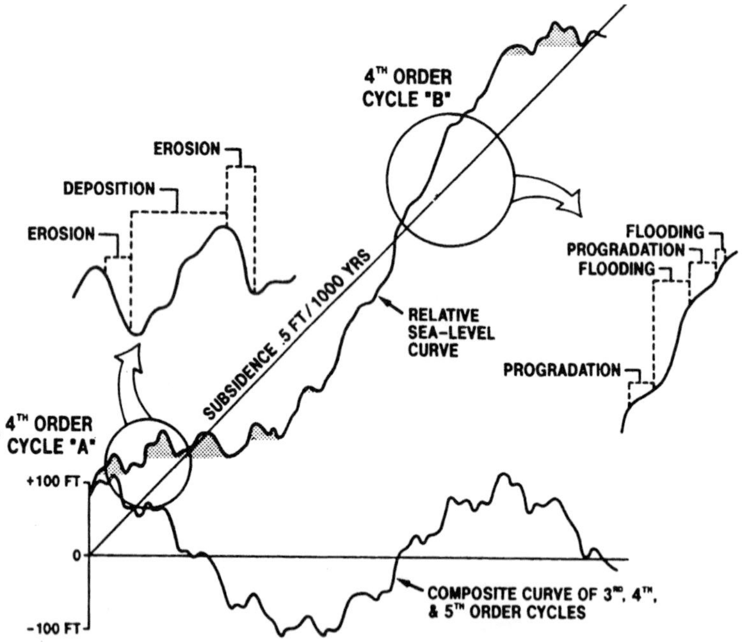

Fig. 13.2. The creation and reduction of accommodation space in response to sea-level change and tectonism. The composite sea-level curve (*bottom*) indicates the integration of three rates of sea-level change. Accommodation space is generated when the rate of sea-level rise outpaces the rate of subsidence (*dotted areas* in relative sea-level curve). This diagram was adapted by Nummedal et al. (1993) from Van Wagoner et al. (1990) as a basis for discussing fourth-order marine cycles, but it can be applied to the development of fluvial cycles, as follows: In coastal-plain settings accommodation space is directly related to the balance between subsidence and sea-level rise, as shown in this diagram. Where tectonism is the dominant control, the composite sea-level curve could be regarded as a curve tracking the rise and fall of the source terrane as a result of uplift and erosion. Accommodation space then represents the space between the existing fluvial profile and the new, higher profile required to restore equilibrium (the same *dotted area* of the curve)

The problem with the last sentence in this quotation is discussed in Sect. 11.2.2. Apart from that, the quote succinctly explains the importance of the equilibrium profile. Accommodation space in fluvial environments can be defined as the space between an existing graded profile and a new graded profile to which the river adjusts in reponse to changes in base level, tectonic movements, or hydraulic conditions (discharge, sediment supply). Accommodation space is positive, and sedimentation occurs, under the following conditions (Fig. 13.2):

1. At the coast, when the graded profile moves upward in response to a rise in sea level. Most sedimentation occurs near the mouth of the river, and there may be little change in accommodation space in upper reaches.
2. At the coast, when the graded profile moves upward in response to coastal regression and progradation, under conditions of sea-level stillstand. This effect is small, and in the case of progradation by lobate deltas it is limited by the tendency for avulsive delta abandonment when the profile of the main channel is raised by aggradation.
3. Inland, when the graded profile moves upward in response to uplift of the source area, for example along a basin-margin fault. New accommodation space is generated adjacent to the zone of uplift and, given adequate sediment supply, progradation may shift the equilibrium profile basinward.
4. Throughout the profile when, because of tectonic or climatic changes in the source area, there is an increase in discharge and/or sediment supply.

Accommodation space is negative, and incision occurs, under the reverse of these conditions: lowering of sea level, reduction in source-area elevation, e.g., by regional erosion, and reduction in discharge or sediment supply.

All these allogenic controls fluctuate constantly in all sedimentary basins. Sea level has been rising and falling eustatically at least throughout the Phanerozoic (see review by Miall 1990, Chap. 8). Tectonic subsidence and uplift, driven by crustal stresses and subcrustal thermal changes, is in a continual state of dynamic balance with uplift driven by erosional downcutting of uplifted areas and consequent isostatic buoyancy. Climatic changes, driven by orbital forcing and by the drift of continental plates through latitudinal climatic zones, bring about changes in discharge and sediment yield. All these changes have been occurring throughout geologic time over a wide range of time scales, ranging from a few thousand years to hundreds of millions of years. Several of these processes of change may be occurring simultaneously at more than one time scale. Few rivers are ever, therefore, in a state of true equilibrium. The concept of the graded profile is a geomorphic one, which is appropriate for short time scales (tens to hundreds of years), but over geologic time scales the concept of dynamic equilibrium is more appropriate and useful. Local conditions must always be taken into account. For example, along much of the present-day coastal plain of South Island, New Zealand, a slow eustatic rise and shelf subsidence are underway. This is causing a transgression, but one that is being recorded not by deposition, but by valley incision and erosional shoreface retreat, because of high wave energies (Leckie 1994).

For marine shoreline and nearshore deposits predictions of sequence architecture in response to sea-level change are easy to make. When sea level rises the coast is flooded, and transgression takes place. Sea-level fall is accompanied by regression and progradation. Such predictions were the basis for the first sequence models in the modern sense, those of Frazier (1974), followed by Vail et al. (1977). However, in the case of rivers, the response to change is more complex. For example, as discussed in Sect. 11.2.2, a lowering of sea level and the extension of a river profile out to a new mouth, lower out and further down on the continental shelf, may lead to incision, progradation, or simple changes in sinuosity with neither enhanced aggradation or erosion, depending on a variety of factors, of which the slope of the newly exposed shelf is one of the most critical. The development of sequence models for fluvial systems has therefore lagged behind that for marine environments, and the subject could be said to have arrived at a state of maturity only recently, with the publication of the review by Shanley and McCabe (1994).

In the case of tectonic control, the effects of uplift, subsidence, and erosion at a faulted basin margin have long been known [see the discussion of the important paper by Heward (1978a) in Sect. 11.2.1.]. Only in recent years, however, have we begun to appreciate the complexities of tectonism as a sedimentary control. For example, in foreland basins it now seems likely that regional crustal shortening and loading on the one hand, and the movement of individual thrust or nappe plates, on the other hand, may generate pulsed, irregular subsidence on both long and short time scales at the same time (Sects. 11.3.6, 11.4.6). Uplift and subsidence may also be

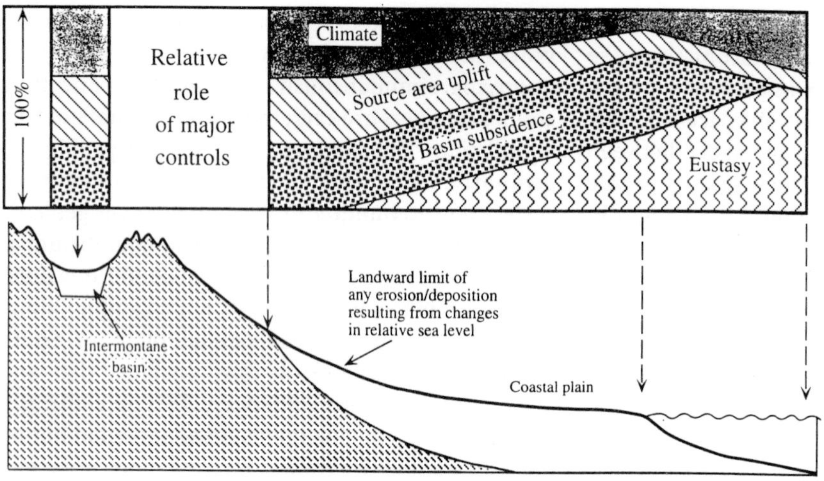

Fig. 13.3. The relative importance of the major controls on fluvial sequence architecture, showing how the balance of control varies between source area and shoreline. Intermontane basins (*left*) are, of course, unaffected by sea-level change. (Shanley and McCabe 1994, reprinted by permission)

thermal in origin, related to volcanic activity or plutonism (Dickinson 1981b; Dorsey and Burns 1994). Major changes in sediment supply (Sect. 11.3.4) and regional paleoslopes are likely to be reflected in the resulting stratigraphy. Patterns of basin subsidence and the sedimentary response are complex in these various tectonic settings and have yet to be fully resolved. Studies of the sequence stratigraphy of nonmarine deposits can play an important part in this research.

In general the importance of base-level change diminishes upstream, as discussed by Miall (1991c) and Shanley and McCabe (1994). In large rivers, such as the Mississippi, the evidence from the Quaternary record indicates that sea-level changes affect aggradation and degradation as far upstream as the region of Natchez, Mississippi, about 220 km upstream from the present mouth (Saucier, in Autin et al. 1991). Farther upstream than this, source-area effects, including changes in discharge and sediment supply, resulting from tectonism and climate change are much more important. In the Colorado River of Texas base-level influence extends about 90 km upstream, beyond which point the river has been affected primarily by the climate changes of the Late Cenozoic glaciations (Blum 1994; see Chap. 12). Based on his detailed work on the Gulf Coast rivers, Blum (1994, p. 275) stated: "At some point upstream rivers become completely independent of higher-order relative changes in base level, and are responding to a tectonically controlled long-term average base level of erosion." Figure 13.3 illustrates a conceptual estimate of the relative importance of these processes in relationship to position within a fluvial system.

It is ironic that interpretations of some fluvial cycles may be coming full circle. As summarized in Sect. 2.3.6.2, in the 1940s some fluvial fining-upward cycles were interpreted as the result of pulsed subsidence of a sedimentary basin. For example, Bersier (1948) expressed this interpretation in his study of the Cenozoic molasse cycles of the Alpine foreland basin. With the advent of facies models in the 1960s, the emphasis shifted, and cyclic sedimentary processes were attributed almost entirely to autogenic controls (Chap. 2). The importance of tectonism was downplayed in the enthusiasm for such processes as delta-lobe switching and fluvial meander migration. We are now returning to a recognition of the importance of allogenic controls and, as pointed out by Shanley and McCabe (1994), there is a danger that the importance of autogenic processes may be forgotten in the new enthusiasm for allogenic mechanisms, particularly eustasy. This researcher shares the uneasiness expressed by Shanley and McCabe (1994) regarding the interpretation of all observed cyclicity in terms of sequence concepts based on allogenic controls, such as in the study of some Gulf Coast deposits by Mitchum and Van Wagoner (1991). Nevertheless, the appearance of the new sequence concepts, and a more sophisticated understanding of tectonic subsidence processes, have raised anew the possibility that many cyclic successions currently attributed to autogenic processes may in fact be allogenic in origin. This may be the case particularly in foreland basins, where the evidence for short-term, irregular or pulsed subsidence driven by thrust-sheet loading, and for rapid uplift and erosional unroofing, is now abundant (see Sect. 11.4.6).

We are beginning to realize that more than one process may control sequence development. These processes can act over different time scales and may vary in importance from source area to shoreline. The result can be a complex stratigraphy. This is exemplified by the Castlegate Sandstone of Utah, which is one of the units discussed in the following sections.

13.3 Main Components of the Fluvial Sequence Model

This discussion uses the standard sequence terminology, as set out by Vail (1987) and Van Wagoner et al. (1987) and as emended by Hunt and Tucker (1992). Their definition of sequences, including the placement of sequence boundaries at subaerial erosion surfaces formed during base-level lowstand, is preferred to that of Galloway (1989a), who recommended use of the maximum flooding surface as the sequence boundary. However, this latter work contains many important ideas and concepts that are discussed later in this chapter, and in at least one recent study (Gibling and Bird 1994) it was found easier to follow Galloway's approach to the definition of sequences, because of the difficulty of correlating sequence boundaries associated with discontinuous paleovalley fills [a difficulty anticipated by Galloway (1989a)]. Reference is made here to the two main sequence models for fluvial deposits (Figs. 13.4, 13.5) published by Shanley and McCabe (1991, 1993, 1994) and Wright and Marriott (1993). The main photographic illustrations are drawn from the work of the writer and his students Shuji Yoshida and Andrew Willis on the Castlegate Sandstone of Utah (Fig. 9.18). Standard abbreviations used in many sequence diagrams are as follows: *SB*, sequence boundary; *LST* or *LWST*, lowstand systems tract; *TST*, transgressive systems tract; *MFS*, maximum flooding surface; *HST*, highstand systems tract; *FSST*, falling-stage systems tract [=forced-regressive-wedge systems tract of Hunt and Tucker (1992)]. A composite sequence model showing the relationships between the sequence terminology, allostratigraphic terms, and the bounding-surface hierarchy of this book is illustrated in Fig. 13.6.

13.3.1 Sequence Boundary

In the original work of Vail et al. (1977), sequence boundaries were defined as the regional unconformities that separate the stratigraphic record into successions of conformable, genetically related strata. It was recognized that in some areas, such as zones of especially rapid subsidence, or in the deep sea, below the level to which sea level could fall eustatically, an unconformity may pass laterally into a "correlative conformity." In later work (Vail and Todd 1981; Vail et al. 1984) the differences between unconformable and conformable sequence boundaries were clarified, and two types were defined.

A *type 1 unconformity* develops where sea-level fall is rapid, more rapid than tectonic subsidence. The coastline may move out to near the shelf edge, and extensive subaerial erosion takes place, with the development of incised fluvial valleys on the shelf and the deepening of submarine canyons on the continental slope. Clastic detritus is transported down these fluvial and canyon systems to the base of

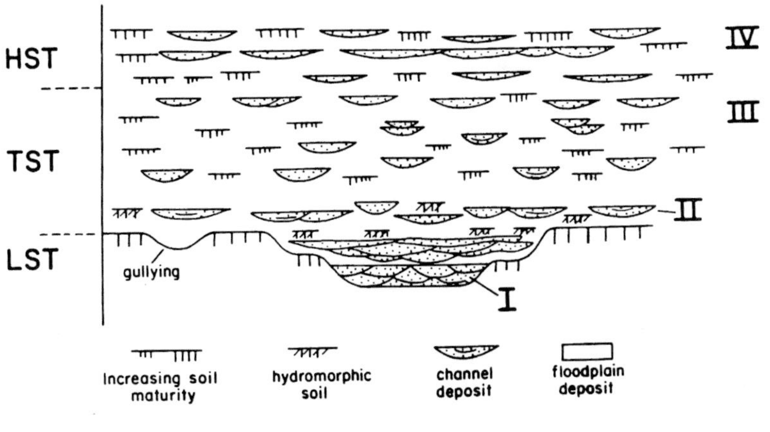

Fig. 13.4. The fluvial sequence model of Wright and Marriott (1993).

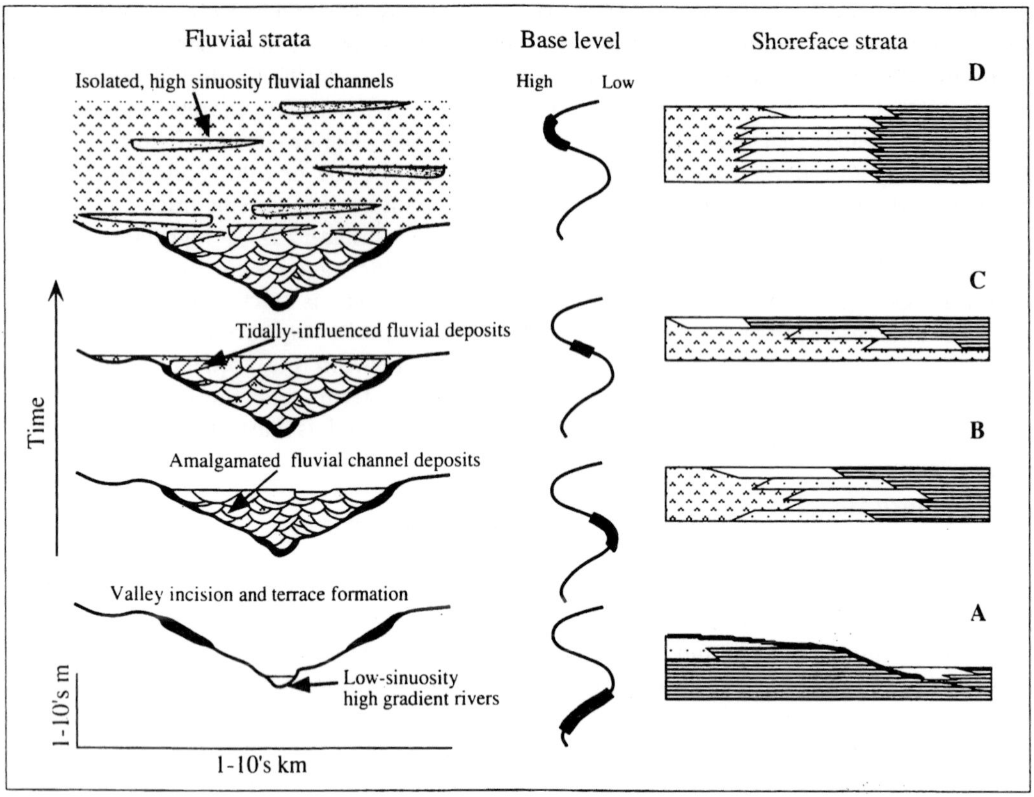

Fig. 13.5. The fluvial sequence model of Shanley and McCabe (1991, 1993, 1994, reprinted by permission), showing the relationship between shoreface and fluvial architecture and base-level change. *A* Falling stage systems tract, with development of incised valley and fluvial terraces; *B* lowstand systems tract; *C* tidal influence indicates the beginning of the transgressive systems tract; *D* highstand systems tract

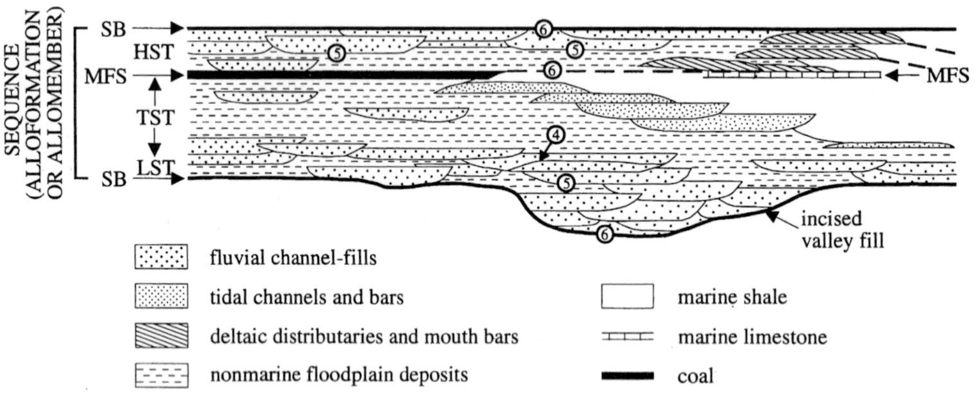

Fig. 13.6. Composite sequence model amalgamating the concepts of Wright and Marriott (1993), Shanley and McCabe (1994), and Gibling and Bird (1994), showing the relationship between sequence terminology (abbreviations explained in text), the bounding-surface hierarchy of this book (*numerals within circles*), and allostratigraphic terms

the continental slope, forming extensive falling-stage and lowstand systems tracts. At type 1 unconformities facies belts undergo a substantial basinward shift. Highstand deposits below the unconformity may be deeply eroded.

A *type 2 unconformity* develops when relative sea level falls slowly, resulting in a gradual seaward shift of facies tracts but only minor subaerial exposure and erosion. According to Vail et al. (1987, 1991), a shelf-margin systems tract is formed under these conditions. Type 2 unconformities are much more difficult to indentify in seismic and outcrop records because they are not characterized by deep erosion or major facies shifts.

On extensional continental margins, where the nonmarine environment occupies the proximal part of the basin, close to the hinge line, subsidence rates are relatively slow, and type 1 unconformities are common. The Quaternary record of modern continental margins has been characterized by rapid glacioeustatic fluctuations, and the record of fluvial deposition is condensed, fragmentary, and cut by numerous erosional unconformities (e.g., Atlantic Coastal Plain: Pazzaglia 1993), except where the hinterland provides an unusually large sediment supply, as along the Gulf Coast (Galloway 1989b; Sect. 11.3.4).

In foreland basins the pattern of subsidence and sedimentation is quite different, and Posamentier and Allen (1993) suggested that the margins of such basins could be subdivided into two zones, reflecting the relative rates of subsidence and base-level change (Fig. 13.7). Drawing on the work of Jervey (1988), they defined the position of an equilibrium point, the point on the basin margin marking a balance between the rate of subsidence and the rate of sea-level change. As the sea rises and falls and the rate of sea-level change fluctuates against a background of steady tectonic subsidence, this point moves in and out across the basin margin, as shown in Fig. 13.7. Landward of the equilbrium point, an area termed zone A, is a region in which subsidence is always faster than the rate of sea-level change and sedimentation is continuous. The sequence boundary here should be of type 2. Seaward of the equilibrium point sea-level change is at times the faster process, and sequence boundaries will be type 1. Posamentier and Allen (1993) developed a simple stratigraphic model that incorporates these ideas (Fig. 13.8).

This analysis invokes only two processes, regional foreland-basin subsidence and eustasy. However, much detailed stratigraphic and modeling work on foreland basins is demonstrating that tectonic and sedimentary processes there may be much more complex (e.g., Jordan and Flemings 1991; Plint et al. 1993). Rates of tectonic uplift and subsidence in convergent tectonic settings can be very high, as much as 10 m/ka over intervals as long as a few hundred thousand years (Burbank and Raynolds

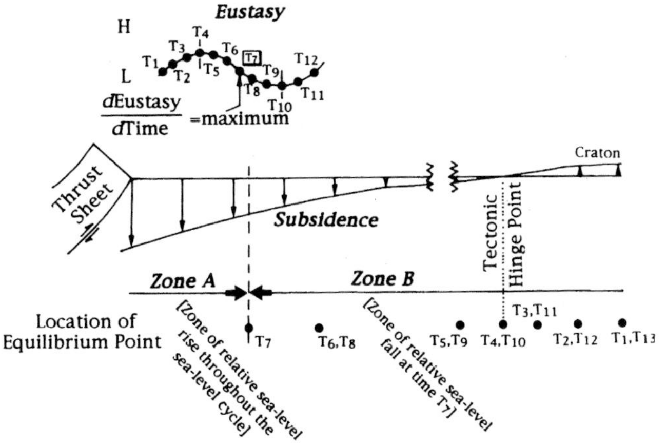

Fig. 13.7. Schematic subsidence profile across a foreland basin, showing the rate of subsidence along a dip-oriented profile. The position of the equilibrium point is shown at time T7, with other positions given by the *numbered dots* below the profile. The equilibrium profile marks the boundary between *zones A* and *B*, as discussed in the text. (Posamentier and Allen 1993)

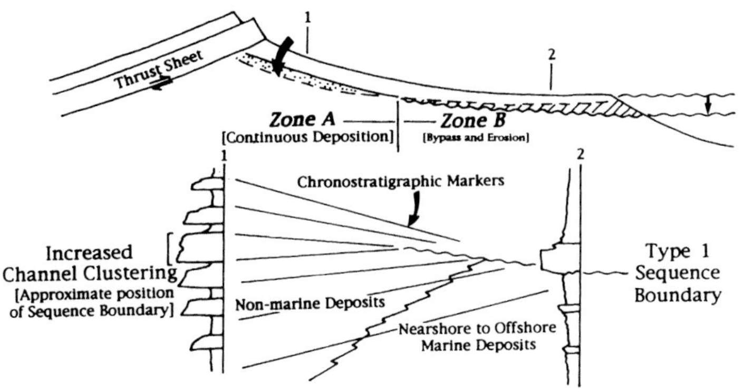

Fig. 13.8. Theoretical correlation between stratigraphic sections located in *zones A* and *B* of a foreland basin. In *zone B* the sequence boundary is expressed as an unconformity, possibly with an incised valley and a lowstand fluvial deposit. Continuous subsidence and sedimentation may occur in *zone A*, leading only to clustering of channel deposits at the time the unconformity develops in *zone B*. (Posamentier and Allen 1993)

Fig. 13.9. Sharp, and probably erosional sequence boundary between the Blackhawk Formation and overlying Castlegate Sandstone at Price Canyon, Utah, indicated by *arrow*. (Miall 1993; Olsen et al. 1995; Yoshida et al., in press)

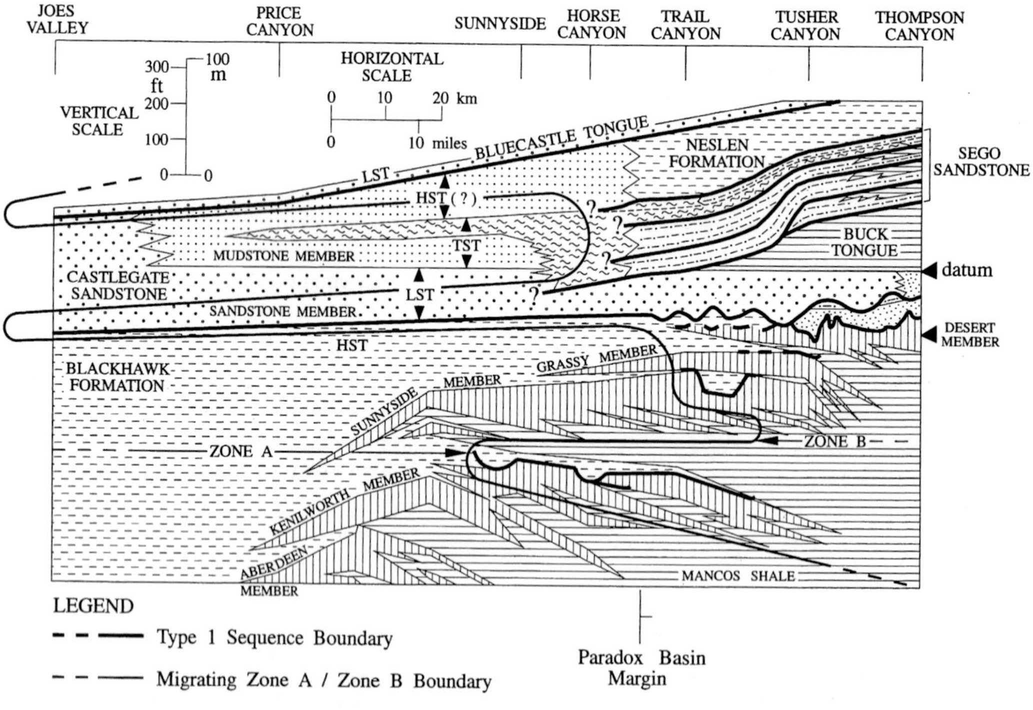

Fig. 13.10. Schematic synthesis of the sequence stratigraphy of the upper Mesa Verde Group in the Book Cliffs, Utah, showing the interpreted migration of the zone A/ zone B boundary of Posamentier and Allen (1993), and the distribution of type 1 sequence boundaries. This architecture is interpreted as the product of the superimposition of high-frequency (< 1 million years cyclicity) base-level changes of probable tectonic origin on a long-term (5 million years) cycle of flexural subsidence. Additional stratigraphic details of the Castlegate Sandstone and related units are shown in Fig. 9.18. (Yoshida et al., in press)

1988; Fortuin and de Smet 1991), possibly even exceeding the rates of sea-level change induced by glacioeustasy. Long-term rates of vertical tectonic movement reflecting regional plate-tectonic processes may be punctuated by short-term rates of very rapid subsidence, reflecting movements on individual thrust sheets or nappes. Therefore, there may not be a zone A in some foreland basins. In fact, erosional sequence boundaries are present in all the studies of nonmarine sequences in foreland basins published to date (e.g., Fig. 13.9), suggesting an absence of zone A. It is not yet known whether the channel clustering effect suggested in Fig. 13.8 actually occurs.

Figure 13.10 illustrates an interpretation of the sequence architecture of the Castlegate Sandstone and associated units in the Book Cliffs, based on the work of Yoshida et al. (in press; Fig. 9.18). It is suggested that the basin was affected by a pattern of long-term flexural subsidence which was greatest in proximal regions (sections toward the left side of the diagram) and decreased distally, and which fluctuated in rate over a time scale of a few millions of years. Superimposed on this was a pattern of higher-frequency base-level fluctuation which may have been tectonic or eustatic in origin. Erosional (type-1) sequence boundaries in distal areas help to define the position of zone B of the foreland basin, but, as shown in Fig. 13.10, the lateral extent of these erosion surfaces varies considerably. The boundary between zones A and B appears to have moved proximally and distally. During deposition of the Lower Castlegate Sandstone and the Bluecastle Sandstone the long-term rate of subsidence was slow and zone A was absent, or was restricted to the most proximal areas west of the project area.

Away from the main meander belt, in interfluve areas, type 2 unconformities may be characterized by well-developed paleosols, which require stability of the land surface and low sediment influx for their development (Sect. 7.4.2). Posamentier and Allen (1993) suggested the Boulder Creek Formation of British Columbia as a possible example. In this unit 15 paleosols are associated with a major regional unconformity (Leckie et al. 1989). Superimposed paleosols also mark unconformity surfaces in Eocene-Oligocene deposits of South Dakota and have been interpreted as indicating two episodes of either base-level fall and rise, or changes in sediment supply relating to volcanism or climate change (Evans and Terry 1994). Other examples of paleosols marking sequence-boundary disconformities have been described by Hanneman and Wideman (1991) and Gibling and Bird (1994). Hanneman et al. (1994) discussed the identification of sequence-bounding paleosols in the subsurface using seismic and well-log data (see Sects. 9.5.2, 9.5.4). In the Breathitt Group of eastern Kentucky interfluves are underlain by carbonaceous siltstones, underclays, and rooted surfaces (Aitken and Flint 1995). There is little to distinguish such deposits from floodplain deposits of the transgressive and highstand systems tracts, and Aitken and Flint (1995) were able to designate sequence boundaries in the interfluve areas only by physically tracing them from the base of incised valleys.

Tandon and Gibling (1994) argued for a climatic control on paleosols at sequence boundaries in the late Paleozoic cyclothems they studied in Nova Scotia. They interpreted calcareous paleosols as indicating seasonally dry climates, contrasting with the humid climates indicated by coal-bearing deposits of the highstand systems tracts. These cyclothems are, of course, ascribed to glacioeustatic controls (Crowell 1978). During major episodes of continental glaciation large-scale shifts in the earth's climatic belts occur, and it is to be expected that the effects of eustasy will be overlayered with the sedimentary effects of climate change. In this case, a northern hemisphere seasonal dryness is correlated with southern hemisphere glacial coldness (sea-level lowstand), and a northern hemisphere warm, humid climate is correlated with a southern hemisphere warm interglacial climate (sea-level highstand). As discussed in Sect. 12.12.2, with reference to the late Cenozoic record, the complexities of these linked changes in the controlling mechanisms are only just beginning to be properly understood.

Throughout the earlier studies of sequence stratigraphy the assumption has been made that the regional unconformities that constitute sequence boundaries have chronostratigraphic significance and can be used for regional, even global, correlation. A discussion of this premise is beyond the scope of this book. It should be noted, however, that it has long been known that the processes that generate nonmarine erosion surfaces are not instantaneous or regionally synchronous in their effect. For example, an erosion surface generated by a fall in base level develops by headward erosion as the longitudinal river profiles adjust to the new lower level. A change in slope from the old to the new, lower slope may generate a "nickpoint", which migrates landward with time (Butcher 1990; see Sect. 11.2.2). The time required for this migration is on the order of at least a few thousand years, which is a significant length of time in the context of high-frequency cycles, such as those generated by glacioeustasy. The response of fluvial systems to tectonic and climatic changes of all kinds may be characterized by a significant lag time (Shanley and McCabe 1994), which is likely to increase with distance from the source of the change and may therefore introduce significant complexities in the regional correlation of nonmarine sequences.

Within successions that are entirely nonmarine, especially within inland basins, where correlation with marine units cannot be attempted, the definition of sequences and the recognition of sequence boundaries may be a challenging task. The response of fluvial systems to tectonic and climatic changes is complex, and the resulting sequences may not display the simple succession of depositional-systems tracts that are now becoming familiar and perhaps even standardized features of the shallow-marine sedimentary record. The following examples of nonmarine sequence boundaries illustrate this point.

The adjustment of longitudinal fluvial profiles to tectonic and climatic change may not be complete if the ability of the system to respond to change is outpaced by the rate of change. The existence of river terraces flanking major alluvial valleys is a good example of this. Terrace complexes are present in most modern river valleys and have commonly been attributed to base-level changes following the last ice age. However, in a detailed review of the alluvial valley of the Colorado River, Texas, Blum (1992) showed that postglacial terraces represent periods of increased alluviation in response to climatically driven increases in sediment supply. Each alluvial level underwent dissection, leaving terrace remnants, at times of reduced sediment supply (Fig. 13.11). The nature of the climatic forcing process that controls this pattern is discussed in Sect. 12.12.2. Blum (1994) demonstrated that nowhere within coastal fluvial systems is there a single erosion surface that can be related to lowstand erosion. Such surfaces are continually modified by channel scour, even during transgression, because episodes of channel incision may reflect climatically controlled

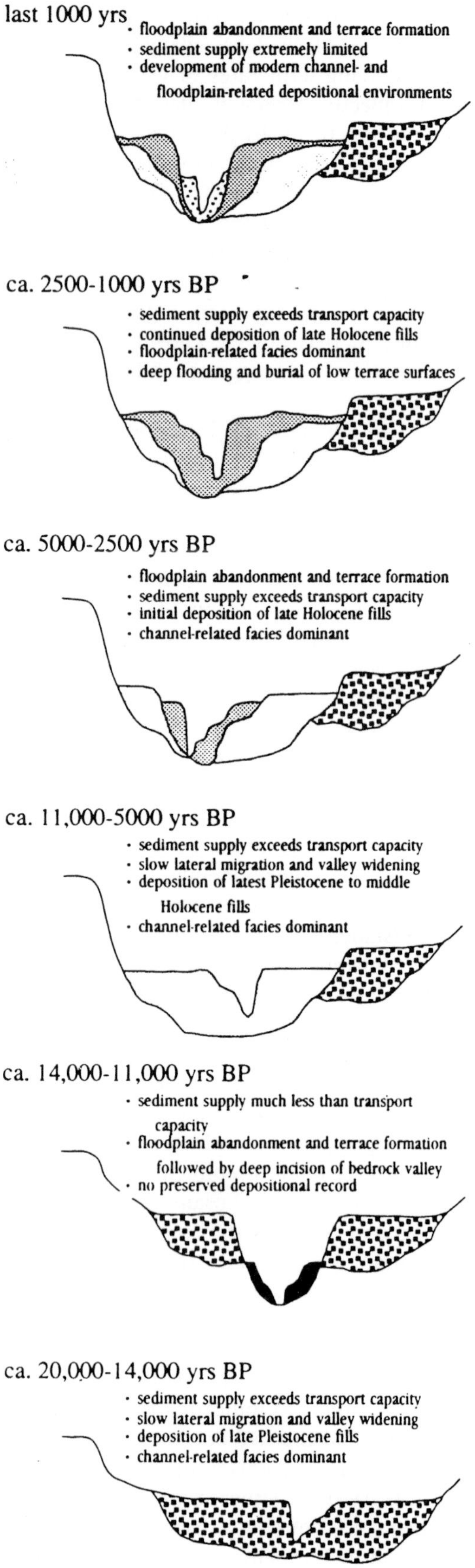

Fig. 13.11. History of alluvial incision and aggradation in response to postglacial climate change, Colorado River, Texas (Shanley and McCabe 1994, modified from Blum 1992, reprinted by permission). As discussed in Chap. 12, the major phase of valley incision within the 14–11 ka period corelates to the beginning of the last postglacial transgression

times of low sediment load, which are not synchronous with changes in base level. This is particularly evident landward of the limit of base-level influence. Postglacial terraces within inland river valleys reveal a history of alternating aggradation and channel incision, reflecting climate changes, all of which occurred during the last postglacial rise in sea level (Blum 1994). The implications of this have yet to be resolved for inland basins where aggradation occurs (because of tectonic subsidence), rather than incision and terrace formation. However, it would seem to suggest that no simple relationship between major bounding surfaces and base-level change should be expected.

Patterns of terrace fragments and incised valleys are probably associated with many nonmarine sequence boundaries in the geological record, where they may be related to tectonic, eustatic, or climatic processes. For example, the Mannville Group (Lower Cretaceous) of Alberta is characterized by several regional unconformity surfaces with incised valleys, many of them containing producing oil fields (Hopkins 1981, 1985; Chap. 15). Stratigraphic separation of the various channel and terrace fragments, which also include Jurassic remnants, is a difficult task in the subsurface.

In another part of the Alberta Basin careful observation and interpretation made it possible to demonstrate the importance of fluvial incision and sediment transport during falling base level, even in the absence of any recognizable preserved fluvial deposits. Conglomerates of the Upper Cretaceous Cardium Formation are interpreted as shoreface deposits (Plint et al. 1986), reworked from original fluvial channel deposits by wave processes during transgression (Arnott 1992). Depressions in the transgressive surface, which here corresponds to the sequence boundary, are interpreted as incised fluvial valleys widened by marine processes, which also reworked all remaining fluvial deposits.

Good outcrops in Petrified Forest National Park of Arizona allowed Kraus and Middleton (1987) to map a major intraformational disconformity surface within the Upper Triassic Chinle Formation. At least two periods of incision and aggradation are present, comparable in geometry to those shown in Fig. 13.11 but probably of different origin. Kraus and Middleton (1987) attributed the erosion to lowering of base level, which led to gullying of a mature alluvial landscape characterized by well-developed paleosols. The fill of the incised valley is mud dominated. It contains several generations of less mature paleosols, suggesting more rapid alluviation during episodes of raised base level.

Fluvial deposits are, by their very nature, characterized by multiple channel scours. One of the difficulties that must be tackled in a sequence analysis of entirely nonmarine deposits is the distinction of sequence boundary erosion surfaces from autogenic channel scours. Major changes in fluvial style or grain size across the boundary may be indicative, as in the sequences described by Steel and Ryseth (1990; see Figs. 9.21–9.23) and Olsen et al. (1995). Rogers (1994) recognized two major erosion surfaces in a nonmarine succession in Montana based on erosional relief, changes in grain size, and the presence of coarse channel-lag deposits with evidence of pervasive oxidation. However, such sedimentological changes may not occur. A much more subtle sequence boundary was recognized by Bromley (1991b) within a monotonous succession of braid-bar sandstones in the Kayenta Formation (Jurassic) of western Colorado. Cut-and-fill erosional features along this surface are unusually steep sided, suggesting that the sand was at least partially consolidated at the time of incision. Petrographic examination of the sandstone immediately below this surface revealed limited grain-to-grain contacts and abundant euhedral, syntaxial quartz overgrowths, suggesting cementation prior to compaction, in contrast to other sandstones within this deposit. Cementation can probably be attributed to silcrete development on an exposed desert surface. Paleocurrent directions are markedly different above and below this surface. Bromley (1991b) classified the surface as sixth-order in rank (in the classification system used in this book) and attributed it to tectonic adjustments of the alluvial plain. In this particular case salt tectonics, originating in the Paleozoic evaporites of the underlying Paradox Basin, was thought to have caused shifts in the paleoslope and changes in fluvial dispersal, resulting in lengthy exposure of the alluvial plain in the area under study.

In another study of a monotonous braided sheet sandstone unit López-Gómez and Arche (1993) were able to distinguish sixth-order surfaces on the basis of detailed mapping of large cliff sections. They subdivided their Triassic unit in Spain into six sequences bounded by these surfaces and related the development of the sequences to periodic basin-margin fault movement.

These studies by Kraus and Middleton (1987), Bromley (1991b), López-Gómez and Arche (1993), and Rogers (1994) identified sequence boundaries located entirely within nonmarine deposits and revealed the subtle nature of such boundaries. They reflect major tectonic, and possibly climatic changes that may have no relationship to changes in sea level.

13.3.2 Lowstand Systems Tract

The lowstand systems tract is defined as the succession of beds between the sequence boundary and the first widespread transgressive surface. Most sequence models, including those developed for nonmarine deposits (Figs. 13.4, 13.5), indicate that lowstand deposits occupy incised valleys and are laterally limited in extent. This may indeed be the case; for example, studies by Hopkins (1981, 1985), Wood and Hopkins (1989, 1992), Shanley and McCabe (1991, 1993), Legarreta and Uliana (1991), Kvale and Vondra (1993), Gibling and Bird (1994), Gibling and Wightman (1994), and Aitken and Flint (1995) all described laterally limited lowstand paleovalley fills. A recent book on incised-valley systems (Dalrymple et al. 1994) contains numerous examples of fluvial to estuarine lowstand deposits. Ribbon sandstone bodies occupying linear, commonly sinuous paleovalleys incised into unconformities are an important type of fluvial oil and gas reservoir (Harms 1966; Wood and Hopkins 1989, 1992; Weimer 1986; Chaps. 14, 15). In Ordos basin,

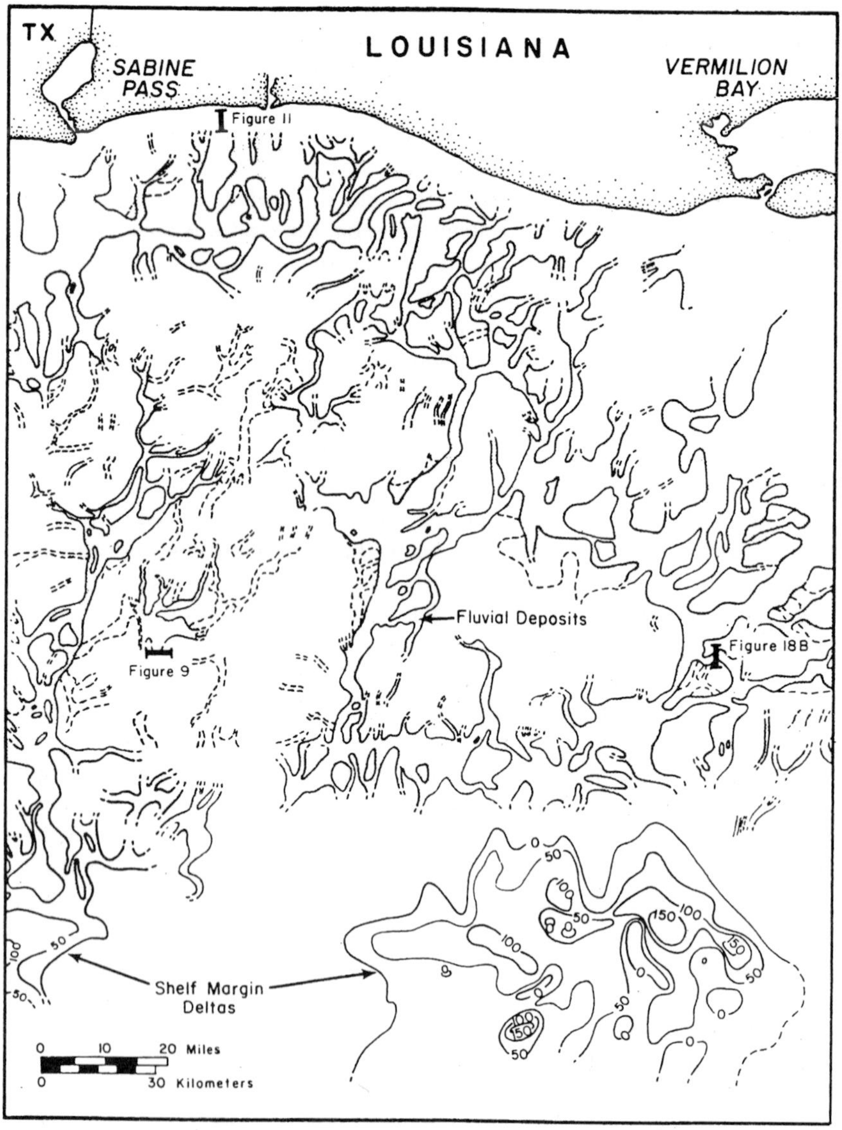

Fig. 13.12. Network of paleovalleys forming a lowstand systems tract developed during glacioeustatic sea-level low, Gul Coast (Suter et al. 1987). See also Fig. 9.12

China, incised valleys up to 300 m deep have been mapped (Moore et al. 1986; Song Guochu 1988). Most of the world's large rivers occupy the fill of canyons incised into the continental margin during glacial sea-level lowstands (e.g., Fisk 1944).

Individual paleovalleys can commonly be traced along the valley axis for many kilometers in the subsurface, as in the case of the Mannville Group valleys of Alberta (Wood and Hopkins 1989, 1992; Fig. 15.1) and those filled with the "J" sandstone in the Denver Basin (Harms 1966). Paleovalleys are associated with considerable stratigraphic relief and lateral facies changes, and this may make for difficulties in correlation between valley fills in both the surface and the subsurface (e.g., Gibling and Bird 1994). As Galloway (1989a) pointed out, this is one good reason for defining sequence boundaries in nonmarine deposits at the nonmarine equivalent of the maximum flooding surface, a proposal followed by Gibling and Bird (1994; see Sect. 13.3.4). In many cases, extensive networks of channels may be present on unconformity surfaces, in which individual channels are commonly incised tens of meters into the substrate (Hopkins 1981; Aubry 1989). In some cases the channel fills merge laterally to form extensive sheet sandstone bodies separated by interfluves, such as the lowstand deposits of the Gulf Coast Shelf off Texas and Louisiana (Suter et al. 1987). Here, the subsurface data base is adequate for the mapping of individual valleys (Fig. 13.12).

In some cases cases the mobile channel belts in which the lowstand deposits are formed merge laterally, to form broad sheets that are continuous across strike and lack mappable interfluves. The lower Castlegate Sandstone of Utah (Fig. 9.18) is a good example (Fouch et al. 1983; Yoshida et al., in press). Van Wagoner et al. (1991) mapped a network of incised channels at the base of the Castlegate Sandstone that define the initial braidplain system (Fig. 13.13), but aggradation resulted in these channels merging laterally, and most of the lower Castlegate Sandstone consists of amalgamated bar and channel deposits (Fig. 13.14) in which individual channel forms can rarely be traced for more than a few tens of meters (Miall 1994). The Lower Cretaceous Mesa Rica Sandstone of northeastern New Mexico is a similar lowstand sheet sandstone consisting of amalgamated channel deposits (Holbrook and Dunbar 1992).

The Castlegate Sandstone and the overlying and underlying beds east of Green River, Utah, comprise a good example of a mixed marine-nonmarine succession in which fluvial deposits characteristically form the lowstand systems tract and are not well represented elsewhere in the succession (Fig. 9.18). Another good example is the Early Jurassic succession of Poland and Sweden (Pienkowski 1991; Fig. 13.15). Highstand deposits might be expected to include deltaic and fluvial deposits, as shown in some standard sequence models (Fig. 13.1), but commonly insufficient time is available for coastal progradation during the highstand, or the deposits are removed by the erosion that generates the next sequence boundary.

Another of the prevailing assumptions about lowstand fluvial deposits is that they are likely to

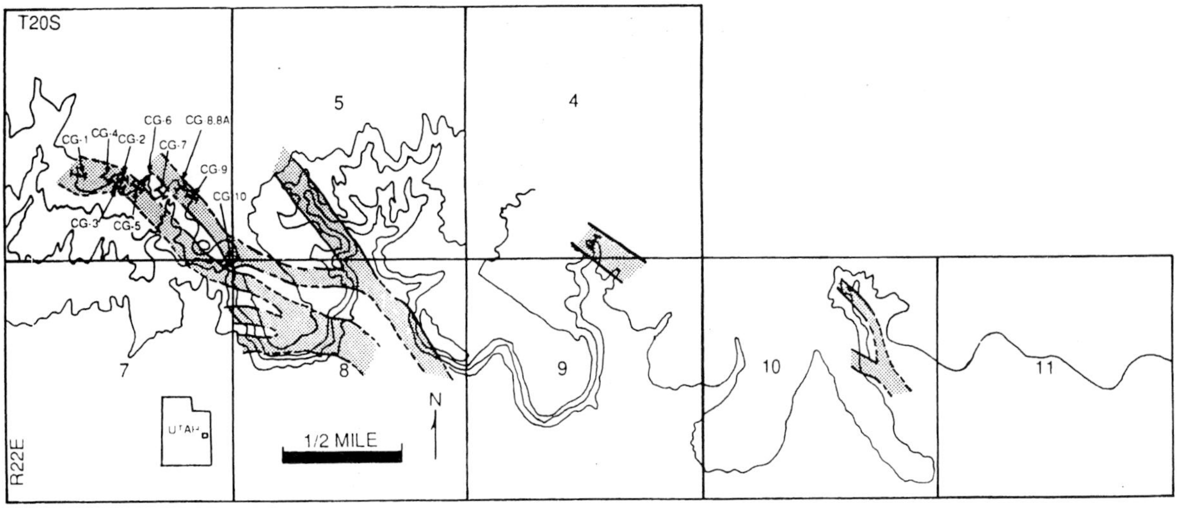

Fig. 13.13. Network of incised channels mapped at the base of the Castlegate Sandstone, near Green River, Utah. (Van Wagoner et al. 1991, reprinted by permission)

Fig. 13.14. Panorama of Castlegate Sandstone, Price Canyon, Utah, showing upper and lower sequence boundaries (*SB*) and the subdivision into Lower and Upper units (*Kcl, Kcu*), corresponding to LST and TST-HST deposits

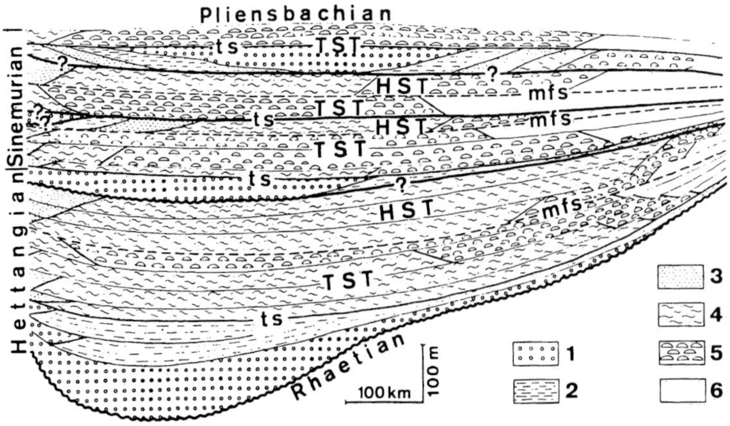

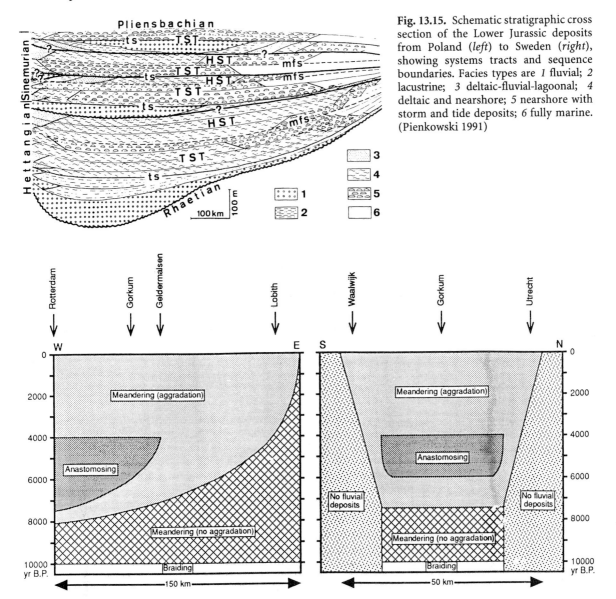

Fig. 13.15. Schematic stratigraphic cross section of the Lower Jurassic deposits from Poland (*left*) to Sweden (*right*), showing systems tracts and sequence boundaries. Facies types are *1* fluvial; *2* lacustrine; *3* deltaic-fluvial-lagoonal; *4* deltaic and nearshore; *5* nearshore with storm and tide deposits; *6* fully marine. (Pienkowski 1991)

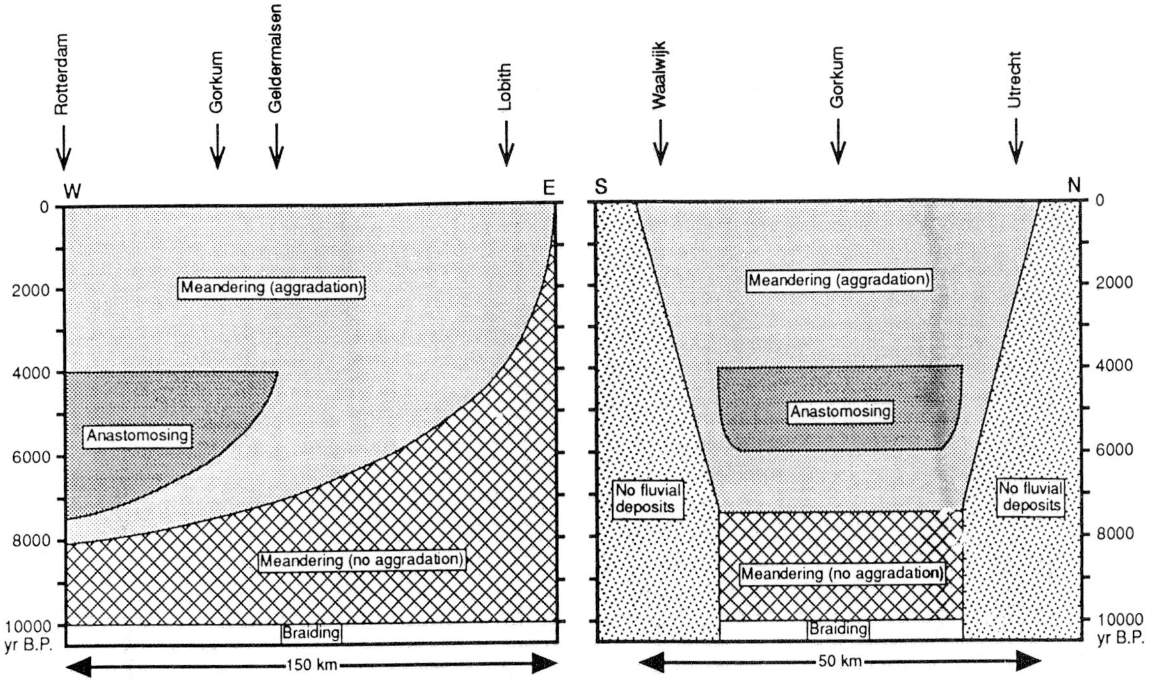

Fig. 13.16. Time-space model showing the Holocene development of channel patterns in the Rhine-Meuse delta. *W-E* diagram is longitudinal, *S-N* diagram is a transect across the alluvial valley. (Törnqvist 1993)

consist of amalgamated channel deposits exhibiting high net-to-gross sandstone ratios, in constrast to the deposits of the transgressive and highstand systems tracts, which current models postulate should consist of more isolated channel bodies encased within thicker floodplain fines (Figs. 13.4, 13.5). This assumption has a long history; it may stem from the work of Fisk (1944) on the Mississippi River valley, as repeated by many later workers (e.g., Weimer 1986). The oldest lowstand deposits, resting on the incised valley floor, consist primarily of a braided sandstone sheet. These pass upward into transgressive deposits in which a greater proportion of floodbasin fine deposits are preserved, and the channels exhibit a greater sinuosity. Most of the examples of nonmarine sequences referred to in this chapter exhibit this pattern (see in particular Gibling and Wightman 1994), but the nature of the sedimentary record depends entirely on the balance between subsidence, base-level change, and sediment supply. Where accommodation space is generated more rapidly than it can be filled by sediment, incised valleys will be invaded by arms of the sea, and the environment will be estuarine, not fluvial (Zaitlin et al.

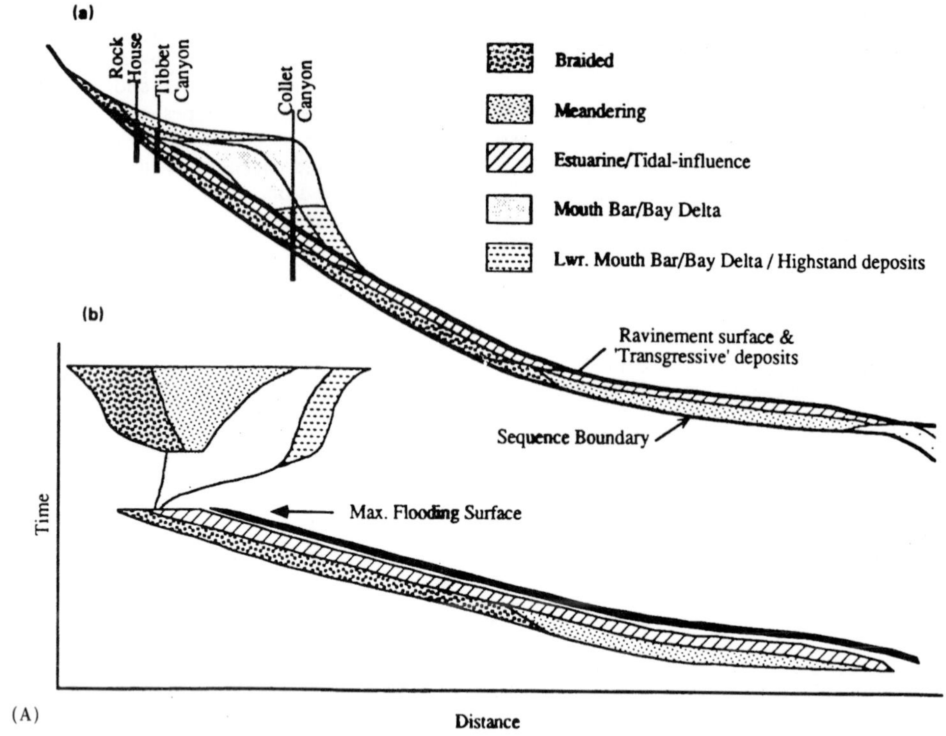

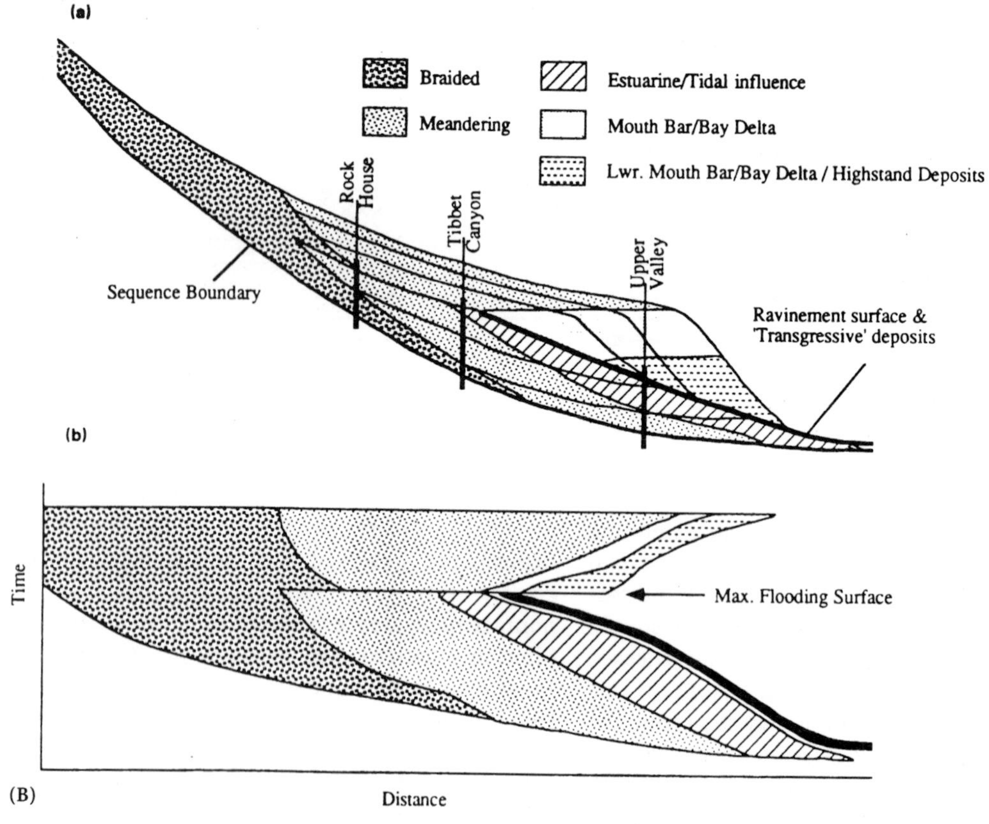

Fig. 13.17A,B. Longitudinal stratigraphic cross section (*a*) and chronostratigraphic diagram (*b*)(Wheeler chart) of fluvial-coastal sequences developed during conditions of **A** rapid rise of base level and **B** slow rise of base level. Based on studies of Upper Cretaceous sequences in southern Utah. (Shanley et al. 1992)

1994). Such deposits are better classified with the transgressive systems tract and are discussed in the next section.

13.3.3 Transgressive Systems Tract

The transgressive systems tract is defined as the beds between the first widespread transgressive surface below and the maximum flooding surface above (Posamentier and Vail 1988; Zaitlin et al. 1994). In practice, there is considerable variation in the composition of the lower part of stratigraphic sequences. As noted above, in some cases little or no fluvial deposit may be preserved at the base of the sequence, although there may be evidence for their former presence, as in the example discussed by Arnott (1992). In this case a marine transgressive systems tract rests directly on the sequence boundary. Incised valleys may be filled with fluvial or estuarine strata, depending on the balance between the rate of generation of accommodation space and the rate of sediment supply (Zaitlin et al. 1994). There may be an upward transition from fluvial to estuarine deposits corresponding to the boundary between the LST and the TST, although this is likely to be diachronous, as the rate of generation of accommodation space increases, and the coastline undergoes transgression and retrogression (Fig. 13.17). Alternatively, there may be simply a change in fluvial style, as suggested in some hypothetical fluvial sequence models (Fig. 13.4). In the Holocene deposits of the Rhine-Meuse delta Törnqvist (1993, 1994) and Törnqvist et al. (1993) demonstrated that a rapid rise in base level led to the development of an anastomosed fluvial style with a high avulsion frequency, whereas a lower rate resulted in the formation of a meandering channel system (Fig. 13.16).

The thickness of the transgressive systems tract depends on the balance between sediment supply and the rate of generation of accommodation space (Fig. 13.17). A rapid rise in base level may result in a very thin TST. In the case of the Castlegate Sandstone of Utah, there is virtually no recognizable TST in distal regions, east of Green River, where the marine shales of the Buck Tongue overlie the LST sandstones with a sharp contact (Fig. 9.18), whereas in updip areas, such as near Sunnyside, much of the Upper Castlegate Sandstone is assigned to the transgressive systems tract (Fig. 9.18). In the modern Canturbury Plains of New Zealand, transgression, resulting from tectonic subsidence and eustatic sea-level rise, is leading to shoreface erosion and the incision of river valleys that drain into the retreating coastline. This is occurring because of high coastal wave energies, and is resulting in an extension of the marine ravinement surface inland up the major river valleys (Leckie 1994).

The key to the recognition of the transgressive systems tract and the maximum flooding surface may be the recognition of marine influence in otherwise nonmarine successions (Shanley and McCabe 1994). The evidence may consist of tidal or wave-formed sedimentary structures or marine faunas (Figs. 13.5, 13.18). The landward extent of this marine influence depends on the balance of such factors as wave power, tide range, sediment supply, and subsidence rate. A detailed description of tidal features in transgressive deposits of Upper Cretaceous age in southern Utah is given by Shanley et al. (1992), who mapped tidal facies some 65 km inland of coeval shoreline deposits. Similar features are present in the Castlegate Sandstone of central Utah and are illustrated here. The most prominent sedimentary indicators of tidal influence are sand waves (Allen 1980), tidal bedding (Reineck and Wunderlich 1968), and inclined heterolithic strata (Thomas et al. 1987), examples of which are illustrated in Fig. 13.18. Tidal sand waves are commonly characterized by sigmoidal bedding and rhythmically spaced mud drapes, which are distinctive structures that may be recognizable even in small outcrops or core.

Marine influence may extend inland for tens of kilometers, but a limit is reached at the time of highest base level, when the maximum flooding surface is formed. This marine limit may be well within the depositional basin, especially in foreland basins, where such proximal deposits have a high preservation potential. Updip from the marine limit, the transgressive systems tract may be recognizable only by a change in fluvial style. The study by Shanley et al. (1992) in southern Utah (Fig. 13.17) and that of the Castlegate Sandstone (Olsen et al. 1995; and as reported here: Fig. 9.18) both show upward changes in fluvial style. In Price Canyon the Castlegate Sandstone can be subdivided into two members (Olsen et al. 1995). There is a lower sandstone member, which consists of a succession of superimposed braided sandstone sheets with very little in the way of preserved overbank deposits [e.g., Castlegate-A location of Miall (1993)] and an upper member containing thicker and more extensive mudstone deposits (Fig. 13.14). There is little evidence of a change in channel style [contrary to the interpretations of Olsen et al. (1995)], and the change may simply reflect an increase in the rate of generation of

A

Fig. 13.18A,B. Examples of tidal sedimentary structures in transgressive systems tracts within dominantly fluvial sequences, Castlegate Sandstone, near Price, Utah. **A** Inclined heterolithic stratification (*IHS*). *SB* indicates the sequence boundary at the top of the Castlegate Sandstone; **B** tidal bedding

accommodation space from the LST to the TST, resulting in greater vertical separation of channel bodies (a lower net-to-gross sandstone ratio). It is not clear whether any of the upper Castlegate Sandstone corresponds to a highstand systems tract. Present evidence suggests that the HST of the Castlegate sandstone is not well preserved.

In the absence of marine-influenced deposits, the distribution of coal may provide a useful indication of transgressive to highstand depositional conditions. Coal is formed in various low-energy environments, notably alluvial backswamps and back-barrier lagoons, but, as McCabe (1984) has pointed out, in many such environments a continuous clastic influx (from overbank flooding or barrier washover) inhibits the development of true coals, and carbonaceous shales are the result. The existence of widespread, ash-free coals therefore indicates that some additional process has occurred. McCabe (1984) emphasized the importance of raised mires, which are organic swamps in which the rapid rate of organic growth and accumulation outpaces clastic sedimentation. An additional factor may be the reduction in clastic influx into backswamp areas associated with certain regional allogenic conditions. Many coals are associated with transgressions, and this has long formed part of the interpretation of the classic Carboniferous cyclothems of Europe and North America (e.g., Moore 1964; Wilson 1975; Crowell 1978; Gibling and Bird 1994). In these deposits coals typically occur above the lowstand fluvial sandstones and below marine shales and limestones. During the transgressive phase of sequence formation the rate of generation of accommodation space is at its highest, while clastic depositional systems are undergoing flooding and retrogradation. These are ideal conditions for coal development, and the thickest and most areally extensive coals are typically formed at such times (Ryer 1984; Hamilton and Tadros 1994). As noted above, fluvial deposits comprising the TST commonly contain a higher proportion of floodplain deposits than those formed during lowstand or highstand, and so, given appropriate conditions for restricting clastic influx, the more

Fig. 13.18B

widespread distribution of coal in this part of a base-level cycle is to be expected.

Still further updip from the coastal plain, fluvial style may remain constant through the LST, TST, and HST. The effects of base-level change are likely to be minimal at a distance of 100 km or more upstream from the river mouth at lowstand (Fig. 13.3), and in a foreland basin the more rapid subsidence rate prevailing in proximal regions is likely to be the dominant control. Shanley et al. (1992) indicated a continuous succession of braided fluvial deposits through their sequences (Fig. 13.17), and the westernmost (most proximal) exposures yet examined in the Castlegate Sandstone (Franczyk and Pitman 1991) reveal that the type of fluvial facies that characterizes the lower Castlegate at Price Canyon extends through the entire thickness of the unit (Fig. 9.18).

The onlap of sequence boundaries by the transgressive systems tract is one of the major lines of architectural evidence used by Vail et al. (1977) for the definition of sequences in the subsurface, especially when using seismic data. However, it is considered very unlikely that simple transgressive onlap will be observed in fluvial sequences. It is unlikely, in the first place, that in a fluvial system any depositional response to changes in accommodation space will generate a single, simple bedding surface that could be traced for long distances on a seismic record. Stratigraphic architecture is complex in fluvial systems and may be difficult to trace for very far on seismic records (Sect. 9.5.4.). Modern seismic acquisition and processing methods are capable of yielding a much higher degree of resolution than they did when the models of Vail et al. (1977) were first published, and Cartwright et al. (1993) have pointed out the difficulty of "forcing through" seismic correlations in many clastic settings that are characterized by rapid lateral facies change. In the second place, the response to the addition of fluvial accommodation space during transgression may be to generate a "backwater effect", whereby the river responds to a rising relative base level by the aggradation of a wedge of sediment that gradually tapers upstream to zero thickness over tens of kilometers, yielding no clear-cut onlap geometry (Fig. 13.17; the effects of base-level change on rivers are discussed in detail in Sect. 11.2.2).

13.3.4 Equivalent of Maximum Flooding Surface

The maximum flooding surface (MFS) is represented by the maximum landward extent of marine deposits; it is immediately followed by the highstand systems tract (Vail et al. 1977; Vail 1987). In marine deposits the surface may be represented by a condensed section (Loutit et al. 1988). Typically, highstand deposits such as regressive delta and coastal barrier-strandplain systems prograde seaward, developing a downlap relationship to the MFS. This downlap architecture is a distinctive feature of seismic cross sections (Vail et al. 1977; Vail 1987). Galloway (1989a) argued that the MFS is commonly a more readily mappable surface than a subaerial erosion surface, and he recommended using the MFS as the sequence boundary. To distinguish such sequences from those of Vail and his colleagues he termed them *genetic stratigraphic sequences.* This usage has been followed by some recent workers. For example, Gibling and Bird (1994), who studied the coal-bearing cyclothems of the Sydney Basin, Nova Scotia, found that coal beds which form at or close to the time of maximum flooding are much easier to correlate in the subsurface than the unconformity surfaces marking the lowstand. The latter are cut by laterally impersistent valley-fill deposits and may juxtapose similar facies, making correlation difficult. A similar approach was recommended by Hamilton and Tadros (1994). Gibling and Bird (1994) placed the boundaries of their cyclothems (sequences) at the base of the maximum flooding surface (Fig. 13.19). The most well-developed coals occur in the TST, while the maximum flooding surface itself is typically represented by carbonaceous limestones and shales with limited faunas, deposited in restricted bay environments.

Interfluve areas may be characterized by slow rates of aggradation and the development of extensive paleosols. During times of base-level lowstand water tables are at their lowest, and such soils are likely to be oxidized types (Shanley and McCabe 1994; Gibling and Bird 1994). During the transgressive phase, and culminating in the MFS, the water table rises, and soils may undergo a change from oxidized types to those formed in reducing conditions (see Sect. 7.4.2 for paleosol description and classification). However, such changes may be confused with those resulting from changes in climate, which are to be expected in cycles caused by glacioeustatic mechanisms (see Sect. 12.12.2).

As in the case of the transgressive systems tract, well inland from the coastal plain, beyond the reach of marine processes, there may be no facies indicators of maximum flooding. In these proximal zones, basin-margin influences, such as rapid rates of subsidence resulting from extensional faulting or crustal loading, are the dominant control on sedimentary style (Fig. 13.3).

13.3.5 Highstand Systems Tract

The highstand systems tract develops when base-level rise slows down and the rate of generation of accommodation space decreases to a minimum (Figs. 13.2, 13.5). There are two possible depositional scenarios for this phase of sequence development. Retrogradation of the river systems during transgression will have led to reduced slopes and to a low-

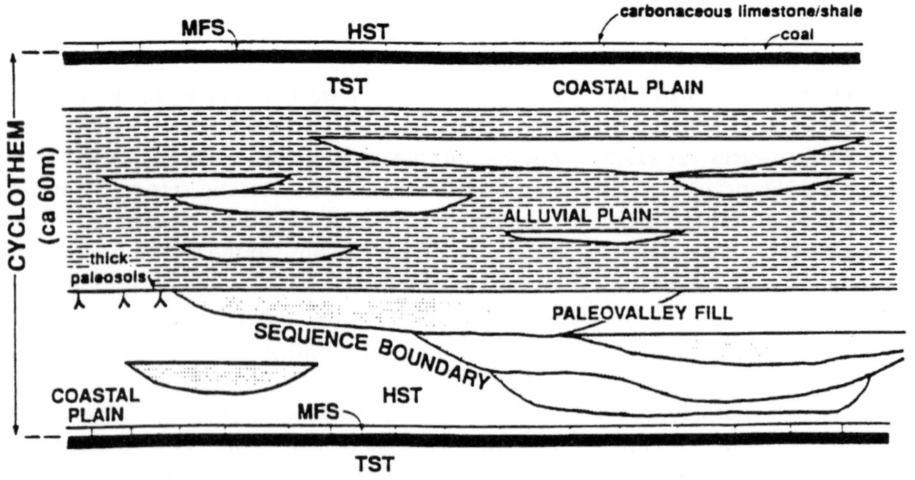

Fig. 13.19. Stratigraphic model for Carboniferous cyclothems of Sydney Basin, Nova Scotia. (Gibling and Bird 1994)

energy landscape undergoing slow accumulation of floodplain deposits, limited channel aggradation, and closely-spaced, well-developed soil profiles (Shanley and McCabe 1994). Given no change in source-area conditions, however, the sediment supply into the basin will continue, and vigorous channel systems will eventually be re-established. Under these conditions, channel bodies will form that show reduced vertical separation relative to the TST, leading to lateral amalgamation of sandstone units and high net-to-gross sandstone ratios (Wright and Marriott 1993; Olsen et al. 1995; Fig. 13.4). Basinward progradation of coastal depositional systems leads to downlap of deltaic and barrier-strandplain deposits onto the maximum flooding surface (Fig. 13.17). A good nonmarine example of this was described by Ray (1982), who mapped the progradation of an alluvial plain and deltaic system into lake deposits (Fig. 9.44).

It seems likely that the HST will be poorly represented in most nonmarine basins, because the highstand is immediately followed by the next cycle of falling base level, which may result in the removal of much or all of the just-formed HST deposits by subaerial erosion. A minor increase in the sand-shale ratio immediately below the sequence boundary may be the only indication of the highstand phase, as in the Castlegate Sandstone of Utah (Olsen et al. 1995; Fig. 9.18).

Care must be taken to evaluate all the evidence in interpreting such data as net-to-gross sandstone ratios. Changes in this parameter may not always be attributable to changes in the rate of generation of accommodation space. G.A. Smith (1994) described a case where an increase in the proportion of channel sandstones in a section seems to be related not to changes in the rate of generation of accommodation space, but to increased sediment runoff resulting from increased rainfall. In the case of sequences driven by orbital forcing mechanisms, where both base-level change and climate change may be involved, unraveling the complexity of causes and effects is likely to be a continuing challenge. Some aspects of this problem are discussed in Sect. 12.12.2. In the model of Shanley and McCabe (1994) a greater degree of channel amalgamation is shown in the TST than in the HST (Fig. 13.5), the opposite of that shown in the model of Wright and Marriott (Fig. 13.4). Shanley and McCabe (1994) suggested that where rising base level is the main control on the rate and style of channel stacking, the rate of generation of accommodation will be small during transgression in inland areas while the coastline is still distant, and will increase only when transgression has brought the coastline farther inland, where the base-level rise on the lower reaches of the river produces a more rapid increase in accommodation. In this model the rate of generation of accommodation is greater during the highstand than during transgression and results in low net-to-gross sandstone ratios. However, this line of reasoning omits the influence of upstream factors, and must therefore not be followed dogmatically. One must also be cautious in using systems-tract terminology derived from marine processes for the labeling of nonmarine events. There may be a considerable lag in the transmission of a transgression upstream to inland positions (Fig. 13.17). The inland reaches of the river will not "know" that a transgression is occurring, and it is questionable, therefore, whether the deposits formed during the initial stages of the marine transgression should be included with the TST.

13.3.6 Falling-stage Systems Tract

This systems tract has not yet been identified in any ancient fluvial deposits. In nonmarine settings a fall in base level is a time when the generation of accommodation space is negative, and erosion occurs. This constrasts with shoreline environments, in which "sharp-based shoreface bodies" form "stranded parasequences" that have a high preservation potential (Plint 1988; Hunt and Tucker 1992; Posamentier et al. 1992).

Terraces flanking incised valleys may correspond to the FSST. The origin of fluvial terraces is complex, however. Blum (1992; Fig. 13.11) argued that climate changes may have been more important than base-level changes in the development of postglacial terraces of the Colorado River, Texas (this topic is discussed at greater length in Chap. 12). As argued by Arnott (1992), detritus generated by fluvial erosion during the falling stage may provide an important sediment supply for shoreline sedimentation, even when all traces of the river system are removed by the falling-stage erosion itself.

13.4 Time Scales of Nonmarine Sequences and Their Causes

Stratigraphic sequences occur over time scales ranging through five orders of magnitude, from tens of thousands of years to hundreds of millions of years

Table 13.1. Stratigraphic cycles and their causes

Type[a]	Terminology	Duration, (million years)	Probable causes
First-order	– –	200–400	Major eustatic cycles caused by formation and breakup of supercontinents
Second-order	supercycle (Vail et al. 1977); sequence (Sloss 1963)	10–100	1. Eustatic cycles induced by volume changes in global midocean spreading centers 2. Regional extensional downwarp and crustal loading
Third-order	mesothem (Ramsbottom 1979); megacyclothem (Heckel 1986)	1–10	Regional cycles caused by intraplate stresses. Most are probably not of global extent
Fourth-order	Cyclothem (Wanless and Weller 1932); major cycle (Heckel 1986)	0.2–0.5	1. Milankovitch glacioeustatic cycles, astronomical forcing 2. Regional cycles caused by flexural loading, especially in foreland basins
Fifth-order	minor cycle (Heckel 1986)	0.01–0.2	1. Milankovitch glacioeustatic cycles, astronomical forcing 2. Regional cycles caused by flexural loading, especially in foreland basins

[a] Hierarchy modified from Vail et al. (1977).

(Vail et al. 1977; Miall 1990, Chap. 8). They have been conveniently subdivided into a fivefold hierarchy (Table 13.1), but the subdivision into the five groups is very approximate and classifies sequences together that probably have very different origins. Examples of nonmarine stratigraphic units that can be classified into these various ranks, and the corresponding "groups" used in this book, are listed in Table 9.1 and are described and illustrated in Chap. 9. The purpose of this section is to provide a summary of these five types of sequence, together with a brief discussion of the various allogenic processes responsible for their formation, processes that are discussed extensively in Chaps. 11 and 12. The focus here is on nonmarine deposits, marine sequences having received extensive discussion elsewhere in the geological literature.

13.4.1 First-order Cycles

First-order cycles are related to the formation and breakup of supercontinents over periods of several hundreds of millions of years. Worsley et al. (1984, 1986) described two Phanerozoic cycles related to the breakup and dispersal of Pangea, and Hoffman (1991) developed a tentative plate-tectonic model that extended the cycle back into the Proterozoic. The most important stratigraphic effect of these long-term plate-tectonic movements is the variation in the long-term rate and global extent of sea-floor spreading. Lengthy, active spreading centers are thermally elevated and displace ocean waters onto the continents. The breakup and dispersion of supercontinents therefore tends to be accompanied by high sea levels, such as those that characterized the earth during the rapid opening of the Atlantic, Indian, and Southern Oceans during the Cretaceous. These cycles have only a marginal, long-term relevance to fluvial sedimentation, however, and are not discussed further, except for the following point: Ager (1981) noted the widespread distribution of certain facies at certain geological times, and this can be attributed to control by these long-term global cycles. For example, nonmarine deposits were particularly widespread during the Permian and Triassic (the New Red Sandstone of Europe). At this time Pangea was beginning to undergo extension, and rift basins developed in virtually all continents. Well-known examples include the Newark rift basins flanking the Atlantic margins of eastern North America (Fig. 11.40) and northwest Africa, the Karoo rifts of southern Africa, and numerous rift systems in Europe (e.g., the Buntsandstein deposits of Germany).

13.4.2 Second-order Cycles

Second-order cycles are the original interregional cycles of Sloss (1963). There is increasingly convincing evidence that many of these are global in scope and are generated by cycles of eustatic sea-level change lasting several tens of millions of years (Miall

1990, Chap. 8). The primary mechanism is probably variations in global sea-floor spreading rates superimposed on the first-order cycle discussed above. Such variations reflect continental-scale adjustments to spreading patterns in response to plate rifting and collision events. The mechanism was first suggested by Hallam (1963), and Pitman (1978) is credited with the theoretical research that lead to general acceptance of the idea.

Cycles of base-level change can also be generated by crustal movements at continental margins. On extensional margins crustal thinning and post-rift cooling lead to subsidence on a time scale of tens of millions of years (McKenzie 1978; Watts 1981), and rift-fill and extensional-margins deposits may include significant thicknesses of fluvial deposits (Sect. 11.4.2). Clastic wedges are also generated by convergent and strike-slip tectonics, especially in foreland basins (Sect. 11.4.6), where wedges of nonmarine "molasse" spanning several millions of years to more than 10 million years are common.

13.4.3 Third-order Cycles

The nonmarine sequences and sequence models described earlier in this chapter are based mainly on examples of third- and fourth-order cycles that have been described in the outcrop or subsurface.

It is this writer's view that most, perhaps even all third-order cycles are tectonic in origin (Miall 1991c; Yoshida et al., in press). Regional tectonic episodes localized along major faults generate basin-margin uplifts and cycles of basin subsidence on time scales of several millions of years. Tectonism is triggered by the regional response to plate-tectonic movements and may include minor changes in sea-floor spreading vectors and the adjustment of major fault blocks and terrane fragments on the continents. Stresses are transmitted laterally by intraplate stress (Sect. 11.3.5), and in this way may cause the development of synchronous cycles of uplift and subsidence over areas as large as entire continental plates. Several examples are given in Sect. 9.4 (group 10 deposits of Table 9.1), and the plate-tectonic settings of these various types of sequence are discussed in Sect. 11.4.

13.4.4 Fourth- and Fifth-order Cycles

Fourth- and fifth-order cycles correspond to groups 9 and 8, respectively, in Table 9.1, where several nonmarine examples are listed. It seems increasingly likely that many fourth- and fifth-order cycles in tectonically active basins, such as foreland basins, are caused by short-term episodes of crustal loading (Peper et al. 1992; Plint et al. 1993; Yoshida et al., in press). If this is the case, these cycles are likely to be localized in extent and confined to individual basins. The mechanics of basin movement and consequent changes in sediment delivery, depositional slope, and accumulation rates are the subject of much theoretical study at the present time (e.g., Jordan and Flemings 1991).

Other cycles are driven by orbital forcing of climate. Minor climatic cycles may simply lead to variations in discharge within a river system, whereas major cycles may cause regional or hemispheric continental glaciation and consequent changes in base level. This has long been thought to have been the cause of the classic cyclothems of Europe and North America. Minor cycles may have occurred at any time, whereas those involving glacioeustatic base-level change probably are confined to the Neogene, the Late Paleozoic (the time of the Gondwana glaciation), and various times in the Precambrian. An additional discussion of orbital forcing is given in Sect. 12.12, where examples of minor and major cycles are described.

13.5 Discussion

Sequence stratigraphy is a subject undergoing rapid evolution at this time, and there is a danger that this chapter in the book, in particular, may be out of date before it is published. Although a consensus seems to be emerging regarding the application of sequence concepts to the nonmarine record, very few well-documented examples of nonmarine sequences have yet been described. There is room for a considerable amount of additional research in this area. Quantitative modeling is also required, of the type discussed in Sect. 11.3.6, to more fully explore the relationship between tectonic processes, basin movements, sediment supply, sediment delivery, and facies changes within sedimentary basins.

The example of the Castlegate Sandstone (Fig. 9.18) illustrates some of the difficulties in applying sequence concepts to fluvial deposits (Yoshida et al., in press). Where marine deposits interfinger with the fluvial succession, as in the area east of Trail Canyon, recognition of sequence boundaries, flooding surfaces, and systems tracts is relatively straightforward (Van Wagoner et al. 1990, 1991). The fact that some of these features are difficult to trace updip, away from the coastal plain, raises an important question regarding the sensitivity of fluvial systems to base-level change. In the most proximal exposures of the Castlegate Sandstone (Fig. 9.18; Joes Valley, Price Canyon) there appears to be only a single sequence, of probable third order, according to the chronostratigraphic reconstruction of Fouch et al. (1983). These beds appear to correlate with a succession of higher-order sequences in the coastal plain area (Trail Canyon and points east). The updip merging of these higher order sequences into a single undifferentiated succession may be interpreted in one of at least two ways: (a) Fluvial systems are relatively insensitive to base-level change, the effects of which disappear updip; or (b) the Castlegate stratigraphy may be controlled by more than one process, such as third-order tectonism (fluvial rejuvenation, changes in sediment supply), especially in updip regions, and a fourth-order cycle of base-level change, of tectonic or eustatic origin, which is most pronounced in coastal areas. Local complications arise in this particular case, because the downdip areas (Trail Canyon and areas to the east) are underlain by the Late Paleozoic evaporites of Paradox Basin, and the effects of contemporaneous salt tectonics may have been significant. This should serve to remind us that in each real-life case study, the lessons to be learned from theoretical or numerical models must be tempered by the practicalities of the given basin setting, with its inherited crustal heterogeneity and history of subsidence and tectonism.

Chapter 14

Stratigraphic and Tectonic Controls on the Distribution and Architecture of Fluvial Oil and Gas Reservoirs

14.1 Introduction

Significant volumes of oil and gas are trapped in fluvial sandstones. Major reservoirs include the Statfjord Formation (Triassic-Jurassic) of the North Sea Basin, the Sadlerochit Group (Permian-Triassic) of Prudhoe Bay on the Alaskan North Slope, the Lower Cretaceous reservoirs of the giant Daqing field of the Songliao Basin, China, the heavy-oil sands of the Cretaceous Athabasca and related deposits in Alberta, Canada, and numerous large to small fields in mature areas such as the Alberta Basin (Mannville fields, Cretaceous-age reservoirs), the southern Midcontinent (Cherokee fields, Pennsylvanian), and Gulf Coast (Cretaceous Tuscaloosa and Travis Peak fields, and Oligocene Frio fields). The tectonic setting and reservoir geometries of these and other fluvial reservoirs are varied, but this chapter demonstrates that oil and gas fields in fluvial sandstones fall into three broad classes, on the basis of two major descriptive criteria. An extensive review of fields around the world is drawn upon to illustrate this classification.

The two criteria are: (a) geometry and origin of the depositional system and (b) geometry of the reservoir bodies (Table 14.1). One of the major purposes of this chapter is to develop a threefold descriptive grouping of fluvial reservoirs based on these criteria. The three reservoir types tend to occur in particular tectonic settings, but the tectonic control is varied enough that this cannot form part of the classification. This chapter deals with only part of what constitutes a complete "play". Other aspects including source rocks, migration history, trap timing, etc. (White 1980, 1988) are not discussed here.

Some of the ideas regarding the various styles of fluvial traps were described briefly by Cant (1982). Blackbourn (1984) discussed the facies variability of fluvial deposits but not the details of petroleum occurrence or reservoir types. Davies et al. (1991, 1993) offered two models for fluvial channel-fill reservoirs and suggested that their classification has a broad applicability, but they did not develop this idea further. Conybeare (1976) described numerous fluvial oil and gas fields but did not offer any useful ideas regarding classification.

Detailed published information on the facies architecture of producing fluvial reservoirs is available for many fields in North America and northwest Europe, and for some in Australia. Most of this paper is based on data from these areas. For other parts of the world much less information is available, in part

Table 14.1. Criteria for classification of petroleum reservoirs in fluvial sandstones

Criterion	Classes	Comments
Geometry of depo. system	1. Clastic wedge 2. Paleovalley	Active tectonism Low base level
Geometry of reservoir	1. Sheet 2. Ribbon or lens	Mainly braided systems Meandering, anastomosed
Tectonic setting[a]	1. Rift basin 2. Extensional-margin basin 3. Backarc basin 4. Retroarc foreland basin 5. Forearc basin 6. Strike-slip basin 7. Peripheral foreland basin	Assoc. with lake beds May be volcaniclastic Assoc. with collision Assoc. with collision

[a] Tectonic setting is included for discussion purposes but does not form a primary part of the classification.

because petroleum exploration and development are still in relatively early stages, and the details of reservoir characteristics are not yet well known (or have not been published). This applies to much of Africa (Gondwanan and Cenozoic rifts), Andean South America, southeast Asia, India, and China. In all these areas nonmarine reservoirs are significant and may even be the most important reservoir type.

The classification proposed here, with the accompanying extensive referencing to published examples, is offered as an aid to exploration and production. The numerous published case studies cited here may provide useful analogs for future field studies.

14.2 The Geometry of Fluvial Reservoirs

14.2.1 Geometry and Origin of Depositional Systems

There are two broad architectural styles of importance for petroleum trapping, the clastic wedge, and the paleovalley fill (Table 14.1, Fig. 14.1).

14.2.1.1 Clastic Wedges

Major clastic wedges and thick sandstone sheets occur wherever tectonic conditions maintain steep paleoslopes and significant relief. Such wedges are therefore commonly associated with rift basins and foreland basins, which are generally bounded by upfaulted mountainous source areas (Figs. 14.1A,B, 14.2). Clastic wedges may be hundreds of meters to a few thousand meters thick, and may extend along strike for tens to hundreds of kilometers. They typically pass downslope into major delta or strandplain systems (Fig. 14.1A) that may themselves be significant petroleum reservoirs. Alternatively, they may grade into lacustrine deposits (Fig. 14.1B). Whether the distal margins are marine or lacustrine, base-level change is a significant control on the distribution and geometry of the fluvial wedge. Tongues or sheets of fluvial deposits, formed during times of low sea level or during periods of slow subsidence, may be interbedded with fine-grained sediments formed in subaqueous settings (lake basin, marine shelf, or interdeltaic bay). Autogenic switching of delta dis-

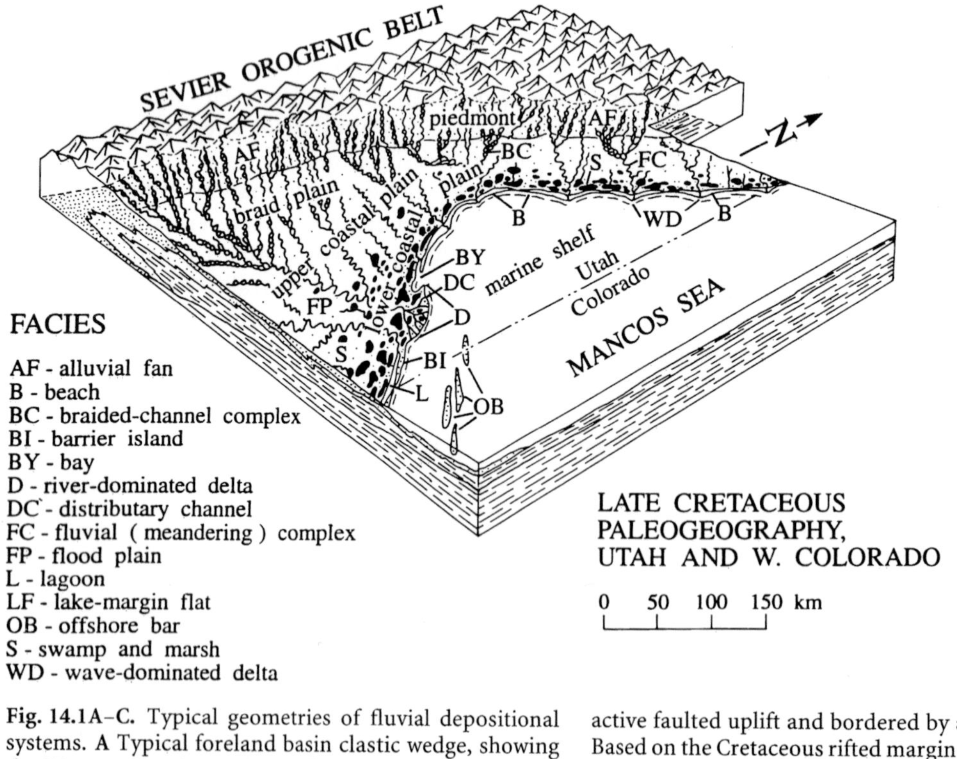

A

Fig. 14.1A–C. Typical geometries of fluvial depositional systems. **A** Typical foreland basin clastic wedge, showing the Western Interior Basin, Utah–Colorado, during accumulation of the Mesa Verde Group (Upper Cretaceous) (Cole and Friberg 1989). **B** Clastic wedge bounded by active faulted uplift and bordered by an arid lake system. Based on the Cretaceous rifted margin of Brazil (Mello and Maxwell 1990; modified from Eugster and Hardie 1975). **C** A typical paleovalley system. Early Cretaceous paleogeography of west-central Alberta. (Rosenthal 1988)

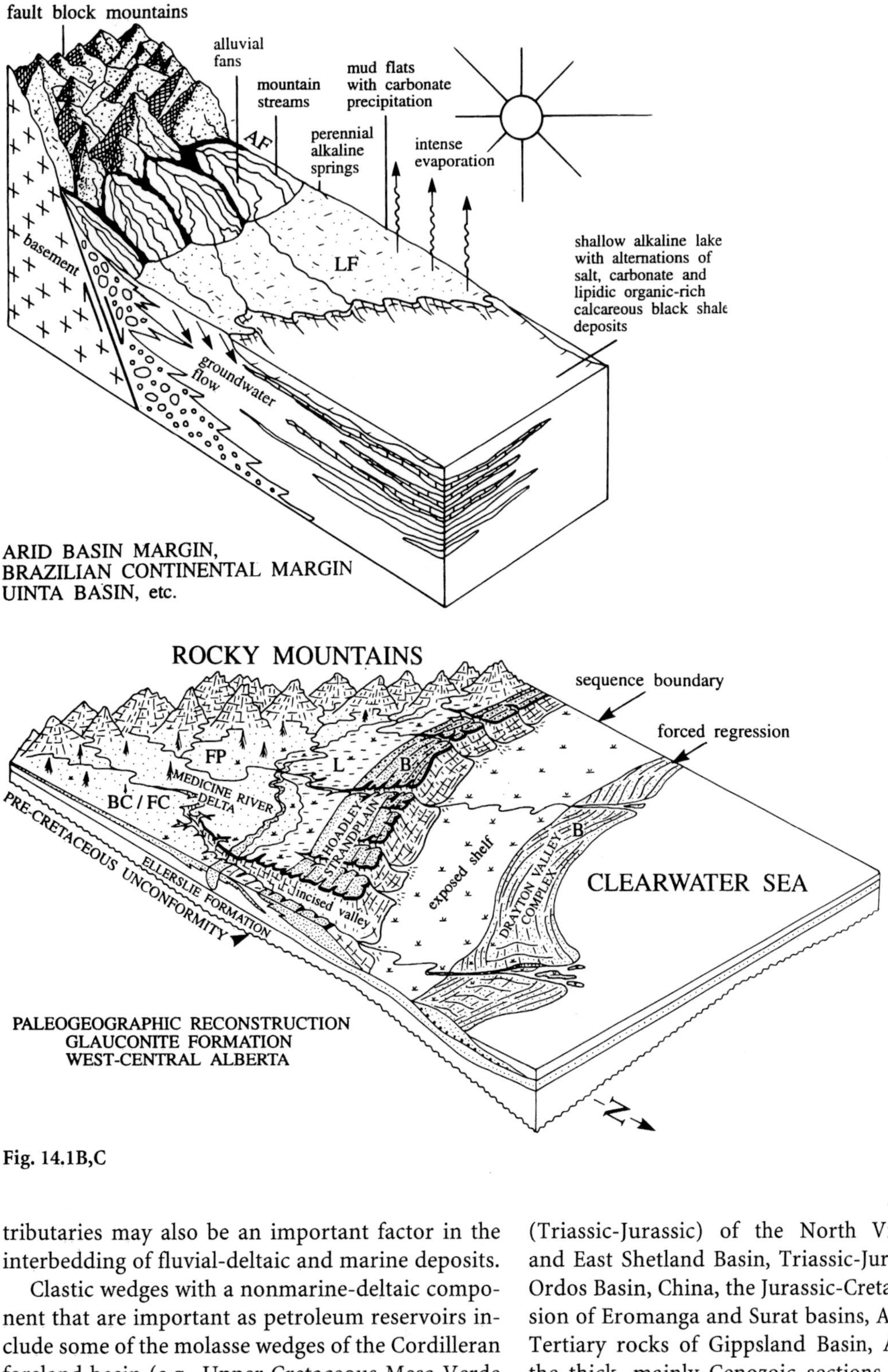

Fig. 14.1B,C

tributaries may also be an important factor in the interbedding of fluvial-deltaic and marine deposits.

Clastic wedges with a nonmarine-deltaic component that are important as petroleum reservoirs include some of the molasse wedges of the Cordilleran foreland basin (e.g., Upper Cretaceous Mesa Verde Group of the Colorado Plateau; Fig. 14.2), the Cherokee Group (Pennsylvanian) of the Arkoma Basin, the Sadlerochit Group (Permian-Triassic) and Kekiktuk Formation (Mississippian) of the North Slope of Alaska, the Statfjord Formation and Brent Group (Triassic-Jurassic) of the North Viking Graben and East Shetland Basin, Triassic-Jurassic rocks of Ordos Basin, China, the Jurassic-Cretaceous succession of Eromanga and Surat basins, Australia, early Tertiary rocks of Gippsland Basin, Australia, and the thick, mainly Cenozoic sections of the Himalayan and Andean foreland basins. Along the US Gulf Coast, clastic wedges of fluvial and fluvial delta-plain origin include the Denkman Sandstone (Upper Jurassic), the Travis Peak formation (Lower Cretaceous), the Woodbine-Tuscaloosa Formation

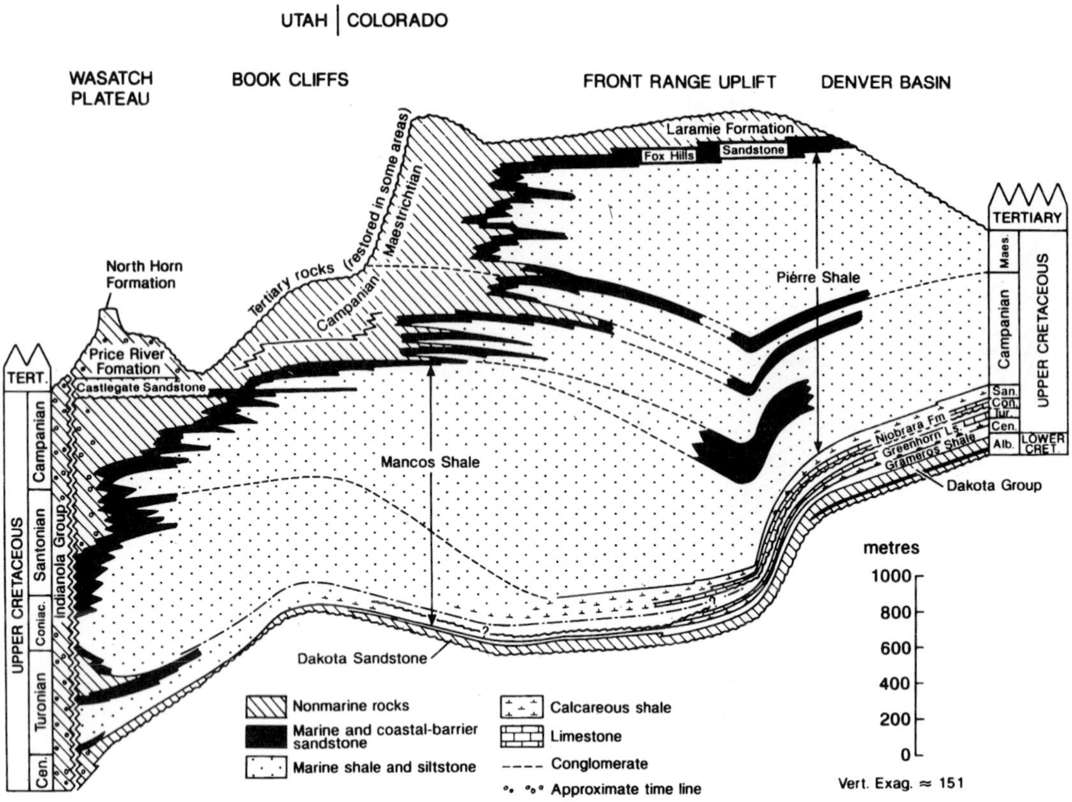

Fig. 14.2. Stratigraphic cross section of Cretaceous-Tertiary rocks from central Utah to northeastern Colorado. Clastic wedges of this type are very common in retroarc and peripheral foreland basins. This example is from the foreland basin of the North American Western Interior. Note interfingering of marine and nonmarine strata. (Molenaar and Rice 1988)

(Upper Cretaceous), the Wilcox Group (Paleocene-Eocene), and the Frio Formation (Oligocene) (references to all these examples are provided later in this chapter). In all of these cases separating deltaic production from strictly fluvial production is a difficult and sometimes an arbitrary exercise. Wherever fluvial reservoir sandstones are interbedded with fine-grained marine sediments deposited in interdeltaic or interdistributary bays, they have been omitted from this review. The geology of such fields is distinctly different from that of purely fluvial production. For this reason, some major producing regions such as the Niger Delta and many of the major clastic units of the US Gulf Coast are not discussed.

Fluvial deposits that are interbedded with lacustrine facies include the Green River deposits of the Green River and Uinta basins of the United States (Paleocene-Eocene), the Tulare Formation in South Belridge field, California (Pleistocene), the Bahia Supergroup of the Recôncavo Basin (Jurassic-Lower Cretaceous), and the fill of most of the Chinese basins (Songliao, Jiangsu, Ordos, Bohai Bay).

14.2.1.2 Paleovalley Fills

A quite different style of fluvial depositional system is the paleovalley fill (Fig. 14.1C). Such valleys are typically a few kilometers wide, a few tens of meters deep, and may be traced for a few tens of kilometers. These valleys are formed by subaerial channel incision during times of falling base level, and are filled during subsequent lowstand to trangressive stages (Sect. 13.3.2). They may therefore be associated with a major regional or interregional unconformity. They may be filled entirely with fluvial deposits, or in part with estuarine sediments, depending on local conditions of paleoslope, sediment supply, and rate of base-level change (Zaitlin et al. 1994). Fluvial depositional systems of this type occur primarily in areas of low regional slope, such as cratonic basins, extensional continental margins, and foreland basins distant from the mountainous source area [the "B" zone of Posamentier and Allen (1993)]. For example, the Quaternary continental shelf of the Gulf Coast is crossed by numerous incised valley systems

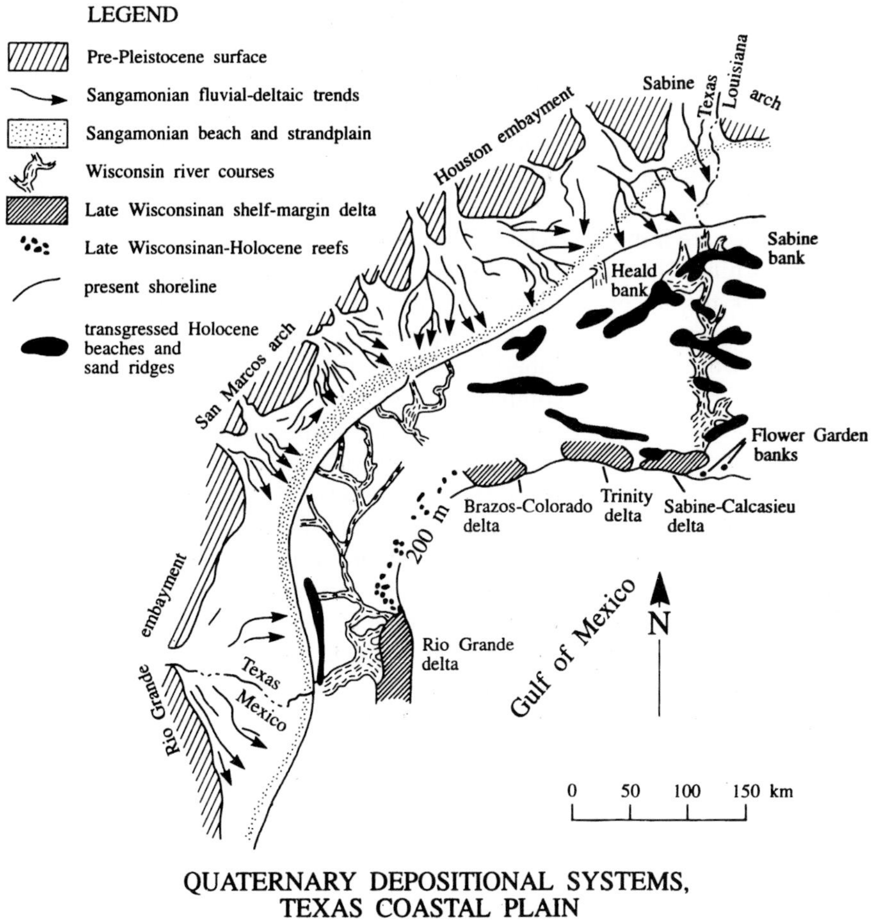

Fig. 14.3. Late Pleistocene and Holocene depositional systems of the Texas coast, showing location of incised fluvial systems and lowstand deltas formed during glacial phases. (Simplifed from Morton and Price 1987)

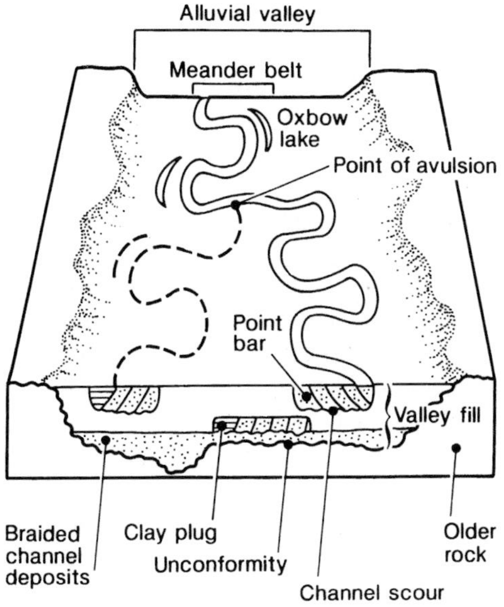

Fig. 14.4. Schematic block diagram of the valley-fill deposits of the Mississippi River. The basal bounding discontinuity of the valley formed during the Pleistocene lowstand of sea level, when the river incised bedrock to adjust itself to a lower base level. As sea level rose following glaciation, the valley filled first with coarse sands deposited in a braided-stream environment and then with lower-energy meandering-stream deposits. Many ancient valley-fill alluvial successions formed in this way. In some, the uppermost deposits are estuarine to marine in depositional environment and may be followed by widespread transgressive-marine blanket deposits. (Weimer 1986; based on the work of H. Fisk)

formed during glacial lowstands of sea level (Figs. 14.3, 14.4).

Several important petroleum plays in the Cordilleran foreland basin produce from paleovalleys incised into regional unconformities generated during the Early Cretaceous (Mannville fields, Alberta; "J" and "D" sandstones, Denver Basin). Deep, fluvial, incised valley fills of Lower Jurassic age are important reservoirs in the Ordos Basin of China. There is much ongoing debate about whether the unconformities in the Western Interior Basin are tectonic or eustatic in origin (Weimer 1986; Dolson et al. 1991).

14.2.2 Geometry of Reservoir Bodies

The second criterion for the classification of fluvial petroleum reservoirs is that of the geometry of the reservoir body. Two similar classifications have been used. Friend (1983) suggested that fluvial sandstones occur primarily in two modes, the laterally extensive sheet and the more laterally restricted ribbon or lens. Galloway (1981) adopted the geomorphic classification of Schumm (1977) in his review of Gulf Coast fluvial systems, and this approach has been used by several other workers (e.g., Davies et al. 1991, 1993). The classification suggests three architectural modes, (a) multilateral channel deposits formed by bedload rivers, with high net-to-gross sandstone ratios and good interconnectedness; (b) multistory, mixed sandstone-shale deposits with intermediate values for interconnectedness, formed in mixed-load fluvial systems; and (c) multistory ribbon bodies with low net-to-gross ratios and poor interconnectedness, formed in suspended-load rivers. The first and third types are similar to the two classes of Friend (1983), which are the subdivisions used here.

14.2.2.1 Sheet Sandstones

Deposits of this type are commonly formed by steep-gradient, bed-load systems such as braided rivers, where channels comb across broad areas of the valley floor. Sheet sandstones typically form laterally extensive reservoir bodies that may require a structural component to generate a petroleum trap. Excellent examples are parts of the Permian-Triassic Sadlerochit Group of Prudhoe Bay, Alaska (Lawton et al. 1987; Atkinson et al. 1990), the braided-fluvial zone 2 of the Kekiktuk Formation in Endicott field, Alaska (Melvin 1993), the Lower Cretaceous Sarir Sandstone of Libya (Clifford et al. 1980), the unnamed Cambrian sandstone of the Oued Mya basin, Algeria (Balducchi and Pommier 1970), the Sherwood sandstone (Triassic) at South Morecambe field in the Irish Sea (Stuart and Cowan 1991; G. Cowan 1993), and many of the Hutton Sandstone fields of Eromanga Basin, Australia (Moore et al. 1986; Wecker 1989). In most of these examples broad sheets are limited at their margins by lateral pinchout or channel incision. The Peco oilfield of Alberta produces from a single braided channel complex 20 m deep and up to 3 km wide enclosed in impermeable floodplain deposits (Gardiner et al. 1990; Putnam 1993). The Travis Peak formation of East Texas includes several separate channel belts up to 10 km wide (Tye 1991).

Sheet sandstones can also accumulate where lower-slope and lower-energy rivers, such as some meandering systems, develop in areas of low subsidence rate, so that channel deposits are incised into and superimposed upon each other and the preservation potential of fine-grained floodplain deposits is low (Miall 1980). Examples include at least some of the Permian fluvial reservoirs of Cooper Basin, Australia (Stuart et al. 1988; Stanmore and Johnstone 1988).

Slightly different is the style of the lake-margin fluvial reservoirs in the Green River Formation of Red Wash field, Uinta Basin. Here, individual producing sandstones are extensive, but much less so than the area of the entire field. Isopachs of the producing sandstones suggest that they include thin, fluvially dominated, deltaic sheets (Chatfield 1972; Castle 1990). Examples of fields producing from sheet-like reservoir bodies are described in Sect. 15.3.

14.2.2.2 Sandstone Ribbons and Lenses

These form by the accumulation of sandstone bodies within channel systems that are isolated within the fine-grained sediments of the overbank environment. Point bars are the best known of such deposits; they form distinctive lens-shaped reservoir bodies, many of which preserve the fine-grained fill of the abandoned channel around the margin of the lens (Berg 1968; Busch 1974; Werren et al. 1990; Davies et al. 1991, 1993; Chapin and Mayer 1991). Not all braided fluvial systems form sandstones with a sheet geometry. In Wytch Farm oilfield, England, sandstones interpreted as braided channel deposits and "sheet-flood" units are lenticular, and form part of a heterogeneous clastic succession (Dranfield et al. 1987). According to Bentham et al. (1993), braided systems consisting of channel sandstone bodies interbedded with significant proportions of floodplain fine-grained deposits constitute a distinctive type of fluvial assemblage that is much more important in the geological record than has hitherto been recognized.

Many fluvial sand bodies are mapped and described as ribbons and may represent the fill of high- or low-sinuosity channels. Examples are described by Davies et al. (1991) and Yinan et al. (1987). Hastings (1990) illustrated a complex of linear sandbodies forming a low-sinuosity meander belt in the Tyler Formation of North Dakota. Many may, however, represent the amalgamation of point bars or other bar deposits along the channel trend, producing an overall ribbon geometry (Ebanks and Weber 1982). Where sufficient well control is available, such ribbons may be broken down into their individual bar units. Examples of these various conditions are described later. Ribbon and lens reservoirs may form purely stratigraphic traps. The ribbon shape of these sandstones should not be confused with the ribbon shape of many paleovalley-fill sandstones, which are typically much wider and thicker and represent the fill of entire, incised meander belts.

The Cherokee Group (Pennsylvanian) of Oklahoma, Kansas, and Missouri has provided many classic examples of shoestring or ribbon sandstones. Examples of producing sandstones in this group, particularly the Bartlesville Sandstone (Desmoinesian), have been known since the 1920s and have been illustrated in many classic textbooks (e.g., Levorsen 1967).

Another classic area of production from fields consisting largely of stratigraphically trapped petroleum in ribbons and lenses is the Gulf Coast. Major fluvial producers there are (a) the Upper Cretaceous Lower Tuscaloosa Sandstone of southern Mississippi and Louisiana, which constitutes a transgressive, retrograde, fluvial-deltaic depositional systems tract (Karges 1962; Berg and Cook 1968; Stancliffe and Adams 1986; Hamilton and Cameron 1986; Hamlin and Cameron 1987; Wiygul and Young 1987; Hersch 1987; Klicman et al. 1988; Hogg 1988; Miller and Groth 1990; Werren et al. 1990; Garrison and Chancellor 1991); (b) the Oligocene Frio Formation of southwest Texas, another fluvial-deltaic system which is an important gas producer (Galloway et al. 1982; Jirik 1990; Kerr 1990; Kerr and Jirik 1990); and (c) the Lower Cretaceous Donovan/Sligo Formation in Texas. The Citronelle field, which produces from this unit, is an excellent example, illustrated later (Eaves 1976).

Most of the producing zones in the Songliao and Ordos basins, China, are lenticular or ribbon-shaped in geometry, and a detailed knowledge of their subsurface distribution and facies is required for the development of efficient production programs (Jin et al. 1985; Moore et al. 1986; Yinan et al. 1987). Likewise, fluvial-deltaic channel-and-bar sands of the Oligocene Barail Sands of northeast India provide numerous ribbon-shaped and lenticular reservoir bodies in combined structural-stratigraphic traps (Murty 1983; Desikachar 1984). Tertiary production in the Green River Basin (McDonald 1976) also provides excellent examples of this type of reservoir geometry. It may be very difficult to produce accurate maps of these ribbon sandstones. A tight well spacing is required, as otherwise spurious correlations may result. This is discussed further below. Examples of fields producing from ribbons and lenses are described in Sect. 15.4.

14.2.2.3 Stratigraphic Variations in Reservoir Geometry

In many fields the geometry of the reservoirs varies stratigraphically, reflecting allogenic influences on fluvial style or autogenic migration of the fluvial system. For example, in South Belridge field, California, the Tulare Formation consists of a braid-delta complex that prograded into a lake. Lower Tulare reservoir sands are lenticular or ribbon bodies deposited in meandering fluvial systems, whereas Upper Tulare sandstones were deposited in wider, more proximal, braided channels and are more laterally continuous (Miller et al. 1990). Similar stratigraphic variations have been described in Prudhoe Bay and Endicott fields, Alaska (Lawton et al. 1987; Melvin 1993). In the Gullfaks field, in the Norwegian sector of the East Shetlands Basin, the Statfjord Formation is divided into three members comprising at least six reservoir units, reflecting stratigraphic variations in fluvial style (Petterson et al. 1990). In the nearby Brent field, eight reservoir units have been mapped in the Statfjord Formation (Struijk and Green 1991). Sequence concepts which may explain these variations are discussed below.

14.3 Tectonic Setting of Fluvial Reservoirs

The thickest and most widespread fluvial deposits accumulate in a few specific tectonic settings. These group into two main categories – extensional basins, and basins associated with plate collision, including foreland basins, forearc basins, and strike-slip basins (Fig. 11.37). Petroleum plays in fluvial sandstones reflect this distribution, most of the major plays occurring in rift basins and foreland basins, with a few major plays in extensional-margin basins and forearc basins. This localization of the petroleum is not suprising, because it is only in these specific types of tectonic setting that fluvial deposits are volumetrically important enough to constitute a distinctive, major category of reservoir body. Fluvial successions up to 8 km thick occur in some of these basins.

The following paragraphs describe the tectonic setting of selected producing units. Other examples of the tectonic setting of fluvial deposits are given in Sect. 11.4.

14.3.1 Retroarc (Backarc) Foreland Basins

The Jurassic-Tertiary foreland basin adjacent to the Cordilleran Orogen of North America includes many petroleum-bearing fluvial sandstones. Foreland basins associated with the Paleozoic Appalachian Orogen and its southward extension (e.g., Arkoma Basin) are also important, although in the case of the Arkoma Basin, the major producing unit, the Pennsylvanian Cherokee Group, was derived from cratonic uplifts to the north of the basin (Rascoe and Adler 1983), and not from uplifts associated with the fold-thrust belt adjacent to the basin, which is usually the case for foreland-basin deposits.

Cenozoic fluvial deposits in the foreland basins flanking the Andes of Colombia (Llanos Orientales and Putumayo basins) are also productive (Simlote et al. 1985; Salazar 1990; Stabler 1990). For example, the giant Caño Limón oil field of Colombia produces mainly from Eocene and Oligocene channel sands in fluvial-deltaic complexes (Gabela 1990; Cleveland and Molina 1990).

The Eromanga and Surat basins of Australia are Jurassic-Cretaceous foreland basins that contain significant fluvial production. However, the Hutton sandstone, which is the major reservoir unit, was derived from uplifts on the cratonic sides of the basins (Moore et al. 1986; Watts 1987).

14.3.2 Backarc Basins

The Songliao Basin of eastern China, where the giant Daqing field is reservoired in Lower Cretaceous fluvial deposits, is a possible backarc basin, although it may have been influenced by escape tectonics (Burke and Sengör 1986). The Jiangsu Basin, to the south, has a similar tectonic origin, and includes some early Tertiary fluvial reservoirs forming parts of lacustrine delta complexes (Moore et al. 1986). The North China Basin, which includes the important producing region of Bohai Bay, is also a Cenozoic backarc basin and contains fluvial-deltaic wedges and lacustrine deposits forming numerous structural and stratigraphic trap types (Shaui Defu et al. 1988; Matsuzawa 1988; Zhao Xueping et al. 1988).

Some of the productive Andean basins of Argentina, notably Cuyo and San Jorge basins, were backarc basins during much of the Cretaceous, at the time the main reservoir units were deposited (Mpodozis and Ramos 1990; Stabler 1990).

14.3.3 Forearc Basins

Cook Inlet, Alaska, is a forearc basin with significant production from the upper Tertiary Kenai Group, a predominantly fluvial succession more than 7 km thick (Kirschner and Lyon 1973; Hayes et al. 1976; Boss et al. 1976; Figs. 11.51, 11.52).

14.3.4 Collision-Related Basins

The nonmarine Cenozoic basins of western China, including Tarim, Djunggar, Jiuxi, and Qaidam basins, all include important production from fluvial sandstones [e.g., Oligocene-Miocene reservoirs in Jiuxi Basin: Wang and Coward (1993)]. They are basins associated with plate collision – their main features were developed during the Himalayan orogeny, and they have the characteristics of foreland or strike-slip basins (Figs. 11.65, 11.66). Many show lateral block movements, a characteristic of escape tectonics (Burke and Sengör 1986).

The Ordos Basin in China is a foreland basin of Triassic-Jurassic age which has important petroleum production from fluvial reservoirs (Moore et al. 1986). Cenozoic fluvial deposits in the peripheral foreland basins flanking the Himalayan Mountains of India are also productive (Murty 1983; Desikachar 1984).

Cenozoic basins in southeast Asia were generated by tectonism associated with the India-Asia collision (Tapponier et al. 1986; Fig. 11.65). These include rift and transtensional basins in offshore Malaysia, Thailand, Cambodia, Vietnam, and the South China Sea and some onshore basins in these same countries [grouped together as the Sundaland rifts by Derksen and McLean-Hodgson (1988)]. Fluvial-deltaic progradation from adjacent contemporaneous uplifts provided numerous potential reservoir units, and offshore exploration is now discovering many important oil and gas fields in these basins (Lee 1982; Crostella 1983; Roberts 1988; Derksen and McLean-Hodgson 1988; Blanche 1990). For example, in Pattani trough, offshore Thailand, a Tertiary-Quaternary fluvial-deltaic section up to 8 km thick is present and contains important gas accumulations (Blanche 1990). In the Phitsanulok Basin of central Thailand, 8 $m^3 \times 10^6$ of recoverable oil have been discovered to date from an 8 km Oligocene-Lower Miocene fluvial-lacustrine succession (Derksen and McLean-Hodgson 1988).

14.3.5 Basins in Continental-Transform Settings

Few producing fluvial reservoirs are located in this type of tectonic setting. The South Belridge field, California, is an example. The reservoir is a Pliocene fluvial-deltaic unit derived from uplifts associated with the San Andreas fault and trapped in associated anticlinal structures (Miller et al. 1990).

14.3.6 Rift Basins

Several major fluvial exploration plays occur in rift basins, including East Shetland Basin and Viking Graben, North Sea (Triassic-Tertiary; Parsley 1984; Figs. 11.44, 11.45), and the geologically very similar Gippsland Basin, Australia (Late Cretaceous-Tertiary; Rahmanian et al. 1990; Figs. 11.46–11.48), as well as the Recóncavo Basin of Brazil (Jurassic-Cretaceous) (Figueiredo et al. 1994).

Triassic rift-fill fluvial sediments host a significant oil field at Wytch Farm in southern Britain (Dranfield et al. 1987) and gas off the west coast, at Morecambe field (Stuart and Cowan 1991; G. Cowan 1993; Meadows and Beach 1993).

The rift basins of Africa, including the Karoo rifts and those originating during the breakup of Gondwana in the Mesozoic-Cenozoic, all contain significant fluvial clastic units that have yielded oil and gas, but most of these basins are relatively unexplored and the full potential remains to be determined (Genik 1993; Winn et al. 1993; Derksen and McLean-Hodgson 1988).

14.3.7 Basins on Extensional Continental Margins

Fluvial reservoirs are very important in some extensional, craton-margin basins. The Gulf Coast of the United States (Fig. 11.42) is among the most well-known of such basins, as it has produced huge volumes of oil and gas from terrigenous clastic reservoirs (Tyler et al. 1984). Most of the reservoirs are marine in origin, but significant production has also been obtained from fluvial units forming part of inland alluvial systems or delta plains distant from the coast. Nonmarine stratigraphic units include parts of the Denkman Sandstone (Upper Jurassic), Travis Peak Formation (Lower Cretaceous), Woodbine-Tuscaloosa Formation (Upper Cretaceous), the Wilcox Group (Paleocene-Eocene), and the Frio Formation (Oligocene).

The Triassic-Tertiary extensional continental margins of western and northwestern Australia have yielded significant gas production from fluvial reservoirs (e.g., Rankin field: Beston 1986; Posaner and Goldthorpe 1986; Warris 1988).

The Upper Paleozoic to Triassic reservoirs of the North Slope of Alaska were deposited on the south-facing margins of a craton that was rifted and dispersed during the Cretaceous when the Arctic Ocean basin was generated (Morgridge and Smith 1972). Other important extensional-margin basins occur in North Africa (Sirte Basin, of Cretaceous age, and Cambrian Oued Mya Basin; e.g., Clifford et al. 1980).

14.3.8 Intracratonic Basins

Intracratonic basins do not constitute a major class of basin from the point of view of nonmarine petroleum fields, but a few significant discoveries have been made in this setting. The Morrow Sandstone (Mississipian) of eastern Colorado and western Kansas produces from paleovalleys crossing an intracratonic basin at the margin of the Anadarko basin. The valleys were incised during glacio-eustatic falls in sea level. The Clifford-Siaana trend in eastern Colorado, which is still undergoing active exploration, is estimated to contain up to 9.5 $m^3 \times 10^6$ of recoverable oil (Bowen et al. 1993).

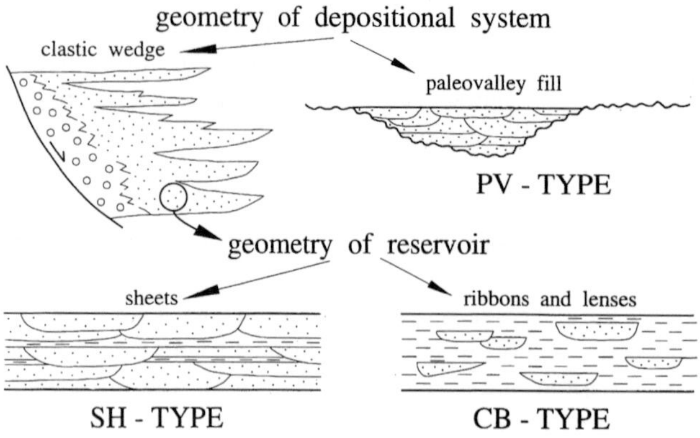

Fig. 14.5. Classification of nonmarine reservoirs according to the geometry of the depositional system and the geometry of the reservoir body

Table 14.2. The three major types of petroleum reservoir in fluvial sandstones

Type	Code	Depositional system	Reservoir geometry	Most typical tectonic setting
Paleovalley	PV	Incised valley	Ribbon	Foreland basin
Sheet	SH	Clastic wedge	Sheet	Rift basin, extensional margin
Channel and bar	CB	Clastic wedge	Multiple lenses, ribbons	Foreland basin, extensional margin

Table 14.3. Examples of petroleum production from fluvial sandstones in North America

Location	Tectonic setting	Stratigraphy	Age	Reservoir type	Reserves oil ($m^3 \times 10^6$)	Reserves gas ($m^3 \times 10^9$)
Alberta Basin	Retroarc foreland	Mannville Group	Cretaceous	PV PV-CB	305	962
Alberta Basin	Retroarc foreland	Athabasca Oil Sands	Cretaceous	CB	144000 (HO)	
Denver Basin	Retroarc foreland	"J" and "D" Ss	Cretaceous	PV	127	
Uinta Basin	Retroarc foreland	Mesa Verde, Green River	Cretaceous Tertiary	CB	80	422
Green River Basin	Retroarc foreland	Hoback, Green River	Tertiary	CB	12	70
Arkoma Basin	Retroarc foreland	Cherokee Group	Penn.	CB	436 +320(HO)	
East Texas Basin	Extensional margin	Woodbine	Cretaceous	CB	252	
Texas Gulf Coast	Extensional margin	Frio	Oligocene	CB	59	
Prudhoe Bay Alaska[a]	Extensional margin	Sadlerochit Formation	Perm-Tr	SH	1525	732
Endicott Alaska[a]	Extensional margin	Kekiktuk	Miss.	SH	56	22
Cook Inlet Alaska	Forearc basin	Kenai Group	Tertiary	SH,CB	238	99

HO Heavy oil
[a] Reserve values based on one field.

Table 14.4. Examples of petroleum production from fluvial sandstones in Europe, Africa, Asia, Australia, and South America

Location	Tectonic setting	Stratigraphy	Age	Reservoir type	Reserves oil ($m^3 \times 10^6$)	Reserves gas ($m^3 \times 10^9$)
N. Viking Graben	Rift basin	Statfjord, Brent	Triassic-Jurassic	SH-CB	2170	
Sirte B Libya[a]	Extensional margin	Sarir Sandstone	Cretaceous	SH	240	
Oued Mya B Algeria[a]	Extensional margin	unnamed	Cambrian	SH	397	
Songliao B China	Rift basin (backarc?)	various	Cretaceous	CB	1460	
Sundaland S. E. Asia	Collisional rifts	various	Tertiary	SH?	3813	1943
Gippsland B Australia	Rift basin	Latrobe Group	Tertiary	CB	665	234
Eromanga B Australia	Retroarc foreland	Hutton Ss	Jurasic	SH	40	
Cooper B Australia	Intracratonic rift	Gidgealpa Gp	Permian	SH?	–	87
Recôncavo B Brazil	rift basin	Sergi Formation	U. Jurassic	SH	302	28

B Basin; *HO* heavy oil
[a] Reserve values based on one field.

The Main Consolidated field of the Illinois Basin is of similar age and origin (Howard and Whitaker 1990).

An intracratonic Gondwana basin located in what is now Oman contains the Permian Gharif Formation, a fluvial coastal-plain deposit up to 2.5 km thick that contains more than 30 producing fields (Focke and van Popta 1989).

The Cooper Basin, Australia's major onshore petroleum producer, is an intracratonic basin with several largely nonmarine reservoir units of Permian-Triassic age that have yielded significant gas production (Cosgrove 1987; Heath 1989).

14.4 Styles of Fluvial Reservoir

The stratigraphic criteria discussed in this chapter can be grouped to define three major types of nonmarine reservoir, as shown in Fig. 14.5 and Table 14.2. Each of the three types of reservoir occurs in more than one tectonic setting; therefore, this does not form part of the classification, although some typical tectonic settings are listed in Table 14.2. Selected examples of producing units are listed in Tables 14.3 and 14.4.

The primary criterion used in this classification is the type of depositional system – clastic wedge versus incised paleovalley. The second criterion is reservoir geometry. Incised-valley fills are ribbon bodies, but they may be broad in scale; clastic wedges consist of either sheet bodies or multiple lenses and ribbons enclosed in impermeable fine-grained sediment.

Weber and van Geuns (1990) classified reservoir heterogeneity into three basic types, all of which are represented by fluvial deposits. Some braided systems and sheet-flood deposits are examples of their "layercake" reservoirs, which are laterally extensive and can be treated as tabular units for mapping and modeling purposes. "Jigsaw-puzzle" reservoirs consist of separate sand bodies such as channel sands, that fit together without major gaps but with occasional low-permeability barriers and baffles. Many braided and coarse-grained meander-belt fluvial systems are of this type. "Labyrinth" reservoirs consist of complex arrangements of lenses and pods which are poorly interconnected and difficult to map. Fine-grained meandering and anastomosing systems are commonly of this type.

14.4.1 Paleovalley Bodies (PV Type)

Paleovalley bodies are distinguished by the ribbon or shoestring shape of the reservoir and by their association with regional unconformities. The reservoirs are typically tens of kilometers in length, up to a few kilometers in width, and several tens of meters in thickness (Sect. 15.2). In Ordos basin incised valleys up to 300 m deep have been mapped (Moore et al. 1986; Song Guochu 1988). Paleovalleys define the former courses of incised meander belts and may be virtually straight, or more or less sinous (e.g., Harms 1966; Weimer 1986). The fill may be entirely fluvial, or it may contain an estuarine component (e.g., Wood and Hopkins 1989, 1992; Howard and Whitaker 1990). Internally, the reservoir may be homogeneous, or it may contain important facies variations, such as the channel-and-bar geometries described below under the heading of the CB type of reservoir. Some traps are therefore combination PV-CB types. Examples include some of the Mannville fields of Alberta (Little Bow field, Figs. 14.6, 14.7; Sect. 15.2.1), Cut Bank field of Montana (Shelton 1967; Sect. 15.2.2), and Blackwell field, Texas (Bloomer 1977).

PV-type fields can occur only where the paleovalley fill is incised into impermeable strata, and they also may require a top seal and a structural component to generate a trap. For example, the paleovalley may be oriented parallel to strike, and

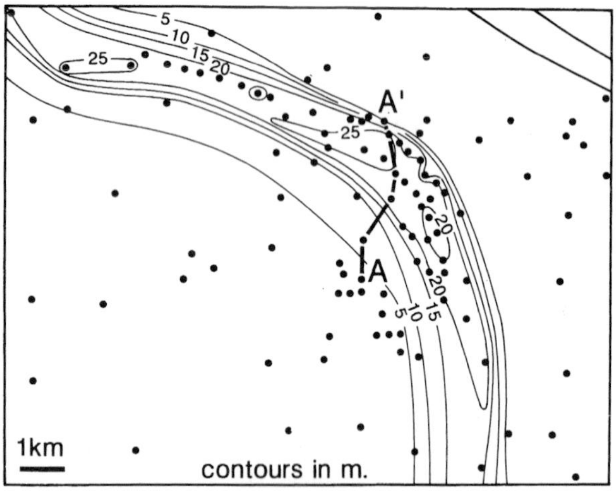

Fig. 14.6. Isopach map (contours in meters) of valley-fill sandstone, Little Bow Field, Mannville Group, Alberta. Cross section A-A' is shown in Fig. 14.7. (Wood and Hopkins 1989)

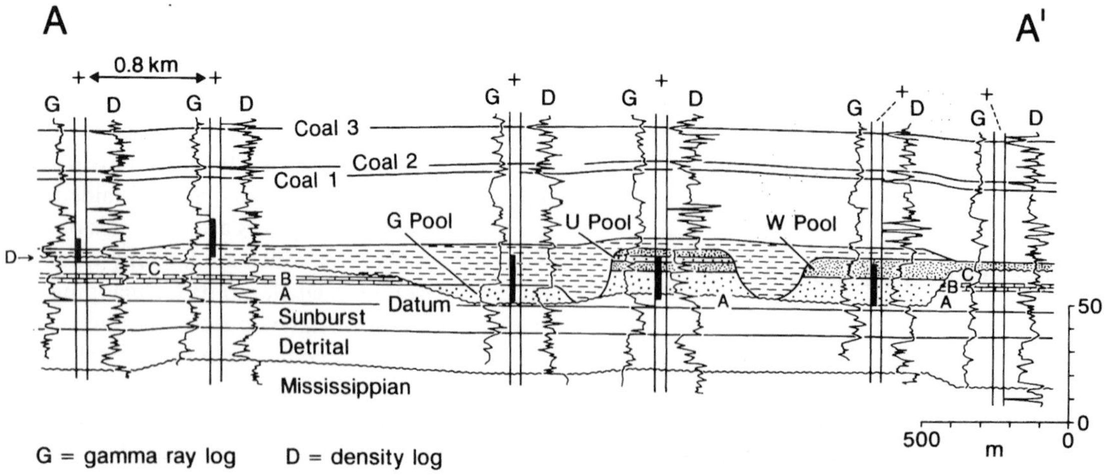

Fig. 14.7. Stratigraphic cross sections across the valley fill at Little Bow field. Note the presence of three separate pools in the valley fill, separated by shale-filled channels. The location of the cross section is given in Fig. 14.6. (Wood and Hopkins 1989)

the trap occurs where bends in the paleovalley include a convexity facing updip. South Ceres field, Oklahoma (Lyons and Dobrin 1972; Sect. 15.2.4), Recluse field, in Powder River Basin (Woncik 1972), and Midland field, Kentucky (Reynolds and Vincent 1972) are good examples. Another type of trap is where the paleovalley is folded over an anticline. Glenrock field, Wyoming (Curry and Curry 1972), and Lane field, Colorado (Harms 1966), are examples. Cheyenne Valley field, Oklahoma (Withrow 1968), occurs where a paleovalley is oriented oblique to regional dip but porosity decreases updip, providing a trapping mechanism.

A subtle paleogeomorphic trapping mechanism may be created by the erosional topography over a bar deposit. Such is the case in the Bellshill Lake and Hughenden fields, Alberta (Martin 1966; Conybeare 1976), in which the actual trap occurs where penecontemporaneous fluvial erosion sculpted the top of the sandstone body into a series of terraces.

Nearly all PV-type fields documented to date occur in foreland-basin settings, including the those in the Cretaceous of the Western Interior Basin (Mannville fields, Alberta; "J" and "D" Sandstone, Denver Basin; South Glenrock and Recluse fields, Powder River Basin, Wyoming), Arkoma Basin (South Ceres and Cheyenne Valley fields), and those in Ordos Basin, China. Paleovalleys are, of course, common in other structural settings, but most do not appear to be associated with significant petroleum trapped in fluvial reservoirs. For example, a major unconformity with incised valleys underlies the Tuscaloosa Sandstone of Mississippi, in the extensional-margin setting of the Gulf Coast (Hogg 1988; Garrison and Chancellor 1991), but fluvial production is located in a fluvial-deltaic complex that overlies the valley fill. Incised valley complexes are an important component of the Wilcox Group in Texas, but they are mainly mud filled (Devine and Wheeler 1989). There is believed to be minor fluvial sandstone reservoir potential in the base of some of the channels, but production from them is not known to this writer. Similarly, the Latrobe Group in the Gippsland Basin contains incised valley fills that are largely mud dominated, but some isolated channel-sand deposits are present which may have future potential as exploration targets, given improvements in seismic methods (Mebberson 1989; Rahmanian et al. 1990). Exceptions to this generalization include the Jiyang Depression of Bohai Bay basin, China, and Mississippian-Pennsylvanian fields of the U.S. Midcontinent. The former is an extensional backarc basin, in which sandstone pinchouts within erosional valleys constitute one of four types of "lithologic" trap, providing about 3% of total oil reserves (Shaui Defu 1988). The U.S. Midcontinent fields include the Morrow Sandstone trend of Colorado and Kansas (Bowen et al. 1993) and Main Consolidated field in the Illinois Basin (Howard and Whitaker 1990).

14.4.2 Sheet Bodies (SH Type)

Sheet sandstones reflect high source-area relief and steep paleoslopes, with the development in most cases of a broad braidplain. Such conditions do not occur in any particular tectonic setting, although they might be expected to be rare within cratonic areas, except in association with an active local

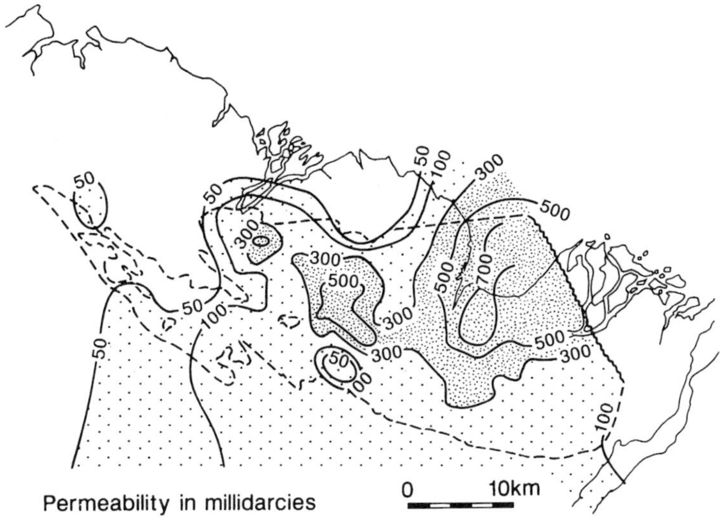

Fig. 14.8. Isopermeability map of the braided channel complex in the Sadlerochit Formation, Alaska North Slope, in millidarcies. The Prudhoe Bay field is outlined by the *dashed line*. The *wavy line* indicates the faulted margin of the field. High permeability values are yielded by the coarse, texturally more mature deposits formed in proximal reaches of the complex. The crudely lobate shape of the area of high permeability corresponds to the distribution of the sand sheet, in the form of a large, fan-shaped braided delta complex. (Wadman et al. 1979)

structure. The large fields of the North Viking Graben (e.g., Kirk 1980; Parsley 1984; Sect. 15.3.3) and the North Slope of Alaska (Morgridge and Smith 1972; Wadman et al. 1979; Bloch et al. 1990; Fig. 14.8; Sect. 15.3.1) include good examples of SH-type reservoirs, as does Messla field, Libya (Clifford et al. 1980; Sect. 15.3.2). All these are located in extensional-margin tectonic settings, the North Viking Graben being a major rift structure. The Hutton Sandstone of Eromanga Basin, Australia, forms the reservoir in many structurally trapped SH-type fields. The basin is a foreland basin, but the sandstone was derived from structural upwarps on the cratonic side of the basin (Moore et al. 1986; Watts 1987; John and Almond 1987; Wecker 1989).

These fields require a structural trapping mechanism because, by definition, the reservoir consists of a sheetlike porous unit providing little or no constraints on fluid movement. Most of the cases referred to here consist of large faulted anticlines in which the reservoir sandstone is overlain by an impermeable shale. The Hassi Messaoud field of Algeria produces from a breached anticline, in which the reservoir sandstone has been eroded from the crest of the anticline (Balducchi and Pommier 1970).

A different type of structural SH-type field is exemplified by Red Earth field, Alberta, which produces from the basal Paleozoic Granite Wash sandstone. The reservoir is a sheetlike unit, and the trap is provided by drape over paleotopographic highs, reflecting the original erosional topography that was covered by Granite Wash sedimentation (Hunter 1966).

The detailed geological and engineering information required for production modeling may result in the definition of considerable internal heterogeneity within reservoir bodies that, at an early exploration stage, might appear fairly uniform. Fields that are originally classified as SH type might later be better considered CB type, at least in terms of the style of research required to further develop them. The considerable amount of production modeling carried out on the braided "sheet" of the Sadlerochit Formation at Prudhoe Bay field (Wadman et al. 1979; Geehan et al. 1986; Atkinson et al. 1990), the Kekiktuk Formation of Endicott field (Woidneck et al. 1987; Melvin 1993), and the Statfjord Formation of Statfjord field (MacDonald and Halland 1993; Martin 1993) are good examples. Recent studies, particularly of North Sea fields, have stimulated a considerable body of work in the area of computer modeling of fluvial systems and comparisons with outcrop analogs (e.g., Buller et al. 1990; Miall and Tyler 1991; Ashton 1992; Flint and Bryant 1993; Martin 1993), the details of which are beyond the scope of this chapter (some aspects are discussed in Sect. 10.4.3). Similarly, developments in the use of 3-D seismic mapping for channel definition are proceeding at a rapid pace (e.g., A.R. Brown 1991; D. Brown 1993; Weber 1993). Techniques of "surveillance geology", including pressure tests and fluid-flow monitoring, may prove to be extremely powerful tools for subsurface mapping as field development proceeds (Sect. 9.5.9).

14.4.3 Channel-and-Bar Bodies (CB Type)

CB-type fields are characterized by the small size of the individual reservoir body, although in some cases there may be hundreds to thousands of individual reservoir units which, if all filled with petroleum (as they are in some cases), can add up to significant, even supergiant accumulations (Figs. 14.9, 14.10). Development of these fields may be

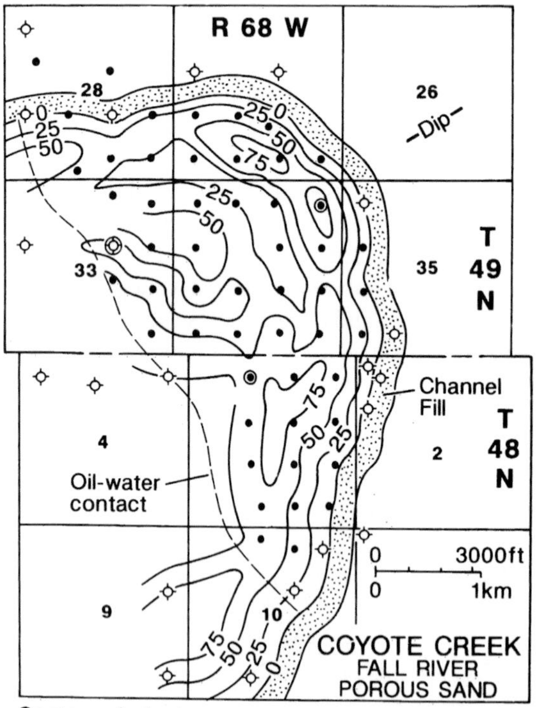

Fig. 14.9. Isopach map (contours in feet) of the Coyote Creek field, Wyoming. Regional dip is toward the west. Oil is trapped against the updip flank of the point bar, which is bordered by the impermeable shale of the abandoned channel. Detailed wireline-log cross sections show that the point-bar lens is probably a composite of more than one bar, with the mapped channel representing the final channel position at the time of abandonment. (Berg 1968)

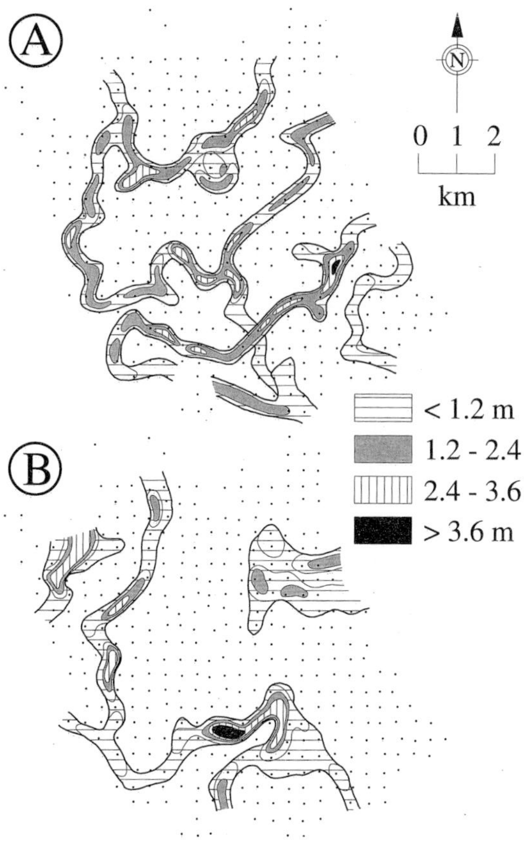

Fig. 14.10. Isopach maps of typical sandstone bodies, Citronelle field, Alabama. Contours in meters. Vertical channel interconnectedness has created continuous fluid-flow paths. Very dense well spacing is required to map such narrow ribbon sandstone bodies. Dots indicate wells. (Eaves 1976)

extremely difficult. They constitute examples of what Weber and Van Geuns (1990) termed the "labyrinth" and "jigsaw" types of reservoir, reflecting the tortuous flow paths that the stratigraphy imposes on the fluids. Examples include Citronelle field, Texas (Eaves 1976), and Daqing field, China (Jin et al. 1985; Yinan et al. 1987; Sect. 15.4.2). Examples where significant accumulations occur in numerous small to medium-sized fields include those of the Cherokee Group of Kansas-Oklahoma and the Lower Tuscaloosa sandstone of Mississippi-Lousiana (Sect. 15.4.1), references to which are given above. Traps that are entirely stratigraphic in character are typical of all these examples, although initial discoveries in the Cherokee and Tuscaloosa were all in domal structures, and most fields occur in areas of gentle regional dip, which affects migration routes and trap configuration.

Prospecting for and definition of this type of trap, and the development of production models, require a detailed knowledge of depositional style. For example, many small fields occur within point-bar lenses, particularly where the point bar is flanked on its updip side by the shale fill of the channel, formed when the channel was abandoned and point-bar sedimentation ceased. In such cases, subsurface prediction for field extension and development may require recognition of the scale and orientation of the channel and its associated bars. Good examples are Miller and Coyote Creek fields, Wyoming (Berg 1968, Fig. 10), Mallalieu and Little Creek fields, Mississippi (Busch 1974; Berg 1986; Werren et al. 1990; Sect. 15.4.1), Berwick field, Mississippi (Garrison and Chancellor 1991), Peoria field, Colorado (Chapin and Mayer 1991), and Eastburn field, Missouri (Ebanks and Weber 1982). In Citronelle (Fig. 14.10) and Daqing fields, where multiple reservoir horizons occur, several types of reservoir geometry are present, indicating subtle local variations in fluvial style.

The classic shoestring or ribbon sandstones of the Cherokee Group (Bartlesville Sandstone) of Kansas, Oklahoma and Missouri are a different category of CB-type fluvial oil field. Here the reservoir body is lenticular rather than lens-shaped (e.g., Charles 1941). These bodies may represent the fill of relatively straight channels, as suggested by the regional isopach mapping of Visher et al. (1971). Such channels are common in some delta-plain environments (e.g., modern Mississippi delta), as is interpreted to be the case for the Cherokee Group. However, some of the ribbons may in fact consist of laterally coalesced bar lenses. The very detailed drilling carried out in the area of the Eastburn field, Missouri, suggests that this is the case there (Ebanks and Weber 1982).

Seismic-stratigraphic mapping is becoming increasingly important in the delineation of small structural-stratigraphic traps in many CB-type fluvial oil plays. This technique has formed a key to the location of most of the Tuscaloosa fields in Mississippi (e.g., Hersch 1987; Garrison and Chancellor 1991) and is becoming of increasing importance in the Gippsland, Surat, Cooper, and other Australian basins, which are beginning to enter the mature exploration phase (e.g., Clark and Thomas 1988; Stanmore and Johnstone 1988; Elliott 1989).

An important aspect of production modeling may be to ascertain the degree of interconnectedness of the reservoir units, and the definition of separate flow units within the field, as discussed in Sect. 10.4.

14.5 Conclusions

The potential for future discoveries of SH- and CB-type fields in clastic wedges would seem to be greatest in the collision-related basins of southeast Asia, where clastic terrigenous sediment supply during the Cenozoic was very high and fluvial conditions were widespread. Some giant discoveries may be expected.

Considerable potential remains for discovery of smaller fields in mature areas, where detailed subsurface information provides the basis for calibration of seismic-reflection data. Seismic mapping of channels, sedimentological techniques for interpreting fluvial channel-and-bar geometries, and geostatistical techniques for predicting their scale are essential. Developments along the U.S. Gulf Coast, in the North American Western Interior basin, and in inland Australia are at this stage.

Much of the best work now being carried out on the integration of geological and engineering skills for enhanced production purposes is being applied to fluvial oil fields, such as those in the North Sea basin, where the expense of offshore operations requires the use of the most sophisticated prediction and modeling tools. Such work is bringing petroleum geologists back to the field to collect ever more detailed data on the characteristics of outrop analogs, and is providing scope for specialized numerical modeling of sand body geometries.

Chapter 15

Case Studies of Oil and Gas Fields in Fluvial Reservoirs

15.1 Introduction

Selected examples of petroleum plays in fluvial sandstones are summarized in Tables 14.3, 14.4. Some of the major oil fields from these plays are summarized in Tables 15.1 and 15.2. A large number of additional fields are described briefly in the literature cited in Chap. 14 (particularly in the SE Asia and Australia areas), but reservoir and production data are inadequate for complete tabulation. Gas fields have not been tabulated for this book. The purpose of this chapter is to provide a suite of documented case studies of selected examples of these fields, in order to illustrate the three main types of reservoir configuration identified in Chap. 14.

15.2 Paleovalley Fields (PV Type)

15.2.1 Little Bow Area, Alberta

The Mannville Sandstone (Lower Cretaceous) is a fluvial, lacustrine, coastal deltaic and barrier, and open marine unit that extends throughout central and southern Alberta. This play is located within the Alberta Basin, a large foreland basin that extends throughout Western Canada and was formed by terrane collision and suturing activity between the Jurassic and the mid-Cenozoic (Sect. 11.4.6). The Mannville and underlying Jurassic deposits contain many small oil fields trapped in fluvial and deltaic channel sandstones in central and southern Alberta. Many are formed in paleovalleys. Wood and Hopkins (1992) identified eight levels of valley cut-and-fill in the Glauconitic Sandstone Member of the Upper Mannville. The Mannville Sandstone is in part correlative with the McMurray Formation (Athabasca Oil Sands) of northern Alberta, the Cutbank Sandstone of Montana, and the "J" and "D" sandstones of the Denver Basin, all of which contain similar PV-type fields. The group consists of several formations and members and contains several internal, regional unconformities. The internal stratigraphy of the unit is very complex, because one or more of the members have locally been removed by intra-Mannville erosion. Recognition and correlation of the members is difficult, because they are present only in the subsurface, and detailed stratigraphic work must rely on precise lithostratigraphic correlation of closely spaced wireline logs and petrographic analysis of the sandstones. The unit and its component formations/members have been assigned a large number of alternative local names, reflecting the fact that much of the outcrop and subsurface stratigraphic work represents local studies carried out in isolation from one another. A thorough regional reevaluation of the Mannville incorporating concepts of sequence stratigraphy still needs to be carried out [the most recent synthesis is that by Hayes et al. (1994)].

The Alberta Basin generally is characterized by a gentle dip to the southwest, toward the fold-thrust belt of the Rocky Mountains. In the southeast corner of the province is the buried Sweetgrass Arch; regionally the beds dip to the north and northwest away from this subtle feature. These dips impose an overriding structural control on all stratigraphic traps, in that it is necessary for there to be an updip seal on the east side of each trap. In most cases in the Mannville Group, this is provided by the juxtaposition of porous Mannville strata filling a paleovalley against impermeable beds of the host rocks (generally older Mannville beds).

In southern Alberta (Fig. 15.1) the Glauconitic Sandstone of the Upper Mannville Group contains numerous sandstone ribbons that are interpreted as valley fills (Hopkins et al. 1982; Hopkins 1985). These rocks exemplify a problem common in the subsurface analysis of fluvial channels associated with marginal-marine rocks: Are the channels contemporaneous with the host strata, or are they younger deposits formed following a period of base-level lowering and incision? The presence of a deeply scoured contact between the channel and the host rocks does not demonstrate that the channels are

Table 15.1. Summary of oil fields in North America

Name	Location	Basin name	Tectonic setting	Basin code[a]	Trap type	Reservoir	Age	Type[b]	Area (km^2)	Ultimate Recovery ($m^3 \times 10^6$)	References
Bantry	Alberta	Alberta	Foreland	4.3	Strat	Mannville	Cretaceous	PV	41	10.5	Farshori and Hopkins (1989)
Bellshill Lk	Alberta	Alberta	Foreland	4.3	Strat	Mannville	Cretaceous	PV-CB	15	12.4	Martin (1966); Conybeare (1976)
Medicine River	Alberta	Alberta	Foreland	4.3	Strat	J2 Ss	Jurassic	PV	7	1.3	Hopkins (1981)
Little Bow	Alberta	Alberta	Foreland	4.3	Strat	Mannville	Cretaceous	PV-CB	10	2.6	Hopkins et al. (1982); Wood and Hopkins (1989)
Peco	Alberta	Alberta	Foreland	4.3	Struct/strat	Belly River	Cretaceous	SH	19	0.5	Gardiner et al. (1990)
Cut Bank	Montana	Rocky Mt.	Foreland	4.3	Strat	Kootenai	Cretaceous	PV-CB	930	4.8	Shelton (1967)
S. Glenrock	Wyoming	Powder River	Foreland	4.3	Struct/strat	L. Muddy	Cretaceous	PV	36	11.9[c]	Curry and Curry (1972)
Recluse	Wyoming	Powder River	Foreland	4.3	Strat	Muddy	Cretaceous	PV	36	10.0	Forgotson and Stark (1972)
Coyote Creek	Wyoming	Powder River	Foreland	4.3	Strat	Fall R.	Cretaceous	CB	8	3.0	Stapp (1967); Berg (1968)
Lane	Colorado	Denver	Foreland	4.3	Strat	"J" Ss	Cretaceous	PV	25	?	Harms (1966)
Zenith	Colorado	Denver	Foreland	4.3	Strat	"D" Ss	Cretaceous	PV	9	?	Sonnenberg (1987)
Red Wash	Utah	Uinta	Foreland	4.3	Struct/strat	Green R.	Eocene	CB	185	9.0	Chatfield (1972); Pitman et al. (1982)
Altamont-Bluebell	Utah	Uinta	Foreland	4.3	Struct/strat	Green R.	Eocene	CB	907	39.7	Lucas and Drexler (1976)
McDonald Draw	Wyoming	Green River	Foreland	4.3	Struct/strat	Hoback	Paleocene	CB	4	1.1	McDonald (1976)
South Belridge	California	San Joaquin	Transform fault	3.1	Struct	Tulare	Pleistocene	CB-SH	50	190.7	Miller et al. (1990)
Citronelle	Alabama	Gulf Coast	Ex. margin	1.2.2	Struct/strat	Donovan	Cretaceous	CB	66	19.8	Eaves (1976)
Mallalieu	Mississippi	Gulf Coast	Ex. margin	1.2.2	Struct/strat	Tuscaloosa	Cretaceous	CB	12	> 5.6	Berg (1986)
Little Creek	Mississippi	Gulf Coast	Ex. margin	1.2.2	Struct/strat	Tuscaloosa	Cretaceous	CB	47	11.0	Busch (1974); Werren et al. (1990)
Berwick	Mississippi	Gulf Coast	Ex. margin	1.2.2	Strat	Tuscaloosa	Cretaceous	CB	~ 5	0.6	Garrison and Chancellor (1991)
Cushing	Oklahoma	Arkoma	Periph. frland.	4.1	Anticline	Bartlesville	Penn.	CB	72	> 74.8[c]	Weirich (1929); Visher et al. (1971)
Bush City	Kansas	Arkoma	Periph. frland.	4.1	Strat	Squirrel	Penn.	CB	8	1.0	Charles (1941)
Cheyenne Valley	Oklahoma	Arkoma	Periph. frland.	4.1	Strat	Red Fork	Penn.	PV	9	1.0	Withrow (1968)
South Ceres	Oklahoma	Arkoma	Periph. frland.	4.1	Strat	Red Fork	Penn.	PV	80	1.6	Lyons and Dobrin (1972)
Blackwell	Texas	Midland	Ex. margin	1.2.2	Struct/strat	Cook	Permian	PV-CB	1	> 0.25	Bloomer (1977)
Big Wall	Montana	–	Ex. margin	1.2.2	Struct/strat	Tyler	Penn.	PV	10	0.9	Kranzler (1966)
Rocky Ridge	North Dakota	Williston	Cratonic	5	Struct/strat	Tyler	Penn.	PV-CB	21	0.5	Hastings (1990)
Main Consolidated	Illinois	Illinois	Cratonic	5	Struct/strat	Caseyville	Penn.	PV	3	0.2	Howard and Whitaker (1990)
Clifford-Siaana	Colorado	Oklahoma	Cratonic	5	Strat	Morrow	Miss	PV	80	9.5	Bowen et al. (1993)
Red Earth	Alberta	Alberta	Ex. margin	1.2.2	Struct	Granite Wsh	Devonian	SH	29	3.5	Hunter (1966); Conybeare (1976)
Endicott	N. Slope, Alaska	North Slope	Ex. margin	1.2.2	Anticline	Kekiktuk	Miss	SH	3580	56.0	Melvin 1987; Woidneck et al. (1987)
Prudhoe Bay	N. Slope, Alaska	North Slope	Ex. margin	1.2.2	Anticline	Sadlerochit	Perm-Tr	SH	506	1700.0[c]	Morgridge and Smith (1972); Eckelmann et al. (1975) Atkinson et al. (1990)
Swanson River	Cook Inlet, Alaska	Cook Inlet	Forearc	2.2	Anticline	Kenai	Tertiary	SH	?	40.0	Kirschner and Lyons (1973)
Middle Grnd. Shoal	Cook Inlet, Alaska	Cook Inlet	Forearc	2.2	Anticline	Kenai	Tertiary	CB	16	> 13.8	Kirschner and Lyons (1973); Boss et al. (1976)

[a] Basin type, from Table 11.2.

[b] Trap type, from Table 14.2: *PV*, paleovalley; *CB*, channel-bar form; *SH*, sheet.

[c] Data for all pools, fluvial portion not separated; >=only cumulative production data are available. Some fields have significant gas reserves; these are not listed in this table.

Table 15.2. Summary of fluvial oil fields in the remainder of the world

Name	Location	Basin name	Tectonic setting	Basin code[a]	Trap type	Reservoir	Age	Type[b]	Area (km^2)	Ultimate Recovery ($m^3 \times 10^6$)	References
Statfjord	North Sea	North Viking Graben	Rift	1.1	Anticline	Statfjord	Tr-Jur	SH	36	120.0	Kirk (1980)
Ninian	North Sea	North Viking Graben	Rift	1.1	Anticline	Brent	Jurassic	SH	580	190.0[c]	Albright et al. (1980)
Brent	North Sea	North Viking Graben	Rift	1.1	Anticline	Statfjord+ Brent	Tr-Jur	SH?	52	318.0	Bowen (1975)
Snorre	North Sea	North Viking Graben	Rift	1.1	Struct	Statfjord	Tr-Jur	CB	30	115.0	Stanley et al. (1990); Martin (1993)
Beatrice	North Sea	Moray Firth	Rift	1.1	Anticline	Unnamed	Jurassic	CB?	17	26.0	Linsley et al. (1980)
Wytch Farm	S. England	W. Approaches Trough	Rift	1.1	Struct	Sherwood Ss	Triassic	CB	34	55.6	Dranfield et al. (1987); Martin (1993)
Messla	Libya	Sirte	Rift	1.1	Struct	Sarir	Cretaceous	SH	200	240.0	Clifford et al. (1980)
Sarir C-Main	Libya	Sirte	Rift	1.1	Struct	Sarir	Cretaceous	SH	800	1271.3	Sanford (1970); Martin (1993)
Hassi Messaoud	Algeria	Oued Mya	Craton	5	Anticline	Unnamed	Cambrian	SH	1300	397.0	Balducchi and Pommier (1970)
Daqing	China	Songliao	Rift	1.1 or 2.3?	Struct/strat	Various	Cretaceous	CB	2800	1460.0	Tang (1982); Jin et al. (1985); Yang (1985)
Halibut	Australia	Gippsland	Rift	1.1	Anticline	Latrobe	Eocene	CB	27	> 4.0	Franklin and Clifton (1971); Rahmanian et al. (1990)
Jackson	Australia	Eromanga	Foreland	2.4	Struct	Hutton	Jurassic	SH	21	14.3	Wecker (1989)
Agua Grande	Brazil	Recôncavo	Rift	1.1	Anticline	Bahia	Jur-Cret	SH	80	41.0	Ghignone and Andrade (1970)
Aracas	Brazil	Recôncavo	Rift	1.1	?	Sergi	Jurassic	SH?	8	18.0	Nascimento et al. (1982); Martin (1993)
Vacas Muertas	Argentina	Cuyo	Foreland	2.4	Struct	Barrancas	Cretaceous	SH	~84	25.0	Simlote et al. (1985); Martin (1993)

[a] Basin type, from Table 11.2.
[b] Trap type, from Table 14.2: *PV*, paleovalley; *CB*, channel-bar form; *SH*, sheet.
[c] Data for all pools, fluvial portion not separated; >=only cumulative production data are available. Some fields have significant gas reserves; these are not listed in this table.

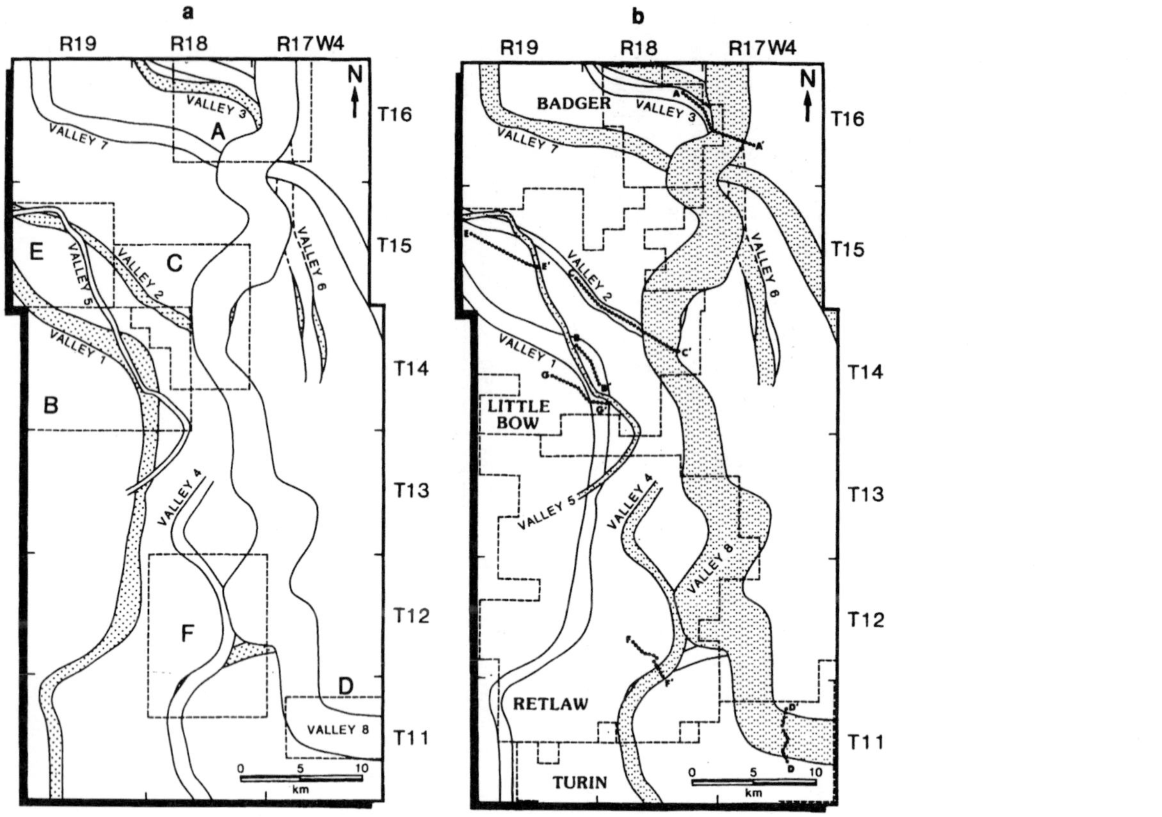

Fig. 15.1. Channels in the Glauconitic Sandstone, Upper Mannville Group, southern Alberta. (Wood and Hopkins 1992, reprinted by permission)

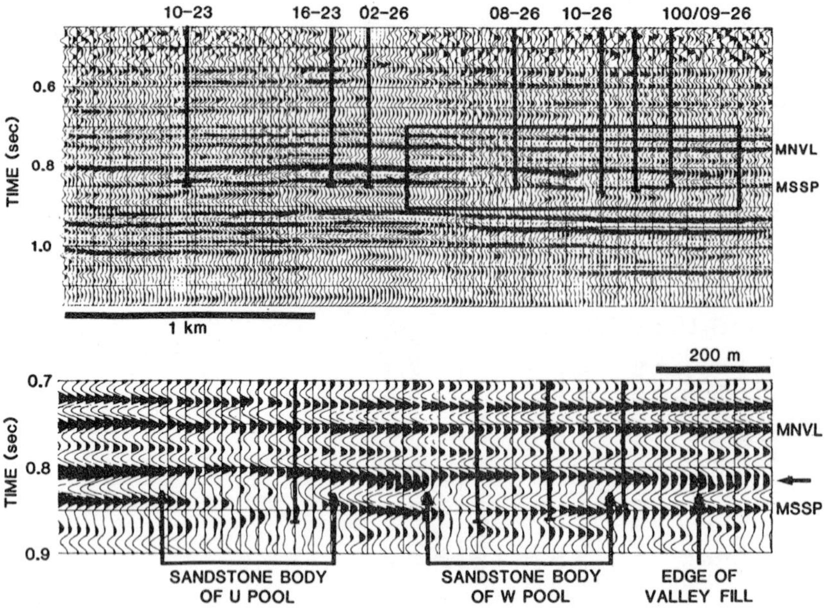

Fig. 15.2. North-south seismic section across Little Bow field. Location is close to that of the section in Fig. 14.7. (Wood and Hopkins 1989, reprinted by permission)

younger, because fluvial channels are by their very nature erosive. Modern delta distributaries are known to scour for several meters below base level into associated delta-front deposits. The scale of the incised channel may be suggestive. Individual active channels, such as delta distributaries, are typically a few hundred meters wide, whereas paleovalleys may be significantly larger (a few kilometers wide) and contain composite fills indicating several periods of incision and infilling. A close comparison of facies and petrography between the fluvial channels and the host strata might suggest a close genetic association (e.g., delta distributaries within an active delta) but does not conclusively prove such a relationship. Consideration of the succession in a regional sequence-stratigraphic context is probably the most reliable method of interpretation.

Eight separate valley-fill levels have been identified in the Little Bow area (Fig. 15.1). Initially, these were interpreted by Hopkins et al. (1982) as being delta distributaries, but later work has indicated that an estuarine interpretation is more realistic (Wood and Hopkins 1989, 1992). Some of the valley fills have been traced for more than 60 km. They are 2–2.5 km wide and up to 30 m deep (Fig. 15.1) They contain three main facies: sandstone, sandstone-shale interbeds, and shale. The shale contains evidence of marine trace fossils, and the sandstone includes sand-mud couplets interpreted as tidal in origin. Wood and Hopkins (1989) compared the facies and distribution of the beds to those of the modern estuarine sediments of the Gironde Estuary, France. They suggested that the valley filled by fluvial and subsequent estuarine sedimentation following a period of base-level lowering and downcutting.

Details of one of the valley fills at Little Bow (valley 2 in Fig. 15.1b), at the bend in the channel, are shown in the detailed map of Fig. 14.6, and Fig. 14.7 provides a stratigraphic cross section through the Little Bow field. Three separate pools (G, U, and W) have been delineated. Proof of their separation is based on two lines of evidence: (a) The shale caprock that separates the pool has been penetrated by cores in several wells and has been shown to occupy the entire stratigraphic interval of the valley fill. (b) Each pool has different pressures and oil-gas contact elevations (Wood and Hopkins 1989). The pools may also be separated on the basis of seismic evidence, as shown in Fig. 15.2. The pools each represent major sand banks that formed within the estuary. They may have been contemporaneous, or they may represent separate periods of incision and valley filling. Wood and Hopkins (1989) preferred the first of these interpretations and pointed to the simultaneous, juxtaposed development of sand bars and mud banks in the Gironde Estuary. Figure 9.72 illustrates fluid-flow paths through a cross section of this field, inferred from data obtained during waterflooding of the field.

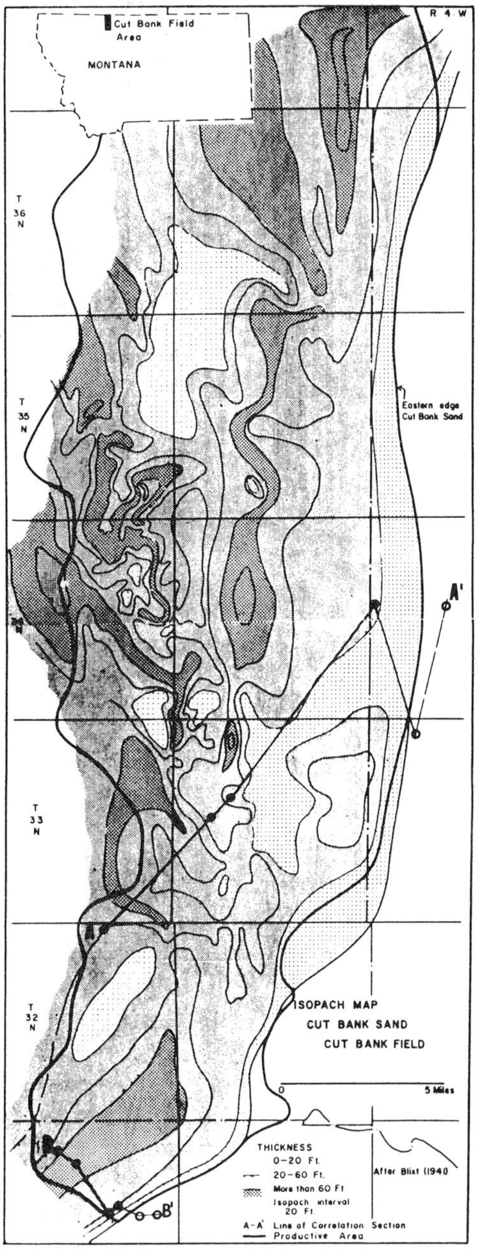

Fig. 15.3. Isopach map of Cut Bank Sandstone in Cut Bank field area. Producing area is *outlined*. Note sinuous ribbon pattern of thick (> 60 ft) sandstone distribution. (Shelton 1967, reprinted by permission)

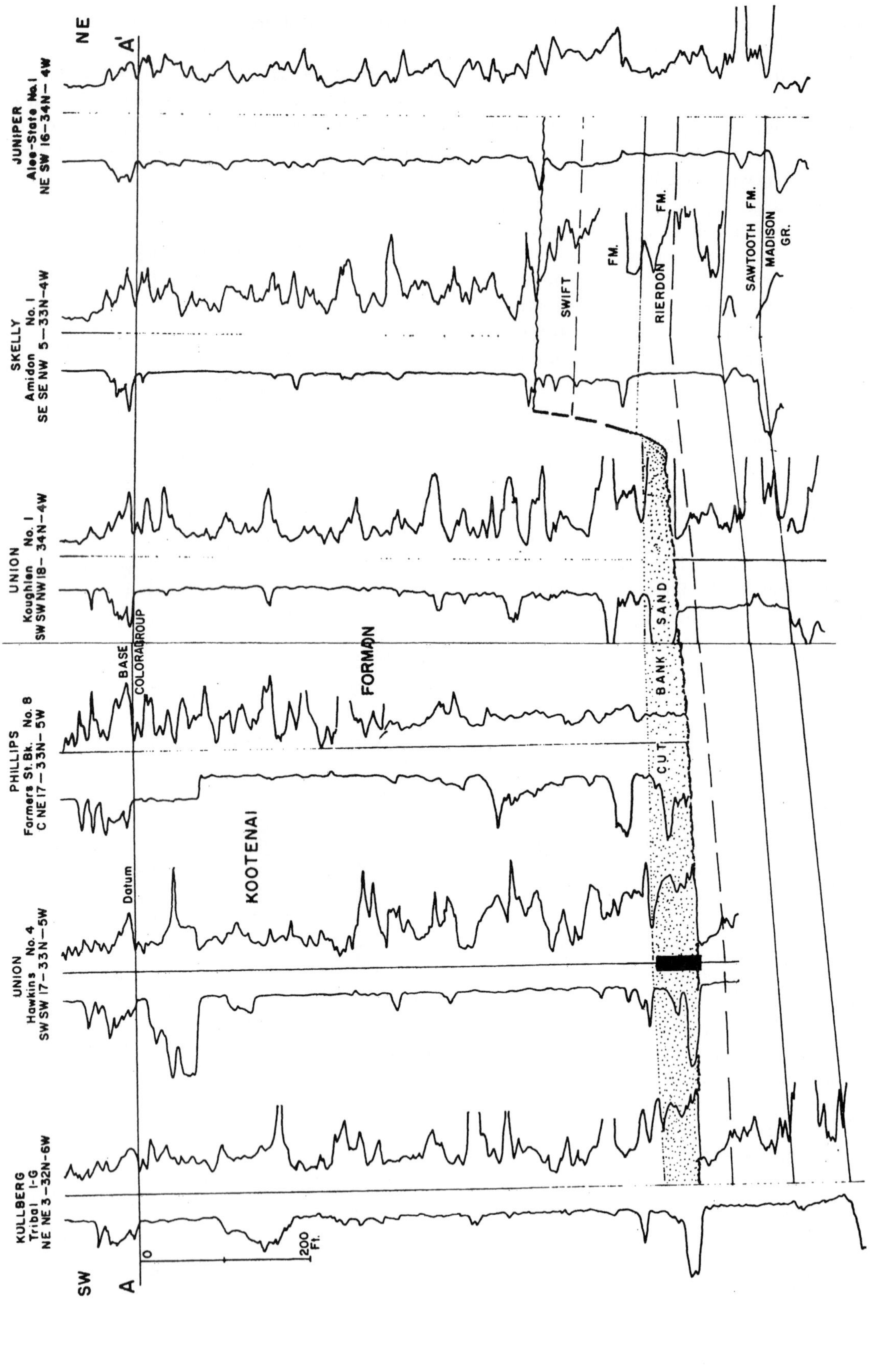

Fig. 15.4. Cross section through Cut Bank field. Line of section is shown on Fig. 15.3. (Shelton 1967, reprinted by permission)

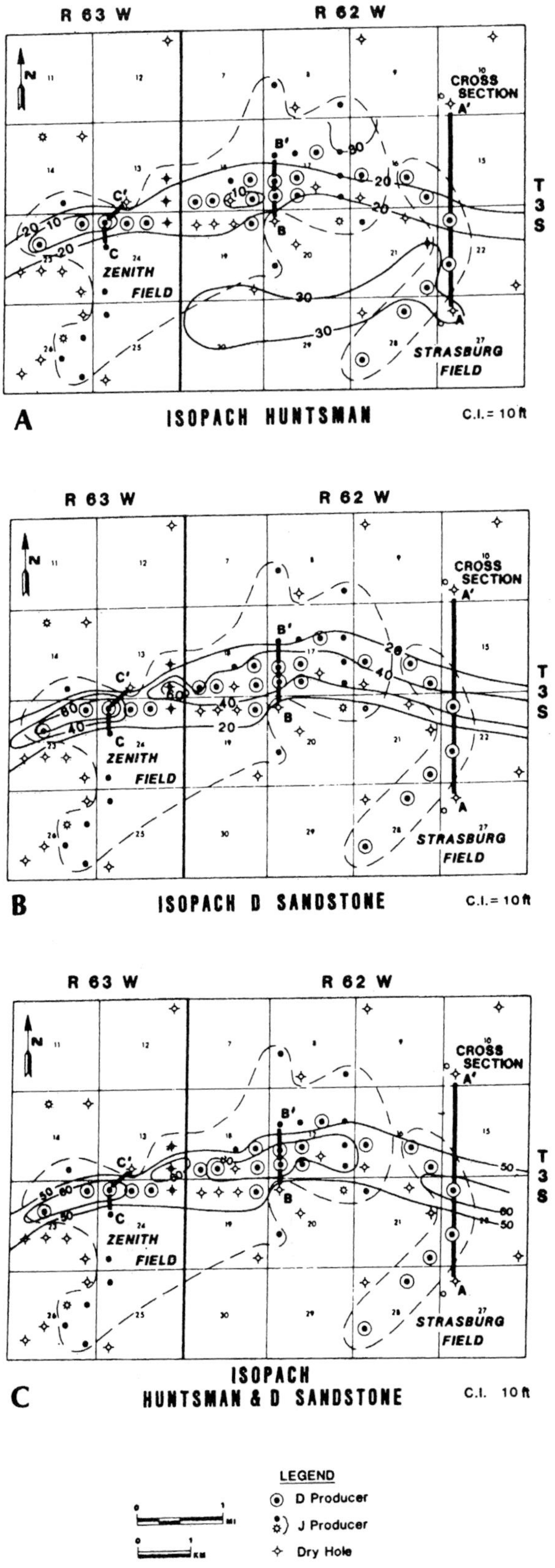

Fig. 15.5. Isopachs of the Huntsman Shale and "D" sandstone. The Huntsman is thin along the paleovalley trend, as a result of valley incision. When combined, both units show a linear trend of increased thickness, probably because these units fill a paleostructural depression. (Sonnenberg 1987, reprinted by permission)

15.2.2 Cut Bank Sandstone, Montana

The Cut Bank field is located in northernmost Montana, adjacent to the border with Alberta. The geological setting is that of the Rocky Mountain foreland basin, a southward extension of the Alberta Basin. The Cut Bank field produces from meandering channels filling a broad paleovalley at the base of the Kootenai Formation, which is the Montana equivalent of the Lower Mannville Group of Alberta. The producing beds are incised into Jurassic strata. The field is located on the west flank of the Sweetgrass Arch. Regional drainage was to the north. This paleovalley is considered to be a southward extension of the major drainage systems that have been identified in Alberta [the Spirit River Channel of McLean and Wall (1981); see Fig. 11.25].

The valley is more than 50 km wide (Blixt 1941; Conybeare 1976). The Cut Bank Sandstone is up to 70 m thick in outcrop, to the west of the field, but within the field itself it has a maximum thickness of 25 m. The dominant lithology is medium-grained, cross-bedded sandstone showing an upward decrease in grain size. An isopach map of the sandstone in the field area is shown in Fig. 15.3. Here the thickness contours suggest a pattern of sinous sandstone ribbons up to 5 km wide. According to Conybeare (1976), individual channels are 450–1200 m wide, but the evidence for this is not presented. Shelton (1967) and Conybeare (1976) interpreted these beds as point-bar deposits within a broad meandering river system.

A cross section through the area confirms the incision of the valley into Jurassic strata which provide the updip seal to the field (Fig. 15.4). Well logs through the Cut Bank Sandstone are serrated in character and do not contain picks correlatable from well to well. This line of evidence, coupled with the overall thickness and areal extent of the valley fill, suggests that the fluvial succession is a composite one, consisting of multiple, multistory meander-belt sandstones.

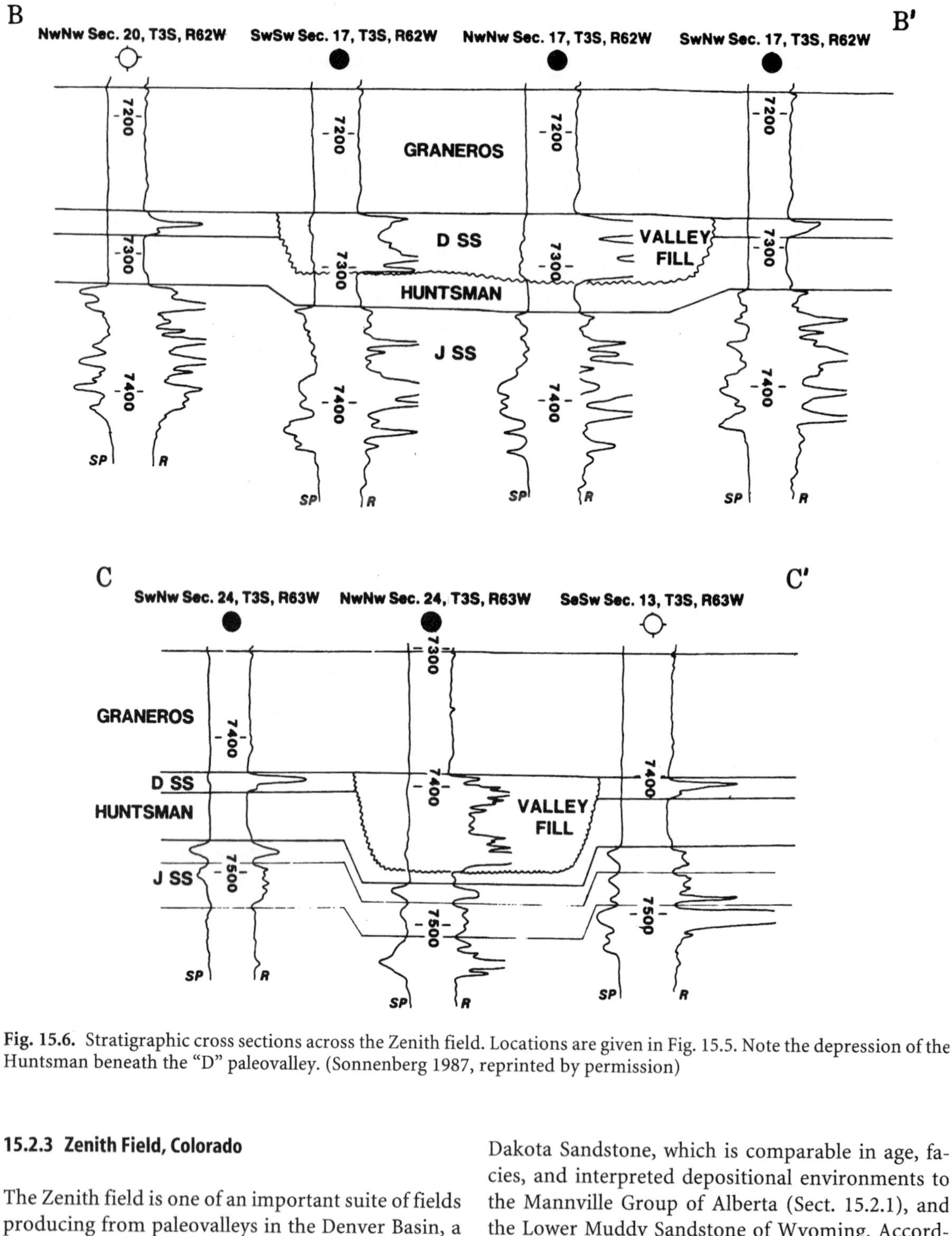

Fig. 15.6. Stratigraphic cross sections across the Zenith field. Locations are given in Fig. 15.5. Note the depression of the Huntsman beneath the "D" paleovalley. (Sonnenberg 1987, reprinted by permission)

15.2.3 Zenith Field, Colorado

The Zenith field is one of an important suite of fields producing from paleovalleys in the Denver Basin, a basin formed by deformation of the Rocky Mountain foreland basin during the Late Cretaceous-Cenozoic Laramide orogeny (Weimer 1986). The major producing trends in the Denver Basin are the valley-fill deposits of the "J" and "D" Sandstones (Harms 1966). These units are part of the Lower Cretaceous Dakota Sandstone, which is comparable in age, facies, and interpreted depositional environments to the Mannville Group of Alberta (Sect. 15.2.1), and the Lower Muddy Sandstone of Wyoming. According to McCaslin (1983), the ultimate recovery for the Denver Basin is expected to be 127 $m^3 \times 10^6$, most of which will be derived from stratigraphic traps in the "J" and "D" sandstones.

The valley fills incise a regional unconformity, just as the producing sandstones of the Mannville do

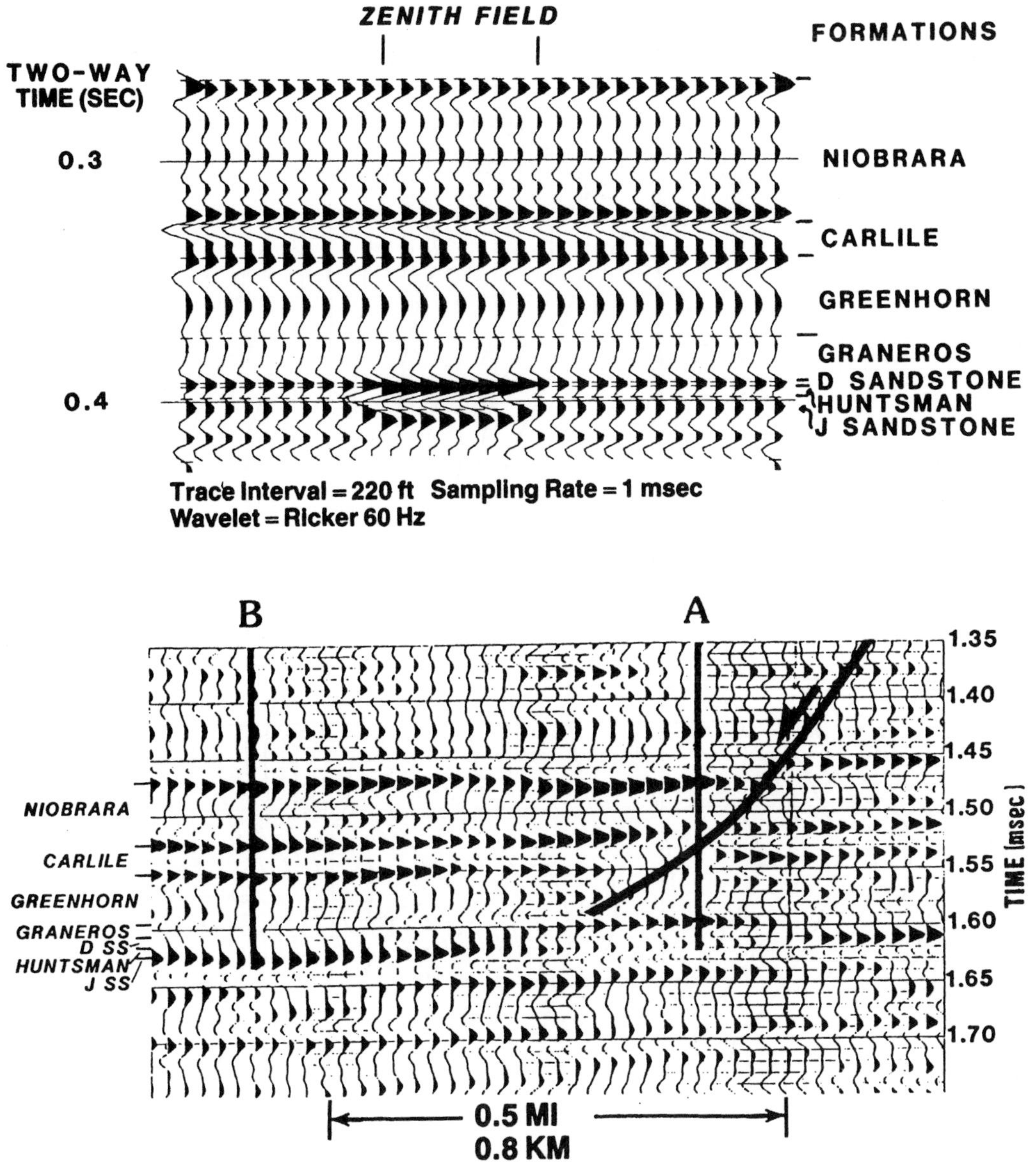

Fig. 15.7. Synthetic seismic section (*above*) and actual cross section (*below*) through the Zenith field. (Sonnenberg 1987, reprinted by permission)

(indeed, the comparisons between the "J" and "D" sandstones and the Mannville in terms of their petroleum geology are quite striking). Weimer (1986) showed that the age of the regional unconformity underlying the "J" sandstone is 97 Ma (late Albian; Fig. 11.70). The valley fill cuts down into a regressive unit [J1 and J2 units of Harms (1966); Fort Collins Member of Weimer (1986)] consisting of fine-grained sandstones deposited in deltaic and shoreline environments. The valley fill itself consists of fluvial and estuarine strata, indicating increasing marine influence upward through the section. these beds are themselves then overlain by the widespread Mowry Shale, deposited during a regional transgression. The Mowry Shale is, in turn, overlain by the Huntsman Shale, and this is incised by paleovalleys, which are part of a regional unconformity developed at the end of the Albian, at about 95 Ma (Weimer 1986; Sonnenberg 1987; Fig. 11.70). The "D" sandstone infills these valleys, with a fluvial-estuarine succession similar to that of the "J" sandstone, and is followed by another widespread shale unit.

The Denver Basin is an asymmetric trough. Important secondary structures provide a structural component to the traps in the "J" and "D" sandstones. A series of broad, NE-SW trending arches

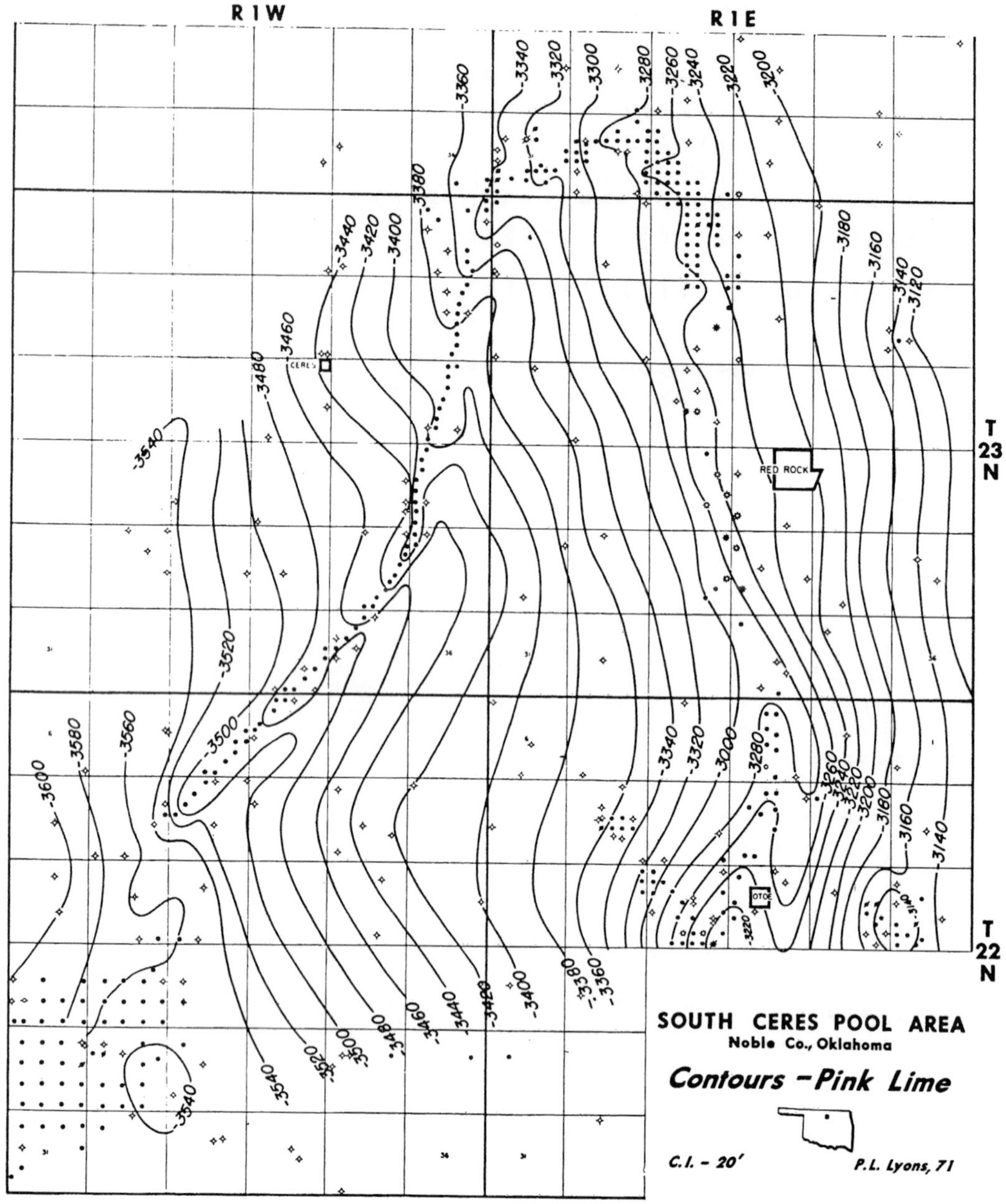

Fig. 15.8. Structure contour map, South Ceres Pool area. Contours are drawn on a marker overlying the Red Fork sandstone. Note the "nosing" of contours over the pool, as a result of structural drape. (Lyons and Dobrin 1972, reprinted by permission)

cross the north end of the Denver Basin. These developed in response to recurrent movement on buried Precambrian lineaments throughout the Cretaceous. Weimer (1986) grouped them into the broad structural feature termed the Transcontinental Arch. During the major late Albian regression, this arch became an exposed drainage divide, with valleys incised across it, draining to the north and south.

The Zenith field is located near the center of the Denver Basin, some 45 km east of Denver. It produces primarily from the "D" Sandstone. Isopach maps of the "D" sandstone and the underlying Huntsman Shale are shown in Fig. 15.5, and these delineate an E-W trending paleovalley 518–790 m wide. The "D" Sandstone ranges from 3.6 to 20.4 m in thickness in the Zenith field area. Cross sections through the field show that the "D" sandstone is incised into the underlying Huntsman Shale, and that the base of the shale is also depressed beneath the field (Fig. 15.6).

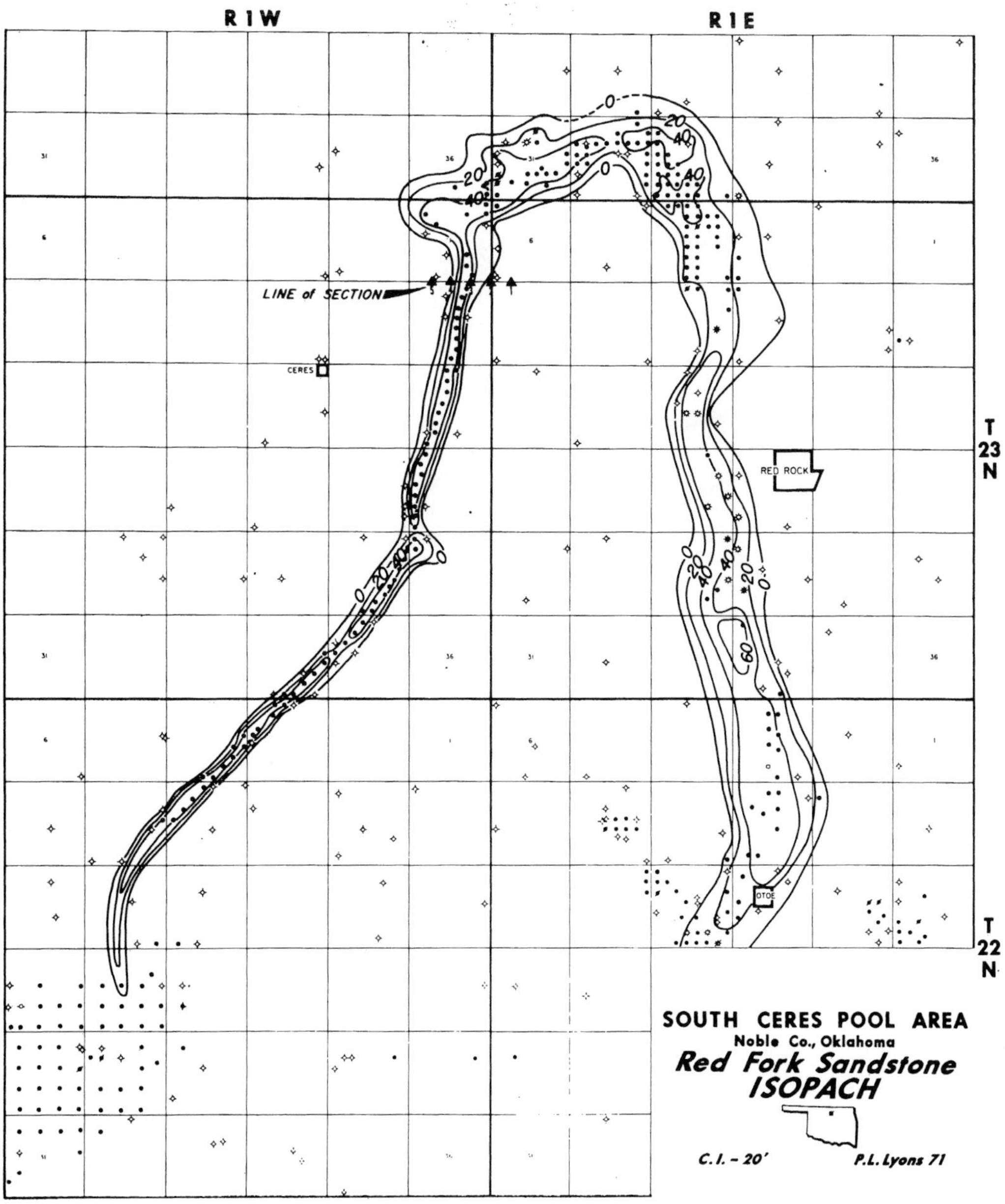

Fig. 15.9. Isopach map of Red Fork Sandstone, South Ceres Pool, Oklahoma. (Lyons and Dobrin 1972, reprinted by permission)

The "D" sandstone trend could be intepreted as a deltaic distributary comparable in age to the beds into which it is incised, but Sonnenberg (1987) rejected this intepretation on several grounds. First, the contact between the valley fill and the host strata is sharp. No gradational facies change can be mapped. Second, the top surface of the valley-fill and adjacent strata is flat, whereas a prograding distributary of this size would be expected to develop pronounced levees. Sonnenberg's (1987) interpretation of the depressed Huntsman underlying the "D" sandstone is that both units were deposited in a subtle structural depression developed during the Laramide reactivation of basement fault blocks.

Sonnenberg (1987) demonstrated that "D" sandstone paleovalleys could probably be mapped using seismic techniques. He developed a synthetic seismic model by determining reflection coefficients from a sonic well-log, and then building a seismic cross section using a paleovalley up to 15.8 m deep. Figure 15.7 compares the synthetic section with an actual seismic cross section through the field. In the

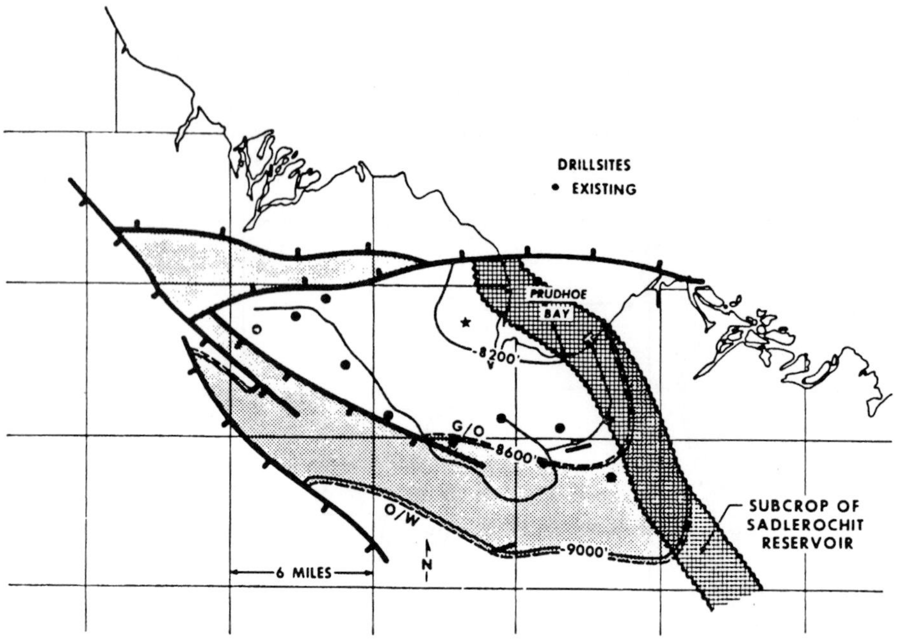

Fig. 15.10. Structure contour map of Prudhoe Bay field. (Morgridge and Smith 1972)

actual section the "D" sandstone is 17.6 m and 4.8 m thick in wells A and B, respectively. Note the slight velocity pull-up beneath the "D" sandstone, and the presence of a listric fault that does not offset the "D" sandstone.

15.2.4 South Ceres Field, Oklahoma

South Ceres field is a classic example of a paleovalley field; it was discovered by seismic methods in the 1940s, a time when seismic acquisition and processing techniques were very primitive by today's standards. The field is located on the north flank of the Anadarko Basin, a peripheral foreland basin that developed during the early and midle Paleozoic (Rascoe and Adler 1983). The Cherokee Group, of Desmoinesian age, is a major producing unit deposited on the north flank of this basin and the adjacent Arkoma Basin in Missouri, Kansas, and Oklahoma. It contains many small to giant fields, many discovered in the early part of this century. The South Ceres field is one of a suite of fields in which the reservoir is the Red Fork Sandstone, one of the component fluvial-deltaic units of the Cherokee. A structure contour map and an isopach map of the producing interval are shown in Figs. 15.8 and 15.9. Figure 9.45 illustrates a seismic cross section through the field. The giant U-shape of the sandstone isopach reflects the sinuosity of the incised valley rather than the sinuosity of individual channels.

15.3 Sheet Reservoirs (SH Type)

15.3.1 Prudhoe Bay Field, Alaska

Prudhoe Bay field and Endicott field are located on the North Slope of Alaska, adjacent to the Beaufort Sea. At present this area can be classified as a foreland basin (Colville Trough) that developed in response to Cretaceous tectonism in the Brooks Range area, to the south. However, during the late Paleozoic and early Mesozoic, when the reservoir units were deposited, the tectonic environment of the region was quite different. The most widely accepted hypothesis for the evolution of northern Alaska interprets the North Slope as a fragment of the continental margin of North America that was formerly continguous with Arctic Canada, but was rotated anticlockwise away from Canada during the Cretaceous (Embry 1990b). In this tectonic model the North Slope formed part of an extensional continental margin that now faces south in the subsurface beneath Mesozoic cover.

The dominant structure in the region of the Endicott and Prudhoe Bay fields is Barrow Arch, a

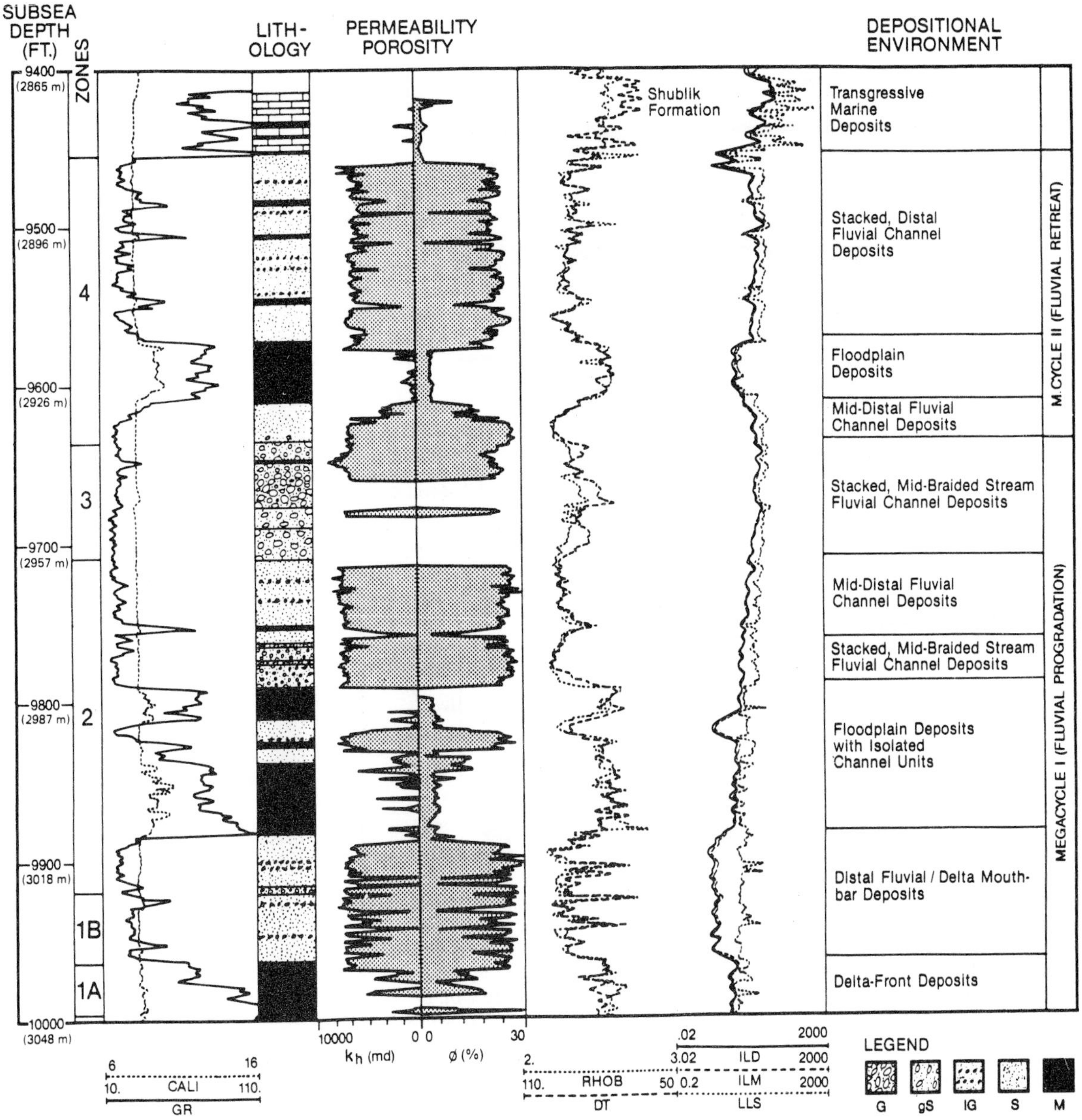

Fig. 15.11. Typical suite of logs through the Ivishak Member of the Sadlerochit Formation. The zonation of the Ivishak Member is based on petrophysical properties and can be mapped throughout the field. *G* Conglomerate; *gS* pebbly sandstone; *IG* intraformational conglomerate; *S* sandstone; *M* shale. (Atkinson et al. 1990)

buried structural high, trending WNW-ESE. This arch was an emergent landmass and a sediment source for the main reservoirs in the oil fields, including the Kekiktuk and Sadlerochit Formations. It seems likely that Barrow Arch was an intracratonic uplift that was active throughout much of the Paleozoic, and that it functioned as part of a rift system when the Canada Basin (Arctic Ocean) opened to the north. Beginning in the Jurassic, it then began to undergo subsidence beneath the sediments of Colville Trough.

The Prudhoe Bay structure is a large faulted anticline (Fig. 15.10). There are several producing horizons in this field, including the Lisburne Group and the Sadlerochit Formation, but the most important is the latter. This unit is truncated beneath a Cretaceous shale on the flanks of Barrow Arch. The shale acts as the seal. The northern limit of the field is

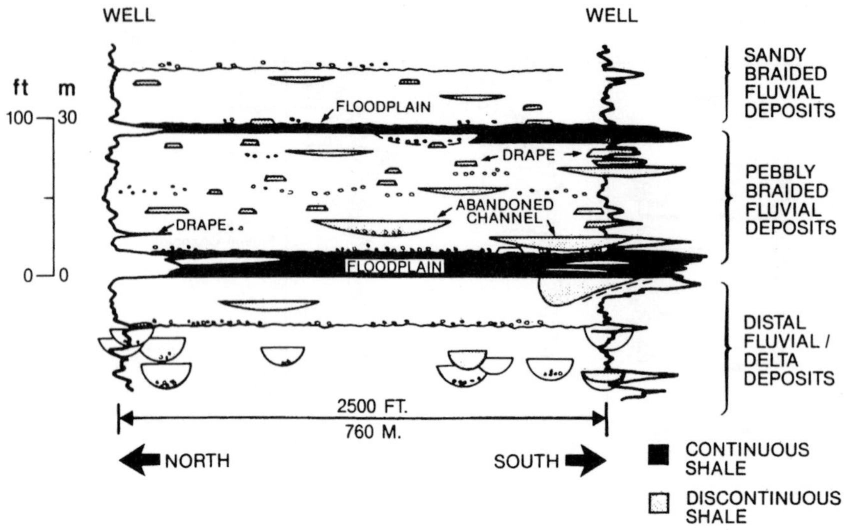

Fig. 15.12

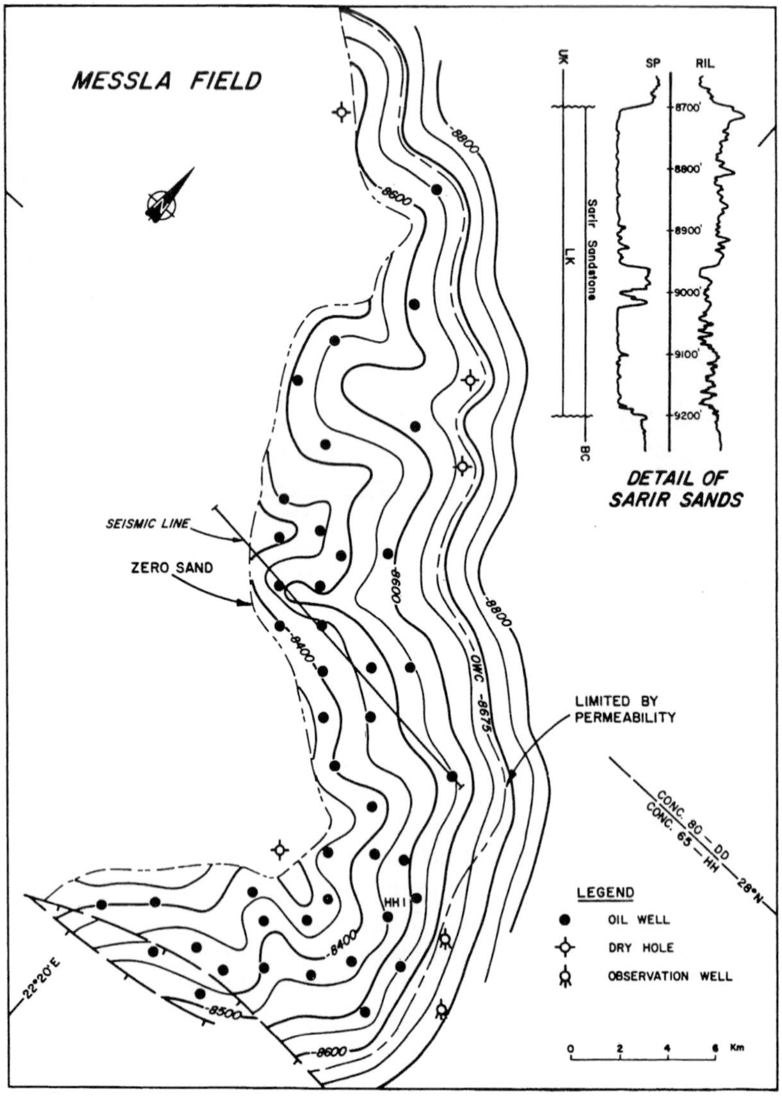

Fig. 15.13

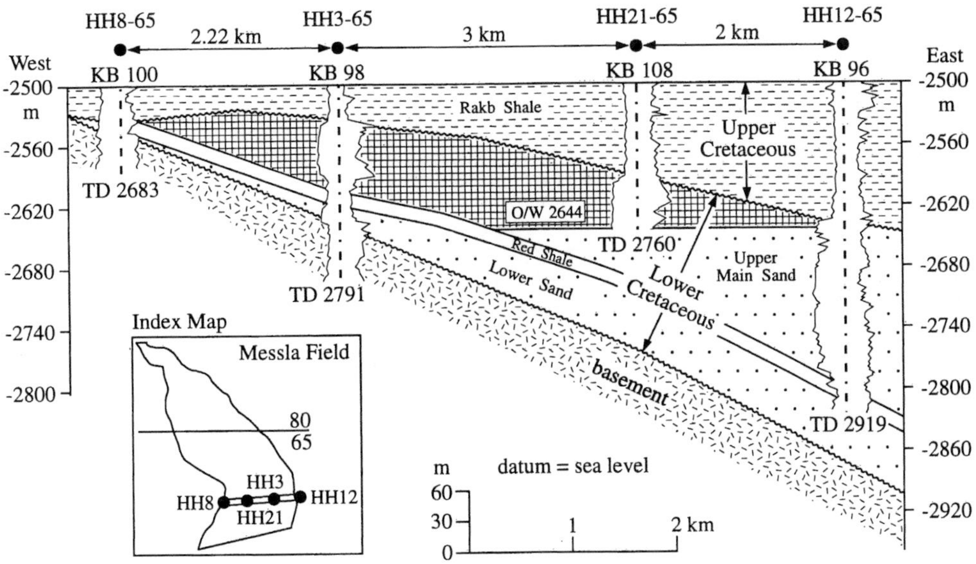

Fig. 15.14. Structural cross section through southern part of Messla field, showing regional stratigraphy and petrophysical log character of the Sarir Sandstone, its truncation by the Rakb Shale, and the position of the oil/water contact. (Clifford et al. 1980, reprinted by permission)

along a fault which drops the reservoir rock to the north.

The Sadlerochit Formation consists of a southward-prograding deltaic complex, in which the main producing horizons constitute a thick braided channel complex of the upper delta plain (Morgridge and Smith 1972; Eckelmann et al. 1975; Fig. 15.11). Braided stream deposits are estimated to constitute 55% of the total rock volume of the Ivishak Member of the Sadlerochit Formation. The lower portion of the interval consists of repeated, stacked channel sequences 3–6 m thick, each grading upward from conglomerate to medium-grained sandstone. In the upper part of the Ivishak Member conglomerates are rare.

Very detailed reservoir studies have been undertaken on the Sadlerochit Formation to permit efficient enhanced recovery (Eckelmann et al. 1975; Wadman et al. 1979; Geehan et al. 1986; Atkinson et al. 1990). A permeability map for the braided steam complex [zone 4 of Wadman et al. (1979)] is shown in Fig. 14.8. Note the variation from a high of more than 700 millidarcy in the north to less than 50 md in the south. The contours are thought to outline, in a crude sense, the distribution of the braided upper delta-plain complex. Figure 15.12 is a reconstructed interpretation of the depositional environment and schematic architecture of zone 2 of the Ivishak Member, based on log response and core data. In terms of primary production, the reservoir units can be regarded as homogeneous (typical SH-type geometries), but enhanced recovery work, including surveillance geology, has detected a considerable degree of internal heterogeneity of the reservoir, based particularly on the presence of shale beds of varying lateral persistence (Geehan et al. 1986; Atkinson et al. 1990). This is typical of many braided "sheet" reservoirs (Martin 1993).

15.3.2 Messla Field, Libya

Messla oil field is located in the southeast part of Sirte Basin, in central Libya. The Sirte Basin is a Late Mesozoic-Tertiary rift basin, resulting from extension of the north Africa craton (Clifford et al. 1980). More than 7600 m of Cretaceous-Tertiary sediment are present, most derived from land areas to the south.

Fig. 15.12. Interpreted depositional environment and generalized depositional architecture of zone 2 of the Ivishak Member at Prudhoe Bay. (Atkinson et al. 1990; based on Geehan et al. 1986)

Fig. 15.13. Structure contour map, top Sarir Sandstone, Messla field, Libya. East-trending lows on this surface may be the result of subaerial erosion preceding the Late Cretaceous transgression. (Clifford et al. 1980, reprinted by permission)

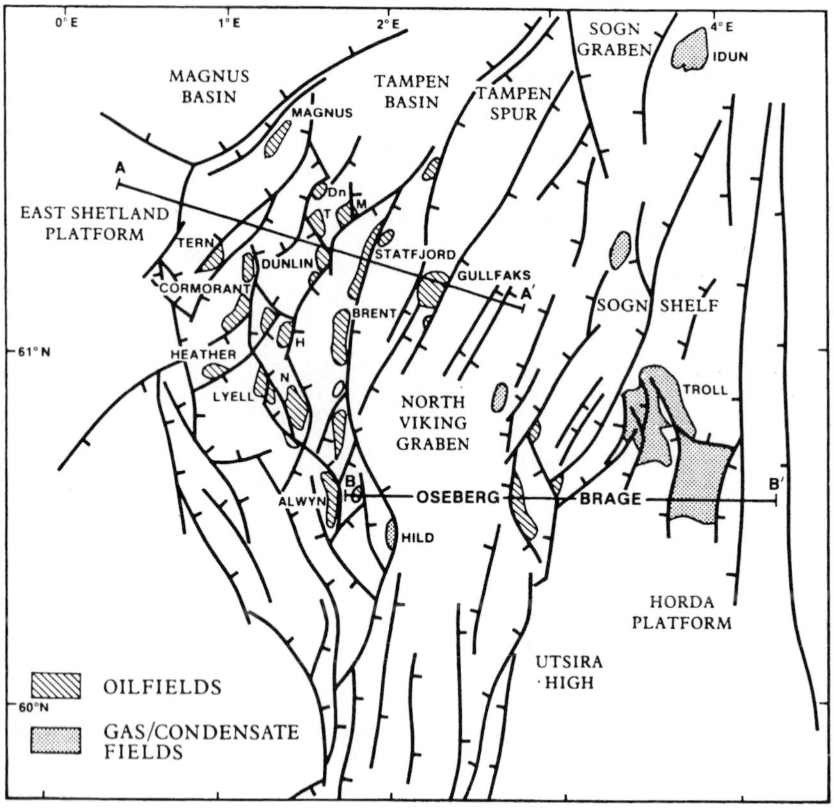

Fig. 15.15. Major fault trends and location of main oil fields in the north Viking Graben and East Shetland Basin area. (Parsley 1984)

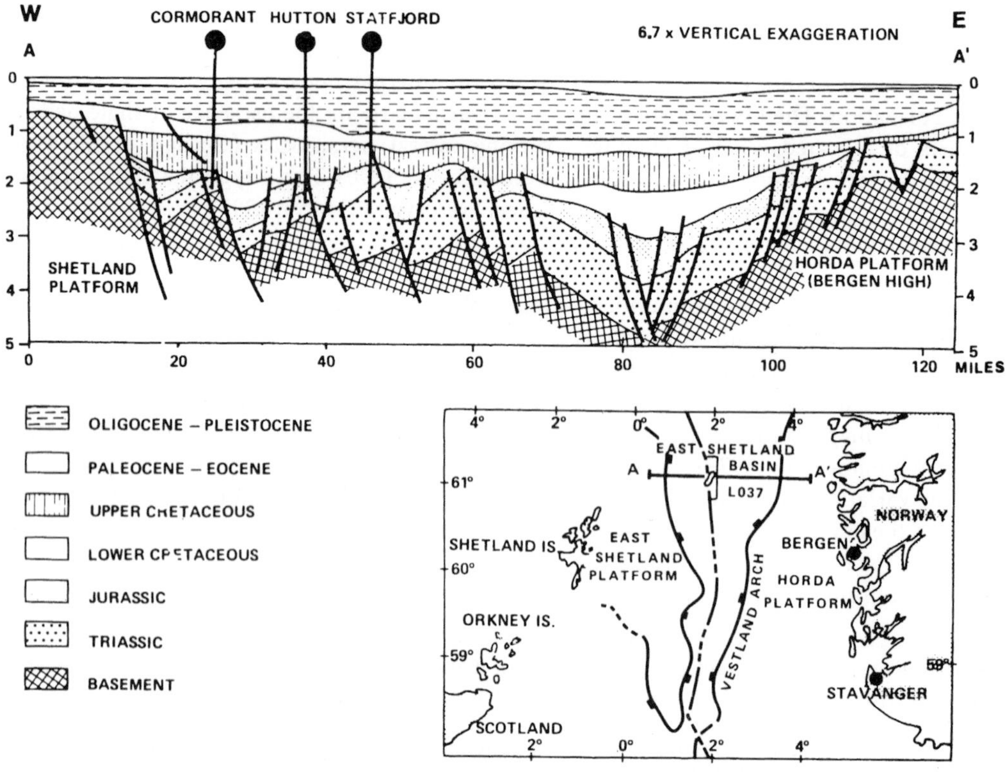

Fig. 15.16. Typical structural cross section through the north Viking Graben, showing position of Statfjord and other fields, on tilted fault blocks. (Kirk 1980, reprinted by permission)

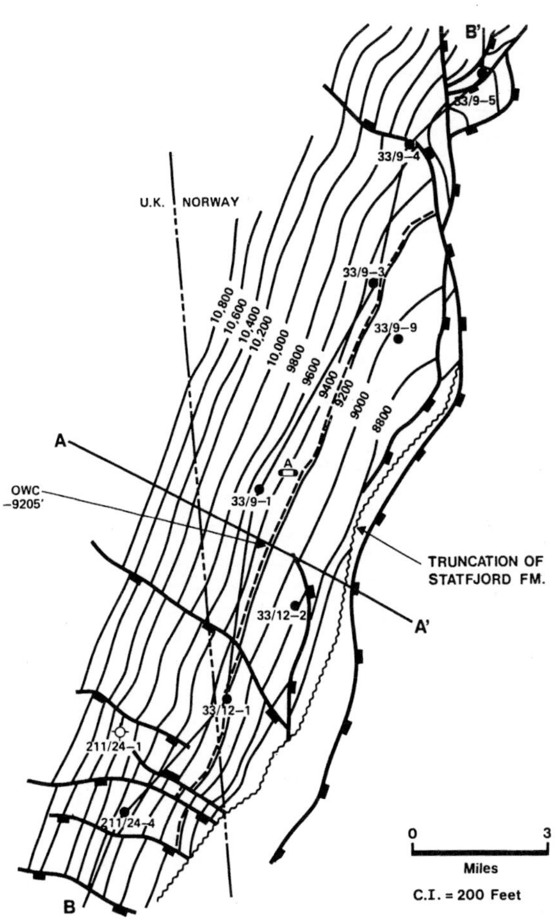

Fig. 15.17. Structure contour map, top of Statfjord Formation, Statfjord field, North Sea. (Kirk 1980, reprinted by permission)

Rift faulting began in the Early Cretaceous, as evidenced by basin-fill sediments of that age, which consist of poorly sorted nonmarine sandstones and siltstones. This succession is overlain by the Sarir Sandstone, a clean fluvial sandstone deposited by northward-flowing river systems. The beds were tilted and partially eroded during an episode of intracratonic tectonism in the mid-Cretaceous; then the basin subsided once again, and a marine transgression from Tethys, to the north, led to the deposition of evaporites, shales, and carbonates of Upper Cretaceous to Miocene age. Tilting of the area occurred during later extensional tectonism in the Tertiary (Clifford et al. 1980).

In the Messla field area the result of this basin evolution was the development of an east-dipping wedge of Sarir Sandstone, overlain and truncated to the west by Upper Cretaceous shales. In the field area the beds dip northeast at less than 1°. Faulting within the field area is minimal, except at the southern end, where the field is truncated by extensional faults, with downthrows to the south (Fig. 15.13).

The Messla Sandstone includes two sandstone members, separated by an extensive tight unit named the Red Shale. The upper sandstone, called the Main Pay, is more than 150 m thick in the south, but was truncated to the north and west by intra-Cretaceous erosion (Fig. 15.14). The Main Pay is a lithologically homogeneous unit of fine- to medium-grained, cross-bedded sandstone, interpreted to have been deposited predominantly in braided stream environments. The underlying Red Shale is a playa-lake or flood-basin deposit. Almost all the oil production is from the Main Pay unit. Petrophysical properties and production characteristics indicate a remarkable lack of heterogeneity in this reservoir. The beds probably represent the constant channeling and reworking of relatively well-sorted sands within an areally extensive braided fluvial system, in which overbank or floodplain areas had minimal preservation potential.

15.3.3 Statfjord Field, North Sea

The North Sea is an intracratonic basin which developed by a complex and lengthy series of extensional movements beginning in the late Paleozoic (Ziegler 1988). Fluvial-deltaic sandstones of Upper Triassic to Middle Jurassic age comprise 1 of 11

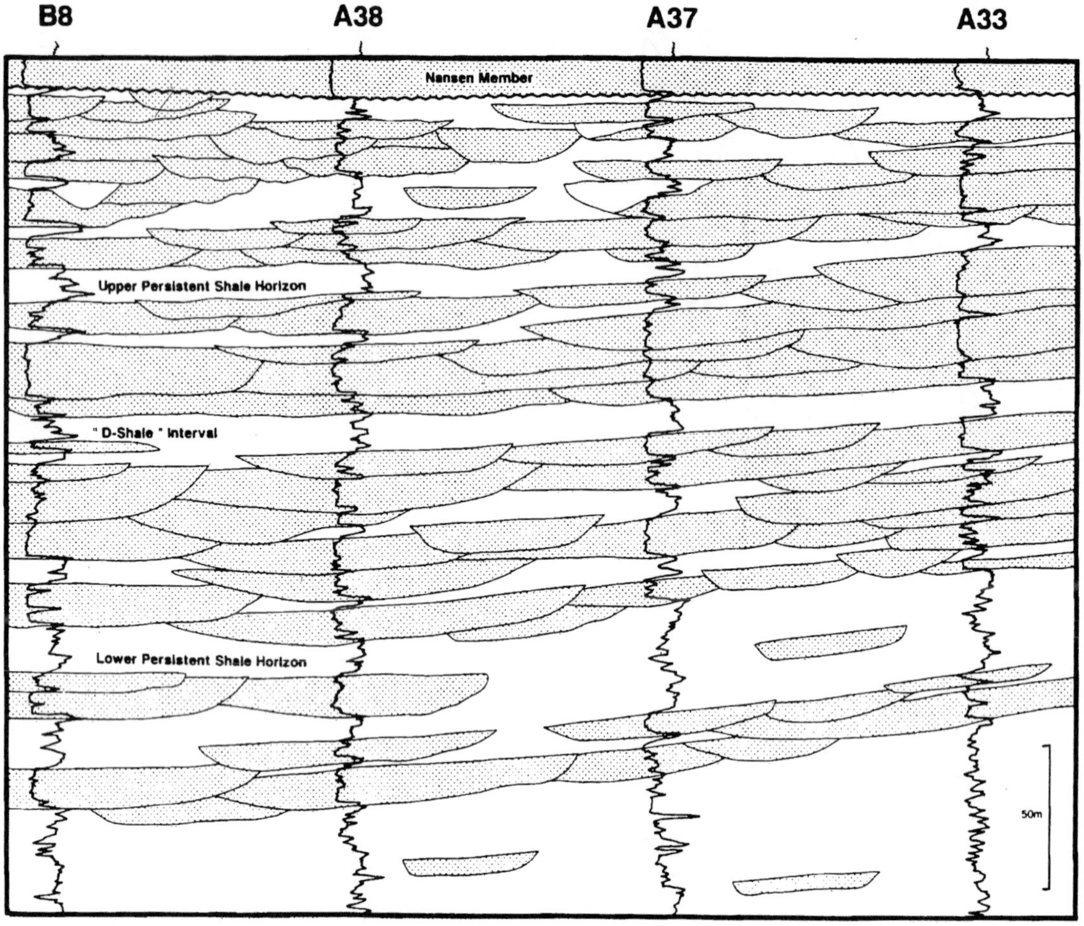

Fig. 15.18. Conceptual model of the large-scale architecture of the Statfjord Formation, Statfjord field. Well spacing is approximately 1 km. The section is oriented north-south, parallel to assumed drainage direction. (MacDonald and Halland 1993, reprinted by permission)

major petroleum plays in this area (Parsley 1984). Several giant oil fields contain significant reserves in these fluvial sandstones, including Brent, Snorre, Ninian, and Statfjord in the north Viking Graben and Beatrice in the Moray Firth Basin.

The Triassic and Jurassic rocks of the North Sea were developed during a phase of repeated rift faulting, related to the incipient opening of the Atlantic Ocean (Ziegler 1988). The Viking-Central Graben was the major structure to develop at this time (Fig. 11.44). Regional subsidence that accompanied rifting activated regional drainages, leading to the development of major fluvial units, the Cormorant and Statfjord Formations (Rhaetic-Hettangian) (Fisher 1984). With continued subsidence marine transgression occurred, and the largely deltaic Brent Group (Aalenian-Bajocian) was deposited (Brown 1984). Faulting and folding occurred in mid- to Late Jurassic time, and the fault blocks were tilted, uplifted, and eroded (Figs. 15.15, 15.16). Subsequent Late Jurassic (Kimmeridgian) and Cretaceous transgressions covered the eroded fault blocks with shales, including the Kimmeridge Shale, a prolific source rock. The shales also provide the seals for these north Viking Graben oil fields.

Petroleum occurs in two main fluvial units, the Statfjord Formation, the sandstones of which are relatively widely distributed and characterized by good porosity, and the Brent Group, where this unit includes sandstones of fluvial, rather than deltaic origin. The Cormorant Formation typically is thin and impersistent, and displays poor reservoir characteristics, although it has yielded some oil and gas shows, notably in the Beryl field (Fisher 1984).

Petroleum occurrences in Triassic-Jurassic strata of the northern North Sea are trapped in fault-bounded anticlinal structures (Fig. 15.17). Stratigraphic trapping mechanisms are secondary. The

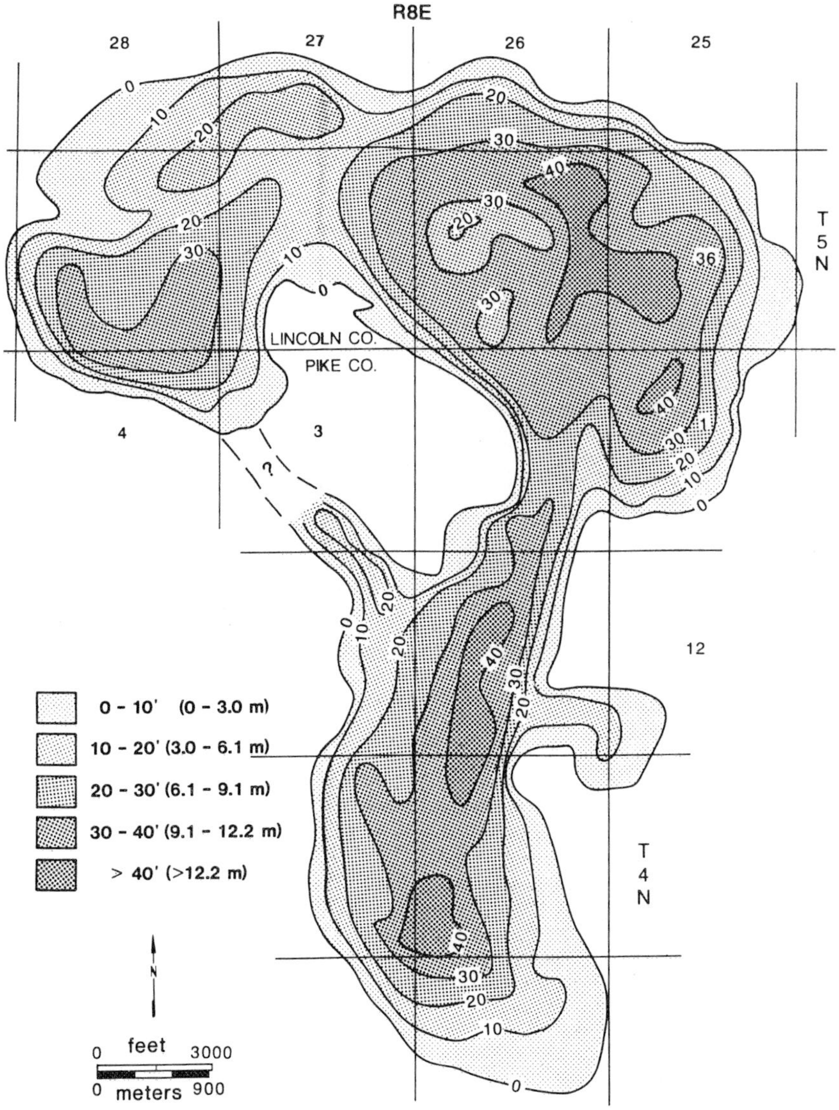

Fig. 15.19. Isopach of the upper producing unit (the Q sandstone) at Little Creek field. The three main pay areas constitute three separate point bars formed inside the meander bends of a highly sinuous river. (Werren et al. 1990)

major exploration method for these fields has been seismic structural mapping. The reservoirs are sheet-like in their overall geometry (Fig. 15.18), but considerable modeling work has now been done on this and nearby fields in attempts to provide detailed descriptions of the heterogeneities within the channel sheets (Stanley et al. 1990; Nybråten et al. 1990; Høimyr et al. 1993; MacDonald and Halland 1993; Martin 1993). This is particularly important for enhanced recovery of mature fields such as Statfjord, which is now being produced by high-pressure miscible gas drive, because of the potential for gas breakthroughs.

15.4 Channel-and-Bar Reservoirs (CB Type)

15.4.1 Little Creek Field, Mississippi

Little Creek field is a classic example of a field defined largely by the stratigraphic configuration of a local depositional system. It was one of the first to be explored using facies concepts, in the late 1950s (Werren et al. 1990), and has been widely figured in textbooks (e.g., Busch 1974).

The field is located within the Upper Cretaceous Tuscaloosa Sandstone, one of numerous productive

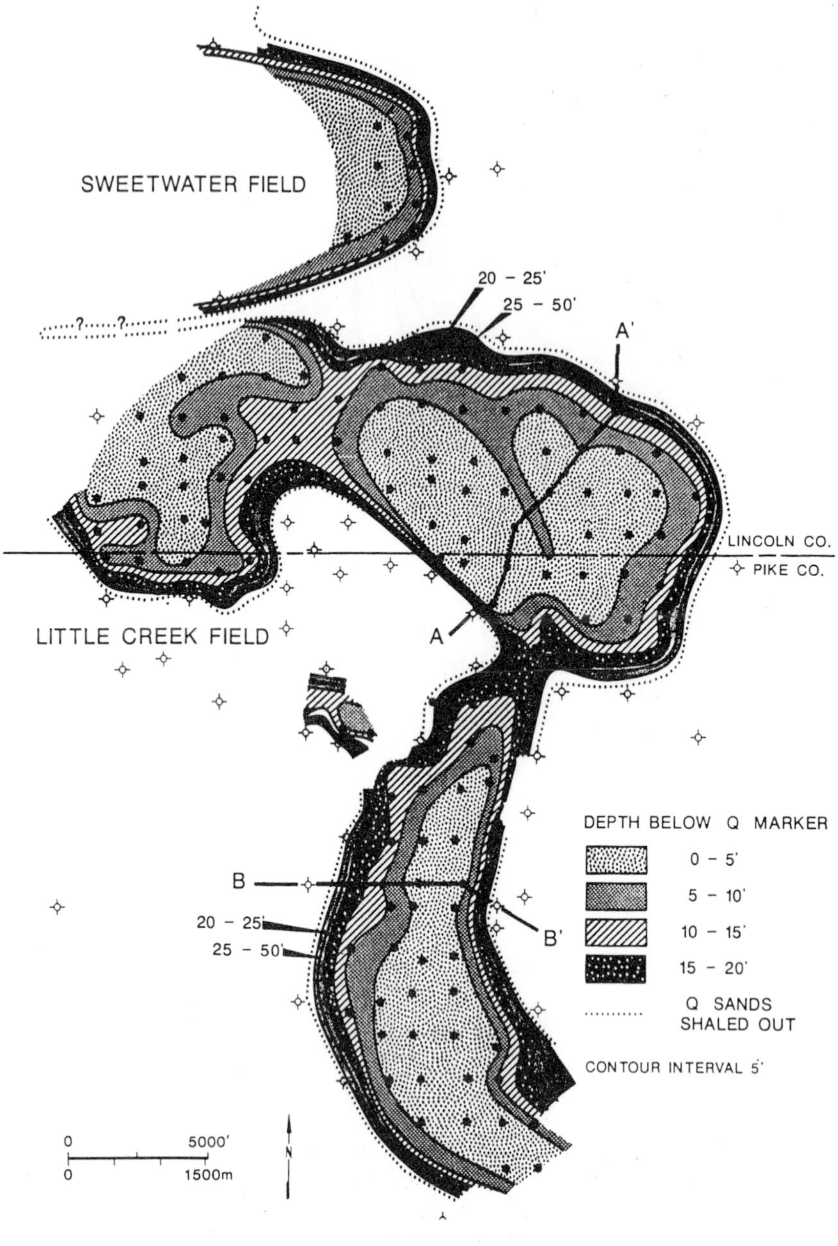

Fig. 15.20. Isopach of fine-grained sandstone, siltstone, and mudstone, constituting the fill of the abandoned channel overlying the Q sandstone at Little Creek field and underlying a regional marker that defines the top of the fluvial interval. (Werren et al. 1990)

sandstone units comprising the Mesozoic-Cenozoic clastic wedge of the Gulf Coast (Sect. 11.4.2.2). The field consists of point-bar lenses occupying the crestal region of a low-relief anticline. Floodplain deposits provide lateral seals, and the point-bar sandstones are overlain by transgressive tidal deposits that constitute the top seal. Two major sandstones are present, reaching maximum thicknesses of 16.8 and 9.1 m, respectively. Figure 15.19 is an isopach of the upper producing sandstone (Fig. 9.39 is an older version of this map). The configuration of the top surface of this point bar was formed by abandonment, and the shape of this surface is shown by Figs. 15.20 and 15.21. These diagrams indicate a northeast-dipping point-bar surface prior to abandonment along the line of section A–A', and a westward-dipping surface in the opposite-facing point bar along the line of section B–B'.

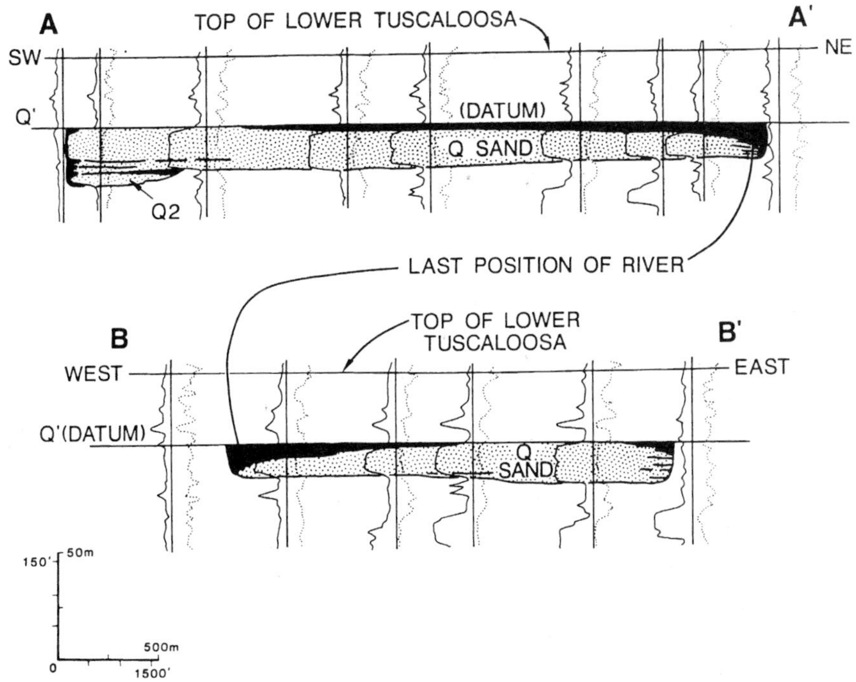

Fig. 15.21. Cross sections through the Q sandstone and overlying beds at Little Creek field. The thickness of the unit between the Q datum and the Q sandstone is shown in Fig. 15.20. (Werren et al. 1990)

15.4.2 Daqing Field, China

Daqing field is the largest in China, and one of the few giant fields in the world producing oil of non-marine origin (Yang 1985). It is located in Songliao Basin, an extensional basin in northeast China (Fig. 15.22), which first appeared as a small rift structure in the Jurassic and underwent active enlargement, and faulting and folding, during the Cretaceous and Cenozoic. A major period of subsidence occurred during the Early Cretaceous, when the main reservoir rocks were formed in a complex of fluvial, deltaic, and lacustrine environments. A lake extended though the basin in the Late Cretaceous, and an episode of folding, leading to the development of the Daqing structure, occurred before the end of the Cretaceous (Tang 1982; Liu 1986).

The tectonic interpretation of the Songliao Basin is uncertain. Klimetz (1983) constructed a model of regional subduction and microplate collision in which the Songliao Basin is interpreted as a backarc rift. However, Burke and Sengör (1986) suggested that, rather than extension, there may have been convergence between Asia and a North China block, with the Songliao basin developing by lateral movement along strike-slip faults, a process termed "escape tectonics". The field is located within a complex anticlinal structure, which extends NNE-SSW along the center of the basin. Reserves are now estimated at 1460 $m^3 \times 10^6$ (Soeparjadi et al. 1986).

Production is from four nonmarine Lower Cretaceous formations, the Quanton, Qingshankou, Yaojia, and Nenjiang formations, in order from base to top (Tang 1982). These units developed as a series of fluvial-deltaic complexes that prograded into the basin from the north. The total reservoir thickness is approximately 3 km, but this includes thousands of separate oil-producing bodies. When waterflooding was initiated, at an early stage of field development, the field was divided into as many as 50 flow units to permit separate production and pressure maintenance (Jin et al. 1985). Individual producing sandstones are less than 10 m thick.

Eight different types of sand bodies have been distinguished (Jin et al. 1985). These include fluvial meander-belt and various deltaic and shoreline facies. Meander-belt sandstones are the thickest (4–10 m). They range from 2 to 4.5 km in width, and individual units have been traced for more than 3 km. Waterflooding experience has shown that the meander-belt sandstones are typically "bottom-flooding"; that is, injected water advances rapidly along the bottom of the zone, because this is where the sandstones are the coarsest and best sorted, with the best permeabilities (Jin et al. 1985).

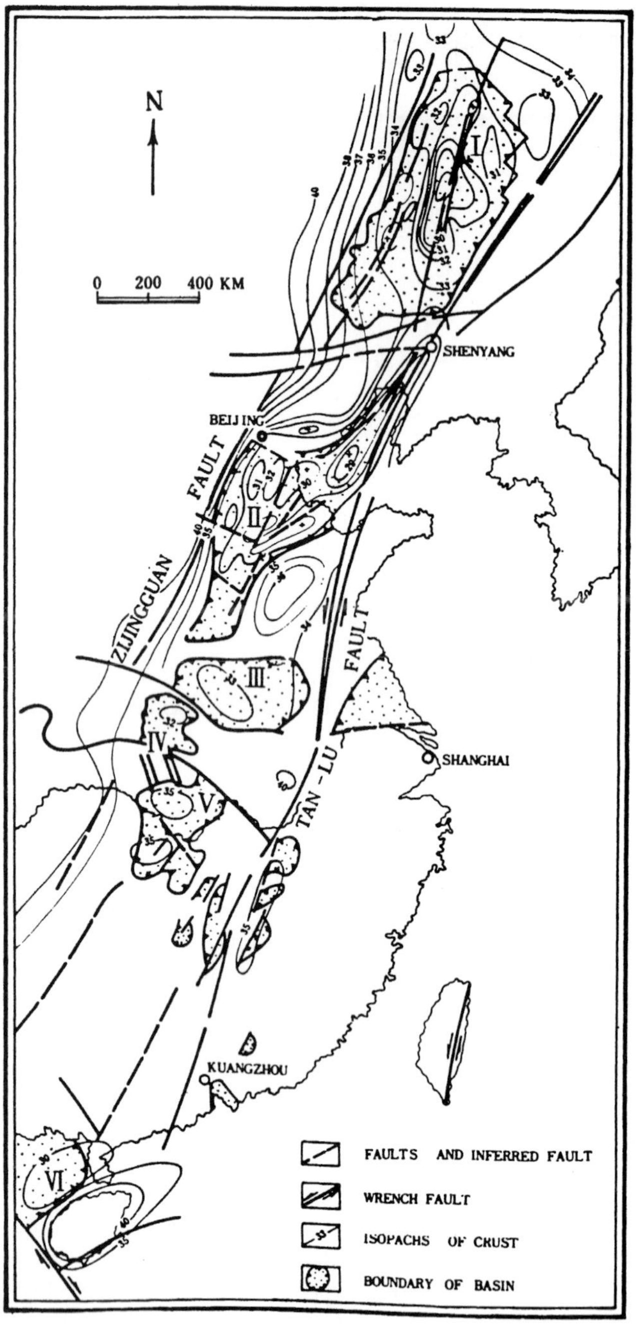

Fig. 15.22. Structural setting of Songliao Basin, east China. The Songliao Basin is lettered *I* at the north end of the map area. It is bounded by extensional faults oriented NNE-SSW, forming part of the Cathaysian Rift System. (Liu 1986, reprinted by permission)

An example of a meander-belt sandstone body is illustrated in Fig. 15.23. This unit consists of a fining-upward succession of medium- to fine-grained, cross-bedded sandstone, with a basal gravel lag. It is interpreted as a succession of amalgamated point-bar deposits (Yinan et al. 1987). The meander belt is 800–1000 m wide, with a width/depth ration of 130–170. The complete point-bar sequence is estimated from core studies to be 5–7 m thick. Given a 5°–10° dip on the point-bar accretion surface, this indicates a channel width of 90–120 m. The meander belt is therefore 10–15 times the width of the channel, which is a typical ratio for meander belts of highly sinuous rivers. Yinan et al. (1987) also described braided sandstone units from Daqing oil field.

Reservoir connectedness is considered by Jin et al. (1985). They illustrated an example of a section containing three flow units: a meandering-channel unit, consisting of fine sandstone, overlying a

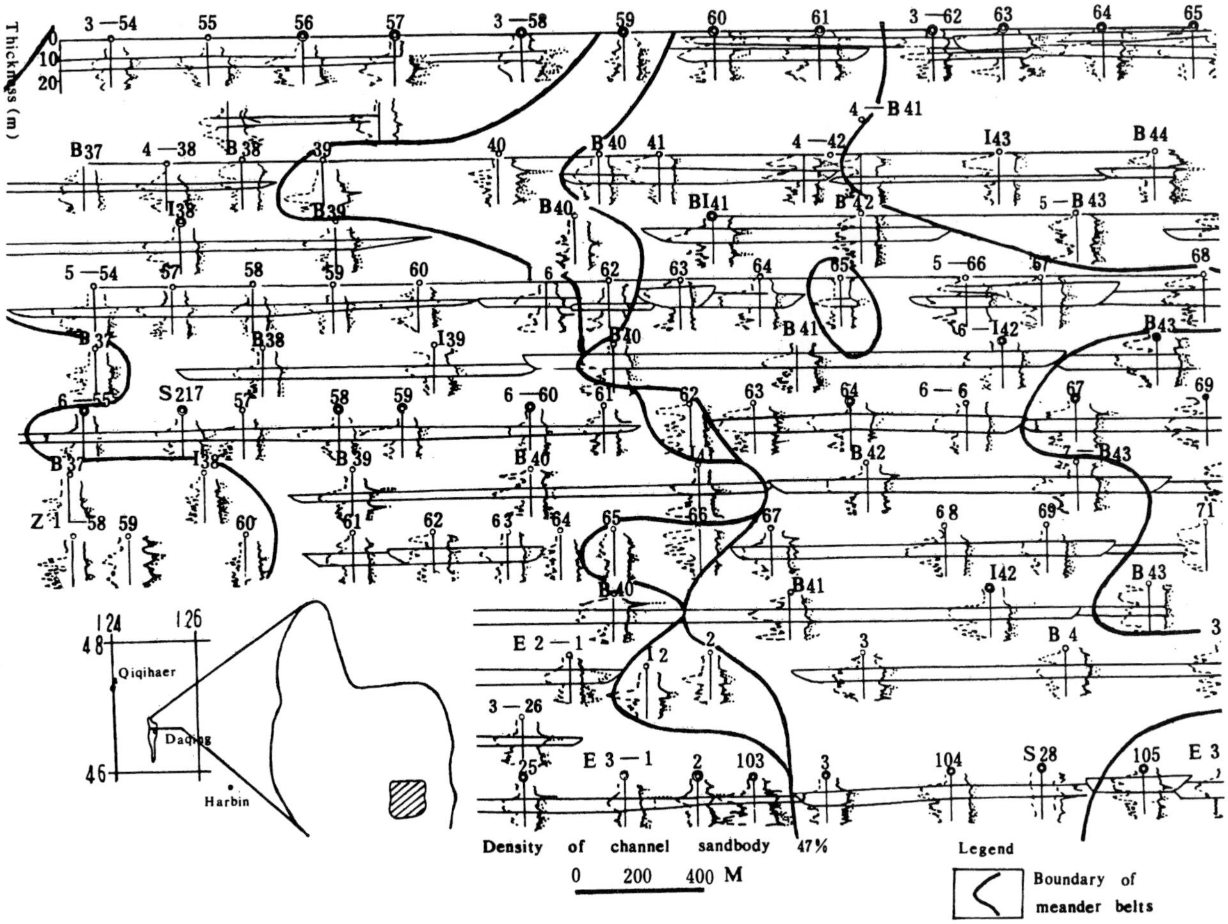

Fig. 15.23. Geometry of a meander-belt sandstone body, Daqing oil field. (Yinan et al. 1987)

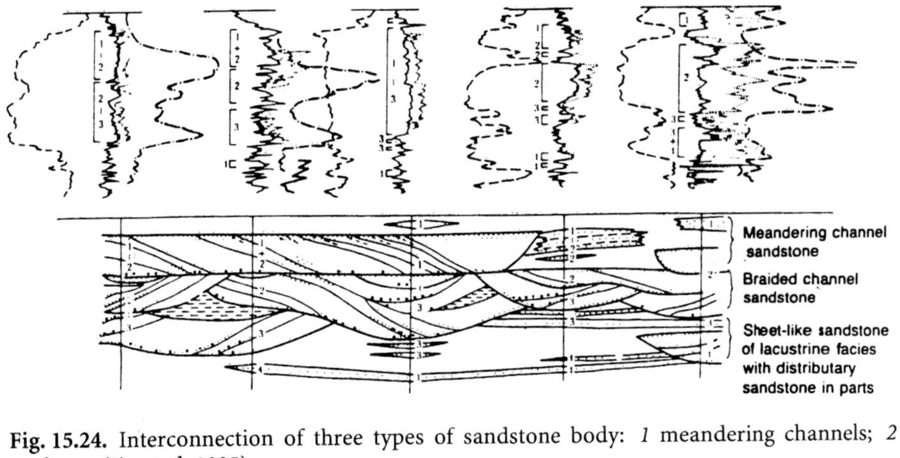

Fig. 15.24. Interconnection of three types of sandstone body: *1* meandering channels; *2* braided channels; *3* core analyses. (Jin et al. 1985)

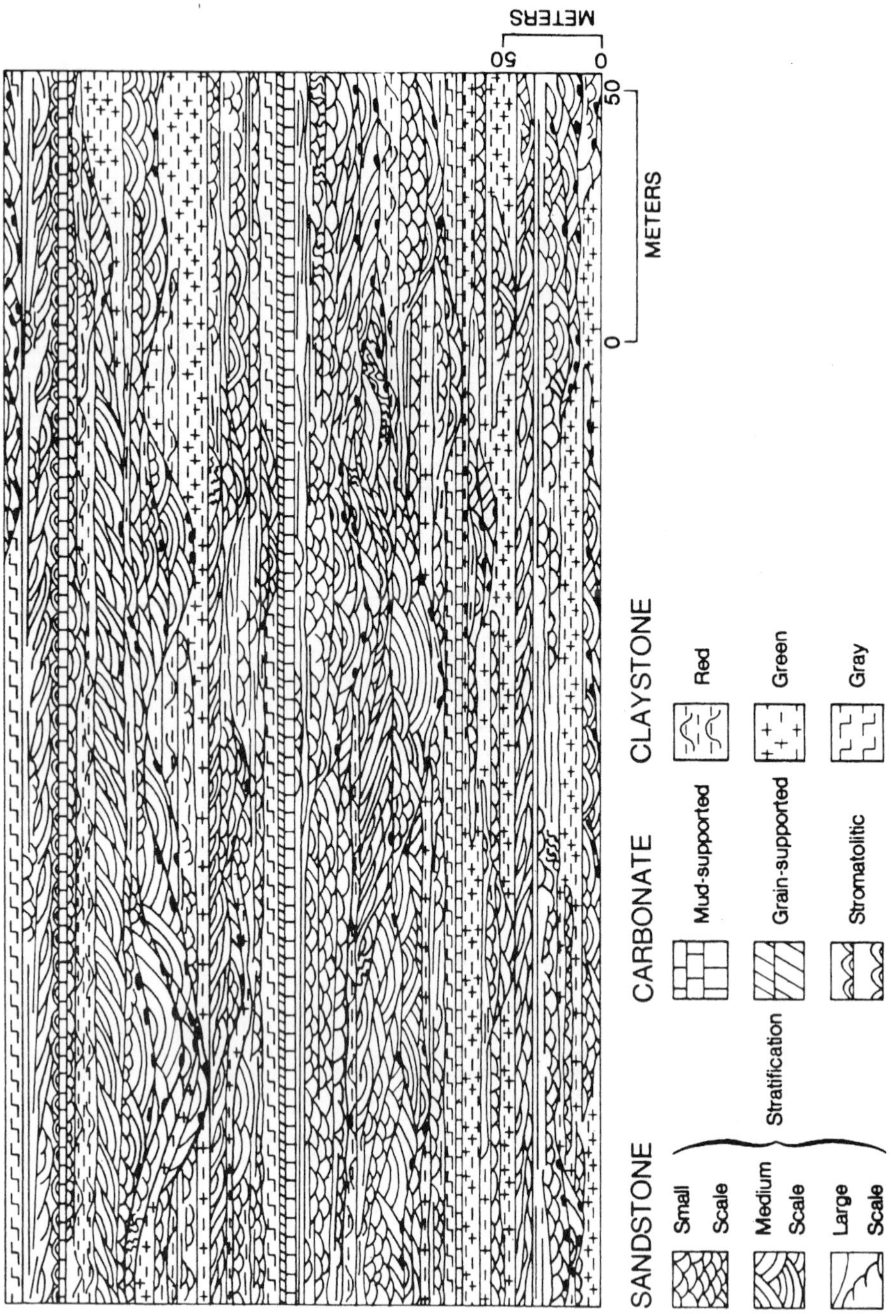

Fig. 15.25. Schematic stratigraphic cross section showing interfingering of fluvial sandstones and lacustrine carbonates and claystones in Uinta Basin. (Pitman et al. 1982, reprinted by permission)

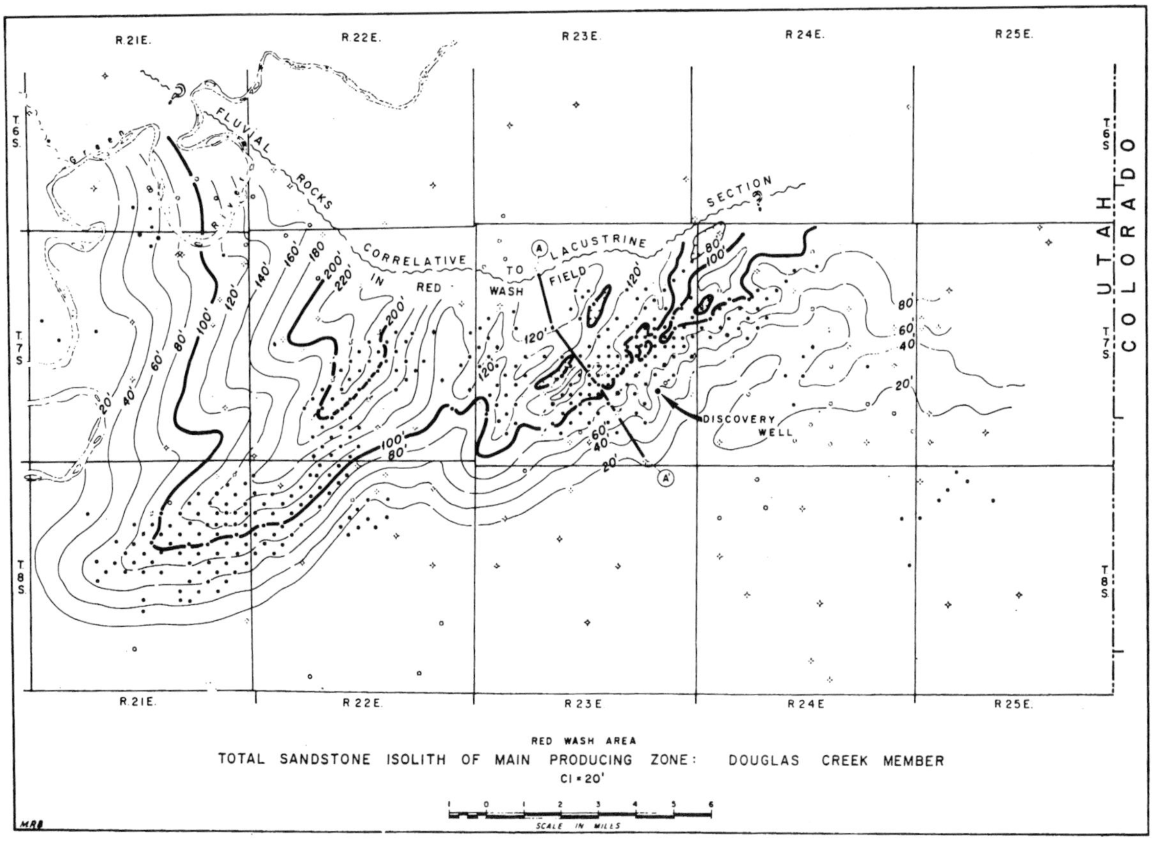

Fig. 15.26. Total sandstone isolith of producing zones in Red Wash field, Utah. Shape of sandstone thickness contours suggests a coastal plain or deltaic complex thickening toward the source area on the northern margin of the basin. In detail individual sandstone lenses occupy only a small part of this area. (Chatfield 1972, reprinted by permission)

braided, coarse-grained sandstone with a high-relief scoured base, which in turn overlies a succession of thin, sheetlike lacustrine sand bodies (Fig. 15.24). Permeability barriers, such as shales draping third- and fourth-order surfaces, may be as little as 10–40 cm thick but may effectively separate sand bodies into distinct flow zones.

15.4.3 Red Wash Field, Utah

During the late Maaastrichtian and early Paleocene, Laramide tectonism in the Rocky Mountain foreland basin created new uplifts in the Colorado Plateau area, and the foreland basin was partitioned (Lawton 1986b; Fouch et al. 1983; Dickinson et al. 1988; Fig. 11.26). The Uinta Basin, situated in northeastern Utah and westernmost Colorado, is one of the intermontane basins generated by this tectonic episode. In northern Uinta Basin, where there is significant oil and gas production, rocks of Paleocene-Eocene age disconformably overly Cretaceous clastic strata of the foreland-basin clastic wedge. Cenozoic deposits were formed in alluvial and lacustrine depositional systems, with the main sediment source located in Uinta Uplift, to the north of the basin.

More than 80 $m^3 \times 10^6$ of recoverable oil and 422 000 $m^3 \times 10^6$ of recoverable gas have been discovered in the Upper Cretaceous and Paleocene-Eocene rocks of Uinta Basin (Fouch 1983). Most of this production is from lenticular fluvial sandstone bodies that were deposited in coastal plain depositional systems. In the Paleocene-Eocene the Uinta Basin was part of a large lacustrine system. Sandstone bodies were formed as channel and point-bar lenses encased within widespread, low-energy floodplain and marine/lacustrine bay-fill deposits (Fig. 15.25) and shaling out to the south. These coastal complexes have been referred to as deltas (e.g., Fouch 1983), but they do not show the thick (hundreds of meters), prograding, coarsening-upward character of typical major delta systems and are interpreted as

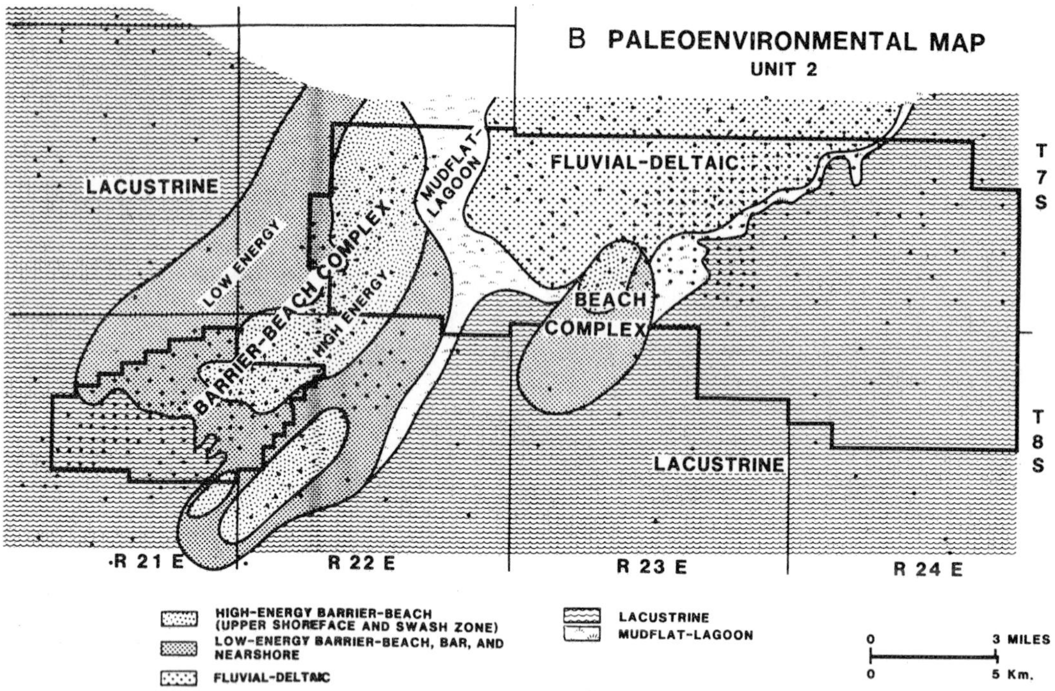

Fig. 15.27. Paleoenvironmental map of one of the producing units in the Red Wash field. Compare with Fig. 15.26. (Castle 1990, reprinted by permission)

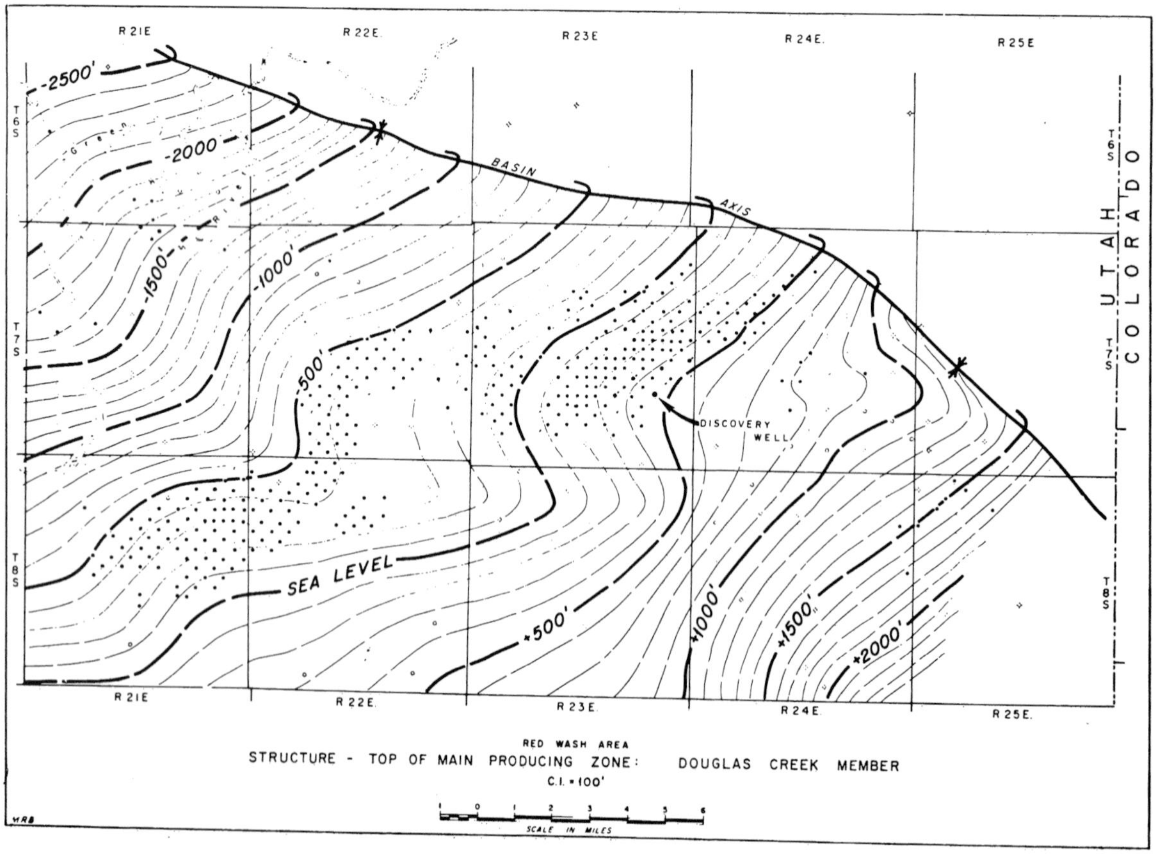

Fig. 15.28. Structure contour map of top of producing zone, Red Wash field, Utah. (Chatfield 1972, reprinted by permission)

thin coastal-plain alluvial sheet sandstone bodies. Figure 15.26 shows an old isopach map of the producing zones, which suggests a lobate, deltaic form to the deposits. However, modern, more detailed facies analysis has yielded the map shown in Fig. 15.27, in which a complex range of interfingering depositional systems has been identified. Traps are present where the shale-out takes place updip within Laramide structures (Chatfield 1972; Pitman et al. 1982; Fouch 1983; Fig. 15.28). Reservoir quality in most of the fields has been enhanced by the development of fractures and secondary porosity, and the traps therefore also have a diagenetic component to them (Keighin and Fouch 1981).

Chapter 16

Future Research Trends

In this brief concluding chapter some attempt is made to assess the present state of knowledge of the geology of fluvial deposits, and to predict the kinds of research developments that might be expected in the next decade.

Facies Analysis. Although some workers still prefer to establish new facies classifications with each new field study, the classification and coding system described in Chaps. 4 and 5 are by now virtually standard methodologies. Room remains for the recognition and better description of facies that are rarely formed or preserved in the sedimentary record, such as those formed under very high shear stresses. There continues to be a need for additional case studies of ancient units to serve as a basis for a wide variety of interpretive studies.

In-channel Depositional Processes. Observations of natural sand-bed and gravel-bed rivers and flume observations and models have provided a thorough understanding of the processes of formation of most bedforms and macroforms. Work in this area continues, with the use of flume experiments, flow visualization techniques, and numerical models.

The Architecture of Channel-and-Bar Deposits. The development of the architectural-element method of description has facilitated the documentation of the complex three-dimensional nature of the channel-and-bar deposits that characterize many modern and ancient fluvial systems. Classifications of architectural elements and bounding surfaces have been developed, but there is scope for considerable development and refinement. This is a case where work on large outcrops of ancient deposits has forced a reexamination of many concepts derived from studies of modern rivers that were restricted to data derived from shallow trenches dug above the water table. For example, it is only relatively recently that the accretionary geometry of most bar complexes in braided systems has been recognized. The application of high-resolution seismic reflection techniques and ground-penetrating radar promises to provide much valuable information on the three-dimensional structure of both modern and ancient fluvial systems in the near future, to further improve our data base of architectural information.

The Architecture of Overbank Regions. There is room for considerable additional work on the regional mapping of floodplain units, using catena and pedofacies concepts. Such work is needed in order to understand the long-term dynamics of channel avulsion and flooding mechanisms, as well as floodplain construction.

Fluvial Styles. Most sedimentological studies of modern river systems have been carried out in temperate, arctic, and hot arid regions. There is a need for comparable research in tropical humid regions, such as the rivers of the African and South American rain forests. Very few detailed studies of the ancient record have been completed, and a wide variety of architectural case studies is needed to complement our knowledge of modern rivers.

Stratigraphic Architecture. One of the most interesting areas of research in recent years has been the application of high-resolution chronostratigraphic techniques, particularly magnetostratigraphy, to the study of alluvial stratigraphies. The availability of correlation frameworks of this type permits much more detailed reconstructions of regional sedimentary controls, particularly tectonism, and allows for the quantitative examination of the rates of such processes as basin accumulation rates and channel avulsion frequency. Such data are essential for the modeling of the autogenic and tectonic control of alluvial stratigraphy, many studies of which are now appearing, particularly as an aid to the reconstruction and interpretation of fluvial petroleum reservoirs.

Sequence Stratigraphy. Basic sequence models for fluvial systems are now available and are being applied to a wide array of case examples. Some of the most interesting work underway at the time of writing is the attempt to determine the relative importance of climate change and base-level change in the development of Quaternary fluvial valley fills. An alternation between valley incision and channel aggradation has been documented that appears to be a complex response to changes in discharge regime and sediment supply associated with changes in rainfall and vegetation cover. Not only is such work of importance in the understanding of the recent (postglacial) history of the earth, but it will also provide models for the interpretation of earlier periods in earth history when orbital-forcing mechanisms may have been responsible for regular or episodic climate change.

Quantitative Basin Models. One of the most active areas of research in basin analysis is the development of numerical forward and inverse models of basin subsidence and fill. Much attention has been paid to the tectonic and eustatic control of stratigraphic architecture of shallow-marine and continental-margin deposits, and recent work has also focused on the nature of the balance between subsidence and sediment supply at tectonically active nonmarine basin margins. There is scope for much additional work on the bringing together of models describing the physics of sedimentation, models describing the autogenic construction of alluvial plains, and models that examine large-scale basin architecture and incorporate differential subsidence and base-level change.

Petroleum Geology. Patterns of fluvial reservoir development are now relatively well understood. Our knowledge of fluvial styles, sequence stratigraphy, and the tectonic control of fluvial depositional systems has provided a range of tools for the discovery and mapping of fluvial reservoirs. Many additional discoveries remain to be made, particularly in frontier areas. At the exploitation stage, seismic mapping of channel bodies and numerical modeling of channel architecture are both areas of active research. Dipmeter data, cross-well seismic tomography, horizontal drilling (two areas of research not discussed in this book), and surveillance methods are all being used in conjunction with outcrop analogue studies as input to architectural models that then form the basis for detailed flow studies. Such work is essential, both for primary production and for enhanced recovery from old fields.

Hydrogeology. Few workers have applied architectural concepts to problems in hydrogeology, and this is an area that is likely to become considerably more important in the next few years. Geophysical techniques for imaging the shallow subsurface, particularly ground-penetrating radar, can provide a wealth of architectural detail on the structure of aquifers and aquitards, which can now be more accurately interpreted using sedimentological knowledge derived from architectural techniques. Such knowledge is likely to be of particular value in the study of the flow of contaminants through surficial sediments, such as the alluvial floodplains upon which much of the developed world's industrial plant has been constructed. Architectural details can be used to constrain mathematical flow models in order to provide precise, site-specific information on fluid transport paths.

References

Abdullatif OM (1989) Channel-fill and sheet-flood facies sequences in the ephemeral terminal River Gash, Kassala, Sudan. Sediment Geol 63: 171–184

Abrahams MJ, Chadwick OH (1994) Tectonic and climatic implications of alluvial fans sequences along the Batinah coast, Oman. J Geol Soc (Lond) 151: 51–58

Ager DV (1981) The nature of the stratigraphical record, 2nd edn. Halsted, Somerset

Ager DV (1993) The new catastrophism: the importance of the rare event in geological history. Cambridge University Press, Cambridge

Ahmad R, Scatena FN, Gupta A (1993) Morphology and sedimentation in Caribbean montane streams: examples from Jamaica and Puerto Rico. Sediment Geol 85: 157–169

Aitken JD (1966) Middle Cambrian to Middle Ordovician cyclic sedimentation, southern Rocky Mountains, Canada. Bull Can Petrol Geol 14: 405–441

Aitken JD (1978) Revised models for depositional grand cycles, Cambrian of the southern Rocky Mountains, Canada. Bull Can Petrol Geol 26: 515–542

Aitken JF, Flint SS (1995) The application of high-resolution sequence stratigraphy to fluvial systems: a case study from the Upper Carboniferous Breathitt Group, eastern Kentucky, USA. Sedimentology 42: 3–30

Albright WA, Turner WL, Williamson KR (1980) Ninian field, UK sector, North Sea. In: Halbouty MT (ed) Giant oil and gas fields of the decade: 1968–1978. Am Assoc Petrol Geol Mem 30: 173–193

Alexander J (1986) Idealized flow models to predict alluvial sandstone body distribution in the Middle Jurassic Yorkshire Basin. Mar Petrol Geol 3: 298–305

Alexander J (1992) Nature and origin of a laterally extensive alluvial sandstone body in the Middle Scalby Formation. J Geol Soc (Lond) 149: 431–441

Alexander J, Gawthorpe RL (1993) The complex nature of a Jurassic multistorey, alluvial sandstone body, Whitby, North Yorkshire. In: North CP, Prosser DJ (eds) Characterization of fluvial and aeolian reservoirs. Geological Society, London, pp 123–142 (Spec Publ 73)

Alexander J, Leeder MR (1987) Active tectonic control on alluvial architecture. In: Ethridge FG, Flores RM, Harvey MD (eds) Recent developments in fluvial sedimentology. Soc Econ Paleontol Mineral Spec Publ 39: 243–252

Alexander J, Bridge JS, Leeder MR, Collier EL, Gawthorpe RL (1994) Holocene meander-belt evolution in an active extensional basin, southwestern Montana. J Sediment Res B64: 542–559

Algeo TJ, Wilkinson BH (1988) Periodicity of mesoscale phanerozoic sedimentary cycles and the role of Milankovitch orbital modulation. J Geol 96: 313–322

Allen JRL (1962) Petrology, origin and deposition of the highest Lower Old Red Sandstone of Shropshire, England. J Sediment Petrol 32: 657–697

Allen JRL (1963a) The classification of cross-stratified units, with notes on their origin. Sedimentology 2: 93–114

Allen JRL (1963b) Henry Clifton Sorby and the sedimentary structures of sands and sandstones in relation to flow conditions. Geol Mijnbouw 42: 223–228

Allen JRL (1964) Studies in fluviatile sedimentation: six cyclothems from the Lower Old Red Sandstone, Anglo-Welsh basin. Sedimentology 3: 163–198

Allen JRL (1965a) A review of the origin and characteristics of recent alluvial sediments. Sedimentology 5: 89–191

Allen JRL (1965b) The sedimentation and palaeogeography of the Old Red Sandstone of Anglesey, North Wales. Yorkshire Geol Soc Proc 35: 139–185

Allen JRL (1966) On bed forms and paleocurrents. Sedimentology 6: 153–190

Allen JRL (1967) Notes of some fundamentals of paleocurrent analysis, with reference to preservation potential and sources of variance. Sedimentology 9: 75–88

Allen JRL (1968) Current ripples. North Holland, Amsterdam

Allen JRL (1970a) Physical processes of sedimentation. Allen and Unwin, London

Allen JRL (1970b) Studies in fluviatile sedimentation: a comparison of fining-upward cyclothems, with special reference to coarse-member composition and interpretation. J Sediment Petrol 40: 298–323

Allen JRL (1973) Phase differences between bed configurations and flow in natural environments, and their geological relevance. Sedimentology 20: 323–329

Allen JRL (1974a) Reaction, relaxation and lag in natural sedimentary systems: general principles, examples and lessons. Earth Sci Rev 10: 263–342

Allen JRL (1974b) Studies in fluviatile sedimentation: implications of pedogenic carbonate units, Lower Old Red Sandstone, Anglo-Welsh outcrop. Geol J 9: 181–208

Allen JRL (1977) The plan shape of current ripples in relation to flow conditions. Sedimentology 24: 53–62

Allen JRL (1978) Studies in fluviatile sedimentation: an exploratory quantitative model for the architecture of avulsion-controlled suites. Sediment Geol 21: 129–147

Allen JRL (1980) Sand waves: a model of origin and internal structure. Sediment Geol 26: 281–328

Allen JRL (1983a) Studies in fluviatile sedimentation: bars, bar complexes and sandstone sheets (low-sinuosity braided streams) in the Brownstones (L. Devonian), Welsh Borders. Sediment Geol 33: 237–293

Allen JRL (1983b) River bedforms: progress and problems. In: Collinson JD, Lewin J (eds) Modern and ancient

fluvial systems. Int Assoc Sedimentol Spec Publ 6: 19–33
Allen JRL (1983c) Gravel overpassing on humpback bars supplied with mixed sediment: examples from the Lower Old Red Sandstone, southern Britain. Sedimentology 30: 285–294
Allen JRL (1984) Sedimentary structures: their character and physical basis. Elsevier, Amsterdam, 663 p (Developments in sedimentology, vol 30)
Allen JRL, Williams BPJ (1982) The architecture of an alluvial suite: rocks between the Townsend Tuff and Pickard bay Tuff Beds (Early Devonian), southwest Wales. Philos Transact R Soc Lond [Biol] 297: 51–89
Allen MB, Windley BF, Zhang C (1991) Active alluvial systems in the Korla Basin, Tien Shan, northwest China, sedimentation in a complex foreland basin. Geol Mag 128: 661–666
Allen PA, Allen JR (1990) Basin analysis: principles and applications. Blackwell Scientific Publications, Oxford
Allen PA, Matter A (1982) Oligocene meandering stream sedimentation in the eastern Ebro basin, Spain. Eclog Geol Helv 75: 33–50
Alonso Zarza MA, Wright VP, Calvo JP, García del Cura MA (1992) Soil-landscape and climatic relationships in the Middle Miocene of the Madrid Basin: Sedimentology 39: 17–35
Ambrose WA, Tyler N, Parsley MJ (1991) Facies heterogeneity, pay continuity, and infill potential in barrier-island, fluvial, and submarine-fan reservoirs: examples from the Texas Gulf Coast and Midland Basin. In: Miall AD, Tyler N (eds) The three-dimensional facies architecture of terrigenous clastic sediments, and its implications for hydrocarbon discovery and recovery. Soc Econ Paleontol Mineral Concepts Models Ser 3: 13–21
Anadón P, Cabrera L, Colombo F, Marzo M, Riba O (1986) Syntectonic intraformational unconformities in alluvial fan deposits, eastern Ebro basin margins (NE Spain). In: Allen PA, Homewood P (eds) Foreland basins. Int Assoc Sedimentol Spec Publ 8: 259–271
Anderson B (1961) The Rufiji basin, Tanganyika, vol 7: soils of the main irrigable area. FAO report to Government of Tanganyika
Anstey NA (1980) Seismic exploration for sandstone reservoirs. International Human Resources Development Corporation, Boston
Arche A (1983) Coarse-grained meander lobe deposits in the Jarama River, Madrid, Spain. In: Collinson JD, Lewin J (eds) Modern and ancient fluvial systems. Int Assoc Sedimentol Spec Publ 6: 313–321
Arnott RWC (1992) The role of fluvial processes during deposition of the (Cardium) Carrott Creek/Cyn-Pem conglomerates. Bull Can Petrol Geol 40: 356–362
Ashley GM (1990) Classification of large-scale subaqueous bedforms: a new look at an old problem. J Sediment Petrol 60: 160–172
Ashley GM, Shaw J, Smith ND (1985) Glacial sedimentary environments. Soc Econ Paleontol Mineral Short Course 16
Ashmore PE (1991) How do gravel-bed rivers braid? Can J Earth Sci 28: 326–341
Ashmore PE (1993) Anabranch confluence kinetics and sedimentation processes in gravel-bed streams. In: Best JL, Bristow CS (eds) Braided rivers. Geol Soc Lond Spec Publ 75: 129–146
Ashton M (ed) (1992) Advances in reservoir geology. Geol Soc Lond Spec Publ 69: 240
Atkinson CD (1986) Tectonic control on alluvial sedimentation as revealed by an ancient catena in the Capella Formation (Eocene) of northern Spain. In: Wright VP (ed) Paleosols: their recognition and interpretation. Blackwell Scientific, Oxford, pp 139–179
Atkinson CD, McGowen JH, Bloch S, Lundell LL, Trumbly PN (1990) Braidplain and deltaic reservoir, Prudhoe Bay field, Alaska. In: Barwis JH, McPherson JG, Studlick RJ (eds) Sandstone petroleum reservoirs. Springer, Berlin Heidelberg New York, pp 7–29
Aubry WM (1989) Mid-Cretaceous incision related to eustasy, southern Colorado Plateau. Geol Soc Am Bull 101: 443–449
Autin WJ (1992) Use of alloformations for definition of Holocene meander belts in the middle Amite River, southeastern Louisiana. Geol Soc Am Bull 104: 233–241
Autin WJ, Burns SF, Miller BJ, Sucier RT, Snead JI (1991) Quaternary geology of the Lower Mississippi valley. In: Morrison RB (ed) The geology of North America, vol K-2: Quaternary nonglacial geology conterminous US. Geological Society of America, pp 547–581
Bachman SB, Lewis SD, Schweller WJ (1983) Evolution of a forearc basin, Luzon Central Valley, Philippines. Am Assoc Petrol Geol Bull 67: 1143–1162
Badgley C, Tauxe L (1990) Paleomagnetic stratigraphy and time in sediments: studies in alluvial Siwalik rocks of Pakistan. J Geol 98: 457–477
Bailey EB (1962) Charles Lyell. Nelson, London
Bailey EB (1967) James Hutton – the founder of modern geology. Elsevier, Amsterdam
Baker VR (1973) Paleohydrology and sedimentology of Lake Missoula flooding in eastern Washington. Geological Society of America, Spec Pap 144
Baker VR (1978a) Adjustment of fluvial systems to climate and source terrain in tropical and subtropical environments. In: Miall AD (ed) Fluvial sedimentology. Can Soc Petrol Geol Mem 5: 211–230
Baker VR (1978b) Large-scale erosional and depositional features of the Channeled Scabland. In: Baker VR, Nummedal D (eds) The channeled scabland. National Aeronautics and Space Administration, Washington, pp 81–115
Balducchi A, Pommier G (1970) Cambrian oil field of Hassi Messaoud, Algeria. In: Halbouty MT (ed) Geology of giant petroleum fields. Am Assoc Petrol Geol Mem 14: 477–488
Ballance PF (1988) The Huriwai braidplain delta of New Zealand: a Late Jurassic, coarse-grained, volcanic-fed depositional system in Gondwana forearc basin. In: Nemec W, Steel RJ (eds) Fan deltas: sedimentology and tectonic settings. Blackie, Glasgow, pp 430–444
Ballance PF, Reading HG (eds) (1980) Sedimentation in oblique-slip mobile zones. Int Assoc Sedimentol Spec Publ 4: 265
Bally AW (1984) Structural styles and the evolution of sedimentary basins. American Association of Petroleum Geologists Short Course, Tulsa
Bally AW, Gordy PL, Stewart GA (1966) Structure, seismic data and orogenic evolution of southern Canadian Rocky Mountains. Bull Can Petrol Geol 14: 337–381

Bally AW, Snelson S (1980) Realms of subsidence. In: Miall AD (ed) Facts and principles of world petroleum occurrence. Can Soc Petrol Geol Mem 6: 9–94
Baltzer F, Purser BH (1990) Modern alluvial fan and deltaic sedimentation in a foreland tectonic setting: the lower Mesopotamian Plain and the Arabian Gulf. Sediment Geol 67: 175–197
Banks NL (1973) The origin and significance of some down current dipping cross stratified sets. J Sediment Petrol 43: 423–427
Bardaji T, Dabrio CJ, Goy JL, Somoza L, Zazo C (eds) (1990) Pleistocene fan deltas in southeasern Iberian peninsula: sedmentary controls and sea-level changes. In: Colella A, Prior DB (eds) Coarse-grained deltas. Int Assoc Sedimentol Spec Publ 10: 129–151
Barrell J (1912) Criteria for the recognition of ancient delta deposits. Geol Soc Am Bull 23: 377–446
Barrell J (1913/1914) The Upper Devonian delta of the Appalachian geosyncline. Am J Sci 36: 429–472; 37: 87–109, 229–253
Barrell J (1917) Rhythms and the measurement of geological time. Geol Soc Am Bull 28: 745–904
Barrell J (1925) Marine and terrestrial conglomerates. Geol Soc Am Bull 36: 279–342
Barry RG, Chorley RJ (1992) Atmosphere, weather and climate, 6th edn. Routledge, London
Bartow JA (1987) Cenozoic nonmarine sedimentation in the San Joaquin Basin, central California. In: Ingersoll RV, Ernst WG (eds) Cenozoic basin development of coastal California, Rubey, vol VI. Prentice-Hall, Englewood Cliffs, pp 146–171
Basu A (1985) Influence of climate and relief on compositions of sands released at source areas. In: Zuffa GG (ed) Provenance of arenites. Reidel, Dordrecht, pp 1–18
Bates RL, Jackson JA (eds) (1987) Glossary of geology, 3rd edn. American Geological Institute, Alexandria
Beaty CB (1963) Origin of alluvial fans, White Mountains, California and Nevada. Assoc Am Geogr Ann 53: 516–535
Beaty CB (1990) Anatomy of a White Mountain debris flow - the making of an alluvial fan. In: Rachocki AH, Church M (eds) Alluvial fans: a field approach. Wiley, Chichester, pp 69–89
Beaumont C (1981) Foreland basins. Geophys J R Astron Soc 65: 291–329
Beer JA (1990) Steady sedimentation and lithologic completeness, Bermejo basin, Argentina. J Geol 98: 501–517
Beerbower JR (1964) Cyclothems and cyclic depositional mechanisms in alluvial plain sedimentation. Geol Surv Kansas Bull 169 (1): 31–42
Begin ZB (1981) Development of longitudinal profiles of alluvial channels in response to baselevel lowering. Earth Surface Proc Landforms 6: 49–68
Begin ZB, Meyer DF, Schumm SA (1980) Knickpoint migration in alluvial channels due to baselevel lowering. Am Soc Civil Eng J Waterway Port Coast Ocean Div 106: 369–388
Behrensmeyer AK (1987) Miocene fluvial facies and vertebrate taphonomy in northern Pakistan. In: Ethridge FG, Flores RM, Harvey MD (eds) Recent developments in fluvial sedimentology. Soc Econ Paleontol Mineral Spec Publ 39: 169–176
Behrensmeyer AK, Tauxe L (1982) Isochronous fluvial systems in Miocene deposits of northern Pakistan. Sedimentology 29: 331–352
Belt ES (1993) Tectonically induced sediment diversion and the origin of thick, widespread coal beds (Appalachian and Williston Basins, USA). In: Frostick LE, Steel RJ (eds) Tectonic controls and signatures in sedimentary successions. Int Assoc Sedimentol Spec Publ 20: 337–397
Ben-Avraham Z, Emery KO (1973) Structural framework of Sunda Shelf. Am Assoc Petrol Geol Bull 57: 2323–2366
Bendix J (1992) Fluvial adjustments on varied timescales in Bear Creek Arroyo, Utah, USA. Z Geomorph NF 36: 141–163
Benkhelil J (1982) Benue trough and Benue chain. Geol Mag 119: 155–168
Bentham PA, Burbank DW, Puigdefábregas C (1992) Temporal and spatial controls on the alluvial architecture of an axial drainage system: late Eocene Escanilla Formation, southern Pyrenean foreland basin, Spain. Basin Res 4: 335–352
Bentham PA, Talling PJ, Burbank DW (1993) Braided stream and flood-plain deposition in a rapidly aggrading basin: the Escanilla formation, Spanish Pyrenees. In: Best JL, Bristow CS (eds) Braided rivers. Geol Soc Lond Spec Publ 75: 177–194
Berg RR (1968) Point-bar origin of Fall River Sandstone reservoirs. Am Assoc Petrol Geol Bull 52: 2116–2122
Berg RR (1986) Reservoir sandstones. Prentice Hall, Englewood Cliffs, New Jersey
Berg RR, Cook BC (1968) Petrography and origin of Lower Tuscaloosa sandstones, Mallalieu field, Lincoln County, Mississippi. Trans Gulf Coast Assoc Geol Soc 18: 242–255
Berger A, Imbrie J, Hays J, Kukla G, Saltzman B (eds) (1984) Milankovitch and climate. NATO ASI Ser D. Reidel, Dordrecht, 2 vols
Bernard HA, Major CJ (1963) Recent meander belt deposits of the Brazos River; an alluvial "sand" model (abstract). Am Assoc Petrol Geol Bull 47: 350
Bernard HA, Leblanc RJ, Major CJ (1962) Recent and pleistocene geology of southeast Texas. In: Rainwater EH, Zingula RP (eds) Geology of the Gulf Coast and central Texas. Geological Society of America guidebook for 1962 annual meeting, pp 175–224
Berné S, Auffret J-P, Walker P (1988) Internal structure of subtidal sandwaves revealed by high-resolution seismic reflection. Sedimentology 35: 5–20
Berné S, Durand J, Weber O (1991), Architecture of modern subtidal dunes (sand waves), Bay of Bourgneuf, France. In: Miall AD, Tyler N (eds) The three-dimensional facies architecture of terrigenous clastic sediments, and its implications for hydrocarbon discovery and recovery. Soc Econ Paleontol Mineral Conc Sedimentol Paleontol 3: 245–260
Bersier A (1948) Les sédimentation rhythmiques synorogéniques dans l'avant-fosse molassique alpine. 18th international geological congress, part IV, pp 83–93
Bersier A (1958) Séquences détritiques et divagations fluviales. Eclog Geol Helv 51: 854–893
Besly BM (1988) Palaeogeographic implications of late Westphalian to early Permian red-beds, central England. In: Besly BM, Kelling G (eds) Sedimentation in a synorogenic basin complex: the Upper Carboniferous of northwest Europe. Blackie, Glasgow, pp 200–221
Belsy BM, Fielding CR (1989) Palaeosols in Westphalian coal-bearing and red-bed sequences central and north-

ern England. Palaeogeogr Palaeoclimatol Palaeoecol 70: 303–330

Besly BM, Kelling G (eds) (1988) Sedimentation in a synorogenic basin complex: the upper Carboniferous of northwest Europe. Blackie, Glasgow, 276 p

Best JL (1987) Flow dynamics at river channel confluences: implications for sediment transport and bed morphology. In: Ethridge FG, Flores RM (eds) Recent and ancient nonmarine depositional environments. Soc Econ Paleontol Mineral Spec Publ 31: 27–35

Best JL (1988) Sediment transport and bed morphology at river channel confluences. Sedimentology 35: 481–498

Best JL, Brayshaw AC (1985) Flow separation – a physical process for the concentration of heavy minerals within alluvial channels. J Geol Soc Lond 142: 747–755

Best JL, Bristow CS (eds) (1993) Braided rivers. Geol Soc Lond Spec Publ 75: 419

Best JL, Bristow CS, Roy AG (1989) The morphology of river channel confluences: scales and dynamics, program and abstracts of the 4th international conference on fluvial sedimentology, Barcelona, Sitges, p 75

Beston NB (1986) Reservoir geological modelling of the North Rankin field, northwest Australia. Aust Petrol Explor Assoc J 26: 375–389

Beutner EC, Flueckinger LA, Gard TM (1967) Bedding geometry in a Pennsylvanian channel sandstone. Geol Soc Am Bull 78: 911–916

Bhattacharyya DP, Lorenz JC (1983) Different depositional settings of the Nubian lithofacies in Libya and southern Egypt. In: Collinson JD, Lewin J (eds) Modern and ancient fluvial systems. Int Assoc Sedimentol Spec Publ 6: 435–448

Bhattacharya J (1991) Regional to sub-regional facies architecture of river-dominated deltas, Upper Cretaceous Dunvegan Formation, Alberta subsurface. In: Miall AD, Tyler N (eds) The three-dimensional facies architecture of terrigenous clastic sediments and its implications for hydrocarbon discovery and recovery. Soc Econ Paleontol Mineral Conc Sedimentol Paleontol 3: 189–206

Biddle KT, Christie-Blick N (eds) (1985) Strike-slip deformation, basin formation and sedimentation. Soc Econ Paleontol Mineral Spec Publ 37: 386

Biddle KT, Uliana MA, Mitchum RM Jr, Fitzgerald MG, Wright RC (1986) The stratigraphic and structural evolution of the central and eastern Magallanes Basin, southern South America. In: Allen PA, Homewood P (eds) Foreland basins. Int Assoc Sedimentol Spec Publ 8: 41–61

Billi P, Magi M, Sagri M (1987) Coarse-grained low-sinuosity river deposits: examples from Plio-Pleistocene Valdarno Basin, Italy. In: Ethridge FG, Flores RM, Harvey MD (eds) Recent developments in fluvial sedimentology. Soc Econ Paleontol Mineral Spec Publ 39: 197–203

Blackwelder E (1928) Mudflow as a geologic agent in semiarid mountains. Geol Soc Am Bull 39: 465–484

Blackbourn GA (1984) Sedimentary facies variation and hydrocarbon reservoirs in continental sediments – a predictive model. J Petrol Geol 7: 67–76

Blair TC (1987) Sedimentary processes, vertical stratification sequences, and geomorphology of the Roaring River alluvial fan, Rocky Mountain National Park. J Sediment Petrol 57: 1–18

Blair TC, Baker FG, Turner JB (1991) Cenozoic fluvial-facies architecture and aquifer heterogeneity, Oroville, California, superfund site and vicinity. In: Miall AD, Tyler N (eds) The three-dimensional facies architecture of terrigenous clastic sediments and its implications for hydrocarbon discovery and recovery. Soc Econ Paleontol Mineral Conc Sedimentol Paleontol 3: 148–159

Blair TC, Bilodeau WL (1988) Development of tectonic cyclothems in rift, pull-apart, and foreland basins: sedimentary response to episodic tectonism. Geology 16: 517–520

Blair TC, McPherson JG (1992) The Trollheim alluvial fan and facies model revisited. Geol Soc Am Bull 104: 762–769

Blair TC, McPherson JG (1994) Alluvial fans and their natural distinction from rivers based on morphology, hydraulic processes, sedimentary processes, and facies assemblages. J Sediment Res A64: 450–489

Blakey RC, Gubitosa R (1984) Controls of sandstone body geometry and architecture in the Chinle Formation (Upper Triassic), Colorado Plateau. Sediment Geol 38: 51–86

Blanche JB (1990) An overview of the exploration history and hydrocarbon potential of Cambodia and Laos. SEAPEX Proc 9: 89–99

Blissenbach E (1954) Geology of alluvial fans in semi-arid regions. Geol Soc Am Bull 65: 175–190

Blixt JE (1941) Cut Bank oil and gas field, Glacier County, Montana. In: Levorsen AI (ed) Stratigraphic type oil fields. American Association of Petroleum Geologists, Tulsa, pp 327–381

Bloch S, McGowen JH, Duncan JR, Brizzolara DW (1990) Porosity prediction, prior to drilling. In: Sandstones of the Kekiktuk Formation (Mississippian), North Slope of Alaska. Am Assoc Petrol Geol Bull 74: 1371–1385

Blodgett RH, Stanley KO (1980) Stratification, bedforms, and discharge relations of the Platte braided river system, Nebraska. J Sediment Petrol 50: 139–148

Bloomer RR (1977) Depositional environments of a reservoir sandstone in west-central Texas. Am Assoc Petrol Geol Bull 61: 344–359

Bluck BJ (1967) Deposition of some Upper Old Red Sandstone conglomerates in the Clyde area: a study in the significance of bedding. Scott J Geol 3: 139–167

Bluck BJ (1971) Sedimentation in the meandering River Endrick. Scott J Geol 7: 93–138

Bluck BJ (1979) Structure of coarse grained braided stream alluvium. Trans R Soc Edinb 70: 181–221

Bluck BJ (1980) Structure, generation and preservation of upward fining, braided stream cycles in the Old Red Sandstone of Scotland. Trans R Soc Edinb 71: 29–46

Bluck BJ (1991) Terrane provenance and amalgamation: examples from the Caledonides. In: Dewey JF, Gass IG, Curry GB, Harris NBW, Sengör AMC (eds) Allochthonous terranes. Cambridge University Press, Cambridge, pp 143–153

Blum MD (1992) Modern depositional environments and recent alluvial history of the lower Colorado River, Gulf Coastal Plain, Texas, PhD dissertation, University of Texas, Austin, Texas, 304 p

Blum MD (1994) Genesis and architecture of incised valley fill sequences: a late Quaternary example from the Colorado River, Gulf Coastal Plain of Texas. In: Weimer

P, Posamentier HW (eds) Siliclastic sequence stratigraphy: recent developments and applications. Am Assoc Petrol Geol Mem 58: 259–283

Blum MD, Price DM (1994) Glacio-eustatic and climatic controls on Quaternary alluvial plain deposition, Texas coastal plain. Gulf Coast Assoc Geol Soc Trans 44: 1–9

Boothroyd JC, Ashley GM (1975) Process, bar morphology, and sedimentary structures on braided outwash fans, northeastern Gulf of Alaska. In: Jopling AV, McDonald BC (eds) Glacio-fluvial and glaciolacustrine sedimentation. Soc Econ Paleontol Mineral Spec Paper 23: 193–222

Boothroyd JC, Nummedal D (1978) Proglacial braided outwash: a model for humid alluvial-fan deposits. In: Miall AD (ed) Fluvial sedimentology. Can Soc Petrol Geol Mem 5: 641–668

Boss RF, Lennon RB, Wilson BW (1976) Middle Ground Shoal field, Alaska. In: Braunstein J (ed) North American oil and gas fields. Am Assoc Petrol Geol Mem 24: 1–22

Botvinkina LN, Feofilova AP, Yablokov VS (1954) Alluvial deposits for the Middle Carboniferous coal-bearing beds of the Donets Basin. Trans Inst Geol Nauk Akad Nauk SSSR Geol Ser 151: 30–89

Bowen DW, Weimer P, Scott AJ (1993) The relative success of siliciclastic sequence stratigraphic concepts in exploration: examples from incised valley fill and turbidite systems reservoirs. In: Weimer P, Posamentier HW (eds) Siliciclastic sequence stratigraphy. Am Assoc Petrol Geol Mem 58: 15–42

Bowen JM (1975) The Brent oil-field. In: Woodland AW (ed) Petroleum and the continental shelf of north-west Europe, vol 1: geology. Wiley, New York, pp 353–361

Bown TM, Kraus MJ (1981) Lower Eocene alluvial paleosols (Wilwood Formation, northwest Wyoming, USA) and their significance for paleoecology, paleoclimatology, and basin analysis. Palaeogeogr Palaeoclimatol Palaeoecol 34: 1–30

Bown TM, Kraus MJ (1987) Integration of channel and floodplain suites, I. Development of sequence and lateral relations of alluvial paleosols. J Sediment Petrol 57: 587–601

Brenner RL, Swift DJP, Gaynor GC (1985) Re-evaluation of coquinoid sandstone depositional model, Upper Jurassic of central Wyoming and south-central Montana. Sedimentology 32: 363–372

Brice JC (1964) Channel patterns and terraces of the Loup Rivers in Nebraska. US Geol Survey, Prof Pap 422-D

Bridge JS (1975) Computer simulation of sedimentation in meandering streams. Sedimentology 22: 3–44

Bridge JS (1984) Large scale facies sequences in alluvial overbank environment. J Sediment Petrol 54: 583–588

Bridge JS (1985) Paleochannel patterns inferred from alluvial deposits: a critical evaluation. J Sediment Petrol 55: 579–589

Bridge JS (1993a) Description and interpretation of fluvial deposits: a critical perspective. Sedimentology 40: 801–810

Bridge JS (1993b) The interaction between channel geometry, water flow, sediment transport and deposition in braided rivers. In: Best JL, Bristow CS (eds) Braided rivers. Geological Society, London, pp 13–71 (Spec Publ 75)

Bridge JS, Diemer JA (1983) Quantitative interpretation of an evolving ancient river system. Sedimentology 30: 599–623

Bridge JS, Gordon EA (1985) The Catskill magnafacies of New York State. In: Flores RM, Harvey MD (eds) Field guide to modern and ancient fluvial systems in the United States. 3rd Int Sedimentology Conf, Fort Collins, Colorado, pp 3–17

Bridge JS, Jarvis J (1976) Flow and sedimentary processes in the meandering River South Esk, Glen Clova, Scotland. Earth Surface Proc 1: 303–336

Bridge JS, Leeder MR (1979) A simulation model of alluvial stratigraphy. Sedimentology 26: 617–644

Bridge JS, Mackey SD (1993a) A revised alluvial stratigraphy model. In: Marzo M, Puigdefábregas C (eds) Alluvial sedimentation. Int Assoc Sedimentol Spec Publ 17: 319–336

Bridge JS, Mackey SD (1993b) A theoretical study of fluvial sandstone body dimensions. In: Flint SS, Bryant ID (eds) The geological modelling of hydrocarbon reservoirs and outcrop analogues. Int Assoc Sedimentol Spec Publ 15: 213–236

Bridge JS, Smith ND, Trent F, Gabel SL, Bernstein P (1986) Sedimentology and morphology of a low-sinuosity river: Calamus River, Nebraska Sand Hills. Sedimentology 33: 851–870

Brierley GJ (1991) Floodplain sedimentology of the Squamish River, British Columbia: relevance of element analysis. Sedimentology 38: 735–750

Brierley GJ (1993) Facies analysis of fine-grained fluvial systems: a constructivist approach: keynote address, 5th international conference on fluvial sedimentology, Brisbane, Australia, July 1993, keynote addresses and abstracts, pp K5–K13

Brierley GJ, Hickin EJ (1991) Channel planform as a non-controlling factor in fluvial sedimentology: the case of the Squamish River floodplain, British Columbia. Sediment Geol 75: 67–83

Brierley GJ, Liu K, Crook KAW (1993) Sedimentology of coarse-grained alluvial fans in the Markham Valley, Papua New Guinea. Sediment Geol 86: 297–324

Brinkman R (1933) Über Kreuzschichtung im deutschen Buntsandsteinbecken. Nachrichten von der Gesellschaft der Wissenschaften zu Göttingen, Math Phys Kl Fachgruppe IV, no 32

Bristow CS (1987) Brahmaputra river: channel migration and deposition. In: Ethridge FG, Flores RM, Harvey MD (eds) Recent developments in fluvial sedimentology. Soc Econ Paleontol Mineral Spec Publ 39: 63–74

Bristow CS (1988) Controls on the sedimentation of the Rough Rock Group (Namurian) from the Pennine Basin of northern England. In: Besly BM, Kelling G (eds) Sedimentation in a synorogenic basin complex: the Upper Carboniferous of northwest Europe. Blackie, Glasgow, pp 114–131

Bristow CS (1993a) Sedimentary structures exposed in bar tops in the Brahmaputra River, Bangladesh. In: Best JL, Bristow CS (eds) Braided rivers. Geol Soc Lond Spec Publ 75: 277–289

Bristow CS (1993b) Sedimentology of the Rough Rock: a carboniferous braided sheet sandstone in northern England. In: Best JL, Bristow CS (eds) Braided rivers. Geol Soc Lond Spec Publ 75: 291–304

Bristow CS, Best JL (1993) Braided rivers: perspectives and problems. In: Best JL, Bristow CS (eds) Braided rivers. Geol Soc Lond Spec Publ 75: 1–11
Bristow CS, Best JL, Roy AG (1993) Morphology and facies models of channel confluences. In: Marzo M, Puigdefábregas C (eds) Alluvial sedimentation. Int Assoc Sedimentol Spec Publ 17: 91–100
Broadhurst FM (1988) Seasons and tides in the Westphalian. In: Besly BM, Kelling G (eds) Sedimentation in a synorogenic basin complex: the Upper Carboniferous of northwest Europe. Blackie, Glasgow, pp 264–272
Bromley MH (1991a) Variations in fluvial style as revealed by architectural elements, Kayenta Formation, Mesa Creek, Colorado, USA: evidence for both ephemeral and perennial fluvial processes. In: Miall AD, Tyler N (eds) The three-dimensional facies architecture of terrigenous clastic sediments and its implications for hydrocarbon discovery and recovery. Soc Econ Paleontol Mineral Conc Sedimentol Paleontol 3: 94–102
Bromley MH (1991b) Architectural features of the Kayenta Formation (Lower Jurassic), Colorado Plateau, USA: relationship to salt tectonics in the Paradox Basin. Sediment Geol 73: 77–99
Brookfield ME (1977) The origin of bounding surfaces in ancient aeolian sandstones. Sedimentology 24: 303–332
Brookfield ME (1992) The paleorivers of central Asia: the interrelationship of Cenozoic tectonism, erosion and sedimentation. 29th Int Geol Congr, Kyoto, Japan, Abstracts, vol 2, p 292
Brookfield ME (1993) The interrelations of post-collision tectonics and sedimentation in central Asia. In: Frostick LE, Steel RJ (eds) Tectonic controls and signatures in sedimentary successions. Int Assoc Sedimentol Spec Publ 20: 13–35
Brown AR (1985) The role of horizontal seismic sections in stratigraphic intepretation. In: Berg OR, Woolverton DG (eds) Seismic stratigraphy II. Am Assoc Petrol Geol Mem 39: 37–47
Brown AR (1991) Interpretation of three-dimensional seismic data, 3rd edn. Am Assoc Petrol Geol Mem 42: 341
Brown D (1993) Make room! It's 3-D for your PC. American Association of Petroleum Geologists Explorer, Tulsa, June 1993, p 14
Brown LF Jr, Fisher WL (1977) Seismic-stratigraphic interpretation of depositional systems: examples from Brazilian rift and pull-apart basins. In: Payton CE (ed) Seismic stratigraphy – applications to hydrocarbon exploration. Am Assoc Petrol Geol Mem 26: 213–248
Brown S (1984) Jurassic. In: Glennie KW (ed) Introduction to the petroleum geology of the North Sea. Blackwell Scientific, Oxford, pp 103–131
Brush LM Jr (1958) Study of stratification in a large laboratory flume (abstract). Geol Soc Am Bull 69: 1542
Bryant ID (1983) Facies sequences associated with some braided river deposits of late Pleistocene age from southern Britain. In: Collinson JD, Lewin J (eds) Modern and ancient fluvial systems. Int Assoc Sedimentol Spec Publ 6: 267–275
Bucher WH (1919) On ripples and related sedimentary surface forms and their paleogeographic interpretation. Am J Sci 47: 149–210, 241–269
Buck SG (1983) The Saiplaas Quartzite Member: a braided system of gold and uranium bearing channel placers within the Proterozoic Witwatersrand Supergroup of South Africa. In: Collinson JD, Lewin J (eds) Modern and ancient fluvial systems. Int Assoc Sedimentol Spec Publ 6: 549–562
Bull WB (1963) Alluvial-fan deposits in western Fresno County, California. J Geol 71: 243–251
Bull WB (1964) Alluvial fans and near surface subsidence in western Fresno County, California. US geological survey professional paper 437-A
Bull WB (1972) Recognition of alluvial fan deposits in the stratigraphic record. In: Rigby JK, Hamblin WK (eds) Recognition of ancient sedimentary environments. Soc Econ Paleontol Mineral Spec Publ 16: 63–83
Bull WB (1977) The alluvial fan environment. Prog Phys Geogr 1: 222–270
Bull WB (1991) Geomorphic responses to climate change. Oxford University Press, New York
Buller AT, Berg E, Hjelmeland O, Kleppe J, Torsaeter O, Aasen JO (eds) (1990) North Sea oil and gas reservoirs II. Graham and Trotman, London, 453 p
Burbank DW (1992) Causes of recent Himalayan uplift deduced from deposited patterns in the Ganges basin. Nature 357: 680–683
Burbank DW, Beck RA (1991) Models of aggradation versus progradation in the Himalayan foreland. Geol Rundsch 80: 623–638
Burbank DW, Raynolds RGH (1984) Sequential late Cenozoic structural disruption of the northern Himalayan foredeep. Nature 311: 114–118
Burbank DW, Raynolds RGH, Johnson GD (1986) Late Cenozoic tectonics and sedimentation in the northwestern Himalayan foredeep: II. Eastern limb of the Northwest Syntaxis and regional synthesis. In: Allen PA, Homewood P (eds) Foreland basins. Int Assoc Sedimentol Spec Publ 8: 293–306
Burbank DW, Beck RA, Raynolds RGH, Hobbs R, Tahirkheli RAK (1988a) Thrusting and gravel progradation in foreland basins: a test of post-thrusting gravel dispersal. Geology 16: 1143–1146
Burbank DW, Raynolds RGH (1988b) Stratigraphic keys to the timing of thrusting in terrestrial foreland basins: applications to the northwestern Himalaya. In: Kleinspehn KL, Paola C (eds) New perspectives in basin analysis: Springer, Berlin Heidelberg New York, pp 331–351
Burbank DW, Beck RA, Raynolds RGH (1989) Reply to comment by Heller et al. on "Thrusting and gravel progradation in foreland basins: a test of post-thrusting gravel dispersal". Geology 17: 960–961
Burbank DW, Vergés J, Muñoz JA, Bentham P (1992) Coeval hindward- and forward-imbricating thrusting in the south-central Pyrenees, Spain: timing and rates of shortening and deposition. Geol Soc Am Bull 104: 3–17
Burchfiel BC, Royden L (1982) Carpathian fold and thrust belt and its relation to Pannonian and other basins. Am Assoc Petrol Geol Bull 66: 1179–1195
Bürgisser HM (1984) A unique mass flow marker bed in a Miocene streamflow molasse sequence, Switzerland. In: Koster EH, Steel RJ (eds) Sedimentology of gravels and conglomerates. Can Soc Petrol Geol Mem 10: 147–163
Burke K, Sengör C (1986) Tectonic escape in the evolution of the continental crust. In: Barazangi M, Brown L (eds) Reflection seismology: the continental crust. Am Geophys Union Geodyn Ser 14: 41–53

Burnett AW, Schumm SA (1983) Alluvial-river response to neotectonic deformation in Louisiana and Mississippi. Science 222: 49–50

Busby CJ, Ingersoll RV (1995) Tectonics of sedimentary basins. Blackwell Science, Oxford

Busch DA (1974) Stratigraphic traps in sandstones – exploration techniques. Am Assoc Petrol Geol Mem 21: 174

Butcher SW (1990) The nickpoint concept and its implications regarding onlap to the stratigraphic record. In: Cross TA (ed) Quantitative dynamic stratigraphy. Prentice-Hall, Englewood Cliffs, pp 375–385

Cairncross B (1980) Anastomosing river deposits: paleoenvironmental control on coal quality and distribution, northern Karoo Basin. Trans Geol Soc South Afr 83: 327–332

Cameron GIF, Collinson JD, Rider MH, Li Xu Analogue dipmeter logs through a prograding deltaic sandbody. In: Ashton M (ed) Advances in reservoir geology. Geological Society, London, pp 195–217 (Spec Publ 69)

Campbell CV (1976) Reservoir geometry of a fluvial sheet sandstone. Am Assoc Petrol Geol Bull 60: 1009–1020

Campbell JE, Hendry HE (1987) Anatomy of a gravelly meander lobe in the Saskatchewan River, near Nipawin, Canada. In: Ethridge FG, Flores RM, Harvey MD (eds) Recent developments in fluvial sedimentology. Soc Econ Paleontol Mineral Spec Publ 39: 179–189

Cant DJ (1976) Braided stream sedimentation in the South Saskatchewan River. PhD Thesis, McMaster University, Hamilton

Cant DJ (1978) Development of a facies model for sandy braided river sedimentation: comparison of the South Saskatchewan River and Battery Point Formation. In: Miall AD (ed) Fluvial sedimentology. Can Soc Petrol Geol Mem 5: 627–640

Cant DJ (1982) Fluvial facies models. In: Scholle PA, Spearing D (eds) Sandstone depositional environments. Am Assoc Petrol Geol Mem 31: 115–137

Cant DJ, Stockmal GS (1989) The Alberta foreland basin: relationship between stratigraphy and terrane-accretion events. Can J Earth Sci 26: 1964–1975

Cant DJ, Walker RG (1976) Development of a braided-fluvial facies model for the Devonian Battery Point Sandstone, Quebec. Can J Earth Sci 13: 102–119

Cant DJ, Walker RG (1978) Fluvial processes and facies sequences in the sandy braided South Saskatchewan River, Canada. Sedimentology 25: 625–648

Carey WC, Keller MD (1957) Systematic changes in the beds of alluvial rivers. Am Soc Civil Eng Proc 83 (HY4): 1–24

Carling PA (1990) Particle over-passing on depth-limited gravel bars. Sedimentology 37: 345–355

Carling PA, Glaister MS (1987) Rapid deposition of sand and gravel mixtures downstream of a negative step: the role of matrix-infilling and particle-overpassing in the process of bar-front accretion. J Geol Soc Lond 144: 543–551

Carlston CW (1965) The relation of free meander geometry to stream discharge and its geomorphic implications. Am J Sci 263: 864–885

Carson MA (1984a) The meandering-braided river threshold: a reappraisal. J Hydrol 73: 315–334

Carson MA (1984b) Observations on the meandering-braided river transition, the Canterbury Plains, New Zealand. N Z Geogr 40 (1): 12–17

Carson (1984c) Observations on the meandering-braided river transition, the Canterbury Plains, New Zealand. N Z Geogr 40 (2): 89–99

Carson MA (1986) Characteristics of high-energy "meandering rivers" the Canterbury Plains, New Zealand. Geol Soc Am Bull 97: 886–895

Cartwright JA, Haddock RC, Pinheiro LM (1993) The lateral extent of sequence boundaries. In: Williams GD, Dobbs A (eds) Tectonics and seismic sequence stratigraphy. Geol Soc Lond Spec Publ 71: 15–34

Casshyap SM, Tewari RC (1982) Facies analysis and paleogeographic implications of a Late Paleozoic glacial outwash deposit, Bihar, India. J Sediment Petrol 52: 1243–1256

Castle JW (1990) Sedimentation in Eocene Lake Uinta (Lower Green River Formation), northeastern Uinta Basin, Utah. In: Katz BJ (ed) Lacustrine basin exploration. Am Assoc Petrol Geol Mem 50: 243–263

Cavazza W (1989) Sedimentation pattern of a rift-filling unit, Tesuque Formastion (Miocene), Española Basin, Rio Grande Rift, New Mexico. J Sediment Petrol 59: 287–296

Cecil CB (1990) Paleoclimate controls on stratigraphic repetition of chemical and siliciclastic rocks. Geology 18: 533–536

Cerveny PF, Naeser ND, Zeitler PK, Naeser CW, Johnson NM (1988) History of uplift and relief of the Himalaya during the past 18 million years: evidence from fission-track ages of detrital zircons from sandstones of the Siwalik Group. In: Kleinspehn KL, Paola C (eds) New perspectives in basin analysis. Springer, Berlin Heidelberg New York, pp 43–61

Chamberlin TC, Salisbury RD (1909) Geology: processes and their results, second edition. Murray, London

Chapin MA, Mayer DF (1991) Constructing a three-dimensional rock-property model of fluvial sandstones in the Peoria field, Colorado. In: Miall AD, Tyler N (eds) The three-dimensional facies architecture of terrigenous clastic sediments, and its implications for hydrocarbon discovery and recovery. Soc Econ Paleontol Mineral Conc Models Ser 3: 160–171

Charles HH (1941) Bush City oil field, Anderson County, Kansas. In: Levorsen AI (ed) Stratigraphic type oil fields. American Association of Petroleum Geologists, Tulsa, pp 43–56

Chatfield J (1972) Case history of Red Wash field, Uintah County, Utah. In: King RE (ed) Stratigraphic oil and gas fields. Am Assoc Petrol Geol Mem 16: 342–353

Chawner WD (1935) Alluvial fan flooding, the Montrose, California, flood of 1934. Geogr Rev 25: 225–263

Chow N, James NP (1987) Cambrian grand cycles: a northern Appalachian perspective. Geol Soc Am Bull 98: 418–429

Church M (1983) Pattern of instability in a wandering gravel bed channel. In: Collinson JD, Lewin J (eds) Modern and ancient fluvial systems. Int Assoc Sedimentol Spec Publ 6: 169–180

Church M, Gilbert R (1975) Proglacial fluvial and lacustrine environments. In: Jopling AV, McDonald BC (eds) Glaciofluvial and glaciolacustrine sedimentation. Soc Econ Paleontol Mineral Spec Publ 23: 22–100

Church M, Rood K (1983) Catalogue of alluvial river channel regime data. Department of Geography, University of British Columbia, 99 p

Church M, Ryder JM (1972) Paraglacial sedimentation: a consideration of fluvial processes conditioned by glaciation. Geol Soc Am Bull 83: 3059–3072

Clark ABS, Thomas BM (1988) The intra-Latrobe playa: a case history from the Basker/Manta Block (Vic/P19), Gippsland Basin. Aust Petrol Explor Assoc J 28: 100–112

Clark SL (1987) Seismic stratigraphy of Early Pennsylvanian Morrowan sandstones, Minneola Complex, Ford and Clark Counties, Kansas. Am Assoc Petrol Geol Bull 71: 1329–1341

Clement WA (1977) A case history of geoseismic modeling of basal Morrow-Springer Sandstones, Watonga-Chickasha trend: Geary, Oklahoma – T13N, R10 W. In: Payton CE (ed) Seismic stratigraphy – applications to hydrocarbon exploration. Am Assoc Petrol Geol Mem 26: 451–476

Clemmensen LB, Tirsgaard H (1990) Sand-drift surfaces: a neglected type of bounding surface. Geology 18: 1142–1145

Clemmensen LB, Øxnevad IEI, De Boer PL (1994) Climatic controls on ancient desert sedimentation: some late Palaeozoic and Mesozoic examples from NW Europe and the Western Interior of the USA. In: De Boer PL, Smith DG (eds) Orbital forcing and cyclic sequences. Int Assoc Sedimentol Spec Publ 19: 439–457

Cleveland MN, Molina J (1990) Deltaic reservoirs of the Caño Limón field, Colombia, South America. In: Barwis JH, McPherson JG, Studlick RJ (eds) Sandstone petroleum reservoirs: Springer, Berlin Heidelberg New York, p 281–315

Clifford HJ, Grund R, Musrati H (1980) Geology of a stratigraphic giant: Messla oil field, Libya. In: Halbouty MT (ed) Giant oil and gas fields of the decade: 1968–1978. Am Assoc Petrol Geol Mem 30: 507–524

Cloetingh S (1988) Intraplate stress: a new element in basin analysis. In: Kleinspehn K, Paola C (eds) New Perspectives in basin analysis. Springer, Berlin Heidelberg New York, pp 205–230

Cloetingh S, Kooi H (1990) Intraplate stresses: a new perspective on QDS and Vail's third-order cycles. In: Cross TA (ed) Quantitative dynamic stratigraphy. Prentice-Hall, Englewood Cliffs, pp 127–148

Cloetingh S, McQueen H, Lambeck K (1985) On a tectonic mechanism for regional sea-level variations. Earth Planet Sci Lett 75: 157–166

Cluzel D, Cadet J-P, Lapierre H (1990) Geodynamics of the Ogcheon Belt (South Korea). Tectonophysics 183: 41–56

Cole RD, Friberg JF (1989) Stratigraphy and sedimentation of the Book Cliffs, Utah. In: Nummedal D, Wright R (eds) Cretaceous shelf sandstones and shelf depositional sequences, Western Interior Basin, Utah, Colorado and New Mexico. 28th international geological congress, Washington DC, Field Trip Guidebook T119, pp 13–24

Colella A, Prior DB (eds) (1990) Coarse-grained deltas. Int Assoc Sedimentol Spec Publ 10: 357

Coleman JM (1969) Brahmaputra river: channel processes and sedimentation. Sediment Geol 3: 129–239

Collinson JD (1970) Bedforms of the Tana river, Norway. Geogr Ann 52A: 31–55

Collinson JD (1971a) Current vector dispersion in a river of fluctuating discharge. Geol Mijnbouw 50: 671–678

Collinson JD (1971b) Some effects of ice on a river bed. J Sediment Petrol 41: 557–564

Collinson JD (1978) Vertical sequence and sand body shape in alluvial sequences. In: Miall AD (ed) Fluvial sedimentology. Can Soc Petrol Geol Mem 5: 577–586

Collinson JD (1986) Alluvial sediments. In: Reading HG (ed) Sedimentary environments and facies. Blackwell Scientific, Oxford, pp 20–62

Collinson JD, Lewin J (eds) (1983) Modern and ancient fluvial systems. Int Assoc Sedimentol Spec Publ 6: 575

Collinson JD, Thompson DB (1982) Sedimentary structures. Allen and Unwin, London, 194 p

Conaghan PJ, Jones JG (1975) The Hawkesbury Sandstone and the Brahmaputra: a depositional model for continental sheet sandstones. J Geol Soc Aust 22: 275–283

Conybeare CEB (1976) Geomorphology of oil and gas fields in sandstone bodies. Elsevier Scientific, Amsterdam (Developments in petroleum science, vol 4)

Conybeare CEB, Crook KAW (1968) Manual of sedimentary structures. Aust Bur Miner Resources Geol Geophys Bull 102: 327

Conybeare WD, Phillips W (1824) Outline of the geology of England and Wales. Am J Sci 7: 203

Cornish V (1899) On kinematology. The study of waves and waves structures of the atmosphere, hydrosphere and lithosphere. Geogr J 13: 624–626

Cosgrove JL (1987) South-west Queensland gas – a resource for the future. Austral Petrol Explor Assoc J 27: 245–262

Costa JE (1974) Stratigraphic, morphologic, and pedogenic evidence of large floods in humid environments. Geology 2: 301–303

Costello WR, Walker RG (1972) Pleistocene sedimentology, Credit River, southern Ontario: a new component of the braided river model. J Sediment Petrol 42: 389–400

Cotter E (1971) Sedimentary structures and the interpretation of paleoflow characteristics of the Ferron Sandstone (Upper Cretaceous), Utah. J Sediment Petrol 41: 129–138

Cotter E (1978) The evolution of fluvial style, with special reference to the central Appalachian Paleozoic. In: Miall AD (ed) Fluvial sedimentology. Can Soc Petrol Geol Mem 5: 361–383

Cotter E, Graham JR (1991) Coastal plain sedimentation in the late Devonian of southern Ireland: hummocky cross-stratification in fluvial deposits. Sediment Geol 72: 201–224

Covey M (1986) The evolution of foreland basins to steady state: the foreland basin of the Banda Orogen. In: Allen PA, Homewood P (eds) Foreland basins. Int Assoc Sedimentol (Spec Publ 8), pp 77-90

Cowan EJ (1991) The large-scale architecture of the fluvial Westwater Canyon Member, Morrison Formation (Jurassic), San Juan Basin, New Mexico. In: Miall AD, Tyler N (eds) The three-dimensional facies architecture of terrigenous clastic sediments, and its implications for hydrocarbon discovery and recovery. Soc Econ Paleontol Mineral Conc Sedimentol Paleontol 3: 80–93

Cowan EJ (1993) Longitudinal fluvial drainage patterns within a foreland basin-fill: Permo-Triassic Sydney basin, Australia. Sediment Geol 85: 557–577

Cowan G (1993) Identification and significance of aeolian deposits within the dominantly fluvial Sherwood Sandstone Group of the East Irish Sea Basin, UK. In: North CP, Prosser DJ (eds) Characterization of fluvial and aeolian reservoirs. Geol Soc Lond Spec Publ 73: 231–245

Coward MP (1990) the Precambrian, Caledonian and Variscan framework to NW Europe. In: Hardman RFP, Brooks J (eds) Tectonic events responsible for Britain's oil and gas reserves. Geol Soc Lond Special Publ 55: 1–34

Crews SG, Ethridge FG (1993) Laramide tectonics and humid alluvial fan sedimentation, NE Uinta Uplift, Utah and Wyoming. J Sediment Petrol 63: 420–436

Crook KAW (1989) Suturing history of an allochthonous terrane at a modern plate boundary traced by flysch-to-molasse transitions. Sediment Geol 61: 49–80

Cross TA (1986) Tectonic controls of foreland basin subsidence and Laramide style deformation, western United States. In: Allen PA, Homewood P (eds) Foreland basins. Int Assoc Sedimentol Spec Publ 8: 15–39

Cross TA (ed) (1990) Quantitative dynamic stratigraphy. Prentice Hall, Englewood Cliffs, New Jersey, 625 p

Crostella A (1983) Malacca Strait wrench fault controlled Laland and Mengkapan oil fields. SEAPEX Proc 6: 24–34

Crowell JC (1974a) Origin of late Cenozoic basins in southern California. In: Dickinson WR (ed) Tectonics and sedimentation. Soc Econ Paleontol Mineral Spec Publ 22: 190–204

Crowell JC (1974b) Sedimentation along the San Andreas Fault, California. In: Dott RH, Shaver RH (eds) Modern and ancient geosynclinal sedimentation. Soc Econ Paleontol Mineral Spec Publ 19: 292–303

Crowell JC (1978) Gondwanan glaciation, cyclothems, continental positioning, and climate change. Am J Sci 278: 1345–1372

Crowell JC, Link MH (eds) (1982) Geologic history of Ridge Basin, southern California. Pacific Section, Society of Economic Paleontologists and Mineralogists, Tulsa, Oklahoma

Crowley KD (1983) Large-scale bed configurations (macroforms), Platte River basin, Colorado and Nebraska: primary structures and formative processes. Geol Soc Am Bull 94: 117–133

Crumeyrolle P, Rubino J-L, Clauzon G (1991) Miocene depositional sequences within a tectonically controlled transgressive-regressive cycle. In: Macdonald DIM (ed) (1991) Sedimentation, tectonics and eustasy: sea-level changes at active margins. Int Assoc Sedimentol Spec Publ 12: 373–390

Cuevas Gozalo MC, Martinius AW (1993) Outcrop database for the geological characterization of fluvial reservoirs: an example from distal fluvial fan deposits in the Loranca Basin, Spain. In: North CP, Prosser DJ (ed) Characterization of fluvial and aeolian reservoirs. Geol Soc Lond Spec Publ 73: 79–94

Cullingford RA, Davidson DA, Lewin J (1980) Timescales in geomorphology. Wiley, Chichester

Curray JR (1956) The analysis of two-dimensional orientation data. J Geol 64: 117–131

Curry WH, Curry WH III (1972) South Glenrock oil field, Wyoming: prediscovery thinking and postdiscovery description. In: King RE (ed) Stratigraphic oil and gas fields. Am Assoc Petrol Geol Mem 16: 415–427

Dahlstrom CDA (1970) Structural geology in the eastern margin of the Canadian Rocky Mountains. Bull Can Petrol Geol 18: 332–406

Dalrymple RW (1984) Morphology and internal structure of sandwaves in the Bay of Fundy. Sedimentology 31: 365–382

Dalrymple RW, Boyd R, Zaitlin BA (eds) (1994) Incised-valley systems: origin and sedimentary sequences. Soc Econ Paleontol Mineral Spec Publ 51

Dalrymple RW, Makino Y (1989) Description and genesis of tidal bedding in the Cobequid Bay-Salmon River Estuary, Bay of Fundy, Canada. In: Taira A, Masuda F (eds) Sedimentary facies in the active plate margin. Terra Scientific, Tokyo, pp 151–177

Damanti JF (1993) Geomorphic and structural controls on facies patterns and sediment composition in a modern foreland basin. In: Marzo M, Puigdefábregas C (eds) Alluvial sedimentation. Int Assoc Sedimentol Spec Publ 17: 221–233

Dana JD (1850a) On denudation in the Pacific. Am J Sci 9: 48–62

Dana JD (1850b) On the degradation of the rocks of New South Wales and formation of valleys. Am J Sci 9: 289–294

Dana JD (1862) Manual of geology. Ivison, Blakeman and Taylor, New York

Darby DA, Whittecar GR, Barringer RA, Garrett JR (1990) Alluvial lithofacies recognition in a humid-tropical setting. Sediment Geol 67: 161–174

Darwin GH (1883) On the formation of ripple marks. R Soc Lond Proc 36: 18–43

Davies DK (1966) Sedimentary structures and subfacies of Mississippi River point bar. J Geol 74: 234–239

Davies DK, Williams BPJ, Vessell RK (1991) Reservoir models for meandering and straight fluvial channels: examples from the Travis Peak Formation, East Texas. Trans Gulf Coast Assoc Geol Soc 41: 152–174

Davies DK, Williams BPJ, Vessell RK (1993) Dimensions and quality of reservoirs originating in low and high sinuosity channel systems, Lower Cretaceous Travis Peak Formation, East Texas, USA. In: North CP, Prosser DJ (eds) Characterization of fluvial and aeolian reservoirs. Geol Soc Lond Spec Publ 73: 95–121

Davis JL, Annan AP (1986) High resolution sounding using ground-probing radar. Geosci Can 13: 205–208

Davis JL, Annan AP (1989) Ground-penetrating radar for high-resolution mapping of soil and rock stratigraphy. Geophys Prospect 37: 531–551

Davis WM (1898a) The Triassic formations of Connecticut. US Geol Surv Annu Rep 18 (2): 1–192

Davis WM (1898b) Physical geography. Ginn, Boston, 432 p

Davis WM (1899) The geographical cycle. Geogr J 14: 481–504

Davis WM (1900) The fresh-water Tertiary formations of the Rocky Mountains region. Proc Am Acad Arts Sci 35: 345–373

Dawson MR, Bryant ID (1987) Three-dimensional facies geometry in Pleistocene outwash sediments, Worcestershire, UK. In: Ethridge FG, Flores RM, Harvey MD (eds) Recent developments in fluvial sedimentology. Soc Econ Paleontol Mineral Spec Publ 39: 191–196

De Boer PL, Smith DG (eds) (1994) Orbital forcing and cyclic sequences. Int Assoc Sedimentol Spec Publ 19: 559

De Boer PL, Smith DG (eds) (1994b) Orbital forcing and cyclic sequences. In: De Boer PL, Smith DG (eds) Orbital forcing and cyclic sequences. Int Assoc Sedimentol Spec Publ 19: 1–14

De Boer PL, Pragt JSJ, Oost AP (1991) Vertically persistent facies boundaries along growth anticlines and climate-controlled sedimentation in the thrust-sheet-top South Pyrenean Tremp-Graus Foreland Basin. Basin Res 3: 63–78

DeCelles PG, Tolson RB, Graham SA, Smith GA, Ingersoll RV, White J, Schmidt CJ, Rice R, Moxon I, Lemke L, Handschy JW, Follow MF, Edwards DP, Cavazza W, Caldwell M, Bargar E (1987) Laramide thrust-generated alluvial-fan sedimentation, Sphinx Conglomerate, southwestern Montana. Am Assoc Petrol Geol Bull 71: 135–155

DeCelles PG, Gray MB, Ridgway KD, Cole RB, Pivnik DA, Pequera N, Srivastava P (1991) Controls on synorogenic alluvial-fan architecture, Beartooth Conglomerate (Paleocene), Wyoming and Montana. Sedimentology 38: 567–590

DeLuca JL, Eriksson KA (1989) Controls on synchronous ephemeral- and perennial-river sediments in the middle sandstone member of the Triassic Chinle Formation, northeastern New Mexico. Sediment Geol 61: 155–175

Denny CS (1967) Fans and pediments. Am J Sci 265: 81–105

Derbyshire E, Owen LA (1990) Quaternary alluvial fans in the Karakoram Mountains. In: Rachocki AH, Church M (eds) Alluvial fans: a field approach. Wiley, Chichester, pp 27–53

Derksen SJ, McLean-Hodgson J (1988) Hydrocarbon potential and structural style of continental rifts: examples from East Africa and southeast Asia. SEAPEX 8: 47–62

Desikachar SV (1984) Exploration plays in north east Indian hydrocarbon bearing province. Petrol Asia J: 51–59

Desloges JR, Church M (1987) Channel and floodplain facies in a wandering gravel-bed river. In: Ethridge FG, Flores RM, Harvey MD (eds) Recent developments in fluvial sedimentology. Soc Econ Paleontol Mineral Spec Publ 39: 99–109

Devine PE, Wheeler DM (1989) Correlation, interpretation, and exploration potential of Lower Wilcox valley-fill sequences, Colorado and Lavaca counties, Texas. Trans Gulf Coast Assoc Geol Soc 39: 57–74

Dewey JF (1977) Suture zone complexities: a review. Tectonophysics 40: 53–67

Dewey JF (1982) Plate tectonics and the evolution of the British Isles. J Geol Soc Lond 139: 371–414

Dewey JF, Hempton MR, Kidd WSF, Saroglu F, Sengör AMC (1986) Shortening of continental lithosphere: the neotectonics of eastern Anatolia – a young collision zone. In: Coward MP, Ries AC (eds) Collision tectonics. Geol Soc Lond Spec Publ 19: 3–36

Díaz-Molina M (1993) Geometry and lateral accretion patterns in meander loops: examples from the Upper Oligocene-Lower Miocene, Loranca Basin, Spain. In: Marzo M, Puigdefábregas C (eds) Alluvial sedimentation. Int Assoc Sedimentol Spec Publ 17: 115–131

Dickinson WR (1974) Plate tectonics and sedimentation. In: Dickinson WR (ed) Tectonics and sedimentation. Soc Econ Paleontol Mineral Spec Publ 22: 1–27

Dickinson WR (1981a) Plate tectonics and the continental margin of California. In: Ernst WG (ed) The geotectonic development of California. Prentice-Hall, Englewood Cliffs, New Jersey, pp 1–28

Dickinson WR (1981b) Plate tectonic evolution of the southern Cordillera. Arizona Geol Soc Dig 14: 113–135

Dickinson WR, Klute MA, Hayes MJ, Janecke SU, Lundin ER, McKittrick MA, Olivares MD (1988) Paleogeographic and paleotectonic setting of Laramide sedimentary basins in the central Rocky Mountain region. Geol Soc Am Bull 100: 1023–1039

Dickinson WR, Seely DR (1979) Structure and stratigraphy of forearc regions. Am Assoc Petrol Geol Bull 63: 2–31

Diemer JA, Belt ES (1991) Sedimentology and paleohydraulics of the meandering river systems of the Fort Union Formation, southeastern Montana. Sediment Geol 75: 85–108

Diessel CFK (1992) Coal-bearing depositional systems. Springer, Berlin Heidelberg New York

Dixon EEL (1921) The geology of the South Wales Coalfield, part XIII: the country around Tenby. Geological Survey of Great Britain, Memoir

Dobrin MB (1977) Seismic exploration for stratigraphic traps. In: Payton CE (ed) Seismic stratigraphy – applications to hydrocarbon exploration. Am Assoc Petrol Geol Mem 26: 329–351

Doeglas DJ (1962) The structure of sedimentary deposits of braided rivers. Sedimentology 1: 167–190

Dolson J, Muller D, Evetts MJ, Stein JA (1991) Regional paleotopographic trends and production, Muddy Sandstone (Lower Cretaceous), central and northern Rocky Mountains. Am Assoc Petrol Geol Bull 75: 409–435

Dorsey RJ, Burns B (1994) Regional stratigraphy, sedimentology, and tectonic significance of Oligocene-Miocene sedimentary and volcanic rocks, northern Baja Calfironia, Mexico. Sediment Geol 88: 231–251

Dott RH Jr, Bourgeois J (1982) Hummocky stratification: significance of its variable bedding sequences. Geol Soc Am Bull 93: 663–680

Dott RH Jr, Bourgeois J (1983) Hummocky stratification: significance of its variable bedding sequences: reply to discussion by RG Walker et al. Geol Soc Am Bull 94: 1245–1251

Dott RH Jr, Byers CW, Fielder GW, Stenzel SR, Winfree KE (1986) Aeolian to marine transition in Cambro-Ordovician cratonic sheet sandstones of the northern Mississippi valley, USA. Sedimentology 33: 345–368

Doyle JD, Sweet ML (1995) Three-dimensional distribution of lithofacies, bounding surfaces, porosity, and permeability in a fluvial sandstone – Gypsy sandstone of northern Oklahoma. Am Assoc Petrol Geol Bull 79: 70–96

Dranfield P, Begg SH, Carter RR (1987) Wytch Farm oilfield: reservoir characterization of the Triassisc Sherwood Sandstone for input to reservoir simulation studies. In: Brooks J, Glennie KW (eds) Petroleum geology of northwest Europe. Graham and Trotman, London, pp 149–160

Drew F (1873) Alluvial and lacustrine deposits and glacial records of the Upper-Indus Basin. Q J Geol Soc Lond 29: 441–471

Dreyer T (1990) Sand body dimensions and infill sequences of stable, humid-climate delta plain channels.

In: Buller AT, Berg E, Hjelmeland O, Kleppe J, Torsaeter O, Aasen JO (eds) North Sea oil and gas reservoirs II. Graham and Trotman, London, pp 337–351

Dreyer T (1993a) Geometry and facies of large-scale flow units in fluvial-dominated fan-delta-front sequences. In: Ashton M (ed) Advances in reservoir geology. Geol Soc Lond Spec Publ 69: 135–174

Dreyer T (1993b) Quantified fluvial architecture in ephemeral stream deposits of the Esplugafreda Formation (Palaeocene), Tremp-Graus Basin, northern Spain. In: Marzo M, Puigdefábregas C (eds) Alluvial sedimentation. Int Assoc Sedimentol Spec Publ 17: 337–362

Dreyer T, Fält L-M, Høy T, Knarud R, Steel RJ, Cuevas J-L (1993) Sedimentary architecture of field analogues for reservoir information (SAFARI): a case study of the fluvial Escanilla Formation, Spanish Pyrenees. In: Flint SS, Bryant ID (eds) The geological modelling of hydrocarbon reservoirs and outcrop analogues. Int Assoc Sedimentol Spec Publ 15: 57–80

Dubiel RF (1991) Architectural-facies analysis of nonmarine depositional systems in the Upper Triassic Chinle Formation, southeastern Utah. In: Miall AD, Tyler N (eds) The three-dimensional facies architecture of terrigenous clastic sediments, and its implications for hydrocarbon discovery and recovery. Soc Econ Paleontol Mineral Conc Sedimentol Paleontol 3: 103–110

Dubiel RF, Parrish JT, Parrish JM, Good SC (1991) The Pangaean megamonsoon – evidence from the upper Triassic Chinle Formation, Colorado Plateau. Palaios 6: 347–370

Du Boys PFD (1879) Le Rhône et le rivier a lit affouillable. Ann Ponts Chaussees 18 (5): 141–195

Dueck RN, Paauwe EFW (1994) The use of borehole imaging techniques in the exploration for stratigraphic traps: an example from the Middle Devonian Gilwood channels in north-central Alberta. Bull Can Petrol Geol 42: 137–154

Dueholm KS, Olsen T (1993) Reservoir analog studies using multimodal photogrammetry: a new tool for the petroleum industry. Am Assoc Petrol Geol Bull 77: 2023–2031

Dumont JF (1993) Lake patterns as related to neotectonics in subsiding basins: the example of the Ucamara Depression, Peru. Tectonophysics 222: 69–78

Dumont JF, Fournier M (1994) Geodynamic environment of Quaternary morphostructures of the Subandean foreland basins of Peru and Bolivia: characteristics and study methods. Q Int 21: 129–142

Dunbar CO, Rodgers J (1957) Principles of stratigraphy. Wiley, New York

Dury GH (1964) Principles of underfit streams. US Geological Survey professional paper 452-A

Eaves E (1976) Citronelle oil field, Mobile county, Alabama. In: Braunstein J (ed) North American oil and gas fields. Am Assoc Petrol Geol Mem 24: 259–275

Ebanks WJ Jr, Weber JF (1982) Development of a shallow heavy-oil deposit in Missouri. Oil Gas J Sept 27: 222–234

Eberth DA, Miall AD (1991) Stratigraphy, sedimentology and evolution of a vertebrate-bearing, braided to anastomosed fluvial system, Cutler Formation (Permian-Pennsylvanian), north-central New Mexico. Sediment Geol 72: 225–252

Eckelmann WR, Dewitt RJ, Fisher WL (1975) Prediction of fluvial-deltaic reservoir geometry, Prudhoe Bay field, Alaska. In: Proceedings of the 9th world petroleum congress, vol 2: geology

Eicher DL (1969) Paleobathymetry of the Cretaceous Greenhorn Sea in eastern Colorado. Am Assoc Petrol Geol Bull 53: 1075–1090

Einsele G, Liu B, Dürr S, Frisch W, Liu G, Luterbacher HP, Ratschbacher L, Ricken W, Wendt J, Wetzel A, Yu G, Zheng H (1994) The Xigaze forearc basin: evolution and facies architecture (Cretaceous, Tibet). Sediment Geol 90: 1–32

Einstein HA (1950) The bed-load function for sediment transportation in open channel flows. US Department of Agriculture, Washington DC (Soil conservation service technical bulletin 1026)

Eisbacher GH (1985) Pericollisional strike-slip faults and synorogenic basins, Canadian Cordillera. In: Biddle KT, Christie-Blick N (eds) Strike-slip deformation, basin formation and sedimentation. Soc Econ Paleontol Mineral Spec Publ 37: 265–280

Eisbacher GH, Carrigy MA, Campbell RB (1974) Paleodrainage pattern and late orogenic basins of the Canadian Cordillera. In: Dickinson WR (ed) Tectonics and sedimentation. Soc Econ Paleontol Mineral Spec Publ 22: 143–166

Ekes C (1993) Bedload-transported pedogenic mud aggregates in the Lower Old Red Sandstone in southwest Wales. J Geol Soc Lond 150: 469–472

Elliott L (1989) The Surat and Bowen basins. Aust Petrol Explor Assoc J 29: 398–416

Embry AF (1990a) A tectonic origin for third-order depositional sequences in extensional basins – implications for basin modelling. In: Cross TA (ed) Quantitative dynamic stratigraphy. Prentice-Hall, Englewood Cliffs, pp 491–501

Embry AF (1990b) Geological and geophysical evidence in support of the hypothesis of anticlockwise rotation of northern Alaska. Mar Geol 93: 317–329

Embry AF (1991) Mesozoic history of the Arctic Islands. In: Trettin HP (ed) Geology of the Innuitian Orogen and Arctic Platform of Canada and Greenland. Geological Survey of Canada. Geol Can 3: 369–433

Epry C (1913) Ripple marks. Annual report of the Smithsonian Institution, pp 307–318

Ethridge FG, Flores RM (eds) (1981) Recent and ancient nonmarine depositional environments: models for exploration. Soc Econ Paleontol Mineral Spec Publ 31: 349

Ethridge FG, Jackson TJ, Youngberg AD (1981) Floodbasin sequence of a fine-grained meander belt subsystem: the coal-bearing Lower Wasatch and Upper Fort Union Formations, southern Powder River Basin, Wyoming. In: Ethridge FG, Flores RM (eds) Recent and ancient nonmarine depositional environments, models for exploration. Soc Econ Paleontol Mineral Spec Publ 31: 191–209

Ethridge FG, Flores RM, Harvey MD (eds) (1987) Recent developments in fluvial sedimentology. Soc Econ Paleontol Mineral Spec Publ 39

Ethridge FG, Schumm SA (1978) Reconstructing paleochannel morphologic and flow characteristics: methodology, limitations and assessment. In: Miall AD (ed) Fluvial sedimentology. Can Soc Petrol Geol Mem 5: 703–721

Eugster HP, Hardie LA (1975) Sedimentation in an ancient playa-lake complex: the Wilkins Peak Member of the Green River Formation of Wyoming. Geol Soc Am Bull 86: 319–334

Evans JE (1991) Facies relationships, alluvial architecture, and paleohydrology of a Paleogene humid-tropical alluvial fan system: Chumstick Formation, Washington State, USA. J Sediment Petrol 61: 732–755

Evans JE, Terry DO Jr (1994) The significance of incision and fluvial sedimentation in the Basal White River Group (Eocene-Oligocene), Badlands of South Dakota, USA. Sediment Geol 90: 137–152

Everest R (1832) A quantitative study of stream transportation. J Asiatic Soc Bengal 1: 238–240

Eyles N (1993) Earth's glacial record and its tectonic setting. Earth Sci Rev 35: 1–248

Eyles N, Eyles CH (1992) Glacial depositional systems. In: Walker RG, James NP (eds) Facies models, response to sea level change. Geological Association of Canada, St. John's Newfoundland, pp 73–100

Eyles N, Eyles CH, Miall AD (1983) Lithofacies types and vertical profile models; an alternative approach to the description and environmental interpretation of glacial diamict and diamictite sequences. Sedimentology 30: 393–410

Fairhead JD (1986) Geophysical controls on sedimentation within the African Rift System. In: Frostick LE, Renaut RW, Reid I, Tiercelin JJ (eds) Sedimentation in the African rifts. Geol Soc Lond Spec Publ 25: 19–27

Farrell KM (1987) Sedimentology and facies architecture of overbank deposits of the Mississippi River, False River region, Louisiana. In: Ethridge FG, Flores RM, Harvey MD (eds) Recent developments in fluvial sedimentology. Soc Econ Paleontol Mineral Spec Publ 39: 111–120

Farshori MZ, Hopkins JC (1989) Sedimentology and petroleum geology of fluvial and shoreline deposits of the Lower Cretaceous Sunburst Sandstone Member, Mannville Group, southern Alberta. Bull Can Petrol Geol 37: 371–388

Fenneman NM (1906) Floodplains produced without floods. Am Geogr Soc Bull 38: 89–91

Ferguson RI (1993) Understanding braiding processes in gravel-bed braided rivers: progress and unsolved problems. In: Best JL, Bristow CS (eds) Braided rivers. Geol Soc Lond Spec Publ 75: 73–87

Ferguson RI, Werritty A (1983) Bar development and channel changes in the gravelly River Feshie, Scotland. In: Collinson JD, Lewin J (eds) Modern and ancient fluvial systems. Int Assoc Sedimentol Spec Publ 6: 181–193

Ferm JC, Cavaroc VV Jr (1968) A nonmarine sedimentary model for the Allegheny Rocks of West Virginia. In: Klein G, De V (ed) Late Paleozoic and Mesozoic continental sedimentation, northeastern North America. Geol Soc Am Spec Paper 106: 1–20

Ferm JC, Staub JR (1984) Depositional controls of mineable coal bodies. In: Rahmani RA, Flores RM (eds) Sedimentology of coal and coal-bearing sequences. Int Assoc Sedimentol Spec Publ 7: 275–289

Fernández J, Bluck BJ, Viseras C (1993) The effects of fluctuating base level on the structure of alluvial fan and associated fan delta deposits: an example from the Tertiary of the Betic Cordillera, Spain. Sedimentology 40: 879–893

Fielding CR (1984) A coal depositional model for the Durham Coal Measures of NE England. J Geol Soc Lond 141: 919–932

Fielding CR (1986) Fluvial channel and overbank deposits from the Westphalian of the Durham Coalfield, NE England. Sedimentology 33: 119–140

Fielding CR (ed) (1993a) Current research in fluvial sedimentology. Sediment Geol 85: 1–656 (special issue)

Fielding CR (1993b) A review of recent research in fluvial sedimentology. Sediment Geol 85: 3–14

Fielding CR, Crane RC (1987) An application of statistical modelling to the prediction of hydrocarbon recovery factors in fluvial reservoir sequences. In: Ethridge FG, Flores RM, Harvey MD (eds) Recent developments in fluvial sedimentology. Soc Econ Paleontol Mineral Spec Publ 39: 321–327

Fielding CR, Falkner AJ, Scott SG (1993) Fluvial response to foreland basin overfilling; the Late Permian Rangal coal measures in the Bowen Basin, Queensland, Australia. Sediment Geol 85: 475–497

Figueiredo AMF (1994) Recôncavo Basin, Brazil: a prolific intracontinental rift basin. In: Landon SM (ed) Interior rift basins. Am Assoc Petrol Geol Mem 59: 157–203

Finch J (1823) Geological essay on the Tertiary formations in America. Am J Sci 7: 31–43

Fischer AG (1986) Climatic rhythms recorded in strata. Annu Rev Earth Planet Sci 14: 351–376

Fischer AG, De Boer PL, Premoli Silva I (1990) Cyclostratigraphy. In: Ginsburg RN, Beaudoin B (eds) Cretaceous resources, events and rhythms: Background and plans for research. Kluwer Academic, Dordrecht, pp 139–172

Fisher MJ (1984) Triassic. In: Glennie KW (ed) Introduction to the petroleum geology of the North Sea. Blackwell Scientific, Oxford, pp 85–101

Fisher WL, McGowen JH (1967) Depositional systems in the Wilcox Group of Texas and their relationship to occurrence of oil and gas. Trans Gulf Coast Assoc Geol Soc 17: 105–125

Fisk HN (1944) Geological investigation of the alluvial valley of the lower Mississippi River. Mississippi River Commission, Vicksburg, Mississippi

Fisk HN (1947) Fine-grained alluvial deposits and their effect on Mississippi River activity. Mississippi River Commission, Vicksburg, Mississippi

Fisk HN (1960) Recent Mississippi River sedimentation and peat accumulation: Compte rendu 4th congres l'avancement des etudes de stratigraphie et de geologie du Carbonifere, Heerlen 1958, vol 1, pp 187–199

Fisk HN (1952) Geological investigations of the Atchafalaya basin and the problem of Mississippi River diversion. US Army Corps of Engineers, Waterways Experimental Station, Vicksburg, Misissippi (2 vols)

Flach PD, Mossop GD (1985) Depositional environments of Lower Cretaceous McMurray Formation, Athabasca Oil Sands, Alberta. Am Assoc Petrol Geol Bull 69: 1195–1207

Flemming BW (1988) Zur Klassifikation subaquatischer, strömungstransversaler Transportkörper. Boch Geol Geotechn Arb 29: 44–47

Flint S (1985) Alluvial fan and playa sedimentation in an Andean arid closed basin: the Pacencia Group, Antofagasta Province, Chile. J Geol Soc Lond 142: 533–546

Flint SS, Bryant ID (eds) (1993) The geological modelling of hydrocarbon reservoirs and outcrop analogues. Int Assoc Sedimentol Spec Publ 15: 269

Flores RM (1981) Coal deposition in fluvial paleoenvironments of the Paleocene Tongue River Member of the Fort Union Formation, Powder River area, Powder River basin, Wyoming and Montana. In: Ethridge FG, Flores RM (eds) Recent and ancient nonmarine depositional environments: models for exploration. Soc Econ Paleontol Mineral Spec Publ 31: 161–190

Flores RM (ed) (1983a) Fluvial systems, their economic and field applications. American Association of Petroleum Geologists Field Seminar, Tulsa

Flores RM (1983b) Basin analysis of coal-rich Tertiary fluvial deposits, northern Powder River Basin, Montana and Wyoming. In: Collinson JD, Lewin J (eds) Modern and ancient fluvial systems. Int Assoc Sedimentol Spec Publ 6: 501–515

Flores RM (1984) Comparative analysis of coal accumulation in Cretaceous alluvial deposits, southern United States Rocky Mountain basins. In: Stott DF, Glass DJ (eds) The Mesozoic of Middle North America. Can Soc Petrol Geol Mem 9: 373–385

Flores RM, Ethridge FG, Miall AD, Galloway WE, Fouch TD (1985) Recognition of fluvial depositional systems and their resource potential. Soc Econ Paleontol Mineral Short Course 19: 290

Flores RM, Hanley JH (1984) Anastomosed and associated coal-bearing fluvial deposits: upper tongue river member, Paleocene Fort Union Formation, northern Powder River Basin, Wyoming, USA. In: Rahmani RA, Flores RM (eds) Sedimentology of coal and coal-bearing sequences. Int Assoc Sedimentol Spec Publ 7: 85–103

Focke JW, van Popta J (1989) Reservoir evaluation of the Permian Gharif Formation, Sultanate of Oman. Society of Petroleum Engineers paper 17978, pp 517–528

Folk RL (1966) A review of grain-size parameters. Sedimentology 6: 73–93

Folk RL, Ward WC (1957) Brazos River bar, a study in the significance of grain size parameters. J Sediment Petrol 27: 3–26

Forbes DL (1983) Morphology and sedimentology of a sinuous gravel-bed channel system, lower Babbage River, Yukon coastal plain, Canada. In: Collinson JD, Lewin J (eds) Modern and ancient fluvial systems. Int Assoc Sedimentol Spec Publ 6: 195–206

Forgotson JM, Stark PH (1972) Well-data files and the computer, a case history from northern Rocky Mountains. Am Assoc Petrol Geol Bull 56: 1114–1127

Fortuin AR, de Smet MEM (1991) Rates and magnitudes of late Cenozoic vertical movements in the Indonesian Banda Arc and the distinction of eustatic effects. In: Macdonald DIM (ed) Sedimentation, tectonics and eustasy: sea-level changes at active margins. Int Assoc Sedimentol Spec Publ 12: 79–89

Fouch TD (1983) Oil- and gas-bearing Upper Cretaceous and Paleogene fluvial rocks in central and northeast Utah. In: Flores RM (ed) Fluvial systems: their economic and field applications. American Association of Petroleum Geologists Field Seminar, Tulsa, pp 312–343

Fouch TD, Lawton TF, Nichols DJ, Cashion WB, Cobban WA (1983) Patterns and timing of synorogenic sedimentation in Upper Cretaceous rocks of central and northeast Utah. In: Reynolds M, Dolly E (eds) Mesozoic paleogeography of west-central United States. Society of Economic Paleontologists and Mineralogists, Rocky Mountain section, symposium, vol 2, pp 305–334

Frakes LA (1979) Climates throughout geologic time. Elsevier, Amsterdam

Franczyk KJ, Pitman JK (1991) Latest Cretaceous nonmarine depositional systems in the Wasatch Plateau area: reflections of foreland to intermontane basin transitions. In: Chidsey TC Jr (ed) Geology of east-central Utah. Utah Geol Assoc Publ 19: 77–93

Franklin EH, Clifton BB (1971) Halibut field, Southeastern Australia. Am Assoc Petrol Geol Bull 55: 1262–1279

Fraser AJ, Gawthorpe RL (1990) Tectono-stratigraphic development and hydrocarbon habitat of the Carboniferous in northern England. In: Hardman RFP, Brooks J (eds) Tectonic events responsible for Britain's oil and gas reserves. Geol Soc Lond Spec Publ 55: 49–86

Fraser GS, DeCelles PG (1992) Geomorphic controls on sediment accumulation and margins of foreland basins. Basin Res 4: 233–252

Frazier DE (1967) Recent deltaic deposits of the Mississippi River: their development and chronology. Trans Gulf Coast Assoc Geol Soc 17: 287–315

Frazier DE (1974) Depositional episodes: their relationship to the Quaternary stratigraphic framework in the northwestern portion of the Gulf Basin. Bureau of Economic Geology, University of Texas, geological circular 74–1

Friedkin JF (1945) A laboratory study of the meandering of alluvial rivers. Mississippi River Commission, Vicksburg

Friedman GM (1967) Dynamic processes and statistical parameters compared for size frequency distribution of beach and river sands. J Sediment Petrol 37: 327–354

Friedman GM (1971) Distinction between dune, beach and river sands from their textural characteristics. J Sediment Petrol 31: 514–529

Friend PF (1978) Distinctive features of some ancient river systems. In: Miall AD (ed) Fluvial sedimentology. Can Soc Petrol Geol Mem 5: 531–542

Friend PF (1983) Towards the field classification of alluvial architecture or sequence. In: Collinson JD, Lewin J (eds) Modern and ancient fluvial systems. Int Assoc Sedimentol Spec Publ 6: 345–354

Friend PF (1985) Molasse basins of Europe: a tectonic assessment. Trans R Soc Edinb 76: 451–462

Friend PF, Alexander-Marrack PD, Nicholson J, Yeats AK (1976) Devonian sediments of east Greenland I. Meddelelser om Gronland, vol 206, no 1

Friend PF, Johnson NM, McRae LE (1989) Time-level plots and accumulation patterns of sediment sequences. Geol Mag 126: 491–498

Friend PF, Moody-Stuart M (1972) Sedimentation of the Wood Bay Formation (Devonian) of Spitsbergen: regional analysis of a late orogenic basin. Norsk Polarinstitutt Skrifter no 157

Friend PF, Sinha R (1993) Braiding and meandering parameters. In: Best JL, Bristow CS (eds) Braided rivers. Geol Soc Lond Spec Publ 75: 105–111
Friend PF, Slater MJ, Williams RC (1979) Vertical and lateral building of river sandstone bodies, Ebro Basin, Spain. J Geol Soc Lond 136: 39–46
Frostick LE, Reid I (1977) The origin of horizontal laminae in ephemeral stream channel fill. Sedimentology 24: 1–10
Frostick LE, Reid I (1989) Climatic versus tectonic controls of fan sequences: lessons from the Dead Sea, Israel. J Geol Soc Lond 146: 527–538
Frostick LE, Renaut RW, Reid I, Tiercelin JJ (eds) (1986) Sedimentation in the African rifts. Geol Soc Lond Spec Publ 25: 382
Fulford MM, Busby CJ (1993) Tectonic controls on nonmarine sedimentation in a Cretaceous fore-arc basin, Baja California, Mexico. In: Frostick LE, Steel RJ (eds) Tectonic controls and signatures in sedimentary successions. Int Assoc Sedimentol Spec Publ 20: 301–333
Fulthorpe CS (1991) Geological controls on seismic sequence resolution. Geology 19: 61–65
Gabela VH (1990) Exploration and geologic framework of the Caño Limón oil filed, Llanos Orientales de Colombia. In: Ericksen GE, Pinochet MTC, Reinemund JA (eds) Geology of the Andes and its relation to hydrocarbon and mineral resources, Circum-Pacific Council for Energy and Mineral Resources. Earth Sci Ser 11: 363–382
Gagliano SM, Van Beek JL (1970) Geologic and geomorphic aspects of deltaic processes, Mississippi delta system. In: Flores RM (ed) Hydrologic and geologic studies of coastal Louisiana. Centre for Wetland Resources, Louisiana State University, report 1
Galay VJ, Kellerhals R, Bray DI (1973) Diversity of river types in Canada. In: Fluvial processes and sedimentation: proceedings of the Hydrology Symposium, Edmonton, National Research Council, Canada, pp 217–250
Galloway WE (1975) Process framework for describing the morphologic and stratigraphic evolution of the deltaic depositional systems. In: Broussard ML (ed) Deltas, models for exploration. Houston Geological Society, Houston, pp 87–98
Galloway WE (1980) Deposition and early hydrologic evolution of Westwater Canyon wet alluvial-fan systems. New Mexico Bureau Mines Miner Resourc Mem 38: 59–69
Galloway WE (1981) Depositional architecture of Cenozoic Gulf Coastal plain fluvial systems. In: Ethridge FG, Flores RM (eds) Recent and ancient nonmarine depositional environments: models for exploration. Soc Econ Paleontol Mineral Spec Publ 31: 127–155
Galloway WE (1989a) Genetic stratigraphic sequences in basin analysis I: architecture and genesis of flooding-surface bounded depositional units. Am Assoc Petrol Geol Bull 73: 125–142
Galloway WE (1989b) Genetic stratigraphic sequences in basin analysis II: application to northwest Gulf of Mexico Cenozoic basin. Am Assoc Petrol Geol Bull 73: 143–154
Galloway WE, Hobday DK (1983) Terrigenous clastic depositional systems. Springer, Berlin Heidelberg New York
Galloway WE, Hobday DK (1995) Terigenous clastic depositional systems, 2nd edition. Springer, Berlin Heidelberg New York (in press)
Galloway WE, Hobday DK, Magara K (1982) Frio Formation of the Texas Gulf Coast Basin – depositional systems, structural framework, and hydrocarbon origin, migration, distribution, and exploration potential. University of Texas at Austin, Bureau of Economic Geology, report of investigations 122
Galloway WE, Williams TA (1991) Sediment accumulation rates in time and space: Paleogene genetic stratigraphic sequences of the northwestern Gulf of Mexico. Geology 19: 986–989
Gansser A (1964) Geology of the Himalayas. Interscience, London
Gardiner S, Thomas DV, Bowering ED, McMinn LS (1990) A braided fluvial reservoir, Peco field, Alberta, Canada. In: Barwis JH, McPherson JG, Studlick RJ (eds) Sandstone petroleum reservoirs: Springer, Berlin Heidelberg New York, pp 31–56
Garrison RK, Chancellor R (1991) Berwick field: the geologic half of the seismic stratigraphic story in the Lower Tuscaloosa Mississippi. Trans Gulf Coast Assoc Geol Soc 41: 299–307
Gawthorpe RL, Colella A (1990) Tectonic controls on coarse-grained delta depositional systems in rift basins. In: Colella A, Prior DB (eds) Coarse-grained deltas. Int Assoc Sedimentol Spec Publ 10: 113–127
Gawthorpe RL, Collier REL, Alexander J, Bridge JS, Leeder MR (1993) Ground penetrating radar: application to sandbody geometry and heterogeneity studies. In: North CP, Prosser DJ (eds) Characterization of fluvial and aeolian reservoirs. Geol Soc Lond Spec Publ 73: 421–432
Geddes A (1960) The alluvial morphology of the Indo-Gangetic Plains. Trans Inst Br Geogr 28: 253–276
Geehan GW, Lawton TF, Sakurai S, Klob H, Clifton TR, Inman KF, Nitzberg KE (1986) Geologic prediction of shale continuity, Prudhoe Bay field. In: Lake LW, Carroll HB Jr (eds) Reservoir characterization. Academic, Orlando, pp 63–82
Geikie A (1882) Text book of geology. Macmillan, London
Geikie A (1905) The founders of Geology (republished in 1962 by Dover, New York)
Genik GJ (1993) Petroleum geology of Cretaceous-Tertiary rift basins in Niger, Chad, and Central African Republic. Am Assoc Petrol Geol Bull 77: 1405–1434
George GT, Berry JK (1993) A new lithostratigraphy and depositional model for the Upper Rotliegend of the UK sector of the southern North Sea. In: North CP, Prosser DJ (ed) Characterization of fluvial and aeolian reservoirs. Geol Soc Lond Spec Publ 73: 291–319
Germanoski D, Schumm SA (1993) Changes in braided river morphology resulting from aggradation and degradation. J Geol 101: 451–466
Gersib GA, McCabe PJ (1981) Continental coal-bearing sediments of the Port Hood Formation (Carboniferous), Cape Linzee, Nova Scotia, Canada. In: Ethridge FG, Flores RM (eds) Recent and ancient nonmarine depositional environments: models for exploration. Soc Econ Paleontol Mineral Spec Publ 31: 95–108
Ghignone JI, de Andrade G (1970, General geology and major oil fields of Reconcavo Basin, Brazil. In:

Halbouty MT (ed) Geology of giant petroleum fields. Am Assoc Petrol Geol Mem 14: 337–358
Gibling MR, Bird DJ (1994) Late Carboniferous cyclothems and alluvial paleovalleys in the Sydney Basin, Nova Scotia. Geol Soc Am Bull 106: 105–117
Gibling MR, Rust BR (1990) Ribbon sandstones in the Pennsylvanian Waddens Cove Formation, Sydney Basin, Atlantic Canada: the influence of siliceous duricrusts on channel-body geometry. Sedimentology 37: 45–65
Gibling MR, Rust BR (1993) Alluvial ridge-and-swale topography: a case study from the Morien Group of Atlantic Canada. In: Marzo M, Puigdefábregas C (eds) Alluvial sedimentation. Int Assoc Sedimentol Spec Publ 17: 133–150
Gibling MR, Wightman WG (1994) Palaeovalleys and proterozoan assemblages in a Late Carboniferous cyclothem, Sydney Basin, Nova Scotia. Sedimentology 41: 699–719
Gibling MR, Calder JH, Ryan R, Can de Poll HW, Yeo GW (1992) Late Carboniferous and Early Permian drainage patterns in Atlantic Canada. Can J Earth Sci 29: 338–352
Gilbert GK (1880) Land, sculpture, geology of the Henry Mountains. US Geographical and Geological Survey of the Rocky Mountain Region, 2nd edn, 160 p
Gilbert GK (1884) Ripple marks. Science 3: 375–376
Gilbert GK (1899) Ripple marks and cross-bedding. Geol Soc Am Bull 10: 135–140
Gilbert GK (1914) The transportation of debris by running water. US Geological Survey professional paper 86
Glennie KW (1970) Desert sedimentary environments. Elsevier, Amsterdam (Developments in sedimentology 14)
Glennie KW (1972) Permian Rotliegendes of northwest Europe interpreted in light of modern desert sedimentation studies. Am Assoc Petrol Geol Bull 56: 1048–1071
Glennie KW (1983) Lower Permian Rotliegend desert sedimentation in the North Sea area. In: Brookfield ME, Ahlbrandt TS (eds) Eolian sediments and processes. Dev Sedimentol 38: 521–541
Glennie KW (1987) Desert sedimentary environments, present and past – a summary. Sediment Geol 50: 135–165
Gloppen TG, Steel RJ (1981) The deposits, internal structure and geometry in six alluvial fan-fan delta bodies (Devonian – Norway) – a study in the significance of bedding sequences in conglomerates. In: Ethridge FG, Flores RM (eds) Recent and ancient nonmarine depositional environments: models for exploration. Soc Econ Paleontol Mineral Spec Publ 31: 49–69
Godin P (1991) Fining-upward cycles in the sandy-braided river deposits of the Westwater Canyon Member (Upper Jurassic), Morrison Formation, New Mexico. Sediment Geol 70: 61–82
Gohain K, Parkash B (1990) Morphology of the Kosi megafan. In: Rachocki AH, Church M (eds) Alluvial fans: a field approach. Wiley, New York, pp 151–178
Gole CV, Chitale SV (1966) Inland delta building activity of Kosi River. Journal of the Hydraulics Division. Proc Am Soc Civil Eng 92 (HY2): 111–126
Golin V, Smyth M (1986) Depositional environments and hydrocarbon potential of the Evergreen Formation, ATP 145P, Surat Basin, Queensland. Aust Petrol Expl Assoc J 26: 156–171
Gordon I, Heller PL (1993) Evaluating major controls on basinal stratigraphy, Pine Valley, Nevada: implications for syntectonic deposition. Geol Soc Am Bull 105: 47–55
Gomez B, Naff RL, Hubbell DW (1989) Temporal variations in bedload transport rates associated with the migration of bedforms. Earth Surface Proc Landforms 14: 135–156
González E (1990) Hydrocarbon resources in the coastal zone of Chile. In: Ericksen GE, Pinochet MTC, Reinemund JA (eds) Geology of the Andes and its relation to hydrocarbon and mineral resources, Circum-Pacific Council for Energy and Mineral Resources. Earth Sci Ser 11: 383–404
Goodwin PW, Anderson EJ (1985) Punctuated aggradational cycles: a general hypothesis of episodic stratigraphic accumulation. J Geol 93: 515–533
Gould HR (1970) The Mississippi Delta complex. In: Morgan JP (ed) Deltaic sedimentation: modern and ancient. Soc Econ Paleontol Mineral Spec Publ 15: 3–30
Grabau AW (1906) Types of sedimentary overlap. Geol Soc Am Bull 17: 567–636
Grabau AW (1907) Types of cross-bedding and their stratigraphic significance. Science 25: 295–296
Grabau AW (1913a) Principles of stratigraphy. Seiler, New York, 1185 p
Grabau AW (1913b) Early Paleozoic delta deposits of North America. Geol Soc Am Bull 24: 399–528
Grabau AW (1917) Problems of the interpretation of sedimentary rocks. Geol Soc Am Bull 28: 735–744
Gradzinski R, Gagol J, Slaczka A (1979) The Tumlin Sandstone (Holy Cross Mts., central Poland): Lower Triassic deposits of aeolian dunes and interdune areas. Acta Geol Polon 29: 151–175
Graham SA, Dickinson WR, Ingersoll RV (1975) Himalayan-Bengal model for flysch dispersal in the Appalachian-Ouachita system. Geol Soc Am Bull 86: 273–286
Griffith WM (1927) A theory of silt and scour. Institute of Civil Engineers Proceedings, Tulsa, Oklahoma, pp 223–314
Guccione MJ (1993) Grain-size distribution of overbank sediment and its use to locate channel positions. In: Marzo M, Puigdefábregas C (eds) Alluvial sedimentation. Int Assoc Sedimentol Spec Publ 17: 185–194
Gurnis M (1990) Bounds on global dynamic topography from Phanerozoic flooding of continental platforms. Nature 344: 754–756
Gurnis M (1992) Long-term controls on eustatic and epeirogenic motions by mantle convection. GSA Today 2: 141–157
Gustavson TC (1974) Sedimentation on gravel outwash fans, Malaspina Glacier foreland, Alaska. J Sediment Petrol 44: 374–389
Gustavson TC (1978) Bed forms and stratification types of modern gravel meander lobes, Nueces River, Texas. Sedimentology 25: 401–426
Hack JT (1957) Studies of longitudinal stream profiles in Virginia and Maryland. US Geological Survey professional paper 294-B
Hacquebard PA, Donaldson JR (1969) Carboniferous coal deposition associated with flood-plain and limnic environments in Nova Scotia. In: Dapples EC, Hopkins ME (eds) Environments of coal deposition. Geol Soc Am Spec Paper 114: 143–191

Hahmann P (1912) Die Bildung von Sandduenen bei gleichmäßiger Strömung. Ann Phys 637–676
Hallam A (1963) Major epeirogenic and eustatic changes since the Cretaceous and their possible relationship to crustal structure. Am J Sci 261: 397–423
Hallam A (1984) Continental humid and arid zones during the Jurassic and Cretaceous. Palaeogeogr Palaeoclimatol Palaeoecol 47: 195–223
Hamblin AP, Rust BR (1989) Tectono-sedimentary analysis of alternate-polarity half-graben basin-fill successions: Late Carboniferous Horton Group, Cape Breton Island, Nova Scotia. Basin Res 2: 239–255
Hambrey MJ, Harland WB (1981) Criteria for the identification of glacigenic deposits. In: Hambrey MJ, Harland WB (eds) Earth's Pre-Pleistocene glacial record. Cambridge University Press, Cambridge, pp 14–27
Hamilton DS, Galloway WE (1989) New exploration techniques in the analysis of diagenetically complex reservoir sandstones, Sydney Basin, NSW. Aust Petrol Explor Assoc J 29: 235–257
Hamilton DS, Tadros NZ (1994) Utility of coal seams as genetic stratigraphic sequence boundaries in nonmarine basins: an example from the Gunnedah Basin, Australia. Am Assoc Petrol Geol Bull 78: 267–286
Hamilton W (1979) Tectonics of the Indonesian region. US Geological Survey professional paper 1078
Hamilton W (1985) Subduction, magmatic arcs, and foreland deformation. In: Howell DG (ed) Tectonostratigraphic terranes of the circum-Pacific region. Circum-Pacific Council for Energy and Mineral Resources. Earth Sci Ser 1: 259–262
Hamilton W (1987) Crustal extension in the Basin and Range province, southwestern United States. In: Coward MP, Dewey JF, Hancock PL (eds) Continental extensional tectonics. Geol Soc Lond Spec Publ 28: 155–176
Hamilton WS, Cameron CP (1986) Facies relationships and depositional environments of Lower Tuscaloosa Formation reservoir sandstones in the McComb field area, southwest Mississippi. Trans Gulf Coast Assoc Geol Soc 36: 141
Hamlin KH, Cameron CP (1987) Sandstone petrology and diagenesis of Lower Tuscaloosa Formation reservoirs in the McComb and Little Creek field areas, southwest Mississippi. Trans Gulf Coast Assoc Geol Soc 37: 95–104
Hanneman DL, Wideman CJ (1991) Sequence stratigraphy of Cenozoic continental rocks, southwestern Montana. Geol Soc Am Bull 103: 1335–1345
Hanneman DL, Wideman CJ, Halvorsen JW (1994) Calcic paleosols: their use in subsurface stratigraphy. Am Assoc Petrol Geol Bull 78: 1360–1371
Happ SC, Rittenhouse G, Dobson GC (1940) Some principles of accelerated stream valley sedimentation. US Department of Agriculture, technical bulletin 695
Harms JC (1966) Stratigraphic traps in valley fill, western Nebraska. Am Assoc Petrol Geol Bull 50: 2119–2149
Haq BU, Hardenbol J, Vail PR (1987) Chronology of fluctuating sea levels since the Triassic (250 million years ago to present). Science 235: 1156–1167
Haq BU, Hardenbol J, Vail PR (1988) Mesozoic and Cenozoic chronostratigraphy and cycles of sea-level change. In: Wilgus CK, Hastings BS, Kendall CGSC, Posamentier HW, Ross CA, Van Wagoner JC (eds) Sea-level research: an integrated approach. Soc Econ Paleontol Mineral Spec Publ 42: 71–108
Hardie LA, Smoot JP, Eugster HP (1978) Saline lakes and their deposits: a sedimentological approach. In: Matter A, Tucker ME (eds) Modern and ancient lake sediments. Int Assoc Sedimentol Spec Publ 2: 7–41
Harms JC (1966) Stratigraphic traps in valley fill, western Nebraska. Am Assoc Petrol Geol Bull 50: 2119–2149
Harms JC, Fahnestock RK (1965) Stratification, bed forms, and flow phenomena (with an example from the Rio Grande). In: Middleton GV (ed) Primary sedimentary structures and their hydrodynamic interpretation. Soc Econ Paleontol Mineral Spec Publ 12: 84–115
Harms JC, Mackenzie DB, McCubbin DG (1963) Stratification in modern sands of the Red River, Louisiana. J Geol 71: 566–580
Harms JC, Southard JB, Spearing DR, Walker RG (1975) Depositional environments as interpreted from primary sedimentary structures and stratification sequences. Soc Econ Paleontol Mineral Short Course 2
Harms JC, Southard JB, Walker RG (1982) Structures and sequences in clastic rocks. Soc Econ Paleontol Mineral Short Course 9
Harris AL, Fettes DJ (eds) (1988) The Caledonian-Appalachian orogen. Geological Society, London (Spec Pap 38)
Harris PT (1988) Large-scale bedforms as indicators of mutually evasive sand transport and the sequential infilling of wide-mouthed estuaries. Sediment Geol 57: 273–298
Hartley AJ (1993) Sedimentological response of an alluvial system to source area tectonism: the Seilao Member of the Late Cretaceous to Eocene Purilactis Formation of northern Chile. In: Marzo M, Puigdefábregas C (eds) Alluvial sedimentation. Int Assoc Sedimentol Spec Publ 17: 489–500
Harvey AM (1990) Factors influencing Quaternary alluvial fan development in southeast Spain. In: Rachocki AH, Church M (eds) Alluvial fans: a field approach. Wiley, Chichester, pp 247–269
Hastings JO Jr (1990) Coarse-grained meander-belt reservoirs, Rocky Ridge field, North Dakota. In: Barwis JH, McPherson JG, Studlick RJ (eds) Sandstone petroleum reservoirs. Springer, Berlin Heidelberg New York, pp 57–84
Haszeldine RS (1983a) Fluvial bars reconstructed from a deep, straight channel, Upper Carboniferous coalfield of northeast England. J Sediment Petrol 53: 1233–1248
Haszeldine RS (1983b) Descending tabular cross-bed sets and bounding surfaces from a fluvial channel in the Upper Carboniferous coalfield of north-east England. In: Collinson JD, Lewin J (eds) Modern and ancient fluvial systems. Int Assoc Sedimentol Spec Publ 6: 449–456
Hayden HH (1821) Geological essays, or an enquiry into some of the geological phenomena to be found in various parts of America and elsewhere. Am J Sci 3: 47–57
Hayes BJR, Christopher JE, Rosenthal L, Los G, McKercher B (1994) Cretaceous Mannville Group of the Western Canada Sedimentary Basin. In: Mossop GD, Shetsen I (eds) Geological atlas of the Western Canada Sedimentary Basin. Canadian Society of Petroleum Geologists, Calgary, Alberta, pp 317–334
Hayes JB, Harms JC, Wilson T Jr (1976) Contrasts between braided and meandering stream deposits, Beluga and

Sterling formations (Tertiary), Cook Inlet, Alaska. In: Miller TP (ed) Recent and ancient sedimentary environments in Alaska. Alaska Geological Society, Anchorage, Alaska, pp J1-J27

Hayes MO, Kana TW (eds) (1976) Terrigenous clastic depositional environments. Coastal Research Division, Department of Geology, University of South Carolina, technical report 11-CRD

Heath R (1989) Exploration in the Cooper Basin. Aust Petrol Explor Assoc J 29: 366–378

Heckel PH (1986) Sea-level curve for Pennsylvanian eustatic marine transgressive-regressive depositional cycles along midcontinent outcrop belt, North America. Geology 14: 330–334

Hein FJ, Walker RG (1977) Bar evolution and development of stratification in the gravelly, braided, Kicking Horse River, British Columbia. Can J Earth Sci 14: 562–570

Heller PL, Gordon I (1994) Evaluating major controls on basinal stratigraphy, Pine Valley, Nevada: implications for syntectonic deposition: reply to Discussion. Geol Soc Am Bull 106: 156–157

Heller PL, Paola C (1989) The paradox of Lower Cretaceous gravels and the initiation of thrusting in the Sevier orogenic belt, United States Western Interior. Geol Soc Am Bull 101: 864–875

Heller PL, Paola C (1992) The large-scale dynamics of grain-size variation in alluvial basins, 2: application to syntectonic conglomerate. Basin Res 4: 91–102

Heller PL, Angevine CL, Winslow NS, Paola C (1988) Two-phase stratigraphic model of foreland-basin sequences. Geology 16: 501–504

Heller PL, Angevine CL, Paola C (1989) Comment on "Thrusting and gravel progradation in foreland basins: a test of post-thrusting gravel dispersal". Geology 17: 959–960

Heller PL, Beekman F, Angevine CL, Cloetingh SAPL (1993) Cause of tectonic reactivation and subtle uplifts in the Rocky Mountain region and its effect on the stratigraphic record. Geology 21: 1003–1006

Hempton MR, Dunne LA (1984) Sedimentation in pull-apart basins: active examples in eastern Turkey. J Geol 92: 513–530

Herries RD (1993) Constrasting styles of fluvial-aeolian interaction at a downwind erg margin: Jurassic Kayenta-Navajo transition, northeastern Arizona. In: North CP, Prosser DJ (ed) Characterization of fluvial and aeolian reservoirs. Geol Soc Lond Spec Publ 73: 199–218

Hersch JB (1987) Exploration methods – Lower Tuscaloosa trend, southwest Mississippi. Trans Gulf Coast Assoc Geol Soc 37: 105–112

Heward AP (1978a) Alluvial fan sequence and megasequence models: with examples from Westphalian D – Stephanian B coalfields, northern Spain. In: Miall AD (ed) Fluvial sedimentology. Can Soc Petrol Geol Mem 5: 669–702

Heward AP (1978b) Alluvial fan and lacustrine sediments from the Stephanian A and B (La Magdalena, Ciñera – Matallana and Sabero) coalfields, northern Spain. Sedimentology 25: 451–488

Hickin EJ (1969) A newly identified process of point bar formation in natural streams. Am J Sci 267: 999–1010

Hickin EJ (1983) River channel changes: retrospect and prospect. In: Collinson JD, Lewin J (eds) Modern and ancient fluvial systems. Int Assoc Sedimentol Spec Publ 6: 61–83

Hider A (1882) Appendix D, Report of assistant engineer Arthur Hider upon obervations at Lake Providence, November, 1879 to November, 1880. Report of the Mississippi River Commission, pp 80–98

Higham N (1963) A very scientific gentleman: the major achievements of Henry Clifton Sorby. Pergamon, Oxford

Hirst JPP (1991) Variations in alluvial architecture across the Oligo-Miocene Huesca fluvial system, Ebro basin, Spain. In: Miall AD, Tyler N (eds) The three-dimensional facies architecture of terrigenous clastic sediments, and its implications for hydrocarbon discovery and recovery. Soc Econ Paleontol Mineral Conc Sedimentol Paleontol 3: 111–121

Hirst JPP, Blackstock CR, Tyson S (1993) Stochastic modelling of fluvial sandstone bodies. In: Flint SS, Bryant ID (eds) The geological modelling of hydrocarbon reservoirs and outcrop analogues. Int Assoc Sedimentol Spec Publ 15: 237–252

Hjulström F (1935) Studies in the morphological activity of rivers as illustrated by the River Fyris. Geol Inst Uppsala Bull 25: 221–528

Hobbs WH (1906) Gaudix formation of Granada, Spain. Geol Soc Am Bull 17: 285–294

Hobday DK (1978) Fluvial deposits of the Ecca and Beaufort Groups in the eastern Karoo Basin, South Africa. In: Miall AD (ed) Fluvial sedimentology. Canadian Society of Petroleum Geologists Memoir 5, pp 413–429

Hobday DK, Woodruff CM Jr, McBride MW (1981) Paleotopographic and structural controls on non-marine sedimentation of the Lower Cretaceous Antlers Formation and correlatives, North Texas and southeastern Oklahoma. In: Ethridge FG, Flores RM (eds) Recent and ancient nonmarine depositional environments: models for exploration. Soc Econ Paleontol Mineral Spec Publ 31: 71–87

Hoey TB (1992) Temporal variations in bedload transport rates and sediment storage in gravel-bed rivers. Prog Phys Geogr 16: 319–338

Hoffman PF (1991) Did the breakout of Laurentia turn Gondwanaland inside-out? Science 252: 1409–1412

Hoffman PF, Grotzinger JP (1993) Orographic precipitation, erosional unloading, and tectonic style. Geology 21: 195–198

Hogg MD (1988) Newtonia field: a model for mid-dip Lower Tuscaloosa retrograde deltaic sedimentation. Trans Gulf Coast Assoc Geol Soc 38: 461–471

Høimyr Ø, Kleppe A, Nystuen JP (1993) Effects of heterogeneities in a braided stream channel sandsbody on the simulation of oil recovery: a case study from the Lower Jurassic Statfjord Formation, Snorre field, North Sea. In: Ashton M (ed) Advances in reservoir geology. Geol Soc Lond Spec Publ 69: 105–134

Holbrook JM, Dunbar RW (1992) Depositional history of Lower Cretaceous strata in northeastern New Mexico: implications for regional tectonics and depositional sequences. Geol Soc Am Bull 104: 802–813

Holdsworth BK, Collinson JD (1988) Millstone Grit cyclicity revisited. In: Besly BM, Kelling G (eds) Sedimentation in a synorogenic basin complex: the Upper Carboniferous of northwest Europe. Blackie, Glasgow, pp 132–152

Holmes A (1965) Principles of physical geology, rev edn. Nelson, London
Holmes DA (1968) The recent history of the Indus. Geogr J 134: 367–382
Homewood P, Allen PA, Williams GD (1986) Dynamics of the Molasse basin of western Switzerland. In: Allen PA, Homewood P (eds) Foreland basins. Int Assoc Sedimentol Spec Publ 8: 199–217
Hooke JM (1986) The significance of mid-channel bars in an active meandering river. Sedimentology 33: 839–850
Hooke R, Le B (1967) Processes on arid-region alluvial fans. J Geol 75: 438–460
Hoover HC, Hoover LH (1950) Georgius Agricola: de re metallica (translated from the first Latin Edition of 1556). Dover, New York
Hopkins JC (1981) Sedimentology of quartzose sandstones of Lower Mannville and associated units, Medicine River area, central Alberta. Bull Can Petrol Geol 29: 12–41
Hopkins JC (1985) Channel-fill deposits formed by aggradation in deeply scoured, superimposed distributaries of the Lower Kootenai Formation. J Sediment Petrol 55: 42–52
Hopkins JC, Hermanson SW, Lawton DC (1982) Morphology of channel-sand bodies in the Glauconitic Sandstone member (Upper Mannville), Little Bow area, Alberta. Bull Can Petrol Geol 30: 274–285
Hopkins JC, Wood JM, Krause FF (1991) Waterflood response of reservoirs in an estuarine valley fill: Upper Mannville G, U, and W pools, Little Bow field, Alberta. Am Assoc Petrol Geol Bull 75: 1064–1088
Horne JC, Ferm JC (1976) Carboniferous depositional environments in the Pocahontas Basin, eastern Kentucky and southern West Virginia, Guidebook. Department of Geology, University of South Carolina
Horne JC, Ferm JC, Caruccio FT, Baganz BP (1978) Depositional models in coal exploration and mine planning in Appalachian region. Am Assoc Petrol Geol Bull 62: 2379–2411
Hossack JR (1984) The geometry of listric growth faults in the Devonian basins of Sunnfjord, W Norway. J Geol Soc Lond 141: 629–638
Houbolt JJHC (1968) Recent sediments in the southern Bight of the North Sea. Geologie Mijnbouw 47: 245–273
Howell DG (1989) Tectonics of suspect terranes. Chapman and Hall, London
Hoyt JH, Henry VJ Jr (1967) Influence of island migration on barrier-island sedimentation. Geol Soc Am Bull 78: 77–86
Hsü KJ, Li J, Chen H, Wang Q, Sun S, Sengör AMC (1990) Tectonics of South China: key to understanding West Pacific geology. Tectonophysics 183: 9–39
Hubert JF, Filipov AJ (1989) Debris-flow deposits in alluvial fans on the west flank of the White Mountains, Owens Valley, California, USA. Sediment Geol 61: 177–206
Hubert JF, Reed AA, Carey PJ (1976) Paleogeography of the East Berlin Formation, Newark Group, Connecticut Valley. Am J Sci 276: 1183–1207
Huggenberger P (1993) Radar facies: recognition of facies patterns and heterogeneities within Pleistocene Rhine gravels, NE Switzerland. In: Best JL, Bristow CS (eds) Braided rivers. Geol Soc Lond Spec Publ 75: 163–176
Huggett RJ (1991) Climate, earth processes and earth history. Springer, Berlin Heidelberg New York
Humphreys AA, Abbot HL (1861) Report upon the physics and hydraulics of the Mississippi. US Army Corps of Topographical Engineers, professional paper 4
Hunt AR (1882) On the formation of ripple marks. R Soc Lond Proc 34: 1–19
Hunt D, Tucker ME (1992) Stranded parasequences and the forced regressive wedge systems tract: deposition during base-level fall. Sediment Geol 81: 1–9
Hunter GM (1966) Red earth field, A and C pools. Oilfields of Alberta: supplement. Alberta Society of Petroleum Geologists, Calgary, Alberta, pp 85–86 (Spec Publ)
Hunter RE, Richmond BM (1988) Daily cycles in coastal dunes. Sediment Geol 55: 43–67
Hunter RE, Rubin DM (1983) Interpreting cyclic cross-bedding, with an example from the Navajo Sandstone. In: Brookfield ME, Ahlbrandt TS (eds) Eolian sediments and processes. Elsevier, Amsterdam. Dev Sedimentol 38: 429–454
Hutchison CS (1989) Geological evolution of south-east Asia. Oxford Monogr Geol Geophys 13
Hutton J (1795) Theory of the earth, 2 vols (reprinted 1959). Engelmann, Wheldon and Wesley, Weinheim
Imbrie J (1985) A theoretical framework for the Pleistocene ice ages. J Geol Soc Lond 142: 417–432
Ingersoll RV (1988) Tectonics of sedimentary basins. Geol Soc Am Bull 100: 1704–1719
Inman DL (1949) Sorting of sediments in the light of fluid mechanics. J Sediment Petrol 19: 51–70
Inman DL (1952) Measures for describing size distribution of sediments. J Sediment Petrol 22: 125–145
Iseya F, Ikeda H (1989) Sedimentation in coarse-grained sand-bedded meanders: distinctive deposition of suspended sediment. In: Taira A, Masuda F (eds) Sedimentary facies in the active plate margin. Terra Scientific, Tokyo, pp 81–112
Ito M, Masuda F (1986) Evolution of clastic piles in an arc-arc collision zone. Late Cenozoic depositional history around the Tanzawa Mountains, Central Honshu, Japan. Sediment Geol 49: 223–259
Iwaniw E (1984) Lower Cantabrian basin margin deposits in NE León, Spain - a model for valley-fill sedimentation in a tectonically active, humid climatic setting. Sedimentology 31: 91–110
Jackson JR (1834) Hints on the subject of geographical arrangement and nomenclature. R Geogr Soc J 4: 72–88
Jackson RG II (1975) Hierarchical attributes and a unifying model of bed forms composed of cohesionless material and produced by shearing flow. Geol Soc Am Bull 86: 1523–1533
Jackson RG II (1976a) Large scale ripples of the lower Wabash River. Sedimentology 23: 593–632
Jackson RG II (1976b) Depositional model of point bars in the lower Wabash River. J Sediment Petrol 46: 579–594
Jackson RG II (1976c) Sedimentological and fluid dynamic implications of the turbulent bursting phenomenon in geophysical flows. J Fluid Mech 77: 531–560
Jackson RG II (1976d) Velocity - bed-form - texture patterns of meander bends in the lower Wabash River of Illinois and Indiana. Geol Soc Am Bull 86: 1511–1522
Jackson RG II (1978) Preliminary evaluation of lithofacies models for meandering alluvial streams. In: Miall AD

(ed) Fluvial sedimentology. Can Soc Petrol Geol Mem 5: 543–576
Jackson RG II (1981) Sedimentology of muddy fine-grained channel deposits in meandering streams of the American middle west. J Sediment Petrol 51: 1169–1192
Jamieson TF (1860) On the drift and rolled gravel of the North of Scotland. Q J Geol Soc Lond 16: 347–371
Jeffreys H (1929) On the transport of sediment by streams. Cambr Philos Soc Proc 25: 272–276
Jervey MT (1988) Quantitative geological modeling of siliciclastic rock sequences and their seismic expression. In: Wilgus CK, Hastings BS, Kendall CGSC, Posamentier HW, Ross CA, Van Wagoner JC (eds) Sea level research – an integrated approach. Soc Econ Paleontol Mineral Spec Publ 42: 47–69
Jerzykiewicz T, Sweet AR (1988) Sedimentological and palynological evidence of regional climate changes in the Campanian to Paleocene sediments of the Rocky Mountain Foothills, Canada. Sediment Geol 59: 29–76
Jin Y, Liu D, Luo C (1985) Development of Daqing oil field by waterflooding. J Petrol Technol 37: 269–274
Jirik LA (1990) Reservoir heterogeneity in Middle Frio fluvial sandstones: case studies in Seeligson field, Jim Wells county, Texas. Trans Gulf Coast Assoc Geol Soc 49: 335–351
Jo HR, Rhee CW, Ryang WH (1994) Evaluating major controls on basinal stratigraphy, Pine Valley, Nevada: implications for syntectonic deposition: Discussion. Geol Soc Am Bull 106: 155–156
Johansson CE (1963) Orientation of pebbles in running water. A laboratory study. Geogr Ann 45: 85–112
John BH, Almond CS (1987) Lithostratigraphy of the Lower Eromanga Basin sequence in south-west Queensland. Aust Petrol Explor Assoc J 27: 196–214
Johnson AM (1970) Physical processes in geology. Freeman, San Francisco
Johnson D (1932) Streams and their significance. J Geol 40: 481–496
Johnson GD (1977) Paleopedology of *Ramapithecus*-bearing sediments, North India. Geol Rundsch 66: 192–216
Johnson GD, Vondra CF (1972) Siwalik sediments in a portion of the Punjab re-entrant: the sequence at Haritalyangar Disrict, Bilaspur, HP. Himal Geol 3: 118–144
Johnson GD, Raynolds RGH, Burbank DW (1986) Late Cenozoic tectonics and sedimentation in the north-western Himalayan foredeep: I. Thrust ramping and associated deformation in the Potwar region. In: Allen PA, Homewood P (eds) Foreland basins. Int Assoc Sedimentol Spec Publ 8: 273–291
Johnson NM, Stix J, Tauxe L, Cerveny PF, Tahirkheli RAK (1985) Paleomagnetic chronology, fluvial processes, and tectonic implications of the Siwalik deposits near Chinji Village, Pakistan. J Geol 93: 27–40
Johnson NM, Jordan TE, Johnsson PA, Naeser CW (1986) Magnetic polarity stratigraphy, age and tectonic setting of fluvial sediments in an eastern Andean foreland basin, San Juan Province, Argentina. In: Allen PA, Homewood P (eds) Foreland basins. Int Assoc Sedimentol Spec Publ 8: 63–75
Johnsson MJ, Basu A (1993) Processes controlling the composition of clastic sediments. Geol Soc Am Spec Paper 284
Jolley EJ, Turner P, Williams GD, Hartley AJ, Flint S (1990) Sedimentological response of an alluvial system to Neogene thrust tectonics, Atacama Desert, northern Chile. J Geol Soc Lond 147: 769–784
Jones BG, Rust BR (1983) Massive sandstone facies in the Hawkesbury Sandstone, a Triassic fluvial deposit near Sydney, Australia. J Sediment Petrol 53: 1249–1260
Jones CM (1977) Effects of varying discharge regimes on bed-form sedimentary structures in modern rivers. Geology 5: 567–570
Jones CM, McCabe PJ (1980) Erosion surfaces within giant fluvial cross-beds of the Carboniferous in northern England. J Sediment Petrol 50: 613–620
Jopling AV (1963) Hydraulic studies on the origin of bedding. Sedimentology 2: 115–121
Jopling AV (1965) Hydraulic factors controlling the shape of laminae in laboratory deltas. J Sediment Petrol 35: 777–791
Jopling AV (1975) Early studies on stratified drift. In: Jopling AV, McDonald BC (eds) Glaciofluvial and glaciolacustrine sedimentation. Soc Econ Paleontol Mineral Spec Publ 23: 4–21
Jopling AV, Walker RG (1968) Morphology and origin of ripple-drift cross-lamination, with examples from the Pleistocene of Massachusetts. J Sediment Petrol 38: 971–984
Jordan DW, Pryor WA (1992) Hierarchical levels of heterogeneity in a Mississippi River meander belt and application to reservoir systems. Am Assoc Petrol Geol Bull 76: 1601–1624
Jordan TE (1981) Thrust loads and foreland basin evolution, Cretaceous western United States. Am Assoc Petrol Geol Bull 65: 2506–2520
Jordan TE (1995) Retroarc foreland and related basins. In: Busby CJ, Ingersoll RV (eds) Tectonics of sedimentary basins. Blackwell Scientific, Oxford: 331–362
Jordan TE, Flemings PB (1990) From geodynamic models to basin fill – a stratigraphic perspective. In: Cross TA (ed) Quantitative dynamic stratigraphy. Prentice-Hall, Englewood Cliffs, pp 149–163
Jordan TE, Flemings PB (1991) Large-scale stratigraphic architecture, eustatic variation, and unsteady tectonism: a theoretical evaluation. J Geophys Res 96B: 6681–6699
Jordan TE, Flemings PB, Beer JA (1988) Dating thrust-fault activity by use of foreland-basin strata. In: Kleinspehn KL, Paola C (eds) New perspectives in basin analysis. Springer, Berlin Heidelberg New York, pp 307–330
Karges HE (1962) Significance of Lower Tuscaloosa sand patterns in southwest Mississippi. Trans Gulf Coast Assoc Geol Soc 12: 171–173
Karlsen DA, Larter S (1989) A rapid correlation method for petroleum population mapping within individual petroleum reservoirs: application to petroleum reservoir description. In: Collinson JD (ed) Correlation in hydrocarbon exploration. Graham and Trotman, London, pp 77–85
Katz BJ (ed) (1990) Lacustrine basin exploration: case studies and modern analogs. Am Assoc Petrol Geol Mem 50
Kauffman EG (1969) Cretaceous marine cycles of the Western Interior. Mountain Geol 6: 227–245

Kauffman EG (1984) Paleobiogeography and evolutionary response dynamic in the Cretaceous Western Interior Seaway of North America. In: Westermann EG (ed) Jurassic-Cretaceous biochronology and paleogeography of North America. Geol Assoc Can Spec Paper 27: 273–306

Keighin CW, Fouch TD (1981) Depositional environments and diagenesis of some nonmarine Upper Cretaceous reservoir rocks, Uinta Basin, Utah. In: Ethridge FG, Flores RM (eds) Recent and ancient nonmarine depositional environments: models for exploration. Soc Econ Paleontol Mineral Spec Publ 31: 109–125

Kelly SB (1993) Cyclical discharge variations recorded in alluvial sediments: an example from the Devonian of southwest Ireland. In: North CP, Prosser DJ (eds) Characterization of fluvial and aeolian reservoirs. Geol Soc Lond Spec Publ 73: 157–166

Kelly SB, Olsen HO (1993) Terminal fans – a review with reference to Devonian examples. Sediment Geol 85: 339–374

Kennedy JF (1963) The mechanics of dunes and antidunes in erodible-bed channels. J Fluid Mech 16: 521–544

Kennedy RG (1895) The prevention of silting in irrigation canals. Inst Civil Eng 119: 281–290

Kerr DR (1990) Reservoir heterogeneity in the Middle Frio Formation: case studies in Stratton and Agua Dulce fields, Nueces County, Texas. Trans Gulf Coast Assoc Geol Soc 49: 363–372

Kerr DR, Jirik LA (1990) Fluvial architecture and reservoir compartmentalization in the Oligocene Middle Frio Formation, south Texas. Trans Gulf Coast Assoc Geol Soc 49: 374–380

Kindle EM (1911) Cross-bedding and absence of fossils considered as criteria of continental deposits. Am J Sci 32: 225–230

Kindle EM (1917) Recent and fossil ripple marks. Geological Survey of Canada, museum bulletin 25

King PB (1977) The evolution of North America, rev edn. Princeton University Press, Princeton

King WSH (1916) The nature and formation of sand ripples and dunes. Geogr J 46: 189–209

Kirk M (1983) Bar developments in a fluvial sandstone (Westphalian "A"), Scotland. Sedimentology 30: 727–742

Kirk RH (1980) Statfjord field: a North Sea giant. In: Halbouty MT (ed) Giant oil and gas fields of the decade: 1968–1978. Am Assoc Petrol Geol Mem 30: 95–116

Kirschbaum MA, McCabe PJ (1992) Controls on the accumulation of coal and on the development of anastomosed fluvial systems in the Cretaceous Dakota formation of southern Utah. Sedimentology 39: 581–598

Kirschner CE, Lyon CA (1973) Stratigraphic and tectonic development of Cook Inlet petroleum province. In: Pitcher MG (ed) Arctic geology. Am Assoc Petrol Geol Mem 19: 396–407

Klein G deV (1987) Current aspects of basin analysis. Sediment Geol 50: 95–118

Klein G deV, Willard DA (1989) Origin of the Pennsylvanian coal-bearing cyclothems of North America. Geology 17: 152–155

Kleinspehn KL (1985) Cretaceous sedimentation and tectonics, Tyaughton-Methow Basin, southwestern British Columbia. Can J Earth Sci 22: 154–174

Kleinspehn KL (1988) Sedimentary basins in the context of allochthonous terranes. In: Kleinspehn KL, Paola C (eds) New perspectives in basin analysis: Springer, Berlin Heidelberg New York, pp 295–305

Klicman DP, Cameron CP, Meylan MA (1988) Petrology and depositional environments of Lower Tuscaloosa Formation (Upper Cretaceous) sandstones in the North Hustler and Thompson field areas, southwest Mississippi. Trans Gulf Coast Assoc Geol Soc 38: 47–58

Klimetz MP (1983) Speculations on the Mesozoic plate tectonic evolution of eastern China. Tectonics 2: 139–166

Klitgord KD, Schouten H (1986) Plate kinematics of the central Atlantic. In: Vogt PR, Tucholke BE (eds) The Western North Atlantic region. Geological Society of America, Boulder, Colorado, pp 351–378 (Geology of North America, vol M)

Klovan JE (1966) The use of factor analysis in determining depositional environments from grain-size distributions. J Sediment Petrol 36: 115–125

Kluth CF, Coney PJ (1981) Plate tectonics of the ancestral Rocky Mountains. Geology 9: 10–15

Knight SH (1929) The Fountain and Casper Formations of the Laramie Basin. University of Wyoming Publications in Science. Geology 1: 1–82

Knoll MD, Rea J, Knight R, Grimm K, Beckie R (1994) Architectural-element analysis of ground penetrating radar data: a multidisciplinary approach to aquifer characterization. Symposium on the application of geophysics to environmental and engineering problems, Boston

Kochel RC (1990) Humid fans of the Appalachian Mountains. In: Rachocki AH, Church M (eds) Alluvial fans: a field approach. Wiley, Chichester, pp 109–129

Kochel RC, Johnson RA (1984) Geomorphology and sedimentology of humid-temperate alluvial fans, central Virginia. In: Koster EH, Steel RJ (eds) Sedimentology of gravels and conglomerates. Can Soc Petrol Geol Mem 10: 109–122

Kocurek G (1981) Significance of interdune deposits and bounding surfaces in aeolian dune sands. Sedimentology 28: 753–780

Kocurek G (1988) First-order and super bounding surfaces in eolian sequences – bounding surfaces revisited. Sediment Geol 56: 193–206

Kocurek G, Hunter RE (1986) Origin of polygonal fractures in sand, uppermost Navajo and Page sandstones, Page, Arizona. J Sediment Petrol 56: 895–904

Kocurek G, Knight J, Havholm K (1991) Outcrop and semi-regional three-dimensional architecture and reconstruction of a portion of the eolian Page Sandstone (Jurassic). In: Miall AD, Tyler N (eds) The three-dimensional facies architecture of terrigenous clastic sediments, and its implications for hydrocarbon discovery and recovery. Soc of Econ Paleontol and Mineral Conc Sedimentol Paleontol 3: 25–43

Kolb CR, Van Lopik JR (1966) Depositional environment of the Mississippi River deltaic plain – southeastern Louisiana. In: Shirley ML (ed) Deltas in their geologic framework. Houston Geological Society, Houston, Texas, pp 17–61

Kominz MA, Bond GC (1991) Unusually large subsidence and sea-level events during middle Paleozoic time: new evidence supporting mantle convection models for supercontinent assembly. Geology 19: 56–60

Koster EH (1978) Transverse ribs: their characteristics, origin and paleohydraulic significance. In: Miall AD (ed) Fluvial sedimentology. Can Soc Petrol Geol Mem 5: 161–186

Koster EH, Steel RJ (eds) (1984) Sedimentology of gravels and conglomerates. Can Soc Petrol Geol Mem 10

Kranzler I (1966) Origin of oil in lower member of Tyler Formation of central Montana. Am Assoc Petrol Geol Bull 50: 2245–2259

Kraus MJ (1987) Integration of channel and floodplain suites, II. Vertical relations of alluvial paleosols. J Sediment Petrol 57: 602–612

Kraus MJ (1992) Mesozoic and tertiary paleosols. In: Martini IP, Chesworth W (eds) Weathering, soils and paleosols. Elsevier, Amsterdam, pp 525–542 (Developments in earth surface processes, no 2)

Kraus MJ, Bown TM (1988) Pedofacies analysis; a new approach to reconstructing ancient fluvial sequences. Geol Soc Am Spec Paper 216: 143–152

Kraus MJ, Bown TM (1993) Short-term sediment accumulation rates determined from Eocene alluvial paleosols. Geology 21: 743–746

Kraus MJ, Middleton LT (1987) Dissected paleotopography and base-level changes in a Triassic fluvial sequence. Geology 15: 18–21

Krumbein WC (1934) Size frequency of sediments. J Sediment Petrol 4: 65–77

Krumbein WC (1941) Measurement and geologic significance of shape and roundness of sedimentary particles. J Sediment Petrol 11: 64–72

Kuenen PH (1957) Longitudinal filling of oblong sedimentary basins. Geol Mijnbouw 18: 189–195

Kuenzi WD, Horst OH, McGehee RV (1979) Effect of volcanic activity on fluvial-deltaic sedimentation in a modern arc-trench gap, southwestern Guatemala. Geol Soc Am Bull 90 (1): 827–838

Kvale EP, Vondra CF (1993) Effects of relative sea-level changes and local tectonics on a Lower Cretaceous fluvial to transitional marine sequence. Bighorn Basin, Wyoming, USA. In: Marzo M, Puigdefábregas C (eds) Alluvial sedimentation. Int Assoc Sedimentol (Spec Publ 17), pp 383–399

Lajoie J, Stix J (1992) Volcaniclastic rocks. In: Walker RG, James NP (eds) Facies models, response to sea level change. Geological Association of Canada, St. John's, Newfoundland, pp 101–118

Lane EW (1935) Stable channels in erodible materials. Trans Am Soc Civil Eng 63: 123–142

Lane EW (1955) The importance of fluvial morphology in hydraulic engineering. Am Soc Civil Eng Proc 81 (745): 1–17

Lane EW (1957) A study of the shape of channels formed by natural streams flowing in erodible material. US Army Corps of Engineers, Missouri River Division, Omaha, Nebraska (Sediment series 9)

Lang SC (1993) Evolution of Devonian alluvial systems in an oblique-slip mobile zone – an example from the Broken River Province, northeastern Australia. Sediment Geol 85: 501–535

Langbein WB, Leopold LB (1966) River meanders – theory of minimum variance. US Geological Survey professional paper 422-H

Langbein WB, Schumm SA (1958) Yield of sediment in relation to mean annual precipitation. Trans Am Geophys Union 39: 1076–1084

Langford RP (1989) Fluvial-aeolian interactions, part I: modern systems. Sedimentology 36: 1023–1035

Langford RP, Bracken B (1987) Medano Creek, Colorado, a model for upper-flow-regime fluvial deposition. J Sediment Petrol 57: 863–870

Langford RP, Chan MA (1989) Fluvial-aeolian interactions, part II, ancient systems. Sedimentology 36: 1037–1051

Larsen V, Steel RJ (1978) The sedimentary history of a debris flow-dominated, Devonian alluvial fan – a study of textural inversion. Sedimentology 25: 37–59

Lash GG (1990) The Shochary Ridge sequence, southeastern Pennsylvania – a possible Ordovician piggyback basin fill. Sediment Geol 68: 39–53

Lawrence DA, Williams BPJ (1987) Evolution of drainage systems in response to Acadian deformation: the Devonian Battery Point Formation, eastern Canada. In: Ethridge FG, Flores RM, Harvey MD (eds) Recent developments in fluvial sedimentology. Soc Econ Paleontol Mineral Spec Publ 39: 287–300

Lawson AC (1913) The petrographic designation of alluvial fan formations. Univ Calif Publ Dep G 7: 325–334

Lawton TF (1985) Style and timing of frontal structures, thrust belt, central Utah. Am Assoc Petrol Geol Bull 69: 1145–1159

Lawton TF (1986a) Compositional trends within a clastic wedge adjacent to a fold-thrust belt: Indianola Group, central Utah, USA. In: Allen PA, Homewood P (eds) Foreland basins. Int Assoc Sedimentol Spec Publ 8: 411–423

Lawton TF (1986b) Fluvial systems of the Upper Cretaceous Mesaverde Group and Paleocene North Horn Formation, central Utah: a record of transition from thin-skinned to thick-skinned in the foreland region. In: Peterson JA (ed) Paleotectonics and sedimentation in the Rocky Mountain region, United States. Am Assoc Petrol Geol Mem 41: 423–442

Lawton TF, Geehan GW, Voorhees BJ (1987) Lithofacies and depositional environments of the Ivishak Formation, Prudhoe Bay field. In: Tailleur IL, Weimer PJ (eds) Alaskan North Slope geology. Alaska Geol Soc Econ Paleontol Mineral Pac Sect 1: 61–76

Leblanc Smith G, Eriksson KA (1979) A fluvioglacial and glaciolacustrine deltaic depositional model for Permo-Carboniferous coals of the northeastern Karoo Basin, South Africa. Palaeogeogr Palaeoclimatol Palaeoecol 27: 67–84

Leckie DA (1994) Canterbury Plains, New Zealand – implications for sequence stratigraphic models. Am Assoc Petrol Geol Bull 78: 1240–1256

Leckie DA, Fox C, Tarnocai C (1989) Multiple paleosols of the late Albian Boulder Creek Formation, British Columbia, Canada. Sedimentology 36: 307–323

Leddy JO, Ashworth PJ, Best JL (1993) Mechanisms of anabranch avulsion within gravel-bed braided rivers: observations from a scaled physical model. In: Best JL, Bristow CS (eds) Braided rivers. Geol Soc Lond Spec Publ 75: 119–127

Lee RA (1982) Petroleum geology of the Malacca Strait contract area (Central Sumatra Basin). Proceedings of

the 11th annual convention of the Indonesian Petroleum Association, vol 1, pp 243–263
Leeder MR (1973) Fluviatile fining-upward cycles and the magnitude of palaeochannels. Geol Mag 110: 265–276
Leeder MR (1975) Pedogenic carbonates and flood sediment accumulation rates: a quantitative model for arid-zone lithofacies. Geol Mag 112: 257–270
Leeder MR (1978) A quantitative stratigraphic model for alluvium, with special reference to channel deposit density and interconnectedness. In: Miall AD (ed) Fluvial sedimentology. Can Soc Petrol Geol Mem 5: 587–596
Leeder MR (1982) Upper Paleozoic basins of the British Isles – Caledonide inheritance versus Hercynian plate margin processes. J Geol Soc Lond 139: 479–491
Leeder MR (1983) On the interactions between turbulent flow, sediment transport and bedform mechanics in channelized flows. In: Collinson JD, Lewin J (eds) Modern and ancient fluvial systems. Int Assoc Sedimentol Spec Publ 6: 5–18
Leeder MR (1988) Recent developments in Carboniferous geology: a critical review with implications for the British Isles and N.W. Europe. Proc Geol Assoc 99: 73–100
Leeder MR (1993) Tectonic controls upon drainage basin development, river channel migration and alluvial architecture: implications for hydrocarbon reservoir development and characterization. In: North CP, Prosser DJ (eds) Characterization of fluvial and aeolian reservoirs. Geol Soc Lond Spec Publ 73: 7–22
Leeder MR, Alexander J (1987) The origin and tectonic significance of asymmetric meander belts. Sedimentology 34: 217–226
Leeder MR, Gawthorpe RL (1987) Sedimentary models for extensional tilt-block/half-graben basins. In: Coward MP, Dewey JF, Hancock PL (eds) Continental extension tectonics. Geol Soc Lond Spec Publ 28: 139–152
Leeder MJ, Hardman M (1990) Carboniferous geology of the southern North Sea basin and controls on hydrocarbon prospectivity. In: Hardman RFP, Brooks J (eds) Tectonic events responsible for Britain's oil and gas reserves. Geol Soc Lond Spec Publ 55: 87–105
Leeder MR, Jackson JA (1993) The interaction between normal faulting and drainage in active extensional basins, with examples from the western United States and central Greece. Basin Res 5: 79–102
Leeder MR, Seger MJ, Stark CP (1991) Sedimentation and tectonic geomorphology adjacent to major active and inactive normal faults, southern Greece. J Geol Soc Lond 148: 331–343
Lees GM (1955) Recent earth movements in the Middle East. Geol Rundsch 43: 221–226
Legarreta L, Gulisano CA (1989) Analisis estratgrafico sequencial de la cuenca neuquina (Triasico superior-Terciario inferior). Cuencas Sediment Argent 6: 221–243
Legarreta L, Uliana MA (1991) Jurassic-Cretaceous marine oscillations and geometry of back-arc basin fill, central Argentine Andes. In: Macdonald DIM (ed) Sedimentation, tectonics and eustasy. Int Assoc Sedimentol (Spec Publ 12), pp 429–450
Leliavsky S (1955) An introduction to fluid hydraulics. Constable, London (reprinted 1966 by Dover, New York)
Leopold LB, Bull WB (1979) Base level, aggradation, and grade. Proc Am Philos Soc 123: 168–202
Leopold LB, Maddock T Jr (1953) The hydraulic geometry of stream channels and some physiographic implications. US Geological Survey professional paper 252
Leopold LB, Wolman MG (1957) River channel patterns; braided, meandering, and straight. US Geological Survey professional paper 282-B
Leopold LB, Wolman MG (1960) River meanders. Geol Soc Am Bull 71: 769–794
Leopold LB, Wolman MG, Miller JP (1964) Fluvial processes in geomorphology. Freeman, San Francisco
Le Roux JP (1992) Determining the channel sinuosity of ancient fluvial systems from paleocurrent data. J Sediment Petrol 62, p. 283–291
Le Roux JP (1994) The angular deviation of paleocurrent directions as applied to the calculation of channel sinuosities. J Sediment Res A64: 86–87
Levey RA (1978) Bed-form distribution and internal stratification of coarse-grained point bars, Upper Congaree River, SC. In: Miall AD (ed) Fluvial sedimentology. Can Soc Petrol Geol Mem 5: 105–127
Levorsen AI (1967) Geology of petroleum, second edition. Freeman, San Francisco
Lewis GW, Lewin J (1983) Alluvial cutoffs in Wales and the Borderlands. In: Collinson JD, Lewin J (eds) Modern and ancient fluvial systems. Int Assoc Sedimentol Spec Publ 6: 145–154
Lewin J (1976) Initiation of bed forms and meanders in coarse-grained sediment. Geol Soc Am Bull 87: 281–285
Li S, Finlayson B (1993) Flood management on the lower Yellow River: hydrological and geomorphological perspectives. Sediment Geol 85: 285–296
Link MH (1984) Fluvial facies of the Miocene Ridge Route Formation, Ridge Basin, California. Sediment Geol 38: 263–286
Link MH, Osborne RH (1978) Lacustrine facies in the Pliocene Ridge Basin Group: Ridge Basin, California. In: Matter A, Tucker ME (eds) Modern and ancient lake sediments. Int Assoc Sedimentol Spec Publ 2: 169–187
Linsley PN, Potter HC, McNab G, Racher D (1980) The Beatrice field, inner Moray Firth, U.K. North Sea. In: Halbouty MT (ed) Giant oil and gas fields of the decade: 1968–1978. Am Assoc Petrol Geol Mem 30: 117–129
Liu H (1986) Geodynamic scenario and structural styles of Mesozoic and Cenozoic basins in China. Am Assoc Petrol Geol Bull 70: 377–395
Long DGF (1978) Proterozoic stream deposits: some problems of recognition and interpretation of ancient sandy fluvial systems. In: Miall AD (ed) Fluvial sedimentology. Can Soc Petrol Geol Mem 5: 313–342
Long DGF (1981) Dextral strike-slip faults in the Canadian Cordillera and depositional environments of related fresh-water intermontane coal basins. In: Miall AD (ed) Sedimentation and tectonics in alluvial basins. Geol Assoc Can Spec Paper 23: 153–186
López-Gómez J, Arche A (1993) Architecture of the Cañizar fluvial sheet sandstones, Early Triassic, Iberian Ranges, eastern Spain. In: Marzo M, Puigdefábregas C (eds) Alluvial sedimentation. Int Assoc Sedimentol Spec Publ 17: 363–381
Lorenz JC, Heinze DM, Clark JA, Searls CA (1985) Determination of widths of meander-belt sandstone reser-

voirs from vertical downhole data, Mesaverde Group, Piceance Creek Basin, Colorado. Am Assoc Petrol Geol Bull 69: 710–721

Lorenz JC, Warpinski NR, Branagan PT (1991) Subsurface characterization of Mesaverde reservoirs in Colorado: geophysical and reservoir-engineering checks on predictive sedimentology. In: Miall AD, Tyler N (eds) The three-dimensional facies architecture of terrigenous clastic sediments, and its implications for hydrocarbon discovery and recovery. Soc Econ Paleontol Mineral Conc Sedimentol Paleontol 3: 57–79

Loutit TS, Hardenbol J, Vail PR, Baum GR (1988) Condensed sections: the key to age dating and correlation of continental margin sequences. In: Wilgus CK, Hastings BS, Kendall CGSC, Posamentier HW, Ross CA, Van Wagoner JC (eds) Sea-level research: an integrated approach. Soc Econ Paleontol Mineral Spec Publ 42: 183–213

Lucas PT, Drexler JM (1976) Altamont-Bluebell – a major, naturally fractured stratigraphic trap, Uinta Basin, Utah. In: Braunstein J (ed) North American oil and gas fields. Am Assoc Petrol Geol Mem 24: 121–135

Lucchitta I, Suneson N (1981) Flash flood in Arizona – observations and their application to the identification of flash-flood deposits in the geologic record. Geology 9: 414–418

Lundberg N, Dorsey RJ (1988) Synorogenic sedimentation and subsidence in a Plio-Pleistocene collisional basin, eastern Taiwan. In: Kleinspehn K, Paola C (eds) New perspectives in basin analysis: Springer, Berlin Heidelberg New York, pp 265–280

Lustig LK (1965) Clastic sedimentation in Deep Springs Valley, California. US Geological Survey professional paper 352-F

Luterbacher HP, Eichenseer H, Betzler C, Van den Hurk AM (1991) Carbonate-siliciclastic depositional systems in the Paleogene of the south Pyrenean foreland basin: a sequence-stratigraphic approach. In: Macdonald DIM (ed) Sedimentation, tectonics and eustasy: sea-level changes at active margins. Int Assoc Sedimentol Spec Publ 12: 391–407

Luttrell PR (1993) Basinwide sedimentation and the continuum of paleoflow in an ancient river system: Kayenta Formation (Lower Jurassic), central portion Colorado Plateau. Sediment Geol 85: 411–434

Luyendyk BP, Hornafius JS (1987) Neogene crustal rotations, fault slip, and basin development in southern California. In: Ingersoll RV, Ernst WG (eds) Cenozoic basin development of coastal California (Rubey vol 6). Prentice-Hall, Englewood Cliffs, pp 259–283

Lyell C (1830–1833) Principles of geology, 3 vols. Murray, London (reprinted by Johnson, New York, 1969)

Lyons PL, Dobrin MB (1972) Seismic exploration for stratigraphic traps. In: King RA (ed) Stratigraphic oil and gas fields. Am Assoc Petrol Geol Mem 16: 225–243

MacDonald AC, Halland EK (1993) Sedimentology and shale modeling of a sandstone-rich fluvial reservoir: Upper Statfjord Formation, Statfjord field, North Sea. Am Assoc Petrol Geol Bull 77: 1016–1040

Mack GH, James WC (1994) Paleoclimate and the global distribution of paleosols. J Geol 102: 360–366

Mack GH, James WC, Monger HC (1993) Classification of paleosols. Geol Soc Am Bull 105: 129-136

MacKay D (1945) Ancient river beds and dead cities. Antiquity 19: 135–144

Mackin JH (1937) Erosional history of the Big Horn Basin, Wyoming. Geol Soc Am Bull 48: 813–894

Mackin JH (1948) Concept of the graded river. Geol Soc Am Bull 59: 463–512

Maizels J (1989) Sedimentology, paleoflow dynamics and flood history of jökulhlaup deposits: paleohydrology of Holocene sediment sequences in southern Iceland sandur deposits. J Sediment Petrol 59: 204–223

Maizels J (1990) Long-term paleochannel evolution during episodic growth of an exhumed Plio-Pleistocene alluvial fan, Oman. In: Rachocki AH, Church M (eds) Alluvial fans: a field approach. Wiley, Chichester, pp 271–304

Maizels J (1993) Lithofacies variations within sandur deposits: the role of runoff regime, flow dynamics and sediment suply characteristics. Sediment Geol 85: 299–325

Manspeizer W (1985) The Dead Sea Rift: impact of climate and tectonism on Pleistocene and Holocene sedimentation. In: Biddle KT, Christie-Blick N (eds) Strike-slip deformation, basin formation and sedimentation. Soc Econ Paleontol Mineral Spec Publ 37: 143–158

Marple RT, Talwani P (1993) Evidence of possible tectonic upwarping along the South Carolina coastal plain from an examination of river morphology and elevation data. Geology 21: 651–654

Marriott SB, Wright VP (1993) Palaeosols as indicators of geomorphic stability in two Old Red Sandstone alluvial suites, South Wales. J Geol Soc Lond 150: 1109–1120

Martin JH (1993) A review of braided fluvial hydrocarbon reservoirs: the petroleum engineer's perspective. In: Best JL, Bristow CS (eds) Braided rivers. Geol Soc Lond Spec Publ 75: 333–367

Martin R (1966) Paleogeomorphology and its application to exploration for oil and gas (with example from western Canada). Am Assoc Petrol Geol Bull 50: 2277–2311

Martini IP, Chesworth W (eds) (1992) Weathering, soils and paleosols. Elsevier, Amsterdam (Developments in earth surface processes, no 2)

Martini IP, Kwong JK, Sadura S (1993) Sediment rafting and cold climate fluvial deposits: Albany River, Ontario. In: Marzo M, Puigdefábregas C (eds) Alluvial sedimentation. Int Assoc Sedimentol Spec Publ 17: 63–76

Martinsen OJ, Martinsen RS, Steidtmann JR (1993) Mesaverde Group (Upper Cretaceous), southeastern Wyoming: allostratigraphy versus sequence stratigraphy in a tectonically active area. Am Assoc Petrol Geol Bull 77: 1351–1373

Marzo M, Puigdefábregas C (eds) (1993) Alluvial sedimentation. Int Assoc Sedimentol Spec Publ 17

Massari F (1983) Tabular cross-bedding in Messinian fluvial channel conglomerates, southern Alps, Italy. In: Collinson JD, Lewin J (eds) Modern and ancient fluvial systems. Int Assoc Sedimentol Spec Publ 6: 287–300

Massari F, Mellere D, Doglioni C (1993) Cyclicity in non-marine foreland-basin sedimentary fill: the Messinian conglomerate-bearing succession of the Venetian Alps (Italy). In: Marzo M, Puigdefábregas C (eds) Alluvial sedimentation. Int Assoc Sedimentol Spec Publ 17: 501–520

Mastalerz K, Wojewoda J (1993) Alluvial-fan sedimentation along an active strike-slip fault: Plio-Pleistocene pre-Kaczawa fan, SW Poland. In: Marzo M, Puigdefábregas C (eds) Alluvial sedimentation. Int Assoc Sedimentol Spec Publ 17: 293–304

Mather AE (1993) Basin inversion: some consequences for drainage evolution and alluvial architecture. Sedimentology 40: 1069–1089

Mathisen ME, Vondra CF (1983) The fluvial and pyroclastic deposits of the Cagayan Basin, northern Luzon, Philippines – an example of non-marine volcaniclastic sedimentation in an interarc basin. Sedimentology 30: 369–392

Matsuzawa A (1988) Oil and gas exploration in Bohai Bay, China. In: Wagner HC, Wagner LC, Wang FFH, Wong FL (eds) Petroleum resources of China and related subjects. Circum Pac Council Energy Miner Res Earth Sci Ser 10: 263–275

May SR, Ehman KD, Gray GG, Crowell JC (1993) A new angle on the tectonic evolution of the Ridge Basin, a "strike-slip" basin in southern California. Geol Soc Am Bull 105: 1357–1372

McCabe PJ (1975) The sedimentology and stratigraphy of the Kinderscout Grit Group (Namurian, R) between Wharfedale and Longdendale. PhD Thesis, University of Keele

McCabe PJ (1977) Deep distributary channels and giant bedforms in the Upper Carboniferous of the central Pennines, northern England. Sedimentology 24: 271–290

McCabe PJ (1984) Depositional environments of coal and coal-bearing strata. In: Rahmani RA, Flores RM (eds) Sedimentology of coal and coal-bearing sequences. Int Assoc Sedimentol Spec Publ 7: 13–42

McCabe PJ, Jones CM (1977) Formation of reactivation surfaces within superimposed deltas and bedforms. J Sediment Petrol 47: 707–715

McCabe PJ, Parrish JT (1992) Tectonic and climatic controls on the distribution and quality of Cretaceous coals. In: McCabe PJ, Parrish JT (eds) Controls on the distribution and quality of Cretaceous coals. Geol Soc Am Spec Paper 267: 1–15

McCarthy TS, Ellery WN, Stanistreet IG (1992) Avulsion mechanisms on the Okavango fan, Botswana: the control of a fluvial system by vegetation. Sedimentology 39: 779–795

McCaslin J (1983) Sohio to test Denver basin's Arbuckle. Oil Gas J 81: 87–88

McClay KR, Norton MG, Coney P, Davis GH (1986) Collapse of the Caledonian orogen and the Old Red Sandstone. Nature 323: 147–149

McDonald BC, Lewis CP (1973) Geomorphologic and sedimentologic processes of rivers and coasts, Yukon coastal plain. Task Force on northern Oil Development, Information Canada Catalogue no R72-12173

McDonald RE (1976) Big Piney-La Barge producing complex, Sublette and Lincoln conties, Wyoming. In: Braunstein J (ed) North American oil and gas fields. Am Assoc Petrol Geol Mem 24: 91–120

McDonnell KL (1978) Transition matrices and the depositional environments of a fluvial sequence. J Sediment Petrol 48: 43–48

McDougall JW (1989) Tectonically-induced diversion of the Indus River west of the Salt Range, Pakistan. Palaeogeogr Palaeoclimatol Palaeoecol 71: 301–307

McGee WJ (1897) Sheetflood erosion. Geol Soc Am Bull 8: 87–112

McGowen JH, Garner LE (1970) Physiographic features and stratification types of coarse-grained point bars; modern and ancient examples. Sedimentology 14: 77–112

McKee BA, Nittrouer CA, Demaster DJ (1983) Concepts of sediment deposition and accumulation applied to the continental shelf near the mouth of the Yangtze River. Geology 11: 631–633

McKee ED (1938) Original structures in Colorado River flood deposits of Grand Canyon. J Sediment Petrol 8: 77–83

McKee ED (1939) Some types of bedding in the Colorado River delta. J Geol 47: 64–81

McKee ED (1957) Flume experiments on the production of stratification and cross-stratification. J Sediment Petrol 27: 129–134

McKee ED, Weir GW (1953) Terminology for stratification and cross-stratification in sedimentary rocks. Geol Soc Am Bull 64: 381–389

McKee ED, Crosby EJ, Berryhill HL Jr (1967) Flood deposits, Bijou Creek, Colorado, June 1965. J Sediment Petrol 37: 829–851

McKenzie DP (1978) Some remarks on the development of sedimentary basins. Earth Planet Sci Lett 40: 25–32

McLean JR (1977) The Cadomin Formation: stratigraphy, sedimentology, and tectonic implications. Bull Can Petrol Geol 25: 792–827

McLean JR, Jerzykiewicz T (1978) Cyclicity, tectonics and coal: some aspects of fluvial sedimentology in the Brazeau-Paskapoo Formations, Coal Valley area, Alberta, Canada. In: Miall AD (ed) Fluvial sedimentology. Can Soc Petrol Geol Mem 5: 441–468

McLean JR, Wall JH (1981) The Early Cretaceous Moosebar Sea in Alberta. Bull Can Petrol Geol 29: 334–377

McMillan NJ (1973) Shelves of Labrador Sea and Baffin Bay, Canada. In: McCrossan RG (ed) The future petroleum provinces of Canada – their geology and potential. Can Soc Petrol Geol Mem 1: 473–517

McPherson JG, Blair TC (1993) Alluvial fans: fluvial or not? Keynote address, 5th international conference on fluvial sedimentology, Brisbane, Australia, July 1993, Keynote addresses and abstracts, pp K33-K41

McPherson JG, Shanmugam G, Moiola RJ (1987) Fan-deltas and braid deltas: varieties of coarse-grained deltas. Geol Soc Am Bull 99: 331–340

Meadows NS, Beach A (1993) Structural and climatic controls on facies distribution in a mixed fluvial and aeolian reservoir: the Triassic Sherwood Sandstone in the Irish Sea. In: North CP, Prosser DJ (eds) Characterization of fluvial and aeolian reservoirs. Geol Soc Lond Spec Publ 73: 247–264

Mebberson AJ (1989) The future for exploration in the Gippsland Basin. Aust Petrol Explor Assoc J 29: 431–439

Medlicott HB, Blanford WT (1879) Manual of the geology of India, part II

Mello MR, Maxwell JR (1990) Organic geochemical and biological marker characterization of source rocks and oils derived from lacustrine environments in the Brazilian continental margin. In: Katz BJ (ed) Lacustrine basin exploration. Am Assoc Petrol Geol Mem 50: 77–97

Melton FA (1936) An empirical classification of floodplain streams. Geogr Rev 26: 593–609
Melvin J (1987) Fluvio-paludal deposits in the Lower Kekiktuk Formation (Mississippian), Endicott Field, northeast Alaska. In: Ethridge FG, Flores RM, Harvey MD (eds) Recent developments in fluvial sedimentology. Soc Econ Paleontol Mineral Spec Publ 39: 343–352
Melvin J (1993) Evolving fluvial style in the Kekiktuk Formation (Mississippian), Endicott field area, Alaska: base level response to contemporaneous tectonism. AAPG Bull 77: 1723–1744
Mertz KA Jr, Hubert JF (1990) Cycles of sand-flat sandstone and playa-lacustrine mudstone in the Triassic-Jurassic Blomidon redbeds, Fundy rift basin, Nova Scotia; implications for tectonics and climatic controls. Can J Earth Sci 27: 442–451
Miall AD (1970a) Devonian alluvial fans, Prince of Wales Island, Arctic Canada. J Sediment Petrol 40: 556–571
Miall AD (1970b) Continental-marine transition in the Devonian of Prince of Wales Island, Northwest Territories. Can J Earth Sci 7: 125–144
Miall AD (1973) Markov chain analysis applied to an ancient alluvial plain succession. Sedimentology 20: 347–364
Miall AD (1974) Paleocurrent analysis of alluvial sediments: a discussion of directional variance and vector magnitude. J Sediment Petrol 44: 1174–1185
Miall AD (1976) Paleocurrent and paleohydrologic analysis of some vertical profiles through a Cretaceous braided stream deposit, Banks Island, Arctic Canada. Sedimentology 23: 459–484
Miall AD (1977) A review of the braided river depositional environment. Earth Sci Rev 13: 1–62
Miall AD (ed) (1978a) Fluvial sedimentology. Can Soc Petrol Geol Mem 5
Miall AD (1978b) Fluvial sedimentology: an historical review. In: Miall AD (ed) Fluvial sedimentology. Can Soc Petrol Geol Mem 5: 1–47
Miall AD (1978c) Lithofacies types and vertical profile models in braided river deposits: a summary. In: Miall AD (ed) Fluvial sedimentology. Can Soc Petrol Geol Mem 5: 597–604
Miall AD (1978d) Tectonic setting and syndepositional deformation of molasse and other nonmarine-paralic sedimentary basins. Can J Earth Sci 15: 1613–1632
Miall AD (1979a) Tertiary fluvial sediments in the Lake Hazen intermontane basin, Ellesmere Island, Arctic Canada. Geological Survey of Canada paper 79-9
Miall AD (1979b) Mesozoic and tertiary geology of Banks Island, Arctic Canada: the history of an unstable craton margin. Geol Surv Can Mem 387
Miall AD (1980) Cyclicity and the facies model concept in geology. Bull Can Petrol Geol 28: 59–80
Miall AD (1981a) Analysis of fluvial depositional systems. Am Assoc Petrol Geol Educ Course Notes Ser 20
Miall AD (1981b) Alluvial sedimentary basins: tectonic setting and basin architecture. In: Miall AD (ed) Sedimentation and tectonics in alluvial basins. Geol Assoc Can Spec Paper 23: 1–33
Miall AD (1983a) Basin analysis of fluvial sediments. In: Collinson JD, Lewin JL (eds) Modern and ancient fluvial systems. Int Assoc Sedimentol Spec Publ 6: 279–286
Miall AD (1983b) Glaciofluvial transport and deposition. In: Eyles N (ed) Glacial geology: an introduction for engineers and earth scientists: Pergamon, New York, pp 168–183
Miall AD (1984a) Principles of sedimentary basin analysis. Springer, Berlin Heidelberg New York
Miall AD (1984b) Variations in fluvial style in the Lower Cenozoic synorogenic sediments of the Canadian Arctic Islands. Sediment Geol 38: 499–523
Miall AD (1985) Architectural-element analysis: a new method of facies analysis applied to fluvial deposits. Earth Sci Rev 22: 261–308
Miall AD (1986) Effects of Caledonian tectonism in Arctic Canada. Geology 14: 904–907
Miall AD (1987) Recent developments in the study of fluvial facies models. In: Ethridge FG, Flores RM (eds) Recent developments in fluvial sedimentology. Soc Econ Paleontol Mineral Spec Publ 39: 1–9
Miall AD (1988a) Reservoir heterogeneities in fluvial sandstones: lessons from outcrop studies. Am Assoc Petrol Geol Bull 72: 682–697
Miall AD (1988b) Facies architecture in clastic sedimentary basins. In: Kleinspehn K, Paola C (eds) New perspectives in basin analysis. Springer, Berlin Heidelberg New York, pp 67–81
Miall AD (1988c) Architectural elements and bounding surfaces in fluvial deposits: anatomy of the Kayenta Formation (Lower Jurassic), southwest Colorado. Sediment Geol 55: 233–262
Miall AD (1989) Architectural elements and bounding surfaces in channelized clastic deposits: notes on comparisons between fluvial and turbidite systems. In: Taira A, Masuda F (eds) Sedimentary facies in the active plate margin. Terra Scientific, Tokyo, pp 3–15
Miall AD (1990) Principles of sedimentary basin analysis, 2nd edn. Springer, Berlin Heidelberg New York
Miall AD (1991a) Hierarchies of architectural units in clastic rocks, and their relationship to sedimentation rate. In: Miall AD, Tyler N (eds) The three-dimensional facies architecture of terrigenous clastic sediments, and its implications for hydrocarbon discovery and recovery. Soc Econ Paleontol Mineral Conc Sedimentol Paleontol 3: 6–12
Miall AD (1991b) Sedimentology of a sequence boundary within the nonmarine Torrivio Member, Gallup Sandstone (Cretaceous), San Juan Basin, New Mexico. In: Miall AD, Tyler N (eds) The three-dimensional facies architecture of terrigenous clastic sediments, and its implications for hydrocarbon discovery and recovery. Soc Eco Paleontol Mineral Conc Sedimentol Paleontol 3: 224–232
Miall AD (1991c) Stratigraphic sequences and their chronostratigraphic correlation. J Sediment Petrol 61: 497–505
Miall AD (1991d) Late Cretaceous and Tertiary basin development and sedimentation, Arctic Islands. In: Trettin HP (ed) Geology of the Innuitian Orogen and Arctic Platform of Canada and Greenland, chap 15. Geol Surv Can Geol Can 3: 437–458
Miall AD (1992a) Alluvial deposits. In: Walker RG, James NP (eds) Facies models: response to sea level change. Geological Association of Canada, St. John's, Newfoundland, pp 119–142
Miall AD (1992b) The Exxon global cycle chart: an event for every occasion? Geology 20: 787–790

Miall AD (1993) The architecture of fluvial-deltaic sequences in the Upper Mesaverde Group (Upper Cretaceous), Book Cliffs, Utah. In: Best JL, Bristow CS (eds) Braided rivers. Geol Soc Lond Spec Publ 75: 305–332
Miall AD (1994) Reconstructing fluvial macroform architecture from two-dimensional outcrops: examples from the Castlegate Sandstone, Book Cliffs, Utah. J Sediment Res B64: 146–158
Miall AD (1995a) Description and interpretation of fluvial deposits: a critical perspective: discussion. Sedimentology 42: 379–384
Miall AD (1995b) Collision-related foreland basins. In: Busby-Spera C, Ingersoll RV (eds) Tectonics of sedimentary basins. Blackwell Scientific, Oxford: 393–424
Miall AD, Gibling MR (1978) The Siluro-Devonian clastic wedge of Somerset Island, Arctic Canada, and some regional paleogeographic implications. Sediment Geol 21: 85–127
Miall AD, Smith ND (1989) Rivers and their deposits. Society of Economic Paleontologists and Mineralogists, Tulsa, Oklahoma, slide set 4
Miall AD, Tyler N (eds) (1991) The three-dimensional facies architecture of terrigenous clastic sediments, and its implications for hydrocarbon discovery and recovery. Soc Econ Paleontol Mineral Conc Mod Ser 3
Miall AD, Kerr JW, Gibling MR (1978) The Somerset Island Formation: an Upper Silurian to Lower Devonian intertidal/supratidal succession, Boothia Uplift region, Arctic Canada. Can J Earth Sci 15: 181–189
Middleton GV (ed) (1965) Primary sedimentary structures and their hydrodynamic interpretation. Soc Econ Paleontol Mineral Spec Publ 12
Middleton GV (1973) Johannes Walther's law of the correlation of facies. Geol Soc Am Bull 84: 979–988
Middleton GV (1976) Hydraulic interpretation of sand size distributions. J Geol 84: 405–426
Middleton GV (1977) Introduction – progress in hydraulic interpretation of sedimentary structures. In: Middleton GV (ed) Sedimentary process: hydraulic interpretation of primary sedimentary structures. Soc Econ Paleontol Mineral Repr Ser 3: 1–15
Middleton GV (1978) Sedimentology - history. In: Fairbridge RW, Bourgeois J (eds) The encyclopedia of sedimentology. Dowden, Hutchinson and Ross, Stroudsberg, pp 707–712
Middleton GV, Hampton MA (1976) Subaqueous sediment transport and deposition by sediment gravity flows. In: Stanley DJ, Swift DJP (eds) Marine sediment transport and environmental management. Wiley, New York, pp 197–218
Middleton GV, Southard JB (1977) Mechanics of sediment movement. Society of Economic Paleontologists and Mineralogists short course 3, Tulsa, Oklahoma
Middleton LT, Blakey RC (1983) Processes and controls on the intertonguing of the Kayenta and Navajo Formations, northern Arizona: Eolian-fluvial interactions. In: Brookfield ME, Ahlbrandt TS (eds) Eolian sediments and processes. Elsevier, Amsterdam, pp 613–634
Mike K (1975) Utilization of the analysis of ancient river beds for the detection of Holocene crustal movements. Tectonophysics 29: 359–368
Miller DD, McPherson JG, Covington TE (1990) Fluviodeltaic reservoir, South Belridge field, San Joaquin Valley, California. In: Barwis JH, McPherson JG, Studlick RJ (eds) Sandstone petroleum reservoirs. Springer, Berlin Heidelberg New York, pp 109–130
Miller RS, Groth JL (1990) Depositional environment and reservoir properties of the Lower Tuscaloosa "B" Sandstone Baywood field, St. Helena Parish, Louisiana. Trans Gulf Coast Assoc Geol Soc 40: 601–605
Milne G (1935) Some suggested units for classification and mapping, particularly for East African soils. Soil Res Berl 4: 183–198
Minter WEL (1978) A sedimentological synthesis of placer gold, uranium and pyrite concentrations in Proterozoic Witwatersrand sediments. In: Miall AD (ed) Fluvial sedimentology. Can Soc Petrol Geol Mem 5: 801–829
Mitchell AHG, McKerrow WS (1975) Analogous evolution of the Burma Orogen and the Scottish Caledonides. Geol Soc Am Bull 86: 305–315
Mitchum RM Jr, Van Wagoner JC (1991) High-frequency sequences and their stacking patterns: sequence-stratigraphic evidence of high-frequency eustatic cycles. Sediment Geol 70: 131–160
Mitrovica JX, Beaumont C, Jarvis GT (1989) Tilting of continental interiors by the dynamical effects of subduction. Tectonics 8: 1079–1094
Mjøs R, Walderhaug O, Prestholm E (1993) Crevasse splay sandstone geometries in the Middle Jurassic Ravenscar Group of Yorkshire, UK. In: Marzo M, Puigdefábregas C (eds) Alluvial sedimentation. Int Assoc Sedimentol Spec Publ 17: 167–184
Molenaar CM, Rice DD (1988) Cretaceous rocks of the Western Interior Basin. In: Sloss LL (ed) Sedimentary cover – North American Craton. US Geol Soc Am Geol North Am D-2: 77–82
Mollard JD (1973) Airphoto interpretation of fluvial features. In: Fluvial processes and sedimentation. Proceedings of the Hydrology Symposium, Edmonton, National Research Council, Canada, pp 341–380
Molnar P, England P (1990) Late Cenozoic uplift of mountain ranges and global climatic change: chicken or egg? Nature 346: 29–34
Molnar P, Tapponnier P (1975) Cenozoic tectonics of Asia: effects of a continental collision. Science 189: 419–426
Moody-Stuart M (1966) High and low sinuosity stream deposits, with examples from the Devonian of Spitzbergen. J Sediment Petrol 36: 1102–1117
Moore PD (1987) Ecological and hydrological aspects of peat formation. In: Scott AC (ed) Coal and coal-bearing strata: recent advances. Geol Soc Lond Spec Publ 32: 7–15
Moore PS, Hobday DK, Mai H, Sun ZC (1986) Comparison of selected non-marine petroleum-bearing basins in Australia and China. Aust Petrol Explor Assoc J 26: 285–309
Moore RC (1964) Paleoecological aspects of Kansas Pennsylvanian and Permian cyclothems. In: Merriam DF (ed) Symposium on cyclic sedimentation. Kansas Geol Surv Bull 169: 287–380
Moorman BJ, Judge AS, Smith DG (1991) Examining fluvial sediments using ground penetrating radar in British Columbia. Geol Surv Can Paper 91 (1A): 31–36
Morgan JP, McIntire WG (1959) Quaternary geology of the Bengal Basin, East Pakistan and India. Geol Soc Am Bull 70: 319–342
Morgan KH (1993) Development, sedimentation and economic potential of palaeoriver systems of the Yilgarn

Craton of Western Australia. Sediment Geol 85: 637–656

Morgridge DL, Smith WB (1972) Geology and discovery of Prudhoe Bay field, eastern Arctic Slope, Alaska. In: King RE (ed) Stratigraphic oil and gas fields. Am Assoc Petrol Geol Mem 16: 489–501

Morison SR, Hein FJ (1987) Sedimentology of the Whie Channel gravels, Klondike area, Yukon Territory: fluvial deposits of a confined valley. In: Ethridge FG, Flores RM, Harvey MD (eds) Recent developments in fluvial sedimentology. Soc Econ Paleontol Mineral Spec Publ 39: 205–216

Morley CK (1986) A classification of thrust fronts. Am Assoc Petrol Geol Bull 70: 12–35

Morton RA, Price WA (1987) Late Quaternary sea-level fluctuations and sedimentary phases of the Texas coastal plain and shelf. In: Nummedal D, Pilkey OH, Howard JD (eds) Sea-level fluctuation and coastal evolution. Soc Econ Paleontol Mineral Spec Publ 41: 181–198

Mosley MP (1976) An experimental study of channel confluences. J Geol 84: 535–562

Mosley MP, Schumm SA (1976) Stream junctions – a probable location for bedrock placers. Econ Geol 72: 691–697

Mossop GD, Flach PD (1983) Deep channel sedimentation in the Lower Cretaceous McMurray Formation, Athabasca Oil Sands, Alberta. Sedimentology 30: 493–509

Mossop GD, Shetsen I (1994) Geological atlas of the Western Canada Sedimentary Basin. Canadian Society of Petroleum Geologists, Calgary, Alberta

Mpodozis C, Ramos V (1990) The Andes of Chile and Argentina. In: Ericksen GE, Pinochet MTC, Reinemund JA (eds) Geology of the Andes and its relation to hydrocarbon and mineral resources. Circum Pac Council Energy Miner Res Earth Sci Ser 11: 59–90

Mukerji AB (1976) Terminal fans of inland streams in Sutlej-Yamuna Plain, India. Z Geomorphol 20: 190–204

Mulder TJ, Burbank DW (1993) The impact of incipient uplift on patterns of fluvial deposition: an example from the Salt Range, northwest Himalayan foreland, Pakistan. In: Marzo M, Puigdefábregas C (eds) Alluvial sedimentation. Int Assoc Sedimentol Spec Publ 17: 521–539

Muñoz A, Ramos A, Sánchez-Moya Y, Sopeña A (1992) Evolving fluvial architecture during a marine transgression: Upper Buntsandstein, Triassic, central Spain. Sediment Geol 75: 257–281

Muñoz A, Ramos A, Sánchez-Moya Y, Sopeña A (1992) Evolving fluvial architecture during a marine transgression: Upper Buntsandstein, Triassic, central Spain. Sediment Geol 75: 257–281

Murty KN (1983) Geology and hydrocarbon prospects of Assam Shelf – recent advances and present status: Petroleum Asia Journal, Nov 1983, pp 1–14

Muto T (1987) Coastal fan processes controlled by sea level changes: a Quaternary example from the Tenryugawa systesm, Pacific coast of central Japan. J Geol 95: 716–724

Muto T (1988) Stratigraphical patterns of coastal-fan sedimentation adjacent to high-gradient submarine slopes affected by sea-level changes. In: Nemec W, Steel RJ (eds) Fan deltas: sedimentology and tectonic settings. Blackie, Glasgow, pp 84–90

Mutti E, Normark WR (1987) Comparing examples of modern and ancient turbidite systems: problems and concepts. In: Leggett JK, Zuffa GG (eds) Marine clastic sedimentology: concepts and case studies. Graham and Trotman, London, pp 1–38

Muwais W, Smith DG (1990) Types of channel-fills interpreted from dipmeter logs in the McMurray Formation, northeast Alberta. Bull Can Petrol Geol 38: 53–63

Nadler CT, Schumm SA (1981) Metamorphosis of South Platte and Arkansas rivers, eastern Colorado. Phys Geogr 2: 95–115

Nadon G (1991) Architectural element analysis of a foreland basin clastic wedge. PhD Thesis, University of Toronto

Nadon G (1994) The genesis and recognition of anastomosed fluvial deposits: data from the St Mary River Formation, southwestern Alberta. J Sediment Res B64: 451–463

Nami M, Leeder MR (1978) Changing channel morphology and magnitude in the Scalby Formation (M. Jurassic) of Yorkshire, England. In: Miall AD (ed) Fluvial sedimentology. Can Soc Petrol Geol Mem 5: 431–440

Nanson GC (1980) Point bar and floodplain formation of the meandering Beatton River, northeastern British Columbia, Canada. Sedimentology 27: 3–30

Nanson GC, Croke JC (1992) A genetic classification of floodplains. Geomorphology 4: 459–486

Nanson GC, Page K (1983) Lateral accretion of fine-grained concave benches on meandering streams. In: Collinson JD, Lewin J (eds) Modern and ancient fluvial systems. Int Assoc Sedimentol Spec Publ 6: 133–144

Nanson GC, Rust BR (1985) Channel patterns and mud deposits in an arid zone river: Cooper Creek, central Australia, 3rd international sedimentology conference, Fort Collins, abstracts, p 31

Nanson GC, Rust BR, Taylor G (1986) Coexistent mud braids in an arid-zone river: Cooper Creek, central Australia. Geology 14: 175–178

Nanz RH Jr (1954) Genesis of Oligocene sandstone reservoir, Seeligson field, Jim Wells and Kleberg Counties, Texas. Am Assoc Petrol Geol Bull 38: 96–117

Nascimento OS, Bornemann E, Jobim LDC, Carvalho MD, Pimentel AM, Bonet EJ, Rodriques EB, Sandoval JRL, Lassandro V, Rodrigues TC, Hocott CR (1982) Aracas field-reservoir heterogeneities and secondary recovery performance (abstract). Am Assoc Petrol Geol Bull 66: 612

Neidell NS, Beard JH (1985) Seismic visibility of stratigraphic objectives. Society of Petroleum Engineers paper 14175

Nemec W (1988) Coal correlations and intrabasinal subsidence: a new analytical perspective. In: Kleinspehn K, Paola C (eds) New perspectives in basin analysis. Springer, Berlin Heidelberg New York, pp 161–188

Nemec W, Muszynski A (1982) Volcaniclastic alluvial aprons in the Tertiary of Sofia district (Bulgaria). Ann Soc Geol Polon 52: 239–303

Nemec W, Postma G (1993) Quaternary alluvial fans in southwestern Crete: sedimentation processes and geomorphic evolution. In: Marzo M, Puigdefábregas C (eds) Alluvial sedimentation. Int Assoc Sedimentol Spec Publ 17: 235–276

Nemec W, Steel RJ (1984) Alluvial and coastal conglomerates: their significant features and some comments on gravelly mass-flow deposits. In: Koster EH, Steel RJ (eds) Sedimentology of gravels and conglomerates. Can Soc Petrol Geol Mem 10: 1–31

Nemec W, Steel RJ (eds) (1988) Fan deltas: sedimentology and tectonic settings. Blackie, Glasgow

Nichols GJ (1987) Structural controls on fluvial distributary systems – the Luna system, northern Spain. In: Ethridge FG, Flores RM, Harvey MD (eds) Recent developments in fluvial sedimentology. Soc Econ Paleontol Mineral Spec Publ 39: 269–277

Nijman W, Puigdefábregas C (1978) Coarse-grained point bar structure in a molasse-type fluvial system, Eocene Castisent sandstone Formation, south Pyrenean Basin. In: Miall AD (ed) Fluvial sedimentology. Can Soc Petrol Geol Mem 5: 487–510

Nilsen TH (1968) The relationship of sedimentation to tectonics in the Solund Devonian district of southwestern Norway. Universitetsforlaget, Oslo (Norges geologiske undersökelse, no 259)

Nilsen TH (ed) (1984) Fluvial sedimentation and related tectonic framework, western North America. Sediment Geol 38 (special issue)

Nilsen TH (1985) Modern and ancient alluvial fan deposits. Van Nostrand Reinhold, New York (Hutchinson Ross Benchmark, vol 87)

Nilsen TH (1987) Paleogene tectonics and sedimentation of coastal California. In: Ingersoll RV, Ernst WG (eds) Cenozoic basin development of coastal California, Rubey, vol VI. Prentice-Hall, Englewood Cliffs, pp 81–123

Nilsen T, McLaughlin RJ (1985) Comparison of tectonic framework and depositional patterns of the Hornelen strike-slip basin of Norway ad the Ridge and Little Sulphur Creek strike-slip basins of California. In: Biddle KT, Christie-Blick N (eds) Strike-slip deformation, basin formation and sedimentation. Soc Econ Paleontol Mineral Spec Publ 37: 79–103

North American Commission on Stratigraphic Nomenclature (1983) North American Stratigraphic Code. Am Assoc Petrol Geol Bull 67: 841–875

Nummedal D, Swift DJP (1987) Transgressive stratigraphy at sequence-bounding unconformities: some principles derived from Holocene and Cretaceous examples. In: Nummedal D, Pilkey OH, Howard JD (eds) Sea-level fluctuation and coastal evolution. Soc Econ Paleontol Mineral Spec Publ 41: 241–260

Nummedal D, Pilkey OH, Howard JD (eds) (1987) Sea-level fluctuation and coastal evolution. Soc Econ Paleontol Mineral Spec Publ 41

Nummedal D, Riley GW, Temple PL (1993) High-resolution sequence architecture: a chronostratigraphic model based on equilibrium profile studies. In: Posamentier HW, Summerhayes CP, Haq BU, Allen GP (eds) Sequence stratigraphy and facies associations. Int Assoc Sedimentol Spec Publ 18: 55–68

Nybråten G, Skolem E, Østby K (1990) Reservoir simulation of the Snorre field. In: Buller AT, Berg E, Hjelmeland O, Kleppe J, Torsaeter O, Aasen JO (eds) North Sea oil and gas reservoirs II. Graham and Trotman, London, pp 103–114

Nystuen JP, Knarud R, Jorde K, Stanley KO (1989) Correlation of Triassic to Jurassic sequences, Snorre Field and adjacent areas, northern North Sea. In: Collinson JD (ed) Correlation in hydrocarbon exploration. Graham and Trotman, London, pp 273–289

Okolo SA (1983) Fluvial distributary channels in the Fletcher Bank Grit (Namurian R2b), at Ramsbottom, Lancashire, England. In: Collinson JD, Lewin J (eds) Modern and ancient fluvial systems. Int Assoc Sedimentol Spec Publ 6: 421–433

Olsen H (1988) The architecture of a sandy braided-meandering river system: an example from the Lower Triassic Solling Formation (M. Buntsandstein) in W-Germany. Geol Rundsch 77: 797–814

Olsen H (1989) Sandstone-body structures and ephemeral stream processes in the Dinosaur Canyon Member, Moenave Formation (Lower Jurassic), Utah, USA. Sediment Geol 61: 207–221

Olsen H (1990) Astronomical forcing of meandering river behaviour: Milankovitch cycles in Devonian of East Greenland. Palaeogeogr Palaeoclimatol Palaeoecol 79: 99–115

Olsen H (1994) Orbital forcing on continental depositional systems – lacustrine and fluvial cyclicity in the Devonian of East Greenland. In: De Boer PL, Smith DG (eds) Orbital forcing and cyclic sequences. Int Assoc Sedimentol Spec Publ 19: 429–438

Olsen H, Larsen PH (1993) Structural and climatic controls on fluvial depositional systems: Devonian, north-east Greenland. In: Marzo M, Puigdefábregas C (eds) Alluvial sedimentation. Int Assoc Sedimentol Spec Publ 17: 401–423

Olsen PE (1990) Tectonic, climatic, and biotic modulation of lacustrine ecosystems – examples from Newark Supergroup of eastern North America. In: Katz BJ (ed) Lacustrine basin exploration: case studies and modern analogs. Am Assoc Petrol Geol Mem 50: 209–224

Olsen T (1993) Large fluvial systems: the Atane Formation, a fluvio-deltaic example from the Upper Cretaceous of central West Greenland. Sediment Geol 85: 457–473

Olsen T, Steel RJ, Høgseth K, Skar T, Røe S-L (1995) Sequential architecture in a fluvial succession: sequence stratigraphy in the Upper Cretaceous Mesaverde Group, Price, Utah. J Sediment Res B65: 265–280

Olson JS, Potter PE (1954) Variance components of cross-bedding directions in some basal Pennsylvanian sandstones of the eastern Interior Basin: statistical methods. J Geol 62: 26–49

Ono Y (1990) Alluvial fans in Japan and South Korea. In: Rachocki AH, Church M (eds) Alluvial fans: a field approach. Wiley, Chichester, pp 91–107

Oomkens E (1970) Depositional sequences and sand distribution in the postglacial Rhône delta complex. In: Morgan JP (ed) Deltaic sedimentation: modern and ancient. Soc Econ Paleontol Mineral Spec Publ 15: 198–212

Oomkens E, Terwindt JHJ (1960) Inshore estuarine sediments in the Haringvliet (The Netherlands). Geol Mijnbouw 39: 701–710

Ordóñez S, García del Cura MA (1983) Recent and Tertiary fluvial carbonates in central Spain. In: Collinson JD, Lewin J (eds) Modern and ancient fluvial systems. Int Assoc Sedimentol Spec Publ 6: 485–497

Ore HT (1964) Some criteria for recognition of braided stream deposits. Wyoming Contrib Geol 3: 1–14

Ori GG (1979) Barre di meandre nelle alluvioni ghiaiose del fiume Reno (Bologna). Bull Soc Geol Ital 98: 35–54

Ori GG (1982) Braided to meandering channel patterns in humid-region alluvial fan deposits, River Reno, Po Plain (northern Italy). Sediment Geol 31: 231–248

Ori GG (1993) Continental depositional systems of the Quaternary of the Po Plain (northern Italy). Sediment Geol 83: 1–14

Ori GG, Friend PF (1984) Sedimentary basins formed and carried piggyback on active thrust sheets. Geology 12: 475–478

Orton GJ (1988) A spectrum of Middle Ordovician fan deltas and braidplain deltas, North Wales: a consequence of varying fluvial clastic input. In: Nemec W, Steel RJ (eds) Fan deltas: sedimentology and tectonic setting: Blackie, London, pp 23–49

Ouchi S (1985) Response of alluvial rivers to slow active tectonic movement. Geol Soc Am Bull 96: 504–515

Ouyang J, Ma Z, He D, Tian H (1988) Sedimentary facies recognition in China with well-logging data. In: Wagner HC, Wagner LC, Wang FFH, Wong FL (eds) Petroleum resources of China and related subjects. Circum-Pacific Council for Energy and Mineral Resources. Earth Sci Ser 10: 467–474

Owens JS (1908) Experiments on the transporting power of sea currents. Geogr J 31: 415–420

Paola C (1988) Subsidence and gravel transport in alluvial basins. In: Kleinspehn K, Paola C (eds) New perspectives in basin analysis. Springer, Berlin Heidelberg New York, pp 231–243

Paola C (1990) A simple basin-filling model for coarse-grained alluvial systems. In: Cross TA (ed) Quantitative dynamic stratigraphy. Prentice-Hall, Englewood Cliffs, pp 363–374

Paola C, Heller PL, Angevine CL (1992) The large-scale dynamics of grain-size variation in alluvial basins, 1: theory. Basin Res 4: 73–90

Padgett V, Ehrlich R (1976) Paleohydrologic analysis of a Late Carboniferous fluvial system, southern Morocco. Geol Soc Am Bull 87: 1101–1104

Parkash B, Awasthi AK, Gohain K (1983) Lithofacies of the Markanda terminal fan, Kurukshetra district, Haryana, India. In: Collinson JD, Lewin J (eds) Modern and ancient fluvial systems. Int Assoc Sedimentol Spec Publ 6: 337–344

Parrish JT (1993) Climate of the supercontinent Pangea. J Geol 101: 215–233

Parrish JT, Barron EJ (1986) Paleoclimates and economic geology. Soc Econ Paleontol Mineral Short Course 18

Parrish JT, Peterson F (1988) Wind directions predicted from global circulation models and wind directions determined from eolian sandstones of the western United States – a comparison. Sediment Geol 56: 261–282

Parsley AJ (1984) North Sea hydrocarbon plays. In: Glennie KW (ed) Introduction to the petroleum geology of the North Sea. Blackwell Scientific, Oxford, pp 205–230

Passega R (1957) Texture as a characteristic of clastic deposition. Am Assoc Petrol Geol Bull 41: 1952–1984

Payton CE (ed) (1977) Seismic stratigraphy – applications to hydrocarbon exploration. Am Assoc Petrol Geol Mem 26

Pazzaglia FJ (1993) Stratigraphy, petrography, and correlation of late Cenozoic middle Atlantic Coastal Plain deposits: implications for late-stage passive-margin geologic evolution. Geol Soc Am Bull 105: 1617–1634

Peale AC (1879) Report on the geology of the Green River District. In: Hayden FV (ed) US geological and geographical survey of the territories, 9th annual report. Washington DC

Pelletier BR (1958) Pocono paleocurrents in Pennsylvania and Maryland. Geol Soc Am Bull 69: 1033–1064

Peper T, Beekman F, Cloetingh S (1992) Consequences of thrusting and intraplate stress fluctuations for vertical motions in foreland basins and peripheral areas. Geophys J Int 111: 104–126

Perlmutter MA, Matthews MD (1989) Global cyclostratigraphy – a model. In: Cross TA (ed) Quantitative dynamic stratigraphy. Prentice Hall, Englewood Cliffs, New Jersey, pp 33–260

Peterson F (1984) Fluvial sedimentation on a quivering craton: influence of slight crustal movements on fluvial processes, Upper Jurassic Morrison Formation, western Colorado Plateau. Sediment Geol 38: 21–50

Petterson O, Storli A, Ljosland E, Massie I (1990) The Gullfaks field: Geology and reservoir development. In: Buller AT, Berg E, Hjelmeland O, Kleppe J, Torsaeter O, Aasen JO (eds) North Sea oil and gas reservoirs II. Graham and Trotman, London, pp 67–90

Pettijohn FJ (1949) Sedimentary rocks. Harper and Row, New York

Pettijohn FJ (1957) Sedimentary rocks, 2nd edn. Harper, New York

Pettijohn FJ (1962) Paleocurrents and paleogeography. Am Assoc Petrol Geol Bull 46: 1468–1493

Pettijohn FJ, Potter PE (1964) Atlas and glossary of primary sedimentary structures. Springer, Berlin Heidelberg New York

Pettijohn FJ, Potter PE, Siever R (1972) Sand and sandstone. Springer, Berlin Heidelberg New York

Pfiffner OA (1986) Evolution of the north Alpine foreland basin in the central Alps. In: Allen PA, Homewood P (eds) Foreland basins. Int Assoc Sedimentol Spec Publ 8: 219–228

Pienkowski G (1991) Eustatically-controlled sedimentation in the Hettangian-Sinemurian (Early Jurassic) of Poland and Sweden. Sedimentology 38: 503–518

Pierson TC (1980) Erosion and deposition by debris flows at Mt. Thomas, New Zealand. Earth Surface Proc 5: 1952–1984

Pirson SJ (1970) Geologic well log analysis. Gulf, Houston

Pirson SJ (1977) Geologic well log analysis, 2nd edn. Gulf, Houston

Pitman JK, Fouch TD, Goldhaber MB (1982) Evolution of some Tertiary unconventional reservoir rocks, Uinta Basin, Utah. Am Assoc Petrol Geol Bull 66: 1581–1596

Pitman WC III (1978) Relationship between eustacy and stratigraphic sequences of passive margins. Geol Soc Am Bull 89: 1389–1403

Pitman WC III (1986) Effects of sea level change on basin stratigraphy. Am Assoc Petrol Geol Bull 70: 1762

Pitman WC III, Golovchenko X (1988) Sea-level changes and their effect on the stratigraphy of Atlantic-type margins. In: Sheridan RE, Grow JA (eds) The geology of North America, vol I-2, The Atlantic continental margin, United States, Geological Society of America, Boulder, Colorado, pp 429–436

Platt NH, Keller B (1992) Distal alluvial deposits in a foreland basin setting – the Lower Freshwater Molasse (Lower Miocene), Switzerland: sedimentology, architecture and palaeosols. Sedimentology 39: 545–565

Platt NH, Wright VP (1992) Palustrine carbonates and the Florida Everglades: towards an exposure index for the fresh-water environment. J Sediment Petrol 62: 1058–1071
Playfair J (1802) Illustrations of the Huttonian theory of the Earth. Cadell and Davies, London (Reprinted by Dover, New York, 1964)
Plint AG (1983) Sandy fluvial point-bar sediments from the middle Eocene of Dorset, England. In: Collinson JD, Lewin J (eds) Modern and ancient fluvial systems. Int Assoc Sedimentol Spec Publ 6: 355–368
Plint AG (1988) Sharp-based shoreface sequences and "offshore bars" in the Cardium Formation of Alberta: their relationship to relative changes in sea level. In: Wilgus CK, Hastings BS, Kendall CGSC, Posamentier HW, Ross CA, Van Wagoner JC (eds) Sea-level research: an integrated approach. Soc Econ Paleontol Mineral Spec Publ 42: 357–370
Plint AG (1990) An allostratigraphic correlation of the Muskiki and Marshybanks formations (Coniacian-Santonian) in the Foothills and subsurface of the Alberta basin. Bull Can Petrol Geol 38: 288–306
Plint AG, Eyles N, Eyles CH, Walker RG (1992) Control of sea level change. In: Walker RG, James NP (eds) Facies models: response to sea level change. Geological Association of Canada, St. John's, Newfoundland, pp 15–25
Plint AG, Hart BS, Donaldson WS (1993) Lithospheric flexure as a control on stratal geometry and facies distribution in Upper Cretaceous rocks of the Alberta foreland basin. Basin Res 5: 69–77
Plint AG, Walker RG, Bergman KM (1986) Cardium Formation 6. Stratigraphic framework of the Cardium in subsurface. Bull Can Petrol Geol 34: 213–225
Poag CW, Sevon WD (1989) A record of Appalachian denudation in postrift Mesozoic and Cenozoic sedimentary deposits of the US Middle Atlantic continental margin. Geomorphology 2: 119–157
Posamentier HW, Allen GP (1993) Siliciclastic sequence stratigraphic patterns in foreland ramp-type basins. Geology 21: 455–458
Posamentier HW, Allen GP (in press) Siliciclastic sequence stratigraphy
Posamentier HW, James DP (1993) An overview of sequence-stratigraphic concepts: uses and abuses. In: Posamentier HW, Summerhayes CP, Haq BU, Allen GP (eds) Sequence stratigraphy and facies associations. Int Assoc Sedimentol Spec Publ 18: 3–18
Posamentier HW, Vail PR (1988) Eustatic controls on clastic deposition II – sequence and systems tracts models. In: Wilgus CK, Hastings BS, Kendall CGSC, Posamentier HW, Ross CA, Van Wagoner JC (eds) Sea-level research: an integrated approach. Soc Econ Paleontol Mineral Spec Publ 42: 125–154
Posamentier HW, Weimer P (1993) Siliciclastic sequence stratigraphy and petroleum geology – where to from here? Am Assoc Petrol Geol Bull 77: 731–742
Posamentier HW, Jervey MT, Vail PR (1988) Eustatic controls on clastic deposition I – conceptual framework. In: Wilgus CK, Hastings BS, Kendall CGSC, Posamentier HW, Ross CA, Van Wagoner JC (eds) Sea-level research: an integrated approach. Soc Econ Paleontol Mineral Spec Publ 42: 109–124
Posamentier HW, Allen GP, James DP, Tesson M (1992) Forced regressions in a sequence stratigraphic framework: concepts, examples, and exploration significance. Am Assoc Petrol Geol Bull 76: 1687–1709
Posaner EM, Goldthorpe WH (1986) The development and early performance of the North Rankin field. Aust Petrol Explor Assoc J 26: 420–427
Potter PE (1955) The petrology and origin of the Lafayette Gravel, part I: mineralogy and petrology. J Geol 63: 1–38
Potter PE (1959) Facies models conference. Science 129: 1292–1294
Potter PE (1978) Significance and origin of big rivers. J Geol 86: 13–33
Potter PE, Pettijohn FJ (1963) Paleocurrents and basin analysis. Academic, New York (2nd edn: 1977, Springer, Berlin Heidelberg New York)
Powell JW (1875) Exploration of the Colorado River of the West. Washington (see also US 43rd congress, 1st session, H miscellaneous documents 265, 1874)
Pratt BR, Miall AD (1993) Anatomy of a bioclastic grainstone megashoal (Middle Silurian, southern Ontario) revealed by ground-penetrating radar. Geology 21: 223–226
Prosser S (1993) Rift-related depositional systems and their seismic expression. In: Williams GD, Dobb A (eds) Tectonics and seismic sequence stratigraphy. Geol Soc Lond Spec Publ 71: 35–66
Pryor WA (1960) Cretaceous sedimentation in upper Mississippi embayment. Am Assoc Petrol Geol Bull 44: 1473–1504
Puigdefábregas C (1973) Miocene point-bar deposits in the Ebro Basin, northern Spain. Sedimentology 20: 133–144
Puigdefábregas C, Van Vliet A (1978) Meandering stream deposits from the Tertiary of the southern Pyrenees. In: Miall AD (ed) Fluvial sedimentology. Can Soc Petrol Geol Mem 5: 469–485
Puigdefábregas C, Munoz JA, Marzo M (1986) Thrust belt development in the eastern Pyrenees and related depositional sequences in the southern foreland basin. In: Allen PA, Homewood P (eds) Foreland basins. Int Assoc Sedimentol Spec Publ 8: 229–246
Putnam PE (1982a) Fluvial channel sandstones within upper Mannville (Albian) of Lloydminster area, Canada – geometry, petrography, and paleogeographic implications. Am Assoc Petrol Geol Bull 66: 436–459
Putnam PE (1982b) Aspects of the petroleum geology of the Lloydminster heavy oil fields, Alberta and Saskatchewan. Bull Can Petrol Geol 30: 81–111
Putnam PE (1983) Fluvial deposits and hydrocarbon accumulations: examples from the Lloydminster area, Canada. In: Collinson JD, Lewin J (eds) Modern and ancient fluvial systems. Int Assoc Sedimentol Spec Publ 6: 517–532
Putnam PE (1993) A multidisciplinary analysis of Belly River-Brazeau (Campanian) fluvial channel reservoirs in west-central Alberta, Canada. Bull Can Petrol Geol 41: 186–217
Putnam PE, Oliver TA (1980) Stratigraphic traps in channel sandstones in the Upper Mannville (Albian) of east-central Alberta. Bull Can Petrol Geol 28: 489–508
Qian N (1990) Fluvial processes in the lower Yellow River after levee breaching at Tongwaxiang in 1855. Int J Sediment Res 5: 1–14

Rachocki AH, Church M (eds) (1990) Alluvial fans: a field approach. Wiley, New York

Rahmani RA, Flores RM (eds) (1984) Sedimentology of coal and coal-bearing sequences. Int Assoc Sedimentol Spec Publ 7

Rahmani RA, Lerbekmo JF (1975) Heavy mineral analysis of Upper Cretaceous and Paleocene sandstones in Alberta and adjacent areas of Saskatchewan. In: Caldwell WGE (ed) The Cretaceous system in the Western Interior of North America. Geol Assoc Can Spec Paper 13: 607–632

Rahmanian VD, Moore PS, Mudge WJ, Spring DE (1990) Sequence stratigraphy and the habitat of hydrocarbons, Gippsland Basin, Australia. In: Brooks J (ed) Classic petroleum provinces. Geol Soc Lond Spec Publ 50: 525–541

Ramos A, Sopeña A (1983) Gravel bars in low-sinuosity streams (Permian and Triassic, central Spain). In: Collinson JD, Lewin J (eds) Modern and ancient fluvial systems. Int Assoc Sedimentol Spec Publ 6: 301–312

Ramos A, Sopeña A, Perez-Arlucea M (1986) Evolution of Buntsandstein fluvial sedimentation in the northwest Iberian Ranges (Central Spain). J Sediment Petrol 56: 862–875

Ramsbottom WHC (1979) Rates of transgression and regression in the Carboniferous of NW Europe. J Geol Soc Lond 136: 147–153

Räsänen M, Neller R, Salo J, Jungner H (1992) Recent and ancient fluvial depositional systems in the Amazon foreland basin, Peru. Geol Mag 129: 293–306

Räsänen M, Salo JS, Kalliola RJ (1987) Fluvial perturbance in the Western Amazon Basin: regulation by long-term sub-Andean tectonics. Science 238: 1398–1401

Rascoe BR Jr, Adler FJ (1983) Permo-Carboniferous hydrocarbon accumulations, Mid-Continent, USA. Am Assoc Petrol Geol Bull 67: 979–1001

Ray RR (1982) Seismic stratigraphic interpretation of the Fort Union Formation, western Wind River Basin: example of subtle trap exploration in a nonmarine sequence. In: Halbouty MT (ed) The deliberate search for the subtle trap. Am Assoc Petrol Geol Mem 32: 169–180

Reade TM (1884) Ripple marks in drift in Shropshire and Cheshire. Q J Geol Soc Lond 40: 267–269

Reading HG (ed) (1986) Sedimentary environments and facies, 2nd edn. Blackwell, Oxford

Reiche P (1938) An analysis of cross-lamination: the Coconino Sandstone. J Geol 46.905–932

Reineck HE, Wunderlich R (1968) Classification and origin of flaser and lenticular bedding. Sedimentology 11: 99–104

Reinhardt J, Sigleo WR (eds) (1988) Paleosols and weathering through geologic time: principles and applications. Geol Soc Am Spec Paper 216

Reinfelds I, Nanson G (1993) Formation of braided river floodplains, Waimakariri River, New Zealand. Sedimentology 40: 1113–1127

Retallack GJ (1981) Fossil soils: indicators of ancient terrestrial environments. In: Niklas KJ (ed) Paleobotany, paleoecology and evolution, vol 1. Praeger, New York, pp 55–102

Retallack GJ (1984) Completeness of the rock and fossil record: some estimates using fossil soils. Paleobiology 10: 59–78

Retallack GJ (1986) Fossil soils as grounds for intepreting long-term controls on ancient rivers. J Sediment Petrol 56: 1–18

Retallack GJ (1988) Field recognition of paleosols. In: Reinhardt J, Sigleo WR (eds) Paleosols and weathering through geologic time: Principles and applications. Geol Soc Am Spec Paper 216: 1–20

Reynolds DW, Vincent JK (1972) Midland gas field, western Kentucky. In: King RE (ed) Stratigraphic oil and gas fields. Am Assoc Petrol Geol Mem 16: 585–598

Rhee CW, Ryand WH, Chough SK (1993) Contrasting development patterns of crevasse channel deposits in Cretaceous alluvial successions, Korea. Sediment Geol 85: 401–410

Riba O (1976) Syntectonic unconformities of the Alto Cardener, Spanish Pyrenees, a genetic interpretation. Sediment Geol 15: 213–233

Ricci Lucchi F (1986) The Oligocene to Recent foreland basins of the northern Apennines. In: Allen PA, Homewood P (eds) Foreland basins. Int Assoc Sedimentol Spec Publ 8: 105–139

Richards K, Chandra S, Friend PF (1993) Avulsive channel systems: characteristics and examples. In: Best JL, Bristow CS (eds) Braided rivers. Geol Soc Lond Spec Publ 75: 195–203

Rider MH (1990) Gamma-ray log shape used as a facies indicator: critical analysis of an oversimplified methodology. In: Hurst A, Lovell MA, Morton AC (eds) Geological applications of wireline logs. Geol Soc Lond Spec Publ 48: 27–37

Ridgway KD, DeCelles PG (1993) Stream-dominated alluvial fan and lacustrine depositional systems in Cenozoic strike-slip basins, Denali fault system, Yukon Territory, Canada. Sedimentology 40: 645–666

Rieu EV (1953) Homer: The Iliad. Methuen, London

Rivenaes JC (1992) Application of a dual-lithology, depth dependent diffusion equation in stratigraphic simulation. Basin Res 4: 133–146

Roberts DG (1988) Basin evolution and hydrocarbon exploration in the South China Sea. In: Wagner HC, Wagner LC, Wang FFH, Wong FL (eds) Petroleum resources of China and related subjects. Circum-Pacific Council for Energy and Mineral Resources. Earth Sci Ser 10: 157–177

Roberts HH, Adams RD, Cunningham RHW (1980) Evolution of sand-dominant subaerial phase, Atchafalaya Delta. Am Assoc Petrol Geol Bull 64: 264–279

Robertson JD (1991) Reservoir management using 3-D seismic data. In: Lake LW, Carroll HB Jr, Wesson TC (eds) Reservoir characterization II. Academic, San Diego, pp 340–354

Robinson JE (1981) Well spacing and the identification of subsurface drainage systems. Bull Can Petrol Geol 29: 250–258

Rodine JD, Johnson AM (1976) The ability of debris, heavily freighted with coarse clastic materials to flow on gentle slopes. Sedimentology 23: 213–234

Rodolfo KS (1975) The Irrawaddy delta: tertiary setting and modern offshore sedimentation. In: Broussard ML (ed) Deltas, models for exploration: Houston Geological Society, Houston, pp 339–356

Røe S-L, Hermansen M (1993) Processes and products of large, late Precambrian sandy rivers in northern

Norway. In: Marzo M, Puigdefábregas C (eds) Alluvial sedimentation. Int Assoc Sedimentol Spec Publ 17: 151–166

Rogers RR (1994) Nature and origin of through-going discontinuities in nonmarine foreland basin strata, Upper Cretaceous, Montana: implications for sequence analysis. Geology 22: 1119–1122

Rosen MR (1994) The importance of groundwater in playas: a review of playa classifications and the sedimentology and hydrology of playas. In: Rosen MR (ed) Paleoclimate and basin evolution of playa systems. Geol Soc Am Spec Paper 289: 1–18

Rosenthal L (1988) Wave dominated shorelines and incised channel trends: Lower Cretaceous Glauconite Formation, west-central Alberta. In: James DP, Leckie DA (eds) 1988, Sequences, stratigraphy, sedimentology: surface and subsurface. Can Soc Petrol Geol Mem 15: 207–230

Ross WC (1990) Modeling base-level dynamics as a control on basin-fill geometries and facies distribution: a conceptual framework. In: Cross TA (ed) Quantitative dynamic stratigraphy. Prentice-Hall, Englewood Cliffs, pp 387–399

Rouse H (1939) Experiments on the mechanics of sediment suspension. 5th international congress on applied mechanics, Cambridge, Massachusetts, pp 550–554

Royden LH (1985) The Vienna Basin: a thin-skinned pull-apart basin. In: Biddle KT, Christie-Blick N (eds) Strike-slip deformation, basin formation and sedimentation. Soc Econ Paleontol Mineral Spec Publ 37: 319–338

Royden L, Horvath F, Rumpler J (1983) Evolution of the Pannonian Basin system 1. Tectonics 2: 63–90

Rubey WW (1938) The force required to move particles on a stream bed. US Geol Surv Profess Paper 189-E: 121–141

Rubey WW, Bass NW (1925) The geology of Russell County, Kansas. Kansas State Geol Surv Bull 10: 1–86

Ruddiman WF, Kutzbach JE (1990) Late Cenozoic plateau uplift and climate change. R Soc Edinb Trans 81: 301–314

Ruddiman WF, Prell WL, Raymo ME (1989) History of Late Cenozoic uplift in southern Asia and the American west: rationale for general circulation modeling experiments. J Geophys Res 94: 18379–18391

Russell RJ (1954) Alluvial morphology of Anatolian rivers. Ann Assoc Am Geogr 44: 363–391

Rust BR (1972) Structure and process in a braided river. Sedimentology 18: 221–245

Rust BR (1975) Fabric and structure in glaciofluvial gravels. In: Jopling AV, McDonald BC (eds) Glaciofluvial and glaciolacustrine sedimentation. Soc Econ Paleontol Mineral Spec Publ 23.238–248

Rust BR (1978a) A classification of alluvial channel systems. In: Fluvial sedimentology. In: Miall AD (ed) Fluvial sedimentology. Can Soc Petrol Geol Mem 5: 187–198

Rust BR (1978b) Depositional models for braided alluvium. In: Miall AD (ed) Fluvial sedimentology. Can Soc Petrol Geol Mem 5: 605–625

Rust BR (1981) Sedimentation in an arid-zone anastomosing fluvial system: Cooper's Creek, central Australia. J Sediment Petrol 51: 745–755

Rust BR (1984) Proximal braidplain deposits in the Middle Devonian Malbaie Formation of eastern Gaspé, Quebec, Canada. Sedimentology 31: 675–695

Rust BR, Gostin VA (1981) Fossil transverse ribs in Holocene alluvial fan deposits, Depot Creek, South Australia. J Sediment Petrol 51: 441–444

Rust BR, Jones BG (1987) The Hawkesbury Sandstone south of Sydney, Australia: Triassic analogue for the deposit of a large braided river. J Sediment Petrol 57: 222–233

Rust BR, Koster EH (1984) Coarse alluvial deposits. In: Walker RG (ed) Facies models, 2nd edn. Geosci Can Reprint Ser 1: 53–69

Rust BR, Legun AS (1983) Modern anastomosing-fluvial deposits in arid central Australia, and a Carboniferous analogue in New Brunswick, Canada. In: Collinson JD, Lewin J (eds) Modern and ancient fluvial systems. Int Assoc Sedimentol Spec Publ 6: 385–392

Rust RB, Nanson GC (1989) Bedload transport of mud as pedogenic aggregates in modern and ancient rivers. Sedimentology 36: 291–306

Rust BR, Gibling MR, Legun AS (1984) Coal deposition in an anastomosed fluvial system: the Pennsylvanian Cumberland Group south of Joggins, Nova Scotia, Canada. In: Rahmani RA, Flores RM (eds) Sedimentology of coal and coal-bearing sequences. Int Assoc Sedimentol Spec Publ 7: 105–120

Ryer TA (1984) Transgressive-regressive cycles and the occurrence of coal in some Upper Cretaceous strata of Utah, USA. In: Rahmani RA, Flores RM (eds) Sedimentology of coal and coal-bearing sequences. Int Assoc Sedimentol Spec Publ 7: 217–227

Ryer TA, Langer AW (1980) Thickness change involved in the peat-to-coal transformation for a bituminous coal of Cretaceous age in central Utah. J Sediment Petrol 50: 987–992

Ryseth A (1989) Correlation of depositional patterns in the Ness Formation, Osberg area. In: Collinson JD (ed) Correlation in hydrocarbon exploration. Graham and Trotman, London, pp 313–326

Sadler PM (1981) Sedimentation rates and the completeness of stratigraphic sections. J Geol 89: 569–584

Sagoe K-MO, Visher GS (1977) Population breaks in grain-size distributions of sand – a theoretical model. J Sediment Petrol 47: 285–310

Sahu BK (1964) Depositional mechanisms from the size analysis of clastic sediments. J Sediment Petrol 34: 73–83

Salazar RB (1990) Hydrocarbon resources in the sub-Andean basins of Colombia. In: Eriksen GE, Pinochet MTC, Reinemund JA (eds) Geology of the Andes and its relation to hydrocarbon and mineral resources. Circum-Pacific Council for Energy and Mineral Resources. Earth Sci Ser 11: 345–362

Saleeby JS (1983) Accretionary tectonics of the North American Cordillera. Annu Rev Earth Planet Sci 15: 45–73

Salter T (1993) Fluvial scour and incision: models for their influence on the development of realistic reservoir geometries. In: North CP, Prosser DJ (eds) Characterization of fluvial and aeolian reservoirs. Geol Soc Lond Spec Publ 73: 33–51

Sanford BV, Thompson FJ, McFall FJ (1985) Plate tectonics – a possible controlling mechanism in the development of hydrocarbon traps in southwestern Ontario. Bull Can Petrol Geol 33: 52–71

Sanford RM (1970) Sarir oil field, Libya – desert surprise. In: Halbouty MT (ed) Geology of giant petroleum fields. Am Assoc Petrol Geol Mem 14: 449–476

Saucier AE (1976) Tectonic influence on uraniferous trends in the Late Jurassic Morrison Formation. In: Woodward LA, Northrop SA (eds) Tectonics and mineral resources of southwestern North America. New Mex Geol Soc Spec Publ 6: 151–157

Saucier RT (1974) Quaternary geology of the lower Mississippi Valley. Arkansas Archaeol Surv Res Ser 6

Saunderson HC, Lockett FPJ (1983) Flume experiments on bedforms and structures at the dune-plane bed transition. In: Collinson JD, Lewin J (eds) Modern and ancient fluvial systems. Int Assoc Sedimentol Spec Publ 6: 49–58

Schermer E, Howell DG, Jones DL (1984) The origin of allochthonous terranes: perspectives on the growth and shaping of continents. Annu Rev Earth Planet Sci 12: 107–131

Schlumberger Limited (1970) Fundamental of dipmeter interpretation. Schlumberger, New York

Schultz AW (1984) Subaerial debris flow deposition in the Upper Paleozoic Cutler Formation, western Colorado. J Sediment Petrol 54: 759–772

Schumm SA (1963) A tentative classification of alluvial river channels. US Geol Surv Circ 477

Schumm SA (1968a) Speculations concerning paleohydrologic controls of terrestrial sedimentation. Geol Soc Am Bull 79: 1573–1588

Schumm SA (1968b) River adjustment to altered hydrologic regimen – Murrumbidgee River and paleochannels, Australia. US Geol Surv Profess Paper 598

Schumm SA (1969) River metamorphosis. J Hydr Div Am Soc Civil Eng 95: 255–273

Schumm SA (1972a) River morphology. Benchmark papers in geology. Dowden, Hutchinson and Ross, Stroudsburg, Pennsylvania

Schumm SA (1972b) Fluvial paleochannels. In: Rigby JK, Hamblin WK (eds) Recognition of ancient sedimentary environments. Soc Econ Paleontol Mineral Spec Publ 16: 98–107

Schumm SA (1977) The fluvial system. Wiley, New York

Schumm SA (1979) Geomorphic thresholds: the concept and its applications. Trans Inst Br Geogr 4: 485–515

Schumm SA (1981) Evolution and response of the fluvial system, sedimentological implications. In: Ethridge FG, Flores RM (eds) Recent and ancient nonmarine depositional environments: models for exploration. Soc Econ Paleontol Mineral Spec Publ 31: 19–29

Schumm SA (1985a) Explanation and extrapolation in geomorphology: seven reasons for geologic uncertainty. Trans Jpn Geomorphol Union 6: 1–18

Schumm SA (1985b) Patterns of alluvial rivers. Annu Rev Earth Planet Sci 13: 5–27

Schumm SA (1988) Variability of the fluvial system in space and time. In: Rosswall T, Woodmansee RG, Risser PG (eds) Scales and global change. Wiley, New York, pp 225–250

Schumm SA (1993) River response to baselevel change: implications for sequence stratigraphy. J Geol 101: 279–294

Schumm SA, Brakenridge GR (1987) River responses. In: Ruddiman WF, Wright HE Jr (eds) North America and adjacent oceans during the last deglaciation. Geol Soc Am Geol North Am K-3: 221–240

Schumm SA, Khan HR (1972) Experimental study of channel patterns. Geol Soc Am Bull 83: 1755–1770

Schumm SA, Lichty RW (1963) Channel widening and flood plain construction along Cimarron River in southwestern Kansas. US Geol Surv Profess Paper 352-D: 71-88

Schumm SA, Mosley MP, Weaver WE (1987) Experimental fluvial geomorphology. Wiley, New York

Schwartz DE (1978) Hydrology and current orientation analysis of a braided to meandering transition; the Red River in Oklahoma and Texas, USA. In: Miall AD (ed) Fluvial sedimentology. Can Soc Petrol Geol Mem 5: 231–256

Seeber L, Gornitz V (1983) River profiles along Himalayan arc as indicators of active tectonics. Tectonophysics 92: 335–367

Sengör AMC (1976) Collision of irregular continental margins: implications for foreland deformation of Alpine-type orogens. Geology 4: 779–782

Sengör AMC, Görür D, Saroglu F (1985) Strike-slip faulting and related basin formation in zones of tectonic escape: Turkey as a case study. In: Biddle KT, Christie-Blick N (eds) Strike-slip deformation, basin formation and sedimentation. Soc Econ Paleontol Mineral Spec Publ 37: 227–264

Seranne M, Seguret M (1987) The Devonian basins of western Norway: tectonics and kinematics of an extending crust. In: Coward MP, Dewey JF, Hancock PL (eds) Continental extension tectonics. Geol Soc Lond Spec Publ 28: 537–548

Serra O (1989) Formation MicroScanner imager interpretation. Schlumberger Educational Services, Schlumberger

Shanley KW, McCabe PJ (1991) Predicting facies architecture through sequence stratigraphy – an example from the Kaiparowits Plateau, Utah. Geology 19: 742–745

Shanley KW, McCabe PJ (1993) Alluvial architecture in a sequence stratigraphic framework: a case history from the Upper Cretaceous of southern Utah, USA. In: Flint SS, Bryant ID (eds) The geological modelling of hydrocarbon reservoirs and outcrop analogues. Int Assoc Sedimentol Spec Publ 15: 21–56

Shanley KW, McCabe PJ (1994) Perspectives on the sequence stratigraphy of continental strata. Am Assoc Petrol Geol Bull 78: 544–568

Shanley KW, McCabe PJ, Hettinger RD (1992) Tidal influences in Cretaceous fluvial strata from Utah, USA: a key to sequence stratigraphic interpretation. Sedimentology 39: 905–930

Shannon PM, Naylor D (1989) Petroleum basin studies. Graham and Trotman, London

Shantzer EV (1951) Alluvium of plains rivers in a temperate zone and its significance for understanding the laws governing the structure and formation of alluvial suites. Trans Inst Geol Nauk Akad Nauk SSSR Geol Ser 135

Sharp RP, Nobles LH (1953) Mudflows in 1941 at Wrightwood, southern California. Geol Soc Am Bull 64: 547–560

Shaui D, Qian K, Song Y, Ge R (1988) Stratigraphic-lithologic oil and gas pools in the Jiyang Depression, China.

In: Wagner HC, Wagner LC, Wang FFH, Wong FL (eds) Petroleum resources of China and related subjects. Circum-Pacific Council for Energy and Mineral Resources. Earth Sci Ser 10: 329–344

Shaw J, Kellerhals R (1977) Paleohydraulic interpretation of antidune bedforms with applications to antidunes in gravel. J Sediment Petrol 47: 257–266

Shelton JW (1967) Stratigraphic models and general criteria for recognition of alluvial, barrier-bar, and turbidity-current sand deposits. Am Assoc Petrol Geol Bull 51: 2441–2461

Shelton JW, Noble RL (1974) Depositional features of braided-meandering stream. Am Assoc Petrol Geol Bull 58: 742–749

Shen HW (1978) Sediment transport models. In: Miall AD (ed) Fluvial sedimentology. Can Soc Petrol Geol Mem 5: 49–60

Shepherd RG (1987) Lateral accretion surfaces in ephemeral-stream point bars, Rio Puerco, New Mexico. In: Ethridge FG, Flores RM, Harvey MD (eds) Recent developments in fluvial sedimentology. Soc Econ Paleontol Mineral Spec Publ 39: 93–98

Shields A (1936) Anwendung der Ähnlichkeitsmechanik und der Turbulenzforschung auf die Geschiebebewegung. Mitteil Preuss Versuch Aust Wasserbau Schiffbau Berl 26

Shlemon RJ (1975) Subaqueous delta formation – Atchafalaya Bay, Lousiana. In: Broussard ML (ed) Deltas. Houston Geological Society, Texas, pp 209–222

Shultz AW (1984) Subaerial debris-flow deposition in the Upper Paleozoic Cutler Formation, western Colorado. J Sediment Petrol 54: 759–772

Shultz EH (1982) The chronosome and supersome: terms proposed for low-rank chronostratigraphic units. Bull Can Petrol Geol 30: 29–33

Shurr GW (1984) Geometry of shelf-sandstone bodies in the Shannon Sandstone of southeastern Montana. In: Tillman RW, Siemers CT (eds) Siliciclastic shelf sediments. Soc Econ Paleontol Mineral Spec Publ 34: 63–83

Shuster MW, Steidtmann JR (1987) Fluvial-sandstone architecture and thrust-induced subsidence, northern Green River Basin, Wyoming. In: Ethridge FG, Flores RM, Harvey MD (eds) Recent developments in fluvial sedimentology. Soc Econ Paleontol Mineral Spec Publ 39: 279–285

Siegenthaler C, Huggenberger P (1993) Pleistocene Rhine gravel: deposits of a braided river system with dominant pool preservation. In: Best JL, Bristow CS (eds) Braided rivers. Geol Soc Lond Spec Publ 75: 147–162

Simlote VN, Ebanks WJ Jr, Eslinger AV, Harpole KJ (1985) Synergistic evaluation of a complex conglomerate reservoir for EOR, Barrancas Formation, Argentina. J Petrol Technol 37: 295–305

Simons DB, Richardson EV (1961) Forms of bed roughness in alluvial channels. Am Soc Civil Eng Proc 87 (HY3): 87–105

Simons DB, Richardson EV, Nordin CF Jr (1965) Sedimentary structures generated by flow in alluvial channels. In: Middleton GV (ed) Primary sedimentary structures and their hydrodynamic interpretation. Soc Econ Paleontol Mineral Spec Publ 12: 34–52

Sinclair HD, Allen PA (1992) Vertical versus horizontal motions in the Alpine orogenic wedge: stratigraphic response in the foreland basin. Basin Res 4: 215–232

Singh H, Parkash B, Gohain K (1993) Facies analysis of the Kosi megafan deposits. Sediment Geol 85: 87–113

Singh IB, Kumar S (1974) Mega- and giant ripples in the Ganga, Yamuna, and Son Rivers, Uttar Pradesh, India. Sediment Geol 12: 53–66

Sinha R, Friend PF (1994) River systems and their sediment flux, Indo-Gangetic plains, northern Bihar, India. Sedimentology 41: 825–845

Sloss LL (1962) Stratigraphic models in exploration. Am Assoc Petrol Geol Bull 46: 1050–1057

Sloss LL (1963) Sequences in the cratonic interior of North America. Geol Soc Am Bull 74: 93–113

Sloss LL (1979) Global sea level changes: a view from the craton. In: Watkins JS, Montadert L, Dickerson PW (eds) Geological and geophysical investigations of continental margins. Am Assoc Petrol Geol Mem 29: 461–468

Smith DG (1973) Aggradation of the Alexandra-North Saskatchewan River, Banff Park, Alberta. In: Morisawa M (ed) Fluvial geomorphology. Proceedings of the 4th annual geomorphology symposium, publications in geomorphology. SUNY-Binghamton, New York, pp 201–219

Smith DG (1976) Effect of vegetation on lateral migration of anastomosed channels of a glacial meltwater river. Geol Soc Am Bull 87: 857–860

Smith DG (1983) Anastomosed fluvial deposits: modern examples from western Canada. In: Collinson JD, Lewin J (eds) Modern and ancient fluvial systems. Int Assoc Sedimentol Spec Publ 6: 155–168

Smith DG (1986) Anastomosing river deposits, sedimentation rates and basin subsidence, Magdalena River, northwestern Colombia, South America. Sediment Geol 46: 177–196

Smith DG (1987) Meandering river point bar lithofacies models: modern and ancient examples compared. In: Ethridge FG, Flores RM, Harvey MD (eds) Recent developments in fluvial sedimentology. Soc Econ Paleontol Mineral Spec Publ 39: 83–91

Smith DG, Locking T (1989) Anastomosed river flood duration and depositional rates in the upper Columbia River, BC, Canada; implications for peat accumulation and interpretation of ancient coal deposits. In: 4th international conference on fluvial sedimentology, Barcelona, Oct 1989, programme and abstracts, p 223

Smith DG, Putnam PE (1980) Anastomosed river deposits: modern and ancient examples in Alberta, Canada. Can J Earth Sci 17: 1396–1406

Smith DG, Smith ND (1980) Sedimentation in anastomosed river systems: examples from alluvial valleys near Banff, Alberta. J Sediment Petrol 50: 157–164

Smith GA (1987) Sedimentology of volcanism-induced aggradation in fluvial basins: examples from the Pacific northwest, USA. In: Ethridge FG, Flores RM, Harvey MD (eds) Recent developments in fluvial sedimentology. Soc Econ Paleontol Mineral Spec Publ 39: 217–228

Smith GA (1994) Climatic influences on continental deposition during late-stage filling of an extensional basin, southeastern Arizona. Geol Soc Am Bull 106: 1212–1228

Smith ND (1970) The braided stream depositional environment: comparison of the Platte River with some Silurian clastic rocks, north central Appalachians. Geol Soc Am Bull 81: 2993–3014

Smith ND (1971) Transverse bars and braiding in the lower Platte River, Nebraska. Geol Soc Am Bull 82: 3407–3420

Smith ND (1972) Some sedimentological aspects of planar cross-stratification in a sandy braided river. J Sediment Petrol 42: 624–634

Smith ND (1974) Sedimentology and bar formation in the upper Kicking Horse River, a braided outwash stream. J Geol 82: 205–224

Smith ND (1978) Some comments on terminology for bars in shallow rivers. In: Miall AD (ed) Fluvial sedimentology. Can Soc Petrol Geol Mem 5: 85–88

Smith ND, Smith DG (1984) William River, an outstanding example of channel widening and braiding caused by bed-load addition. Geology 12: 78–82

Smith ND, Cross TA, Dufficy JP, Clough SR (1989) Anatomy of an avulsion. Sedimentology 36: 1–23

Smith RMH (1987) Morphology and depositional history of exhumed Permian point bars in the southwest Karoo, South Africa. J Sediment Petrol 57: 19–29

Smith SA (1987) Gravel counterpoint bars: examples from the River Tywi, South Wales. In: Ethridge FG, Flores RM, Harvey MD (eds) Recent developments in fluvial sedimentology. Soc Econ Paleontol Mineral Spec Publ 39: 75–81

Smith SA (1990) The sedimentology and accretionary style of an ancient gravel-bed stream: the Budleigh Salterton Pebble Beds (Lower Triassic), southwest England. Sediment Geol 67: 199–219

Smoot JP (1983) Depositional subenvironments in an arid closed basin; the Wilkins Peak Member of the Green River Formation (Eocene), Wyoming, USA. Sedimentology 30: 801–827

Sneed ED, Folk RL (1958) Pebbles in the lower Colorado River, Texas, a study in particle morphogenesis. J Geol 66: 114–150

Sneh A (1983) Desert stream sequences in the Sinai Peninsula. J Sediment Petrol 53: 1271–1280

Sneider R, Massell W, Mathis R, Loren D, Wichmann P (1991) The integration of geology, geophysics, petrophysics and petroleum engineering in reservoir delineation, description and management, Proceedings of the 1st Arche conference, Oct 1990. American Association of Petroleum Geologists, Tulsa, Oklahoma

Soegaard K (1990) Fan-delta and braid-delta systems in Pennsylvanian Sandia Formation, Taos Trough, northern New Mexico: depositional and tectonic implications. Geol Soc Am Bull 102: 1325–1343

Soegaard K (1992) Architectural elements of fan-delta complex in Pennsylvanian Sandia Formation, Taos Trough, Northern New Mexico. In: Miall AD, Tyler N (eds) The three-dimensional facies architecture of terrigenous clastic sediments, and its implications for hydrocarbon discovery and recovery. Soc Econ Paleontol Mineral Conc Models Ser, pp 217–223

Soeparjadi RA, Valachi LZ, Sosromihardjo S (1986) Oil and gas developments in Far East in 1985. Am Assoc Petrol Geol Bull 70: 1479–1565

Song G (1988) Jurassic paleo-landform oil fields in the Ordos Basin, North China. In: Wagner HC, Wagner LC, Wang FFH, Wong FL (eds) Petroleum resources of China and related subjects. Circum-Pacific Council for Energy and Mineral Resources. Earth Sci Ser 10: 371–386

Sonnenberg SA (1987) Tectonic, sedimentary, and seismic models for D sandstone, Zenith field area, Denver Basin, Colorado. Am Assoc Petrol Geol Bull 71: 1366–1377

Sorby HC (1852) On the oscillation of the currents drifting the sandstone beds of the southeast of Northumberland, and on their general direction in the coalfield in the neighbourhood of Edinburgh. Proc West Yorkshire Geol Soc 3: 232–240

Sorby HC (1859) On the structures produced by the currents present during the deposition of stratified rocks. Geologist 2: 137–147

Sorby HC (1908) On the application of quantitative methods to the study of the structure and history of rocks. Q J Geol Soc Lond 64: 171–233

Southard JB (1971) Representation of bed configurations in depth-velocity-size diagrams. J Sediment Petrol 41: 903–915

Southard JB, Boguchwal LA, Romea RD (1980) Test of scale modelling of sediment transport in steady unidirectional flow. Earth Surface Proc 5: 17–23

Southard JB, Smith ND, Kuhnle RA (1984) Chutes and lobes: newly identified elements of braiding in shallow gravelly streams. In: Koster EH, Steel RJ (eds) Sedimentology of gravels and conglomerates. Can Soc Petrol Geol Mem 10: 51–59

Spencer DW (1963) The interpretation of grain size distribution curves of clastic sediments. J Sediment Petrol 33: 180–190

Spieker EM (1946) Late Mesozoic and Early Cenozoic history of central Utah. US Geol Surv Profess Paper 205-D: 117–161

Spurr JE (1894a) False bedding in stratified drift deposits: American Geologist, v 13, p 43–47

Spurr JE (1894b) Oscillation and single current ripple marks. Am Geol 13: 201–206

Srivastava P, Parkash B, Sehgal JL, Kumar S (1994) Role of neotectonics and climate in development of the Holocene geomorphology and soils of the Gangetic Plains between the Ramganga and Rapti rivers. Sediment Geol 94: 129–151

Srivastava SP, Tapscott CR (1986) Plate kinematics of the North Atlantic. In: Vogt PR, Tucholke BE (eds) The Western North Atlantic region. The Geology of North America vol M. Geological Society of America, Boulder, Colorado, pp 379–404

Stabler CL (1990) Andean hydrocarbon resources – an overview. In: Eriksen GE, Pinochet MTC, Reinemund JA (eds) Geology of the Andes and its relation to hydrocarbon and mineral resources. Circum-Pacific Council for Energy and Mineral Resources. Earth Sci Ser 11: 431–438

Stamp LD (1925) Seasonal rhythms in the Tertiary sediments of Burma. Geol Mag 62: 515–528

Stancliffe RJ, Adams ER (1986) Lower Tuscaloosa fluvial channel styles at Liberty field, Amite County, Mississippi. Trans Gulf Coast Assoc Geol Soc 36: 305–313

Stanistreet IG (1993) Ancient and modern examples of tectonic escape basins: the Archaean Witwatersrand Basin compared with the Cenozoic Maracaibo Basin. In: Frostick LE, Steel RJ (eds) Tectonic controls and signatures in sedimentary successions. Int Assoc Sedimentol Spec Publ 20: 363–376

Stanistreet IG, McCarthy TS (1993) The Okavango fan and the classification of subaerial fan systems. Sediment Geol 85: 115–133

Stanistreet IG, Cairncross B, McCarthy TS (1993) Low sinuosity and meandering bedload rivers of the Okavango fan: channel confinement by vegetated levées without fine sediment. Sediment Geol 85: 135–156

Stanley KO (1976) Sandstone petrofacies in the Cenozoic High Plains sequence, eastern Wyoming and Nebraska. Geol Soc Am Bull 87: 297–309

Stanley KO, Jorde K, Raestad N, Stockbridge CP (1990) Stochastic modeling of reservoir sand bodies for input to reservoir simulation, Snorre field, northern North Sea, Norway. In: Buller AT, Berg E, Hjelmeland O, Kleppe J, Torsaeter O, Aasen JO (eds) North Sea oil and gas reservoirs I: Graham and Trotman, London, pp 91–101

Stanley KO, Wayne WJ (1972) Epeirogenic and climatic controls of Early Pleistocene fluvial sediment dispersal in Nebraska. Geol Soc Am Bull 83: 3675–3690

Stanmore PJ, Johnstone EM (1988) The search for stratigraphic traps in the southern Patchawarra Trough, South Australia. Aust Petrol Explor Assoc J 28: 156–165

Stapp RW (1967) Relationship of Lower Cretaceous depositional environment to oil accumulation, northeastern Powder River Basin, Wyoming. Am Assoc Petrol Geol Bull 51: 2044–2055

Statham I (1976) Debris flows on vegetated screes in the Black Mountain, Carmarthenshire. Earth Surface Proc 1: 173–180

Stear WM (1983) Morphological characteristics of ephemeral stream channel and overbank splay sandstone bodies in the Permian Lower Beaufort Group, Karoo Basin, South Africa. In: Collinson JD, Lewin J (eds) Modern and ancient fluvial systems. Int Assoc Sedimentol Spec Publ 6: 405–420

Stear WM (1985) Comparison of the bedform distribution and dynamics of modern and ancient sandy ephemeral flood deposits in the southwestern Karoo region, South Africa. Sediment Geol 45: 209–230

Steel RJ (1974) New Red Sandstone floodplain and piedmont sedimentation in the Hebridian Province, Scotland. J Sediment Petrol 44: 336–357

Steel RJ, Aasheim SM (1978) Alluvial sand deposition in a rapidly subsiding basin (Devonian, Norway). In: Miall AD (ed) Fluvial sedimentology. Can Soc Petrol Geol Mem 5: 385–412

Steel RJ, Gloppen TG (1980) Late Caledonian (Devonian) basin formation, western Norway: signs of strike-slip tectonics during infilling. In: Ballance PF, Reading HG (eds) Sedimentation in oblique-slip mobile zones. Int Assoc Sedimentol Spec Publ 4: 79–103

Steel RJ, Ryseth A (1990) The Triassic – Early Jurassic succession in the northern North Sea: megasequence stratigraphy and intra-Triassic tectonics. In: Hardman RFP, Brooks J (eds) Tectonic events responsible for Britain's oil and gas reserves. Geol Soc Lond Spec Publ 55: 139–168

Steel RJ, Dalland A, Kalgraff K, Larsen V (1981) The Central Tertiary Basin of Spitsbergen: sedimentary development of a sheared-margin basin. In: Kerr JW, Fergusson AJ (eds) Geology of the North Atlantic borderlands. Can Soc Petrol Geol Mem 7: 647–664

Steel RJ, Gjelberg J, Helland-Hansen W, Kleinspehn K, Nottvedt A, Rye-Larsen M (1985) The Tertiary strike-slip basins and orogenic belt of Spitsbergen. In: Biddle KT, Christie-Blick N (eds) Strike-slip deformation, basin formation and sedimentation. Soc Econ Paleontol Mineral Spec Publ 37: 339–359

Steel RJ, Maehl S, Nilsen H, Roe SL, Spinnangr Å (1977) Coarsening-upward cycles in the alluvium of Hornelen Basin (Devonian), Norway. Sedimentary response to tectonic events. Geol Soc Am Bull 88: 1124–1134

Stephens M (1994) Architectural element analysis within the Kayenta Formation (Lower Jurassic) using ground-probing radar and sedimentological profiling, southwestern Colorado. Sediment Geol 90: 179–211

Stewart DJ (1983) Possible suspended-load channel deposits from the Wealden Group (Lower Cretaceous) of southern England. In: Collinson JD, Lewin J (eds) Modern and ancient fluvial systems. Int Assoc Sedimentol Spec Publ 6: 369–384

Stewart JH (1978) Basin-range structure in western North America: a review. In: Smith RB, Eaton GP (eds) Cenozoic tectonics and regional geophysics of the Western Cordillera. Geol Soc Am Mem 152: 1–31

Stockmal GS, Beaumont C (1987) Geodynamic models of convergent margin tectonics: the southern Canadian Cordillera and the Swiss Alps. In: Beaumont C, Tankard AJ (eds) Sedimentary basins and basin-forming mechanisms. Can Soc Petrol Geol Mem 12: 393–411

Stockmal GS, Beaumont C, Boutilier R (1986) Geodynamic models of convergent margin tectonics: transition from rifted margin to overthrust belt and consequences for foreland-basin development. Am Assoc Petrol Geol Bull 70: 181–190

Stockmal GS, Cant DJ, Bell JS (1992) Relationship of the stratigraphy of the Western Canada foreland basin to Cordilleran tectonics: insights from geodynamic models. In: Macqueen RW, Leckie DA (eds) Foreland basin and fold belts. Am Assoc Petrol Geol Mem 55: 107–124

Stollhofen H, Stanistreet IG (1994) Interaction between bimodal volcanism, fluvial sedimentation and basin development in the Permo-Carboniferous Saar-Nahe Basin (south-west Germany). Basin Res 6: 245–267

Stott DF, Aitken JD (eds) (1993) Sedimentary cover of the craton in Canada. Geological Survey of Canada, Ottawa, Canada (Geology of Canada no 5)

Struijk AP, Green RT (1991) The Brent field, Block 211/29, UK North Sea. In: Abbotts IL (ed) United Kingdom oil and gas fields, 25 year Commemorative volume. Geological Society, London, pp 63–72

Stuart IA, Cowan G (1991) The South Morecambe field, Blocks 110/2a, 110/3a, 110/8a, UK East Irish Sea. In: Abbotts IL (ed) United Kingdom oil and gas fields, 25 year Commemorative volume. Geological Society, London, pp 527–541

Stuart WJ, Kennedy S, Thomas AD (1988) The influence of structural growth and other factors on the configuration of fluviatile sandstones, Permian Cooper Basin. Aust Petrol Explor Assoc J 28: 255–266

Summerson CH (ed) (1976) Sorby on sedimentology: a collection of papers from 1851 to 1908 by Henry Clifton Sorby. Geological Milestones I. Comparative sedimentology laboratory, University of Miami, Miami

Sundborg Å (1956) The River Klarälven, a study of fluvial processes. Geogr Ann 38: 125–316

Surell A (1841) Etude sur les torrents des Hautes-Alpes. Dunod, Paris

Surell A (1870) Etude sur les torrents des Hautes-Alpes, 2nd edn Dunod, Paris

Suter JR, Berryhill HL, Penland S (1987) Late Quaternary sea-level fluctuations and depositional sequences, southwest Louisiana continental shelf. In: Nummedal D, Pilkey OH, Howard JD (eds) Sea-level fluctuation and coastal evolution. Soc Econ Paleontol Mineral Spec Publ 41: 199–219

Swift DJP, Rice DD (1984) Sand bodies on muddy shelves: a model for sedimentation in the Western Interior Cretaceous seaway, North America. In: Tillman RW, Siemers CT (eds) Siliciclastic shelf sediments. Soc Econ Paleontol Mineral Spec Publ 34: 43–62

Swift DJP, Hudelson PM, Brenner RL, Thompson P (1987) Shelf construction in a foreland basin: storm beds, shelf sandbodies, and shelf-slope depositional sequences in the Upper Cretaceous Mesaverde Group, Book Cliffs, Utah. Sedimentology 34: 423–457

Talbot MR, Holm K, Williams MAJ (1994) Sedimentation in low-gradient desert margin systems: a comparison of the Late Triassic of northwest Somerset (England) and the late Quaternary of east-central Australia. In: Rosen MR (ed) Paleoclimate and basin evolution of playa systems. Geol Soc Am Spec Paper 289: 97–117

Tandon SK, Gibling MR (1994) Calcrete and coal in late Carboniferous cyclothems of Nova Scotia, Canada: climate and sea-level changes linked. Geology 22: 755–758

Tang Z (1982) Tectonic features of oil and gas basins in eastern part of China. Am Assoc Petrol Geol Bull 66: 509–521

Tankard AJ, Jackson MPA, Eriksson KA, Hobday DK, Hunter DR, Minter WEL (1982) Crustal evolution of Southern Africa. Springer, Berlin Heidelberg New York

Tanner WF (1955) Paleogeographic reconstructions from cross-bedding studies. Am Assoc Petrol Geol Bull 39: 2471–2483

Tapponnier P, Peltzer G, Armijo R (1986) On the mechanics of the collision between India and Asia. In: Coward MP, Ries AC (eds) Collision tectonics. Geol Soc Lond Spec Publ 19: 115–157

Taylor G, Woodyer KD (1978) Bank deposition in suspended-load streams. In: Miall AD (ed) Fluvial sedimentology. Can Soc Petrol Geol Mem 5: 257–275

Teng LS (1990) Geotectonic evolution of late Cenozoic arc-continent collision in Taiwan. Tectonophysics 183: 57–76

Thakur GC (1991) Waterflood surveillance techniques – a reservoir management approach. J Petrol Technol 43: 1180–1192

Theriault P, Desrochers A (1993) Carboniferous calcretes in the Canadian Arctic. Sedimentology 40: 449–466

Thomas RG, Smith DG, Wood JM, Visser J, Calverley-Range EA, Koster EH (1987) Inclined heterolithic stratification – terminology, description, interpretation and significance. Sediment Geol 53: 123–179

Thorne CR, Russell APG, Alam MK (1993) Planform pattern and channel evolution of the Brahmaputra River, Bangladesh. In: Best JL, Bristow CS (eds) Braided rivers. Geol Soc Lond Spec Publ 75: 257–276

Tipper JC (1989) Computer modelling of seismic facies: implications for seismic and sequence stratigraphy. In: Collinson JD (ed) Correlation in hydrocarbon exploration. Graham and Trotman, London, pp 45–51

Titheridge DG (1993) The influence of half-graben syndepositional tilting on thickness variation and seam splitting in the Brunner Coal Measures, New Zealand. Sediment Geol 87: 195–213

Todd SP, Went DJ (1991) Lateral migration of sand-bed rivers: examples from the Devonian Glashabeg Formation, SW Ireland and the Cambrian Alderney Sandstone Formation, Channel Islands. Sedimentology 38: 997–1020

Tolman CF (1909) Erosion and deposition in the southern Arizona bolson region. J Geol 17: 136–163

Törnqvist TE (1993) Holocene alternation of meandering and anastomosing fluvial systems in the Rhine-Mesue delta (central Netherlands) controlled by sea-level rise and subsoil erodibility. J Sediment Petrol 63: 683–693

Törnqvist TE (1994) Middle and late Holocene avulsion history of the River Rhine (Rhine-Meuse delta, Netherlands). Geology 22: 711–714

Törnqvist TE, van Ree MHM, Faessen ELJH (1993) Longitudinal facies architectural changes of a Middle Holocene anastomosing distributary system (Rhine-Meuse delta, central Netherlands). Sediment Geol 85: 203–219

Trewin NH (1993a) Mixed aeolian sandsheet and fluvial deposits in the Tumblagooda Sandstone, Western Australia. In: North CP, Prosser DJ (eds) Characterization of fluvial and aeolian reservoirs. Geol Soc Lond Spec Publ 73: 219–230

Trewin NH (1993b) Controls on fluvial deposition in mixed fluvial and aeolian facies within the Tumblagooda Sandstone (Late Silurian) of Western Australia. Sediment Geol 85: 387–400

Trifonov VG (1978) Late Quaternary tectonic movements of western and central Asia. Geol Soc Am Bull 89: 1059–1072

Trowbridge AC (1911) The terrestrial deposits of Owens Valley, California. J Geol 19: 706–747

Tunbridge IP (1981) Sandy high-energy flood sedimentation – some criteria for recognition, with an example from the Devonian of SW England. Sediment Geol 28: 79–96

Tunbridge IP (1984) Facies models for a sandy ephemeral stream and clay playa complex; the Middle Devonian Trentishoe Formation of North Devon, UK. Sedimentology 31: 697–716

Turner BR (1978) Sedimentary patterns of uranium mineralisation in the Beaufort Group of the southern Karoo (Gondwana) Basin, South Africa. In: Miall AD (ed) Fluvial sedimentology. Can Soc Petrol Geol Mem 5: 831–848

Turner BR (1983) Braidplain deposition of the Upper Triassic Molteno Formation in the main Karoo (Gondwana) Basin, South Africa. Sedimentology 30: 77–89

Turner BR, Munro M (1987) Channel formation and migration by mass-flow processes in the Lower Carboniferous fluviatile Fell Sandstone Group, northeast England. Sedimentology 34: 1107–1122

Turner P (1980) Continental red beds. Elsevier Scientific, Amsterdam (Developments in sedimentology, vol 29)

Twenhofel WH (1932) Treatise on sedimentation, 2nd edn. Williams and Wilkins, New York

Tye RS (1991) Fluvial-sandstone reservoirs of the Travis Peak Formation, East Texas Basin. In: Miall AD, Tyler, N (eds) The three-dimensional facies architecture of terrigenous clastic sediments, and its implications for hydrocarbon discovery and recovery. Soc Econ Paleontol Mineral Conc Models Ser 3: 172–188

Tyler N, Ethridge FG (1983) Fluvial architecture of Jurassic uranium-bearing sandstones, Colorado Plateau, western United States. In: Collinson JD, Lewin J (eds) Modern and ancient fluvial systems. Int Assoc Sedimentol Spec Publ 6: 533–547

Tyler N, Galloway WE, Garrett CM Jr, Ewing TE (1984) Oil accumulation, production characteristics, and targets for additional recovery in major oil reservoirs of Texas. University of Texas, Bureau of Economic Geology (Geological circular 84-2)

Udden JA (1914) Mechanical composition of clastic sediments. Geol Soc Am Bull 25: 655–744

Vail PR (1987) Seismic stratigraphy interpretation using sequence stratigraphy, part 1: seismic stratigraphy interpretation procedure. In: Bally AW (ed) Atlas of seismic stratigraphy. Am Assoc Petrol Geol Stud Geol 27 (1): 1–10

Vail PR, Todd RG (1981) Northern North Sea Jurassic unconformities, chronostratigraphy and sea-level changes from seismic stratigraphy. In: Illing LV, Hobson GD (eds) Petroleum geology of the continental shelf of northwest Europe. Institute of Petroleum, London, pp 216–235

Vail PR, Mitchum RM Jr, Todd RG, Widmier JM, Thompson S III, Sangree JB, Bubb JN, Hatlelid WG (1977) Seismic stratigraphy and global changes of sea-level. In: Payton CE (ed) Seismic stratigraphy – applications to hydrocarbon exploration. Am Assoc Petrol Geol Mem 26: 49–212

Vail PR, Hardenbol J, Todd RG (1984) Jurassic unconformities, chronostratigraphy and sea-level changes from seismic stratigraphy and biostratigraphy. In: Schlee JS (ed) Interregional unconformities and hydrocarbon exploration. Am Assoc Petrol Geol Mem 36: 129–144

Vail PR, Audemard F, Bowman SA, Eisner PN, Perez-Crus C (1991) The stratigraphic signatures of tectonics, eustasy and sedimentology – an overview. In: Einsele G, Ricken W, Seilacher A (eds) Cycles and events in stratigraphy. Springer, Berlin Heidelberg New York, pp 617–659

Vandenberghe J (1993) Changing fluvial processes under changing periglacial conditions. Z Geomorph NF 88: 17–28

Vandenberghe J, Kasse C, Bohnke S, Kozarski S (1994) Climate-related river activity at the Weichselian-Holocene transition: a comparative study of the Warta and Maas rivers. Terra Nova 6: 476–485

van der Lingen GJ (1982) Development of the North Island subduction system, New Zealand. In: Leggett JK (ed) Trench-forearc geology. Geol Soc Lond Spec Publ 10: 259–272

Van Houten FB (1973) Origin of red beds: a review – 1961–1972. Annu Rev Earth Planet Sci 1: 39–61

Van Houten FB (1981) The Odyssey of Molasse. In: Miall AD (ed) Sedimentation and tectonics in alluvial basins. Geol Assoc Can Spec Paper 23: 35–48

Vanoni VA (1946) Transportation of suspended sediment by water. Am Assoc Civil Eng Trans 111: 67–102

Van Overmeeren RA, Staal JH (1976) Floodfan sedimentation and gravitational anomalies in the Salar de Punta Negra, northern Chile. Geol Rundsch 65: 195–211

Van Straaten LMJU (1954) Sedimentology of Recent tidal flat deposits and the Psammites du Condroz (Devonian). Geol Mijnbouw 16: 25–47

Van Wagoner JC, Mitchum RM Jr, Posamentier HW, Vail PR (1987) Seismic stratigraphy interpretation using sequence stratigraphy, Part 2: key definitions of sequence stratigraphy. In: Bally AW (ed) Atlas of seismic stratigraphy. Am Assoc Petrol Geol Stud Geol 27 (1): 11–14

Van Wagoner JC, Mitchum RM, Campion KM, Rahmanian VD (1990) Siliciclastic sequence stratigraphy in well logs, cores, and outcrops. Am Assoc Petrol Geol Methods Explor Ser 7

Van Wagoner JC, Nummedal D, Jones CR, Taylor DR, Jennette DC, Riley GW (1991) Sequence stratigraphy applications to shelf sandstone reservoirs. American Association of Petroleum Geologists Field Conference Guidebook, Tulsa

Veevers JJ (ed) (1984) Phanerozoic earth history of Australia. Oxford University Press, Oxford

Verdier AC, Oki T, Suardy A (1980) Geology of the Handil field (East Kalimantan, Indonesia). In: Halbouty MT (ed) Giant oil and gas fields of the decade: 1968–1978. Am Assoc Petrol Geol Mem 30: 399–421

Vessell RK, Davies DK (1981) Nonmarine sedimentation in an active fore arc basin. In: Ethridge FG, Flores RM (eds) Recent and ancient nonmarine depositional environments: models for exploration. Soc Econ Paleontol Mineral Spec Publ 31: 31–45

Vincent P, Gartner JE, Attali G (1977) Geodip: an approach to detailed dip determination using correlation by pattern recognition. 6th formation evaluation symposium, Canadian Well Logging Society, paper L

Vincent P, Gartner JE, Attali G (1979) Geodip: an approach to detailed dip determination using correlation by pattern recognition. J Petrol Technol 31: 232–240

Viseras C, Fernández J (1994) Channel migration patterns and related sequences in some alluvial fan valleys. Sediment Geol 88: 201–217

Visher GS (1965a) Fluvial processes as interpreted from ancient and recent fluvial deposits. In: Middleton GV (ed) Primary sedimentary structures and their hydrodynamic interpretation. Soc Econ Paleontol Mineral Spec Publ 12: 84–115

Visher GS (1965b) Use of vertical profile in environmental reconstruction. Am Assoc Petrol Geol Bull 49: 41–61

Visher GS (1969) Grain size distributions and depositional processes. J Sediment Petrol 39: 1074–1106

Visher GS, Saitta S, Phares RS (1971) Pennsylvanian delta patterns and petroleum occurrences in eastern Oklahoma. Am Assoc Petrol Geol Bull 55: 1206–1230

Visser JNJ, Dukas BA (1979) Upward-fining fluviatile megacycles in the Beaufort Group, north of Graaff-Reinet, Cape Province. Trans Geol Soc South Afr 82: 149–154

Visser MJ (1980) Neap-spring cycles reflected in Holocene subtidal large-scale bedform deposits: a preliminary note. Geology 8: 543–546

von Hoff KEA (1822–1824) Geschichte der durch Überlieferung nachgewiesenen natnrlichen Veränderungen der Erdoberflache, 2 vols. Perthes, Gotha

von Zittel KA (1901) History of geology and paleontology to the end of the nineteenth century (translated by MM Ogilvie-Gordon). Scott, London

Vos RG, Tankard AJ (1981) Braided fluvial sedimentation in the Lower Paleozoic Cape Basin, South Africa. Sediment Geol 29: 171–193

Wadman DH, Lamprecht DE, Mrosovsky I (1979) Joint geologic/engineering analysis of the Sadlerochit reservoir, Prudhoe Bay field. J Petrol Technol 31: 933–940

Walker RG (1976) Facies models 3: sandy fluvial systems. Geosci Can 3: 101–109

Walker RG (ed) (1979) Facies models. Geological Association of Canada, St. John's, Newfoundland (Geosci Can Reprint Ser 1)

Walker RG (ed) (1984) Facies models, 2nd edn. Geological Association of Canada, St. John's, Newfoundland (Geosci Can Reprint Ser 1)

Walker RG (1990) Perspective – facies modelling and sequence stratigraphy. J Sediment Petrol 60: 777–786

Walker RG, Cant DJ (1979) Sandy fluvial systems. In: Walker RG (ed) Facies models. Geosci Can Reprint Ser 1: 23–31

Walker RG, Cant DJ (1984) Sandy fluvial systems. In: Walker, R. G (ed) Facies models, 2nd edn. Geosci Can Reprint Ser 1: 71–89

Walker RG, James NP (eds) (1992) Facies models: response to sea level change. Geological Association of Canada, St. John's, Newfoundland

Walker TR (1967) Formation of red beds in modern and ancient deserts. Geol Soc Am Bull 78: 353–368

Walling DE, Webb BW (1983) Patterns of sediment yield. In: Gregory KJ (ed) Background to paleohydrology. Wiley, Chichester, pp 69–100

Walther J (1893/1894) Einleitung in die Geologie als historische Wissenschaft. Fischer, Jena, 3 vols

Walton AW, Bouquet DJ, Evenson RA, Rofheart DH, Woody MD (1986) Characterization of sandstone reservoirs in the Cherokee Group (Pennsylvanian, Desmoinesian) of southeastern Kansas. In: Lake LW, Carroll HB Jr (eds) Reservoir characterization. Academic, Orlando, pp 39–62

Wang QM, Coward MP (1993) The Jiuxi Basin, Hexi corridor, NW China: foreland structural features and hydrocarbon potential. J Petrol Geol 16: 169–182

Wanless HR, Weller JM (1932) Correlation and extent of Pennsylvanian cyclothems. Geol Soc Am Bull 43: 1003–1016

Wanless HR, Tyrrell KM, Tedesco LP, Dravis JJ (1988) Tidal-flat sedimentation from Hurricane Kate, Caicos platform, British West Indies. J Sediment Petrol 58: 724–738

Warris BJ (1988) The geology of the Mount Horner oilfield, Perth Basin, Western Australia. Aust Petrol Explor Assoc J 28: 88–99

Warwick PD, Flores RM (1987) Evolution of fluvial styles in the Eocene Wasatch Formation, Powder River Basin, Wyoming. In: Ethridge FG, Flores RM (eds) Recent and ancient nonmarine depositional environments: models for exploration. Soc Econ Paleontol Mineral Spec Publ 31: 303–310

Wasson RJ (1977a) Catchment processes and the evolution of alluvial fans in the lower Derwent valley, Tasmania. Z Geomorphol 21: 147–168

Wasson RJ (1977b) Last-glacial alluvial fan sedimentation in the Lower Derwent valley, Tasmania. Sedimentology 24: 781–799

Waters KH, Rice JW (1975) Some statistical and probabilistic techniques to optimize the search for stratigraphic traps using seismic data. Ninth World Petroleum Congress Proceedings, Tokyo (Panel Discussion 9, paper 1)

Watts AB (1981) The US Atlantic margin: subsidence history, crustal structure and thermal evolution. American Association of Petroleum Geologists (Education Course Notes Series 19)

Watts KJ (1987) The Hutton Sandstone-Birkhead Formation transition, ATP 269P(1), Eromanga Basin. Aust Petrol Explor Assoc J 27: 215–228

Weber KJ (1993) The use of 3-D seismic in reservoir geological modelling. In: Flint SS, Bryant ID (eds) The geological modelling of hydrocarbon reservoirs and outcrop analogues. Int Assoc Sedimentol Spec Publ 15: 181–188

Weber KJ, Van Geuns LC (1990) Framework for constructing clastic reservoir simulation models. J Petrol Technol 42: 1248–1253, 1296–1297

Wecker HRB (1989) The Eromanga Basin. Aust Petrol Explor Assoc J 29: 379–397

Weimer P, Posamentier HW (1993) Recent developments and applications in siliciclastic sequence stratigraphy. In: Weimer P, Posamentier HW (eds) Siliciclastic sequence stratigraphy. Am Assoc Petrol Geol Mem 58: 3–12

Weimer RJ (1960) Upper Cretaceous stratigraphy, Rocky Mountain area. Am Assoc Petrol Geol Bull 44: 1–20

Weimer RJ (1970) Rates of deltaic sedimentation and intrabasin deformation, Upper Cretaceous of Rocky Mountain region. In: Morgan JP (ed) Deltaic sedimentation, modern and ancient. Soc Econ Paleontol Mineral Spec Publ 15: 270–292

Weimer RJ (1986) Relationship of unconformities, tectonics, and sea level change in the Cretaceous of the Western Interior, United States. In: Peterson JA (ed) Paleotectonics and sedimentation in the Rocky Mountain region, United States. Am Assoc Petrol Geol Mem 41: 397–422

Weirich TE (1929) Cushing oil and gas field, Creek County, Oklahoma. In: Structure of typical American oil fields, vol 2. American Association of Petroleum Geologists, Tulsa, pp 396–406

Wells NA (1983) Transient streams in sand-poor red beds: Early Middle Eocene Kuldana Formation of northern Pakistan. In: Collinson JD, Lewin J (eds) Modern and ancient fluvial systems. Int Assoc Sedimentol Spec Publ 6: 393–403

Wells NA, Dorr JA Jr (1987) Shifting of the Kosi River, northern India. Geology 15: 204–207

Wells SG, Harvey AM (1987) Sedimentologic and geomorphic variations in storm-generated alluvial fans, Howgill fells, northwest England. Geol Soc Am Bull 98: 182–198

Wentworth CK (1922) A scale of grade and class terms for clastic sediments. J Geol 30: 377–392

Werren EG, Shew RD, Adams ER, Stancliffe RJ (1990) Meander-belt reservoir geology, mid-dip Tuscaloosa, Little Creek field, Mississippi. In: Barwis JH, McPherson JG, Studlick RJ (eds) Sandstone petroleum reservoirs. Springer, Berlin Heidelberg New York, pp 85–107

Wescott WA (1990) The Yallahs fan delta: a coastal fan in a humid tropical climate. In: Rachocki AH, Church M (eds) Alluvial fans: a field approach. Wiley, Chichester, pp 213–225

Wescott WA (1993) Geomorphic thresholds and complex response of fluvial systems – some implications for sequence stratigraphy. Am Assoc Petrol Geol Bull 77: 1208–1218

Weston PJ, Alexander J (1993) Computer modelling of flow lines over deformed surfaces: the implications for prediction of alluvial facies distribution. In: Marzo M, Puigdefábregas C (eds) Alluvial sedimentation. Int Assoc Sedimentol Spec Publ 17: 211–217

White DA (1980) Assessing oil and gas plays in facies-cycle wedges. Am Assoc Petrol Geol Bull 64: 1158–1178

White DA (1988) Oil and gas play maps in exploration and assessment. Am Assoc Petrol Geol Bull 72: 944–949

Wightman DM, Tilley BJ, Last BM (1981) Stratigraphic traps in channels sandstones in the upper Mannville (Albian) of east-central Alberta: discussion. Bull Can Petrol Geol 29: 622–625

Wightman DM, Pemberton SG, Singh C (1987) Depositional modelling of the Upper Manville (Lower Cretaceous), east central Alberta: implications for the recognition of brackish water deposits. In: Tillman RW, Weber KJ (eds) Reservoir sedimentology. Soc Econ Paleontol Mineral Spec Publ 40: 189–220

Wilgus CK, Hastings BS, Kendall CGSC, Posamentier HW, Ross CA, Van Wagoner JC (eds) (1988) Sea-level changes: an integrated approach. Soc Econ Paleontol Mineral Spec Publ 42

Williams GE (1971) Flood deposits of the sandbed ephemeral streams of central Australia. Sedimentology 17: 1–40

Williams H, Soek HF (1993) Predicting reservoir sandbody orientation from dipmeter data: the use of sedimentary dip profiles from outcrop studies. In: Flint SS, Bryant ID (eds) The geological modelling of hydrocarbon reservoirs and outcrop analogues. Int Assoc Sedimentol Spec Publ 15: 143–156

Williams PF, Rust BR (1969) The sedimentology of a braided river. J Sediment Petrol 39: 649–679

Willis BJ (1989) Palaeochannel reconstruction from point bar deposits: a three-dimensional perspective. Sedimentology 36: 757–766

Willis BJ (1993a) Interpretation of bedding geometry within ancient point-bar deposits. In: Marzo M, Puigdefábregas C (eds) Alluvial sedimentation. Int Assoc Sedimentol Spec Publ 17: 101–114

Willis BJ (1993b) Ancient river systems in the Himalayan foredeep, Chinji Village area, northern Pakistan. Sediment Geol 88: 1–76

Willis BJ (1993c) Evolution of Miocene fluvial systems in the Himalayan foredeep through a two-kilometer-thick succession in northern Pakistan. Sediment Geol 88: 77–121

Willis BJ, Behrensmeyer AK (1994) Architecture of Miocene overbank deposits in northern Pakistan. J Sediment Res B64: 60–67

Wilson JL (1975) Carbonate facies in geologic history. Springer, Berlin Heidelberg New York

Wilson LG (1972) Charles Lyell, the years to 1841: the revolution in geology. Yale University Press, New Haven, Connecticut

Wilson RCL, Williams CA (1979) Oceanic transform structures and the development of Atlantic continental margin sedimentary basins – a review. J Geol Soc Lond 136: 311–320

Winder CG (1965) Alluvial cone construction by alpine mudflow in a humid temperate region. Can J Earth Sci 2: 270–277

Wing SL (1984) Relation of paleovegetation to geometry and cyclicity of some fluvial carbonaceous deposits. J Sediment Petrol 54: 52–66

Winker CD (1982) Cenozoic shelf margins, northwest Gulf of Mexico basin. Trans Gulf Coast Assoc Geol Soc 32: 427–448

Winkley BR (1977) Manmade cutoffs on the lower Mississippi River; conception, construction and river response. US Army Corps of Engineers, Vicksburg, Mississippi

Winn RD Jr, Steinmetz JC, Kerekgyarto WL (1993) Stratigraphy and rifting history of the Mesozoic-Cenozoic Anza Rift, Kenya. Am Assoc Petrol Geol Bull 77: 1989–2005

Winsemann J, Seyfried H (1991) Response of deep-water fore-arc systems to sea-level changes, tectonic activity and volcaniclastic input in Central America. In: Macdonald DIM (ed) Sedimentation, tectonics and eustasy: sea-level changes at active margins. Int Assoc Sedimentol Spec Publ 12: 273–292

Wise DU, Belt ES, Lyons PC (1991) Clastic diversion by fold salients and blind thrust ridges in coal-swamp development. Geology 19: 514–517

Withrow PC (1968) Depositional environments of Pennsylvanian Red Fork Sandstone in northeastern Anadarko Basin, Oklahoma. Am Assoc Petrol Geol Bull 52: 1638–1654

Wiygul GJ, Young A (1987) A subsurface study of the Lower Tuscaloosa Formation at Olive field, Pike and Amite Counties, Mississippi. Trans Gulf Coast Assoc Geol Soc 37: 295–302

Wizevich MC (1992a) Photomosaics of outcrops: useful photographic techniques. In: Miall AD, Tyler N (eds) The three-dimensional facies architecture of terrigenous clastic sediments, and its implications for hydrocarbon discovery and recovery. Soc Econ Paleontol Mineral Conc Sedimentol Paleontol 3: 22–24

Wizevich MC (1992b) Sedimentology of Pennsylvanian quartzose sandstones of the Lee Formation, central Appalachian Basin: fluvial interpretation based on lateral profile analysis. Sediment Geol 78: 1–47

Wizevich MC (1993) Depositional controls in a bedload-dominated fluvial system: internal architecture of the Lee Formation, Kentucky. Sediment Geol 85: 537–556

Woidneck K, Behrman P, Soule C, Wu J (1987) Reservoir description of the Endicott field, North Slope, Alaska. In: Tailleur I, Weimer P (eds) Alaska North Slope geology. Society of Economic Paleontologists and Mineralogists Pacific Section, Los Angeles, California, and Alaska Geological Society, Anchorage, Alaska, pp 43–59

Wolman MG, Leopold LB (1957) River flood plains: some observations on their formation. US Geological Survey professional paper 282-C

Woncik J (1972) Recluse field, Campbell County, Wyoming. In: King RA (ed) Stratigraphic oil and gas fields. Am Assoc Petrol Geol Mem 16: 376–382

Wood JM (1989) Alluvial architecture of the Upper Cretaceous Judith River Formation, Dinosaur Provincial Park, Alberta, Canada. Bull Can Petrol Geol 37: 169–181

Wood JM, Hopkins JC (1989) Reservoir sandstone bodies in estuarine valley fill: Lower Cretaceous Glauconitic Member, Little Bow Field, Alberta, Canada. Am Assoc Petrol Geol Bull 73: 1361–1382

Wood JM, Hopkins JC (1992) Traps associated with paleovalleys and interfluves in an unconformity bounded sequence: Lower Cretaceous Glauconitic Member, southern Alberta, Canada. Am Assoc Petrol Geol Bull 76: 904–926

Wood LJ, Ethridge FG, Schumm SA (1993) An experimental study of the influence of subaqueous shelf angles on coastal plain and shelf deposits. In: Weimer P, Posamentier HW (eds) Siliciclastic sequence stratigraphy. Am Assoc Petrol Geol Mem 58: 381–391

Woodland AW, Evans WB (1964) The geology of South Wales Coalfield, part IV: the country around Pontypridd and Maesteg, 3rd edn. Geological Survey of Great Britain Memoir, London

Wopfner H, Callen R, Harris WK (1974) The Lower Tertiary Eyre Formation of the southwestern Great Artesian Basin. J Geol Soc Aust 21: 17–51

Worrall DM, Snelson S (1989) Evolution of the northern Gulf of Mexico, with emphasis on Cenozoic growth faulting and the role of salt. In: Bally AW, Palmer AR (eds) The geology of North America – an overview. Geol Soc Am Geol North Am A: 97–138

Worsley TW, Nance D, Moody JB (1984) Global tectonics and eustasy for the past 2 billion years. Mar Geol 58: 373–400

Worsley TW, Nance D, Moody JB (1986) Tectonic cycles and the history of the earth's biogeochemical and paleoceanographic record. Paleoceanography 1: 233–263

Wright MD (1959) The formation of cross-bedding by a meandering or braided stream. J Sediment Petrol 29: 610–615

Wright VP (ed) (1986) Paleosols: their recognition and interpretation. Blackwell Scientific, Oxford

Wright VP (1990) Estimating rates of calcrete formation and sediment accretion in ancient alluvial deposits. Geol Mag 127: 273–276

Wright VP (1992) Paleopedology, stratigraphic relationships and empirical models. In: Martini IP, Chesworth W (eds) Weathering, soils and paleosols: developments in earth surface processes, no 2. Elsevier, Amsterdam, pp 475–499

Wright VP, Marriott SB (1993) The sequence stratigraphy of fluvial depositional systems: the role of floodplain sediment storage. Sediment Geol 86: 203–210

Yang C-S, Nio S-D (1989) An ebb-tide delta depositional model – a comparison between the modern Eastern Scheldt tidal basin (southwest Netherlands) and the Lower Eocene Roda Sandstone in the southern Pyrenees (Spain). Sediment Geol 64: 175–196

Yang C-S, Nio S-D (1993) Application of high-resolution sequence stratigraphy to the Upper Rotliegend in the Netherlands offshore. In: Weimer P, Posamentier HW (eds) Siliciclastic sequence stratigraphy. Am Assoc Petrol Geol Mem 58: 285–316

Yang W (1985) Daqing oil field, People's Republic of China: a giant field with oil of nonmarine origin. Am Assoc Petrol Geol Bull 69: 1101–1111

Yinan Q, Peihua X, Jingsiu X (1987) Fluvial sandstone bodies as hydrocarbon reservoirs in lake basins. In: Ethridge FG, Flores RM, Harvey MD (eds) Recent developments in fluvial sedimentology. Soc Econ Paleontol Mineral Spec Publ 39: 329–342

Yoshida S, Willis A, Miall AD (in press) Tectonic control of nested sequence architectures in the Castlegate Sandstone (Upper Cretaceous), Book Cliffs, Utah. J Sediment Res

Zaitlin BA, Dalrymple RW, Boyd R (1994) The stratigraphic organization of incised-valley systems associated with relative sea-level changes. In: Dalrymple RW, Boyd R, Zaitlin BA (eds) Incised-valley systems: origin and sedimentary sequences. Soc Econ Paleontol Mineral Spec Publ 51: 45–60

Zhao X, Zhang F, Yang H (1988) Seismic-reflection studies in the Huanghua Depression, Bohai Bay Basin, China. In: Wagner HC, Wagner LC, Wang FFH, Wong FL (eds) Petroleum resources of China and related subjects. Circum-Pacific Council for Energy and Mineral Resources. Earth Sci Ser 10: 277–296

Ziegler AM, Raymond AL, Gierlowski TC, Horrell MA, Rowley DB, Lottes AL (1987) Coal, climate and terrestrial productivity: the present and early Cretaceous compared. In: Scott AC (ed) Coal and coal-bearing strata: recent advances. Geol Soc Lond Spec Publ 32: 25–49

Ziegler PA (1988) Evolution of the Arctic-North Atlantic and the western Tethys. Am Assoc Petrol Geol Mem 43

Author Index

Subject Index